# TRAFFIC ENGINEERING TABLES: THE COMPLETE PRACTICAL ENCYCLOPEDIA

Written by J. Jewett; J. Shrago; J. Gilliland; B. Yomtov

Telephony Publishing Corporation
Chicago, Illinois

Edited and Illustrated by L. Saryusz-Dembowski
TRAFFIC ENGINEERING TABLES: THE COMPLETE PRACTICAL ENCYCLOPEDIA


Library of Congress Catalogue Card Number: 80-50601

Dedicated to C. West Churchman
Professor of Business Administration
University of California at Berkeley

## ACKNOWLEDGEMENTS

This book is the product of half a decade of work by the authors and would never have come into existence without the support and interest of hundreds of persons: some academic colleagues, some editors of publications, some personal acquaintances, but mostly telecommunications professionals with urgent problems to solve.

This work began with Jim Jewett's 'Integrative Applied Projects' course at Vanderbilt University's Owen Graduate School of Management in 1974. The objective of the course was to address actual management problems in the business community. And, in 1974, a problem was supplied by William Tallent, then Director of Management Information Systems at Genesco, who cited a need to improve the quality of long distance telephone service, save money, or both.

A student team, led by Bruce Johnson, clarified objectives, surveyed existing approaches (known techniques resided primarily inside telephone company research laboratories), and designed project stages to collect data, conduct tests, and revise ideas. Other members of the team included Jackie Shrago, Juliane Nelson, and Jim Gordon. Dissatisfied with the inaccuracies of existing telephone company techniques, Jackie Shrago and Bruce Johnson developed many of the ideas that later became known as the Jewett-Shrago Extended Erlang B. Simulation and other mathematical approaches were used to answer Genesco's demand for a more cost-effective telephone service, resulting in a savings of approximately $30,000 as well as the installation of a WATSBOX in the spring of 1975.

Members of the Genesco staff who were particularly helpful included Jim Poteet, Mel Dunn, John Garrett, Brady Olmstead, Dennis Harris, and Bill Jackson. The project received recognition in *Business Week* and local newspapers. In addition, the project motivated Jackie Shrago and Bruce Johnson to begin a software firm after their graduation.

Publication of the initial work came about because of the encouragement from Jerry Goldstone, editor of *Business Communications Review*. The first article was published in October, 1975. Many letters, both critical and encouraging, quickly arrived from telecommunications professionals. These letters became the driving force behind our continued research in the field and our efforts to reveal our findings to the telecommunications community. Many of the authors of those first letters later supplied data and case illustrations valuable to the development of this book.

Probably the greatest single contribution to this work was made by Ramses Mina, now with the Federal Communications Commission and author of *Telephony's Introduction to Traffic Engineering*. He unselfishly has been available for the sounding of ideas, for researching materials, and providing moral support. There are others who have acted in a similar capacity: a few include John Bridges of Satellite Business Systems; David Health of Cornell University; John Fike of Southern Methodist University; Arvind Kumar, a Ph.D. student from the University of Texas; Jerry Finnefrock of SPC; Al Peyser of TDX Systems; Terry Easton of Golden Gate College; and Bob Cooper. Help also came from Bell Laboratories, A.T.&T., Bell operating companies, GTE, and United Telephone. Specifically, this book was aided by the work of Wilkinson, Neal, Steve Katz, Henry Jacobsen, Jonathon Brown, Jim Allen, Mike Blackwell, Jerry Tlapa, Norm Wingard, Bob Lucian, Ted Crater, Elliot Berdy, Larry Schultz, and Paul Mullen. Although these individuals have not entirely endorsed and agreed with the research results, their interest and input have strengthened it considerably.

There are still others who deserve mention for their stimulation and support. Student researchers of particular merit include Joan Holland, now with A.T.&T.; Joe Baker, now with United Telecommunications; Neal Schunemann, now with TCS Communications Consultants; George Pallas; and Steve Vestal. We are grateful that we have had the privilege of working with these people and are glad to see them begin telecommunications careers.

Special note should also be made of those who created the environment where such inquiry could take place. Samuel B. Richmond, dean of the Owen Graduate School of Management, has provided this stimulus. The acting dean before him, James V. Davis, also was quite helpful. Research support at times was furnished by the

Owen Graduate School of Management Associates, business professionals who contribute to Vanderbilt's Owen School in many ways, including support for faculty research.

Other corporate professionals who have assisted this work include Lloyd Isaacs of Bank of America; Wayne Mueller and John Skiragis of Chrysler; Bernard Overreynder of Xerox; Charles Baker of ICA; Bob Bennis of Westinghouse; Joe Manoni, John Manzione, and Kevin Johnson of Merrill Lynch; Bruce Mueller of Upjohn; Gus Bender of Travelers Insurance Company; Dick Jenifer of CNA Insurance; Dick Burtis of Corning Glass; Lance Ede of General Motors; John Oldfield of Transunion; Dave Waldon, Pete Guile, Ed Novak, and Andy Schneider of Carrier Corporation; Frank Watson of Massachusetts Mutual; Roy Hoffman of Standard Oil of Indiana; Jerry Palkovich of Honeywell; Al Gordon, Dave Bittner, and Dale Mullen of Johns-Manville; Phil Fennel of Pullman; and Chuck Schweis of Ciba-Geigy. Some other individuals include Jim Ruttenberg, Joe Massey, Dick Kuehne, Barry Crowley, Jane Johnson, Leo Geoller, Mike Hill, and Richard Mates. Note should also be made of individuals in the interconnect industry: Pat Ryan and Taylor Roberts of I.T.&T.; Dan Hosage, Steve James, Jim Carreker, and Earl Stamen of Datapoint Corporation; Jerry Van Pelt of Action Communications; John Burger and Ernie St. Germaine of General Dynamics; Ed Jungerman and Fred Doidge of Northern Telecom; and Win Bogel and Roger Fetterman of Bell Northern Research.

Many telecommunications professional associations have invited us to present our work and have been most helpful in setting up a forum for the exchange of ideas: International Communications Association (ICA), Communication Management Association (CMA), New England Telecommunications Association (NETA), Connecticut Telecommunications Association (CTA), Tri-Valley Telecommunications Association, South Eastern Telecommunications Association (SETA), Chicago Industrial Communication Association (CICA), and the four chapters of Telecommunications Association (TCA) — Phoenix, Los Angeles, San Francisco, and Seattle.

We would like to recognize the Telco Research Corporation staff: Linda Vacca, Rosalind Barr, Alan Wand, Ann Ewen, and Ann Carter; for their administrative support in connection with the research and publication of this book. In addition we would like to thank Telco's Board of Directors: Phillip Pfeffer, President of Ingram Book Company; William Tallent, President of Coleraine Clothes, Inc.; and Alan Peyser, Executive Vice President of TDX Systems; for their vigorous endorsement of our efforts. And, finally, special thanks to Karen, Eric, and Susan Jewett, and Pat and Dan Burton for their understanding and encouragement.

The list could continue. In fact, the authors have received generous support for our ideas, our applied research, and this book; and we are deeply grateful to all of these individuals.

Jim Jewett
December, 1979
Owen Graduate School of Management
Vanderbilt University
Nashville, Tennessee

Jackie Shrago
Telco Research Corporation
Nashville, Tennessee

# Contents

# PREFACE

This book was written for several reasons. First, there is no complete design resource available that is organized from today's management point of view. By this, we mean a set of tables that accurately predicts cost and grade of service and that can be easily used by a manager who is not a technician or mathematician. As a result, this book is a blend of classical network design tables: Poisson, Erlang B, and Erlang C; and recent developments in caller retrial behavior, queuing techniques, and 'smooth' (VMR less than 1) and 'peaked' (VMR greater than 1) traffic. All the tables are organized in the same manner — by HOURS FIRST ATTEMPT (i.e., caller demand). The tables include trunk-by-trunk usage detail as well as variance-to-mean ratio (VMR) information for CONNECTED and OVERFLOW traffic. Such new detail in tables is critical for today's telephone company and management decisions.

The second reason for the writing of this book involves the errors and omissions in existing tables. Confusion and poor management decisions result from good tables not being available or from the unwillingness of research organizations to reveal their techniques and results. In particular, the tables in this book make the A.T.&T. Neal-Wilkinson tables for 'peakedness' and the A.T.&T. Jacobsen Retrial Tables easy to apply. The Erlang-Engset table in this book corrects the errors found in other published sources.

In addition, this book is a response to the many requests received for such tables over the last few years. Conducting seminars for thousands each year has revealed such a demand.

Finally, this work was published in an effort to diminish the time required to gather all the resources necessary to perform network analyses with these techniques. One table alone required 14.5 hours of computer time! Although some of the tables are complex and expensive to develop, consolidating these materials in one unified volume is worth the effort.

With these tables, telecommunications professionals having minimal technical expertise can more quickly grasp network design and obtain the benefits of better service and/or lower cost. Furthermore, managers can now readily analyze vendor proposals for realism and accuracy.

# SECTION 1
# CHOOSING THE RIGHT TABLE

# SECTION 1:
## CHOOSING THE RIGHT TABLE
### 1.1 Historical Use and Limitations of Network Tables

Tables have been used for many years by both telephone companies and private corporations to design networks. For the last two decades, the most popular tables in the U. S. have been the Poisson tables; while in most other countries, Erlang B tables have been used. Other than A.T.&T., Japan's telephone company was the last to use Poisson tables — it converted to Erlang B in 1965. As of the printing of this book, A.T.&T. is still officially using Poisson, although its well-known inaccuracies have caused many network designers within A.T.&T. to use other techniques.

Historically, network tables have been printed as 'capacity' tables, organized by desired grade of service. Therefore, the user has had to establish that criterion before using the tables. Then, by finding the capacity load in the table nearest to the actual load, the user could determine the number of circuits needed.

There are significant limitations to those traditional tables: no individual circuit usage is available to 1) verify each technique's predictions, and 2) evaluate usage-sensitive tariffed circuits. The traditional tables also do not clearly recognize the difference between 'first-offered' load and 'carried' load. Let's look at these limitations in more detail.

The function of any traffic engineering technique is to predict the actual behavior of telephone systems, specifically the grade of service for a given traffic load and number of circuits. From this information, the cost should be easy to calculate. The two pieces of management information required — cost and grade of service — should be readily available. It is helpful to verify a technique's accuracy by considering more than simply the grade of service prediction. This is best done by additionally using the technique to predict carried traffic and trunk-by-trunk usages. These values, as well as projected blockage, can be compared to the actual system behavior. If discrepancies appear, the managers should try to determine the source of the problem: Are the assumptions of the techniques violated? Is there an error in the technique? Were some circuits unavailable during the data collection period? As a result, carried traffic and trunk-by-trunk usage predictions are quite desirable for verification.

As mentioned above, costs should be easy to calculate. From a telephone company perspective, cost is not often a consideration. Much to A.T.&T.'s credit, its philosophy has been to put in as many circuits as required to provide excellent service. This has not been true of telephone companies in other countries, nor is it necessarily a sensible policy for corporate management. Trunk-by-trunk usage information has not been necessary from a telephone company perspective because usage on an additional trunk has no usage-sensitive cost associated with it. However, telephone customers must consider these circuit usage costs. With usage-sensitive tariffs (especially measured-time WATS), individual (trunk-by-trunk) circuit usages are needed to calculate cost. Hence, circuit detail information is needed both to verify a technique and to evaluate the cost of circuits with usage-sensitive tariffs.

The third limitation of traditional tables is that they do not make effective use of commonly available data. This is because they are based on first-attempt traffic and do not consider total offered or carried traffic. To determine the number of required circuits, we must first estimate traffic demand. If no circuits are in place, this should be the first-attempt load on which traditional tables are based. However, once circuits are in place, demand differs from this estimate. This is because some callers are blocked on their first attempt and may try again. Thus, a given quantity of first-attempt traffic generates a somewhat larger quantity of total offered traffic. Usually, not all traffic can be accommodated on the available circuits, even with retries. Therefore, carried traffic differs, in turn, from both first-attempt and total offered traffic. In general, data available from circuits in place reflects either total offered or, more often, carried traffic. Using this data with traditional tables without allowing for retrials can lead to serious errors. The tables in this book do consider both total offered and carried traffic. While not a complete solution to the data problem, they do put the user in position to make better-informed decisions.

Formulas to translate carried into first-attempt load and first-attempt load into total attempt (offered) load are (load is always defined as hours of traffic — number of calls times average length of completed calls, with the total converted to hours):

$$F = C(1\text{-}BR)/(1\text{-}B) = C + OV + D$$
$$O = F/(1\text{-}BR) = F + T$$
$$T = OBR$$
$$C = (1\text{-}B)O$$

F = First Attempt Load
T = Retrial Load from Blocked Callers Retrying
C = Carried Load
O = Total Offered Load
B = Blockage(0-1.0)
R = Portion of Blocked Callers Who Retry (0-1.0)
D = Traffic which tries a circuit, is blocked and does not retry (dead attempts x any call length of completed calls)
OV = Traffic which tries a circuit, is blocked and uses an alternate set of circuits, often DDD circuits; expressed in hours

## 1.2 Uniqueness of these Network Tables

Unlike the traditional 'capacity' tables, grade of service decision options ranging from one-half of one percent (.5%) blockage (also referred to as P005) to a maximum of 35% (P35) blockage are listed for any traffic load. In addition, usage of each circuit in the group is provided in each table. To our knowledge, this is the first time the trunk-by-trunk usage has been published for the Poisson, Erlang B, Erlang-Engset, Erlang C, and the Equivalent Random Theory tables. It now allows managers to use much more fully the network design information from the techniques, and the information that the historical telephone company perspective has not found necessary until recently.

All of the tables are published here in the same format so that even the infrequent user will find them easy to understand. Once the user becomes familiar with the logic of one table, he will have little difficulty reading the others.

Finally, this comprehensive set of tables covers the full range of today's voice telecommunication situations — both the traditional techniques that have been widely used for the last 60 years and newer ones developed to respond to the changes in today's network economics of private corporations.

## 1.3 Explanation of Types of Traffic

In this section, we begin by defining a traffic characteristic, which plays an important role in choosing a table. At first glance, it may appear that the amount of traffic is all that is needed to choose a table. For instance, it seems that perhaps the only item that must be known is that an average of twenty-five four minute calls is placed during a one-hour period. All the tables are organized so that once the amount of traffic is known, we will know where to look in each table. However, the amount of traffic does not determine which table to select. We will use some examples to help explain.

For these examples, the 'amount' of traffic is fixed — in each case an average of 60 calls are placed during a one-hour period. The average call length is four minutes. Hence, there is an average of four hours, or Erlangs (144 CCSs or 240 minutes), of traffic demand.

In the first case, there are four people in a sales office who are constantly placing calls. Since each person makes one call virtually right after another, it is easy to see that there is a need for one circuit per person. Anything less would be inadequate; anything more would be unnecessary. Hence, four circuits are required.

We might think of this situation another way. We might picture the circuits coming out of the office so that we can see how many are in use at any one time. Clearly, we would see the four circuits almost constantly in use. This pattern would be very steady, with little fluctuation.

Now let's turn to the second example. There are 60 people placing calls. These are highly competitive sales personnel watching a ticker tape for information. An average of 60 calls is still the traffic demand per hour. However, the pattern of calls is very different. Very few calls are made by each sales person. However, every once in a while the ticker tape contains some critical information that sets almost everyone calling. Upon seeing a particularly interesting announcement, each of the 60 brokers makes a call to buy, sell, or tell someone. Again, to provide adequate circuits for the calling pattern means that 60 circuits are needed, one circuit for each person.

The traffic pattern here dictates very different circuit requirements. If we watched the 60 available circuits to see when they are in use, we would see them almost always vacant. Once in a great while, they would all suddenly be used, then, in a few minutes, unused again. The calling pattern has violent fluctuations.

The third example lies between these two extremes. Here, an office of 200 people makes the same average of 60 calls per hour. In this group, there will be times when more than four people will be placing a call simultaneously. However, it is almost certain that as many as 60 people will never need to make calls at the same instant. In fact, 99 1/2% of the time, there will be ten or fewer people placing calls. As a result, only ten circuits need to be available.

It is obvious that circuit requirements vary enormously in these three cases. The amount of traffic alone is not enough information to determine the required circuits. Something must also be known about the traffic pattern. The first example with four people has very steady calling. This traffic is called "smooth", and the circuits have very high and steady utilization. The second calling pattern is termed "peaked" — referring to the violent fluctuation. The third calling pattern is normal, and is referred to as "random" traffic.

Although the amount of traffic is the same in the three examples, whatever technique is used must clearly allow for the calling pattern (smooth, peaked, or random) if it is to determine required circuits accurately. A statistic can be calculated to tell us just how smooth, peaked, or random the calling pattern is. This measure is called the statistical variance of the traffic. It is quite complicated. Generally, we assume the pattern is random or normal unless there is other information. Sections 2 and 3 contain tables for normal or random traffic.

As a result of the concept of statistical variance, each call-arrival pattern can be measured to determine the correct number of needed circuits. The call-arrival 'rate' is translated into the "variance" of the traffic. While it is not our purpose to give a detailed explanation of statistical properties, it must be noted that the concept of variance satisfactorily describes the fluctuation of the traffic throughout the hour for network design purposes.

In each of the three examples here, the average arrival rate was 1 call per minute (an average of 60 calls per hour). In the first two examples, the minute-by-minute dispersion of calls is significantly different from the hourly average. This dispersion or "variance" is divided by the average traffic amount or "mean" to give another important network design statistic, called the traffic variance-to-mean ratio, (VMR). The traffic VMR succinctly summarizes the arrival pattern. Random or normal traffic always has a VMR = 1.0; smooth traffic always has a VMR less than 1.0; and peaked traffic always has a VMR greater than 1.0. Graphically, this measure shows how steady or spiked (peaked) the traffic pattern is within the hour:

In each of the three examples, the average traffic demand requiring circuits was the same within the hour. The variance, fluctuation of the calls, generated different circuit requirements to assure callers of good service.

In Figure 1, traffic patterns are characterized as follows:

| Source of Traffic | Type Traffic | VMR (variance-to-mean ratio) |
|---|---|---|
| Many calling stations with at least a ratio of 8 callers per circuit. | Random | 1 |
| Few callers, a ratio of less than 8 callers per circuit | Smooth | < 1 usually .1 to .9 |
| A tandem network connection through another group of circuits (called a "multiple connect' call). | Smooth | < 1 usually .6 to .9 |
| Overflow from another busy circuit group. | Peaked | > 1 usually 1.5 to 2.5 |
| Callers blocked redial (retry) quickly in attempting to place calls | Peaked | > 1 usually 1.3 - 3.0 |
| Traffic varies significantly from day to day (like 30% of a firm's traffic occurring on Monday) | Peaked | > 1 usually 1.5 - 4.0 |

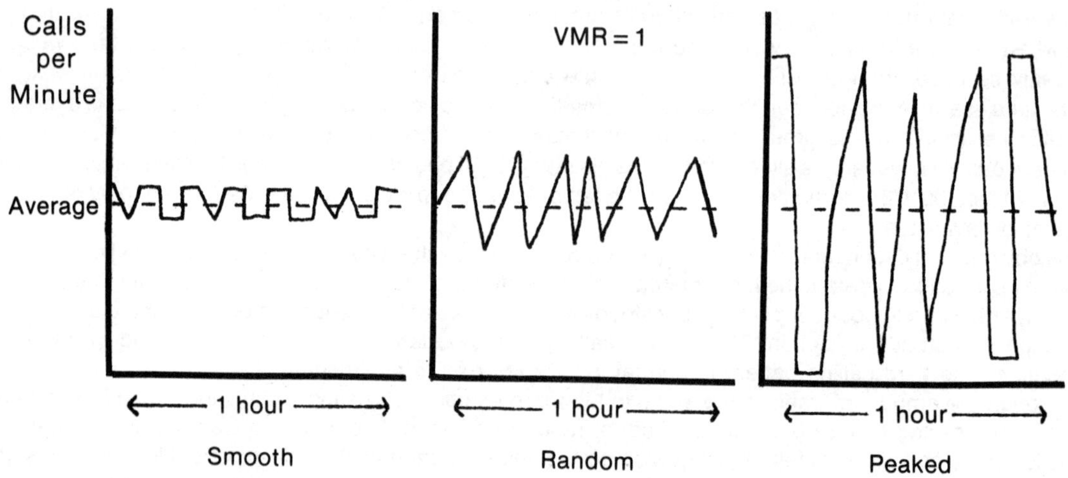


FIG. 1 COMPARISON OF TRAFFIC VARIANCE

These three figures have the same average traffic — only the call pattern (call fluctuation) varies.

Sections 2 and 3 of this book provide techniques that assume the traffic is normal or random (VMR = 1.0). Section 4 provides techniques for non-random situations. Next, we will describe when each table should be used.

### 1.4 The Selection of a Network Table Through a Decision Tree

Having described some of the ways traffic engineering tables have been used in the past, as well as some situations in which they could be helpful, we would like to identify the underlying assumptions and the best applications for each table. This section develops a decision tree of questions and then provides a list of the tables and their assumptions for reference purposes.

We begin by assuming that one or several traffic amounts are known — busy hour, average hour, or traffic for each hour of the day. The traffic amount may be First Attempt traffic. It may be Total Offered traffic. It may be the carried traffic on an existing circuit group. It may be the carried traffic on a part of an existing circuit group — like knowing the measured time WATS usage hours but not the full WATS in a WATS Band. Furthermore, the blocked traffic may be known — either as an amount, hours or minutes, or as a percentage, i.e., a peg count.

QUESTION 1: Is the traffic pattern normal, i.e., random?

We can easily separate traffic engineering situations into two categories: non-random and random. For most cases, such as those for WATS at a single location, the traffic is 'assumed' to be random, and the reader can proceed to Question 2.

The underlying question in this separation is whether or not the traffic is random in its arrival pattern. If it is, then we know that many of the more commonly available tables will accurately describe the situation. As described in Section 1.3, if the traffic is smooth or peaked, then it is not random.

Situations where there are at least 8 callers for each circuit are usually assumed to be random. Frequently, it may be necessary to determine the number of circuits by assuming random traffic and then to return to this question to discover that it is non-random (less than 8 callers in each trunk).

In a complex multi-location network, the traffic may be largely non-random. For instance, the engineering of circuits between tandem switching nodes (usually referred to as inter-machine trunks or IMTs) and circuits providing off-network access via WATS, FX, or Local Off-Network Access Lines (LONALS) may involve non-random traffic. In fact, the only circuits in a multi-location network that do not usually require a non-random assumption are the access lines.

For a multi-location network, you must have more information on how the calls are handled before deciding whether traffic is non-random. We will cover this situation at several points with additional questions.

QUESTION 2: If the traffic is random, is there queuing or no queuing?

Queuing describes a situation where callers wait if all circuits are in use when the call first attempts completion. The queue may be 'manual' with an operator listing the calling party and calling them back when a circuit is available. Or, more commonly today, the call is held by a computer-controlled telephone switching system. In this case, the caller may be required to hold the telephone instrument off-hook for the duration of the queue. This is referred to as a "camp-on", "hold-on", or "short-term" queue. Alternatively, the computer switch may allow the caller to hang-up and be reconnected when a line is available. This is referred to as "call-back", "on-hook", or "long-term" queuing. Many switches today allow for either. This will be discussed in more depth under Question 5.

QUESTION 3: If the traffic is random and there is queuing, is there a "maximum wait time." for those callers who must wait?

If the answer is NO — in other words, if all of the callers will wait indefinitely (i.e., whatever it takes) until a line becomes available — then the ERLANG C table should be used. In many firms, there is no other way to complete a long distance phone call; so, the wait is indefinite. This situation is most frequently associated with an operator-handled queue. ERLANG C is also used to engineer data networks.

QUESTION 4: If the traffic is random, there is queuing, and queued callers have a maximum wait time, how long is the maximum wait?

When a maximum queuing time has been designed (1, 5, 10 minutes), there is usually the ability to "overflow" callers to Direct Distance Dialing (or another route) if a line has not become available during the waiting period.

The maximum wait time is predetermined in computerized switches with queuing capabilities. The time for "camp-on" or "off-hook" queues is usually 30 seconds to 90 seconds. In this case, use the Jewett EQUIVALENT QUEUE EXTENDED ERLANG B for a 1 minute maximum wait (EQEEB-1 MIN).

Other options for maximum wait times are also available. We have included two other tables for the specific maximum wait times of 5 minutes and 10 minutes. If the maximum wait is 5 or 10 minutes, use Jewett EQUIVALENT QUEUE EXTENDED ERLANG B-5 or 10 minutes, respectively. Both assume callers will wait in queue if necessary up to the maximum time specified, and if a line is still not available, will automatically be connected via Direct Distance Dialing (DDD). While we typically think of overflow to another route as DDD, an overflow call could also use another route, like a wide-area WATS service.

In a queuing situation where: (a) all callers do not have the same predetermined maximum queue wait, or (b) there is no automatic overflow to DDD, adjustments must be made in using these EQEEB tables.

Where some callers have a short maximum queue and others have a longer maximum queue, it is appropriate to use a weighted average of the maximum queue times. For example, if the executives and sales force make one-half of the calls and have a maximum wait of 60 seconds, while the other half of the calls wait as long as 10 minutes, we have a weighted maximum queue length of 5.5 minutes $(.5)(1) + (.5)(10) = 5.5$ minutes. Use the Jewett EQUIVALENT QUEUE EXTENDED ERLANG B — 5 minutes for this situation as an approximation.

Many situations may not come as close as this example in approximating the 5 minute table. You may have to use two tables. Note the range of solution both in terms of total number of circuits and circuit usage, and then decide. The important distinction here is not how many people have a short or long term queue but the volume of calls (traffic) which are subject to one queue or another. If an executive group represents only 5% or 10% of the call volume and they have no queue, while everyone else has a one minute queue, use the one minute EQEEB table. On the other hand, if 75% of the employees in a manufacturing firm make almost no calls and have an unlimited wait, while the other 25% have no queue and make 90% of the calls, then you do not really have queuing and should go on to Question 5.

You may also have caller retries at the end of queue. Where there is no automatic overflow to DDD, it is harder to calculate how the table use should be adjusted. If there is significant recalling (retries) after callers wait the maximum time, the effective maximum wait may be considerably longer than that predetermined in the computerized switch. Callers may wait in queue more than once! Especially in situations where (a) short term queues are implemented, or (b) people have a choice before using DDD, or (c) they are charged more for a DDD call, they are likely to retry later and, with some likelihood, may enter the queue two, and even three, or more times. We have determined that when 75% of the callers wait one minute then retry, that is equivalent to a 5 minute maximum queue, and accurate results can be produced by using the Jewett EQUIVALENT QUEUE EXTENDED ERLANG B - 5 minutes maximum wait table. If you are engineering for a similar situation, an alternative is to compare the actual line usage on the existing circuits (and overflow) with the same information from the queuing tables to determine which table is the best predictor.

QUESTION 5: If the traffic is random and there is no queuing, do callers retry when all lines are in use?

If no retries occur, then either all blocked callers automatically use another route to complete the call, or the blocked calls disappear entirely. The blocked caller alternate route is usually Direct Distance Dialing (DDD). For this call routing pattern, use ERLANG B.

In today's electronic switch environment, 'least cost routing' or 'automatic route selection' require engineering FX and WATS trunk groups to get the maximum dollar benefit. The primary trunk group engineering should usually use Erlang B so long as all blocked calls are automatically routed to a secondary route (such as a WATS Band or DDD). To engineer the secondary route, see Question 8. If no alternate route is available, see Question 6.

QUESTION 6: If the traffic is random, there is no queuing but there are retrials, how quickly, following a busy signal, do the callers retry?

The answer to this question requires judgment since detailed data on retries is not usually available. The 'wait before retry' is, however, important because the retrial speed can dramatically affect blockage and, thus, the number of circuits required.

The retrial speed is usually quantified in relation to the length of holding time (or call length) for a completed call. The retrial speed question can be translated as: Within how many holding times will the retrial, on an average, be made? The following table clarifies which table to use in different situations:

| Speed of Average Retrial After Busy Signal | Applicable Table |
|---|---|
| 3-5 Holding Times | Jewett-Shrago Extended Erlang B (EEB) — "Slow" Retrial |
| 1 Holding Time | Jewett-Shrago "Fast" Retrial |
| 1/3-1/2 Holding Time | Jewett-Shrago "Very Fast" Retrial |

If callers in a 'least cost routing' system are all barred (toll restricted) from using DDD, then recalling is the only option. There is no choice. Once you have determined how 'fast' they retry, use the table with 100% blocked callers retrying and 0% overflowing. On the other hand, if the callers may use DDD, select the set of tables with the appropriate retrial speed and use the 50% blocked and 50% overflow tables. The choice between tables with different percentages of recalling, 100% or 50%, is usually straightforward. The estimate of recalling is usually based on judgment. Interpolation between tables is usually not required. The tables are not very sensitive to major changes between 25% and 75% recalling.

Jacobsen Tables used by AT&T, while approaching retrials slightly differently, are similar to the Jewett-Shrago Extended Erlang B, Fast Retrial, and Very Fast Retrial Tables.

This completes the series of questions and techniques for normal or random traffic.

QUESTION 7: If the situation is non-random, is the traffic 'smooth'?

In one of the examples described in Section 1.3, a few callers, 4, were making all the long distance calls, and the traffic was described as smooth. As mentioned, the degree of smoothness is defined measuring the first-attempt traffic variance-to-mean ratio (VMR). But fortunately, to engineer in smooth traffic situations, you need only know the number of sources of the traffic. The Erlang-Engset table is used for smooth traffic. The

Erlang-Engset table printed in this book is organized to show different numbers of sources for each amount of first attempt traffic. Each number of sources has a different VMR. This is reflected in the number of required circuits.

If the ratio of number of sources to the number of circuits is 8 or more to 1, the traffic is usually assumed to be random. The results are equivalent to the Erlang B table. As the smooth VMR approaches 1, results are not shown in the Erlang-Engset table since these results are in the Erlang B table.

There are also some other important situations where the traffic is smooth, i.e., has a VMR less than 1. If the traffic is for a tandem network connection and consists primarily of traffic already carried on one trunk group that is offered to a second trunk group, then the traffic may be smooth. ERLANG-ENGSET should be used. For example, if calls seizing access line groups also seize a WATS line group at a tandem switch and there is little other traffic seizing the WATS group, then the WATS traffic is probably smooth. This is especially true if the access line groups have high blockage. In this case, the number of sources would represent the number of access lines and not the number of callers. If there is a large number of access lines compared to the number of WATS lines (more than 8 to 1), then we are back to a random traffic situation again. Low blockage on the access lines would also create random traffic.

Two formulas are quite useful in the analysis of smooth traffic and in using the Erlang-Engset table. These formulas help due to the fact that when using the tables the user usually has first attempt traffic information and either (1) number of sources; or (2) the variance-to-mean ratio (VMR) of first attempt traffic. The number of sources is usually known when there are few callers (stations) placing the calls. The variance-to-mean ratio (VMR) is usually known when the smooth traffic being studied consists, in part, of traffic already connected on other circuits groups. The two formulas are given below. In using each formula, you begin with the information on the right side, using the information in the computation to derive the third parameter. All three parameters, hours first attempt traffic, VMR of first attempt traffic, and number of sources, are required to use the Erlang-Engset table.

$$VMR = 1 - (A/S)$$
$$S = A/(1 - VMR)$$

A-Hours First Attempt Traffic Load
S-Number of Sources
VMR-Variance-to-Mean Ratio of First Attempt Traffic

QUESTION 8: If the situation is non-random and the traffic is not smooth, is the traffic peaked?

Peaked traffic was described in Section 1.3 as erratic traffic, with variance-to-mean ratios greater than 1.0. This pattern occurs when traffic is route advanced or alternately routed from a "primary" circuit group to a "secondary" one. Again, the variance-to-mean ratio (VMR) is the measure used to describe this characteristic. EQUIVALENT RANDOM THEORY tables are used for peaked traffic. Four tables of Equivalent Random Theory are provided here. They are used for traffic VMRs equal to 1.5, 2, 3, and 4, respectively. Choosing the correct table depends on determining the erratic nature of the first attempt traffic.

In route advancement situations, if there is high blockage on the primary route, the traffic attempting the second group will be peaked and require the Equivalent Random Theory tables. Of course, if the route advanced peaked traffic is mixed with a large amount of random traffic, then the VMR of the combined traffic may again be close to 1.0. In this case, assume the traffic is random.

Peakedness is usually of critical importance when engineering the intermachine trunks in a multi-node network and may also be important in a tandem tie line network. In the latter case, where high blockage occurs on one route and callers have an alternate path, the traffic on one link can greatly impact the VMR of offered traffic on an alternate route.

Neal-Wilkinson Tables used by A.T.&T. have been specifically designed to adjust for peakedness and day-to-day traffic variation. These tables utilize Equivalent Random Theory and generate results similar to the tables presented in this book.

QUESTION 9: If the situation is non-random and is neither peaked or smooth, then can you specify the retrial and other traffic characteristics?

In any other situation, such as those involving an Automatic Call Distributor (ACD) or retrials which are not independent of each other (as happens in some multi-location network problems), SIMULATION is required. This technique requires a computer to act out complex behavior because the interactions of the traffic characteristics are so complex that no known formulas can describe the traffic implications accurately.

Summary List of Questions

Question 1 Traffic Normal (Random) or Not?
Question 2 Queuing or No Queuing?
Question 3 Maximum Wait or Not?
Question 4 How Long is Maximum Wait?
Question 5 Do Blocked Callers Retry?
Question 6 Speed of Retries?
Question 7 Smooth Traffic or Not?
Question 8 Peaked Traffic or Not?
Question 9 Retrial and Other Characteristics for Simulation?

In deciding on the applicability of each table, it is important to know when an underlying assumption is violated. The following table summarizes the information that is required for maximum accuracy in traffic engineering.

| Technique (Table) | Traffic Characteristics: Arrival Pattern of First Attempt Traffic | Do Blocked Callers Retry? | Number of Callers | Do Blocked Callers Queue? |
|---|---|---|---|---|
| Erlang B | Random Traffic VMR = 1 | No | Infinite | No |
| *Jewett-Shrago Extended Erlang B (Slow Retrials) | Random; Retrials are random | Yes | Infinite | No |
| *Jewett-Shrago Fast Retrials | Random; Retrials are peaked | Yes | Infinite | No |
| *Jewett-Shrago Very Fast Retrials | Random; Retrials are very peaked | Yes | Infinite | No |
| Jewett Equivalent Queue Extended Erlang B: | | | | |
| 1 Minute | Random | No | Infinite | Yes up to Max Wait (1 Minute) |
| 5 Minute | Random | No | Infinite | Yes up to Max Wait (5 Minute) |
| 10 Minute | Random | No | Infinite | Yes up to Max Wait (10 Minute) |

| | | | | |
|---|---|---|---|---|
| Erlang C | Random | No | Infinite | Yes, Infinite Wait |
| Erlang-Engset | Smooth VMR $< 1$ | No | Specified as "Number of Sources" in table | No |
| Equivalent Random Theory: | | | | |
| **VMR = 1.5 | Peaked VMR $> 1$ | No | Infinite | No |
| **VMR = 2 | Peaked VMR $> 1$ | No | Infinite | No |
| **VMR = 3 | Peaked VMR | No | Infinite | No |
| **VMR = 4 | Peaked VMR | No | Infinite | No |

Simulation - - - - - - - - - - No Restrictions - - - - - - - - - -
* Similar to A.T.&T. Jacobsen Retrial tables
** Similar to A.T.&T. Neal-Wilkinson tables

There are some additional assumptions which apply to all of the techniques.

1. CALL PROCESSING OR CALL SET-UP TIME. WATS and Direct Distance Dialing are charged for time beginning when the parties are connected. All outward circuits, however, are often in use at other times as well. In other words, there is time when the line is in use but for which you are not billed. This extra use includes dialing time, network connection, ringing time, busy signals, and 'no answer' at the distant end. However, since the line is 'in use', this 'non-billed' time must be included in the first attempt traffic load in order to use the tables properly.

2. ROTARY LINE SEIZING. The tables accurately predict trunk-by-trunk usage for systems with rotary hunt line seizing. All the tables assume that the first circuit is seized before the second, which is seized before the third, etc. The tables can be used where rotary line seizing is not used. While many electronic switches seize in a different hunt sequence, the average trunk usage for other hunt sequences can be directly calculated from the tables. The 'average' trunk usage of all circuits in a group will be the same as the 'average' usage for the number of circuits in the table. If in a WATS group, you have two full-time circuits and four measured-time circuits, find the place in the tables where the carried load for six circuits matches your actual usage. The average of the usage on the first two circuits should approximate the usage on each of your full-time circuits. The average of the usage on the last four circuits should approximate the usage on each of your measured circuits. By using the table usage, you can determine whether or not you have the right mix of full and measured WATS circuits. Thus, the trunk-by-trunk usages from the tables can be used in all situations no matter how the lines are hunted (seized).

With the questions in this section, we have tried to describe the precise application of each table. There are many situations where it is not clear exactly which table is appropriate or where the use of a table involves some error. Since the manual effort to use one or even two or three tables is usually not significant, you can expect to obtain a range of information which provides rich insight. Remembering that traffic engineering must still be considered an art rather than a science will assist you in realizing that an approximation derived from the tables can be quite valuable and may serve as a check for reasonableness of results generated by a computer technique.

There is no single table which incorporates the complex combinations of behavior where route advance alternatives, bulk data, recalling, and queuing are ALL utilized. There is no single table for peaked retrials from a finite number of sources where route advancement occurs. And yet, these situations occur frequently in parts of multi-node networks. There is no single table for determining multi-stage queuing or priority classes of service with different queue lengths. It is in these situations where tables can provide an approximation. Computerized algorithms which incorporate these techniques are required for maximum accuracy.

## 1.5 Table Column Headings

The first column heading for all tables is called HOURS FIRST ATTEMPT. This is the raw telecommunications need expressed in hours (sometimes called Erlangs). Minutes or CCSs can be easily converted to hours: 60

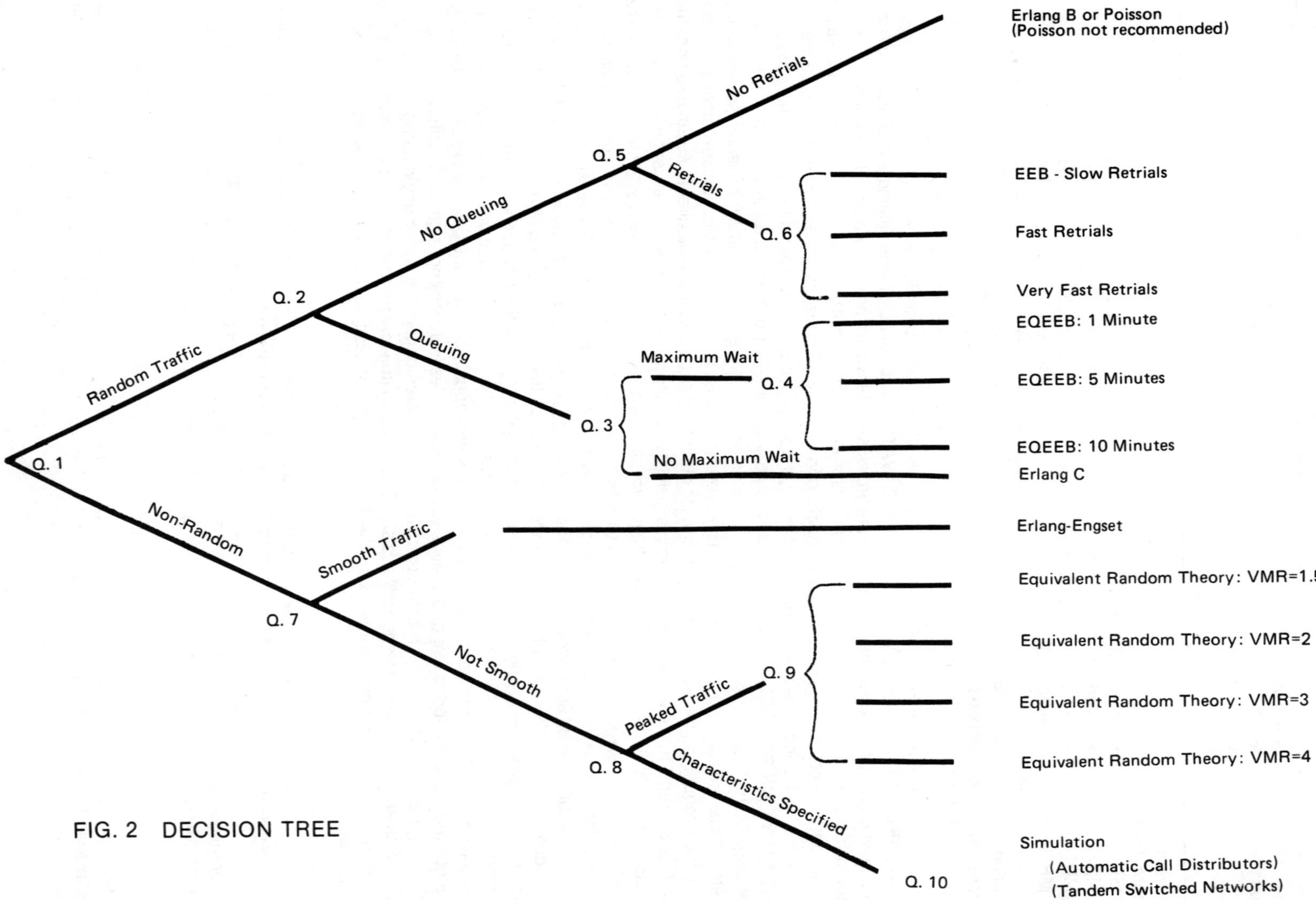


FIG. 2 DECISION TREE

minutes in one hour, 36 CCS in one hour. Each table is also 'organized' by HOURS FIRST ATTEMPT, starting with .200 Hours (12 minutes) and increasing to 30.000 Hours — generally 1-45 circuits.

All of the tables, except one, state that the variance-to-mean ratio (VMR) of the HOURS FIRST ATTEMPT is the same throughout the table.

| | | |
|---|---|---|
| VMR = 1 | : | Poisson |
| | | Erlang B |
| | | Jewett-Shrago Extended Erlang B (Slow Retrials) |
| | | Jewett-Shrago Fast Retrial Tables |
| | | Jewett-Shrago Very Fast Retrial Tables |
| | | Jewett Equivalent Queue Extended Erlang B |
| | | Erlang C. |
| VMR = 1.5 | : | Equivalent Random Theory |
| VMR = 2 | : | Equivalent Random Theory |
| VMR = 3 | : | Equivalent Random Theory |
| VMR = 4 | : | Equivalent Random Theory |

The exception is the Erlang-Engset table where the HOURS FIRST ATTEMPT VMR varies (less than 1.0) with the NUMBER SOURCES in the table. As a result, the Erlang-Engset table has two additional column headings the other tables do not have: FIRST ATTEMPT VMR, and NUMBER SOURCES. Note that the value of the FIRST ATTEMPT VMR fluctuates throughout the Erlang-Engset table. It varies depending upon both the HOURS FIRST ATTEMPT and NUMBER SOURCES.

One additional note, HOURS FIRST ATTEMPT is generally not 'total offered' traffic. The two are identical only in situations where there are no retrials of blocked calls. Thus, for the slow, fast, and very fast retrial tables, offered traffic would be the sum of HOURS FIRST ATTEMPT and HOURS RECALL/RETRY. The amount of additional offered traffic created by retries is given in the HOURS RECALL/RETRY column. This represents an approximation of the retries multiplied by the average call length of completed calls. By taking a ratio of this number to the HOURS FIRST ATTEMPT, you can approximate the portion of calls which are retries. For all the other tables, HOURS FIRST ATTEMPT is 'total offered' traffic.

The next column on each table is HOURS CONNECT. This heading must be explained with the 'LINE' column heading. The 'LINE' column shows the number of circuits available. The HOURS CONNECT column shows how much of the HOURS FIRST ATTEMPT is expected to be connected on the number of available lines.

The other disposition of HOURS FIRST ATTEMPT Is to HOURS OVFL, or hours of overflow traffic. As a result, HOURS OVFL and HOURS CONNECT will always sum to HOURS FIRST ATTEMPT. The actual overflow of the HOURS OVFL for the queuing tables occurs after the specified maximum wait without a line becoming available.

Next, each table gives trunk-by-trunk usage detail for the connected traffic. These amounts are in the ten right hand columns labelled HOURS CONNECTED TRAFFIC ON LINES 1-10. An amount is given for each line, i.e., three amounts for three lines, five amounts for five lines, etc. If more than ten lines are available, the eleventh amount is given under the column '1' on a second table row. Usages for up to approximately 50 lines are provided by the tables. The sum of the line usages is always HOURS CONNECT.

For purposes of studying overflow, route advancement, and connecting traffic on multiple circuit groups, the variance-to-mean ratios (VMR) are given for both connected and overflow traffic, if available. These two column headings are labelled CONNECT VMR and OVFL VMR. Values for these are not provided for POISSON, the queuing, and Erlang-Engset tables.

The connected traffic and overflow traffic VMRs are quite useful in determining the variances of these traffic loads. They are given below:

Variance of Overflow = (OVFL VMR) × (HOURS OVFL)
Variance of Connected Traffic = (CONNECT VMR) × (HOURS CONNECT)

Another useful relationship should be mentioned at this point. This relationship concerns the situation where there are two or more types of traffic, each with a different VMR, which comprise the HOURS FIRST ATTEMPT traffic.

Assume there are three kinds of traffic, hours being represented by A, B, and C. Then the HOURS FIRST ATTEMPT traffic, F, is the sum: $F = A + B + C$. This relationship has to do with the determination of the VMR of this traffic when each traffic segment has a separate VMR. Assume each of the three traffic amounts has a different VMR represented by X, Y, and Z. To obtain the VMR of the total HOURS FIRST ATTEMPT, F, we must first know the variance of this traffic, which is the sum of the three variances of the separate traffic amounts, A, B, and C. We know the variance of A is simply the traffic, A, times the VMR of A, X. Hence, the variance of A is (A)x(X). Similar logic results in the calculation of the total variance of F being $V = (AX) + (BY) + (CZ)$. The VMR of the first attempt traffic then becomes $VMR = V/F$. Two or more different traffic segments can be handled in this manner.

The grade of service columns are immediately before the line usage columns. For all the tables which assume blocked callers are not queued, the grade of service column is PERCENT BLOCKAGE. This is simply the percent of the total offered traffic which is blocked. The blockage in each 'non-queuing' table varies generally from .5%-35% (P005-P35). However, the Erlang B table has blockages as high as 75% for route advancement studies.

For queuing, the grade of service has two columns: PORTION DELAYED and DELAY CALL LENGTH FACTOR. The PORTION DELAY is that portion (between 0 to 1.0) of traffic that experiences some delay. To determine the "average delay" of "delayed callers" the DELAY CALL LENGTH FACTOR is multiplied by the average length of call. The PORTION DELAY represents only the portion who are queued and connected on the circuits. The portion who are queued and overflow must be added separately. To obtain this portion, divide the amount of HOURS OVERFLOW by the HOURS FIRST ATTEMPT. Add this portion to the PORTION DELAYED for the total queued portion. This number is an approximation.

# SECTION 2
# RANDOM TRAFFIC WITH NO QUEUING OF BLOCKED CALLERS

*POISSON TABLE: BLOCKAGE REPRESENTS ALL TRUNKS BUSY

| HOURS FIRST ATTEMPT VMR = 1.0 | HOURS CON-NECT | HOURS RECALL /RETRY | HOURS OVFL | PERCENT BLOCKED (GRADE) | LINE | HOURS CONNECTED TRAFFIC ON LINES | | | | | | | | | |
|---|---|---|---|---|---|---|---|---|---|---|---|---|---|---|---|
| | | | | | | 1 | 2 | 3 | 4 | 5 | 6 | 7 | 8 | 9 | 10 |
| .200 | .181 | .000 | .019 | 18.1% | 1 | .181 | | | | | | | | | |
| .200 | .199 | .000 | .001 | 1.8% | 2 | .181 | .018 | | | | | | | | |
| .400 | .330 | .000 | .070 | 33.0% | 1 | .330 | | | | | | | | | |
| .400 | .391 | .000 | .009 | 6.2% | 2 | .330 | .062 | | | | | | | | |
| .400 | .399 | .000 | .001 | .8% | 3 | .330 | .062 | .008 | | | | | | | |
| .600 | .451 | .000 | .149 | 45.1% | 1 | .451 | | | | | | | | | |
| .600 | .573 | .000 | .027 | 12.2% | 2 | .451 | .122 | | | | | | | | |
| .600 | .596 | .000 | .004 | 2.3% | 3 | .451 | .122 | .023 | | | | | | | |
| .800 | .742 | .000 | .058 | 19.1% | 2 | .551 | .191 | | | | | | | | |
| .800 | .789 | .000 | .011 | 4.7% | 3 | .551 | .191 | .047 | | | | | | | |
| .800 | .798 | .000 | .002 | .9% | 4 | .551 | .191 | .047 | .009 | | | | | | |
| 1.000 | .896 | .000 | .104 | 26.4% | 2 | .632 | .264 | | | | | | | | |
| 1.000 | .977 | .000 | .023 | 8.0% | 3 | .632 | .264 | .080 | | | | | | | |
| 1.000 | .996 | .000 | .004 | 1.9% | 4 | .632 | .264 | .080 | .019 | | | | | | |
| 1.200 | 1.036 | .000 | .164 | 33.7% | 2 | .699 | .337 | | | | | | | | |
| 1.200 | 1.157 | .000 | .043 | 12.1% | 3 | .699 | .337 | .121 | | | | | | | |
| 1.200 | 1.190 | .000 | .010 | 3.4% | 4 | .699 | .337 | .121 | .034 | | | | | | |
| 1.200 | 1.198 | .000 | .002 | .8% | 5 | .699 | .337 | .121 | .034 | .008 | | | | | |
| 1.400 | 1.162 | .000 | .238 | 40.8% | 2 | .753 | .408 | | | | | | | | |
| 1.400 | 1.328 | .000 | .072 | 16.7% | 3 | .753 | .408 | .167 | | | | | | | |
| 1.400 | 1.382 | .000 | .018 | 5.4% | 4 | .753 | .408 | .167 | .054 | | | | | | |
| 1.400 | 1.396 | .000 | .004 | 1.4% | 5 | .753 | .408 | .167 | .054 | .014 | | | | | |
| 1.600 | 1.273 | .000 | .327 | 47.5% | 2 | .798 | .079 | | | | | | | | |
| 1.600 | 1.490 | .000 | .110 | 21.7% | 3 | .798 | .475 | .217 | | | | | | | |
| 1.600 | 1.569 | .000 | .031 | 7.9% | 4 | .798 | .475 | .217 | .079 | | | | | | |
| 1.600 | 1.592 | .000 | .008 | 2.4% | 5 | .798 | .475 | .217 | .079 | .024 | | | | | |
| 1.600 | 1.598 | .000 | .002 | .6% | 6 | .798 | .475 | .217 | .079 | .024 | .006 | | | | |
| 1.800 | 1.641 | .000 | .159 | 26.9% | 3 | .835 | .537 | .269 | | | | | | | |
| 1.800 | 1.750 | .000 | .050 | 10.9% | 4 | .835 | .537 | .269 | .109 | | | | | | |
| 1.800 | 1.786 | .000 | .014 | 3.6% | 5 | .835 | .537 | .269 | .109 | .036 | | | | | |
| 1.800 | 1.797 | .000 | .003 | 1.0% | 6 | .835 | .537 | .269 | .109 | .036 | .010 | | | | |
| 2.000 | 1.782 | .000 | .218 | 32.3% | 3 | .865 | .594 | .323 | | | | | | | |
| 2.000 | 1.925 | .000 | .075 | 14.3% | 4 | .865 | .594 | .323 | .143 | | | | | | |
| 2.000 | 1.978 | .000 | .022 | 5.3% | 5 | .865 | .594 | .323 | .143 | .053 | | | | | |
| 2.000 | 1.994 | .000 | .006 | 1.7% | 6 | .865 | .594 | .323 | .143 | .053 | .017 | | | | |
| 2.200 | 1.912 | .000 | .288 | 37.7% | 3 | .889 | .645 | .377 | | | | | | | |
| 2.200 | 2.093 | .000 | .107 | 18.1% | 4 | .889 | .645 | .377 | .181 | | | | | | |
| 2.200 | 2.165 | .000 | .035 | 7.2% | 5 | .889 | .645 | .377 | .181 | .072 | | | | | |
| 2.200 | 2.190 | .000 | .010 | 2.5% | 6 | .889 | .645 | .377 | .181 | .072 | .025 | | | | |
| 2.200 | 2.197 | .000 | .003 | .7% | 7 | .889 | .645 | .377 | .181 | .072 | .025 | .007 | | | |

*NOT RECOMMENDED FOR BLOCKAGE MORE THAN 5%, ALSO UNDERSTATES MEASURED TIME USAGE

*POISSON TABLE: BLOCKAGE REPRESENTS ALL TRUNKS BUSY

| HOURS FIRST ATTEMPT VMR = 1.0 | HOURS CON-NECT | HOURS RECALL /RETRY | HOURS OVFL | PERCENT BLOCKED (GRADE) | LINE | HOURS CONNECTED TRAFFIC ON LINES | | | | | | | | | |
|---|---|---|---|---|---|---|---|---|---|---|---|---|---|---|---|
| | | | | | | 1 | 2 | 3 | 4 | 5 | 6 | 7 | 8 | 9 | 10 |
| 2.400 | 2.031 | .000 | .369 | 43.0% | 3 | .909 | .692 | .430 | | | | | | | |
| 2.400 | 2.252 | .000 | .148 | 22.1% | 4 | .909 | .692 | .430 | .221 | | | | | | |
| 2.400 | 2.348 | .000 | .052 | 9.6% | 5 | .909 | .692 | .430 | .221 | .096 | | | | | |
| 2.400 | 2.384 | .000 | .016 | 3.6% | 6 | .909 | .692 | .430 | .221 | .096 | .036 | | | | |
| 2.400 | 2.396 | .000 | .004 | 1.2% | 7 | .909 | .692 | .430 | .221 | .096 | .036 | .012 | | | |
| 2.600 | 2.140 | .000 | .462 | 48.2% | 3 | .926 | .733 | .482 | | | | | | | |
| 2.600 | 2.404 | .000 | .196 | 26.4% | 4 | .926 | .733 | .482 | .264 | | | | | | |
| 2.600 | 2.526 | .000 | .074 | 12.3% | 5 | .926 | .733 | .482 | .264 | .123 | | | | | |
| 2.600 | 2.576 | .000 | .024 | 4.9% | 6 | .926 | .733 | .482 | .264 | .123 | .049 | | | | |
| 2.600 | 2.593 | .000 | .007 | 1.7% | 7 | .926 | .733 | .482 | .264 | .123 | .049 | .017 | | | |
| 2.600 | 2.598 | .000 | .002 | .5% | 8 | .926 | .733 | .482 | .264 | .123 | .049 | .017 | .005 | | |
| 2.800 | 2.547 | .000 | .253 | 30.8% | 4 | .939 | .769 | .531 | .308 | | | | | | |
| 2.800 | 2.699 | .000 | .101 | 15.2% | 5 | .939 | .769 | .531 | .308 | .152 | | | | | |
| 2.800 | 2.764 | .000 | .036 | 6.5% | 6 | .939 | .769 | .531 | .308 | .152 | .065 | | | | |
| 2.800 | 2.789 | .000 | .011 | 2.4% | 7 | .939 | .769 | .531 | .308 | .152 | .065 | .024 | | | |
| 2.800 | 2.797 | .000 | .003 | .8% | 8 | .939 | .769 | .531 | .308 | .152 | .065 | .024 | | | |
| 3.000 | 2.681 | .000 | .319 | 35.3% | 4 | .950 | .801 | .577 | .353 | | | | | | |
| 3.000 | 2.865 | .000 | .135 | 18.5% | 5 | .950 | .801 | .577 | .353 | .185 | | | | | |
| 3.000 | 2.949 | .000 | .051 | 8.4% | 6 | .950 | .801 | .577 | .353 | .185 | .084 | | | | |
| 3.000 | 2.983 | .000 | .017 | 3.4% | 7 | .950 | .801 | .577 | .353 | .185 | .084 | .034 | | | |
| 3.000 | 2.995 | .000 | .005 | 1.2% | 8 | .950 | .801 | .577 | .353 | .185 | .084 | .034 | .012 | | |
| 3.200 | 2.806 | .000 | .394 | 39.7% | 4 | .959 | .829 | .620 | .397 | | | | | | |
| 3.200 | 3.025 | .000 | .175 | 21.9% | 5 | .959 | .829 | .620 | .397 | .219 | | | | | |
| 3.200 | 3.130 | .000 | .070 | 10.5% | 6 | .959 | .829 | .620 | .397 | .219 | .105 | | | | |
| 3.200 | 3.175 | .000 | .025 | 4.5% | 7 | .959 | .829 | .620 | .397 | .219 | .105 | .045 | | | |
| 3.200 | 3.192 | .000 | .008 | 1.7% | 8 | .959 | .829 | .620 | .397 | .219 | .105 | .045 | .017 | | |
| 3.200 | 3.198 | .000 | .002 | .6% | 9 | .959 | .829 | .620 | .397 | .219 | .105 | .045 | .017 | .006 | |
| 3.400 | 2.922 | .000 | .478 | 44.2% | 4 | .967 | .853 | .660 | .442 | | | | | | |
| 3.400 | 3.178 | .000 | .222 | 25.6% | 5 | .967 | .853 | .660 | .442 | .256 | | | | | |
| 3.400 | 3.307 | .000 | .093 | 12.9% | 6 | .967 | .853 | .660 | .442 | .256 | .129 | | | | |
| 3.400 | 3.365 | .000 | .035 | 5.8% | 7 | .967 | .853 | .660 | .442 | .256 | .129 | .058 | | | |
| 3.400 | 3.388 | .000 | .012 | 2.3% | 8 | .967 | .853 | .660 | .442 | .256 | .129 | .058 | .023 | | |
| 3.400 | 3.396 | .000 | .004 | .8% | 9 | .967 | .853 | .660 | .442 | .256 | .129 | .058 | .023 | .008 | |
| 3.600 | 3.029 | .000 | .571 | 48.5% | 4 | .973 | .874 | .697 | .485 | | | | | | |
| 3.600 | 3.323 | .000 | .277 | 29.4% | 5 | .973 | .874 | .697 | .485 | .294 | | | | | |
| 3.600 | 3.478 | .000 | .122 | 15.6% | 6 | .973 | .874 | .697 | .485 | .294 | .156 | | | | |
| 3.600 | 3.552 | .000 | .048 | 7.3% | 7 | .973 | .874 | .697 | .485 | .294 | .156 | .073 | | | |
| 3.600 | 3.583 | .000 | .017 | 3.1% | 8 | .973 | .874 | .697 | .485 | .294 | .156 | .073 | .031 | | |
| 3.600 | 3.594 | .000 | .006 | 1.2% | 9 | .973 | .874 | .697 | .485 | .294 | .156 | .073 | .031 | .012 | |
| 3.800 | 3.460 | .000 | .340 | 33.2% | 5 | .978 | .893 | .731 | .527 | .332 | | | | | |

*NOT RECOMMENDED FOR BLOCKAGE MORE THAN 5%, ALSO UNDERSTATES MEASURED TIME USAGE

*POISSON TABLE: BLOCKAGE REPRESENTS ALL TRUNKS BUSY

| HOURS FIRST ATTEMPT VMR = 1.0 | HOURS CON-NECT | HOURS RECALL /RETRY | HOURS OVFL | PERCENT BLOCKED (GRADE) | LINE | HOURS CONNECTED TRAFFIC ON LINES 1 | 2 | 3 | 4 | 5 | 6 | 7 | 8 | 9 | 10 |
|---|---|---|---|---|---|---|---|---|---|---|---|---|---|---|---|
| 3.800 | 3.644 | .000 | .156 | 18.4% | 6 | .978 | .893 | .731 | .527 | .332 | .184 | | | | |
| 3.800 | 3.735 | .000 | .065 | 9.1% | 7 | .978 | .893 | .731 | .527 | .332 | .184 | .091 | | | |
| 3.800 | 3.775 | .000 | .025 | 4.0% | 8 | .978 | .893 | .731 | .527 | .332 | .184 | .091 | .040 | | |
| 3.800 | 3.791 | .000 | .009 | 1.6% | 9 | .978 | .893 | .731 | .527 | .332 | .184 | .091 | .040 | .016 | |
| 3.800 | 3.797 | .000 | .003 | .6% | 10 | .978 | .893 | .731 | .527 | .332 | .184 | .091 | .040 | .016 | .006 |
| 4.000 | 3.590 | .000 | .410 | 37.1% | 5 | .982 | .762 | .567 | .371 | | | | | | |
| 4.000 | 3.805 | .000 | .195 | 21.5% | 6 | .982 | .908 | .762 | .567 | .371 | .215 | | | | |
| 4.000 | 3.915 | .000 | .085 | 11.1% | 7 | .982 | .908 | .762 | .567 | .371 | .215 | .111 | | | |
| 4.000 | 3.966 | .000 | .034 | 5.1% | 8 | .982 | .908 | .762 | .567 | .371 | .215 | .111 | .051 | | |
| 4.000 | 3.988 | .000 | .012 | 2.1% | 9 | .982 | .908 | .762 | .567 | .371 | .215 | .111 | .051 | .021 | |
| 4.000 | 3.996 | .000 | .004 | .8% | 10 | .982 | .908 | .762 | .567 | .371 | .215 | .111 | .051 | .021 | .008 |
| 4.200 | 3.712 | .000 | .488 | 41.0% | 5 | .985 | .922 | .790 | .605 | .410 | | | | | |
| 4.200 | 3.958 | .000 | .242 | 24.7% | 6 | .985 | .922 | .790 | .605 | .410 | .247 | | | | |
| 4.200 | 4.091 | .000 | .109 | 13.3% | 7 | .985 | .922 | .790 | .605 | .410 | .247 | .133 | | | |
| 4.200 | 4.155 | .000 | .045 | 6.4% | 8 | .985 | .922 | .790 | .605 | .410 | .247 | .133 | .064 | | |
| 4.200 | 4.183 | .000 | .017 | 2.8% | 9 | .985 | .922 | .790 | .605 | .410 | .247 | .133 | .064 | .028 | |
| 4.200 | 4.194 | .000 | .006 | 1.1% | 10 | .985 | .922 | .790 | .605 | .410 | .247 | .133 | .064 | .028 | .011 |
| 4.400 | 3.826 | .000 | .574 | 44.9% | 5 | .988 | .934 | .815 | .641 | .449 | | | | | |
| 4.400 | 4.106 | .000 | .294 | 28.0% | 6 | .988 | .934 | .815 | .641 | .449 | .280 | | | | |
| 4.400 | 4.262 | .000 | .138 | 15.6% | 7 | .988 | .934 | .815 | .641 | .449 | .280 | .156 | | | |
| 4.400 | 4.341 | .000 | .059 | 7.9% | 8 | .988 | .934 | .815 | .641 | .449 | .280 | .156 | .079 | | |
| 4.400 | 4.376 | .000 | .024 | 3.6% | 9 | .988 | .934 | .815 | .641 | .449 | .280 | .156 | .079 | .036 | |
| 4.400 | 4.391 | .000 | .009 | 1.5% | 10 | .988 | .934 | .815 | .641 | .449 | .280 | .156 | .079 | .036 | .015 |
| 4.400 | 4.397 | .000 | .003 | .6% | 11 | .988 | .934 | .815 | .641 | .449 | .280 | .156 | .079 | .036 | .015 |
| | | | | | | .006 | | | | | | | | | |
| 4.600 | 3.932 | .000 | .668 | 48.7% | 5 | .990 | .944 | .837 | .674 | .487 | | | | | |
| 4.600 | 4.246 | .000 | .354 | 31.4% | 6 | .990 | .944 | .837 | .674 | .487 | .314 | | | | |
| 4.600 | 4.428 | .000 | .172 | 18.2% | 7 | .990 | .944 | .837 | .674 | .487 | .314 | .182 | | | |
| 4.600 | 4.523 | .000 | .077 | 9.5% | 8 | .990 | .944 | .837 | .674 | .487 | .314 | .182 | .095 | | |
| 4.600 | 4.568 | .000 | .032 | 4.5% | 9 | .990 | .944 | .837 | .674 | .487 | .314 | .182 | .095 | .045 | |
| 4.600 | 4.588 | .000 | .012 | 2.0% | 10 | .990 | .944 | .837 | .674 | .487 | .314 | .182 | .095 | .045 | .020 |
| 4.600 | 4.596 | .000 | .004 | .8% | 11 | .990 | .944 | .837 | .674 | .487 | .314 | .182 | .095 | .045 | .020 |
| | | | | | | .008 | | | | | | | | | |
| 4.800 | 4.380 | .000 | .420 | 34.9% | 6 | .992 | .952 | .857 | .706 | .524 | .349 | | | | |
| 4.800 | 4.589 | .000 | .211 | 20.9% | 7 | .992 | .952 | .857 | .706 | .524 | .349 | .209 | | | |
| 4.800 | 4.703 | .000 | .097 | 11.3% | 8 | .992 | .952 | .857 | .706 | .524 | .349 | .209 | .113 | | |
| 4.800 | 4.758 | .000 | .042 | 5.6% | 9 | .992 | .952 | .857 | .706 | .524 | .349 | .209 | .113 | .056 | |
| 4.800 | 4.783 | .000 | .017 | 2.5% | 10 | .992 | .952 | .857 | .706 | .524 | .349 | .209 | .113 | .056 | .025 |
| 4.800 | 4.794 | .000 | .006 | 1.0% | 11 | .992 | .952 | .857 | .706 | .524 | .349 | .209 | .113 | .056 | .025 |
| | | | | | | .010 | | | | | | | | | |

*NOT RECOMMENDED FOR BLOCKAGE MORE THAN 5%, ALSO UNDERSTATES MEASURED TIME USAGE

## *POISSON TABLE: BLOCKAGE REPRESENTS ALL TRUNKS BUSY

| HOURS FIRST ATTEMPT VMR = 1.0 | HOURS CON-NECT | HOURS RECALL /RETRY | HOURS OVFL | PERCENT BLOCKED (GRADE) | LINE | HOURS CONNECTED TRAFFIC ON LINES | | | | | | | | | |
|---|---|---|---|---|---|---|---|---|---|---|---|---|---|---|---|
| | | | | | | 1 | 2 | 3 | 4 | 5 | 6 | 7 | 8 | 9 | 10 |
| 5.000 | 4.507 | .000 | .493 | 38.4% | 6 | .993 | .960 | .875 | .735 | .560 | .384 | | | | |
| 5.000 | 4.745 | .000 | .255 | 23.8% | 7 | .993 | .960 | .875 | .735 | .560 | .384 | .238 | | | |
| 5.000 | 4.878 | .000 | .122 | 13.3% | 8 | .993 | .960 | .875 | .735 | .560 | .384 | .238 | .133 | | |
| 5.000 | 4.946 | .000 | .054 | 6.8% | 9 | .993 | .960 | .875 | .735 | .560 | .384 | .238 | .133 | .068 | |
| 5.000 | 4.978 | .000 | .022 | 3.2% | 10 | .993 | .960 | .875 | .735 | .560 | .384 | .238 | .133 | .068 | .032 |
| 5.000 | 4.992 | .000 | .008 | 1.4% | 11 | .993 | .960 | .875 | .735 | .560 | .384 | .238 | .133 | .068 | .032 |
| | | | | | | .014 | | | | | | | | | |
| 5.000 | 4.997 | .000 | .003 | .5% | 12 | .993 | .960 | .875 | .735 | .560 | .384 | .238 | .133 | .068 | .032 |
| | | | | | | .014 | .005 | | | | | | | | |
| 5.500 | 4.793 | .000 | .707 | 47.1% | 6 | .996 | .973 | .912 | .798 | .642 | .471 | | | | |
| 5.500 | 5.107 | .000 | .393 | 31.4% | 7 | .996 | .973 | .912 | .798 | .642 | .471 | .314 | | | |
| 5.500 | 5.297 | .000 | .203 | 19.1% | 8 | .996 | .973 | .912 | .798 | .642 | .471 | .314 | .191 | | |
| 5.500 | 5.403 | .000 | .097 | 10.6% | 9 | .996 | .973 | .912 | .798 | .642 | .471 | .314 | .191 | .106 | |
| 5.500 | 5.457 | .000 | .043 | 5.4% | 10 | .996 | .973 | .912 | .798 | .642 | .471 | .314 | .191 | .106 | .054 |
| 5.500 | 5.482 | .000 | .018 | 2.5% | 11 | .996 | .973 | .912 | .798 | .642 | .471 | .314 | .191 | .106 | .054 |
| | | | | | | .025 | | | | | | | | | |
| 5.500 | 5.493 | .000 | .007 | 1.1% | 12 | .996 | .973 | .912 | .798 | .642 | .471 | .314 | .191 | .106 | .054 |
| | | | | | | .025 | .011 | | | | | | | | |
| 6.000 | 5.430 | .000 | .570 | 39.4% | 7 | .998 | .983 | .849 | .715 | .554 | .394 | | | | |
| 6.000 | 5.686 | .000 | .314 | 25.6% | 8 | .998 | .983 | .938 | .849 | .715 | .554 | .394 | .256 | | |
| 6.000 | 5.839 | .000 | .161 | 15.3% | 9 | .998 | .983 | .849 | .715 | .554 | .394 | .256 | .153 | | |
| 6.000 | 5.923 | .000 | .077 | 8.4% | 10 | .998 | .983 | .938 | .849 | .715 | .554 | .394 | .256 | .153 | .084 |
| 6.000 | 5.965 | .000 | .035 | 4.3% | 11 | .998 | .983 | .938 | .849 | .715 | .554 | .394 | .256 | .153 | .084 |
| | | | | | | .043 | | | | | | | | | |
| 6.000 | 5.985 | .000 | .015 | 2.0% | 12 | .998 | .983 | .938 | .849 | .715 | .554 | .394 | .256 | .153 | .084 |
| | | | | | | .043 | .020 | | | | | | | | |
| 6.000 | 5.994 | .000 | .006 | .9% | 13 | .998 | .983 | .938 | .849 | .715 | .554 | .394 | .256 | .153 | .084 |
| | | | | | | .043 | .020 | .009 | | | | | | | |
| 6.500 | 5.713 | .000 | .787 | 47.3% | 7 | .998 | .989 | .957 | .888 | .776 | .631 | .473 | | | |
| 6.500 | 6.040 | .000 | .460 | 32.7% | 8 | .998 | .989 | .957 | .888 | .776 | .631 | .473 | .327 | | |
| 6.500 | 6.249 | .000 | .251 | 20.8% | 9 | .998 | .989 | .957 | .888 | .776 | .631 | .473 | .327 | .208 | |
| 6.500 | 6.371 | .000 | .129 | 12.3% | 10 | .998 | .989 | .957 | .888 | .776 | .631 | .473 | .327 | .208 | .123 |
| 6.500 | 6.438 | .000 | .062 | 6.7% | 11 | .998 | .989 | .957 | .888 | .776 | .631 | .473 | .327 | .208 | .123 |
| | | | | | | .067 | | | | | | | | | |
| 6.500 | 6.472 | .000 | .028 | 3.4% | 12 | .998 | .989 | .957 | .888 | .776 | .631 | .473 | .327 | .208 | .123 |
| | | | | | | .067 | .034 | | | | | | | | |
| 6.500 | 6.488 | .000 | .012 | 1.6% | 13 | .998 | .989 | .957 | .888 | .776 | .631 | .473 | .327 | .208 | .123 |
| | | | | | | .067 | .034 | .016 | | | | | | | |
| 6.500 | 6.495 | .000 | .005 | .7% | 14 | .998 | .989 | .957 | .888 | .776 | .631 | .473 | .327 | .208 | .123 |
| | | | | | | .067 | .034 | .016 | .007 | | | | | | |

*NOT RECOMMENDED FOR BLOCKAGE MORE THAN 5%, ALSO UNDERSTATES MEASURED TIME USAGE

*POISSON TABLE: BLOCKAGE REPRESENTS ALL TRUNKS BUSY

| HOURS FIRST ATTEMPT VMR = 1.0 | HOURS CON-NECT | HOURS RECALL /RETRY | HOURS OVFL | PERCENT BLOCKED (GRADE) | LINE | HOURS CONNECTED TRAFFIC ON LINES 1 | 2 | 3 | 4 | 5 | 6 | 7 | 8 | 9 | 10 |
|---|---|---|---|---|---|---|---|---|---|---|---|---|---|---|---|
| 7.000 | 6.358 | .000 | .642 | 40.1% | 8 | .999 | .993 | .970 | .918 | .827 | .699 | .550 | .401 | | |
| 7.000 | 6.629 | .000 | .371 | 27.1% | 9 | .999 | .993 | .970 | .918 | .827 | .699 | .550 | .401 | .271 | |
| 7.000 | 6.799 | .000 | .201 | 17.0% | 10 | .999 | .993 | .970 | .918 | .827 | .699 | .550 | .401 | .271 | .170 |
| 7.000 | 6.897 | .000 | .103 | 9.9% | 11 | .999 | .993 | .970 | .918 | .827 | .699 | .550 | .401 | .271 | .170 |
| | | | | | | .099 | | | | | | | | | |
| 7.000 | 6.951 | .000 | .049 | 5.3% | 12 | .999 | .993 | .970 | .918 | .827 | .699 | .550 | .401 | .271 | .170 |
| | | | | | | .099 | .053 | | | | | | | | |
| 7.000 | 6.978 | .000 | .022 | 2.7% | 13 | .999 | .993 | .970 | .918 | .827 | .699 | .550 | .401 | .271 | .170 |
| | | | | | | .099 | .053 | .027 | | | | | | | |
| 7.000 | 6.990 | .000 | .010 | 1.3% | 14 | .999 | .993 | .970 | .918 | .827 | .699 | .550 | .401 | .271 | .170 |
| | | | | | | .099 | .053 | .027 | .013 | | | | | | |
| 7.000 | 6.996 | .000 | .004 | .6% | 15 | .999 | .993 | .970 | .918 | .827 | .699 | .550 | .401 | .271 | .170 |
| | | | | | | .099 | .053 | .027 | .013 | .006 | | | | | |
| 7.500 | 6.639 | .000 | .861 | 47.5% | 8 | .999 | .995 | .980 | .941 | .868 | .759 | .622 | .475 | | |
| 7.500 | 6.977 | .000 | .523 | 33.8% | 9 | .999 | .995 | .980 | .941 | .868 | .759 | .622 | .475 | .338 | |
| 7.500 | 7.201 | .000 | .299 | 22.4% | 10 | .999 | .995 | .980 | .941 | .868 | .759 | .622 | .475 | .338 | .224 |
| 7.500 | 7.338 | .000 | .162 | 13.8% | 11 | .999 | .995 | .980 | .941 | .868 | .759 | .622 | .475 | .338 | .224 |
| | | | | | | .138 | | | | | | | | | |
| 7.500 | 7.418 | .000 | .082 | 7.9% | 12 | .999 | .995 | .980 | .941 | .868 | .759 | .622 | .475 | .338 | .224 |
| | | | | | | .138 | .079 | | | | | | | | |
| 7.500 | 7.460 | .000 | .040 | 4.3% | 13 | .999 | .995 | .980 | .941 | .868 | .759 | .622 | .475 | .338 | .224 |
| | | | | | | .138 | .079 | .043 | | | | | | | |
| 7.500 | 7.482 | .000 | .018 | 2.2% | 14 | .999 | .995 | .980 | .941 | .868 | .759 | .622 | .475 | .338 | .224 |
| | | | | | | .138 | .079 | .043 | .022 | | | | | | |
| 7.500 | 7.492 | .000 | .008 | 1.0% | 15 | .999 | .995 | .980 | .941 | .868 | .759 | .622 | .475 | .338 | .224 |
| | | | | | | .138 | .079 | .043 | .022 | .010 | | | | | |
| 8.000 | 7.291 | .000 | .709 | 40.7% | 9 | .999 | .997 | .986 | .958 | .900 | .809 | .687 | .547 | .407 | |
| 8.000 | 7.574 | .000 | .426 | 28.3% | 10 | .999 | .997 | .986 | .958 | .900 | .809 | .687 | .547 | .407 | .283 |
| 8.000 | 7.758 | .000 | .242 | 18.4% | 11 | .999 | .997 | .986 | .958 | .900 | .809 | .687 | .547 | .407 | .283 |
| | | | | | | .184 | | | | | | | | | |
| 8.000 | 7.870 | .000 | .130 | 11.2% | 12 | .999 | .997 | .986 | .958 | .900 | .809 | .687 | .547 | .407 | .283 |
| | | | | | | .184 | .112 | | | | | | | | |
| 8.000 | 7.934 | .000 | .066 | 6.4% | 13 | .999 | .997 | .986 | .958 | .900 | .809 | .687 | .547 | .407 | .283 |
| | | | | | | .184 | .112 | .064 | | | | | | | |
| 8.000 | 7.968 | .000 | .032 | 3.4% | 14 | .000 | .007 | .086 | .958 | .900 | .009 | .087 | .547 | .407 | .283 |
| | | | | | | .184 | .112 | .064 | .034 | | | | | | |
| 8.000 | 7.985 | .000 | .015 | 1.7% | 15 | .999 | .997 | .986 | .958 | .900 | .809 | .687 | .547 | .407 | .283 |
| | | | | | | .184 | .112 | .064 | .034 | .017 | | | | | |
| 8.000 | 7.994 | .000 | .006 | .8% | 16 | .999 | .997 | .986 | .958 | .900 | .809 | .687 | .547 | .407 | .283 |
| | | | | | | .184 | .112 | .064 | .034 | .017 | .008 | | | | |

*NOT RECOMMENDED FOR BLOCKAGE MORE THAN 5%, ALSO UNDERSTATES MEASURED TIME USAGE

*POISSON TABLE: BLOCKAGE REPRESENTS ALL TRUNKS BUSY

| HOURS FIRST ATTEMPT VMR=1.0 | HOURS CON-NECT | HOURS RECALL /RETRY | HOURS OVFL | PERCENT BLOCKED (GRADE) | LINE | HOURS CONNECTED TRAFFIC ON LINES 1 | 2 | 3 | 4 | 5 | 6 | 7 | 8 | 9 | 10 |
|---|---|---|---|---|---|---|---|---|---|---|---|---|---|---|---|
| 8.500 | 7.570 | .000 | .930 | 47.7% | 9 | .999 | .998 | .991 | .970 | .926 | .850 | .744 | .614 | .477 | |
| 8.500 | 7.917 | .000 | .583 | 34.7% | 10 | .999 | .998 | .991 | .970 | .926 | .850 | .744 | .614 | .477 | .347 |
| 8.500 | 8.153 | .000 | .347 | 23.7% | 11 | .999 | .998 | .991 | .970 | .926 | .850 | .744 | .614 | .477 | .347 |
| | | | | | | .237 | | | | | | | | | |
| 8.500 | 8.305 | .000 | .195 | 15.1% | 12 | .999 | .998 | .991 | .970 | .926 | .850 | .744 | .614 | .477 | .347 |
| | | | | | | .237 | .151 | | | | | | | | |
| 8.500 | 8.396 | .000 | .104 | 9.1% | 13 | .999 | .998 | .991 | .970 | .926 | .850 | .744 | .614 | .477 | .347 |
| | | | | | | .237 | .151 | .091 | | | | | | | |
| 8.500 | 8.447 | .000 | .053 | 5.1% | 14 | .999 | .998 | .991 | .970 | .926 | .850 | .744 | .614 | .477 | .347 |
| | | | | | | .237 | .151 | .091 | .051 | | | | | | |
| 8.500 | 8.474 | .000 | .026 | 2.7% | 15 | .999 | .998 | .991 | .970 | .926 | .850 | .744 | .614 | .477 | .347 |
| | | | | | | .237 | .151 | .091 | .051 | .027 | | | | | |
| 8.500 | 8.488 | .000 | .012 | 1.4% | 16 | .999 | .998 | .991 | .970 | .926 | .850 | .744 | .614 | .477 | .347 |
| | | | | | | .237 | .151 | .091 | .051 | .027 | .014 | | | | |
| 8.500 | 8.495 | .000 | .005 | .7% | 17 | .999 | .998 | .991 | .970 | .926 | .850 | .744 | .614 | .477 | .347 |
| | | | | | | .237 | .151 | .091 | .051 | .027 | .014 | .007 | | | |
| 9.000 | 8.227 | .000 | .773 | 41.3% | 10 | .999 | .999 | .994 | .979 | .945 | .884 | .793 | .676 | .544 | .413 |
| 9.000 | 8.521 | .000 | .479 | 29.4% | 11 | .999 | .999 | .994 | .979 | .945 | .884 | .793 | .676 | .544 | .413 |
| | | | | | | .294 | | | | | | | | | |
| 9.000 | 8.718 | .000 | .282 | 19.7% | 12 | .999 | .999 | .994 | .979 | .945 | .884 | .793 | .676 | .544 | .413 |
| | | | | | | .294 | .197 | | | | | | | | |
| 9.000 | 8.842 | .000 | .158 | 12.4% | 13 | .999 | .999 | .994 | .979 | .945 | .884 | .793 | .676 | .544 | .413 |
| | | | | | | .294 | .197 | .124 | | | | | | | |
| 9.000 | 8.916 | .000 | .084 | 7.4% | 14 | .999 | .999 | .994 | .979 | .945 | .884 | .793 | .676 | .544 | .413 |
| | | | | | | .294 | .197 | .124 | .074 | | | | | | |
| 9.000 | 8.957 | .000 | .043 | 4.1% | 15 | .999 | .999 | .994 | .979 | .945 | .884 | .793 | .676 | .544 | .413 |
| | | | | | | .294 | .197 | .124 | .074 | .041 | | | | | |
| 9.000 | 8.979 | .000 | .021 | 2.2% | 16 | .999 | .999 | .994 | .979 | .945 | .884 | .793 | .676 | .544 | .413 |
| | | | | | | .294 | .197 | .124 | .074 | .041 | .022 | | | | |
| 9.000 | 8.990 | .000 | .010 | 1.1% | 17 | .999 | .999 | .994 | .979 | .945 | .884 | .793 | .676 | .544 | .413 |
| | | | | | | .294 | .197 | .124 | .074 | .041 | .022 | .011 | | | |
| 9.000 | 8.996 | .000 | .004 | .5% | 18 | .999 | .999 | .994 | .979 | .945 | .884 | .793 | .676 | .544 | .413 |
| | | | | | | .294 | .197 | .124 | .074 | .041 | .022 | .011 | .005 | | |
| 9.500 | 8.504 | .000 | .996 | 47.8% | 10 | .999 | .999 | .996 | .985 | .960 | .911 | .835 | .731 | .608 | .478 |
| 9.500 | 8.859 | .000 | .641 | 35.5% | 11 | .999 | .999 | .996 | .985 | .960 | .911 | .835 | .731 | .608 | .478 |
| | | | | | | .355 | | | | | | | | | |
| 9.500 | 9.107 | .000 | .393 | 24.8% | 12 | .999 | .999 | .996 | .985 | .960 | .911 | .835 | .731 | .608 | .478 |
| | | | | | | .355 | .248 | | | | | | | | |
| 9.500 | 9.270 | .000 | .230 | 16.4% | 13 | .999 | .999 | .996 | .985 | .960 | .911 | .835 | .731 | .608 | .478 |
| | | | | | | .355 | .248 | .164 | | | | | | | |

*NOT RECOMMENDED FOR BLOCKAGE MORE THAN 5%, ALSO UNDERSTATES MEASURED TIME USAGE

*POISSON TABLE: BLOCKAGE REPRESENTS ALL TRUNKS BUSY

| HOURS FIRST ATTEMPT VMR=1.0 | HOURS CON-NECT | HOURS RECALL /RETRY | HOURS OVFL | PERCENT BLOCKED (GRADE) | LINE | HOURS CONNECTED TRAFFIC ON LINES 1 | 2 | 3 | 4 | 5 | 6 | 7 | 8 | 9 | 10 |
|---|---|---|---|---|---|---|---|---|---|---|---|---|---|---|---|
| 9.500 | 9.372 | .000 | .128 | 10.2% | 14 | .999 | .999 | .996 | .985 | .960 | .911 | .835 | .731 | .608 | .478 |
| | | | | | | .355 | .248 | .164 | .102 | | | | | | |
| 9.500 | 9.432 | .000 | .068 | 6.0% | 15 | .999 | .999 | .996 | .985 | .960 | .911 | .835 | .731 | .608 | .478 |
| | | | | | | .355 | .248 | .164 | .102 | .060 | | | | | |
| 9.500 | 9.466 | .000 | .034 | 3.3% | 16 | .999 | .999 | .996 | .985 | .960 | .911 | .835 | .731 | .608 | .478 |
| | | | | | | .355 | .248 | .164 | .102 | .060 | .033 | | | | |
| 9.500 | 9.483 | .000 | .017 | 1.8% | 17 | .999 | .999 | .996 | .985 | .960 | .911 | .835 | .731 | .608 | .478 |
| | | | | | | .355 | .248 | .164 | .102 | .060 | .033 | .018 | | | |
| 9.500 | 9.492 | .000 | .008 | .9% | 18 | .999 | .999 | .996 | .985 | .960 | .911 | .835 | .731 | .608 | .478 |
| | | | | | | .355 | .248 | .164 | .102 | .060 | .033 | .018 | .009 | | |
| 10.000 | 9.166 | .000 | .834 | 41.7% | 11 | .999 | .999 | .997 | .990 | .971 | .933 | .870 | .780 | .667 | .542 |
| | | | | | | .417 | | | | | | | | | |
| 10.000 | 9.469 | .000 | .531 | 30.3% | 12 | .999 | .999 | .997 | .990 | .971 | .933 | .870 | .780 | .667 | .542 |
| | | | | | | .417 | .303 | | | | | | | | |
| 10.000 | 9.678 | .000 | .322 | 20.8% | 13 | .999 | .999 | .997 | .990 | .971 | .933 | .870 | .780 | .667 | .542 |
| | | | | | | .417 | .303 | .208 | | | | | | | |
| 10.000 | 9.813 | .000 | .187 | 13.6% | 14 | .999 | .999 | .997 | .990 | .971 | .933 | .870 | .780 | .667 | .542 |
| | | | | | | .417 | .303 | .208 | .136 | | | | | | |
| 10.000 | 9.897 | .000 | .103 | 8.3% | 15 | .999 | .999 | .997 | .990 | .971 | .933 | .870 | .780 | .667 | .542 |
| | | | | | | .417 | .303 | .208 | .136 | .083 | | | | | |
| 10.000 | 9.945 | .000 | .055 | 4.9% | 16 | .999 | .999 | .997 | .990 | .971 | .933 | .870 | .780 | .667 | .542 |
| | | | | | | .417 | .303 | .208 | .136 | .083 | .049 | | | | |
| 10.000 | 9.972 | .000 | .028 | 2.7% | 17 | .999 | .999 | .997 | .990 | .971 | .933 | .870 | .780 | .667 | .542 |
| | | | | | | .417 | .303 | .208 | .136 | .083 | .049 | .027 | | | |
| 10.000 | 9.987 | .000 | .013 | 1.4% | 18 | .999 | .999 | .997 | .990 | .971 | .933 | .870 | .780 | .667 | .542 |
| | | | | | | .417 | .303 | .208 | .136 | .083 | .049 | .027 | .014 | | |
| 10.000 | 9.994 | .000 | .006 | .7% | 19 | .999 | .999 | .997 | .990 | .971 | .933 | .870 | .780 | .667 | .542 |
| | | | | | | .417 | .303 | .208 | .136 | .083 | .049 | .027 | .014 | .007 | |
| 11.000 | 10.108 | .000 | .892 | 42.1% | 12 | .999 | .999 | .999 | .995 | .985 | .962 | .921 | .857 | .768 | .659 |
| | | | | | | .540 | .421 | | | | | | | | |
| 11.000 | 0.419 | .000 | .581 | 31.1% | 13 | .999 | .999 | .999 | .995 | .985 | .962 | .921 | .857 | .768 | .659 |
| | | | | | | .540 | .421 | .311 | | | | | | | |
| 11.000 | 10.638 | .000 | .362 | 21.9% | 14 | .999 | .999 | .999 | .995 | .985 | .962 | .921 | .857 | .768 | .659 |
| | | | | | | .540 | .421 | .311 | .219 | | | | | | |
| 11.000 | 10.784 | .000 | .216 | 14.6% | 15 | .999 | .999 | .999 | .995 | .985 | .962 | .921 | .857 | .768 | .659 |
| | | | | | | .540 | .421 | .311 | .219 | .146 | | | | | |
| 11.000 | 10.876 | .000 | .124 | 9.3% | 16 | .999 | .999 | .999 | .995 | .985 | .962 | .921 | .857 | .768 | .659 |
| | | | | | | .540 | .421 | .311 | .219 | .146 | .093 | | | | |
| 11.000 | 10.932 | .000 | .068 | 5.6% | 17 | .999 | .999 | .999 | .995 | .985 | .962 | .921 | .857 | .768 | .659 |
| | | | | | | .540 | .421 | .311 | .219 | .146 | .093 | .056 | | | |

*NOT RECOMMENDED FOR BLOCKAGE MORE THAN 5%, ALSO UNDERSTATES MEASURED TIME USAGE

*POISSON TABLE: BLOCKAGE REPRESENTS ALL TRUNKS BUSY

| HOURS FIRST ATTEMPT VMR = 1.0 | HOURS CON-NECT | HOURS RECALL /RETRY | HOURS OVFL | PERCENT BLOCKED (GRADE) | LINE | HOURS CONNECTED TRAFFIC ON LINES 1 | 2 | 3 | 4 | 5 | 6 | 7 | 8 | 9 | 10 |
|---|---|---|---|---|---|---|---|---|---|---|---|---|---|---|---|
| 11.000 | 10.964 | .000 | .036 | 3.2% | 18 | .999 | .999 | .999 | .995 | .985 | .962 | .921 | .857 | .768 | .659 |
| | | | | | | .540 | .421 | .311 | .219 | .146 | .093 | .056 | .032 | | |
| 11.000 | 10.982 | .000 | .018 | 1.8% | 19 | .999 | .999 | .999 | .995 | .985 | .962 | .921 | .857 | .768 | .659 |
| | | | | | | .540 | .421 | .311 | .219 | .146 | .093 | .056 | .032 | .018 | |
| 11.000 | 10.991 | .000 | .009 | .9% | 20 | .999 | .999 | .999 | .995 | .985 | .962 | .921 | .857 | .768 | .659 |
| | | | | | | .540 | .421 | .311 | .219 | .146 | .093 | .056 | .032 | .018 | .009 |
| 12.000 | 11.052 | .000 | .948 | 42.4% | 13 | .999 | .999 | .999 | .998 | .992 | .980 | .954 | .910 | .845 | .758 |
| | | | | | | .653 | .538 | .424 | | | | | | | |
| 12.000 | 11.370 | .000 | .630 | 31.8% | 14 | .999 | .999 | .999 | .998 | .992 | .980 | .954 | .910 | .845 | .758 |
| | | | | | | .653 | .538 | .424 | .318 | | | | | | |
| 12.000 | 11.598 | .000 | .402 | 22.8% | 15 | .999 | .999 | .999 | .998 | .992 | .980 | .954 | .910 | .845 | .758 |
| | | | | | | .653 | .538 | .424 | .318 | .228 | | | | | |
| 12.000 | 11.754 | .000 | .246 | 15.6% | 16 | .999 | .999 | .999 | .998 | .992 | .980 | .954 | .910 | .845 | .758 |
| | | | | | | .653 | .538 | .424 | .318 | .228 | .156 | | | | |
| 12.000 | 11.855 | .000 | .145 | 10.1% | 17 | .999 | .999 | .999 | .998 | .992 | .980 | .954 | .910 | .845 | .758 |
| | | | | | | .653 | .538 | .424 | .318 | .228 | .156 | .101 | | | |
| 12.000 | 11.918 | .000 | .082 | 6.3% | 18 | .999 | .999 | .999 | .998 | .992 | .980 | .954 | .910 | .845 | .758 |
| | | | | | | .653 | .538 | .424 | .318 | .228 | .156 | .101 | .063 | | |
| 12.000 | 11.955 | .000 | .045 | 3.7% | 19 | .999 | .999 | .999 | .998 | .992 | .980 | .954 | .910 | .845 | .758 |
| | | | | | | .653 | .538 | .424 | .318 | .228 | .156 | .101 | .063 | .037 | |
| 12.000 | 11.977 | .000 | .023 | 2.1% | 20 | .999 | .999 | .999 | .998 | .992 | .980 | .954 | .910 | .845 | .758 |
| | | | | | | .653 | .538 | .424 | .318 | .228 | .156 | .101 | .063 | .037 | .021 |
| 12.000 | 11.988 | .000 | .012 | 1.2% | 21 | .999 | .999 | .999 | .998 | .992 | .980 | .954 | .910 | .845 | .758 |
| | | | | | | .653 | .538 | .424 | .318 | .228 | .156 | .101 | .063 | .037 | .021 |
| | | | | | | .012 | | | | | | | | | |
| 12.000 | 11.994 | .000 | .006 | .6% | 22 | .999 | .999 | .999 | .998 | .992 | .980 | .954 | .910 | .845 | .758 |
| | | | | | | .653 | .538 | .424 | .318 | .228 | .156 | .101 | .063 | .037 | .021 |
| | | | | | | .012 | .006 | | | | | | | | |
| 13.000 | 11.998 | .000 | 1.002 | 42.7% | 14 | .999 | .999 | .999 | .999 | .996 | .989 | .974 | .946 | .900 | .834 |
| | | | | | | .748 | .647 | .537 | .427 | | | | | | |
| 13.000 | 12.323 | .000 | .677 | 32.5% | 15 | .999 | .999 | .999 | .999 | .996 | .989 | .974 | .946 | .900 | .834 |
| | | | | | | .748 | .647 | .537 | .427 | .325 | | | | | |
| 13.000 | 12.559 | .000 | .441 | 23.6% | 16 | .999 | .999 | .999 | .999 | .996 | .989 | .974 | .946 | .900 | .834 |
| | | | | | | .748 | .647 | .537 | .427 | .325 | .236 | | | | |
| 13.000 | 12.724 | .000 | .276 | 16.5% | 17 | .999 | .999 | .999 | .999 | .996 | .989 | .974 | .946 | .900 | .834 |
| | | | | | | .748 | .647 | .537 | .427 | .325 | .236 | .165 | | | |
| 13.000 | 12.833 | .000 | .167 | 11.0% | 18 | .999 | .999 | .999 | .999 | .996 | .989 | .974 | .946 | .900 | .834 |
| | | | | | | .748 | .647 | .537 | .427 | .325 | .236 | .165 | .110 | | |
| 13.000 | 12.903 | .000 | .097 | 7.0% | 19 | .999 | .999 | .999 | .999 | .996 | .989 | .974 | .946 | .900 | .834 |
| | | | | | | .748 | .647 | .537 | .427 | .325 | .236 | .165 | .110 | .070 | |

*NOT RECOMMENDED FOR BLOCKAGE MORE THAN 5%, ALSO UNDERSTATES MEASURED TIME USAGE

*POISSON TABLE: BLOCKAGE REPRESENTS ALL TRUNKS BUSY

| HOURS FIRST ATTEMPT VMR = 1.0 | HOURS CON-NECT | HOURS RECALL /RETRY | HOURS OVFL | PERCENT BLOCKED (GRADE) | LINE | HOURS CONNECTED TRAFFIC ON LINES | | | | | | | | | |
|---|---|---|---|---|---|---|---|---|---|---|---|---|---|---|---|
| | | | | | | 1 | 2 | 3 | 4 | 5 | 6 | 7 | 8 | 9 | 10 |
| 13.000 | 12.946 | .000 | .054 | 4.3% | 20 | .999 | .999 | .999 | .999 | .996 | .989 | .974 | .946 | .900 | .834 |
| | | | | | | .748 | .647 | .537 | .427 | .325 | .236 | .165 | .110 | .070 | .043 |
| 13.000 | 12.971 | .000 | .029 | 2.5% | 21 | .999 | .999 | .999 | .999 | .996 | .989 | .974 | .946 | .900 | .834 |
| | | | | | | .748 | .647 | .537 | .427 | .325 | .236 | .165 | .110 | .070 | .043 |
| | | | | | | .025 | | | | | | | | | |
| 13.000 | 12.985 | .000 | .015 | 1.4% | 22 | .999 | .999 | .999 | .999 | .996 | .989 | .974 | .946 | .900 | .834 |
| | | | | | | .748 | .647 | .537 | .427 | .325 | .236 | .165 | .110 | .070 | .043 |
| | | | | | | .025 | .014 | | | | | | | | |
| 13.000 | 12.992 | .000 | .008 | .8% | 23 | .999 | .999 | .999 | .999 | .996 | .989 | .974 | .946 | .900 | .834 |
| | | | | | | .748 | .647 | .537 | .427 | .325 | .236 | .165 | .110 | .070 | .043 |
| | | | | | | .025 | .014 | .008 | | | | | | | |
| 14.000 | 12.946 | .000 | 1.054 | 43.0% | 15 | .999 | .999 | .999 | .999 | .998 | .994 | .986 | .968 | .938 | .891 |
| | | | | | | .824 | .740 | .642 | .536 | .430 | | | | | |
| 14.000 | 13.276 | .000 | .724 | 33.1% | 16 | .999 | .999 | .999 | .999 | .998 | .994 | .986 | .968 | .938 | .891 |
| | | | | | | .824 | .740 | .642 | .536 | .430 | .331 | | | | |
| 14.000 | 13.520 | .000 | .480 | 24.4% | 17 | .999 | .999 | .999 | .999 | .998 | .994 | .986 | .968 | .938 | .891 |
| | | | | | | .824 | .740 | .642 | .536 | .430 | .331 | .244 | | | |
| 14.000 | 13.693 | .000 | .307 | 17.3% | 18 | .999 | .999 | .999 | .999 | .998 | .994 | .986 | .968 | .938 | .891 |
| | | | | | | .824 | .740 | .642 | .536 | .430 | .331 | .244 | .173 | | |
| 14.000 | 13.811 | .000 | .189 | 11.7% | 19 | .999 | .999 | .999 | .999 | .998 | .994 | .986 | .968 | .938 | .891 |
| | | | | | | .824 | .740 | .642 | .536 | .430 | .331 | .244 | .173 | .117 | |
| 14.000 | 13.887 | .000 | .113 | 7.7% | 20 | .999 | .999 | .999 | .999 | .998 | .994 | .986 | .968 | .938 | .891 |
| | | | | | | .824 | .740 | .642 | .536 | .430 | .331 | .244 | .173 | .117 | .077 |
| 14.000 | 13.935 | .000 | .065 | 4.8% | 21 | .999 | .999 | .999 | .999 | .998 | .994 | .986 | .968 | .938 | .891 |
| | | | | | | .824 | .740 | .642 | .536 | .430 | .331 | .244 | .173 | .117 | .077 |
| | | | | | | .048 | | | | | | | | | |
| 14.000 | 13.964 | .000 | .036 | 2.9% | 22 | .999 | .999 | .999 | .999 | .998 | .994 | .986 | .968 | .938 | .891 |
| | | | | | | .824 | .740 | .642 | .536 | .430 | .331 | .244 | .173 | .117 | .077 |
| | | | | | | .048 | .029 | | | | | | | | |
| 14.000 | 13.981 | .000 | .019 | 1.7% | 23 | .999 | .999 | .999 | .999 | .998 | .994 | .986 | .968 | .938 | .891 |
| | | | | | | .824 | .740 | .642 | .536 | .430 | .331 | .244 | .173 | .117 | .077 |
| | | | | | | .048 | .029 | .017 | | | | | | | |
| 14.000 | 13.990 | .000 | .010 | .9% | 24 | .999 | .999 | .999 | .999 | .998 | .994 | .986 | .968 | .938 | .891 |
| | | | | | | .824 | .740 | .642 | .536 | .430 | .331 | .244 | .173 | .117 | .077 |
| | | | | | | .048 | .029 | .017 | .009 | | | | | | |
| 14.000 | 13.995 | .000 | .005 | .5% | 25 | .999 | .999 | .999 | .999 | .998 | .994 | .986 | .968 | .938 | .891 |
| | | | | | | .824 | .740 | .642 | .536 | .430 | .331 | .244 | .173 | .117 | .077 |
| | | | | | | .048 | .029 | .017 | .009 | .005 | | | | | |
| 15.000 | 13.895 | .000 | 1.105 | 43.2% | 16 | .999 | .999 | .999 | .999 | .999 | .997 | .992 | .982 | .963 | .930 |
| | | | | | | .882 | .816 | .732 | .637 | .534 | .432 | | | | |

*NOT RECOMMENDED FOR BLOCKAGE MORE THAN 5%, ALSO UNDERSTATES MEASURED TIME USAGE

*POISSON TABLE: BLOCKAGE REPRESENTS ALL TRUNKS BUSY

| HOURS FIRST ATTEMPT VMR = 1.0 | HOURS CON-NECT | HOURS RECALL /RETRY | HOURS OVFL | PERCENT BLOCKED (GRADE) | LINE | HOURS CONNECTED TRAFFIC ON LINES 1 | 2 | 3 | 4 | 5 | 6 | 7 | 8 | 9 | 10 |
|---|---|---|---|---|---|---|---|---|---|---|---|---|---|---|---|
| 15.000 | 14.231 | .000 | .769 | 33.6% | 17 | .999 | .999 | .999 | .999 | .999 | .997 | .992 | .982 | .963 | .930 |
| | | | | | | .882 | .815 | .732 | .637 | .534 | .432 | .336 | | | |
| 15.000 | 14.482 | .000 | .518 | 25.1% | 18 | .999 | .999 | .999 | .999 | .999 | .997 | .992 | .982 | .963 | .930 |
| | | | | | | .882 | .815 | .732 | .637 | .534 | .432 | .336 | .251 | | |
| 15.000 | 14.663 | .000 | .337 | 18.1% | 19 | .999 | .999 | .999 | .999 | .999 | .997 | .992 | .982 | .963 | .930 |
| | | | | | | .882 | .815 | .732 | .637 | .534 | .432 | .336 | .251 | .181 | |
| 15.000 | 14.788 | .000 | .212 | 12.5% | 20 | .999 | .999 | .999 | .999 | .999 | .997 | .992 | .982 | .963 | .930 |
| | | | | | | .882 | .815 | .732 | .637 | .534 | .432 | .336 | .251 | .181 | .125 |
| 15.000 | 14.871 | .000 | .129 | 8.3% | 21 | .999 | .999 | .999 | .999 | .999 | .997 | .992 | .982 | .963 | .930 |
| | | | | | | .882 | .815 | .732 | .637 | .534 | .432 | .336 | .251 | .181 | .125 |
| | | | | | | .083 | | | | | | | | | |
| 15.000 | 14.924 | .000 | .076 | 5.3% | 22 | .999 | .999 | .999 | .999 | .999 | .997 | .992 | .982 | .963 | .930 |
| | | | | | | .882 | .815 | .732 | .637 | .534 | .432 | .336 | .251 | .181 | .125 |
| | | | | | | .083 | .053 | | | | | | | | |
| 15.000 | 14.957 | .000 | .043 | 3.3% | 23 | .999 | .999 | .999 | .999 | .999 | .997 | .992 | .982 | .963 | .930 |
| | | | | | | .882 | .815 | .732 | .637 | .534 | .432 | .336 | .251 | .181 | .125 |
| | | | | | | .083 | .053 | .033 | | | | | | | |
| 15.000 | 14.976 | .000 | .024 | 1.9% | 24 | .999 | .999 | .999 | .999 | .999 | .997 | .992 | .982 | .963 | .930 |
| | | | | | | .882 | .815 | .732 | .637 | .534 | .432 | .336 | .251 | .181 | .125 |
| | | | | | | .083 | .053 | .033 | .019 | | | | | | |
| 15.000 | 14.987 | .000 | .013 | 1.1% | 25 | .999 | .999 | .999 | .999 | .999 | .997 | .992 | .982 | .963 | .930 |
| | | | | | | .882 | .815 | .732 | .637 | .534 | .432 | .336 | .251 | .181 | .125 |
| | | | | | | .083 | .053 | .033 | .019 | .011 | | | | | |
| 15.000 | 14.993 | .000 | .007 | .6% | 26 | .999 | .999 | .999 | .999 | .999 | .997 | .992 | .982 | .963 | .930 |
| | | | | | | .882 | .815 | .732 | .637 | .534 | .432 | .336 | .251 | .181 | .125 |
| | | | | | | .083 | .053 | .033 | .019 | .011 | .006 | | | | |
| 16.000 | 14.847 | .000 | 1.153 | 43.4% | 17 | .999 | .999 | .999 | .999 | .999 | 999 | .996 | .990 | .978 | .957 |
| | | | | | | .923 | .873 | .807 | .725 | .632 | .533 | .434 | | | |
| 16.000 | 15.187 | .000 | .813 | 34.1% | 18 | .999 | .999 | .999 | .999 | .999 | .999 | .996 | .990 | .978 | .957 |
| | | | | | | .923 | .873 | .807 | .725 | .632 | .533 | .434 | .341 | | |
| 16.000 | 15.445 | .000 | .555 | 25.8% | 19 | .999 | .999 | .999 | .999 | .999 | .999 | .996 | .990 | .978 | .957 |
| | | | | | | .923 | .873 | .807 | .725 | .632 | .533 | .434 | .341 | .258 | |
| 16.000 | 15.633 | .000 | .367 | 18.8% | 20 | .999 | .999 | .999 | .999 | .999 | .999 | .996 | .990 | .978 | .957 |
| | | | | | | .923 | .873 | .807 | .725 | .632 | .533 | .434 | .341 | .258 | .188 |
| 16.000 | 15.764 | .000 | .236 | 13.2% | 21 | .999 | .999 | .999 | .999 | .999 | .999 | .996 | .990 | .978 | .957 |
| | | | | | | .923 | .873 | .807 | .725 | .632 | .533 | .434 | .341 | .258 | .188 |
| | | | | | | .132 | | | | | | | | | |
| 16.000 | 15.854 | .000 | .146 | 8.9% | 22 | .999 | .999 | .999 | .999 | .999 | .999 | .996 | .990 | .978 | .957 |
| | | | | | | .923 | .873 | .807 | .725 | .632 | .533 | .434 | .341 | .258 | .188 |
| | | | | | | .132 | .089 | | | | | | | | |

*NOT RECOMMENDED FOR BLOCKAGE MORE THAN 5%, ALSO UNDERSTATES MEASURED TIME USAGE

*POISSON TABLE: BLOCKAGE REPRESENTS ALL TRUNKS BUSY

| HOURS FIRST ATTEMPT VMR = 1.0 | HOURS CON-NECT | HOURS RECALL /RETRY | HOURS OVFL | PERCENT BLOCKED (GRADE) | LINE | HOURS CONNECTED TRAFFIC ON LINES 1 | 2 | 3 | 4 | 5 | 6 | 7 | 8 | 9 | 10 |
|---|---|---|---|---|---|---|---|---|---|---|---|---|---|---|---|
| 16.000 | 15.912 | .000 | .088 | 5.8% | 23 | .999 | .999 | .999 | .999 | .999 | .999 | .996 | .990 | .978 | .957 |
| | | | | | | .923 | .873 | .807 | .725 | .632 | .533 | .434 | .341 | .258 | .188 |
| | | | | | | .132 | .089 | .058 | | | | | | | |
| 16.000 | 15.949 | .000 | .051 | 3.7% | 24 | .999 | .999 | .999 | .999 | .999 | .999 | .996 | .990 | .978 | .957 |
| | | | | | | .923 | .873 | .807 | .725 | .632 | .533 | .434 | .341 | .258 | .188 |
| | | | | | | .132 | .089 | .058 | .037 | | | | | | |
| 16.000 | 15.971 | .000 | .029 | 2.2% | 25 | .999 | .999 | .999 | .999 | .999 | .999 | .996 | .990 | .978 | .957 |
| | | | | | | .923 | .873 | .807 | .725 | .632 | .533 | .434 | .341 | .258 | .188 |
| | | | | | | .132 | .089 | .058 | .037 | .022 | | | | | |
| 16.000 | 15.984 | .000 | .016 | 1.3% | 26 | .999 | .999 | .999 | .999 | .999 | .999 | .996 | .990 | .978 | .957 |
| | | | | | | .923 | .873 | .807 | .725 | .632 | .533 | .434 | .341 | .258 | .188 |
| | | | | | | .132 | .089 | .058 | .037 | .022 | .013 | | | | |
| 16.000 | 15.991 | .000 | .009 | .7% | 27 | .999 | .999 | .999 | .999 | .999 | .999 | .996 | .990 | .978 | .957 |
| | | | | | | .923 | .873 | .807 | .725 | .632 | .533 | .434 | .341 | .258 | .188 |
| | | | | | | .132 | .089 | .058 | .037 | .022 | .013 | .007 | | | |
| | | | | | | | | | | | | | | | |
| 17.000 | 15.799 | .000 | 1.201 | 43.6% | 18 | .999 | .999 | .999 | .999 | .999 | .999 | .998 | .995 | .987 | .974 |
| | | | | | | .951 | .915 | .865 | .799 | .719 | .629 | .532 | .436 | | |
| 17.000 | 16.144 | .000 | .856 | 34.5% | 19 | .999 | .999 | .999 | .999 | .999 | .999 | .998 | .995 | .987 | .974 |
| | | | | | | .951 | .915 | .865 | .799 | .719 | .629 | .532 | .436 | .345 | |
| 17.000 | 16.408 | .000 | .592 | 26.4% | 20 | .999 | .999 | .999 | .999 | .999 | .999 | .998 | .995 | .987 | .974 |
| | | | | | | .951 | .915 | .865 | .799 | .719 | .629 | .532 | .436 | .345 | .264 |
| 17.000 | 16.602 | .000 | .398 | 19.5% | 21 | .999 | .999 | .999 | .999 | .999 | .999 | .998 | .995 | .987 | .974 |
| | | | | | | .951 | .915 | .865 | .799 | .719 | .629 | .532 | .436 | .345 | .264 |
| | | | | | | .195 | | | | | | | | | |
| 17.000 | 16.741 | .000 | .259 | 13.9% | 22 | .999 | .999 | .999 | .999 | .999 | .999 | .998 | .995 | .987 | .974 |
| | | | | | | .951 | .915 | .865 | .799 | .719 | .629 | .532 | .436 | .345 | .264 |
| | | | | | | .195 | .139 | | | | | | | | |
| 17.000 | 16.836 | .000 | .164 | 9.5% | 23 | .999 | .999 | .999 | .999 | .999 | .999 | .998 | .995 | .987 | .974 |
| | | | | | | .951 | .915 | .865 | .799 | .719 | .629 | .532 | .436 | .345 | .264 |
| | | | | | | .195 | .139 | .095 | | | | | | | |
| 17.000 | 16.899 | .000 | .101 | 6.3% | 24 | .999 | .999 | .999 | .999 | .999 | .999 | .998 | .995 | .987 | .974 |
| | | | | | | .951 | .915 | .865 | .799 | .719 | .629 | .532 | .436 | .345 | .264 |
| | | | | | | .195 | .139 | .095 | .063 | | | | | | |
| 17.000 | 16.940 | .000 | .060 | 4.1% | 25 | .999 | .999 | .999 | .999 | .999 | .999 | .998 | .995 | .987 | .974 |
| | | | | | | .951 | .915 | .865 | .799 | .719 | .629 | .532 | .436 | .345 | .264 |
| | | | | | | .195 | .139 | .095 | .063 | .041 | | | | | |
| 17.000 | 16.965 | .000 | .035 | 2.5% | 26 | .999 | .999 | .999 | .999 | .999 | .999 | .998 | .995 | .987 | .974 |
| | | | | | | .951 | .915 | .865 | .799 | .719 | .629 | .532 | .436 | .345 | .264 |
| | | | | | | .195 | .139 | .095 | .063 | .041 | .025 | | | | |
| 17.000 | 16.981 | .000 | .019 | 1.5% | 27 | .999 | .999 | .999 | .999 | .999 | .999 | .998 | .995 | .987 | .974 |

*NOT RECOMMENDED FOR BLOCKAGE MORE THAN 5%, ALSO UNDERSTATES MEASURED TIME USAGE

*POISSON TABLE: BLOCKAGE REPRESENTS ALL TRUNKS BUSY

| HOURS FIRST ATTEMPT VMR = 1.0 | HOURS CON-NECT | HOURS RECALL /RETRY | HOURS OVFL | PERCENT BLOCKED (GRADE) | LINE | HOURS CONNECTED TRAFFIC ON LINES 1 | 2 | 3 | 4 | 5 | 6 | 7 | 8 | 9 | 10 |
|---|---|---|---|---|---|---|---|---|---|---|---|---|---|---|---|
| | | | | | | .951 | .915 | .865 | .799 | .719 | .629 | .532 | .436 | .345 | .264 |
| | | | | | | .195 | .139 | .095 | .063 | .041 | .025 | .015 | | | |
| 17.000 | 16.989 | .000 | .011 | .9% | 28 | .999 | .999 | .999 | .999 | .999 | .999 | .998 | .995 | .987 | .974 |
| | | | | | | .195 | .139 | .095 | .063 | .041 | .025 | .015 | .009 | | |
| 18.000 | 16.753 | .000 | 1.247 | 43.8% | 19 | .999 | .999 | .999 | .999 | .999 | .999 | .999 | .997 | .993 | .985 |
| | | | | | | .970 | .945 | .908 | .857 | .792 | .713 | .625 | .531 | .438 | |
| 18.000 | 17.102 | .000 | .898 | 34.9% | 20 | .999 | .999 | .999 | .999 | .999 | .999 | .999 | .997 | .993 | .985 |
| | | | | | | .970 | .945 | .908 | .857 | .792 | .713 | .625 | .531 | .438 | .349 |
| 18.000 | 17.371 | .000 | .629 | 26.9% | 21 | .999 | .999 | .999 | .999 | .999 | .999 | .999 | .997 | .993 | .985 |
| | | | | | | .970 | .945 | .908 | .857 | .792 | .713 | .625 | .531 | .438 | .349 |
| | | | | | | .269 | | | | | | | | | |
| 18.000 | 17.572 | .000 | .428 | 20.1% | 22 | .999 | .999 | .999 | .999 | .999 | .999 | .999 | .997 | .993 | .985 |
| | | | | | | .970 | .945 | .908 | .857 | .792 | .713 | .625 | .531 | .438 | .349 |
| | | | | | | .269 | .201 | | | | | | | | |
| 18.000 | 17.717 | .000 | .283 | 14.5% | 23 | .999 | .999 | .999 | .999 | .999 | .999 | .999 | .997 | .993 | .985 |
| | | | | | | .970 | .945 | .908 | .857 | .792 | .713 | .625 | .531 | .438 | .349 |
| | | | | | | .269 | .201 | .145 | | | | | | | |
| 18.000 | 17.818 | .000 | .182 | 10.1% | 24 | .999 | .999 | .999 | .999 | .999 | .999 | .999 | .997 | .993 | .985 |
| | | | | | | .970 | .945 | .908 | .857 | .792 | .713 | .625 | .531 | .438 | .349 |
| | | | | | | .269 | .201 | .145 | .101 | | | | | | |
| 18.000 | 17.887 | .000 | .113 | 6.8% | 25 | .999 | .999 | .999 | .999 | .999 | .999 | .999 | .997 | .993 | .985 |
| | | | | | | .970 | .945 | .908 | .857 | .792 | .713 | .625 | .531 | .438 | .349 |
| | | | | | | .269 | .201 | .145 | .101 | .068 | | | | | |
| 18.000 | 17.931 | .000 | .069 | 4.5% | 26 | .999 | .999 | .999 | .999 | .999 | .999 | .999 | .997 | .993 | .985 |
| | | | | | | .970 | .945 | .908 | .857 | .792 | .713 | .625 | .531 | .438 | .349 |
| | | | | | | .269 | .201 | .145 | .101 | .068 | .045 | | | | |
| 18.000 | 17.959 | .000 | .041 | 2.8% | 27 | .999 | .999 | .999 | .999 | .999 | .999 | .999 | .997 | .993 | .985 |
| | | | | | | .970 | .945 | .908 | .857 | .792 | .713 | .625 | .531 | .438 | .349 |
| | | | | | | .269 | .201 | .145 | .101 | .068 | .045 | .028 | | | |
| 18.000 | 17.977 | .000 | .023 | 1.7% | 28 | .999 | .999 | .999 | .999 | .999 | .999 | .999 | .997 | .993 | .985 |
| | | | | | | .970 | .945 | .908 | .857 | .792 | .713 | .625 | .531 | .438 | .349 |
| | | | | | | .269 | .201 | .145 | .101 | .068 | .045 | .028 | .017 | | |
| 18.000 | 17.987 | .000 | .013 | 1.0% | 29 | .999 | .999 | .999 | .999 | .999 | .999 | .999 | .997 | .993 | .985 |
| | | | | | | .970 | .945 | .908 | .857 | .792 | .713 | .625 | .531 | .438 | .349 |
| | | | | | | .269 | .201 | .145 | .101 | .068 | .045 | .028 | .017 | .010 | |
| 18.000 | 17.993 | .000 | .007 | .6% | 30 | .999 | .999 | .999 | .999 | .999 | .999 | .999 | .997 | .993 | .985 |
| | | | | | | .970 | .945 | .908 | .857 | .792 | .713 | .625 | .531 | .438 | .349 |
| | | | | | | .269 | .201 | .145 | .101 | .068 | .045 | .028 | .017 | .010 | .006 |
| 19.000 | 17.708 | .000 | 1.292 | 43.9% | 20 | .999 | .999 | .999 | .999 | .999 | .999 | .999 | .998 | .996 | .991 |

*NOT RECOMMENDED FOR BLOCKAGE MORE THAN 5%, ALSO UNDERSTATES MEASURED TIME USAGE

*POISSON TABLE: BLOCKAGE REPRESENTS ALL TRUNKS BUSY

| HOURS FIRST ATTEMPT VMR = 1.0 | HOURS CON-NECT | HOURS RECALL /RETRY | HOURS OVFL | PERCENT BLOCKED (GRADE) | LINE | HOURS CONNECTED TRAFFIC ON LINES 1 | 2 | 3 | 4 | 5 | 6 | 7 | 8 | 9 | 10 |
|---|---|---|---|---|---|---|---|---|---|---|---|---|---|---|---|
| | | | | | | .982 | .965 | .939 | .902 | .850 | .785 | .708 | .622 | .531 | .439 |
| 19.000 | 18.061 | .000 | .939 | 35.3% | 21 | .999 | .999 | .999 | .999 | .999 | .999 | .999 | .998 | .996 | .991 |
| | | | | | | .982 | .965 | .939 | .902 | .850 | .785 | .708 | .622 | .531 | .439 |
| | | | | | | .353 | | | | | | | | | |
| 19.000 | 18.335 | .000 | .665 | 27.5% | 22 | .999 | .999 | .999 | .999 | .999 | .999 | .999 | .998 | .996 | .991 |
| | | | | | | .982 | .965 | .939 | .902 | .850 | .785 | .708 | .622 | .531 | .439 |
| | | | | | | .353 | .275 | | | | | | | | |
| 19.000 | 18.542 | .000 | .458 | 20.7% | 23 | .999 | .999 | .999 | .999 | .999 | .999 | .999 | .998 | .996 | .991 |
| | | | | | | .982 | .965 | .939 | .902 | .850 | .785 | .708 | .622 | .531 | .439 |
| | | | | | | .353 | .275 | .207 | | | | | | | |
| 19.000 | 18.693 | .000 | .307 | 15.1% | 24 | .999 | .999 | .999 | .999 | .999 | .999 | .999 | .998 | .996 | .991 |
| | | | | | | .982 | .965 | .939 | .902 | .850 | .785 | .708 | .622 | .531 | .439 |
| | | | | | | .353 | .275 | .207 | .151 | | | | | | |
| 19.000 | 18.800 | .000 | .200 | 10.7% | 25 | .999 | .999 | .999 | .999 | .999 | .999 | .999 | .998 | .996 | .991 |
| | | | | | | .982 | .965 | .939 | .902 | .850 | .785 | .708 | .622 | .531 | .439 |
| | | | | | | .353 | .275 | .207 | .151 | .107 | | | | | |
| 19.000 | 18.873 | .000 | .127 | 7.3% | 26 | .999 | .999 | .999 | .999 | .999 | .999 | .999 | .998 | .996 | .991 |
| | | | | | | .982 | .965 | .939 | .902 | .850 | .785 | .708 | .622 | .531 | .439 |
| | | | | | | .353 | .275 | .207 | .151 | .107 | .073 | | | | |
| 19.000 | 18.922 | .000 | .078 | 4.9% | 27 | .999 | .999 | .999 | .999 | .999 | .999 | .999 | .998 | .996 | .991 |
| | | | | | | .982 | .965 | .939 | .902 | .850 | .785 | .708 | .622 | .531 | .439 |
| | | | | | | .353 | .275 | .207 | .151 | .107 | .073 | .049 | | | |
| 19.000 | 18.953 | .000 | .047 | 3.1% | 28 | .999 | .999 | .999 | .999 | .999 | .999 | .999 | .998 | .996 | .991 |
| | | | | | | .982 | .965 | .939 | .902 | .850 | .785 | .708 | .622 | .531 | .439 |
| | | | | | | .353 | .275 | .207 | .151 | .107 | .073 | .049 | .031 | | |
| 19.000 | 18.972 | .000 | .028 | 2.0% | 29 | .999 | .999 | .999 | .999 | .999 | .999 | .999 | .998 | .996 | .991 |
| | | | | | | .982 | .965 | .939 | .902 | .850 | .785 | .708 | .622 | .531 | .439 |
| | | | | | | .353 | .275 | .207 | .151 | .107 | .073 | .049 | .031 | .020 | |
| 19.000 | 18.984 | .000 | .016 | 1.2% | 30 | .999 | .999 | .999 | .999 | .999 | .999 | .999 | .998 | .996 | .991 |
| | | | | | | .982 | .965 | .939 | .902 | .850 | .785 | .708 | .622 | .531 | .439 |
| | | | | | | .353 | .275 | .207 | .151 | .107 | .073 | .049 | .031 | .020 | .012 |
| 19.000 | 18.991 | .000 | .009 | .7% | 31 | .999 | .999 | .999 | .999 | .999 | .999 | .999 | .998 | .996 | .991 |
| | | | | | | .982 | .965 | .939 | .902 | .850 | .785 | .708 | .622 | .531 | .439 |
| | | | | | | .353 | .275 | .207 | .151 | .107 | .073 | .049 | .031 | .020 | .012 |
| | | | | | | .007 | | | | | | | | | |
| 20.000 | 18.664 | .000 | 1.336 | 44.1% | 21 | .999 | .999 | .999 | .999 | .999 | .999 | .999 | .999 | .998 | .995 |
| | | | | | | .989 | .979 | .961 | .934 | .895 | .843 | .779 | .703 | .619 | .530 |
| | | | | | | .441 | | | | | | | | | |
| 20.000 | 19.021 | .000 | .979 | 35.6% | 22 | .999 | .999 | .999 | .999 | .999 | .999 | .999 | .999 | .998 | .995 |
| | | | | | | .989 | .979 | .961 | .934 | .895 | .843 | .779 | .703 | .619 | .530 |

*NOT RECOMMENDED FOR BLOCKAGE MORE THAN 5%, ALSO UNDERSTATES MEASURED TIME USAGE

*POISSON TABLE: BLOCKAGE REPRESENTS ALL TRUNKS BUSY

| HOURS FIRST ATTEMPT VMR = 1.0 | HOURS CON-NECT | HOURS RECALL /RETRY | HOURS OVFL | PERCENT BLOCKED (GRADE) | LINE | HOURS CONNECTED TRAFFIC ON LINES 1 | 2 | 3 | 4 | 5 | 6 | 7 | 8 | 9 | 10 |
|---|---|---|---|---|---|---|---|---|---|---|---|---|---|---|---|
| | | | | | | .441 | .356 | | | | | | | | |
| 20.000 | 19.300 | .000 | .700 | 27.9% | 23 | .999 | .999 | .999 | .999 | .999 | .999 | .999 | .999 | .998 | .995 |
| | | | | | | .989 | .979 | .961 | .934 | .895 | .843 | .779 | .703 | .619 | .530 |
| | | | | | | .441 | .356 | .279 | | | | | | | |
| 20.000 | 19.512 | .000 | .488 | 21.3% | 24 | .999 | .999 | .999 | .999 | .999 | .999 | .999 | .999 | .998 | .995 |
| | | | | | | .989 | .979 | .961 | .934 | .895 | .843 | .779 | .703 | .619 | .530 |
| | | | | | | .441 | .356 | .279 | .213 | | | | | | |
| 20.000 | 19.669 | .000 | .331 | 15.7% | 25 | .999 | .999 | .999 | .999 | .999 | .999 | .999 | .999 | .998 | .995 |
| | | | | | | .989 | .979 | .961 | .934 | .895 | .843 | .779 | .703 | .619 | .530 |
| | | | | | | .441 | .356 | .279 | .213 | .157 | | | | | |
| 20.000 | 19.781 | .000 | .219 | 11.2% | 26 | .999 | .999 | .999 | .999 | .999 | .999 | .999 | .999 | .998 | .995 |
| | | | | | | .989 | .979 | .961 | .934 | .895 | .843 | .779 | .703 | .619 | .530 |
| | | | | | | .441 | .356 | .279 | .213 | .157 | .112 | | | | |
| 20.000 | 19.859 | .000 | .141 | 7.8% | 27 | .999 | .999 | .999 | .999 | .999 | .999 | .999 | .999 | .998 | .995 |
| | | | | | | .989 | .979 | .961 | .934 | .895 | .843 | .779 | .703 | .619 | .530 |
| | | | | | | .441 | .356 | .279 | .213 | .157 | .112 | .078 | | | |
| 20.000 | 19.912 | .000 | .088 | 5.2% | 28 | .999 | .999 | .999 | .999 | .999 | .999 | .999 | .999 | .998 | .995 |
| | | | | | | .989 | .979 | .961 | .934 | .895 | .843 | .779 | .703 | .619 | .530 |
| | | | | | | .441 | .356 | .279 | .213 | .157 | .112 | .078 | .052 | | |
| 20.000 | 19.946 | .000 | .054 | 3.4% | 29 | .999 | .999 | .999 | .999 | .999 | .999 | .999 | .999 | .998 | .995 |
| | | | | | | .989 | .979 | .961 | .934 | .895 | .843 | .779 | .703 | .619 | .530 |
| | | | | | | .441 | .356 | .279 | .213 | .157 | .112 | .078 | .052 | .034 | |
| 20.000 | 19.968 | .000 | .032 | 2.2% | 30 | .999 | .999 | .999 | .999 | .999 | .999 | .999 | .999 | .998 | .995 |
| | | | | | | .989 | .979 | .961 | .934 | .895 | .843 | .779 | .703 | .619 | .530 |
| | | | | | | .441 | .356 | .279 | .213 | .157 | .112 | .078 | .052 | .034 | .022 |
| 20.000 | 19.981 | .000 | .019 | 1.3% | 31 | .999 | .999 | .999 | .999 | .999 | .999 | .999 | .999 | .998 | .995 |
| | | | | | | .989 | .979 | .961 | .934 | .895 | .843 | .779 | .703 | .619 | .530 |
| | | | | | | .441 | .356 | .279 | .213 | .157 | .112 | .078 | .052 | .034 | .022 |
| | | | | | | .013 | | | | | | | | | |
| 20.000 | 19.989 | .000 | .011 | .8% | 32 | .999 | .999 | .999 | .999 | .999 | .999 | .999 | .999 | .998 | .995 |
| | | | | | | .989 | .979 | .961 | .934 | .895 | .843 | .779 | .703 | .619 | .530 |
| | | | | | | .441 | .356 | .279 | .213 | .157 | .112 | .078 | .052 | .034 | .022 |
| | | | | | | .013 | | | | | | | | | |
| 22.000 | 20.579 | .000 | 1.421 | 44.4% | 23 | .999 | .999 | .999 | .999 | .999 | .999 | .999 | .999 | .999 | .998 |
| | | | | | | .996 | .992 | .985 | .972 | .952 | .923 | .883 | .831 | .768 | .694 |
| | | | | | | .613 | .528 | .444 | | | | | | | |
| 22.000 | 20.942 | .000 | 1.058 | 36.3% | 24 | .999 | .999 | .999 | .999 | .999 | .999 | .999 | .999 | .999 | .998 |
| | | | | | | .996 | .992 | .985 | .972 | .952 | .923 | .883 | .831 | .768 | .694 |
| | | | | | | .613 | .528 | .444 | .363 | | | | | | |
| 22.000 | 21.230 | .000 | .770 | 28.8% | 25 | .999 | .999 | .999 | .999 | .999 | .999 | .999 | .999 | .999 | .998 |

*NOT RECOMMENDED FOR BLOCKAGE MORE THAN 5%, ALSO UNDERSTATES MEASURED TIME USAGE

*POISSON TABLE: BLOCKAGE REPRESENTS ALL TRUNKS BUSY

| HOURS FIRST ATTEMPT VMR = 1.0 | HOURS CON-NECT | HOURS RECALL /RETRY | HOURS OVFL | PERCENT BLOCKED (GRADE) | LINE | HOURS CONNECTED TRAFFIC ON LINES 1 | 2 | 3 | 4 | 5 | 6 | 7 | 8 | 9 | 10 |
|---|---|---|---|---|---|---|---|---|---|---|---|---|---|---|---|
| | | | | | | .996 | .992 | .985 | .972 | .952 | .923 | .883 | .831 | .768 | .694 |
| | | | | | | .613 | .528 | .444 | .363 | .288 | | | | | |
| 22.000 | 21.453 | .000 | .547 | 22.3% | 26 | .999 | .999 | .999 | .999 | .999 | .999 | .999 | .999 | .999 | .998 |
| | | | | | | .996 | .992 | .985 | .972 | .952 | .923 | .883 | .831 | .768 | .694 |
| | | | | | | .613 | .528 | .444 | .363 | .288 | .223 | | | | |
| 22.000 | 21.621 | .000 | .379 | 16.8% | 27 | .999 | .999 | .999 | .999 | .999 | .999 | .999 | .999 | .999 | .998 |
| | | | | | | .996 | .992 | .985 | .972 | .952 | .923 | .883 | .831 | .768 | .694 |
| | | | | | | .613 | .528 | .444 | .363 | .288 | .223 | .168 | | | |
| 22.000 | 21.743 | .000 | .257 | 12.3% | 28 | .999 | .999 | .999 | .999 | .999 | .999 | .999 | .999 | .999 | .998 |
| | | | | | | .996 | .992 | .985 | .972 | .952 | .923 | .883 | .831 | .768 | .694 |
| | | | | | | .613 | .528 | .444 | .363 | .288 | .223 | .168 | .123 | | |
| 22.000 | 21.830 | .000 | .170 | 8.7% | 29 | .999 | .999 | .999 | .999 | .999 | .999 | .999 | .999 | .999 | .998 |
| | | | | | | .996 | .992 | .985 | .972 | .952 | .923 | .883 | .831 | .768 | .694 |
| | | | | | | .613 | .528 | .444 | .363 | .288 | .223 | .168 | .123 | .087 | |
| 22.000 | 21.891 | .000 | .109 | 6.0% | 30 | .999 | .999 | .999 | .999 | .999 | .999 | .999 | .999 | .999 | .998 |
| | | | | | | .996 | .992 | .985 | .972 | .952 | .923 | .883 | .831 | .768 | .694 |
| | | | | | | .613 | .528 | .444 | .363 | .288 | .223 | .168 | .123 | .087 | .060 |
| 22.000 | 21.931 | .000 | .069 | 4.1% | 31 | .999 | .999 | .999 | .999 | .999 | .999 | .999 | .999 | .999 | .998 |
| | | | | | | .996 | .992 | .985 | .972 | .952 | .923 | .883 | .831 | .768 | .694 |
| | | | | | | .613 | .528 | .444 | .363 | .288 | .223 | .168 | .123 | .087 | .060 |
| | | | | | | .041 | | | | | | | | | |
| 22.000 | 21.958 | .000 | .042 | 2.7% | 32 | .999 | .999 | .999 | .999 | .999 | .999 | .999 | .999 | .999 | .998 |
| | | | | | | .996 | .992 | .985 | .972 | .952 | .923 | .883 | .831 | .768 | .694 |
| | | | | | | .613 | .528 | .444 | .363 | .288 | .223 | .168 | .123 | .087 | .060 |
| | | | | | | .041 | .027 | | | | | | | | |
| 22.000 | 21.975 | .000 | .025 | 1.7% | 33 | .999 | .999 | .999 | .999 | .999 | .999 | .999 | .999 | .999 | .998 |
| | | | | | | .996 | .992 | .985 | .972 | .952 | .923 | .883 | .831 | .768 | .694 |
| | | | | | | .613 | .528 | .444 | .363 | .288 | .223 | .168 | .123 | .087 | .060 |
| | | | | | | .041 | .027 | .017 | | | | | | | |
| 22.000 | 21.985 | .000 | .015 | 1.1% | 34 | .999 | .999 | .999 | .999 | .999 | .999 | .999 | .999 | .999 | .998 |
| | | | | | | .996 | .992 | .985 | .972 | .952 | .923 | .883 | .831 | .768 | .694 |
| | | | | | | .613 | .528 | .444 | .363 | .288 | .223 | .168 | .123 | .087 | .060 |
| | | | | | | .041 | .027 | .017 | .011 | | | | | | |
| 22.000 | 21.991 | .000 | .009 | .6% | 35 | .999 | .999 | .999 | .999 | .999 | .999 | .999 | .999 | .999 | .998 |
| | | | | | | .996 | .992 | .985 | .972 | .952 | .923 | .883 | .831 | .768 | .694 |
| | | | | | | .613 | .528 | .444 | .363 | .288 | .223 | .168 | .123 | .087 | .060 |
| | | | | | | .041 | .027 | .017 | .011 | .006 | | | | | |
| 24.000 | 22.498 | .000 | 1.502 | 44.0% | 25 | .999 | .999 | .999 | .999 | .999 | .999 | .999 | .999 | .999 | .999 |
| | | | | | | .999 | .997 | .995 | .989 | .980 | .966 | .944 | .913 | .872 | .820 |
| | | | | | | .757 | .686 | .608 | .527 | .446 | | | | | |

*NOT RECOMMENDED FOR BLOCKAGE MORE THAN 5%, ALSO UNDERSTATES MEASURED TIME USAGE

*POISSON TABLE: BLOCKAGE REPRESENTS ALL TRUNKS BUSY

| HOURS FIRST ATTEMPT VMR = 1.0 | HOURS CON-NECT | HOURS RECALL /RETRY | HOURS OVFL | PERCENT BLOCKED (GRADE) | LINE | HOURS CONNECTED TRAFFIC ON LINES 1 | 2 | 3 | 4 | 5 | 6 | 7 | 8 | 9 | 10 |
|---|---|---|---|---|---|---|---|---|---|---|---|---|---|---|---|
| 24.000 | 22.866 | .000 | 1.134 | 36.8% | 26 | .999 | .999 | .999 | .999 | .999 | .999 | .999 | .999 | .999 | .999 |
| | | | | | | .999 | .997 | .995 | .989 | .980 | .966 | .944 | .913 | .872 | .820 |
| | | | | | | .757 | .686 | .608 | .527 | .446 | .368 | | | | |
| 24.000 | 23.163 | .000 | .837 | 29.6% | 27 | .999 | .999 | .999 | .999 | .999 | .999 | .999 | .999 | .999 | .999 |
| | | | | | | .999 | .997 | .995 | .989 | .980 | .966 | .944 | .913 | .872 | .820 |
| | | | | | | .757 | .686 | .608 | .527 | .446 | .368 | .296 | | | |
| 24.000 | 23.395 | .000 | .605 | 23.2% | 28 | .999 | .999 | .999 | .999 | .999 | .999 | .999 | .999 | .999 | .999 |
| | | | | | | .999 | .997 | .995 | .989 | .980 | .966 | .944 | .913 | .872 | .820 |
| | | | | | | .757 | .686 | .608 | .527 | .446 | .368 | .296 | .232 | | |
| 24.000 | 23.572 | .000 | .428 | 17.7% | 29 | .999 | .999 | .999 | .999 | .999 | .999 | .999 | .999 | .999 | .999 |
| | | | | | | .999 | .997 | .995 | .989 | .980 | .966 | .944 | .913 | .872 | .820 |
| | | | | | | .757 | .686 | .608 | .527 | .446 | .368 | .296 | .232 | .177 | |
| 24.000 | 23.704 | .000 | .296 | 13.2% | 30 | .999 | .999 | .999 | .999 | .999 | .999 | .999 | .999 | .999 | .999 |
| | | | | | | .999 | .997 | .995 | .989 | .980 | .966 | .944 | .913 | .872 | .820 |
| | | | | | | .757 | .686 | .608 | .527 | .446 | .368 | .296 | .232 | .177 | .132 |
| 24.000 | 23.800 | .000 | .200 | 9.6% | 31 | .999 | .999 | .999 | .999 | .999 | .999 | .999 | .999 | .999 | .999 |
| | | | | | | .999 | .997 | .995 | .989 | .980 | .966 | .944 | .913 | .872 | .820 |
| | | | | | | .757 | .686 | .608 | .527 | .446 | .368 | .296 | .232 | .177 | .132 |
| | | | | | | .096 | | | | | | | | | |
| 24.000 | 23.868 | .000 | .132 | 6.8% | 32 | .999 | .999 | .999 | .999 | .999 | .999 | .999 | .999 | .999 | .999 |
| | | | | | | .999 | .997 | .995 | .989 | .980 | .966 | .944 | .913 | .872 | .820 |
| | | | | | | .757 | .686 | .608 | .527 | .446 | .368 | .296 | .232 | .177 | .132 |
| | | | | | | .096 | .068 | | | | | | | | |
| 24.000 | 23.915 | .000 | .085 | 4.7% | 33 | .999 | .999 | .999 | .999 | .999 | .999 | .999 | .999 | .999 | .999 |
| | | | | | | .999 | .997 | .995 | .989 | .980 | .966 | .944 | .913 | .872 | .820 |
| | | | | | | .757 | .686 | .608 | .527 | .446 | .368 | .296 | .232 | .177 | .132 |
| | | | | | | .096 | .068 | .047 | | | | | | | |
| 24.000 | 23.946 | .000 | .054 | 3.1% | 34 | .999 | .999 | .999 | .999 | .999 | .999 | .999 | .999 | .999 | .999 |
| | | | | | | .999 | .997 | .995 | .989 | .980 | .966 | .944 | .913 | .872 | .820 |
| | | | | | | .757 | .686 | .608 | .527 | .446 | .368 | .296 | .232 | .177 | .132 |
| | | | | | | .096 | .068 | .047 | .031 | | | | | | |
| 24.000 | 23.967 | .000 | .033 | 2.1% | 35 | .999 | .999 | .999 | .999 | .999 | .999 | .999 | .999 | .999 | .999 |
| | | | | | | .999 | .997 | .995 | .989 | .980 | .966 | .944 | .913 | .872 | .820 |
| | | | | | | .757 | .686 | .608 | .527 | .446 | .368 | .296 | .232 | .177 | .132 |
| | | | | | | .096 | .068 | .047 | .031 | .021 | | | | | |
| 24.000 | 23.980 | .000 | .020 | 1.3% | 36 | .999 | .999 | .999 | .999 | .999 | .999 | .999 | .999 | .999 | .999 |
| | | | | | | .999 | .997 | .995 | .989 | .980 | .966 | .944 | .913 | .872 | .820 |
| | | | | | | .757 | .686 | .608 | .527 | .446 | .368 | .296 | .232 | .177 | .132 |
| | | | | | | .096 | .068 | .047 | .031 | .021 | .013 | | | | |
| 24.000 | 23.988 | .000 | .012 | .8% | 37 | .999 | .999 | .999 | .999 | .999 | .999 | .999 | .999 | .999 | .999 |

*NOT RECOMMENDED FOR BLOCKAGE MORE THAN 5%, ALSO UNDERSTATES MEASURED TIME USAGE

*POISSON TABLE: BLOCKAGE REPRESENTS ALL TRUNKS BUSY

| HOURS FIRST ATTEMPT VMR=1.0 | HOURS CON-NECT | HOURS RECALL /RETRY | HOURS OVFL | PERCENT BLOCKED (GRADE) | LINE | HOURS CONNECTED TRAFFIC ON LINES 1 | 2 | 3 | 4 | 5 | 6 | 7 | 8 | 9 | 10 |
|---|---|---|---|---|---|---|---|---|---|---|---|---|---|---|---|
| | | | | | | .999 | .997 | .995 | .989 | .980 | .966 | .944 | .913 | .872 | .820 |
| | | | | | | .757 | .686 | .608 | .527 | .446 | .368 | .296 | .232 | .177 | .132 |
| | | | | | | .096 | .068 | .047 | .031 | .021 | .013 | .008 | | | |
| 24.000 | 23.993 | .000 | .007 | .5% | 38 | .999 | .999 | .999 | .999 | .999 | .999 | .999 | .999 | .999 | .999 |
| | | | | | | .999 | .997 | .995 | .989 | .980 | .966 | .944 | .913 | .872 | .820 |
| | | | | | | .757 | .686 | .608 | .527 | .446 | .368 | .296 | .232 | .177 | .132 |
| | | | | | | .096 | .068 | .047 | .031 | .021 | .013 | .008 | .005 | | |
| 26.000 | 24.420 | .000 | 1.580 | 44.8% | 27 | .999 | .999 | .999 | .999 | .999 | .999 | .999 | .999 | .999 | .999 |
| | | | | | | .999 | .999 | .998 | .996 | .992 | .986 | .975 | .959 | .935 | .903 |
| | | | | | | .861 | .810 | .748 | .679 | .604 | .526 | .448 | | | |
| 26.000 | 24.793 | .000 | 1.207 | 37.3% | 28 | .999 | .999 | .999 | .999 | .999 | .999 | .999 | .999 | .999 | .999 |
| | | | | | | .999 | .999 | .998 | .996 | .992 | .986 | .975 | .959 | .935 | .903 |
| | | | | | | .861 | .810 | .748 | .679 | .604 | .526 | .448 | .373 | | |
| 26.000 | 25.097 | .000 | .903 | 30.3% | 29 | .999 | .999 | .999 | .999 | .999 | .999 | .999 | .999 | .999 | .999 |
| | | | | | | .999 | .999 | .998 | .996 | .992 | .986 | .975 | .959 | .935 | .903 |
| | | | | | | .861 | .810 | .748 | .679 | .604 | .526 | .448 | .373 | .303 | |
| 26.000 | 25.337 | .000 | .663 | 24.1% | 30 | .999 | .999 | .999 | .999 | .999 | .999 | .999 | .999 | .999 | .999 |
| | | | | | | .999 | .999 | .998 | .996 | .992 | .986 | .975 | .959 | .935 | .903 |
| | | | | | | .861 | .810 | .748 | .679 | .604 | .526 | .448 | .373 | .303 | .241 |
| 26.000 | 25.524 | .000 | .476 | 18.7% | 31 | .999 | .999 | .999 | .999 | .999 | .999 | .999 | .999 | .999 | .999 |
| | | | | | | .999 | .999 | .998 | .996 | .992 | .986 | .975 | .959 | .935 | .903 |
| | | | | | | .861 | .810 | .748 | .679 | .604 | .526 | .448 | .373 | .303 | .241 |
| | | | | | | .187 | | | | | | | | | |
| 26.000 | 25.665 | .000 | .335 | 14.1% | 32 | .999 | .999 | .999 | .999 | .999 | .999 | .999 | .999 | .999 | .999 |
| | | | | | | .999 | .999 | .998 | .996 | .992 | .986 | .975 | .959 | .935 | .903 |
| | | | | | | .861 | .810 | .748 | .679 | .604 | .526 | .448 | .373 | .303 | .241 |
| | | | | | | .187 | .141 | | | | | | | | |
| 26.000 | 25.769 | .000 | .231 | 10.4% | 33 | .999 | .999 | .999 | .999 | .999 | .999 | .999 | .999 | .999 | .999 |
| | | | | | | .999 | .999 | .998 | .996 | .992 | .986 | .975 | .959 | .935 | .903 |
| | | | | | | .861 | .810 | .748 | .679 | .604 | .526 | .448 | .373 | .303 | .241 |
| | | | | | | .187 | .141 | .104 | | | | | | | |
| 26.000 | 25.844 | .000 | .156 | 7.5% | 34 | .999 | .999 | .999 | .999 | .999 | .999 | .999 | .999 | .999 | .999 |
| | | | | | | .999 | .999 | .998 | .996 | .992 | .986 | .975 | .959 | .935 | .903 |
| | | | | | | .861 | .810 | .748 | .679 | .604 | .526 | .448 | .373 | .303 | .241 |
| | | | | | | .187 | .141 | .104 | .075 | | | | | | |
| 26.000 | 25.897 | .000 | .103 | 5.3% | 35 | .999 | .999 | .999 | .999 | .999 | .999 | .999 | .999 | .999 | .999 |
| | | | | | | .999 | .999 | .998 | .996 | .992 | .986 | .975 | .959 | .935 | .903 |
| | | | | | | .861 | .810 | .748 | .679 | .604 | .526 | .448 | .373 | .303 | .241 |
| | | | | | | .187 | .141 | .104 | .075 | .053 | | | | | |
| 26.000 | 25.933 | .000 | .067 | 3.6% | 36 | .999 | .999 | .999 | .999 | .999 | .999 | .999 | .999 | .999 | .999 |

*NOT RECOMMENDED FOR BLOCKAGE MORE THAN 5%, ALSO UNDERSTATES MEASURED TIME USAGE

*POISSON TABLE: BLOCKAGE REPRESENTS ALL TRUNKS BUSY

| HOURS FIRST ATTEMPT VMR = 1.0 | HOURS CON-NECT | HOURS RECALL /RETRY | HOURS OVFL | PERCENT BLOCKED (GRADE) | LINE | HOURS CONNECTED TRAFFIC ON LINES 1 | 2 | 3 | 4 | 5 | 6 | 7 | 8 | 9 | 10 |
|---|---|---|---|---|---|---|---|---|---|---|---|---|---|---|---|
| | | | | | | .999 | .999 | .998 | .996 | .992 | .986 | .975 | .959 | .935 | .903 |
| | | | | | | .861 | .810 | .748 | .679 | .604 | .526 | .448 | .373 | .303 | .241 |
| | | | | | | .187 | .141 | .104 | .075 | .053 | .036 | | | | |
| 26.000 | 25.958 | .000 | .042 | 2.4% | 37 | .999 | .999 | .999 | .999 | .999 | .999 | .999 | .999 | .999 | .999 |
| | | | | | | .999 | .999 | .998 | .996 | .992 | .986 | .975 | .959 | .935 | .903 |
| | | | | | | .861 | .810 | .748 | .679 | .604 | .526 | .448 | .373 | .303 | .241 |
| | | | | | | .187 | .141 | .104 | .075 | .053 | .036 | .024 | | | |
| 26.000 | 25.974 | .000 | .026 | 1.6% | 38 | .999 | .999 | .999 | .999 | .999 | .999 | .999 | .999 | .999 | .999 |
| | | | | | | .999 | .999 | .998 | .996 | .992 | .986 | .975 | .959 | .935 | .903 |
| | | | | | | .861 | .810 | .748 | .679 | .604 | .526 | .448 | .373 | .303 | .241 |
| | | | | | | .187 | .141 | .104 | .075 | .053 | .036 | .024 | .016 | | |
| 26.000 | 25.984 | .000 | .016 | 1.0% | 39 | .999 | .999 | .999 | .999 | .999 | .999 | .999 | .999 | .999 | .999 |
| | | | | | | .999 | .999 | .998 | .996 | .992 | .986 | .975 | .959 | .935 | .903 |
| | | | | | | .861 | .810 | .748 | .679 | .604 | .526 | .448 | .373 | .303 | .241 |
| | | | | | | .187 | .141 | .104 | .075 | .053 | .036 | .024 | .016 | .010 | |
| 26.000 | 25.991 | .000 | .009 | .6% | 40 | .999 | .999 | .999 | .999 | .999 | .999 | .999 | .999 | .999 | .999 |
| | | | | | | .999 | .999 | .998 | .996 | .992 | .986 | .975 | .959 | .935 | .903 |
| | | | | | | .861 | .810 | .748 | .679 | .604 | .526 | .448 | .373 | .303 | .241 |
| | | | | | | .187 | .141 | .104 | .075 | .053 | .036 | .024 | .016 | .010 | .006 |
| 28.000 | 26.345 | .000 | 1.655 | 45.0% | 29 | .999 | .999 | .999 | .999 | .999 | .999 | .999 | .999 | .999 | .999 |
| | | | | | | .999 | .999 | .999 | .999 | .997 | .995 | .990 | .982 | .970 | .952 |
| | | | | | | .927 | .894 | .852 | .800 | .740 | .673 | .600 | .525 | .450 | |
| 28.000 | 26.723 | .000 | 1.277 | 37.7% | 30 | .999 | .999 | .999 | .999 | .999 | .999 | .999 | .999 | .999 | .999 |
| | | | | | | .999 | .999 | .999 | .999 | .997 | .995 | .990 | .982 | .970 | .952 |
| | | | | | | .927 | .894 | .852 | .800 | .740 | .673 | .600 | .525 | .450 | .377 |
| 28.000 | 27.032 | .000 | .968 | 31.0% | 31 | .999 | .999 | .999 | .999 | .999 | .999 | .999 | .999 | .999 | .999 |
| | | | | | | .999 | .999 | .999 | .999 | .997 | .995 | .990 | .982 | .970 | .952 |
| | | | | | | .927 | .894 | .852 | .800 | .740 | .673 | .600 | .525 | .450 | .377 |
| | | | | | | .310 | | | | | | | | | |
| 28.000 | 27.281 | .000 | .719 | 24.8% | 32 | .999 | .999 | .999 | .999 | .999 | .999 | .999 | .999 | .999 | .999 |
| | | | | | | .999 | .999 | .999 | .999 | .997 | .995 | .990 | .982 | .970 | .952 |
| | | | | | | .927 | .894 | .852 | .800 | .740 | .673 | .600 | .525 | .450 | .377 |
| | | | | | | .310 | .248 | | | | | | | | |
| 28.000 | 27.476 | .000 | .524 | 19.5% | 33 | .999 | .999 | .999 | .999 | .999 | .999 | .999 | .999 | .999 | .999 |
| | | | | | | .999 | .999 | .999 | .999 | .997 | .995 | .990 | .982 | .970 | .952 |
| | | | | | | .927 | .894 | .852 | .800 | .740 | .673 | .600 | .525 | .450 | .377 |
| | | | | | | 310 | .248 | .195 | | | | | | | |
| 28.000 | 27.625 | .000 | .375 | 15.0% | 34 | .999 | .999 | .999 | .999 | .999 | .999 | .999 | .999 | .999 | .999 |
| | | | | | | .999 | .999 | .999 | .999 | .997 | .995 | .990 | .982 | .970 | .952 |
| | | | | | | .927 | .894 | .852 | .800 | .740 | .673 | .600 | .525 | .450 | .377 |

*NOT RECOMMENDED FOR BLOCKAGE MORE THAN 5%, ALSO UNDERSTATES MEASURED TIME USAGE

*POISSON TABLE: BLOCKAGE REPRESENTS ALL TRUNKS BUSY

| HOURS FIRST ATTEMPT VMR = 1.0 | HOURS CON-NECT | HOURS RECALL /RETRY | HOURS OVFL | PERCENT BLOCKED (GRADE) | LINE | HOURS CONNECTED TRAFFIC ON LINES 1 | 2 | 3 | 4 | 5 | 6 | 7 | 8 | 9 | 10 |
|---|---|---|---|---|---|---|---|---|---|---|---|---|---|---|---|
| | | | | | | .310 | .248 | .195 | .150 | | | | | | |
| 28.000 | 27.737 | .000 | .263 | 11.2% | 35 | .999 | .999 | .999 | .999 | .999 | .999 | .999 | .999 | .999 | .999 |
| | | | | | | .999 | .999 | .999 | .999 | .997 | .995 | .990 | .982 | .970 | .952 |
| | | | | | | .927 | .894 | .852 | .800 | .740 | .673 | .600 | .525 | .450 | .377 |
| | | | | | | .310 | .248 | .195 | .150 | .112 | | | | | |
| 28.000 | 27.819 | .000 | .181 | 8.2% | 36 | .999 | .999 | .999 | .999 | .999 | .999 | .999 | .999 | .999 | .999 |
| | | | | | | .999 | .999 | .999 | .999 | .997 | .995 | .990 | .982 | .970 | .952 |
| | | | | | | .927 | .894 | .852 | .800 | .740 | .673 | .600 | .525 | .450 | .377 |
| | | | | | | .310 | .248 | .195 | .150 | .112 | .082 | | | | |
| 28.000 | 27.878 | .000 | .122 | 5.9% | 37 | .999 | .999 | .999 | .999 | .999 | .999 | .999 | .999 | .999 | .999 |
| | | | | | | .999 | .999 | .999 | .999 | .997 | .995 | .990 | .982 | .970 | .952 |
| | | | | | | .927 | .894 | .852 | .800 | .740 | .673 | .600 | .525 | .450 | .377 |
| | | | | | | .310 | .248 | .195 | .150 | .112 | .082 | .059 | | | |
| 28.000 | 27.920 | .000 | .080 | 4.1% | 38 | .999 | .999 | .999 | .999 | .999 | .999 | .999 | .999 | .999 | .999 |
| | | | | | | .999 | .999 | .999 | .999 | .997 | .995 | .990 | .982 | .970 | .952 |
| | | | | | | .927 | .894 | .852 | .800 | .740 | .673 | .600 | .525 | .450 | .377 |
| | | | | | | .310 | .248 | .195 | .150 | .112 | .082 | .059 | .041 | | |
| 28.000 | 27.948 | .000 | .052 | 2.8% | 39 | .999 | .999 | .999 | .999 | .999 | .999 | .999 | .999 | .999 | .999 |
| | | | | | | .999 | .999 | .999 | .999 | .997 | .995 | .990 | .982 | .970 | .952 |
| | | | | | | .927 | .894 | .852 | .800 | .740 | .673 | .600 | .525 | .450 | .377 |
| | | | | | | .310 | .248 | .195 | .150 | .112 | .082 | .059 | .041 | .028 | |
| 28.000 | 27.967 | .000 | .033 | 1.9% | 40 | .999 | .999 | .999 | .999 | .999 | .999 | .999 | .999 | .999 | .999 |
| | | | | | | .999 | .999 | .999 | .999 | .997 | .995 | .990 | .982 | .970 | .952 |
| | | | | | | .927 | .894 | .852 | .800 | .740 | .673 | .600 | .525 | .450 | .377 |
| | | | | | | .310 | .248 | .195 | .150 | .112 | .082 | .059 | .041 | .028 | .019 |
| 28.000 | 27.979 | .000 | .021 | 1.2% | 41 | .999 | .999 | .999 | .999 | .999 | .999 | .999 | .999 | .999 | .999 |
| | | | | | | .999 | .999 | .999 | .999 | .997 | .995 | .990 | .982 | .970 | .952 |
| | | | | | | .927 | .894 | .852 | .800 | .740 | .673 | .600 | .525 | .450 | .377 |
| | | | | | | .310 | .248 | .195 | .150 | .112 | .082 | .059 | .041 | .028 | .019 |
| | | | | | | .012 | | | | | | | | | |
| 28.000 | 27.987 | .000 | .013 | .8% | 42 | .999 | .999 | .999 | .999 | .999 | .999 | .999 | .999 | .999 | .999 |
| | | | | | | .999 | .999 | .999 | .999 | .997 | .995 | .990 | .982 | .970 | .952 |
| | | | | | | .927 | .894 | .852 | .800 | .740 | .673 | .600 | .525 | .450 | .377 |
| | | | | | | .310 | .248 | .195 | .150 | .112 | .082 | .059 | .041 | .028 | .019 |
| | | | | | | .012 | .008 | | | | | | | | |
| 28.000 | 27.992 | .000 | .008 | .5% | 43 | .999 | .999 | .999 | .999 | .999 | .999 | .999 | .999 | .999 | .999 |
| | | | | | | .999 | .999 | .999 | .999 | .997 | .995 | .990 | .982 | .970 | .952 |
| | | | | | | .927 | .894 | .852 | .800 | .740 | .673 | .600 | .525 | .450 | .377 |
| | | | | | | .310 | .248 | .195 | .150 | .112 | .082 | .059 | .041 | .028 | .019 |
| | | | | | | .012 | .008 | .005 | | | | | | | |

*NOT RECOMMENDED FOR BLOCKAGE MORE THAN 5%, ALSO UNDERSTATES MEASURED TIME USAGE

*POISSON TABLE: BLOCKAGE REPRESENTS ALL TRUNKS BUSY

| HOURS FIRST ATTEMPT VMR = 1.0 | HOURS CON-NECT | HOURS RECALL /RETRY | HOURS OVFL | PERCENT BLOCKED (GRADE) | LINE | HOURS CONNECTED TRAFFIC ON LINES 1 | 2 | 3 | 4 | 5 | 6 | 7 | 8 | 9 | 10 |
|---|---|---|---|---|---|---|---|---|---|---|---|---|---|---|---|
| 30.000 | 28.273 | .000 | 1.727 | 45.2% | 31 | .999 | .999 | .999 | .999 | .999 | .999 | .999 | .999 | .999 | .999 |
| | | | | | | .999 | .999 | .999 | .999 | .999 | .998 | .996 | .993 | .987 | .978 |
| | | | | | | .965 | .946 | .919 | .885 | .843 | .792 | .733 | .667 | .597 | .524 |
| | | | | | | .452 | | | | | | | | | |
| 30.000 | 28.654 | .000 | 1.346 | 38.1% | 32 | .999 | .999 | .999 | .999 | .999 | .999 | .999 | .999 | .999 | .999 |
| | | | | | | .999 | .999 | .999 | .999 | .999 | .998 | .996 | .993 | .987 | .978 |
| | | | | | | .965 | .946 | .919 | .885 | .843 | .792 | .733 | .667 | .597 | .524 |
| | | | | | | .452 | .381 | | | | | | | | |
| 30.000 | 28.969 | .000 | 1.031 | 31.5% | 33 | .999 | .999 | .999 | .999 | .999 | .999 | .999 | .999 | .999 | .999 |
| | | | | | | .999 | .999 | .999 | .999 | .999 | .998 | .996 | .993 | .987 | .978 |
| | | | | | | .965 | .946 | .919 | .885 | .843 | .792 | .733 | .667 | .597 | .524 |
| | | | | | | .452 | .381 | .315 | | | | | | | |
| 30.000 | 29.225 | .000 | .775 | 25.6% | 34 | .999 | .999 | .999 | .999 | .999 | .999 | .999 | .999 | .999 | .999 |
| | | | | | | .999 | .999 | .999 | .999 | .999 | .998 | .996 | .993 | .987 | .978 |
| | | | | | | .965 | .946 | .919 | .885 | .843 | .792 | .733 | .667 | .597 | .524 |
| | | | | | | .452 | .381 | .315 | .256 | | | | | | |
| 30.000 | 29.428 | .000 | .572 | 20.3% | 35 | .999 | .999 | .999 | .999 | .999 | .999 | .999 | .999 | .999 | .999 |
| | | | | | | .999 | .999 | .999 | .999 | .999 | .998 | .996 | .993 | .987 | .978 |
| | | | | | | .965 | .946 | .919 | .885 | .843 | .792 | .733 | .667 | .597 | .524 |
| | | | | | | .452 | .381 | .315 | .256 | .203 | | | | | |
| 30.000 | 29.585 | .000 | .415 | 15.7% | 36 | .999 | .999 | .999 | .999 | .999 | .999 | .999 | .999 | .999 | .999 |
| | | | | | | .999 | .999 | .999 | .999 | .999 | .998 | .996 | .993 | .987 | .978 |
| | | | | | | .965 | .946 | .919 | .885 | .843 | .792 | .733 | .667 | .597 | .524 |
| | | | | | | .452 | .381 | .315 | .256 | .203 | .157 | | | | |
| 30.000 | 29.705 | .000 | .295 | 12.0% | 37 | .999 | .999 | .999 | .999 | .999 | .999 | .999 | .999 | .999 | .999 |
| | | | | | | .999 | .999 | .999 | .999 | .999 | .998 | .996 | .993 | .987 | .978 |
| | | | | | | .965 | .946 | .919 | .885 | .843 | .792 | .733 | .667 | .597 | .524 |
| | | | | | | .452 | .381 | .315 | .256 | .203 | .157 | .120 | | | |
| 30.000 | 29.794 | .000 | .206 | 8.9% | 38 | .999 | .999 | .999 | .999 | .999 | .999 | .999 | .999 | .999 | .999 |
| | | | | | | .999 | .999 | .999 | .999 | .999 | .998 | .996 | .993 | .987 | .978 |
| | | | | | | .965 | .946 | .919 | .885 | .843 | .792 | .733 | .667 | .597 | .524 |
| | | | | | | .452 | .381 | .315 | .256 | .203 | .157 | .120 | .089 | | |
| 30.000 | 29.859 | .000 | .141 | 6.5% | 39 | .999 | .999 | .999 | .999 | .999 | .999 | .999 | .999 | .999 | .999 |
| | | | | | | .999 | .999 | .999 | .999 | .999 | .998 | .996 | .993 | .987 | .978 |
| | | | | | | .965 | .946 | .919 | .885 | .843 | .792 | .733 | .667 | .597 | .524 |
| | | | | | | .452 | .381 | .315 | .256 | .203 | .157 | .120 | .089 | .065 | |
| 30.000 | 29.905 | .000 | .095 | 4.6% | 40 | .999 | .999 | .999 | .999 | .999 | .999 | .999 | .999 | .999 | .999 |
| | | | | | | .999 | .999 | .999 | .999 | .999 | .998 | .996 | .993 | .987 | .978 |
| | | | | | | .965 | .946 | .919 | .885 | .843 | .792 | .733 | .667 | .597 | .524 |
| | | | | | | .452 | .381 | .315 | .256 | .203 | .157 | .120 | .089 | .065 | .046 |

*NOT RECOMMENDED FOR BLOCKAGE MORE THAN 5%, ALSO UNDERSTATES MEASURED TIME USAGE

*POISSON TABLE: BLOCKAGE REPRESENTS ALL TRUNKS BUSY

| HOURS FIRST ATTEMPT VMR = 1.0 | HOURS CON-NECT | HOURS RECALL /RETRY | HOURS OVFL | PERCENT BLOCKED (GRADE) | LINE | HOURS CONNECTED TRAFFIC ON LINES 1 | 2 | 3 | 4 | 5 | 6 | 7 | 8 | 9 | 10 |
|---|---|---|---|---|---|---|---|---|---|---|---|---|---|---|---|
| 30.000 | 29.937 | .000 | .063 | 3.2% | 41 | .999 | .999 | .999 | .999 | .999 | .999 | .999 | .999 | .999 | .999 |
| | | | | | | .999 | .999 | .999 | .999 | .999 | .998 | .996 | .993 | .987 | .978 |
| | | | | | | .965 | .946 | .919 | .885 | .843 | .792 | .733 | .667 | .597 | .524 |
| | | | | | | .452 | .381 | .315 | .256 | .203 | .157 | .120 | .089 | .065 | .046 |
| | | | | | | .032 | | | | | | | | | |
| 30.000 | 29.959 | .000 | .041 | 2.2% | 42 | .999 | .999 | .999 | .999 | .999 | .999 | .999 | .999 | .999 | .999 |
| | | | | | | .999 | .999 | .999 | .999 | .999 | .998 | .996 | .993 | .987 | .978 |
| | | | | | | .965 | .946 | .919 | .885 | .843 | .792 | .733 | .667 | .597 | .524 |
| | | | | | | .452 | .381 | .315 | .256 | .203 | .157 | .120 | .089 | .065 | .046 |
| | | | | | | .032 | .022 | | | | | | | | |
| 30.000 | 29.974 | .000 | .026 | 1.5% | 43 | .999 | .999 | .999 | .999 | .999 | .999 | .999 | .999 | .999 | .999 |
| | | | | | | .999 | .999 | .999 | .999 | .999 | .998 | .996 | .993 | .987 | .978 |
| | | | | | | .965 | .946 | .919 | .885 | .843 | .792 | .733 | .667 | .597 | .524 |
| | | | | | | .452 | .381 | .315 | .256 | .203 | .157 | .120 | .089 | .065 | .046 |
| | | | | | | .032 | .022 | .015 | | | | | | | |
| 30.000 | 29.984 | .000 | .016 | 1.0% | 44 | .999 | .999 | .999 | .999 | .999 | .999 | .999 | .999 | .999 | .999 |
| | | | | | | .999 | .999 | .999 | .999 | .999 | .998 | .996 | .993 | .987 | .978 |
| | | | | | | .965 | .946 | .919 | .885 | .843 | .792 | .733 | .667 | .597 | .524 |
| | | | | | | .452 | .381 | .315 | .256 | .203 | .157 | .120 | .089 | .065 | .046 |
| | | | | | | .032 | .022 | .015 | .010 | | | | | | |
| 30.000 | 29.990 | .000 | .010 | .6% | 45 | .999 | .999 | .999 | .999 | .999 | .999 | .999 | .999 | .999 | .999 |
| | | | | | | .999 | .999 | .999 | .999 | .999 | .998 | .996 | .993 | .987 | .978 |
| | | | | | | .965 | .946 | .919 | .885 | .843 | .792 | .733 | .667 | .597 | .524 |
| | | | | | | .452 | .381 | .315 | .256 | .203 | .157 | .120 | .089 | .065 | .046 |
| | | | | | | .032 | .022 | .015 | .010 | .006 | | | | | |

*NOT RECOMMENDED FOR BLOCKAGE MORE THAN 5%, ALSO UNDERSTATES MEASURED TIME USAGE

## ERLANG B: 100% BLOCKED CALLERS ROUTE ADVANCE, OVERFLOW, OR DISAPPEAR

| HOURS FIRST ATTEMPT VMR = 1.0 | HOURS CON-NECT | HOURS RECALL /RETRY | HOURS OVFL | CON-NECT VMR | OVFL VMR | PERCENT BLOCKED (GRADE) | LINE | HOURS CONNECTED TRAFFIC ON LINES 1 | 2 | 3 | 4 | 5 | 6 | 7 | 8 | 9 | 10 |
|---|---|---|---|---|---|---|---|---|---|---|---|---|---|---|---|---|---|
| .200 | .167 | .000 | .033 | .83 | 1.08 | 16.7% | 1 | .167 | | | | | | | | | |
| .200 | .197 | .000 | .003 | .97 | 1.07 | 1.6% | 2 | .167 | .030 | | | | | | | | |
| .400 | .286 | .000 | .114 | .71 | 1.12 | 28.6% | 1 | .286 | | | | | | | | | |
| .400 | .378 | .000 | .022 | .91 | 1.13 | 5.4% | 2 | .286 | .093 | | | | | | | | |
| .400 | .397 | .000 | .003 | .98 | 1.11 | .7% | 3 | .286 | .093 | .019 | | | | | | | |
| .600 | .375 | .000 | .225 | .63 | 1.14 | 37.5% | 1 | .375 | | | | | | | | | |
| .600 | .539 | .000 | .061 | .84 | 1.18 | 10.1% | 2 | .375 | .164 | | | | | | | | |
| .600 | .588 | .000 | .012 | .95 | 1.16 | 2.0% | 3 | .375 | .164 | .049 | | | | | | | |
| .800 | .444 | .000 | .356 | .56 | 1.16 | 44.4% | 1 | .444 | | | | | | | | | |
| .800 | .679 | .000 | .121 | .77 | 1.22 | 15.1% | 2 | .444 | .235 | | | | | | | | |
| .800 | .769 | .000 | .031 | .91 | 1.22 | 3.9% | 3 | .444 | .235 | .090 | | | | | | | |
| .800 | .794 | .000 | .006 | .98 | 1.18 | .8% | 4 | .444 | .235 | .090 | .025 | | | | | | |
| 1.000 | .500 | .000 | .500 | .50 | 1.17 | 50.0% | 1 | .500 | | | | | | | | | |
| 1.000 | .800 | .000 | .200 | .70 | 1.25 | 20.0% | 2 | .500 | .300 | | | | | | | | |
| 1.000 | .938 | .000 | .063 | .86 | 1.26 | 6.3% | 3 | .500 | .300 | .137 | | | | | | | |
| 1.000 | .985 | .000 | .015 | .95 | 1.23 | 1.5% | 4 | .500 | .300 | .137 | .047 | | | | | | |
| 1.200 | .545 | .000 | .655 | .45 | 1.17 | 54.5% | 1 | .545 | | | | | | | | | |
| 1.200 | .904 | .000 | .296 | .64 | 1.28 | 24.7% | 2 | .545 | .359 | | | | | | | | |
| 1.200 | 1.092 | .000 | .108 | .81 | 1.30 | 9.0% | 3 | .545 | .359 | .188 | | | | | | | |
| 1.200 | 1.169 | .000 | .031 | .92 | 1.28 | 2.6% | 4 | .545 | .359 | .188 | .076 | | | | | | |
| 1.200 | 1.192 | .000 | .008 | .98 | 1.24 | .6% | 5 | .545 | .359 | .188 | .076 | .024 | | | | | |
| 1.400 | .583 | .000 | .817 | .42 | 1.17 | 58.3% | 1 | .583 | | | | | | | | | |
| 1.400 | .994 | .000 | .406 | .59 | 1.29 | 29.0% | 2 | .583 | .411 | | | | | | | | |
| 1.400 | 1.233 | .000 | .167 | .76 | 1.34 | 11.9% | 3 | .583 | .411 | .239 | | | | | | | |
| 1.400 | 1.344 | .000 | .056 | .89 | 1.33 | 4.0% | 4 | .583 | .411 | .239 | .111 | | | | | | |
| 1.400 | 1.384 | .000 | .016 | .96 | 1.29 | 1.1% | 5 | .583 | .411 | .239 | .111 | .041 | | | | | |
| 1.600 | .615 | .000 | .985 | .38 | 1.17 | 61.5% | 1 | .615 | | | | | | | | | |
| 1.600 | 1.072 | .000 | .528 | .54 | 1.30 | 33.0% | 2 | .615 | .457 | | | | | | | | |
| 1.600 | 1.361 | .000 | .239 | .71 | 1.37 | 15.0% | 3 | .615 | .457 | .288 | | | | | | | |
| 1.600 | 1.510 | .000 | .090 | .85 | 1.37 | 5.6% | 4 | .615 | .457 | .288 | .149 | | | | | | |
| 1.600 | 1.572 | .000 | .028 | .94 | 1.33 | 1.8% | 5 | .615 | .457 | .288 | .149 | .062 | | | | | |
| 1.800 | .643 | .000 | 1.157 | .36 | 1.17 | 64.3% | 1 | .643 | | | | | | | | | |
| 1.800 | 1.140 | .000 | .660 | .50 | 1.31 | 36.7% | 2 | .643 | .497 | | | | | | | | |
| 1.800 | 1.476 | .000 | .324 | .66 | 1.39 | 18.0% | 3 | .643 | .497 | .335 | | | | | | | |
| 1.800 | 1.665 | .000 | .135 | .81 | 1.40 | 7.5% | 4 | .643 | .497 | .335 | .189 | | | | | | |
| 1.800 | 1.753 | .000 | .047 | .91 | 1.38 | 2.6% | 5 | .643 | .497 | .335 | .189 | .088 | | | | | |
| 1.800 | 1.786 | .000 | .014 | .97 | 1.33 | .8% | 6 | .643 | .497 | .335 | .189 | .088 | .033 | | | | |
| 2.000 | .667 | .000 | 1.333 | .33 | 1.17 | 66.7% | 1 | .667 | | | | | | | | | |
| 2.000 | 1.200 | .000 | .800 | .47 | 1.31 | 40.0% | 2 | .667 | .533 | | | | | | | | |
| 2.000 | 1.579 | .000 | .421 | .62 | 1.41 | 21.1% | 3 | .667 | .533 | .379 | | | | | | | |

## ERLANG B: 100% BLOCKED CALLERS ROUTE ADVANCE, OVERFLOW, OR DISAPPEAR

| HOURS FIRST ATTEMPT VMR = 1.0 | HOURS CON-NECT | HOURS RECALL /RETRY | HOURS OVFL | CON-NECT VMR | OVFL VMR | PERCENT BLOCKED (GRADE) | LINE | HOURS CONNECTED TRAFFIC ON LINES | | | | | | | | | |
|---|---|---|---|---|---|---|---|---|---|---|---|---|---|---|---|---|---|
| | | | | | | | | 1 | 2 | 3 | 4 | 5 | 6 | 7 | 8 | 9 | 10 |
| 2.000 | 1.810 | .000 | .190 | .77 | 1.44 | 9.5% | 4 | .667 | .533 | .379 | .231 | | | | | | |
| 2.000 | 1.927 | .000 | .073 | .88 | 1.42 | 3.7% | 5 | .667 | .533 | .379 | .231 | .117 | | | | | |
| 2.000 | 1.976 | .000 | .024 | .95 | 1.37 | 1.2% | 6 | .667 | .533 | .379 | .231 | .117 | .049 | | | | |
| 2.200 | .688 | .000 | 1.512 | .31 | 1.16 | 68.7% | 1 | .688 | | | | | | | | | |
| 2.200 | 1.253 | .000 | .947 | .43 | 1.31 | 43.1% | 2 | .688 | .565 | | | | | | | | |
| 2.200 | 1.672 | .000 | .528 | .58 | 1.42 | 24.0% | 3 | .688 | .565 | .419 | | | | | | | |
| 2.200 | 1.943 | .000 | .257 | .73 | 1.46 | 11.7% | 4 | .688 | .565 | .419 | .271 | | | | | | |
| 2.200 | 2.093 | .000 | .107 | .85 | 1.46 | 4.9% | 5 | .688 | .565 | .419 | .271 | .149 | | | | | |
| 2.200 | 2.161 | .000 | .039 | .93 | 1.42 | 1.8% | 6 | .688 | .565 | .419 | .271 | .149 | .069 | | | | |
| 2.200 | 2.188 | .000 | .012 | .97 | 1.37 | .5% | 7 | .688 | .565 | .419 | .271 | .149 | .069 | .027 | | | |
| 2.400 | .706 | .000 | 1.694 | .29 | 1.16 | 70.6% | 1 | .706 | | | | | | | | | |
| 2.400 | 1.299 | .000 | 1.101 | .41 | 1.31 | 45.9% | 2 | .706 | .593 | | | | | | | | |
| 2.400 | 1.756 | .000 | .644 | .54 | 1.43 | 26.8% | 3 | .706 | .593 | .456 | | | | | | | |
| 2.400 | 2.067 | .000 | .333 | .69 | 1.49 | 13.9% | 4 | .706 | .593 | .456 | .311 | | | | | | |
| 2.400 | 2.250 | .000 | .150 | .82 | 1.49 | 6.2% | 5 | .706 | .593 | .456 | .311 | .183 | | | | | |
| 2.400 | 2.342 | .000 | .058 | .91 | 1.46 | 2.4% | 6 | .706 | .593 | .456 | .311 | .183 | .091 | | | | |
| 2.400 | 2.380 | .000 | .020 | .96 | 1.41 | .8% | 7 | .706 | .593 | .456 | .311 | .183 | .091 | .039 | | | |
| 2.600 | .722 | .000 | 1.878 | .28 | 1.16 | 72.2% | 1 | .722 | | | | | | | | | |
| 2.600 | 1.341 | .000 | 1.259 | .38 | 1.31 | 48.4% | 2 | .722 | .619 | | | | | | | | |
| 2.600 | 1.831 | .000 | .769 | .51 | 1.43 | 29.6% | 3 | .722 | .619 | .490 | | | | | | | |
| 2.600 | 2.181 | .000 | .419 | .65 | 1.50 | 16.1% | 4 | .722 | .619 | .490 | .350 | | | | | | |
| 2.600 | 2.399 | .000 | .201 | .78 | 1.52 | 7.7% | 5 | .722 | .619 | .490 | .350 | .218 | | | | | |
| 2.600 | 2.516 | .000 | .084 | .88 | 1.50 | 3.2% | 6 | .722 | .619 | .490 | .350 | .218 | .117 | | | | |
| 2.600 | 2.569 | .000 | .031 | .95 | 1.45 | 1.2% | 7 | .722 | .619 | .490 | .350 | .218 | .117 | .053 | | | |
| 2.800 | .737 | .000 | 2.063 | .26 | 1.15 | 73.7% | 1 | .737 | | | | | | | | | |
| 2.800 | 1.378 | .000 | 1.422 | .36 | 1.30 | 50.8% | 2 | .737 | .641 | | | | | | | | |
| 2.800 | 1.900 | .000 | .900 | .48 | 1.43 | 32.2% | 3 | .737 | .641 | .521 | | | | | | | |
| 2.800 | 2.286 | .000 | .514 | .61 | 1.52 | 18.4% | 4 | .737 | .641 | .521 | .386 | | | | | | |
| 2.800 | 2.539 | .000 | .261 | .75 | 1.55 | 9.3% | 5 | .737 | .641 | .521 | .386 | .253 | | | | | |
| 2.800 | 2.683 | .000 | .117 | .86 | 1.53 | 4.2% | 6 | .737 | .641 | .521 | .386 | .253 | .144 | | | | |
| 2.800 | 2.754 | .000 | .046 | .93 | 1.49 | 1.6% | 7 | .737 | .641 | .521 | .386 | .253 | .144 | .071 | | | |
| 2.800 | 2.784 | .000 | .016 | .97 | 1.43 | .6% | 8 | .737 | .641 | .521 | .386 | .253 | .144 | .071 | .030 | | |
| 3.000 | .750 | .000 | 2.250 | .25 | 1.15 | 75.0% | 1 | .750 | | | | | | | | | |
| 3.000 | 1.412 | .000 | 1.588 | .34 | 1.30 | 52.9% | 2 | .750 | .662 | | | | | | | | |
| 3.000 | 1.962 | .000 | 1.038 | .45 | 1.43 | 34.6% | 3 | .750 | .662 | .550 | | | | | | | |
| 3.000 | 2.382 | .000 | .618 | .58 | 1.53 | 20.6% | 4 | .750 | .662 | .550 | .420 | | | | | | |
| 3.000 | 2.670 | .000 | .330 | .71 | 1.57 | 11.0% | 5 | .750 | .662 | .550 | .420 | .288 | | | | | |
| 3.000 | 2.844 | .000 | .156 | .83 | 1.57 | 5.2% | 6 | .750 | .662 | .550 | .420 | .288 | .174 | | | | |
| 3.000 | 2.934 | .000 | .066 | .91 | 1.53 | 2.2% | 7 | .750 | .662 | .550 | .420 | .288 | .174 | .091 | | | |
| 3.000 | 2.976 | .000 | .024 | .96 | 1.47 | .8% | 8 | .750 | .662 | .550 | .420 | .288 | .174 | .091 | .041 | | |

# ERLANG B: 100% BLOCKED CALLERS ROUTE ADVANCE, OVERFLOW, OR DISAPPEAR

| HOURS FIRST ATTEMPT VMR=1.0 | HOURS CON-NECT | HOURS RECALL /RETRY | HOURS OVFL | CON-NECT VMR | OVFL VMR | PERCENT BLOCKED (GRADE) | LINE | HOURS CONNECTED TRAFFIC ON LINES 1 | 2 | 3 | 4 | 5 | 6 | 7 | 8 | 9 | 10 |
|---|---|---|---|---|---|---|---|---|---|---|---|---|---|---|---|---|---|
| 3.200 | 1.442 | .000 | 1.758 | .32 | 1.30 | 54.9% | 2 | .762 | .680 | | | | | | | | |
| 3.200 | 2.018 | .000 | 1.182 | .42 | 1.43 | 36.9% | 3 | .762 | .680 | .576 | | | | | | | |
| 3.200 | 2.470 | .000 | .730 | .55 | 1.53 | 22.8% | 4 | .762 | .680 | .576 | .452 | | | | | | |
| 3.200 | 2.792 | .000 | .408 | .68 | 1.59 | 12.7% | 5 | .762 | .680 | .576 | .452 | .322 | | | | | |
| 3.200 | 2.996 | .000 | .204 | .80 | 1.60 | 6.4% | 6 | .762 | .680 | .576 | .452 | .322 | .204 | | | | |
| 3.200 | 3.110 | .000 | .090 | .89 | 1.56 | 2.8% | 7 | .762 | .680 | .576 | .452 | .322 | .204 | .113 | | | |
| 3.200 | 3.164 | .000 | .036 | .95 | 1.51 | 1.1% | 8 | .762 | .680 | .576 | .452 | .322 | .204 | .113 | .055 | | |
| 3.400 | 1.470 | .000 | 1.930 | .30 | 1.29 | 56.8% | 2 | .773 | .697 | | | | | | | | |
| 3.400 | 2.069 | .000 | 1.331 | .40 | 1.43 | 39.2% | 3 | .773 | .697 | .599 | | | | | | | |
| 3.400 | 2.551 | .000 | .849 | .52 | 1.54 | 25.0% | 4 | .773 | .697 | .599 | .482 | | | | | | |
| 3.400 | 2.906 | .000 | .494 | .64 | 1.61 | 14.5% | 5 | .773 | .697 | .599 | .482 | .355 | | | | | |
| 3.400 | 3.142 | .000 | .258 | .76 | 1.62 | 7.6% | 6 | .773 | .697 | .599 | .482 | .355 | .235 | | | | |
| 3.400 | 3.279 | .000 | .121 | .86 | 1.60 | 3.6% | 7 | .773 | .697 | .599 | .482 | .355 | .235 | .137 | | | |
| 3.400 | 3.349 | .000 | .051 | .93 | 1.55 | 1.5% | 8 | .773 | .697 | .599 | .482 | .355 | .235 | .137 | .070 | | |
| 3.400 | 3.381 | .000 | .019 | .97 | 1.49 | .6% | 9 | .773 | .697 | .599 | .482 | .355 | .235 | .137 | .070 | .032 | |
| 3.600 | 1.495 | .000 | 2.105 | .29 | 1.29 | 58.5% | 2 | .783 | .712 | | | | | | | | |
| 3.600 | 2.115 | .000 | 1.485 | .38 | 1.43 | 41.2% | 3 | .783 | .712 | .621 | | | | | | | |
| 3.600 | 2.626 | .000 | .974 | .49 | 1.54 | 27.1% | 4 | .783 | .712 | .621 | .510 | | | | | | |
| 3.600 | 3.013 | .000 | .587 | .61 | 1.62 | 16.3% | 5 | .783 | .712 | .621 | .510 | .387 | | | | | |
| 3.600 | 3.279 | .000 | .321 | .73 | 1.65 | 8.9% | 6 | .783 | .712 | .621 | .510 | .387 | .266 | | | | |
| 3.600 | 3.442 | .000 | .158 | .84 | 1.63 | 4.4% | 7 | .783 | .712 | .621 | .510 | .387 | .266 | .163 | | | |
| 3.600 | 3.530 | .000 | .070 | .91 | 1.59 | 1.9% | 8 | .783 | .712 | .621 | .510 | .387 | .266 | .163 | .088 | | |
| 3.600 | 3.572 | .000 | .028 | .96 | 1.53 | .8% | 9 | .783 | .712 | .621 | .510 | .387 | .266 | .163 | .088 | .042 | |
| 3.800 | 1.517 | .000 | 2.283 | .27 | 1.28 | 60.1% | 2 | .792 | .726 | | | | | | | | |
| 3.800 | 2.158 | .000 | 1.642 | .36 | 1.42 | 43.2% | 3 | .792 | .726 | .641 | | | | | | | |
| 3.800 | 2.694 | .000 | 1.106 | .46 | 1.54 | 29.1% | 4 | .792 | .726 | .641 | .536 | | | | | | |
| 3.800 | 3.112 | .000 | .688 | .58 | 1.63 | 18.1% | 5 | .792 | .726 | .641 | .536 | .418 | | | | | |
| 3.800 | 3.409 | .000 | .391 | .70 | 1.67 | 10.3% | 6 | .792 | .726 | .641 | .536 | .418 | .297 | | | | |
| 3.800 | 3.599 | .000 | .201 | .81 | 1.66 | 5.3% | 7 | .792 | .726 | .641 | .536 | .418 | .297 | .190 | | | |
| 3.800 | 3.707 | .000 | .093 | .89 | 1.62 | 2.5% | 8 | .792 | .726 | .641 | .536 | .418 | .297 | .190 | .108 | | |
| 3.800 | 3.761 | .000 | .039 | .95 | 1.57 | 1.0% | 9 | .792 | .726 | .641 | .536 | .418 | .297 | .190 | .108 | .054 | |
| 4.000 | 1.538 | .000 | 2.462 | .26 | 1.28 | 61.5% | 2 | .800 | .738 | | | | | | | | |
| 4.000 | 2.197 | .000 | 1.803 | .34 | 1.42 | 45.1% | 3 | .800 | .738 | .659 | | | | | | | |
| 4.000 | 2.757 | .000 | 1.243 | .44 | 1.54 | 31.1% | 4 | .800 | .738 | .659 | .560 | | | | | | |
| 4.000 | 3.204 | .000 | .796 | .55 | 1.63 | 19.9% | 5 | .800 | .738 | .659 | .560 | .446 | | | | | |
| 4.000 | 3.531 | .000 | .469 | .67 | 1.68 | 11.7% | 6 | .800 | .738 | .659 | .560 | .446 | .328 | | | | |
| 4.000 | 3.749 | .000 | .251 | .78 | 1.69 | 6.3% | 7 | .800 | .738 | .659 | .560 | .446 | .328 | .218 | | | |
| 4.000 | 3.878 | .000 | .122 | .87 | 1.66 | 3.0% | 8 | .800 | .738 | .659 | .560 | .446 | .328 | .218 | .129 | | |
| 4.000 | 3.947 | .000 | .053 | .93 | 1.61 | 1.3% | 9 | .800 | .738 | .659 | .560 | .446 | .328 | .218 | .129 | .068 | |
| 4.000 | 3.979 | .000 | .021 | .97 | 1.55 | .5% | 10 | .800 | .738 | .659 | .560 | .446 | .328 | .218 | .129 | .068 | .032 |

# ERLANG B: 100% BLOCKED CALLERS ROUTE ADVANCE, OVERFLOW, OR DISAPPEAR

| HOURS FIRST ATTEMPT VMR = 1.0 | HOURS CON-NECT | HOURS RECALL /RETRY | HOURS OVFL | CON-NECT VMR | OVFL VMR | PERCENT BLOCKED (GRADE) | LINE | HOURS CONNECTED TRAFFIC ON LINES | | | | | | | | | |
|---|---|---|---|---|---|---|---|---|---|---|---|---|---|---|---|---|---|
| | | | | | | | | 1 | 2 | 3 | 4 | 5 | 6 | 7 | 8 | 9 | 10 |
| 4.200 | 1.558 | .000 | 2.642 | .25 | 1.27 | 62.9% | 2 | .808 | .750 | | | | | | | | |
| 4.200 | 2.233 | .000 | 1.967 | .32 | 1.41 | 46.8% | 3 | .808 | .750 | .675 | | | | | | | |
| 4.200 | 2.816 | .000 | 1.384 | .42 | 1.54 | 33.0% | 4 | .808 | .750 | .675 | .582 | | | | | | |
| 4.200 | 3.289 | .000 | .911 | .53 | 1.64 | 21.7% | 5 | .808 | .750 | .675 | .582 | .474 | | | | | |
| 4.200 | 3.646 | .000 | .554 | .64 | 1.70 | 13.2% | 6 | .808 | .750 | .675 | .582 | .474 | .357 | | | | |
| 4.200 | 3.892 | .000 | .308 | .75 | 1.71 | 7.3% | 7 | .808 | .750 | .675 | .582 | .474 | .357 | .246 | | | |
| 4.200 | 4.044 | .000 | .156 | .85 | 1.69 | 3.7% | 8 | .808 | .750 | .675 | .582 | .474 | .357 | .246 | .152 | | |
| 4.200 | 4.129 | .000 | .071 | .92 | 1.64 | 1.7% | 9 | .808 | .750 | .675 | .582 | .474 | .357 | .246 | .152 | .084 | |
| 4.200 | 4.170 | .000 | .030 | .96 | 1.59 | .7% | 10 | .808 | .750 | .675 | .582 | .474 | .357 | .246 | .152 | .084 | |
| 4.400 | 1.576 | .000 | 2.824 | .24 | 1.26 | 64.2% | 2 | .815 | .761 | | | | | | | | |
| 4.400 | 2.266 | .000 | 2.134 | .31 | 1.40 | 48.5% | 3 | .815 | .761 | .691 | | | | | | | |
| 4.400 | 2.869 | .000 | 1.531 | .40 | 1.53 | 34.8% | 4 | .815 | .761 | .691 | .603 | | | | | | |
| 4.400 | 3.369 | .000 | 1.031 | .50 | 1.64 | 23.4% | 5 | .815 | .761 | .691 | .603 | .499 | | | | | |
| 4.400 | 3.755 | .000 | .645 | .61 | 1.71 | 14.7% | 6 | .815 | .761 | .691 | .603 | .499 | .386 | | | | |
| 4.400 | 4.029 | .000 | .371 | .73 | 1.74 | 8.4% | 7 | .815 | .761 | .691 | .603 | .499 | .386 | .274 | | | |
| 4.400 | 4.205 | .000 | .195 | .82 | 1.72 | 4.4% | 8 | .815 | .761 | .691 | .603 | .499 | .386 | .274 | .176 | | |
| 4.400 | 4.307 | .000 | .093 | .90 | 1.68 | 2.1% | 9 | .815 | .761 | .691 | .603 | .499 | .386 | .274 | .176 | .102 | |
| 4.400 | 4.359 | .000 | .041 | .95 | 1.62 | .9% | 10 | .815 | .761 | .691 | .603 | .499 | .386 | .274 | .176 | .102 | .053 |
| 4.600 | 1.592 | .000 | 3.008 | .23 | 1.26 | 65.4% | 2 | .821 | .771 | | | | | | | | |
| 4.600 | 2.297 | .000 | 2.303 | .30 | 1.40 | 50.1% | 3 | .821 | .771 | .705 | | | | | | | |
| 4.600 | 2.919 | .000 | 1.681 | .38 | 1.53 | 36.5% | 4 | .821 | .771 | .705 | .622 | | | | | | |
| 4.600 | 3.443 | .000 | 1.157 | .48 | 1.64 | 25.2% | 5 | .821 | .771 | .705 | .622 | .523 | | | | | |
| 4.600 | 3.856 | .000 | .744 | .59 | 1.72 | 16.2% | 6 | .821 | .771 | .705 | .622 | .523 | .413 | | | | |
| 4.600 | 4.158 | .000 | .442 | .70 | 1.76 | 9.6% | 7 | .821 | .771 | .705 | .622 | .523 | .413 | .302 | | | |
| 4.600 | 4.359 | .000 | .241 | .80 | 1.75 | 5.2% | 8 | .821 | .771 | .705 | .622 | .523 | .413 | .302 | .201 | | |
| 4.600 | 4.480 | .000 | .120 | .88 | 1.71 | 2.0% | 9 | .821 | .771 | .705 | .622 | .523 | .413 | .302 | .201 | .121 | |
| 4.600 | 4.546 | .000 | .054 | .93 | 1.66 | 1.2% | 10 | .821 | .771 | .705 | .622 | .523 | .413 | .302 | .201 | .121 | .065 |
| 4.800 | 1.607 | .000 | 3.193 | .22 | 1.25 | 66.5% | 2 | .828 | .780 | | | | | | | | |
| 4.800 | 2.325 | .000 | 2.475 | .28 | 1.39 | 51.6% | 3 | .828 | .780 | .718 | | | | | | | |
| 4.800 | 2.965 | .000 | 1.835 | .36 | 1.52 | 38.2% | 4 | .828 | .780 | .718 | .640 | | | | | | |
| 4.800 | 3.512 | .000 | 1.288 | .45 | 1.64 | 26.8% | 5 | .828 | .780 | .718 | .640 | .546 | | | | | |
| 4.800 | 3.951 | .000 | .849 | .56 | 1.73 | 17.7% | 6 | .828 | .780 | .718 | .640 | .546 | .440 | | | | |
| 4.800 | 4.281 | .000 | .519 | .67 | 1.77 | 10.8% | 7 | .828 | .780 | .718 | .640 | .546 | .440 | .330 | | | |
| 4.800 | 4.508 | .000 | .292 | .77 | 1.78 | 6.1% | 8 | .828 | .780 | .718 | .640 | .546 | .440 | .330 | .227 | | |
| 4.800 | 4.649 | .000 | .151 | .86 | 1.75 | 3.1% | 9 | .828 | .780 | .718 | .640 | .546 | .440 | .330 | .227 | .141 | |
| 4.800 | 4.729 | .000 | .071 | .92 | 1.69 | 1.5% | 10 | .828 | .780 | .718 | .640 | .546 | .440 | .330 | .227 | .141 | .080 |
| 4.800 | 4.769 | .000 | .031 | .96 | 1.63 | .6% | 11 | .828 | .780 | .718 | .640 | .546 | .440 | .330 | .227 | .141 | .080 |
| | | | | | | | | .040 | | | | | | | | | |
| 5.000 | 1.622 | .000 | 3.378 | .21 | 1.25 | 67.6% | 2 | .833 | .788 | | | | | | | | |
| 5.000 | 2.352 | .000 | 2.648 | .27 | 1.39 | 53.0% | 3 | .833 | .788 | .730 | | | | | | | |

# ERLANG B: 100% BLOCKED CALLERS ROUTE ADVANCE, OVERFLOW, OR DISAPPEAR

| HOURS FIRST ATTEMPT VMR=1.0 | HOURS CON-NECT | HOURS RECALL /RETRY | HOURS OVFL | CON-NECT VMR | OVFL VMR | PERCENT BLOCKED (GRADE) | LINE | HOURS CONNECTED TRAFFIC ON LINES 1 | 2 | 3 | 4 | 5 | 6 | 7 | 8 | 9 | 10 |
|---|---|---|---|---|---|---|---|---|---|---|---|---|---|---|---|---|---|
| 5.000 | 3.008 | .000 | 1.992 | .34 | 1.52 | 39.8% | 4 | .833 | .788 | .730 | .657 | | | | | | |
| 5.000 | 3.576 | .000 | 1.424 | .43 | 1.64 | 28.5% | 5 | .833 | .788 | .730 | .657 | .567 | | | | | |
| 5.000 | 4.041 | .000 | .959 | .53 | 1.73 | 19.2% | 6 | .833 | .788 | .730 | .657 | .567 | .465 | | | | |
| 5.000 | 4.397 | .000 | .603 | .64 | 1.79 | 12.1% | 7 | .833 | .788 | .730 | .657 | .567 | .465 | .357 | | | |
| 5.000 | 4.650 | .000 | .350 | .75 | 1.80 | 7.0% | 8 | .833 | .788 | .730 | .657 | .567 | .465 | .357 | .252 | | |
| 5.000 | 4.813 | .000 | .187 | .84 | 1.78 | 3.7% | 9 | .833 | .788 | .730 | .657 | .567 | .465 | .357 | .252 | .163 | |
| 5.000 | 4.908 | .000 | .092 | .90 | 1.73 | 1.8% | .10 | .833 | .788 | .730 | .657 | .567 | .465 | .357 | .252 | .163 | .095 |
| 5.000 | 4.959 | .000 | .041 | .95 | 1.67 | .8% | 11 | .833 | .788 | .730 | .657 | .567 | .465 | .357 | .252 | .163 | .095 |
| | | | | | | | | .050 | | | | | | | | | |
| 5.500 | 1.653 | .000 | 3.847 | .19 | 1.24 | 69.9% | 2 | .846 | .807 | | | | | | | | |
| 5.500 | 2.410 | .000 | 3.090 | .24 | 1.37 | 56.2% | 3 | .846 | .807 | .757 | | | | | | | |
| 5.500 | 3.103 | .000 | 2.397 | .31 | 1.50 | 43.6% | 4 | .846 | .807 | .757 | .693 | | | | | | |
| 5.500 | 3.718 | .000 | 1.782 | .39 | 1.63 | 32.4% | 5 | .846 | .807 | .757 | .693 | .615 | | | | | |
| 5.500 | 4.240 | .000 | 1.260 | .48 | 1.73 | 22.9% | 6 | .846 | .807 | .757 | .693 | .615 | .523 | | | | |
| 5.500 | 4.661 | .000 | .839 | .58 | 1.81 | 15.3% | 7 | .846 | .807 | .757 | .693 | .615 | .523 | .421 | | | |
| 5.500 | 4.978 | .000 | .522 | .68 | 1.85 | 9.5% | 8 | .846 | .807 | .757 | .693 | .615 | .523 | .421 | .317 | | |
| 5.500 | 5.199 | .000 | .301 | .78 | 1.84 | 5.5% | 9 | .846 | .807 | .757 | .693 | .615 | .523 | .421 | .317 | .220 | |
| 5.500 | 5.339 | .000 | .161 | .86 | 1.81 | 2.9% | 10 | .846 | .807 | .757 | .693 | .615 | .523 | .421 | .317 | .220 | .141 |
| 5.500 | 5.421 | .000 | .079 | .92 | 1.76 | 1.4% | 11 | .846 | .807 | .757 | .693 | .615 | .523 | .421 | .317 | .220 | .141 |
| | | | | | | | | .082 | | | | | | | | | |
| 5.500 | 5.464 | .000 | .036 | .96 | 1.69 | .7% | 12 | .846 | .807 | .757 | .693 | .615 | .523 | .421 | .317 | .220 | .141 |
| | | | | | | | | .082 | .043 | | | | | | | | |
| 6.000 | 1.680 | .000 | 4.320 | .18 | 1.23 | 72.0% | 2 | .857 | .823 | | | | | | | | |
| 6.000 | 2.459 | .000 | 3.541 | .22 | 1.35 | 59.0% | 3 | .857 | .823 | .779 | | | | | | | |
| 6.000 | 3.183 | .000 | 2.817 | .28 | 1.48 | 47.0% | 4 | .857 | .823 | .779 | .724 | | | | | | |
| 6.000 | 3.838 | .000 | 2.162 | .35 | 1.61 | 36.0% | 5 | .857 | .823 | .779 | .724 | .655 | | | | | |
| 6.000 | 4.410 | .000 | 1.590 | .43 | 1.73 | 26.5% | 6 | .857 | .823 | .779 | .724 | .655 | .573 | | | | |
| 6.000 | 4.890 | .000 | 1.110 | .52 | 1.82 | 18.5% | 7 | .857 | .823 | .779 | .724 | .655 | .573 | .479 | | | |
| 6.000 | 5.269 | .000 | .731 | .62 | 1.88 | 12.2% | 8 | .857 | .823 | .779 | .724 | .655 | .573 | .479 | .379 | | |
| 6.000 | 5.549 | .000 | .451 | .72 | 1.90 | 7.5% | 9 | .857 | .823 | .779 | .724 | .655 | .573 | .479 | .379 | .280 | |
| 6.000 | 5.741 | .000 | .259 | .81 | 1.88 | 4.3% | 10 | .857 | .823 | .779 | .724 | .655 | .573 | .479 | .379 | .280 | .192 |
| 6.000 | 5.862 | .000 | .138 | .88 | 1.84 | 2.3% | 11 | .857 | .823 | .779 | .724 | .655 | .573 | .479 | .379 | .280 | .192 |
| | | | | | | | | .121 | | | | | | | | | |
| 6.000 | 5.932 | .000 | .068 | .93 | 1.78 | 1.1% | 12 | .857 | .823 | .779 | .724 | .655 | .573 | .479 | .379 | .280 | .192 |
| | | | | | | | | .121 | .070 | | | | | | | | |
| 6.000 | 5.969 | .000 | .031 | .96 | 1.72 | .5% | 13 | .857 | .823 | .779 | .724 | .655 | .573 | .479 | .379 | .280 | .192 |
| | | | | | | | | .121 | .070 | .037 | | | | | | | |
| 6.500 | 1.703 | .000 | 4.797 | .16 | 1.21 | 73.8% | 2 | .867 | .836 | | | | | | | | |
| 6.500 | 2.501 | .000 | 3.999 | .20 | 1.34 | 61.5% | 3 | .867 | .836 | .798 | | | | | | | |
| 6.500 | 3.250 | .000 | 3.250 | .25 | 1.47 | 50.0% | 4 | .867 | .836 | .798 | .749 | | | | | | |

# ERLANG B: 100% BLOCKED CALLERS ROUTE ADVANCE, OVERFLOW, OR DISAPPEAR

| HOURS FIRST ATTEMPT VMR=1.0 | HOURS CON-NECT | HOURS RECALL /RETRY | HOURS OVFL | CON-NECT VMR | OVFL VMR | PERCENT BLOCKED (GRADE) | LINE | HOURS CONNECTED TRAFFIC ON LINES 1 | 2 | 3 | 4 | 5 | 6 | 7 | 8 | 9 | 10 |
|---|---|---|---|---|---|---|---|---|---|---|---|---|---|---|---|---|---|
| 6.500 | 3.940 | .000 | 2.560 | .31 | 1.59 | 39.4% | 5 | .867 | .836 | .798 | .749 | .689 | | | | | |
| 6.500 | 4.556 | .000 | 1.944 | .38 | 1.72 | 29.9% | 6 | .867 | .836 | .798 | .749 | .689 | .616 | | | | |
| 6.500 | 5.087 | .000 | 1.413 | .47 | 1.82 | 21.7% | 7 | .867 | .836 | .798 | .749 | .689 | .616 | .531 | | | |
| 6.500 | 5.624 | .000 | .976 | .56 | 1.89 | 15.0% | 8 | .867 | .836 | .798 | .749 | .689 | .616 | .531 | .437 | | |
| 6.500 | 5.864 | .000 | .636 | .66 | 1.94 | 9.8% | 9 | .867 | .836 | .798 | .836 | .798 | .749 | .689 | .616 | .531 | .437 |
| .340 | | | | | | | | | | | | | | | | | |
| 6.500 | 6.111 | .000 | .389 | .75 | 1.94 | 6.0% | 10 | .867 | .836 | .798 | .749 | .689 | .616 | .531 | .437 | .340 | .247 |
| 6.500 | 6.278 | .000 | .222 | .83 | 1.91 | 3.4% | 11 | .867 | .836 | .798 | .749 | .689 | .616 | .531 | .437 | .340 | .247 |
| | | | | | | | | .167 | | | | | | | | | |
| 6.500 | 6.382 | .000 | .118 | .90 | 1.86 | 1.8% | 12 | .867 | .836 | .798 | .749 | .689 | .616 | .531 | .437 | .340 | .247 |
| | | | | | | | | .167 | .104 | | | | | | | | |
| 6.500 | 6.442 | .000 | .058 | .94 | 1.80 | .9% | 13 | .867 | .936 | .798 | .749 | .689 | .616 | .531 | .437 | .340 | .247 |
| | | | | | | | | .167 | .104 | .059 | | | | | | | |
| 6.500 | 6.442 | .000 | .058 | .94 | 1.80 | .9% | 13 | .867 | .836 | .798 | .749 | .689 | .616 | .531 | .437 | .340 | .247 |
| | | | | | | | | .167 | .104 | .059 | | | | | | | |
| 7.000 | 2.537 | .000 | 4.463 | .19 | 1.32 | 63.8% | 3 | .875 | .848 | .814 | | | | | | | |
| 7.000 | 3.309 | .000 | 3.691 | .23 | 1.45 | 52.7% | 4 | .875 | .848 | .814 | .771 | | | | | | |
| 7.000 | 4.027 | .000 | 2.973 | .28 | 1.57 | 42.5% | 5 | .875 | .848 | .814 | .771 | .718 | | | | | |
| 7.000 | 4.681 | .000 | 2.319 | .35 | 1.70 | 33.1% | 6 | .875 | .848 | .814 | .771 | .718 | .654 | | | | |
| 7.000 | 5.258 | .000 | 1.742 | .42 | 1.81 | 24.9% | 7 | .875 | .848 | .814 | .771 | .718 | .654 | .577 | | | |
| 7.000 | 5.748 | .000 | 1.252 | .51 | 1.90 | 17.9% | 8 | .875 | .848 | .814 | .771 | .718 | .654 | .577 | .490 | | |
| 7.000 | 6.145 | .000 | .855 | .60 | 1.96 | 12.2% | 9 | .875 | .848 | .814 | .771 | .718 | .654 | .577 | .490 | .397 | |
| 7.000 | 6.449 | .000 | .551 | .70 | 1.99 | 7.9% | 10 | .875 | .848 | .814 | .771 | .718 | .654 | .577 | .397 | .304 | |
| 7.000 | 6.666 | .000 | .334 | .78 | 1.98 | 4.8% | 11 | .875 | .848 | .814 | .771 | .718 | .654 | .577 | .490 | .397 | .304 |
| | | | | | | | | .217 | .144 | | | | | | | | |
| 7.000 | 6.899 | .000 | .101 | .91 | 1.89 | 1.4% | 13 | .875 | .848 | .814 | .771 | .718 | .654 | .577 | .490 | .397 | .304 |
| | | | | | | | | .217 | .144 | .080 | | | | | | | |
| 7.000 | 6.950 | .000 | .050 | .95 | 1.82 | .7% | 14 | .875 | .848 | .814 | .771 | .718 | .654 | .577 | .490 | .397 | .304 |
| | | | | | | | | .217 | .144 | .089 | .051 | | | | | | |
| 7.500 | 2.569 | .000 | 4.931 | .17 | 1.31 | 65.8% | 3 | .882 | .858 | .828 | | | | | | | |
| 7.500 | 3.359 | .000 | 4.141 | .21 | 1.43 | 55.2% | 4 | .882 | .858 | .828 | .790 | | | | | | |
| 7.500 | 4.102 | .000 | 3.398 | .26 | 1.55 | 45.3% | 5 | .882 | .858 | .828 | .790 | .743 | | | | | |
| 7.500 | 4.788 | .000 | 2.712 | .31 | 1.68 | 36.2% | 6 | .882 | .858 | .828 | .790 | .743 | .686 | | | | |
| 7.500 | 5.406 | .000 | 2.094 | .38 | 1.80 | 27.9% | 7 | .882 | .858 | .828 | .790 | .743 | .686 | .617 | | | |
| 7.500 | 5.944 | .000 | 1.556 | .46 | 1.90 | 20.7% | 8 | .882 | .858 | .828 | .790 | .743 | .686 | .617 | .538 | | |
| 7.500 | 6.395 | .000 | 1.105 | .55 | 1.97 | 14.7% | 9 | .882 | .858 | .828 | .790 | .743 | .686 | .617 | .538 | .450 | |
| 7.500 | 6.753 | .000 | .747 | .64 | 2.02 | 10.0% | 10 | .882 | .858 | .828 | .790 | .743 | .686 | .617 | .538 | .450 | .359 |
| 7.500 | 7.023 | .000 | .477 | .73 | 2.03 | 6.4% | 11 | .882 | .858 | .828 | .790 | .743 | .686 | .617 | .538 | .450 | .359 |
| | | | | | | | | .270 | | | | | | | | | |
| 7.500 | 7.213 | .000 | .287 | .81 | 2.01 | 3.8% | 12 | .882 | .858 | .828 | .790 | .743 | .686 | .617 | .538 | .450 | .359 |
| | | | | | | | | .270 | .190 | | | | | | | | |

# ERLANG B: 100% BLOCKED CALLERS ROUTE ADVANCE, OVERFLOW, OR DISAPPEAR

| HOURS FIRST ATTEMPT VMR=1.0 | HOURS CONNECT | HOURS RECALL /RETRY | HOURS OVFL | CONNECT VMR | OVFL VMR | PERCENT BLOCKED (GRADE) | LINE | HOURS CONNECTED TRAFFIC ON LINES 1 | 2 | 3 | 4 | 5 | 6 | 7 | 8 | 9 | 10 |
|---|---|---|---|---|---|---|---|---|---|---|---|---|---|---|---|---|---|
| 7.500 | 7.338 | .000 | .162 | .88 | 1.96 | 2.2% | 13 | .882 | .858 | .828 | .790 | .743 | .686 | .617 | .538 | .450 | .359 |
| | | | | | | | | .270 | .190 | .125 | | | | | | | |
| 7.500 | 7.414 | .000 | .086 | .92 | 1.90 | 1.1% | 14 | .882 | .858 | .828 | .790 | .743 | .686 | .617 | .538 | .450 | .359 |
| | | | | | | | | .270 | .190 | .125 | .076 | | | | | | |
| 7.500 | 7.457 | .000 | .043 | .96 | 1.84 | .6% | 15 | .882 | .858 | .828 | .790 | .743 | .686 | .617 | .538 | .450 | .359 |
| | | | | | | | | .270 | .190 | .125 | .076 | .043 | | | | | |
| 8.000 | 2.596 | .000 | 5.404 | .16 | 1.30 | 67.5% | 3 | .889 | .867 | .840 | | | | | | | |
| 8.000 | 3.403 | .000 | 4.597 | .19 | 1.41 | 57.5% | 4 | .889 | .867 | .840 | .807 | | | | | | |
| 8.000 | 4.168 | .000 | 3.832 | .23 | 1.53 | 47.9% | 5 | .889 | .867 | .840 | .807 | .765 | | | | | |
| 8.000 | 4.882 | .000 | 3.118 | .29 | 1.66 | 39.0% | 6 | .889 | .867 | .840 | .807 | .765 | .714 | | | | |
| 8.000 | 5.535 | .000 | 2.465 | .35 | 1.78 | 30.8% | 7 | .889 | .867 | .840 | .807 | .765 | .714 | .653 | | | |
| 8.000 | 6.115 | .000 | 1.885 | .42 | 1.89 | 23.6% | 8 | .889 | .867 | .840 | .807 | .765 | .714 | .653 | .581 | | |
| 8.000 | 6.615 | .000 | 1.385 | .50 | 1.98 | 17.3% | 9 | .889 | .867 | .840 | .807 | .765 | .714 | .653 | .581 | .499 | |
| 8.000 | 7.027 | .000 | .973 | .59 | 2.04 | 12.2% | 10 | .889 | .867 | .840 | .807 | .765 | .714 | .653 | .581 | .499 | .412 |
| 8.000 | 7.350 | .000 | .650 | .68 | 2.07 | 8.1% | 11 | .889 | .867 | .840 | .807 | .765 | .714 | .653 | .581 | .499 | .412 |
| | | | | | | | | .323 | | | | | | | | | |
| 8.000 | 7.589 | .000 | .411 | .76 | 2.07 | 5.1% | 12 | .889 | .867 | .840 | .807 | .765 | .714 | .653 | .581 | .499 | .412 |
| | | | | | | | | .323 | .239 | | | | | | | | |
| 8.000 | 7.755 | .000 | .245 | .83 | 2.04 | 3.1% | 13 | .889 | .867 | .840 | .807 | .765 | .714 | .653 | .581 | .499 | .412 |
| | | | | | | | | .323 | .239 | .166 | | | | | | | |
| 8.000 | 7.862 | .000 | .138 | .89 | 1.98 | 1.7% | 14 | .889 | .867 | .840 | .807 | .765 | .714 | .653 | .581 | .499 | .412 |
| | | | | | | | | .323 | .239 | .166 | .108 | | | | | | |
| 8.000 | 7.927 | .000 | .073 | .94 | 1.92 | .9% | 15 | .889 | .867 | .840 | .807 | .765 | .714 | .653 | .581 | .499 | .412 |
| | | | | | | | | .323 | .239 | .166 | .108 | .065 | | | | | |
| 8.500 | 2.621 | .000 | 5.879 | .15 | 1.28 | 69.2% | 3 | .895 | .875 | .851 | | | | | | | |
| 8.500 | 3.442 | .000 | 5.058 | .18 | 1.40 | 59.5% | 4 | .895 | .875 | .851 | .821 | | | | | | |
| 8.500 | 4.225 | .000 | 4.275 | .22 | 1.51 | 50.3% | 5 | .895 | .875 | .851 | .821 | .784 | | | | | |
| 8.500 | 4.964 | .000 | 3.536 | .26 | 1.64 | 41.6% | 6 | .895 | .875 | .851 | .821 | .784 | .738 | | | | |
| 8.500 | 5.647 | .000 | 2.853 | .32 | 1.76 | 33.6% | 7 | .895 | .875 | .851 | .821 | .784 | .738 | .683 | | | |
| 8.500 | 6.266 | .000 | 2.234 | .38 | 1.87 | 26.3% | 8 | .895 | .875 | .851 | .821 | .784 | .738 | .683 | .618 | | |
| 8.500 | 6.809 | .000 | 1.691 | .46 | 1.97 | 19.9% | 9 | .895 | .875 | .851 | .821 | .784 | .738 | .683 | .618 | .544 | |
| 8.500 | 7.271 | .000 | 1.229 | .54 | 2.05 | 14.5% | 10 | .895 | .875 | .851 | .821 | .784 | .738 | .683 | .618 | .544 | .461 |
| 8.500 | 7.646 | .000 | .854 | .63 | 2.10 | 10.1% | 11 | .895 | .875 | .851 | .821 | .784 | .738 | .683 | .618 | .544 | .461 |
| | | | | | | | | .375 | | | | | | | | | |
| 8.500 | 7.935 | .000 | .565 | .71 | 2.11 | 6.6% | 12 | .895 | .875 | .851 | .821 | .784 | .738 | .683 | .618 | .544 | .461 |
| | | | | | | | | .375 | .289 | | | | | | | | |
| 8.500 | 8.146 | .000 | .354 | .79 | 2.10 | 4.2% | 13 | .895 | .875 | .851 | .821 | .784 | .738 | .683 | .618 | .544 | .461 |
| | | | | | | | | .375 | .289 | .211 | | | | | | | |
| 8.500 | 8.290 | .000 | .210 | .86 | 2.06 | 2.5% | 14 | .895 | .875 | .851 | .821 | .784 | .738 | .683 | .618 | .544 | .461 |
| | | | | | | | | .375 | .289 | .211 | .144 | | | | | | |

## ERLANG B: 100% BLOCKED CALLERS ROUTE ADVANCE, OVERFLOW, OR DISAPPEAR

| HOURS FIRST ATTEMPT VMR = 1.0 | HOURS CON-NECT | HOURS RECALL /RETRY | HOURS OVFL | CON-NECT VMR | OVFL VMR | PERCENT BLOCKED (GRADE) | LINE | HOURS CONNECTED TRAFFIC ON LINES 1 | 2 | 3 | 4 | 5 | 6 | 7 | 8 | 9 | 10 |
|---|---|---|---|---|---|---|---|---|---|---|---|---|---|---|---|---|---|
| 8.500 | 8.383 | .000 | .117 | .91 | 2.00 | 1.4% | 15 | .895 | .875 | .851 | .821 | .784 | .738 | .683 | .618 | .544 | .461 |
| | | | | | | | | .375 | .289 | .211 | .144 | .092 | | | | | |
| 8.500 | 8.438 | .000 | .062 | .94 | 1.93 | .7% | 16 | .895 | .875 | .851 | .821 | .784 | .738 | .683 | .618 | .544 | .461 |
| | | | | | | | | .375 | .289 | .211 | .144 | .092 | .055 | | | | |
| 9.000 | 2.642 | .000 | 6.358 | .14 | 1.27 | 70.6% | 3 | .900 | .882 | .860 | | | | | | | |
| 9.000 | 3.476 | .000 | 5.524 | .17 | 1.38 | 61.4% | 4 | .900 | .882 | .860 | .833 | | | | | | |
| 9.000 | 4.276 | .000 | 4.724 | .20 | 1.50 | 52.5% | 5 | .900 | .882 | .860 | .833 | .800 | .760 | | | | |
| 9.000 | 5.035 | .000 | 3.965 | .24 | 1.62 | 44.1% | 6 | .900 | .882 | .860 | .833 | .800 | .760 | | | | |
| 9.000 | 5.746 | .000 | 3.254 | .29 | 1.74 | 36.2% | 7 | .900 | .882 | .860 | .833 | .800 | .760 | .710 | | | |
| 9.000 | 6.398 | .000 | 2.602 | .35 | 1.86 | 28.9% | 8 | .900 | .882 | .860 | .833 | .800 | .760 | .710 | .652 | | |
| 9.000 | 6.981 | .000 | 2.019 | .42 | 1.96 | 22.4% | 9 | .900 | .882 | .860 | .833 | .800 | .760 | .710 | .652 | .584 | |
| 9.000 | 7.488 | .000 | 1.512 | .49 | 2.05 | 16.8% | 10 | .900 | .882 | .860 | .833 | .800 | .760 | .710 | .652 | .584 | .507 |
| 9.000 | 7.913 | .000 | 1.087 | .58 | 2.11 | 12.1% | 11 | .900 | .882 | .860 | .833 | .800 | .760 | .710 | .652 | .584 | .507 |
| | | | | | | | | .424 | | | | | | | | | |
| 9.000 | 8.252 | .000 | .748 | .66 | 2.15 | 8.3% | 12 | .900 | .882 | .860 | .833 | .800 | .760 | .710 | .652 | .584 | .507 |
| | | | | | | | | .424 | .340 | | | | | | | | |
| 9.000 | 8.510 | .000 | .490 | .74 | 2.15 | 5.4% | 13 | .900 | .882 | .860 | .833 | .800 | .760 | .710 | .652 | .584 | .507 |
| | | | | | | | | .424 | .340 | .258 | | | | | | | |
| 9.000 | 8.696 | .000 | .304 | .81 | 2.12 | 3.4% | 14 | .900 | .882 | .860 | .833 | .800 | .760 | .710 | .652 | .584 | .507 |
| | | | | | | | | .424 | .340 | .258 | .185 | | | | | | |
| 9.000 | 8.821 | .000 | .179 | .87 | 2.07 | 2.0% | 15 | .900 | .882 | .860 | .833 | .800 | .760 | .710 | .652 | .584 | .507 |
| | | | | | | | | .424 | .340 | .258 | .185 | .125 | | | | | |
| 9.000 | 8.901 | .000 | .099 | .92 | 2.01 | 1.1% | 16 | .900 | .882 | .860 | .833 | .800 | .760 | .710 | .652 | .584 | .507 |
| | | | | | | | | .424 | .340 | .258 | .185 | .125 | .079 | | | | |
| 9.000 | 8.948 | .000 | .052 | .95 | 1.94 | .6% | 17 | .900 | .882 | .860 | .833 | .800 | .760 | .710 | .652 | .584 | .507 |
| | | | | | | | | .424 | .340 | .258 | .185 | .125 | .079 | .047 | | | |
| 9.500 | 2.662 | .000 | 6.838 | .13 | 1.26 | 72.0% | 3 | .905 | .888 | .869 | | | | | | | |
| 9.500 | 3.506 | .000 | 5.994 | .16 | 1.37 | 63.1% | 4 | .905 | .888 | .869 | .844 | | | | | | |
| 9.500 | 4.321 | .000 | 5.179 | .19 | 1.48 | 54.5% | 5 | .905 | .888 | .869 | .844 | .814 | | | | | |
| 9.500 | 5.099 | .000 | 4.401 | .22 | 1.60 | 46.3% | 6 | .905 | .888 | .869 | .844 | .814 | .778 | | | | |
| 9.500 | 5.833 | .000 | 3.667 | .27 | 1.72 | 38.6% | 7 | .905 | .888 | .869 | .844 | .814 | .778 | .734 | | | |
| 9.500 | 6.514 | .000 | 2.986 | .32 | 1.84 | 31.4% | 8 | .905 | .888 | .869 | .844 | .814 | .778 | .734 | .681 | | |
| 9.500 | 7.133 | .000 | 2.367 | .38 | 1.95 | 24.9% | 9 | .905 | .888 | .869 | .844 | .814 | .778 | .734 | .681 | .619 | |
| 9.500 | 7.682 | .000 | 1.818 | .45 | 2.04 | 19.1% | 10 | .888 | .869 | .844 | .814 | .888 | .869 | .844 | .814 | .778 | .734 |
| .681 | .619 | .540 | | | | | | | | | | | | | | | |
| 9.500 | 8.153 | .000 | 1.347 | .53 | 2.12 | 14.2% | 11 | .905 | .888 | .869 | .844 | .814 | .888 | 0 | 8.541 | .000 | .959 |
| .61 | 2.17 | | | | | | | | | | | | | | | | |
| | | 10.1% | 12 | .905 | .888 | .869.844 | | .814 | .778 | .734 | .681 | .610 | .549 | | | | |
| | | | | | | | | .471 | .388 | | | | | | | | |
| 9.500 | 8.847 | .000 | .653 | .69 | 2.10 | 6.9% | 13 | .905 | .888 | .869 | .844 | .814 | .778 | .734 | .681 | .619 | .549 |
| | | | | | | | | .471 | .388 | .306 | | | | | | | |

## ERLANG B: 100% BLOCKED CALLERS ROUTE ADVANCE, OVERFLOW, OR DISAPPEAR

| HOURS FIRST ATTEMPT VMR = 1.0 | HOURS CON-NECT | HOURS RECALL /RETRY | HOURS OVFL | CON-NECT VMR | OVFL VMR | PERCENT BLOCKED (GRADE) | LINE | HOURS CONNECTED TRAFFIC ON LINES 1 | 2 | 3 | 4 | 5 | 6 | 7 | 8 | 9 | 10 |
|---|---|---|---|---|---|---|---|---|---|---|---|---|---|---|---|---|---|
| 9.500 | 9.077 | .000 | .423 | .77 | 2.18 | 4.5% | 14 | .905 | .888 | .869 | .844 | .814 | .778 | .734 | .681 | .619 | .549 |
| | | | | | | | | .471 | .388 | .306 | .230 | | | | | | |
| 9.500 | 9.230 | .000 | .261 | .84 | 2.14 | 2.7% | 15 | .905 | .888 | .869 | .844 | .814 | .778 | .734 | .681 | .619 | .549 |
| | | | | | | | | .471 | .388 | .306 | .230 | .163 | | | | | |
| 9.500 | 9.348 | .000 | .152 | .89 | 2.09 | 1.6% | 16 | .905 | .888 | .869 | .844 | .814 | .778 | .734 | .681 | .619 | .549 |
| | | | | | | | | .471 | .388 | .306 | .230 | .163 | .108 | | | | |
| 9.500 | 9.416 | .000 | .084 | .93 | 2.02 | .9% | 17 | .905 | .888 | .869 | .844 | .814 | .778 | .734 | .681 | .619 | .549 |
| | | | | | | | | .471 | .388 | .306 | .230 | .163 | .108 | .068 | | | |
| 10.000 | 2.679 | .000 | 7.321 | .12 | 1.25 | 73.2% | 3 | .909 | .894 | .876 | | | | | | | |
| 10.000 | 3.533 | .000 | 6.467 | .15 | 1.35 | 64.7% | 4 | .909 | .894 | .876 | .854 | | | | | | |
| 10.000 | 4.360 | .000 | 5.640 | .17 | 1.46 | 56.4% | 5 | .909 | .894 | .876 | .854 | .827 | | | | | |
| 10.000 | 5.155 | .000 | 4.845 | .21 | 1.57 | 48.5% | 6 | .909 | .894 | .876 | .854 | .827 | .794 | | | | |
| 10.000 | 5.910 | .000 | 4.090 | .25 | 1.69 | 40.9% | 7 | .909 | .894 | .876 | .854 | .827 | .794 | .755 | | | |
| 10.000 | 6.617 | .000 | 3.383 | .29 | 1.81 | 33.8% | 8 | .909 | .894 | .876 | .854 | .827 | .794 | .755 | .707 | | |
| 10.000 | 7.268 | .000 | 2.732 | .35 | 1.93 | 27.3% | 9 | .909 | .894 | .876 | .854 | .827 | .794 | .755 | .707 | .651 | |
| 10.000 | 7.854 | .000 | 2.146 | .41 | 2.03 | 21.5% | 10 | .909 | .894 | .876 | .854 | .827 | .794 | .755 | .707 | .651 | .586 |
| 10.000 | 8.368 | .000 | 1.632 | .49 | 2.12 | 16.3% | 11 | .909 | .894 | .876 | .854 | .827 | .794 | .755 | .707 | .651 | .586 |
| | | | | | | | | .514 | | | | | | | | | |
| 10.000 | 8.803 | .000 | 1.197 | .57 | 2.19 | 12.0% | 12 | .909 | .894 | .876 | .854 | .827 | .794 | .755 | .707 | .651 | .586 |
| | | | | | | | | .514 | .435 | | | | | | | | |
| 10.000 | 9.157 | .000 | .843 | .65 | 2.22 | 8.4% | 13 | .909 | .894 | .876 | .854 | .827 | .794 | .755 | .707 | .651 | .586 |
| | | | | | | | | .514 | .435 | .354 | | | | | | | |
| 10.000 | 9.432 | .000 | .568 | .72 | 2.23 | 5.7% | 14 | .909 | .894 | .876 | .854 | .827 | .794 | .755 | .707 | .651 | .586 |
| | | | | | | | | .514 | .435 | .354 | .275 | | | | | | |
| 10.000 | 9.635 | .000 | .365 | .80 | 2.21 | 3.6% | 15 | .909 | .894 | .876 | .854 | .827 | .794 | .755 | .707 | .651 | .586 |
| | | | | | | | | .514 | .435 | .354 | .275 | .203 | | | | | |
| 10.000 | 9.777 | .000 | .223 | .86 | 2.16 | 2.2% | 16 | .909 | .894 | .876 | .854 | .827 | .794 | .755 | .707 | .651 | .586 |
| | | | | | | | | .514 | .435 | .354 | .275 | .203 | .142 | | | | |
| 10.000 | 9.871 | .000 | .129 | .91 | 2.10 | 1.3% | 17 | .909 | .894 | .876 | .854 | .827 | .794 | .755 | .707 | .651 | .586 |
| | | | | | | | | .514 | .435 | .354 | .275 | .203 | .142 | .094 | | | |
| 10.000 | 9.929 | .000 | .071 | .94 | 2.03 | .7% | 18 | .909 | .894 | .876 | .854 | .827 | .794 | .755 | .707 | .651 | .586 |
| | | | | | | | | .514 | .435 | .354 | .275 | .203 | .142 | .094 | .058 | | |
| 11.000 | 3.580 | .000 | 7.420 | .13 | 1.33 | 67.5% | 4 | .917 | .904 | .889 | .870 | | | | | | |
| 11.000 | 4.428 | .000 | 6.572 | .15 | 1.43 | 59.7% | 5 | .917 | .904 | .889 | .870 | .848 | | | | | |
| 11.000 | 5.250 | .000 | 5.750 | .18 | 1.54 | 52.3% | 6 | .917 | .904 | .889 | .870 | .848 | .822 | | | | |
| 11.000 | 6.039 | .000 | 4.961 | .21 | 1.65 | 45.1% | 7 | .917 | .904 | .889 | .870 | .848 | .822 | .789 | | | |
| 11.000 | 6.790 | .000 | 4.210 | .25 | 1.77 | 38.3% | 8 | .917 | .904 | .889 | .870 | .848 | .822 | .789 | .751 | | |
| 11.000 | 7.494 | .000 | 3.506 | .30 | 1.88 | 31.9% | 9 | .917 | .904 | .889 | .870 | .848 | .822 | .789 | .751 | .704 | |
| 11.000 | 8.145 | .000 | 2.855 | .35 | 2.00 | 26.0% | 10 | .917 | .904 | .889 | .870 | .848 | .822 | .789 | .751 | .704 | .650 |
| 11.000 | 8.733 | .000 | 2.267 | .41 | 2.10 | 20.6% | 11 | .917 | .904 | .889 | .870 | .848 | .822 | .789 | .751 | .704 | .650 |

## ERLANG B: 100% BLOCKED CALLERS ROUTE ADVANCE, OVERFLOW, OR DISAPPEAR

| HOURS FIRST ATTEMPT VMR=1.0 | HOURS CON-NECT | HOURS RECALL /RETRY | HOURS OVFL | CON-NECT VMR | OVFL VMR | PERCENT BLOCKED (GRADE) | LINE | HOURS CONNECTED TRAFFIC ON LINES 1 | 2 | 3 | 4 | 5 | 6 | 7 | 8 | 9 | 10 |
|---|---|---|---|---|---|---|---|---|---|---|---|---|---|---|---|---|---|
| | | | | | | | | .588 | | | | | | | | | |
| 11.000 | 9.252 | .000 | 1.748 | .48 | 2.19 | 15.9% | 12 | .917 | .904 | .889 | .870 | .848 | .822 | .789 | .751 | .704 | .650 |
| | | | | | | | | .588 | .519 | | | | | | | | |
| 11.000 | 9.696 | .000 | 1.304 | .56 | 2.25 | 11.9% | 13 | .917 | .904 | .889 | .870 | .848 | .822 | .789 | .751 | .704 | .650 |
| | | | | | | | | .588 | .519 | .444 | | | | | | | |
| 11.000 | 10.063 | .000 | .937 | .63 | 2.29 | 8.5% | 14 | .917 | .904 | .889 | .870 | .848 | .822 | .789 | .751 | .704 | .650 |
| | | | | | | | | .588 | .519 | .444 | .367 | | | | | | |
| 11.000 | 10.353 | .000 | .647 | .71 | 2.30 | 5.9% | 15 | .917 | .904 | .889 | .870 | .848 | .822 | .789 | .751 | .704 | .650 |
| | | | | | | | | .588 | .519 | .444 | .367 | .290 | | | | | |
| 11.000 | 10.573 | .000 | .427 | .78 | 2.28 | 3.9% | 16 | .917 | .904 | .889 | .870 | .848 | .822 | .789 | .751 | .704 | .650 |
| | | | | | | | | .588 | .519 | .444 | .367 | .290 | .219 | | | | |
| 11.000 | 10.730 | .000 | .270 | .84 | 2.24 | 2.5% | 17 | .917 | .904 | .889 | .870 | .848 | .822 | .789 | .751 | .704 | .650 |
| | | | | | | | | .588 | .519 | .444 | .367 | .290 | .219 | .158 | | | |
| 11.000 | 10.838 | .000 | .162 | .89 | 2.19 | 1.5% | 18 | .917 | .904 | .889 | .870 | .848 | .822 | .789 | .751 | .704 | .650 |
| | | | | | | | | .588 | .519 | .444 | .367 | .290 | .219 | .158 | .107 | | |
| 11.000 | 10.907 | .000 | .093 | .93 | 2.12 | .8% | 19 | .917 | .904 | .889 | .870 | .848 | .822 | .789 | .751 | .704 | .650 |
| | | | | | | | | .588 | .519 | .444 | .367 | .290 | .219 | .158 | .107 | .069 | |
| 12.000 | 3.618 | .000 | 8.382 | .12 | 1.30 | 69.8% | 4 | .923 | .912 | .899 | .884 | | | | | | |
| 12.000 | 4.484 | .000 | 7.516 | .13 | 1.40 | 62.6% | 5 | .923 | .912 | .899 | .884 | .865 | | | | | |
| 12.000 | 5.327 | .000 | 6.673 | .16 | 1.50 | 55.6% | 6 | .923 | .912 | .899 | .884 | .865 | .843 | | | | |
| 12.000 | 6.143 | .000 | 5.857 | .18 | 1.61 | 48.8% | 7 | .923 | .912 | .899 | .884 | .865 | .843 | .817 | | | |
| 12.000 | 6.928 | .000 | 5.072 | .22 | 1.72 | 42.3% | 8 | .923 | .912 | .899 | .884 | .865 | .843 | .817 | .785 | | |
| 12.000 | 7.675 | .000 | 4.325 | .25 | 1.84 | 36.0% | 9 | .923 | .912 | .899 | .884 | .865 | .843 | .817 | .785 | .747 | |
| 12.000 | 8.377 | .000 | 3.623 | .30 | 1.95 | 30.2% | 10 | .923 | .912 | .899 | .884 | .865 | .843 | .817 | .785 | .747 | .702 |
| 12.000 | 9.027 | .000 | 2.973 | .35 | 2.06 | 24.8% | 11 | .923 | .912 | .899 | .884 | .865 | .843 | .817 | .785 | .747 | .702 |
| | | | | | | | | .650 | | | | | | | | | |
| 12.000 | 9.617 | .000 | 2.383 | .41 | 2.16 | 10.0% | 12 | .923 | .912 | .899 | .884 | .865 | .843 | .817 | .785 | .747 | .702 |
| | | | | | | | | .650 | .590 | | | | | | | | |
| 12.000 | 10.141 | .000 | 1.859 | .48 | 2.25 | 15.5% | 13 | .923 | .912 | .899 | .884 | .865 | .843 | .817 | .785 | .747 | .702 |
| | | | | | | | | .650 | .590 | .524 | | | | | | | |
| 12.000 | 10.593 | .000 | 1.407 | .55 | 2.32 | 11.7% | 14 | .923 | .912 | .899 | .884 | .865 | .843 | .817 | .785 | .747 | .702 |
| | | | | | | | | .650 | .590 | .524 | .452 | | | | | | |
| 12.000 | 10.971 | .000 | 1.029 | .62 | 2.36 | 8.6% | 15 | .923 | .912 | .899 | .884 | .865 | .843 | .817 | .785 | .747 | .702 |
| | | | | | | | | .650 | .590 | .524 | .452 | .378 | | | | | |
| 12.000 | 11.275 | .000 | .725 | .70 | 2.37 | 6.0% | 16 | .923 | .912 | .899 | .884 | .865 | .843 | .817 | .785 | .747 | .702 |
| | | | | | | | | .650 | .590 | .524 | .452 | .378 | .304 | | | | |
| 12.000 | 11.509 | .000 | .491 | .77 | 2.36 | 4.1% | 17 | .923 | .912 | .899 | .884 | .865 | .843 | .817 | .785 | .747 | .702 |
| | | | | | | | | .650 | .590 | .524 | .452 | .378 | .304 | .234 | | | |
| 12.000 | 11.681 | .000 | .319 | .83 | 2.32 | 2.7% | 18 | .923 | .912 | .899 | .884 | .865 | .843 | .817 | .785 | .747 | .702 |
| | | | | | | | | .650 | .590 | .524 | .452 | .378 | .304 | .234 | .172 | | |

# ERLANG B: 100% BLOCKED CALLERS ROUTE ADVANCE, OVERFLOW, OR DISAPPEAR

| HOURS FIRST ATTEMPT VMR = 1.0 | HOURS CON- NECT | HOURS RECALL /RETRY | HOURS OVFL | CON- NECT VMR | OVFL VMR | PERCENT BLOCKED (GRADE) | LINE | HOURS CONNECTED TRAFFIC ON LINES 1 | 2 | 3 | 4 | 5 | 6 | 7 | 8 | 9 | 10 |
|---|---|---|---|---|---|---|---|---|---|---|---|---|---|---|---|---|---|
| 12.000 | 11.802 | .000 | .198 | .88 | 2.27 | 1.6% | 19 | .923 | .912 | .899 | .884 | .865 | .843 | .817 | .785 | .747 | .702 |
| | | | | | | | | .650 | .590 | .524 | .452 | .378 | .304 | .234 | .172 | .121 | |
| 12.000 | 11.882 | .000 | .118 | .92 | 2.20 | 1.0% | 20 | .923 | .912 | .899 | .884 | .865 | .843 | .817 | .785 | .747 | .702 |
| | | | | | | | | .650 | .590 | .524 | .452 | .378 | .304 | .234 | .172 | .121 | .080 |
| 12.000 | 11.933 | .000 | .067 | .95 | 2.13 | .6% | 21 | .923 | .912 | .899 | .884 | .865 | .843 | .817 | .785 | .747 | .702 |
| | | | | | | | | .650 | .590 | .524 | .452 | .378 | .304 | .234 | .172 | .121 | .080 |
| | | | | | | | | .051 | | | | | | | | | |
| 13.000 | 3.651 | .000 | 9.349 | .11 | 1.28 | 71.9% | 4 | .929 | .919 | .908 | .895 | | | | | | |
| 13.000 | 4.530 | .000 | 8.470 | .12 | 1.37 | 65.2% | 5 | .929 | .919 | .908 | .895 | .879 | | | | | |
| 13.000 | 5.390 | .000 | 7.610 | .14 | 1.47 | 58.5% | 6 | .929 | .919 | .908 | .895 | .879 | .861 | | | | |
| 13.000 | 6.229 | .000 | 6.771 | .16 | 1.57 | 52.1% | 7 | .929 | .919 | .908 | .895 | .879 | .861 | .838 | | | |
| 13.000 | 7.041 | .000 | 5.959 | .19 | 1.68 | 45.8% | 8 | .929 | .919 | .908 | .895 | .879 | .861 | .838 | .812 | | |
| 13.000 | 7.821 | .000 | 5.179 | .22 | 1.79 | 39.8% | 9 | .929 | .919 | .908 | .895 | .879 | .861 | .838 | .812 | .781 | |
| 13.000 | 8.565 | .000 | 4,435 | .26 | 1.90 | 34.1% | 10 | .929 | .919 | .908 | .895 | .879 | .861 | .838 | .812 | .781 | .743 |
| 13.000 | 9.264 | .000 | 3.736 | .30 | 2.02 | 28.7% | 11 | .929 | .919 | .908 | .895 | .879 | .861 | .838 | .812 | .781 | .743 |
| | | | | | | | | .700 | | | | | | | | | |
| 13.000 | 9.914 | .000 | 3.086 | .35 | 2.13 | 23.7% | 12 | .929 | .919 | .908 | .895 | .879 | .861 | .838 | .812 | .781 | .743 |
| | | | | | | | | .700 | .649 | | | | | | | | |
| 13.000 | 10.506 | .000 | 2.494 | .41 | 2.23 | 19.2% | 13 | .929 | .919 | .908 | .895 | .879 | .861 | .838 | .812 | .781 | .743 |
| | | | | | | | | .700 | .649 | .592 | | | | | | | |
| 13.000 | 11.034 | .000 | 1.966 | .47 | 2.31 | 15.1% | 14 | .929 | .919 | .908 | .895 | .879 | .861 | .838 | .812 | .781 | .743 |
| | | | | | | | | .700 | .649 | .592 | .528 | | | | | | |
| 13.000 | 11.494 | .000 | 1.506 | .54 | 2.38 | 11.6% | 15 | .929 | .919 | .908 | .895 | .879 | .861 | .838 | .812 | .781 | .743 |
| | | | | | | | | .700 | .649 | .592 | .528 | .459 | | | | | |
| 13.000 | 11.881 | .000 | 1.119 | .61 | 2.42 | 8.6% | 16 | .929 | .919 | .908 | .895 | .879 | .861 | .838 | .812 | .781 | .743 |
| | | | | | | | | .700 | .649 | .592 | .528 | .459 | .388 | | | | |
| 13.000 | 12.197 | .000 | .803 | .68 | 2.44 | 6.2% | 17 | .929 | .919 | .908 | .895 | .879 | .861 | .838 | .812 | .781 | .743 |
| | | | | | | | | .700 | .649 | .592 | .528 | .459 | .388 | .316 | | | |
| 13.000 | 12.445 | .000 | .555 | .75 | 2.43 | 4.3% | 18 | .929 | .919 | .908 | .895 | .879 | .861 | .838 | .812 | .781 | .743 |
| | | | | | | | | .700 | .649 | .592 | .528 | .459 | .388 | .316 | .248 | | |
| 13.000 | 12.631 | .000 | .369 | .81 | 2.40 | 2.8% | 19 | .929 | .919 | .908 | .895 | .879 | .861 | .838 | .812 | .781 | .743 |
| | | | | | | | | .700 | .649 | .592 | .528 | .459 | .388 | .316 | .248 | .186 | |
| 13.000 | 12.765 | .000 | .235 | .87 | 2.34 | 1.8% | 20 | .929 | .919 | .908 | .895 | .879 | .861 | .838 | .812 | .781 | .743 |
| | | | | | | | | .700 | .649 | .592 | .528 | .459 | .388 | .316 | .248 | .186 | .133 |
| 13.000 | 12.856 | .000 | .144 | .91 | 2.28 | 1.1% | 21 | .929 | .919 | .908 | .895 | .879 | .861 | .838 | .812 | .781 | .743 |
| | | | | | | | | .700 | .649 | .592 | .528 | .459 | .388 | .316 | .248 | .186 | .133 |
| | | | | | | | | .091 | | | | | | | | | |
| 13.000 | 12.915 | .000 | .085 | .94 | 2.20 | .7% | 22 | .929 | .919 | .908 | .895 | .879 | .861 | .838 | .812 | .781 | .743 |
| | | | | | | | | .700 | .649 | .592 | .528 | .459 | .388 | .316 | .248 | .186 | .133 |
| | | | | | | | | .091 | .060 | | | | | | | | |

## ERLANG B: 100% BLOCKED CALLERS ROUTE ADVANCE, OVERFLOW, OR DISAPPEAR

| HOURS FIRST ATTEMPT VMR = 1.0 | HOURS CON-NECT | HOURS RECALL /RETRY | HOURS OVFL | CON-NECT VMR | OVFL VMR | PERCENT BLOCKED (GRADE) | LINE | HOURS CONNECTED TRAFFIC ON LINES | | | | | | | | | |
|---|---|---|---|---|---|---|---|---|---|---|---|---|---|---|---|---|---|
| | | | | | | | | 1 | 2 | 3 | 4 | 5 | 6 | 7 | 8 | 9 | 10 |
| 14.000 | 3.678 | .000 | 10.322 | .10 | 1.27 | 73.7% | 4 | .933 | .925 | .915 | .904 | | | | | | |
| 14.000 | 4.569 | .000 | 9.431 | .11 | 1.35 | 67.4% | 5 | .933 | .925 | .915 | .904 | .891 | | | | | |
| 14.000 | 5.443 | .000 | 8.557 | .13 | 1.44 | 61.1% | 6 | .933 | .925 | .915 | .904 | .891 | .875 | | | | |
| 14.000 | 6.300 | .000 | 7.700 | .14 | 1.53 | 55.0% | 7 | .933 | .925 | .915 | .904 | .891 | .875 | .856 | | | |
| 14.000 | 7.134 | .000 | 6.866 | .17 | 1.63 | 49.0% | 8 | .933 | .925 | .915 | .904 | .891 | .875 | .856 | .834 | | |
| 14.000 | 7.941 | .000 | 6.059 | .19 | 1.74 | 43.3% | 9 | .933 | .925 | .915 | .904 | .891 | .875 | .856 | .834 | .808 | |
| 14.000 | 8.718 | .000 | 5.282 | .22 | 1.85 | 37.7% | 10 | .933 | .925 | .915 | .904 | .891 | .875 | .856 | .834 | .808 | .777 |
| 14.000 | 9.458 | .000 | 4.542 | .26 | 1.97 | 32.4% | 11 | .933 | .925 | .915 | .904 | .891 | .875 | .856 | .834 | .808 | .777 |
| | | | | | | | | .740 | | | | | | | | | |
| 14.000 | 10.156 | .000 | 3.844 | .30 | 2.08 | 27.5% | 12 | .933 | .925 | .915 | .904 | .891 | .875 | .856 | .834 | .808 | .777 |
| | | | | | | | | .740 | .698 | | | | | | | | |
| 14.000 | 10.805 | .000 | 3.195 | .35 | 2.19 | 22.8% | 13 | .933 | .925 | .915 | .904 | .891 | .875 | .856 | .834 | .808 | .777 |
| | | | | | | | | .740 | .698 | .649 | | | | | | | |
| 14.000 | 11.399 | .000 | 2.601 | .41 | 2.29 | 18.6% | 14 | .933 | .925 | .915 | .904 | .891 | .875 | .856 | .834 | .808 | .777 |
| | | | | | | | | .740 | .698 | .649 | .594 | | | | | | |
| 14.000 | 11.931 | .000 | 2.069 | .47 | 2.37 | 14.8% | 15 | .933 | .925 | .915 | .904 | .891 | .875 | .856 | .834 | .808 | .777 |
| | | | | | | | | .740 | .698 | .649 | .594 | .532 | | | | | |
| 14.000 | 12.397 | .000 | 1.603 | .53 | 2.44 | 11.5% | 16 | .933 | .925 | .915 | .904 | .891 | .875 | .856 | .834 | .808 | .777 |
| | | | | | | | | .740 | .698 | .649 | .594 | .532 | .466 | | | | |
| 14.000 | 12.794 | .000 | 1.206 | .60 | 2.48 | 8.6% | 17 | .933 | .925 | .915 | .904 | .891 | .875 | .856 | .834 | .808 | .777 |
| | | | | | | | | .740 | .698 | .649 | .594 | .532 | .466 | .397 | | | |
| 14.000 | 13.121 | .000 | .879 | .67 | 2.50 | 6.3% | 18 | .933 | .925 | .915 | .904 | .891 | .875 | .856 | .834 | .808 | .777 |
| | | | | | | | | .740 | .698 | .649 | .594 | .532 | .466 | .397 | .327 | | |
| 14.000 | 13.381 | .000 | .619 | .74 | 2.50 | 4.4% | 19 | .933 | .925 | .915 | .904 | .891 | .875 | .856 | .834 | .808 | .777 |
| | | | | | | | | .740 | .698 | .649 | .594 | .532 | .466 | .397 | .327 | .260 | |
| 14.000 | 13.580 | .000 | .420 | .80 | 2.47 | 3.0% | 20 | .933 | .925 | .915 | .904 | .891 | .875 | .856 | .834 | .808 | .777 |
| | | | | | | | | .740 | .698 | .649 | .594 | .532 | .466 | .397 | .327 | .260 | .199 |
| 14.000 | 13.725 | .000 | .275 | .85 | 2.42 | 2.0% | 21 | .933 | .925 | .915 | .904 | .891 | .875 | .856 | .834 | .808 | .777 |
| | | | | | | | | .740 | .698 | .649 | .594 | .532 | .466 | .397 | .327 | .260 | .199 |
| | | | | | | | | .146 | | | | | | | | | |
| 14.000 | 13.827 | .000 | .173 | .90 | 2.35 | 1.2% | 22 | .933 | .925 | .915 | .904 | .891 | .875 | .856 | .834 | .808 | .777 |
| | | | | | | | | .740 | .698 | .649 | .594 | .532 | .466 | .397 | .327 | .260 | .199 |
| | | | | | | | | .146 | .102 | | | | | | | | |
| 14.000 | 13.896 | .000 | .104 | .93 | 2.28 | .7% | 23 | .933 | .925 | .915 | .904 | .891 | .875 | .856 | .834 | .808 | .777 |
| | | | | | | | | .740 | .698 | .649 | .594 | .532 | .466 | .397 | .327 | .260 | .199 |
| | | | | | | | | .146 | .102 | .068 | | | | | | | |
| 15.000 | 4.602 | .000 | 10.398 | .10 | 1.33 | 69.3% | 5 | .938 | .930 | .922 | .912 | .900 | | | | | |
| 15.000 | 5.488 | .000 | 9.512 | .11 | 1.41 | 63.4% | 6 | .938 | .930 | .922 | .912 | .900 | .887 | | | | |
| 15.000 | 6.359 | .000 | 8.641 | .13 | 1.50 | 57.6% | 7 | .938 | .930 | .922 | .912 | .900 | .887 | .871 | | | |
| 15.000 | 7.211 | .000 | 7.789 | .15 | 1.60 | 51.9% | 8 | .938 | .930 | .922 | .912 | .900 | .887 | .871 | .852 | | |

## ERLANG B: 100% BLOCKED CALLERS ROUTE ADVANCE, OVERFLOW, OR DISAPPEAR

| HOURS FIRST ATTEMPT VMR = 1.0 | HOURS CON-NECT | HOURS RECALL /RETRY | HOURS OVFL | CON-NECT VMR | OVFL VMR | PERCENT BLOCKED (GRADE) | LINE | HOURS CONNECTED TRAFFIC ON LINES | | | | | | | | | |
|---|---|---|---|---|---|---|---|---|---|---|---|---|---|---|---|---|---|
| | | | | | | | | 1 | 2 | 3 | 4 | 5 | 6 | 7 | 8 | 9 | 10 |
| 15.000 | 8.041 | .000 | 6.959 | .17 | 1.70 | 46.4% | 9 | .938 | .930 | .922 | .912 | .900 | .887 | .871 | .852 | .830 | |
| 15.000 | 8.845 | .000 | 6.155 | .20 | 1.81 | 41.0% | 10 | .938 | .930 | .922 | .912 | .900 | .887 | .871 | .852 | .830 | .804 |
| 15.000 | 9.618 | .000 | 5.382 | .23 | 1.92 | 35.9% | 11 | .938 | .930 | .922 | .912 | .900 | .887 | .871 | .852 | .830 | .804 |
| | | | | | | | | .773 | | | | | | | | | |
| 15.000 | 10.356 | .000 | 4.644 | .26 | 2.03 | 31.0% | 12 | .938 | .930 | .922 | .912 | .900 | .887 | .871 | .852 | .830 | .804 |
| | | | | | | | | .773 | .737 | | | | | | | | |
| 15.000 | 11.052 | .000 | 3.948 | .30 | 2.14 | 26.3% | 13 | .938 | .930 | .922 | .912 | .900 | .887 | .871 | .852 | .830 | .804 |
| | | | | | | | | .773 | .737 | .696 | | | | | | | |
| 15.000 | 11.700 | .000 | 3.300 | .35 | 2.25 | 22.0% | 14 | .938 | .930 | .922 | .912 | .900 | .887 | .871 | .852 | .830 | .804 |
| | | | | | | | | .773 | .737 | .696 | .649 | | | | | | |
| 15.000 | 12.295 | .000 | 2.705 | .41 | 2.34 | 18.0% | 15 | .938 | .930 | .922 | .912 | .900 | .887 | .871 | .852 | .830 | .804 |
| | | | | | | | | .773 | .737 | .696 | .649 | .595 | | | | | |
| 15.000 | 12.831 | .000 | 2.169 | .46 | 2.43 | 14.5% | 16 | .938 | .930 | .922 | .912 | .900 | .887 | .871 | .852 | .830 | .804 |
| | | | | | | | | .773 | .737 | .696 | .649 | .595 | .536 | | | | |
| 15.000 | 13.303 | .000 | 1.697 | .53 | 2.50 | 11.3% | 17 | .938 | .930 | .922 | .912 | .900 | .887 | .871 | .852 | .830 | .804 |
| | | | | | | | | .773 | .737 | .696 | .649 | .595 | .536 | .472 | | | |
| 15.000 | 13.707 | .000 | 1.293 | .60 | 2.54 | 8.6% | 18 | .938 | .930 | .922 | .912 | .900 | .887 | .871 | .852 | .830 | .804 |
| | | | | | | | | .773 | .737 | .696 | .649 | .595 | .536 | .472 | .405 | | |
| 15.000 | 14.045 | .000 | .955 | .66 | 2.56 | 6.4% | 19 | .938 | .930 | .922 | .912 | .900 | .887 | .871 | .852 | .830 | .804 |
| | | | | | | | | .773 | .737 | .696 | .649 | .595 | .536 | .472 | .405 | .337 | |
| 15.000 | 14.316 | .000 | .684 | .73 | 2.56 | 4.6% | 20 | .938 | .930 | .922 | .912 | .900 | .887 | .871 | .852 | .830 | .804 |
| | | | | | | | | .773 | .737 | .696 | .649 | .595 | .536 | .472 | .405 | .337 | .272 |
| 15.000 | 14.527 | .000 | .473 | .79 | 2.53 | 3.2% | 21 | .938 | .930 | .922 | .912 | .900 | .887 | .871 | .852 | .830 | .804 |
| | | | | | | | | .773 | .737 | .696 | .649 | .595 | .536 | .472 | .405 | .337 | .272 |
| | | | | | | | | .211 | | | | | | | | | |
| 15.000 | 14.684 | .000 | .316 | .84 | 2.49 | 2.1% | 22 | .938 | .930 | .922 | .912 | .900 | .887 | .871 | .852 | .830 | .804 |
| | | | | | | | | .773 | .737 | .696 | .649 | .595 | .536 | .472 | .405 | .337 | .272 |
| | | | | | | | | .211 | .157 | | | | | | | | |
| 15.000 | 14.797 | .000 | .203 | .89 | 2.43 | 1.4% | 23 | .938 | .930 | .922 | .912 | .900 | .887 | .871 | .852 | .830 | .804 |
| | | | | | | | | .773 | .737 | .696 | .649 | .595 | .536 | .472 | .405 | .337 | .272 |
| | | | | | | | | .211 | .157 | .113 | | | | | | | |
| 15.000 | 14.874 | .000 | .126 | .92 | 2.36 | .8% | 24 | .938 | .930 | .922 | .912 | .900 | .887 | .871 | .852 | .830 | .804 |
| | | | | | | | | .773 | .737 | .696 | .649 | .595 | .536 | .472 | .405 | .337 | .272 |
| | | | | | | | | .211 | .157 | .113 | .077 | | | | | | |
| 15.000 | 14.925 | .000 | .075 | .95 | 2.28 | .5% | 25 | .938 | .930 | .922 | .912 | .900 | .887 | .871 | .852 | .830 | .804 |
| | | | | | | | | .773 | .737 | .696 | .649 | .595 | .536 | .472 | .405 | .337 | .272 |
| | | | | | | | | .211 | .157 | .113 | .077 | .051 | | | | | |
| 16.000 | 4.630 | .000 | 11.370 | .09 | 1.31 | 71.1% | 5 | .941 | .935 | .927 | .919 | .908 | | | | | |
| 16.000 | 5.527 | .000 | 10.473 | .10 | 1.39 | 65.5% | 6 | .941 | .935 | .927 | .919 | .908 | .897 | | | | |
| 16.000 | 6.410 | .000 | 9.590 | .12 | 1.47 | 59.9% | 7 | .941 | .935 | .927 | .919 | .908 | .897 | .883 | | | |

## ERLANG B: 100% BLOCKED CALLERS ROUTE ADVANCE, OVERFLOW, OR DISAPPEAR

| HOURS FIRST ATTEMPT VMR = 1.0 | HOURS CON-NECT | HOURS RECALL /RETRY | HOURS OVFL | CON-NECT VMR | OVFL VMR | PERCENT BLOCKED (GRADE) | LINE | HOURS CONNECTED TRAFFIC ON LINES 1 | 2 | 3 | 4 | 5 | 6 | 7 | 8 | 9 | 10 |
|---|---|---|---|---|---|---|---|---|---|---|---|---|---|---|---|---|---|
| 16.000 | 7.277 | .000 | 8.723 | .13 | 1.56 | 54.5% | 8 | .941 | .935 | .927 | .919 | .908 | .897 | .883 | .867 | | |
| 16.000 | 8.125 | .000 | 7.875 | .15 | 1.66 | 49.2% | 9 | .941 | .935 | .927 | .919 | .908 | .897 | .883 | .867 | .848 | |
| 16.000 | 8.951 | .000 | 7.049 | .17 | 1.76 | 44.1% | 10 | .941 | .935 | .927 | .919 | .908 | .897 | .883 | .867 | .848 | .826 |
| 16.000 | 9.751 | .000 | 6.249 | .20 | 1.87 | 39.1% | 11 | .941 | .935 | .927 | .919 | .908 | .897 | .883 | .867 | .848 | .826 |
| | | | | | | | | .800 | | | | | | | | | |
| 16.000 | 10.521 | .000 | 5.479 | .23 | 1.98 | 34.2% | 12 | .941 | .935 | .927 | .919 | .908 | .897 | .883 | .867 | .848 | .826 |
| | | | | | | | | .800 | .770 | | | | | | | | |
| 16.000 | 11.256 | .000 | 4.744 | .27 | 2.09 | 29.6% | 13 | .941 | .935 | .927 | .919 | .908 | .897 | .883 | .867 | .848 | .826 |
| | | | | | | | | .800 | .770 | .735 | | | | | | | |
| 16.000 | 11.951 | .000 | 4.049 | .31 | 2.20 | 25.3% | 14 | .941 | .935 | .927 | .919 | .908 | .897 | .883 | .867 | .848 | .826 |
| | | | | | | | | .800 | .770 | .735 | .694 | | | | | | |
| 16.000 | 12.599 | .000 | 3.401 | .35 | 2.30 | 21.3% | 15 | .941 | .935 | .927 | .919 | .908 | .897 | .883 | .867 | .848 | .826 |
| | | | | | | | | .800 | .770 | .735 | .694 | .648 | | | | | |
| 16.000 | 13.195 | .000 | 2.805 | .40 | 2.40 | 17.5% | 16 | .941 | .935 | .927 | .919 | .908 | .897 | .883 | .867 | .848 | .826 |
| | | | | | | | | .800 | .770 | .735 | .694 | .648 | .596 | | | | |
| 16.000 | 13.734 | .000 | 2.266 | .46 | 2.48 | 14.2% | 17 | .941 | .935 | .927 | .919 | .908 | .897 | .883 | .867 | .848 | .826 |
| | | | | | | | | .800 | .770 | .735 | .694 | .648 | .596 | .539 | | | |
| 16.000 | 14.211 | .000 | 1.789 | .52 | 2.55 | 11.2% | 18 | .941 | .935 | .927 | .919 | .908 | .897 | .883 | .867 | .848 | .826 |
| | | | | | | | | .800 | .770 | .735 | .694 | .648 | .596 | .539 | .477 | | |
| 16.000 | 14.623 | .000 | 1.377 | .59 | 2.60 | 8.6% | 19 | .941 | .935 | .927 | .919 | .908 | .897 | .883 | .867 | .848 | .826 |
| | | | | | | | | .800 | .770 | .735 | .694 | .648 | .596 | .539 | .477 | .412 | |
| 16.000 | 14.969 | .000 | 1.031 | .65 | 2.62 | 6.4% | 20 | .941 | .935 | .927 | .919 | .908 | .897 | .883 | .867 | .848 | .826 |
| | | | | | | | | .800 | .770 | .735 | .694 | .648 | .596 | .539 | .477 | .412 | .346 |
| 16.000 | 15.252 | .000 | .748 | .72 | 2.62 | 4.7% | 21 | .941 | .935 | .927 | .919 | .908 | .897 | .883 | .867 | .848 | .826 |
| | | | | | | | | .800 | .770 | .735 | .694 | .648 | .596 | .539 | .477 | .412 | .346 |
| | | | | | | | | .282 | | | | | | | | | |
| 16.000 | 15.474 | .000 | .526 | .78 | 2.60 | 3.3% | 22 | .941 | .935 | .927 | .919 | .908 | .897 | .883 | .867 | .848 | .826 |
| | | | | | | | | .000 | .770 | .735 | .694 | .648 | .596 | .539 | .477 | .412 | .346 |
| | | | | | | | | .282 | .222 | | | | | | | | |
| 16.000 | 15.642 | .000 | .358 | .83 | 2.56 | 2.2% | 23 | .941 | .935 | .927 | .919 | .908 | .897 | .883 | .867 | .848 | .826 |
| | | | | | | | | .800 | .770 | .735 | .694 | .648 | .596 | .539 | .477 | .412 | .346 |
| | | | | | | | | .282 | .222 | .168 | | | | | | | |
| 16.000 | 15.765 | .000 | .235 | .88 | 2.50 | 1.5% | 24 | .941 | .935 | .927 | .919 | .908 | .897 | .883 | .867 | .848 | .826 |
| | | | | | | | | .800 | .770 | .735 | .694 | .648 | .596 | .539 | .477 | .412 | .346 |
| | | | | | | | | .202 | .222 | .168 | .123 | | | | | | |
| 16.000 | 15.851 | .000 | .149 | .91 | 2.43 | .9% | 25 | .941 | .935 | .927 | .919 | .908 | .897 | .883 | .867 | .848 | .826 |
| | | | | | | | | .800 | .770 | .735 | .694 | .648 | .596 | .539 | .477 | .412 | .346 |
| | | | | | | | | .282 | .222 | .168 | .123 | .086 | | | | | |
| 16.000 | 15.909 | .000 | .091 | .94 | 2.35 | .6% | 26 | .941 | .935 | .927 | .919 | .908 | .897 | .883 | .867 | .848 | .826 |
| | | | | | | | | .800 | .770 | .735 | .694 | .648 | .596 | .539 | .477 | .412 | .340 |

## ERLANG B: 100% BLOCKED CALLERS ROUTE ADVANCE, OVERFLOW, OR DISAPPEAR

| HOURS FIRST ATTEMPT VMR = 1.0 | HOURS CONNECT | HOURS RECALL /RETRY | HOURS OVFL | CONNECT VMR | OVFL VMR | PERCENT BLOCKED (GRADE) | LINE | HOURS CONNECTED TRAFFIC ON LINES 1 | 2 | 3 | 4 | 5 | 6 | 7 | 8 | 9 | 10 |
|---|---|---|---|---|---|---|---|---|---|---|---|---|---|---|---|---|---|
| | | | | | | | | .282 | .222 | .168 | .123 | .086 | .058 | | | | |
| 17.000 | 4.655 | .000 | 12.345 | .08 | 1.29 | 72.6% | 5 | .944 | .939 | .932 | .924 | .915 | | | | | |
| 17.000 | 5.560 | .000 | 11.440 | .09 | 1.37 | 67.3% | 6 | .944 | .939 | .932 | .924 | .915 | .905 | | | | |
| 17.000 | 6.453 | .000 | 10.547 | .11 | 1.45 | 62.0% | 7 | .944 | .939 | .932 | .924 | .915 | .905 | .893 | | | |
| 17.000 | 7.333 | .000 | 9.667 | .12 | 1.53 | 56.9% | 8 | .944 | .939 | .932 | .924 | .915 | .905 | .893 | .879 | | |
| 17.000 | 8.196 | .000 | 8.804 | .14 | 1.62 | 51.8% | 9 | .944 | .939 | .932 | .924 | .915 | .905 | .893 | .879 | .863 | |
| 17.000 | 9.041 | .000 | 7.959 | .16 | 1.72 | 46.8% | 10 | .944 | .939 | .932 | .924 | .915 | .905 | .893 | .879 | .863 | .845 |
| 17.000 | 9.863 | .000 | 7.137 | .18 | 1.82 | 42.0% | 11 | .944 | .939 | .932 | .924 | .915 | .905 | .893 | .879 | .863 | .845 |
| | | | | | | | | .823 | | | | | | | | | |
| 17.000 | 10.660 | .000 | 6.340 | .20 | 1.93 | 37.3% | 12 | .944 | .939 | .932 | .924 | .915 | .905 | .893 | .879 | .863 | .845 |
| | | | | | | | | .823 | .797 | | | | | | | | |
| 17.000 | 11.427 | .000 | 5.573 | .23 | 2.03 | 32.8% | 13 | .944 | .939 | .932 | .924 | .915 | .905 | .893 | .879 | .863 | .845 |
| | | | | | | | | .823 | .797 | .767 | | | | | | | |
| 17.000 | 12.160 | .000 | 4.840 | .27 | 2.15 | 28.5% | 14 | .944 | .939 | .932 | .924 | .915 | .905 | .893 | .879 | .863 | .845 |
| | | | | | | | | .823 | .797 | .767 | .733 | | | | | | |
| 17.000 | 12.853 | .000 | 4.147 | .31 | 2.25 | 24.4% | 15 | .944 | .939 | .932 | .924 | .915 | .905 | .893 | .879 | .863 | .845 |
| | | | | | | | | .823 | .797 | .767 | .733 | .693 | | | | | |
| 17.000 | 13.501 | .000 | 3.499 | .35 | 2.36 | 20.6% | 16 | .944 | .939 | .932 | .924 | .915 | .905 | .893 | .879 | .863 | .845 |
| | | | | | | | | .823 | .797 | .767 | .733 | .693 | .648 | | | | |
| 17.000 | 14.098 | .000 | 2.902 | .40 | 2.45 | 17.1% | 17 | .944 | .939 | .932 | .924 | .915 | .905 | .893 | .879 | .863 | .845 |
| | | | | | | | | .823 | .797 | .767 | .733 | .693 | .648 | .597 | | | |
| 17.000 | 14.640 | .000 | 2.360 | .46 | 2.54 | 13.9% | 18 | .944 | .939 | .932 | .924 | .915 | .905 | .893 | .879 | .863 | .845 |
| | | | | | | | | .823 | .797 | .767 | .733 | .693 | .648 | .597 | .542 | | |
| 17.000 | 15.122 | .000 | 1.878 | .52 | 2.61 | 11.0% | 19 | .944 | .939 | .932 | .924 | .915 | .905 | .893 | .879 | .863 | .845 |
| | | | | | | | | .823 | .797 | .767 | .733 | .693 | .648 | .597 | .542 | .482 | |
| 17.000 | 15.540 | .000 | 1.460 | .58 | 2.65 | 8.6% | 20 | .944 | .939 | .932 | .924 | .915 | .905 | .893 | .879 | .863 | .845 |
| | | | | | | | | .823 | .797 | .767 | .733 | .693 | .648 | .597 | .542 | .482 | .419 |
| 17.000 | 15.895 | .000 | 1.105 | .65 | 2.68 | 6.5% | 21 | .944 | .939 | .932 | .924 | .915 | .905 | .893 | .879 | .863 | .845 |
| | | | | | | | | .823 | .797 | .767 | .733 | .693 | .648 | .597 | .542 | .482 | .419 |
| | | | | | | | | .355 | | | | | | | | | |
| 17.000 | 16.187 | .000 | .813 | .71 | 2.68 | 4.8% | 22 | .944 | .939 | .932 | .924 | .915 | .905 | .893 | .879 | .863 | .845 |
| | | | | | | | | .823 | .797 | .767 | .733 | .693 | .648 | .597 | .542 | .482 | .419 |
| | | | | | | | | .355 | .292 | | | | | | | | |
| 17.000 | 16.420 | .000 | .580 | .77 | 2.66 | 3.4% | 23 | .944 | .939 | .932 | .924 | .915 | .905 | .893 | .879 | .863 | .845 |
| | | | | | | | | .823 | .797 | .767 | .733 | .693 | .648 | .597 | .542 | .482 | .419 |
| | | | | | | | | .355 | .292 | .233 | | | | | | | |
| 17.000 | 16.599 | .000 | .401 | .82 | 2.62 | 2.4% | 24 | .944 | .939 | .932 | .924 | .915 | .905 | .893 | .879 | .863 | .845 |
| | | | | | | | | .823 | .797 | .767 | .733 | .693 | .648 | .597 | .542 | .482 | .419 |
| | | | | | | | | .355 | .292 | .233 | .179 | | | | | | |
| 17.000 | 16.731 | .000 | .269 | .87 | 2.57 | 1.6% | 25 | .944 | .939 | .932 | .924 | .915 | .905 | .893 | .879 | .863 | .845 |

## ERLANG B: 100% BLOCKED CALLERS ROUTE ADVANCE, OVERFLOW, OR DISAPPEAR

| HOURS FIRST ATTEMPT VMR = 1.0 | HOURS CON-NECT | HOURS RECALL /RETRY | HOURS OVFL | CON-NECT VMR | VFL VMR | PERCENT BLOCKED (GRADE) | LINE | HOURS CONNECTED TRAFFIC ON LINES 1 | 2 | 3 | 4 | 5 | 6 | 7 | 8 | 9 | 10 |
|---|---|---|---|---|---|---|---|---|---|---|---|---|---|---|---|---|---|
| | | | | | | | | .823 | .797 | .767 | .733 | .693 | .648 | .597 | .542 | .482 | .419 |
| | | | | | | | | .355 | .292 | .233 | .179 | .133 | | | | | |
| 17.000 | 16.826 | .000 | .174 | .91 | 2.50 | 1.0% | 26 | .944 | .939 | .932 | .924 | .915 | .905 | .893 | .879 | .863 | .845 |
| | | | | | | | | .823 | .797 | .767 | .733 | .693 | .648 | .597 | .542 | .482 | .419 |
| | | | | | | | | .355 | .292 | .233 | .179 | .133 | .095 | | | | |
| 17.000 | 16.891 | .000 | .109 | .93 | 2.42 | .6% | 27 | .944 | .939 | .932 | .924 | .915 | .905 | .893 | .879 | .863 | .845 |
| | | | | | | | | .823 | .797 | .767 | .733 | .693 | .648 | .597 | .542 | .482 | .419 |
| | | | | | | | | .355 | .292 | .233 | .179 | .133 | .095 | .065 | | | |
| 18.000 | 4.677 | .000 | 13.323 | .08 | 1.28 | 74.0% | 5 | .947 | .942 | .936 | .929 | .922 | | | | | |
| 18.000 | 5.589 | .000 | 12.411 | .09 | 1.35 | 68.9% | 6 | .947 | .942 | .936 | .929 | .922 | .913 | | | | |
| 18.000 | 6.491 | .000 | 11.509 | .10 | 1.42 | 63.9% | 7 | .947 | .942 | .936 | .929 | .922 | .913 | .902 | | | |
| 18.000 | 7.381 | .000 | 10.619 | .11 | 1.50 | 59.0% | 8 | .947 | .942 | .936 | .929 | .922 | .913 | .902 | .890 | | |
| 18.000 | 8.257 | .000 | 9.743 | .12 | 1.59 | 54.1% | 9 | .947 | .942 | .936 | .929 | .922 | .913 | .902 | .890 | .876 | |
| 18.000 | 9.117 | .000 | 8.883 | .14 | 1.68 | 49.3% | 10 | .947 | .942 | .936 | .929 | .922 | .913 | .902 | .890 | .876 | .860 |
| 18.000 | 9.958 | .000 | 8.042 | .16 | 1.78 | 44.7% | 11 | .947 | .942 | .936 | .929 | .922 | .913 | .902 | .890 | .876 | .860 |
| | | | | | | | | .841 | | | | | | | | | |
| 18.000 | 10.778 | .000 | 7.222 | .18 | 1.88 | 40.1% | 12 | .947 | .942 | .936 | .929 | .922 | .913 | .902 | .890 | .876 | .860 |
| | | | | | | | | .841 | .819 | | | | | | | | |
| 18.000 | 11.571 | .000 | 6.429 | .21 | 1.98 | 35.7% | 13 | .947 | .942 | .936 | .929 | .922 | .913 | .902 | .890 | .876 | .860 |
| | | | | | | | | .841 | .819 | .794 | | | | | | | |
| 18.000 | 12.336 | .000 | 5.664 | .24 | 2.09 | 31.5% | 14 | .947 | .942 | .936 | .929 | .922 | .913 | .902 | .890 | .876 | .860 |
| | | | | | | | | .841 | .819 | .794 | .764 | | | | | | |
| 18.000 | 13.066 | .000 | 4.934 | .27 | 2.20 | 27.4% | 15 | .947 | .942 | .936 | .929 | .922 | .913 | .902 | .890 | .876 | .860 |
| | | | | | | | | .841 | .819 | .794 | .764 | .730 | | | | | |
| 18.000 | 13.757 | .000 | 4.243 | .31 | 2.31 | 23.6% | 16 | .947 | .942 | .936 | .929 | .922 | .913 | .902 | .890 | .876 | .860 |
| | | | | | | | | .841 | .819 | .794 | .764 | .730 | .692 | | | | |
| 18.000 | 14.405 | .000 | 3.595 | .35 | 2.41 | 20.0% | 17 | .947 | .942 | .936 | .929 | .922 | .913 | .902 | .890 | .876 | .860 |
| | | | | | | | | .841 | .819 | .794 | .764 | .730 | .692 | .648 | | | |
| 18.000 | 15.004 | .000 | 2.996 | .40 | 2.51 | 16.6% | 18 | .947 | .942 | .936 | .929 | .922 | .913 | .902 | .890 | .876 | .860 |
| | | | | | | | | .841 | .819 | .794 | .764 | .730 | .692 | .648 | .598 | | |
| 18.000 | 15.548 | .000 | 2.452 | .46 | 2.59 | 13.6% | 19 | .947 | .942 | .936 | .929 | .922 | .913 | .902 | .890 | .876 | .860 |
| | | | | | | | | .841 | .819 | .794 | .764 | .730 | .692 | .648 | .598 | .544 | |
| 18.000 | 16.034 | .000 | 1.966 | .51 | 2.66 | 10.9% | 20 | .947 | .942 | .936 | .929 | .922 | .913 | .902 | .890 | .876 | .860 |
| | | | | | | | | .841 | .819 | .794 | .764 | .730 | .692 | .648 | .598 | .544 | .486 |
| 18.000 | 16.459 | .000 | 1.541 | .57 | 2.71 | 8.6% | 21 | .947 | .942 | .936 | .929 | .922 | .913 | .902 | .890 | .876 | .860 |
| | | | | | | | | .841 | .819 | .794 | .764 | .730 | .692 | .648 | .598 | .544 | .486 |
| | | | | | | | | .425 | | | | | | | | | |
| 18.000 | 16.822 | .000 | 1.178 | .64 | 2.74 | 6.5% | 22 | .947 | .942 | .936 | .929 | .922 | .913 | .902 | .890 | .876 | .860 |
| | | | | | | | | .841 | .819 | .794 | .764 | .730 | .692 | .648 | .598 | .544 | .486 |
| | | | | | | | | .425 | .363 | | | | | | | | |

# ERLANG B: 100% BLOCKED CALLERS ROUTE ADVANCE, OVERFLOW, OR DISAPPEAR

| HOURS FIRST ATTEMPT VMR = 1.0 | HOURS CON-NECT | HOURS RECALL /RETRY | HOURS OVFL | CON-NECT VMR | OVFL VMR | PERCENT BLOCKED (GRADE) | LINE | HOURS CONNECTED TRAFFIC ON LINES 1 | 2 | 3 | 4 | 5 | 6 | 7 | 8 | 9 | 10 |
|---|---|---|---|---|---|---|---|---|---|---|---|---|---|---|---|---|---|
| 18.000 | 17.123 | .000 | .877 | .70 | 2.74 | 4.9% | 23 | .947 | .942 | .936 | .929 | .922 | .913 | .902 | .890 | .876 | .860 |
| | | | | | | | | .841 | .819 | .794 | .764 | .730 | .692 | .648 | .598 | .544 | .486 |
| | | | | | | | | .425 | .363 | .301 | | | | | | | |
| 18.000 | 17.365 | .000 | .635 | .76 | 2.72 | 3.5% | 24 | .947 | .942 | .936 | .929 | .922 | .913 | .902 | .890 | .876 | .860 |
| | | | | | | | | .841 | .819 | .794 | .764 | .730 | .692 | .648 | .598 | .544 | .486 |
| | | | | | | | | .425 | .363 | .301 | .242 | | | | | | |
| 18.000 | 17.554 | .000 | .446 | .81 | 2.69 | 2.5% | 25 | .947 | .942 | .936 | .929 | .922 | .913 | .902 | .890 | .876 | .860 |
| | | | | | | | | .841 | .819 | .794 | .764 | .730 | .692 | .648 | .598 | .544 | .486 |
| | | | | | | | | .425 | .363 | .301 | .242 | .189 | | | | | |
| 18.000 | 17.697 | .000 | .303 | .86 | 2.63 | 1.7% | 26 | .947 | .942 | .936 | .929 | .922 | .913 | .902 | .890 | .876 | .860 |
| | | | | | | | | .841 | .819 | .794 | .764 | .730 | .692 | .648 | .598 | .544 | .486 |
| | | | | | | | | .425 | .363 | .301 | .242 | .189 | .142 | | | | |
| 18.000 | 17.800 | .000 | .200 | .90 | 2.56 | 1.1% | 27 | .947 | .942 | .936 | .929 | .922 | .913 | .902 | .890 | .876 | .860 |
| | | | | | | | | .841 | .819 | .794 | .764 | .730 | .692 | .648 | .598 | .544 | .486 |
| | | | | | | | | .425 | .363 | .301 | .242 | .189 | .142 | .103 | | | |
| 18.000 | 17.872 | .000 | .128 | .93 | 2.49 | .7% | 28 | .947 | .942 | .936 | .929 | .922 | .913 | .902 | .890 | .876 | .860 |
| | | | | | | | | .841 | .819 | .794 | .764 | .730 | .692 | .648 | .598 | .544 | .486 |
| | | | | | | | | .425 | .363 | .301 | .242 | .189 | .142 | .103 | .072 | | |
| 19.000 | 5.615 | .000 | 13.385 | .08 | 1.33 | 70.4% | 6 | .950 | .945 | .940 | .934 | .927 | .919 | | | | |
| 19.000 | 6.524 | .000 | 12.476 | .09 | 1.40 | 65.7% | 7 | .950 | .945 | .940 | .934 | .927 | .919 | .910 | | | |
| 19.000 | 7.423 | .000 | 11.577 | .10 | 1.47 | 60.9% | 8 | .950 | .945 | .940 | .934 | .927 | .919 | .910 | .899 | | |
| 19.000 | 8.310 | .000 | 10.690 | .11 | 1.56 | 56.3% | 9 | .950 | .945 | .940 | .934 | .927 | .919 | .910 | .899 | .887 | |
| 19.000 | 9.183 | .000 | 9.817 | .13 | 1.64 | 51.7% | 10 | .950 | .945 | .940 | .934 | .927 | .919 | .910 | .899 | .887 | .873 |
| 19.000 | 10.040 | .000 | 8.960 | .14 | 1.73 | 47.2% | 11 | .950 | .945 | .940 | .934 | .927 | .919 | .910 | .899 | .887 | .873 |
| | | | | | | | | .857 | | | | | | | | | |
| 19.000 | 10.878 | .000 | 8.122 | .16 | 1.83 | 42.7% | 12 | .950 | .945 | .940 | .934 | .927 | .919 | .910 | .899 | .887 | .873 |
| | | | | | | | | .857 | .838 | | | | | | | | |
| 19.000 | 11.694 | .000 | 7.306 | .18 | 1.93 | 38.5% | 13 | .950 | .945 | .940 | .934 | .927 | .919 | .910 | .899 | .887 | .873 |
| | | | | | | | | .857 | .838 | .816 | | | | | | | |
| 19.000 | 12.485 | .000 | 6.515 | .21 | 2.04 | 34.3% | 14 | .950 | .945 | .940 | .934 | .927 | .919 | .910 | .899 | .887 | .873 |
| | | | | | | | | .857 | .838 | .816 | .791 | | | | | | |
| 19.000 | 13.246 | .000 | 5.754 | .24 | 2.15 | 30.3% | 15 | .950 | .945 | .940 | .934 | .927 | .919 | .910 | .899 | .887 | .873 |
| | | | | | | | | .857 | .838 | .816 | .791 | .762 | | | | | |
| 19.000 | 13.975 | .000 | 5.025 | .27 | 2.26 | 26.4% | 16 | .950 | .945 | .940 | .934 | .927 | .919 | .910 | .899 | .887 | .873 |
| | | | | | | | | .857 | .838 | .816 | .791 | .762 | .728 | | | | |
| 19.000 | 14.665 | .000 | 4.335 | .31 | 2.36 | 22.8% | 17 | .950 | .945 | .940 | .934 | .927 | .919 | .910 | .899 | .887 | .873 |
| | | | | | | | | .857 | .838 | .816 | .791 | .762 | .728 | .690 | | | |
| 19.000 | 15.312 | .000 | 3.688 | .35 | 2.46 | 19.4% | 18 | .950 | .945 | .940 | .934 | .927 | .919 | .910 | .899 | .887 | .873 |
| | | | | | | | | .857 | .838 | .816 | .791 | .762 | .728 | .690 | .647 | | |
| 19.000 | 15.912 | .000 | 3.088 | .40 | 2.56 | 16.3% | 19 | .950 | .945 | .940 | .934 | .927 | .919 | .910 | .899 | .887 | .873 |

# ERLANG B: 100% BLOCKED CALLERS ROUTE ADVANCE, OVERFLOW, OR DISAPPEAR

| HOURS FIRST ATTEMPT VMR = 1.0 | HOURS CON-NECT | HOURS RECALL /RETRY | HOURS OVFL | CON-NECT VMR | OVFL VMR | PERCENT BLOCKED (GRADE) | LINE | HOURS CONNECTED TRAFFIC ON LINES 1 | 2 | 3 | 4 | 5 | 6 | 7 | 8 | 9 | 10 |
|---|---|---|---|---|---|---|---|---|---|---|---|---|---|---|---|---|---|
| | | | | | | | | .857 | .838 | .816 | .791 | .762 | .728 | .690 | .647 | .599 | |
| 19.000 | 16.459 | .000 | 2.541 | .45 | 2.64 | 13.4% | 20 | .950 | .945 | .940 | .934 | .927 | .919 | .910 | .899 | .887 | .873 |
| | | | | | | | | .857 | .838 | .816 | .791 | .762 | .728 | .690 | .647 | .599 | .547 |
| 19.000 | 16.949 | .000 | 2.051 | .51 | 2.71 | 10.8% | 21 | .950 | .945 | .940 | .934 | .927 | .919 | .910 | .899 | .887 | .873 |
| | | | | | | | | .857 | .838 | .816 | .791 | .762 | .728 | .690 | .647 | .599 | .547 |
| | | | | | | | | .490 | | | | | | | | | |
| 19.000 | 17.380 | .000 | 1.620 | .57 | 2.76 | 8.5% | 22 | .950 | .945 | .940 | .934 | .927 | .919 | .910 | .899 | .887 | .873 |
| | | | | | | | | .857 | .838 | .816 | .791 | .762 | .728 | .690 | .647 | .599 | .547 |
| | | | | | | | | .490 | .431 | | | | | | | | |
| 19.000 | 17.750 | .000 | 1.250 | .63 | 2.79 | 6.6% | 23 | .950 | .945 | .940 | .934 | .927 | .919 | .910 | .899 | .887 | .873 |
| | | | | | | | | .857 | .838 | .816 | .791 | .762 | .728 | .690 | .647 | .599 | .547 |
| | | | | | | | | .490 | .431 | .370 | | | | | | | |
| 19.000 | 18.059 | .000 | .941 | .69 | 2.80 | 5.0% | 24 | .950 | .945 | .940 | .934 | .927 | .919 | .910 | .899 | .887 | .873 |
| | | | | | | | | .857 | .838 | .816 | .791 | .762 | .728 | .690 | .647 | .599 | .547 |
| | | | | | | | | .490 | .431 | .370 | .310 | | | | | | |
| 19.000 | 18.311 | .000 | .689 | .75 | 2.78 | 3.6% | 25 | .950 | .945 | .940 | .934 | .927 | .919 | .910 | .899 | .887 | .873 |
| | | | | | | | | .857 | .838 | .816 | .791 | .762 | .728 | .690 | .647 | .599 | .547 |
| | | | | | | | | .490 | .431 | .370 | .310 | .252 | | | | | |
| 19.000 | 18.509 | .000 | .491 | .80 | 2.75 | 2.6% | 26 | .950 | .945 | .940 | .934 | .927 | .919 | .910 | .899 | .887 | .873 |
| | | | | | | | | .857 | .838 | .816 | .791 | .762 | .728 | .690 | .647 | .599 | .547 |
| | | | | | | | | .490 | .431 | .370 | .310 | .252 | .199 | | | | |
| 19.000 | 18.661 | .000 | .339 | .85 | 2.70 | 1.8% | 27 | .950 | .945 | .940 | .934 | .927 | .919 | .910 | .899 | .887 | .873 |
| | | | | | | | | .857 | .838 | .816 | .791 | .762 | .728 | .690 | .647 | .599 | .547 |
| | | | | | | | | .490 | .431 | .370 | .310 | .252 | .199 | .152 | | | |
| 19.000 | 18.773 | .000 | .227 | .89 | 2.63 | 1.2% | 28 | .950 | .945 | .940 | .934 | .927 | .919 | .910 | .899 | .887 | .873 |
| | | | | | | | | .857 | .838 | .816 | .791 | .762 | .728 | .690 | .647 | .599 | .547 |
| | | | | | | | | .490 | .431 | .370 | .310 | .252 | .199 | .152 | .112 | | |
| 19.000 | 18.852 | .000 | .148 | .92 | 2.56 | .8% | 29 | .950 | .945 | .940 | .934 | .927 | .919 | .910 | .899 | .887 | .873 |
| | | | | | | | | .857 | .838 | .816 | .791 | .762 | .728 | .690 | .647 | .599 | .547 |
| | | | | | | | | .490 | .431 | .370 | .310 | .199 | .152 | .112 | .080 | | |
| 20.000 | 5.637 | .000 | 14.363 | .08 | 1.31 | 71.8% | 6 | .952 | .948 | .943 | .938 | .931 | .924 | | | | |
| 20.000 | 6.553 | .000 | 13.447 | .08 | 1.38 | 67.2% | 7 | .952 | .948 | .943 | .938 | .931 | .924 | .916 | | | |
| 20.000 | 7.460 | .000 | 12.540 | .09 | 1.45 | 62.7% | 8 | .952 | .948 | .943 | .938 | .931 | .924 | .916 | .907 | | |
| 20.000 | 8.357 | .000 | 11.643 | .10 | 1.53 | 58.2% | 9 | .952 | .948 | .943 | .938 | .931 | .924 | .916 | .907 | .896 | |
| 20.000 | 9.241 | .000 | 10.759 | .12 | 1.61 | 53.8% | 10 | .952 | .948 | .943 | .938 | .931 | .924 | .916 | .907 | .896 | .884 |
| 20.000 | 10.111 | .000 | 9.889 | .13 | 1.70 | 49.4% | 11 | .952 | .948 | .943 | .938 | .931 | .924 | .916 | .907 | .896 | .884 |
| | | | | | | | | .870 | | | | | | | | | |
| 20.000 | 10.964 | .000 | 9.036 | .15 | 1.79 | 45.2% | 12 | .952 | .948 | .943 | .938 | .931 | .924 | .916 | .907 | .896 | .884 |
| | | | | | | | | .870 | .854 | | | | | | | | |
| 20.000 | 11.799 | .000 | 8.201 | .17 | 1.89 | 41.0% | 13 | .952 | .948 | .943 | .938 | .931 | .924 | .916 | .907 | .896 | .884 |

## ERLANG B: 100% BLOCKED CALLERS ROUTE ADVANCE, OVERFLOW, OR DISAPPEAR

| HOURS FIRST ATTEMPT VMR=1.0 | HOURS CON-NECT | HOURS RECALL /RETRY | HOURS OVFL | CON-NECT VMR | OVFL VMR | PERCENT BLOCKED (GRADE) | LINE | HOURS CONNECTED TRAFFIC ON LINES 1 | 2 | 3 | 4 | 5 | 6 | 7 | 8 | 9 | 10 |
|---|---|---|---|---|---|---|---|---|---|---|---|---|---|---|---|---|---|
| | | | | | | | | .870 | .854 | .835 | | | | | | | |
| 20.000 | 12.612 | .000 | 7.388 | .19 | 1.99 | 36.9% | 14 | .952 | .948 | .943 | .938 | .931 | .924 | .916 | .907 | .896 | .884 |
| | | | | | | | | .870 | .854 | .835 | .813 | | | | | | |
| 20.000 | 13.400 | .000 | 6.600 | .21 | 2.09 | 33.0% | 15 | .952 | .948 | .943 | .938 | .931 | .924 | .916 | .907 | .896 | .884 |
| | | | | | | | | .870 | .854 | .835 | .813 | .788 | | | | | |
| 20.000 | 14.159 | .000 | 5.841 | .24 | 2.20 | 29.2% | 16 | .952 | .948 | .943 | .938 | .931 | .924 | .916 | .907 | .896 | .884 |
| | | | | | | | | .870 | .854 | .835 | .813 | .788 | .759 | | | | |
| 20.000 | 14.886 | .000 | 5.114 | .27 | 2.31 | 25.6% | 17 | .952 | .948 | .943 | .938 | .931 | .924 | .916 | .907 | .896 | .884 |
| | | | | | | | | .870 | .854 | .835 | .813 | .788 | .759 | .726 | | | |
| 20.000 | 15.575 | .000 | 4.425 | .31 | 2.41 | 22.1% | 18 | .952 | .948 | .943 | .938 | .931 | .924 | .916 | .907 | .896 | .884 |
| | | | | | | | | .870 | .854 | .835 | .813 | .788 | .759 | .726 | .689 | | |
| 20.000 | 16.222 | .000 | 3.778 | .35 | 2.52 | 18.9% | 19 | .952 | .948 | .943 | .938 | .931 | .924 | .916 | .907 | .896 | .884 |
| | | | | | | | | .870 | .854 | .835 | .813 | .788 | .759 | .726 | .689 | .647 | |
| 20.000 | 16.822 | .000 | 3.178 | .40 | 2.61 | 15.9% | 20 | .952 | .948 | .943 | .938 | .931 | .924 | .916 | .907 | .896 | .884 |
| | | | | | | | | .870 | .854 | .835 | .813 | .788 | .759 | .726 | .689 | .647 | .600 |
| 20.000 | 17.371 | .000 | 2.629 | .45 | 2.69 | 13.1% | 21 | .952 | .948 | .943 | .938 | .931 | .924 | .916 | .907 | .896 | .884 |
| | | | | | | | | .870 | .854 | .835 | .813 | .788 | .759 | .726 | .689 | .647 | .600 |
| | | | | | | | | .549 | | | | | | | | | |
| 20.000 | 17.865 | .000 | 2.135 | .51 | 2.76 | 10.7% | 22 | .952 | .948 | .943 | .938 | .931 | .924 | .916 | .907 | .896 | .884 |
| | | | | | | | | .870 | .854 | .835 | .813 | .788 | .759 | .726 | .689 | .647 | .600 |
| | | | | | | | | .549 | .494 | | | | | | | | |
| 20.000 | 18.301 | .000 | 1.699 | .56 | 2.81 | 8.5% | 23 | .952 | .948 | .943 | .938 | .931 | .924 | .916 | .907 | .896 | .884 |
| | | | | | | | | .870 | .854 | .835 | .813 | .788 | .759 | .726 | .689 | .647 | .600 |
| | | | | | | | | .549 | .494 | .436 | | | | | | | |
| 20.000 | 18.678 | .000 | 1.322 | .62 | 2.84 | 6.6% | 24 | .952 | .948 | .943 | .938 | .931 | .924 | .916 | .907 | .896 | .884 |
| | | | | | | | | .870 | .854 | .835 | .813 | .788 | .759 | .726 | .689 | .647 | .600 |
| | | | | | | | | .549 | .494 | .436 | .377 | | | | | | |
| 20.000 | 18.996 | .000 | 1.004 | .68 | 2.85 | 5.0% | 25 | .952 | .948 | .943 | .938 | .931 | .924 | .916 | .907 | .896 | .884 |
| | | | | | | | | .870 | .854 | .835 | .813 | .788 | .759 | .726 | .689 | .647 | .600 |
| | | | | | | | | .549 | .494 | .436 | .377 | .317 | | | | | |
| 20.000 | 19.256 | .000 | .744 | .74 | 2.84 | 3.7% | 26 | .952 | .948 | .943 | .938 | .931 | .924 | .916 | .907 | .896 | .884 |
| | | | | | | | | .870 | .854 | .835 | .813 | .788 | .759 | .726 | .689 | .647 | .600 |
| | | | | | | | | .549 | .494 | .436 | .377 | .317 | .261 | | | | |
| 20.000 | 19.464 | .000 | .536 | .79 | 2.81 | 2.7% | 27 | .952 | .948 | .943 | .938 | .931 | .924 | .916 | .907 | .896 | .884 |
| | | | | | | | | .870 | .854 | .835 | .813 | .788 | .759 | .726 | .689 | .647 | .600 |
| | | | | | | | | .549 | .494 | .436 | .377 | .317 | .261 | .208 | | | |
| 20.000 | 19.624 | .000 | .376 | .84 | 2.76 | 1.9% | 28 | .952 | .948 | .943 | .938 | .931 | .924 | .916 | .907 | .896 | .884 |
| | | | | | | | | .870 | .854 | .835 | .813 | .788 | .759 | .726 | .689 | .647 | .600 |
| | | | | | | | | .549 | .494 | .436 | .377 | .317 | .261 | .208 | .160 | | |
| 20.000 | 19.744 | .000 | .256 | .88 | 2.69 | 1.3% | 29 | .952 | .948 | .943 | .938 | .931 | .924 | .916 | .907 | .896 | .884 |

## ERLANG B: 100% BLOCKED CALLERS ROUTE ADVANCE, OVERFLOW, OR DISAPPEAR

| HOURS FIRST ATTEMPT VMR = 1.0 | HOURS CON- NECT | HOURS RECALL /RETRY | HOURS OVFL | CON- NECT VMR | OVFL VMR | PERCENT BLOCKED (GRADE) | LINE | 1 | 2 | 3 | 4 | 5 | 6 | 7 | 8 | 9 | 10 |
|---|---|---|---|---|---|---|---|---|---|---|---|---|---|---|---|---|---|
| | | | | | | | | .870 | .854 | .835 | .813 | .788 | .759 | .726 | .689 | .647 | .600 |
| | | | | | | | | .549 | .494 | .436 | .377 | .317 | .261 | .208 | .160 | .120 | |
| 20.000 | 19.831 | .000 | .169 | .91 | 2.62 | .8% | 30 | .952 | .948 | .943 | .938 | .931 | .924 | .916 | .907 | .896 | .884 |
| | | | | | | | | .870 | .854 | .835 | .813 | .788 | .759 | .726 | .689 | .647 | .600 |
| | | | | | | | | .549 | .494 | .436 | .377 | .317 | .261 | .208 | .160 | .120 | .087 |
| 20.000 | 19.891 | .000 | .109 | .94 | 2.54 | .5% | 31 | .952 | .948 | .943 | .938 | .931 | .924 | .916 | .907 | .896 | .884 |
| | | | | | | | | .870 | .854 | .835 | .813 | .788 | .759 | .726 | .689 | .647 | .600 |
| | | | | | | | | .549 | .494 | .436 | .377 | .317 | .261 | .208 | .160 | .120 | .087 |
| | | | | | | | | .061 | | | | | | | | | |
| 22.000 | 5.675 | .000 | 16.325 | .07 | 1.28 | 74.2% | 6 | .957 | .953 | .949 | .944 | .939 | .934 | | | | |
| 22.000 | 6.602 | .000 | 15.398 | .07 | 1.34 | 70.0% | 7 | .957 | .953 | .949 | .944 | .939 | .934 | .927 | | | |
| 22.000 | 7.522 | .000 | 14.478 | .08 | 1.41 | 65.8% | 8 | .957 | .953 | .949 | .944 | .939 | .934 | .927 | .920 | | |
| 22.000 | 8.433 | .000 | 13.567 | .09 | 1.48 | 61.7% | 9 | .957 | .953 | .949 | .944 | .939 | .934 | .927 | .920 | .911 | |
| 22.000 | 9.335 | .000 | 12.665 | .10 | 1.55 | 57.6% | 10 | .957 | .953 | .949 | .944 | .939 | .934 | .927 | .920 | .911 | .902 |
| 22.000 | 10.226 | .000 | 11.774 | .11 | 1.63 | 53.5% | 11 | .957 | .953 | .949 | .944 | .939 | .934 | .927 | .920 | .911 | .902 |
| | | | | | | | | .891 | | | | | | | | | |
| 22.000 | 11.105 | .000 | 10.895 | .12 | 1.71 | 49.5% | 12 | .957 | .953 | .949 | .944 | .939 | .934 | .927 | .920 | .911 | .902 |
| | | | | | | | | .891 | .878 | | | | | | | | |
| 22.000 | 11.969 | .000 | 10.031 | .14 | 1.80 | 45.6% | 13 | .957 | .953 | .949 | .944 | .939 | .934 | .927 | .920 | .911 | .902 |
| | | | | | | | | .891 | .878 | .864 | | | | | | | |
| 22.000 | 12.817 | .000 | 9.183 | .15 | 1.89 | 41.7% | 14 | .957 | .953 | .949 | .944 | .939 | .934 | .927 | .920 | .911 | .902 |
| | | | | | | | | .891 | .878 | .864 | .848 | | | | | | |
| 22.000 | 13.646 | .000 | 8.354 | .17 | 1.99 | 38.0% | 15 | .957 | .953 | .949 | .944 | .939 | .934 | .927 | .920 | .911 | .902 |
| | | | | | | | | .891 | .878 | .864 | .848 | .829 | | | | | |
| 22.000 | 14.453 | .000 | 7.547 | .19 | 2.09 | 34.3% | 16 | .957 | .953 | .949 | .944 | .939 | .934 | .927 | .920 | .911 | .902 |
| | | | | | | | | .891 | .878 | .864 | .848 | .829 | .808 | | | | |
| 22.000 | 15.236 | .000 | 6.764 | .22 | 2.20 | 30.7% | 17 | .957 | .953 | .949 | .944 | .939 | .934 | .927 | .920 | .911 | .902 |
| | | | | | | | | .891 | .878 | .864 | .848 | .829 | .808 | .783 | | | |
| 22.000 | 15.991 | .000 | 6.009 | .25 | 2.30 | 27.3% | 18 | .957 | .953 | .949 | .944 | .939 | .934 | .927 | .920 | .911 | .902 |
| | | | | | | | | .891 | .878 | .864 | .848 | .829 | .808 | .783 | .755 | | |
| 22.000 | 16.714 | .000 | 5.286 | .28 | 2.41 | 24.0% | 19 | .957 | .953 | .949 | .944 | .939 | .934 | .927 | .920 | .911 | .902 |
| | | | | | | | | .891 | .878 | .864 | .848 | .829 | .808 | .783 | .755 | .723 | |
| 22.000 | 17.401 | .000 | 4.599 | .31 | 2.51 | 20.9% | 20 | .957 | .953 | .949 | .944 | .939 | .934 | .927 | .920 | .911 | .902 |
| | | | | | | | | .891 | .878 | .864 | .848 | .829 | .808 | .783 | .755 | .723 | .687 |
| 22.000 | 18.048 | .000 | 3.952 | .35 | 2.61 | 18.0% | 21 | .957 | .953 | .949 | .944 | .939 | .934 | .927 | .920 | .911 | .902 |
| | | | | | | | | .891 | .878 | .864 | .848 | .829 | .808 | .783 | .755 | .723 | .687 |
| | | | | | | | | .647 | | | | | | | | | |
| 22.000 | 18.650 | .000 | 3.350 | .40 | 2.71 | 15.2% | 22 | .957 | .953 | .949 | .944 | .939 | .934 | .927 | .920 | .911 | .902 |
| | | | | | | | | .891 | .878 | .864 | .848 | .829 | .808 | .783 | .755 | .723 | .687 |
| | | | | | | | | .647 | .602 | | | | | | | | |

# ERLANG B: 100% BLOCKED CALLERS ROUTE ADVANCE, OVERFLOW, OR DISAPPEAR

| HOURS FIRST ATTEMPT VMR = 1.0 | HOURS CON-NECT | HOURS RECALL /RETRY | HOURS OVFL | CON-NECT VMR | OVFL VMR | PERCENT BLOCKED (GRADE) | LINE | HOURS CONNECTED TRAFFIC ON LINES 1 | 2 | 3 | 4 | 5 | 6 | 7 | 8 | 9 | 10 |
|---|---|---|---|---|---|---|---|---|---|---|---|---|---|---|---|---|---|
| 22.000 | 19.203 | .000 | 2.797 | .45 | 2.79 | 12.7% | 23 | .957 | .953 | .949 | .944 | .939 | .934 | .927 | .920 | .911 | .902 |
| | | | | | | | | .891 | .878 | .864 | .848 | .829 | .808 | .783 | .755 | .723 | .687 |
| | | | | | | | | .647 | .602 | .553 | | | | | | | |
| 22.000 | 19.703 | .000 | 2.297 | .50 | 2.86 | 10.4% | 24 | .957 | .953 | .949 | .944 | .939 | .934 | .927 | .920 | .911 | .902 |
| | | | | | | | | .891 | .878 | .864 | .848 | .829 | .808 | .783 | .755 | .723 | .687 |
| | | | | | | | | .647 | .602 | .553 | .501 | | | | | | |
| 22.000 | 20.149 | .000 | 1.851 | .55 | 2.91 | 8.4% | 25 | .957 | .953 | .949 | .944 | .939 | .934 | .927 | .920 | .911 | .902 |
| | | | | | | | | .891 | .878 | .864 | .848 | .829 | .808 | .783 | .755 | .723 | .687 |
| | | | | | | | | .647 | .602 | .553 | .501 | .446 | | | | | |
| 22.000 | 20.538 | .000 | 1.462 | .61 | 2.94 | 6.6% | 26 | .957 | .953 | .949 | .944 | .939 | .934 | .927 | .920 | .911 | .902 |
| | | | | | | | | .891 | .878 | .864 | .848 | .829 | .808 | .783 | .755 | .723 | .687 |
| | | | | | | | | .647 | .602 | .553 | .501 | .446 | .389 | | | | |
| 22.000 | 20.870 | .000 | 1.130 | .67 | 2.96 | 5.1% | 27 | .957 | .953 | .949 | .944 | .939 | .934 | .927 | .920 | .911 | .902 |
| | | | | | | | | .891 | .878 | .864 | .848 | .829 | .808 | .783 | .755 | .723 | .687 |
| | | | | | | | | .647 | .602 | .553 | .501 | .446 | .389 | .332 | | | |
| 22.000 | 21.146 | .000 | .854 | .72 | 2.95 | 3.9% | 28 | .957 | .953 | .949 | .944 | .939 | .934 | .927 | .920 | .911 | .902 |
| | | | | | | | | .891 | .878 | .864 | .848 | .829 | .808 | .783 | .755 | .723 | .687 |
| | | | | | | | | .647 | .602 | .553 | .501 | .446 | .389 | .332 | .277 | | |
| 22.000 | 21.371 | .000 | .629 | .78 | 2.92 | 2.9% | 29 | .957 | .953 | .949 | .944 | .939 | .934 | .927 | .920 | .911 | .902 |
| | | | | | | | | .891 | .878 | .864 | .848 | .829 | .808 | .783 | .755 | .723 | .687 |
| | | | | | | | | .647 | .602 | .553 | .501 | .446 | .389 | .332 | .277 | .225 | |
| 22.000 | 21.548 | .000 | .452 | .82 | 2.88 | 2.1% | 30 | .957 | .953 | .949 | .944 | .939 | .934 | .927 | .920 | .911 | .902 |
| | | | | | | | | .891 | .878 | .864 | .848 | .829 | .808 | .783 | .755 | .723 | .687 |
| | | | | | | | | .647 | .602 | .553 | .501 | .446 | .389 | .332 | .277 | .225 | .177 |
| 22.000 | 21.684 | .000 | .316 | .86 | 2.82 | 1.4% | 31 | .957 | .953 | .949 | .944 | .939 | .934 | .927 | .920 | .911 | .902 |
| | | | | | | | | .891 | .878 | .864 | .848 | .829 | .808 | .783 | .755 | .723 | .687 |
| | | | | | | | | .647 | .602 | .553 | .501 | .446 | .389 | .332 | .277 | .225 | .177 |
| | | | | | | | | .136 | | | | | | | | | |
| 22.000 | 21.785 | .000 | .215 | .90 | 2.75 | 1.0% | 32 | .957 | .953 | .949 | .944 | .939 | .934 | .927 | .920 | .911 | .902 |
| | | | | | | | | .891 | .878 | .864 | .848 | .829 | .808 | .783 | .755 | .723 | .687 |
| | | | | | | | | .647 | .602 | .553 | .501 | .446 | .389 | .332 | .277 | .225 | .177 |
| | | | | | | | | .136 | .101 | | | | | | | | |
| 22.000 | 21.858 | .000 | .142 | .93 | 2.67 | .6% | 33 | .957 | .953 | .949 | .944 | .939 | .934 | .927 | .920 | .911 | .902 |
| | | | | | | | | .891 | .878 | .864 | .848 | .829 | .808 | .783 | .755 | .723 | .687 |
| | | | | | | | | .647 | .602 | .553 | .501 | .446 | .389 | .332 | .277 | .225 | .177 |
| | | | | | | | | .136 | .101 | .073 | | | | | | | |
| 24.000 | 6.642 | .000 | 17.358 | .06 | 1.32 | 72.3% | 7 | .960 | .957 | .954 | .950 | .946 | .941 | .936 | | | |
| 24.000 | 7.572 | .000 | 16.428 | .07 | 1.37 | 68.5% | 8 | .960 | .957 | .954 | .950 | .946 | .941 | .936 | .930 | | |
| 24.000 | 8.494 | .000 | 15.506 | .08 | 1.44 | 64.6% | 9 | .960 | .957 | .954 | .950 | .946 | .941 | .936 | .930 | .923 | |
| 24.000 | 9.410 | .000 | 14.590 | .08 | 1.50 | 60.8% | 10 | .960 | .957 | .954 | .950 | .946 | .941 | .936 | .930 | .923 | .915 |

## ERLANG B: 100% BLOCKED CALLERS ROUTE ADVANCE, OVERFLOW, OR DISAPPEAR

| HOURS FIRST ATTEMPT VMR = 1.0 | HOURS CON-NECT | HOURS RECALL /RETRY | HOURS OVFL | CON-NECT VMR | OVFL VMR | PERCENT BLOCKED (GRADE) | LINE | HOURS CONNECTED TRAFFIC ON LINES 1 | 2 | 3 | 4 | 5 | 6 | 7 | 8 | 9 | 10 |
|---|---|---|---|---|---|---|---|---|---|---|---|---|---|---|---|---|---|
| 24.000 | 10.316 | .000 | 13.684 | .09 | 1.57 | 57.0% | 11 | .960 | .957 | .954 | .950 | .946 | .941 | .936 | .930 | .923 | .915 |
| | | | | | | | | .907 | | | | | | | | | |
| 24.000 | 11.213 | .000 | 12.787 | .10 | 1.65 | 53.3% | 12 | .960 | .957 | .954 | .950 | .946 | .941 | .936 | .930 | .923 | .915 |
| | | | | | | | | .907 | .897 | | | | | | | | |
| 24.000 | 12.099 | .000 | 11.901 | .11 | 1.73 | 49.6% | 13 | .960 | .957 | .954 | .950 | .946 | .941 | .936 | .930 | .923 | .915 |
| | | | | | | | | .907 | .897 | .886 | | | | | | | |
| 24.000 | 12.973 | .000 | 11.027 | .13 | 1.81 | 45.9% | 14 | .960 | .957 | .954 | .950 | .946 | .941 | .936 | .930 | .923 | .915 |
| | | | | | | | | .907 | .897 | .886 | .873 | | | | | | |
| 24.000 | 13.832 | .000 | 10.168 | .14 | 1.90 | 42.4% | 15 | .960 | .957 | .954 | .950 | .946 | .941 | .936 | .930 | .923 | .915 |
| | | | | | | | | .907 | .897 | .886 | .873 | .859 | | | | | |
| 24.000 | 14.674 | .000 | 9.326 | .16 | 1.99 | 38.9% | 16 | .960 | .957 | .954 | .950 | .946 | .941 | .936 | .930 | .923 | .915 |
| | | | | | | | | .907 | .897 | .886 | .873 | .859 | .843 | | | | |
| 24.000 | 15.498 | .000 | 8.502 | .18 | 2.09 | 35.4% | 17 | .960 | .957 | .954 | .950 | .946 | .941 | .936 | .930 | .923 | .915 |
| | | | | | | | | .907 | .897 | .886 | .873 | .859 | .843 | .824 | | | |
| 24.000 | 16.301 | .000 | 7.699 | .20 | 2.19 | 32.1% | 18 | .960 | .957 | .954 | .950 | .946 | .941 | .936 | .930 | .923 | .915 |
| | | | | | | | | .907 | .897 | .886 | .873 | .859 | .843 | .824 | .803 | | |
| 24.000 | 17.079 | .000 | 6.921 | .22 | 2.30 | 28.8% | 19 | .960 | .957 | .954 | .950 | .946 | .941 | .936 | .930 | .923 | .915 |
| | | | | | | | | .907 | .897 | .886 | .873 | .859 | .843 | .824 | .803 | .778 | |
| 24.000 | 17.830 | .000 | 6.170 | .25 | 2.40 | 25.7% | 20 | .960 | .957 | .954 | .950 | .946 | .941 | .936 | .930 | .923 | .915 |
| | | | | | | | | .907 | .897 | .886 | .873 | .859 | .843 | .824 | .803 | .778 | .751 |
| 24.000 | 18.550 | .000 | 5.450 | .28 | 2.51 | 22.7% | 21 | .960 | .957 | .954 | .950 | .946 | .941 | .936 | .930 | .923 | .915 |
| | | | | | | | | .907 | .897 | .886 | .873 | .859 | .843 | .824 | .803 | .778 | .751 |
| | | | | | | | | .720 | | | | | | | | | |
| 24.000 | 19.235 | .000 | 4.765 | .31 | 2.61 | 19.9% | 22 | .960 | .957 | .954 | .950 | .946 | .941 | .936 | .930 | .923 | .915 |
| | | | | | | | | .907 | .897 | .886 | .873 | .859 | .843 | .824 | .803 | .778 | .751 |
| | | | | | | | | .720 | .685 | | | | | | | | |
| 24.000 | 19.881 | .000 | 4.119 | .35 | 2.71 | 17.2% | 23 | .960 | .957 | .954 | .950 | .946 | .941 | .936 | .930 | .923 | .915 |
| | | | | | | | | .907 | .897 | .886 | .873 | .859 | .843 | .824 | .803 | .778 | .751 |
| | | | | | | | | .720 | .685 | .646 | | | | | | | |
| 24.000 | 20.484 | .000 | 3.516 | .40 | 2.80 | 14.6% | 24 | .960 | .957 | .954 | .950 | .946 | .941 | .936 | .930 | .923 | .915 |
| | | | | | | | | .907 | .897 | .886 | .873 | .859 | .843 | .824 | .803 | .778 | .751 |
| | | | | | | | | .720 | .685 | .646 | .603 | | | | | | |
| 24.000 | 21.041 | .000 | 2.959 | .44 | 2.88 | 12.3% | 25 | .960 | .957 | .954 | .950 | .946 | .941 | .936 | .930 | .923 | .915 |
| | | | | | | | | .907 | .897 | .886 | .873 | .859 | .843 | .824 | .803 | .778 | .751 |
| | | | | | | | | .720 | .685 | .646 | .603 | .557 | | | | | |
| 24.000 | 21.548 | .000 | 2.452 | .49 | 2.95 | 10.2% | 26 | .960 | .957 | .954 | .950 | .946 | .941 | .936 | .930 | .923 | .915 |
| | | | | | | | | .907 | .897 | .886 | .873 | .859 | .843 | .824 | .803 | .778 | .751 |
| | | | | | | | | .720 | .685 | .646 | .603 | .557 | .507 | | | | |
| 24.000 | 22.002 | .000 | 1.998 | .55 | 3.00 | 8.3% | 27 | .960 | .957 | .954 | .950 | .946 | .941 | .936 | .930 | .923 | .915 |
| | | | | | | | | .907 | .897 | .886 | .873 | .859 | .843 | .824 | .803 | .778 | .751 |

# ERLANG B: 100% BLOCKED CALLERS ROUTE ADVANCE, OVERFLOW, OR DISAPPEAR

| HOURS FIRST ATTEMPT VMR = 1.0 | HOURS CON-NECT | HOURS RECALL /RETRY | HOURS OVFL | CON-NECT VMR | OVFL VMR | PERCENT BLOCKED (GRADE) | LINE | HOURS CONNECTED TRAFFIC ON LINES 1 | 2 | 3 | 4 | 5 | 6 | 7 | 8 | 9 | 10 |
|---|---|---|---|---|---|---|---|---|---|---|---|---|---|---|---|---|---|
| | | | | | | | | .720 | .685 | .646 | .603 | .557 | .507 | .454 | | | |
| 24.000 | 22.401 | .000 | 1.599 | .60 | 3.04 | 6.7% | 28 | .960 | .957 | .954 | .950 | .946 | .941 | .936 | .930 | .923 | .915 |
| | | | | | | | | .907 | .897 | .886 | .873 | .859 | .843 | .824 | .803 | .778 | .751 |
| | | | | | | | | .720 | .685 | .646 | .603 | .557 | .507 | .454 | .400 | | |
| 24.000 | 22.746 | .000 | 1.254 | .66 | 3.05 | 5.2% | 29 | .960 | .957 | .954 | .950 | .946 | .941 | .936 | .930 | .923 | .915 |
| | | | | | | | | .907 | .897 | .886 | .873 | .859 | .843 | .824 | .803 | .778 | .751 |
| | | | | | | | | .720 | .685 | .646 | .603 | .557 | .507 | .454 | .400 | .345 | |
| 24.000 | 23.037 | .000 | .963 | .71 | 3.05 | 4.0% | 30 | .960 | .957 | .954 | .950 | .946 | .941 | .936 | .930 | .923 | .915 |
| | | | | | | | | .907 | .897 | .886 | .873 | .859 | .843 | .824 | .803 | .778 | .751 |
| | | | | | | | | .720 | .685 | .646 | .603 | .557 | .507 | .454 | .400 | .345 | .291 |
| 24.000 | 23.277 | .000 | .723 | .76 | 3.03 | 3.0% | 31 | .960 | .957 | .954 | .950 | .946 | .941 | .936 | .930 | .923 | .915 |
| | | | | | | | | .907 | .897 | .886 | .873 | .859 | .843 | .824 | .803 | .778 | .751 |
| | | | | | | | | .720 | .685 | .646 | .603 | .557 | .507 | .454 | .400 | .345 | .291 |
| | | | | | | | | .240 | | | | | | | | | |
| 24.000 | 23.470 | .000 | .530 | .81 | 2.99 | 2.2% | 32 | .960 | .957 | .954 | .950 | .946 | .941 | .936 | .930 | .923 | .915 |
| | | | | | | | | .907 | .897 | .886 | .873 | .859 | .843 | .824 | .803 | .778 | .751 |
| | | | | | | | | .720 | .685 | .646 | .603 | .557 | .507 | .454 | .400 | .345 | .291 |
| | | | | | | | | .240 | .193 | | | | | | | | |
| 24.000 | 23.620 | .000 | .380 | .85 | 2.93 | 1.6% | 33 | .960 | .957 | .954 | .950 | .946 | .941 | .936 | .930 | .923 | .915 |
| | | | | | | | | .907 | .897 | .886 | .873 | .859 | .843 | .824 | .803 | .778 | .751 |
| | | | | | | | | .720 | .685 | .646 | .603 | .557 | .507 | .454 | .400 | .345 | .291 |
| | | | | | | | | .240 | .193 | .151 | | | | | | | |
| 24.000 | 23.735 | .000 | .265 | .89 | 2.87 | 1.1% | 34 | .960 | .957 | .954 | .950 | .946 | .941 | .936 | .930 | .923 | .915 |
| | | | | | | | | .907 | .897 | .886 | .873 | .859 | .843 | .824 | .803 | .778 | .751 |
| | | | | | | | | .720 | .685 | .646 | .603 | .557 | .507 | .454 | .400 | .345 | .291 |
| | | | | | | | | .240 | .193 | .151 | .115 | | | | | | |
| 24.000 | 23.820 | .000 | .180 | .92 | 2.79 | .8% | 35 | .960 | .957 | .954 | .950 | .946 | .941 | .936 | .930 | .923 | .915 |
| | | | | | | | | .907 | .897 | .886 | .873 | .859 | .843 | .824 | .803 | .778 | .751 |
| | | | | | | | | .720 | .685 | .646 | .603 | .557 | .507 | .454 | .400 | .345 | .291 |
| | | | | | | | | .240 | .193 | .151 | .115 | .085 | | | | | |
| 26.000 | 6.675 | .000 | 19.325 | .06 | 1.29 | 74.3% | 7 | .963 | .960 | .957 | .954 | .951 | .947 | .942 | | | |
| 26.000 | 7.612 | .000 | 18.388 | .06 | 1.34 | 70.7% | 8 | .963 | .960 | .957 | .954 | .951 | .947 | .942 | .937 | | |
| 26.000 | 8.544 | .000 | 17.456 | .07 | 1.40 | 67.1% | 9 | .963 | .960 | .957 | .954 | .951 | .947 | .942 | .937 | .932 | |
| 26.000 | 9.470 | .000 | 16.530 | .07 | 1.46 | 63.6% | 10 | .963 | .960 | .957 | .954 | .951 | .947 | .942 | .937 | .932 | .926 |
| 26.000 | 10.389 | .000 | 15.611 | .08 | 1.52 | 60.0% | 11 | .963 | .960 | .957 | .954 | .951 | .947 | .942 | .937 | .932 | .926 |
| | | | | | | | | .919 | | | | | | | | | |
| 26.000 | 11.300 | .000 | 14.700 | .09 | 1.59 | 56.5% | 12 | .963 | .960 | .957 | .954 | .951 | .947 | .942 | .937 | .932 | .926 |
| | | | | | | | | .919 | .911 | | | | | | | | |
| 26.000 | 12.202 | .000 | 13.798 | .10 | 1.66 | 53.1% | 13 | .963 | .960 | .957 | .954 | .951 | .947 | .942 | .937 | .932 | .926 |
| | | | | | | | | .919 | .911 | .902 | | | | | | | |

## ERLANG B: 100% BLOCKED CALLERS ROUTE ADVANCE, OVERFLOW, OR DISAPPEAR

| HOURS FIRST ATTEMPT VMR = 1.0 | HOURS CON-NECT | HOURS RECALL /RETRY | HOURS OVFL | CON-NECT VMR | OVFL VMR | PERCENT BLOCKED (GRADE) | LINE | HOURS CONNECTED TRAFFIC ON LINES 1 | 2 | 3 | 4 | 5 | 6 | 7 | 8 | 9 | 10 |
|---|---|---|---|---|---|---|---|---|---|---|---|---|---|---|---|---|---|
| 26.000 | 13.094 | .000 | 12.906 | .11 | 1.74 | 49.6% | 14 | .963 | .960 | .957 | .954 | .951 | .947 | .942 | .937 | .932 | .926 |
| | | | | | | | | .919 | .911 | .902 | .892 | | | | | | |
| 26.000 | 13.976 | .000 | 12.024 | .12 | 1.82 | 46.2% | 15 | .963 | .960 | .957 | .954 | .951 | .947 | .942 | .937 | .932 | .926 |
| | | | | | | | | .919 | .911 | .902 | .892 | .881 | | | | | |
| 26.000 | 14.844 | .000 | 11.156 | .13 | 1.91 | 42.9% | 16 | .963 | .960 | .957 | .954 | .951 | .947 | .942 | .937 | .932 | .926 |
| | | | | | | | | .919 | .911 | .902 | .892 | .881 | .869 | | | | |
| 26.000 | 15.698 | .000 | 10.302 | .15 | 1.99 | 39.6% | 17 | .963 | .960 | .957 | .954 | .951 | .947 | .942 | .937 | .932 | .926 |
| | | | | | | | | .919 | .911 | .902 | .892 | .881 | .869 | .854 | | | |
| 26.000 | 16.536 | .000 | 9.464 | .16 | 2.09 | 36.4% | 18 | .963 | .960 | .957 | .954 | .951 | .947 | .942 | .937 | .932 | .926 |
| | | | | | | | | .919 | .911 | .902 | .892 | .881 | .869 | .854 | .838 | | |
| 26.000 | 17.355 | .000 | 8.645 | .18 | 2.19 | 33.2% | 19 | .963 | .960 | .957 | .954 | .951 | .947 | .942 | .937 | .932 | .926 |
| | | | | | | | | .919 | .911 | .902 | .892 | .881 | .869 | .854 | .838 | .819 | |
| 26.000 | 18.153 | .000 | 7.847 | .20 | 2.29 | 30.2% | 20 | .963 | .960 | .957 | .954 | .951 | .947 | .942 | .937 | .932 | .926 |
| | | | | | | | | .919 | .911 | .902 | .892 | .881 | .869 | .854 | .838 | .819 | .798 |
| 26.000 | 18.928 | .000 | 7.072 | .23 | 2.39 | 27.2% | 21 | .963 | .960 | .957 | .954 | .951 | .947 | .942 | .937 | .932 | .926 |
| | | | | | | | | .919 | .911 | .902 | .892 | .881 | .869 | .854 | .838 | .819 | .798 |
| | | | | | | | | .774 | | | | | | | | | |
| 26.000 | 19.675 | .000 | 6.325 | .25 | 2.49 | 24.3% | 22 | .963 | .960 | .957 | .954 | .951 | .947 | .942 | .937 | .932 | .926 |
| | | | | | | | | .919 | .911 | .902 | .892 | .881 | .869 | .854 | .838 | .819 | .798 |
| | | | | | | | | .774 | .747 | | | | | | | | |
| 26.000 | 20.392 | .000 | 5.608 | .28 | 2.60 | 21.6% | 23 | .963 | .960 | .957 | .954 | .951 | .947 | .942 | .937 | .932 | .926 |
| | | | | | | | | .919 | .911 | .902 | .892 | .881 | .869 | .854 | .838 | .819 | .798 |
| | | | | | | | | .774 | .747 | .717 | | | | | | | |
| 26.000 | 21.076 | .000 | 4.924 | .32 | 2.70 | 18.9% | 24 | .963 | .960 | .957 | .954 | .951 | .947 | .942 | .937 | .932 | .926 |
| | | | | | | | | .919 | .911 | .902 | .892 | .881 | .869 | .854 | .838 | .819 | .798 |
| | | | | | | | | .774 | .747 | .717 | .683 | | | | | | |
| 26.000 | 21.721 | .000 | 4.279 | .35 | 2.80 | 16.5% | 25 | .963 | .960 | .957 | .954 | .951 | .947 | .942 | .037 | .932 | .926 |
| | | | | | | | | .910 | .011 | .902 | .892 | .881 | .869 | .854 | .838 | .819 | .798 |
| | | | | | | | | .774 | .747 | .717 | .683 | .646 | | | | | |
| 26.000 | 22.326 | .000 | 3.674 | .40 | 2.89 | 14.1% | 26 | .963 | .960 | .957 | .954 | .951 | .947 | .942 | .937 | .932 | .926 |
| | | | | | | | | .919 | .911 | .902 | .892 | .881 | .869 | .854 | .838 | .819 | .798 |
| | | | | | | | | .774 | .747 | .717 | .683 | .646 | .605 | | | | |
| 26.000 | 22.886 | .000 | 3.114 | .44 | 2.97 | 12.0% | 27 | .963 | .960 | .957 | .954 | .951 | .947 | .942 | .937 | .932 | .926 |
| | | | | | | | | .919 | .911 | .902 | .892 | .881 | .869 | .854 | .838 | .819 | .798 |
| | | | | | | | | .771 | .747 | .717 | .683 | .646 | .605 | .560 | | | |
| 26.000 | 23.398 | .000 | 2.602 | .49 | 3.04 | 10.0% | 28 | .963 | .960 | .957 | .954 | .951 | .947 | .942 | .937 | .932 | .926 |
| | | | | | | | | .919 | .911 | .902 | .892 | .881 | .869 | .854 | .838 | .819 | .798 |
| | | | | | | | | .774 | .747 | .717 | .683 | .646 | .605 | .560 | .512 | | |
| 26.000 | 23.859 | .000 | 2.141 | .54 | 3.09 | 8.2% | 29 | .963 | .960 | .957 | .954 | .951 | .947 | .942 | .937 | .932 | .926 |
| | | | | | | | | .919 | .911 | .902 | .892 | .881 | .869 | .854 | .838 | .819 | .798 |

# ERLANG B: 100% BLOCKED CALLERS ROUTE ADVANCE, OVERFLOW, OR DISAPPEAR

| HOURS FIRST ATTEMPT VMR = 1.0 | HOURS CON-NECT | HOURS RECALL /RETRY | HOURS OVFL | CON-NECT VMR | OVFL VMR | PERCENT BLOCKED (GRADE) | LINE | HOURS CONNECTED TRAFFIC ON LINES 1 | 2 | 3 | 4 | 5 | 6 | 7 | 8 | 9 | 10 |
|---|---|---|---|---|---|---|---|---|---|---|---|---|---|---|---|---|---|
| | | | | | | | | .774 | .747 | .717 | .683 | .646 | .605 | .560 | .512 | .461 | |
| 26.000 | 24.268 | .000 | 1.732 | .59 | 3.13 | 6.7% | 30 | .963 | .960 | .957 | .954 | .951 | .947 | .942 | .937 | .932 | .926 |
| | | | | | | | | .919 | .911 | .902 | .892 | .881 | .869 | .854 | .838 | .819 | .798 |
| | | | | | | | | .774 | .747 | .717 | .683 | .646 | .605 | .560 | .512 | .461 | .409 |
| 26.000 | 24.624 | .000 | 1.376 | .64 | 3.15 | 5.3% | 31 | .963 | .960 | .957 | .954 | .951 | .947 | .942 | .937 | .932 | .926 |
| | | | | | | | | .919 | .911 | .902 | .892 | .881 | .869 | .854 | .838 | .819 | .798 |
| | | | | | | | | .774 | .747 | .717 | .683 | .646 | .605 | .560 | .512 | .461 | .409 |
| | | | | | | | | .356 | | | | | | | | | |
| 26.000 | 24.928 | .000 | 1.072 | .70 | 3.15 | 4.1% | 32 | .963 | .960 | .957 | .954 | .951 | .947 | .942 | .937 | .932 | .926 |
| | | | | | | | | .919 | .911 | .902 | .892 | .881 | .869 | .854 | .838 | .819 | .798 |
| | | | | | | | | .774 | .747 | .717 | .683 | .646 | .605 | .560 | .512 | .461 | .409 |
| | | | | | | | | .356 | .304 | | | | | | | | |
| 26.000 | 25.182 | .000 | .818 | .75 | 3.13 | 3.1% | 33 | .963 | .960 | .957 | .954 | .951 | .947 | .942 | .937 | .932 | .926 |
| | | | | | | | | .919 | .911 | .902 | .892 | .881 | .869 | .854 | .838 | .819 | .798 |
| | | | | | | | | .774 | .747 | .717 | .683 | .646 | .605 | .560 | .512 | .461 | .409 |
| | | | | | | | | .356 | .304 | .254 | | | | | | | |
| 26.000 | 25.389 | .000 | .611 | .79 | 3.09 | 2.3% | 34 | .963 | .960 | .957 | .954 | .951 | .947 | .942 | .937 | .932 | .926 |
| | | | | | | | | .919 | .911 | .902 | .892 | .881 | .869 | .854 | .838 | .819 | .798 |
| | | | | | | | | .774 | .747 | .717 | .683 | .646 | .605 | .560 | .512 | .461 | .409 |
| | | | | | | | | .356 | .304 | .254 | .207 | | | | | | |
| 26.000 | 25.554 | .000 | .446 | .84 | 3.04 | 1.7% | 35 | .963 | .960 | .957 | .954 | .951 | .947 | .942 | .937 | .932 | .926 |
| | | | | | | | | .919 | .911 | .902 | .892 | .881 | .869 | .854 | .838 | .819 | .798 |
| | | | | | | | | .774 | .747 | .717 | .683 | .646 | .605 | .560 | .512 | .461 | .409 |
| | | | | | | | | .356 | .304 | .254 | .207 | .165 | | | | | |
| 26.000 | 25.682 | .000 | .318 | .87 | 2.98 | 1.2% | 36 | .963 | .960 | .957 | .954 | .951 | .947 | .942 | .937 | .932 | .926 |
| | | | | | | | | .919 | .911 | .902 | .892 | .881 | .869 | .854 | .838 | .819 | .798 |
| | | | | | | | | .774 | .747 | .717 | .683 | .646 | .605 | .560 | .512 | .461 | .409 |
| | | | | | | | | .356 | .304 | .254 | .207 | .165 | .128 | | | | |
| 26.000 | 25.778 | .000 | .222 | .90 | 2.91 | .9% | 37 | .963 | .960 | .957 | .954 | .951 | .947 | .942 | .937 | .932 | .926 |
| | | | | | | | | .919 | .911 | .902 | .892 | .881 | .869 | .854 | .838 | .819 | .798 |
| | | | | | | | | .774 | .747 | .717 | .683 | .646 | .605 | .560 | .512 | .461 | .409 |
| | | | | | | | | .356 | .304 | .254 | .207 | .165 | .128 | .096 | | | |
| 26.000 | 25.849 | .000 | .151 | .93 | 2.83 | .6% | 38 | .963 | .960 | .957 | .954 | .951 | .947 | .942 | .937 | .932 | .926 |
| | | | | | | | | .919 | .911 | .902 | .892 | .881 | .869 | .854 | .838 | .819 | .798 |
| | | | | | | | | .774 | .747 | .717 | .683 | .646 | .605 | .560 | .512 | .461 | .409 |
| | | | | | | | | .356 | .304 | .254 | .207 | .165 | .128 | .096 | | | |
| 28.000 | 7.646 | .000 | 20.354 | .06 | 1.32 | 72.7% | 8 | .966 | .963 | .961 | .958 | .955 | .952 | .948 | .944 | | |
| 28.000 | 8.585 | .000 | 19.415 | .06 | 1.37 | 69.3% | 9 | .966 | .963 | .961 | .958 | .955 | .952 | .948 | .944 | .939 | |
| 28.000 | 9.519 | .000 | 18.481 | .07 | 1.42 | 66.0% | 10 | .966 | .963 | .961 | .958 | .955 | .952 | .948 | .944 | .939 | .934 |
| 28.000 | 10.447 | .000 | 17.553 | .07 | 1.48 | 62.7% | 11 | .966 | .963 | .961 | .958 | .955 | .952 | .948 | .944 | .939 | .934 |

## ERLANG B: 100% BLOCKED CALLERS ROUTE ADVANCE, OVERFLOW, OR DISAPPEAR

| HOURS FIRST ATTEMPT VMR=1.0 | HOURS CON- NECT | HOURS RECALL /RETRY | HOURS OVFL | CON- NECT VMR | OVFL VMR | PERCENT BLOCKED (GRADE) | LINE | HOURS CONNECTED TRAFFIC ON LINES 1 | 2 | 3 | 4 | 5 | 6 | 7 | 8 | 9 | 10 |
|---|---|---|---|---|---|---|---|---|---|---|---|---|---|---|---|---|---|
| | | | | | | | | .928 | | | | | | | | | |
| 28.000 | 11.370 | .000 | 16.630 | .08 | 1.54 | 59.4% | 12 | .966 | .963 | .961 | .958 | .955 | .952 | .948 | .944 | .939 | .934 |
| | | | | | | | | .928 | .922 | | | | | | | | |
| 28.000 | 12.285 | .000 | 15.715 | .08 | 1.61 | 56.1% | 13 | .966 | .963 | .961 | .958 | .955 | .952 | .948 | .944 | .939 | .934 |
| | | | | | | | | .928 | .922 | .915 | | | | | | | |
| 28.000 | 13.192 | .000 | 14.808 | .09 | 1.68 | 52.9% | 14 | .966 | .963 | .961 | .958 | .955 | .952 | .948 | .944 | .939 | .934 |
| | | | | | | | | .928 | .922 | .915 | .907 | | | | | | |
| 28.000 | 14.090 | .000 | 13.910 | .10 | 1.75 | 49.7% | 15 | .966 | .963 | .961 | .958 | .955 | .952 | .948 | .944 | .939 | .934 |
| | | | | | | | | .928 | .922 | .915 | .907 | .898 | | | | | |
| 28.000 | 14.978 | .000 | 13.022 | .11 | 1.83 | 46.5% | 16 | .966 | .963 | .961 | .958 | .955 | .952 | .948 | .944 | .939 | .934 |
| | | | | | | | | .928 | .922 | .915 | .907 | .898 | .888 | | | | |
| 28.000 | 15.855 | .000 | 12.145 | .12 | 1.91 | 43.4% | 17 | .966 | .963 | .961 | .958 | .955 | .952 | .948 | .944 | .939 | .934 |
| | | | | | | | | .928 | .922 | .915 | .907 | .898 | .888 | .877 | | | |
| 28.000 | 16.719 | .000 | 11.281 | .14 | 2.00 | 40.3% | 18 | .966 | .963 | .961 | .958 | .955 | .952 | .948 | .944 | .939 | .934 |
| | | | | | | | | .928 | .922 | .915 | .907 | .898 | .888 | .877 | .864 | | |
| 28.000 | 17.569 | .000 | 10.431 | .15 | 2.09 | 37.3% | 19 | .966 | .963 | .961 | .958 | .955 | .952 | .948 | .944 | .939 | .934 |
| | | | | | | | | .928 | .922 | .915 | .907 | .898 | .888 | .877 | .864 | .850 | |
| 28.000 | 18.402 | .000 | 9.598 | .17 | 2.18 | 34.3% | 20 | .966 | .963 | .961 | .958 | .955 | .952 | .948 | .944 | .939 | .934 |
| | | | | | | | | .928 | .922 | .915 | .907 | .898 | .888 | .877 | .864 | .850 | .833 |
| 28.000 | 19.217 | .000 | 8.783 | .19 | 2.28 | 31.4% | 21 | .966 | .963 | .961 | .958 | .955 | .952 | .948 | .944 | .939 | .934 |
| | | | | | | | | .928 | .922 | .915 | .907 | .898 | .888 | .877 | .864 | .850 | .833 |
| | | | | | | | | .815 | | | | | | | | | |
| 28.000 | 20.011 | .000 | 7.989 | .21 | 2.38 | 28.5% | 22 | .966 | .963 | .961 | .958 | .955 | .952 | .948 | .944 | .939 | .934 |
| | | | | | | | | .928 | .922 | .915 | .907 | .898 | .888 | .877 | .864 | .850 | .833 |
| | | | | | | | | .815 | .794 | | | | | | | | |
| 28.000 | 20.782 | .000 | 7.218 | .23 | 2.48 | 25.8% | 23 | .966 | .963 | .961 | .958 | .955 | .952 | .948 | .944 | .939 | .934 |
| | | | | | | | | .928 | .922 | .915 | .907 | .898 | .888 | .877 | .864 | .850 | .833 |
| | | | | | | | | .815 | .794 | .771 | | | | | | | |
| 28.000 | 21.526 | .000 | 6.474 | .26 | 2.59 | 23.1% | 24 | .966 | .963 | .961 | .958 | .955 | .952 | .948 | .944 | .939 | .934 |
| | | | | | | | | .928 | .922 | .915 | .907 | .898 | .888 | .877 | .864 | .850 | .833 |
| | | | | | | | | .815 | .794 | .771 | .744 | | | | | | |
| 28.000 | 22.240 | .000 | 5.760 | .29 | 2.69 | 20.6% | 25 | .966 | .963 | .961 | .958 | .955 | .952 | .948 | .944 | .939 | .934 |
| | | | | | | | | .928 | .922 | .915 | .907 | .898 | .888 | .877 | .864 | .850 | .833 |
| | | | | | | | | .815 | .794 | .771 | .744 | .715 | | | | | |
| 28.000 | 22.922 | .000 | 5.078 | .32 | 2.79 | 18.1% | 26 | .966 | .963 | .961 | .958 | .955 | .952 | .948 | .944 | .939 | .934 |
| | | | | | | | | .928 | .922 | .915 | .907 | .898 | .888 | .877 | .864 | .850 | .833 |
| | | | | | | | | .815 | .794 | .771 | .744 | .715 | .682 | | | | |
| 28.000 | 23.568 | .000 | 4.432 | .35 | 2.89 | 15.8% | 27 | .966 | .963 | .961 | .958 | .955 | .952 | .948 | .944 | .939 | .934 |
| | | | | | | | | .928 | .922 | .915 | .907 | .898 | .888 | .877 | .864 | .850 | .833 |
| | | | | | | | | .815 | .794 | .771 | .744 | .715 | .682 | .645 | | | |

## ERLANG B: 100% BLOCKED CALLERS ROUTE ADVANCE, OVERFLOW, OR DISAPPEAR

| HOURS FIRST ATTEMPT VMR = 1.0 | HOURS CON-NECT | HOURS RECALL /RETRY | HOURS OVFL | CON-NECT VMR | OVFL VMR | PERCENT BLOCKED (GRADE) | LINE | HOURS CONNECTED TRAFFIC ON LINES 1 | 2 | 3 | 4 | 5 | 6 | 7 | 8 | 9 | 10 |
|---|---|---|---|---|---|---|---|---|---|---|---|---|---|---|---|---|---|
| 28.000 | 24.173 | .000 | 3.827 | .39 | 2.97 | 13.7% | 28 | .966 | .963 | .961 | .958 | .955 | .952 | .948 | .944 | .939 | .934 |
| | | | | | | | | .928 | .922 | .915 | .907 | .898 | .888 | .877 | .864 | .850 | .833 |
| | | | | | | | | .815 | .794 | .771 | .744 | .715 | .682 | .645 | .606 | | |
| 28.000 | 24.736 | .000 | 3.264 | .44 | 3.06 | 11.7% | 29 | .966 | .963 | .961 | .958 | .955 | .952 | .948 | .944 | .939 | .934 |
| | | | | | | | | .928 | .922 | .915 | .907 | .898 | .888 | .877 | .864 | .850 | .833 |
| | | | | | | | | .815 | .794 | .771 | .744 | .715 | .682 | .645 | .606 | .563 | |
| 28.000 | 25.253 | .000 | 2.747 | .48 | 3.12 | 9.8% | 30 | .966 | .963 | .961 | .958 | .955 | .952 | .948 | .944 | .939 | .934 |
| | | | | | | | | .928 | .922 | .915 | .907 | .898 | .888 | .877 | .864 | .850 | .833 |
| | | | | | | | | .815 | .794 | .771 | .744 | .715 | .682 | .645 | .606 | .563 | .517 |
| 28.000 | 25.720 | .000 | 2.280 | .53 | 3.18 | 8.1% | 31 | .966 | .963 | .961 | .958 | .955 | .952 | .948 | .944 | .939 | .934 |
| | | | | | | | | .928 | .922 | .915 | .907 | .898 | .888 | .877 | .864 | .850 | .833 |
| | | | | | | | | .815 | .794 | .771 | .744 | .715 | .682 | .645 | .606 | .563 | .517 |
| | | | | | | | | .468 | | | | | | | | | |
| 28.000 | 26.138 | .000 | 1.862 | .58 | 3.22 | 6.6% | 32 | .966 | .963 | .961 | .958 | .955 | .952 | .948 | .944 | .939 | .934 |
| | | | | | | | | .928 | .922 | .915 | .907 | .898 | .888 | .877 | .864 | .850 | .833 |
| | | | | | | | | .815 | .794 | .771 | .744 | .715 | .682 | .645 | .606 | .563 | .517 |
| | | | | | | | | .468 | .418 | | | | | | | | |
| 28.000 | 26.505 | .000 | 1.495 | .63 | 3.24 | 5.3% | 33 | .966 | .963 | .961 | .958 | .955 | .952 | .948 | .944 | .939 | .934 |
| | | | | | | | | .928 | .922 | .915 | .907 | .898 | .888 | .877 | .864 | .850 | .833 |
| | | | | | | | | .815 | .794 | .771 | .744 | .715 | .682 | .645 | .606 | .563 | .517 |
| | | | | | | | | .468 | .418 | .366 | | | | | | | |
| 28.000 | 26.820 | .000 | 1.180 | .68 | 3.24 | 4.2% | 34 | .966 | .963 | .961 | .958 | .955 | .952 | .948 | .944 | .939 | .934 |
| | | | | | | | | .928 | .922 | .915 | .907 | .898 | .888 | .877 | .864 | .850 | .833 |
| | | | | | | | | .815 | .794 | .771 | .744 | .715 | .682 | .645 | .606 | .563 | .517 |
| | | | | | | | | .468 | .418 | .366 | .316 | | | | | | |
| 28.000 | 27.087 | .000 | .913 | .73 | 3.23 | 3.3% | 35 | .966 | .963 | .961 | .958 | .955 | .952 | .948 | .944 | .939 | .934 |
| | | | | | | | | .928 | .922 | .915 | .907 | .898 | .888 | .877 | .864 | .850 | .833 |
| | | | | | | | | .815 | .794 | .771 | .744 | .715 | .682 | .645 | .606 | .563 | .517 |
| | | | | | | | | .468 | .418 | .366 | .316 | .267 | | | | | |
| 28.000 | 27.307 | .000 | .693 | .78 | 3.20 | 2.5% | 36 | .966 | .963 | .961 | .958 | .955 | .952 | .948 | .944 | .939 | .934 |
| | | | | | | | | .928 | .922 | .915 | .907 | .898 | .888 | .877 | .864 | .850 | .833 |
| | | | | | | | | .815 | .794 | .771 | .744 | .715 | .682 | .645 | .606 | .563 | .517 |
| | | | | | | | | .468 | .418 | .366 | .316 | .267 | .220 | | | | |
| 28.000 | 27.486 | .000 | .514 | .82 | 3.15 | 1.8% | 37 | .966 | .963 | .961 | .958 | .955 | .952 | .948 | .944 | .939 | .934 |
| | | | | | | | | .928 | .922 | .915 | .907 | .898 | .888 | .877 | .864 | .850 | .833 |
| | | | | | | | | .815 | .794 | .771 | .744 | .715 | .682 | .645 | .606 | .563 | .517 |
| | | | | | | | | .468 | .418 | .366 | .316 | .267 | .220 | .178 | | | |
| 28.000 | 27.626 | .000 | .374 | .86 | 3.09 | 1.3% | 38 | .966 | .963 | .961 | .958 | .955 | .952 | .948 | .944 | .939 | .934 |
| | | | | | | | | .928 | .922 | .915 | .907 | .898 | .888 | .877 | .864 | .850 | .833 |
| | | | | | | | | .815 | .794 | .771 | .744 | .715 | .682 | .645 | .606 | .563 | .517 |

# ERLANG B: 100% BLOCKED CALLERS ROUTE ADVANCE, OVERFLOW, OR DISAPPEAR

| HOURS FIRST ATTEMPT VMR = 1.0 | HOURS CON- NECT | HOURS RECALL /RETRY | HOURS OVFL | CON- NECT VMR | OVFL VMR | PERCENT BLOCKED (GRADE) | LINE | HOURS CONNECTED TRAFFIC ON LINES 1 | 2 | 3 | 4 | 5 | 6 | 7 | 8 | 9 | 10 |
|---|---|---|---|---|---|---|---|---|---|---|---|---|---|---|---|---|---|
| | | | | | | | | .468 | .418 | .366 | .316 | .267 | .220 | .178 | .140 | | |
| 28.000 | 27.734 | .000 | .266 | .89 | 3.02 | .9% | 39 | .966 | .963 | .961 | .958 | .955 | .952 | .948 | .944 | .939 | .934 |
| | | | | | | | | .928 | .922 | .915 | .907 | .898 | .888 | .877 | .864 | .850 | .833 |
| | | | | | | | | .815 | .794 | .771 | .744 | .715 | .682 | .645 | .606 | .563 | .517 |
| | | | | | | | | .468 | .418 | .366 | .316 | .267 | .220 | .178 | .140 | .108 | |
| 28.000 | 27.815 | .000 | .185 | .92 | 2.94 | .7% | 40 | .966 | .963 | .961 | .958 | .955 | .952 | .948 | .944 | .939 | .934 |
| | | | | | | | | .928 | .922 | .915 | .907 | .898 | .888 | .877 | .864 | .850 | .833 |
| | | | | | | | | .815 | .794 | .771 | .744 | .715 | .682 | .645 | .606 | .563 | .517 |
| | | | | | | | | .468 | .418 | .366 | .316 | .267 | .220 | .178 | .140 | .108 | .081 |
| 30.000 | 7.674 | .000 | 22.326 | .05 | 1.30 | 74.4% | 8 | .968 | .966 | .964 | .961 | .959 | .956 | .952 | .949 | | |
| 30.000 | 8.619 | .000 | 21.381 | .05 | 1.34 | 71.3% | 9 | .968 | .966 | .964 | .961 | .959 | .956 | .952 | .949 | .945 | |
| 30.000 | 9.560 | .000 | 20.440 | .06 | 1.39 | 68.1% | 10 | .968 | .966 | .964 | .961 | .959 | .956 | .952 | .949 | .945 | .941 |
| 30.000 | 10.496 | .000 | 19.504 | .06 | 1.45 | 65.0% | 11 | .968 | .966 | .964 | .961 | .959 | .956 | .952 | .949 | .945 | .941 |
| | | | | | | | | .936 | | | | | | | | | |
| 30.000 | 11.427 | .000 | 18.573 | .07 | 1.50 | 61.9% | 12 | .968 | .966 | .964 | .961 | .959 | .956 | .952 | .949 | .945 | .941 |
| | | | | | | | | .936 | .931 | | | | | | | | |
| 30.000 | 12.352 | .000 | 17.648 | .07 | 1.56 | 58.8% | 13 | .968 | .966 | .964 | .961 | .959 | .956 | .952 | .949 | .945 | .941 |
| | | | | | | | | .936 | .931 | .925 | | | | | | | |
| 30.000 | 13.271 | .000 | 16.729 | .08 | 1.62 | 55.8% | 14 | .968 | .966 | .964 | .961 | .959 | .956 | .952 | .949 | .945 | .941 |
| | | | | | | | | .936 | .931 | .925 | .919 | | | | | | |
| 30.000 | 14.183 | .000 | 15.817 | .09 | 1.69 | 52.7% | 15 | .968 | .966 | .964 | .961 | .959 | .956 | .952 | .949 | .945 | .941 |
| | | | | | | | | .936 | .931 | .925 | .919 | .912 | | | | | |
| 30.000 | 15.086 | .000 | 14.914 | .10 | 1.76 | 49.7% | 16 | .968 | .966 | .964 | .961 | .959 | .956 | .952 | .949 | .945 | .941 |
| | | | | | | | | .936 | .931 | .925 | .919 | .912 | .903 | | | | |
| 30.000 | 15.981 | .000 | 14.019 | .11 | 1.84 | 46.7% | 17 | .968 | .966 | .964 | .961 | .959 | .956 | .952 | .949 | .945 | .941 |
| | | | | | | | | .936 | .931 | .925 | .919 | .912 | .903 | .894 | | | |
| 30.000 | 16.865 | .000 | 13.135 | .12 | 1.91 | 43.8% | 18 | .968 | .966 | .964 | .961 | .959 | .956 | .952 | .949 | .945 | .941 |
| | | | | | | | | .936 | .931 | .925 | .919 | .912 | .903 | .894 | .884 | | |
| 30.000 | 17.738 | .000 | 12.262 | .13 | 2.00 | 40.9% | 19 | .968 | .966 | .964 | .961 | .959 | .956 | .952 | .949 | .945 | .941 |
| | | | | | | | | .936 | .931 | .925 | .919 | .912 | .903 | .894 | .884 | .873 | |
| 30.000 | 18.597 | .000 | 11.403 | .14 | 2.08 | 38.0% | 20 | .968 | .966 | .964 | .961 | .959 | .956 | .952 | .949 | .945 | .941 |
| | | | | | | | | .936 | .931 | .925 | .919 | .912 | .903 | .894 | .884 | .873 | .860 |
| 30.000 | 19.443 | .000 | 10.557 | .15 | 2.18 | 35.2% | 21 | .968 | .966 | .964 | .961 | .959 | .956 | .952 | .949 | .945 | .941 |
| | | | | | | | | .936 | .931 | .925 | .919 | .912 | .903 | .894 | .884 | .873 | .860 |
| | | | | | | | | .845 | | | | | | | | | |
| 30.000 | 20.272 | .000 | 9.728 | .17 | 2.27 | 32.4% | 22 | .968 | .966 | .964 | .961 | .959 | .956 | .952 | .949 | .945 | .941 |
| | | | | | | | | .936 | .931 | .925 | .919 | .912 | .903 | .894 | .884 | .873 | .860 |
| | | | | | | | | .845 | .829 | | | | | | | | |
| 30.000 | 21.083 | .000 | 8.917 | .19 | 2.37 | 29.7% | 23 | .968 | .966 | .964 | .961 | .959 | .956 | .952 | .949 | .945 | .941 |
| | | | | | | | | .936 | .931 | .925 | .919 | .912 | .903 | .894 | .884 | .873 | .860 |

## ERLANG B: 100% BLOCKED CALLERS ROUTE ADVANCE, OVERFLOW, OR DISAPPEAR

| HOURS FIRST ATTEMPT VMR=1.0 | HOURS CON-NECT | HOURS RECALL /RETRY | HOURS OVFL | CON-NECT VMR | OVFL VMR | PERCENT BLOCKED (GRADE) | LINE | HOURS CONNECTED TRAFFIC ON LINES 1 | 2 | 3 | 4 | 5 | 6 | 7 | 8 | 9 | 10 |
|---|---|---|---|---|---|---|---|---|---|---|---|---|---|---|---|---|---|
| | | | | | | | | .845 | .829 | .811 | | | | | | | |
| 30.000 | 21.873 | .000 | 8.127 | .21 | 2.47 | 27.1% | 24 | .968 | .966 | .964 | .961 | .959 | .956 | .952 | .949 | .945 | .941 |
| | | | | | | | | .936 | .931 | .925 | .919 | .912 | .903 | .894 | .884 | .873 | .860 |
| | | | | | | | | .845 | .829 | .811 | .790 | | | | | | |
| 30.000 | 22.640 | .000 | 7.360 | .23 | 2.57 | 24.5% | 25 | .968 | .966 | .964 | .961 | .959 | .956 | .952 | .949 | .945 | .941 |
| | | | | | | | | .936 | .931 | .925 | .919 | .912 | .903 | .894 | .884 | .873 | .860 |
| | | | | | | | | .845 | .829 | .811 | .790 | .767 | | | | | |
| 30.000 | 23.381 | .000 | 6.619 | .26 | 2.67 | 22.1% | 26 | .968 | .966 | .964 | .961 | .959 | .956 | .952 | .949 | .945 | .941 |
| | | | | | | | | .936 | .931 | .925 | .919 | .912 | .903 | .894 | .884 | .873 | .860 |
| | | | | | | | | .845 | .829 | .811 | .790 | .767 | .741 | | | | |
| 30.000 | 24.094 | .000 | 5.906 | .29 | 2.77 | 19.7% | 27 | .968 | .966 | .964 | .961 | .959 | .956 | .952 | .949 | .945 | .941 |
| | | | | | | | | .936 | .931 | .925 | .919 | .912 | .903 | .894 | .884 | .873 | .860 |
| | | | | | | | | .845 | .829 | .811 | .790 | .767 | .741 | .712 | | | |
| 30.000 | 24.774 | .000 | 5.226 | .32 | 2.87 | 17.4% | 28 | .968 | .966 | .964 | .961 | .959 | .956 | .952 | .949 | .945 | .941 |
| | | | | | | | | .936 | .931 | .925 | .919 | .912 | .903 | .894 | .884 | .873 | .860 |
| | | | | | | | | .845 | .829 | .811 | .790 | .767 | .741 | .712 | .680 | | |
| 30.000 | 25.419 | .000 | 4.581 | .35 | 2.97 | 15.3% | 29 | .968 | .966 | .964 | .961 | .959 | .956 | .952 | .949 | .945 | .941 |
| | | | | | | | | .936 | .931 | .925 | .919 | .912 | .903 | .894 | .884 | .873 | .860 |
| | | | | | | | | .845 | .829 | .811 | .790 | .767 | .741 | .712 | .680 | .645 | |
| 30.000 | 26.026 | .000 | 3.974 | .39 | 3.06 | 13.2% | 30 | .968 | .966 | .964 | .961 | .959 | .956 | .952 | .949 | .945 | .941 |
| | | | | | | | | .936 | .931 | .925 | .919 | .912 | .903 | .894 | .884 | .873 | .860 |
| | | | | | | | | .845 | .829 | .811 | .790 | .767 | .741 | .712 | .680 | .645 | .607 |
| 30.000 | 26.591 | .000 | 3.409 | .43 | 3.14 | 11.4% | 31 | .968 | .966 | .964 | .961 | .959 | .956 | .952 | .949 | .945 | .941 |
| | | | | | | | | .936 | .931 | .925 | .919 | .912 | .903 | .894 | .884 | .873 | .860 |
| | | | | | | | | .845 | .829 | .811 | .790 | .767 | .741 | .712 | .680 | .645 | .607 |
| | | | | | | | | .565 | | | | | | | | | |
| 30.000 | 27.112 | .000 | 2.888 | .48 | 3.21 | 9.6% | 32 | .968 | .966 | .964 | .961 | .959 | .956 | .952 | .949 | .945 | .941 |
| | | | | | | | | .936 | .931 | .925 | .919 | .912 | .903 | .894 | .884 | .873 | .860 |
| | | | | | | | | .845 | .829 | .811 | .790 | .767 | .741 | .712 | .680 | .645 | .607 |
| | | | | | | | | .565 | .521 | | | | | | | | |
| 30.000 | 27.586 | .000 | 2.414 | .53 | 3.26 | 8.0% | 33 | .968 | .966 | .964 | .961 | .959 | .956 | .952 | .949 | .945 | .941 |
| | | | | | | | | .936 | .931 | .925 | .919 | .912 | .903 | .894 | .884 | .873 | .860 |
| | | | | | | | | .845 | .829 | .811 | .790 | .767 | .741 | .712 | .680 | .645 | .607 |
| | | | | | | | | .565 | .521 | .474 | | | | | | | |
| 30.000 | 28.011 | .000 | 1.989 | .57 | 3.30 | 6.6% | 34 | .968 | .966 | .964 | .961 | .959 | .956 | .952 | .949 | .945 | .941 |
| | | | | | | | | .936 | .931 | .925 | .919 | .912 | .903 | .894 | .884 | .873 | .860 |
| | | | | | | | | .845 | .829 | .811 | .790 | .767 | .741 | .712 | .680 | .645 | .607 |
| | | | | | | | | .565 | .521 | .474 | .425 | | | | | | |
| 30.000 | 28.387 | .000 | 1.613 | .62 | 3.33 | 5.4% | 35 | .968 | .966 | .964 | .961 | .959 | .956 | .952 | .949 | .945 | .941 |
| | | | | | | | | .936 | .931 | .925 | .919 | .912 | .903 | .894 | .884 | .873 | .860 |

# ERLANG B: 100% BLOCKED CALLERS ROUTE ADVANCE, OVERFLOW, OR DISAPPEAR

| HOURS FIRST ATTEMPT VMR = 1.0 | HOURS CON-NECT | HOURS RECALL /RETRY | HOURS OVFL | CON-NECT VMR | OVFL VMR | PERCENT BLOCKED (GRADE) | LINE | HOURS CONNECTED TRAFFIC ON LINES 1 | 2 | 3 | 4 | 5 | 6 | 7 | 8 | 9 | 10 |
|---|---|---|---|---|---|---|---|---|---|---|---|---|---|---|---|---|---|
| | | | | | | | | .845 | .829 | .811 | .790 | .767 | .741 | .712 | .680 | .645 | .607 |
| | | | | | | | | .565 | .521 | .474 | .425 | .376 | | | | | |
| 30.000 | 28.713 | .000 | 1.287 | .67 | 3.33 | 4.3% | 36 | .968 | .966 | .964 | .961 | .959 | .956 | .952 | .949 | .945 | .941 |
| | | | | | | | | .936 | .931 | .925 | .919 | .912 | .903 | .894 | .884 | .873 | .860 |
| | | | | | | | | .845 | .829 | .811 | .790 | .767 | .741 | .712 | .680 | .645 | .607 |
| | | | | | | | | .565 | .521 | .474 | .425 | .376 | .327 | | | | |
| 30.000 | 28.992 | .000 | 1.008 | .72 | 3.32 | 3.4% | 37 | .968 | .966 | .964 | .961 | .959 | .956 | .952 | .949 | .945 | .941 |
| | | | | | | | | .936 | .931 | .925 | .919 | .912 | .903 | .894 | .884 | .873 | .860 |
| | | | | | | | | .845 | .829 | .811 | .790 | .767 | .741 | .712 | .680 | .645 | .607 |
| | | | | | | | | .565 | .521 | .474 | .425 | .376 | .327 | .278 | | | |
| 30.000 | 29.225 | .000 | .775 | .77 | 3.29 | 2.6% | 38 | .968 | .966 | .964 | .961 | .959 | .956 | .952 | .949 | .945 | .941 |
| | | | | | | | | .936 | .931 | .925 | .919 | .912 | .903 | .894 | .884 | .873 | .860 |
| | | | | | | | | .845 | .829 | .811 | .790 | .767 | .741 | .712 | .680 | .645 | .607 |
| | | | | | | | | .565 | .521 | .474 | .425 | .376 | .327 | .278 | .233 | | |
| 30.000 | 29.415 | .000 | .585 | .81 | 3.25 | 1.9% | 39 | .968 | .966 | .964 | .961 | .959 | .956 | .952 | .949 | .945 | .941 |
| | | | | | | | | .936 | .931 | .925 | .919 | .912 | .903 | .894 | .884 | .873 | .860 |
| | | | | | | | | .845 | .829 | .811 | .790 | .767 | .741 | .712 | .680 | .645 | .607 |
| | | | | | | | | .565 | .521 | .474 | .425 | .376 | .327 | .278 | .233 | .191 | |
| 30.000 | 29.568 | .000 | .432 | .85 | 3.19 | 1.4% | 40 | .968 | .966 | .964 | .961 | .959 | .956 | .952 | .949 | .945 | .941 |
| | | | | | | | | .936 | .931 | .925 | .919 | .912 | .903 | .894 | .884 | .873 | .860 |
| | | | | | | | | .845 | .829 | .811 | .790 | .767 | .741 | .712 | .680 | .645 | .607 |
| | | | | | | | | .565 | .521 | .474 | .425 | .376 | .327 | .278 | .233 | .191 | .153 |
| 30.000 | 29.687 | .000 | .313 | .88 | 3.12 | 1.0% | 41 | .968 | .966 | .964 | .961 | .959 | .956 | .952 | .949 | .945 | .941 |
| | | | | | | | | .936 | .931 | .925 | .919 | .912 | .903 | .894 | .884 | .873 | .860 |
| | | | | | | | | .845 | .829 | .811 | .790 | .767 | .741 | .712 | .680 | .645 | .607 |
| | | | | | | | | .565 | .521 | .474 | .425 | .376 | .327 | .278 | .233 | .191 | .153 |
| | | | | | | | | .119 | | | | | | | | | |
| 30.000 | 29.778 | .000 | .222 | .91 | 3.05 | 7% | 42 | .968 | .066 | .064 | .901 | .959 | .956 | .952 | .949 | .945 | .941 |
| | | | | | | | | .936 | .931 | .925 | .919 | .912 | .903 | .894 | .884 | .873 | .860 |
| | | | | | | | | .845 | .829 | .811 | .790 | .767 | .741 | .712 | .680 | .645 | .607 |
| | | | | | | | | .565 | .521 | .474 | .425 | .376 | .327 | .278 | .233 | .191 | .153 |
| | | | | | | | | .119 | .091 | | | | | | | | |
| 30.000 | 29.846 | .000 | .154 | .93 | 2.97 | .5% | 43 | .968 | .966 | .964 | .961 | .959 | .956 | .952 | .949 | .945 | .941 |
| | | | | | | | | .936 | .931 | .925 | .919 | .912 | .903 | .894 | .884 | .873 | .860 |
| | | | | | | | | .845 | .820 | .811 | .790 | .707 | .741 | .712 | .680 | .645 | .607 |
| | | | | | | | | .565 | .521 | .474 | .425 | .376 | .327 | .278 | .233 | .191 | .153 |
| | | | | | | | | .119 | .091 | .068 | | | | | | | |

*SLOW RETRIALS: JEWETT-SHRAGO EXTENDED ERLANG B - 50% BLOCKED CALLERS RETRY, 50% OVERFLOW (RETRIES 3-5 HOLDING TIMES LATER)

| HOURS FIRST ATTEMPT VMR = 1.0 | HOURS CON-NECT | HOURS RECALL /RETRY | HOURS OVFL | CON-NECT VMR | OVFL VMR | PERCENT BLOCKED (GRADE) | LINE | HOURS CONNECTED TRAFFIC ON LINES | | | | | | | | | |
|---|---|---|---|---|---|---|---|---|---|---|---|---|---|---|---|---|---|
| | | | | | | | | 1 | 2 | 3 | 4 | 5 | 6 | 7 | 8 | 9 | 10 |
| .200 | .180 | .020 | .020 | .82 | 1.08 | 18.0% | 1 | .180 | | | | | | | | | |
| .200 | .198 | .002 | .002 | .97 | 1.07 | 1.7% | 2 | .168 | .030 | | | | | | | | |
| .400 | .323 | .077 | .077 | .68 | 1.13 | 32.3% | 1 | .323 | | | | | | | | | |
| .400 | .388 | .012 | .012 | .90 | 1.13 | 5.7% | 2 | .292 | .097 | | | | | | | | |
| .400 | .399 | .001 | .001 | .98 | 1.11 | .7% | 3 | .286 | .093 | .019 | | | | | | | |
| .600 | .565 | .035 | .035 | .82 | 1.19 | 11.0% | 2 | .388 | .177 | | | | | | | | |
| .600 | .594 | .006 | .006 | .95 | 1.17 | 2.0% | 3 | .377 | .167 | .050 | | | | | | | |
| .800 | .726 | .074 | .074 | .74 | 1.24 | 16.9% | 2 | .466 | .260 | | | | | | | | |
| .800 | .783 | .017 | .017 | .91 | 1.22 | 4.0% | 3 | .450 | .240 | .094 | | | | | | | |
| .800 | .797 | .003 | .003 | .97 | 1.18 | .8% | 4 | .445 | .236 | .090 | .025 | | | | | | |
| 1.000 | .870 | .130 | .130 | .66 | 1.27 | 23.1% | 2 | .531 | .339 | | | | | | | | |
| 1.000 | .965 | .035 | .035 | .85 | 1.27 | 6.7% | 3 | .509 | .311 | .146 | | | | | | | |
| 1.000 | .992 | .008 | .008 | .95 | 1.24 | 1.6% | 4 | .502 | .302 | .139 | .048 | | | | | | |
| 1.200 | .996 | .204 | .204 | .59 | 1.29 | 29.1% | 2 | .584 | .412 | | | | | | | | |
| 1.200 | 1.138 | .062 | .062 | .80 | 1.32 | 9.9% | 3 | .558 | .376 | .204 | | | | | | | |
| 1.200 | 1.183 | .017 | .017 | .92 | 1.29 | 2.7% | 4 | .549 | .363 | .192 | .079 | | | | | | |
| 1.200 | 1.196 | .004 | .004 | .98 | 1.24 | .6% | 5 | .546 | .360 | .189 | .077 | .024 | | | | | |
| 1.400 | 1.106 | .294 | .294 | .52 | 1.31 | 34.8% | 2 | .629 | .477 | | | | | | | | |
| 1.400 | 1.299 | .101 | .101 | .74 | 1.35 | 13.4% | 3 | .600 | .435 | .264 | | | | | | | |
| 1.400 | 1.370 | .030 | .030 | .88 | 1.33 | 4.2% | 4 | .589 | .418 | .247 | .116 | | | | | | |
| 1.400 | 1.392 | .008 | .008 | .96 | 1.29 | 1.1% | 5 | .585 | .413 | .241 | .112 | .041 | | | | | |
| 1.600 | 1.449 | .151 | .151 | .68 | 1.38 | 17.3% | 3 | .637 | .488 | 324 | | | | | | | |
| 1.600 | 1.550 | .050 | .050 | .84 | 1.38 | 6.1% | 4 | .623 | .467 | .300 | .159 | | | | | | |
| 1.600 | 1.585 | .015 | .015 | .94 | 1.34 | 1.8% | 5 | .618 | .460 | .292 | .152 | .064 | | | | | |
| 1.800 | 1.586 | .214 | .214 | .62 | 1.41 | 21.3% | 3 | .668 | .536 | .382 | | | | | | | |
| 1.800 | 1.722 | .078 | .078 | .79 | 1.42 | 8.3% | 4 | .653 | .512 | .353 | .205 | | | | | | |
| 1.800 | 1.775 | .025 | .025 | .91 | 1.38 | 2.8% | 5 | .646 | .502 | .341 | .195 | .091 | | | | | |
| 1.800 | 1.793 | .007 | .007 | .97 | 1.33 | .8% | 6 | .644 | .499 | .337 | .191 | .089 | .034 | | | | |
| 2.000 | 1.711 | .289 | .289 | .56 | 1.42 | 25.3% | 3 | .696 | .578 | .436 | | | | | | | |
| 2.000 | 1.887 | .113 | .113 | .75 | 1.45 | 10.7% | 4 | .679 | .552 | .402 | .254 | | | | | | |
| 2.000 | 1.960 | .040 | .040 | .88 | 1.43 | 3.9% | 5 | .671 | .540 | .387 | .239 | .123 | | | | | |
| 2.000 | 1.988 | .012 | .012 | .95 | 1.38 | 1.2% | 6 | .668 | .535 | .382 | .233 | .119 | .050 | | | | |
| 2.200 | 1.823 | .377 | .377 | .51 | 1.43 | 29.3% | 3 | .720 | .616 | .487 | | | | | | | |
| 2.200 | 2.042 | .158 | .158 | .70 | 1.48 | 13.4% | 4 | .702 | .588 | .449 | .303 | | | | | | |
| 2.200 | 2.140 | .060 | .060 | .84 | 1.47 | 5.3% | 5 | .693 | .574 | .431 | .283 | .159 | | | | | |
| 2.200 | 2.180 | .020 | .020 | .93 | 1.42 | 1.8% | 6 | .689 | .568 | .423 | .276 | .153 | .071 | | | | |
| 2.200 | 2.194 | .006 | .006 | .97 | 1.37 | .6% | 7 | .688 | .566 | .421 | .273 | .150 | .069 | .027 | | | |
| 2.400 | 1.924 | .476 | .476 | .47 | 1.43 | 33.1% | 3 | .742 | .649 | .533 | | | | | | | |
| 2.400 | 2.188 | .212 | .212 | .65 | 1.50 | 16.3% | 4 | .723 | .620 | .492 | .352 | | | | | | |
| 2.400 | 2.315 | .085 | .085 | .80 | 1.50 | 6.9% | 5 | .713 | .605 | .471 | .328 | .198 | | | | | |

*SIMILAR TO A.T.&T. JACOBSEN RETRIAL TABLES

## *SLOW RETRIALS: JEWETT-SHRAGO EXTENDED ERLANG B - 50% BLOCKED CALLERS RETRY, 50% OVERFLOW (RETRIES 3-5 HOLDING TIMES LATER)

| HOURS FIRST ATTEMPT VMR = 1.0 | HOURS CON- NECT | HOURS RECALL /RETRY | HOURS OVFL | CON- NECT VMR | OVFL VMR | PERCENT BLOCKED (GRADE) | LINE | HOURS CONNECTED TRAFFIC ON LINES 1 | 2 | 3 | 4 | 5 | 6 | 7 | 8 | 9 | 10 |
|---|---|---|---|---|---|---|---|---|---|---|---|---|---|---|---|---|---|
| 2.400 | 2.369 | .031 | .031 | .90 | 1.46 | 2.6% | 6 | .709 | .598 | .462 | .317 | .188 | .095 | | | | |
| 2.400 | 2.390 | .010 | .010 | .96 | 1.41 | .8% | 7 | .707 | .595 | .458 | .313 | .185 | .093 | .039 | | | |
| 2.600 | 2.323 | .277 | .277 | .60 | 1.52 | 19.2% | 4 | .742 | .649 | .533 | .399 | | | | | | |
| 2.600 | 2.482 | .118 | .118 | .76 | 1.54 | 8.7% | 5 | .731 | .632 | .509 | .371 | .239 | | | | | |
| 2.600 | 2.554 | .046 | .046 | .88 | 1.50 | 3.4% | 6 | .726 | .624 | .498 | .358 | .226 | .123 | | | | |
| 2.600 | 2.584 | .016 | .016 | .95 | 1.45 | 1.2% | 7 | .723 | .621 | .493 | .353 | .221 | .119 | .055 | | | |
| 2.800 | 2.449 | .351 | .351 | .56 | 1.53 | 22.3% | 4 | .759 | .676 | .570 | .445 | | | | | | |
| 2.800 | 2.643 | .157 | .157 | .72 | 1.57 | 10.6% | 5 | .747 | .658 | .544 | .413 | .281 | | | | | |
| 2.800 | 2.736 | .064 | .064 | .85 | 1.54 | 4.5% | 6 | .741 | .648 | .531 | .397 | .265 | .154 | | | | |
| 2.800 | 2.776 | .024 | .024 | .93 | 1.49 | 1.7% | 7 | .739 | .644 | .525 | .390 | .257 | .148 | .073 | | | |
| 2.800 | 2.792 | .008 | .008 | .97 | 1.44 | .6% | 8 | .737 | .642 | .523 | .387 | .255 | .146 | .072 | .030 | | |
| 3.000 | 2.565 | .435 | .435 | .51 | 1.54 | 25.3% | 4 | .775 | .700 | .603 | .487 | | | | | | |
| 3.000 | 2.795 | .205 | .205 | .68 | 1.59 | 12.8% | 5 | .762 | .681 | .576 | .453 | .323 | | | | | |
| 3.000 | 2.912 | .088 | .088 | .81 | 1.58 | 5.7% | 6 | .755 | .670 | .561 | .435 | .303 | .187 | | | | |
| 3.000 | 2.965 | .035 | .035 | .91 | 1.53 | 2.3% | 7 | .752 | .665 | .554 | .426 | .294 | .179 | .095 | | | |
| 3.000 | 2.987 | .013 | .013 | .96 | 1.48 | .8% | 8 | .751 | .663 | .551 | .422 | .290 | .176 | .092 | .042 | | |
| 3.200 | 2.671 | .529 | .529 | .47 | 1.54 | 28.4% | 4 | .789 | .721 | .634 | .527 | | | | | | |
| 3.200 | 2.940 | .260 | .260 | .63 | 1.61 | 15.1% | 5 | .776 | .702 | .606 | .491 | .365 | | | | | |
| 3.200 | 3.083 | .117 | .117 | .78 | 1.61 | 7.1% | 6 | .768 | .690 | .590 | .470 | .342 | .222 | | | | |
| 3.200 | 3.151 | .049 | .049 | .88 | 1.57 | 3.0% | 7 | .765 | .684 | .582 | .460 | .331 | .212 | .119 | | | |
| 3.200 | 3.182 | .018 | .018 | .94 | 1.52 | 1.1% | 8 | .763 | .682 | .578 | .455 | .325 | .207 | .115 | .056 | | |
| 3.400 | 2.707 | .833 | .633 | .44 | 1.54 | 31.4% | 4 | .801 | .740 | .662 | .504 | | | | | | |
| 3.400 | 3.075 | .325 | .325 | .59 | 1.62 | 17.4% | 5 | .788 | .721 | .633 | .527 | .406 | | | | | |
| 3.400 | 3.247 | .153 | .153 | .74 | 1.64 | 8.6% | 6 | .780 | .709 | .616 | .504 | .380 | .259 | | | | |
| 3.400 | 3.334 | .066 | .066 | .85 | 1.61 | 3.8% | 7 | .776 | .702 | .607 | .492 | .366 | .245 | .146 | | | |
| 3.400 | 3.374 | .026 | .026 | .93 | 1.56 | 1.5% | 8 | .774 | .699 | .602 | .486 | .360 | .239 | .141 | .073 | | |
| 3.400 | 3.390 | .010 | .010 | .97 | 1.50 | .6% | 9 | .773 | .698 | .600 | .484 | .357 | .237 | .139 | .071 | .032 | |
| 3.600 | 2.855 | .745 | .745 | .40 | 1.54 | 34.3% | 4 | .813 | .758 | .687 | .598 | | | | | | |
| 3.600 | 3.203 | .397 | .397 | .55 | 1.63 | 19.9% | 5 | .800 | .738 | .659 | .560 | .446 | | | | | |
| 3.600 | 3.406 | .194 | .194 | .70 | 1.67 | 10.2% | 6 | .701 | .725 | .040 | .535 | .417 | .296 | | | | |
| 3.600 | 3.512 | .088 | .088 | .83 | 1.65 | 4.8% | 7 | .787 | .718 | .630 | .522 | .401 | .280 | .175 | | | |
| 3.600 | 3.563 | .037 | .037 | .91 | 1.60 | 2.0% | 8 | .784 | .715 | .625 | .515 | .393 | .272 | .168 | .092 | | |
| 3.600 | 3.586 | .014 | .014 | .96 | 1.54 | .8% | 9 | .783 | .713 | .622 | .512 | .389 | .268 | .165 | .089 | .043 | |
| 3.800 | 3.321 | .479 | .479 | .52 | 1.64 | 22.4% | 5 | .811 | .754 | .682 | .601 | .404 | | | | | |
| 3.800 | 3.557 | .243 | .243 | .67 | 1.69 | 12.0% | 6 | .802 | .741 | .662 | .565 | .452 | .334 | | | | |
| 3.800 | 3.686 | .114 | .114 | .79 | 1.68 | 5.8% | 7 | .797 | .733 | .651 | .550 | .434 | .315 | .206 | | | |
| 3.800 | 3.750 | .050 | .050 | .89 | 1.63 | 2.6% | 8 | .794 | .729 | .645 | .542 | .425 | .305 | .197 | .113 | | |
| 3.800 | 3.780 | .020 | .020 | .94 | 1.57 | 1.1% | 9 | .793 | .727 | .642 | .539 | .421 | .300 | .193 | .110 | .056 | |
| 4.000 | 3.432 | .568 | .568 | .48 | 1.64 | 24.9% | 5 | .820 | .769 | .703 | .619 | .520 | | | | | |
| 4.000 | 3.701 | .299 | .299 | .63 | 1.70 | 13.9% | 6 | .811 | .756 | .683 | .593 | .487 | .372 | | | | |

*SIMILAR TO A.T.&T. JACOBSEN RETRIAL TABLES

*SLOW RETRIALS: JEWETT-SHRAGO EXTENDED ERLANG B - 50% BLOCKED CALLERS RETRY, 50% OVERFLOW (RETRIES 3-5 HOLDING TIMES LATER)

| HOURS FIRST ATTEMPT VMR = 1.0 | HOURS CON-NECT | HOURS RECALL /RETRY | HOURS OVFL | CON-NECT VMR | OVFL VMR | PERCENT BLOCKED (GRADE) | LINE | HOURS CONNECTED TRAFFIC ON LINES 1 | 2 | 3 | 4 | 5 | 6 | 7 | 8 | 9 | 10 |
|---|---|---|---|---|---|---|---|---|---|---|---|---|---|---|---|---|---|
| 4.000 | 3.854 | .146 | .146 | .76 | 1.71 | 7.0% | 7 | .806 | .747 | .671 | .577 | .466 | .349 | .238 | | | |
| 4.000 | 3.934 | .066 | .066 | .86 | 1.67 | 3.3% | 8 | .803 | .742 | .664 | .568 | .456 | .337 | .227 | .137 | | |
| 4.000 | 3.972 | .028 | .028 | .93 | 1.61 | 1.4% | 9 | .801 | .740 | .661 | .563 | .450 | .332 | .222 | .132 | .070 | |
| 4.000 | 3.989 | .011 | .011 | .97 | 1.55 | .5% | 10 | .800 | .739 | .660 | .561 | .448 | .329 | .219 | .131 | .069 | .033 |
| 4.200 | 3.534 | .666 | .666 | .45 | 1.64 | 27.4% | 5 | .830 | .783 | .722 | .646 | .553 | | | | | |
| 4.200 | 3.838 | .362 | .362 | .59 | 1.72 | 15.9% | 6 | .820 | .769 | .702 | .619 | .519 | .408 | | | | |
| 4.200 | 4.017 | .183 | .183 | .73 | 1.73 | 8.3% | 7 | .814 | .760 | .689 | .601 | .497 | .383 | .272 | | | |
| 4.200 | 4.114 | .086 | .086 | .84 | 1.71 | 4.0% | 8 | .811 | .755 | .682 | .591 | .485 | .370 | .258 | .162 | | |
| 4.200 | 4.162 | .038 | .038 | .91 | 1.65 | 1.8% | 9 | .809 | .752 | .678 | .586 | .479 | .363 | .251 | .157 | .087 | |
| 4.200 | 4.185 | .015 | .015 | .96 | 1.59 | .7% | 10 | .808 | .751 | .677 | .584 | .476 | .359 | .248 | .154 | .085 | .042 |
| 4.400 | 3.628 | .772 | .772 | .42 | 1.64 | 29.9% | 5 | .838 | .795 | .740 | .670 | .585 | | | | | |
| 4.400 | 3.967 | .433 | .433 | .56 | 1.73 | 17.9% | 6 | .829 | .781 | .720 | .643 | .550 | .444 | | | | |
| 4.400 | 4.174 | .226 | .226 | .69 | 1.76 | 9.8% | 7 | .822 | .772 | .707 | .625 | .526 | .417 | .306 | | | |
| 4.400 | 4.290 | .110 | .110 | .81 | 1.74 | 4.9% | 8 | .819 | .766 | .699 | .614 | .513 | .401 | .289 | .190 | | |
| 4.400 | 4.350 | .050 | .050 | .89 | 1.69 | 2.2% | 9 | .817 | .763 | .694 | .608 | .505 | .393 | .281 | .182 | .106 | |
| 4.400 | 4.379 | .021 | .021 | .95 | 1.63 | 1.0% | 10 | .816 | .762 | .692 | .605 | .502 | .389 | .277 | .179 | .104 | .054 |
| 4.600 | 3.714 | .886 | .886 | .39 | 1.63 | 32.3% | 5 | .846 | .807 | .756 | .692 | .614 | | | | | |
| 4.600 | 4.088 | .512 | .512 | .52 | 1.73 | 20.0% | 6 | .836 | .793 | .736 | .665 | .579 | .479 | | | | |
| 4.600 | 4.325 | .275 | .275 | .66 | 1.78 | 11.3% | 7 | .830 | .783 | .723 | .646 | .554 | .449 | .340 | | | |
| 4.600 | 4.462 | .138 | .138 | .78 | 1.77 | 5.8% | 8 | .826 | .777 | .714 | .635 | .539 | .432 | .321 | .219 | | |
| 4.600 | 4.535 | .065 | .065 | .87 | 1.72 | 2.8% | 9 | .824 | .774 | .709 | .628 | .531 | .422 | .311 | .209 | .127 | |
| 4.600 | 4.572 | .028 | .028 | .93 | 1.66 | 1.2% | 10 | .822 | .772 | .707 | .625 | .527 | .417 | .306 | .205 | .124 | .067 |
| 4.600 | 4.588 | .012 | .012 | .97 | 1.60 | .5% | 11 | .822 | .771 | .706 | .623 | .525 | .415 | .304 | .203 | .122 | .066 |
| | | | | | | | | .032 | | | | | | | | | |
| 4.800 | 3.794 | 1.006 | 1.006 | .36 | 1.62 | 34.7% | 5 | .853 | .817 | .771 | .712 | .640 | | | | | |
| 4.800 | 4.202 | .598 | .598 | .49 | 1.73 | 22.1% | 6 | .844 | .803 | .752 | .686 | .606 | .512 | | | | |
| 4.800 | 4.470 | .330 | .330 | .63 | 1.79 | 12.9% | 7 | .837 | .794 | .737 | .667 | .580 | .481 | .374 | | | |
| 4.800 | 4.629 | .171 | .171 | .75 | 1.80 | 6.9% | 8 | .833 | .787 | .728 | .654 | .564 | .461 | .353 | .249 | | |
| 4.800 | 4.717 | .083 | .083 | .85 | 1.76 | 3.4% | 9 | .830 | .783 | .723 | .647 | .555 | .450 | .341 | .237 | .150 | |
| 4.800 | 4.763 | .037 | .037 | .92 | 1.70 | 1.5% | 10 | .829 | .781 | .720 | .643 | .550 | .445 | .335 | .231 | .145 | .082 |
| 4.800 | 4.784 | .016 | .016 | .96 | 1.64 | .7% | 11 | .828 | .781 | .719 | .641 | .548 | .442 | .332 | .229 | .143 | .081 |
| | | | | | | | | .041 | | | | | | | | | |
| 5.000 | 4.309 | .691 | .691 | .46 | 1.73 | 24.3% | 6 | .851 | .813 | .766 | .705 | .631 | .543 | | | | |
| 5.000 | 4.608 | .392 | .392 | .59 | 1.80 | 14.6% | 7 | .844 | .803 | .751 | .686 | .605 | .511 | .407 | | | |
| 5.000 | 4.791 | .209 | .209 | .72 | 1.82 | 8.0% | 8 | .839 | .797 | .742 | .673 | .588 | .490 | .384 | .279 | | |
| 5.000 | 4.896 | .104 | .104 | .83 | 1.79 | 4.1% | 9 | .836 | .792 | .736 | .665 | .578 | .478 | .370 | .266 | .175 | |
| 5.000 | 4.951 | .049 | .049 | .90 | 1.74 | 1.9% | 10 | .835 | .790 | .733 | .660 | .572 | .471 | .363 | .259 | .168 | .099 |
| 5.000 | 4.979 | .021 | .021 | .95 | 1.67 | .9% | 11 | .834 | .789 | .731 | .658 | .570 | .468 | .359 | .255 | .165 | .097 |
| | | | | | | | | .052 | | | | | | | | | |
| 5.500 | 4.544 | .956 | .956 | .39 | 1.72 | 29.6% | 6 | .866 | .835 | .796 | .747 | .686 | .613 | | | | |

*SIMILAR TO A.T.&T. JACOBSEN RETRIAL TABLES

# *SLOW RETRIALS: JEWETT-SHRAGO EXTENDED ERLANG B - 50% BLOCKED CALLERS RETRY, 50% OVERFLOW (RETRIES 3-5 HOLDING TIMES LATER)

| HOURS FIRST ATTEMPT VMR = 1.0 | HOURS CON-NECT | HOURS RECALL /RETRY | HOURS OVFL | CON-NECT VMR | OVFL VMR | PERCENT BLOCKED (GRADE) | LINE | HOURS CONNECTED TRAFFIC ON LINES 1 | 2 | 3 | 4 | 5 | 6 | 7 | 8 | 9 | 10 |
|---|---|---|---|---|---|---|---|---|---|---|---|---|---|---|---|---|---|
| 5.500 | 4.922 | .578 | .578 | .51 | 1.82 | 19.0% | 7 | .859 | .825 | .782 | .728 | .661 | .580 | .488 | | | |
| 5.500 | 5.173 | .327 | .327 | .64 | 1.87 | 11.2% | 8 | .854 | .818 | .772 | .714 | .642 | .556 | .460 | .358 | | |
| 5.500 | 5.325 | .175 | .175 | .76 | 1.86 | 6.2% | 9 | .850 | .813 | .765 | .704 | .630 | .541 | .442 | .339 | .241 | |
| 5.500 | 5.412 | .088 | .088 | .85 | 1.82 | 3.1% | 10 | .848 | .810 | .761 | .699 | .622 | .532 | .432 | .328 | .231 | .149 |
| 5.500 | 5.458 | .042 | .042 | .92 | 1.76 | 1.5% | 11 | .847 | .808 | .759 | .696 | .618 | .527 | .426 | .322 | .225 | .145 |
| | | | | | | | | .085 | | | | | | | | | |
| 5.500 | 5.481 | .019 | .019 | .96 | 1.70 | .7% | 12 | .847 | .808 | .758 | .694 | .616 | .525 | .423 | .319 | .223 | .142 |
| | | | | | | | | .083 | .044 | | | | | | | | |
| 6.000 | 4.739 | 1.261 | 1.261 | .33 | 1.69 | 34.7% | 6 | .879 | .854 | .822 | .782 | .732 | .671 | | | | |
| 6.000 | 5.195 | .805 | .805 | .44 | 1.81 | 23.7% | 7 | .872 | .844 | .808 | .763 | .708 | .640 | .560 | | | |
| 6.000 | 5.517 | .483 | .483 | .56 | 1.89 | 14.9% | 8 | .866 | .836 | .797 | .749 | .688 | .615 | .530 | .435 | | |
| 6.000 | 5.726 | .274 | .274 | .69 | 1.92 | 8.7% | 9 | .863 | .831 | .790 | .738 | .674 | .597 | .508 | .411 | .313 | |
| 6.000 | 5.853 | .147 | .147 | .79 | 1.90 | 4.8% | 10 | .860 | .827 | .785 | .732 | .666 | .586 | .495 | .397 | .298 | .208 |
| 6.000 | 5.926 | .074 | .074 | .87 | 1.85 | 2.4% | 11 | .859 | .825 | .782 | .728 | .660 | .580 | .487 | .388 | .289 | .200 |
| | | | | | | | | .127 | | | | | | | | | |
| 6.000 | 5.964 | .036 | .036 | .93 | 1.79 | 1.2% | 12 | .858 | .824 | .780 | .726 | .658 | .576 | .483 | .383 | .285 | .196 |
| | | | | | | | | .124 | .072 | | | | | | | | |
| 6.000 | 5.984 | .016 | .016 | .96 | 1.72 | .5% | 13 | .858 | .823 | .780 | .725 | .656 | .574 | .481 | .381 | .282 | .194 |
| | | | | | | | | .122 | .071 | .037 | | | | | | | |
| 6.500 | 5.426 | 1.074 | 1.074 | .38 | 1.79 | 28.4% | 7 | .883 | .860 | .830 | .793 | .747 | .690 | .623 | | | |
| 6.500 | 5.821 | .679 | .679 | .49 | 1.90 | 18.9% | 8 | .878 | .852 | .819 | .779 | .728 | .666 | .592 | .508 | | |
| 6.500 | 6.095 | .405 | .405 | .61 | 1.96 | 11.7% | 9 | .874 | .846 | .811 | .768 | .713 | .647 | .569 | .481 | .386 | |
| 6.500 | 6.270 | .230 | .230 | .73 | 1.96 | 6.8% | 10 | .871 | .842 | .806 | .760 | .703 | .634 | .553 | .462 | .367 | .273 |
| 6.500 | 6.377 | .123 | .123 | .82 | 1.93 | 3.7% | 11 | .869 | .839 | .802 | .755 | .697 | .626 | .543 | .451 | .354 | .281 |
| | | | | | | | | .170 | | | | | | | | | |
| 6.500 | 6.437 | .063 | .063 | .89 | 1.87 | 1.9% | 12 | .868 | .838 | .800 | .752 | .693 | .621 | .537 | .444 | .347 | .254 |
| | | | | | | | | .173 | .109 | | | | | | | | |
| 6.500 | 6.470 | .030 | .030 | .94 | 1.81 | .9% | 13 | .867 | .837 | .799 | .751 | .691 | .619 | .534 | .441 | .343 | .251 |
| | | | | | | | | .170 | .106 | .061 | | | | | | | |
| 7.000 | 5.621 | 1.379 | 1.379 | .32 | 1.76 | 32.9% | 7 | .893 | .873 | .848 | .818 | .779 | .733 | .676 | | | |
| 7.000 | 6.087 | .913 | .913 | .43 | 1.89 | 23.1% | 8 | .888 | .866 | .830 | .804 | .761 | .709 | .647 | .574 | | |
| 7.000 | 6.428 | .572 | .572 | .54 | 1.98 | 15.1% | 9 | .883 | .860 | .830 | .793 | .747 | .690 | .623 | .545 | .458 | |
| 7.000 | 6.660 | .340 | .340 | .66 | 2.01 | 8.3% | 10 | .880 | .855 | .824 | .785 | .736 | .676 | .605 | .523 | .434 | .341 |
| 7.000 | 6.807 | .193 | .193 | .76 | 2.00 | 5.4% | 11 | .878 | .852 | .820 | .779 | .728 | .667 | .593 | .509 | .418 | .325 |
| | | | | | | | | .237 | | | | | | | | | |
| 7.000 | 6.896 | .104 | .104 | .85 | 1.96 | 2.9% | 12 | .877 | .850 | .817 | .776 | .724 | .661 | .586 | .501 | .408 | .316 |
| | | | | | | | | .228 | .154 | | | | | | | | |
| 7.000 | 6.947 | .053 | .053 | .91 | 1.89 | 1.5% | 13 | .876 | .840 | .816 | .774 | .721 | .657 | .582 | .496 | .403 | .309 |
| | | | | | | | | .223 | .149 | .092 | | | | | | | |
| 7.000 | 6.974 | .026 | .026 | .95 | 1.82 | .7% | 14 | .875 | .849 | .815 | .772 | .720 | .656 | .579 | .493 | .400 | .306 |

*SIMILAR TO A.T.¢T. JACOBSEN RETRIAL TABLES

*SLOW RETRIALS: JEWETT-SHRAGO EXTENDED ERLANG B - 50% BLOCKED CALLERS RETRY, 50% OVERFLOW (RETRIES 3-5 HOLDING TIMES LATER)

| HOURS FIRST ATTEMPT VMR = 1.0 | HOURS CONNECT | HOURS RECALL /RETRY | HOURS OVFL | CONNECT VMR | OVFL VMR | PERCENT BLOCKED (GRADE) | LINE | HOURS CONNECTED TRAFFIC ON LINES 1 | 2 | 3 | 4 | 5 | 6 | 7 | 8 | 9 | 10 |
|---|---|---|---|---|---|---|---|---|---|---|---|---|---|---|---|---|---|
| | | | | | | | | .220 | .147 | .091 | .052 | | | | | | |
| 7.500 | 6.316 | 1.184 | 1.184 | .37 | 1.87 | 27.3% | 8 | .897 | .878 | .854 | .826 | .790 | .746 | .694 | .631 | | |
| 7.500 | 6.725 | .775 | .775 | .48 | 1.98 | 18.7% | 9 | .892 | .872 | .846 | .815 | .776 | .728 | .670 | .602 | .524 | |
| 7.500 | 7.018 | .482 | .482 | .59 | 2.04 | 12.1% | 10 | .889 | .867 | .840 | .806 | .764 | .713 | .652 | .579 | .498 | .410 |
| 7.500 | 7.214 | .286 | .286 | .70 | 2.05 | 7.3% | 11 | .886 | .864 | .835 | .800 | .756 | .703 | .638 | .563 | .479 | .390 |
| | | | | | | | | .300 | | | | | | | | | |
| 7.500 | 7.338 | .162 | .162 | .79 | 2.03 | 4.2% | 12 | .885 | .861 | .832 | .796 | .751 | .696 | .629 | .552 | .467 | .376 |
| | | | | | | | | .287 | .206 | | | | | | | | |
| 7.500 | 7.413 | .087 | .087 | .87 | 1.98 | 2.3% | 13 | .884 | .860 | .830 | .793 | .747 | .691 | .624 | .546 | .459 | .368 |
| | | | | | | | | .279 | .198 | .132 | | | | | | | |
| 7.500 | 7.455 | .045 | .045 | .92 | 1.91 | 1.2% | 14 | .883 | .859 | .829 | .792 | .745 | .689 | .621 | .542 | .455 | .364 |
| | | | | | | | | .275 | .194 | .128 | .079 | | | | | | |
| 7.500 | 7.478 | .022 | .022 | .96 | 1.84 | .6% | 15 | .883 | .859 | .829 | .791 | .744 | .687 | .619 | .540 | .453 | .361 |
| | | | | | | | | .272 | .192 | .126 | .077 | .044 | | | | | |
| 8.000 | 6.512 | 1.488 | 1.488 | .32 | 1.84 | 31.4% | 8 | .905 | .888 | .868 | .844 | .814 | .778 | .733 | .681 | | |
| 8.000 | 6.986 | 1.014 | 1.014 | .42 | 1.96 | 22.5% | 9 | .900 | .882 | .861 | .834 | .801 | .760 | .711 | .653 | .585 | |
| 8.000 | 7.342 | .658 | .658 | .52 | 2.05 | 15.2% | 10 | .897 | .877 | .854 | .825 | .789 | .745 | .692 | .629 | .557 | .476 |
| 8.000 | 7.593 | .407 | .407 | .63 | 2.09 | 9.7% | 11 | .894 | .874 | .849 | .818 | .780 | .734 | .678 | .612 | .536 | .452 |
| | | | | | | | | .365 | | | | | | | | | |
| 8.000 | 7.759 | .241 | .241 | .74 | 2.09 | 5.8% | 12 | .892 | .871 | .846 | .814 | .774 | .726 | .668 | .600 | .521 | .436 |
| | | | | | | | | .348 | .263 | | | | | | | | |
| 8.000 | 7.864 | .136 | .136 | .82 | 2.05 | 3.3% | 13 | .891 | .869 | .843 | .811 | .770 | .721 | .661 | .591 | .512 | .426 |
| | | | | | | | | .337 | .253 | .178 | | | | | | | |
| 8.000 | 7.927 | .073 | .073 | .89 | 1.99 | 1.8% | 14 | .890 | .868 | .842 | .809 | .768 | .718 | .657 | .587 | .506 | .419 |
| | | | | | | | | .331 | .246 | .172 | .113 | | | | | | |
| 8.000 | 7.962 | .038 | .038 | .93 | 1.92 | .9% | 15 | .889 | .868 | .841 | .808 | .767 | .716 | .655 | .584 | .503 | .416 |
| | | | | | | | | .327 | .243 | .169 | .110 | .067 | | | | | |
| 8.500 | 7.213 | 1.287 | 1.287 | .36 | 1.94 | 26.3% | 9 | .907 | .892 | .873 | .850 | .822 | .788 | .746 | .697 | .638 | |
| 8.500 | 7.633 | .867 | .867 | .46 | 2.05 | 18.5% | 10 | .904 | .887 | .867 | .842 | .811 | .773 | .728 | .674 | .610 | .538 |
| 8.500 | 7.942 | .558 | .558 | .57 | 2.12 | 12.3% | 11 | .901 | .883 | .861 | .835 | .802 | .762 | .713 | .655 | .588 | .512 |
| | | | | | | | | .430 | | | | | | | | | |
| 8.500 | 8.156 | .344 | .344 | .68 | 2.14 | 7.8% | 12 | .898 | .880 | .857 | .830 | .795 | .753 | .702 | .642 | .572 | .493 |
| | | | | | | | | .409 | .324 | | | | | | | | |
| 8.500 | 8.297 | .203 | .203 | .77 | 2.12 | 4.7% | 13 | .897 | .878 | .855 | .826 | .791 | .747 | .695 | .633 | .561 | .480 |
| | | | | | | | | .395 | .310 | .230 | | | | | | | |
| 8.500 | 8.385 | .115 | .115 | .85 | 2.07 | 2.7% | 14 | .896 | .877 | .853 | .824 | .788 | .743 | .690 | .627 | .553 | .472 |
| | | | | | | | | .386 | .301 | .222 | .153 | | | | | | |
| 8.500 | 8.438 | .062 | .062 | .90 | 2.01 | 1.4% | 15 | .895 | .876 | .852 | .822 | .786 | .741 | .687 | .623 | .549 | .467 |
| | | | | | | | | .381 | .296 | .217 | .149 | .096 | | | | | |
| 8.500 | 8.468 | .032 | .032 | .94 | 1.94 | .7% | 16 | .895 | .876 | .851 | .822 | .785 | .740 | .685 | .621 | .547 | .464 |

*SIMILAR TO A.T.&T. JACOBSEN RETRIAL TABLES

*SLOW RETRIALS: JEWETT-SHRAGO EXTENDED ERLANG B - 50% BLOCKED CALLERS RETRY, 50% OVERFLOW (RETRIES 3-5 HOLDING TIMES LATER)

| HOURS FIRST ATTEMPT VMR = 1.0 | HOURS CONNECT | HOURS RECALL /RETRY | HOURS OVFL | CONNECT VMR | OVFL VMR | PERCENT BLOCKED (GRADE) | LINE | HOURS CONNECTED TRAFFIC ON LINES 1 | 2 | 3 | 4 | 5 | 6 | 7 | 8 | 9 | 10 |
|---|---|---|---|---|---|---|---|---|---|---|---|---|---|---|---|---|---|
| | | | | | | | | .378 | .293 | .214 | .147 | .094 | .057 | | | | |
| 9.000 | 7.408 | 1.592 | 1.592 | .32 | 1.90 | 30.1% | 9 | .914 | .900 | .884 | .864 | .840 | .811 | .776 | .734 | .684 | |
| 9.000 | 7.890 | 1.110 | 1.110 | .41 | 2.03 | 22.0% | 10 | .910 | .895 | .878 | .856 | .830 | .798 | .759 | .713 | .658 | .594 |
| 9.000 | 8.259 | .741 | .741 | .51 | 2.12 | 15.2% | 11 | .907 | .891 | .872 | .849 | .821 | .786 | .744 | .694 | .635 | .567 |
| | | | | | | | | .492 | | | | | | | | | |
| 9.000 | 8.527 | .473 | .473 | .61 | 2.17 | 10.0% | 12 | .905 | .888 | .868 | .844 | .814 | .777 | .733 | .680 | .618 | .546 |
| | | | | | | | | .468 | .386 | | | | | | | | |
| 9.000 | 8.710 | .290 | .290 | .71 | 2.18 | 6.3% | 13 | .903 | .886 | .865 | .840 | .809 | .771 | .724 | .669 | .605 | .532 |
| | | | | | | | | .452 | .368 | .286 | | | | | | | |
| 9.000 | 8.829 | .171 | .171 | .80 | 2.14 | 3.7% | 14 | .902 | .884 | .863 | .837 | .805 | .766 | .719 | .662 | .596 | .522 |
| | | | | | | | | .440 | .356 | .275 | .200 | | | | | | |
| 9.000 | 8.904 | .096 | .096 | .87 | 2.09 | 2.1% | 15 | .901 | .883 | .862 | .836 | .803 | .763 | .715 | .658 | .591 | .515 |
| | | | | | | | | .433 | .349 | .267 | .194 | .132 | | | | | |
| 9.000 | 8.948 | .052 | .052 | .92 | 2.02 | 1.2% | 16 | .901 | .883 | .861 | .834 | .802 | .762 | .713 | .655 | .588 | .512 |
| | | | | | | | | .429 | .345 | .263 | .190 | .129 | .082 | | | | |
| 9.000 | 8.973 | .027 | .027 | .95 | 1.95 | .6% | 17 | .900 | .883 | .861 | .834 | .801 | .761 | .712 | .654 | .586 | .509 |
| | | | | | | | | .427 | .342 | .261 | .188 | .127 | .081 | .048 | | | |
| 9.500 | 7.576 | 1.924 | 1.924 | .28 | 1.86 | 33.7% | 9 | .920 | .908 | .894 | .877 | .856 | .831 | .802 | .766 | .724 | |
| 9.500 | 8.115 | 1.385 | 1.385 | .36 | 2.00 | 25.4% | 10 | .916 | .903 | .888 | .869 | .846 | .819 | .786 | .746 | .699 | .644 |
| 9.500 | 8.544 | .956 | .956 | .45 | 2.11 | 18.3% | 11 | .913 | .899 | .882 | .862 | .837 | .808 | .772 | .728 | .677 | .617 |
| | | | | | | | | .549 | | | | | | | | | |
| 9.500 | 8.868 | .632 | .632 | .55 | 2.19 | 12.5% | 12 | .910 | .896 | .878 | .856 | .830 | .798 | .760 | .714 | .659 | .596 |
| | | | | | | | | .524 | .447 | | | | | | | | |
| 9.500 | 9.098 | .402 | .402 | .66 | 2.22 | 8.1% | 13 | .908 | .893 | .875 | .852 | .825 | .791 | 751 | .702 | .645 | .579 |
| | | | | | | | | .505 | .426 | .045 | | | | | | | |
| 9.500 | 9.255 | .245 | .245 | .75 | 2.21 | 5.0% | 14 | .907 | .891 | .872 | .849 | .821 | .786 | .745 | .694 | .635 | .568 |
| | | | | | | | | .492 | .412 | .330 | .252 | | | | | | |
| 9.500 | 9.356 | .144 | .144 | .83 | 2.16 | 3.0% | 15 | .906 | .890 | .871 | .847 | .818 | .783 | .740 | .689 | .629 | .560 |
| | | | | | | | | .483 | .402 | .320 | .243 | .174 | | | | | |
| 9.500 | 9.419 | .081 | .081 | .89 | 2.10 | 1.7% | 16 | .906 | .889 | .870 | .846 | .817 | .781 | .738 | .686 | .625 | .555 |
| | | | | | | | | .478 | .396 | .314 | .237 | .169 | .114 | | | | |
| 9.500 | 9.456 | .044 | .044 | .93 | 2.03 | 9% | 17 | .905 | .880 | .869 | .845 | .816 | .780 | .736 | .684 | .622 | .552 |
| | | | | | | | | .475 | .393 | .311 | .233 | .166 | .111 | .070 | | | |
| 10.000 | 8.310 | 1.690 | 1.690 | .31 | 1.97 | 28.9% | 10 | .921 | .910 | .896 | .880 | .860 | .837 | .809 | .775 | .735 | .687 |
| 10.000 | 8.798 | 1.202 | 1.202 | .40 | 2.09 | 21.5% | 11 | .918 | .906 | .891 | .873 | .852 | .826 | .795 | .758 | .714 | .662 |
| | | | | | | | | .602 | | | | | | | | | |
| 10.000 | 9.179 | .821 | .821 | .50 | 2.19 | 15.2% | 12 | .915 | .902 | .887 | .868 | .845 | .817 | .784 | .744 | .696 | .640 |
| | | | | | | | | .576 | .505 | | | | | | | | |
| 10.000 | 9.461 | .539 | .539 | .60 | 2.24 | 10.2% | 13 | .913 | .900 | .883 | .863 | .839 | .810 | .774 | .732 | .681 | .623 |
| | | | | | | | | .556 | .482 | .404 | | | | | | | |

*SIMILAR TO A.T.¢T. JACOBSEN RETRIAL TABLES

*SLOW RETRIALS: JEWETT-SHRAGO EXTENDED ERLANG B - 50% BLOCKED CALLERS RETRY, 50% OVERFLOW (RETRIES 3-5 HOLDING TIMES LATER)

| HOURS FIRST ATTEMPT VMR = 1.0 | HOURS CON-NECT | HOURS RECALL /RETRY | HOURS OVFL | CON-NECT VMR | OVFL VMR | PERCENT BLOCKED (GRADE) | LINE | HOURS CONNECTED TRAFFIC ON LINES 1 | 2 | 3 | 4 | 5 | 6 | 7 | 8 | 9 | 10 |
|---|---|---|---|---|---|---|---|---|---|---|---|---|---|---|---|---|---|
| 10.000 | 9.659 | .341 | .341 | .69 | 2.25 | 6.6% | 14 | .912 | .898 | .881 | .860 | .835 | .804 | .767 | .723 | .671 | .610 |
| | | | | | | | | .541 | .465 | .386 | .307 | | | | | | |
| 10.000 | 9.792 | .208 | .208 | .78 | 2.23 | 4.1% | 15 | .911 | .896 | .879 | .858 | .832 | .801 | .763 | .717 | .663 | .601 |
| | | | | | | | | .530 | .453 | .373 | .294 | .221 | | | | | |
| 10.000 | 9.878 | .122 | .122 | .85 | 2.18 | 2.4% | 16 | .910 | .896 | .878 | .856 | .830 | .798 | .759 | .713 | .658 | .595 |
| | | | | | | | | .523 | .446 | .365 | .286 | .213 | .151 | | | | |
| 10.000 | 9.932 | .068 | .068 | .90 | 2.11 | 1.4% | 17 | .910 | .895 | .877 | .855 | .829 | .796 | .757 | .711 | .655 | .591 |
| | | | | | | | | .519 | .441 | .360 | .281 | .209 | .147 | .097 | | | |
| 10.000 | 9.963 | .037 | .037 | .94 | 2.04 | .7% | 18 | .909 | .895 | .877 | .855 | .828 | .795 | .756 | .709 | .653 | .589 |
| | | | | | | | | .517 | .438 | .357 | .279 | .206 | .145 | .096 | .060 | | |
| 11.000 | 9.217 | 1.783 | 1.783 | .31 | 2.03 | 27.9% | 11 | .927 | .918 | .906 | .893 | .876 | .857 | .834 | .807 | .774 | .735 |
| | | | | | | | | .690 | | | | | | | | | |
| 11.000 | 9.710 | 1.290 | 1.290 | .39 | 2.15 | 21.0% | 12 | .925 | .914 | .902 | .887 | .870 | .849 | .823 | .793 | .757 | .715 |
| | | | | | | | | .665 | .609 | | | | | | | | |
| 11.000 | 10.101 | .899 | .899 | .48 | 2.25 | 15.1% | 13 | .923 | .911 | .898 | .883 | .864 | .841 | .814 | .782 | .743 | .697 |
| | | | | | | | | .644 | .584 | .516 | | | | | | | |
| 11.000 | 10.395 | .605 | .605 | .58 | 2.31 | 10.4% | 14 | .921 | .909 | .895 | .879 | .859 | .835 | .807 | .772 | .731 | .683 |
| | | | | | | | | .627 | .564 | .494 | .419 | | | | | | |
| 11.000 | 10.608 | .392 | .392 | .68 | 2.33 | 6.9% | 15 | .919 | .907 | .893 | .876 | .855 | .831 | .801 | .765 | .722 | .672 |
| | | | | | | | | .614 | .549 | .477 | .401 | .325 | | | | | |
| 11.000 | 10.754 | .246 | .246 | .76 | 2.31 | 4.4% | 16 | .918 | .906 | .892 | .874 | .853 | .827 | .797 | .760 | .716 | .664 |
| | | | | | | | | .605 | .538 | .465 | .388 | .312 | .240 | | | | |
| 11.000 | 10.851 | .149 | .149 | .83 | 2.26 | 2.7% | 17 | .918 | .905 | .891 | .873 | .851 | .825 | .794 | .756 | .711 | .659 |
| | | | | | | | | .598 | .531 | .457 | .380 | .303 | .232 | .168 | | | |
| 11.000 | 10.913 | .087 | .087 | .89 | 2.20 | 1.6% | 18 | .917 | .905 | .890 | .872 | .850 | .824 | .792 | .754 | .709 | .655 |
| | | | | | | | | .594 | .526 | .452 | .374 | .298 | .227 | .164 | .112 | | |
| 11.000 | 10.951 | .049 | .049 | .93 | 2.12 | .9% | 19 | .917 | .904 | .889 | .871 | .849 | .823 | .791 | .752 | .707 | .653 |
| | | | | | | | | .592 | .523 | .448 | .371 | .295 | .223 | .161 | .110 | .071 | |
| 12.000 | 9.537 | 2.463 | 2.463 | .24 | 1.94 | 34.1% | 11 | .935 | .928 | .918 | .908 | .895 | .881 | .863 | .843 | .819 | .790 |
| | | | | | | | | .756 | | | | | | | | | |
| 12.000 | 10.128 | 1.872 | 1.872 | .31 | 2.09 | 27.0% | 12 | .933 | .924 | .915 | .903 | .889 | .873 | .854 | .831 | .805 | .773 |
| | | | | | | | | .736 | .692 | | | | | | | | |
| 12.000 | 10.625 | 1.375 | 1.375 | .39 | 2.21 | 20.6% | 13 | .930 | .922 | .911 | .899 | .884 | .866 | .845 | .821 | .791 | .757 |
| | | | | | | | | .716 | .669 | .615 | | | | | | | |
| 12.000 | 11.025 | .975 | .975 | .47 | 2.31 | 15.0% | 14 | .928 | .919 | .908 | .895 | .879 | .860 | .838 | .811 | .780 | .742 |
| | | | | | | | | .699 | .648 | .591 | .527 | | | | | | |
| 12.000 | 11.331 | .669 | .669 | .57 | 2.38 | 10.6% | 15 | .927 | .917 | .905 | .891 | .875 | .855 | .832 | .804 | .770 | .731 |
| | | | | | | | | .684 | .631 | .571 | .504 | .433 | | | | | |
| 12.000 | 11.556 | .444 | .444 | .66 | 2.40 | 7.1% | 16 | .926 | .915 | .903 | .889 | .872 | .851 | .827 | .798 | .763 | .721 |
| | | | | | | | | .673 | .618 | .556 | .487 | .415 | .341 | | | | |

*SIMILAR TO A.T.&T. JACOBSEN RETRIAL TABLES

*SLOW RETRIALS: JEWETT-SHRAGO EXTENDED ERLANG B - 50% BLOCKED CALLERS RETRY, 50% OVERFLOW (RETRIES 3-5 HOLDING TIMES LATER)

| HOURS FIRST ATTEMPT VMR=1.0 | HOURS CON-NECT | HOURS RECALL /RETRY | HOURS OVFL | CON-NECT VMR | OVFL VMR | PERCENT BLOCKED (GRADE) | LINE | HOURS CONNECTED TRAFFIC ON LINES 1 | 2 | 3 | 4 | 5 | 6 | 7 | 8 | 9 | 10 |
|---|---|---|---|---|---|---|---|---|---|---|---|---|---|---|---|---|---|
| 12.000 | 11.715 | .285 | .285 | .74 | 2.38 | 4.6% | 17 | .925 | .914 | .902 | .887 | .870 | .849 | .823 | .793 | .757 | .715 |
| | | | | | | | | .665 | .608 | .545 | .475 | .402 | .328 | .257 | | | |
| 12.000 | 11.823 | .177 | .177 | .81 | 2.34 | 2.9% | 18 | .924 | .914 | .901 | .886 | .868 | .847 | .821 | .790 | .753 | .710 |
| | | | | | | | | .660 | .602 | .537 | .467 | .393 | .319 | .248 | .185 | | |
| 12.000 | 11.894 | .106 | .106 | .87 | 2.28 | 1.8% | 19 | .924 | .913 | .900 | .885 | .867 | .845 | .819 | .788 | .751 | .707 |
| | | | | | | | | .656 | .597 | .532 | .461 | .387 | .313 | .243 | .180 | .127 | |
| 12.000 | 11.938 | .062 | .062 | .92 | 2.21 | 1.0% | 20 | .923 | .913 | .900 | .885 | .866 | .844 | .818 | .787 | .749 | .705 |
| | | | | | | | | .653 | .594 | .529 | .457 | .383 | .309 | .239 | .177 | .124 | .083 |
| 12.000 | 11.966 | .034 | .034 | .95 | 2.13 | .6% | 21 | .923 | .913 | .900 | .884 | .866 | .844 | .817 | .786 | .748 | .704 |
| | | | | | | | | .652 | .593 | .527 | .455 | .381 | .307 | .237 | .175 | .123 | .082 |
| | | | | | | | | .052 | | | | | | | | | |
| 13.000 | 10.451 | 2.549 | 2.549 | .24 | 2.00 | 32.8% | 12 | .940 | .933 | .925 | .916 | .905 | .892 | .878 | .861 | .840 | .817 |
| | | | | | | | | .789 | .756 | | | | | | | | |
| 13.000 | 11.043 | 1.957 | 1.957 | .31 | 2.14 | 26.2% | 13 | .937 | .930 | .922 | .912 | .900 | .886 | .870 | .851 | .829 | .803 |
| | | | | | | | | .772 | .736 | .694 | | | | | | | |
| 13.000 | 11.544 | 1.456 | 1.456 | .38 | 2.27 | 20.1% | 14 | .935 | .928 | .918 | .908 | .895 | .881 | .863 | .843 | .818 | .790 |
| | | | | | | | | .756 | .717 | .672 | .620 | | | | | | |
| 13.000 | 11.951 | 1.049 | 1.049 | .46 | 2.37 | 14.9% | 15 | .934 | .925 | .916 | .905 | .891 | .876 | .857 | .835 | .809 | .778 |
| | | | | | | | | .742 | .700 | .652 | .597 | .536 | | | | | |
| 13.000 | 12.268 | .732 | .732 | .55 | 2.44 | 10.7% | 16 | .932 | .924 | .914 | .902 | .888 | .871 | .852 | .829 | .801 | .768 |
| | | | | | | | | .730 | .686 | .635 | .577 | .514 | .446 | | | | |
| 13.000 | 12.504 | .496 | .496 | .64 | 2.46 | 7.3% | 17 | .931 | .922 | .912 | .900 | .885 | .868 | .848 | .823 | .795 | .761 |
| | | | | | | | | .721 | .675 | .622 | .562 | .497 | .427 | .356 | | | |
| 13.000 | 12.675 | .325 | .325 | .73 | 2.46 | 4.9% | 18 | .930 | .921 | .911 | .898 | .883 | .866 | .845 | .820 | .790 | .755 |
| | | | | | | | | .714 | .666 | .612 | .551 | .484 | .414 | .343 | .273 | | |
| 13.000 | 12.793 | .207 | .207 | .80 | 2.42 | 3.1% | 19 | .930 | .920 | .910 | .897 | .882 | .864 | .842 | .817 | .787 | .751 |
| | | | | | | | | .709 | .660 | .605 | .543 | .475 | .404 | .333 | .264 | .201 | |
| 13.000 | 12.873 | .127 | .127 | .86 | 2.36 | 1.9% | 20 | .929 | .920 | .909 | .896 | .881 | .863 | .841 | .815 | .784 | .748 |
| | | | | | | | | .705 | .656 | .600 | .537 | .469 | .398 | .326 | .258 | .195 | .141 |
| 13.000 | 12.924 | .076 | .076 | .91 | 2.29 | 1.2% | 21 | .929 | .920 | .909 | .896 | .880 | .862 | .840 | .814 | .783 | .746 |
| | | | | | | | | .703 | .653 | .597 | .534 | .465 | .394 | .322 | .254 | .191 | .138 |
| | | | | | | | | .095 | | | | | | | | | |
| 13.000 | 12.956 | .044 | .044 | .94 | 2.21 | .7% | 22 | .929 | .919 | .908 | .895 | .880 | .861 | .839 | .813 | .782 | .745 |
| | | | | | | | | .702 | .652 | .595 | .531 | .463 | .391 | .320 | .251 | .189 | .136 |
| | | | | | | | | .093 | .061 | | | | | | | | |
| 14.000 | 11.369 | 2.631 | 2.631 | .24 | 2.05 | 31.6% | 13 | .943 | .937 | .930 | .922 | .913 | .902 | .890 | .875 | .858 | .838 |
| | | | | | | | | .815 | .788 | .756 | | | | | | | |
| 14.000 | 11.960 | 2.040 | 2.040 | .30 | 2.20 | 25.4% | 14 | .941 | .935 | .927 | .919 | .909 | .897 | .883 | .868 | .849 | .827 |
| | | | | | | | | .801 | .771 | .736 | .696 | | | | | | |
| 14.000 | 12.465 | 1.535 | 1.535 | .38 | 2.32 | 19.8% | 15 | .940 | .933 | .925 | .916 | .905 | .892 | .878 | .860 | .840 | .816 |

*SIMILAR TO A.T.&T. JACOBSEN RETRIAL TABLES

## *SLOW RETRIALS: JEWETT-SHRAGO EXTENDED ERLANG B - 50% BLOCKED CALLERS RETRY, 50% OVERFLOW (RETRIES 3-5 HOLDING TIMES LATER)

| HOURS FIRST ATTEMPT VMR = 1.0 | HOURS CON-NECT | HOURS RECALL /RETRY | HOURS OVFL | CON-NECT VMR | OVFL VMR | PERCENT BLOCKED (GRADE) | LINE | HOURS CONNECTED TRAFFIC ON LINES 1 | 2 | 3 | 4 | 5 | 6 | 7 | 8 | 9 | 10 |
|---|---|---|---|---|---|---|---|---|---|---|---|---|---|---|---|---|---|
| | | | | | | | | .788 | .756 | .718 | .674 | .625 | | | | | |
| 14.000 | 12.879 | 1.121 | 1.121 | .46 | 2.43 | 14.8% | 16 | .938 | .931 | .922 | .913 | .901 | .888 | .872 | .854 | .832 | .807 |
| | | | | | | | | .777 | .742 | .701 | .655 | .602 | .544 | | | | |
| 14.000 | 13.205 | .795 | .795 | .54 | 2.50 | 10.7% | 17 | .937 | .929 | .921 | .910 | .898 | .885 | .868 | .849 | .826 | .799 |
| | | | | | | | | .767 | .730 | .687 | .638 | .583 | .522 | .457 | | | |
| 14.000 | 13.453 | .547 | .547 | .63 | 2.53 | 7.5% | 18 | .936 | .928 | .919 | .909 | .896 | .882 | .865 | .844 | .820 | .792 |
| | | | | | | | | .759 | .721 | .676 | .625 | .568 | .505 | .439 | .370 | | |
| 14.000 | 13.634 | .366 | .366 | .71 | 2.52 | 5.1% | 19 | .935 | .927 | .918 | .907 | .894 | .880 | .862 | .841 | .816 | .787 |
| | | | | | | | | .753 | .713 | .667 | .615 | .556 | .492 | .425 | .356 | .288 | |
| 14.000 | 13.762 | .238 | .238 | .78 | 2.49 | 3.3% | 20 | .934 | .926 | .917 | .906 | .893 | .878 | .860 | .839 | .813 | .784 |
| | | | | | | | | .749 | .708 | .661 | .608 | .548 | .483 | .415 | .346 | .278 | .216 |
| 14.000 | 13.850 | .150 | .150 | .85 | 2.44 | 2.1% | 21 | .934 | .926 | .916 | .905 | .892 | .877 | .859 | .837 | .811 | .781 |
| | | | | | | | | .746 | .704 | .657 | .602 | .542 | .477 | .408 | .339 | .272 | .209 |
| | | | | | | | | .155 | | | | | | | | | |
| 14.000 | 13.908 | .092 | .092 | .89 | 2.37 | 1.3% | 22 | .934 | .926 | .916 | .905 | .892 | .876 | .858 | .836 | .810 | .779 |
| | | | | | | | | .744 | .702 | .654 | .599 | .538 | .473 | .404 | .334 | .267 | .205 |
| | | | | | | | | .151 | .107 | | | | | | | | |
| 14.000 | 13.946 | .054 | .054 | .93 | 2.29 | .8% | 23 | .934 | .925 | .916 | .905 | .891 | .876 | .857 | .835 | .809 | .778 |
| | | | | | | | | .742 | .700 | .652 | .597 | .536 | .470 | .401 | .331 | .264 | .203 |
| | | | | | | | | .149 | .105 | .070 | | | | | | | |
| 15.000 | 12.289 | 2.711 | 2.711 | .24 | 2.11 | 30.6% | 14 | .947 | .941 | .935 | .928 | .920 | .911 | .900 | .887 | .873 | .856 |
| | | | | | | | | .836 | .813 | .787 | .756 | | | | | | |
| 15.000 | 12.881 | 2.119 | 2.119 | .30 | 2.25 | 24.8% | 15 | .945 | .939 | .933 | .925 | .916 | .906 | .894 | .881 | .865 | .847 |
| | | | | | | | | .825 | .800 | .771 | .737 | .698 | | | | | |
| 15.000 | 13.389 | 1.611 | 1.611 | .37 | 2.38 | 19.4% | 16 | .943 | .937 | .930 | .922 | .913 | .902 | .890 | .875 | .858 | .838 |
| | | | | | | | | .814 | .787 | .755 | .719 | .676 | .629 | | | | |
| 15.000 | 13.809 | 1.191 | 1.191 | .45 | 2.48 | 14.7% | 17 | .942 | .936 | .928 | .920 | .910 | .898 | .885 | .870 | .851 | .830 |
| | | | | | | | | .805 | .776 | .742 | .702 | .657 | .607 | .551 | | | |
| 15.000 | 14.144 | .856 | .856 | .53 | 2.55 | 10.8% | 18 | .941 | .934 | .926 | .918 | .907 | .895 | .881 | .865 | .846 | .823 |
| | | | | | | | | .797 | .766 | .730 | .688 | .641 | .588 | .530 | .467 | | |
| 15.000 | 14.401 | .599 | .599 | .62 | 2.59 | 7.7% | 19 | .940 | .933 | .925 | .916 | .905 | .893 | .878 | .861 | .841 | .818 |
| | | | | | | | | .790 | .758 | .720 | .677 | .628 | .573 | .513 | .449 | .382 | |
| 15.000 | 14.593 | .407 | .407 | .70 | 2.59 | 5.3% | 20 | .939 | .932 | .924 | .915 | .904 | .891 | .876 | .858 | .838 | .813 |
| | | | | | | | | .785 | .752 | .713 | .668 | .618 | .561 | .500 | .435 | .368 | .302 |
| 15.000 | 14.730 | .270 | .270 | .77 | 2.56 | 3.5% | 21 | .939 | .931 | .923 | .914 | .903 | .890 | .874 | .856 | .835 | .810 |
| | | | | | | | | .781 | .747 | .707 | .662 | .610 | .553 | .491 | .425 | .358 | .292 |
| | | | | | | | | .230 | | | | | | | | | |
| 15.000 | 14.827 | .173 | .173 | .83 | 2.51 | 2.3% | 22 | .938 | .931 | .923 | .913 | .902 | .889 | .873 | .855 | .833 | .808 |
| | | | | | | | | .778 | .744 | .703 | .657 | .605 | .547 | .484 | .418 | .350 | .284 |
| | | | | | | | | .223 | .168 | | | | | | | | |

*SIMILAR TO A.T.&T. JACOBSEN RETRIAL TABLES

*SLOW RETRIALS: JEWETT-SHRAGO EXTENDED ERLANG B - 50% BLOCKED CALLERS RETRY, 50% OVERFLOW (RETRIES 3-5 HOLDING TIMES LATER)

| HOURS FIRST ATTEMPT VMR=1.0 | HOURS CON-NECT | HOURS RECALL /RETRY | HOURS OVFL | CON-NECT VMR | OVFL VMR | PERCENT BLOCKED (GRADE) | LINE | HOURS CONNECTED TRAFFIC ON LINES 1 | 2 | 3 | 4 | 5 | 6 | 7 | 8 | 9 | 10 |
|---|---|---|---|---|---|---|---|---|---|---|---|---|---|---|---|---|---|
| 15.000 | 14.892 | .108 | .108 | .88 | 2.44 | 1.4% | 23 | .938 | .931 | .922 | .913 | .901 | .888 | .872 | .854 | .832 | .806 |
| | | | | | | | | .776 | .741 | .701 | .654 | .601 | .543 | .479 | .413 | .345 | .280 |
| | | | | | | | | .218 | .164 | .118 | | | | | | | |
| 15.000 | 14.934 | .066 | .066 | .92 | 2.37 | .9% | 24 | .938 | .931 | .922 | .912 | .901 | .887 | .872 | .853 | .831 | .805 |
| | | | | | | | | .775 | .740 | .699 | .652 | .599 | .540 | .476 | .410 | .342 | .276 |
| | | | | | | | | .215 | .161 | .116 | .080 | | | | | | |
| 15.000 | 14.961 | .039 | .039 | .95 | 2.28 | .5% | 25 | .938 | .930 | .922 | .912 | .901 | .887 | .871 | .853 | .831 | .805 |
| | | | | | | | | .774 | .739 | .698 | .651 | .597 | .538 | .474 | .408 | .340 | .274 |
| | | | | | | | | .213 | .160 | .115 | .079 | .052 | | | | | |
| 16.000 | 13.212 | 2.788 | 2.788 | .24 | 2.16 | 29.7% | 15 | .950 | .945 | .939 | .933 | .926 | .918 | .908 | .897 | .885 | .870 |
| | | | | | | | | .854 | .834 | .812 | .786 | .756 | | | | | |
| 16.000 | 13.804 | 2.196 | 2.196 | .30 | 2.30 | 24.1% | 16 | .948 | .943 | .937 | .930 | .923 | .914 | .904 | .892 | .878 | .863 |
| | | | | | | | | .844 | .823 | .798 | .770 | .737 | .699 | | | | |
| 16.000 | 14.314 | 1.686 | 1.686 | .37 | 2.43 | 19.1% | 17 | .947 | .941 | .935 | .928 | .920 | .910 | .900 | .887 | .872 | .855 |
| | | | | | | | | .836 | .813 | .786 | .755 | .719 | .679 | .633 | | | |
| 16.000 | 14.740 | 1.260 | 1.260 | .44 | 2.53 | 14.6% | 18 | .945 | .940 | .933 | .926 | .917 | .907 | .896 | .882 | .867 | .849 |
| | | | | | | | | .828 | .803 | .774 | .741 | .703 | .660 | .611 | .557 | | |
| 16.000 | 15.083 | .917 | .917 | .52 | 2.61 | 10.8% | 19 | .944 | .938 | .932 | .924 | .915 | .905 | .893 | .879 | .862 | .843 |
| | | | | | | | | .821 | .795 | .765 | .730 | .689 | .644 | .593 | .537 | .476 | |
| 16.000 | 15.350 | .650 | .650 | .61 | 2.65 | 7.8% | 20 | .943 | .937 | .930 | .922 | .913 | .902 | .890 | .875 | .858 | .838 |
| | | | | | | | | .815 | .788 | .756 | .720 | .678 | .631 | .578 | .520 | .458 | .394 |
| 16.000 | 15.551 | .449 | .449 | .69 | 2.65 | 5.5% | 21 | .943 | .937 | .929 | .921 | .912 | .901 | .888 | .873 | .855 | .835 |
| | | | | | | | | .811 | .783 | .750 | .712 | .669 | .620 | .566 | .507 | .444 | .379 |
| | | | | | | | | .315 | | | | | | | | | |
| 16.000 | 15.698 | .302 | .302 | .76 | 2.63 | 3.7% | 22 | .942 | .936 | .929 | .920 | .911 | .899 | .886 | .871 | .853 | .832 |
| | | | | | | | | .807 | .779 | .745 | .707 | .663 | .613 | .557 | .497 | .434 | .369 |
| | | | | | | | | .304 | .243 | | | | | | | | |
| 16.000 | 15.802 | .198 | .198 | .82 | 2.58 | 2.4% | 23 | .942 | .936 | .928 | .920 | .910 | .899 | .885 | .870 | .851 | .830 |
| | | | | | | | | .805 | .776 | .742 | .703 | .658 | .607 | .551 | .490 | .426 | .361 |
| | | | | | | | | .297 | .236 | .181 | | | | | | | |
| 16.000 | 15.874 | .126 | .126 | .87 | 2.51 | 1.6% | 24 | .942 | .935 | .928 | .919 | .909 | .898 | .884 | .869 | .850 | .829 |
| | | | | | | | | .803 | .774 | .739 | .700 | .654 | .603 | .547 | .486 | .421 | .356 |
| | | | | | | | | .291 | .231 | .176 | .129 | | | | | | |
| 16.000 | 15.922 | .078 | .078 | .91 | 2.44 | 1.0% | 25 | .941 | .935 | .928 | .919 | .909 | .897 | .884 | .868 | .849 | .828 |
| | | | | | | | | .802 | .772 | .738 | .698 | .652 | .601 | .544 | .482 | .418 | .352 |
| | | | | | | | | .288 | .227 | .173 | .127 | .089 | | | | | |
| 16.000 | 15.953 | .047 | .047 | .94 | 2.36 | .6% | 26 | .941 | .935 | .928 | .919 | .909 | .897 | .884 | .868 | .849 | .827 |
| | | | | | | | | .801 | .771 | .737 | .696 | .650 | .599 | .542 | .480 | .415 | .350 |
| | | | | | | | | .285 | .225 | .171 | .125 | .088 | .059 | | | | |
| 17.000 | 13.473 | 3.527 | 3.527 | .20 | 2.06 | 34.4% | 15 | .954 | .949 | .945 | .940 | .934 | .927 | .919 | .911 | .901 | .889 |

*SIMILAR TO A.T.&T. JACOBSEN RETRIAL TABLES

*SLOW RETRIALS: JEWETT-SHRAGO EXTENDED ERLANG B - 50% BLOCKED CALLERS RETRY, 50% OVERFLOW (RETRIES 3-5 HOLDING TIMES LATER)

| HOURS FIRST ATTEMPT VMR=1.0 | HOURS CON-NECT | HOURS RECALL /RETRY | HOURS OVFL | CON-NECT VMR | OVFL VMR | PERCENT BLOCKED (GRADE) | LINE | HOURS CONNECTED TRAFFIC ON LINES 1 | 2 | 3 | 4 | 5 | 6 | 7 | 8 | 9 | 10 |
|---|---|---|---|---|---|---|---|---|---|---|---|---|---|---|---|---|---|
| | | | | | | | | .876 | .861 | .843 | .823 | .800 | | | | | |
| 17.000 | 14.137 | 2.863 | 2.863 | .24 | 2.21 | 28.8% | 16 | .952 | .948 | .943 | .937 | .931 | .924 | .915 | .906 | .895 | .883 |
| | | | | | | | | .868 | .852 | .832 | .810 | .785 | .755 | | | | |
| 17.000 | 14.729 | 2.271 | 2.271 | .30 | 2.35 | 23.6% | 17 | .951 | .946 | .941 | .935 | .928 | .920 | .912 | .901 | .890 | .876 |
| | | | | | | | | .861 | .842 | .821 | .797 | .769 | .737 | .701 | | | |
| 17.000 | 15.242 | 1.758 | 1.758 | .36 | 2.48 | 18.7% | 18 | .949 | .945 | .939 | .933 | .926 | .917 | .908 | .897 | .885 | .870 |
| | | | | | | | | .853 | .834 | .811 | .785 | .755 | .720 | .681 | .636 | | |
| 17.000 | 15.674 | 1.326 | 1.326 | .44 | 2.58 | 14.5% | 19 | .948 | .943 | .937 | .931 | .923 | .915 | .905 | .893 | .880 | .864 |
| | | | | | | | | .846 | .826 | .801 | .773 | .741 | .704 | .662 | .615 | .563 | |
| 17.000 | 16.024 | .976 | .976 | .52 | 2.66 | 10.9% | 20 | .947 | .942 | .936 | .929 | .921 | .912 | .902 | .890 | .876 | .860 |
| | | | | | | | | .841 | .819 | .793 | .764 | .730 | .691 | .646 | .597 | .543 | .485 |
| 17.000 | 16.299 | .701 | .701 | .60 | 2.70 | 7.9% | 21 | .947 | .941 | .935 | .928 | .920 | .911 | .900 | .887 | .873 | .856 |
| | | | | | | | | .836 | .813 | .786 | .755 | .720 | .679 | .633 | .582 | .526 | .467 |
| | | | | | | | | .404 | | | | | | | | | |
| 17.000 | 16.509 | .491 | .491 | .67 | 2.71 | 5.6% | 22 | .946 | .940 | .934 | .927 | .919 | .909 | .898 | .885 | .870 | .852 |
| | | | | | | | | .832 | .808 | .781 | .749 | .712 | .670 | .623 | .571 | .513 | .453 |
| | | | | | | | | .390 | .327 | | | | | | | | |
| 17.000 | 16.664 | .336 | .336 | .74 | 2.69 | 3.9% | 23 | .946 | .940 | .933 | .926 | .918 | .908 | .896 | .883 | .868 | .850 |
| | | | | | | | | .829 | .805 | .777 | .744 | .706 | .663 | .615 | .562 | .504 | .442 |
| | | | | | | | | .379 | .316 | .255 | | | | | | | |
| 17.000 | 16.776 | .224 | .224 | .81 | 2.65 | 2.6% | 24 | .945 | .939 | .933 | .926 | .917 | .907 | .895 | .882 | .866 | .848 |
| | | | | | | | | .827 | .802 | .773 | .740 | .702 | .658 | .609 | .555 | .496 | .434 |
| | | | | | | | | .371 | .308 | .248 | .193 | | | | | | |
| 17.000 | 16.855 | .145 | .145 | .86 | 2.58 | 1.7% | 25 | .945 | .939 | .933 | .925 | .916 | .906 | .895 | .881 | .865 | .847 |
| | | | | | | | | .825 | .800 | .771 | .738 | .699 | .655 | .605 | .550 | .491 | .429 |
| | | | | | | | | .365 | .302 | .242 | .188 | .140 | | | | | |
| 17.000 | 16.908 | .092 | .092 | .90 | 2.51 | 1.1% | 26 | .945 | .939 | .932 | .925 | .916 | .906 | .894 | .881 | .865 | .846 |
| | | | | | | | | .824 | .799 | .770 | .736 | .697 | .652 | .602 | .547 | .488 | .425 |
| | | | | | | | | .361 | .298 | .239 | .185 | .138 | .099 | | | | |
| 17.000 | 16.944 | .056 | .056 | .93 | 2.43 | .7% | 27 | .945 | .939 | .932 | .925 | .916 | .906 | .894 | .880 | .864 | .845 |
| | | | | | | | | .824 | .798 | .769 | .734 | .695 | .651 | .600 | .545 | .486 | .423 |
| | | | | | | | | .359 | .296 | .236 | .182 | .136 | .097 | .067 | | | |
| 18.000 | 14.402 | 3.598 | 3.598 | .20 | 2.11 | 33.3% | 16 | .956 | .952 | .948 | .943 | .938 | .932 | .925 | .917 | .909 | .899 |
| | | | | | | | | .887 | .874 | .859 | .842 | .822 | .799 | | | | |
| 18.000 | 15.064 | 2.936 | 2.936 | .24 | 2.26 | 28.0% | 17 | .954 | .951 | .946 | .941 | .935 | .929 | .922 | .913 | .904 | .893 |
| | | | | | | | | .881 | .866 | .850 | .831 | .809 | .784 | .755 | | | |
| 18.000 | 15.657 | 2.343 | 2.343 | .30 | 2.40 | 23.0% | 18 | .953 | .949 | .944 | .939 | .933 | .926 | .918 | .909 | .899 | .888 |
| | | | | | | | | .874 | .858 | .841 | .820 | .796 | .769 | .737 | .702 | | |
| 18.000 | 16.172 | 1.828 | 1.828 | .36 | 2.52 | 18.4% | 19 | .952 | .948 | .943 | .937 | .931 | .923 | .915 | .906 | .895 | .882 |
| | | | | | | | | .868 | .851 | .832 | .809 | .784 | .754 | .721 | .682 | .639 | |

*SIMILAR TO A.T.&T. JACOBSEN RETRIAL TABLES

# *SLOW RETRIALS: JEWETT-SHRAGO EXTENDED ERLANG B - 50% BLOCKED CALLERS RETRY, 50% OVERFLOW (RETRIES 3-5 HOLDING TIMES LATER)

| HOURS FIRST ATTEMPT VMR = 1.0 | HOURS CON-NECT | HOURS RECALL /RETRY | HOURS OVFL | CON-NECT VMR | OVFL VMR | PERCENT BLOCKED (GRADE) | LINE | HOURS CONNECTED TRAFFIC ON LINES 1 | 2 | 3 | 4 | 5 | 6 | 7 | 8 | 9 | 10 |
|---|---|---|---|---|---|---|---|---|---|---|---|---|---|---|---|---|---|
| 18.000 | 16.608 | 1.392 | 1.392 | .43 | 2.63 | 14.4% | 20 | .951 | .946 | .941 | .935 | .929 | .921 | .912 | .902 | .891 | .878 |
| | | | | | | | | .862 | .844 | .824 | .800 | .773 | .741 | .705 | .664 | .619 | .569 |
| 18.000 | 16.965 | 1.035 | 1.035 | .51 | 2.71 | 10.9% | 21 | .950 | .945 | .940 | .934 | .927 | .919 | .910 | .899 | .887 | .873 |
| | | | | | | | | .857 | .838 | .817 | .792 | .763 | .729 | .692 | .649 | .601 | .549 |
| | | | | | | | | .492 | | | | | | | | | |
| 18.000 | 17.249 | .751 | .751 | .59 | 2.76 | 8.0% | 22 | .949 | .945 | .939 | .933 | .926 | .917 | .908 | .897 | .885 | .870 |
| | | | | | | | | .853 | .834 | .811 | .785 | .754 | .720 | .680 | .636 | .586 | .532 |
| | | | | | | | | .475 | .414 | | | | | | | | |
| 18.000 | 17.467 | .533 | .533 | .66 | 2.77 | 5.8% | 23 | .949 | .944 | .938 | .932 | .924 | .916 | .906 | .895 | .882 | .867 |
| | | | | | | | | .850 | .830 | .806 | .779 | .748 | .712 | .671 | .625 | .575 | .519 |
| | | | | | | | | .461 | .399 | .338 | | | | | | | |
| 18.000 | 17.631 | .369 | .369 | .73 | 2.75 | 4.0% | 24 | .948 | .943 | .938 | .931 | .924 | .915 | .905 | .894 | .880 | .865 |
| | | | | | | | | .847 | .826 | .802 | .775 | .743 | .706 | .664 | .617 | .566 | .509 |
| | | | | | | | | .450 | .388 | .327 | .267 | | | | | | |
| 18.000 | 17.750 | .250 | .250 | .80 | 2.71 | 2.7% | 25 | .948 | .943 | .937 | .931 | .923 | .914 | .904 | .893 | .879 | .863 |
| | | | | | | | | .845 | .824 | .800 | .771 | .739 | .701 | .659 | .611 | .559 | .502 |
| | | | | | | | | .442 | .380 | .318 | .259 | .204 | | | | | |
| 18.000 | 17.835 | .165 | .165 | .85 | 2.65 | 1.8% | 26 | .948 | .943 | .937 | .930 | .922 | .914 | .903 | .892 | .878 | .862 |
| | | | | | | | | .844 | .823 | .798 | .769 | .736 | .698 | .655 | .607 | .554 | .497 |
| | | | | | | | | .436 | .374 | .312 | .253 | .199 | .151 | | | | |
| 18.000 | 17.894 | .106 | .106 | .89 | 2.58 | 1.2% | 27 | .948 | .942 | .937 | .930 | .922 | .913 | .903 | .891 | .877 | .861 |
| | | | | | | | | .843 | .821 | .796 | .767 | .734 | .696 | .652 | .604 | .551 | .493 |
| | | | | | | | | .432 | .370 | .308 | .249 | .195 | .148 | .108 | | | |
| 18.000 | 17.934 | .066 | .066 | .93 | 2.50 | .7% | 28 | .948 | .942 | .936 | .930 | .922 | .913 | .903 | .891 | .877 | .861 |
| | | | | | | | | .842 | .821 | .795 | .766 | .733 | .694 | .651 | .602 | .548 | .490 |
| | | | | | | | | .430 | .367 | .306 | .247 | .193 | .146 | .100 | .075 | | |
| 19.000 | 15.333 | 3.667 | 3.667 | .20 | 2.16 | 32.4% | 17 | .958 | .954 | .951 | .946 | .942 | .936 | .930 | .923 | .916 | .907 |
| | | | | | | | | .897 | .885 | .872 | .857 | .840 | .821 | .798 | | | |
| 19.000 | 15.993 | 3.007 | 3.007 | .25 | 2.30 | 27.3% | 18 | .957 | .953 | .949 | .944 | .939 | .934 | .927 | .920 | .911 | .902 |
| | | | | | | | | .891 | .879 | .864 | .848 | .829 | .808 | .783 | .755 | | |
| 19.000 | 16.586 | 2.414 | 2.414 | .30 | 2.44 | 22.5% | 19 | .955 | .952 | .947 | .943 | .937 | .931 | .924 | .916 | .907 | .897 |
| | | | | | | | | .885 | .872 | .857 | .839 | .818 | .795 | .768 | .738 | .703 | |
| 19.000 | 17.104 | 1.896 | 1.896 | .36 | 2.57 | 18.1% | 20 | .954 | .950 | .948 | .941 | .935 | .929 | .921 | .913 | .904 | .893 |
| | | | | | | | | .880 | .866 | .849 | .830 | .808 | .783 | .754 | .721 | .684 | .642 |
| 19.000 | 17.544 | 1.456 | 1.456 | .43 | 2.68 | 14.2% | 21 | .953 | .949 | .945 | .939 | .933 | .927 | .919 | .910 | .900 | .889 |
| | | | | | | | | .875 | .860 | .842 | .822 | .799 | .772 | .741 | .706 | .666 | .622 |
| | | | | | | | | .574 | | | | | | | | | |
| 19.000 | 17.908 | 1.092 | 1.092 | .50 | 2.76 | 10.9% | 22 | .953 | .948 | .944 | .938 | .932 | .925 | .917 | .908 | .897 | .885 |
| | | | | | | | | .871 | .855 | .836 | .815 | .790 | .762 | .729 | .693 | .651 | .605 |
| | | | | | | | | .551 | .500 | | | | | | | | |

*SIMILAR TO A.T.&T. JACOBSEN RETRIAL TABLES

*SLOW RETRIALS: JEWETT-SHRAGO EXTENDED ERLANG B - 50% BLOCKED CALLERS RETRY, 50% OVERFLOW (RETRIES 3-5 HOLDING TIMES LATER)

| HOURS FIRST ATTEMPT VMR = 1.0 | HOURS CON-NECT | HOURS RECALL /RETRY | HOURS OVFL | CON-NECT VMR | OVFL VMR | PERCENT BLOCKED (GRADE) | LINE | HOURS CONNECTED TRAFFIC ON LINES 1 | 2 | 3 | 4 | 5 | 6 | 7 | 8 | 9 | 10 |
|---|---|---|---|---|---|---|---|---|---|---|---|---|---|---|---|---|---|
| 19.000 | 18.198 | .802 | .802 | .58 | 2.81 | 8.1% | 23 | .952 | .948 | .943 | .937 | .931 | .923 | .915 | .906 | .895 | .882 |
| | | | | | | | | .867 | .851 | .831 | .809 | .783 | .754 | .720 | .681 | .638 | .590 |
| | | | | | | | | .538 | .482 | .423 | | | | | | | |
| 19.000 | 18.425 | .575 | .575 | .65 | 2.83 | 5.9% | 24 | .951 | .947 | .942 | .936 | .930 | .922 | .914 | .904 | .893 | .880 |
| | | | | | | | | .865 | .847 | .827 | .804 | .777 | .747 | .712 | .672 | .628 | .578 |
| | | | | | | | | .525 | .468 | .408 | .348 | | | | | | |
| 19.000 | 18.596 | .404 | .404 | .72 | 2.81 | 4.2% | 25 | .951 | .946 | .941 | .935 | .929 | .921 | .912 | .902 | .891 | .878 |
| | | | | | | | | .862 | .845 | .824 | .800 | .773 | .741 | .706 | .665 | .619 | .569 |
| | | | | | | | | .515 | .457 | .397 | .337 | .278 | | | | | |
| 19.000 | 18.723 | .277 | .277 | .78 | 2.78 | 2.9% | 26 | .951 | .946 | .941 | .935 | .928 | .920 | .912 | .901 | .890 | .876 |
| | | | | | | | | .861 | .842 | .822 | .797 | .769 | .737 | .701 | .660 | .613 | .562 |
| | | | | | | | | .507 | .449 | .389 | .328 | .270 | .215 | | | | |
| 19.000 | 18.815 | .185 | .185 | .84 | 2.72 | 1.9% | 27 | .951 | .946 | .941 | .935 | .928 | .920 | .911 | .901 | .889 | .875 |
| | | | | | | | | .859 | .841 | .820 | .795 | .767 | .734 | .697 | .656 | .609 | .557 |
| | | | | | | | | .502 | .443 | .382 | .322 | .264 | .210 | .161 | | | |
| 19.000 | 18.879 | .121 | .121 | .88 | 2.65 | 1.3% | 28 | .950 | .946 | .940 | .934 | .927 | .920 | .911 | .900 | .888 | .874 |
| | | | | | | | | .858 | .840 | .818 | .794 | .765 | .732 | .695 | .653 | .606 | .554 |
| | | | | | | | | .498 | .439 | .378 | .318 | .260 | .206 | .158 | .117 | | |
| 19.000 | 18.923 | .077 | .077 | .92 | 2.57 | .8% | 29 | .950 | .945 | .940 | .934 | .927 | .919 | .910 | .900 | .888 | .874 |
| | | | | | | | | .858 | .839 | .818 | .793 | .764 | .731 | .693 | .651 | .603 | .551 |
| | | | | | | | | .495 | .436 | .375 | .315 | .257 | .203 | .156 | .115 | .082 | |
| 19.000 | 18.952 | .048 | .048 | .94 | 2.48 | .5% | 30 | .950 | .945 | .940 | .934 | .927 | .919 | .910 | .900 | .888 | .874 |
| | | | | | | | | .857 | .839 | .817 | .792 | .763 | .730 | .692 | .649 | .602 | .550 |
| | | | | | | | | .493 | .434 | .373 | .313 | .255 | .201 | .154 | .114 | .081 | .056 |
| 20.000 | 16.265 | 3.735 | 3.735 | .20 | 2.21 | 31.5% | 18 | .960 | .956 | .953 | .949 | .945 | .940 | .935 | .928 | .922 | .914 |
| | | | | | | | | .905 | .895 | .883 | .870 | .856 | .839 | .819 | .797 | | |
| 20.000 | 16.924 | 3.076 | 3.076 | .25 | 2.35 | 26.7% | 19 | .959 | .955 | .952 | .947 | .943 | .938 | .932 | .925 | .918 | .910 |
| | | | | | | | | .900 | .889 | .877 | .863 | .846 | .828 | .807 | .782 | .755 | |
| 20.000 | 17.518 | 2.482 | 2.482 | .30 | 2.49 | 22.1% | 20 | .957 | .954 | .950 | .946 | .941 | .935 | .929 | .922 | .914 | .905 |
| | | | | | | | | .895 | .883 | .870 | .855 | .837 | .817 | .794 | .768 | .738 | .704 |
| 20.000 | 18.037 | 1.963 | 1.963 | .36 | 2.62 | 17.9% | 21 | .956 | .953 | .949 | .944 | .939 | .933 | .927 | .920 | .911 | .902 |
| | | | | | | | | .891 | .878 | .864 | .847 | .828 | .807 | .782 | .754 | .722 | .686 |
| | | | | | | | | .645 | | | | | | | | | |
| 20.000 | 18.482 | 1.518 | 1.518 | .42 | 2.72 | 14.1% | 22 | .956 | .952 | .948 | .943 | .938 | .932 | .925 | .917 | .908 | .898 |
| | | | | | | | | .886 | .873 | .858 | .840 | .820 | .797 | .771 | .741 | .707 | .668 |
| | | | | | | | | .625 | .578 | | | | | | | | |
| 20.000 | 18.851 | 1.149 | 1.149 | .49 | 2.81 | 10.9% | 23 | .955 | .951 | .947 | .942 | .936 | .930 | .923 | .915 | .905 | .895 |
| | | | | | | | | .883 | .869 | .853 | .834 | .813 | .789 | .761 | .729 | .694 | .653 |
| | | | | | | | | .608 | .559 | .506 | | | | | | | |
| 20.000 | 19.149 | .851 | .851 | .57 | 2.86 | 8.2% | 24 | .954 | .950 | .946 | .941 | .935 | .929 | .921 | .913 | .903 | .892 |

*SIMILAR TO A.T.&T. JACOBSEN RETRIAL TABLES

*SLOW RETRIALS: JEWETT-SHRAGO EXTENDED ERLANG B - 50% BLOCKED CALLERS RETRY, 50% OVERFLOW (RETRIES 3-5 HOLDING TIMES LATER)

| HOURS FIRST ATTEMPT VMR=1.0 | HOURS CON-NECT | HOURS RECALL /RETRY | HOURS OVFL | CON-NECT VMR | OVFL VMR | PERCENT BLOCKED (GRADE) | LINE | HOURS CONNECTED TRAFFIC ON LINES 1 | 2 | 3 | 4 | 5 | 6 | 7 | 8 | 9 | 10 |
|---|---|---|---|---|---|---|---|---|---|---|---|---|---|---|---|---|---|
| | | | | | | | | .880 | .865 | .848 | .829 | .807 | .782 | .753 | .720 | .682 | .640 |
| | | | | | | | | .594 | .543 | .489 | .432 | | | | | | |
| 20.000 | 19.382 | .618 | .618 | .64 | 2.88 | 6.0% | 25 | .954 | .950 | .945 | .940 | .934 | .927 | .920 | .911 | .901 | .890 |
| | | | | | | | | .877 | .862 | .845 | .825 | .802 | .776 | .746 | .712 | .673 | .630 |
| | | | | | | | | .582 | .530 | .475 | .417 | .358 | | | | | |
| 20.000 | 19.562 | .438 | .438 | .71 | 2.87 | 4.3% | 26 | .953 | .949 | .945 | .939 | .933 | .927 | .919 | .910 | .900 | .888 |
| | | | | | | | | .875 | .860 | .842 | .822 | .798 | .771 | .740 | .705 | .666 | .621 |
| | | | | | | | | .573 | .520 | .464 | .405 | .346 | .289 | | | | |
| 20.000 | 19.696 | .304 | .304 | .77 | 2.84 | 3.0% | 27 | .953 | .949 | .944 | .939 | .933 | .926 | .918 | .909 | .899 | .887 |
| | | | | | | | | .874 | .858 | .840 | .819 | .795 | .768 | .736 | .701 | .660 | .615 |
| | | | | | | | | .566 | .512 | .455 | .397 | .338 | .280 | .226 | | | |
| 20.000 | 19.793 | .207 | .207 | .83 | 2.78 | 2.0% | 28 | .953 | .949 | .944 | .939 | .932 | .925 | .918 | .908 | .898 | .886 |
| | | | | | | | | .872 | .857 | .838 | .817 | .793 | .765 | .733 | .697 | .656 | .610 |
| | | | | | | | | .560 | .506 | .449 | .390 | .331 | .274 | .220 | .171 | | |
| 20.000 | 19.863 | .137 | .137 | .87 | 2.71 | 1.4% | 29 | .953 | .948 | .944 | .938 | .932 | .925 | .917 | .908 | .897 | .885 |
| | | | | | | | | .872 | .856 | .837 | .816 | .791 | .763 | .731 | .694 | .653 | .607 |
| | | | | | | | | .557 | .502 | .445 | .386 | .327 | .269 | .216 | .168 | .126 | |
| 20.000 | 19.911 | .089 | .089 | .91 | 2.63 | .9% | 30 | .953 | .948 | .943 | .938 | .932 | .925 | .917 | .908 | .897 | .885 |
| | | | | | | | | .871 | .855 | .836 | .815 | .790 | .762 | .729 | .692 | .651 | .605 |
| | | | | | | | | .554 | .499 | .442 | .383 | .323 | .266 | .213 | .165 | .124 | .090 |
| 20.000 | 19.944 | .056 | .056 | .94 | 2.55 | .6% | 31 | .953 | .948 | .943 | .938 | .932 | .925 | .917 | .907 | .897 | .885 |
| | | | | | | | | .871 | .854 | .836 | .814 | .789 | .761 | .728 | .691 | .649 | .603 |
| | | | | | | | | .552 | .497 | .440 | .380 | .321 | .264 | .211 | .163 | .122 | .089 |
| | | | | | | | | .062 | | | | | | | | | |
| 22.000 | 17.424 | 4.576 | 4.576 | .17 | 2.16 | 34.4% | 19 | .964 | .961 | .958 | .955 | .952 | .948 | .944 | .939 | .934 | .928 |
| | | | | | | | | .922 | .915 | .900 | .897 | .887 | .875 | .861 | .846 | .829 | |
| 22.000 | 18.135 | 3.865 | 3.865 | .20 | 2.30 | 29.9% | 20 | .963 | .960 | .957 | .954 | .950 | .946 | .942 | .937 | .931 | .925 |
| | | | | | | | | .918 | .910 | .901 | .891 | .880 | .867 | .852 | .836 | .817 | .795 |
| 22.000 | 18.792 | 3.208 | 3.208 | .25 | 2.44 | 25.5% | 21 | .962 | .959 | .956 | .953 | .949 | .945 | .940 | .935 | .929 | .922 |
| | | | | | | | | .914 | .906 | .896 | .886 | .873 | .859 | .843 | .825 | .805 | .781 |
| | | | | | | | | .755 | | | | | | | | | |
| 22.000 | 19.385 | 2.615 | 2.615 | .29 | 2.58 | 21.2% | 22 | .961 | .958 | .955 | .951 | .947 | .943 | .938 | .932 | .926 | .919 |
| | | | | | | | | .911 | .902 | .891 | .880 | .867 | .851 | .834 | .815 | .792 | .767 |
| | | | | | | | | .738 | .706 | | | | | | | | |
| 22.000 | 19.908 | 2.092 | 2.092 | .35 | 2.70 | 17.4% | 23 | .900 | .957 | .954 | .950 | .946 | .941 | .936 | .930 | .923 | .916 |
| | | | | | | | | .907 | .898 | .887 | .874 | .860 | .844 | .826 | .804 | .781 | .753 |
| | | | | | | | | .723 | .688 | .650 | | | | | | | |
| 22.000 | 20.360 | 1.640 | 1.640 | .41 | 2.81 | 13.9% | 24 | .959 | .956 | .953 | .949 | .945 | .940 | .934 | .928 | .921 | .913 |
| | | | | | | | | .904 | .894 | .882 | .869 | .854 | .837 | .817 | .795 | .770 | .741 |
| | | | | | | | | .708 | .672 | .631 | .587 | | | | | | |

*SIMILAR TO A.T.&T. JACOBSEN RETRIAL TABLES

*SLOW RETRIALS: JEWETT-SHRAGO EXTENDED ERLANG B - 50% BLOCKED CALLERS RETRY, 50% OVERFLOW (RETRIES 3-5 HOLDING TIMES LATER)

| HOURS FIRST ATTEMPT VMR = 1.0 | HOURS CON-NECT | HOURS RECALL /RETRY | HOURS OVFL | CON-NECT VMR | OVFL VMR | PERCENT BLOCKED (GRADE) | LINE | HOURS CONNECTED TRAFFIC ON LINES 1 | 2 | 3 | 4 | 5 | 6 | 7 | 8 | 9 | 10 |
|---|---|---|---|---|---|---|---|---|---|---|---|---|---|---|---|---|---|
| 22.000 | 20.740 | 1.260 | 1.260 | .48 | 2.90 | 10.8% | 25 | .959 | .956 | .952 | .948 | .943 | .938 | .933 | .926 | .919 | .911 |
| | | | | | | | | .901 | .891 | .879 | .865 | .849 | .831 | .810 | .787 | .760 | .729 |
| | | | | | | | | .695 | .657 | .615 | .568 | .518 | | | | | |
| 22.000 | 21.050 | .950 | .950 | .55 | 2.96 | 8.3% | 26 | .958 | .955 | .951 | .947 | .942 | .937 | .931 | .925 | .917 | .909 |
| | | | | | | | | .899 | .888 | .875 | .861 | .844 | .826 | .804 | .779 | .751 | .720 |
| | | | | | | | | .684 | .644 | .600 | .552 | .501 | .447 | | | | |
| 22.000 | 21.298 | .702 | .702 | .62 | 2.98 | 6.2% | 27 | .958 | .954 | .951 | .946 | .942 | .936 | .930 | .923 | .916 | .907 |
| | | | | | | | | .897 | .886 | .873 | .858 | .841 | .821 | .799 | .773 | .744 | .712 |
| | | | | | | | | .675 | .634 | .589 | .539 | .487 | .432 | .376 | | | |
| 22.000 | 21.492 | .508 | .508 | .69 | 2.98 | 4.5% | 28 | .958 | .954 | .950 | .946 | .941 | .936 | .929 | .922 | .915 | .906 |
| | | | | | | | | .895 | .884 | .870 | .855 | .838 | .818 | .795 | .768 | .739 | .705 |
| | | | | | | | | .667 | .625 | .579 | .529 | .476 | .420 | .364 | .308 | | |
| 22.000 | 21.639 | .361 | .361 | .75 | 2.95 | 3.2% | 29 | .957 | .954 | .950 | .945 | .941 | .935 | .929 | .922 | .914 | .905 |
| | | | | | | | | .894 | .882 | .869 | .853 | .835 | .815 | .791 | .765 | .734 | .700 |
| | | | | | | | | .661 | .619 | .572 | .521 | .467 | .411 | .355 | .299 | .246 | |
| 22.000 | 21.749 | .251 | .251 | .81 | 2.90 | 2.3% | 30 | .957 | .953 | .950 | .945 | .940 | .935 | .928 | .921 | .913 | .904 |
| | | | | | | | | .893 | .881 | .867 | .851 | .833 | .813 | .789 | .762 | .731 | .696 |
| | | | | | | | | .657 | .614 | .566 | .515 | .461 | .405 | .348 | .292 | .239 | .191 |
| 22.000 | 21.829 | .171 | .171 | .86 | 2.84 | 1.5% | 31 | .957 | .953 | .949 | .945 | .940 | .934 | .928 | .921 | .912 | .903 |
| | | | | | | | | .892 | .880 | .866 | .850 | .832 | .811 | .787 | .760 | .728 | .693 |
| | | | | | | | | .654 | .610 | .562 | .510 | .456 | .399 | .343 | .287 | .234 | .186 |
| | | | | | | | | .144 | | | | | | | | | |
| 22.000 | 21.886 | .114 | .114 | .89 | 2.76 | 1.0% | 32 | .957 | .953 | .949 | .945 | .940 | .934 | .928 | .920 | .912 | .903 |
| | | | | | | | | .892 | .880 | .866 | .850 | .831 | .810 | .786 | .758 | .727 | .691 |
| | | | | | | | | .651 | .607 | .559 | .507 | .452 | .396 | .339 | .284 | .231 | .183 |
| | | | | | | | | .141 | .105 | | | | | | | | |
| 22.000 | 21.926 | .074 | .074 | .93 | 2.68 | .7% | 33 | .957 | .953 | .949 | .945 | .940 | .934 | .927 | .920 | .912 | .902 |
| | | | | | | | | .892 | .879 | .865 | .849 | .830 | .809 | .785 | .757 | .725 | .690 |
| | | | | | | | | .650 | .605 | .557 | .505 | .450 | .393 | .337 | .281 | .229 | .181 |
| | | | | | | | | .139 | .104 | .075 | | | | | | | |
| 24.000 | 19.303 | 4.697 | 4.697 | .17 | 2.24 | 32.7% | 21 | .966 | .964 | .962 | .959 | .956 | .953 | .950 | .946 | .941 | .937 |
| | | | | | | | | .931 | .926 | .919 | .912 | .903 | .894 | .884 | .872 | .858 | .843 |
| | | | | | | | | .826 | | | | | | | | | |
| 24.000 | 20.011 | 3.989 | 3.989 | .21 | 2.38 | 28.5% | 22 | .966 | .963 | .961 | .958 | .955 | .952 | .948 | .944 | .939 | .934 |
| | | | | | | | | .929 | .922 | .915 | .907 | .898 | .888 | .877 | .864 | .850 | .833 |
| | | | | | | | | .815 | .794 | | | | | | | | |
| 24.000 | 20.665 | 3.335 | 3.335 | .25 | 2.52 | 24.4% | 23 | .965 | .962 | .960 | .957 | .954 | .950 | .946 | .942 | .937 | .932 |
| | | | | | | | | .926 | .919 | .911 | .903 | .893 | .882 | .870 | .856 | .840 | .823 |
| | | | | | | | | .803 | .780 | .754 | | | | | | | |
| 24.000 | 21.258 | 2.742 | 2.742 | .29 | 2.66 | 20.5% | 24 | .964 | .962 | .959 | .956 | .952 | .949 | .945 | .940 | .935 | .929 |

*SIMILAR TO A.T.&T. JACOBSEN RETRIAL TABLES

## *SLOW RETRIALS: JEWETT-SHRAGO EXTENDED ERLANG B - 50% BLOCKED CALLERS RETRY, 50% OVERFLOW (RETRIES 3-5 HOLDING TIMES LATER)

| HOURS FIRST ATTEMPT VMR = 1.0 | HOURS CON-NECT | HOURS RECALL /RETRY | HOURS OVFL | CON-NECT VMR | OVFL VMR | PERCENT BLOCKED (GRADE) | LINE | HOURS CONNECTED TRAFFIC ON LINES 1 | 2 | 3 | 4 | 5 | 6 | 7 | 8 | 9 | 10 |
|---|---|---|---|---|---|---|---|---|---|---|---|---|---|---|---|---|---|
| | | | | | | | | .923 | .916 | .908 | .898 | .888 | .877 | .863 | .848 | .832 | .812 |
| | | | | | | | | .791 | .766 | .739 | .708 | | | | | | |
| 24.000 | 21.785 | 2.215 | 2.215 | .35 | 2.79 | 16.9% | 25 | .963 | .961 | .958 | .955 | .951 | .947 | .943 | .938 | .933 | .927 |
| | | | | | | | | .920 | .912 | .904 | .894 | .883 | .871 | .857 | .841 | .823 | .802 |
| | | | | | | | | .779 | .753 | .724 | .691 | .654 | | | | | |
| 24.000 | 22.243 | 1.757 | 1.757 | .41 | 2.90 | 13.6% | 26 | .963 | .960 | .957 | .954 | .950 | .946 | .942 | .937 | .931 | .925 |
| | | | | | | | | .918 | .910 | .901 | .890 | .879 | .866 | .851 | .834 | .815 | .793 |
| | | | | | | | | .769 | .741 | .710 | .675 | .636 | .594 | | | | |
| 24.000 | 22.632 | 1.368 | 1.368 | .47 | 2.99 | 10.8% | 27 | .962 | .959 | .956 | .953 | .949 | .945 | .940 | .935 | .929 | .923 |
| | | | | | | | | .915 | .907 | .898 | .887 | .875 | .861 | .846 | .828 | .808 | .785 |
| | | | | | | | | .759 | .730 | .697 | .660 | .620 | .576 | .528 | | | |
| 24.000 | 22.953 | 1.047 | 1.047 | .54 | 3.05 | 8.4% | 28 | .962 | .959 | .956 | .952 | .948 | .944 | .939 | .934 | .928 | .921 |
| | | | | | | | | .913 | .905 | .895 | .884 | .871 | .857 | .841 | .822 | .801 | .777 |
| | | | | | | | | .750 | .720 | .686 | .648 | .606 | .561 | .512 | .460 | | |
| 24.000 | 23.214 | .786 | .786 | .61 | 3.08 | 6.3% | 29 | .961 | .958 | .955 | .952 | .948 | .943 | .938 | .933 | .927 | .920 |
| | | | | | | | | .912 | .903 | .893 | .882 | .869 | .854 | .837 | .818 | .796 | .771 |
| | | | | | | | | .743 | .712 | .676 | .637 | .594 | .548 | .498 | .446 | .392 | |
| 24.000 | 23.421 | .579 | .579 | .67 | 3.08 | 4.7% | 30 | .961 | .958 | .955 | .951 | .947 | .943 | .938 | .932 | .926 | .919 |
| | | | | | | | | .911 | .901 | .891 | .879 | .866 | .851 | .834 | .814 | .792 | .766 |
| | | | | | | | | .737 | .705 | .669 | .629 | .585 | .537 | .487 | .434 | .380 | .326 |
| 24.000 | 23.581 | .419 | .419 | .74 | 3.06 | 3.4% | 31 | .961 | .958 | .954 | .951 | .947 | .942 | .937 | .931 | .925 | .918 |
| | | | | | | | | .910 | .900 | .890 | .878 | .864 | .849 | .831 | .811 | .788 | .762 |
| | | | | | | | | .733 | .700 | .663 | .622 | .577 | .529 | .478 | .424 | .370 | .316 |
| | | | | | | | | .264 | | | | | | | | | |
| 24.000 | 23.702 | .298 | .298 | .79 | 3.02 | 2.5% | 32 | .960 | .957 | .954 | .950 | .946 | .942 | .937 | .931 | .924 | .917 |
| | | | | | | | | .909 | .899 | .889 | .877 | .863 | .847 | .829 | .809 | .785 | .759 |
| | | | | | | | | .729 | .695 | .658 | .617 | .571 | .523 | .471 | .417 | .363 | .309 |
| | | | | | | | | .257 | .208 | | | | | | | | |
| 24.000 | 23.793 | .207 | .207 | .84 | 2.96 | 1.7% | 33 | .960 | .957 | .954 | .950 | .946 | .942 | .936 | .930 | .924 | .917 |
| | | | | | | | | .908 | .899 | .888 | .876 | .862 | .846 | .827 | .807 | .783 | .756 |
| | | | | | | | | .726 | .692 | .654 | .613 | .567 | .518 | .466 | .412 | .357 | .303 |
| | | | | | | | | .252 | .204 | .160 | | | | | | | |
| 24.000 | 23.859 | .141 | .141 | .88 | 2.88 | 1.2% | 34 | .960 | .957 | .954 | .950 | .946 | .941 | .936 | .930 | .924 | .916 |
| | | | | | | | | .908 | .898 | .887 | .875 | .861 | .845 | .826 | .805 | .782 | .755 |
| | | | | | | | | .724 | .690 | .652 | .610 | .564 | .514 | .462 | .408 | .353 | .299 |
| | | | | | | | | .248 | .200 | .157 | .120 | | | | | | |
| 24.000 | 23.905 | .095 | .095 | .91 | 2.80 | .8% | 35 | .960 | .957 | .954 | .950 | .946 | .941 | .936 | .930 | .923 | .916 |
| | | | | | | | | .907 | .898 | .887 | .874 | .860 | .844 | .826 | .805 | .781 | .753 |
| | | | | | | | | .723 | .688 | .650 | .608 | .561 | .512 | .459 | .405 | .350 | .297 |
| | | | | | | | | .245 | .198 | .155 | .118 | .088 | | | | | |

*SIMILAR TO A.T.&T. JACOBSEN RETRIAL TABLES

# *SLOW RETRIALS: JEWETT-SHRAGO EXTENDED ERLANG B - 50% BLOCKED CALLERS RETRY, 50% OVERFLOW (RETRIES 3-5 HOLDING TIMES LATER)

| HOURS FIRST ATTEMPT VMR = 1.0 | HOURS CON-NECT | HOURS RECALL /RETRY | HOURS OVFL | CON-NECT VMR | OVFL VMR | PERCENT BLOCKED (GRADE) | LINE | HOURS CONNECTED TRAFFIC ON LINES 1 | 2 | 3 | 4 | 5 | 6 | 7 | 8 | 9 | 10 |
|---|---|---|---|---|---|---|---|---|---|---|---|---|---|---|---|---|---|
| 24.000 | 23.938 | .062 | .062 | .94 | 2.72 | .5% | 36 | .960 | .957 | .954 | .950 | .946 | .941 | .936 | .930 | .923 | .916 |
| | | | | | | | | .907 | .898 | .887 | .874 | .860 | .844 | .825 | .804 | .780 | .753 |
| | | | | | | | | .722 | .687 | .649 | .606 | .560 | .510 | .457 | .403 | .348 | .295 |
| | | | | | | | | .243 | .196 | .154 | .117 | .087 | .062 | | | | |
| 26.000 | 21.187 | 4.813 | 4.813 | .18 | 2.32 | 31.2% | 23 | .969 | .967 | .965 | .962 | .960 | .957 | .954 | .951 | .947 | .943 |
| | | | | | | | | .939 | .934 | .929 | .923 | .916 | .909 | .900 | .891 | .881 | .869 |
| | | | | | | | | .856 | .841 | .824 | | | | | | | |
| 26.000 | 21.892 | 4.108 | 4.108 | .21 | 2.46 | 27.3% | 24 | .968 | .966 | .964 | .961 | .959 | .956 | .953 | .949 | .946 | .941 |
| | | | | | | | | .937 | .932 | .926 | .919 | .912 | .904 | .895 | .885 | .874 | .861 |
| | | | | | | | | .847 | .831 | .813 | .792 | | | | | | |
| 26.000 | 22.543 | 3.457 | 3.457 | .25 | 2.60 | 23.5% | 25 | .967 | .965 | .963 | .960 | .958 | .955 | .951 | .948 | .944 | .939 |
| | | | | | | | | .934 | .929 | .923 | .916 | .908 | .900 | .890 | .879 | .867 | .853 |
| | | | | | | | | .838 | .821 | .801 | .779 | .754 | | | | | |
| 26.000 | 23.137 | 2.863 | 2.863 | .29 | 2.74 | 19.8% | 26 | .967 | .964 | .962 | .959 | .957 | .953 | .950 | .946 | .942 | .937 |
| | | | | | | | | .932 | .926 | .920 | .913 | .904 | .895 | .885 | .873 | .860 | .846 |
| | | | | | | | | .829 | .810 | .789 | .766 | .739 | .709 | | | | |
| 26.000 | 23.666 | 2.334 | 2.334 | .34 | 2.87 | 16.5% | 27 | .966 | .964 | .961 | .959 | .956 | .952 | .949 | .945 | .940 | .935 |
| | | | | | | | | .930 | .924 | .917 | .909 | .901 | .891 | .880 | .868 | .854 | .838 |
| | | | | | | | | .821 | .801 | .778 | .753 | .724 | .693 | .658 | | | |
| 26.000 | 24.130 | 1.870 | 1.870 | .40 | 2.98 | 13.4% | 28 | .965 | .963 | .961 | .958 | .955 | .951 | .948 | .943 | .939 | .934 |
| | | | | | | | | .928 | .922 | .914 | .906 | .897 | .887 | .876 | .863 | .848 | .831 |
| | | | | | | | | .813 | .791 | .768 | .741 | .711 | .677 | .641 | .600 | | |
| 26.000 | 24.527 | 1.473 | 1.473 | .46 | 3.07 | 10.7% | 29 | .965 | .963 | .960 | .957 | .954 | .950 | .947 | .942 | .937 | .932 |
| | | | | | | | | .926 | .920 | .912 | .904 | .894 | .884 | .872 | .858 | .842 | .825 |
| | | | | | | | | .805 | .783 | .758 | .730 | .698 | .663 | .625 | .583 | .538 | |
| 26.000 | 24.859 | 1.141 | 1.141 | .53 | 3.14 | 8.4% | 30 | .964 | .962 | .959 | .956 | .953 | .950 | .946 | .941 | .936 | .931 |
| | | | | | | | | .925 | .918 | .910 | .901 | .892 | .880 | .868 | .854 | .838 | .819 |
| | | | | | | | | .799 | .776 | .749 | .720 | .687 | .651 | .611 | .568 | .521 | .472 |
| 26.000 | 25.131 | .869 | .869 | .59 | 3.17 | 6.5% | 31 | .964 | .962 | .959 | .956 | .953 | .949 | .945 | .940 | .935 | .930 |
| | | | | | | | | .923 | .916 | .908 | .899 | .889 | .878 | .865 | .850 | .834 | .815 |
| | | | | | | | | .793 | .769 | .742 | .712 | .678 | .641 | .599 | .555 | .508 | .458 |
| | | | | | | | | .406 | | | | | | | | | |
| 26.000 | 25.349 | .651 | .651 | .66 | 3.18 | 4.9% | 32 | .964 | .961 | .959 | .956 | .952 | .948 | .944 | .940 | .934 | .929 |
| | | | | | | | | .922 | .915 | .907 | .898 | .887 | .876 | .862 | .847 | .830 | .811 |
| | | | | | | | | .789 | .764 | .736 | .705 | .670 | .632 | .590 | .545 | .496 | .446 |
| | | | | | | | | .394 | .341 | | | | | | | | |
| 26.000 | 25.521 | .479 | .479 | .72 | 3.17 | 3.6% | 33 | .964 | .961 | .958 | .955 | .952 | .948 | .944 | .939 | .934 | .928 |
| | | | | | | | | .921 | .914 | .906 | .896 | .886 | .874 | .860 | .845 | .827 | .808 |
| | | | | | | | | .785 | .760 | .731 | .699 | .664 | .625 | .582 | .536 | .487 | .436 |
| | | | | | | | | .384 | .332 | .281 | | | | | | | |

*SIMILAR TO A.T.&T. JACOBSEN RETRIAL TABLES

*SLOW RETRIALS: JEWETT-SHRAGO EXTENDED ERLANG B - 50% BLOCKED CALLERS RETRY, 50% OVERFLOW (RETRIES 3-5 HOLDING TIMES LATER)

| HOURS FIRST ATTEMPT VMR = 1.0 | HOURS CON-NECT | HOURS RECALL /RETRY | HOURS OVFL | CON-NECT VMR | OVFL VMR | PERCENT BLOCKED (GRADE) | LINE | HOURS CONNECTED TRAFFIC ON LINES 1 | 2 | 3 | 4 | 5 | 6 | 7 | 8 | 9 | 10 |
|---|---|---|---|---|---|---|---|---|---|---|---|---|---|---|---|---|---|
| 26.000 | 25.654 | .346 | .346 | .77 | 3.13 | 2.6% | 34 | .963 | .961 | .958 | .955 | .951 | .948 | .943 | .939 | .933 | .927 |
| | | | | | | | | .921 | .913 | .905 | .895 | .885 | .872 | .859 | .843 | .825 | .805 |
| | | | | | | | | .782 | .756 | .727 | .695 | .659 | .619 | .576 | .530 | .480 | .429 |
| | | | | | | | | .376 | .324 | .273 | .225 | | | | | | |
| 26.000 | 25.754 | .246 | .246 | .82 | 3.07 | 1.9% | 35 | .963 | .961 | .958 | .955 | .951 | .947 | .943 | .938 | .933 | .927 |
| | | | | | | | | .920 | .913 | .904 | .895 | .884 | .871 | .857 | .841 | .823 | .803 |
| | | | | | | | | .780 | .754 | .725 | .692 | .655 | .615 | .571 | .524 | .475 | .423 |
| | | | | | | | | .370 | .318 | .267 | .220 | .176 | | | | | |
| 26.000 | 25.829 | .171 | .171 | .87 | 3.00 | 1.3% | 36 | .963 | .961 | .958 | .955 | .951 | .947 | .943 | .938 | .933 | .927 |
| | | | | | | | | .920 | .912 | .904 | .894 | .883 | .870 | .856 | .840 | .822 | .802 |
| | | | | | | | | .778 | .752 | .722 | .689 | .652 | .612 | .568 | .521 | .471 | .419 |
| | | | | | | | | .366 | .314 | .263 | .216 | .173 | .135 | | | | |
| 26.000 | 25.883 | .117 | .117 | .90 | 2.92 | .9% | 37 | .963 | .961 | .958 | .954 | .951 | .947 | .943 | .938 | .932 | .926 |
| | | | | | | | | .920 | .912 | .903 | .893 | .882 | .870 | .856 | .840 | .821 | .800 |
| | | | | | | | | .777 | .750 | .721 | .687 | .650 | .610 | .565 | .518 | .468 | .416 |
| | | | | | | | | .363 | .311 | .260 | .213 | .170 | .133 | .101 | | | |
| 26.000 | 25.922 | .078 | .078 | .93 | 2.84 | .6% | 38 | .963 | .960 | .958 | .954 | .951 | .947 | .943 | .938 | .932 | .926 |
| | | | | | | | | .919 | .912 | .903 | .893 | .882 | .869 | .855 | .839 | .821 | .800 |
| | | | | | | | | .776 | .749 | .720 | .686 | .649 | .608 | .564 | .516 | .466 | .414 |
| | | | | | | | | .361 | .309 | .258 | .211 | .168 | .131 | .099 | .073 | | |
| 28.000 | 22.331 | 5.669 | 5.669 | .15 | 2.27 | 33.7% | 24 | .971 | .970 | .968 | .966 | .964 | .962 | .959 | .957 | .954 | .951 |
| | | | | | | | | .947 | .943 | .939 | .934 | .929 | .924 | .917 | .910 | .903 | .894 |
| | | | | | | | | .885 | .874 | .862 | .848 | | | | | | |
| 28.000 | 23.075 | 4.925 | 4.925 | .18 | 2.40 | 29.9% | 25 | .971 | .969 | .967 | .965 | .963 | .961 | .958 | .955 | .952 | .949 |
| | | | | | | | | .945 | .941 | .937 | .932 | .926 | .920 | .913 | .906 | .898 | .888 |
| | | | | | | | | .878 | .000 | .853 | .839 | .822 | | | | | |
| 20.000 | 23.776 | 4.224 | 4.224 | .21 | 2.54 | 26.2% | 26 | .970 | .968 | .966 | .964 | .962 | .960 | .957 | .954 | .951 | .947 |
| | | | | | | | | .943 | .939 | .934 | .929 | .923 | .917 | .910 | .901 | .893 | .882 |
| | | | | | | | | .871 | .859 | .845 | .829 | .811 | .791 | | | | |
| 28.000 | 24.425 | 3.575 | 3.575 | .25 | 2.68 | 22.6% | 27 | .969 | .968 | .966 | .963 | .961 | .958 | .956 | .953 | .949 | .945 |
| | | | | | | | | .941 | .937 | .932 | .926 | .920 | .913 | .905 | .897 | .887 | .877 |
| | | | | | | | | .864 | .851 | .836 | .819 | .799 | .778 | .754 | | | |
| 28.000 | 25.019 | 2.981 | 2.981 | .29 | 2.82 | 19.2% | 28 | .960 | .907 | .965 | .963 | .960 | .957 | .955 | .951 | .948 | .944 |
| | | | | | | | | .940 | .935 | .929 | .924 | .917 | .910 | .902 | .892 | .882 | .871 |
| | | | | | | | | .858 | .843 | .827 | .809 | .788 | .765 | .739 | .711 | | |
| 28.000 | 25.551 | 2.449 | 2.449 | .34 | 2.95 | 16.1% | 29 | .968 | .966 | .964 | .962 | .959 | .957 | .953 | .950 | .940 | .942 |
| | | | | | | | | .938 | .933 | .927 | .921 | .914 | .906 | .898 | .888 | .877 | .865 |
| | | | | | | | | .851 | .836 | .818 | .799 | .777 | .753 | .725 | .695 | .661 | |
| 28.000 | 26.020 | 1 980 | 1.080 | .39 | 3.06 | 13.2% | 30 | .968 | .966 | .964 | .961 | .959 | .956 | .952 | .949 | .945 | .941 |
| | | | | | | | | .936 | .931 | .925 | .919 | .911 | .903 | .894 | .884 | .873 | .860 |

*SIMILAR TO A.T.¢T. JACOBSEN RETRIAL TABLES

# *SLOW RETRIALS: JEWETT-SHRAGO EXTENDED ERLANG B - 50% BLOCKED CALLERS RETRY, 50% OVERFLOW (RETRIES 3-5 HOLDING TIMES LATER)

| HOURS FIRST ATTEMPT VMR=1.0 | HOURS CON-NECT | HOURS RECALL /RETRY | HOURS OVFL | CON-NECT VMR | OVFL VMR | PERCENT BLOCKED (GRADE) | LINE | HOURS CONNECTED TRAFFIC ON LINES 1 | 2 | 3 | 4 | 5 | 6 | 7 | 8 | 9 | 10 |
|---|---|---|---|---|---|---|---|---|---|---|---|---|---|---|---|---|---|
| | | | | | | | | .845 | .829 | .810 | .790 | .767 | .741 | .712 | .680 | .644 | .606 |
| 28.000 | 26.425 | 1.575 | 1.575 | .45 | 3.15 | 10.7% | 31 | .967 | .965 | .963 | .961 | .958 | .955 | .952 | .948 | .944 | .940 |
| | | | | | | | | .935 | .929 | .923 | .917 | .909 | .901 | .891 | .880 | .868 | .855 |
| | | | | | | | | .840 | .823 | .803 | .781 | .757 | .730 | .700 | .666 | .629 | .589 |
| | | | | | | | | .546 | | | | | | | | | |
| 28.000 | 26.766 | 1.234 | 1.234 | .52 | 3.22 | 8.4% | 32 | .967 | .965 | .963 | .960 | .957 | .954 | .951 | .947 | .943 | .939 |
| | | | | | | | | .933 | .928 | .922 | .915 | .907 | .898 | .888 | .877 | .865 | .851 |
| | | | | | | | | .835 | .817 | .797 | .774 | .749 | .720 | .689 | .654 | .616 | .574 |
| | | | | | | | | .530 | .483 | | | | | | | | |
| 28.000 | 27.049 | .951 | .951 | .58 | 3.26 | 6.6% | 33 | .967 | .965 | .962 | .960 | .957 | .954 | .950 | .946 | .942 | .938 |
| | | | | | | | | .932 | .927 | .920 | .913 | .905 | .896 | .886 | .874 | .861 | .847 |
| | | | | | | | | .830 | .812 | .791 | .768 | .741 | .712 | .679 | .643 | .604 | .562 |
| | | | | | | | | .516 | .468 | .419 | | | | | | | |
| 28.000 | 27.278 | .722 | .722 | .64 | 3.28 | 5.0% | 34 | .966 | .964 | .962 | .959 | .956 | .953 | .950 | .946 | .942 | .937 |
| | | | | | | | | .932 | .926 | .919 | .912 | .903 | .894 | .884 | .872 | .859 | .844 |
| | | | | | | | | .827 | .808 | .786 | .762 | .735 | .705 | .672 | .635 | .595 | .551 |
| | | | | | | | | .505 | .456 | .406 | .356 | | | | | | |
| 28.000 | 27.461 | .539 | .539 | .70 | 3.26 | 3.8% | 35 | .966 | .964 | .962 | .959 | .956 | .953 | .949 | .945 | .941 | .936 |
| | | | | | | | | .931 | .925 | .918 | .911 | .902 | .893 | .882 | .870 | .857 | .841 |
| | | | | | | | | .824 | .804 | .783 | .758 | .730 | .699 | .665 | .628 | .587 | .543 |
| | | | | | | | | .496 | .447 | .396 | .346 | .296 | | | | | |
| 28.000 | 27.604 | .396 | .396 | .76 | 3.23 | 2.8% | 36 | .966 | .964 | .961 | .959 | .956 | .952 | .949 | .945 | .941 | .936 |
| | | | | | | | | .930 | .924 | .917 | .910 | .901 | .892 | .881 | .869 | .855 | .839 |
| | | | | | | | | .822 | .802 | .779 | .754 | .726 | .695 | .660 | .622 | .580 | .536 |
| | | | | | | | | .488 | .439 | .388 | .338 | .288 | .241 | | | | |
| 28.000 | 27.714 | .286 | .286 | .81 | 3.18 | 2.0% | 37 | .966 | .964 | .961 | .959 | .956 | .952 | .949 | .945 | .940 | .935 |
| | | | | | | | | .930 | .924 | .917 | .909 | .900 | .891 | .880 | .867 | .853 | .838 |
| | | | | | | | | .820 | .800 | .777 | .752 | .723 | .691 | .656 | .617 | .576 | .530 |
| | | | | | | | | .483 | .433 | .382 | .332 | .282 | .235 | .191 | | | |
| 28.000 | 27.797 | .203 | .203 | .85 | 3.11 | 1.4% | 38 | .966 | .964 | .961 | .958 | .955 | .952 | .948 | .944 | .940 | .935 |
| | | | | | | | | .929 | .923 | .916 | .909 | .900 | .890 | .879 | .866 | .852 | .836 |
| | | | | | | | | .818 | .798 | .775 | .749 | .721 | .689 | .653 | .614 | .572 | .526 |
| | | | | | | | | .478 | .429 | .378 | .327 | .278 | .231 | .188 | .149 | | |
| 28.000 | 27.859 | .141 | .141 | .89 | 3.03 | 1.0% | 39 | .966 | .963 | .961 | .958 | .955 | .952 | .948 | .944 | .940 | .935 |
| | | | | | | | | .929 | .923 | .916 | .908 | .899 | .889 | .878 | .866 | .851 | .835 |
| | | | | | | | | .817 | .797 | .774 | .748 | .719 | .687 | .651 | .612 | .569 | .523 |
| | | | | | | | | .475 | .425 | .374 | .324 | .274 | .228 | .185 | .146 | .113 | |
| 28.000 | 27.903 | .097 | .097 | .92 | 2.95 | .7% | 40 | .966 | .963 | .961 | .958 | .955 | .952 | .948 | .944 | .940 | .935 |
| | | | | | | | | .929 | .923 | .916 | .908 | .899 | .889 | .878 | .865 | .851 | .835 |
| | | | | | | | | .817 | .796 | .773 | .747 | .718 | .685 | .649 | .610 | .567 | .521 |

*SIMILAR TO A.T.&T. JACOBSEN RETRIAL TABLES

*SLOW RETRIALS: JEWETT-SHRAGO EXTENDED ERLANG B - 50% BLOCKED CALLERS RETRY, 50% OVERFLOW (RETRIES 3-5 HOLDING TIMES LATER)

| HOURS FIRST ATTEMPT VMR = 1.0 | HOURS CON-NECT | HOURS RECALL /RETRY | HOURS OVFL | CON-NECT VMR | OVFL VMR | PERCENT BLOCKED (GRADE) | LINE | HOURS CONNECTED TRAFFIC ON LINES 1 | 2 | 3 | 4 | 5 | 6 | 7 | 8 | 9 | 10 |
|---|---|---|---|---|---|---|---|---|---|---|---|---|---|---|---|---|---|
| | | | | | | | | .473 | .423 | .372 | .321 | .272 | .225 | .183 | .144 | .111 | .084 |
| 30.000 | 24.227 | 5.773 | 5.773 | .15 | 2.34 | 32.3% | 26 | .973 | .971 | .970 | .968 | .966 | .964 | .962 | .960 | .958 | .955 |
| | | | | | | | | .952 | .949 | .945 | .941 | .937 | .932 | .927 | .921 | .915 | .908 |
| | | | | | | | | .900 | .892 | .882 | .871 | .859 | .846 | | | | |
| 30.000 | 24.967 | 5.033 | 5.033 | .18 | 2.48 | 28.7% | 27 | .972 | .971 | .969 | .967 | .966 | .964 | .961 | .959 | .956 | .953 |
| | | | | | | | | .950 | .947 | .943 | .939 | .934 | .929 | .924 | .918 | .911 | .903 |
| | | | | | | | | .895 | .886 | .875 | .864 | .851 | .837 | .821 | | | |
| 30.000 | 25.665 | 4.335 | 4.335 | .21 | 2.62 | 25.3% | 28 | .972 | .970 | .969 | .967 | .965 | .963 | .960 | .958 | .955 | .952 |
| | | | | | | | | .949 | .945 | .941 | .937 | .932 | .927 | .921 | .914 | .907 | .899 |
| | | | | | | | | .890 | .880 | .869 | .856 | .842 | .827 | .809 | .790 | | |
| 30.000 | 26.312 | 3.688 | 3.688 | .25 | 2.76 | 21.9% | 29 | .971 | .970 | .968 | .966 | .964 | .962 | .959 | .957 | .954 | .951 |
| | | | | | | | | .947 | .943 | .939 | .935 | .929 | .924 | .918 | .911 | .903 | .894 |
| | | | | | | | | .885 | .874 | .862 | .849 | .834 | .817 | .798 | .777 | .754 | |
| 30.000 | 26.906 | 3.094 | 3.094 | .29 | 2.89 | 18.7% | 30 | .971 | .969 | .967 | .965 | .963 | .961 | .958 | .956 | .953 | .949 |
| | | | | | | | | .946 | .942 | .937 | .932 | .927 | .921 | .914 | .907 | .899 | .890 |
| | | | | | | | | .880 | .868 | .855 | .841 | .825 | .807 | .787 | .765 | .740 | .712 |
| 30.000 | 27.441 | 2.559 | 2.559 | .34 | 3.02 | 15.7% | 31 | .970 | .969 | .967 | .965 | .962 | .960 | .957 | .955 | .951 | .948 |
| | | | | | | | | .944 | .940 | .935 | .930 | .925 | .918 | .911 | .904 | .895 | .885 |
| | | | | | | | | .875 | .862 | .849 | .834 | .817 | .797 | .776 | .752 | .726 | .697 |
| | | | | | | | | .664 | | | | | | | | | |
| 30.000 | 27.914 | 2.086 | 2.086 | .39 | 3.13 | 13.0% | 32 | .970 | .968 | .966 | .964 | .962 | .959 | .957 | .954 | .950 | .947 |
| | | | | | | | | .943 | .939 | .934 | .928 | .923 | .916 | .909 | .901 | .891 | .881 |
| | | | | | | | | .870 | .857 | .843 | .827 | .809 | .788 | .766 | .741 | .713 | .682 |
| | | | | | | | | .648 | .611 | | | | | | | | |
| 30.000 | 28.325 | 1.675 | 1.675 | .45 | 3.23 | 10.6% | 33 | .969 | .968 | .966 | .964 | .961 | .959 | .956 | .953 | .949 | .946 |
| | | | | | | | | .942 | .937 | .932 | .927 | .921 | .914 | .906 | .898 | .888 | .877 |
| | | | | | | | | .866 | .852 | .837 | .820 | .801 | .780 | .756 | .730 | .701 | .668 |
| | | | | | | | | .633 | .595 | .553 | | | | | | | |
| 30.000 | 28.674 | 1.326 | 1.326 | .51 | 3.30 | 8.5% | 34 | .969 | .967 | .965 | .963 | .961 | .958 | .955 | .952 | .949 | .945 |
| | | | | | | | | .941 | .936 | .931 | .925 | .919 | .912 | .904 | .895 | .885 | .874 |
| | | | | | | | | .862 | .848 | .832 | .815 | .795 | .773 | .748 | .721 | .690 | .657 |
| | | | | | | | | .620 | .580 | .538 | .493 | | | | | | |
| 30.000 | 28.967 | 1.033 | 1.033 | .57 | 3.35 | 6.7% | 35 | .969 | .967 | .965 | .963 | .960 | .958 | .955 | .951 | .948 | .944 |
| | | | | | | | | .940 | .935 | .930 | .924 | .917 | .910 | .902 | .893 | .883 | .871 |
| | | | | | | | | .858 | .844 | .828 | .810 | .789 | .766 | .741 | .712 | .681 | .646 |
| | | | | | | | | .608 | .568 | .524 | .478 | .430 | | | | | |
| 30.000 | 29.207 | .793 | .793 | .63 | 3.36 | 5.2% | 36 | .969 | .967 | .965 | .962 | .960 | .957 | .954 | .951 | .947 | .943 |
| | | | | | | | | .939 | .934 | .929 | .923 | .916 | .909 | .900 | .891 | .880 | .869 |
| | | | | | | | | .856 | .841 | .824 | .805 | .784 | .761 | .734 | .705 | .673 | .637 |
| | | | | | | | | .599 | .557 | .513 | .466 | .410 | .309 | | | | |

*SIMILAR TO A.T.&T. JACOBSEN RETRIAL TABLES

*SLOW RETRIALS: JEWETT-SHRAGO EXTENDED ERLANG B - 50% BLOCKED CALLERS RETRY, 50% OVERFLOW (RETRIES 3-5 HOLDING TIMES LATER)

| HOURS FIRST ATTEMPT VMR=1.0 | HOURS CON-NECT | HOURS RECALL /RETRY | HOURS OVFL | CON-NECT VMR | OVFL VMR | PERCENT BLOCKED (GRADE) | LINE | HOURS CONNECTED TRAFFIC ON LINES 1 | 2 | 3 | 4 | 5 | 6 | 7 | 8 | 9 | 10 |
|---|---|---|---|---|---|---|---|---|---|---|---|---|---|---|---|---|---|
| 30.000 | 29.400 | .600 | .600 | .69 | 3.36 | 3.9% | 37 | .968 | .966 | .964 | .962 | .960 | .957 | .954 | .950 | .947 | .943 |
| | | | | | | | | .938 | .933 | .928 | .922 | .915 | .907 | .899 | .889 | .879 | .867 |
| | | | | | | | | .853 | .838 | .821 | .802 | .780 | .756 | .729 | .699 | .666 | .630 |
| | | | | | | | | .591 | .548 | .503 | .456 | .408 | .359 | .310 | | | |
| 30.000 | 29.553 | .447 | .447 | .74 | 3.33 | 2.9% | 38 | .968 | .966 | .964 | .962 | .959 | .956 | .953 | .950 | .946 | .942 |
| | | | | | | | | .938 | .933 | .927 | .921 | .914 | .906 | .898 | .888 | .877 | .865 |
| | | | | | | | | .851 | .836 | .818 | .799 | .777 | .752 | .725 | .695 | .661 | .624 |
| | | | | | | | | .584 | .542 | .496 | .449 | .400 | .351 | .302 | .255 | | |
| 30.000 | 29.672 | .328 | .328 | .79 | 3.28 | 2.2% | 39 | .968 | .966 | .964 | .962 | .959 | .956 | .953 | .950 | .946 | .942 |
| | | | | | | | | .937 | .932 | .927 | .920 | .913 | .906 | .897 | .887 | .876 | .864 |
| | | | | | | | | .850 | .834 | .816 | .797 | .774 | .750 | .722 | .691 | .657 | .620 |
| | | | | | | | | .579 | .536 | .490 | .442 | .393 | .344 | .296 | .249 | .206 | |
| 30.000 | 29.764 | .236 | .236 | .84 | 3.22 | 1.6% | 40 | .968 | .966 | .964 | .962 | .959 | .956 | .953 | .950 | .946 | .942 |
| | | | | | | | | .937 | .932 | .926 | .920 | .913 | .905 | .896 | .886 | .875 | .863 |
| | | | | | | | | .849 | .833 | .815 | .795 | .772 | .747 | .719 | .688 | .654 | .616 |
| | | | | | | | | .575 | .532 | .486 | .438 | .388 | .339 | .291 | .245 | .202 | .162 |
| 30.000 | 29.833 | .167 | .167 | .87 | 3.14 | 1.1% | 41 | .968 | .966 | .964 | .961 | .959 | .956 | .953 | .949 | .946 | .941 |
| | | | | | | | | .937 | .932 | .926 | .920 | .913 | .905 | .896 | .886 | .874 | .862 |
| | | | | | | | | .848 | .832 | .814 | .794 | .771 | .746 | .717 | .686 | .651 | .613 |
| | | | | | | | | .572 | .529 | .482 | .434 | .385 | .336 | .287 | .241 | .198 | .160 |
| | | | | | | | | .125 | | | | | | | | | |
| 30.000 | 29.883 | .117 | .117 | .91 | 3.06 | .8% | 42 | .968 | .966 | .964 | .961 | .959 | .956 | .953 | .949 | .945 | .941 |
| | | | | | | | | .937 | .931 | .926 | .919 | .912 | .904 | .895 | .885 | .874 | .861 |
| | | | | | | | | .847 | .831 | .813 | .793 | .770 | .744 | .716 | .684 | .649 | .611 |
| | | | | | | | | .570 | .526 | .480 | .431 | .382 | .333 | .285 | .239 | .196 | .157 |
| | | | | | | | | .124 | .095 | | | | | | | | |
| 30.000 | 29.920 | .080 | .080 | .93 | 2.98 | .5% | 43 | .968 | .966 | .964 | .961 | .959 | .956 | .953 | .949 | .945 | .941 |
| | | | | | | | | .937 | .931 | .926 | .919 | .912 | .904 | .895 | .885 | .874 | .861 |
| | | | | | | | | .847 | .830 | .812 | .792 | .769 | .743 | .715 | .683 | .648 | .610 |
| | | | | | | | | .569 | .524 | .478 | .429 | .380 | .331 | .283 | .237 | .194 | .156 |
| | | | | | | | | .122 | .093 | .070 | | | | | | | |

*SIMILAR TO A.T.&T. JACOBSEN RETRIAL TABLES

*SLOW RETRIALS: JEWETT-SHRAGO EXTENDED ERLANG B - 100% BLOCKED CALLERS RETRY, 0% OVERFLOW (RETRIES 3-5 HOLDING TIMES LATER)

| HOURS FIRST ATTEMPT VMR=1.0 | HOURS CON-NECT | HOURS RECALL /RETRY | HOURS OVFL | CON-NECT VMR | OVFL VMR | PERCENT BLOCKED (GRADE) | LINE | HOURS CONNECTED TRAFFIC ON LINES 1 | 2 | 3 | 4 | 5 | 6 | 7 | 8 | 9 | 10 |
|---|---|---|---|---|---|---|---|---|---|---|---|---|---|---|---|---|---|
| .200 | .200 | .050 | .000 | .80 | 1.09 | 20.0% | 1 | .200 | | | | | | | | | |
| .200 | .200 | .003 | .000 | .97 | 1.07 | 1.7% | 2 | .169 | .031 | | | | | | | | |
| .400 | .400 | .025 | .000 | .90 | 1.14 | 6.0% | 2 | .298 | .102 | | | | | | | | |
| .400 | .400 | .003 | .000 | .98 | 1.11 | .7% | 3 | .287 | .094 | .019 | | | | | | | |
| .600 | .600 | .083 | .000 | .81 | 1.20 | 12.2% | 2 | .406 | .194 | | | | | | | | |
| .600 | .600 | .013 | .000 | .95 | 1.17 | 2.1% | 3 | .380 | .169 | .051 | | | | | | | |
| .800 | .800 | .200 | .000 | .70 | 1.25 | 20.0% | 2 | .500 | .300 | | | | | | | | |
| .800 | .800 | .036 | .000 | .90 | 1.23 | 4.3% | 3 | .455 | .247 | .098 | | | | | | | |
| .800 | .800 | .006 | .000 | .97 | 1.19 | .8% | 4 | .446 | .237 | .091 | .025 | | | | | | |
| 1.000 | 1.000 | .414 | .000 | .59 | 1.29 | 29.3% | 2 | .586 | .414 | | | | | | | | |
| 1.000 | 1.000 | .079 | .000 | .84 | 1.28 | 7.3% | 3 | .519 | .324 | .157 | | | | | | | |
| 1.000 | 1.000 | .016 | .000 | .95 | 1.24 | 1.6% | 4 | .504 | .305 | .142 | .049 | | | | | | |
| 1.200 | 1.200 | .151 | .000 | .77 | 1.33 | 11.2% | 3 | .575 | .399 | .227 | | | | | | | |
| 1.200 | 1.200 | .035 | .000 | .92 | 1.29 | 2.8% | 4 | .553 | .368 | .197 | .082 | | | | | | |
| 1.200 | 1.200 | .008 | .000 | .98 | 1.24 | .6% | 5 | .547 | .361 | .190 | .078 | .025 | | | | | |
| 1.400 | 1.400 | .266 | .000 | .70 | 1.37 | 16.0% | 3 | .625 | .471 | .304 | | | | | | | |
| 1.400 | 1.400 | .066 | .000 | .88 | 1.34 | 4.5% | 4 | .595 | .427 | .256 | .123 | | | | | | |
| 1.400 | 1.400 | .016 | .000 | .96 | 1.29 | 1.2% | 5 | .586 | .415 | .243 | .114 | .042 | | | | | |
| 1.600 | 1.600 | .443 | .000 | .61 | 1.41 | 21.7% | 3 | .671 | .541 | .388 | | | | | | | |
| 1.600 | 1.600 | .115 | .000 | .83 | 1.39 | 6.7% | 4 | .632 | .481 | .316 | .172 | | | | | | |
| 1.600 | 1.600 | .031 | .000 | .93 | 1.34 | 1.9% | 5 | .620 | .463 | .296 | .155 | .066 | | | | | |
| 1.800 | 1.800 | .714 | .000 | .52 | 1.43 | 28.4% | 3 | .715 | .608 | .476 | | | | | | | |
| 1.800 | 1.800 | .186 | .000 | .77 | 1.43 | 9.4% | 4 | .665 | .531 | .376 | .228 | | | | | | |
| 1.800 | 1.800 | .054 | .000 | .90 | 1.39 | 2.9% | 5 | .650 | .507 | .347 | .200 | .095 | | | | | |
| 1.800 | 1.800 | .015 | .000 | .97 | 1.33 | .8% | 6 | .645 | .500 | .339 | .192 | .090 | .034 | | | | |
| 2.000 | 2.000 | .289 | .000 | .71 | 1.47 | 12.6% | 4 | .696 | .578 | .436 | .209 | | | | | | |
| 2.000 | 2.000 | .087 | .000 | .87 | 1.43 | 4.2% | 5 | .676 | .548 | .397 | .248 | .131 | | | | | |
| 2.000 | 2.000 | .026 | .000 | .95 | 1.38 | 1.3% | 6 | .670 | .538 | .384 | .236 | .121 | .052 | | | | |
| 2.200 | 2.200 | .435 | .000 | .64 | 1.51 | 16.5% | 4 | .725 | .623 | .496 | .356 | | | | | | |
| 2.200 | 2.200 | .135 | .000 | .83 | 1.48 | 5.8% | 5 | .700 | .585 | .445 | .298 | .172 | | | | | |
| 2.200 | 2.200 | .042 | .000 | .93 | 1.42 | 1.9% | 6 | .692 | .571 | .427 | .280 | .156 | .073 | | | | |
| 2.200 | 2.200 | .012 | .000 | .97 | 1.37 | .6% | 7 | .689 | .567 | .422 | .274 | .151 | .070 | .027 | | | |
| 2.400 | 2.400 | .640 | .000 | .57 | 1.53 | 21.1% | 4 | .753 | .666 | .555 | .427 | | | | | | |
| 2.400 | 2.400 | .201 | .000 | .78 | 1.52 | 7.7% | 5 | .722 | .619 | .491 | .350 | .218 | | | | | |
| 2.400 | 2.400 | .066 | .000 | .90 | 1.47 | 2.7% | 6 | .712 | .602 | .468 | .324 | .195 | .099 | | | | |
| 2.400 | 2.400 | .021 | .000 | .96 | 1.41 | .9% | 7 | .708 | .596 | .460 | .315 | .187 | .094 | .040 | | | |
| 2.600 | 2.600 | .929 | .000 | .50 | 1.54 | 26.3% | 4 | .779 | .707 | .613 | .500 | | | | | | |
| 2.600 | 2.600 | .292 | .000 | .73 | 1.56 | 10.1% | 5 | .743 | .651 | .535 | .402 | .269 | | | | | |
| 2.600 | 2.600 | .100 | .000 | .87 | 1.51 | 3.7% | 6 | .730 | .630 | .506 | .368 | .236 | .130 | | | | |
| 2.600 | 2.600 | .033 | .000 | .94 | 1.45 | 1.3% | 7 | .725 | .623 | .496 | .356 | .224 | .121 | .056 | | | |

*SIMILAR TO A.T.&T. JACOBSEN RETRIAL TABLES

## *SLOW RETRIALS: JEWETT-SHRAGO EXTENDED ERLANG B - 100% BLOCKED CALLERS RETRY, 0% OVERFLOW (RETRIES 3-5 HOLDING TIMES LATER)

| HOURS FIRST ATTEMPT VMR=1.0 | HOURS CON-NECT | HOURS RECALL /RETRY | HOURS OVFL | CON-NECT VMR | OVFL VMR | PERCENT BLOCKED (GRADE) | LINE | HOURS CONNECTED TRAFFIC ON LINES 1 | 2 | 3 | 4 | 5 | 6 | 7 | 8 | 9 | 10 |
|---|---|---|---|---|---|---|---|---|---|---|---|---|---|---|---|---|---|
| 2.800 | 2.800 | 1.344 | .000 | .42 | 1.54 | 32.4% | 4 | .806 | .747 | .671 | .576 | | | | | | |
| 2.800 | 2.800 | .413 | .000 | .68 | 1.59 | 12.9% | 5 | .763 | .681 | .577 | .454 | .325 | | | | | |
| 2.800 | 2.800 | .145 | .000 | .83 | 1.56 | 4.9% | 6 | .747 | .656 | .542 | .411 | .279 | .165 | | | | |
| 2.800 | 2.800 | .050 | .000 | .92 | 1.50 | 1.8% | 7 | .740 | .647 | .529 | .395 | .262 | .152 | .076 | | | |
| 2.800 | 2.800 | .017 | .000 | .97 | 1.44 | .6% | 8 | .738 | .643 | .524 | .389 | .256 | .147 | .072 | .031 | | |
| 3.000 | 3.000 | .575 | .000 | .62 | 1.62 | 16.1% | 5 | .781 | .710 | .618 | .507 | .383 | | | | | |
| 3.000 | 3.000 | .205 | .000 | .80 | 1.60 | 6.4% | 6 | .762 | .681 | .576 | .453 | .323 | .205 | | | | |
| 3.000 | 3.000 | .074 | .000 | .90 | 1.54 | 2.4% | 7 | .755 | .669 | .560 | .432 | .301 | .185 | .099 | | | |
| 3.000 | 3.000 | .026 | .000 | .96 | 1.48 | .8% | 8 | .752 | .664 | .553 | .424 | .293 | .178 | .094 | .043 | | |
| 3.200 | 3.200 | .791 | .000 | .55 | 1.63 | 19.8% | 5 | .800 | .738 | .658 | .559 | .445 | | | | | |
| 3.200 | 3.200 | .284 | .000 | .75 | 1.63 | 8.1% | 6 | .777 | .703 | .608 | .494 | .369 | .248 | | | | |
| 3.200 | 3.200 | .106 | .000 | .87 | 1.58 | 3.2% | 7 | .768 | .689 | .588 | .468 | .340 | .220 | .126 | | | |
| 3.200 | 3.200 | .038 | .000 | .94 | 1.52 | 1.2% | 8 | .764 | .683 | .580 | .458 | .329 | .210 | .118 | .058 | | |
| 3.400 | 3.400 | 1.082 | .000 | .49 | 1.64 | 24.1% | 5 | .818 | .765 | .697 | .611 | .509 | | | | | |
| 3.400 | 3.400 | .386 | .000 | .71 | 1.67 | 10.2% | 6 | .791 | .725 | .639 | .534 | .416 | .295 | | | | |
| 3.400 | 3.400 | .148 | .000 | .84 | 1.62 | 4.2% | 7 | .780 | .708 | .615 | .503 | .379 | .258 | .156 | | | |
| 3.400 | 3.400 | .056 | .000 | .92 | 1.56 | 1.6% | 8 | .776 | .701 | .605 | .490 | .364 | .244 | .144 | .075 | | |
| 3.400 | 3.400 | .020 | .000 | .97 | 1.50 | .6% | 9 | .774 | .698 | .601 | .485 | .359 | .238 | .140 | .072 | .033 | |
| 3.600 | 3.600 | 1.479 | .000 | .42 | 1.64 | 29.1% | 5 | .836 | .792 | .735 | .663 | .575 | | | | | |
| 3.600 | 3.600 | .518 | .000 | .65 | 1.69 | 12.6% | 6 | .805 | .745 | .669 | .573 | .463 | .345 | | | | |
| 3.600 | 3.600 | .201 | .000 | .81 | 1.66 | 5.3% | 7 | .792 | .726 | .641 | .536 | .418 | .297 | .190 | | | |
| 3.600 | 3.600 | .078 | .000 | .90 | 1.60 | 2.1% | 8 | .786 | .718 | .629 | .521 | .399 | .278 | .173 | .096 | | |
| 3.600 | 3.600 | .029 | .000 | .96 | 1.54 | .8% | 9 | .784 | .714 | .624 | .514 | .392 | .271 | .167 | .091 | .044 | |
| 3.800 | 3.800 | 2.033 | .000 | .36 | 1.62 | 34.9% | 5 | .854 | .818 | .772 | .714 | .642 | | | | | |
| 3.800 | 3.800 | .687 | .000 | .60 | 1.71 | 15.3% | 6 | .818 | .765 | .697 | .612 | .510 | .398 | | | | |
| 3.800 | 3.800 | .270 | .000 | .77 | 1.70 | 6.6% | 7 | .803 | .743 | .665 | .568 | .456 | .338 | .227 | | | |
| 3.800 | 3.800 | .108 | .000 | .88 | 1.64 | 2.8% | 8 | .796 | .733 | .651 | .549 | .433 | .314 | .205 | .119 | | |
| 3.800 | 3.800 | .042 | .000 | .94 | 1.58 | 1.1% | 9 | .793 | .729 | .645 | .541 | .424 | .304 | .196 | .112 | .057 | |
| 4.000 | 4.000 | .907 | .000 | .55 | 1.73 | 18.5% | 6 | .831 | .784 | .725 | .649 | .558 | .454 | | | | |
| 4.000 | 4.000 | .357 | .000 | .73 | 1.73 | 8.2% | 7 | .813 | .759 | .688 | .599 | .494 | .380 | .268 | | | |
| 4.000 | 4.000 | .146 | .000 | .85 | 1.68 | 3.5% | 8 | .806 | .747 | .671 | .577 | .466 | .349 | .238 | .146 | | |
| 4.000 | 4.000 | .058 | .000 | .93 | 1.62 | 1.4% | 9 | .802 | .742 | .664 | .567 | .455 | .336 | .226 | .136 | .073 | |
| 4.000 | 4.000 | .022 | .000 | .97 | 1.55 | .5% | 10 | .801 | .740 | .661 | .563 | .450 | .331 | .221 | .132 | .070 | .033 |
| 4.200 | 4.200 | 1.191 | .000 | .49 | 1.73 | 22.1% | 6 | .844 | .803 | .751 | .686 | .605 | .511 | | | | |
| 4.200 | 4.200 | .467 | .000 | .69 | 1.76 | 10.0% | 7 | .824 | .774 | .709 | .628 | .531 | .422 | .311 | | | |
| 4.200 | 4.200 | .194 | .000 | .82 | 1.72 | 4.4% | 8 | .815 | .760 | .690 | .602 | .499 | .385 | .273 | .175 | | |
| 4.200 | 4.200 | .080 | .000 | .91 | 1.66 | 1.9% | 9 | .811 | .754 | .682 | .591 | .484 | .369 | .257 | .162 | .091 | |
| 4.200 | 4.200 | .031 | .000 | .96 | 1.59 | .7% | 10 | .809 | .752 | .678 | .586 | .478 | .362 | .250 | .156 | .087 | .043 |
| 4.400 | 4.400 | 1.566 | .000 | .43 | 1.73 | 26.2% | 6 | .856 | .822 | .778 | .722 | .652 | .540 | | | | |
| 4.400 | 4.400 | .604 | .000 | .64 | 1.79 | 12.1% | 7 | .834 | .789 | .730 | .657 | .568 | .466 | .357 | | | |

*SIMILAR TO A.T.&T. JACOBSEN RETRIAL TABLES

## *SLOW RETRIALS: JEWETT-SHRAGO EXTENDED ERLANG B - 100% BLOCKED CALLERS RETRY, 0% OVERFLOW (RETRIES 3-5 HOLDING TIMES LATER)

| HOURS FIRST ATTEMPT VMR=1.0 | HOURS CON-NECT | HOURS RECALL /RETRY | HOURS OVFL | CON-NECT VMR | OVFL VMR | PERCENT BLOCKED (GRADE) | LINE | HOURS CONNECTED TRAFFIC ON LINES 1 | 2 | 3 | 4 | 5 | 6 | 7 | 8 | 9 | 10 |
|---|---|---|---|---|---|---|---|---|---|---|---|---|---|---|---|---|---|
| 4.400 | 4.400 | .254 | .000 | .79 | 1.76 | 5.5% | 8 | .823 | .773 | .709 | .627 | .530 | .421 | .309 | .208 | | |
| 4.400 | 4.400 | .107 | .000 | .89 | 1.70 | 2.4% | 9 | .818 | .766 | .698 | .614 | .512 | .401 | .289 | .189 | .112 | |
| 4.400 | 4.400 | .043 | .000 | .94 | 1.63 | 1.0% | 10 | .816 | .763 | .694 | .607 | .505 | .392 | .280 | .182 | .106 | .055 |
| 4.600 | 4.600 | 2.067 | .000 | .37 | 1.71 | 31.0% | 6 | .870 | .841 | .804 | .757 | .699 | .629 | | | | |
| 4.600 | 4.600 | .777 | .000 | .59 | 1.80 | 14.5% | 7 | .843 | .803 | .751 | .685 | .604 | .509 | .406 | | | |
| 4.600 | 4.600 | .329 | .000 | .76 | 1.79 | 6.7% | 8 | .831 | .785 | .726 | .651 | .560 | .456 | .347 | .243 | | |
| 4.600 | 4.600 | .141 | .000 | .86 | 1.74 | 3.0% | 9 | .826 | .777 | .714 | .635 | .540 | .432 | .322 | .219 | .135 | |
| 4.600 | 4.600 | .059 | .000 | .93 | 1.67 | 1.3% | 10 | .823 | .773 | .709 | .628 | .530 | .421 | .310 | .209 | .127 | .069 |
| 4.600 | 4.600 | .024 | .000 | .97 | 1.60 | .5% | 11 | .822 | .772 | .706 | .624 | .526 | .417 | .305 | .204 | .123 | .067 |
| | | | | | | | | .033 | | | | | | | | | |
| 4.800 | 4.800 | .994 | .000 | .54 | 1.82 | 17.2% | 7 | .853 | .817 | .770 | .712 | .639 | .553 | .456 | | | |
| 4.800 | 4.800 | .421 | .000 | .72 | 1.82 | 8.1% | 8 | .839 | .797 | .742 | .674 | .589 | .492 | .386 | .281 | | |
| 4.800 | 4.800 | .184 | .000 | .84 | 1.77 | 3.7% | 9 | .833 | .788 | .729 | .655 | .566 | .463 | .355 | .250 | .161 | |
| 4.800 | 4.800 | .079 | .000 | .91 | 1.71 | 1.6% | 10 | .830 | .783 | .723 | .647 | .555 | .450 | .340 | .237 | .150 | .086 |
| 4.800 | 4.800 | .033 | .000 | .96 | 1.64 | .7% | 11 | .829 | .781 | .720 | .643 | .550 | .444 | .334 | .231 | .145 | .082 |
| | | | | | | | | .042 | | | | | | | | | |
| 5.000 | 5.000 | 1.270 | .000 | .49 | 1.82 | 20.3% | 7 | .862 | .830 | .790 | .738 | .674 | .597 | .508 | | | |
| 5.000 | 5.000 | .535 | .000 | .68 | 1.85 | 9.7% | 8 | .847 | .808 | .758 | .695 | .618 | .526 | .425 | .321 | | |
| 5.000 | 5.000 | .237 | .000 | .81 | 1.81 | 4.5% | 9 | .840 | .798 | .743 | .675 | .591 | .493 | .388 | .283 | .190 | |
| 5.000 | 5.000 | .104 | .000 | .90 | 1.75 | 2.0% | 10 | .836 | .792 | .736 | .665 | .578 | .478 | .370 | .266 | .175 | .104 |
| 5.000 | 5.000 | .044 | .000 | .95 | 1.68 | .9% | 11 | .835 | .790 | .733 | .660 | .572 | .470 | .362 | .258 | .168 | .099 |
| | | | | | | | | .053 | | | | | | | | | |
| 5.500 | 5.500 | 2.357 | .000 | .36 | 1.79 | 30.0% | 7 | .887 | .865 | .837 | .802 | .759 | .708 | .643 | | | |
| 5.500 | 5.500 | .949 | .000 | .57 | 1.89 | 14.7% | 8 | .866 | .835 | .796 | .747 | .686 | .612 | .526 | .431 | | |
| 5.500 | 5.500 | .427 | .000 | .73 | 1.89 | 7.2% | 9 | .856 | .821 | .776 | .719 | .649 | .566 | .471 | .370 | .272 | |
| 5.500 | 5.500 | .196 | .000 | .84 | 1.84 | 3.4% | 10 | .851 | .814 | .766 | .706 | .631 | .543 | .444 | .342 | .244 | .160 |
| 5.500 | 5.500 | .088 | .000 | .91 | 1.77 | 1.6% | 11 | .848 | .810 | .761 | .699 | .622 | .532 | .432 | .328 | .231 | .149 |
| | | | | | | | | .088 | | | | | | | | | |
| 5.500 | 5.500 | .038 | .000 | .96 | 1.70 | .7% | 12 | .847 | .808 | .759 | .696 | .618 | .527 | .426 | .322 | .225 | .144 |
| | | | | | | | | .084 | .045 | | | | | | | | |
| 6.000 | 6.000 | 1.655 | .000 | .45 | 1.90 | 21.6% | 8 | .885 | .861 | .832 | .798 | .750 | .695 | .629 | .552 | | |
| 6.000 | 6.000 | .734 | .000 | .63 | 1.95 | 10.9% | 9 | .871 | .842 | .806 | .760 | .703 | .634 | .553 | .463 | .367 | |
| 6.000 | 6.000 | .345 | .000 | .77 | 1.92 | 5.4% | 10 | .864 | .832 | .792 | .742 | .679 | .603 | .516 | .420 | .322 | .230 |
| 6.000 | 6.000 | .162 | .000 | .86 | 1.86 | 2.6% | 11 | .860 | .828 | .786 | .732 | .667 | .588 | .497 | .398 | .300 | .210 |
| | | | | | | | | .135 | | | | | | | | | |
| 6.000 | 6.000 | .074 | .000 | .93 | 1.79 | 1.2% | 12 | .859 | .825 | .782 | .728 | .660 | .580 | .487 | .388 | .289 | .200 |
| | | | | | | | | .127 | .074 | | | | | | | | |
| 6.000 | 6.000 | .033 | .000 | .96 | 1.72 | .5% | 13 | .858 | .824 | .780 | .725 | .657 | .576 | .483 | .383 | .284 | .196 |
| | | | | | | | | .124 | .072 | .038 | | | | | | | |
| 6.500 | 6.500 | 2.935 | .000 | .32 | 1.84 | 31.1% | 8 | .904 | .888 | .868 | .843 | .813 | .776 | .731 | .678 | | |

*SIMILAR TO A.T.&T. JACOBSEN RETRIAL TABLES

*SLOW RETRIALS: JEWETT-SHRAGO EXTENDED ERLANG B - 100% BLOCKED CALLERS RETRY, 0% OVERFLOW (RETRIES 3-5 HOLDING TIMES LATER)

| HOURS FIRST ATTEMPT VMR = 1.0 | HOURS CON-NECT | HOURS RECALL /RETRY | HOURS OVFL | CON-NECT VMR | OVFL VMR | PERCENT BLOCKED (GRADE) | LINE | HOURS CONNECTED TRAFFIC ON LINES 1 | 2 | 3 | 4 | 5 | 6 | 7 | 8 | 9 | 10 |
|---|---|---|---|---|---|---|---|---|---|---|---|---|---|---|---|---|---|
| 6.500 | 6.500 | 1.231 | .000 | .53 | 1.98 | 15.9% | 9 | .886 | .863 | .834 | .798 | .754 | .700 | .634 | .558 | .474 | |
| 6.500 | 6.500 | .580 | .000 | .69 | 1.99 | 8.2% | 10 | .876 | .850 | .817 | .775 | .723 | .659 | .584 | .498 | .406 | .313 |
| 6.500 | 6.500 | .281 | .000 | .81 | 1.95 | 4.1% | 11 | .872 | .843 | .807 | .762 | .706 | .638 | .558 | .468 | .372 | .279 |
| | | | | | | | | .195 | | | | | | | | | |
| 6.500 | 6.500 | .135 | .000 | .89 | 1.89 | 2.0% | 12 | .869 | .840 | .803 | .756 | .698 | .627 | .544 | .452 | .356 | .262 |
| | | | | | | | | .180 | .114 | | | | | | | | |
| 6.500 | 6.500 | .063 | .000 | .94 | 1.81 | 1.0% | 13 | .868 | .838 | .800 | .752 | .693 | .621 | .537 | .444 | .347 | .254 |
| | | | | | | | | .173 | .109 | .063 | | | | | | | |
| 7.000 | 7.000 | 2.057 | .000 | .41 | 1.96 | 22.7% | 9 | .901 | .883 | .861 | .835 | .802 | .762 | .713 | .655 | .588 | |
| 7.000 | 7.000 | .948 | .000 | .59 | 2.04 | 11.9% | 10 | .888 | .866 | .839 | .805 | .763 | .711 | .649 | .577 | .495 | .406 |
| 7.000 | 7.000 | .466 | .000 | .73 | 2.03 | 6.2% | 11 | .882 | .858 | .827 | .789 | .742 | .684 | .615 | .535 | .447 | .355 |
| | | | | | | | | .266 | | | | | | | | | |
| 7.000 | 7.000 | .231 | .000 | .83 | 1.97 | 3.2% | 12 | .879 | .853 | .821 | .781 | .730 | .669 | .596 | .513 | .422 | .329 |
| | | | | | | | | .241 | .165 | | | | | | | | |
| 7.000 | 7.000 | .113 | .000 | .90 | 1.90 | 1.6% | 13 | .877 | .851 | .817 | .776 | .724 | .661 | .587 | .502 | .409 | .316 |
| | | | | | | | | .229 | .154 | .096 | | | | | | | |
| 7.000 | 7.000 | .053 | .000 | .95 | 1.83 | .8% | 14 | .876 | .849 | .816 | .774 | .721 | .657 | .582 | .496 | .403 | .309 |
| | | | | | | | | .223 | .149 | .092 | .053 | | | | | | |
| 7.500 | 7.500 | 3.526 | .000 | .29 | 1.88 | 32.0% | 9 | .917 | .904 | .889 | .871 | .849 | .822 | .790 | .752 | .706 | |
| 7.500 | 7.500 | 1.528 | .000 | .49 | 2.05 | 16.9% | 10 | .900 | .883 | .861 | .834 | .801 | .761 | .712 | .654 | .586 | .509 |
| 7.500 | 7.500 | .747 | .000 | .65 | 2.09 | 9.1% | 11 | .892 | .871 | .846 | .814 | .775 | .727 | .668 | .600 | .522 | .437 |
| | | | | | | | | .349 | | | | | | | | | |
| 7.500 | 7.500 | .378 | .000 | .77 | 2.05 | 4.8% | 12 | .887 | .865 | .837 | .803 | .760 | .708 | .645 | .571 | .488 | .399 |
| | | | | | | | | .310 | .227 | | | | | | | | |
| 7.500 | 7.500 | .191 | .000 | .86 | 1.99 | 2.5% | 13 | .885 | .862 | .833 | .797 | .752 | .697 | .631 | .555 | .470 | .379 |
| | | | | | | | | .290 | .208 | .140 | | | | | | | |
| 7.500 | 7.500 | .094 | .000 | .92 | 1.92 | 1.2% | 14 | .884 | .860 | .831 | .794 | .748 | .692 | .624 | .547 | .460 | .369 |
| | | | | | | | | .280 | .199 | .132 | .082 | | | | | | |
| 7.500 | 7.500 | .045 | .000 | .96 | 1.84 | .6% | 15 | .883 | .859 | .829 | .792 | .745 | .689 | .621 | .542 | .455 | .364 |
| | | | | | | | | .275 | .194 | .128 | .079 | .045 | | | | | |
| 8.000 | 8.000 | 2.472 | .000 | .38 | 2.02 | 23.6% | 10 | .913 | .899 | .882 | .862 | .838 | .808 | .772 | .729 | .678 | .618 |
| 8.000 | 8.000 | 1.175 | .000 | .56 | 2.12 | 12.8% | 11 | .902 | .885 | .863 | .837 | .805 | .766 | .719 | .663 | .597 | .522 |
| | | | | | | | | .441 | | | | | | | | | |
| 8.000 | 8.000 | .598 | .000 | .70 | 2.12 | 7.0% | 12 | .896 | .877 | .853 | .823 | .787 | .743 | .689 | .625 | .552 | .471 |
| | | | | | | | | .385 | .299 | | | | | | | | |
| 8.000 | 8.000 | .309 | .000 | .81 | 2.08 | 3.7% | 13 | .893 | .872 | .847 | .816 | .777 | .730 | .672 | .605 | .527 | .443 |
| | | | | | | | | .355 | .270 | .193 | | | | | | | |
| 8.000 | 8.000 | .158 | .000 | .88 | 2.01 | 1.9% | 14 | .891 | .870 | .844 | .811 | .771 | .722 | .663 | .593 | .514 | .428 |
| | | | | | | | | .340 | .255 | .180 | .119 | | | | | | |
| 8.000 | 8.000 | .079 | .000 | .93 | 1.93 | 1.0% | 15 | .890 | .869 | .842 | .809 | .768 | .718 | .658 | .587 | .507 | .420 |

*SIMILAR TO A.T.&T. JACOBSEN RETRIAL TABLES

*SLOW RETRIALS: JEWETT-SHRAGO EXTENDED ERLANG B - 100% BLOCKED CALLERS RETRY, 0% OVERFLOW (RETRIES 3-5 HOLDING TIMES LATER)

| HOURS FIRST ATTEMPT VMR=1.0 | HOURS CON-NECT | HOURS RECALL /RETRY | HOURS OVFL | CON-NECT VMR | OVFL VMR | PERCENT BLOCKED (GRADE) | LINE | HOURS CONNECTED TRAFFIC ON LINES 1 | 2 | 3 | 4 | 5 | 6 | 7 | 8 | 9 | 10 |
|---|---|---|---|---|---|---|---|---|---|---|---|---|---|---|---|---|---|
| | | | | | | | | .331 | .247 | .173 | .113 | .069 | | | | | |
| 8.500 | 8.500 | 4.129 | .000 | .27 | 1.92 | 32.7% | 10 | .927 | .917 | .905 | .891 | .874 | .855 | .831 | .803 | .769 | .729 |
| 8.500 | 8.500 | 1.836 | .000 | .46 | 2.12 | 17.8% | 11 | .912 | .898 | .881 | .860 | .835 | .804 | .767 | .723 | .670 | .609 |
| | | | | | | | | .540 | | | | | | | | | |
| 8.500 | 8.500 | .926 | .000 | .62 | 2.17 | 9.8% | 12 | .904 | .888 | .867 | .843 | .812 | .775 | .731 | .677 | .614 | .543 |
| | | | | | | | | .464 | .381 | | | | | | | | |
| 8.500 | 8.500 | .485 | .000 | .74 | 2.15 | 5.4% | 13 | .900 | .882 | .860 | .833 | .800 | .759 | .710 | .651 | .583 | .506 |
| | | | | | | | | .423 | .338 | .257 | | | | | | | |
| 8.500 | 8.500 | .255 | .000 | .84 | 2.09 | 2.9% | 14 | .898 | .879 | .856 | .827 | .792 | .750 | .698 | .636 | .565 | .485 |
| | | | | | | | | .400 | .315 | .235 | .165 | | | | | | |
| 8.500 | 8.500 | .132 | .000 | .90 | 2.02 | 1.5% | 15 | .896 | .877 | .853 | .824 | .788 | .744 | .691 | .628 | .555 | .474 |
| | | | | | | | | .388 | .303 | .223 | .155 | .101 | | | | | |
| 8.500 | 8.500 | .066 | .000 | .94 | 1.94 | .8% | 16 | .895 | .876 | .852 | .823 | .786 | .741 | .687 | .623 | .549 | .468 |
| | | | | | | | | .381 | .296 | .217 | .150 | .097 | .058 | | | | |
| 9.000 | 9.000 | 2.897 | .000 | .36 | 2.07 | 24.4% | 11 | .923 | .911 | .898 | .883 | .864 | .841 | .814 | .782 | .743 | .697 |
| | | | | | | | | .644 | | | | | | | | | |
| 9.000 | 9.000 | 1.413 | .000 | .53 | 2.19 | 13.6% | 12 | .912 | .899 | .882 | .861 | .837 | .806 | .770 | .726 | .675 | .614 |
| | | | | | | | | .546 | .471 | | | | | | | | |
| 9.000 | 9.000 | .741 | .000 | .67 | 2.21 | 7.6% | 13 | .907 | .891 | .872 | .849 | .821 | .786 | .744 | .694 | .635 | .567 |
| | | | | | | | | .492 | .411 | .329 | | | | | | | |
| 9.000 | 9.000 | .396 | .000 | .78 | 2.17 | 4.2% | 14 | .904 | .887 | .867 | .842 | .812 | .774 | .729 | .675 | .612 | .540 |
| | | | | | | | | .461 | .378 | .296 | .220 | | | | | | |
| 9.000 | 9.000 | .211 | .000 | .86 | 2.11 | 2.3% | 15 | .902 | .885 | .864 | .838 | .806 | .768 | .721 | .665 | .599 | .525 |
| | | | | | | | | .444 | .360 | .278 | .204 | .140 | | | | | |
| 9.000 | 9.000 | .110 | .000 | .91 | 2.03 | 1.2% | 16 | .901 | .884 | .862 | .836 | .803 | .764 | .716 | .659 | .592 | .516 |
| | | | | | | | | .435 | .350 | .269 | .195 | .133 | .085 | | | | |
| 9.000 | 9.000 | .055 | .000 | .95 | 1.95 | .6% | 17 | .901 | .883 | .861 | .835 | .802 | .762 | .713 | .655 | .588 | .512 |
| | | | | | | | | .430 | .345 | .264 | .190 | .129 | .082 | .049 | | | |
| 9.500 | 9.500 | 4.739 | .000 | .25 | 1.95 | 33.3% | 11 | .934 | .926 | .917 | .906 | .893 | .878 | .860 | .839 | .813 | .784 |
| | | | | | | | | .749 | | | | | | | | | |
| 9.500 | 9.500 | 2.154 | .000 | .43 | 2.17 | 18.5% | 12 | .921 | .910 | .896 | .880 | .860 | .836 | .808 | .774 | .733 | .685 |
| | | | | | | | | .630 | .567 | | | | | | | | |
| 9.500 | 9.500 | 1.114 | .000 | .59 | 2.24 | 10.5% | 13 | .914 | .900 | .884 | .865 | .841 | .812 | .777 | .735 | .685 | .627 |
| | | | | | | | | .561 | .488 | .410 | | | | | | | |
| 9.500 | 9.500 | .600 | .000 | .72 | 2.24 | 5.9% | 14 | .910 | .895 | .878 | .856 | .829 | .797 | .759 | .712 | .657 | .593 |
| | | | | | | | | .522 | .444 | .363 | .284 | | | | | | |
| 9.500 | 9.500 | .326 | .000 | .81 | 2.19 | 3.3% | 15 | .908 | .892 | .874 | .851 | .823 | .789 | .748 | .699 | .640 | .574 |
| | | | | | | | | .499 | .419 | .338 | .259 | .189 | | | | | |
| 9.500 | 9.500 | .175 | .000 | .88 | 2.12 | 1.8% | 16 | .906 | .891 | .871 | .848 | .819 | .784 | .742 | .691 | .631 | .562 |
| | | | | | | | | .486 | .405 | .323 | .245 | .176 | .120 | | | | |

*SIMILAR TO A.T.&T. JACOBSEN RETRIAL TABLES

# *SLOW RETRIALS: JEWETT-SHRAGO EXTENDED ERLANG B - 100% BLOCKED CALLERS RETRY, 0% OVERFLOW (RETRIES 3-5 HOLDING TIMES LATER)

| HOURS FIRST ATTEMPT VMR=1.0 | HOURS CON-NECT | HOURS RECALL /RETRY | HOURS OVFL | CON-NECT VMR | OVFL VMR | PERCENT BLOCKED (GRADE) | LINE | HOURS CONNECTED TRAFFIC ON LINES 1 | 2 | 3 | 4 | 5 | 6 | 7 | 8 | 9 | 10 |
|---|---|---|---|---|---|---|---|---|---|---|---|---|---|---|---|---|---|
| 9.500 | 9.500 | .091 | .000 | .93 | 2.04 | 1.0% | 17 | .906 | .890 | .870 | .846 | .817 | .781 | .738 | .686 | .625 | .556 |
| | | | | | | | | .479 | .397 | .315 | .238 | .170 | .114 | .072 | | | |
| 10.000 | 10.000 | 3.330 | .000 | .33 | 2.11 | 25.0% | 12 | .930 | .921 | .911 | .898 | .883 | .866 | .845 | .820 | .790 | .755 |
| | | | | | | | | .714 | .667 | | | | | | | | |
| 10.000 | 10.000 | 1.660 | .000 | .50 | 2.25 | 14.2% | 13 | .921 | .910 | .896 | .880 | .860 | .836 | .808 | .774 | .733 | .686 |
| | | | | | | | | .630 | .568 | .498 | | | | | | | |
| 10.000 | 10.000 | .892 | .000 | .64 | 2.29 | 8.2% | 14 | .916 | .903 | .888 | .869 | .846 | .819 | .786 | .746 | .699 | .644 |
| | | | | | | | | .581 | .511 | .435 | .357 | | | | | | |
| 10.000 | 10.000 | .491 | .000 | .75 | 2.26 | 4.7% | 15 | .913 | .899 | .883 | .863 | .838 | .809 | .773 | .730 | .679 | .619 |
| | | | | | | | | .552 | .478 | .399 | .320 | .245 | | | | | |
| 10.000 | 10.000 | .269 | .000 | .84 | 2.20 | 2.6% | 16 | .911 | .897 | .880 | .859 | .833 | .802 | .765 | .720 | .667 | .605 |
| | | | | | | | | .535 | .459 | .379 | .300 | .226 | .162 | | | | |
| 10.000 | 10.000 | .145 | .000 | .90 | 2.12 | 1.4% | 17 | .910 | .896 | .878 | .857 | .831 | .799 | .760 | .714 | .660 | .596 |
| | | | | | | | | .525 | .448 | .368 | .289 | .216 | .152 | .102 | | | |
| 10.000 | 10.000 | .076 | .000 | .94 | 2.04 | .8% | 18 | .910 | .895 | .877 | .855 | .829 | .797 | .758 | .711 | .656 | .592 |
| | | | | | | | | .520 | .442 | .361 | .282 | .210 | .147 | .098 | .061 | | |
| 11.000 | 11.000 | 3.771 | .000 | .31 | 2.15 | 25.5% | 13 | .937 | .929 | .920 | .910 | .898 | .884 | .868 | .848 | .825 | .798 |
| | | | | | | | | .766 | .729 | .686 | | | | | | | |
| 11.000 | 11.000 | 1.914 | .000 | .48 | 2.31 | 14.8% | 14 | .928 | .919 | .907 | .894 | .878 | .859 | .837 | .810 | .778 | .740 |
| | | | | | | | | .696 | .645 | .587 | .522 | | | | | | |
| 11.000 | 11.000 | 1.050 | .000 | .62 | 2.36 | 8.7% | 15 | .923 | .913 | .900 | .884 | .866 | .844 | .818 | .786 | .749 | .704 |
| | | | | | | | | .653 | .594 | .528 | .456 | .382 | | | | | |
| 11.000 | 11.000 | .591 | .000 | .73 | 2.34 | 5.1% | 16 | .921 | .909 | .895 | .879 | .859 | .835 | .806 | .772 | .731 | .682 |
| | | | | | | | | .626 | .563 | .493 | .418 | .342 | .269 | | | | |
| 11.000 | 11.000 | .334 | .000 | .82 | 2.29 | 2.9% | 17 | .919 | .907 | .893 | .875 | .854 | .829 | .799 | .763 | .720 | .669 |
| | | | | | | | | .610 | .544 | .472 | .396 | .320 | .247 | .182 | | | |
| 11.000 | 11.000 | .186 | .000 | .88 | 2.21 | 1.7% | 18 | .918 | .906 | .891 | .873 | .852 | .826 | .795 | .757 | .713 | .661 |
| | | | | | | | | .601 | .533 | .460 | .383 | .307 | .235 | .171 | .118 | | |
| 11.000 | 11.000 | .101 | .000 | .93 | 2.13 | .9% | 19 | .917 | .905 | .890 | .872 | .850 | .824 | .792 | .754 | .709 | .656 |
| | | | | | | | | .595 | .527 | .453 | .376 | .299 | .228 | .165 | .113 | .074 | |
| 12.000 | 12.000 | 4.217 | .000 | .30 | 2.19 | 26.0% | 14 | .942 | .936 | .928 | .920 | .910 | .899 | .885 | .870 | .852 | .830 |
| | | | | | | | | .805 | .776 | .742 | .703 | | | | | | |
| 12.000 | 12.000 | 2.174 | .000 | .46 | 2.37 | 15.3% | 15 | .934 | .926 | .917 | .906 | .893 | .877 | .859 | .837 | .812 | .782 |
| | | | | | | | | .747 | .705 | .658 | .604 | .544 | | | | | |
| 12.000 | 12.000 | 1.215 | .000 | .59 | 2.43 | 9.2% | 16 | .930 | .921 | .910 | .897 | .882 | .864 | .843 | .817 | .787 | .751 |
| | | | | | | | | .709 | .661 | .605 | .543 | .476 | .405 | | | | |
| 12.000 | 12.000 | .699 | .000 | .71 | 2.42 | 5.5% | 17 | .927 | .917 | .906 | .892 | .875 | .856 | .832 | .804 | .771 | .732 |
| | | | | | | | | .686 | .633 | .573 | .507 | .436 | .363 | .291 | | | |
| 12.000 | 12.000 | .403 | .000 | .80 | 2.37 | 3.3% | 18 | .925 | .915 | .903 | .889 | .871 | .851 | .826 | .796 | .761 | .720 |
| | | | | | | | | .671 | .616 | .553 | .484 | .412 | .338 | .267 | .202 | | |

*SIMILAR TO A.T.&T. JACOBSEN RETRIAL TABLES

# *SLOW RETRIALS: JEWETT-SHRAGO EXTENDED ERLANG B - 100% BLOCKED CALLERS RETRY, 0% OVERFLOW (RETRIES 3-5 HOLDING TIMES LATER)

| HOURS FIRST ATTEMPT VMR = 1.0 | HOURS CON-NECT | HOURS RECALL /RETRY | HOURS OVFL | CON-NECT VMR | VFL VMR | PERCENT BLOCKED (GRADE) | LINE | HOURS CONNECTED TRAFFIC ON LINES 1 | 2 | 3 | 4 | 5 | 6 | 7 | 8 | 9 | 10 |
|---|---|---|---|---|---|---|---|---|---|---|---|---|---|---|---|---|---|
| 12.000 | 12.000 | .231 | .000 | .87 | 2.30 | 1.9% | 19 | .924 | .914 | .901 | .887 | .869 | .848 | .822 | .792 | .755 | .712 |
| | | | | | | | | .662 | .605 | .541 | .471 | .397 | .323 | .253 | .189 | .135 | |
| 12.000 | 12.000 | .129 | .000 | .91 | 2.22 | 1.1% | 20 | .924 | .913 | .901 | .885 | .867 | .846 | .820 | .789 | .752 | .708 |
| | | | | | | | | .657 | .599 | .533 | .463 | .389 | .315 | .245 | .182 | .128 | .086 |
| 12.000 | 12.000 | .071 | .000 | .95 | 2.14 | .6% | 21 | .924 | .913 | .900 | .885 | .866 | .845 | .818 | .787 | .749 | .705 |
| | | | | | | | | .654 | .595 | .529 | .458 | .384 | .310 | .240 | .177 | .125 | .084 |
| | | | | | | | | .053 | | | | | | | | | |
| 13.000 | 13.000 | 4.668 | .000 | .28 | 2.22 | 26.4% | 15 | .946 | .941 | .935 | .928 | .920 | .910 | .899 | .887 | .872 | .855 |
| | | | | | | | | .835 | .812 | .785 | .754 | .719 | | | | | |
| 13.000 | 13.000 | 2.441 | .000 | .44 | 2.42 | 15.8% | 16 | .939 | .932 | .924 | .915 | .904 | .891 | .876 | .859 | .838 | .814 |
| | | | | | | | | .786 | .753 | .714 | .670 | .620 | .563 | | | | |
| 13.000 | 13.000 | 1.385 | .000 | .57 | 2.49 | 9.6% | 17 | .935 | .927 | .918 | .907 | .895 | .880 | .862 | .841 | .817 | .788 |
| | | | | | | | | .754 | .714 | .668 | .616 | .558 | .494 | .426 | | | |
| 13.000 | 13.000 | .811 | .000 | .69 | 2.49 | 5.9% | 18 | .933 | .924 | .914 | .902 | .889 | .872 | .853 | .830 | .803 | .771 |
| | | | | | | | | .733 | .689 | .639 | .582 | .519 | .452 | .382 | .312 | | |
| 13.000 | 13.000 | .478 | .000 | .78 | 2.45 | 3.5% | 19 | .931 | .922 | .912 | .900 | .885 | .868 | .847 | .823 | .794 | .760 |
| | | | | | | | | .720 | .674 | .621 | .561 | .495 | .426 | .355 | .285 | .221 | |
| 13.000 | 13.000 | .280 | .000 | .85 | 2.38 | 2.1% | 20 | .930 | .921 | .910 | .898 | .883 | .865 | .844 | .819 | .789 | .753 |
| | | | | | | | | .712 | .664 | .609 | .548 | .481 | .410 | .339 | .270 | .206 | .151 |
| 13.000 | 13.000 | .161 | .000 | .90 | 2.30 | 1.2% | 21 | .929 | .920 | .909 | .896 | .881 | .863 | .842 | .816 | .785 | .749 |
| | | | | | | | | .707 | .658 | .602 | .540 | .472 | .401 | .329 | .260 | .197 | .143 |
| | | | | | | | | .099 | | | | | | | | | |
| 13.000 | 13.000 | .091 | .000 | .94 | 2.22 | .7% | 22 | .929 | .920 | .909 | .896 | .880 | .862 | .840 | .814 | .783 | .747 |
| | | | | | | | | .704 | .654 | .598 | .535 | .466 | .395 | .323 | .255 | .192 | .139 |
| | | | | | | | | .096 | .063 | | | | | | | | |
| 14.000 | 14.000 | 5.122 | .000 | .27 | 2.25 | 26.8% | 16 | .950 | .946 | .940 | .934 | .927 | .920 | .911 | .900 | .888 | .874 |
| | | | | | | | | .858 | .840 | .819 | .794 | .765 | .732 | | | | |
| 14.000 | 14.000 | 2.712 | .000 | .42 | 2.46 | 16.2% | 17 | .944 | .938 | .931 | .923 | .914 | .903 | .891 | .876 | .859 | .840 |
| | | | | | | | | .817 | .790 | .758 | .722 | .681 | .634 | .581 | | | |
| 14.000 | 14.000 | 1.560 | .000 | .55 | 2.55 | 10.0% | 18 | .940 | .933 | .925 | .916 | .905 | .893 | .878 | .861 | .841 | .817 |
| | | | | | | | | .789 | .757 | .719 | .675 | .626 | .571 | .510 | .446 | | |
| 14.000 | 14.000 | .928 | .000 | .67 | 2.56 | 6.2% | 19 | .937 | .930 | .921 | .911 | .900 | .886 | .870 | .851 | .828 | .802 |
| | | | | | | | | .771 | .735 | .693 | .645 | .591 | .531 | .467 | .399 | .331 | |
| 14.000 | 14.000 | .557 | .000 | .76 | 2.52 | 3.8% | 20 | .936 | .928 | .919 | .909 | .896 | .882 | .865 | .844 | .821 | .793 |
| | | | | | | | | .759 | .721 | .676 | .625 | .568 | .506 | .439 | .371 | .303 | .239 |
| 14.000 | 14.000 | .333 | .000 | .83 | 2.46 | 2.3% | 21 | .935 | .927 | .918 | .907 | .894 | .879 | .861 | .840 | .816 | .786 |
| | | | | | | | | .752 | .712 | .666 | .613 | .554 | .490 | .422 | .353 | .286 | .222 |
| | | | | | | | | .166 | | | | | | | | | |
| 14.000 | 14.000 | .196 | .000 | 89 | 2.38 | 1.4% | 22 | .934 | .928 | .917 | .906 | .893 | .877 | .859 | .838 | .812 | .782 |
| | | | | | | | | .747 | .706 | .659 | .605 | .545 | .480 | .412 | .343 | .275 | .213 |

*SIMILAR TO A.T.&T. JACOBSEN RETRIAL TABLES

*SLOW RETRIALS: JEWETT-SHRAGO EXTENDED ERLANG B - 100% BLOCKED CALLERS RETRY, 0% OVERFLOW (RETRIES 3-5 HOLDING TIMES LATER)

| HOURS FIRST ATTEMPT VMR=1.0 | HOURS CON-NECT | HOURS RECALL /RETRY | HOURS OVFL | CON-NECT VMR | OVFL VMR | PERCENT BLOCKED (GRADE) | LINE | HOURS CONNECTED TRAFFIC ON LINES 1 | 2 | 3 | 4 | 5 | 6 | 7 | 8 | 9 | 10 |
|---|---|---|---|---|---|---|---|---|---|---|---|---|---|---|---|---|---|
| 14.000 | 14.000 | .113 | .000 | .93 | 2.30 | .8% | 23 | .934 | .926 | .916 | .905 | .892 | .876 | .858 | .836 | .810 | .780 |
| | | | | | | | | .744 | .703 | .655 | .600 | .540 | .474 | .405 | .336 | .269 | .207 |
| | | | | | | | | .153 | .108 | .073 | | | | | | | |
| 15.000 | 15.000 | 5.582 | .000 | .26 | 2.28 | 27.1% | 17 | .954 | .950 | .945 | .940 | .934 | .927 | .920 | .911 | .901 | .890 |
| | | | | | | | | .877 | .862 | .844 | .824 | .801 | .775 | .745 | | | |
| 15.000 | 15.000 | 2.987 | .000 | .40 | 2.51 | 16.6% | 18 | .947 | .942 | .936 | .929 | .921 | .912 | .902 | .890 | .876 | .860 |
| | | | | | | | | .841 | .819 | .793 | .764 | .730 | .691 | .647 | .598 | | |
| 15.000 | 15.000 | 1.741 | .000 | .54 | 2.61 | 10.4% | 19 | .944 | .938 | .931 | .923 | .914 | .903 | .891 | .876 | .860 | .840 |
| | | | | | | | | .817 | .790 | .759 | .723 | .682 | .635 | .583 | .526 | .464 | |
| 15.000 | 15.000 | 1.050 | .000 | .65 | 2.62 | 6.5% | 20 | .941 | .935 | .927 | .919 | .909 | .897 | .884 | .868 | .849 | .827 |
| | | | | | | | | .801 | .771 | .737 | .696 | .651 | .599 | .542 | .480 | .416 | .350 |
| 15.000 | 15.000 | .640 | .000 | .74 | 2.59 | 4.1% | 21 | .940 | .933 | .925 | .916 | .906 | .893 | .879 | .862 | .842 | .819 |
| | | | | | | | | .791 | .759 | .722 | .679 | .630 | .575 | .516 | .452 | .386 | .319 |
| | | | | | | | | .256 | | | | | | | | | |
| 15.000 | 15.000 | .389 | .000 | .82 | 2.53 | 2.5% | 22 | .939 | .932 | .924 | .915 | .904 | .891 | .876 | .858 | .837 | .813 |
| | | | | | | | | .784 | .751 | .712 | .667 | .617 | .560 | .499 | .434 | .367 | .301 |
| | | | | | | | | .238 | .182 | | | | | | | | |
| 15.000 | 15.000 | .234 | .000 | .88 | 2.46 | 1.5% | 23 | .938 | .931 | .923 | .914 | .902 | .889 | .874 | .856 | .834 | .809 |
| | | | | | | | | .780 | .746 | .706 | .660 | .608 | .551 | .488 | .422 | .355 | .289 |
| | | | | | | | | .227 | .172 | .125 | | | | | | | |
| 15.000 | 15.000 | .138 | .000 | .92 | 2.38 | .9% | 24 | .938 | .931 | .923 | .913 | .902 | .888 | .873 | .854 | .833 | .807 |
| | | | | | | | | .777 | .742 | .702 | .655 | .603 | .545 | .481 | .415 | .348 | .282 |
| | | | | | | | | .220 | .166 | .120 | .083 | | | | | | |
| 15.000 | 15.000 | .080 | .000 | .95 | 2.29 | .5% | 25 | .938 | .931 | .922 | .912 | .901 | .888 | .872 | .853 | .831 | .806 |
| | | | | | | | | .776 | .740 | .699 | .653 | .600 | .541 | .477 | .411 | .343 | .277 |
| | | | | | | | | .216 | .162 | .117 | .080 | .053 | | | | | |
| 16.000 | 16.000 | 6.043 | .000 | .24 | 2.30 | 27.4% | 18 | .957 | .953 | .949 | .945 | .939 | .934 | .927 | .920 | .912 | .902 |
| | | | | | | | | .891 | .879 | .865 | .849 | .830 | .808 | .784 | .756 | | |
| 16.000 | 16.000 | 3.266 | .000 | .39 | 2.55 | 17.0% | 19 | .951 | .946 | .941 | .935 | .928 | .920 | .912 | .901 | .890 | .876 |
| | | | | | | | | .860 | .842 | .821 | .797 | .769 | .737 | .700 | .659 | .613 | |
| 16.000 | 16.000 | 1.924 | .000 | .52 | 2.66 | 10.7% | 20 | .947 | .942 | .936 | .929 | .921 | .912 | .902 | .889 | .875 | .859 |
| | | | | | | | | .840 | .818 | .792 | .762 | .728 | .688 | .644 | .594 | .540 | .481 |
| 16.000 | 16.000 | 1.175 | .000 | .63 | 2.69 | 6.8% | 21 | .945 | .939 | .933 | .925 | .917 | .907 | .895 | .882 | .866 | .847 |
| | | | | | | | | .826 | .801 | .772 | .739 | .700 | .656 | .607 | .552 | .493 | .431 |
| | | | | | | | | .367 | | | | | | | | | |
| 16.000 | 16.000 | .726 | .000 | .73 | 2.66 | 4.3% | 22 | .944 | .938 | .931 | .923 | .914 | .903 | .891 | .876 | .860 | .840 |
| | | | | | | | | .817 | .790 | .759 | .723 | .681 | .635 | .582 | .525 | .463 | .399 |
| | | | | | | | | .335 | .272 | | | | | | | | |
| 16.000 | 16.000 | .449 | .000 | .80 | 2.61 | 2.7% | 23 | .943 | .937 | .929 | .921 | .912 | .901 | .888 | .873 | .855 | .835 |

*SIMILAR TO A.T.&T. JACOBSEN RETRIAL TABLES

*SLOW RETRIALS: JEWETT-SHRAGO EXTENDED ERLANG B - 100% BLOCKED CALLERS RETRY, 0% OVERFLOW (RETRIES 3-5 HOLDING TIMES LATER)

| HOURS FIRST ATTEMPT VMR=1.0 | HOURS CON-NECT | HOURS RECALL /RETRY | HOURS OVFL | CON-NECT VMR | OVFL VMR | PERCENT BLOCKED (GRADE) | LINE | HOURS CONNECTED TRAFFIC ON LINES 1 | 2 | 3 | 4 | 5 | 6 | 7 | 8 | 9 | 10 |
|---|---|---|---|---|---|---|---|---|---|---|---|---|---|---|---|---|---|
| | | | | | | | | .811 | .783 | .750 | .712 | .669 | .620 | .566 | .507 | .444 | .379 |
| | | | | | | | | .315 | .253 | .196 | | | | | | | |
| 16.000 | 16.000 | .275 | .000 | .86 | 2.53 | 1.7% | 24 | .942 | .936 | .929 | .920 | .911 | .899 | .886 | .871 | .853 | .832 |
| | | | | | | | | .807 | .778 | .744 | .706 | .661 | .611 | .556 | .496 | .432 | .367 |
| | | | | | | | | .302 | .241 | .185 | .137 | | | | | | |
| 16.000 | 16.000 | .165 | .000 | .91 | 2.45 | 1.0% | 25 | .942 | .935 | .928 | .920 | .910 | .898 | .885 | .869 | .851 | .829 |
| | | | | | | | | .804 | .775 | .741 | .701 | .656 | .605 | .549 | .488 | .424 | .358 |
| | | | | | | | | .294 | .233 | .178 | .131 | .093 | | | | | |
| 16.000 | 16.000 | .098 | .000 | .94 | 2.37 | .6% | 26 | .942 | .935 | .928 | .919 | .909 | .898 | .884 | .868 | .850 | .828 |
| | | | | | | | | .803 | .773 | .738 | .698 | .653 | .602 | .545 | .484 | .419 | .354 |
| | | | | | | | | .289 | .229 | .174 | .128 | .090 | .061 | | | | |
| 17.000 | 17.000 | 6.508 | .000 | .23 | 2.32 | 27.7% | 19 | .959 | .956 | .953 | .949 | .944 | .939 | .934 | .927 | .920 | .912 |
| | | | | | | | | .903 | .893 | .881 | .868 | .852 | .835 | .815 | .792 | .766 | |
| 17.000 | 17.000 | 3.549 | .000 | .37 | 2.59 | 17.3% | 20 | .954 | .950 | .945 | .940 | .934 | .927 | .920 | .911 | .901 | .889 |
| | | | | | | | | .876 | .861 | .844 | .824 | .801 | .774 | .744 | .709 | .670 | .627 |
| 17.000 | 17.000 | 2.112 | .000 | .50 | 2.71 | 11.1% | 21 | .950 | .946 | .940 | .934 | .927 | .920 | .911 | .900 | .888 | .874 |
| | | | | | | | | .858 | .840 | .818 | .793 | .765 | .732 | .695 | .652 | .605 | .553 |
| | | | | | | | | .497 | | | | | | | | | |
| 17.000 | 17.000 | 1.304 | .000 | .62 | 2.75 | 7.1% | 22 | .948 | .943 | .937 | .931 | .923 | .915 | .905 | .893 | .880 | .864 |
| | | | | | | | | .846 | .825 | .801 | .773 | .740 | .703 | .661 | .614 | .562 | .505 |
| | | | | | | | | .445 | .384 | | | | | | | | |
| 17.000 | 17.000 | .816 | .000 | .71 | 2.73 | 4.6% | 23 | .947 | .942 | .936 | .929 | .921 | .911 | .901 | .888 | .874 | .857 |
| | | | | | | | | .838 | .815 | .789 | .759 | .724 | .684 | .639 | .589 | .533 | .474 |
| | | | | | | | | .412 | .350 | .288 | | | | | | | |
| 17.000 | 17.000 | .512 | .000 | .79 | 2.68 | 2.9% | 24 | .946 | .941 | .934 | .927 | .919 | .909 | .898 | .885 | .870 | .853 |
| | | | | | | | | .832 | .809 | .781 | .750 | .713 | .671 | .624 | .572 | .515 | .454 |
| | | | | | | | | .391 | .328 | .267 | .211 | | | | | | |
| 17.000 | 17.000 | .318 | .000 | .85 | 2.61 | 1.8% | 25 | .945 | .940 | .933 | .926 | .918 | .908 | .896 | .883 | .868 | .850 |
| | | | | | | | | .829 | .804 | .776 | .743 | .706 | .663 | .614 | .561 | .502 | .441 |
| | | | | | | | | .378 | .314 | .254 | .198 | .150 | | | | | |
| 17.000 | 17.000 | .195 | .000 | .90 | 2.52 | 1.1% | 26 | .945 | .939 | .933 | .925 | .917 | .907 | .895 | .882 | .866 | .848 |
| | | | | | | | | .826 | .802 | .773 | .739 | .701 | .657 | .608 | .553 | .495 | .432 |
| | | | | | | | | .369 | .306 | .246 | .191 | .143 | .103 | | | | |
| 17.000 | 17.000 | .117 | .000 | .93 | 2.44 | .7% | 27 | .945 | .939 | .933 | .925 | .916 | .906 | .894 | .881 | .865 | .846 |
| | | | | | | | | .825 | .800 | .770 | .737 | .698 | .653 | .604 | .549 | .490 | .427 |
| | | | | | | | | .363 | .300 | .240 | .186 | .139 | .100 | .069 | | | |
| 18.000 | 18.000 | 6.975 | .000 | .22 | 2.35 | 27.9% | 20 | .962 | .959 | .956 | .952 | .948 | .944 | .939 | .934 | .928 | .921 |
| | | | | | | | | .913 | .904 | .895 | .883 | .871 | .856 | .840 | .821 | .800 | .776 |
| 18.000 | 18.000 | 3.835 | .000 | .36 | 2.62 | 17.6% | 21 | .958 | .953 | .949 | .944 | .939 | .933 | .926 | .919 | .910 | .901 |
| | | | | | | | | .889 | .877 | .862 | .845 | .826 | .804 | .779 | .750 | .718 | .681 |

*SIMILAR TO A.T.&T. JACOBSEN RETRIAL TABLES

# *SLOW RETRIALS: JEWETT-SHRAGO EXTENDED ERLANG B - 100% BLOCKED CALLERS RETRY, 0% OVERFLOW (RETRIES 3-5 HOLDING TIMES LATER)

| HOURS FIRST ATTEMPT VMR = 1.0 | HOURS CON-NECT | HOURS RECALL /RETRY | HOURS OVFL | CON-NECT VMR | OVFL VMR | PERCENT BLOCKED (GRADE) | LINE | HOURS CONNECTED TRAFFIC ON LINES 1 | 2 | 3 | 4 | 5 | 6 | 7 | 8 | 9 | 10 |
|---|---|---|---|---|---|---|---|---|---|---|---|---|---|---|---|---|---|
| 18.000 | 18.000 | 2.303 | .000 | .49 | 2.76 | 11.3% | 22 | .953 | .949 | .944 | .939 | .933 | .926 | .918 | .909 | .899 | .887 |
| | | | | | | | | .874 | .858 | .840 | .819 | .795 | .768 | .736 | .700 | .660 | .615 |
| | | | | | | | | .566 | .512 | | | | | | | | |
| 18.000 | 18.000 | 1.437 | .000 | .60 | 2.80 | 7.4% | 23 | .951 | .947 | .941 | .936 | .929 | .921 | .913 | .903 | .891 | .878 |
| | | | | | | | | .863 | .845 | .825 | .801 | .774 | .742 | .707 | .666 | .621 | .571 |
| | | | | | | | | .517 | .459 | .399 | | | | | | | |
| 18.000 | 18.000 | .910 | .000 | .70 | 2.79 | 4.8% | 24 | .950 | .945 | .940 | .933 | .926 | .918 | .909 | .898 | .886 | .872 |
| | | | | | | | | .855 | .836 | .814 | .789 | .759 | .725 | .687 | .643 | .595 | .542 |
| | | | | | | | | .485 | .425 | .364 | .303 | | | | | | |
| 18.000 | 18.000 | .577 | .000 | .78 | 2.74 | 3.1% | 25 | .949 | .944 | .938 | .932 | .925 | .916 | .907 | .896 | .883 | .868 |
| | | | | | | | | .850 | .830 | .807 | .780 | .749 | .714 | .673 | .628 | .577 | .522 |
| | | | | | | | | .463 | .402 | .341 | .281 | .225 | | | | | |
| 18.000 | 18.000 | .364 | .000 | .84 | 2.68 | 2.0% | 26 | .948 | .943 | .938 | .931 | .924 | .915 | .905 | .894 | .880 | .865 |
| | | | | | | | | .847 | .826 | .802 | .774 | .742 | .706 | .664 | .617 | .565 | .509 |
| | | | | | | | | .449 | .388 | .326 | .267 | .211 | .162 | | | | |
| 18.000 | 18.000 | .227 | .000 | .89 | 2.60 | 1.2% | 27 | .948 | .943 | .937 | .930 | .923 | .914 | .904 | .892 | .879 | .863 |
| | | | | | | | | .845 | .824 | .799 | .771 | .738 | .700 | .658 | .610 | .558 | .501 |
| | | | | | | | | .440 | .378 | .317 | .257 | .203 | .154 | .113 | | | |
| 18.000 | 18.000 | .139 | .000 | .92 | 2.51 | .8% | 28 | .948 | .943 | .937 | .930 | .922 | .913 | .903 | .891 | .878 | .862 |
| | | | | | | | | .843 | .822 | .797 | .768 | .735 | .697 | .654 | .606 | .552 | .495 |
| | | | | | | | | .434 | .372 | .311 | .252 | .197 | .150 | .109 | .077 | | |
| 19.000 | 19.000 | 7.443 | .000 | .22 | 2.37 | 28.1% | 21 | .964 | .961 | .958 | .955 | .952 | .948 | .944 | .939 | .934 | .928 |
| | | | | | | | | .921 | .914 | .906 | .896 | .885 | .873 | .860 | .844 | .827 | .807 |
| | | | | | | | | .784 | | | | | | | | | |
| 19.000 | 19.000 | 4.122 | .000 | .35 | 2.66 | 17.8% | 22 | .959 | .955 | .952 | .948 | .943 | .938 | .932 | .926 | .918 | .910 |
| | | | | | | | | .900 | .890 | .877 | .863 | .847 | .829 | .808 | .784 | .756 | .725 |
| | | | | | | | | .690 | .651 | | | | | | | | |
| 19.000 | 19.000 | 2.497 | .000 | .47 | 2.80 | 11.6% | 23 | .956 | .952 | .948 | .943 | .938 | .931 | .925 | .917 | .908 | .898 |
| | | | | | | | | .886 | .873 | .858 | .840 | .820 | .797 | .770 | .740 | .706 | .667 |
| | | | | | | | | .625 | .577 | .526 | | | | | | | |
| 19.000 | 19.000 | 1.572 | .000 | .59 | 2.86 | 7.6% | 24 | .954 | .950 | .945 | .940 | .934 | .927 | .920 | .911 | .901 | .890 |
| | | | | | | | | .877 | .862 | .844 | .824 | .801 | .775 | .745 | .710 | .671 | .628 |
| | | | | | | | | .580 | .527 | .472 | .414 | | | | | | |
| 19.000 | 19.000 | 1.006 | .000 | .68 | 2.85 | 5.0% | 25 | .952 | .948 | .943 | .938 | .932 | .924 | .916 | .907 | .896 | .884 |
| | | | | | | | | .870 | .854 | .835 | .813 | .788 | .759 | .727 | .689 | .647 | .601 |
| | | | | | | | | .549 | .494 | .436 | .377 | .318 | | | | | |
| 19.000 | 19.000 | .646 | .000 | .76 | 2.81 | 3.3% | 26 | .952 | .947 | .942 | .936 | .930 | .923 | .914 | .904 | .893 | .880 |
| | | | | | | | | .865 | .848 | .829 | .806 | .779 | .749 | .714 | .675 | .631 | .582 |
| | | | | | | | | .529 | .472 | .413 | .353 | .294 | .238 | | | | |

*SIMILAR TO A.T.&T. JACOBSEN RETRIAL TABLES

*SLOW RETRIALS: JEWETT-SHRAGO EXTENDED ERLANG B - 100% BLOCKED CALLERS RETRY, 0% OVERFLOW (RETRIES 3-5 HOLDING TIMES LATER)

| HOURS FIRST ATTEMPT VMR = 1.0 | HOURS CON-NECT | HOURS RECALL /RETRY | HOURS OVFL | CON-NECT VMR | OVFL VMR | PERCENT BLOCKED (GRADE) | LINE | HOURS CONNECTED TRAFFIC ON LINES 1 | 2 | 3 | 4 | 5 | 6 | 7 | 8 | 9 | 10 |
|---|---|---|---|---|---|---|---|---|---|---|---|---|---|---|---|---|---|
| 19.000 | 19.000 | .413 | .000 | .83 | 2.74 | 2.1% | 27 | .951 | .946 | .941 | .935 | .929 | .921 | .913 | .903 | .891 | .878 |
| | | | | | | | | .862 | .845 | .824 | .800 | .773 | .742 | .706 | .665 | .620 | .570 |
| | | | | | | | | .515 | .458 | .398 | .337 | .279 | .224 | .174 | | | |
| 19.000 | 19.000 | .261 | .000 | .88 | 2.67 | 1.4% | 28 | .951 | .946 | .941 | .935 | .928 | .920 | .912 | .901 | .890 | .876 |
| | | | | | | | | .860 | .842 | .821 | .797 | .769 | .737 | .700 | .659 | .613 | .561 |
| | | | | | | | | .506 | .448 | .388 | .327 | .269 | .214 | .165 | .124 | | |
| 19.000 | 19.000 | .162 | .000 | .91 | 2.58 | .8% | 29 | .950 | .946 | .941 | .935 | .928 | .920 | .911 | .901 | .889 | .875 |
| | | | | | | | | .859 | .841 | .819 | .795 | .766 | .734 | .697 | .655 | .608 | .556 |
| | | | | | | | | .500 | .441 | .381 | .320 | .262 | .208 | .160 | .119 | .085 | |
| 19.000 | 19.000 | .099 | .000 | .94 | 2.49 | .5% | 30 | .950 | .946 | .940 | .934 | .927 | .919 | .910 | .900 | .888 | .874 |
| | | | | | | | | .858 | .840 | .818 | .793 | .765 | .732 | .694 | .652 | .604 | .552 |
| | | | | | | | | .496 | .437 | .377 | .316 | .258 | .204 | .157 | .116 | .083 | .057 |
| 20.000 | 20.000 | 7.915 | .000 | .21 | 2.38 | 28.4% | 22 | .965 | .963 | .961 | .958 | .955 | .951 | .948 | .944 | .939 | .934 |
| | | | | | | | | .928 | .922 | .915 | .907 | .898 | .888 | .876 | .863 | .849 | .832 |
| | | | | | | | | .813 | .792 | | | | | | | | |
| 20.000 | 20.000 | 4.413 | .000 | .34 | 2.69 | 18.1% | 23 | .961 | .958 | .954 | .951 | .947 | .942 | .937 | .931 | .925 | .918 |
| | | | | | | | | .910 | .900 | .890 | .878 | .864 | .849 | .831 | .811 | .788 | .762 |
| | | | | | | | | .732 | .699 | .662 | | | | | | | |
| 20.000 | 20.000 | 2.693 | .000 | .46 | 2.84 | 11.9% | 24 | .958 | .954 | .951 | .946 | .942 | .936 | .930 | .923 | .916 | .907 |
| | | | | | | | | .897 | .886 | .872 | .858 | .841 | .821 | .799 | .773 | .744 | .711 |
| | | | | | | | | .674 | .633 | .588 | .539 | | | | | | |
| 20.000 | 20.000 | 1.710 | .000 | .57 | 2.91 | 7.9% | 25 | .956 | .952 | .948 | .944 | .938 | .932 | .926 | .918 | .909 | .900 |
| | | | | | | | | .888 | .875 | .861 | .844 | .824 | .802 | .776 | .747 | .713 | .676 |
| | | | | | | | | .634 | .588 | .538 | .484 | .428 | | | | | |
| 20.000 | 20.000 | 1.105 | .000 | .67 | 2.91 | 5.2% | 26 | .955 | .951 | .947 | .942 | .936 | .930 | .923 | .914 | .905 | .895 |
| | | | | | | | | .882 | .868 | .852 | .834 | .812 | .788 | .760 | .728 | .692 | .651 |
| | | | | | | | | .606 | .557 | .504 | .447 | .390 | .332 | | | | |
| 20.000 | 20.000 | .717 | .000 | .75 | 2.87 | 3.5% | 27 | .954 | .950 | .945 | .940 | .935 | .928 | .920 | .912 | .902 | .891 |
| | | | | | | | | .878 | .863 | .846 | .827 | .804 | .778 | .749 | .715 | .677 | .634 |
| | | | | | | | | .587 | .536 | .481 | .423 | .365 | .307 | .251 | | | |
| 20.000 | 20.000 | .464 | .000 | .81 | 2.81 | 2.3% | 28 | .953 | .949 | .945 | .939 | .933 | .927 | .919 | .910 | .900 | .889 |
| | | | | | | | | .875 | .860 | .842 | .822 | .799 | .772 | .741 | .706 | .667 | .623 |
| | | | | | | | | .574 | .521 | .465 | .407 | .348 | .290 | .236 | .186 | | |
| 20.000 | 20.000 | .297 | .000 | .87 | 2.73 | 1.5% | 29 | .953 | .949 | .944 | .939 | .933 | .926 | .918 | .909 | .899 | .887 |
| | | | | | | | | .873 | .850 | .840 | .819 | .795 | .768 | .736 | .700 | .660 | .615 |
| | | | | | | | | .565 | .512 | .455 | .396 | .337 | .280 | .225 | .176 | .134 | |
| 20.000 | 20.000 | .187 | .000 | .91 | 2.65 | .9% | 30 | .953 | .949 | .944 | .938 | .932 | .925 | .917 | .908 | .898 | .886 |
| | | | | | | | | .872 | .856 | .838 | .817 | .792 | .765 | .733 | .696 | .655 | .610 |
| | | | | | | | | .559 | .505 | .448 | .389 | .330 | .272 | .219 | .170 | .128 | .094 |
| 20.000 | 20.000 | .116 | .000 | .94 | 2.56 | .6% | 31 | .953 | .948 | .944 | .938 | .932 | .925 | .917 | .908 | .897 | .885 |

*SIMILAR TO A.T.&T. JACOBSEN RETRIAL TABLES

## *SLOW RETRIALS: JEWETT-SHRAGO EXTENDED ERLANG B - 100% BLOCKED CALLERS RETRY, 0% OVERFLOW (RETRIES 3-5 HOLDING TIMES LATER)

| HOURS FIRST ATTEMPT VMR=1.0 | HOURS CON-NECT | HOURS RECALL /RETRY | HOURS OVFL | CON-NECT VMR | OVFL VMR | PERCENT BLOCKED (GRADE) | LINE | HOURS CONNECTED TRAFFIC ON LINES 1 | 2 | 3 | 4 | 5 | 6 | 7 | 8 | 9 | 10 |
|---|---|---|---|---|---|---|---|---|---|---|---|---|---|---|---|---|---|
| | | | | | | | | .871 | .855 | .837 | .815 | .791 | .763 | .730 | .694 | .652 | .606 |
| | | | | | | | | .556 | .501 | .443 | .384 | .325 | .268 | .215 | .167 | .125 | .091 |
| | | | | | | | | .064 | | | | | | | | | |
| 22.000 | 22.000 | 8.861 | .000 | .19 | 2.42 | 28.7% | 24 | .969 | .967 | .965 | .962 | .960 | .957 | .954 | .951 | .947 | .944 |
| | | | | | | | | .939 | .934 | .929 | .923 | .916 | .909 | .901 | .891 | .881 | .869 |
| | | | | | | | | .856 | .842 | .825 | .806 | | | | | | |
| 22.000 | 22.000 | 5.001 | .000 | .32 | 2.75 | 18.5% | 25 | .964 | .962 | .959 | .956 | .953 | .949 | .945 | .941 | .936 | .930 |
| | | | | | | | | .924 | .917 | .909 | .900 | .890 | .879 | .866 | .852 | .836 | .817 |
| | | | | | | | | .796 | .772 | .746 | .716 | .683 | | | | | |
| 22.000 | 22.000 | 3.094 | .000 | .44 | 2.92 | 12.3% | 26 | .962 | .959 | .956 | .952 | .949 | .944 | .939 | .934 | .928 | .921 |
| | | | | | | | | .914 | .905 | .895 | .884 | .872 | .858 | .842 | .823 | .802 | .778 |
| | | | | | | | | .752 | .721 | .687 | .650 | .608 | .563 | | | | |
| 22.000 | 22.000 | 1.995 | .000 | .55 | 3.00 | 8.3% | 27 | .960 | .957 | .954 | .950 | .946 | .941 | .936 | .930 | .923 | .915 |
| | | | | | | | | .907 | .897 | .886 | .873 | .859 | .843 | .824 | .803 | .778 | .751 |
| | | | | | | | | .720 | .685 | .646 | .603 | .556 | .506 | .454 | | | |
| 22.000 | 22.000 | 1.311 | .000 | .64 | 3.02 | 5.6% | 28 | .959 | .956 | .952 | .948 | .944 | .939 | .933 | .926 | .919 | .911 |
| | | | | | | | | .902 | .891 | .879 | .865 | .850 | .832 | .811 | .788 | .761 | .731 |
| | | | | | | | | .697 | .659 | .617 | .571 | .521 | .468 | .413 | .358 | | |
| 22.000 | 22.000 | .867 | .000 | .72 | 2.99 | 3.8% | 29 | .958 | .955 | .951 | .947 | .942 | .937 | .931 | .924 | .917 | .908 |
| | | | | | | | | .898 | .887 | .874 | .860 | .843 | .824 | .802 | .777 | .749 | .717 |
| | | | | | | | | .681 | .641 | .596 | .548 | .496 | .442 | .386 | .330 | .276 | |
| 22.000 | 22.000 | .572 | .000 | .79 | 2.94 | 2.5% | 30 | .958 | .954 | .950 | .946 | .941 | .936 | .930 | .923 | .915 | .906 |
| | | | | | | | | .896 | .884 | .871 | .856 | .839 | .819 | .796 | .770 | .741 | .707 |
| | | | | | | | | .670 | .628 | .582 | .532 | .480 | .424 | .368 | .312 | .258 | .208 |
| 22.000 | 22.000 | .375 | .000 | .85 | 2.86 | 1.7% | 31 | .957 | .954 | .950 | .945 | .941 | .935 | .929 | .922 | .914 | .905 |
| | | | | | | | | .894 | .882 | .869 | .853 | .835 | .815 | .792 | .765 | .735 | .700 |
| | | | | | | | | .662 | .619 | .572 | .522 | .468 | .412 | .356 | .300 | .246 | .197 |
| | | | | | | | | .153 | | | | | | | | | |
| 22.000 | 22.000 | .243 | .000 | .89 | 2.78 | 1.1% | 32 | .957 | .953 | .950 | .945 | .940 | .935 | .928 | .921 | .913 | .904 |
| | | | | | | | | .893 | .881 | .867 | .851 | .833 | .812 | .789 | .761 | .731 | .696 |
| | | | | | | | | .657 | .613 | .566 | .514 | .460 | .404 | .347 | .292 | .239 | .190 |
| | | | | | | | | .147 | .110 | | | | | | | | |
| 22.000 | 22.000 | .155 | .000 | .92 | 2.69 | .7% | 33 | .957 | .953 | .949 | .945 | .940 | .934 | .928 | .921 | .912 | .903 |
| | | | | | | | | .892 | .880 | .866 | .850 | .832 | .811 | .787 | .759 | .728 | .693 |
| | | | | | | | | .653 | .609 | .561 | .510 | .455 | .398 | .342 | .286 | .234 | .185 |
| | | | | | | | | .143 | .107 | .078 | | | | | | | |
| 24.000 | 24.000 | 9.814 | .000 | .18 | 2.45 | 29.0% | 26 | .971 | .970 | .968 | .966 | .964 | .962 | .959 | .957 | .954 | .951 |
| | | | | | | | | .947 | .944 | .940 | .935 | .930 | .924 | .918 | .911 | .904 | .895 |
| | | | | | | | | .886 | .875 | .863 | .850 | .835 | .819 | | | | |
| 24.000 | 24.000 | 5.598 | .000 | .30 | 2.80 | 18.9% | 27 | .967 | .965 | .963 | .961 | .958 | .955 | .952 | .948 | .944 | .940 |

*SIMILAR TO A.T.&T. JACOBSEN RETRIAL TABLES

*SLOW RETRIALS: JEWETT-SHRAGO EXTENDED ERLANG B - 100% BLOCKED CALLERS RETRY, 0% OVERFLOW (RETRIES 3-5 HOLDING TIMES LATER)

| HOURS FIRST ATTEMPT VMR = 1.0 | HOURS CON-NECT | HOURS RECALL /RETRY | HOURS OVFL | CON-NECT VMR | OVFL VMR | PERCENT BLOCKED (GRADE) | LINE | HOURS CONNECTED TRAFFIC ON LINES 1 | 2 | 3 | 4 | 5 | 6 | 7 | 8 | 9 | 10 |
|---|---|---|---|---|---|---|---|---|---|---|---|---|---|---|---|---|---|
| | | | | | | | | .935 | .929 | .923 | .917 | .909 | .901 | .891 | .881 | .869 | .855 |
| | | | | | | | | .840 | .823 | .804 | .782 | .758 | .731 | .700 | | | |
| 24.000 | 24.000 | 3.504 | .000 | .42 | 3.00 | 12.7% | 28 | .965 | .963 | .960 | .957 | .954 | .950 | .947 | .942 | .938 | .932 |
| | | | | | | | | .926 | .920 | .912 | .904 | .894 | .884 | .872 | .858 | .843 | .826 |
| | | | | | | | | .806 | .784 | .759 | .731 | .699 | .664 | .626 | .584 | | |
| 24.000 | 24.000 | 2.289 | .000 | .52 | 3.09 | 8.7% | 29 | .963 | .961 | .958 | .955 | .951 | .947 | .943 | .938 | .933 | .927 |
| | | | | | | | | .920 | .913 | .904 | .895 | .884 | .872 | .858 | .842 | .824 | .804 |
| | | | | | | | | .781 | .755 | .726 | .693 | .657 | .617 | .573 | .527 | .477 | |
| 24.000 | 24.000 | 1.526 | .000 | .62 | 3.12 | 6.0% | 30 | .962 | .960 | .957 | .953 | .950 | .945 | .941 | .936 | .930 | .924 |
| | | | | | | | | .916 | .908 | .899 | .888 | .877 | .863 | .848 | .830 | .811 | .788 |
| | | | | | | | | .763 | .734 | .702 | .666 | .627 | .583 | .536 | .487 | .435 | .382 |
| 24.000 | 24.000 | 1.025 | .000 | .70 | 3.10 | 4.1% | 31 | .962 | .959 | .956 | .952 | .948 | .944 | .939 | .934 | .928 | .921 |
| | | | | | | | | .913 | .905 | .095 | .884 | .871 | .857 | .841 | .822 | .801 | .777 |
| | | | | | | | | .750 | .719 | .685 | .647 | .605 | .559 | .511 | .459 | .406 | .352 |
| | | | | | | | | .299 | | | | | | | | | |
| 24.000 | 24.000 | .689 | .000 | .77 | 3.05 | 2.8% | 32 | .961 | .958 | .955 | .951 | .947 | .943 | .938 | .933 | .926 | .919 |
| | | | | | | | | .911 | .902 | .892 | .881 | .867 | .852 | .835 | .816 | .794 | .769 |
| | | | | | | | | .740 | .709 | .673 | .633 | .590 | .543 | .493 | .440 | .386 | .332 |
| | | | | | | | | .279 | .230 | | | | | | | | |
| 24.000 | 24.000 | .460 | .000 | .83 | 2.99 | 1.9% | 33 | .961 | .958 | .955 | .951 | .947 | .942 | .937 | .932 | .925 | .918 |
| | | | | | | | | .910 | .901 | .890 | .878 | .865 | .849 | .832 | .812 | .789 | .763 |
| | | | | | | | | .734 | .701 | .664 | .624 | .579 | .531 | .480 | .427 | .372 | .318 |
| | | | | | | | | .266 | .217 | .173 | | | | | | | |
| 24.000 | 24.000 | .305 | .000 | .87 | 2.90 | 1.3% | 34 | .961 | .958 | .954 | .951 | .946 | .942 | .937 | .931 | .924 | .917 |
| | | | | | | | | .909 | .899 | .889 | .877 | .863 | .847 | .829 | .809 | .785 | .759 |
| | | | | | | | | .729 | .696 | .658 | .617 | .572 | .523 | .471 | .418 | .363 | .300 |
| | | | | | | | | .257 | .209 | 165 | .127 | | | | | | |
| 24.000 | 24.000 | .199 | .000 | .91 | 2.82 | .8% | 35 | .960 | .957 | .954 | .950 | .946 | .942 | .936 | .930 | .924 | .916 |
| | | | | | | | | .908 | .899 | .888 | .876 | .862 | .846 | .827 | .807 | .783 | .756 |
| | | | | | | | | .726 | .692 | .654 | .612 | .567 | .517 | .465 | .411 | .357 | .303 |
| | | | | | | | | .251 | .203 | .160 | .123 | .091 | | | | | |
| 24.000 | 24.000 | .128 | .000 | .94 | 2.73 | .5% | 36 | .960 | .957 | .954 | .950 | .946 | .941 | .936 | .930 | .924 | .916 |
| | | | | | | | | .908 | .898 | .887 | .875 | .861 | .845 | .826 | .805 | .781 | .754 |
| | | | | | | | | .724 | .690 | .651 | .609 | .563 | .514 | .461 | .407 | .352 | .299 |
| | | | | | | | | .247 | .199 | .157 | .120 | .089 | .064 | | | | |
| 26.000 | 26.000 | 10.770 | .000 | .17 | 2.48 | 29.3% | 28 | .974 | .972 | .971 | .969 | .967 | .966 | .964 | .962 | .959 | .957 |
| | | | | | | | | .954 | .051 | .948 | .944 | .940 | .936 | .931 | .926 | .920 | .914 |
| | | | | | | | | .907 | .899 | .890 | .880 | .870 | .858 | .844 | .830 | | |
| 26.000 | 26.000 | 6.199 | .000 | .28 | 2.85 | 19.3% | 29 | .970 | .968 | .966 | .964 | .962 | .959 | .957 | .954 | .951 | .947 |
| | | | | | | | | .943 | .939 | .934 | .929 | .923 | .017 | .909 | .901 | .892 | .882 |

*SIMILAR TO A.T.&T. JACOBSEN RETRIAL TABLES

*SLOW RETRIALS: JEWETT-SHRAGO EXTENDED ERLANG B - 100% BLOCKED CALLERS RETRY, 0% OVERFLOW (RETRIES 3-5 HOLDING TIMES LATER)

| HOURS FIRST ATTEMPT VMR=1.0 | HOURS CON-NECT | HOURS RECALL /RETRY | HOURS OVFL | CON-NECT VMR | OVFL VMR | PERCENT BLOCKED (GRADE) | LINE | HOURS CONNECTED TRAFFIC ON LINES 1 | 2 | 3 | 4 | 5 | 6 | 7 | 8 | 9 | 10 |
|---|---|---|---|---|---|---|---|---|---|---|---|---|---|---|---|---|---|
| | | | | | | | | .871 | .858 | .844 | .828 | .811 | .791 | .768 | .744 | .716 | |
| 26.000 | 26.000 | 3.919 | .000 | .40 | 3.06 | 13.1% | 30 | .968 | .966 | .964 | .961 | .958 | .956 | .952 | .949 | .945 | .941 |
| | | | | | | | | .936 | .931 | .925 | .918 | .911 | .903 | .894 | .884 | .872 | .859 |
| | | | | | | | | .844 | .828 | .809 | .789 | .765 | .739 | .710 | .678 | .642 | .603 |
| 26.000 | 26.000 | 2.589 | .000 | .50 | 3.17 | 9.1% | 31 | .966 | .964 | .962 | .959 | .956 | .953 | .949 | .945 | .941 | .936 |
| | | | | | | | | .931 | .925 | .918 | .911 | .903 | .893 | .883 | .871 | .857 | .842 |
| | | | | | | | | .825 | .805 | .784 | .759 | .732 | .701 | .667 | .630 | .589 | .545 |
| | | | | | | | | .498 | | | | | | | | | |
| 26.000 | 26.000 | 1.747 | .000 | .60 | 3.21 | 6.3% | 32 | .965 | .963 | .960 | .958 | .954 | .951 | .947 | .943 | .938 | .933 |
| | | | | | | | | .927 | .921 | .914 | .906 | .896 | .886 | .874 | .861 | .846 | .829 |
| | | | | | | | | .810 | .789 | .765 | .737 | .707 | .673 | .636 | .595 | .551 | .504 |
| | | | | | | | | .454 | .404 | | | | | | | | |
| 26.000 | 26.000 | 1.191 | .000 | .68 | 3.21 | 4.4% | 33 | .965 | .962 | .959 | .957 | .953 | .950 | .946 | .941 | .936 | .931 |
| | | | | | | | | .925 | .918 | .910 | .902 | .892 | .881 | .868 | .854 | .838 | .820 |
| | | | | | | | | .800 | .777 | .751 | .722 | .689 | .653 | .613 | .570 | .524 | .475 |
| | | | | | | | | .424 | .372 | .321 | | | | | | | |
| 26.000 | 26.000 | .813 | .000 | .75 | 3.17 | 3.0% | 34 | .964 | .962 | .959 | .956 | .952 | .949 | .945 | .940 | .935 | .929 |
| | | | | | | | | .923 | .916 | .908 | .899 | .889 | .877 | .864 | .849 | .833 | .814 |
| | | | | | | | | .792 | .768 | .741 | .710 | .676 | .638 | .597 | .552 | .505 | .455 |
| | | | | | | | | .403 | .351 | .299 | .250 | | | | | | |
| 26.000 | 26.000 | .553 | .000 | .81 | 3.10 | 2.1% | 35 | .964 | .961 | .958 | .955 | .952 | .948 | .944 | .939 | .934 | .928 |
| | | | | | | | | .922 | .914 | .906 | .897 | .886 | .875 | .861 | .846 | .829 | .809 |
| | | | | | | | | .787 | .762 | .733 | .702 | .667 | .628 | .586 | .540 | .491 | .440 |
| | | | | | | | | .388 | .336 | .285 | .236 | .191 | | | | | |
| 26.000 | 26.000 | .373 | .000 | .86 | 3.02 | 1.4% | 36 | .964 | .961 | .958 | .955 | .952 | .948 | .943 | .939 | .933 | .928 |
| | | | | | | | | .921 | .913 | .905 | .896 | .885 | .873 | .859 | .843 | .826 | .805 |
| | | | | | | | | .783 | .757 | .728 | .696 | .660 | .620 | .577 | .531 | .482 | .430 |
| | | | | | | | | .378 | .325 | .275 | .227 | .183 | .143 | | | | |
| 26.000 | 26.000 | .249 | .000 | .89 | 2.94 | .9% | 37 | .963 | .961 | .958 | .955 | .951 | .947 | .943 | .938 | .933 | .927 |
| | | | | | | | | .920 | .913 | .904 | .895 | .884 | .871 | .857 | .841 | .823 | .803 |
| | | | | | | | | .780 | .754 | .725 | .692 | .655 | .615 | .572 | .525 | .475 | .423 |
| | | | | | | | | .371 | .318 | .268 | .220 | .177 | .138 | .105 | | | |
| 26.000 | 26.000 | .164 | .000 | .92 | 2.85 | .6% | 38 | .963 | .961 | .958 | .955 | .951 | .947 | .943 | .938 | .933 | .927 |
| | | | | | | | | .920 | .912 | .904 | .894 | .883 | .870 | .856 | .840 | .822 | .801 |
| | | | | | | | | .778 | .752 | .722 | .689 | .652 | .612 | .568 | .520 | .470 | .418 |
| | | | | | | | | .366 | .313 | .263 | .216 | .172 | .134 | .102 | .075 | | |
| 28.000 | 28.000 | 11.732 | .000 | .16 | 2.50 | 29.5% | 30 | .976 | .974 | .973 | .972 | .970 | .969 | .967 | .965 | .963 | .961 |
| | | | | | | | | .959 | .957 | .954 | .951 | .948 | .944 | .941 | .937 | .932 | .927 |
| | | | | | | | | .922 | .916 | .909 | .902 | .894 | .885 | .875 | .865 | .853 | .839 |
| 28.000 | 28.000 | 6.806 | .000 | .27 | 2.89 | 19.6% | 31 | .972 | .971 | .969 | .967 | .965 | .963 | .961 | .959 | .956 | .953 |

*SIMILAR TO A.T.&T. JACOBSEN RETRIAL TABLES

*SLOW RETRIALS: JEWETT-SHRAGO EXTENDED ERLANG B - 100% BLOCKED CALLERS RETRY, 0% OVERFLOW (RETRIES 3-5 HOLDING TIMES LATER)

| HOURS FIRST ATTEMPT VMR = 1.0 | HOURS CON-NECT | HOURS RECALL /RETRY | HOURS OVFL | CON-NECT VMR | OVFL VMR | PERCENT BLOCKED (GRADE) | LINE | HOURS CONNECTED TRAFFIC ON LINES 1 | 2 | 3 | 4 | 5 | 6 | 7 | 8 | 9 | 10 |
|---|---|---|---|---|---|---|---|---|---|---|---|---|---|---|---|---|---|
| | | | | | | | | .950 | .946 | .942 | .938 | .934 | .929 | .923 | .917 | .910 | .902 |
| | | | | | | | | .894 | .884 | .873 | .862 | .848 | .834 | .817 | .799 | .778 | .755 |
| | | | | | | | | .730 | | | | | | | | | |
| 28.000 | 28.000 | 4.342 | .000 | .38 | 3.12 | 13.4% | 32 | .970 | .968 | .966 | .964 | .962 | .960 | .957 | .954 | .951 | .947 |
| | | | | | | | | .944 | .939 | .935 | .929 | .924 | .917 | .910 | .902 | .893 | .884 |
| | | | | | | | | .872 | .860 | .846 | .830 | .813 | .793 | .771 | .747 | .720 | .690 |
| | | | | | | | | .657 | .621 | | | | | | | | |
| 28.000 | 28.000 | 2.898 | .000 | .48 | 3.25 | 9.4% | 33 | .969 | .967 | .965 | .962 | .960 | .957 | .954 | .951 | .948 | .944 |
| | | | | | | | | .939 | .934 | .929 | .923 | .917 | .909 | .901 | .892 | .881 | .870 |
| | | | | | | | | .857 | .842 | .826 | .807 | .786 | .763 | .737 | .708 | .676 | .641 |
| | | | | | | | | .603 | .562 | .518 | | | | | | | |
| 28.000 | 28.000 | 1.978 | .000 | .58 | 3.30 | 6.6% | 34 | .968 | .966 | .964 | .961 | .959 | .956 | .952 | .949 | .945 | .941 |
| | | | | | | | | .936 | .931 | .925 | .919 | .911 | .903 | .894 | .884 | .873 | .860 |
| | | | | | | | | .845 | .829 | .810 | .790 | .767 | .741 | .712 | .680 | .644 | .606 |
| | | | | | | | | .564 | .520 | .473 | .424 | | | | | | |
| 28.000 | 28.000 | 1.364 | .000 | .66 | 3.31 | 4.6% | 35 | .967 | .965 | .963 | .960 | .957 | .954 | .951 | .948 | .943 | .939 |
| | | | | | | | | .934 | .928 | .922 | .915 | .908 | .899 | .889 | .878 | .866 | .852 |
| | | | | | | | | .837 | .819 | .799 | .777 | .752 | .724 | .693 | .659 | .621 | .580 |
| | | | | | | | | .536 | .489 | .441 | .391 | .341 | | | | | |
| 28.000 | 28.000 | .944 | .000 | .73 | 3.27 | 3.3% | 36 | .967 | .964 | .962 | .960 | .957 | .954 | .950 | .946 | .942 | .938 |
| | | | | | | | | .932 | .927 | .920 | .913 | .905 | .896 | .886 | .874 | .861 | .847 |
| | | | | | | | | .830 | .812 | .791 | .767 | .741 | .712 | .679 | .643 | .604 | .561 |
| | | | | | | | | .516 | .468 | .418 | .368 | .318 | .270 | | | | |
| 28.000 | 28.000 | .651 | .000 | .79 | 3.21 | 2.3% | 37 | .966 | .964 | .962 | .959 | .956 | .953 | .949 | .946 | .941 | .937 |
| | | | | | | | | .931 | .925 | .919 | .911 | .903 | .894 | .883 | .871 | .858 | .843 |
| | | | | | | | | .826 | .807 | .785 | .761 | .733 | .703 | .669 | .632 | .592 | .548 |
| | | | | | | | | .501 | .450 | .403 | .352 | .302 | .254 | .209 | | | |
| 28.000 | 28.000 | .446 | .000 | .84 | 3.14 | 1.6% | 38 | .966 | .964 | .961 | .959 | .956 | .953 | .949 | .945 | .941 | .936 |
| | | | | | | | | .930 | .924 | .918 | .910 | .902 | .892 | .881 | .869 | .855 | .840 |
| | | | | | | | | .822 | .803 | .781 | .756 | .728 | .696 | .662 | .624 | .583 | .538 |
| | | | | | | | | .491 | .442 | .391 | .341 | .291 | .243 | .199 | .159 | | |
| 28.000 | 28.000 | .303 | .000 | .88 | 3.06 | 1.1% | 39 | .966 | .964 | .961 | .958 | .956 | .952 | .949 | .945 | .940 | .935 |
| | | | | | | | | .930 | .924 | .917 | .909 | .900 | .891 | .880 | .868 | .854 | .838 |
| | | | | | | | | .820 | .800 | .777 | .752 | .724 | .692 | .657 | .618 | .576 | .531 |
| | | | | | | | | .484 | .434 | .383 | .333 | .283 | .236 | .192 | .153 | .119 | |
| 28.000 | 28.000 | .203 | .000 | .91 | 2.07 | .7% | 40 | .966 | .964 | .961 | .958 | .955 | .952 | .948 | .944 | .940 | .935 |
| | | | | | | | | .929 | .923 | .916 | .909 | .900 | .890 | .879 | .866 | .852 | .836 |
| | | | | | | | | .818 | .798 | .775 | .749 | .721 | .689 | .653 | .614 | .572 | .526 |
| | | | | | | | | .478 | .429 | .378 | .327 | .278 | .231 | .188 | .149 | .115 | .007 |
| 30.000 | 30.000 | 12.695 | .000 | .15 | 2.52 | 29.7% | 32 | .977 | .076 | .975 | .974 | .973 | .971 | .970 | .968 | .967 | .965 |

*SIMILAR TO A.T.&T. JACOBSEN RETRIAL TABLES

*SLOW RETRIALS: JEWETT-SHRAGO EXTENDED ERLANG B - 100% BLOCKED CALLERS RETRY, 0% OVERFLOW (RETRIES 3-5 HOLDING TIMES LATER)

| HOURS FIRST ATTEMPT VMR = 1.0 | HOURS CON-NECT | HOURS RECALL /RETRY | HOURS OVFL | CON-NECT VMR | OVFL VMR | PERCENT BLOCKED (GRADE) | LINE | HOURS CONNECTED TRAFFIC ON LINES 1 | 2 | 3 | 4 | 5 | 6 | 7 | 8 | 9 | 10 |
|---|---|---|---|---|---|---|---|---|---|---|---|---|---|---|---|---|---|
| | | | | | | | | .963 | .961 | .959 | .957 | .954 | .951 | .948 | .945 | .941 | .937 |
| | | | | | | | | .933 | .929 | .923 | .918 | .912 | .905 | .898 | .890 | .881 | .871 |
| | | | | | | | | .860 | .848 | | | | | | | | |
| 30.000 | 30.000 | 7.418 | .000 | .26 | 2.93 | 19.8% | 33 | .974 | .973 | .971 | .970 | .968 | .966 | .964 | .962 | .960 | .958 |
| | | | | | | | | .955 | .952 | .949 | .946 | .942 | .938 | .933 | .928 | .923 | .917 |
| | | | | | | | | .910 | .903 | .895 | .886 | .876 | .865 | .852 | .838 | .823 | .806 |
| | | | | | | | | .787 | .766 | .743 | | | | | | | |
| 30.000 | 30.000 | 4.771 | .000 | .36 | 3.18 | 13.7% | 34 | .972 | .971 | .969 | .967 | .965 | .963 | .961 | .958 | .956 | .953 |
| | | | | | | | | .950 | .946 | .942 | .938 | .934 | .928 | .923 | .916 | .910 | .902 |
| | | | | | | | | .893 | .884 | .873 | .861 | .848 | .833 | .816 | .798 | .777 | .754 |
| | | | | | | | | .729 | .701 | .670 | .637 | | | | | | |
| 30.000 | 30.000 | 3.211 | .000 | .46 | 3.32 | 9.7% | 35 | .971 | .969 | .967 | .965 | .963 | .961 | .959 | .956 | .953 | .950 |
| | | | | | | | | .946 | .942 | .938 | .933 | .928 | .922 | .915 | .908 | .900 | .891 |
| | | | | | | | | .881 | .869 | .857 | .843 | .827 | .809 | .789 | .767 | .743 | .715 |
| | | | | | | | | .685 | .652 | .616 | .577 | .536 | | | | | |
| 30.000 | 30.000 | 2.213 | .000 | .56 | 3.39 | 6.9% | 36 | .970 | .968 | .966 | .964 | .962 | .960 | .957 | .954 | .951 | .947 |
| | | | | | | | | .943 | .939 | .934 | .929 | .923 | .917 | .909 | .901 | .892 | .882 |
| | | | | | | | | .871 | .859 | .844 | .829 | .811 | .791 | .769 | .744 | .716 | .686 |
| | | | | | | | | .652 | .616 | .576 | .534 | .489 | .443 | | | | |
| 30.000 | 30.000 | 1.543 | .000 | .64 | 3.40 | 4.9% | 37 | .969 | .968 | .966 | .963 | .961 | .958 | .956 | .953 | .949 | .945 |
| | | | | | | | | .941 | .937 | .932 | .926 | .920 | .913 | .905 | .897 | .887 | .876 |
| | | | | | | | | .864 | .851 | .835 | .818 | .799 | .777 | .753 | .727 | .697 | .664 |
| | | | | | | | | .628 | .589 | .547 | .503 | .456 | .409 | .360 | | | |
| 30.000 | 30.000 | 1.080 | .000 | .71 | 3.37 | 3.5% | 38 | .969 | .967 | .965 | .963 | .960 | .958 | .955 | .952 | .948 | .944 |
| | | | | | | | | .940 | .935 | .930 | .924 | .918 | .910 | .902 | .893 | .883 | .872 |
| | | | | | | | | .859 | .845 | .828 | .810 | .790 | .767 | .742 | .714 | .682 | .648 |
| | | | | | | | | .610 | .570 | .526 | .480 | .433 | .384 | .336 | .288 | | |
| 30.000 | 30.000 | .755 | .000 | .77 | 3.32 | 2.5% | 39 | .969 | .967 | .965 | .962 | .960 | .957 | .954 | .951 | .947 | .943 |
| | | | | | | | | .939 | .934 | .929 | .923 | .916 | .908 | .900 | .891 | .880 | .868 |
| | | | | | | | | .855 | .840 | .823 | .805 | .783 | .760 | .733 | .704 | .672 | .636 |
| | | | | | | | | .597 | .555 | .511 | .464 | .416 | .367 | .318 | .271 | .226 | |
| 30.000 | 30.000 | .525 | .000 | .83 | 3.25 | 1.7% | 40 | .968 | .966 | .964 | .962 | .959 | .957 | .954 | .950 | .947 | .943 |
| | | | | | | | | .938 | .933 | .928 | .921 | .915 | .907 | .898 | .889 | .878 | .866 |
| | | | | | | | | .852 | .837 | .820 | .800 | .779 | .754 | .727 | .697 | .664 | .627 |
| | | | | | | | | .588 | .545 | .500 | .452 | .404 | .355 | .306 | .259 | .215 | .175 |
| 30.000 | 30.000 | .362 | .000 | .87 | 3.17 | 1.2% | 41 | .968 | .966 | .964 | .962 | .959 | .956 | .953 | .950 | .946 | .942 |
| | | | | | | | | .937 | .932 | .927 | .921 | .914 | .906 | .897 | .887 | .876 | .864 |
| | | | | | | | | .850 | .835 | .817 | .797 | .775 | .750 | .723 | .692 | .658 | .621 |
| | | | | | | | | .581 | .538 | .492 | .444 | .395 | .346 | .298 | .251 | .208 | .168 |
| | | | | | | | | .133 | | | | | | | | | |

*SIMILAR TO A.T.&T. JACOBSEN RETRIAL TABLES

*SLOW RETRIALS: JEWETT-SHRAGO EXTENDED ERLANG B - 100% BLOCKED CALLERS RETRY, 0% OVERFLOW (RETRIES 3-5 HOLDING TIMES LATER)

| HOURS FIRST ATTEMPT VMR=1.0 | HOURS CON-NECT | HOURS RECALL /RETRY | HOURS OVFL | CON-NECT VMR | OVFL VMR | PERCENT BLOCKED (GRADE) | LINE | HOURS CONNECTED TRAFFIC ON LINES 1 | 2 | 3 | 4 | 5 | 6 | 7 | 8 | 9 | 10 |
|---|---|---|---|---|---|---|---|---|---|---|---|---|---|---|---|---|---|
| 30.000 | 30.000 | .247 | .000 | .90 | 3.08 | .8% | 42 | .968 | .966 | .964 | .962 | .959 | .956 | .953 | .950 | .946 | .942 |
| | | | | | | | | .937 | .932 | .926 | .920 | .913 | .905 | .896 | .886 | .875 | .863 |
| | | | | | | | | .849 | .833 | .815 | .795 | .773 | .748 | .719 | .688 | .654 | .617 |
| | | | | | | | | .576 | .532 | .486 | .438 | .389 | .340 | .291 | .245 | .202 | .163 |
| | | | | | | | | .128 | .099 | | | | | | | | |
| 30.000 | 30.000 | .166 | .000 | .93 | 2.99 | .6% | 43 | .968 | .966 | .964 | .961 | .959 | .956 | .953 | .949 | .946 | .941 |
| | | | | | | | | .937 | .932 | .926 | .920 | .913 | .905 | .896 | .886 | .874 | .862 |
| | | | | | | | | .848 | .832 | .814 | .794 | .771 | .745 | .717 | .686 | .651 | .613 |
| | | | | | | | | .572 | .529 | .482 | .434 | .385 | .335 | .287 | .241 | .198 | .159 |
| | | | | | | | | .125 | .096 | .072 | | | | | | | |

*SIMILAR TO A.T.&T. JACOBSEN RETRIAL TABLES

*FAST RETRIALS: JEWETT-SHRAGO FAST RETRIAL TABLES - 50% BLOCKED CALLERS RETRY, 50% OVERFLOW (RETRIES 1 HOLDING TIMES LATER)

| HOURS FIRST ATTEMPT VMR=1.0 | HOURS CON-NECT | HOURS RECALL /RETRY | HOURS OVFL | CON-NECT VMR | OVFL VMR | PERCENT BLOCKED (GRADE) | LINE | HOURS CONNECTED TRAFFIC ON LINES 1 | 2 | 3 | 4 | 5 | 6 | 7 | 8 | 9 | 10 |
|---|---|---|---|---|---|---|---|---|---|---|---|---|---|---|---|---|---|
| .200 | .180 | .020 | .020 | .82 | 1.08 | 18.0% | 1 | .180 | | | | | | | | | |
| .200 | .198 | .002 | .002 | .97 | 1.07 | 1.7% | 2 | .168 | .030 | | | | | | | | |
| .400 | .323 | .077 | .077 | .68 | 1.13 | 32.3% | 1 | .323 | | | | | | | | | |
| .400 | .388 | .012 | .012 | .90 | 1.13 | 5.7% | 2 | .292 | .097 | | | | | | | | |
| .400 | .399 | .001 | .001 | .98 | 1.11 | .7% | 3 | .286 | .093 | .019 | | | | | | | |
| .600 | .565 | .035 | .035 | .82 | 1.19 | 11.0% | 2 | .388 | .177 | | | | | | | | |
| .600 | .594 | .006 | .006 | .95 | 1.17 | 2.0% | 3 | .377 | .167 | .050 | | | | | | | |
| .800 | .726 | .074 | .074 | .74 | 1.24 | 16.9% | 2 | .466 | .260 | | | | | | | | |
| .800 | .783 | .017 | .017 | .91 | 1.22 | 4.0% | 3 | .450 | .240 | .094 | | | | | | | |
| .800 | .797 | .003 | .003 | .97 | 1.18 | .8% | 4 | .445 | .236 | .090 | .025 | | | | | | |
| 1.000 | .870 | .130 | .130 | .66 | 1.27 | 23.1% | 2 | .531 | .339 | | | | | | | | |
| 1.000 | .965 | .035 | .035 | .85 | 1.27 | 6.7% | 3 | .509 | .311 | .146 | | | | | | | |
| 1.000 | .992 | .008 | .008 | .95 | 1.24 | 1.6% | 4 | .502 | .302 | .139 | .048 | | | | | | |
| 1.200 | .996 | .204 | .204 | .59 | 1.29 | 29.1% | 2 | .584 | .412 | | | | | | | | |
| 1.200 | 1.138 | .062 | .062 | .80 | 1.32 | 9.9% | 3 | .558 | .376 | .204 | | | | | | | |
| 1.200 | 1.183 | .017 | .017 | .92 | 1.29 | 2.7% | 4 | .549 | .363 | .192 | .079 | | | | | | |
| 1.200 | 1.196 | .004 | .004 | .98 | 1.24 | .6% | 5 | .546 | .360 | .189 | .077 | .024 | | | | | |
| 1.400 | 1.106 | .294 | .294 | .52 | 1.31 | 34.8% | 2 | .629 | .477 | | | | | | | | |
| 1.400 | 1.299 | .101 | .101 | .74 | 1.35 | 13.4% | 3 | .600 | .435 | .264 | | | | | | | |
| 1.400 | 1.370 | .030 | .030 | .88 | 1.33 | 4.2% | 4 | .589 | .418 | .247 | .116 | | | | | | |
| 1.400 | 1.392 | .008 | .008 | .96 | 1.29 | 1.1% | 5 | .585 | .413 | .241 | .112 | .041 | | | | | |
| 1.600 | 1.449 | .151 | .151 | .68 | 1.38 | 17.3% | 3 | .637 | .488 | .324 | | | | | | | |
| 1.600 | 1.550 | .050 | .050 | .84 | 1.38 | 6.1% | 4 | .623 | .467 | .300 | .159 | | | | | | |
| 1.600 | 1.585 | .015 | .015 | .94 | 1.34 | 1.8% | 5 | .618 | .460 | .292 | .152 | .064 | | | | | |
| 1.800 | 1.586 | .214 | .214 | .62 | 1.41 | 21.3% | 3 | .668 | .536 | .382 | | | | | | | |
| 1.800 | 1.722 | .078 | .078 | .79 | 1.42 | 8.3% | 4 | .653 | .512 | .353 | .205 | | | | | | |
| 1.800 | 1.775 | .025 | .025 | .91 | 1.38 | 2.8% | 5 | .646 | .502 | .341 | .195 | .091 | | | | | |
| 1.800 | 1.793 | .007 | .007 | .97 | 1.33 | .8% | 6 | .644 | .499 | .337 | .191 | .089 | .034 | | | | |
| 2.000 | 1.711 | .289 | .289 | .56 | 1.42 | 25.3% | 3 | .696 | .578 | .436 | | | | | | | |
| 2.000 | 1.887 | .113 | .113 | .75 | 1.45 | 10.7% | 4 | .679 | .552 | .402 | .254 | | | | | | |
| 2.000 | 1.960 | .040 | .040 | .88 | 1.43 | 3.9% | 5 | .671 | .540 | .387 | .239 | .123 | | | | | |
| 2.000 | 1.988 | .012 | .012 | .95 | 1.38 | 1.2% | 6 | .668 | .535 | .382 | .233 | .119 | .050 | | | | |
| 2.200 | 1.823 | .377 | .377 | .51 | 1.43 | 29.3% | 3 | .720 | .616 | .487 | | | | | | | |
| 2.200 | 2.042 | .158 | .158 | .70 | 1.48 | 13.4% | 4 | .702 | .588 | .449 | .303 | | | | | | |
| 2.200 | 2.140 | .060 | .060 | .84 | 1.47 | 5.3% | 5 | .693 | .574 | .431 | .283 | .159 | | | | | |
| 2.200 | 2.180 | .020 | .020 | .93 | 1.42 | 1.8% | 6 | .689 | .568 | .423 | .276 | .153 | .071 | | | | |
| 2.200 | 2.194 | .006 | .006 | .97 | 1.37 | .6% | 7 | .688 | .566 | .421 | .273 | .150 | .069 | .027 | | | |
| 2.400 | 2.188 | .212 | .212 | .65 | 1.50 | 16.3% | 4 | .723 | .620 | .492 | .352 | | | | | | |
| 2.400 | 2.315 | .085 | .085 | .80 | 1.50 | 6.9% | 5 | .713 | .605 | .471 | .328 | .198 | | | | | |
| 2.400 | 2.369 | .031 | .031 | .90 | 1.46 | 2.6% | 6 | .709 | .598 | .462 | .317 | .188 | .095 | | | | |

*SIMILAR TO A.T.&T. JACOBSEN RETRIAL TABLES

## *FAST RETRIALS: JEWETT-SHRAGO FAST RETRIAL TABLES - 50% BLOCKED CALLERS RETRY, 50% OVERFLOW (RETRIES 1 HOLDING TIMES LATER)

| HOURS FIRST ATTEMPT VMR=1.0 | HOURS CON-NECT | HOURS RECALL /RETRY | HOURS OVFL | CON-NECT VMR | OVFL VMR | PERCENT BLOCKED (GRADE) | LINE | HOURS CONNECTED TRAFFIC ON LINES 1 | 2 | 3 | 4 | 5 | 6 | 7 | 8 | 9 | 10 |
|---|---|---|---|---|---|---|---|---|---|---|---|---|---|---|---|---|---|
| 2.400 | 2.390 | .010 | .010 | .96 | 1.41 | .8% | 7 | .707 | .595 | .458 | .313 | .185 | .093 | .039 | | | |
| 2.600 | 2.323 | .277 | .277 | .60 | 1.52 | 19.2% | 4 | .742 | .649 | .533 | .399 | | | | | | |
| 2.600 | 2.482 | .118 | .118 | .76 | 1.54 | 8.7% | 5 | .731 | .632 | .509 | .371 | .239 | | | | | |
| 2.600 | 2.554 | .046 | .046 | .88 | 1.50 | 3.4% | 6 | .726 | .624 | .498 | .358 | .226 | .123 | | | | |
| 2.600 | 2.584 | .016 | .016 | .95 | 1.45 | 1.2% | 7 | .723 | .621 | .493 | .353 | .221 | .119 | .055 | | | |
| 2.800 | 2.449 | .351 | .351 | .56 | 1.53 | 22.3% | 4 | .759 | .676 | .570 | .445 | | | | | | |
| 2.800 | 2.643 | .157 | .157 | .72 | 1.57 | 10.6% | 5 | .747 | .658 | .544 | .413 | .281 | | | | | |
| 2.800 | 2.736 | .064 | .064 | .85 | 1.54 | 4.5% | 6 | .741 | .648 | .531 | .397 | .265 | .154 | | | | |
| 2.800 | 2.776 | .024 | .024 | .93 | 1.49 | 1.7% | 7 | .739 | .644 | .525 | .390 | .257 | .148 | .073 | | | |
| 2.800 | 2.792 | .008 | .008 | .97 | 1.44 | .6% | 8 | .737 | .642 | .523 | .387 | .255 | .146 | .072 | .030 | | |
| 3.000 | 2.517 | .483 | .483 | .51 | 1.64 | 27.8% | 4 | .755 | .683 | .592 | .486 | | | | | | |
| 3.000 | 2.795 | .205 | .205 | .68 | 1.59 | 12.8% | 5 | .762 | .681 | .576 | .453 | .323 | | | | | |
| 3.000 | 2.912 | .088 | .088 | .81 | 1.58 | 5.7% | 6 | .755 | .670 | .561 | .435 | .303 | .187 | | | | |
| 3.000 | 2.965 | .035 | .035 | .91 | 1.53 | 2.3% | 7 | .752 | .665 | .554 | .426 | .294 | .179 | .095 | | | |
| 3.000 | 2.987 | .013 | .013 | .96 | 1.48 | .8% | 8 | .751 | .663 | .551 | .422 | .290 | .176 | .092 | .042 | | |
| 3.200 | 2.621 | .579 | .579 | .48 | 1.64 | 30.6% | 4 | .771 | .705 | .622 | .523 | | | | | | |
| 3.200 | 2.940 | .260 | .260 | .63 | 1.61 | 15.1% | 5 | .776 | .702 | .606 | .491 | .365 | | | | | |
| 3.200 | 3.083 | .117 | .117 | .78 | 1.61 | 7.1% | 6 | .768 | .690 | .590 | .470 | .342 | .222 | | | | |
| 3.200 | 3.151 | .049 | .049 | .88 | 1.57 | 3.0% | 7 | .765 | .684 | .582 | .460 | .331 | .212 | .119 | | | |
| 3.200 | 3.182 | .018 | .018 | .94 | 1.52 | 1.1% | 8 | .763 | .682 | .578 | .455 | .325 | .207 | .115 | .056 | | |
| 3.400 | 2.718 | .682 | .682 | .44 | 1.64 | 33.4% | 4 | .785 | .725 | .050 | .558 | | | | | | |
| 3.400 | 3.075 | .325 | .325 | .60 | 1.62 | 17.4% | 5 | .788 | .721 | .633 | .527 | .406 | | | | | |
| 3.400 | 3.247 | .153 | .153 | .74 | 1.64 | 8.6% | 6 | .780 | .709 | .616 | .504 | .380 | .259 | | | | |
| 3.400 | 3.334 | .066 | .066 | .85 | 1.61 | 3.8% | 7 | .776 | .702 | .607 | .492 | .366 | .245 | .146 | | | |
| 3.400 | 3.374 | .026 | .026 | .93 | 1.56 | 1.5% | 8 | .774 | .699 | .602 | .486 | .360 | .239 | .141 | .073 | | |
| 3.400 | 3.390 | .010 | .010 | .97 | 1.50 | .6% | 9 | .773 | .698 | .600 | .484 | .357 | .237 | .139 | .071 | .032 | |
| 3.600 | 3.156 | .444 | .444 | .55 | 1.73 | 22.0% | 5 | .783 | .723 | .646 | .554 | .449 | | | | | |
| 3.600 | 3.406 | .194 | .194 | .70 | 1.67 | 10.2% | 6 | .791 | .725 | .640 | .535 | .417 | .296 | | | | |
| 3.600 | 3.512 | .088 | .088 | .83 | 1.65 | 4.8% | 7 | .787 | .718 | .630 | .522 | .401 | .280 | .175 | | | |
| 3.600 | 3.563 | .037 | .037 | .91 | 1.60 | 2.0% | 8 | .784 | .715 | .625 | .515 | .393 | .272 | .168 | .092 | | |
| 3.600 | 3.586 | .014 | .014 | .96 | 1.54 | .8% | 9 | .783 | .713 | .622 | .512 | .389 | .268 | .165 | .089 | .043 | |
| 3.800 | 3.273 | .527 | .527 | .52 | 1.73 | 24.4% | 5 | .795 | .739 | .660 | .504 | .485 | | | | | |
| 3.000 | 3.557 | .243 | .243 | .67 | 1.69 | 12.0% | 6 | .802 | .741 | .662 | .565 | .452 | .334 | | | | |
| 3.800 | 3.686 | .114 | .114 | .79 | 1.68 | 5.8% | 7 | .797 | .733 | .651 | .550 | .434 | .315 | .206 | | | |
| 3.800 | 3.750 | .050 | .050 | .89 | 1.63 | 2.6% | 8 | .794 | .729 | .645 | .542 | .425 | .305 | .197 | .113 | | |
| 3.800 | 3.780 | .020 | .020 | .94 | 1.57 | 1.1% | 9 | .793 | .727 | .642 | .539 | .421 | .300 | .193 | .110 | .056 | |
| 4.000 | 3.382 | .618 | .618 | .48 | 1.73 | 26.8% | 5 | .806 | .755 | .691 | .612 | .519 | | | | | |
| 4.000 | 3.701 | .299 | .299 | .63 | 1.70 | 13.9% | 6 | .811 | .756 | .683 | .593 | .487 | .372 | | | | |
| 4.000 | 3.854 | .146 | .146 | .76 | 1.71 | 7.0% | 7 | .806 | .747 | .671 | .577 | .466 | .349 | .238 | | | |
| 4.000 | 3.934 | .066 | .066 | .86 | 1.67 | 3.3% | 8 | .803 | .742 | .664 | .568 | .456 | .337 | .227 | .137 | | |

*SIMILAR TO A.T.&T. JACOBSEN RETRIAL TABLES

*FAST RETRIALS: JEWETT-SHRAGO FAST RETRIAL TABLES - 50% BLOCKED CALLERS RETRY, 50% OVERFLOW (RETRIES 1 HOLDING TIMES LATER)

| HOURS FIRST ATTEMPT VMR = 1.0 | HOURS CON-NECT | HOURS RECALL /RETRY | HOURS OVFL | CON-NECT VMR | OVFL VMR | PERCENT BLOCKED (GRADE) | LINE | HOURS CONNECTED TRAFFIC ON LINES 1 | 2 | 3 | 4 | 5 | 6 | 7 | 8 | 9 | 10 |
|---|---|---|---|---|---|---|---|---|---|---|---|---|---|---|---|---|---|
| 4.000 | 3.972 | .028 | .028 | .93 | 1.61 | 1.4% | 9 | .801 | .740 | .661 | .563 | .450 | .332 | .222 | .132 | .070 | |
| 4.000 | 3.989 | .011 | .011 | .97 | 1.55 | .5% | 10 | .800 | .739 | .660 | .561 | .448 | .329 | .219 | .131 | .069 | .033 |
| 4.200 | 3.483 | .717 | .717 | .45 | 1.73 | 29.2% | 5 | .816 | .769 | .710 | .637 | .551 | | | | | |
| 4.200 | 3.838 | .362 | .362 | .59 | 1.72 | 15.9% | 6 | .820 | .769 | .702 | .619 | .519 | .408 | | | | |
| 4.200 | 4.017 | .183 | .183 | .73 | 1.73 | 8.3% | 7 | .814 | .760 | .689 | .601 | .497 | .383 | .272 | | | |
| 4.200 | 4.114 | .086 | .086 | .84 | 1.71 | 4.0% | 8 | .811 | .755 | .682 | .591 | .485 | .370 | .258 | .162 | | |
| 4.200 | 4.162 | .038 | .038 | .91 | 1.65 | 1.8% | 9 | .809 | .752 | .678 | .586 | .479 | .363 | .251 | .157 | .087 | |
| 4.200 | 4.185 | .015 | .015 | .96 | 1.59 | .7% | 10 | .808 | .751 | .677 | .584 | .476 | .359 | .248 | .154 | .085 | .042 |
| 4.400 | 3.577 | .823 | .823 | .42 | 1.73 | 31.5% | 5 | .825 | .782 | .728 | .661 | .580 | | | | | |
| 4.400 | 3.920 | .480 | .480 | .55 | 1.81 | 19.7% | 6 | .815 | .767 | .708 | .634 | .547 | .449 | | | | |
| 4.400 | 4.174 | .226 | .226 | .69 | 1.76 | 9.8% | 7 | .822 | .772 | .707 | .625 | .526 | .417 | .306 | | | |
| 4.400 | 4.290 | .110 | .110 | .81 | 1.74 | 4.9% | 8 | .819 | .766 | .699 | .614 | .513 | .401 | .289 | .190 | | |
| 4.400 | 4.350 | .050 | .050 | .89 | 1.69 | 2.2% | 9 | .817 | .763 | .694 | .608 | .505 | .393 | .281 | .182 | .106 | |
| 4.400 | 4.379 | .021 | .021 | .95 | 1.63 | 1.0% | 10 | .816 | .762 | .692 | .605 | .502 | .389 | .277 | .179 | .104 | .054 |
| 4.600 | 3.664 | .936 | .936 | .39 | 1.72 | 33.8% | 5 | .834 | .794 | .745 | .683 | .608 | | | | | |
| 4.600 | 4.040 | .560 | .560 | .52 | 1.82 | 21.7% | 6 | .823 | .780 | .725 | .656 | .575 | .481 | | | | |
| 4.600 | 4.325 | .275 | .275 | .66 | 1.78 | 11.3% | 7 | .830 | .783 | .723 | .646 | .554 | .449 | .340 | | | |
| 4.600 | 4.462 | .138 | .138 | .78 | 1.77 | 5.8% | 8 | .826 | .777 | .714 | .635 | .539 | .432 | .321 | .219 | | |
| 4.600 | 4.535 | .065 | .065 | .87 | 1.72 | 2.8% | 9 | .824 | .774 | .709 | .628 | .531 | .422 | .311 | .209 | .127 | |
| 4.600 | 4.572 | .028 | .028 | .93 | 1.66 | 1.2% | 10 | .822 | .772 | .707 | .625 | .527 | .417 | .306 | .205 | .124 | .067 |
| 4.600 | 4.588 | .012 | .012 | .97 | 1.60 | .5% | 11 | .822 | .771 | .706 | .623 | .525 | .415 | .304 | .203 | .122 | .066 |
| | | | | | | | | .032 | | | | | | | | | |
| 4.800 | 4.153 | .647 | .647 | .49 | 1.82 | 23.8% | 6 | .831 | .791 | .740 | .677 | .601 | .512 | | | | |
| 4.800 | 4.470 | .330 | .330 | .63 | 1.79 | 12.9% | 7 | .837 | .794 | .737 | .667 | .580 | .481 | .374 | | | |
| 4.800 | 4.629 | .171 | .171 | .75 | 1.80 | 6.9% | 8 | .833 | .787 | .728 | .654 | .564 | .461 | .353 | .249 | | |
| 4.800 | 4.717 | .083 | .083 | .85 | 1.76 | 3.4% | 9 | .830 | .783 | .723 | .647 | .555 | .450 | .341 | .237 | .150 | |
| 4.800 | 4.763 | .037 | .037 | .92 | 1.70 | 1.5% | 10 | .829 | .781 | .720 | .643 | .550 | .445 | .335 | .231 | .145 | .082 |
| 4.800 | 4.784 | .016 | .016 | .96 | 1.64 | .7% | 11 | .828 | .781 | .719 | .641 | .548 | .442 | .332 | .229 | .143 | .081 |
| | | | | | | | | .041 | | | | | | | | | |
| 5.000 | 4.258 | .742 | .742 | .46 | 1.82 | 25.8% | 6 | .839 | .802 | .755 | .696 | .625 | .542 | | | | |
| 5.000 | 4.563 | .437 | .437 | .59 | 1.89 | 16.1% | 7 | .831 | .791 | .740 | .676 | .600 | .511 | .414 | | | |
| 5.000 | 4.791 | .209 | .209 | .72 | 1.82 | 8.0% | 8 | .839 | .797 | .742 | .673 | .588 | .490 | .384 | .279 | | |
| 5.000 | 4.896 | .104 | .104 | .83 | 1.79 | 4.1% | 9 | .836 | .792 | .736 | .665 | .578 | .478 | .370 | .266 | .175 | |
| 5.000 | 4.951 | .049 | .049 | .90 | 1.74 | 1.9% | 10 | .835 | .790 | .733 | .660 | .572 | .471 | .363 | .259 | .168 | .099 |
| 5.000 | 4.979 | .021 | .021 | .95 | 1.67 | .9% | 11 | .834 | .789 | .731 | .658 | .570 | .468 | .359 | .255 | .165 | .097 |
| | | | | | | | | .052 | | | | | | | | | |
| 5.500 | 4.494 | 1.006 | 1.006 | .39 | 1.80 | 30.9% | 6 | .856 | .825 | .786 | .738 | .679 | .609 | | | | |
| 5.500 | 4.874 | .626 | .626 | .51 | 1.90 | 20.4% | 7 | .848 | .814 | .771 | .718 | .654 | .577 | .490 | | | |
| 5.500 | 5.173 | .327 | .327 | .64 | 1.87 | 11.2% | 8 | .854 | .818 | .772 | .714 | .642 | .556 | .460 | .358 | | |
| 5.500 | 5.325 | .175 | .175 | .76 | 1.86 | 6.2% | 9 | .850 | .813 | .765 | .704 | .630 | .541 | .442 | .339 | .241 | |

*SIMILAR TO A.T.&T. JACOBSEN RETRIAL TABLES

*FAST RETRIALS: JEWETT-SHRAGO FAST RETRIAL TABLES - 50% BLOCKED CALLERS RETRY, 50% OVERFLOW (RETRIES 1 HOLDING TIMES LATER)

| HOURS FIRST ATTEMPT VMR = 1.0 | HOURS CON-NECT | HOURS RECALL /RETRY | HOURS OVFL | CON-NECT VMR | OVFL VMR | PERCENT BLOCKED (GRADE) | LINE | HOURS CONNECTED TRAFFIC ON LINES 1 | 2 | 3 | 4 | 5 | 6 | 7 | 8 | 9 | 10 | | | | |
|---|---|---|---|---|---|---|---|---|---|---|---|---|---|---|---|---|---|---|---|---|---|
| 5.500 | 5.412 | .088 | .088 | .85 | 1.82 | 3.1% | 10 | .848 | .810 | .761 | .699 | .622 | .532 | .432 | .328 | .231 | .149 | | | | |
| 5.500 | 5.458 | .042 | .042 | .92 | 1.76 | 1.5% | 11 | .847 | .808 | .759 | .696 | .618 | .527 | .426 | .322 | .225 | .145 | .085 | | | |
| 5.500 | 5.481 | .019 | .019 | .96 | 1.70 | .7% | 12 | .847 | .808 | .758 | .694 | .616 | .525 | .423 | .319 | .223 | .142 | .083 | .044 | | |
| 6.000 | 5.144 | .856 | .856 | .44 | 1.89 | 25.0% | 7 | .863 | .834 | .798 | .754 | .700 | .635 | .559 | | | | | | | |
| 6.000 | 5.471 | .529 | .529 | .56 | 1.97 | 16.2% | 8 | .857 | .826 | .787 | .739 | .681 | .611 | .530 | .441 | | | | | | |
| 6.000 | 5.726 | .274 | .274 | .69 | 1.92 | 8.7% | 9 | .863 | .831 | .790 | .738 | .674 | .597 | .508 | .411 | .313 | | | | | |
| 6.000 | 5.853 | .147 | .147 | .79 | 1.90 | 4.8% | 10 | .860 | .827 | .785 | .732 | .666 | .586 | .495 | .397 | .298 | .208 | | | | |
| 6.000 | 5.926 | .074 | .074 | .87 | 1.85 | 2.4% | 11 | .859 | .825 | .782 | .728 | .660 | .580 | .487 | .388 | .289 | .200 | .127 | | | |
| 6.000 | 5.964 | .036 | .036 | .93 | 1.79 | 1.2% | 12 | .858 | .824 | .780 | .726 | .658 | .576 | .483 | .383 | .285 | .196 | .124 | .072 | | |
| 6.000 | 5.984 | .016 | .016 | .96 | 1.72 | .5% | 13 | .858 | .823 | .780 | .725 | .656 | .574 | .481 | .381 | .282 | .194 | .122 | .071 | .037 | |
| 6.500 | 5.330 | 1.170 | 1.170 | .38 | 1.95 | 30.5% | 7 | .868 | .843 | .813 | .777 | .732 | .679 | .617 | | | | | | | |
| 6.500 | 5.773 | .727 | .727 | .49 | 1.98 | 20.1% | 8 | .869 | .843 | .810 | .770 | .720 | .660 | .590 | .510 | | | | | | |
| 6.500 | 6.053 | .447 | .447 | .61 | 2.03 | 12.9% | 9 | .864 | .836 | .801 | .758 | .705 | .641 | .567 | .484 | .395 | | | | | |
| 6.500 | 6.270 | .230 | .230 | .73 | 1.96 | 6.8% | 10 | .871 | .842 | .806 | .760 | .703 | .634 | .553 | .462 | .367 | .273 | | | | |
| 6.500 | 6.377 | .123 | .123 | .82 | 1.93 | 3.7% | 11 | .869 | .839 | .802 | .755 | .697 | .626 | .543 | .451 | .354 | .261 | .179 | | | |
| 6.500 | 6.437 | .063 | .063 | .89 | 1.87 | 1.9% | 12 | .868 | .838 | .800 | .752 | .693 | .621 | .537 | .444 | .347 | .254 | .173 | .109 | | |
| 6.500 | 6.470 | .030 | .030 | .94 | 1.81 | .9% | 13 | .867 | .837 | .799 | .751 | .691 | .019 | .534 | .441 | .343 | .251 | .170 | .106 | .061 | |
| 7.000 | 5.529 | 1.471 | 1.471 | .33 | 1.92 | 34.7% | 7 | .880 | .859 | .834 | .803 | .765 | .720 | .667 | | | | | | | |
| 7.000 | 6.037 | .963 | .963 | .43 | 1.97 | 24.2% | 8 | .880 | .858 | .830 | .796 | .754 | .703 | .643 | .573 | | | | | | |
| 7.000 | 6.382 | .618 | .618 | .54 | 2.05 | 16.2% | 9 | .875 | .851 | .821 | .784 | .739 | .684 | .619 | .545 | .463 | | | | | |
| 7.000 | 6.622 | .378 | .378 | .65 | 2.09 | 10.3% | 10 | .872 | .846 | .815 | .775 | .728 | .670 | .602 | .524 | .439 | .351 | | | | |
| 7.000 | 6.807 | .193 | .193 | .76 | 2.00 | 5.4% | 11 | .878 | .852 | .820 | .779 | .728 | .667 | .593 | .509 | .418 | .325 | .237 | | | |
| 7.000 | 6.896 | .104 | .104 | .85 | 1.96 | 2.9% | 12 | .877 | .850 | .817 | .776 | .724 | .661 | .586 | .501 | .408 | .315 | .228 | .154 | | |
| 7.000 | 6.947 | .053 | .053 | .91 | 1.89 | 1.5% | 13 | .876 | .849 | .816 | .774 | .721 | .657 | .582 | .496 | .403 | .309 | .223 | .149 | .092 | |
| 7.000 | 6.974 | .026 | .026 | .95 | 1.82 | .7% | 14 | .875 | .849 | .815 | .772 | .720 | .656 | .579 | .493 | .400 | .306 | .220 | .147 | .091 | .052 |
| 7.500 | 6.221 | 1.279 | 1.279 | .37 | 2.01 | 29.1% | 8 | .884 | .864 | .840 | .811 | .776 | .734 | .684 | .626 | | | | | | |
| 7.500 | 6.677 | .823 | .823 | .47 | 2.05 | 19.8% | 9 | .885 | .864 | .838 | .807 | .768 | .721 | .666 | .600 | .526 | | | | | |
| 7.500 | 6.975 | .525 | .525 | .58 | 2.11 | 13.1% | 10 | .881 | .859 | .831 | .798 | .756 | .707 | .647 | .578 | .500 | .417 | | | | |

*SIMILAR TO A.T.&T. JACOBSEN RETRIAL TABLES

# *FAST RETRIALS: JEWETT-SHRAGO FAST RETRIAL TABLES - 50% BLOCKED CALLERS RETRY, 50% OVERFLOW (RETRIES 1 HOLDING TIMES LATER)

| HOURS FIRST ATTEMPT VMR=1.0 | HOURS CON-NECT | HOURS RECALL /RETRY | HOURS OVFL | CON-NECT VMR | OVFL VMR | PERCENT BLOCKED (GRADE) | LINE | HOURS CONNECTED TRAFFIC ON LINES 1 | 2 | 3 | 4 | 5 | 6 | 7 | 8 | 9 | 10 |
|---|---|---|---|---|---|---|---|---|---|---|---|---|---|---|---|---|---|
| 7.500 | 7.214 | .286 | .286 | .70 | 2.05 | 7.3% | 11 | .886 | .864 | .835 | .800 | .756 | .703 | .638 | .563 | .479 | .390 |
| | | | | | | | | .300 | | | | | | | | | |
| 7.500 | 7.338 | .162 | .162 | .79 | 2.03 | 4.2% | 12 | .885 | .861 | .832 | .796 | .751 | .696 | .629 | .552 | .467 | .376 |
| | | | | | | | | .287 | .206 | | | | | | | | |
| 7.500 | 7.413 | .087 | .087 | .87 | 1.98 | 2.3% | 13 | .884 | .860 | .830 | .793 | .747 | .691 | .624 | .546 | .459 | .368 |
| | | | | | | | | .279 | .198 | .132 | | | | | | | |
| 7.500 | 7.455 | .045 | .045 | .92 | 1.91 | 1.2% | 14 | .883 | .859 | .829 | .792 | .745 | .689 | .621 | .542 | .455 | .364 |
| | | | | | | | | .275 | .194 | .128 | .079 | | | | | | |
| 7.500 | 7.478 | .022 | .022 | .96 | 1.84 | .6% | 15 | .883 | .859 | .829 | .791 | .744 | .687 | .619 | .540 | .453 | .361 |
| | | | | | | | | .272 | .192 | .126 | .077 | .044 | | | | | |
| 8.000 | 6.419 | 1.581 | 1.581 | .33 | 1.98 | 33.0% | 8 | .893 | .876 | .856 | .831 | .801 | .765 | .723 | .673 | | |
| 8.000 | 6.891 | 1.109 | 1.109 | .42 | 2.10 | 24.4% | 9 | .888 | .869 | .847 | .820 | .787 | .748 | .701 | .646 | .584 | |
| 8.000 | 7.296 | .704 | .704 | .52 | 2.12 | 16.2% | 10 | .890 | .870 | .846 | .817 | .782 | .739 | .687 | .627 | .557 | .480 |
| 8.000 | 7.554 | .446 | .446 | .63 | 2.17 | 10.6% | 11 | .887 | .866 | .841 | .810 | .773 | .727 | .673 | .609 | .537 | .457 |
| | | | | | | | | .374 | | | | | | | | | |
| 8.000 | 7.759 | .241 | .241 | .74 | 2.09 | 5.8% | 12 | .892 | .871 | .846 | .814 | .774 | .726 | .668 | .600 | .521 | .436 |
| | | | | | | | | .348 | .263 | | | | | | | | |
| 8.000 | 7.864 | .136 | .136 | .82 | 2.05 | 3.3% | 13 | .891 | .869 | .843 | .811 | .770 | .721 | .661 | .591 | .512 | .426 |
| | | | | | | | | .337 | .253 | .178 | | | | | | | |
| 8.000 | 7.927 | .073 | .073 | .89 | 1.99 | 1.8% | 14 | .890 | .868 | .842 | .809 | .768 | .718 | .657 | .587 | .506 | .419 |
| | | | | | | | | .331 | .246 | .172 | .113 | | | | | | |
| 8.000 | 7.962 | .038 | .038 | .93 | 1.92 | .9% | 15 | .889 | .868 | .841 | .808 | .767 | .716 | .655 | .584 | .503 | .416 |
| | | | | | | | | .327 | .243 | .169 | .110 | .067 | | | | | |
| 8.500 | 7.118 | 1.382 | 1.382 | .37 | 2.07 | 28.0% | 9 | .896 | .880 | .861 | .837 | .809 | .776 | .736 | .688 | .634 | |
| 8.500 | 7.585 | .915 | .915 | .46 | 2.12 | 19.4% | 10 | .898 | .881 | .860 | .835 | .804 | .767 | .723 | .670 | .609 | .540 |
| 8.500 | 7.899 | .601 | .601 | .56 | 2.19 | 13.2% | 11 | .894 | .876 | .854 | .827 | .795 | .755 | .708 | .652 | .587 | .514 |
| | | | | | | | | .436 | | | | | | | | | |
| 8.500 | 8.121 | .379 | .379 | .67 | 2.21 | 8.5% | 12 | .892 | .873 | .850 | .822 | .788 | .746 | .697 | .638 | .571 | .496 |
| | | | | | | | | .415 | .334 | | | | | | | | |
| 8.500 | 8.297 | .203 | .203 | .77 | 2.12 | 4.7% | 13 | .897 | .878 | .855 | .826 | .791 | .747 | .695 | .633 | .561 | .480 |
| | | | | | | | | .395 | .310 | .230 | | | | | | | |
| 8.500 | 8.385 | .115 | .115 | .85 | 2.07 | 2.7% | 14 | .896 | .877 | .853 | .824 | .788 | .743 | .690 | .627 | .553 | .472 |
| | | | | | | | | .386 | .301 | .222 | .153 | | | | | | |
| 8.500 | 8.438 | .062 | .062 | .90 | 2.01 | 1.4% | 15 | .895 | .876 | .852 | .822 | .786 | .741 | .687 | .623 | .549 | .467 |
| | | | | | | | | .381 | .296 | .217 | .149 | .096 | | | | | |
| 8.500 | 8.468 | .032 | .032 | .94 | 1.94 | .7% | 16 | .895 | .876 | .851 | .822 | .785 | .740 | .685 | .621 | .547 | .464 |
| | | | | | | | | .378 | .293 | .214 | .147 | .094 | .057 | | | | |
| 9.000 | 7.316 | 1.684 | 1.684 | .32 | 2.04 | 31.5% | 9 | .904 | .890 | .873 | .853 | .829 | .800 | .765 | .725 | .677 | |
| 9.000 | 7.795 | 1.205 | 1.205 | .41 | 2.16 | 23.6% | 10 | .900 | .884 | .866 | .844 | .818 | .786 | .748 | .704 | .652 | .593 |
| 9.000 | 8.213 | .787 | .787 | .50 | 2.19 | 16.1% | 11 | .901 | .885 | .866 | .842 | .814 | .780 | .739 | .690 | .633 | .568 |

*SIMILAR TO A.T.&T. JACOBSEN RETRIAL TABLES

*FAST RETRIALS: JEWETT-SHRAGO FAST RETRIAL TABLES - 50% BLOCKED CALLERS RETRY, 50% OVERFLOW (RETRIES 1 HOLDING TIMES LATER)

| HOURS FIRST ATTEMPT VMR = 1.0 | HOURS CON-NECT | HOURS RECALL /RETRY | HOURS OVFL | CON-NECT VMR | OVFL VMR | PERCENT BLOCKED (GRADE) | LINE | HOURS CONNECTED TRAFFIC ON LINES 1 | 2 | 3 | 4 | 5 | 6 | 7 | 8 | 9 | 10 |
|---|---|---|---|---|---|---|---|---|---|---|---|---|---|---|---|---|---|
| | | | | | | | | .495 | | | | | | | | | |
| 9.000 | 8.487 | .513 | .513 | .61 | 2.24 | 10.8% | 12 | .899 | .882 | .861 | .837 | .807 | .770 | .727 | .675 | .615 | .547 |
| | | | | | | | | .472 | .393 | | | | | | | | |
| 9.000 | 8.710 | .290 | .290 | .71 | 2.18 | 6.3% | 13 | .903 | .886 | .865 | .840 | .809 | .771 | .724 | .669 | .605 | .532 |
| | | | | | | | | .452 | .368 | .286 | | | | | | | |
| 9.000 | 8.829 | .171 | .171 | .80 | 2.14 | 3.7% | 14 | .902 | .884 | .863 | .837 | .805 | .766 | .719 | .662 | .596 | .522 |
| | | | | | | | | .440 | .356 | .275 | .200 | | | | | | |
| 9.000 | 8.904 | .096 | .096 | .87 | 2.09 | 2.1% | 15 | .901 | .883 | .862 | .836 | .803 | .763 | .715 | .658 | .591 | .515 |
| | | | | | | | | .433 | .349 | .267 | .194 | .132 | | | | | |
| 9.000 | 8.948 | .052 | .052 | .92 | 2.02 | 1.2% | 16 | .901 | .883 | .861 | .834 | .802 | .762 | .713 | .655 | .588 | .512 |
| | | | | | | | | .429 | .345 | .263 | :190 | .129 | .082 | | | | |
| 9.000 | 8.973 | .027 | .027 | .95 | 1.95 | .6% | 17 | .900 | .883 | .861 | .834 | .801 | .761 | .712 | .654 | .586 | .509 |
| | | | | | | | | .427 | .342 | .261 | .188 | .127 | .081 | .048 | | | |
| 9.500 | 8.021 | 1.479 | 1.479 | .36 | 2.13 | 26.9% | 10 | .907 | .893 | .877 | .858 | .835 | .808 | .775 | .737 | .692 | .640 |
| 9.500 | 8.452 | 1.048 | 1.048 | .45 | 2.24 | 19.9% | 11 | .903 | .888 | .871 | .850 | .826 | .796 | .761 | .719 | .671 | .615 |
| | | | | | | | | .552 | | | | | | | | | |
| 9.500 | 8.824 | .676 | .676 | .55 | 2.25 | 13.3% | 12 | .905 | .890 | .872 | .850 | .824 | .792 | .754 | .709 | .656 | .595 |
| | | | | | | | | .526 | .452 | | | | | | | | |
| 9.500 | 9.062 | .438 | .438 | .65 | 2.28 | 8.8% | 13 | .903 | .887 | .868 | .846 | .818 | .785 | .745 | .697 | .642 | .578 |
| | | | | | | | | .508 | .432 | .353 | | | | | | | |
| 9.500 | 9.255 | .245 | .245 | .75 | 2.21 | 5.0% | 14 | .907 | .891 | .872 | .849 | .821 | .786 | .745 | .694 | .635 | .568 |
| | | | | | | | | .492 | .412 | .330 | .252 | | | | | | |
| 9.500 | 9.356 | .144 | .144 | .83 | 2.16 | 3.0% | 15 | .906 | .890 | .871 | .847 | .818 | .783 | .740 | .689 | .629 | .560 |
| | | | | | | | | .483 | .402 | .320 | .243 | .174 | | | | | |
| 9.500 | 9.419 | .081 | .081 | .89 | 2.10 | 1.7% | 16 | .906 | .889 | .870 | .846 | .817 | .781 | .738 | .686 | .625 | .555 |
| | | | | | | | | .470 | .398 | .314 | .237 | .169 | .114 | | | | |
| 9.500 | 9.456 | .044 | .044 | .93 | 2.03 | .9% | 17 | .905 | .889 | .869 | .845 | .816 | .780 | .736 | .684 | .622 | .552 |
| | | | | | | | | .475 | .393 | .311 | .233 | .166 | .111 | .070 | | | |
| 10.000 | 8.177 | 1.823 | 1.823 | .32 | 2.16 | 30.8% | 10 | .909 | .897 | .882 | .865 | .845 | .822 | .794 | .761 | .723 | .679 |
| 10.000 | 8.704 | 1.296 | 1.296 | .40 | 2.22 | 22.9% | 11 | .909 | .896 | .881 | .863 | .841 | .815 | .785 | .749 | .706 | .657 |
| | | | | | | | | .601 | | | | | | | | | |
| 10.000 | 9.090 | .910 | .910 | .49 | 2.32 | 16.7% | 12 | .900 | .892 | .876 | .857 | .833 | .806 | .773 | .734 | .689 | .636 |
| | | | | | | | | .576 | .511 | | | | | | | | |
| 10.000 | 9.420 | .580 | .580 | .59 | 2.31 | 11.0% | 13 | .908 | .894 | .877 | .857 | .833 | .804 | .769 | .727 | .678 | .621 |
| | | | | | | | | .556 | .485 | .410 | | | | | | | |
| 10.000 | 9.626 | .374 | .374 | .68 | 2.32 | 7.2% | 14 | .907 | .892 | .875 | .854 | .828 | .798 | .761 | .718 | .667 | .608 |
| | | | | | | | | .541 | .469 | .393 | .316 | | | | | | |
| 10.000 | 9.702 | .208 | .208 | .78 | 2.23 | 4.1% | 15 | .911 | .896 | .879 | .858 | .832 | .801 | .763 | .717 | .663 | .601 |
| | | | | | | | | .530 | .453 | .373 | .294 | .221 | | | | | |
| 10.000 | 9.878 | .122 | .122 | .85 | 2.18 | 2.4% | 16 | .910 | .896 | .878 | .856 | .830 | .798 | .759 | .713 | .658 | .595 |

*SIMILAR TO A.T.&T. JACOBSEN RETRIAL TABLES

*FAST RETRIALS: JEWETT-SHRAGO FAST RETRIAL TABLES - 50% BLOCKED CALLERS RETRY, 50% OVERFLOW (RETRIES 1 HOLDING TIMES LATER)

| HOURS FIRST ATTEMPT VMR = 1.0 | HOURS CON- NECT | HOURS RECALL /RETRY | HOURS OVFL | CON- NECT VMR | OVFL VMR | PERCENT BLOCKED (GRADE) | LINE | HOURS CONNECTED TRAFFIC ON LINES 1 | 2 | 3 | 4 | 5 | 6 | 7 | 8 | 9 | 10 |
|---|---|---|---|---|---|---|---|---|---|---|---|---|---|---|---|---|---|
| | | | | | | | | .523 | .446 | .365 | .286 | .213 | .151 | | | | |
| 10.000 | 9.932 | .068 | .068 | .90 | 2.11 | 1.4% | 17 | .910 | .895 | .877 | .855 | .829 | .796 | .757 | .711 | .655 | .591 |
| | | | | | | | | .519 | .441 | .360 | .281 | .209 | .147 | .097 | | | |
| 10.000 | 9.963 | .037 | .037 | .94 | 2.04 | .7% | 18 | .909 | .895 | .877 | .855 | .828 | .795 | .756 | .709 | .653 | .589 |
| | | | | | | | | .517 | .438 | .357 | .279 | .206 | .145 | .096 | .060 | | |
| 11.000 | 9.085 | 1.915 | 1.915 | .32 | 2.21 | 29.7% | 11 | .917 | .906 | .894 | .880 | .863 | .843 | .820 | .793 | .761 | .725 |
| | | | | | | | | .682 | | | | | | | | | |
| 11.000 | 9.617 | 1.383 | 1.383 | .39 | 2.28 | 22.3% | 12 | .917 | .906 | .893 | .878 | .860 | .839 | .814 | .784 | .749 | .708 |
| | | | | | | | | .661 | .608 | | | | | | | | |
| 11.000 | 10.012 | .988 | .988 | .48 | 2.37 | 16.5% | 13 | .914 | .903 | .889 | .873 | .854 | .831 | .804 | .772 | .735 | .691 |
| | | | | | | | | .641 | .584 | .522 | | | | | | | |
| 11.000 | 10.355 | .645 | .645 | .57 | 2.37 | 11.1% | 14 | .916 | .904 | .890 | .874 | .854 | .830 | .801 | .767 | .727 | .680 |
| | | | | | | | | .625 | .564 | .497 | .425 | | | | | | |
| 11.000 | 10.574 | .426 | .426 | .67 | 2.39 | 7.5% | 15 | .915 | .903 | .888 | .871 | .850 | .825 | .795 | .759 | .718 | .669 |
| | | | | | | | | .612 | .549 | .480 | .407 | .334 | | | | | |
| 11.000 | 10.754 | .246 | .246 | .76 | 2.31 | 4.4% | 16 | .918 | .906 | .892 | .874 | .853 | .827 | .797 | .760 | .716 | .664 |
| | | | | | | | | .605 | .538 | .465 | .388 | .312 | .240 | | | | |
| 11.000 | 10.851 | .149 | .149 | .83 | 2.26 | 2.7% | 17 | .918 | .905 | .891 | .873 | .851 | .825 | .794 | .756 | .711 | .659 |
| | | | | | | | | .598 | .531 | .457 | .380 | .303 | .232 | .168 | | | |
| 11.000 | 10.913 | .087 | .087 | .89 | 2.20 | 1.6% | 18 | .917 | .905 | .890 | .872 | .850 | .824 | .792 | .754 | .709 | .655 |
| | | | | | | | | .594 | .526 | .452 | .374 | .298 | .227 | .164 | .112 | | |
| 11.000 | 10.951 | .049 | .049 | .93 | 2.12 | .9% | 19 | .917 | .904 | .889 | .871 | .849 | .823 | .791 | .752 | .707 | .653 |
| | | | | | | | | .592 | .523 | .448 | .371 | .295 | .223 | .161 | .110 | .071 | |
| 12.000 | 9.997 | 2.003 | 2.003 | .31 | 2.26 | 28.6% | 12 | .923 | .914 | .904 | .892 | .877 | .861 | .841 | .819 | .792 | .762 |
| | | | | | | | | .726 | .685 | | | | | | | | |
| 12.000 | 10.491 | 1.509 | 1.509 | .39 | 2.39 | 22.3% | 13 | .920 | .911 | .900 | .886 | .871 | .853 | .832 | .808 | .779 | .746 |
| | | | | | | | | .707 | .663 | .613 | | | | | | | |
| 12.000 | 10.937 | 1.063 | 1.063 | .47 | 2.43 | 16.3% | 14 | .921 | .911 | .900 | .886 | .870 | .851 | .828 | .802 | .771 | .735 |
| | | | | | | | | .693 | .645 | .591 | .531 | | | | | | |
| 12.000 | 11.290 | .710 | .710 | .56 | 2.44 | 11.2% | 15 | .923 | .913 | .901 | .887 | .870 | .850 | .827 | .799 | .766 | .727 |
| | | | | | | | | .682 | .630 | .571 | .507 | .439 | | | | | |
| 12.000 | 11.522 | .478 | .478 | .65 | 2.46 | 7.7% | 16 | .922 | .911 | .899 | .884 | .867 | .846 | .821 | .792 | .758 | .717 |
| | | | | | | | | .670 | .617 | .556 | .490 | .421 | .349 | | | | |
| 12.000 | 11.687 | .313 | .313 | .73 | 2.45 | 5.1% | 17 | .921 | .910 | .897 | .882 | .864 | .843 | .818 | .788 | .752 | .710 |
| | | | | | | | | .662 | .607 | .545 | .478 | .407 | .336 | .267 | | | |
| 12.000 | 11.823 | .177 | .177 | .81 | 2.34 | 2.9% | 18 | .924 | .914 | .901 | .886 | .868 | .847 | .821 | .790 | .753 | .710 |
| | | | | | | | | .660 | .602 | .537 | .467 | .393 | .319 | .248 | .185 | | |
| 12.000 | 11.894 | .106 | .106 | .87 | 2.28 | 1.8% | 19 | .924 | .913 | .900 | .885 | .867 | .845 | .819 | .788 | .751 | .707 |
| | | | | | | | | .656 | .597 | .532 | .461 | .387 | .313 | .243 | .180 | .127 | |
| 12.000 | 11.938 | .062 | .062 | .92 | 2.21 | 1.0% | 20 | .923 | .913 | .900 | .885 | .866 | .844 | .818 | .787 | .749 | .705 |

*SIMILAR TO A.T.&T. JACOBSEN RETRIAL TABLES

*FAST RETRIALS: JEWETT-SHRAGO FAST RETRIAL TABLES - 50% BLOCKED CALLERS RETRY, 50% OVERFLOW (RETRIES 1 HOLDING TIMES LATER)

| HOURS FIRST ATTEMPT VMR = 1.0 | HOURS CON-NECT | HOURS RECALL /RETRY | HOURS OVFL | CON-NECT VMR | OVFL VMR | PERCENT BLOCKED (GRADE) | LINE | HOURS CONNECTED TRAFFIC ON LINES 1 | 2 | 3 | 4 | 5 | 6 | 7 | 8 | 9 | 10 |
|---|---|---|---|---|---|---|---|---|---|---|---|---|---|---|---|---|---|
| | | | | | | | | .653 | .594 | .529 | .457 | .383 | .309 | .239 | .177 | .124 | .083 |
| 12.000 | 11.966 | .034 | .034 | .95 | 2.13 | .6% | 21 | .923 | .913 | .900 | .884 | .866 | .844 | .817 | .786 | .748 | .704 |
| | | | | | | | | .652 | .593 | .527 | .455 | .381 | .307 | .237 | .175 | .123 | .082 |
| | | | | | | | | .052 | | | | | | | | | |
| 13.000 | 10.293 | 2.707 | 2.707 | .26 | 2.23 | 34.5% | 12 | .929 | .922 | .913 | .903 | .892 | .879 | .863 | .846 | .825 | .802 |
| | | | | | | | | .775 | .744 | | | | | | | | |
| 13.000 | 10.873 | 2.127 | 2.127 | .31 | 2.37 | 28.1% | 13 | .926 | .918 | .909 | .898 | .886 | .872 | .855 | .836 | .814 | .788 |
| | | | | | | | | .759 | .725 | .686 | | | | | | | |
| 13.000 | 11.411 | 1.589 | 1.589 | .38 | 2.44 | 21.8% | 14 | .926 | .918 | .908 | .897 | .884 | .869 | .851 | .830 | .806 | .779 |
| | | | | | | | | .746 | .709 | .667 | .619 | | | | | | |
| 13.000 | 11.863 | 1.137 | 1.137 | .46 | 2.48 | 16.1% | 15 | .927 | .919 | .909 | .897 | .883 | .867 | .848 | .826 | .801 | .770 |
| | | | | | | | | .735 | .695 | .649 | .597 | .540 | | | | | |
| 13.000 | 12.188 | .812 | .812 | .55 | 2.55 | 11.8% | 16 | .926 | .917 | .906 | .894 | .879 | .863 | .843 | .820 | .792 | .760 |
| | | | | | | | | .723 | .681 | .632 | .578 | .519 | .455 | | | | |
| 13.000 | 12.470 | .530 | .530 | .64 | 2.52 | 7.8% | 17 | .928 | .918 | .908 | .895 | .881 | .863 | .843 | .819 | .790 | .756 |
| | | | | | | | | .717 | .672 | .620 | .563 | .499 | .432 | .364 | | | |
| 13.000 | 12.646 | .354 | .354 | .72 | 2.52 | 5.3% | 18 | .927 | .917 | .906 | .894 | .879 | .861 | .840 | .815 | .785 | .750 |
| | | | | | | | | .710 | .663 | .610 | .551 | .487 | .419 | .350 | .282 | | |
| 13.000 | 12.793 | .207 | .207 | .80 | 2.42 | 3.1% | 19 | .930 | .920 | .910 | .897 | .882 | .864 | .842 | .817 | .787 | .751 |
| | | | | | | | | .709 | .660 | .605 | .543 | .475 | .404 | .333 | .264 | .201 | |
| 13.000 | 12.873 | .127 | .127 | .86 | 2.36 | 1.9% | 20 | .929 | .920 | .909 | .896 | .881 | .863 | .841 | .815 | .784 | .748 |
| | | | | | | | | .705 | .656 | .600 | .537 | .469 | .398 | .326 | .258 | .195 | .141 |
| 13.000 | 12.924 | .076 | .076 | .91 | 2.29 | 1.2% | 21 | .929 | .920 | .909 | .896 | .880 | .862 | .840 | .814 | .783 | .746 |
| | | | | | | | | .703 | .653 | .597 | .534 | .465 | .394 | .322 | .254 | .191 | .138 |
| | | | | | | | | .095 | | | | | | | | | |
| 13.000 | 12.956 | .044 | .044 | .94 | 2.21 | .7% | 22 | .929 | .910 | .908 | .895 | .880 | .861 | .839 | .813 | .782 | .745 |
| | | | | | | | | .702 | .652 | .595 | .531 | .463 | .391 | .320 | .251 | .189 | .136 |
| | | | | | | | | .093 | .061 | | | | | | | | |
| 14.000 | 11.175 | 2.825 | 2.825 | .26 | 2.33 | 33.6% | 13 | .932 | .925 | .917 | .908 | .898 | .887 | .874 | .859 | .841 | .821 |
| | | | | | | | | .798 | .772 | .742 | | | | | | | |
| 14.000 | 11.792 | 2.208 | 2.208 | .31 | 2.41 | 27.2% | 14 | .931 | .924 | .916 | .907 | .896 | .884 | .870 | .853 | .835 | .813 |
| | | | | | | | | .788 | .759 | .726 | .609 | | | | | | |
| 14.000 | 12.333 | 1.667 | 1.667 | .38 | 2.49 | 21.3% | 15 | .932 | .924 | .916 | .906 | .895 | .882 | .867 | .849 | .829 | .805 |
| | | | | | | | | .778 | .747 | .711 | .670 | .624 | | | | | |
| 14.000 | 12.792 | 1.208 | 1.208 | .45 | 2.54 | 15.9% | 16 | .932 | .925 | .916 | .906 | .894 | .881 | .865 | .846 | .824 | .799 |
| | | | | | | | | .770 | .736 | .697 | .652 | .602 | .547 | | | | |
| 14.000 | 13.126 | .874 | .874 | .53 | 2.61 | 11.8% | 17 | .931 | .923 | .914 | .903 | .891 | .877 | .860 | .840 | .817 | .701 |
| | | | | | | | | .760 | .724 | .683 | .636 | .584 | .527 | .465 | | | |
| 14.000 | 13.418 | .582 | .582 | .62 | 2.59 | 8.0% | 18 | .933 | .925 | .915 | .905 | .892 | .877 | .860 | .840 | .816 | .788 |
| | | | | | | | | .755 | .717 | .673 | .624 | .568 | .508 | .443 | .377 | | |

*SIMILAR TO A.T.&T. JACOBSEN RETRIAL TABLES

*FAST RETRIALS: JEWETT-SHRAGO FAST RETRIAL TABLES - 50% BLOCKED CALLERS RETRY, 50% OVERFLOW (RETRIES 1 HOLDING TIMES LATER)

| HOURS FIRST ATTEMPT VMR = 1.0 | HOURS CON-NECT | HOURS RECALL /RETRY | HOURS OVFL | CON-NECT VMR | OVFL VMR | PERCENT BLOCKED (GRADE) | LINE | HOURS CONNECTED TRAFFIC ON LINES 1 | 2 | 3 | 4 | 5 | 6 | 7 | 8 | 9 | 10 |
|---|---|---|---|---|---|---|---|---|---|---|---|---|---|---|---|---|---|
| 14.000 | 13.605 | .395 | .395 | .70 | 2.58 | 5.5% | 19 | .932 | .924 | .914 | .903 | .890 | .875 | .857 | .836 | .812 | .783 |
| | | | | | | | | .749 | .710 | .665 | .614 | .557 | .495 | .430 | .363 | .296 | |
| 14.000 | 13.762 | .238 | .238 | .78 | 2.49 | 3.3% | 20 | .934 | .926 | .917 | .906 | .893 | .878 | .860 | .839 | .813 | .784 |
| | | | | | | | | .749 | .708 | .661 | .608 | .548 | .483 | .415 | .346 | .278 | .216 |
| 14.000 | 13.850 | .150 | .150 | .85 | 2.44 | 2.1% | 21 | .934 | .926 | .916 | .905 | .892 | .877 | .859 | .837 | .811 | .781 |
| | | | | | | | | .746 | .704 | .657 | .602 | .542 | .477 | .408 | .339 | .272 | .209 |
| | | | | | | | | .155 | | | | | | | | | |
| 14.000 | 13.908 | .092 | .092 | .89 | 2.37 | 1.3% | 22 | .934 | .926 | .916 | .905 | .892 | .876 | .858 | .836 | .810 | .779 |
| | | | | | | | | .744 | .702 | .654 | .599 | .538 | .473 | .404 | .334 | .267 | .205 |
| | | | | | | | | .151 | .107 | | | | | | | | |
| 14.000 | 13.946 | .054 | .054 | .93 | 2.29 | .8% | 23 | .934 | .925 | .916 | .905 | .891 | .876 | .857 | .835 | .809 | .778 |
| | | | | | | | | .742 | .700 | .652 | .597 | .536 | .470 | .401 | .331 | .264 | .203 |
| | | | | | | | | .149 | .105 | .070 | | | | | | | |
| 15.000 | 12.096 | 2.904 | 2.904 | .26 | 2.37 | 32.4% | 14 | .936 | .930 | .923 | .915 | .907 | .897 | .885 | .872 | .857 | .840 |
| | | | | | | | | .820 | .798 | .772 | .743 | | | | | | |
| 15.000 | 12.714 | 2.286 | 2.286 | .31 | 2.46 | 26.5% | 15 | .936 | .930 | .922 | .914 | .905 | .894 | .882 | .868 | .852 | .833 |
| | | | | | | | | .812 | .787 | .759 | .727 | .691 | | | | | |
| 15.000 | 13.258 | 1.742 | 1.742 | .37 | 2.53 | 20.8% | 16 | .936 | .930 | .922 | .914 | .904 | .893 | .880 | .865 | .847 | .827 |
| | | | | | | | | .804 | .778 | .747 | .712 | .672 | .628 | | | | |
| 15.000 | 13.682 | 1.318 | 1.318 | .44 | 2.64 | 16.1% | 17 | .934 | .928 | .920 | .911 | .900 | .889 | .875 | .859 | .840 | .819 |
| | | | | | | | | .794 | .766 | .733 | .696 | .654 | .607 | .556 | | | |
| 15.000 | 14.065 | .935 | .935 | .53 | 2.66 | 11.7% | 18 | .935 | .929 | .921 | .911 | .901 | .888 | .874 | .857 | .838 | .815 |
| | | | | | | | | .789 | .759 | .724 | .684 | .639 | .589 | .534 | .475 | | |
| 15.000 | 14.366 | .634 | .634 | .61 | 2.64 | 8.1% | 19 | .937 | .930 | .922 | .913 | .902 | .889 | .874 | .857 | .837 | .814 |
| | | | | | | | | .786 | .754 | .717 | .675 | .627 | .574 | .515 | .453 | .389 | |
| 15.000 | 14.563 | .437 | .437 | .69 | 2.65 | 5.7% | 20 | .936 | .929 | .921 | .911 | .900 | .887 | .872 | .854 | .833 | .809 |
| | | | | | | | | .781 | .748 | .710 | .666 | .617 | .562 | .502 | .439 | .374 | .310 |
| 15.000 | 14.707 | .293 | .293 | .76 | 2.62 | 3.8% | 21 | .936 | .928 | .920 | .910 | .899 | .886 | .870 | .852 | .831 | .806 |
| | | | | | | | | .777 | .743 | .704 | .659 | .609 | .553 | .493 | .429 | .364 | .299 |
| | | | | | | | | .238 | | | | | | | | | |
| 15.000 | 14.827 | .173 | .173 | .83 | 2.51 | 2.3% | 22 | .938 | .931 | .923 | .913 | .902 | .889 | .873 | .855 | .833 | .808 |
| | | | | | | | | .778 | .744 | .703 | .657 | .605 | .547 | .484 | .418 | .350 | .284 |
| | | | | | | | | .223 | .168 | | | | | | | | |
| 15.000 | 14.892 | .108 | .108 | .88 | 2.44 | 1.4% | 23 | .938 | .931 | .922 | .913 | .901 | .888 | .872 | .854 | .832 | .806 |
| | | | | | | | | .776 | .741 | .701 | .654 | .601 | .543 | .479 | .413 | .345 | .280 |
| | | | | | | | | .218 | .164 | .118 | | | | | | | |
| 15.000 | 14.934 | .066 | .066 | .92 | 2.37 | .9% | 24 | .938 | .931 | .922 | .912 | .901 | .887 | .872 | .853 | .831 | .805 |
| | | | | | | | | .775 | .740 | .699 | .652 | .599 | .540 | .476 | .410 | .342 | .276 |
| | | | | | | | | .215 | .161 | .116 | .080 | | | | | | |
| 15.000 | 14.961 | .039 | .039 | .95 | 2.28 | .5% | 25 | .938 | .930 | .922 | .912 | .901 | .887 | .871 | .853 | .831 | .805 |

*SIMILAR TO A.T.&T. JACOBSEN RETRIAL TABLES

*FAST RETRIALS: JEWETT-SHRAGO FAST RETRIAL TABLES - 50% BLOCKED CALLERS RETRY, 50% OVERFLOW (RETRIES 1 HOLDING TIMES LATER)

| HOURS FIRST ATTEMPT VMR = 1.0 | HOURS CON-NECT | HOURS RECALL /RETRY | HOURS OVFL | CON-NECT VMR | OVFL VMR | PERCENT BLOCKED (GRADE) | LINE | HOURS CONNECTED TRAFFIC ON LINES 1 | 2 | 3 | 4 | 5 | 6 | 7 | 8 | 9 | 10 |
|---|---|---|---|---|---|---|---|---|---|---|---|---|---|---|---|---|---|
| | | | | | | | | .774 | .739 | .698 | .651 | .597 | .538 | .474 | .408 | .340 | .274 |
| | | | | | | | | .213 | .160 | .115 | .079 | .052 | | | | | |
| 16.000 | 13.019 | 2.981 | 2.981 | .26 | 2.42 | 31.4% | 15 | .940 | .935 | .928 | .922 | .914 | .905 | .895 | .883 | .870 | .856 |
| | | | | | | | | .839 | .819 | .797 | .772 | .744 | | | | | |
| 16.000 | 13.638 | 2.362 | 2.362 | .31 | 2.50 | 25.7% | 16 | .940 | .934 | .928 | .921 | .912 | .903 | .892 | .880 | .866 | .850 |
| | | | | | | | | .832 | .811 | .787 | .759 | .728 | .693 | | | | |
| 16.000 | 14.146 | 1.854 | 1.854 | .37 | 2.63 | 20.8% | 17 | .938 | .932 | .925 | .918 | .909 | .899 | .888 | .875 | .860 | .843 |
| | | | | | | | | .823 | .800 | .774 | .745 | .711 | .674 | .632 | | | |
| 16.000 | 14.615 | 1.385 | 1.385 | .44 | 2.69 | 15.9% | 18 | .939 | .932 | .926 | .918 | .909 | .898 | .886 | .873 | .857 | .839 |
| | | | | | | | | .818 | .793 | .766 | .734 | .698 | .657 | .612 | .562 | | |
| 16.000 | 15.005 | .995 | .995 | .52 | 2.71 | 11.7% | 19 | .939 | .933 | .926 | .918 | .909 | .898 | .886 | .872 | .855 | .836 |
| | | | | | | | | .814 | .788 | .758 | .725 | .686 | .642 | .594 | .540 | .483 | |
| 16.000 | 15.315 | .685 | .685 | .60 | 2.70 | 8.2% | 20 | .941 | .935 | .928 | .919 | .910 | .899 | .886 | .872 | .855 | .835 |
| | | | | | | | | .811 | .784 | .753 | .717 | .676 | .630 | .578 | .522 | .462 | .400 |
| 16.000 | 15.521 | .479 | .479 | .68 | 2.71 | 5.8% | 21 | .940 | .934 | .927 | .918 | .908 | .897 | .884 | .869 | .851 | .831 |
| | | | | | | | | .807 | .779 | .747 | .710 | .667 | .620 | .567 | .509 | .448 | .385 |
| | | | | | | | | .322 | | | | | | | | | |
| 16.000 | 15.673 | .327 | .327 | .75 | 2.69 | 4.0% | 22 | .940 | .933 | .926 | .917 | .907 | .896 | .883 | .867 | .849 | .828 |
| | | | | | | | | .803 | .775 | .742 | .704 | .660 | .612 | .558 | .500 | .438 | .374 |
| | | | | | | | | .311 | .251 | | | | | | | | |
| 16.000 | 15.802 | .198 | .198 | .82 | 2.58 | 2.4% | 23 | .942 | .936 | .928 | .920 | .910 | .899 | .885 | .870 | .851 | .830 |
| | | | | | | | | .805 | .776 | .742 | .703 | .658 | .607 | .551 | .490 | .426 | .361 |
| | | | | | | | | .297 | .236 | .181 | | | | | | | |
| 16.000 | 15.874 | .126 | .126 | .87 | 2.51 | 1.6% | 24 | .942 | .935 | .928 | .919 | .909 | .898 | .884 | .869 | .850 | .829 |
| | | | | | | | | .803 | .774 | .739 | 700 | .654 | .603 | .547 | .486 | .421 | .356 |
| | | | | | | | | .201 | .201 | .178 | .129 | | | | | | |
| 16.000 | 15.922 | .078 | .078 | .91 | 2.44 | 1.0% | 25 | .941 | .935 | .928 | .919 | .909 | .897 | .884 | .868 | .849 | .828 |
| | | | | | | | | .802 | .772 | .738 | .698 | .652 | .601 | .544 | .482 | .418 | .352 |
| | | | | | | | | .288 | .227 | .173 | .127 | .089 | | | | | |
| 16.000 | 15.953 | .047 | .047 | .94 | 2.36 | .6% | 26 | .941 | .935 | .928 | .919 | .909 | .897 | .884 | .868 | .849 | .827 |
| | | | | | | | | .801 | .771 | .737 | .696 | .650 | .599 | .542 | .480 | .415 | .350 |
| | | | | | | | | .285 | .225 | .171 | .125 | .088 | .059 | | | | |
| 17.000 | 13.946 | 3.054 | 3.054 | .26 | 2.46 | 30.5% | 16 | .943 | .938 | .933 | .927 | .920 | .912 | .903 | .893 | .882 | 869 |
| | | | | | | | | .054 | .837 | .818 | .797 | .772 | .744 | | | | |
| 17.000 | 14.527 | 2.473 | 2.473 | .31 | 2.60 | 25.4% | 17 | .942 | .936 | .931 | .924 | .917 | .908 | .899 | .888 | .876 | .862 |
| | | | | | | | | .846 | .828 | .807 | .784 | .757 | .727 | .693 | | | |
| 17.000 | 15.076 | 1.924 | 1.924 | .36 | 2.67 | 20.3% | 18 | .942 | .936 | .930 | .924 | .916 | .907 | .807 | .886 | .873 | .858 |
| | | | | | | | | .841 | .822 | .799 | .774 | .745 | .713 | .676 | .635 | | |
| 17.000 | 15.549 | 1.451 | 1.451 | .43 | 2.73 | 15.7% | 19 | .942 | .937 | .931 | .924 | .916 | .907 | .896 | .884 | .871 | .855 |
| | | | | | | | | .837 | .816 | .793 | .765 | .734 | .699 | .660 | .616 | .567 | |

*SIMILAR TO A.T.&T. JACOBSEN RETRIAL TABLES

*FAST RETRIALS: JEWETT-SHRAGO FAST RETRIAL TABLES - 50% BLOCKED CALLERS RETRY, 50% OVERFLOW (RETRIES 1 HOLDING TIMES LATER)

| HOURS FIRST ATTEMPT VMR=1.0 | HOURS CON-NECT | HOURS RECALL /RETRY | HOURS OVFL | CON-NECT VMR | OVFL VMR | PERCENT BLOCKED (GRADE) | LINE | HOURS CONNECTED TRAFFIC ON LINES | | | | | | | | | |
|---|---|---|---|---|---|---|---|---|---|---|---|---|---|---|---|---|---|
| | | | | | | | | 1 | 2 | 3 | 4 | 5 | 6 | 7 | 8 | 9 | 10 |
| 17.000 | 15.945 | 1.055 | 1.055 | .51 | 2.76 | 11.7% | 20 | .943 | .938 | .931 | .924 | .916 | .907 | .896 | .884 | .869 | .853 |
| | | | | | | | | .834 | .812 | .787 | .758 | .725 | .687 | .645 | .598 | .547 | .491 |
| 17.000 | 16.229 | .771 | .771 | .59 | 2.81 | 8.7% | 21 | .942 | .936 | .930 | .923 | .914 | .904 | .893 | .881 | .866 | .849 |
| | | | | | | | | .829 | .806 | .780 | .750 | .715 | .676 | .632 | .583 | .530 | .474 |
| | | | | | | | | .415 | | | | | | | | | |
| 17.000 | 16.479 | .521 | .521 | .67 | 2.77 | 6.0% | 22 | .944 | .938 | .931 | .924 | .916 | .906 | .895 | .881 | .866 | .849 |
| | | | | | | | | .828 | .805 | .777 | .746 | .710 | .668 | .622 | .571 | .516 | .456 |
| | | | | | | | | .395 | .334 | | | | | | | | |
| 17.000 | 16.640 | .360 | .360 | .74 | 2.75 | 4.2% | 23 | .943 | .937 | .931 | .923 | .915 | .905 | .893 | .880 | .864 | .846 |
| | | | | | | | | .825 | .801 | .773 | .741 | .704 | .661 | .614 | .562 | .506 | .446 |
| | | | | | | | | .384 | .322 | .263 | | | | | | | |
| 17.000 | 16.776 | .224 | .224 | .81 | 2.65 | 2.6% | 24 | .945 | .939 | .933 | .926 | .917 | .907 | .895 | .882 | .866 | .848 |
| | | | | | | | | .827 | .802 | .773 | .740 | .702 | .658 | .609 | .555 | .496 | .434 |
| | | | | | | | | .371 | .308 | .248 | .193 | | | | | | |
| 17.000 | 16.855 | .145 | .145 | .86 | 2.58 | 1.7% | 25 | .945 | .939 | .933 | .925 | .916 | .906 | .895 | .881 | .865 | .847 |
| | | | | | | | | .825 | .800 | .771 | .738 | .699 | .655 | .605 | .550 | .491 | .429 |
| | | | | | | | | .365 | .302 | .242 | .188 | .140 | | | | | |
| 17.000 | 16.908 | .092 | .092 | .90 | 2.51 | 1.1% | 26 | .945 | .939 | .932 | .925 | .916 | .906 | .894 | .881 | .865 | .846 |
| | | | | | | | | .824 | .799 | .770 | .736 | .697 | .652 | .602 | .547 | .488 | .425 |
| | | | | | | | | .361 | .298 | .239 | .185 | .138 | .099 | | | | |
| 17.000 | 16.944 | .056 | .056 | .93 | 2.43 | .7% | 27 | .945 | .939 | .932 | .925 | .916 | .906 | .894 | .880 | .864 | .845 |
| | | | | | | | | .824 | .798 | .769 | .734 | .695 | .651 | .600 | .545 | .486 | .423 |
| | | | | | | | | .359 | .296 | .236 | .182 | .136 | .097 | .067 | | | |
| 18.000 | 14.195 | 3.805 | 3.805 | .22 | 2.40 | 34.9% | 16 | .947 | .942 | .938 | .932 | .927 | .920 | .913 | .904 | .895 | .884 |
| | | | | | | | | .872 | .859 | .844 | .826 | .807 | .785 | | | | |
| 18.000 | 14.839 | 3.161 | 3.161 | .26 | 2.55 | 29.9% | 17 | .945 | .940 | .935 | .930 | .923 | .916 | .909 | .900 | .889 | .878 |
| | | | | | | | | .865 | .851 | .834 | .815 | .794 | .770 | .743 | | | |
| 18.000 | 15.456 | 2.544 | 2.544 | .30 | 2.64 | 24.8% | 18 | .945 | .940 | .935 | .929 | .922 | .915 | .907 | .897 | .887 | .874 |
| | | | | | | | | .861 | .845 | .827 | .807 | .784 | .758 | .728 | .695 | | |
| 18.000 | 16.007 | 1.993 | 1.993 | .36 | 2.72 | 19.9% | 19 | .945 | .940 | .935 | .929 | .922 | .914 | .905 | .895 | .884 | .871 |
| | | | | | | | | .856 | .840 | .821 | .799 | .774 | .746 | .714 | .678 | .638 | |
| 18.000 | 16.485 | 1.515 | 1.515 | .43 | 2.78 | 15.5% | 20 | .945 | .941 | .935 | .929 | .922 | .914 | .905 | .894 | .882 | .869 |
| | | | | | | | | .853 | .835 | .815 | .792 | .765 | .735 | .700 | .662 | .619 | .572 |
| 18.000 | 16.887 | 1.113 | 1.113 | .50 | 2.81 | 11.6% | 21 | .946 | .941 | .936 | .929 | .922 | .914 | .904 | .894 | .881 | .867 |
| | | | | | | | | .851 | .832 | .811 | .786 | .757 | .725 | .689 | .648 | .602 | .552 |
| | | | | | | | | .499 | | | | | | | | | |
| 18.000 | 17.178 | .822 | .822 | .58 | 2.86 | 8.7% | 22 | .945 | .940 | .935 | .928 | .920 | .912 | .902 | .891 | .878 | .864 |
| | | | | | | | | .847 | .827 | .804 | .779 | .749 | .715 | .677 | .635 | .588 | .536 |
| | | | | | | | | .481 | .424 | | | | | | | | |
| 18.000 | 17.436 | .564 | .564 | .66 | 2.82 | 6.1% | 23 | .947 | .942 | .936 | .929 | .922 | .913 | .903 | .89 | .879 | .864 |

*SIMILAR TO A.T.&T. JACOBSEN RETRIAL TABLES

*FAST RETRIALS: JEWETT-SHRAGO FAST RETRIAL TABLES - 50% BLOCKED CALLERS RETRY, 50% OVERFLOW (RETRIES 1 HOLDING TIMES LATER)

| HOURS FIRST ATTEMPT VMR = 1.0 | HOURS CON-NECT | HOURS RECALL /RETRY | HOURS OVFL | CON-NECT VMR | OVFL VMR | PERCENT BLOCKED (GRADE) | LINE | HOURS CONNECTED TRAFFIC ON LINES 1 | 2 | 3 | 4 | 5 | 6 | 7 | 8 | 9 | 10 |
|---|---|---|---|---|---|---|---|---|---|---|---|---|---|---|---|---|---|
| | | | | | | | | .846 | .826 | .803 | .776 | .745 | .710 | .670 | .625 | .575 | .521 |
| | | | | | | | | .464 | .404 | .344 | | | | | | | |
| 18.000 | 17.605 | .395 | .395 | .73 | 2.81 | 4.3% | 24 | .946 | .941 | .935 | .928 | .921 | .912 | .902 | .890 | .877 | .861 |
| | | | | | | | | .844 | .823 | .799 | .771 | .740 | .703 | .662 | .617 | .566 | .511 |
| | | | | | | | | .453 | .393 | .333 | .274 | | | | | | |
| 18.000 | 17.730 | .270 | .270 | .79 | 2.77 | 3.0% | 25 | .946 | .941 | .935 | .928 | .920 | .911 | .901 | .889 | .876 | .860 |
| | | | | | | | | .842 | .820 | .796 | .768 | .736 | .699 | .657 | .611 | .559 | .504 |
| | | | | | | | | .445 | .385 | .324 | .266 | .212 | | | | | |
| 18.000 | 17.835 | .165 | .165 | .85 | 2.65 | 1.8% | 26 | .948 | .943 | .937 | .930 | .922 | .914 | .903 | .892 | .878 | .862 |
| | | | | | | | | .844 | .823 | .798 | .769 | .736 | .698 | .655 | .607 | .554 | .497 |
| | | | | | | | | .436 | .374 | .312 | .253 | .199 | .151 | | | | |
| 18.000 | 17.894 | .106 | .106 | .89 | 2.58 | 1.2% | 27 | .948 | .942 | .937 | .930 | .922 | .913 | .903 | .891 | .877 | .861 |
| | | | | | | | | .843 | .821 | .796 | .767 | .734 | .696 | .652 | .604 | .551 | .493 |
| | | | | | | | | .432 | .370 | .308 | .249 | .195 | .148 | .108 | | | |
| 18.000 | 17.934 | .066 | .066 | .93 | 2.50 | .7% | 28 | .948 | .942 | .936 | .930 | .922 | .913 | .903 | .891 | .877 | .861 |
| | | | | | | | | .842 | .821 | .795 | .766 | .733 | .694 | .651 | .602 | .548 | .490 |
| | | | | | | | | .430 | .367 | .306 | .247 | .193 | .146 | .106 | .075 | | |
| 19.000 | 15.095 | 3.905 | 3.905 | .22 | 2.49 | 34.1% | 17 | .948 | .944 | .940 | .935 | .930 | .923 | .917 | .909 | .901 | .891 |
| | | | | | | | | .881 | .869 | .855 | .840 | .823 | .804 | .783 | | | |
| 19.000 | 15.770 | 3.230 | 3.230 | .26 | 2.59 | 29.1% | 18 | .948 | .944 | .939 | .934 | .928 | .922 | .915 | .907 | .898 | .888 |
| | | | | | | | | .877 | .864 | .849 | .833 | .815 | .794 | .770 | .744 | | |
| 19.000 | 16.387 | 2.613 | 2.613 | .30 | 2.68 | 24.2% | 19 | .948 | .944 | .939 | .934 | .928 | .921 | .914 | .905 | .896 | .885 |
| | | | | | | | | .873 | .859 | .844 | .826 | .806 | .783 | .758 | .729 | .697 | |
| 19.000 | 16.940 | 2.060 | 2.060 | .36 | 2.76 | 19.6% | 20 | .948 | .944 | .939 | .933 | .927 | .920 | .912 | .904 | .894 | .882 |
| | | | | | | | | .870 | .855 | .838 | .819 | .798 | .774 | .748 | .715 | .680 | .641 |
| 19.000 | 17.423 | 1.577 | 1.577 | .42 | 2.82 | 15.3% | 21 | .948 | .944 | .939 | .933 | .927 | .920 | .912 | .903 | .892 | .880 |
| | | | | | | | | .807 | .852 | .834 | .814 | .791 | .765 | .735 | .702 | .664 | .623 |
| | | | | | | | | .577 | | | | | | | | | |
| 19.000 | 17.793 | 1.207 | 1.207 | .49 | 2.90 | 12.0% | 22 | .947 | .943 | .938 | .932 | .925 | .918 | .909 | .900 | .889 | .877 |
| | | | | | | | | .862 | .846 | .828 | .806 | .782 | .755 | .724 | .689 | .649 | .606 |
| | | | | | | | | .559 | .508 | | | | | | | | |
| 19.000 | 18.128 | .872 | .872 | .57 | 2.91 | 8.8% | 23 | .948 | .944 | .938 | .933 | .926 | .918 | .910 | .900 | .889 | .876 |
| | | | | | | | | .861 | .845 | .825 | .803 | .778 | .749 | .716 | .679 | .637 | .591 |
| | | | | | | | | .541 | .488 | .432 | | | | | | | |
| 19.000 | 18.394 | .606 | .606 | .65 | 2.88 | 6.2% | 24 | .950 | .945 | .940 | .934 | .927 | .920 | .911 | .901 | .890 | .876 |
| | | | | | | | | .861 | .844 | .824 | .801 | .774 | .744 | .710 | .671 | .627 | .579 |
| | | | | | | | | .527 | .471 | .413 | .354 | | | | | | |
| 19.000 | 18.571 | .429 | .429 | .72 | 2.87 | 4.4% | 25 | .949 | .944 | .939 | .933 | .926 | .919 | .910 | .899 | .888 | .874 |
| | | | | | | | | .859 | .841 | .821 | .797 | .770 | .739 | .703 | .663 | .619 | .570 |
| | | | | | | | | .517 | .460 | .402 | .343 | .285 | | | | | |

*SIMILAR TO A.T.&T. JACOBSEN RETRIAL TABLES

*FAST RETRIALS: JEWETT-SHRAGO FAST RETRIAL TABLES - 50% BLOCKED CALLERS RETRY, 50% OVERFLOW (RETRIES 1 HOLDING TIMES LATER)

| HOURS FIRST ATTEMPT VMR=1.0 | HOURS CONNECT | HOURS RECALL /RETRY | HOURS OVFL | CONNECT VMR | OVFL VMR | PERCENT BLOCKED (GRADE) | LINE | HOURS CONNECTED TRAFFIC ON LINES 1 | 2 | 3 | 4 | 5 | 6 | 7 | 8 | 9 | 10 |
|---|---|---|---|---|---|---|---|---|---|---|---|---|---|---|---|---|---|
| 19.000 | 18.702 | .298 | .298 | .78 | 2.83 | 3.1% | 26 | .949 | .944 | .939 | .933 | .926 | .918 | .909 | .898 | .887 | .873 |
| | | | | | | | | .857 | .839 | .818 | .794 | .766 | .735 | .698 | .658 | .613 | .563 |
| | | | | | | | | .509 | .452 | .393 | .334 | .277 | .223 | | | | |
| 19.000 | 18.815 | .185 | .185 | .84 | 2.72 | 1.9% | 27 | .951 | .946 | .941 | .935 | .928 | .920 | .911 | .901 | .889 | .875 |
| | | | | | | | | .859 | .841 | .820 | .795 | .767 | .734 | .697 | .656 | .609 | .557 |
| | | | | | | | | .502 | .443 | .382 | .322 | .264 | .210 | .161 | | | |
| 19.000 | 18.879 | .121 | .121 | .88 | 2.65 | 1.3% | 28 | .950 | .946 | .940 | .934 | .927 | .920 | .911 | .900 | .888 | .874 |
| | | | | | | | | .858 | .840 | .818 | .794 | .765 | .732 | .695 | .653 | .606 | .554 |
| | | | | | | | | .498 | .439 | .378 | .318 | .260 | .206 | .158 | .117 | | |
| 19.000 | 18.923 | .077 | .077 | .92 | 2.57 | .8% | 29 | .950 | .945 | .940 | .934 | .927 | .919 | .910 | .900 | .888 | .874 |
| | | | | | | | | .858 | .839 | .818 | .793 | .764 | .731 | .693 | .651 | .603 | .551 |
| | | | | | | | | .495 | .436 | .375 | .315 | .257 | .203 | .156 | .115 | .082 | |
| 19.000 | 18.952 | .048 | .048 | .94 | 2.48 | .5% | 30 | .950 | .945 | .940 | .934 | .927 | .919 | .910 | .900 | .888 | .874 |
| | | | | | | | | .857 | .839 | .817 | .792 | .763 | .730 | .692 | .649 | .602 | .550 |
| | | | | | | | | .493 | .434 | .373 | .313 | .255 | .201 | .154 | .114 | .081 | .056 |
| 20.000 | 16.028 | 3.972 | 3.972 | .22 | 2.53 | 33.1% | 18 | .951 | .947 | .943 | .939 | .934 | .928 | .922 | .916 | .908 | .900 |
| | | | | | | | | .890 | .880 | .868 | .854 | .839 | .823 | .804 | .782 | | |
| 20.000 | 16.703 | 3.297 | 3.297 | .26 | 2.63 | 28.3% | 19 | .950 | .947 | .943 | .938 | .933 | .927 | .921 | .914 | .906 | .897 |
| | | | | | | | | .887 | .875 | .863 | .848 | .832 | .814 | .793 | .770 | .744 | |
| 20.000 | 17.285 | 2.715 | 2.715 | .30 | 2.76 | 23.9% | 20 | .949 | .945 | .941 | .936 | .930 | .924 | .918 | .910 | .902 | .892 |
| | | | | | | | | .881 | .869 | .856 | .840 | .823 | .803 | .781 | .756 | .728 | .697 |
| 20.000 | 17.837 | 2.163 | 2.163 | .36 | 2.84 | 19.5% | 21 | .949 | .945 | .941 | .936 | .930 | .924 | .917 | .909 | .900 | .890 |
| | | | | | | | | .879 | .866 | .851 | .835 | .816 | .795 | .771 | .745 | .714 | .681 |
| | | | | | | | | .644 | | | | | | | | | |
| 20.000 | 18.323 | 1.677 | 1.677 | .42 | 2.91 | 15.5% | 22 | .949 | .945 | .941 | .936 | .930 | .923 | .916 | .908 | .899 | .888 |
| | | | | | | | | .876 | .863 | .847 | .830 | .810 | .788 | .762 | .734 | .701 | .666 |
| | | | | | | | | .626 | .583 | | | | | | | | |
| 20.000 | 18.736 | 1.264 | 1.264 | .49 | 2.95 | 11.9% | 23 | .950 | .946 | .941 | .936 | .930 | .924 | .916 | .908 | .898 | .887 |
| | | | | | | | | .875 | .861 | .845 | .826 | .805 | .782 | .755 | .724 | .690 | .652 |
| | | | | | | | | .609 | .563 | .514 | | | | | | | |
| 20.000 | 19.078 | .922 | .922 | .56 | 2.96 | 8.8% | 24 | .951 | .947 | .942 | .937 | .931 | .924 | .916 | .908 | .898 | .887 |
| | | | | | | | | .874 | .859 | .843 | .823 | .802 | .777 | .748 | .716 | .680 | .639 |
| | | | | | | | | .595 | .546 | .494 | .440 | | | | | | |
| 20.000 | 19.351 | .649 | .649 | .64 | 2.93 | 6.3% | 25 | .952 | .948 | .943 | .938 | .932 | .925 | .917 | .909 | .899 | .887 |
| | | | | | | | | .874 | .859 | .842 | .822 | .799 | .773 | .743 | .710 | .672 | .629 |
| | | | | | | | | .583 | .532 | .478 | .421 | .364 | | | | | |
| 20.000 | 19.535 | .465 | .465 | .70 | 2.92 | 4.5% | 26 | .952 | .947 | .943 | .937 | .931 | .924 | .916 | .907 | .897 | .885 |
| | | | | | | | | .872 | .857 | .839 | .819 | .795 | .768 | .738 | .703 | .664 | .621 |
| | | | | | | | | .573 | .522 | .467 | .410 | .352 | .295 | | | | |
| 20.000 | 19.674 | .320 | .326 | .77 | 2.89 | 3.2% | 27 | .951 | .947 | .942 | .937 | .930 | .923 | .915 | .906 | .896 | .884 |

*SIMILAR TO A.T.&T. JACOBSEN RETRIAL TABLES

*FAST RETRIALS: JEWETT-SHRAGO FAST RETRIAL TABLES - 50% BLOCKED CALLERS RETRY, 50% OVERFLOW (RETRIES 1 HOLDING TIMES LATER)

| HOURS FIRST ATTEMPT VMR = 1.0 | HOURS CON-NECT | HOURS RECALL /RETRY | HOURS OVFL | CON-NECT VMR | OVFL VMR | PERCENT BLOCKED (GRADE) | LINE | HOURS CONNECTED TRAFFIC ON LINES | | | | | | | | | |
|---|---|---|---|---|---|---|---|---|---|---|---|---|---|---|---|---|---|
| | | | | | | | | 1 | 2 | 3 | 4 | 5 | 6 | 7 | 8 | 9 | 10 |
| | | | | | | | | .870 | .855 | .837 | .816 | .792 | .765 | .734 | .698 | .659 | .615 |
| | | | | | | | | .566 | .514 | .458 | .401 | .343 | .286 | .233 | | | |
| 20.000 | 19.793 | .207 | .207 | .83 | 2.78 | 2.0% | 28 | .953 | .949 | .944 | .939 | .932 | .925 | .918 | .908 | .898 | .886 |
| | | | | | | | | .872 | .857 | .838 | .817 | .793 | .765 | .733 | .697 | .656 | .610 |
| | | | | | | | | .560 | .506 | .449 | .390 | .331 | .274 | .220 | .171 | | |
| 20.000 | 19.863 | .137 | .137 | .87 | 2.71 | 1.4% | 29 | .953 | .948 | .944 | .938 | .932 | .925 | .917 | .908 | .897 | .885 |
| | | | | | | | | .872 | .856 | .837 | .816 | .791 | .763 | .731 | .694 | .653 | .607 |
| | | | | | | | | .557 | .502 | .445 | .386 | .327 | .269 | .216 | .168 | .126 | |
| 20.000 | 19.911 | .089 | .089 | .91 | 2.63 | .9% | 30 | .953 | .948 | .943 | .938 | .932 | .925 | .917 | .908 | .897 | .885 |
| | | | | | | | | .871 | .855 | .836 | .815 | .790 | .762 | .729 | .692 | .651 | .605 |
| | | | | | | | | .554 | .499 | .442 | .383 | .323 | .266 | .213 | .165 | .124 | .090 |
| 20.000 | 19.944 | .056 | .056 | .94 | 2.55 | .6% | 31 | .953 | .948 | .943 | .938 | .932 | .925 | .917 | .907 | .897 | .885 |
| | | | | | | | | .871 | .854 | .836 | .814 | .789 | .761 | .728 | .691 | .649 | .603 |
| | | | | | | | | .552 | .497 | .440 | .380 | .321 | .264 | .211 | .163 | .122 | .089 |
| | | | | | | | | .062 | | | | | | | | | |
| 22.000 | 17.868 | 4.132 | 4.132 | .22 | 2.64 | 31.6% | 20 | .954 | .951 | .947 | .944 | .940 | .935 | .930 | .924 | .918 | .911 |
| | | | | | | | | .904 | .895 | .886 | .875 | .864 | .851 | .836 | .819 | .801 | .780 |
| 22.000 | 18.539 | 3.461 | 3.461 | .26 | 2.74 | 27.2% | 21 | .954 | .951 | .947 | .943 | .939 | .934 | .929 | .923 | .917 | .909 |
| | | | | | | | | .901 | .892 | .882 | .871 | .858 | .844 | .828 | .811 | .791 | .769 |
| | | | | | | | | .744 | | | | | | | | | |
| 22.000 | 19.155 | 2.845 | 2.845 | .30 | 2.84 | 22.9% | 22 | .954 | .950 | .947 | .943 | .938 | .933 | .928 | .922 | .915 | .907 |
| | | | | | | | | .899 | .889 | .879 | .867 | .853 | .838 | .821 | .802 | .781 | .757 |
| | | | | | | | | .730 | .700 | | | | | | | | |
| 22.000 | 19.713 | 2.287 | 2.287 | .35 | 2.92 | 18.8% | 23 | .954 | .950 | .947 | .943 | .938 | .933 | .927 | .921 | .914 | .906 |
| | | | | | | | | .897 | .887 | .876 | .863 | .849 | .833 | .814 | .794 | .771 | .745 |
| | | | | | | | | .716 | .684 | .649 | | | | | | | |
| 22.000 | 20.204 | 1.796 | 1.796 | .41 | 2.99 | 15.1% | 24 | .954 | .951 | .947 | .943 | .938 | .933 | .927 | .920 | .913 | .905 |
| | | | | | | | | .896 | .885 | .873 | .860 | .845 | .828 | .808 | .787 | .762 | .734 |
| | | | | | | | | .704 | .669 | .632 | .590 | | | | | | |
| 22.000 | 20.627 | 1.373 | 1.373 | .48 | 3.03 | 11.8% | 25 | .955 | .951 | .947 | .943 | .938 | .933 | .927 | .920 | .913 | .904 |
| | | | | | | | | .895 | .884 | .871 | .857 | .842 | .824 | .803 | .780 | .754 | .725 |
| | | | | | | | | .692 | .656 | .616 | .572 | .525 | | | | | |
| 22.000 | 20.980 | 1.020 | 1.020 | .55 | 3.05 | 8.9% | 26 | .955 | .052 | .940 | .944 | .939 | .933 | .927 | .921 | .913 | .904 |
| | | | | | | | | .894 | .883 | .870 | .856 | .839 | .820 | .799 | .775 | .747 | .717 |
| | | | | | | | | .002 | .644 | .601 | .555 | .506 | .454 | | | | |
| 22.000 | 21.236 | .764 | .764 | .61 | 3.08 | 6.7% | 27 | .955 | .951 | .947 | .943 | .938 | .932 | .926 | .919 | .911 | .902 |
| | | | | | | | | .892 | .880 | .867 | .852 | .835 | .816 | .794 | .769 | .740 | .708 |
| | | | | | | | | .673 | .633 | .590 | .543 | .492 | .440 | .386 | | | |
| 22.000 | 21.465 | .535 | .535 | .69 | 3.03 | 4.8% | 28 | .956 | .952 | .948 | .944 | .939 | .934 | .927 | .920 | .912 | .903 |
| | | | | | | | | .893 | .881 | .860 | .052 | .835 | .815 | .792 | .766 | .737 | .703 |

*SIMILAR TO A.T.&T. JACOBSEN RETRIAL TABLES

# *FAST RETRIALS: JEWETT-SHRAGO FAST RETRIAL TABLES - 50% BLOCKED CALLERS RETRY, 50% OVERFLOW (RETRIES 1 HOLDING TIMES LATER)

| HOURS FIRST ATTEMPT VMR=1.0 | HOURS CON-NECT | HOURS RECALL /RETRY | HOURS OVFL | CON-NECT VMR | OVFL VMR | PERCENT BLOCKED (GRADE) | LINE | HOURS CONNECTED TRAFFIC ON LINES 1 | 2 | 3 | 4 | 5 | 6 | 7 | 8 | 9 | 10 |
|---|---|---|---|---|---|---|---|---|---|---|---|---|---|---|---|---|---|
| | | | | | | | | .666 | .625 | .580 | .531 | .479 | .424 | .369 | .314 | | |
| 22.000 | 21.617 | .383 | .383 | .75 | 3.00 | 3.4% | 29 | .956 | .952 | .948 | .944 | .939 | .933 | .927 | .919 | .911 | .902 |
| | | | | | | | | .891 | .879 | .866 | .850 | .832 | .812 | .789 | .762 | .732 | .698 |
| | | | | | | | | .660 | .618 | .572 | .522 | .470 | .415 | .360 | .305 | .252 | |
| 22.000 | 21.731 | .269 | .269 | .80 | 2.96 | 2.4% | 30 | .955 | .952 | .948 | .943 | .938 | .932 | .926 | .919 | .910 | .901 |
| | | | | | | | | .890 | .878 | .864 | .848 | .830 | .810 | .786 | .759 | .728 | .694 |
| | | | | | | | | .656 | .613 | .567 | .516 | .463 | .408 | .353 | .298 | .246 | .197 |
| 22.000 | 21.829 | .171 | .171 | .86 | 2.84 | 1.5% | 31 | .957 | .953 | .949 | .945 | .940 | .934 | .928 | .921 | .912 | .903 |
| | | | | | | | | .892 | .880 | .866 | .850 | .832 | .811 | .787 | .760 | .728 | .693 |
| | | | | | | | | .654 | .610 | .562 | .510 | .456 | .399 | .343 | .287 | .234 | .186 |
| | | | | | | | | .144 | | | | | | | | | |
| 22.000 | 21.886 | .114 | .114 | .89 | 2.76 | 1.0% | 32 | .957 | .953 | .949 | .945 | .940 | .934 | .928 | .920 | .912 | .903 |
| | | | | | | | | .892 | .880 | .866 | .850 | .831 | .810 | .786 | .758 | .727 | .691 |
| | | | | | | | | .651 | .607 | .559 | .507 | .452 | .396 | .339 | .284 | .231 | .183 |
| | | | | | | | | .141 | .105 | | | | | | | | |
| 22.000 | 21.926 | .074 | .074 | .93 | 2.68 | .7% | 33 | .957 | .953 | .949 | .945 | .940 | .934 | .927 | .920 | .912 | .902 |
| | | | | | | | | .892 | .879 | .865 | .849 | .830 | .809 | .785 | .757 | .725 | .690 |
| | | | | | | | | .650 | .605 | .557 | .505 | .450 | .393 | .337 | .281 | .229 | .181 |
| | | | | | | | | .139 | .104 | .075 | | | | | | | |
| 24.000 | 19.030 | 4.970 | 4.970 | .19 | 2.62 | 34.3% | 21 | .958 | .955 | .953 | .950 | .946 | .943 | .939 | .934 | .929 | .924 |
| | | | | | | | | .918 | .912 | .904 | .897 | .888 | .878 | .867 | .855 | .842 | .827 |
| | | | | | | | | .810 | | | | | | | | | |
| 24.000 | 19.715 | 4.285 | 4.285 | .22 | 2.75 | 30.3% | 22 | .957 | .954 | .951 | .948 | .945 | .941 | .936 | .932 | .927 | .921 |
| | | | | | | | | .915 | .908 | .900 | .892 | .882 | .872 | .860 | .847 | .833 | .817 |
| | | | | | | | | .799 | .779 | | | | | | | | |
| 24.000 | 20.382 | 3.618 | 3.618 | .26 | 2.86 | 26.2% | 23 | .957 | .954 | .951 | .948 | .944 | .940 | .935 | .931 | .925 | .919 |
| | | | | | | | | .913 | .905 | .897 | .888 | .878 | .867 | .855 | .841 | .825 | .808 |
| | | | | | | | | .789 | .767 | .743 | | | | | | | |
| 24.000 | 20.999 | 3.001 | 3.001 | .30 | 2.95 | 22.2% | 24 | .957 | .954 | .951 | .947 | .943 | .939 | .935 | .930 | .924 | .918 |
| | | | | | | | | .911 | .903 | .895 | .885 | .875 | .863 | .849 | .835 | .818 | .799 |
| | | | | | | | | .779 | .756 | .730 | .702 | | | | | | |
| 24.000 | 21.557 | 2.443 | 2.443 | .35 | 3.04 | 18.5% | 25 | .957 | .954 | .951 | .947 | .943 | .939 | .934 | .929 | .923 | .917 |
| | | | | | | | | .909 | .901 | .893 | .882 | .871 | .859 | .845 | .829 | .811 | .791 |
| | | | | | | | | .769 | .744 | .717 | .686 | .653 | | | | | |
| 24.000 | 22.053 | 1.947 | 1.947 | .40 | 3.11 | 15.0% | 26 | .957 | .954 | .951 | .947 | .943 | .939 | .934 | .929 | .923 | .916 |
| | | | | | | | | .908 | .900 | .891 | .880 | .868 | .855 | .840 | .824 | .805 | .784 |
| | | | | | | | | .760 | .734 | .704 | .672 | .637 | .598 | | | | |
| 24.000 | 22.485 | 1.515 | 1.515 | .46 | 3.16 | 11.9% | 27 | .957 | .954 | .951 | .947 | .943 | .939 | .934 | .928 | .922 | .915 |
| | | | | | | | | .908 | .899 | .889 | .878 | .866 | .852 | .837 | .819 | .799 | .777 |
| | | | | | | | | .752 | .724 | .693 | .659 | .621 | .580 | .536 | | | |

*SIMILAR TO A.T.&T. JACOBSEN RETRIAL TABLES

*FAST RETRIALS: JEWETT-SHRAGO FAST RETRIAL TABLES - 50% BLOCKED CALLERS RETRY, 50% OVERFLOW (RETRIES 1 HOLDING TIMES LATER)

| HOURS FIRST ATTEMPT VMR = 1.0 | HOURS CONNECT | HOURS RECALL /RETRY | HOURS OVFL | CONNECT VMR | OVFL VMR | PERCENT BLOCKED (GRADE) | LINE | HOURS CONNECTED TRAFFIC ON LINES 1 | 2 | 3 | 4 | 5 | 6 | 7 | 8 | 9 | 10 |
|---|---|---|---|---|---|---|---|---|---|---|---|---|---|---|---|---|---|
| 24.000 | 22.851 | 1.149 | 1.149 | .53 | 3.18 | 9.1% | 28 | .958 | .955 | .952 | .948 | .944 | .939 | .934 | .929 | .922 | .915 |
| | | | | | | | | .907 | .899 | .889 | .877 | .865 | .850 | .834 | .816 | .795 | .771 |
| | | | | | | | | .745 | .716 | .683 | .647 | .607 | .564 | .518 | .470 | | |
| 24.000 | 23.152 | .848 | .848 | .60 | 3.17 | 6.8% | 29 | .959 | .956 | .952 | .949 | .945 | .940 | .935 | .929 | .923 | .916 |
| | | | | | | | | .908 | .899 | .888 | .877 | .864 | .849 | .832 | .813 | .791 | .767 |
| | | | | | | | | .740 | .709 | .675 | .637 | .595 | .550 | .503 | .452 | .401 | |
| 24.000 | 23.366 | .634 | .634 | .66 | 3.18 | 5.1% | 30 | .958 | .955 | .952 | .948 | .944 | .939 | .934 | .928 | .922 | .914 |
| | | | | | | | | .906 | .897 | .886 | .875 | .861 | .846 | .829 | .809 | .787 | .762 |
| | | | | | | | | .734 | .702 | .667 | .628 | .586 | .540 | .491 | .441 | .388 | .336 |
| 24.000 | 23.558 | .442 | .442 | .73 | 3.11 | 3.6% | 31 | .959 | .956 | .953 | .949 | .945 | .940 | .935 | .929 | .923 | .916 |
| | | | | | | | | .907 | .898 | .887 | .875 | .862 | .846 | .828 | .808 | .785 | .760 |
| | | | | | | | | .731 | .698 | .662 | .621 | .578 | .530 | .480 | .428 | .374 | .321 |
| | | | | | | | | .270 | | | | | | | | | |
| 24.000 | 23.683 | .317 | .317 | .79 | 3.07 | 2.6% | 32 | .959 | .956 | .953 | .949 | .945 | .940 | .935 | .929 | .922 | .915 |
| | | | | | | | | .906 | .897 | .886 | .874 | .860 | .844 | .826 | .806 | .783 | .756 |
| | | | | | | | | .727 | .694 | .657 | .616 | .572 | .524 | .473 | .421 | .367 | .314 |
| | | | | | | | | .263 | .215 | | | | | | | | |
| 24.000 | 23.793 | .207 | .207 | .84 | 2.96 | 1.7% | 33 | .960 | .957 | .954 | .950 | .946 | .942 | .936 | .930 | .924 | .917 |
| | | | | | | | | .908 | .899 | .888 | .876 | .862 | .846 | .827 | .807 | .783 | .756 |
| | | | | | | | | .726 | .692 | .654 | .613 | .567 | .518 | .466 | .412 | .357 | .303 |
| | | | | | | | | .252 | .204 | .160 | | | | | | | |
| 24.000 | 23.859 | .141 | .141 | .88 | 2.88 | 1.2% | 34 | .960 | .957 | .954 | .950 | .946 | .941 | .936 | .930 | .924 | .916 |
| | | | | | | | | .908 | .898 | .887 | .875 | .861 | .845 | .826 | .805 | .782 | .755 |
| | | | | | | | | .724 | .690 | .652 | .610 | .564 | .514 | .462 | .408 | .353 | .299 |
| | | | | | | | | .248 | .200 | .157 | .120 | | | | | | |
| 24.000 | 23.905 | .095 | .095 | .91 | 2.80 | .8% | 35 | .960 | .957 | .954 | .950 | .946 | .941 | .936 | .930 | .923 | .916 |
| | | | | | | | | .907 | .898 | .887 | .874 | .860 | .844 | .020 | .805 | .781 | .753 |
| | | | | | | | | .723 | .688 | .650 | .608 | .561 | .512 | .459 | .405 | .350 | .297 |
| | | | | | | | | .245 | .198 | .155 | .118 | .088 | | | | | |
| 24.000 | 23.938 | .062 | .062 | .94 | 2.72 | .5% | 36 | .960 | .957 | .954 | .950 | .946 | .941 | .936 | .930 | .923 | .916 |
| | | | | | | | | .907 | .898 | .887 | .874 | .860 | .844 | .825 | .804 | .780 | .753 |
| | | | | | | | | .722 | .687 | .649 | .606 | .560 | .510 | .457 | .403 | .348 | .295 |
| | | | | | | | | .243 | .196 | .154 | .117 | .087 | .062 | | | | |
| 26.000 | 20.886 | 5.114 | 5.114 | .19 | 2.72 | 32.9% | 23 | .961 | .958 | .956 | .953 | .950 | .947 | .944 | .940 | .936 | .931 |
| | | | | | | | | .926 | .921 | .915 | .908 | .001 | .093 | .884 | .875 | .864 | .852 |
| | | | | | | | | .839 | .824 | .808 | | | | | | | |
| 26.000 | 21.597 | 4.403 | 4.403 | .22 | 2.82 | 29.0% | 24 | .960 | .958 | .055 | .953 | .950 | .946 | .943 | .939 | .935 | .930 |
| | | | | | | | | .925 | .919 | .913 | .906 | .898 | .890 | .880 | .870 | .858 | .846 |
| | | | | | | | | .831 | .816 | .798 | .779 | | | | | | |
| 26.000 | 22.264 | 3.736 | 3.736 | .26 | 2.92 | 25.1% | 25 | .960 | .958 | .955 | .952 | .949 | .946 | .042 | .938 | .933 | .929 |

*SIMILAR TO A.T.&T. JACOBSEN RETRIAL TABLES

## *FAST RETRIALS: JEWETT-SHRAGO FAST RETRIAL TABLES - 50% BLOCKED CALLERS RETRY, 50% OVERFLOW (RETRIES 1 HOLDING TIMES LATER)

| HOURS FIRST ATTEMPT VMR = 1.0 | HOURS CON-NECT | HOURS RECALL /RETRY | HOURS OVFL | CON-NECT VMR | OVFL VMR | PERCENT BLOCKED (GRADE) | LINE | HOURS CONNECTED TRAFFIC ON LINES 1 | 2 | 3 | 4 | 5 | 6 | 7 | 8 | 9 | 10 |
|---|---|---|---|---|---|---|---|---|---|---|---|---|---|---|---|---|---|
| | | | | | | | | .923 | .917 | .911 | .903 | .895 | .886 | .876 | .865 | .853 | .839 |
| | | | | | | | | .824 | .807 | .788 | .767 | .744 | | | | | |
| 26.000 | 22.880 | 3.120 | 3.120 | .30 | 3.02 | 21.4% | 26 | .960 | .958 | .955 | .952 | .949 | .945 | .942 | .937 | .933 | .928 |
| | | | | | | | | .922 | .916 | .909 | .901 | .892 | .883 | .872 | .861 | .848 | .833 |
| | | | | | | | | .817 | .799 | .778 | .756 | .731 | .704 | | | | |
| 26.000 | 23.442 | 2.558 | 2.558 | .34 | 3.11 | 17.9% | 27 | .960 | .958 | .955 | .952 | .949 | .945 | .941 | .937 | .932 | .927 |
| | | | | | | | | .921 | .914 | .907 | .899 | .890 | .880 | .869 | .857 | .843 | .827 |
| | | | | | | | | .810 | .790 | .769 | .745 | .718 | .689 | .657 | | | |
| 26.000 | 23.943 | 2.057 | 2.057 | .40 | 3.18 | 14.7% | 28 | .960 | .958 | .955 | .952 | .949 | .945 | .941 | .937 | .932 | .926 |
| | | | | | | | | .920 | .913 | .906 | .897 | .888 | .878 | .866 | .853 | .838 | .822 |
| | | | | | | | | .803 | .783 | .760 | .734 | .706 | .675 | .641 | .604 | | |
| 26.000 | 24.382 | 1.618 | 1.618 | .46 | 3.23 | 11.7% | 29 | .961 | .958 | .955 | .952 | .949 | .945 | .941 | .936 | .931 | .926 |
| | | | | | | | | .920 | .913 | .905 | .896 | .886 | .876 | .863 | .850 | .834 | .817 |
| | | | | | | | | .798 | .776 | .752 | .725 | .695 | .662 | .626 | .587 | .545 | |
| 26.000 | 24.757 | 1.243 | 1.243 | .52 | 3.26 | 9.1% | 30 | .961 | .959 | .956 | .953 | .949 | .946 | .941 | .937 | .932 | .926 |
| | | | | | | | | .919 | .912 | .904 | .895 | .885 | .874 | .862 | .847 | .831 | .813 |
| | | | | | | | | .793 | .770 | .745 | .716 | .685 | .650 | .612 | .571 | .527 | .481 |
| 26.000 | 25.069 | .931 | .931 | .59 | 3.26 | 6.9% | 31 | .962 | .959 | .957 | .953 | .950 | .946 | .942 | .937 | .932 | .926 |
| | | | | | | | | .920 | .912 | .904 | .895 | .885 | .873 | .860 | .846 | .829 | .810 |
| | | | | | | | | .789 | .766 | .739 | .709 | .676 | .640 | .600 | .558 | .512 | .464 |
| | | | | | | | | .414 | | | | | | | | | |
| 26.000 | 25.295 | .705 | .705 | .65 | 3.27 | 5.3% | 32 | .962 | .959 | .956 | .953 | .949 | .946 | .941 | .936 | .931 | .925 |
| | | | | | | | | .919 | .911 | .903 | .893 | .883 | .871 | .858 | .843 | .825 | .806 |
| | | | | | | | | .785 | .760 | .733 | .702 | .669 | .631 | .591 | .547 | .501 | .452 |
| | | | | | | | | .402 | .351 | | | | | | | | |
| 26.000 | 25.498 | .502 | .502 | .71 | 3.21 | 3.8% | 33 | .962 | .960 | .957 | .954 | .950 | .946 | .942 | .937 | .932 | .926 |
| | | | | | | | | .919 | .912 | .904 | .894 | .883 | .871 | .858 | .842 | .825 | .805 |
| | | | | | | | | .783 | .758 | .730 | .698 | .663 | .625 | .583 | .537 | .489 | .439 |
| | | | | | | | | .388 | .336 | .286 | | | | | | | |
| 26.000 | 25.634 | .366 | .366 | .77 | 3.18 | 2.8% | 34 | .962 | .960 | .957 | .954 | .950 | .946 | .942 | .937 | .932 | .926 |
| | | | | | | | | .919 | .911 | .903 | .893 | .882 | .870 | .856 | .840 | .823 | .803 |
| | | | | | | | | .780 | .754 | .726 | .694 | .658 | .619 | .576 | .531 | .482 | .432 |
| | | | | | | | | .380 | .329 | .279 | .231 | | | | | | |
| 26.000 | 25.738 | .262 | .262 | .82 | 3.12 | 2.0% | 35 | .962 | .959 | .957 | .953 | .950 | .946 | .941 | .937 | .931 | .925 |
| | | | | | | | | .918 | .911 | .902 | .892 | .881 | .869 | .855 | .839 | .821 | .801 |
| | | | | | | | | .778 | .752 | .723 | .690 | .654 | .615 | .572 | .526 | .477 | .426 |
| | | | | | | | | .374 | .323 | .273 | .226 | .182 | | | | | |
| 26.000 | 25.829 | .171 | .171 | .87 | 3.00 | 1.3% | 36 | .963 | .961 | .958 | .955 | .951 | .947 | .943 | .938 | .933 | .927 |
| | | | | | | | | .920 | .912 | .904 | .894 | .883 | .870 | .856 | .840 | .822 | .802 |
| | | | | | | | | .778 | .752 | .722 | .689 | .652 | .612 | .568 | .521 | .471 | .419 |

*SIMILAR TO A.T.&T. JACOBSEN RETRIAL TABLES

*FAST RETRIALS: JEWETT-SHRAGO FAST RETRIAL TABLES - 50% BLOCKED CALLERS RETRY, 50% OVERFLOW (RETRIES 1 HOLDING TIMES LATER)

| HOURS FIRST ATTEMPT VMR = 1.0 | HOURS CON-NECT | HOURS RECALL /RETRY | HOURS OVFL | CON-NECT VMR | OVFL VMR | PERCENT BLOCKED (GRADE) | LINE | HOURS CONNECTED TRAFFIC ON LINES 1 | 2 | 3 | 4 | 5 | 6 | 7 | 8 | 9 | 10 |
|---|---|---|---|---|---|---|---|---|---|---|---|---|---|---|---|---|---|
| | | | | | | | | .366 | .314 | .263 | .216 | .173 | .135 | | | | |
| 26.000 | 25.883 | .117 | .117 | .90 | 2.92 | .9% | 37 | .963 | .961 | .958 | .954 | .951 | .947 | .943 | .938 | .932 | .926 |
| | | | | | | | | .920 | .912 | .903 | .893 | .882 | .870 | .856 | .840 | .821 | .800 |
| | | | | | | | | .777 | .750 | .721 | .687 | .650 | .610 | .565 | .518 | .468 | .416 |
| | | | | | | | | .363 | .311 | .260 | .213 | .170 | .133 | .101 | | | |
| 26.000 | 25.922 | .078 | .078 | .93 | 2.84 | .6% | 38 | .963 | .960 | .958 | .954 | .951 | .947 | .943 | .938 | .932 | .926 |
| | | | | | | | | .919 | .912 | .903 | .893 | .882 | .869 | .855 | .839 | .821 | .800 |
| | | | | | | | | .776 | .749 | .720 | .686 | .649 | .608 | .564 | .516 | .466 | .414 |
| | | | | | | | | .361 | .309 | .258 | .211 | .168 | .131 | .099 | .073 | | |
| 28.000 | 22.748 | 5.252 | 5.252 | .19 | 2.82 | 31.6% | 25 | .963 | .961 | .958 | .956 | .954 | .951 | .948 | .945 | .941 | .937 |
| | | | | | | | | .933 | .928 | .923 | .918 | .912 | .905 | .898 | .890 | .881 | .871 |
| | | | | | | | | .861 | .849 | .836 | .022 | .806 | | | | | |
| 28.000 | 23.454 | 4.546 | 4.546 | .22 | 2.93 | 27.9% | 26 | .962 | .960 | .958 | .956 | .953 | .950 | .947 | .944 | .940 | .936 |
| | | | | | | | | .932 | .927 | .922 | .916 | .909 | .902 | .895 | .886 | .877 | .867 |
| | | | | | | | | .855 | .843 | .829 | .813 | .796 | .777 | | | | |
| 28.000 | 24.119 | 3.881 | 3.881 | .26 | 3.03 | 24.3% | 27 | .962 | .960 | .958 | .955 | .953 | .950 | .947 | .943 | .939 | .935 |
| | | | | | | | | .930 | .925 | .920 | .914 | .907 | .900 | .892 | .883 | .873 | .862 |
| | | | | | | | | .850 | .836 | .821 | .805 | .786 | .766 | .744 | | | |
| 28.000 | 24.734 | 3.266 | 3.266 | .30 | 3.13 | 20.9% | 28 | .962 | .960 | .958 | .955 | .952 | .949 | .946 | .942 | .939 | .934 |
| | | | | | | | | .929 | .924 | .918 | .912 | .905 | .897 | .889 | .879 | .869 | .857 |
| | | | | | | | | .844 | .830 | .814 | .796 | .777 | .755 | .731 | .705 | | |
| 28.000 | 25.297 | 2.703 | 2.703 | .34 | 3.22 | 17.6% | 29 | .962 | .960 | .958 | .955 | .952 | .949 | .946 | .942 | .938 | .934 |
| | | | | | | | | .929 | .923 | .917 | .911 | .903 | .895 | .886 | .876 | .865 | .853 |
| | | | | | | | | .839 | .824 | .807 | .788 | .767 | .744 | .719 | .691 | .660 | |
| 28.000 | 25.802 | 2.198 | 2.198 | .39 | 3.29 | 14.6% | 30 | .963 | .960 | .958 | .955 | .952 | .949 | .946 | .942 | .938 | .933 |
| | | | | | | | | .928 | .922 | .916 | .909 | .902 | .803 | .004 | .874 | .862 | .849 |
| | | | | | | | | .835 | .818 | .801 | .781 | .759 | .734 | .707 | .677 | .645 | .610 |
| 28.000 | 26.247 | 1.753 | 1.753 | .45 | 3.35 | 11.8% | 31 | .963 | .961 | .958 | .955 | .952 | .949 | .946 | .942 | .938 | .933 |
| | | | | | | | | .928 | .922 | .916 | .908 | .901 | .892 | .882 | .871 | .859 | .846 |
| | | | | | | | | .830 | .814 | .795 | .774 | .750 | .725 | .696 | .665 | .630 | .593 |
| | | | | | | | | .554 | | | | | | | | | |
| 28.000 | 26.632 | 1.368 | 1.368 | .51 | 3.38 | 9.3% | 32 | .963 | .961 | .959 | .956 | .953 | .950 | .946 | .942 | .938 | .933 |
| | | | | | | | | .928 | .922 | .915 | .908 | .900 | .801 | .001 | .870 | .857 | .843 |
| | | | | | | | | .827 | .809 | .790 | .768 | .743 | .716 | .686 | .653 | .617 | .578 |
| | | | | | | | | .537 | 493 | | | | | | | | |
| 28.000 | 26.956 | 1.044 | 1.044 | .57 | 3.39 | 7.2% | 33 | .964 | .962 | .959 | .956 | .953 | .950 | .946 | .942 | .938 | .933 |
| | | | | | | | | .928 | .922 | .915 | .908 | .900 | .890 | .880 | .868 | .855 | .841 |
| | | | | | | | | .824 | .806 | .786 | .763 | .737 | .709 | .677 | .643 | .606 | 565 |
| | | | | | | | | .522 | .477 | .429 | | | | | | | |
| 28.000 | 27.223 | .777 | .777 | .64 | 3.36 | 5.4% | 34 | .964 | .962 | .960 | .957 | .954 | .051 | .947 | .943 | .939 | .934 |

*SIMILAR TO A.T.&T. JACOBSEN RETRIAL TABLES

*FAST RETRIALS: JEWETT-SHRAGO FAST RETRIAL TABLES - 50% BLOCKED CALLERS RETRY, 50% OVERFLOW (RETRIES 1 HOLDING TIMES LATER)

| HOURS FIRST ATTEMPT VMR=1.0 | HOURS CON-NECT | HOURS RECALL /RETRY | HOURS OVFL | CON-NECT VMR | OVFL VMR | PERCENT BLOCKED (GRADE) | LINE | HOURS CONNECTED TRAFFIC ON LINES 1 | 2 | 3 | 4 | 5 | 6 | 7 | 8 | 9 | 10 |
|---|---|---|---|---|---|---|---|---|---|---|---|---|---|---|---|---|---|
| | | | | | | | | .928 | .922 | .916 | .908 | .900 | .890 | .880 | .868 | .855 | .840 |
| | | | | | | | | .823 | .804 | .783 | .759 | .732 | .703 | .670 | .634 | .595 | .554 |
| | | | | | | | | .509 | .462 | .414 | .365 | | | | | | |
| 28.000 | 27.437 | .563 | .563 | .70 | 3.31 | 3.9% | 35 | .965 | .963 | .960 | .958 | .955 | .951 | .948 | .944 | .939 | .935 |
| | | | | | | | | .929 | .923 | .916 | .909 | .900 | .891 | .880 | .868 | .854 | .839 |
| | | | | | | | | .822 | .802 | .780 | .756 | .729 | .698 | .664 | .627 | .587 | .544 |
| | | | | | | | | .498 | .450 | .400 | .350 | .301 | | | | | |
| 28.000 | 27.584 | .416 | .416 | .75 | 3.28 | 2.9% | 36 | .965 | .963 | .960 | .957 | .954 | .951 | .947 | .943 | .939 | .934 |
| | | | | | | | | .928 | .922 | .915 | .908 | .899 | .889 | .879 | .866 | .852 | .837 |
| | | | | | | | | .819 | .800 | .777 | .752 | .725 | .694 | .659 | .622 | .581 | .537 |
| | | | | | | | | .490 | .442 | .392 | .342 | .293 | .246 | | | | |
| 28.000 | 27.697 | .303 | .303 | .80 | 3.23 | 2.1% | 37 | .965 | .963 | .960 | .957 | .954 | .951 | .947 | .943 | .939 | .934 |
| | | | | | | | | .928 | .922 | .915 | .907 | .898 | .888 | .877 | .865 | .851 | .835 |
| | | | | | | | | .817 | .797 | .775 | .750 | .721 | .690 | .655 | .617 | .576 | .532 |
| | | | | | | | | .485 | .436 | .386 | .336 | .287 | .240 | .197 | | | |
| 28.000 | 27.797 | .203 | .203 | .85 | 3.11 | 1.4% | 38 | .966 | .964 | .961 | .958 | .955 | .952 | .948 | .944 | .940 | .935 |
| | | | | | | | | .929 | .923 | .916 | .909 | .900 | .890 | .879 | .866 | .852 | .836 |
| | | | | | | | | .818 | .798 | .775 | .749 | .721 | .689 | .653 | .614 | .572 | .526 |
| | | | | | | | | .478 | .429 | .378 | .327 | .278 | .231 | .188 | .149 | | |
| 28.000 | 27.859 | .141 | .141 | .89 | 3.03 | 1.0% | 39 | .966 | .963 | .961 | .958 | .955 | .952 | .948 | .944 | .940 | .935 |
| | | | | | | | | .929 | .923 | .916 | .908 | .899 | .889 | .878 | .866 | .851 | .835 |
| | | | | | | | | .817 | .797 | .774 | .748 | .719 | .687 | .651 | .612 | .569 | .523 |
| | | | | | | | | .475 | .425 | .374 | .324 | .274 | .228 | .185 | .146 | .113 | |
| 28.000 | 27.903 | .097 | .097 | .92 | 2.95 | .7% | 40 | .966 | .963 | .961 | .958 | .955 | .952 | .948 | .944 | .940 | .935 |
| | | | | | | | | .929 | .923 | .916 | .908 | .899 | .889 | .878 | .865 | .851 | .835 |
| | | | | | | | | .817 | .796 | .773 | .747 | .718 | .685 | .649 | .610 | .567 | .521 |
| | | | | | | | | .473 | .423 | .372 | .321 | .272 | .225 | .183 | .144 | .111 | .084 |
| 30.000 | 23.899 | 6.101 | 6.101 | .17 | 2.78 | 33.8% | 26 | .965 | .964 | .962 | .960 | .958 | .955 | .953 | .950 | .947 | .944 |
| | | | | | | | | .941 | .937 | .933 | .928 | .924 | .918 | .913 | .906 | .900 | .892 |
| | | | | | | | | .884 | .875 | .865 | .854 | .842 | .829 | | | | |
| 30.000 | 24.640 | 5.360 | 5.360 | .19 | 2.89 | 30.3% | 27 | .965 | .963 | .961 | .959 | .957 | .955 | .952 | .949 | .946 | .943 |
| | | | | | | | | .939 | .936 | .931 | .927 | .922 | .916 | .910 | .904 | .896 | .888 |
| | | | | | | | | .880 | .870 | .859 | .848 | .835 | .821 | .805 | | | |
| 30.000 | 25.316 | 4.684 | 4.684 | .22 | 3.02 | 27.0% | 28 | .964 | .963 | .961 | .958 | .956 | .954 | .951 | .948 | .945 | .941 |
| | | | | | | | | .937 | .933 | .929 | .924 | .919 | .913 | .906 | .899 | .892 | .883 |
| | | | | | | | | .874 | .864 | .852 | .840 | .826 | .811 | .794 | .776 | | |
| 30.000 | 26.009 | 3.991 | 3.991 | .26 | 3.09 | 23.5% | 29 | .965 | .963 | .961 | .959 | .957 | .954 | .951 | .948 | .945 | .941 |
| | | | | | | | | .938 | .933 | .929 | .924 | .918 | .912 | .905 | .898 | .890 | .881 |
| | | | | | | | | .871 | .860 | .848 | .835 | .820 | .804 | .786 | .766 | .745 | |
| 30.000 | 26.625 | 3.375 | 3.375 | .29 | 3.19 | 20.2% | 30 | .965 | .963 | .961 | .959 | .956 | .954 | .951 | .948 | .944 | .941 |

*SIMILAR TO A.T.&T. JACOBSEN RETRIAL TABLES

*FAST RETRIALS: JEWETT-SHRAGO FAST RETRIAL TABLES - 50% BLOCKED CALLERS RETRY, 50% OVERFLOW (RETRIES 1 HOLDING TIMES LATER)

| HOURS FIRST ATTEMPT VMR = 1.0 | HOURS CON-NECT | HOURS RECALL /RETRY | HOURS OVFL | CON-NECT VMR | OVFL VMR | PERCENT BLOCKED (GRADE) | LINE | HOURS CONNECTED TRAFFIC ON LINES 1 | 2 | 3 | 4 | 5 | 6 | 7 | 8 | 9 | 10 |
|---|---|---|---|---|---|---|---|---|---|---|---|---|---|---|---|---|---|
| | | | | | | | | .937 | .932 | .928 | .922 | .917 | .910 | .903 | .895 | .887 | .877 |
| | | | | | | | | .867 | .856 | .843 | .829 | .813 | .796 | .777 | .755 | .732 | .707 |
| 30.000 | 27.190 | 2.810 | 2.810 | .34 | 3.28 | 17.1% | 31 | .965 | .963 | .961 | .959 | .956 | .954 | .951 | .947 | .944 | .940 |
| | | | | | | | | .936 | .932 | .927 | .921 | .915 | .909 | .901 | .893 | .884 | .874 |
| | | | | | | | | .863 | .851 | .838 | .823 | .806 | .788 | .767 | .745 | .720 | .693 |
| | | | | | | | | .663 | | | | | | | | | |
| 30.000 | 27.699 | 2.301 | 2.301 | .39 | 3.36 | 14.2% | 32 | .965 | .963 | .961 | .959 | .956 | .954 | .951 | .947 | .944 | .940 |
| | | | | | | | | .936 | .931 | .926 | .920 | .914 | .907 | .900 | .891 | .882 | .871 |
| | | | | | | | | .860 | .847 | .833 | .817 | .799 | .780 | .758 | .735 | .708 | .680 |
| | | | | | | | | .648 | .614 | | | | | | | | |
| 30.000 | 28.151 | 1.849 | 1.849 | .44 | 3.42 | 11.6% | 33 | .965 | .964 | .961 | .959 | .956 | .954 | .951 | .947 | .944 | .940 |
| | | | | | | | | .935 | .931 | .925 | .920 | .913 | .906 | .898 | .889 | .880 | .869 |
| | | | | | | | | .857 | .844 | .829 | .812 | .794 | .773 | .750 | .725 | .698 | .667 |
| | | | | | | | | .634 | .599 | .560 | | | | | | | |
| 30.000 | 28.542 | 1.458 | 1.458 | .50 | 3.46 | 9.3% | 34 | .966 | .964 | .962 | .959 | .957 | .954 | .951 | .948 | .944 | .940 |
| | | | | | | | | .935 | .931 | .925 | .919 | .913 | .905 | .897 | .888 | .878 | .867 |
| | | | | | | | | .855 | .841 | .825 | .808 | .788 | .767 | .743 | .717 | .688 | .656 |
| | | | | | | | | .621 | .584 | .544 | .502 | | | | | | |
| 30.000 | 28.875 | 1.125 | 1.125 | .56 | 3.47 | 7.2% | 35 | .966 | .964 | .962 | .960 | .957 | .954 | .951 | .948 | .944 | .940 |
| | | | | | | | | .936 | .931 | .925 | .919 | .913 | .905 | .897 | .887 | .877 | .866 |
| | | | | | | | | .853 | .838 | .822 | .804 | .784 | .762 | .737 | .709 | .679 | .646 |
| | | | | | | | | .610 | .571 | .529 | .486 | .440 | | | | | |
| 30.000 | 29.151 | .849 | .849 | .62 | 3.45 | 5.5% | 36 | .967 | .965 | .963 | .960 | .958 | .955 | .952 | .948 | .945 | .941 |
| | | | | | | | | .936 | .931 | .926 | .920 | .913 | .905 | .897 | .887 | .877 | .865 |
| | | | | | | | | .852 | .837 | .820 | .802 | .781 | .758 | .732 | .703 | .672 | .637 |
| | | | | | | | | .600 | .559 | .516 | 471 | .425 | .377 | | | | |
| 30.000 | 29.352 | .648 | .648 | 68 | 3.44 | 1.2% | 37 | .907 | .905 | .962 | .960 | .957 | .955 | .951 | .948 | .944 | .940 |
| | | | | | | | | .935 | .930 | .925 | .919 | .912 | .904 | .895 | .886 | .875 | .863 |
| | | | | | | | | .849 | .834 | .817 | .798 | .777 | .753 | .727 | .697 | .665 | .630 |
| | | | | | | | | .592 | .551 | .507 | .462 | .415 | .367 | .320 | | | |
| 30.000 | 29.533 | .467 | .467 | .74 | 3.37 | 3.1% | 38 | .967 | .965 | .963 | .961 | .958 | .955 | .952 | .949 | .945 | .941 |
| | | | | | | | | .936 | .931 | .926 | .919 | .912 | .905 | .896 | .886 | .875 | .863 |
| | | | | | | | | .849 | .834 | .816 | .797 | .775 | .751 | .724 | .694 | .660 | .624 |
| | | | | | | | | .505 | .543 | .498 | .451 | .403 | .355 | .307 | .260 | | |
| 30.000 | 29.655 | .345 | .345 | .79 | 3.33 | 2.3% | 39 | .967 | .965 | .963 | .961 | .958 | .955 | .952 | .949 | .945 | .941 |
| | | | | | | | | .936 | .931 | .925 | .919 | .912 | .904 | .895 | .885 | .874 | .862 |
| | | | | | | | | .848 | .832 | .814 | .795 | .772 | .748 | .720 | .690 | .656 | .619 |
| | | | | | | | | .580 | .537 | .492 | .445 | .397 | .348 | .300 | .254 | .211 | |
| 30.000 | 29.750 | .250 | .250 | .83 | 3.27 | 1.7% | 40 | .967 | .965 | .963 | .960 | .958 | .955 | .052 | .940 | .944 | .940 |
| | | | | | | | | .936 | .930 | .925 | .918 | .911 | .903 | .894 | .884 | .873 | .860 |

*SIMILAR TO A.T.&T. JACOBSEN RETRIAL TABLES

# *FAST RETRIALS: JEWETT-SHRAGO FAST RETRIAL TABLES - 50% BLOCKED CALLERS RETRY, 50% OVERFLOW (RETRIES 1 HOLDING TIMES LATER)

| HOURS FIRST ATTEMPT VMR = 1.0 | HOURS CON-NECT | HOURS RECALL /RETRY | HOURS OVFL | CON-NECT VMR | OVFL VMR | PERCENT BLOCKED (GRADE) | LINE | HOURS CONNECTED TRAFFIC ON LINES 1 | 2 | 3 | 4 | 5 | 6 | 7 | 8 | 9 | 10 |
|---|---|---|---|---|---|---|---|---|---|---|---|---|---|---|---|---|---|
| | | | | | | | | .846 | .830 | .813 | .793 | .770 | .745 | .718 | .687 | .653 | .616 |
| | | | | | | | | .576 | .533 | .487 | .440 | .392 | .343 | .295 | .249 | .207 | .167 |
| 30.000 | 29.833 | .167 | .167 | .87 | 3.14 | 1.1% | 41 | .968 | .966 | .964 | .961 | .959 | .956 | .953 | .949 | .946 | .941 |
| | | | | | | | | .937 | .932 | .926 | .920 | .913 | .905 | .896 | .886 | .874 | .862 |
| | | | | | | | | .848 | .832 | .814 | .794 | .771 | .746 | .717 | .686 | .651 | .613 |
| | | | | | | | | .572 | .529 | .482 | .434 | .385 | .336 | .287 | .241 | .198 | .160 |
| | | | | | | | | .125 | | | | | | | | | |
| 30.000 | 29.883 | .117 | .117 | .91 | 3.06 | .8% | 42 | .968 | .966 | .964 | .961 | .959 | .956 | .953 | .949 | .945 | .941 |
| | | | | | | | | .937 | .931 | .926 | .919 | .912 | .904 | .895 | .885 | .874 | .861 |
| | | | | | | | | .847 | .831 | .813 | .793 | .770 | .744 | .716 | .684 | .649 | .611 |
| | | | | | | | | .570 | .526 | .480 | .431 | .382 | .333 | .285 | .239 | .196 | .157 |
| | | | | | | | | .124 | .095 | | | | | | | | |
| 30.000 | 29.920 | .080 | .080 | .93 | 2.98 | .5% | 43 | .968 | .966 | .964 | .961 | .959 | .956 | .953 | .949 | .945 | .941 |
| | | | | | | | | .937 | .931 | .926 | .919 | .912 | .904 | .895 | .885 | .874 | .861 |
| | | | | | | | | .847 | .830 | .812 | .792 | .769 | .743 | .715 | .683 | .648 | .610 |
| | | | | | | | | .569 | .524 | .478 | .429 | .380 | .331 | .283 | .237 | .194 | .156 |
| | | | | | | | | .122 | .093 | .070 | | | | | | | |

*SIMILAR TO A.T.&T. JACOBSEN RETRIAL TABLES

*FAST RETRIALS: JEWETT-SHRAGO FAST RETRIAL TABLES - 100% BLOCKED CALLERS RETRY, 0% OVERFLOW (RETRIES 1 HOLDING TIMES LATER)

| HOURS FIRST ATTEMPT VMR = 1.0 | HOURS CON-NECT | HOURS RECALL /RETRY | HOURS OVFL | CON-NECT VMR | OVFL VMR | PERCENT BLOCKED (GRADE) | LINE | HOURS CONNECTED TRAFFIC ON LINES 1 | 2 | 3 | 4 | 5 | 6 | 7 | 8 | 9 | 10 |
|---|---|---|---|---|---|---|---|---|---|---|---|---|---|---|---|---|---|
| 200 | .200 | .050 | .000 | .80 | 1.09 | 20.0% | 1 | .200 | | | | | | | | | |
| .200 | .200 | .003 | .000 | .97 | 1.07 | 1.7% | 2 | .169 | .031 | | | | | | | | |
| .400 | .400 | .025 | .000 | .90 | 1.14 | 6.0% | 2 | .298 | .102 | | | | | | | | |
| .400 | .400 | .003 | .000 | .98 | 1.11 | .7% | 3 | .287 | .094 | .019 | | | | | | | |
| 600 | .600 | .083 | .000 | .81 | 1.20 | 12.2% | 2 | .406 | .194 | | | | | | | | |
| .600 | .600 | .013 | .000 | .95 | 1.17 | 2.1% | 3 | .380 | .169 | .051 | | | | | | | |
| .800 | .800 | .200 | .000 | .70 | 1.25 | 20.0% | 2 | .500 | .300 | .800 | .800 | .036 | .000 | .90 | 1.23 | 4.3% | 3 |
| .455 | .247 | .098 | | | | | | | | | | | | | | | |
| .800 | .800 | .006 | .000 | .97 | 1.19 | .8% | 4 | .446 | .237 | .091 | .025 | | | | | | |
| 1.000 | 1.000 | .079 | .000 | .84 | 1.28 | 7.3% | 3 | .519 | .324 | .157 | | | | | | | |
| 1.000 | 1.000 | .016 | .000 | .95 | 1.24 | 1.6% | 4 | .504 | .305 | .142 | .049 | | | | | | |
| 1.200 | 1.200 | .151 | .000 | .77 | 1.33 | 11.2% | 3 | .575 | .399 | .227 | | | | | | | |
| 1.200 | 1.200 | .035 | .000 | .92 | 1.29 | 2.8% | 4 | .553 | .368 | .197 | .082 | | | | | | |
| 1.200 | 1.200 | .008 | .000 | .98 | 1.24 | .6% | 5 | .547 | .361 | .190 | .078 | .025 | | | | | |
| 1.400 | 1.400 | .266 | .000 | .70 | 1.37 | 16.0% | 3 | .625 | .471 | .304 | | | | | | | |
| 1.400 | 1.400 | .066 | .000 | .88 | 1.34 | 4.5% | 4 | .595 | .427 | .256 | .123 | | | | | | |
| 1.400 | 1.400 | .016 | .000 | .96 | 1.29 | 1.2% | 5 | .586 | .415 | .243 | .114 | .042 | | | | | |
| 1.600 | 1.600 | .576 | .000 | .59 | 1.52 | 26.5% | 3 | .654 | .539 | .407 | | | | | | | |
| 1.600 | 1.600 | .115 | .000 | .83 | 1.39 | 6.7% | 4 | .632 | .481 | .316 | .172 | | | | | | |
| 1.600 | 1.600 | .031 | .000 | .93 | 1.34 | 1.9% | 5 | 620 | .463 | .206 | .155 | .000 | | | | | |
| 1.800 | 1.800 | .889 | .000 | .51 | 1.54 | 33.1% | 3 | .702 | .606 | .492 | | | | | | | |
| 1.800 | 1.800 | .186 | .000 | .77 | 1.43 | 9.4% | 4 | .665 | .531 | .376 | .228 | | | | | | |
| 1.800 | 1.800 | .054 | .000 | .90 | 1.39 | 2.9% | 5 | .650 | .507 | .347 | .200 | .095 | | | | | |
| 1.800 | 1.800 | .015 | .000 | .97 | 1.33 | .8% | 6 | .645 | .500 | .339 | .192 | .090 | .034 | | | | |
| 2.000 | 2.000 | .289 | .000 | .71 | 1.47 | 12.6% | 4 | 696 | .578 | .436 | .289 | | | | | | |
| 2.000 | 2.000 | .087 | .000 | .87 | 1.43 | 4.2% | 5 | .676 | .548 | .397 | .248 | .131 | | | | | |
| 2.000 | 2.000 | .026 | .000 | .95 | 1.38 | 1.3% | 6 | .670 | .500 | .384 | .236 | .121 | .052 | | | | |
| 2.200 | 2.200 | .556 | .000 | .62 | 1.61 | 20.2% | 4 | .707 | .614 | .501 | .377 | | | | | | |
| 2.200 | 2.200 | .135 | .000 | .83 | 1.48 | 5.8% | 5 | .700 | .585 | .445 | .298 | .172 | | | | | |
| 2.200 | 2.200 | .042 | .000 | .93 | 1.42 | 1.9% | 6 | .692 | .571 | .427 | .280 | .156 | .073 | | | | |
| 2.200 | 2.200 | .012 | .000 | .97 | 1.37 | .6% | 7 | .689 | .567 | .422 | .274 | .151 | .070 | .027 | | | |
| 2.400 | 2.400 | .791 | .000 | .55 | 1.63 | 24.8% | 4 | .738 | .658 | .559 | .445 | | | | | | |
| 2.400 | 2.400 | .201 | .000 | .78 | 1.52 | 7.7% | 5 | .722 | .619 | .491 | .350 | .218 | | | | | |
| 2.400 | 2.400 | .066 | .000 | .90 | 1.47 | 2.7% | 6 | .712 | .602 | .468 | .324 | .195 | .099 | | | | |
| 2.400 | 2.400 | .021 | .000 | .96 | 1.41 | .9% | 7 | .708 | .596 | .460 | .315 | .187 | .094 | .040 | | | |
| 2.600 | 2.600 | 1.286 | .000 | .47 | 1.73 | 33.1% | 4 | .759 | .696 | .618 | .527 | | | | | | |
| 2.600 | 2.600 | .292 | .000 | .73 | 1.56 | 10.1% | 5 | .743 | .651 | .535 | .402 | .269 | | | | | |
| 2.600 | 2.600 | .100 | .000 | .87 | 1.51 | 3.7% | 6 | .730 | .630 | .506 | .368 | .236 | .130 | | | | |
| 2.600 | 2.600 | .033 | .000 | .94 | 1.45 | 1.3% | 7 | .725 | .623 | .496 | .356 | .224 | .121 | .056 | | | |
| 2.800 | 2.800 | .521 | .000 | .65 | 1.69 | 15.7% | 5 | .746 | .669 | .574 | .464 | .340 | | | | | |

*SIMILAR TO A.T.&T. JACOBSEN RETRIAL TABLES

*FAST RETRIALS: JEWETT-SHRAGO FAST RETRIAL TABLES - 100% BLOCKED CALLERS RETRY, 0% OVERFLOW (RETRIES 1 HOLDING TIMES LATER)

| HOURS FIRST ATTEMPT VMR=1.0 | HOURS CON-NECT | HOURS RECALL /RETRY | HOURS OVFL | CON-NECT VMR | OVFL VMR | PERCENT BLOCKED (GRADE) | LINE | HOURS CONNECTED TRAFFIC ON LINES 1 | 2 | 3 | 4 | 5 | 6 | 7 | 8 | 9 | 10 |
|---|---|---|---|---|---|---|---|---|---|---|---|---|---|---|---|---|---|
| 2.800 | 2.800 | .145 | .000 | .83 | 1.56 | 4.9% | 6 | .747 | .656 | .542 | .411 | .279 | .165 | | | | |
| 2.800 | 2.800 | .050 | .000 | .92 | 1.50 | 1.8% | 7 | .740 | .647 | .529 | .395 | .262 | .152 | .076 | | | |
| 2.800 | 2.800 | .017 | .000 | .97 | 1.44 | .6% | 8 | .738 | .643 | .524 | | | | | | | |
| 3.000 | 3.000 | .205 | ..256 | .147 | .072 | .031 | | | | | | | | | | | |
| 3.000 | 3.000 | .706 | .000 | .60 | 1.72 | 19.0% | 5 | .767 | .700 | .615 | .515 | .403 | | | | | |
| 3.000 | 3.000 | .205 | .000 | .80 | 1.60 | 6.4% | 6 | .762 | .681 | .576 | .453 | .323 | .205 | | | | |
| 3.000 | 3.000 | .074 | .000 | .90 | 1.54 | 2.4% | 7 | .755 | .669 | .560 | .432 | .301 | .185 | .099 | | | |
| 3.000 | 3.000 | .026 | .000 | .96 | 1.48 | .8% | 8 | .752 | .664 | .553 | .424 | .293 | .178 | .094 | .043 | | |
| 3.200 | 3.200 | .948 | .000 | .54 | 1.73 | 22.9% | 5 | .788 | .729 | .655 | .565 | .463 | | | | | |
| 3.200 | 3.200 | .284 | .000 | .75 | 1.63 | 8.1% | 6 | .777 | .703 | .608 | .494 | .369 | .248 | | | | |
| 3.200 | 3.200 | .106 | .000 | .87 | 1.58 | 3.2% | 7 | .768 | .689 | .588 | .468 | .340 | .220 | .126 | | | |
| 3.200 | 3.200 | .038 | .000 | .94 | 1.52 | 1.2% | 8 | .764 | .683 | .580 | .458 | .329 | .210 | .118 | .058 | | |
| 3.400 | 3.400 | 1.444 | .000 | .46 | 1.82 | 29.8% | 5 | .800 | .752 | .692 | .620 | .536 | | | | | |
| 3.400 | 3.400 | .482 | .000 | .68 | 1.76 | 12.4% | 6 | .776 | .712 | .632 | .536 | .428 | .317 | | | | |
| 3.400 | 3.400 | .148 | .000 | .84 | 1.62 | 4.2% | 7 | .780 | .708 | .615 | .503 | .379 | .258 | .156 | | | |
| 3.400 | 3.400 | .056 | .000 | .92 | 1.56 | 1.6% | 8 | .776 | .701 | .605 | .490 | .364 | .244 | .144 | .075 | | |
| 3.400 | 3.400 | .020 | .000 | .97 | 1.50 | .6% | 9 | .774 | .698 | .601 | .485 | .359 | .238 | .140 | .072 | .033 | |
| 3.600 | 3.600 | .632 | .000 | .63 | 1.79 | 14.9% | 6 | .791 | .734 | .662 | .574 | .473 | .366 | | | | |
| 3.600 | 3.600 | .201 | .000 | .81 | 1.66 | 5.3% | 7 | .792 | .726 | .641 | .536 | .418 | .297 | .190 | | | |
| 3.600 | 3.600 | .078 | .000 | .90 | 1.60 | 2.1% | 8 | .786 | .718 | .629 | .521 | .399 | .278 | .173 | .096 | | |
| 3.600 | 3.600 | .029 | .000 | .96 | 1.54 | .8% | 9 | .784 | .714 | .624 | .514 | .392 | .271 | .167 | .091 | .044 | |
| 3.800 | 3.800 | .822 | .000 | .58 | 1.81 | 17.8% | 6 | .806 | .755 | .691 | .612 | .519 | .417 | | | | |
| 3.800 | 3.800 | .270 | .000 | .77 | 1.70 | 6.6% | 7 | .803 | .743 | .665 | .568 | .456 | .338 | .227 | | | |
| 3.800 | 3.800 | .108 | .000 | .88 | 1.64 | 2.8% | 8 | .796 | .733 | .651 | .549 | .433 | .314 | .205 | .119 | | |
| 3.800 | 3.800 | .042 | .000 | .94 | 1.58 | 1.1% | 9 | .793 | .729 | .645 | .541 | .424 | .304 | .196 | .112 | .057 | |
| 4.000 | 4.000 | 1.214 | .000 | .52 | 1.90 | 23.3% | 6 | .812 | .769 | .715 | .649 | .571 | .484 | | | | |
| 4.000 | 4.000 | .443 | .000 | .71 | 1.83 | 10.0% | 7 | .799 | .746 | .678 | .595 | .499 | .394 | .289 | | | |
| 4.000 | 4.000 | .146 | .000 | .85 | 1.68 | 3.5% | 8 | .806 | .747 | .671 | .577 | .466 | .349 | .238 | .146 | | |
| 4.000 | 4.000 | .058 | .000 | .93 | 1.62 | 1.4% | 9 | .802 | .742 | .664 | .567 | .455 | .336 | .226 | .136 | .073 | |
| 4.000 | 4.000 | .022 | .000 | .97 | 1.55 | .5% | 10 | .801 | .740 | .661 | .563 | .450 | .331 | .221 | .132 | .070 | .033 |
| 4.200 | 4.200 | 1.713 | .000 | .45 | 1.97 | 29.0% | 6 | .822 | .785 | .740 | .685 | .621 | .547 | | | | |
| 4.200 | 4.200 | .567 | .000 | .67 | 1.85 | 11.9% | 7 | .811 | .762 | .701 | .625 | .535 | .435 | .331 | | | |
| 4.200 | 4.200 | .194 | .000 | .82 | 1.72 | 4.4% | 8 | .815 | .760 | .690 | .602 | .499 | .385 | .273 | .175 | | |
| 4.200 | 4.200 | .080 | .000 | .91 | 1.66 | 1.9% | 9 | .811 | .754 | .682 | .591 | .484 | .369 | .257 | .162 | .091 | |
| 4.200 | 4.200 | .031 | .000 | .96 | 1.59 | .7% | 10 | .809 | .752 | .678 | .586 | .478 | .362 | .250 | .156 | .087 | .043 |
| 4.400 | 4.400 | 2.350 | .000 | .39 | 2.02 | 34.8% | 6 | .834 | .803 | .766 | .721 | .668 | .607 | | | | |
| 4.400 | 4.400 | .721 | .000 | .62 | 1.88 | 14.1% | 7 | .822 | .778 | .722 | .653 | .571 | .477 | .376 | | | |
| 4.400 | 4.400 | .254 | .000 | .79 | 1.76 | 5.5% | 8 | .823 | .773 | .709 | .627 | .530 | .421 | .309 | .208 | | |
| 4.400 | 4.400 | .107 | .000 | .89 | 1.70 | 2.4% | 9 | .818 | .766 | .698 | .614 | .512 | .401 | .289 | .189 | .112 | |
| 4.400 | 4.400 | 043 | .000 | .94 | 1.63 | 1.0% | 10 | .816 | .763 | .694 | .607 | .505 | .392 | .280 | .182 | .106 | .055 |
| 4.600 | 4.600 | .912 | .000 | .58 | 1.89 | 16.5% | 7 | .833 | .794 | .743 | .681 | .606 | .519 | .423 | | | |

*SIMILAR TO A.T.&T. JACOBSEN RETRIAL TABLES

*FAST RETRIALS: JEWETT-SHRAGO FAST RETRIAL TABLES - 100% BLOCKED CALLERS RETRY, 0% OVERFLOW (RETRIES 1 HOLDING TIMES LATER)

| HOURS FIRST ATTEMPT VMR = 1.0 | HOURS CON-NECT | HOURS RECALL /RETRY | HOURS OVFL | CON-NECT VMR | OVFL VMR | PERCENT BLOCKED (GRADE) | LINE | HOURS CONNECTED TRAFFIC ON LINES 1 | 2 | 3 | 4 | 5 | 6 | 7 | 8 | 9 | 10 |
|---|---|---|---|---|---|---|---|---|---|---|---|---|---|---|---|---|---|
| 4.600 | 4.600 | .329 | .000 | .76 | 1.79 | 6.7% | 8 | .831 | .785 | .726 | .651 | .560 | .456 | .347 | .243 | | |
| 4.600 | 4.600 | .141 | .000 | .86 | 1.74 | 3.0% | 9 | .826 | .777 | .714 | .635 | .540 | .432 | .322 | .219 | .135 | |
| 4.600 | 4.600 | .059 | .000 | .93 | 1.67 | 1.3% | 10 | .823 | .773 | .709 | .628 | .530 | .421 | .310 | .209 | .127 | .069 |
| 4.600 | 4.600 | .024 | .000 | .97 | 1.60 | .5% | 11 | .822 | .772 | .706 | .624 | .526 | .417 | .305 | .204 | .123 | .067 |
| | | | | | | | | .033 | | | | | | | | | |
| 4.800 | 4.800 | 1.298 | .000 | .51 | 1.98 | 21.3% | 7 | .837 | .802 | .759 | .706 | .643 | .568 | .485 | | | |
| 4.800 | 4.800 | .510 | .000 | .70 | 1.91 | 9.6% | 8 | .828 | .786 | .733 | .667 | .588 | .498 | .399 | .301 | | |
| 4.800 | 4.800 | .184 | .000 | .84 | 1.77 | 3.7% | 9 | .833 | .788 | .729 | .655 | .566 | .463 | .355 | .250 | .161 | |
| 4.800 | 4.800 | .079 | .000 | .91 | 1.71 | 1.6% | 10 | .830 | .783 | .723 | .647 | .555 | .450 | .340 | .237 | .150 | .086 |
| 4.800 | 4.800 | .033 | .000 | .96 | 1.64 | .7% | 11 | .829 | .781 | .720 | .643 | .550 | .444 | .334 | .231 | .145 | .082 |
| | | | | | | | | .042 | | | | | | | | | |
| 5.000 | 5.000 | 1.779 | 0.000 | .46 | 2.05 | 26.2% | 7 | .843 | .813 | .776 | .731 | .678 | .615 | .544 | | | |
| 5.000 | 5.000 | 0.637 | 0.000 | .66 | 1.94 | 11.3% | 8 | .836 | .798 | .750 | .689 | .617 | .532 | .438 | .340 | | |
| 5.000 | 5.000 | 0.237 | 0.000 | .81 | 1.81 | 4.5% | 9 | .840 | .798 | .743 | .675 | .591 | .493 | .388 | .283 | .190 | |
| 5.000 | 5.000 | 0.104 | 0.000 | .90 | 1.75 | 2.0% | 10 | .836 | .792 | .736 | .665 | .578 | .478 | .370 | .266 | .175 | .104 |
| 5.000 | 5.000 | 0.044 | 0.000 | .95 | 1.68 | 0.9% | 11 | .835 | .790 | .733 | .660 | .572 | .470 | .362 | .258 | .168 | .099 |
| | | | | | | | | .053 | | | | | | | | | |
| 5.500 | 5.500 | 1.228 | .000 | .54 | 2.05 | 18.2% | 8 | .851 | .821 | .784 | .738 | .683 | .618 | .544 | .461 | | |
| 5.500 | 5.500 | .510 | .000 | .71 | 1.98 | 8.5% | 9 | .845 | .811 | .767 | .712 | .645 | .567 | .478 | .384 | .290 | |
| 5.500 | 5.500 | .196 | .000 | .84 | 1.84 | 3.4% | 10 | .851 | .814 | .766 | .706 | .631 | .543 | .444 | .342 | .244 | .160 |
| 5.500 | 5.500 | .088 | .000 | .91 | 1.77 | 1.6% | 11 | .848 | .810 | .761 | .699 | .622 | .532 | .432 | .328 | .231 | .149 |
| | | | | | | | | .088 | | | | | | | | | |
| 5.500 | 5.500 | .038 | .000 | .96 | 1.70 | .7% | 12 | .847 | .808 | .759 | .696 | .618 | .527 | .426 | .322 | .225 | .144 |
| | | | | | | | | .084 | .045 | | | | | | | | |
| 6.000 | 6.000 | 2.385 | .000 | .41 | 2.16 | 28.4% | 8 | .865 | .843 | .817 | .785 | .747 | .702 | .650 | .591 | | |
| 6.000 | 6.000 | .849 | .000 | .62 | 2.03 | 12.4% | 9 | .863 | .834 | .798 | .754 | .700 | .635 | .559 | .474 | .384 | |
| 6.000 | 6.000 | .414 | .000 | .75 | 2.01 | 6.5% | 10 | .854 | .823 | .783 | .733 | .673 | .601 | .519 | .429 | .336 | .248 |
| 6.000 | 6.000 | .162 | .000 | .86 | 1.86 | 2.6% | 11 | .860 | .828 | .786 | .732 | .667 | .588 | .497 | .398 | .300 | .210 |
| | | | | | | | | .135 | | | | | | | | | |
| 6.000 | 6.000 | .074 | .000 | .93 | 1.79 | 1.2% | 12 | .859 | .825 | .782 | .728 | .660 | .580 | .487 | .388 | .289 | .200 |
| | | | | | | | | .127 | .074 | | | | | | | | |
| 6.000 | 6.000 | .033 | .000 | .96 | 1.72 | .5% | 13 | .858 | .824 | .780 | .725 | .657 | .576 | .483 | .383 | .284 | .196 |
| | | | | | | | | .124 | .072 | .038 | | | | | | | |
| 7.000 | 7.000 | 3.192 | .000 | .37 | 2.32 | 31.3% | 9 | .880 | .863 | .843 | .820 | .793 | .761 | .724 | .682 | .634 | |
| 7.000 | 7.000 | 1.198 | .000 | .57 | 2.19 | 14.6% | 10 | .876 | .854 | .827 | .794 | .755 | .707 | .651 | .586 | .514 | .435 |
| 7.000 | 7.000 | .544 | .000 | .72 | 2.11 | 7.2% | 11 | .874 | .850 | .819 | .782 | .736 | .680 | .614 | .539 | .455 | .368 |
| | | | | | | | | .283 | | | | | | | | | |
| 7.000 | 7.000 | .231 | .000 | .83 | 1.97 | 3.2% | 12 | .879 | .853 | .821 | .781 | .730 | .669 | .596 | .513 | .422 | .329 |
| | | | | | | | | .241 | .165 | | | | | | | | |

*SIMILAR TO A.T.&T. JACOBSEN RETRIAL TABLES

## *FAST RETRIALS: JEWETT-SHRAGO FAST RETRIAL TABLES - 100% BLOCKED CALLERS RETRY, 0% OVERFLOW (RETRIES 1 HOLDING TIMES LATER)

| HOURS FIRST ATTEMPT VMR = 1.0 | HOURS CON-NECT | HOURS RECALL /RETRY | HOURS OVFL | CON-NECT VMR | OVFL VMR | PERCENT BLOCKED (GRADE) | LINE | HOURS CONNECTED TRAFFIC ON LINES 1 | 2 | 3 | 4 | 5 | 6 | 7 | 8 | 9 | 10 |
|---|---|---|---|---|---|---|---|---|---|---|---|---|---|---|---|---|---|
| 7.000 | 7.000 | .113 | .000 | .90 | 1.90 | 1.6% | 13 | .877 | .851 | .817 | .776 | .724 | .661 | .587 | .502 | .409 | .316 |
| | | | | | | | | .229 | .154 | .096 | | | | | | | |
| 7.000 | 7.000 | .053 | .000 | .95 | 1.83 | .8% | 14 | .876 | .849 | .816 | .774 | .721 | .657 | .582 | .496 | .403 | .309 |
| | | | | | | | | .223 | .149 | .092 | .053 | | | | | | |
| 7.500 | 7.500 | 2.162 | .000 | .45 | 2.31 | 22.4% | 10 | .883 | .865 | .844 | .820 | .790 | .755 | .714 | .666 | .612 | .551 |
| 7.500 | 7.500 | .851 | .000 | .64 | 2.16 | 10.2% | 11 | .885 | .865 | .839 | .808 | .769 | .723 | .667 | .602 | .529 | .448 |
| | | | | | | | | .365 | | | | | | | | | |
| 7.500 | 7.500 | .444 | .000 | .76 | 2.13 | 5.6% | 12 | .880 | .857 | .830 | .795 | .753 | .702 | .642 | .572 | .493 | .409 |
| | | | | | | | | .324 | .243 | | | | | | | | |
| 7.500 | 7.500 | .191 | .000 | .86 | 1.99 | 2.5% | 13 | .885 | .862 | .833 | .797 | .752 | .697 | .631 | .555 | .470 | .379 |
| | | | | | | | | .290 | .208 | .140 | | | | | | | |
| 7.500 | 7.500 | .094 | .000 | .92 | 1.92 | 1.2% | 14 | .884 | .860 | .831 | .794 | .748 | .692 | .624 | .547 | .460 | .369 |
| | | | | | | | | .280 | .199 | .132 | .082 | | | | | | |
| 7.500 | 7.500 | .045 | .000 | .96 | 1.84 | .6% | 15 | .883 | .859 | .829 | .792 | .745 | .689 | .621 | .542 | .455 | .364 |
| | | | | | | | | .275 | .194 | .128 | .079 | .045 | | | | | |
| 8.000 | 8.000 | 4.028 | .000 | .33 | 2.44 | 33.5% | 10 | .892 | .878 | .863 | .846 | .825 | .802 | .775 | .743 | .708 | .668 |
| 8.000 | 8.000 | 1.573 | .000 | .52 | 2.32 | 16.4% | 11 | .888 | .870 | .849 | .824 | .794 | .758 | .716 | .667 | .610 | .547 |
| | | | | | | | | .477 | | | | | | | | | |
| 8.000 | 8.000 | .685 | .000 | .68 | 2.20 | 7.9% | 12 | .890 | .870 | .846 | .817 | .781 | .738 | .686 | .625 | .556 | .479 |
| | | | | | | | | .397 | .315 | | | | | | | | |
| 8.000 | 8.000 | .309 | .000 | .81 | 2.08 | 3.7% | 13 | .893 | .872 | .847 | .816 | .777 | .730 | .672 | .605 | .527 | .443 |
| | | | | | | | | .355 | .270 | .193 | | | | | | | |
| 8.000 | 8.000 | .158 | .000 | .88 | 2.01 | 1.9% | 14 | .891 | .870 | .844 | .811 | .771 | .722 | .663 | .593 | .514 | .428 |
| | | | | | | | | .340 | .255 | .180 | .119 | | | | | | |
| 8.000 | 8.000 | .079 | .000 | .93 | 1.93 | 1.0% | 15 | .890 | .869 | .842 | .809 | .768 | .718 | .658 | .587 | .507 | .420 |
| | | | | | | | | .331 | .247 | .173 | .113 | .069 | | | | | |
| 8.500 | 8.500 | 2.666 | .000 | .42 | 2.41 | 23.9% | 11 | .895 | .881 | .864 | .845 | .822 | .795 | .764 | .727 | .685 | .638 |
| | | | | | | | | .584 | | | | | | | | | |
| 8.500 | 8.500 | 1.148 | .000 | .59 | 2.31 | 11.9% | 12 | .894 | .877 | .857 | .832 | .803 | .768 | .726 | .676 | .619 | .555 |
| | | | | | | | | .484 | .408 | | | | | | | | |
| 8.500 | 8.500 | .557 | .000 | .73 | 2.22 | 6.2% | 13 | .894 | .876 | .853 | .826 | .793 | .753 | .706 | .649 | .584 | .511 |
| | | | | | | | | .432 | .351 | .272 | | | | | | | |
| 8.500 | 8.500 | .255 | .000 | .04 | 2.00 | 2.9% | 14 | .898 | .879 | .856 | .827 | .792 | .750 | .698 | .636 | .565 | .485 |
| | | | | | | | | .400 | .315 | .235 | .165 | | | | | | |
| 8.500 | 8.500 | .132 | .000 | .90 | 2.02 | 1.5% | 15 | .896 | .877 | .853 | .824 | .788 | .744 | .691 | .628 | .555 | .474 |
| | | | | | | | | .388 | .303 | .223 | .155 | .101 | | | | | |
| 8.500 | 8.500 | .066 | .000 | .94 | 1.94 | .8% | 16 | .895 | .876 | .852 | .823 | .786 | .741 | .687 | .623 | .549 | .468 |
| | | | | | | | | .381 | .296 | .217 | .150 | .097 | .058 | | | | |
| 9.000 | 9.000 | 1.970 | .000 | .49 | 2.43 | 18.0% | 12 | .897 | .883 | .866 | .846 | .823 | .795 | .763 | .725 | .681 | .631 |
| | | | | | | | | .575 | .513 | | | | | | | | |

*SIMILAR TO A.T.&T. JACOBSEN RETRIAL TABLES

*FAST RETRIALS: JEWETT-SHRAGO FAST RETRIAL TABLES - 100% BLOCKED CALLERS RETRY, 0% OVERFLOW (RETRIES 1 HOLDING TIMES LATER)

| HOURS FIRST ATTEMPT VMR = 1.0 | HOURS CON-NECT | HOURS RECALL /RETRY | HOURS OVFL | CON-NECT VMR | OVFL VMR | PERCENT BLOCKED (GRADE) | LINE | HOURS CONNECTED TRAFFIC ON LINES 1 | 2 | 3 | 4 | 5 | 6 | 7 | 8 | 9 | 10 |
|---|---|---|---|---|---|---|---|---|---|---|---|---|---|---|---|---|---|
| 9.000 | 9.000 | .928 | .000 | .64 | 2.35 | 9.3% | 13 | .897 | .881 | .862 | .838 | .811 | .777 | .737 | .691 | .636 | .574 |
| | | | | | | | | .506 | .432 | .357 | | | | | | | |
| 9.000 | 9.000 | .457 | .000 | .77 | 2.25 | 4.8% | 14 | .898 | .881 | .861 | .836 | .805 | .768 | .724 | .672 | .612 | .543 |
| | | | | | | | | .467 | .388 | .309 | .235 | | | | | | |
| 9.000 | 9.000 | .211 | .000 | .86 | 2.11 | 2.3% | 15 | .902 | .885 | .864 | .838 | .806 | .768 | .721 | .665 | .599 | .525 |
| | | | | | | | | .444 | .360 | .278 | .204 | .140 | | | | | |
| 9.000 | 9.000 | .110 | .000 | .91 | 2.03 | 1.2% | 16 | .901 | .884 | .862 | .836 | .803 | .764 | .716 | .659 | .592 | .516 |
| | | | | | | | | .435 | .350 | .269 | .195 | .133 | .085 | | | | |
| 9.000 | 9.000 | .055 | .000 | .95 | 1.95 | .6% | 17 | .901 | .883 | .861 | .835 | .802 | .762 | .713 | .655 | .588 | .512 |
| | | | | | | | | .430 | .345 | .264 | .190 | .129 | .082 | .049 | | | |
| 9.500 | 9.500 | 3.344 | .000 | .38 | 2.54 | 26.0% | 12 | .902 | .891 | .877 | .862 | .844 | .823 | .800 | .772 | .741 | .705 |
| | | | | | | | | .664 | .618 | | | | | | | | |
| 9.500 | 9.500 | 1.466 | .000 | .55 | 2.44 | 13.4% | 13 | .902 | .888 | .871 | .852 | .829 | .801 | .769 | .730 | .686 | .635 |
| | | | | | | | | .577 | .514 | .446 | | | | | | | |
| 9.500 | 9.500 | .679 | .000 | .70 | 2.31 | 6.7% | 14 | .905 | .890 | .872 | .850 | .824 | .792 | .754 | .709 | .656 | .595 |
| | | | | | | | | .526 | .452 | .375 | .299 | | | | | | |
| 9.500 | 9.500 | .377 | .000 | .80 | 2.26 | 3.8% | 15 | .902 | .886 | .867 | .844 | .816 | .783 | .742 | .694 | .638 | .574 |
| | | | | | | | | .503 | .426 | .348 | .272 | .202 | | | | | |
| 9.500 | 9.500 | .175 | .000 | .88 | 2.12 | 1.8% | 16 | .906 | .891 | .871 | .848 | .819 | .784 | .742 | .691 | .631 | .562 |
| | | | | | | | | .486 | .405 | .323 | .245 | .176 | .120 | | | | |
| 9.500 | 9.500 | .091 | .000 | .93 | 2.04 | 1.0% | 17 | .906 | .890 | .870 | .846 | .817 | .781 | .738 | .686 | .625 | .556 |
| | | | | | | | | .479 | .397 | .315 | .238 | .170 | .114 | .072 | | | |
| 10.000 | 10.000 | 2.384 | .000 | .46 | 2.54 | 19.3% | 13 | .906 | .894 | .880 | .864 | .845 | .823 | .798 | .768 | .734 | .695 |
| | | | | | | | | .650 | .600 | .544 | | | | | | | |
| 10.000 | 10.000 | 1.001 | .000 | .82 | 2.42 | 9.8% | 14 | .908 | .801 | .070 | .800 | .837 | .810 | .779 | .741 | .697 | .646 |
| | | | | | | | | .588 | .524 | .455 | .383 | | | | | | |
| 10.000 | 10.000 | .557 | .000 | .74 | 2.33 | 5.3% | 15 | .908 | .894 | .877 | .857 | .832 | .803 | .768 | .726 | .676 | .619 |
| | | | | | | | | .555 | .484 | .408 | .332 | .259 | | | | | |
| 10.000 | 10.000 | .269 | .000 | .84 | 2.20 | 2.6% | 16 | .911 | .897 | .880 | .859 | .833 | .802 | .765 | .720 | .667 | .605 |
| | | | | | | | | .535 | .459 | .379 | .300 | .226 | .162 | | | | |
| 10.000 | 10.000 | .145 | .000 | .90 | 2.12 | 1.4% | 17 | .910 | .896 | .878 | .857 | .831 | .799 | .760 | .714 | .660 | .596 |
| | | | | | | | | .525 | .448 | .308 | .289 | .216 | .152 | .102 | | | |
| 10.000 | 10.000 | .076 | .000 | .94 | 2.04 | .8% | 18 | .910 | .895 | .877 | .855 | .829 | .797 | .758 | .711 | .656 | .592 |
| | | | | | | | | .520 | .442 | .361 | .282 | .210 | .147 | .098 | .061 | | |
| 11.000 | 11.000 | 2.811 | .000 | .43 | 2.63 | 20.4% | 14 | .913 | .903 | .891 | .878 | .863 | .845 | .825 | .801 | .774 | .742 |
| | | | | | | | | .707 | .666 | .621 | .571 | | | | | | |
| 11.000 | 11.000 | 1.364 | .000 | .58 | 2.55 | 11.0% | 15 | .913 | .902 | .888 | .873 | .855 | .833 | .808 | .778 | .743 | .703 |
| | | | | | | | | .656 | .604 | .540 | .483 | .416 | | | | | |
| 11.000 | 11.000 | .663 | .000 | .72 | 2.41 | 5.7% | 16 | .916 | .905 | .891 | .874 | .854 | .830 | .801 | .768 | .727 | .681 |
| | | | | | | | | .627 | .566 | .498 | .427 | .354 | .282 | | | | |

*SIMILAR TO A.T.&T. JACOBSEN RETRIAL TABLES

## *FAST RETRIALS: JEWETT-SHRAGO FAST RETRIAL TABLES - 100% BLOCKED CALLERS RETRY, 0% OVERFLOW (RETRIES 1 HOLDING TIMES LATER)

| HOURS FIRST ATTEMPT VMR = 1.0 | HOURS CON-NECT | HOURS RECALL /RETRY | HOURS OVFL | CON-NECT VMR | OVFL VMR | PERCENT BLOCKED (GRADE) | LINE | HOURS CONNECTED TRAFFIC ON LINES 1 | 2 | 3 | 4 | 5 | 6 | 7 | 8 | 9 | 10 |
|---|---|---|---|---|---|---|---|---|---|---|---|---|---|---|---|---|---|
| 11.000 | 11.000 | .382 | .000 | .81 | 2.36 | 3.4% | 17 | .914 | .902 | .887 | .870 | .849 | .824 | .794 | .758 | .716 | .666 |
| | | | | | | | | .610 | .546 | .477 | .404 | .330 | .259 | .195 | | | |
| 11.000 | 11.000 | .186 | .000 | .88 | 2.21 | 1.7% | 18 | .918 | .906 | .891 | .873 | .852 | .826 | .795 | .757 | .713 | .661 |
| | | | | | | | | .601 | .533 | .460 | .383 | .307 | .235 | .171 | .118 | | |
| 11.000 | 11.000 | .101 | .000 | .93 | 2.13 | .9% | 19 | .917 | .905 | .890 | .872 | .850 | .824 | .792 | .754 | .709 | .656 |
| | | | | | | | | .595 | .527 | .453 | .376 | .299 | .228 | .165 | .113 | .074 | |
| 12.000 | 12.000 | 3.251 | .000 | .41 | 2.71 | 21.3% | 15 | .919 | .910 | .901 | .890 | .877 | .862 | .846 | .826 | .804 | .779 |
| | | | | | | | | .751 | .718 | .681 | .640 | .595 | | | | | |
| 12.000 | 12.000 | 1.544 | .000 | .56 | 2.60 | 11.4% | 16 | .921 | .911 | .900 | .887 | .872 | .854 | .833 | .809 | .780 | .747 |
| | | | | | | | | .709 | .665 | .615 | .560 | .501 | .437 | | | | |
| 12.000 | 12.000 | .851 | .000 | .68 | 2.55 | 6.6% | 17 | .920 | .910 | .898 | .884 | .867 | .847 | .824 | .797 | .765 | .727 |
| | | | | | | | | .684 | .634 | .578 | .517 | .451 | .383 | .315 | | | |
| 12.000 | 12.000 | .456 | .000 | .79 | 2.44 | 3.7% | 18 | .922 | .911 | .899 | .884 | .866 | .846 | .821 | .792 | .757 | .716 |
| | | | | | | | | .669 | .615 | .555 | .489 | .419 | .348 | .278 | .214 | | |
| 12.000 | 12.000 | .231 | .000 | .87 | 2.30 | 1.9% | 19 | .924 | .914 | .901 | .887 | .869 | .848 | .822 | .792 | .755 | .712 |
| | | | | | | | | .662 | .605 | .541 | .471 | .397 | .323 | .253 | .189 | .135 | |
| 12.000 | 12.000 | .129 | .000 | .91 | 2.22 | 1.1% | 20 | .924 | .913 | .901 | .885 | .867 | .846 | .820 | .789 | .752 | .708 |
| | | | | | | | | .657 | .599 | .533 | .463 | .389 | .315 | .245 | .182 | .128 | .086 |
| 12.000 | 12.000 | .071 | .000 | .95 | 2.14 | .6% | 21 | .924 | .913 | .900 | .885 | .866 | .845 | .818 | .787 | .749 | .705 |
| | | | | | | | | .654 | .595 | .529 | .458 | .384 | .310 | .240 | .177 | .125 | .084 |
| | | | | | | | | .053 | | | | | | | | | |
| 13.000 | 13.000 | 3.847 | .000 | .38 | 2.83 | 22.8% | 16 | .923 | .915 | .907 | .898 | .887 | .875 | .861 | .846 | .828 | .808 |
| | | | | | | | | .785 | .759 | .730 | .697 | .661 | .620 | | | | |
| 13.000 | 13.000 | 1.840 | .000 | .53 | 2.71 | 12.4% | 17 | .925 | .916 | .907 | .896 | .883 | .868 | .851 | .831 | .808 | .781 |
| | | | | | | | | .750 | .714 | .674 | .629 | .578 | .524 | .465 | | | |
| 13.000 | 13.000 | .971 | .000 | .67 | 2.62 | 7.0% | 18 | .926 | .918 | .907 | .895 | .881 | .865 | .846 | .823 | .796 | .765 |
| | | | | | | | | .729 | .688 | .641 | .587 | .529 | .466 | .401 | .335 | | |
| 13.000 | 13.000 | .534 | .000 | .77 | 2.51 | 3.9% | 19 | .928 | .918 | .908 | .895 | .881 | .863 | .843 | .819 | .790 | .757 |
| | | | | | | | | .717 | .672 | .621 | .563 | .500 | .433 | .364 | .296 | .232 | |
| 13.000 | 13.000 | .318 | .000 | .84 | 2.45 | 2.4% | 20 | .926 | .917 | .906 | .893 | .878 | .860 | .839 | .814 | .784 | .749 |
| | | | | | | | | .709 | .662 | .608 | .549 | .484 | .416 | .347 | .279 | .217 | .161 |
| 13.000 | 13.000 | .161 | .000 | .90 | 2.30 | 1.2% | 21 | .929 | .920 | .909 | .896 | .881 | .863 | .842 | .816 | .785 | .749 |
| | | | | | | | | .707 | .658 | .602 | .540 | .472 | .401 | .329 | .260 | .197 | .143 |
| | | | | | | | | .099 | | | | | | | | | |
| 13.000 | 13.000 | .091 | .000 | .94 | 2.22 | .7% | 22 | .929 | .920 | .909 | .896 | .880 | .862 | .840 | .814 | .703 | .747 |
| | | | | | | | | .704 | .654 | .598 | .535 | .466 | .395 | .323 | .255 | .192 | .139 |
| | | | | | | | | .096 | .063 | | | | | | | | |
| 14.000 | 14.000 | 4.308 | .000 | .36 | 2.89 | 23.5% | 17 | .928 | .921 | .914 | .906 | .897 | .887 | .875 | .862 | .847 | .831 |
| | | | | | | | | .812 | .790 | .766 | .739 | .709 | .676 | .639 | | | |
| 14.000 | 14.000 | 2.146 | .000 | .51 | 2.81 | 13.3% | 18 | .929 | .921 | .913 | .903 | .892 | .880 | .865 | .849 | .830 | .808 |

*SIMILAR TO A.T.&T. JACOBSEN RETRIAL TABLES

# *FAST RETRIALS: JEWETT-SHRAGO FAST RETRIAL TABLES - 100% BLOCKED CALLERS RETRY, 0% OVERFLOW (RETRIES 1 HOLDING TIMES LATER)

| HOURS FIRST ATTEMPT VMR=1.0 | HOURS CON-NECT | HOURS RECALL /RETRY | HOURS OVFL | CON-NECT VMR | OVFL VMR | PERCENT BLOCKED (GRADE) | LINE | HOURS CONNECTED TRAFFIC ON LINES 1 | 2 | 3 | 4 | 5 | 6 | 7 | 8 | 9 | 10 |
|---|---|---|---|---|---|---|---|---|---|---|---|---|---|---|---|---|---|
| | | | | | | | | .782 | .753 | .720 | .683 | .641 | .595 | .544 | .490 | | |
| 14.000 | 14.000 | 1.096 | .000 | .65 | 2.68 | 7.3% | 19 | .932 | .924 | .915 | .905 | .893 | .879 | .863 | .844 | .822 | .796 |
| | | | | | | | | .766 | .732 | .692 | .647 | .596 | .540 | .480 | .417 | .353 | |
| 14.000 | 14.000 | .616 | .000 | .75 | 2.59 | 4.2% | 20 | .933 | .925 | .916 | .905 | .892 | .878 | .861 | .840 | .817 | .789 |
| | | | | | | | | .756 | .719 | .675 | .626 | .570 | .510 | .446 | .379 | .313 | .250 |
| 14.000 | 14.000 | .374 | .000 | .82 | 2.53 | 2.6% | 21 | .932 | .924 | .914 | .903 | .890 | .875 | .857 | .836 | .811 | .782 |
| | | | | | | | | .748 | .709 | .664 | .613 | .556 | .494 | .428 | .361 | .295 | .233 |
| | | | | | | | | .177 | | | | | | | | | |
| 14.000 | 14.000 | .196 | .000 | .89 | 2.38 | 1.4% | 22 | .934 | .926 | .917 | .906 | .893 | .877 | .859 | .838 | .812 | .782 |
| | | | | | | | | .747 | .706 | .659 | .605 | .545 | .480 | .412 | .343 | .275 | .213 |
| | | | | | | | | .158 | .112 | | | | | | | | |
| 14.000 | 14.000 | .113 | .000 | .93 | 2.30 | .8% | 23 | .934 | .926 | .916 | .905 | .892 | .876 | .858 | .836 | .810 | .780 |
| | | | | | | | | .744 | .703 | .655 | .600 | .540 | .474 | .405 | .336 | .269 | .207 |
| | | | | | | | | .153 | .108 | .073 | | | | | | | |
| 15.000 | 15.000 | 4.923 | .000 | .34 | 2.99 | 24.7% | 18 | .931 | .925 | .919 | .912 | .904 | .895 | .886 | .875 | .862 | .848 |
| | | | | | | | | .833 | .815 | .796 | .774 | .749 | .722 | .692 | .659 | | |
| 15.000 | 15.000 | 2.462 | .000 | .49 | 2.90 | 14.1% | 19 | .932 | .926 | .918 | .910 | .901 | .890 | .877 | .863 | .847 | .829 |
| | | | | | | | | .808 | .784 | .757 | .726 | .692 | .653 | .610 | .563 | .512 | |
| 15.000 | 15.000 | 1.312 | .000 | .62 | 2.79 | 8.0% | 20 | .934 | .928 | .920 | .911 | .900 | .888 | .875 | .859 | .840 | .819 |
| | | | | | | | | .794 | .766 | .733 | .696 | .654 | .607 | .555 | .499 | .440 | .380 |
| 15.000 | 15.000 | .702 | .000 | .73 | 2.66 | 4.5% | 21 | .937 | .930 | .922 | .913 | .902 | .890 | .875 | .858 | .838 | .815 |
| | | | | | | | | .788 | .756 | .720 | .678 | .630 | .578 | .520 | .458 | .394 | .329 |
| | | | | | | | | .267 | | | | | | | | | |
| 15.000 | 15.000 | .433 | .000 | .81 | 2.60 | 2.8% | 22 | .936 | .929 | .921 | .911 | .900 | .887 | .872 | .854 | .833 | .809 |
| | | | | | | | | .781 | .748 | .710 | .666 | .617 | .562 | .502 | .439 | .374 | .309 |
| | | | | | | | | .248 | .192 | | | | | | | | |
| 15.000 | 15.000 | .234 | .000 | .88 | 2.40 | 1.5% | 23 | .938 | .931 | .923 | .914 | .902 | .889 | .874 | .856 | .834 | .809 |
| | | | | | | | | .780 | .746 | .706 | .660 | .608 | .551 | .488 | .422 | .355 | .289 |
| | | | | | | | | .227 | .172 | .125 | | | | | | | |
| 15.000 | 15.000 | .138 | .000 | .92 | 2.38 | .9% | 24 | .938 | .931 | .923 | .913 | .902 | .888 | .873 | .854 | .833 | .807 |
| | | | | | | | | .777 | .742 | .702 | .655 | .603 | .545 | .481 | .415 | .348 | .282 |
| | | | | | | | | .220 | .166 | .120 | .083 | | | | | | |
| 15.000 | 15.000 | .080 | .000 | .95 | 2.29 | .5% | 25 | .938 | .931 | .922 | .912 | .901 | .888 | .872 | .853 | .831 | .806 |
| | | | | | | | | .776 | .740 | .699 | .653 | .600 | .541 | .477 | .411 | .343 | .277 |
| | | | | | | | | .216 | .162 | .117 | .080 | .053 | | | | | |
| 16.000 | 16.000 | 2.666 | .000 | .47 | 2.94 | 14.3% | 20 | .937 | .931 | .925 | .917 | .909 | .900 | .889 | .877 | .864 | .848 |
| | | | | | | | | .830 | .810 | .787 | .761 | .731 | .698 | .661 | .620 | .575 | .527 |
| 16.000 | 16.000 | 1.447 | .000 | .60 | 2.85 | 8.3% | 21 | .939 | .933 | .928 | .918 | .909 | .899 | .887 | .873 | .858 | .840 |
| | | | | | | | | .819 | .795 | .767 | .736 | .700 | .660 | .615 | .565 | .511 | .455 |
| | | | | | | | | .396 | | | | | | | | | |

*SIMILAR TO A.T.&T. JACOBSEN RETRIAL TABLES

## *FAST RETRIALS: JEWETT-SHRAGO FAST RETRIAL TABLES - 100% BLOCKED CALLERS RETRY, 0% OVERFLOW (RETRIES 1 HOLDING TIMES LATER)

| HOURS FIRST ATTEMPT VMR = 1.0 | HOURS CON-NECT | HOURS RECALL /RETRY | HOURS OVFL | CON-NECT VMR | OVFL VMR | PERCENT BLOCKED (GRADE) | LINE | HOURS CONNECTED TRAFFIC ON LINES | | | | | | | | | |
|---|---|---|---|---|---|---|---|---|---|---|---|---|---|---|---|---|---|
| | | | | | | | | 1 | 2 | 3 | 4 | 5 | 6 | 7 | 8 | 9 | 10 |
| 16.000 | 16.000 | .857 | .000 | .71 | 2.78 | 5.1% | 22 | .939 | .933 | .926 | .917 | .908 | .897 | .884 | .870 | .853 | .833 |
| | | | | | | | | .811 | .785 | .754 | .720 | .680 | .636 | .586 | .532 | .475 | .414 |
| | | | | | | | | .353 | .293 | | | | | | | | |
| 16.000 | 16.000 | .495 | .000 | .79 | 2.67 | 3.0% | 23 | .940 | .934 | .927 | .918 | .909 | .897 | .884 | .869 | .852 | .831 |
| | | | | | | | | .807 | .779 | .747 | .710 | .668 | .620 | .568 | .510 | .449 | .386 |
| | | | | | | | | .323 | .262 | .206 | | | | | | | |
| 16.000 | 16.000 | .307 | .000 | .85 | 2.60 | 1.9% | 24 | .940 | .933 | .926 | .917 | .907 | .896 | .882 | .867 | .849 | .828 |
| | | | | | | | | .803 | .774 | .741 | .703 | .660 | .611 | .557 | .498 | .436 | .373 |
| | | | | | | | | .310 | .250 | .194 | .146 | | | | | | |
| 16.000 | 16.000 | .165 | .000 | .91 | 2.45 | 1.0% | 25 | .942 | .935 | .928 | .920 | .910 | .898 | .885 | .869 | .851 | .829 |
| | | | | | | | | .804 | .775 | .741 | .701 | .656 | .605 | .549 | .488 | .424 | .358 |
| | | | | | | | | .294 | .233 | .178 | .131 | .093 | | | | | |
| 16.000 | 16.000 | .098 | .000 | .94 | 2.37 | .6% | 26 | .942 | .935 | .928 | .919 | .909 | .898 | .884 | .868 | .850 | .828 |
| | | | | | | | | .803 | .773 | .738 | .698 | .653 | .602 | .545 | .484 | .419 | .354 |
| | | | | | | | | .289 | .229 | .174 | .128 | .090 | .061 | | | | |
| 17.000 | 17.000 | 6.326 | .000 | .31 | 3.19 | 27.1% | 20 | .936 | .932 | .927 | .921 | .915 | .909 | .902 | .894 | .885 | .875 |
| | | | | | | | | .864 | .852 | .838 | .823 | .807 | .789 | .768 | .746 | .721 | .695 |
| 17.000 | 17.000 | 3.115 | .000 | .45 | 3.06 | 15.5% | 21 | .938 | .933 | .927 | .921 | .914 | .906 | .896 | .886 | .874 | .861 |
| | | | | | | | | .846 | .830 | .810 | .789 | .765 | .738 | .707 | .673 | .636 | .595 |
| | | | | | | | | .551 | | | | | | | | | |
| 17.000 | 17.000 | 1.678 | .000 | .58 | 2.95 | 9.0% | 22 | .941 | .935 | .929 | .922 | .914 | .905 | .895 | .883 | .870 | .855 |
| | | | | | | | | .837 | .817 | .794 | .768 | .738 | .704 | .666 | .624 | .578 | .528 |
| | | | | | | | | .474 | .419 | | | | | | | | |
| 17.000 | 17.000 | .953 | .000 | .69 | 2.84 | 5.3% | 23 | .943 | .937 | .931 | .924 | .915 | .906 | .895 | .882 | .868 | .851 |
| | | | | | | | | .832 | .810 | .784 | .755 | .721 | .683 | .640 | .593 | .541 | .485 |
| | | | | | | | | .426 | .367 | .308 | | | | | | | |
| 17.000 | 17.000 | .561 | .000 | .78 | 2.74 | 3.2% | 24 | .944 | .938 | .932 | .924 | .916 | .906 | .895 | .882 | .867 | .849 |
| | | | | | | | | .829 | .806 | .778 | .747 | .711 | .670 | .624 | .573 | .518 | .459 |
| | | | | | | | | .398 | .336 | .276 | .220 | | | | | | |
| 17.000 | 17.000 | .353 | .000 | .84 | 2.67 | 2.0% | 25 | .943 | .937 | .931 | .923 | .915 | .905 | .893 | .880 | .864 | .846 |
| | | | | | | | | .825 | .801 | .773 | .740 | .703 | .661 | .614 | .562 | .505 | .445 |
| | | | | | | | | .384 | .322 | .262 | .207 | .158 | | | | | |
| 17.000 | 17.000 | .195 | .000 | .90 | 2.52 | 1.1% | 26 | .945 | .939 | .933 | .925 | .917 | .907 | .895 | .882 | .866 | .848 |
| | | | | | | | | .826 | .802 | .773 | .739 | .701 | .657 | .608 | .553 | .495 | .432 |
| | | | | | | | | .369 | .306 | .246 | .191 | .143 | .103 | | | | |
| 17.000 | 17.000 | .117 | .000 | .93 | 2.44 | .7% | 27 | .945 | .939 | .933 | .925 | .916 | .906 | .894 | .881 | .865 | .846 |
| | | | | | | | | .825 | .800 | .770 | .737 | .698 | .653 | .604 | .549 | .490 | .427 |
| | | | | | | | | .363 | .300 | .240 | .186 | .139 | .100 | .069 | | | |
| 18.000 | 18.000 | 6.965 | .000 | .29 | 3.26 | 27.9% | 21 | .939 | .935 | .931 | .926 | .920 | .915 | .908 | .901 | .894 | .885 |
| | | | | | | | | .876 | .865 | .854 | .841 | .827 | .812 | .795 | .776 | .756 | .733 |

*SIMILAR TO A.T.&T. JACOBSEN RETRIAL TABLES

*FAST RETRIALS: JEWETT-SHRAGO FAST RETRIAL TABLES - 100% BLOCKED CALLERS RETRY, 0% OVERFLOW (RETRIES 1 HOLDING TIMES LATER)

| HOURS FIRST ATTEMPT VMR=1.0 | HOURS CON-NECT | HOURS RECALL /RETRY | HOURS OVFL | CON-NECT VMR | OVFL VMR | PERCENT BLOCKED (GRADE) | LINE | HOURS CONNECTED TRAFFIC ON LINES 1 | 2 | 3 | 4 | 5 | 6 | 7 | 8 | 9 | 10 |
|---|---|---|---|---|---|---|---|---|---|---|---|---|---|---|---|---|---|
| 18.000 | 18.000 | 3.453 | .000 | .43 | 3.14 | 16.1% | 22 | .941 | .936 | .931 | .926 | .919 | .912 | .904 | .895 | .885 | .874 |
| | | | | | | | | .861 | .846 | .830 | .812 | .792 | .769 | .743 | .714 | .683 | .648 |
| | | | | | | | | .610 | .568 | | | | | | | | |
| 18.000 | 18.000 | 1.821 | .000 | .57 | 3.00 | 9.2% | 23 | .944 | .940 | .934 | .928 | .921 | .913 | .904 | .894 | .882 | .869 |
| | | | | | | | | .854 | .837 | .817 | .795 | .770 | .741 | .708 | .672 | .631 | .586 |
| | | | | | | | | .538 | .487 | .433 | | | | | | | |
| 18.000 | 18.000 | 1.052 | .000 | .68 | 2.90 | 5.5% | 24 | .946 | .941 | .935 | .929 | .922 | .913 | .904 | .893 | .881 | .866 |
| | | | | | | | | .850 | .831 | .809 | .784 | .756 | .723 | .686 | .645 | .599 | .549 |
| | | | | | | | | .495 | .438 | .380 | .322 | | | | | | |
| 18.000 | 18.000 | .629 | .000 | .77 | 2.80 | 3.4% | 25 | .947 | .942 | .936 | .930 | .922 | .914 | .904 | .893 | .880 | .865 |
| | | | | | | | | .847 | .827 | .804 | .778 | .747 | .712 | .672 | .628 | .579 | .525 |
| | | | | | | | | .468 | .409 | .349 | .290 | .234 | | | | | |
| 18.000 | 18.000 | .401 | .000 | .83 | 2.74 | 2.2% | 26 | .946 | .941 | .935 | .929 | .921 | .912 | .902 | .890 | .877 | .862 |
| | | | | | | | | .844 | .823 | .799 | .771 | .740 | .704 | .663 | .617 | .566 | .512 |
| | | | | | | | | .454 | .394 | .333 | .275 | .220 | .170 | | | | |
| 18.000 | 18.000 | .227 | .000 | .89 | 2.60 | 1.2% | 27 | .948 | .943 | .937 | .930 | .923 | .914 | .904 | .892 | .879 | .863 |
| | | | | | | | | .845 | .824 | .799 | .771 | .738 | .700 | .658 | .610 | .558 | .501 |
| | | | | | | | | .440 | .378 | .317 | .257 | .203 | .154 | .113 | | | |
| 18.000 | 18.000 | .139 | .000 | .92 | 2.51 | .8% | 28 | .948 | .943 | .937 | .930 | .922 | .913 | .903 | .891 | .878 | .862 |
| | | | | | | | | .843 | .822 | .797 | .768 | .735 | .697 | .654 | .606 | .552 | .495 |
| | | | | | | | | .434 | .372 | .311 | .252 | .197 | .150 | .109 | .077 | | |
| 19.000 | 19.000 | 7.755 | .000 | .28 | 3.35 | 29.0% | 22 | .941 | .937 | .933 | .929 | .924 | .919 | .914 | .908 | .901 | .893 |
| | | | | | | | | .885 | .877 | .867 | .856 | .844 | .831 | .817 | .802 | .785 | .766 |
| | | | | | | | | .745 | .723 | | | | | | | | |
| 19.000 | 19.000 | 3.795 | .000 | .42 | 3.20 | 16.6% | 23 | .944 | .939 | .935 | .930 | .924 | .918 | .910 | .903 | .894 | .884 |
| | | | | | | | | .873 | .861 | .847 | .831 | .814 | .701 | .770 | .748 | .721 | .692 |
| | | | | | | | | .659 | .623 | .584 | | | | | | | |
| 19.000 | 19.000 | 2.065 | .000 | .55 | 3.09 | 9.8% | 24 | .946 | .942 | .937 | .931 | .925 | .918 | .910 | .901 | .891 | .880 |
| | | | | | | | | .867 | .852 | .835 | .816 | .795 | .771 | .743 | .712 | .678 | .640 |
| | | | | | | | | .598 | .552 | .504 | .453 | | | | | | |
| 19.000 | 19.000 | 1.226 | .000 | .66 | 3.01 | 6.1% | 25 | .947 | .943 | .938 | .932 | .925 | .918 | .910 | .900 | .889 | .877 |
| | | | | | | | | .863 | .847 | .828 | .807 | .783 | .755 | .724 | .689 | .650 | .607 |
| | | | | | | | | .560 | .509 | .455 | .400 | .344 | | | | | |
| 19.000 | 19.000 | .754 | .000 | .74 | 2.92 | 3.8% | 26 | .948 | .943 | .938 | .932 | .925 | .918 | .909 | .899 | .888 | .875 |
| | | | | | | | | .860 | .843 | .823 | .800 | .775 | .745 | .712 | .674 | .632 | .585 |
| | | | | | | | | .535 | .481 | .425 | .367 | .311 | .256 | | | | |
| 19.000 | 19.000 | .452 | .000 | .82 | 2.80 | 2.3% | 27 | .949 | .944 | .939 | .933 | .926 | .919 | .910 | .900 | .888 | .875 |
| | | | | | | | | .859 | .842 | .821 | .797 | .770 | .739 | .704 | .664 | .620 | .571 |
| | | | | | | | | .518 | .462 | .403 | .344 | .286 | .232 | .182 | | | |

*SIMILAR TO A.T.&T. JACOBSEN RETRIAL TABLES

*FAST RETRIALS: JEWETT-SHRAGO FAST RETRIAL TABLES - 100% BLOCKED CALLERS RETRY, 0% OVERFLOW (RETRIES 1 HOLDING TIMES LATER)

| HOURS FIRST ATTEMPT VMR = 1.0 | HOURS CON-NECT | HOURS RECALL /RETRY | HOURS OVFL | CON-NECT VMR | OVFL VMR | PERCENT BLOCKED (GRADE) | LINE | HOURS CONNECTED TRAFFIC ON LINES 1 | 2 | 3 | 4 | 5 | 6 | 7 | 8 | 9 | 10 |
|---|---|---|---|---|---|---|---|---|---|---|---|---|---|---|---|---|---|
| 19.000 | 19.000 | .289 | .000 | .87 | 2.73 | 1.5% | 28 | .949 | .944 | .939 | .933 | .926 | .918 | .909 | .898 | .886 | .873 |
| | | | | | | | | .857 | .839 | .818 | .794 | .766 | .734 | .698 | .658 | .612 | .562 |
| | | | | | | | | .508 | .451 | .393 | .333 | .276 | .222 | .173 | .131 | | |
| 19.000 | 19.000 | .162 | .000 | .91 | 2.58 | .8% | 29 | .950 | .946 | .941 | .935 | .928 | .920 | .911 | .901 | .889 | .875 |
| | | | | | | | | .859 | .841 | .819 | .795 | .766 | .734 | .697 | .655 | .608 | .556 |
| | | | | | | | | .500 | .441 | .381 | .320 | .262 | .208 | .160 | .119 | .085 | |
| 19.000 | 19.000 | .099 | .000 | .94 | 2.49 | .5% | 30 | .950 | .946 | .940 | .934 | .927 | .919 | .910 | .900 | .888 | .874 |
| | | | | | | | | .858 | .840 | .818 | .793 | .765 | .732 | .694 | .652 | .604 | .552 |
| | | | | | | | | .496 | .437 | .377 | .316 | .258 | .204 | .157 | .116 | .083 | .057 |
| 20.000 | 20.000 | 8.404 | .000 | .27 | 3.41 | 29.6% | 23 | .944 | .940 | .937 | .933 | .929 | .924 | .919 | .913 | .908 | .901 |
| | | | | | | | | .894 | .886 | .878 | .869 | .858 | .847 | .835 | .822 | .807 | .791 |
| | | | | | | | | .774 | .755 | .734 | | | | | | | |
| 20.000 | 20.000 | 4.142 | .000 | .40 | 3.27 | 17.2% | 24 | .946 | .942 | .938 | .933 | .928 | .922 | .916 | .909 | .901 | .893 |
| | | | | | | | | .883 | .873 | .861 | .847 | .832 | .816 | .797 | .776 | .753 | .728 |
| | | | | | | | | .700 | .669 | .635 | .598 | | | | | | |
| 20.000 | 20.000 | 2.314 | .000 | .53 | 3.18 | 10.4% | 25 | .948 | .944 | .939 | .934 | .929 | .923 | .915 | .908 | .899 | .889 |
| | | | | | | | | .878 | .865 | .851 | .834 | .816 | .795 | .772 | .746 | .717 | .684 |
| | | | | | | | | .648 | .609 | .566 | .520 | .471 | | | | | |
| 20.000 | 20.000 | 1.332 | .000 | .64 | 3.06 | 6.2% | 26 | .950 | .946 | .941 | .936 | .930 | .924 | .916 | .908 | .899 | .888 |
| | | | | | | | | .876 | .862 | .846 | .827 | .807 | .783 | .756 | .726 | .692 | .654 |
| | | | | | | | | .612 | .567 | .518 | .466 | .412 | .357 | | | | |
| 20.000 | 20.000 | .830 | .000 | .73 | 2.98 | 4.0% | 27 | .951 | .946 | .942 | .936 | .930 | .924 | .916 | .907 | .897 | .886 |
| | | | | | | | | .873 | .858 | .841 | .822 | .800 | .774 | .746 | .713 | .676 | .635 |
| | | | | | | | | .590 | .542 | .489 | .434 | .378 | .323 | .268 | | | |
| 20.000 | 20.000 | .505 | .000 | .81 | 2.87 | 2.5% | 28 | .952 | .947 | .943 | .937 | .931 | .924 | .917 | .908 | .897 | .886 |
| | | | | | | | | .872 | .857 | .840 | .819 | .796 | .769 | .739 | .705 | .666 | .623 |
| | | | | | | | | .575 | .524 | .469 | .412 | .355 | .298 | .244 | .194 | | |
| 20.000 | 20.000 | .327 | .000 | .86 | 2.79 | 1.6% | 29 | .951 | .947 | .942 | .937 | .930 | .923 | .915 | .906 | .896 | .884 |
| | | | | | | | | .870 | .855 | .837 | .816 | .792 | .765 | .734 | .698 | .659 | .615 |
| | | | | | | | | .566 | .514 | .458 | .401 | .343 | .286 | .233 | .184 | .141 | |
| 20.000 | 20.000 | .187 | .000 | .91 | 2.65 | .9% | 30 | .953 | .949 | .944 | .938 | .932 | .925 | .917 | .908 | .898 | .886 |
| | | | | | | | | .872 | .856 | .838 | .817 | .792 | .765 | .733 | .696 | .655 | .610 |
| | | | | | | | | .559 | .505 | .448 | .389 | .330 | .272 | .219 | .170 | .128 | .094 |
| 20.000 | 20.000 | .116 | .000 | .94 | 2.56 | .6% | 31 | .953 | .948 | .944 | .938 | .932 | .925 | .917 | .908 | .897 | .885 |
| | | | | | | | | .871 | .855 | .837 | .815 | .791 | .763 | .730 | .694 | .652 | .606 |
| | | | | | | | | .556 | .501 | .443 | .384 | .325 | .268 | .215 | .167 | .125 | .091 |
| | | | | | | | | .064 | | | | | | | | | |
| 22.000 | 22.000 | 10.150 | .000 | .24 | 3.58 | 31.6% | 25 | .947 | .944 | .941 | .938 | .934 | .931 | .927 | .922 | .918 | .913 |
| | | | | | | | | .907 | .901 | .895 | .888 | .880 | .872 | .863 | .854 | .843 | .832 |
| | | | | | | | | .819 | .806 | .791 | .775 | .758 | | | | | |

*SIMILAR TO A.T.&T. JACOBSEN RETRIAL TABLES

*FAST RETRIALS: JEWETT-SHRAGO FAST RETRIAL TABLES - 100% BLOCKED CALLERS RETRY, 0% OVERFLOW (RETRIES 1 HOLDING TIMES LATER)

| HOURS FIRST ATTEMPT VMR=1.0 | HOURS CON-NECT | HOURS RECALL /RETRY | HOURS OVFL | CON-NECT VMR | OVFL VMR | PERCENT BLOCKED (GRADE) | LINE | HOURS CONNECTED TRAFFIC ON LINES 1 | 2 | 3 | 4 | 5 | 6 | 7 | 8 | 9 | 10 |
|---|---|---|---|---|---|---|---|---|---|---|---|---|---|---|---|---|---|
| 22.000 | 22.000 | 4.978 | .000 | .37 | 3.41 | 18.5% | 26 | .949 | .946 | .943 | .939 | .935 | .930 | .925 | .919 | .913 | .907 |
| | | | | | | | | .899 | .891 | .882 | .872 | .861 | .848 | .835 | .820 | .803 | .784 |
| | | | | | | | | .764 | .741 | .716 | .689 | .660 | .628 | | | | |
| 22.000 | 22.000 | 2.726 | .000 | .50 | 3.30 | 11.0% | 27 | .952 | .949 | .945 | .941 | .936 | .931 | .926 | .919 | .913 | .905 |
| | | | | | | | | .896 | .886 | .875 | .863 | .850 | .834 | .817 | .797 | .776 | .751 |
| | | | | | | | | .724 | .694 | .661 | .625 | .586 | .544 | .499 | | | |
| 22.000 | 22.000 | 1.629 | .000 | .61 | 3.21 | 6.9% | 28 | .954 | .950 | .946 | .942 | .937 | .932 | .926 | .920 | .912 | .904 |
| | | | | | | | | .894 | .884 | .871 | .858 | .842 | .825 | .805 | .783 | .758 | .730 |
| | | | | | | | | .698 | .663 | .625 | .583 | .538 | .490 | .440 | .388 | | |
| 22.000 | 22.000 | .987 | .000 | .71 | 3.10 | 4.3% | 29 | .955 | .952 | .948 | .944 | .939 | .933 | .927 | .920 | .913 | .904 |
| | | | | | | | | .894 | .883 | .870 | .855 | .839 | .820 | .798 | .774 | .747 | .716 |
| | | | | | | | | .681 | .642 | .600 | .554 | .504 | .452 | .399 | .345 | .292 | |
| 22.000 | 22.000 | .617 | .000 | .78 | 2.99 | 2.7% | 30 | .956 | .953 | .949 | .944 | .939 | .934 | .928 | .921 | .913 | .904 |
| | | | | | | | | .893 | .882 | .869 | .853 | .836 | .816 | .794 | .768 | .739 | .706 |
| | | | | | | | | .669 | .628 | .583 | .535 | .483 | .429 | .374 | .319 | .266 | .216 |
| 22.000 | 22.000 | .408 | .000 | .84 | 2.92 | 1.8% | 31 | .956 | .952 | .948 | .944 | .939 | .933 | .927 | .920 | .911 | .902 |
| | | | | | | | | .892 | .880 | .866 | .850 | .833 | .812 | .789 | .763 | .733 | .699 |
| | | | | | | | | .661 | .619 | .573 | .524 | .471 | .417 | .361 | .306 | .254 | .205 |
| | | | | | | | | .161 | | | | | | | | | |
| 22.000 | 22.000 | .267 | .000 | .88 | 2.84 | 1.2% | 32 | .955 | .952 | .948 | .943 | .938 | .932 | .926 | .919 | .910 | .901 |
| | | | | | | | | .890 | .878 | .864 | .848 | .830 | .810 | .786 | .759 | .728 | .694 |
| | | | | | | | | .656 | .613 | .566 | .516 | .463 | .408 | .352 | .298 | .245 | .197 |
| | | | | | | | | .154 | .117 | | | | | | | | |
| 22.000 | 22.000 | .155 | .000 | .92 | 2.69 | .7% | 33 | .957 | .953 | .949 | .945 | .940 | .934 | .928 | .921 | .912 | .903 |
| | | | | | | | | .892 | .880 | .866 | .850 | .832 | .811 | .787 | .759 | .728 | .693 |
| | | | | | | | | .653 | .609 | .561 | .510 | .455 | .308 | .342 | .286 | .234 | .185 |
| | | | | | | | | .143 | .107 | .070 | | | | | | | |
| 24.000 | 24.0001 | 1.909 | .000 | .22 | 3.72 | 33.2% | 27 | .950 | .948 | .945 | .942 | .939 | .936 | .933 | .930 | .926 | .922 |
| | | | | | | | | .917 | .913 | .908 | .902 | .896 | .890 | .883 | .876 | .868 | .860 |
| | | | | | | | | .850 | .840 | .830 | .818 | .806 | .792 | .777 | | | |
| 24.000 | 24.000 | 5.958 | .000 | .34 | 3.57 | 19.9% | 28 | .952 | .949 | .946 | .943 | .939 | .936 | .931 | .927 | .922 | .917 |
| | | | | | | | | .911 | .904 | .897 | .889 | .881 | .872 | .861 | .850 | .837 | .824 |
| | | | | | | | | .809 | .792 | .774 | .754 | .733 | .709 | .683 | .655 | | |
| 24.000 | 24.000 | 3.258 | .000 | .47 | 3.44 | 12.0% | 29 | .955 | .952 | .949 | .945 | .942 | .937 | .933 | .928 | .922 | .916 |
| | | | | | | | | .909 | .901 | .893 | .883 | .873 | .861 | .848 | .834 | .818 | .800 |
| | | | | | | | | .779 | .757 | .733 | .705 | .675 | .643 | .607 | .569 | .529 | |
| 24.000 | 24.000 | 1.942 | .000 | .58 | 3.34 | 7.5% | 30 | .957 | .954 | .951 | .947 | .943 | .939 | .934 | .929 | .923 | .916 |
| | | | | | | | | .908 | .900 | .891 | .880 | .868 | .855 | .840 | .823 | .805 | .784 |
| | | | | | | | | .760 | .734 | .704 | .672 | .636 | .598 | .556 | .511 | .465 | .416 |
| 24.000 | 24.000 | 1.216 | .000 | .68 | 3.25 | 4.8% | 31 | .958 | .955 | .952 | .948 | .944 | .940 | .935 | .929 | .923 | .916 |

*SIMILAR TO A.T.&T. JACOBSEN RETRIAL TABLES

*FAST RETRIALS: JEWETT-SHRAGO FAST RETRIAL TABLES - 100% BLOCKED CALLERS RETRY, 0% OVERFLOW (RETRIES 1 HOLDING TIMES LATER)

| HOURS FIRST ATTEMPT VMR = 1.0 | HOURS CON-NECT | HOURS RECALL /RETRY | HOURS OVFL | CON-NECT VMR | OVFL VMR | PERCENT BLOCKED (GRADE) | LINE | HOURS CONNECTED TRAFFIC ON LINES 1 | 2 | 3 | 4 | 5 | 6 | 7 | 8 | 9 | 10 |
|---|---|---|---|---|---|---|---|---|---|---|---|---|---|---|---|---|---|
| | | | | | | | | .908 | .899 | .889 | .878 | .865 | .851 | .835 | .817 | .796 | .773 |
| | | | | | | | | .747 | .718 | .685 | .649 | .610 | .567 | .522 | .473 | .423 | .372 |
| | | | | | | | | .322 | | | | | | | | | |
| 24.000 | 24.000 | .785 | .000 | .76 | 3.16 | 3.2% | 32 | .959 | .956 | .952 | .949 | .944 | .940 | .935 | .929 | .923 | .915 |
| | | | | | | | | .907 | .898 | .888 | .876 | .863 | .848 | .831 | .812 | .790 | .765 |
| | | | | | | | | .738 | .707 | .672 | .634 | .593 | .547 | .499 | .449 | .397 | .345 |
| | | | | | | | | .294 | .245 | | | | | | | | |
| 24.000 | 24.000 | .497 | .000 | .82 | 3.04 | 2.0% | 33 | .959 | .956 | .953 | .949 | .945 | .941 | .935 | .930 | .923 | .916 |
| | | | | | | | | .908 | .898 | .888 | .876 | .862 | .847 | .829 | .809 | .787 | .761 |
| | | | | | | | | .732 | .700 | .664 | .624 | .580 | .533 | .483 | .431 | .378 | .325 |
| | | | | | | | | .273 | .224 | .180 | | | | | | | |
| 24.000 | 24.000 | .331 | .000 | .87 | 2.96 | 1.4% | 34 | .959 | .956 | .953 | .949 | .945 | .940 | .935 | .929 | .922 | .915 |
| | | | | | | | | .907 | .897 | .886 | .874 | .860 | .844 | .827 | .806 | .783 | .757 |
| | | | | | | | | .727 | .694 | .657 | .617 | .572 | .525 | .474 | .421 | .368 | .315 |
| | | | | | | | | .264 | .215 | .172 | .133 | | | | | | |
| 24.000 | 24.000 | .199 | .000 | .91 | 2.82 | .8% | 35 | .960 | .957 | .954 | .950 | .946 | .942 | .936 | .930 | .924 | .916 |
| | | | | | | | | .908 | .899 | .888 | .876 | .862 | .846 | .827 | .807 | .783 | .756 |
| | | | | | | | | .726 | .692 | .654 | .612 | .567 | .517 | .465 | .411 | .357 | .303 |
| | | | | | | | | .251 | .203 | .160 | .123 | .091 | | | | | |
| 24.000 | 24.000 | .128 | .000 | .94 | 2.73 | .5% | 36 | .960 | .957 | .954 | .950 | .946 | .941 | .936 | .930 | .924 | .916 |
| | | | | | | | | .908 | .898 | .887 | .875 | .861 | .845 | .826 | .805 | .781 | .754 |
| | | | | | | | | .724 | .690 | .651 | .609 | .563 | .514 | .461 | .407 | .352 | .299 |
| | | | | | | | | .247 | .199 | .157 | .120 | .089 | .064 | | | | |
| 26.00026.0001 | | 3.814 | .000 | .21 | 3.87 | 34.7% | 29 | .952 | .950 | .948 | .946 | .944 | .941 | .938 | .935 | .932 | .929 |
| | | | | | | | | .925 | .922 | .918 | .913 | .909 | .904 | .898 | .893 | .887 | .880 |
| | | | | | | | | .873 | .866 | .857 | .849 | .839 | .829 | .819 | .807 | .795 | |
| 26.000 | 26.000 | 6.827 | .000 | .32 | 3.68 | 20.8% | 30 | .955 | .953 | .950 | .947 | .944 | .941 | .938 | .934 | .930 | .925 |
| | | | | | | | | .920 | .915 | .910 | .903 | .897 | .889 | .881 | .872 | .863 | .852 |
| | | | | | | | | .841 | .828 | .814 | .799 | .783 | .765 | .746 | .724 | .701 | .676 |
| 26.000 | 26.000 | 3.805 | .000 | .45 | 3.57 | 12.8% | 31 | .958 | .955 | .952 | .949 | .946 | .943 | .939 | .935 | .930 | .925 |
| | | | | | | | | .919 | .913 | .906 | .899 | .891 | .882 | .871 | .860 | .848 | .834 |
| | | | | | | | | .819 | .802 | .784 | .763 | .740 | .715 | .688 | .659 | .626 | .592 |
| | | | | | | | | .555 | | | | | | | | | |
| 26.000 | 26.000 | 2.267 | .000 | .56 | 3.47 | 8.0% | 32 | .960 | .957 | .954 | .951 | .948 | .944 | .940 | .936 | .931 | .925 |
| | | | | | | | | .919 | .913 | .905 | .897 | .888 | .877 | .866 | .853 | .839 | .823 |
| | | | | | | | | .805 | .785 | .763 | .738 | .710 | .680 | .647 | .611 | .572 | .531 |
| | | | | | | | | .487 | .442 | | | | | | | | |
| 26.000 | 26.000 | 1.458 | .000 | .65 | 3.39 | 5.3% | 33 | .961 | .958 | .955 | .952 | .949 | .945 | .941 | .936 | .931 | .925 |
| | | | | | | | | .919 | .912 | .904 | .895 | .885 | .874 | .862 | .848 | .832 | .814 |
| | | | | | | | | .795 | .773 | .748 | .720 | .690 | .656 | .620 | .580 | .537 | .492 |

*SIMILAR TO A.T.&T. JACOBSEN RETRIAL TABLES

*FAST RETRIALS: JEWETT-SHRAGO FAST RETRIAL TABLES - 100% BLOCKED CALLERS RETRY, 0% OVERFLOW (RETRIES 1 HOLDING TIMES LATER)

| HOURS FIRST ATTEMPT VMR=1.0 | HOURS CON-NECT | HOURS RECALL /RETRY | HOURS OVFL | CON-NECT VMR | OVFL VMR | PERCENT BLOCKED (GRADE) | LINE | HOURS CONNECTED TRAFFIC ON LINES 1 | 2 | 3 | 4 | 5 | 6 | 7 | 8 | 9 | 10 |
|---|---|---|---|---|---|---|---|---|---|---|---|---|---|---|---|---|---|
| | | | | | | | | .445 | .397 | .349 | | | | | | | |
| 26.000 | 26.000 | .915 | .000 | .74 | 3.26 | 3.4% | 34 | .962 | .959 | .957 | .953 | .950 | .946 | .942 | .937 | .932 | .926 |
| | | | | | | | | .920 | .912 | .904 | .895 | .885 | .873 | .860 | .845 | .829 | .810 |
| | | | | | | | | .789 | .765 | .739 | .709 | .676 | .639 | .600 | .557 | .511 | .463 |
| | | | | | | | | .413 | .363 | .313 | .264 | | | | | | |
| 26.000 | 26.000 | .591 | .000 | .80 | 3.15 | 2.2% | 35 | .963 | .960 | .957 | .954 | .951 | .947 | .942 | .938 | .932 | .927 |
| | | | | | | | | .920 | .913 | .904 | .895 | .884 | .872 | .859 | .844 | .826 | .807 |
| | | | | | | | | .785 | .760 | .732 | .701 | .666 | .628 | .587 | .542 | .494 | .444 |
| | | | | | | | | .393 | .341 | .291 | .243 | .198 | | | | | |
| 26.000 | 26.000 | .402 | .000 | .85 | 3.08 | 1.5% | 36 | .962 | .960 | .957 | .954 | .950 | .946 | .942 | .937 | .932 | .926 |
| | | | | | | | | .919 | .911 | .903 | .893 | .883 | .870 | .857 | .841 | .823 | .803 |
| | | | | | | | | .781 | .755 | .727 | .695 | .659 | .620 | .578 | .532 | .484 | .434 |
| | | | | | | | | .382 | .331 | .280 | .233 | .189 | .150 | | | | |
| 26.000 | 26.000 | .270 | .000 | .89 | 2.99 | 1.0% | 37 | .962 | .960 | .957 | .953 | .950 | .946 | .941 | .937 | .931 | .925 |
| | | | | | | | | .918 | .911 | .902 | .892 | .881 | .869 | .855 | .839 | .821 | .801 |
| | | | | | | | | .778 | .752 | .723 | .690 | .655 | .615 | .572 | .526 | .477 | .427 |
| | | | | | | | | .375 | .323 | .273 | .226 | .183 | .144 | .111 | | | |
| 26.000 | 26.000 | .164 | .000 | .92 | 2.85 | .6% | 38 | .963 | .961 | .958 | .955 | .951 | .947 | .943 | .938 | .933 | .927 |
| | | | | | | | | .920 | .912 | .904 | .894 | .883 | .870 | .856 | .840 | .822 | .801 |
| | | | | | | | | .778 | .752 | .722 | .689 | .652 | .612 | .568 | .520 | .470 | .418 |
| | | | | | | | | .366 | .313 | .263 | .216 | .172 | .134 | .102 | .075 | | |
| 28.000 | 28.000 | 7.963 | .000 | .30 | 3.83 | 22.1% | 32 | .957 | .955 | .952 | .950 | .947 | .945 | .942 | .939 | .935 | .931 |
| | | | | | | | | .927 | .923 | .919 | .914 | .908 | .902 | .896 | .889 | .881 | .873 |
| | | | | | | | | .864 | .855 | .844 | .833 | .821 | .807 | .792 | .776 | .759 | .741 |
| | | | | | | | | .720 | .699 | | | | | | | | |
| 28.000 | 28.000 | 4.474 | .000 | .42 | 3.72 | 13.8% | 33 | .959 | .957 | .955 | .952 | .949 | .946 | .943 | .939 | .936 | .931 |
| | | | | | | | | .927 | .922 | .916 | .910 | .904 | .896 | .888 | .880 | .870 | .859 |
| | | | | | | | | .848 | .835 | .821 | .805 | .788 | .769 | .749 | .726 | .701 | .675 |
| | | | | | | | | .646 | .615 | .581 | | | | | | | |
| 28.000 | 28.000 | 2.690 | .000 | .53 | 3.62 | 8.8% | 34 | .962 | .959 | .957 | .954 | .951 | .948 | .945 | .941 | .937 | .932 |
| | | | | | | | | .927 | .922 | .916 | .909 | .902 | .893 | .884 | .874 | .863 | .851 |
| | | | | | | | | .837 | .822 | .805 | .786 | .765 | .742 | .717 | .089 | 659 | .620 |
| | | | | | | | | .600 | .552 | .512 | .470 | | | | | | |
| 28.000 | 28.000 | 1.713 | .000 | .63 | 3.52 | 5.8% | 35 | .963 | .961 | .958 | .955 | .952 | .949 | .946 | .942 | .037 | .933 |
| | | | | | | | | .927 | 922 | .015 | .908 | .900 | .892 | .882 | .871 | .859 | .845 |
| | | | | | | | | .830 | .813 | .794 | .773 | .749 | .723 | .695 | .663 | .029 | .592 |
| | | | | | | | | .552 | .510 | .465 | .420 | .373 | | | | | |
| 28.000 | 28.000 | 1.106 | .000 | .71 | 3.41 | 3.8% | 36 | .964 | .962 | .959 | .956 | .953 | .950 | .947 | .943 | .938 | .933 |
| | | | | | | | | .928 | .922 | .916 | .908 | .900 | .891 | .881 | .869 | .856 | .842 |
| | | | | | | | | .825 | .807 | 787 | .764 | .739 | .711 | .679 | .645 | .608 | .568 |

*SIMILAR TO A.T.&T. JACOBSEN RETRIAL TABLES

*FAST RETRIALS: JEWETT-SHRAGO FAST RETRIAL TABLES - 100% BLOCKED CALLERS RETRY, 0% OVERFLOW (RETRIES 1 HOLDING TIMES LATER)

| HOURS FIRST ATTEMPT VMR=1.0 | HOURS CON-NECT | HOURS RECALL /RETRY | HOURS OVFL | CON-NECT VMR | OVFL VMR | PERCENT BLOCKED (GRADE) | LINE | HOURS CONNECTED TRAFFIC ON LINES 1 | 2 | 3 | 4 | 5 | 6 | 7 | 8 | 9 | 10 |
|---|---|---|---|---|---|---|---|---|---|---|---|---|---|---|---|---|---|
| | | | | | | | | .525 | .480 | .433 | .385 | .337 | .290 | | | | |
| 28.000 | 28.000 | .734 | .000 | .78 | 3.31 | 2.6% | 37 | .964 | .962 | .960 | .957 | .954 | .951 | .947 | .943 | .938 | .934 |
| | | | | | | | | .928 | .922 | .915 | .908 | .899 | .890 | .879 | .867 | .854 | .839 |
| | | | | | | | | .822 | .803 | .782 | .758 | .731 | .702 | .669 | .633 | .594 | .552 |
| | | | | | | | | .507 | .460 | .411 | .362 | .314 | .267 | .222 | | | |
| 28.000 | 28.000 | .477 | .000 | .83 | 3.19 | 1.7% | 38 | .965 | .963 | .960 | .958 | .955 | .951 | .948 | .944 | .939 | .934 |
| | | | | | | | | .929 | .923 | .916 | .908 | .900 | .890 | .879 | .867 | .853 | .838 |
| | | | | | | | | .820 | .801 | .779 | .754 | .726 | .695 | .661 | .624 | .583 | .540 |
| | | | | | | | | .493 | .445 | .395 | .345 | .296 | .249 | .205 | .165 | | |
| 28.000 | 28.000 | .327 | .000 | .88 | 3.11 | 1.2% | 39 | .965 | .963 | .960 | .957 | .954 | .951 | .947 | .943 | .939 | .934 |
| | | | | | | | | .928 | .922 | .915 | .907 | .898 | .889 | .878 | .865 | .851 | .836 |
| | | | | | | | | .818 | .798 | .775 | .750 | .722 | .691 | .656 | .618 | .577 | .533 |
| | | | | | | | | .486 | .437 | .387 | .337 | .288 | .242 | .198 | .159 | .124 | |
| 28.000 | 28.000 | .203 | .000 | .91 | 2.97 | .7% | 40 | .966 | .964 | .961 | .958 | .955 | .952 | .948 | .944 | .940 | .935 |
| | | | | | | | | .929 | .923 | .916 | .909 | .900 | .890 | .879 | .866 | .852 | .836 |
| | | | | | | | | .818 | .798 | .775 | .749 | .721 | .689 | .653 | .614 | .572 | .526 |
| | | | | | | | | .478 | .429 | .378 | .327 | .278 | .231 | .188 | .149 | .115 | .087 |
| 30.000 | 30.000 | 8.986 | .000 | .28 | 3.94 | 23.0% | 34 | .959 | .957 | .955 | .953 | .951 | .948 | .946 | .943 | .940 | .937 |
| | | | | | | | | .934 | .930 | .926 | .922 | .918 | .913 | .908 | .902 | .896 | .890 |
| | | | | | | | | .882 | .875 | .867 | .858 | .848 | .837 | .826 | .814 | .801 | .786 |
| | | | | | | | | .771 | .754 | .736 | .716 | | | | | | |
| 30.000 | 30.000 | 5.047 | .000 | .40 | 3.82 | 14.4% | 35 | .962 | .960 | .958 | .955 | .953 | .950 | .947 | .944 | .941 | .937 |
| | | | | | | | | .934 | .929 | .925 | .920 | .914 | .909 | .902 | .895 | .887 | .879 |
| | | | | | | | | .870 | .860 | .848 | .836 | .823 | .808 | .792 | .775 | .756 | .735 |
| | | | | | | | | .712 | .687 | .661 | .632 | .601 | | | | | |
| 30.000 | 30.000 | 3.038 | .000 | .51 | 3.72 | 9.2% | 36 | .964 | .962 | .960 | .957 | .955 | .952 | .949 | .946 | .942 | .939 |
| | | | | | | | | .934 | .930 | .925 | .919 | .913 | .907 | .900 | .892 | .883 | .873 |
| | | | | | | | | .862 | .850 | .837 | .822 | .806 | .788 | .768 | .746 | .723 | .696 |
| | | | | | | | | .668 | .637 | .604 | .568 | .530 | .491 | | | | |
| 30.000 | 30.000 | 1.977 | .000 | .60 | 3.65 | 6.2% | 37 | .965 | .963 | .961 | .958 | .956 | .953 | .950 | .947 | .943 | .939 |
| | | | | | | | | .935 | .930 | .925 | .919 | .912 | .905 | .897 | .889 | .879 | .868 |
| | | | | | | | | .856 | .843 | .828 | .812 | .794 | .774 | .751 | .727 | .700 | .670 |
| | | | | | | | | .638 | .603 | .565 | .525 | .484 | .440 | .396 | | | |
| 30.000 | 30.000 | 1.307 | .000 | .69 | 3.55 | 4.2% | 38 | .966 | .964 | .961 | .959 | .956 | .954 | .951 | .947 | .943 | .939 |
| | | | | | | | | .935 | .930 | .925 | .919 | .912 | .904 | .896 | .887 | .877 | .865 |
| | | | | | | | | .853 | .839 | .823 | .805 | .786 | .764 | .739 | .713 | .683 | .651 |
| | | | | | | | | .616 | .578 | .537 | .495 | .450 | .405 | .359 | .313 | | |
| 30.000 | 30.000 | .843 | .000 | .76 | 3.42 | 2.7% | 39 | .967 | .965 | .963 | .960 | .958 | .955 | .952 | .948 | .945 | .941 |
| | | | | | | | | .936 | .931 | .926 | .919 | .913 | .905 | .897 | .887 | .877 | .865 |
| | | | | | | | | .852 | .837 | .820 | .801 | .781 | .757 | .732 | .703 | .672 | .637 |

*SIMILAR TO A.T.&T. JACOBSEN RETRIAL TABLES

*FAST RETRIALS: JEWETT-SHRAGO FAST RETRIAL TABLES - 100% BLOCKED CALLERS RETRY, 0% OVERFLOW (RETRIES 1 HOLDING TIMES LATER)

| HOURS FIRST ATTEMPT VMR = 1.0 | HOURS CONNECT | HOURS RECALL /RETRY | HOURS OVFL | CONNECT VMR | OVFL VMR | PERCENT BLOCKED (GRADE) | LINE | HOURS CONNECTED TRAFFIC ON LINES 1 | 2 | 3 | 4 | 5 | 6 | 7 | 8 | 9 | 10 |
|---|---|---|---|---|---|---|---|---|---|---|---|---|---|---|---|---|---|
| | | | | | | | | .600 | .559 | .516 | .471 | .424 | .377 | .330 | .283 | .239 | |
| 30.000 | 30.000 | .559 | .000 | .82 | 3.30 | 1.8% | 40 | .967 | .965 | .963 | .961 | .958 | .956 | .952 | .949 | .945 | .941 |
| | | | | | | | | .937 | .932 | .926 | .920 | .913 | .905 | .897 | .887 | .876 | .864 |
| | | | | | | | | .850 | .835 | .818 | .799 | .777 | .753 | .726 | .696 | .664 | .628 |
| | | | | | | | | .589 | .547 | .502 | .456 | .408 | .359 | .312 | .265 | .221 | .181 |
| 30.000 | 30.000 | .388 | .000 | .86 | 3.22 | 1.3% | 41 | .967 | .965 | .963 | .961 | .958 | .955 | .952 | .949 | .945 | .941 |
| | | | | | | | | .936 | .931 | .925 | .919 | .912 | .904 | .895 | .885 | .874 | .862 |
| | | | | | | | | .848 | .833 | .815 | .795 | .773 | .749 | .721 | .691 | .658 | .621 |
| | | | | | | | | .581 | .539 | .494 | .447 | .399 | .350 | .303 | .256 | .213 | .173 |
| | | | | | | | | .138 | | | | | | | | | |
| 30.000 | 30.000 | .267 | .000 | .90 | 3.13 | .9% | 42 | .967 | .965 | .963 | .960 | .958 | .955 | .952 | .948 | .945 | .940 |
| | | | | | | | | .936 | .930 | .925 | .918 | .911 | .903 | .894 | .884 | .873 | .861 |
| | | | | | | | | .847 | .831 | .813 | .793 | .771 | .746 | .718 | .687 | .653 | .616 |
| | | | | | | | | .576 | .533 | .488 | .441 | .393 | .344 | .296 | .250 | .207 | .168 |
| | | | | | | | | .134 | .104 | | | | | | | | |
| 30.000 | 30.000 | .166 | .000 | .93 | 2.99 | .6% | 43 | .968 | .966 | .964 | .961 | .959 | .956 | .953 | .949 | .946 | .941 |
| | | | | | | | | .937 | .932 | .926 | .920 | .913 | .905 | .896 | .886 | .874 | .862 |
| | | | | | | | | .848 | .832 | .814 | .794 | .771 | .745 | .717 | .686 | .651 | .613 |
| | | | | | | | | .572 | .529 | .482 | .434 | .385 | .335 | .287 | .241 | .198 | .159 |
| | | | | | | | | .125 | .096 | .072 | | | | | | | |

*SIMILAR TO A.T.&T. JACOBSEN RETRIAL TABLES

## * VERY FAST RETRIALS: JEWETT-SHRAGO VERY FAST RETRIAL TABLES - 50% RETRY, 50% OVERFLOW (RETRIES 1/3 HOLDING TIMES LATER)

| HOURS FIRST ATTEMPT VMR = 1.0 | HOURS CON-NECT | HOURS RECALL /RETRY | HOURS OVFL | CON-NECT VMR | OVFL VMR | PERCENT BLOCKED (GRADE) | LINE | HOURS CONNECTED TRAFFIC ON LINES 1 | 2 | 3 | 4 | 5 | 6 | 7 | 8 | 9 | 10 |
|---|---|---|---|---|---|---|---|---|---|---|---|---|---|---|---|---|---|
| .200 | .168 | .032 | .032 | .83 | 1.19 | 27.4% | 1 | .168 | | | | | | | | | |
| .200 | .198 | .002 | .002 | .97 | 1.07 | 1.7% | 2 | .168 | .030 | | | | | | | | |
| .400 | .388 | .012 | .012 | .90 | 1.13 | 5.7% | 2 | .292 | .097 | | | | | | | | |
| .400 | .399 | .001 | .001 | .98 | 1.11 | .7% | 3 | .286 | .093 | .019 | | | | | | | |
| .600 | .565 | .035 | .035 | .82 | 1.19 | 11.0% | 2 | .388 | .177 | | | | | | | | |
| .600 | .594 | .006 | .006 | .95 | 1.17 | 2.0% | 3 | .377 | .167 | .050 | | | | | | | |
| .800 | .699 | .101 | .101 | .74 | 1.35 | 22.4% | 2 | .435 | .264 | | | | | | | | |
| .800 | .783 | .017 | .017 | .91 | 1.22 | 4.0% | 3 | .450 | .240 | .094 | | | | | | | |
| .800 | .797 | .003 | .003 | .97 | 1.18 | .8% | 4 | .445 | .236 | .090 | .025 | | | | | | |
| 1.000 | .810 | .190 | .190 | .67 | 1.50 | 32.0% | 2 | .476 | .333 | | | | | | | | |
| 1.000 | .965 | .035 | .035 | .85 | 1.27 | 6.7% | 3 | .509 | .311 | .146 | | | | | | | |
| 1.000 | .992 | .008 | .008 | .95 | 1.24 | 1.6% | 4 | .502 | .302 | .139 | .048 | | | | | | |
| 1.200 | 1.138 | .062 | .062 | .80 | 1.32 | 9.9% | 3 | .558 | .376 | .204 | | | | | | | |
| 1.200 | 1.183 | .017 | .017 | .92 | 1.29 | 2.7% | 4 | .549 | .363 | .192 | .079 | | | | | | |
| 1.200 | 1.196 | .004 | .004 | .98 | 1.24 | .6% | 5 | .546 | .360 | .189 | .077 | .024 | | | | | |
| 1.400 | 1.268 | .132 | .132 | .72 | 1.47 | 17.3% | 3 | .568 | .424 | .276 | | | | | | | |
| 1.400 | 1.370 | .030 | .030 | .88 | 1.33 | 4.2% | 4 | .589 | .418 | .247 | .116 | | | | | | |
| 1.400 | 1.392 | .008 | .008 | .96 | 1.29 | 1.1% | 5 | .585 | .413 | .241 | .112 | .041 | | | | | |
| 1.600 | 1.412 | .188 | .188 | .67 | 1.50 | 21.0% | 3 | .607 | .474 | .331 | | | | | | | |
| 1.600 | 1.550 | .050 | .050 | .84 | 1.38 | 6.1% | 4 | .623 | .467 | .300 | .159 | | | | | | |
| 1.600 | 1.585 | .015 | .015 | .94 | 1.34 | 1.8% | 5 | .618 | .460 | .292 | .152 | .064 | | | | | |
| 1.800 | 1.511 | .289 | .289 | .62 | 1.62 | 27.7% | 3 | .619 | .508 | .384 | | | | | | | |
| 1.800 | 1.722 | .078 | .078 | .79 | 1.42 | 8.3% | 4 | .653 | .512 | .353 | .205 | | | | | | |
| 1.800 | 1.775 | .025 | .025 | .91 | 1.38 | 2.8% | 5 | .646 | .502 | .341 | .195 | .091 | | | | | |
| 1.800 | 1.793 | .007 | .007 | .97 | 1.33 | .8% | 6 | .644 | .499 | .337 | .191 | .089 | .034 | | | | |
| 2.000 | 1.598 | .402 | .402 | .57 | 1.72 | 33.5% | 3 | .633 | .537 | .429 | | | | | | | |
| 2.000 | 1.855 | .145 | .145 | .73 | 1.56 | 13.5% | 4 | .651 | .534 | .401 | .269 | | | | | | |
| 2.000 | 1.960 | .040 | .040 | .88 | 1.43 | 3.9% | 5 | .671 | .540 | .387 | .239 | .123 | | | | | |
| 2.000 | 1.988 | .012 | .012 | .95 | 1.38 | 1.2% | 6 | .668 | .535 | .382 | .233 | .119 | .050 | | | | |
| 2.200 | 2.006 | .194 | .194 | .69 | 1.59 | 16.2% | 4 | .676 | .570 | .445 | .315 | | | | | | |
| 2.200 | 2.140 | .060 | .060 | .84 | 1.47 | 5.3% | 5 | .693 | .574 | .431 | .283 | .159 | | | | | |
| 2.200 | 2.180 | .020 | .020 | .93 | 1.42 | 1.8% | 6 | .689 | .568 | .423 | .276 | .153 | .071 | | | | |
| 2.200 | 2.194 | .006 | .006 | .97 | 1.37 | .6% | 7 | .688 | .566 | .421 | .273 | .150 | .069 | .027 | | | |
| 2.400 | 2.112 | .288 | .288 | .64 | 1.70 | 21.4% | 4 | .679 | .588 | .480 | .365 | | | | | | |
| 2.400 | 2.315 | .085 | .085 | .80 | 1.50 | 6.9% | 5 | .713 | .605 | .471 | .328 | .198 | | | | | |
| 2.400 | 2.369 | .031 | .031 | .90 | 1.46 | 2.6% | 6 | .709 | .598 | .462 | .317 | .188 | .095 | | | | |
| 2.400 | 2.390 | .010 | .010 | .96 | 1.41 | .8% | 7 | .707 | .595 | .458 | .313 | .185 | .093 | .039 | | | |
| 2.600 | 2.242 | .358 | .358 | .59 | 1.72 | 24.2% | 4 | .701 | .617 | .517 | .406 | | | | | | |
| 2.600 | 2.451 | .149 | .149 | .74 | 1.64 | 10.8% | 5 | .707 | .613 | .500 | .376 | .255 | | | | | |
| 2.600 | 2.554 | .046 | .046 | .88 | 1.50 | 3.4% | 6 | .726 | .624 | .498 | .358 | .226 | .123 | | | | |

*SIMILAR TO A.T.&T. JACOBSEN RETRIAL TABLES

*VERY FAST RETRIALS: JEWETT-SHRAGO VERY FAST RETRIAL TABLES - 50% RETRY, 50% OVERFLOW (RETRIES 1/3 HOLDING TIMES LATER)

| HOURS FIRST ATTEMPT VMR = 1.0 | HOURS CON-NECT | HOURS RECALL /RETRY | HOURS OVFL | CON-NECT VMR | OVFL VMR | PERCENT BLOCKED (GRADE) | LINE | HOURS CONNECTED TRAFFIC ON LINES 1 | 2 | 3 | 4 | 5 | 6 | 7 | 8 | 9 | 10 |
|---|---|---|---|---|---|---|---|---|---|---|---|---|---|---|---|---|---|
| 2.600 | 2.584 | .016 | .016 | .95 | 1.45 | 1.2% | 7 | .723 | .621 | .493 | .353 | .221 | .119 | .055 | | | |
| 2.800 | 2.327 | .473 | .473 | .55 | 1.81 | 28.9% | 4 | .706 | .632 | .544 | .445 | | | | | | |
| 2.800 | 2.608 | .192 | .192 | .71 | 1.67 | 12.9% | 5 | .725 | .639 | .534 | .415 | .295 | | | | | |
| 2.800 | 2.736 | .064 | .064 | .85 | 1.54 | 4.5% | 6 | .741 | .648 | .531 | .397 | .265 | .154 | | | | |
| 2.800 | 2.776 | .024 | .024 | .93 | 1.49 | 1.7% | 7 | .739 | .644 | .525 | .390 | .257 | .148 | .073 | | | |
| 2.800 | 2.792 | .008 | .008 | .97 | 1.44 | .6% | 8 | .737 | .642 | .523 | .387 | .255 | .146 | .072 | .030 | | |
| 3.000 | 2.405 | .595 | .595 | .52 | 1.90 | 33.1% | 4 | .712 | .646 | .568 | .479 | | | | | | |
| 3.000 | 2.756 | .244 | .244 | .67 | 1.69 | 15.0% | 5 | .741 | .663 | .565 | .453 | .334 | | | | | |
| 3.000 | 2.912 | .088 | .088 | .81 | 1.58 | 5.7% | 6 | .755 | .670 | .561 | .435 | .303 | .187 | | | | |
| 3.000 | 2.965 | .035 | .035 | .91 | 1.53 | 2.3% | 7 | .752 | .665 | .554 | .426 | .294 | .179 | .095 | | | |
| 3.000 | 2.987 | .013 | .013 | .96 | 1.40 | .8% | 8 | .751 | .663 | .551 | .422 | .290 | .176 | .092 | .042 | | |
| 3.200 | 2.860 | .340 | .340 | .62 | 1.79 | 19.2% | 5 | .740 | .670 | .585 | .486 | .379 | | | | | |
| 3.200 | 3.053 | .147 | .147 | .76 | 1.71 | 8.8% | 6 | .747 | .672 | .577 | .467 | .350 | .239 | | | | |
| 3.200 | 3.151 | .049 | .049 | .88 | 1.57 | 3.0% | 7 | .765 | .684 | .582 | .460 | .331 | .212 | .119 | | | |
| 3.200 | 3.182 | .018 | .018 | .94 | 1.52 | 1.1% | 8 | .763 | .682 | .578 | .455 | .325 | .207 | .115 | .056 | | |
| 3.400 | 2.991 | .409 | .409 | .58 | 1.81 | 21.5% | 5 | .755 | .690 | .611 | .518 | .416 | | | | | |
| 3.400 | 3.214 | .186 | .186 | .73 | 1.74 | 10.4% | 6 | .761 | .691 | .603 | .499 | .386 | .274 | | | | |
| 3.400 | 3.334 | .066 | .066 | .85 | 1.61 | 3.8% | 7 | .776 | .702 | .607 | .492 | .366 | .245 | .146 | | | |
| 3.400 | 3.374 | .026 | .026 | .93 | 1.56 | 1.5% | 8 | .774 | .699 | .602 | .486 | .360 | .239 | .141 | .073 | | |
| 3.400 | 3.390 | .010 | .010 | .97 | 1.50 | .6% | 9 | .773 | .698 | .600 | .484 | .357 | .237 | .139 | .071 | .032 | |
| 3.600 | 3.077 | .523 | .523 | .55 | 1.90 | 25.4% | 5 | .756 | .697 | .627 | .544 | .452 | | | | | |
| 3.600 | 3.369 | .231 | .231 | .69 | 1.76 | 12.1% | 6 | .770 | .709 | .627 | .530 | .421 | .309 | | | | |
| 3.600 | 3.512 | .088 | .088 | .83 | 1.65 | 4.8% | 7 | .787 | .718 | .630 | .522 | .401 | .280 | .175 | | | |
| 3.600 | 3.563 | .037 | .037 | .91 | 1.60 | 2.0% | 8 | .784 | .715 | .625 | .515 | .393 | .272 | .168 | .092 | | |
| 3.600 | 3.586 | .014 | .014 | .96 | 1.54 | .8% | 9 | .783 | .713 | .622 | .512 | .389 | .268 | .165 | .089 | .043 | |
| 3.800 | 3.156 | .644 | .644 | .52 | 1.98 | 29.0% | 5 | .758 | .705 | .641 | .567 | .484 | | | | | |
| 3.800 | 3.481 | .319 | .319 | .65 | 1.87 | 15.5% | 6 | .770 | .711 | .639 | .552 | .455 | .353 | | | | |
| 3.800 | 3.658 | .142 | .142 | .78 | 1.77 | 7.2% | 7 | .770 | .716 | .637 | .543 | .436 | .325 | .223 | | | |
| 3.800 | 3.750 | .050 | .050 | .89 | 1.63 | 2.6% | 8 | .794 | .729 | .645 | .542 | .425 | .305 | .197 | .113 | | |
| 3.800 | 3.780 | .020 | .020 | .94 | 1.57 | 1.1% | 9 | .793 | .727 | .642 | .539 | .421 | .300 | .193 | .110 | .056 | |
| 4.000 | 3.229 | .771 | .771 | .40 | 2.05 | 32.3% | 5 | .762 | .713 | .655 | .588 | .511 | | | | | |
| 4.000 | 3.620 | .380 | .380 | .61 | 1.88 | 17.4% | 6 | .782 | .727 | .660 | .579 | .486 | .387 | | | | |
| 4.000 | 3.823 | .177 | .177 | .75 | 1.80 | 8.5% | 7 | .789 | .731 | .657 | .568 | .466 | .358 | .254 | | | |
| 4.000 | 3.934 | .066 | .066 | .86 | 1.67 | 3.3% | 8 | .803 | .742 | .664 | .568 | .456 | .337 | .227 | .137 | | |
| 4.000 | 3.972 | .028 | .028 | .93 | 1.61 | 1.4% | 9 | .801 | .740 | .661 | .563 | .450 | .332 | .222 | .132 | .070 | |
| 4.000 | 3.989 | .011 | .011 | .97 | 1.55 | .5% | 10 | .800 | .739 | .660 | .561 | .448 | .329 | .219 | .131 | .069 | .033 |
| 4.200 | 3.715 | .485 | .485 | .58 | 1.97 | 20.7% | 6 | .781 | .731 | .670 | .597 | .514 | .423 | | | | |
| 4.200 | 3.983 | .217 | .217 | .71 | 1.82 | 9.8% | 7 | .798 | .744 | .676 | .593 | .496 | .390 | .286 | | | |
| 4.200 | 4.114 | .086 | .086 | .84 | 1.71 | 4.0% | 8 | .811 | .755 | .682 | .591 | .485 | .370 | .258 | .162 | | |
| 4.200 | 4.162 | .038 | .038 | .91 | 1.65 | 1.8% | 9 | .809 | .752 | .678 | .586 | .479 | .363 | .251 | .157 | .087 | |

*SIMILAR TO A.T.&T. JACOBSEN RETRIAL TABLES

*VERY FAST RETRIALS: JEWETT-SHRAGO VERY FAST RETRIAL TABLES - 50% RETRY, 50% OVERFLOW (RETRIES 1/3 HOLDING TIMES LATER)

| HOURS FIRST ATTEMPT VMR = 1.0 | HOURS CON-NECT | HOURS RECALL /RETRY | HOURS OVFL | CON-NECT VMR | OVFL VMR | PERCENT BLOCKED (GRADE) | LINE | HOURS CONNECTED TRAFFIC ON LINES 1 | 2 | 3 | 4 | 5 | 6 | 7 | 8 | 9 | 10 |
|---|---|---|---|---|---|---|---|---|---|---|---|---|---|---|---|---|---|
| 4.200 | 4.185 | .015 | .015 | .96 | 1.59 | .7% | 10 | .808 | .751 | .677 | .584 | .476 | .359 | .248 | .154 | .085 | .042 |
| 4.400 | 3.839 | .561 | .561 | .55 | 1.98 | 22.6% | 6 | .791 | .745 | .688 | .620 | .541 | .454 | | | | |
| 4.400 | 4.137 | .263 | .263 | .68 | 1.85 | 11.3% | 7 | .807 | .757 | .694 | .616 | .524 | .422 | .318 | | | |
| 4.400 | 4.265 | .135 | .135 | .79 | 1.83 | 6.0% | 8 | .803 | .751 | .685 | .604 | .509 | .406 | .301 | .206 | | |
| 4.400 | 4.350 | .050 | .050 | .89 | 1.69 | 2.2% | 9 | .817 | .763 | .694 | .608 | .505 | .393 | .281 | .182 | .106 | |
| 4.400 | 4.379 | .021 | .021 | .95 | 1.63 | 1.0% | 10 | .816 | .762 | .692 | .605 | .502 | .389 | .277 | .179 | .104 | .054 |
| 4.600 | 3.920 | .680 | .680 | .52 | 2.05 | 25.8% | 6 | .792 | .749 | .697 | .635 | .563 | .484 | | | | |
| 4.600 | 4.249 | .351 | .351 | .64 | 1.95 | 14.2% | 7 | .803 | .757 | .699 | .629 | .547 | .455 | .359 | | | |
| 4.600 | 4.434 | .166 | .166 | .77 | 1.86 | 7.0% | 8 | .811 | .762 | .701 | .625 | .535 | .435 | .331 | .234 | | |
| 4.600 | 4.535 | .065 | .065 | .87 | 1.72 | 2.8% | 9 | .824 | .774 | .709 | .628 | .531 | .422 | .311 | .209 | .127 | |
| 4.600 | 4.572 | .028 | .028 | .93 | 1.66 | 1.2% | 10 | .822 | .772 | .707 | .625 | .527 | .417 | .306 | .205 | .124 | .067 |
| 4.600 | 4.588 | .012 | .012 | .97 | 1.60 | .5% | 11 | .822 | .771 | .706 | .623 | .525 | .415 | .304 | .203 | .122 | .066 |
| | | | | | | | | .032 | | | | | | | | | |
| 4.800 | 3.995 | .805 | .805 | .49 | 2.12 | 28.7% | 6 | .793 | .753 | .705 | .649 | .584 | .510 | | | | |
| 4.800 | 4.389 | .411 | .411 | .61 | 1.96 | 15.8% | 7 | .812 | .769 | .715 | .649 | .571 | .483 | .389 | | | |
| 4.800 | 4.598 | .202 | .202 | .74 | 1.88 | 8.1% | 8 | .819 | .773 | .715 | .644 | .559 | .463 | .362 | .263 | | |
| 4.800 | 4.717 | .083 | .083 | .85 | 1.76 | 3.4% | 9 | .830 | .783 | .723 | .647 | .555 | .450 | .341 | .237 | .150 | |
| 4.800 | 4.763 | .037 | .037 | .92 | 1.70 | 1.5% | 10 | .829 | .781 | .720 | .643 | .550 | .445 | .335 | .231 | .145 | .082 |
| 4.800 | 4.784 | .016 | .016 | .96 | 1.64 | .7% | 11 | .828 | .781 | .719 | .641 | .548 | .442 | .332 | .229 | .143 | .081 |
| | | | | | | | | .041 | | | | | | | | | |
| 5.000 | 4.065 | .935 | .935 | .47 | 2.18 | 31.5% | 6 | .795 | .758 | .714 | .662 | .602 | .534 | | | | |
| 5.000 | 4.485 | .515 | .515 | .58 | 2.04 | 18.7% | 7 | .810 | .770 | .720 | .660 | .590 | .510 | .424 | | | |
| 5.000 | 4.757 | .243 | .243 | .71 | 1.91 | 9.3% | 8 | .826 | .783 | .729 | .662 | .582 | .490 | .391 | .293 | | |
| 5.000 | 4.896 | .104 | .104 | .83 | 1.79 | 4.1% | 9 | .836 | .792 | .736 | .665 | .578 | .478 | .370 | .266 | .175 | |
| 5.000 | 4.951 | .049 | .049 | .90 | 1.74 | 1.9% | 10 | .835 | .790 | .733 | .660 | .572 | .471 | .363 | .259 | .168 | .099 |
| 5.000 | 4.979 | .021 | .021 | .95 | 1.67 | .9% | 11 | .834 | .789 | .731 | .658 | .570 | .468 | .359 | .255 | .165 | .097 |
| | | | | | | | | .052 | | | | | | | | | |
| 5.500 | 4.751 | .749 | .749 | .51 | 2.12 | 24.0% | 7 | .821 | .787 | .746 | .696 | .637 | .569 | .494 | | | |
| 5.500 | 5.094 | .406 | .406 | .62 | 2.03 | 13.7% | 8 | .832 | .795 | .750 | .695 | .629 | .552 | .466 | .375 | | |
| 5.500 | 5.295 | .205 | .205 | .74 | 1.95 | 7.2% | 9 | .838 | .800 | .753 | .694 | .622 | .538 | .445 | .348 | .255 | |
| 5.500 | 5.412 | .088 | .088 | .85 | 1.82 | 3.1% | 10 | .848 | .810 | .761 | .699 | .622 | .532 | .432 | .328 | .231 | .149 |
| 5.500 | 5.458 | .042 | .042 | .92 | 1.76 | 1.5% | 11 | .847 | .808 | .759 | .696 | .618 | .527 | .426 | .322 | .225 | .145 |
| | | | | | | | | .085 | | | | | | | | | |
| 5.500 | 5.481 | .019 | .019 | .96 | 1.70 | .7% | 12 | .847 | .808 | .758 | .694 | .616 | .525 | .423 | .319 | .223 | .142 |
| | | | | | | | | .083 | .044 | | | | | | | | |
| 6.000 | 4.945 | 1.055 | 1.055 | .45 | 2.24 | 29.9% | 7 | .826 | .797 | .761 | .720 | .672 | .616 | .553 | | | |
| 6.000 | 5.390 | .610 | .610 | .55 | 2.12 | 18.5% | 8 | .839 | .808 | .770 | .723 | .668 | .603 | .529 | .449 | | |
| 6.000 | 5.655 | .345 | .345 | .67 | 2.08 | 10.9% | 9 | .843 | .810 | .769 | .719 | .659 | .589 | .509 | .422 | .334 | |
| 6.000 | 5.827 | .173 | .173 | .78 | 1.98 | 5.6% | 10 | .849 | .816 | .773 | .721 | .657 | .581 | .495 | .402 | .309 | .222 |
| 6.000 | 5.926 | .074 | .074 | .87 | 1.85 | 2.4% | 11 | .859 | .825 | .782 | .728 | .660 | .580 | .487 | .388 | .289 | .200 |

*SIMILAR TO A.T.&T. JACOBSEN RETRIAL TABLES

*VERY FAST RETRIALS: JEWETT-SHRAGO VERY FAST RETRIAL TABLES - 50% RETRY, 50% OVERFLOW (RETRIES 1/3 HOLDING TIMES LATER)

| HOURS FIRST ATTEMPT VMR = 1.0 | HOURS CON-NECT | HOURS RECALL /RETRY | HOURS OVFL | CON-NECT VMR | OVFL VMR | PERCENT BLOCKED (GRADE) | LINE | HOURS CONNECTED TRAFFIC ON LINES 1 | 2 | 3 | 4 | 5 | 6 | 7 | 8 | 9 | 10 |
|---|---|---|---|---|---|---|---|---|---|---|---|---|---|---|---|---|---|
| | | | | | | | | .127 | | | | | | | | | |
| 6.000 | 5.964 | .036 | .036 | .93 | 1.79 | 1.2% | 12 | .858 | .824 | .780 | .726 | .658 | .576 | .483 | .383 | .285 | .196 |
| | | | | | | | | .124 | .072 | | | | | | | | |
| 6.000 | 5.984 | .016 | .016 | .96 | 1.72 | .5% | 13 | .858 | .823 | .780 | .725 | .656 | .574 | .481 | .381 | .282 | .194 |
| | | | | | | | | .122 | .071 | .037 | | | | | | | |
| 6.500 | 5.610 | .890 | .890 | .49 | 2.25 | 24.1% | 8 | .841 | .813 | .781 | .742 | .696 | .642 | .582 | .514 | | |
| 6.500 | 5.976 | .524 | .524 | .59 | 2.18 | 14.9% | 9 | .848 | .820 | .785 | .742 | .692 | .632 | .564 | .488 | .407 | |
| 6.500 | 6.238 | .262 | .262 | .71 | 2.04 | 7.8% | 10 | .861 | .832 | .795 | .750 | .695 | .628 | .551 | .465 | .375 | .285 |
| 6.500 | 6.354 | .146 | .146 | .81 | 2.01 | 4.4% | 11 | .859 | .829 | .791 | .745 | .688 | .620 | .541 | .453 | .362 | .273 |
| | | | | | | | | .193 | | | | | | | | | |
| 6.500 | 6.437 | .063 | .063 | .89 | 1.87 | 1.9% | 12 | .868 | .838 | .800 | .752 | .693 | .621 | .537 | .444 | .347 | .254 |
| | | | | | | | | .173 | .109 | | | | | | | | |
| 6.500 | 6.470 | .030 | .030 | .94 | 1.81 | .9% | 13 | .867 | .837 | .799 | .751 | .691 | .619 | .534 | .441 | .343 | .251 |
| | | | | | | | | .170 | .106 | .061 | | | | | | | |
| 7.000 | 5.801 | 1.199 | 1.199 | .43 | 2.36 | 29.3% | 8 | .844 | .820 | .792 | .759 | .720 | .675 | .624 | .567 | | |
| 7.000 | 6.261 | .739 | .739 | .53 | 2.25 | 19.1% | 9 | .854 | .829 | .799 | .763 | .720 | .669 | .610 | .544 | .472 | |
| 7.000 | 6.585 | .415 | .415 | .64 | 2.16 | 11.2% | 10 | .864 | .838 | .806 | .767 | .720 | .664 | .599 | .524 | .443 | .359 |
| 7.000 | 6.778 | .222 | .222 | .75 | 2.08 | 6.1% | 11 | .869 | .843 | .810 | .769 | .720 | .660 | .590 | .510 | .423 | .335 |
| | | | | | | | | .250 | | | | | | | | | |
| 7.000 | 6.896 | .104 | .104 | .85 | 1.96 | 2.9% | 12 | .877 | .850 | .817 | .776 | .724 | .661 | .586 | .501 | .408 | .315 |
| | | | | | | | | .228 | .154 | | | | | | | | |
| 7.000 | 6.947 | .053 | .053 | .91 | 1.89 | 1.5% | 13 | .876 | .849 | .816 | .774 | .721 | .657 | .582 | .496 | .403 | .309 |
| | | | | | | | | .223 | .149 | .092 | | | | | | | |
| 7.000 | 6.974 | .026 | .026 | .95 | 1.82 | .7% | 14 | .875 | .849 | .815 | .772 | .720 | .656 | .579 | .493 | .400 | .306 |
| | | | | | | | | .220 | .147 | .091 | .052 | | | | | | |
| 7.500 | 5.937 | 1.563 | 1.563 | .39 | 2.50 | 34.5% | 8 | .844 | .823 | .798 | .770 | .737 | .699 | .656 | .608 | | |
| 7.500 | 6.476 | 1.024 | 1.024 | .47 | 2.37 | 24.0% | 9 | .856 | .833 | .807 | .776 | .739 | .696 | .647 | .592 | .530 | |
| 7.500 | 6.896 | .604 | .604 | .57 | 2.25 | 14.9% | 10 | .868 | .845 | .817 | .783 | .743 | .695 | .639 | .575 | .504 | .428 |
| 7.500 | 7.147 | .353 | .353 | .68 | 2.20 | 9.0% | 11 | .871 | .847 | .818 | .783 | .740 | .689 | .629 | .560 | .484 | .403 |
| | | | | | | | | .321 | | | | | | | | | |
| 7.500 | 7.313 | .187 | .187 | .78 | 2.11 | 4.9% | 12 | .876 | .852 | .823 | .787 | .742 | .688 | .624 | .551 | .469 | .383 |
| | | | | | | | | .298 | .219 | | | | | | | | |
| 7.500 | 7.413 | .087 | .087 | .87 | 1.98 | 2.3% | 13 | .884 | .860 | .830 | .793 | .747 | .691 | .624 | .546 | .459 | .368 |
| | | | | | | | | .279 | .198 | .132 | | | | | | | |
| 7.500 | 7.455 | .045 | .045 | .92 | 1.91 | 1.2% | 14 | .883 | .859 | .829 | .792 | .745 | .689 | .621 | .542 | .455 | .364 |
| | | | | | | | | .275 | .194 | .128 | .079 | | | | | | |
| 7.500 | 7.478 | .022 | .022 | .96 | 1.84 | .6% | 15 | .883 | .859 | .829 | .791 | .744 | .687 | .619 | .540 | .453 | .361 |
| | | | | | | | | .272 | .192 | .126 | .077 | .044 | | | | | |
| 8.000 | 6.664 | 1.336 | 1.336 | .42 | 2.47 | 28.6% | 9 | .858 | .838 | .815 | .788 | .756 | .720 | .678 | .631 | .578 | |
| 8.000 | 7.136 | .864 | .864 | .51 | 2.38 | 19.5% | 10 | .867 | .846 | .822 | .792 | .758 | .717 | .670 | .617 | .556 | .490 |

*SIMILAR TO A.T.&T. JACOBSEN RETRIAL TABLES

*VERY FAST RETRIALS: JEWETT-SHRAGO VERY FAST RETRIAL TABLES - 50% RETRY, 50% OVERFLOW (RETRIES 1/3 HOLDING TIMES LATER)

| HOURS FIRST ATTEMPT VMR = 1.0 | HOURS CON-NECT | HOURS RECALL /RETRY | HOURS OVFL | CON-NECT VMR | OVFL VMR | PERCENT BLOCKED (GRADE) | LINE | HOURS CONNECTED TRAFFIC ON LINES 1 | 2 | 3 | 4 | 5 | 6 | 7 | 8 | 9 | 10 |
|---|---|---|---|---|---|---|---|---|---|---|---|---|---|---|---|---|---|
| 8.000 | 7.481 | .519 | .519 | .61 | 2.30 | 12.2% | 11 | .874 | .853 | .827 | .796 | .759 | .715 | .664 | .604 | .537 | .464 |
| | | | | | | | | .387 | | | | | | | | | |
| 8.000 | 7.729 | .271 | .271 | .73 | 2.17 | 6.6% | 12 | .884 | .863 | .837 | .805 | .766 | .719 | .662 | .597 | .522 | .441 |
| | | | | | | | | .357 | .275 | | | | | | | | |
| 8.000 | 7.842 | .158 | .158 | .81 | 2.13 | 3.9% | 13 | .883 | .861 | .835 | .802 | .762 | .713 | .655 | .588 | .512 | .430 |
| | | | | | | | | .345 | .264 | .190 | | | | | | | |
| 8.000 | 7.927 | .073 | .073 | .89 | 1.99 | 1.8% | 14 | .890 | .868 | .842 | .809 | .768 | .718 | .657 | .587 | .506 | .419 |
| | | | | | | | | .331 | .246 | .172 | .113 | | | | | | |
| 8.000 | 7.962 | .038 | .038 | .93 | 1.92 | .9% | 15 | .889 | .868 | .841 | .808 | .767 | .716 | .655 | .584 | .503 | .416 |
| | | | | | | | | .327 | .243 | .169 | .110 | .067 | | | | | |
| 8.500 | 6.800 | 1.700 | 1.700 | .38 | 2.60 | 33.3% | 9 | .858 | .840 | .820 | .796 | .768 | .737 | .701 | .661 | .616 | |
| 8.500 | 7.347 | 1.153 | 1.153 | .46 | 2.48 | 23.9% | 10 | .868 | .849 | .827 | .802 | .772 | .737 | .697 | .651 | .600 | .543 |
| 8.500 | 7.782 | .718 | .718 | .55 | 2.38 | 15.6% | 11 | .877 | .858 | .836 | .809 | .776 | .738 | .694 | .642 | .583 | .519 |
| | | | | | | | | .449 | | | | | | | | | |
| 8.500 | 8.087 | .413 | .413 | .66 | 2.28 | 9.3% | 12 | .886 | .866 | .843 | .815 | .781 | .740 | .691 | .634 | .570 | .498 |
| | | | | | | | | .421 | .342 | | | | | | | | |
| 8.500 | 8.270 | .230 | .230 | .76 | 2.19 | 5.3% | 13 | .890 | .871 | .847 | .818 | .783 | .740 | .689 | .628 | .559 | .483 |
| | | | | | | | | .401 | .319 | .242 | | | | | | | |
| 8.500 | 8.366 | .134 | .134 | .83 | 2.15 | 3.1% | 14 | .889 | .869 | .845 | .815 | .779 | .736 | .683 | .622 | .552 | .474 |
| | | | | | | | | .392 | .310 | .233 | .166 | | | | | | |
| 8.500 | 8.438 | .062 | .062 | .90 | 2.01 | 1.4% | 15 | .895 | .876 | .852 | .822 | .786 | .741 | .687 | .623 | .549 | .467 |
| | | | | | | | | .381 | .296 | .217 | .149 | .096 | | | | | |
| 8.500 | 8.468 | .032 | .032 | .94 | 1.94 | .7% | 16 | .895 | .876 | .851 | .822 | .785 | .740 | .685 | .621 | .547 | .464 |
| | | | | | | | | .378 | .293 | .214 | .147 | .094 | .057 | | | | |
| 9.000 | 7.533 | 1.467 | 1.467 | .41 | 2.57 | 28.0% | 10 | .870 | .853 | .834 | .811 | .785 | .755 | .720 | .681 | .636 | .587 |
| 9.000 | 8.016 | .984 | .984 | .49 | 2.49 | 19.7% | 11 | .877 | .860 | .839 | .815 | .787 | .754 | .716 | .672 | .622 | .567 |
| | | | | | | | | .506 | | | | | | | | | |
| 9.000 | 8.412 | .588 | .588 | .59 | 2.37 | 12.3% | 12 | .888 | .870 | .849 | .824 | .794 | .759 | .717 | .668 | .611 | .548 |
| | | | | | | | | .478 | .406 | | | | | | | | |
| 9.000 | 8.646 | .354 | .354 | .69 | 2.31 | 7.6% | 13 | .891 | .873 | .852 | .826 | .795 | .757 | .713 | .660 | .600 | .533 |
| | | | | | | | | .459 | .382 | .306 | | | | | | | |
| 9.000 | 8.805 | .195 | .195 | .79 | 2.22 | 4.2% | 14 | .895 | .878 | .856 | .830 | .798 | .759 | .712 | .657 | .594 | .522 |
| | | | | | | | | .444 | .364 | .285 | .212 | | | | | | |
| 9.000 | 8.904 | .096 | .096 | .87 | 2.09 | 2.1% | 15 | .901 | .883 | .862 | .836 | .803 | .763 | .715 | .658 | .591 | .515 |
| | | | | | | | | .433 | .349 | .267 | .194 | .132 | | | | | |
| 9.000 | 8.948 | .052 | .052 | .92 | 2.02 | 1.2% | 16 | .901 | .883 | .861 | .834 | .802 | .762 | .713 | .655 | .588 | .512 |
| | | | | | | | | .429 | .345 | .263 | .190 | .129 | .082 | | | | |
| 9.000 | 8.973 | .027 | .027 | .95 | 1.95 | .6% | 17 | .900 | .883 | .861 | .834 | .801 | .761 | .712 | .654 | .586 | .509 |
| | | | | | | | | .427 | .342 | .261 | .188 | .127 | .081 | .048 | | | |
| 9.500 | 7.669 | 1.831 | 1.831 | .38 | 2.69 | 32.3% | 10 | .870 | .854 | .837 | .817 | .794 | .767 | .737 | .704 | .666 | .623 |

*SIMILAR TO A.T.&T. JACOBSEN RETRIAL TABLES

*VERY FAST RETRIALS: JEWETT-SHRAGO VERY FAST RETRIAL TABLES - 50% RETRY, 50% OVERFLOW (RETRIES 1/3 HOLDING TIMES LATER)

| HOURS FIRST ATTEMPT VMR = 1.0 | HOURS CON-NECT | HOURS RECALL /RETRY | HOURS OVFL | CON-NECT VMR | OVFL VMR | PERCENT BLOCKED (GRADE) | LINE | HOURS CONNECTED TRAFFIC ON LINES 1 | 2 | 3 | 4 | 5 | 6 | 7 | 8 | 9 | 10 |
|---|---|---|---|---|---|---|---|---|---|---|---|---|---|---|---|---|---|
| 9.500 | 8.224 | 1.276 | 1.276 | .45 | 2.59 | 23.7% | 11 | .878 | .862 | .844 | .822 | .798 | .769 | .736 | .698 | .655 | .607 |
| | | | | | | | | .554 | | | | | | | | | |
| 9.500 | 8.706 | .794 | .794 | .54 | 2.44 | 15.4% | 12 | .890 | .874 | .856 | .833 | .807 | .776 | .739 | .697 | .647 | .592 |
| | | | | | | | | .530 | .464 | | | | | | | | |
| 9.500 | 8.993 | .507 | .507 | .63 | 2.41 | 10.1% | 13 | .892 | .876 | .857 | .834 | .806 | .773 | .734 | .689 | .636 | .577 |
| | | | | | | | | .511 | .440 | .368 | | | | | | | |
| 9.500 | 9.226 | .274 | .274 | .74 | 2.28 | 5.6% | 14 | .901 | .885 | .866 | .842 | .814 | .779 | .738 | .689 | .632 | .567 |
| | | | | | | | | .494 | .417 | .338 | .263 | | | | | | |
| 9.500 | 9.335 | .165 | .165 | .81 | 2.24 | 3.4% | 15 | .900 | .884 | .864 | .840 | .811 | .776 | .734 | .683 | .625 | .559 |
| | | | | | | | | .485 | .407 | .328 | .253 | .185 | | | | | |
| 9.500 | 9.419 | .081 | .081 | .89 | 2.10 | 1.7% | 16 | .906 | .889 | .870 | .846 | .817 | .781 | .738 | .686 | .625 | .555 |
| | | | | | | | | .478 | .396 | .314 | .237 | .169 | .114 | | | | |
| 9.500 | 9.456 | .044 | .044 | .93 | 2.03 | .9% | 17 | .905 | .889 | .869 | .845 | .816 | .780 | .736 | .684 | .622 | .552 |
| | | | | | | | | .475 | .393 | .311 | .233 | .166 | .111 | .070 | | | |
| 10.000 | 8.408 | 1.592 | 1.592 | .40 | 2.66 | 27.5% | 11 | .880 | .865 | .849 | .830 | .808 | .782 | .753 | .721 | .683 | .641 |
| | | | | | | | | .595 | | | | | | | | | |
| 10.000 | 8.935 | 1.065 | 1.065 | .48 | 2.55 | 19.3% | 12 | .890 | .875 | .858 | .838 | .814 | .787 | .755 | .719 | .677 | .629 |
| | | | | | | | | .576 | .518 | | | | | | | | |
| 10.000 | 9.309 | .691 | .691 | .57 | 2.49 | 12.9% | 13 | .895 | .880 | .862 | .841 | .817 | .788 | .754 | .714 | .668 | .616 |
| | | | | | | | | .557 | .493 | .426 | | | | | | | |
| 10.000 | 9.594 | .406 | .406 | .68 | 2.39 | 7.8% | 14 | .902 | .887 | .869 | .848 | .822 | .792 | .756 | .713 | .663 | .606 |
| | | | | | | | | .542 | .472 | .398 | .325 | | | | | | |
| 10.000 | 9.767 | .233 | .233 | .77 | 2.30 | 4.6% | 15 | .905 | .891 | .873 | .851 | .825 | .794 | .756 | .711 | .659 | .598 |
| | | | | | | | | .531 | .457 | .380 | .304 | .232 | | | | | |
| 10.000 | 9.860 | .140 | .140 | .84 | 2.25 | 2.8% | 16 | .905 | .889 | .871 | .849 | .823 | .791 | .753 | .707 | .654 | .592 |
| | | | | | | | | 523 | .440 | .072 | .295 | .224 | .162 | | | | |
| 10.000 | 9.932 | .068 | .068 | .90 | 2.11 | 1.4% | 17 | .910 | .895 | .877 | .855 | .829 | .796 | .757 | .711 | .655 | .591 |
| | | | | | | | | .519 | .441 | .360 | .281 | .209 | .147 | .097 | | | |
| 10.000 | 9.963 | .037 | .037 | .94 | 2.04 | .7% | 18 | .909 | .895 | .877 | .855 | .828 | .795 | .756 | .709 | .653 | .589 |
| | | | | | | | | .517 | .438 | .357 | .279 | .206 | .145 | .096 | .060 | | |
| 11.000 | 8.667 | 2.333 | 2.333 | .34 | 2.87 | 35.0% | 11 | .880 | .868 | .854 | .838 | .820 | .801 | .778 | .753 | .724 | .693 |
| | | | | | | | | 658 | | | | | | | | | |
| 11.000 | 9.288 | 1.712 | 1.712 | .40 | 2.75 | 26.9% | 12 | .888 | .876 | .861 | .845 | .826 | .805 | .780 | .752 | .721 | .685 |
| | | | | | | | | .646 | .002 | | | | | | | | |
| 11.000 | 9.821 | 1.179 | 1.179 | .47 | 2.65 | 19.4% | 13 | .897 | .884 | .869 | .852 | .832 | .810 | .783 | .753 | .718 | .678 |
| | | | | | | | | .033 | .583 | .529 | | | | | | | |
| 11.000 | 10.243 | .757 | .757 | .56 | 2.55 | 12.9% | 14 | .904 | .892 | .877 | .859 | .839 | .815 | .787 | .754 | .715 | .671 |
| | | | | | | | | .621 | .565 | .504 | .439 | | | | | | |
| 11.000 | 10.541 | .459 | .459 | .66 | 2.46 | 8.0% | 15 | .910 | .898 | .883 | .865 | .844 | .819 | .790 | .754 | .713 | .665 |
| | | | | | | | | .611 | .550 | .483 | .413 | .341 | | | | | |

*SIMILAR TO A.T.&T. JACOBSEN RETRIAL TABLES

*VERY FAST RETRIALS: JEWETT-SHRAGO VERY FAST RETRIAL TABLES - 50% RETRY, 50% OVERFLOW (RETRIES 1/3 HOLDING TIMES LATER)

| HOURS FIRST ATTEMPT VMR = 1.0 | HOURS CON-NECT | HOURS RECALL /RETRY | HOURS OVFL | CON-NECT VMR | OVFL VMR | PERCENT BLOCKED (GRADE) | LINE | HOURS CONNECTED TRAFFIC ON LINES 1 | 2 | 3 | 4 | 5 | 6 | 7 | 8 | 9 | 10 |
|---|---|---|---|---|---|---|---|---|---|---|---|---|---|---|---|---|---|
| 11.000 | 10.728 | .272 | .272 | .75 | 2.38 | 4.8% | 16 | .914 | .901 | .886 | .868 | .847 | .821 | .790 | .754 | .711 | .661 |
| | | | | | | | | .603 | .538 | .468 | .394 | .320 | .250 | | | | |
| 11.000 | 10.832 | .168 | .168 | .82 | 2.33 | 3.0% | 17 | .913 | .900 | .885 | .867 | .845 | .819 | .787 | .750 | .706 | .655 |
| | | | | | | | | .596 | .531 | .460 | .386 | .312 | .242 | .179 | | | |
| 11.000 | 10.913 | .087 | .087 | .89 | 2.20 | 1.6% | 18 | .917 | .905 | .890 | .872 | .850 | .824 | .792 | .754 | .709 | .655 |
| | | | | | | | | .594 | .526 | .452 | .374 | .298 | .227 | .164 | .112 | | |
| 11.000 | 10.951 | .049 | .049 | .93 | 2.12 | .9% | 19 | .917 | .904 | .889 | .871 | .849 | .823 | .791 | .752 | .707 | .653 |
| | | | | | | | | .592 | .523 | .448 | .371 | .295 | .223 | .161 | .110 | .071 | |
| 12.000 | 9.548 | 2.452 | 2.452 | .34 | 2.94 | 33.9% | 12 | .888 | .877 | .865 | .851 | .836 | .819 | .799 | .777 | .753 | .725 |
| | | | | | | | | .695 | .662 | | | | | | | | |
| 12.000 | 10.173 | 1.827 | 1.827 | .39 | 2.84 | 26.4% | 13 | .895 | .884 | .872 | .858 | .841 | .823 | .802 | .778 | .751 | .721 |
| | | | | | | | | .687 | .650 | .608 | | | | | | | |
| 12.000 | 10.711 | 1.289 | 1.289 | .46 | 2.75 | 19.4% | 14 | .903 | .892 | .879 | .864 | .847 | .828 | .806 | .780 | .751 | .717 |
| | | | | | | | | .679 | .637 | .590 | .539 | | | | | | |
| 12.000 | 11.142 | .858 | .858 | .55 | 2.66 | 13.3% | 15 | .909 | .898 | .885 | .870 | .853 | .833 | .809 | .782 | .750 | .713 |
| | | | | | | | | .672 | .625 | .572 | .516 | .455 | | | | | |
| 12.000 | 11.456 | .544 | .544 | .64 | 2.58 | 8.7% | 16 | .914 | .903 | .890 | .875 | .857 | .836 | .812 | .783 | .749 | .710 |
| | | | | | | | | .665 | .614 | .557 | .495 | .430 | .363 | | | | |
| 12.000 | 11.660 | .340 | .340 | .72 | 2.51 | 5.5% | 17 | .917 | .906 | .893 | .877 | .859 | .838 | .813 | .783 | .747 | .706 |
| | | | | | | | | .659 | .605 | .546 | .481 | .412 | .343 | .275 | | | |
| 12.000 | 11.802 | .198 | .198 | .80 | 2.41 | 3.2% | 18 | .920 | .909 | .896 | .881 | .863 | .841 | .815 | .784 | .748 | .705 |
| | | | | | | | | .656 | .600 | .537 | .469 | .398 | .326 | .258 | .195 | | |
| 12.000 | 11.879 | .121 | .121 | .86 | 2.35 | 2.0% | 19 | .919 | .908 | .895 | .880 | .861 | .839 | .813 | .782 | .745 | .702 |
| | | | | | | | | .652 | .595 | .532 | .463 | .392 | .320 | .252 | .189 | .136 | |
| 12.000 | 11.938 | .062 | .062 | .92 | 2.21 | 1.0% | 20 | .923 | .913 | .900 | .885 | .866 | .844 | .818 | .787 | .749 | .705 |
| | | | | | | | | .653 | .594 | .529 | .457 | .383 | .309 | .239 | .177 | .124 | .083 |
| 12.000 | 11.966 | .034 | .034 | .95 | 2.13 | .6% | 21 | .923 | .913 | .900 | .884 | .866 | .844 | .817 | .786 | .748 | .704 |
| | | | | | | | | .652 | .593 | .527 | .455 | .381 | .307 | .237 | .175 | .123 | .082 |
| | | | | | | | | .052 | | | | | | | | | |
| 13.000 | 10.433 | 2.567 | 2.567 | .34 | 3.02 | 33.0% | 13 | .895 | .885 | .875 | .863 | .849 | .834 | .817 | .798 | .776 | .753 |
| | | | | | | | | .726 | .697 | .665 | | | | | | | |
| 13.000 | 11.061 | 1.939 | 1.939 | .39 | 2.92 | 26.0% | 14 | .902 | .892 | .881 | .868 | .854 | .838 | .820 | .800 | .777 | .751 |
| | | | | | | | | .722 | .689 | .653 | .613 | | | | | | |
| 13.000 | 11.637 | 1.363 | 1.363 | .45 | 2.79 | 19.0% | 15 | .911 | .901 | .890 | .877 | .863 | .846 | .827 | .805 | .780 | .752 |
| | | | | | | | | .719 | .682 | .642 | .596 | .547 | | | | | |
| 13.000 | 12.078 | .922 | .922 | .53 | 2.71 | 13.2% | 16 | .916 | .907 | .896 | .883 | .868 | .851 | .831 | .808 | .781 | .750 |
| | | | | | | | | .715 | .674 | .629 | .579 | .524 | .466 | | | | |
| 13.000 | 12.403 | .597 | .597 | .62 | 2.64 | 8.8% | 17 | .921 | .911 | .900 | .887 | .872 | .855 | .834 | .810 | .782 | .749 |
| | | | | | | | | .711 | .667 | .618 | .564 | .504 | .441 | .376 | | | |
| 13.000 | 12.619 | .381 | .381 | .71 | 2.58 | 5.7% | 18 | .923 | .914 | .902 | .889 | .874 | .856 | .835 | .810 | .780 | .746 |

*SIMILAR TO A.T.&T. JACOBSEN RETRIAL TABLES

*VERY FAST RETRIALS: JEWETT-SHRAGO VERY FAST RETRIAL TABLES - 50% RETRY, 50% OVERFLOW (RETRIES 1/3 HOLDING TIMES LATER)

| HOURS FIRST ATTEMPT VMR = 1.0 | HOURS CON-NECT | HOURS RECALL /RETRY | HOURS OVFL | CON-NECT VMR | OVFL VMR | PERCENT BLOCKED (GRADE) | LINE | HOURS CONNECTED TRAFFIC ON LINES 1 | 2 | 3 | 4 | 5 | 6 | 7 | 8 | 9 | 10 |
|---|---|---|---|---|---|---|---|---|---|---|---|---|---|---|---|---|---|
| | | | | | | | | .706 | .661 | .609 | .552 | .489 | .423 | .356 | .290 | | |
| 13.000 | 12.771 | .229 | .229 | .79 | 2.48 | 3.5% | 19 | .926 | .917 | .905 | .892 | .877 | .859 | .837 | .812 | .782 | .746 |
| | | | | | | | | .705 | .657 | .603 | .543 | .478 | .409 | .340 | .273 | .210 | |
| 13.000 | 12.856 | .144 | .144 | .85 | 2.43 | 2.2% | 20 | .925 | .916 | .905 | .891 | .876 | .857 | .836 | .810 | .779 | .743 |
| | | | | | | | | .701 | .653 | .598 | .537 | .472 | .403 | .333 | .266 | .204 | .150 |
| 13.000 | 12.924 | .076 | .076 | .91 | 2.29 | 1.2% | 21 | .929 | .920 | .909 | .896 | .880 | .862 | .840 | .814 | .783 | .746 |
| | | | | | | | | .703 | .653 | .597 | .534 | .465 | .394 | .322 | .254 | .191 | .138 |
| | | | | | | | | .095 | | | | | | | | | |
| 13.000 | 12.956 | .044 | .044 | .94 | 2.21 | .7% | 22 | .929 | .919 | .908 | .895 | .880 | .861 | .839 | .813 | .782 | .745 |
| | | | | | | | | .702 | .652 | .595 | .531 | .463 | .391 | .320 | .251 | .189 | .136 |
| | | | | | | | | .093 | .061 | | | | | | | | |
| 14.000 | 11.323 | 2.677 | 2.677 | .33 | 3.09 | 32.1% | 14 | .901 | .893 | .883 | .872 | .860 | .847 | .832 | .815 | .797 | .776 |
| | | | | | | | | .753 | .727 | .699 | .668 | | | | | | |
| 14.000 | 11.982 | 2.018 | 2.018 | .38 | 2.96 | 25.2% | 15 | .909 | .901 | .891 | .880 | .868 | .854 | .838 | .820 | .800 | .778 |
| | | | | | | | | .752 | .724 | .692 | .657 | .619 | | | | | |
| 14.000 | 12.533 | 1.467 | 1.467 | .45 | 2.88 | 19.0% | 16 | .915 | .906 | .897 | .886 | .873 | .858 | .842 | .823 | .802 | .777 |
| | | | | | | | | .750 | .719 | .684 | .645 | .602 | .555 | | | | |
| 14.000 | 12.981 | 1.019 | 1.019 | .52 | 2.81 | 13.6% | 17 | .920 | .912 | .902 | .891 | .878 | .863 | .846 | .826 | .803 | .777 |
| | | | | | | | | .747 | .713 | .675 | .632 | .585 | .533 | .478 | | | |
| 14.000 | 13.318 | .682 | .682 | .61 | 2.75 | 9.3% | 18 | .924 | .915 | .906 | .894 | .881 | .866 | .848 | .828 | .804 | .777 |
| | | | | | | | | .745 | .709 | .667 | .621 | .570 | .514 | .455 | .393 | | |
| 14.000 | 13.576 | .424 | .424 | .70 | 2.64 | 5.9% | 19 | .929 | .920 | .911 | .899 | .886 | .871 | .853 | .832 | .807 | .779 |
| | | | | | | | | .745 | .707 | .663 | .613 | .557 | .497 | .434 | .369 | .304 | |
| 14.000 | 13.739 | .261 | .261 | .78 | 2.55 | 3.7% | 20 | .931 | .923 | .913 | .902 | .889 | .873 | .855 | .834 | .809 | .779 |
| | | | | | | | | .744 | .704 | .658 | .606 | .548 | .486 | .420 | .352 | .286 | .225 |
| 14.000 | 13.833 | .167 | .167 | .84 | 2.50 | 2.1% | 21 | .931 | .922 | .913 | .901 | .888 | .872 | .854 | .832 | .806 | .776 |
| | | | | | | | | .741 | .700 | .654 | .601 | .543 | .479 | .413 | .345 | .279 | .218 |
| | | | | | | | | .164 | | | | | | | | | |
| 14.000 | 13.908 | .092 | .092 | .89 | 2.37 | 1.3% | 22 | .934 | .926 | .916 | .905 | .892 | .876 | .858 | .836 | .810 | .779 |
| | | | | | | | | .744 | .702 | .654 | .599 | .538 | .473 | .404 | .334 | .267 | .205 |
| | | | | | | | | .151 | .107 | | | | | | | | |
| 14.000 | 13.946 | .054 | .054 | .93 | 2.29 | .8% | 23 | .934 | .925 | .916 | .905 | .891 | .876 | .857 | .835 | .809 | .778 |
| | | | | | | | | .742 | .700 | .652 | .597 | .536 | .470 | .401 | .331 | .264 | .203 |
| | | | | | | | | .149 | .105 | .070 | | | | | | | |
| 15.000 | 12.216 | 2.784 | 2.784 | .33 | 3.16 | 31.3% | 15 | .907 | .899 | .890 | .881 | .870 | .858 | .846 | .830 | .814 | .795 |
| | | | | | | | | .775 | .753 | .728 | .700 | .670 | | | | | |
| 15.000 | 12.877 | 2.123 | 2.123 | .38 | 3.04 | 24.8% | 16 | .914 | .906 | .897 | .888 | .877 | .865 | .851 | .835 | .818 | .798 |
| | | | | | | | | .776 | .752 | .724 | .693 | .660 | .623 | | | | |
| 15.000 | 13.431 | 1.569 | 1.569 | .44 | 2.97 | 18.9% | 17 | .919 | .911 | .903 | .893 | .882 | .869 | .855 | .838 | .820 | .799 |
| | | | | | | | | .775 | .748 | .718 | .685 | .647 | .606 | .562 | | | |

*SIMILAR TO A.T.&T. JACOBSEN RETRIAL TABLES

*VERY FAST RETRIALS: JEWETT-SHRAGO VERY FAST RETRIAL TABLES - 50% RETRY, 50% OVERFLOW (RETRIES 1/3 HOLDING TIMES LATER)

| HOURS FIRST ATTEMPT VMR = 1.0 | HOURS CON-NECT | HOURS RECALL /RETRY | HOURS OVFL | CON-NECT VMR | OVFL VMR | PERCENT BLOCKED (GRADE) | LINE | HOURS CONNECTED TRAFFIC ON LINES 1 | 2 | 3 | 4 | 5 | 6 | 7 | 8 | 9 | 10 |
|---|---|---|---|---|---|---|---|---|---|---|---|---|---|---|---|---|---|
| 15.000 | 13.886 | 1.114 | 1.114 | .51 | 2.91 | 13.8% | 18 | .924 | .916 | .907 | .897 | .886 | .873 | .858 | .841 | .822 | .799 |
| | | | | | | | | .774 | .745 | .713 | .676 | .635 | .590 | .541 | .489 | | |
| 15.000 | 14.265 | .735 | .735 | .60 | 2.80 | 9.3% | 19 | .929 | .922 | .913 | .903 | .892 | .879 | .864 | .846 | .826 | .803 |
| | | | | | | | | .776 | .745 | .709 | .669 | .625 | .575 | .521 | .464 | .404 | |
| 15.000 | 14.505 | .495 | .495 | .68 | 2.76 | 6.4% | 20 | .931 | .923 | .915 | .905 | .893 | .880 | .864 | .846 | .826 | .801 |
| | | | | | | | | .773 | .741 | .704 | .662 | .615 | .563 | .507 | .447 | .385 | .323 |
| 15.000 | 14.683 | .317 | .317 | .75 | 2.68 | 4.1% | 21 | .933 | .925 | .917 | .907 | .895 | .882 | .866 | .848 | .827 | .802 |
| | | | | | | | | .773 | .739 | .701 | .657 | .608 | .554 | .495 | .433 | .369 | .306 |
| | | | | | | | | .246 | | | | | | | | | |
| 15.000 | 14.808 | .192 | .192 | .82 | 2.57 | 2.5% | 22 | .935 | .928 | .919 | .909 | .898 | .884 | .869 | .850 | .829 | .803 |
| | | | | | | | | .774 | .740 | .700 | .655 | .604 | .547 | .486 | .422 | .356 | .292 |
| | | | | | | | | .231 | .177 | | | | | | | | |
| 15.000 | 14.878 | .122 | .122 | .87 | 2.51 | 1.6% | 23 | .935 | .928 | .919 | .909 | .897 | .884 | .868 | .849 | .827 | .802 |
| | | | | | | | | .772 | .737 | .697 | .651 | .600 | .543 | .481 | .417 | .351 | .287 |
| | | | | | | | | .226 | .172 | .126 | | | | | | | |
| 15.000 | 14.934 | .066 | .066 | .92 | 2.37 | .9% | 24 | .938 | .931 | .922 | .912 | .901 | .887 | .872 | .853 | .831 | .805 |
| | | | | | | | | .775 | .740 | .699 | .652 | .599 | .540 | .476 | .410 | .342 | .276 |
| | | | | | | | | .215 | .161 | .116 | .080 | | | | | | |
| 15.000 | 14.961 | .039 | .039 | .95 | 2.28 | .5% | 25 | .938 | .930 | .922 | .912 | .901 | .887 | .871 | .853 | .831 | .805 |
| | | | | | | | | .774 | .739 | .698 | .651 | .597 | .538 | .474 | .408 | .340 | .274 |
| | | | | | | | | .213 | .160 | .115 | .079 | .052 | | | | | |
| 16.000 | 13.113 | 2.887 | 2.887 | .33 | 3.23 | 30.6% | 16 | .912 | .904 | .897 | .888 | .878 | .868 | .856 | .843 | .828 | .812 |
| | | | | | | | | .794 | .775 | .753 | .729 | .702 | .673 | | | | |
| 16.000 | 13.775 | 2.225 | 2.225 | .37 | 3.11 | 24.4% | 17 | .918 | .911 | .903 | .895 | .885 | .874 | .862 | .848 | .833 | .816 |
| | | | | | | | | .796 | .775 | .751 | .724 | .695 | .662 | .627 | | | |
| 16.000 | 14.332 | 1.668 | 1.668 | .43 | 3.05 | 18.9% | 18 | .923 | .916 | .908 | .899 | .889 | .878 | .865 | .851 | .835 | .817 |
| | | | | | | | | .796 | .773 | .747 | .718 | .686 | .650 | .611 | .568 | | |
| 16.000 | 14.826 | 1.174 | 1.174 | .50 | 2.95 | 13.7% | 19 | .929 | .922 | .914 | .905 | .895 | .884 | .871 | .857 | .840 | .821 |
| | | | | | | | | .799 | .774 | .746 | .714 | .678 | .638 | .595 | .547 | .496 | |
| 16.000 | 15.181 | .819 | .819 | .58 | 2.90 | 9.7% | 20 | .932 | .925 | .917 | .908 | .898 | .887 | .874 | .858 | .841 | .821 |
| | | | | | | | | .798 | .772 | .742 | .708 | .670 | .627 | .580 | .529 | .475 | .418 |
| 16.000 | 15.463 | .537 | .537 | .67 | 2.82 | 6.5% | 21 | .935 | .929 | .921 | .912 | .902 | .891 | .877 | .862 | .844 | .823 |
| | | | | | | | | .800 | .772 | .741 | .704 | .664 | .618 | .568 | .513 | .455 | .395 |
| | | | | | | | | .335 | | | | | | | | | |
| 16.000 | 15.649 | .351 | .351 | .74 | 2.74 | 4.3% | 22 | .937 | .930 | .923 | .914 | .904 | .892 | .879 | .863 | .845 | .824 |
| | | | | | | | | .799 | .771 | .738 | .701 | .659 | .611 | .558 | .501 | .441 | .379 |
| | | | | | | | | .318 | .258 | | | | | | | | |
| 16.000 | 15.783 | .217 | .217 | .81 | 2.64 | 2.7% | 23 | .939 | .933 | .925 | .917 | .906 | .895 | .881 | .866 | .847 | .826 |
| | | | | | | | | .801 | .772 | .738 | .699 | .655 | .606 | .551 | .492 | .430 | .366 |
| | | | | | | | | .303 | .243 | .189 | | | | | | | |

*SIMILAR TO A.T.&T. JACOBSEN RETRIAL TABLES

*VERY FAST RETRIALS: JEWETT-SHRAGO VERY FAST RETRIAL TABLES - 50% RETRY, 50% OVERFLOW (RETRIES 1/3 HOLDING TIMES LATER)

| HOURS FIRST ATTEMPT VMR = 1.0 | HOURS CON-NECT | HOURS RECALL /RETRY | HOURS OVFL | CON-NECT VMR | OVFL VMR | PERCENT BLOCKED (GRADE) | LINE | HOURS CONNECTED TRAFFIC ON LINES 1 | 2 | 3 | 4 | 5 | 6 | 7 | 8 | 9 | 10 |
|---|---|---|---|---|---|---|---|---|---|---|---|---|---|---|---|---|---|
| 16.000 | 15.859 | .141 | .141 | .86 | 2.58 | 1.7% | 24 | .939 | .932 | .925 | .916 | .906 | .894 | .880 | .865 | .846 | .824 |
| | | | | | | | | .799 | .770 | .735 | .696 | .652 | .602 | .547 | .487 | .425 | .361 |
| | | | | | | | | .298 | .238 | .184 | .137 | | | | | | |
| 16.000 | 15.922 | .078 | .078 | .91 | 2.44 | 1.0% | 25 | .941 | .935 | .928 | .919 | .909 | .897 | .884 | .868 | .849 | .828 |
| | | | | | | | | .802 | .772 | .738 | .698 | .652 | .601 | .544 | .482 | .418 | .352 |
| | | | | | | | | .288 | .227 | .173 | .127 | .089 | | | | | |
| 16.000 | 15.953 | .047 | .047 | .94 | 2.36 | .6% | 26 | .941 | .935 | .928 | .919 | .909 | .897 | .884 | .868 | .849 | .827 |
| | | | | | | | | .801 | .771 | .737 | .696 | .650 | .599 | .542 | .480 | .415 | .350 |
| | | | | | | | | .285 | .225 | .171 | .125 | .088 | .059 | | | | |
| 17.000 | 14.037 | 2.963 | 2.963 | .32 | 3.26 | 29.7% | 17 | .917 | .911 | .904 | .896 | .887 | .878 | .867 | .856 | .843 | .829 |
| | | | | | | | | .813 | .795 | .776 | .754 | .730 | .704 | .676 | | | |
| 17.000 | 14.675 | 2.325 | 2.325 | .37 | 3.19 | 24.1% | 18 | .922 | .915 | .908 | .901 | .892 | .882 | .871 | .859 | .846 | .830 |
| | | | | | | | | .814 | .795 | .774 | .750 | .724 | .696 | .664 | .630 | | |
| 17.000 | 15.235 | 1.765 | 1.765 | .43 | 3.13 | 18.8% | 19 | .926 | .920 | .913 | .905 | .896 | .886 | .875 | .862 | .848 | .832 |
| | | | | | | | | .814 | .794 | .772 | .746 | .718 | .687 | .652 | .615 | .574 | |
| 17.000 | 15.734 | 1.266 | 1.266 | .49 | 3.04 | 13.9% | 20 | .931 | .925 | .918 | .910 | .901 | .891 | .880 | .867 | .852 | .836 |
| | | | | | | | | .817 | .796 | .771 | .744 | .713 | .679 | .641 | .599 | .554 | .505 |
| 17.000 | 16.129 | .871 | .871 | .57 | 2.95 | 9.7% | 21 | .936 | .930 | .923 | .915 | .906 | .896 | .885 | .872 | .857 | .839 |
| | | | | | | | | .820 | .797 | .771 | .742 | .709 | .672 | .630 | .585 | .535 | .482 |
| | | | | | | | | .427 | | | | | | | | | |
| 17.000 | 16.420 | .580 | .580 | .65 | 2.87 | 6.6% | 22 | .939 | .933 | .927 | .919 | .910 | .900 | .888 | .875 | .860 | .842 |
| | | | | | | | | .821 | .798 | .771 | .740 | .705 | .665 | .621 | .572 | .519 | .463 |
| | | | | | | | | .405 | .346 | | | | | | | | |
| 17.000 | 16.615 | .385 | .385 | .73 | 2.80 | 4.4% | 23 | .941 | .935 | .928 | .920 | .912 | .901 | .890 | .876 | .861 | .843 |
| | | | | | | | | .822 | .797 | .770 | .738 | .701 | .660 | .614 | .563 | .508 | .449 |
| | | | | | | | | .389 | .320 | .270 | | | | | | | |
| 17.000 | 16.757 | .243 | .243 | .80 | 2.71 | 2.8% | 24 | .943 | .937 | .930 | .923 | .914 | .904 | .892 | .878 | .863 | .844 |
| | | | | | | | | .823 | .798 | .770 | .737 | .699 | .656 | .608 | .556 | .498 | .438 |
| | | | | | | | | .376 | .314 | .255 | .201 | | | | | | |
| 17.000 | 16.840 | .160 | .160 | .85 | 2.64 | 1.9% | 25 | .942 | .937 | .930 | .922 | .913 | .903 | .891 | .877 | .861 | .843 |
| | | | | | | | | .821 | .796 | .767 | .734 | .696 | .653 | .604 | .551 | .493 | .432 |
| | | | | | | | | .370 | .309 | .250 | .196 | .148 | | | | | |
| 17.000 | 16.908 | .092 | .092 | .90 | 2.51 | 1.1% | 26 | .945 | .939 | .032 | .925 | .916 | .906 | .894 | .881 | .865 | .846 |
| | | | | | | | | .824 | .799 | .770 | .736 | .697 | .652 | .602 | .547 | .488 | .425 |
| | | | | | | | | .361 | .290 | .239 | .185 | .138 | .099 | | | | |
| 17.000 | 16.944 | .056 | .056 | .93 | 2.43 | .7% | 27 | .945 | .939 | .932 | .925 | .916 | .906 | .894 | .880 | .864 | .845 |
| | | | | | | | | .824 | .798 | .769 | .734 | .695 | .651 | .600 | .545 | .486 | .423 |
| | | | | | | | | .359 | .296 | .236 | .182 | .136 | .097 | .067 | | | |
| 18.000 | 14.219 | 3.781 | 3.781 | .29 | 3.44 | 34.7% | 17 | .915 | .910 | .003 | .090 | .888 | .880 | .871 | .861 | .850 | .837 |
| | | | | | | | | .824 | .809 | .793 | .776 | .756 | .735 | .712 | | | |

*SIMILAR TO A.T.&T. JACOBSEN RETRIAL TABLES

# *VERY FAST RETRIALS: JEWETT-SHRAGO VERY FAST RETRIAL TABLES - 50% RETRY, 50% OVERFLOW (RETRIES 1/3 HOLDING TIMES LATER)

| HOURS FIRST ATTEMPT VMR = 1.0 | HOURS CONNECT | HOURS RECALL /RETRY | HOURS OVFL | CONNECT VMR | OVFL VMR | PERCENT BLOCKED (GRADE) | LINE | HOURS CONNECTED TRAFFIC ON LINES 1 | 2 | 3 | 4 | 5 | 6 | 7 | 8 | 9 | 10 |
|---|---|---|---|---|---|---|---|---|---|---|---|---|---|---|---|---|---|
| 18.000 | 14.939 | 3.061 | 3.061 | .32 | 3.32 | 29.1% | 18 | .921 | .915 | .909 | .902 | .894 | .885 | .876 | .865 | .854 | .841 |
| | | | | | | | | .827 | .811 | .794 | .775 | .754 | .731 | .706 | .678 | | |
| 18.000 | 15.578 | 2.422 | 2.422 | .37 | 3.26 | 23.7% | 19 | .925 | .919 | .913 | .906 | .898 | .889 | .880 | .869 | .857 | .843 |
| | | | | | | | | .828 | .812 | .793 | .773 | .750 | .725 | .697 | .667 | .633 | |
| 18.000 | 16.171 | 1.829 | 1.829 | .42 | 3.17 | 18.4% | 20 | .931 | .925 | .919 | .911 | .904 | .895 | .885 | .874 | .861 | .847 |
| | | | | | | | | .832 | .814 | .794 | .772 | .747 | .719 | .689 | .655 | .618 | .579 |
| 18.000 | 16.643 | 1.357 | 1.357 | .49 | 3.13 | 14.0% | 21 | .934 | .928 | .922 | .915 | .907 | .898 | .888 | .876 | .863 | .849 |
| | | | | | | | | .832 | .814 | .793 | .769 | .743 | .713 | .680 | .643 | .603 | .560 |
| | | | | | | | | .514 | | | | | | | | | |
| 18.000 | 17.047 | .953 | .953 | .56 | 3.05 | 10.1% | 22 | .938 | .932 | .926 | .919 | .911 | .902 | .892 | .880 | .867 | .852 |
| | | | | | | | | .835 | .816 | .793 | .768 | .740 | .708 | .672 | .632 | .589 | .542 |
| | | | | | | | | .491 | .439 | | | | | | | | |
| 18.000 | 17.347 | .653 | .653 | .64 | 2.97 | 7.0% | 23 | .941 | .935 | .929 | .922 | .914 | .905 | .895 | .883 | .870 | .854 |
| | | | | | | | | .837 | .817 | .794 | .767 | .737 | .703 | .665 | .623 | .577 | .526 |
| | | | | | | | | .473 | .417 | .361 | | | | | | | |
| 18.000 | 17.580 | .420 | .420 | .72 | 2.86 | 4.6% | 24 | .944 | .939 | .933 | .926 | .918 | .909 | .899 | .887 | .874 | .858 |
| | | | | | | | | .840 | .819 | .796 | .768 | .737 | .701 | .661 | .616 | .567 | .513 |
| | | | | | | | | .456 | .398 | .339 | .281 | | | | | | |
| 18.000 | 17.709 | .291 | .291 | .78 | 2.83 | 3.2% | 25 | .944 | .938 | .932 | .925 | .917 | .908 | .898 | .886 | .872 | .856 |
| | | | | | | | | .838 | .817 | .793 | .765 | .733 | .696 | .655 | .610 | .560 | .506 |
| | | | | | | | | .448 | .389 | .330 | .273 | .219 | | | | | |
| 18.000 | 17.819 | .181 | .181 | .84 | 2.71 | 2.0% | 26 | .946 | .940 | .934 | .927 | .920 | .911 | .900 | .888 | .874 | .859 |
| | | | | | | | | .840 | .819 | .794 | .765 | .733 | .695 | .653 | .606 | .554 | .498 |
| | | | | | | | | .439 | .379 | .318 | .260 | .206 | .158 | | | | |
| 18.000 | 17.882 | .118 | .118 | .89 | 2.64 | 1.3% | 27 | .945 | .940 | .934 | .927 | .919 | .910 | .900 | .888 | .874 | .858 |
| | | | | | | | | .839 | .817 | .792 | .764 | .731 | .693 | .650 | .603 | .551 | .494 |
| | | | | | | | | .435 | .374 | .314 | .256 | .203 | .155 | .115 | | | |
| 18.000 | 17.934 | .066 | .066 | .93 | 2.50 | .7% | 28 | .948 | .942 | .936 | .930 | .922 | .913 | .903 | .891 | .877 | .861 |
| | | | | | | | | .842 | .821 | .795 | .766 | .733 | .694 | .651 | .602 | .548 | .490 |
| | | | | | | | | .430 | .367 | .306 | .247 | .193 | .146 | .106 | .075 | | |
| 19.000 | 15.123 | 3.877 | 3.877 | .29 | 3.50 | 33.9% | 18 | .919 | .914 | .908 | .902 | .895 | .887 | .878 | .869 | .859 | .848 |
| | | | | | | | | .836 | .823 | .809 | .793 | .775 | .756 | .736 | .713 | | |
| 19.000 | 15.844 | 3.156 | 3.156 | .32 | 3.39 | 28.5% | 19 | .925 | .919 | .913 | .907 | .900 | .892 | .883 | .874 | .864 | .852 |
| | | | | | | | | .839 | .825 | .810 | .793 | .774 | .754 | .731 | .707 | .680 | |
| 19.000 | 16.483 | 2.517 | 2.517 | .36 | 3.33 | 23.4% | 20 | .928 | .923 | .917 | .911 | .904 | .896 | .887 | .877 | .866 | .854 |
| | | | | | | | | .841 | .826 | .810 | .792 | .772 | .749 | .725 | .698 | .669 | .637 |
| 19.000 | 17.078 | 1.922 | 1.922 | .42 | 3.24 | 18.4% | 21 | .933 | .928 | .922 | .916 | .909 | .901 | .892 | .882 | .871 | .858 |
| | | | | | | | | .844 | .829 | .812 | .792 | .770 | .746 | .719 | .690 | .657 | .622 |
| | | | | | | | | .583 | | | | | | | | | |
| 19.000 | 17.587 | 1.413 | 1.413 | .48 | 3.17 | 13.8% | 22 | .938 | .933 | .927 | .921 | .913 | .905 | .896 | .886 | .875 | .862 |

*SIMILAR TO A.T.&T. JACOBSEN RETRIAL TABLES

*VERY FAST RETRIALS: JEWETT-SHRAGO VERY FAST RETRIAL TABLES - 50% RETRY, 50% OVERFLOW (RETRIES 1/3 HOLDING TIMES LATER)

| HOURS FIRST ATTEMPT VMR = 1.0 | HOURS CON-NECT | HOURS RECALL /RETRY | HOURS OVFL | CON-NECT VMR | OVFL VMR | PERCENT BLOCKED (GRADE) | LINE | HOURS CONNECTED TRAFFIC ON LINES 1 | 2 | 3 | 4 | 5 | 6 | 7 | 8 | 9 | 10 |
|---|---|---|---|---|---|---|---|---|---|---|---|---|---|---|---|---|---|
| | | | | | | | | .848 | .831 | .813 | .792 | .769 | .743 | .714 | .682 | .646 | .607 |
| | | | | | | | | .565 | .519 | | | | | | | | |
| 19.000 | 17.996 | 1.004 | 1.004 | .55 | 3.09 | 10.0% | 23 | .941 | .936 | .931 | .925 | .917 | .909 | .900 | .890 | .879 | .865 |
| | | | | | | | | .851 | .834 | .814 | .793 | .768 | .740 | .709 | .674 | .635 | .593 |
| | | | | | | | | .547 | .498 | .446 | | | | | | | |
| 19.000 | 18.304 | .696 | .696 | .63 | 3.03 | 7.1% | 24 | .944 | .939 | .934 | .927 | .920 | .912 | .903 | .893 | .881 | .868 |
| | | | | | | | | .852 | .835 | .815 | .792 | .766 | .737 | .704 | .667 | .626 | .580 |
| | | | | | | | | .532 | .480 | .425 | .370 | | | | | | |
| 19.000 | 18.519 | .481 | .481 | .70 | 2.97 | 4.9% | 25 | .945 | .940 | .935 | .929 | .922 | .913 | .904 | .894 | .882 | .868 |
| | | | | | | | | .853 | .835 | .814 | .791 | .764 | .734 | .699 | .661 | .618 | .571 |
| | | | | | | | | .520 | .466 | .410 | .353 | .298 | | | | | |
| 19.000 | 18.681 | .319 | .319 | .77 | 2.89 | 3.3% | 26 | .947 | .942 | .936 | .930 | .923 | .915 | .906 | .896 | .884 | .870 |
| | | | | | | | | .854 | .836 | .815 | .791 | .763 | .732 | .696 | .656 | .612 | .563 |
| | | | | | | | | .511 | .455 | .397 | .339 | .283 | .230 | | | | |
| 19.000 | 18.798 | .202 | .202 | .83 | 2.78 | 2.1% | 27 | .948 | .944 | .938 | .932 | .925 | .917 | .908 | .898 | .886 | .872 |
| | | | | | | | | .856 | .837 | .816 | .792 | .764 | .731 | .695 | .654 | .608 | .558 |
| | | | | | | | | .503 | .446 | .387 | .328 | .270 | .217 | .169 | | | |
| 19.000 | 18.866 | .134 | .134 | .88 | 2.71 | 1.4% | 28 | .948 | .943 | .938 | .932 | .925 | .917 | .908 | .897 | .885 | .871 |
| | | | | | | | | .855 | .836 | .815 | .790 | .762 | .729 | .692 | .651 | .605 | .554 |
| | | | | | | | | .499 | .442 | .382 | .323 | .266 | .213 | .165 | .124 | | |
| 19.000 | 18.923 | .077 | .077 | .92 | 2.57 | .8% | 29 | .950 | .945 | .940 | .934 | .927 | .919 | .910 | .900 | .888 | .874 |
| | | | | | | | | .858 | .839 | .818 | .793 | .764 | .731 | .693 | .651 | .603 | .551 |
| | | | | | | | | .495 | .436 | .375 | .315 | .257 | .203 | .156 | .115 | .082 | |
| 19.000 | 18.952 | .048 | .048 | .94 | 2.48 | .5% | 30 | .950 | .945 | .940 | .934 | .927 | .919 | .910 | .900 | .888 | .874 |
| | | | | | | | | .857 | .839 | .817 | .792 | .763 | .730 | .692 | .649 | .602 | .550 |
| | | | | | | | | .493 | .434 | .373 | .313 | .255 | .201 | .154 | .114 | .081 | .056 |
| 20.000 | 16.029 | 3.971 | 3.971 | .29 | 3.55 | 33.1% | 19 | .923 | .918 | .912 | .906 | .900 | .893 | .885 | .877 | .868 | .858 |
| | | | | | | | | .847 | .835 | .822 | .808 | .792 | .775 | .757 | .736 | .715 | |
| 20.000 | 16.751 | 3.249 | 3.249 | .32 | 3.45 | 28.0% | 20 | .928 | .923 | .917 | .911 | .905 | .898 | .890 | .882 | .872 | .862 |
| | | | | | | | | .850 | .838 | .824 | .809 | .792 | .774 | .754 | .732 | .708 | .681 |
| 20.000 | 17.391 | 2.609 | 2.609 | .36 | 3.40 | 23.1% | 21 | .931 | .926 | .921 | .915 | .909 | .901 | .893 | .885 | .875 | .864 |
| | | | | | | | | .852 | .839 | .825 | .808 | .791 | .771 | .749 | .725 | .699 | .670 |
| | | | | | | | | .630 | | | | | | | | | |
| 20.000 | 17.987 | 2.013 | 2.013 | .41 | 3.32 | 18.3% | 22 | .936 | .931 | .926 | .920 | .913 | .906 | .898 | .889 | .879 | .868 |
| | | | | | | | | .856 | .842 | .827 | .809 | .790 | .769 | .745 | .719 | .691 | .659 |
| | | | | | | | | .625 | .588 | | | | | | | | |
| 20.000 | 18.499 | 1.501 | 1.501 | .47 | 3.25 | 14.0% | 23 | .940 | .935 | .930 | .924 | .918 | .910 | .902 | .893 | .883 | .872 |
| | | | | | | | | .859 | .845 | .828 | .810 | .790 | .767 | .742 | .714 | .682 | .648 |
| | | | | | | | | .610 | .570 | .526 | | | | | | | |
| 20.000 | 18.915 | 1.085 | 1.085 | .54 | 3.18 | 10.3% | 24 | .943 | .938 | .933 | .928 | .921 | .914 | .906 | .897 | .886 | .875 |

*SIMILAR TO A.T.&T. JACOBSEN RETRIAL TABLES

*VERY FAST RETRIALS: JEWETT-SHRAGO VERY FAST RETRIAL TABLES - 50% RETRY, 50% OVERFLOW (RETRIES 1/3 HOLDING TIMES LATER)

| HOURS FIRST ATTEMPT VMR=1.0 | HOURS CON-NECT | HOURS RECALL /RETRY | HOURS OVFL | CON-NECT VMR | OVFL VMR | PERCENT BLOCKED (GRADE) | LINE | HOURS CONNECTED TRAFFIC ON LINES 1 | 2 | 3 | 4 | 5 | 6 | 7 | 8 | 9 | 10 |
|---|---|---|---|---|---|---|---|---|---|---|---|---|---|---|---|---|---|
| | | | | | | | | .862 | .847 | .830 | .811 | .790 | .765 | .738 | .708 | .674 | .637 |
| | | | | | | | | .597 | .553 | .506 | .456 | | | | | | |
| 20.000 | 19.232 | .768 | .768 | .62 | 3.12 | 7.4% | 25 | .945 | .941 | .936 | .930 | .924 | .916 | .908 | .899 | .889 | .877 |
| | | | | | | | | .863 | .848 | .831 | .811 | .789 | .763 | .735 | .703 | .667 | .628 |
| | | | | | | | | .584 | .538 | .488 | .436 | .383 | | | | | |
| 20.000 | 19.484 | .516 | .516 | .69 | 3.02 | 5.0% | 26 | .948 | .944 | .939 | .933 | .927 | .920 | .911 | .902 | .892 | .880 |
| | | | | | | | | .866 | .851 | .833 | .813 | .789 | .763 | .733 | .699 | .662 | .620 |
| | | | | | | | | .574 | .525 | .472 | .418 | .362 | .307 | | | | |
| 20.000 | 19.653 | .347 | .347 | .76 | 2.94 | 3.4% | 27 | .950 | .945 | .940 | .935 | .928 | .921 | .913 | .904 | .893 | .881 |
| | | | | | | | | .867 | .852 | .833 | .813 | .789 | .762 | .731 | .696 | .657 | .614 |
| | | | | | | | | .566 | .515 | .461 | .405 | .348 | .293 | .240 | | | |
| 20.000 | 19.776 | .224 | .224 | .82 | 2.84 | 2.2% | 28 | .951 | .947 | .942 | .936 | .930 | .923 | .915 | .906 | .895 | .883 |
| | | | | | | | | .869 | .853 | .835 | .814 | .790 | .762 | .730 | .695 | .654 | .610 |
| | | | | | | | | .561 | .508 | .452 | .394 | .336 | .280 | .227 | .178 | | |
| 20.000 | 19.850 | .150 | .150 | .87 | 2.77 | 1.5% | 29 | .951 | .946 | .942 | .936 | .930 | .923 | .914 | .905 | .894 | .882 |
| | | | | | | | | .868 | .852 | .834 | .812 | .788 | .760 | .728 | .692 | .651 | .606 |
| | | | | | | | | .557 | .504 | .447 | .390 | .332 | .275 | .222 | .174 | .133 | |
| 20.000 | 19.911 | .089 | .089 | .91 | 2.63 | .9% | 30 | .953 | .948 | .943 | .938 | .932 | .925 | .917 | .908 | .897 | .885 |
| | | | | | | | | .871 | .855 | .836 | .815 | .790 | .762 | .729 | .692 | .651 | .605 |
| | | | | | | | | .554 | .499 | .442 | .383 | .323 | .266 | .213 | .165 | .124 | .090 |
| 20.000 | 19.944 | .056 | .056 | .94 | 2.55 | .6% | 31 | .953 | .948 | .943 | .938 | .932 | .925 | .917 | .907 | .897 | .885 |
| | | | | | | | | .871 | .854 | .836 | .814 | .789 | .761 | .728 | .691 | .649 | .603 |
| | | | | | | | | .552 | .497 | .440 | .380 | .321 | .264 | .211 | .163 | .122 | .089 |
| | | | | | | | | .062 | | | | | | | | | |
| 22.000 | 17.848 | 4.152 | 4.152 | .28 | 3.66 | 31.7% | 21 | .929 | .925 | .920 | .915 | .910 | .904 | .897 | .890 | .883 | .874 |
| | | | | | | | | .865 | .855 | .845 | .833 | .820 | .806 | .791 | .775 | .757 | .737 |
| | | | | | | | | .716 | | | | | | | | | |
| 22.000 | 18.571 | 3.429 | 3.429 | .32 | 3.57 | 27.0% | 22 | .933 | .929 | .924 | .919 | .914 | .908 | .901 | .894 | .887 | .878 |
| | | | | | | | | .869 | .858 | .847 | .835 | .821 | .807 | .791 | .773 | .753 | .732 |
| | | | | | | | | .709 | .684 | | | | | | | | |
| 22.000 | 19.239 | 2.761 | 2.761 | .36 | 3.49 | 22.3% | 23 | .937 | .933 | .929 | .924 | .918 | .912 | .906 | .899 | .891 | .882 |
| | | | | | | | | .872 | .862 | .850 | .837 | .823 | .807 | .790 | .771 | .750 | .727 |
| | | | | | | | | .702 | .674 | .645 | | | | | | | |
| 22.000 | 19.840 | 2.160 | 2.160 | .40 | 3.42 | 17.9% | 24 | .941 | .937 | .933 | .928 | .922 | .916 | .910 | .903 | .894 | .886 |
| | | | | | | | | .876 | .865 | .853 | .839 | .824 | .807 | .789 | .768 | .746 | .721 |
| | | | | | | | | .693 | .663 | .631 | .596 | | | | | | |
| 22.000 | 20.359 | 1.641 | 1.641 | .46 | 3.36 | 13.9% | 25 | .945 | .941 | .936 | .931 | .926 | .920 | .914 | .906 | .898 | .889 |
| | | | | | | | | .879 | .868 | .855 | .841 | .825 | .807 | .788 | .766 | .741 | .714 |
| | | | | | | | | .685 | .652 | .617 | .578 | .537 | | | | | |
| 22.000 | 20.786 | 1.214 | 1.214 | .53 | 3.30 | 10.5% | 26 | .948 | .944 | .939 | .934 | .929 | .923 | .917 | .909 | .901 | .892 |

*SIMILAR TO A.T.&T. JACOBSEN RETRIAL TABLES

*VERY FAST RETRIALS: JEWETT-SHRAGO VERY FAST RETRIAL TABLES - 50% RETRY, 50% OVERFLOW (RETRIES 1/3 HOLDING TIMES LATER)

| HOURS FIRST ATTEMPT VMR = 1.0 | HOURS CON-NECT | HOURS RECALL /RETRY | HOURS OVFL | CON-NECT VMR | OVFL VMR | PERCENT BLOCKED (GRADE) | LINE | HOURS CONNECTED TRAFFIC ON LINES 1 | 2 | 3 | 4 | 5 | 6 | 7 | 8 | 9 | 10 |
|---|---|---|---|---|---|---|---|---|---|---|---|---|---|---|---|---|---|
| | | | | | | | | .881 | .870 | .857 | .842 | .826 | .807 | .786 | .763 | .737 | .708 |
| | | | | | | | | .676 | .641 | .603 | .562 | .518 | .471 | | | | |
| 22.000 | 21.118 | .882 | .882 | .60 | 3.26 | 7.7% | 27 | .949 | .946 | .941 | .936 | .931 | .925 | .919 | .911 | .903 | .894 |
| | | | | | | | | .883 | .871 | .858 | .843 | .826 | .806 | .785 | .760 | .733 | .703 |
| | | | | | | | | .669 | .632 | .591 | .547 | .501 | .452 | .402 | | | |
| 22.000 | 21.386 | .614 | .614 | .67 | 3.17 | 5.4% | 28 | .952 | .948 | .944 | .939 | .934 | .928 | .921 | .914 | .906 | .896 |
| | | | | | | | | .885 | .873 | .860 | .844 | .827 | .807 | .784 | .759 | .731 | .699 |
| | | | | | | | | .663 | .624 | .581 | .535 | .486 | .435 | .382 | .330 | | |
| 22.000 | 21.595 | .405 | .405 | .74 | 3.05 | 3.6% | 29 | .954 | .951 | .946 | .942 | .937 | .931 | .924 | .917 | .909 | .900 |
| | | | | | | | | .889 | .877 | .863 | .847 | .829 | .809 | .786 | .760 | .730 | .696 |
| | | | | | | | | .659 | .618 | .573 | .524 | .472 | .419 | .364 | .310 | .258 | |
| 22.000 | 21.713 | .287 | .287 | .80 | 3.01 | 2.6% | 30 | .954 | .950 | .946 | .941 | .936 | .930 | .924 | .916 | .908 | .899 |
| | | | | | | | | .888 | .876 | .862 | .846 | .827 | .807 | .783 | .756 | .726 | .692 |
| | | | | | | | | .654 | .613 | .567 | .518 | .466 | .412 | .357 | .303 | .252 | .204 |
| 22.000 | 21.815 | .185 | .185 | .85 | 2.89 | 1.7% | 31 | .955 | .952 | .948 | .943 | .938 | .932 | .926 | .918 | .910 | .900 |
| | | | | | | | | .890 | .877 | .863 | .847 | .829 | .808 | .784 | .757 | .726 | .691 |
| | | | | | | | | .652 | .609 | .562 | .512 | .458 | .403 | .347 | .293 | .241 | .193 |
| | | | | | | | | .150 | | | | | | | | | |
| 22.000 | 21.875 | .125 | .125 | .89 | 2.82 | 1.1% | 32 | .955 | .952 | .947 | .943 | .938 | .932 | .925 | .918 | .910 | .900 |
| | | | | | | | | .889 | .877 | .863 | .846 | .828 | .807 | .783 | .755 | .724 | .689 |
| | | | | | | | | .650 | .606 | .559 | .508 | .455 | .399 | .344 | .289 | .237 | .189 |
| | | | | | | | | .147 | .111 | | | | | | | | |
| 22.000 | 21.926 | .074 | .074 | .93 | 2.68 | .7% | 33 | .957 | .953 | .949 | .945 | .940 | .934 | .927 | .920 | .912 | .902 |
| | | | | | | | | .892 | .879 | .865 | .849 | .830 | .809 | .785 | .757 | .725 | .690 |
| | | | | | | | | .650 | .605 | .557 | .505 | .450 | .393 | .337 | .281 | .229 | .181 |
| | | | | | | | | .139 | .104 | .075 | | | | | | | |
| 24.000 | 18.933 | 5.067 | 5.067 | .26 | 3.81 | 34.9% | 22 | .931 | .927 | .923 | .919 | .914 | .909 | .904 | .898 | .892 | .885 |
| | | | | | | | | .877 | .869 | .861 | .852 | .841 | .831 | .819 | .806 | .792 | .777 |
| | | | | | | | | .761 | .744 | | | | | | | | |
| 24.000 | 19.676 | 4.324 | 4.324 | .28 | 3.77 | 30.5% | 23 | .934 | .930 | .926 | .922 | .917 | .912 | .907 | .901 | .894 | .887 |
| | | | | | | | | .880 | .872 | .863 | .853 | .842 | .831 | .819 | .805 | .790 | .774 |
| | | | | | | | | .757 | .738 | .718 | | | | | | | |
| 24.000 | 20.398 | 3.602 | 3.602 | .31 | 3.69 | 26.1% | 24 | .938 | .934 | .930 | .926 | .921 | .916 | .911 | .905 | .898 | .891 |
| | | | | | | | | .883 | .875 | .866 | .856 | .844 | .832 | .819 | .805 | .789 | .772 |
| | | | | | | | | .753 | .733 | .711 | .687 | | | | | | |
| 24.000 | 21.067 | 2.933 | 2.933 | .35 | 3.62 | 21.8% | 25 | .942 | .938 | .934 | .930 | .925 | .920 | .915 | .908 | .902 | .895 |
| | | | | | | | | .887 | .878 | .868 | .858 | .846 | .834 | .820 | .804 | .788 | .769 |
| | | | | | | | | .749 | .727 | .703 | .677 | .649 | | | | | |
| 24.000 | 21.669 | 2.331 | 2.331 | .40 | 3.56 | 17.7% | 26 | .945 | .941 | .938 | .933 | .929 | .924 | .918 | .912 | .905 | .898 |
| | | | | | | | | .890 | .881 | .871 | .860 | .848 | .835 | .820 | .804 | .786 | .766 |

*SIMILAR TO A.T.&T. JACOBSEN RETRIAL TABLES

*VERY FAST RETRIALS: JEWETT-SHRAGO VERY FAST RETRIAL TABLES - 50% RETRY, 50% OVERFLOW (RETRIES 1/3 HOLDING TIMES LATER)

| HOURS FIRST ATTEMPT VMR=1.0 | HOURS CON-NECT | HOURS RECALL /RETRY | HOURS OVFL | CON-NECT VMR | OVFL VMR | PERCENT BLOCKED (GRADE) | LINE | HOURS CONNECTED TRAFFIC ON LINES 1 | 2 | 3 | 4 | 5 | 6 | 7 | 8 | 9 | 10 |
|---|---|---|---|---|---|---|---|---|---|---|---|---|---|---|---|---|---|
| | | | | | | | | .744 | .721 | .695 | .666 | .636 | .603 | | | | |
| 24.000 | 22.195 | 1.805 | 1.805 | .45 | 3.50 | 14.0% | 27 | .948 | .944 | .941 | .936 | .932 | .927 | .921 | .915 | .908 | .901 |
| | | | | | | | | .893 | .884 | .873 | .862 | .850 | .836 | .820 | .803 | .784 | .763 |
| | | | | | | | | .740 | .714 | .686 | .655 | .622 | .586 | .548 | | | |
| 24.000 | 22.660 | 1.340 | 1.340 | .52 | 3.42 | 10.6% | 28 | .951 | .948 | .944 | .940 | .936 | .931 | .925 | .919 | .912 | .905 |
| | | | | | | | | .897 | .887 | .877 | .865 | .852 | .838 | .822 | .804 | .784 | .761 |
| | | | | | | | | .736 | .709 | .678 | .645 | .609 | .570 | .528 | .485 | | |
| 24.000 | 23.033 | .967 | .967 | .58 | 3.34 | 7.7% | 29 | .954 | .951 | .947 | .943 | .939 | .934 | .928 | .922 | .916 | .908 |
| | | | | | | | | .900 | .890 | .880 | .868 | .855 | .840 | .823 | .804 | .783 | .760 |
| | | | | | | | | .733 | .704 | .671 | .636 | .597 | .555 | .510 | .464 | .415 | |
| 24.000 | 23.314 | .686 | .686 | .65 | 3.27 | 5.6% | 30 | .956 | .953 | .949 | .945 | .941 | .936 | .931 | .925 | .918 | .911 |
| | | | | | | | | .902 | .893 | .882 | .870 | .856 | .841 | .824 | .804 | .782 | .758 |
| | | | | | | | | .730 | .699 | .665 | .628 | .587 | .542 | .496 | .447 | .396 | .346 |
| 24.000 | 23.512 | .488 | .488 | .72 | 3.21 | 4.0% | 31 | .957 | .954 | .950 | .946 | .942 | .937 | .932 | .926 | .919 | .911 |
| | | | | | | | | .903 | .893 | .883 | .870 | .857 | .841 | .823 | .803 | .781 | .755 |
| | | | | | | | | .727 | .695 | .660 | .621 | .578 | .533 | .485 | .434 | .383 | .331 |
| | | | | | | | | .281 | | | | | | | | | |
| 24.000 | 23.665 | .335 | .335 | .78 | 3.12 | 2.8% | 32 | .958 | .955 | .951 | .947 | .943 | .938 | .933 | .927 | .920 | .913 |
| | | | | | | | | .904 | .895 | .884 | .871 | .857 | .842 | .824 | .803 | .780 | .754 |
| | | | | | | | | .725 | .692 | .656 | .616 | .572 | .525 | .475 | .424 | .371 | .319 |
| | | | | | | | | .268 | .221 | | | | | | | | |
| 24.000 | 23.778 | .222 | .222 | .83 | 3.01 | 1.8% | 33 | .959 | .956 | .952 | .949 | .944 | .940 | .934 | .928 | .922 | .914 |
| | | | | | | | | .906 | .896 | .885 | .873 | .859 | .843 | .825 | .804 | .781 | .754 |
| | | | | | | | | .724 | .691 | .653 | .612 | .567 | .519 | .468 | .415 | .362 | .308 |
| | | | | | | | | .257 | .210 | .167 | | | | | | | |
| 24.000 | 23.847 | .153 | .153 | .87 | 2.94 | 1.3% | 34 | .959 | .956 | .952 | .948 | .944 | .939 | .934 | .928 | .921 | .914 |
| | | | | | | | | .905 | .896 | .885 | .872 | .858 | .842 | .823 | .803 | .779 | .752 |
| | | | | | | | | .722 | .688 | .650 | .609 | .564 | .515 | .464 | .411 | .357 | .304 |
| | | | | | | | | .253 | .206 | .163 | .126 | | | | | | |
| 24.000 | 23.897 | .103 | .103 | .91 | 2.86 | .9% | 35 | .959 | .956 | .952 | .948 | .944 | .939 | .934 | .928 | .921 | .914 |
| | | | | | | | | .905 | .895 | .884 | .872 | .857 | .841 | .823 | .802 | .778 | .751 |
| | | | | | | | | .720 | .686 | .649 | .607 | .561 | .513 | .461 | .408 | .354 | .301 |
| | | | | | | | | .251 | .203 | .161 | .124 | .093 | | | | | |
| 24.000 | 23.938 | .062 | .062 | .94 | 2.72 | .5% | 36 | .960 | .957 | .954 | .950 | .946 | .941 | .936 | .930 | .923 | .916 |
| | | | | | | | | .907 | .898 | .887 | .874 | .860 | .844 | .825 | .804 | .780 | .753 |
| | | | | | | | | .722 | .687 | .649 | .606 | .560 | .510 | .457 | .403 | .348 | .295 |
| | | | | | | | | .243 | .196 | .154 | .117 | .087 | .062 | | | | |
| 26.000 | 20.768 | 5.232 | 5.232 | .26 | 3.94 | 33.5% | 24 | .936 | .932 | .929 | .925 | .921 | .917 | .912 | .907 | .902 | .896 |
| | | | | | | | | .889 | .883 | .875 | .867 | .859 | .850 | .840 | .829 | .818 | .805 |
| | | | | | | | | .792 | .777 | .761 | .745 | | | | | | |

*SIMILAR TO A.T.&T. JACOBSEN RETRIAL TABLES

*VERY FAST RETRIALS: JEWETT-SHRAGO VERY FAST RETRIAL TABLES - 50% RETRY, 50% OVERFLOW (RETRIES 1/3 HOLDING TIMES LATER)

| HOURS FIRST ATTEMPT VMR = 1.0 | HOURS CON-NECT | HOURS RECALL /RETRY | HOURS OVFL | CON-NECT VMR | OVFL VMR | PERCENT BLOCKED (GRADE) | LINE | HOURS CONNECTED TRAFFIC ON LINES 1 | 2 | 3 | 4 | 5 | 6 | 7 | 8 | 9 | 10 |
|---|---|---|---|---|---|---|---|---|---|---|---|---|---|---|---|---|---|
| 26.000 | 21.510 | 4.490 | 4.490 | .28 | 3.88 | 29.4% | 25 | .939 | .935 | .932 | .928 | .924 | .920 | .915 | .910 | .904 | .898 |
| | | | | | | | | .892 | .885 | .877 | .869 | .860 | .851 | .840 | .829 | .817 | .804 |
| | | | | | | | | .789 | .774 | .757 | .739 | .719 | | | | | |
| 26.000 | 22.231 | 3.769 | 3.769 | .31 | 3.80 | 25.3% | 26 | .942 | .939 | .935 | .932 | .928 | .923 | .918 | .913 | .908 | .902 |
| | | | | | | | | .895 | .888 | .880 | .872 | .863 | .853 | .842 | .830 | .817 | .803 |
| | | | | | | | | .788 | .771 | .753 | .733 | .712 | .689 | | | | |
| 26.000 | 22.900 | 3.100 | 3.100 | .35 | 3.74 | 21.3% | 27 | .945 | .942 | .939 | .935 | .931 | .927 | .922 | .917 | .911 | .905 |
| | | | | | | | | .898 | .891 | .883 | .874 | .865 | .855 | .843 | .831 | .817 | .802 |
| | | | | | | | | .786 | .768 | .749 | .727 | .704 | .680 | .653 | | | |
| 26.000 | 23.506 | 2.494 | 2.494 | .39 | 3.68 | 17.5% | 28 | .948 | .945 | .942 | .938 | .934 | .930 | .925 | .920 | .914 | .908 |
| | | | | | | | | .901 | .894 | .886 | .877 | .867 | .856 | .844 | .831 | .817 | .801 |
| | | | | | | | | .784 | .764 | .744 | .721 | .696 | .669 | .640 | .609 | | |
| 26.000 | 24.063 | 1.937 | 1.937 | .44 | 3.61 | 13.9% | 29 | .952 | .949 | .945 | .942 | .938 | .933 | .929 | .924 | .918 | .912 |
| | | | | | | | | .905 | .898 | .889 | .880 | .870 | .859 | .847 | .833 | .818 | .801 |
| | | | | | | | | .783 | .762 | .740 | .715 | .688 | .659 | .627 | .593 | .556 | |
| 26.000 | 24.537 | 1.463 | 1.463 | .50 | 3.53 | 10.7% | 30 | .954 | .952 | .948 | .945 | .941 | .937 | .932 | .927 | .922 | .915 |
| | | | | | | | | .909 | .901 | .893 | .883 | .873 | .861 | .849 | .834 | .819 | .801 |
| | | | | | | | | .781 | .760 | .736 | .709 | .680 | .649 | .614 | .577 | .538 | .496 |
| 26.000 | 24.921 | 1.079 | 1.079 | .57 | 3.46 | 8.0% | 31 | .957 | .954 | .951 | .947 | .944 | .940 | .935 | .930 | .924 | .918 |
| | | | | | | | | .911 | .904 | .895 | .886 | .875 | .864 | .850 | .836 | .819 | .801 |
| | | | | | | | | .780 | .757 | .732 | .704 | .673 | .639 | .602 | .563 | .521 | .476 |
| | | | | | | | | .430 | | | | | | | | | |
| 26.000 | 25.215 | .785 | .785 | .64 | 3.40 | 5.9% | 32 | .958 | .956 | .953 | .949 | .946 | .941 | .937 | .932 | .926 | .920 |
| | | | | | | | | .913 | .906 | .897 | .888 | .877 | .865 | .851 | .836 | .819 | .800 |
| | | | | | | | | .779 | .755 | .728 | .699 | .666 | .631 | .592 | .550 | .506 | .460 |
| | | | | | | | | .412 | .364 | | | | | | | | |
| 26.000 | 25.428 | .572 | .572 | .70 | 3.35 | 4.0% | 33 | .959 | .956 | .953 | .950 | .946 | .942 | .938 | .933 | .927 | .921 |
| | | | | | | | | .914 | .906 | .898 | .888 | .877 | .865 | .851 | .836 | .818 | .799 |
| | | | | | | | | .777 | .752 | .725 | .694 | .661 | .624 | .584 | .541 | .495 | .448 |
| | | | | | | | | .399 | .350 | .301 | | | | | | | |
| 26.000 | 25.615 | .385 | .385 | .76 | 3.22 | 2.9% | 34 | .961 | .958 | .956 | .952 | .949 | .945 | .940 | .935 | .930 | .924 |
| | | | | | | | | .917 | .909 | .901 | .891 | .880 | .868 | .854 | .838 | .820 | .800 |
| | | | | | | | | .778 | .752 | .724 | .692 | .657 | .619 | .577 | .532 | .484 | .435 |
| | | | | | | | | .384 | .333 | .284 | .237 | | | | | | |
| 26.000 | 25.722 | .278 | .278 | .81 | 3.17 | 2.1% | 35 | .961 | .958 | .955 | .952 | .948 | .944 | .940 | .935 | .929 | .923 |
| | | | | | | | | .916 | .908 | .900 | .890 | .879 | .866 | .852 | .836 | .818 | .798 |
| | | | | | | | | .775 | .749 | .721 | .689 | .653 | .614 | .572 | .527 | .479 | .429 |
| | | | | | | | | .378 | .327 | .278 | .231 | .188 | | | | | |
| 26.000 | 25.816 | .184 | .184 | .86 | 3.05 | 1.4% | 36 | .962 | .959 | .956 | .953 | .950 | .940 | .941 | .936 | .931 | .925 |
| | | | | | | | | .918 | .910 | .901 | .892 | .880 | .868 | .854 | .838 | .820 | .799 |

*SIMILAR TO A.T.&T. JACOBSEN RETRIAL TABLES

*VERY FAST RETRIALS: JEWETT-SHRAGO VERY FAST RETRIAL TABLES - 50% RETRY, 50% OVERFLOW (RETRIES 1/3 HOLDING TIMES LATER)

| HOURS FIRST ATTEMPT VMR=1.0 | HOURS CON-NECT | HOURS RECALL /RETRY | HOURS OVFL | CON-NECT VMR | OVFL VMR | PERCENT BLOCKED (GRADE) | LINE | HOURS CONNECTED TRAFFIC ON LINES 1 | 2 | 3 | 4 | 5 | 6 | 7 | 8 | 9 | 10 |
|---|---|---|---|---|---|---|---|---|---|---|---|---|---|---|---|---|---|
| | | | | | | | | .776 | .750 | .720 | .688 | .651 | .611 | .568 | .522 | .473 | .422 |
| | | | | | | | | .370 | .318 | .268 | .221 | .178 | .140 | | | | |
| 26.000 | 25.873 | .127 | .127 | .89 | 2.97 | 1.0% | 37 | .962 | .959 | .956 | .953 | .949 | .945 | .941 | .936 | .931 | .924 |
| | | | | | | | | .917 | .910 | .901 | .891 | .880 | .867 | .853 | .837 | .819 | .798 |
| | | | | | | | | .774 | .748 | .719 | .686 | .649 | .609 | .565 | .519 | .470 | .418 |
| | | | | | | | | .367 | .315 | .265 | .219 | .176 | .138 | .105 | | | |
| 26.000 | 25.922 | .078 | .078 | .93 | 2.84 | .6% | 38 | .963 | .960 | .958 | .954 | .951 | .947 | .943 | .938 | .932 | .926 |
| | | | | | | | | .919 | .912 | .903 | .893 | .882 | .869 | .855 | .839 | .821 | .800 |
| | | | | | | | | .776 | .749 | .720 | .686 | .649 | .608 | .564 | .516 | .466 | .414 |
| | | | | | | | | .361 | .309 | .258 | .211 | .168 | .131 | .099 | .073 | | |
| 28.000 | 22.609 | 5.391 | 5.391 | .25 | 4.03 | 32.3% | 26 | .940 | .937 | .934 | .931 | .927 | .923 | .919 | .915 | .910 | .905 |
| | | | | | | | | .900 | .894 | .887 | .881 | .873 | .866 | .857 | .848 | .838 | .828 |
| | | | | | | | | .816 | .804 | .791 | .777 | .762 | .745 | | | | |
| 28.000 | 23.352 | 4.648 | 4.648 | .28 | 3.98 | 28.5% | 27 | .943 | .940 | .937 | .933 | .930 | .926 | .922 | .917 | .913 | .907 |
| | | | | | | | | .902 | .896 | .890 | .883 | .875 | .867 | .858 | .849 | .839 | .827 |
| | | | | | | | | .815 | .803 | .789 | .773 | .757 | .740 | .721 | | | |
| 28.000 | 24.071 | 3.929 | 3.929 | .31 | 3.91 | 24.6% | 28 | .946 | .943 | .940 | .936 | .933 | .929 | .925 | .921 | .916 | .911 |
| | | | | | | | | .905 | .899 | .892 | .885 | .878 | .869 | .860 | .850 | .840 | .828 |
| | | | | | | | | .815 | .801 | .787 | .770 | .753 | .734 | .713 | .691 | | |
| 28.000 | 24.741 | 3.259 | 3.259 | .34 | 3.85 | 20.9% | 29 | .948 | .946 | .943 | .939 | .936 | .932 | .928 | .924 | .919 | .914 |
| | | | | | | | | .908 | .902 | .895 | .888 | .880 | .871 | .862 | .852 | .840 | .828 |
| | | | | | | | | .815 | .800 | .784 | .767 | .748 | .728 | .706 | .682 | .656 | |
| 28.000 | 25.373 | 2.627 | 2.627 | .39 | 3.77 | 17.2% | 30 | .952 | .949 | .946 | .943 | .939 | .936 | .932 | .927 | .922 | .917 |
| | | | | | | | | .912 | .905 | .899 | .891 | .883 | .874 | .864 | .854 | .842 | .829 |
| | | | | | | | | .815 | .800 | .783 | .764 | .744 | .722 | .698 | .672 | .644 | .614 |
| 28.000 | 25.908 | 2.092 | 2.092 | .44 | 3.74 | 13.9% | 31 | .954 | .951 | .948 | .945 | .942 | .938 | .934 | .930 | .925 | .920 |
| | | | | | | | | .914 | .908 | .901 | .893 | .885 | .876 | .866 | .855 | .842 | .829 |
| | | | | | | | | .814 | .798 | .780 | .760 | .739 | .715 | .689 | .661 | .631 | .599 |
| | | | | | | | | .565 | | | | | | | | | |
| 28.000 | 26.388 | 1.612 | 1.612 | .49 | 3.68 | 10.9% | 32 | .956 | .954 | .951 | .948 | .945 | .941 | .937 | .933 | .928 | .923 |
| | | | | | | | | .917 | .911 | .904 | .896 | .888 | .878 | .868 | .856 | .844 | .830 |
| | | | | | | | | .814 | .797 | .778 | .757 | .734 | .709 | .681 | .651 | .619 | .584 |
| | | | | | | | | .547 | .508 | | | | | | | | |
| 28.000 | 26.811 | 1.189 | 1.189 | .56 | 3.58 | 8.1% | 33 | .959 | .957 | .954 | .951 | .948 | .944 | .940 | .936 | .932 | .926 |
| | | | | | | | | .921 | .914 | .907 | .900 | .891 | .882 | .871 | .859 | .846 | .832 |
| | | | | | | | | .816 | .798 | .778 | .755 | .731 | .704 | .674 | .642 | .607 | .570 |
| | | | | | | | | .530 | .488 | .444 | | | | | | | |
| 28.000 | 27.116 | .884 | .884 | .62 | 3.53 | 6.1% | 34 | .961 | .958 | .956 | .953 | .949 | .946 | .942 | .938 | .933 | .928 |
| | | | | | | | | .922 | .916 | .909 | .901 | .893 | .883 | .872 | .860 | .847 | .832 |
| | | | | | | | | .815 | .796 | .776 | .753 | .727 | .699 | .668 | .634 | .597 | .558 |

*SIMILAR TO A.T.&T. JACOBSEN RETRIAL TABLES

*VERY FAST RETRIALS: JEWETT-SHRAGO VERY FAST RETRIAL TABLES - 50% RETRY, 50% OVERFLOW (RETRIES 1/3 HOLDING TIMES LATER)

| HOURS FIRST ATTEMPT VMR = 1.0 | HOURS CON-NECT | HOURS RECALL /RETRY | HOURS OVFL | CON-NECT VMR | OVFL VMR | PERCENT BLOCKED (GRADE) | LINE | HOURS CONNECTED TRAFFIC ON LINES 1 | 2 | 3 | 4 | 5 | 6 | 7 | 8 | 9 | 10 |
|---|---|---|---|---|---|---|---|---|---|---|---|---|---|---|---|---|---|
| | | | | | | | | .516 | .472 | .426 | .380 | | | | | | |
| 28.000 | 27.366 | .634 | .634 | .68 | 3.44 | 4.4% | 35 | .962 | .960 | .957 | .954 | .951 | .948 | .944 | .940 | .935 | .930 |
| | | | | | | | | .924 | .918 | .911 | .903 | .895 | .885 | .874 | .862 | .848 | .833 |
| | | | | | | | | .816 | .796 | .775 | .751 | .724 | .695 | .662 | .627 | .588 | .547 |
| | | | | | | | | .503 | .457 | .410 | .363 | .315 | | | | | |
| 28.000 | 27.544 | .456 | .456 | .74 | 3.37 | 3.2% | 36 | .963 | .961 | .958 | .955 | .952 | .949 | .945 | .941 | .936 | .931 |
| | | | | | | | | .925 | .919 | .912 | .904 | .895 | .885 | .874 | .862 | .848 | .833 |
| | | | | | | | | .815 | .795 | .773 | .749 | .721 | .691 | .658 | .621 | .582 | .539 |
| | | | | | | | | .494 | .447 | .399 | .350 | .303 | .257 | | | | |
| 28.000 | 27.681 | .319 | .319 | .80 | 3.27 | 2.3% | 37 | .964 | .962 | .959 | .956 | .953 | .950 | .946 | .942 | .937 | .932 |
| | | | | | | | | .926 | .920 | .913 | .905 | .896 | .886 | .875 | .863 | .849 | .833 |
| | | | | | | | | .815 | .795 | .773 | .748 | .720 | .689 | .654 | .617 | .576 | .533 |
| | | | | | | | | .487 | .438 | .389 | .340 | .292 | .246 | .202 | | | |
| 28.000 | 27.784 | .216 | .216 | .85 | 3.16 | 1.5% | 38 | .965 | .962 | .960 | .957 | .954 | .951 | .947 | .943 | .938 | .933 |
| | | | | | | | | .928 | .921 | .914 | .906 | .898 | .888 | .877 | .864 | .850 | .834 |
| | | | | | | | | .816 | .796 | .773 | .747 | .719 | .687 | .652 | .614 | .572 | .527 |
| | | | | | | | | .480 | .431 | .381 | .331 | .282 | .236 | .193 | .154 | | |
| 28.000 | 27.848 | .152 | .152 | .88 | 3.09 | 1.1% | 39 | .965 | .962 | .960 | .957 | .954 | .951 | .947 | .943 | .938 | .933 |
| | | | | | | | | .927 | .921 | .914 | .906 | .897 | .887 | .876 | .863 | .849 | .833 |
| | | | | | | | | .815 | .794 | .771 | .746 | .717 | .685 | .650 | .611 | .569 | .524 |
| | | | | | | | | .477 | .428 | .378 | .328 | .279 | .233 | .190 | .151 | .118 | |
| 28.000 | 27.895 | .105 | .105 | .91 | 3.00 | .7% | 40 | .965 | .962 | .960 | .957 | .954 | .950 | .947 | .943 | .938 | .933 |
| | | | | | | | | .927 | .921 | .914 | .906 | .897 | .887 | .876 | .863 | .848 | .832 |
| | | | | | | | | .814 | .793 | .770 | .745 | .716 | .683 | .648 | .609 | .567 | .522 |
| | | | | | | | | .475 | .425 | .375 | .325 | .277 | .230 | .188 | .150 | .116 | .088 |
| 30.000 | 23.662 | 6.338 | 6.338 | .23 | 4.22 | 34.9% | 27 | .941 | .938 | .935 | .932 | .929 | .020 | .922 | .918 | .914 | .910 |
| | | | | | | | | .905 | .000 | .895 | .889 | .883 | .876 | .869 | .862 | .854 | .845 |
| | | | | | | | | .836 | .826 | .816 | .804 | .792 | .779 | .765 | | | |
| 30.000 | 24.456 | 5.544 | 5.544 | .25 | 4.13 | 31.2% | 28 | .944 | .941 | .938 | .935 | .932 | .929 | .925 | .921 | .917 | .913 |
| | | | | | | | | .908 | .903 | .898 | .892 | .886 | .879 | .872 | .864 | .856 | .847 |
| | | | | | | | | .837 | .827 | .815 | .803 | .790 | .777 | .762 | .746 | | |
| 30.000 | 25.198 | 4.802 | 4.802 | .28 | 4.07 | 27.6% | 29 | .946 | .943 | .941 | .938 | .935 | .931 | .928 | .924 | .020 | 915 |
| | | | | | | | | .910 | .905 | .900 | .894 | .087 | .880 | .873 | .865 | .856 | .847 |
| | | | | | | | | .037 | .826 | .814 | .801 | .788 | .773 | .757 | .740 | .722 | |
| 30.000 | 25.917 | 4.083 | 4,083 | .31 | 4.01 | 24.0% | 30 | .949 | .946 | .944 | .941 | .937 | .004 | .931 | .927 | .922 | .918 |
| | | | | | | | | .913 | .908 | .902 | .896 | .890 | .883 | .875 | .867 | .858 | .848 |
| | | | | | | | | .837 | .826 | .813 | .800 | .785 | .770 | .753 | .734 | .715 | .693 |
| 30.000 | 26.610 | 3.390 | 3.390 | .34 | 3.93 | 20.3% | 31 | .952 | .949 | .947 | .944 | .941 | .938 | .934 | .930 | .926 | .922 |
| | | | | | | | | .917 | .912 | .906 | .900 | .893 | .886 | .878 | .869 | .860 | .850 |
| | | | | | | | | .839 | .827 | .814 | .799 | .784 | 767 | .748 | .729 | .707 | .684 |

*SIMILAR TO A.T.&T. JACOBSEN RETRIAL TABLES

*VERY FAST RETRIALS: JEWETT-SHRAGO VERY FAST RETRIAL TABLES - 50% RETRY, 50% OVERFLOW (RETRIES 1/3 HOLDING TIMES LATER)

| HOURS FIRST ATTEMPT VMR = 1.0 | HOURS CON-NECT | HOURS RECALL /RETRY | HOURS OVFL | CON-NECT VMR | OVFL VMR | PERCENT BLOCKED (GRADE) | LINE | HOURS CONNECTED TRAFFIC ON LINES 1 | 2 | 3 | 4 | 5 | 6 | 7 | 8 | 9 | 10 |
|---|---|---|---|---|---|---|---|---|---|---|---|---|---|---|---|---|---|
| | | | | | | | | .659 | | | | | | | | | |
| 30.000 | 27.219 | 2.781 | 2.781 | .38 | 3.89 | 17.0% | 32 | .954 | .952 | .949 | .946 | .943 | .940 | .937 | .933 | .929 | .924 |
| | | | | | | | | .919 | .914 | .908 | .902 | .895 | .888 | .880 | .871 | .861 | .850 |
| | | | | | | | | .839 | .826 | .812 | .797 | .781 | .763 | .743 | .722 | .699 | .674 |
| | | | | | | | | .647 | .619 | | | | | | | | |
| 30.000 | 27.784 | 2.216 | 2.216 | .43 | 3.83 | 13.8% | 33 | .957 | .954 | .952 | .949 | .946 | .943 | .939 | .936 | .932 | .927 |
| | | | | | | | | .922 | .917 | .911 | .905 | .898 | .890 | .882 | .873 | .863 | .852 |
| | | | | | | | | .840 | .827 | .812 | .796 | .779 | .760 | .739 | .716 | .691 | .664 |
| | | | | | | | | .635 | .604 | .571 | | | | | | | |
| 30.000 | 28.271 | 1.729 | 1.729 | .48 | 3.78 | 10.9% | 34 | .959 | .957 | .954 | .952 | .949 | .946 | .942 | .938 | .934 | .930 |
| | | | | | | | | .925 | .920 | .914 | .907 | .900 | .893 | .884 | .875 | .865 | .853 |
| | | | | | | | | .841 | .827 | .812 | .795 | .776 | .756 | .734 | .710 | .683 | .654 |
| | | | | | | | | .623 | .590 | .554 | .517 | | | | | | |
| 30.000 | 28.673 | 1.327 | 1.327 | .54 | 3.73 | 8.5% | 35 | .961 | .959 | .956 | .953 | .951 | .947 | .944 | .940 | .936 | .932 |
| | | | | | | | | .927 | .922 | .916 | .910 | .902 | .895 | .886 | .876 | .866 | .854 |
| | | | | | | | | .841 | .827 | .811 | .794 | .774 | .753 | .729 | .704 | .675 | .645 |
| | | | | | | | | .612 | .576 | .539 | .499 | .458 | | | | | |
| 30.000 | 29.018 | .982 | .982 | .61 | 3.64 | 6.3% | 36 | .963 | .961 | .958 | .956 | .953 | .950 | .946 | .943 | .939 | .934 |
| | | | | | | | | .930 | .924 | .919 | .912 | .905 | .897 | .888 | .879 | .868 | .856 |
| | | | | | | | | .843 | .828 | .812 | .793 | .773 | .751 | .726 | .699 | .669 | .637 |
| | | | | | | | | .602 | .564 | .524 | .482 | .439 | .395 | | | | |
| 30.000 | 29.280 | .720 | .720 | .67 | 3.57 | 4.7% | 37 | .964 | .962 | .960 | .957 | .954 | .951 | .948 | .944 | .941 | .936 |
| | | | | | | | | .931 | .926 | .920 | .914 | .907 | .899 | .890 | .880 | .869 | .857 |
| | | | | | | | | .844 | .828 | .811 | .793 | .772 | .748 | .723 | .694 | .663 | .630 |
| | | | | | | | | .593 | .554 | .512 | .468 | .424 | .378 | .332 | | | |
| 30.000 | 29.491 | .509 | .509 | .73 | 3.46 | 3.3% | 38 | .966 | .964 | .961 | .959 | .956 | .953 | .950 | .946 | .943 | .938 |
| | | | | | | | | .934 | .928 | .923 | .916 | .909 | .901 | .892 | .882 | .871 | .859 |
| | | | | | | | | .845 | .830 | .812 | .793 | .772 | .748 | .721 | .691 | .659 | .624 |
| | | | | | | | | .586 | .545 | .501 | .456 | .409 | .362 | .316 | .270 | | |
| 30.000 | 29.638 | .362 | .362 | .78 | 3.37 | 2.4% | 39 | .966 | .964 | .962 | .960 | .957 | .954 | .951 | .947 | .943 | .939 |
| | | | | | | | | .934 | .929 | .923 | .917 | .910 | .902 | .893 | .883 | .872 | .859 |
| | | | | | | | | .845 | .830 | .812 | .793 | .771 | .746 | .719 | .689 | .655 | .619 |
| | | | | | | | | .580 | .538 | .494 | .447 | .400 | .352 | .305 | .259 | .216 | |
| 30.000 | 29.735 | .265 | .265 | .83 | 3.31 | 1.7% | 40 | .966 | .964 | .962 | .959 | .957 | .954 | .951 | .947 | .943 | .939 |
| | | | | | | | | .934 | .929 | .923 | .917 | .909 | .901 | .892 | .882 | .871 | .858 |
| | | | | | | | | .844 | .828 | .811 | .791 | .768 | .744 | .716 | .686 | .652 | .615 |
| | | | | | | | | .576 | .534 | .489 | .442 | .395 | .347 | .300 | .254 | .212 | .173 |
| 30.000 | 29.821 | .179 | .179 | .87 | 3.19 | 1.2% | 41 | .967 | .965 | .963 | .960 | .958 | .955 | .952 | .948 | .944 | .940 |
| | | | | | | | | .935 | .930 | .924 | .918 | .911 | .903 | .894 | .884 | .872 | .860 |
| | | | | | | | | .845 | .829 | .811 | .791 | .769 | .744 | .716 | .684 | .650 | .613 |

*SIMILAR TO A.T.&T. JACOBSEN RETRIAL TABLES

*VERY FAST RETRIALS: JEWETT-SHRAGO VERY FAST RETRIAL TABLES - 50% RETRY, 50% OVERFLOW (RETRIES 1/3 HOLDING TIMES LATER)

| HOURS FIRST ATTEMPT VMR=1.0 | HOURS CON-NECT | HOURS RECALL /RETRY | HOURS OVFL | CON-NECT VMR | OVFL VMR | PERCENT BLOCKED (GRADE) | LINE | HOURS CONNECTED TRAFFIC ON LINES 1 | 2 | 3 | 4 | 5 | 6 | 7 | 8 | 9 | 10 |
|---|---|---|---|---|---|---|---|---|---|---|---|---|---|---|---|---|---|
| | | | | | | | | .573 | .529 | .484 | .436 | .388 | .339 | .292 | .246 | .203 | .164 |
| | | | | | | | | .130 | | | | | | | | | |
| 30.000 | 29.874 | .126 | .126 | .90 | 3.11 | .8% | 42 | .967 | .965 | .963 | .960 | .958 | .955 | .952 | .948 | .944 | .940 |
| | | | | | | | | .935 | .930 | .924 | .918 | .910 | .902 | .893 | .883 | .872 | .859 |
| | | | | | | | | .845 | .829 | .811 | .790 | .768 | .742 | .714 | .683 | .648 | .611 |
| | | | | | | | | .570 | .527 | .481 | .434 | .385 | .336 | .289 | .243 | .201 | .162 |
| | | | | | | | | .128 | .099 | | | | | | | | |
| 30.000 | 29.920 | .080 | .080 | .93 | 2.98 | .5% | 43 | .968 | .966 | .964 | .961 | .959 | .956 | .953 | .949 | .945 | .941 |
| | | | | | | | | .937 | .931 | .926 | .919 | .912 | .904 | .895 | .885 | .874 | .861 |
| | | | | | | | | .847 | .830 | .812 | .792 | .769 | .743 | .715 | .683 | .648 | .610 |
| | | | | | | | | .569 | .524 | .478 | .429 | .380 | .331 | .283 | .237 | .194 | .156 |
| | | | | | | | | .122 | .093 | .070 | | | | | | | |

*SIMILAR TO A.T.&T. JACOBSEN RETRIAL TABLES

## *VERY FAST RETRIALS: JEWETT-SHRAGO VERY FAST RETRIAL TABLES - 100% RETRY, 0% OVERFLOW (RETRIES 1/3 HOLDING TIME LATER)

| HOURS FIRST ATTEMPT VMR = 1.0 | HOURS CON-NECT | HOURS RECALL /RETRY | HOURS OVFL | CON-NECT VMR | OVFL VMR | PERCENT BLOCKED (GRADE) | LINE | HOURS CONNECTED TRAFFIC ON LINES 1 | 2 | 3 | 4 | 5 | 6 | 7 | 8 | 9 | 10 |
|---|---|---|---|---|---|---|---|---|---|---|---|---|---|---|---|---|---|
| .200 | .200 | .003 | .000 | .97 | 1.07 | 1.7% | 2 | .169 | .031 | | | | | | | | |
| .400 | .400 | .025 | .000 | .90 | 1.14 | 6.0% | 2 | .298 | .102 | | | | | | | | |
| .400 | .400 | .003 | .000 | .98 | 1.11 | .7% | 3 | .287 | .094 | .019 | | | | | | | |
| .600 | .600 | .013 | .000 | .95 | 1.17 | 2.1% | 3 | .380 | .169 | .051 | | | | | | | |
| .800 | .800 | .036 | .000 | .90 | 1.23 | 4.3% | 3 | .455 | .247 | .098 | | | | | | | |
| .800 | .800 | .006 | .000 | .97 | 1.19 | .8% | 4 | .446 | .237 | .091 | .025 | | | | | | |
| 1.000 | 1.000 | .126 | .000 | .82 | 1.40 | 11.2% | 3 | .491 | .327 | .182 | | | | | | | |
| 1.000 | 1.000 | .016 | .000 | .95 | 1.24 | 1.6% | 4 | .504 | .305 | .142 | .049 | | | | | | |
| 1.200 | 1.200 | .035 | .000 | .92 | 1.29 | 2.8% | 4 | .553 | .368 | .197 | .082 | | | | | | |
| 1.200 | 1.200 | .008 | .000 | .98 | 1.24 | .6% | 5 | .547 | .361 | .190 | .078 | .025 | | | | | |
| 1.400 | 1.400 | .066 | .000 | .88 | 1.34 | 4.5% | 4 | .595 | .427 | .256 | .123 | | | | | | |
| 1.400 | 1.400 | .016 | .000 | .96 | 1.29 | 1.2% | 5 | .586 | .415 | .243 | .114 | .042 | | | | | |
| 1.600 | 1.600 | .170 | .000 | .80 | 1.50 | 9.6% | 4 | .604 | .471 | .327 | .197 | | | | | | |
| 1.600 | 1.600 | .031 | .000 | .93 | 1.34 | 1.9% | 5 | .620 | .463 | .296 | .155 | .066 | | | | | |
| 1.800 | 1.800 | .397 | .000 | .72 | 1.74 | 18.1% | 4 | .610 | .509 | .396 | .284 | | | | | | |
| 1.800 | 1.800 | .054 | .000 | .90 | 1.39 | 2.9% | 5 | .650 | .507 | .347 | .200 | .095 | | | | | |
| 1.800 | 1.800 | .015 | .000 | .97 | 1.33 | .8% | 6 | .645 | .500 | .339 | .192 | .090 | .034 | | | | |
| 2.000 | 2.000 | .130 | .000 | .85 | 1.54 | 6.1% | 5 | .649 | .532 | .398 | .266 | .155 | | | | | |
| 2.000 | 2.000 | .026 | .000 | .95 | 1.38 | 1.3% | 6 | .670 | .538 | .384 | .236 | .121 | .052 | | | | |
| 2.200 | 2.200 | .191 | .000 | .80 | 1.59 | 8.0% | 5 | .676 | .569 | .444 | .314 | .196 | | | | | |
| 2.200 | 2.200 | .042 | .000 | .93 | 1.42 | 1.9% | 6 | .692 | .571 | .427 | .280 | .156 | .073 | | | | |
| 2.200 | 2.200 | .012 | .000 | .97 | 1.37 | .6% | 7 | .689 | .567 | .422 | .274 | .151 | .070 | .027 | | | |
| 2.400 | 2.400 | .342 | .000 | .74 | 1.73 | 12.5% | 5 | .684 | .594 | .488 | .373 | .261 | | | | | |
| 2.400 | 2.400 | .066 | .000 | .90 | 1.47 | 2.7% | 6 | .712 | .602 | .468 | .324 | .195 | .099 | | | | |
| 2.400 | 2.400 | .021 | .000 | .96 | 1.41 | .9% | 7 | .708 | .596 | .460 | .315 | .187 | .094 | .040 | | | |
| 2.600 | 2.600 | .143 | .000 | .85 | 1.62 | 5.2% | 6 | .706 | .613 | .499 | .375 | .254 | .153 | | | | |
| 2.600 | 2.600 | .033 | .000 | .94 | 1.45 | 1.3% | 7 | .725 | .623 | .496 | .356 | .224 | .121 | .056 | | | |
| 2.800 | 2.800 | .199 | .000 | .81 | 1.66 | 6.6% | 6 | .725 | .640 | .535 | .416 | .296 | .189 | | | | |
| 2.800 | 2.800 | .050 | .000 | .92 | 1.50 | 1.8% | 7 | .740 | .647 | .529 | .395 | .262 | .152 | .076 | | | |
| 2.800 | 2.800 | .017 | .000 | .97 | 1.44 | .6% | 8 | .738 | .643 | .524 | .389 | .256 | .147 | .072 | .031 | | |
| 3.000 | 3.000 | .337 | .000 | .75 | 1.79 | 10.1% | 6 | .727 | .653 | .563 | .460 | .351 | .247 | | | | |
| 3.000 | 3.000 | .074 | .000 | .90 | 1.54 | 2.4% | 7 | .755 | .669 | .560 | .432 | .301 | .185 | .099 | | | |
| 3.000 | 3.000 | .026 | .000 | .96 | 1.48 | .8% | 8 | .752 | .664 | .553 | .424 | .293 | .178 | .094 | .043 | | |
| 3.200 | 3.200 | .519 | .000 | .70 | 1.91 | 13.9% | 6 | .734 | .669 | .591 | .500 | .402 | .304 | | | | |
| 3.200 | 3.200 | .147 | .000 | .85 | 1.68 | 4.4% | 7 | .747 | .672 | .577 | .467 | .350 | .239 | .147 | | | |
| 3.200 | 3.200 | .038 | .000 | .94 | 1.52 | 1.2% | 8 | .764 | .683 | .580 | .458 | .329 | .210 | .118 | .058 | | |
| 3.400 | 3.400 | .198 | .000 | .82 | 1.72 | 5.5% | 7 | .761 | .692 | .604 | .501 | .388 | .276 | .178 | | | |
| 3.400 | 3.400 | .056 | .000 | .92 | 1.56 | 1.6% | 8 | .776 | .701 | .605 | .490 | .364 | .244 | .144 | .075 | | |
| 3.400 | 3.400 | .020 | .000 | .97 | 1.50 | .6% | 9 | .774 | .698 | .601 | .485 | .359 | .238 | .140 | .072 | .033 | |
| 3.600 | 3.600 | .262 | .000 | .79 | 1.76 | 6.8% | 7 | .775 | .711 | .630 | .533 | .425 | .314 | .212 | | | |

*SIMILAR TO A.T.&T. JACOBSEN RETRIAL TABLES

*VERY FAST RETRIALS: JEWETT-SHRAGO VERY FAST RETRIAL TABLES - 100% RETRY, 0% OVERFLOW (RETRIES 1/3 HOLDING TIME LATER)

| HOURS FIRST ATTEMPT VMR = 1.0 | HOURS CON-NECT | HOURS RECALL /RETRY | HOURS OVFL | CON-NECT VMR | OVFL VMR | PERCENT BLOCKED (GRADE) | LINE | HOURS CONNECTED TRAFFIC ON LINES 1 | 2 | 3 | 4 | 5 | 6 | 7 | 8 | 9 | 10 |
|---|---|---|---|---|---|---|---|---|---|---|---|---|---|---|---|---|---|
| 3.600 | 3.600 | .078 | .000 | .90 | 1.60 | 2.1% | 8 | .786 | .718 | .629 | .521 | .399 | .278 | .173 | .096 | | |
| 3.600 | 3.600 | .029 | .000 | .96 | 1.54 | .8% | 9 | .784 | .714 | .624 | .514 | .392 | .271 | .167 | .091 | .044 | |
| 3.800 | 3.800 | .414 | .000 | .73 | 1.89 | 9.8% | 7 | .774 | .717 | .647 | .562 | .467 | .366 | .267 | | | |
| 3.800 | 3.800 | .147 | .000 | .86 | 1.74 | 3.7% | 8 | .779 | .716 | .638 | .543 | .436 | .326 | .223 | .139 | | |
| 3.800 | 3.800 | .042 | .000 | .94 | 1.58 | 1.1% | 9 | .793 | .729 | .645 | .541 | .424 | .304 | .196 | .112 | .057 | |
| 4.000 | 4.000 | .766 | .000 | .66 | 2.15 | 16.1% | 7 | .761 | .713 | .655 | .587 | .511 | .429 | .344 | | | |
| 4.000 | 4.000 | .192 | .000 | .83 | 1.78 | 4.6% | 8 | .789 | .732 | .659 | .570 | .468 | .360 | .256 | .166 | | |
| 4.000 | 4.000 | .058 | .000 | .93 | 1.62 | 1.4% | 9 | .802 | .742 | .664 | .567 | .455 | .336 | .226 | .136 | .073 | |
| 4.000 | 4.000 | .022 | .000 | .97 | 1.55 | .5% | 10 | .801 | .740 | .661 | .563 | .450 | .331 | .221 | .132 | .070 | .033 |
| 4.200 | 4.200 | .249 | .000 | .80 | 1.82 | 5.6% | 8 | .800 | .746 | .679 | .596 | .499 | .395 | .290 | .196 | | |
| 4.200 | 4.200 | .080 | .000 | .91 | 1.66 | 1.9% | 9 | .811 | .754 | .682 | .591 | .484 | .369 | .257 | .162 | .091 | |
| 4.200 | 4.200 | .031 | .000 | .96 | 1.59 | .7% | 10 | .809 | .752 | .678 | .586 | .478 | .362 | .250 | .156 | .087 | .043 |
| 4.400 | 4.400 | .385 | .000 | .75 | 1.94 | 8.0% | 8 | .797 | .749 | .688 | .615 | .530 | .436 | .338 | .246 | | |
| 4.400 | 4.400 | .107 | .000 | .89 | 1.70 | 2.4% | 9 | .818 | .766 | .698 | .614 | .512 | .401 | .289 | .189 | .112 | |
| 4.400 | 4.400 | .043 | .000 | .94 | 1.63 | 1.0% | 10 | .816 | .763 | .694 | .607 | .505 | .392 | .280 | .182 | .106 | .055 |
| 4.600 | 4.600 | .555 | .000 | .70 | 2.05 | 10.8% | 8 | .798 | .754 | .700 | .635 | .559 | .474 | .384 | .295 | | |
| 4.600 | 4.600 | .184 | .000 | .85 | 1.83 | 3.8% | 9 | .811 | .763 | .702 | .626 | .537 | .437 | .334 | .236 | .154 | |
| 4.600 | 4.600 | .059 | .000 | .93 | 1.67 | 1.3% | 10 | .823 | .773 | .709 | .628 | .530 | .421 | .310 | .209 | .127 | .069 |
| 4.600 | 4.600 | .024 | .000 | .97 | 1.60 | .5% | 11 | .822 | .772 | .706 | .624 | .526 | .417 | .305 | .204 | .123 | .067 |
| | | | | | | | | .033 | | | | | | | | | |
| 4.800 | 4.800 | 1.003 | .000 | .63 | 2.36 | 17.3% | 8 | .783 | .745 | .699 | .647 | .587 | .520 | .447 | .373 | | |
| 4.800 | 4.800 | .234 | .000 | .82 | 1.87 | 4.7% | 9 | .820 | .774 | .717 | .647 | .562 | .467 | .366 | .267 | .180 | |
| 4.800 | 4.800 | .079 | .000 | .91 | 1.71 | 1.6% | 10 | .830 | .783 | .723 | .647 | .555 | .450 | .340 | .237 | .150 | .086 |
| 4.800 | 4.800 | .033 | .000 | .96 | 1.64 | .7% | 11 | .829 | .781 | .720 | .643 | .550 | .444 | .334 | .231 | .145 | .082 |
| | | | | | | | | .042 | | | | | | | | | |
| 5.000 | 5.000 | .355 | .000 | .77 | 1.99 | 6.6% | 9 | .816 | .775 | .723 | .659 | .584 | .498 | .406 | 313 | .228 | |
| 5.000 | 5.000 | .104 | .000 | .90 | 1.75 | 2.0% | 10 | .836 | .792 | .736 | .665 | .578 | .478 | .370 | .266 | .175 | .104 |
| 5.000 | 5.000 | .044 | .000 | 95 | 1.60 | .9% | 11 | .835 | .790 | .733 | .660 | .572 | .470 | .362 | .258 | .168 | .000 |
| | | | | | | | | .053 | | | | | | | | | |
| 5.500 | 5.500 | .833 | .000 | .66 | 2.28 | 13.2% | 9 | .815 | .781 | .740 | .692 | .635 | .571 | .499 | .422 | .344 | |
| 5.500 | 5.500 | .245 | .000 | .82 | 1.93 | 4.3% | 10 | .839 | .802 | .755 | .696 | .625 | .542 | .450 | .353 | .260 | .178 |
| 5.500 | 5.500 | .088 | .000 | .91 | 1.77 | 1.6% | 11 | .848 | .810 | .761 | .699 | .622 | .532 | .432 | .328 | .231 | .149 |
| | | | | | | | | .088 | | | | | | | | | |
| 5.500 | 5.500 | .038 | .000 | .96 | 1.70 | 7% | 12 | .847 | .808 | .759 | .696 | .618 | .527 | .426 | .322 | .225 | .144 |
| | | | | | | | | .084 | .045 | | | | | | | | |
| 6.000 | 6.000 | .552 | .000 | .72 | 2.17 | 8.4% | 10 | .838 | .806 | .767 | .720 | .664 | .599 | .524 | .443 | .360 | .278 |
| 6.000 | 6.000 | .203 | .000 | .85 | 1.95 | 3.3% | 11 | .850 | .816 | .775 | .723 | .659 | .584 | .498 | .406 | .312 | .225 |
| | | | | | | | | .151 | | | | | | | | | |
| 6.000 | 6.000 | .074 | .000 | .93 | 1.79 | 1.2% | 12 | .859 | .825 | .782 | .728 | .660 | .580 | .487 | .388 | .289 | .200 |
| | | | | | | | | .127 | .074 | | | | | | | | |

*SIMILAR TO A.T.&T. JACOBSEN RETRIAL TABLES

*VERY FAST RETRIALS: JEWETT-SHRAGO VERY FAST RETRIAL TABLES - 100% RETRY, 0% OVERFLOW (RETRIES 1/3 HOLDING TIME LATER)

| HOURS FIRST ATTEMPT VMR = 1.0 | HOURS CON-NECT | HOURS RECALL /RETRY | HOURS OVFL | CON-NECT VMR | OVFL VMR | PERCENT BLOCKED (GRADE) | LINE | HOURS CONNECTED TRAFFIC ON LINES 1 | 2 | 3 | 4 | 5 | 6 | 7 | 8 | 9 | 10 |
|---|---|---|---|---|---|---|---|---|---|---|---|---|---|---|---|---|---|
| 6.000 | 6.000 | .033 | .000 | .96 | 1.72 | .5% | 13 | .858 | .824 | .780 | .725 | .657 | .576 | .483 | .383 | .284 | .196 |
| | | | | | | | | .124 | .072 | .038 | | | | | | | |
| 6.500 | 6.500 | .397 | .000 | .77 | 2.12 | 5.8% | 11 | .854 | .825 | .790 | .746 | .693 | .630 | .558 | .477 | .392 | .307 |
| | | | | | | | | .227 | | | | | | | | | |
| 6.500 | 6.500 | .169 | .000 | .87 | 1.97 | 2.5% | 12 | .859 | .829 | .792 | .746 | .689 | .621 | .543 | .456 | .365 | .275 |
| | | | | | | | | .195 | .129 | | | | | | | | |
| 6.500 | 6.500 | .063 | .000 | .94 | 1.81 | 1.0% | 13 | .868 | .838 | .800 | .752 | .693 | .621 | .537 | .444 | .347 | .254 |
| | | | | | | | | .173 | .109 | .063 | | | | | | | |
| 7.000 | 7.000 | .853 | .000 | .67 | 2.39 | 10.9% | 11 | .850 | .825 | .795 | .760 | .718 | .669 | .613 | .549 | .480 | .408 |
| | | | | | | | | .334 | | | | | | | | | |
| 7.000 | 7.000 | .279 | .000 | .82 | 2.06 | 3.8% | 12 | .870 | .844 | .812 | .772 | .723 | .664 | .594 | .515 | .429 | .341 |
| | | | | | | | | .256 | .181 | | | | | | | | |
| 7.000 | 7.000 | .113 | .000 | .90 | 1.90 | 1.6% | 13 | .877 | .851 | .817 | .776 | .724 | .661 | .587 | .502 | .409 | .316 |
| | | | | | | | | .229 | .154 | .096 | | | | | | | |
| 7.000 | 7.000 | .053 | .000 | .95 | 1.83 | .8% | 14 | .876 | .849 | .816 | .774 | .721 | .657 | .582 | .496 | .403 | .309 |
| | | | | | | | | .223 | .149 | .092 | .053 | | | | | | |
| 7.500 | 7.500 | .576 | .000 | .73 | 2.28 | 7.1% | 12 | .867 | .844 | .816 | .782 | .742 | .694 | .637 | .573 | .502 | .425 |
| | | | | | | | | .347 | .271 | | | | | | | | |
| 7.500 | 7.500 | .231 | .000 | .85 | 2.07 | 3.0% | 13 | .877 | .853 | .824 | .788 | .744 | .691 | .627 | .554 | .473 | .388 |
| | | | | | | | | .302 | .223 | .154 | | | | | | | |
| 7.500 | 7.500 | .094 | .000 | .92 | 1.92 | 1.2% | 14 | .884 | .860 | .831 | .794 | .748 | .692 | .624 | .547 | .460 | .369 |
| | | | | | | | | .280 | .199 | .132 | .082 | | | | | | |
| 7.500 | 7.500 | .045 | .000 | .96 | 1.84 | .6% | 15 | .883 | .859 | .829 | .792 | .745 | .689 | .621 | .542 | .455 | .364 |
| | | | | | | | | .275 | .194 | .128 | .079 | .045 | | | | | |
| 8.000 | 8.000 | 1.427 | .000 | .60 | 2.75 | 15.1% | 12 | .851 | .830 | .807 | .780 | .749 | .714 | .673 | .628 | .577 | .522 |
| | | | | | | | | .464 | .403 | | | | | | | | |
| 8.000 | 8.000 | .420 | .000 | .78 | 2.23 | 5.0% | 13 | .879 | .858 | .832 | .801 | .763 | .718 | .664 | .602 | .532 | .455 |
| | | | | | | | | .375 | .296 | .223 | | | | | | | |
| 8.000 | 8.000 | .192 | .000 | .87 | 2.09 | 2.3% | 14 | .883 | .862 | .835 | .803 | .763 | .715 | .658 | .591 | .515 | .433 |
| | | | | | | | | .349 | .267 | .193 | .132 | | | | | | |
| 8.000 | 8.000 | .079 | .000 | .93 | 1.93 | 1.0% | 15 | .890 | .869 | .842 | .809 | .768 | .718 | .658 | .587 | .507 | .420 |
| | | | | | | | | .331 | .247 | .173 | .113 | .069 | | | | | |
| 8.500 | 8.500 | .847 | .000 | .68 | 2.50 | 9.1% | 13 | .874 | .855 | .832 | .806 | .774 | .737 | .694 | .644 | .588 | .526 |
| | | | | | | | | .459 | .390 | .320 | | | | | | | |
| 8.500 | 8.500 | .301 | .000 | .82 | 2.17 | 3.4% | 14 | .891 | .872 | .848 | .820 | .785 | .743 | .692 | .633 | .565 | .489 |
| | | | | | | | | .408 | .326 | .248 | .179 | | | | | | |
| 8.500 | 8.500 | .160 | .000 | .89 | 2.10 | 1.9% | 15 | .889 | .870 | .846 | .816 | .780 | .737 | .685 | .624 | .554 | .476 |
| | | | | | | | | .395 | .313 | .235 | .168 | .113 | | | | | |
| 8.500 | 8.500 | .066 | .000 | .94 | 1.94 | .8% | 16 | .895 | .876 | .852 | .823 | .786 | .741 | .687 | .623 | .549 | .468 |
| | | | | | | | | .381 | .296 | .217 | .150 | .097 | .058 | | | | |

*SIMILAR TO A.T.&T. JACOBSEN RETRIAL TABLES

*VERY FAST RETRIALS: JEWETT-SHRAGO VERY FAST RETRIAL TABLES - 100% RETRY, 0% OVERFLOW (RETRIES 1/3 HOLDING TIME LATER)

| HOURS FIRST ATTEMPT VMR = 1.0 | HOURS CON-NECT | HOURS RECALL /RETRY | HOURS OVFL | CON-NECT VMR | OVFL VMR | PERCENT BLOCKED (GRADE) | LINE | HOURS CONNECTED TRAFFIC ON LINES 1 | 2 | 3 | 4 | 5 | 6 | 7 | 8 | 9 | 10 |
|---|---|---|---|---|---|---|---|---|---|---|---|---|---|---|---|---|---|
| 9.000 | 9.000 | .581 | .000 | .74 | 2.39 | 6.1% | 14 | .888 | .870 | .849 | .824 | .794 | .758 | .716 | .667 | .611 | .547 |
| | | | | | | | | .478 | .405 | .331 | .260 | | | | | | |
| 9.000 | 9.000 | .250 | .000 | .85 | 2.18 | 2.7% | 15 | .896 | .878 | .857 | .831 | .799 | .761 | .715 | .660 | .597 | .526 |
| | | | | | | | | .449 | .369 | .290 | .217 | .154 | | | | | |
| 9.000 | 9.000 | .110 | .000 | .91 | 2.03 | 1.2% | 16 | .901 | .884 | .862 | .836 | .803 | .764 | .716 | .659 | .592 | .516 |
| | | | | | | | | .435 | .350 | .269 | .195 | .133 | .085 | | | | |
| 9.000 | 9.000 | .055 | .000 | .95 | 1.95 | .6% | 17 | .901 | .883 | .861 | .835 | .802 | .762 | .713 | .655 | .588 | .512 |
| | | | | | | | | .430 | .345 | .264 | .190 | .129 | .082 | .049 | | | |
| 9.500 | 9.500 | 1.146 | .000 | .64 | 2.68 | 10.8% | 14 | .881 | .865 | .846 | .825 | .799 | .770 | .736 | .697 | .653 | .603 |
| | | | | | | | | .548 | .489 | .426 | .362 | | | | | | |
| 9.500 | 9.500 | .430 | .000 | .78 | 2.33 | 4.3% | 15 | .897 | .881 | .862 | .838 | .811 | .777 | .738 | .691 | .636 | .575 |
| | | | | | | | | .506 | .433 | .357 | .283 | .215 | | | | | |
| 9.500 | 9.500 | .208 | .000 | .87 | 2.19 | 2.1% | 16 | .901 | .884 | .865 | .841 | .812 | .777 | .735 | .686 | .628 | .562 |
| | | | | | | | | .489 | .411 | .332 | .257 | .189 | .132 | | | | |
| 9.500 | 9.500 | .091 | .000 | .93 | 2.04 | 1.0% | 17 | .906 | .890 | .870 | .846 | .817 | .781 | .738 | .686 | .625 | .556 |
| | | | | | | | | .479 | .397 | .315 | .238 | .170 | .114 | .072 | | | |
| 10.000 | 10.000 | .759 | .000 | .71 | 2.53 | 7.1% | 15 | .895 | .880 | .863 | .842 | .818 | .790 | .756 | .717 | .671 | .619 |
| | | | | | | | | .561 | .498 | .431 | .362 | .294 | | | | | |
| 10.000 | 10.000 | .358 | .000 | .81 | 2.34 | 3.5% | 16 | .901 | .886 | .868 | .847 | .821 | .790 | .754 | .711 | .661 | .603 |
| | | | | | | | | .538 | .468 | .394 | .320 | .250 | .186 | | | | |
| 10.000 | 10.000 | .174 | .000 | .89 | 2.20 | 1.7% | 17 | .905 | .890 | .872 | .850 | .824 | .792 | .754 | .709 | .656 | .595 |
| | | | | | | | | .526 | .452 | .375 | .298 | .227 | .164 | .113 | | | |
| 10.000 | 10.000 | .076 | .000 | .94 | 2.04 | .8% | 18 | .910 | .895 | .877 | .855 | .829 | .797 | .758 | .711 | .656 | .592 |
| | | | | | | | | .520 | .442 | .361 | .282 | .210 | .147 | .098 | .061 | | |
| 11.000 | 11.000 | 1.023 | .000 | .67 | 2.72 | 8.5% | 16 | .899 | .886 | .871 | .854 | .834 | .811 | .784 | .752 | .716 | .675 |
| | | | | | | | | .628 | .577 | .520 | .460 | .397 | .334 | | | | |
| 11.000 | 11.000 | .431 | .000 | .79 | 2.43 | 3.8% | 17 | .910 | .898 | .883 | .865 | .844 | .819 | .789 | .753 | .712 | .664 |
| | | | | | | | | .609 | .548 | .481 | .410 | .339 | .270 | .206 | | | |
| 11.000 | 11.000 | .217 | .000 | .87 | 2.29 | 1.9% | 18 | .913 | .901 | .886 | .867 | .846 | .820 | .789 | .752 | .708 | .658 |
| | | | | | | | | .599 | .534 | .464 | .390 | .316 | .245 | .182 | .129 | | |
| 11.000 | 11.000 | .101 | .000 | .93 | 2.13 | .9% | 19 | .917 | .905 | .890 | .872 | .850 | .824 | .792 | .754 | .709 | .656 |
| | | | | | | | | .595 | .527 | .453 | .376 | .299 | .228 | .165 | .113 | .074 | |
| 12.000 | 12.000 | 1.380 | .000 | .62 | 2.94 | 10.3% | 17 | .900 | .889 | .876 | .862 | .845 | .826 | .804 | .778 | .749 | .717 |
| | | | | | | | | .680 | .638 | .593 | .543 | .490 | .434 | .377 | | | |
| 12.000 | 12.000 | .562 | .000 | .76 | 2.57 | 4.5% | 18 | .914 | .903 | .890 | .875 | .858 | .837 | .812 | .783 | 750 | .711 |
| | | | | | | | | .666 | .615 | .558 | .497 | .431 | .364 | .298 | .236 | | |
| 12.000 | 12.000 | .266 | .000 | .85 | 2.37 | 2.2% | 19 | .920 | .910 | .897 | .882 | .864 | .842 | .817 | .786 | .750 | .708 |
| | | | | | | | | .660 | .604 | .542 | .474 | .404 | .332 | .263 | .200 | .145 | |
| 12.000 | 12.000 | .153 | .000 | .90 | 2.29 | 1.3% | 20 | .920 | .909 | .896 | .880 | .862 | .840 | .814 | .783 | .746 | .703 |
| | | | | | | | | .654 | .597 | .534 | .466 | .394 | .323 | .254 | .192 | .138 | .095 |

*SIMILAR TO A.T.&T. JACOBSEN RETRIAL TABLES

*VERY FAST RETRIALS: JEWETT-SHRAGO VERY FAST RETRIAL TABLES - 100% RETRY, 0% OVERFLOW (RETRIES 1/3 HOLDING TIME LATER)

| HOURS FIRST ATTEMPT VMR=1.0 | HOURS CON-NECT | HOURS RECALL /RETRY | HOURS OVFL | CON-NECT VMR | OVFL VMR | PERCENT BLOCKED (GRADE) | LINE | HOURS CONNECTED TRAFFIC ON LINES 1 | 2 | 3 | 4 | 5 | 6 | 7 | 8 | 9 | 10 |
|---|---|---|---|---|---|---|---|---|---|---|---|---|---|---|---|---|---|
| 12.000 | 12.000 | .071 | .000 | .95 | 2.14 | .6% | 21 | .924 | .913 | .900 | .885 | .866 | .845 | .818 | .787 | .749 | .705 |
| | | | | | | | | .654 | .595 | .529 | .458 | .384 | .310 | .240 | .177 | .125 | .084 |
| | | | | | | | | .053 | | | | | | | | | |
| 13.000 | 13.000 | 1.986 | .000 | .57 | 3.26 | 13.3% | 18 | .897 | .887 | .876 | .863 | .849 | .833 | .815 | .795 | .772 | .746 |
| | | | | | | | | .717 | .685 | .650 | .612 | .570 | .525 | .478 | .428 | | |
| 13.000 | 13.000 | .704 | .000 | .74 | 2.70 | 5.1% | 19 | .918 | .909 | .897 | .884 | .869 | .852 | .831 | .807 | .780 | .748 |
| | | | | | | | | .711 | .668 | .621 | .568 | .511 | .450 | .387 | .324 | .263 | |
| 13.000 | 13.000 | .357 | .000 | .83 | 2.52 | 2.7% | 20 | .923 | .913 | .902 | .889 | .874 | .856 | .834 | .809 | .780 | .745 |
| | | | | | | | | .706 | .660 | .608 | .550 | .488 | .422 | .354 | .288 | .227 | .171 |
| 13.000 | 13.000 | .187 | .000 | .89 | 2.37 | 1.4% | 21 | .926 | .916 | .905 | .892 | .876 | .858 | .836 | .811 | .780 | .745 |
| | | | | | | | | .703 | .655 | .601 | .540 | .475 | .406 | .337 | .269 | .207 | .153 |
| | | | | | | | | .108 | | | | | | | | | |
| 13.000 | 13.000 | .091 | .000 | .94 | 2.22 | .7% | 22 | .929 | .920 | .909 | .896 | .880 | .862 | .840 | .814 | .783 | .747 |
| | | | | | | | | .704 | .654 | .598 | .535 | .466 | .395 | .323 | .255 | .192 | .139 |
| | | | | | | | | .096 | .063 | | | | | | | | |
| 14.000 | 14.000 | .858 | .000 | .71 | 2.82 | 5.8% | 20 | .922 | .913 | .904 | .892 | .879 | .864 | .847 | .827 | .804 | .777 |
| | | | | | | | | .746 | .711 | .672 | .627 | .578 | .524 | .467 | .408 | .347 | .289 |
| 14.000 | 14.000 | .416 | .000 | .81 | 2.59 | 2.9% | 21 | .929 | .920 | .911 | .899 | .886 | .871 | .853 | .832 | .807 | .778 |
| | | | | | | | | .745 | .706 | .662 | .612 | .557 | .497 | .433 | .368 | .303 | .242 |
| | | | | | | | | .187 | | | | | | | | | |
| 14.000 | 14.000 | .224 | .000 | .88 | 2.45 | 1.6% | 22 | .931 | .923 | .913 | .902 | .888 | .873 | .855 | .833 | .808 | .778 |
| | | | | | | | | .743 | .703 | .657 | .604 | .546 | .483 | .417 | .349 | .284 | .222 |
| | | | | | | | | .167 | .121 | | | | | | | | |
| 14.000 | 14.000 | .132 | .000 | .92 | 2.37 | .9% | 23 | .931 | .922 | .912 | .901 | .887 | .872 | .853 | .831 | .806 | .775 |
| | | | | | | | | .740 | .699 | .652 | .599 | .540 | .477 | .410 | .342 | .277 | .216 |
| | | | | | | | | .162 | .116 | .080 | | | | | | | |
| 15.000 | 15.000 | 1.019 | .000 | .69 | 2.94 | 6.4% | 21 | .926 | .918 | .909 | .899 | .888 | .875 | .860 | .843 | .824 | .801 |
| | | | | | | | | .775 | .746 | .713 | .675 | .633 | .587 | .536 | .483 | .426 | .369 |
| | | | | | | | | .312 | | | | | | | | | |
| 15.000 | 15.000 | .523 | .000 | .79 | 2.72 | 3.4% | 22 | .931 | .923 | .915 | .905 | .893 | .880 | .865 | .847 | .826 | .802 |
| | | | | | | | | .774 | .742 | .705 | .663 | .617 | .565 | .508 | .449 | .387 | .325 |
| | | | | | | | | .266 | .211 | | | | | | | | |
| 15.000 | 15.000 | .264 | .000 | .87 | 2.52 | 1.7% | 23 | .936 | .928 | .920 | .910 | .899 | .885 | .870 | .851 | .830 | .805 |
| | | | | | | | | .776 | .742 | .703 | .658 | .608 | .552 | .491 | .427 | .362 | .297 |
| | | | | | | | | .236 | .181 | .134 | | | | | | | |
| 15.000 | 15.000 | .159 | .000 | .91 | 2.44 | 1.0% | 24 | .935 | .928 | .919 | .909 | .898 | .884 | .868 | .850 | .828 | .803 |
| | | | | | | | | .773 | .738 | .699 | .653 | .602 | .545 | .484 | .419 | .354 | .289 |
| | | | | | | | | .229 | .175 | .128 | .090 | | | | | | |
| 15.000 | 15.000 | .080 | .000 | .95 | 2.29 | .5% | 25 | .938 | .931 | .922 | .912 | .901 | .888 | .872 | .853 | .831 | .806 |
| | | | | | | | | .776 | .740 | .699 | .653 | .600 | .541 | .477 | .411 | .343 | .277 |

*SIMILAR TO A.T.&T. JACOBSEN RETRIAL TABLES

*VERY FAST RETRIALS: JEWETT-SHRAGO VERY FAST RETRIAL TABLES - 100% RETRY, 0% OVERFLOW (RETRIES 1/3 HOLDING TIME LATER)

| HOURS FIRST ATTEMPT VMR = 1.0 | HOURS CON-NECT | HOURS RECALL /RETRY | HOURS OVFL | CON-NECT VMR | OVFL VMR | PERCENT BLOCKED (GRADE) | LINE | HOURS CONNECTED TRAFFIC ON LINES 1 | 2 | 3 | 4 | 5 | 6 | 7 | 8 | 9 | 10 |
|---|---|---|---|---|---|---|---|---|---|---|---|---|---|---|---|---|---|
| | | | | | | | | .216 | .162 | .117 | .080 | .053 | | | | | |
| 16.000 | 16.000 | 1.256 | .000 | .66 | 3.10 | 7.3% | 22 | .927 | .920 | .912 | .903 | .893 | .882 | .869 | .854 | .838 | .819 |
| | | | | | | | | .797 | .773 | .745 | .714 | .679 | .640 | .597 | .551 | .501 | .449 |
| | | | | | | | | .395 | .341 | | | | | | | | |
| 16.000 | 16.000 | .591 | .000 | .78 | 2.79 | 3.6% | 23 | .936 | .929 | .921 | .913 | .903 | .891 | .878 | .863 | .845 | .824 |
| | | | | | | | | .801 | .774 | .742 | .706 | .666 | .621 | .571 | .516 | .459 | .399 |
| | | | | | | | | .338 | .280 | .225 | | | | | | | |
| 16.000 | 16.000 | .341 | .000 | .85 | 2.66 | 2.1% | 24 | .937 | .930 | .923 | .914 | .904 | .892 | .879 | .863 | .845 | .824 |
| | | | | | | | | .799 | .771 | .738 | .701 | .658 | .610 | .558 | .501 | .441 | .379 |
| | | | | | | | | .317 | .258 | .203 | .155 | | | | | | |
| 16.000 | 16.000 | .188 | .000 | .90 | 2.52 | 1.2% | 25 | .939 | .933 | .925 | .916 | .906 | .895 | .881 | .865 | .847 | .825 |
| | | | | | | | | .800 | .771 | .737 | .698 | .654 | .605 | .550 | .491 | .428 | .364 |
| | | | | | | | | .301 | .241 | .187 | .140 | .100 | | | | | |
| 16.000 | 16.000 | .098 | .000 | .94 | 2.37 | .6% | 26 | .942 | .935 | .928 | .919 | .909 | .898 | .884 | .868 | .850 | .828 |
| | | | | | | | | .803 | .773 | .738 | .698 | .653 | .602 | .545 | .484 | .419 | .354 |
| | | | | | | | | .289 | .229 | .174 | .128 | .090 | .061 | | | | |
| 17.000 | 17.000 | 1.573 | .000 | .63 | 3.29 | 8.5% | 23 | .927 | .921 | .914 | .906 | .897 | .887 | .875 | .863 | .848 | .832 |
| | | | | | | | | .814 | .793 | .770 | .744 | .715 | .683 | .647 | .608 | .566 | .521 |
| | | | | | | | | .473 | .424 | .374 | | | | | | | |
| 17.000 | 17.000 | .712 | .000 | .75 | 2.91 | 4.0% | 24 | .938 | .931 | .925 | .917 | .908 | .898 | .886 | .873 | .858 | .840 |
| | | | | | | | | .820 | .797 | .771 | .741 | .706 | .668 | .625 | .578 | .527 | .472 |
| | | | | | | | | .415 | .358 | .301 | .247 | | | | | | |
| 17.000 | 17.000 | .389 | .000 | .83 | 2.73 | 2.2% | 25 | .941 | .935 | .928 | .920 | .912 | .901 | .890 | .876 | .861 | .843 |
| | | | | | | | | .822 | .798 | .770 | .738 | .701 | .660 | 614 | .560 | .508 | .449 |
| | | | | | | | | .380 | .329 | .270 | .216 | .167 | | | | | |
| 17.000 | 17.000 | .220 | .000 | .89 | 2.59 | 1.0% | 26 | .943 | .937 | .930 | .922 | .914 | .903 | .892 | .878 | .862 | .844 |
| | | | | | | | | .823 | .798 | .769 | .736 | .698 | .655 | .607 | .554 | .497 | .436 |
| | | | | | | | | .374 | .313 | .254 | .199 | .151 | .111 | | | | |
| 17.000 | 17.000 | .134 | .000 | .92 | 2.50 | .8% | 27 | .942 | .936 | .930 | .922 | .913 | .903 | .891 | .877 | .861 | .842 |
| | | | | | | | | .821 | .796 | .767 | .733 | .695 | .651 | .603 | .549 | .491 | .431 |
| | | | | | | | | .368 | .307 | .248 | .194 | .147 | .107 | .075 | | | |
| 18.000 | 18.000 | 1.905 | .000 | .60 | 3.46 | 0.6% | 24 | .928 | .922 | .915 | .908 | .900 | .891 | .881 | .870 | .857 | .843 |
| | | | | | | | | .828 | .810 | .790 | 769 | .744 | .717 | .687 | .654 | .619 | .580 |
| | | | | | | | | .539 | .495 | .449 | .402 | | | | | | |
| 18.000 | 18.000 | .841 | .000 | .73 | 3.02 | 4.5% | 25 | .939 | .934 | 928 | .021 | .913 | .904 | .894 | .882 | .869 | .854 |
| | | | | | | | | .836 | .817 | .794 | .769 | .740 | .707 | .670 | 630 | .505 | .536 |
| | | | | | | | | .485 | .431 | .375 | .321 | .267 | | | | | |
| 18.000 | 18.000 | .439 | .000 | .82 | 2.80 | 2.4% | 26 | .944 | .939 | .933 | .926 | 918 | .009 | .899 | .887 | .874 | .858 |
| | | | | | | | | .840 | .820 | .796 | .769 | .737 | .702 | .662 | .617 | .568 | .514 |
| | | | | | | | | .458 | .399 | .340 | .282 | .228 | .179 | | | | |

*SIMILAR TO A.T.&T. JACOBSEN RETRIAL TABLES

*VERY FAST RETRIALS: JEWETT-SHRAGO VERY FAST RETRIAL TABLES - 100% RETRY, 0% OVERFLOW RETRIES 1/3 HOLDING TIME LATER)

| HOURS FIRST ATTEMPT VMR=1.0 | HOURS CON-NECT | HOURS RECALL /RETRY | HOURS OVFL | CON-NECT VMR | OVFL VMR | PERCENT BLOCKED (GRADE) | LINE | HOURS CONNECTED TRAFFIC ON LINES 1 | 2 | 3 | 4 | 5 | 6 | 7 | 8 | 9 | 10 |
|---|---|---|---|---|---|---|---|---|---|---|---|---|---|---|---|---|---|
| 18.000 | 18.000 | .253 | .000 | .88 | 2.66 | 1.4% | 27 | .946 | .941 | .935 | .928 | .920 | .911 | .901 | .889 | .875 | .860 |
| | | | | | | | | .841 | .820 | .796 | .767 | .735 | .698 | .656 | .610 | .558 | .503 |
| | | | | | | | | .444 | .384 | .323 | .265 | .211 | .162 | .121 | | | |
| 18.000 | 18.000 | .157 | .000 | .92 | 2.57 | .9% | 28 | .946 | .940 | .934 | .927 | .919 | .910 | .900 | .888 | .874 | .858 |
| | | | | | | | | .840 | .818 | .793 | .765 | .732 | .694 | .652 | .605 | .553 | .497 |
| | | | | | | | | .438 | .377 | .317 | .259 | .205 | .157 | .117 | .083 | | |
| 19.000 | 19.000 | 2.461 | .000 | .56 | 3.73 | 11.5% | 25 | .925 | .920 | .914 | .907 | .900 | .892 | .883 | .873 | .862 | .850 |
| | | | | | | | | .836 | .821 | .805 | .787 | .766 | .744 | .720 | .693 | .664 | .632 |
| | | | | | | | | .598 | .561 | .523 | .482 | .440 | | | | | |
| 19.000 | 19.000 | .975 | .000 | .71 | 3.13 | 4.9% | 26 | .941 | .936 | .931 | .924 | .917 | .909 | .900 | .890 | .878 | .865 |
| | | | | | | | | .850 | .833 | .814 | .792 | .767 | .739 | .708 | .673 | .634 | .591 |
| | | | | | | | | .545 | .496 | .445 | .392 | .339 | .287 | | | | |
| 19.000 | 19.000 | .533 | .000 | .80 | 2.92 | 2.7% | 27 | .945 | .941 | .935 | .929 | .922 | .914 | .905 | .894 | .883 | .869 |
| | | | | | | | | .853 | .836 | .815 | .792 | .765 | .735 | .701 | .663 | .620 | .573 |
| | | | | | | | | .523 | .469 | .413 | .357 | .301 | .248 | .198 | | | |
| 19.000 | 19.000 | .318 | .000 | .86 | 2.78 | 1.6% | 28 | .947 | .942 | .936 | .930 | .923 | .915 | .906 | .895 | .883 | .870 |
| | | | | | | | | .854 | .836 | .815 | .791 | .763 | .732 | .696 | .656 | .612 | .563 |
| | | | | | | | | .510 | .455 | .397 | .339 | .283 | .229 | .181 | .138 | | |
| 19.000 | 19.000 | .182 | .000 | .91 | 2.64 | 1.0% | 29 | .948 | .944 | .938 | .932 | .925 | .917 | .908 | .897 | .885 | .872 |
| | | | | | | | | .856 | .837 | .816 | .791 | .763 | .731 | .694 | .653 | .607 | .557 |
| | | | | | | | | .502 | .445 | .386 | .326 | .269 | .216 | .168 | .126 | .092 | |
| 19.000 | 19.000 | .099 | .000 | .94 | 2.49 | .5% | 30 | .950 | .946 | .940 | .934 | .927 | .919 | .910 | .900 | .888 | .874 |
| | | | | | | | | .858 | .840 | .818 | .793 | .765 | .732 | .694 | .652 | .604 | .552 |
| | | | | | | | | .496 | .437 | .377 | .316 | .258 | .204 | .157 | .116 | .083 | .057 |
| 20.000 | 20.000 | 1.117 | .000 | .69 | 3.23 | 5.3% | 27 | .943 | .939 | .933 | .928 | .921 | .914 | .906 | .897 | .887 | .875 |
| | | | | | | | | .862 | .847 | .830 | .812 | .790 | .766 | .739 | .709 | .675 | .638 |
| | | | | | | | | .598 | .554 | .507 | .458 | .407 | .356 | .306 | | | |
| 20.000 | 20.000 | .632 | .000 | .78 | 3.03 | 3.1% | 28 | .947 | .942 | .937 | .931 | .925 | .918 | .910 | .901 | .890 | .878 |
| | | | | | | | | .865 | .849 | .832 | .812 | .789 | .763 | .734 | .701 | .664 | .623 |
| | | | | | | | | .579 | .531 | .480 | .427 | .372 | .318 | .266 | .217 | | |
| 20.000 | 20.000 | .357 | .000 | .85 | 2.85 | 1.8% | 29 | .950 | .945 | .940 | .935 | .928 | .921 | .913 | .904 | .893 | .881 |
| | | | | | | | | .867 | .852 | .834 | .813 | .789 | .762 | .731 | .697 | .658 | .614 |
| | | | | | | | | .567 | .516 | .462 | .406 | .349 | .293 | .240 | .191 | .148 | |
| 20.000 | 20.000 | .209 | .000 | .90 | 2.71 | 1.0% | 30 | .951 | .947 | .942 | .936 | .930 | .923 | .915 | .906 | .895 | .883 |
| | | | | | | | | .869 | .853 | .835 | .814 | .789 | .762 | .730 | .694 | .654 | .609 |
| | | | | | | | | .560 | .507 | .451 | .393 | .335 | .279 | .226 | .178 | .136 | .100 |
| 20.000 | 20.000 | .131 | .000 | .93 | 2.62 | .7% | 31 | .951 | .946 | .941 | .936 | .930 | .922 | .914 | .905 | .894 | .882 |
| | | | | | | | | .868 | .852 | .833 | .812 | .787 | .759 | .727 | .691 | .650 | .605 |
| | | | | | | | | .556 | .502 | .446 | .388 | .330 | .274 | .221 | .173 | .132 | .097 |
| | | | | | | | | .069 | | | | | | | | | |

*SIMILAR TO A.T.&T. JACOBSEN RETRIAL TABLES

*VERY FAST RETRIALS: JEWETT-SHRAGO VERY FAST RETRIAL TABLES - 100% RETRY, 0% OVERFLOW (RETRIES 1/3 HOLDING TIME LATER)

| HOURS FIRST ATTEMPT VMR = 1.0 | HOURS CON-NECT | HOURS RECALL /RETRY | HOURS OVFL | CON-NECT VMR | OVFL VMR | PERCENT BLOCKED (GRADE) | LINE | HOURS CONNECTED TRAFFIC ON LINES 1 | 2 | 3 | 4 | 5 | 6 | 7 | 8 | 9 | 10 |
|---|---|---|---|---|---|---|---|---|---|---|---|---|---|---|---|---|---|
| 22.000 | 22.000 | 1.540 | .000 | .65 | 3.51 | 6.5% | 29 | .944 | .940 | .936 | .931 | .926 | .920 | .913 | .906 | .897 | .888 |
| | | | | | | | | .878 | .866 | .854 | .839 | .823 | .806 | .786 | .763 | .739 | .712 |
| | | | | | | | | .682 | .649 | .613 | .574 | .533 | .489 | .444 | .398 | .351 | |
| 22.000 | 22.000 | .801 | .000 | .75 | 3.20 | 3.5% | 30 | .951 | .947 | .943 | .938 | .933 | .927 | .920 | .913 | .905 | .895 |
| | | | | | | | | .885 | .873 | .859 | .844 | .827 | .808 | .786 | .761 | .733 | .702 |
| | | | | | | | | .668 | .630 | .589 | .544 | .496 | .446 | .395 | .344 | .293 | .245 |
| 22.000 | 22.000 | .476 | .000 | .83 | 3.03 | 2.1% | 31 | .953 | .949 | .945 | .940 | .935 | .929 | .923 | .915 | .907 | .898 |
| | | | | | | | | .887 | .875 | .861 | .845 | .828 | .807 | .784 | .758 | .729 | .696 |
| | | | | | | | | .660 | .619 | .575 | .528 | .477 | .425 | .371 | .318 | .267 | .219 |
| | | | | | | | | .175 | | | | | | | | | |
| 22.000 | 22.000 | .267 | .000 | .88 | 2.84 | 1.2% | 32 | .955 | .952 | .948 | .943 | .938 | .932 | .926 | .919 | .910 | .901 |
| | | | | | | | | .890 | .878 | .864 | .848 | .830 | .810 | .786 | .759 | .728 | .694 |
| | | | | | | | | .656 | .613 | .566 | .516 | .463 | .408 | .352 | .298 | .245 | .197 |
| | | | | | | | | .154 | .117 | | | | | | | | |
| 22.000 | 22.000 | .172 | .000 | .92 | 2.75 | .8% | 33 | .955 | .952 | .948 | .943 | .938 | .932 | .926 | .918 | .910 | .900 |
| | | | | | | | | .890 | .877 | .863 | .847 | .829 | .808 | .784 | .756 | .725 | .691 |
| | | | | | | | | .652 | .609 | .562 | .511 | .458 | .402 | .347 | .292 | .240 | .192 |
| | | | | | | | | .149 | .113 | .083 | | | | | | | |
| 24.000 | 24.000 | 2.057 | .000 | .61 | 3.79 | 7.9% | 31 | .945 | .942 | .938 | .934 | .929 | .924 | .918 | .912 | .905 | .898 |
| | | | | | | | | .889 | .880 | .870 | .859 | .847 | .833 | .818 | .801 | .783 | .763 |
| | | | | | | | | .740 | .715 | .689 | .659 | .628 | .594 | .557 | .519 | .479 | .437 |
| | | | | | | | | .395 | | | | | | | | | |
| 24.000 | 24.000 | 1.034 | .000 | .72 | 3.40 | 4.1% | 32 | .953 | .950 | .946 | .942 | .937 | .933 | .927 | .921 | .914 | .907 |
| | | | | | | | | .898 | .889 | .878 | .867 | .853 | .838 | .822 | .803 | .782 | .759 |
| | | | | | | | | .733 | .704 | .672 | .637 | .599 | .558 | .514 | .468 | .421 | .373 |
| | | | | | | | | .325 | .278 | | | | | | | | |
| 24.000 | 24.000 | .608 | .000 | .80 | 3.20 | 2.5% | 33 | .956 | .953 | .949 | .945 | .941 | .936 | .930 | .924 | .918 | .910 |
| | | | | | | | | .902 | .892 | .881 | .869 | .855 | .840 | .823 | .803 | .781 | .756 |
| | | | | | | | | .728 | .697 | .662 | .624 | .583 | .539 | .492 | .442 | .392 | .341 |
| | | | | | | | | .292 | .244 | .200 | | | | | | | |
| 24.000 | 24.000 | .359 | .000 | .86 | 3.01 | 1.5% | 34 | .958 | .955 | .951 | .947 | .943 | .938 | .933 | .927 | .920 | .913 |
| | | | | | | | | .904 | .895 | .884 | .872 | .858 | .842 | .824 | .804 | .781 | .755 |
| | | | | | | | | .726 | .693 | .657 | .617 | .573 | .526 | .477 | .425 | .373 | .320 |
| | | | | | | | | .270 | .222 | .178 | .140 | | | | | | |
| 24.000 | 24.000 | .219 | .000 | .90 | 2.87 | .9% | 35 | .959 | .956 | .952 | .949 | .944 | .940 | .934 | .928 | .922 | .914 |
| | | | | | | | | .906 | .896 | .885 | .873 | .859 | .843 | .825 | .804 | .700 | .754 |
| | | | | | | | | .724 | .690 | .053 | .612 | .567 | .519 | .468 | .415 | .361 | .308 |
| | | | | | | | | .257 | .209 | .166 | .129 | .097 | | | | | |
| 24.000 | 24.000 | .142 | .000 | .93 | 2.78 | .6% | 36 | .059 | .956 | .952 | .948 | .944 | .939 | .934 | .928 | .921 | .914 |
| | | | | | | | | .905 | .896 | .885 | .872 | .058 | .842 | .823 | .802 | .779 | .752 |

*SIMILAR TO A.T.&T. JACOBSEN RETRIAL TABLES

*VERY FAST RETRIALS: JEWETT-SHRAGO VERY FAST RETRIAL TABLES - 100% RETRY, 0% OVERFLOW (RETRIES 1/3 HOLDING TIME LATER)

| HOURS FIRST ATTEMPT VMR = 1.0 | HOURS CON-NECT | HOURS RECALL /RETRY | HOURS OVFL | CON-NECT VMR | OVFL VMR | PERCENT BLOCKED (GRADE) | LINE | HOURS CONNECTED TRAFFIC ON LINES 1 | 2 | 3 | 4 | 5 | 6 | 7 | 8 | 9 | 10 |
|---|---|---|---|---|---|---|---|---|---|---|---|---|---|---|---|---|---|
| | | | | | | | | .722 | .688 | .650 | .609 | .563 | .515 | .463 | .410 | .357 | .304 |
| | | | | | | | | .253 | .205 | .163 | .125 | .094 | .069 | | | | |
| 26.000 | 26.000 | 2.869 | .000 | .55 | 4.16 | 9.9% | 33 | .944 | .941 | .937 | .933 | .929 | .925 | .920 | .915 | .909 | .903 |
| | | | | | | | | .896 | .888 | .880 | .871 | .861 | .851 | .839 | .826 | .812 | .797 |
| | | | | | | | | .780 | .762 | .742 | .721 | .698 | .672 | .645 | .616 | .585 | .553 |
| | | | | | | | | .518 | .482 | .446 | | | | | | | |
| 26.000 | 26.000 | 1.284 | .000 | .69 | 3.58 | 4.7% | 34 | .955 | .952 | .949 | .946 | .942 | .938 | .933 | .928 | .922 | .916 |
| | | | | | | | | .909 | .902 | .893 | .884 | .873 | .862 | .849 | .834 | .818 | .800 |
| | | | | | | | | .780 | .758 | .733 | .706 | .676 | .644 | .608 | .570 | .530 | .487 |
| | | | | | | | | .443 | .398 | .352 | .308 | | | | | | |
| 26.000 | 26.000 | .751 | .000 | .78 | 3.35 | 2.8% | 35 | .958 | .956 | .953 | .949 | .945 | .941 | .937 | .932 | .926 | .920 |
| | | | | | | | | .913 | .905 | .897 | .887 | .877 | .864 | .851 | .836 | .819 | .799 |
| | | | | | | | | .778 | .754 | .727 | .698 | .665 | .629 | .591 | .549 | .505 | .458 |
| | | | | | | | | .410 | .362 | .314 | .268 | .224 | | | | | |
| 26.000 | 26.000 | .462 | .000 | .84 | 3.18 | 1.7% | 36 | .960 | .957 | .954 | .951 | .947 | .943 | .939 | .934 | .928 | .922 |
| | | | | | | | | .915 | .908 | .899 | .889 | .878 | .866 | .852 | .836 | .819 | .799 |
| | | | | | | | | .777 | .752 | .724 | .693 | .658 | .621 | .580 | .536 | .489 | .440 |
| | | | | | | | | .391 | .341 | .292 | .245 | .201 | .162 | | | | |
| 26.000 | 26.000 | .293 | .000 | .88 | 3.05 | 1.1% | 37 | .961 | .958 | .955 | .952 | .948 | .944 | .940 | .935 | .929 | .923 |
| | | | | | | | | .916 | .909 | .900 | .890 | .879 | .867 | .853 | .837 | .819 | .798 |
| | | | | | | | | .776 | .750 | .721 | .689 | .654 | .615 | .573 | .527 | .479 | .430 |
| | | | | | | | | .379 | .328 | .279 | .232 | .189 | .150 | .116 | | | |
| 26.000 | 26.000 | .179 | .000 | .92 | 2.90 | .7% | 38 | .962 | .959 | .956 | .953 | .950 | .946 | .941 | .936 | .931 | .925 |
| | | | | | | | | .918 | .910 | .901 | .891 | .880 | .868 | .854 | .838 | .819 | .799 |
| | | | | | | | | .776 | .749 | .720 | .687 | .651 | .611 | .568 | .521 | .472 | .421 |
| | | | | | | | | .370 | .318 | .268 | .221 | .178 | .140 | .107 | .080 | | |
| 28.000 | 28.000 | 1.606 | .000 | .66 | 3.79 | 5.4% | 36 | .956 | .954 | .951 | .948 | .945 | .941 | .937 | .933 | .928 | .923 |
| | | | | | | | | .917 | .911 | .904 | .896 | .888 | .878 | .868 | .856 | .844 | .830 |
| | | | | | | | | .814 | .797 | .778 | .757 | .734 | .709 | .681 | .651 | .619 | .584 |
| | | | | | | | | .547 | .508 | .467 | .425 | .382 | .340 | | | | |
| 28.000 | 28.000 | .906 | .000 | .75 | 3.50 | 3.1% | 37 | .961 | .958 | .956 | .953 | .950 | .946 | .942 | .938 | .933 | .928 |
| | | | | | | | | .922 | .916 | .909 | .901 | .893 | .883 | .872 | .860 | .847 | .832 |
| | | | | | | | | .815 | .797 | .776 | .753 | .728 | .699 | .668 | .635 | .598 | .559 |
| | | | | | | | | .517 | .473 | .427 | .381 | .335 | .290 | .246 | | | |
| 28.000 | 28.000 | .542 | .000 | .82 | 3.29 | 1.9% | 38 | .963 | .961 | .958 | .955 | .952 | .949 | .945 | .941 | .936 | .931 |
| | | | | | | | | .926 | .919 | .912 | .905 | .896 | .886 | .875 | .863 | .849 | .834 |
| | | | | | | | | .816 | .797 | .775 | .751 | .724 | .694 | .661 | .624 | .585 | .543 |
| | | | | | | | | .498 | .451 | .403 | .355 | .307 | .261 | .217 | .177 | | |
| 28.000 | 28.000 | .351 | .000 | .87 | 3.16 | 1.2% | 39 | .964 | .962 | .959 | .956 | .953 | .950 | .946 | .942 | .937 | .932 |
| | | | | | | | | .926 | .920 | .913 | .905 | .897 | .887 | .876 | .863 | .849 | .833 |

*SIMILAR TO A.T.&T. JACOBSEN RETRIAL TABLES

*VERY FAST RETRIALS: JEWETT-SHRAGO VERY FAST RETRIAL TABLES - 100% RETRY, 0% OVERFLOW (RETRIES 1/3 HOLDING TIME LATER)

| HOURS FIRST ATTEMPT VMR = 1.0 | HOURS CON-NECT | HOURS RECALL /RETRY | HOURS OVFL | CON-NECT VMR | OVFL VMR | PERCENT BLOCKED (GRADE) | LINE | HOURS CONNECTED TRAFFIC ON LINES 1 | 2 | 3 | 4 | 5 | 6 | 7 | 8 | 9 | 10 |
|---|---|---|---|---|---|---|---|---|---|---|---|---|---|---|---|---|---|
| | | | | | | | | .816 | .796 | .773 | .748 | .721 | .690 | .655 | .618 | .578 | .534 |
| | | | | | | | | .488 | .440 | .391 | .342 | .293 | .247 | .204 | .165 | .130 | |
| 28.000 | 28.000 | .221 | .000 | .91 | 3.02 | .8% | 40 | .965 | .962 | .960 | .957 | .954 | .951 | .947 | .943 | .938 | .933 |
| | | | | | | | | .928 | .921 | .914 | .907 | .898 | .888 | .877 | .864 | .850 | .834 |
| | | | | | | | | .816 | .796 | .773 | .747 | .719 | .687 | .652 | .614 | .572 | .528 |
| | | | | | | | | .481 | .432 | .382 | .332 | .283 | .236 | .193 | .154 | .121 | .092 |
| 28.000 | 28.000 | .147 | .000 | .93 | 2.93 | .5% | 41 | .965 | .962 | .960 | .957 | .954 | .951 | .947 | .943 | .938 | .933 |
| | | | | | | | | .927 | .921 | .914 | .906 | .897 | .887 | .876 | .863 | .849 | .833 |
| | | | | | | | | .815 | .794 | .771 | .746 | .717 | .685 | .650 | .611 | .569 | .524 |
| | | | | | | | | .477 | .428 | .377 | .327 | .279 | .232 | .190 | .151 | .118 | .089 |
| | | | | | | | | .066 | | | | | | | | | |
| 30.000 | 30.000 | 2.005 | .000 | .63 | 4.02 | 6.3% | 38 | .957 | .955 | .952 | .950 | .947 | .943 | .940 | .936 | .932 | .927 |
| | | | | | | | | .923 | .917 | .911 | .905 | .898 | .890 | .882 | .873 | .863 | .851 |
| | | | | | | | | .839 | .826 | .811 | .794 | .777 | .757 | .735 | .712 | .687 | .659 |
| | | | | | | | | .629 | .598 | .564 | .528 | .491 | .452 | .413 | .373 | | |
| 30.000 | 30.000 | 1.116 | .000 | .73 | 3.68 | 3.6% | 39 | .962 | .960 | .958 | .955 | .952 | .949 | .946 | .942 | .938 | .934 |
| | | | | | | | | .929 | .924 | .918 | .911 | .904 | .896 | .888 | .878 | .867 | .856 |
| | | | | | | | | .842 | .828 | .812 | .794 | .774 | .752 | .728 | .701 | .672 | .640 |
| | | | | | | | | .606 | .569 | .530 | .489 | .447 | .403 | .359 | .315 | .273 | |
| 30.000 | 30.000 | .662 | .000 | .80 | 3.44 | 2.2% | 40 | .965 | .963 | .961 | .958 | .955 | .952 | .949 | .946 | .942 | .937 |
| | | | | | | | | .933 | .927 | .922 | .915 | .908 | .900 | .891 | .882 | .871 | .859 |
| | | | | | | | | .845 | .830 | .813 | .794 | .773 | .749 | .723 | .694 | .663 | .628 |
| | | | | | | | | .591 | .551 | .509 | .465 | .419 | .373 | .326 | .281 | .238 | .198 |
| 30.000 | 30.000 | .441 | .000 | .85 | 3.32 | 1.4% | 41 | .966 | .963 | .961 | .959 | .956 | .953 | .950 | .946 | .942 | .938 |
| | | | | | | | | .933 | .928 | .922 | .916 | .909 | .901 | .892 | .882 | .871 | .858 |
| | | | | | | | | .844 | .829 | .811 | .792 | .770 | .746 | .719 | .689 | .657 | .621 |
| | | | | | | | | .583 | .542 | .498 | .453 | .406 | .359 | .312 | .267 | .224 | .184 |
| | | | | | | | | .149 | | | | | | | | | |
| 30.000 | 30.000 | .286 | .000 | .89 | 3.18 | .9% | 42 | .966 | .964 | .962 | .959 | .957 | .954 | .951 | .947 | .943 | .939 |
| | | | | | | | | .934 | .929 | .923 | .917 | .909 | .901 | .892 | .882 | .871 | .859 |
| | | | | | | | | .844 | .829 | .811 | .791 | .769 | .744 | .717 | .686 | .653 | .616 |
| | | | | | | | | .577 | .535 | .490 | .443 | .396 | .348 | .301 | .255 | .213 | .173 |
| | | | | | | | | .138 | .108 | | | | | | | | |
| 30.000 | 30.000 | .181 | .000 | .92 | 3.04 | .6% | 43 | .967 | .965 | .963 | .960 | .958 | .955 | .952 | .948 | .944 | .940 |
| | | | | | | | | .935 | .930 | .924 | .910 | .911 | .903 | .894 | .884 | .872 | .860 |
| | | | | | | | | .845 | .829 | .811 | .791 | .769 | .744 | .716 | .685 | .650 | .613 |
| | | | | | | | | .573 | .529 | .484 | .436 | .388 | .339 | .292 | .246 | .203 | .165 |
| | | | | | | | | .130 | .101 | .076 | | | | | | | |

*SIMILAR TO A.T.&T. JACOBSEN RETRIAL TABLES

# SECTION 3
# RANDOM TRAFFIC WITH QUEUING

JEWETT EQUIVALENT QUEUE EXTENDED ERLANG B(EQEEB): 1 MINUTE **MAXIMUM** WAIT (IF NOT CONNECTED AFTER 1 MINUTE WAIT, 100% OVERFLOW)

| HOURS FIRST ATTEMPT VMR = 1.0 | HOURS CON-NECT | HOURS OVFL | DELAYED PORTION | DELAY LENGTH CALL FACTOR | LINE | HOURS CONNECTED TRAFFIC ON LINES 1 | 2 | 3 | 4 | 5 | 6 | 7 | 8 | 9 | 10 |
|---|---|---|---|---|---|---|---|---|---|---|---|---|---|---|---|
| .200 | .180 | .020 | .190 | .141 | 1 | .180 | | | | | | | | | |
| .400 | .321 | .079 | .355 | .252 | 1 | .321 | | | | | | | | | |
| .400 | .388 | .012 | .058 | .044 | 2 | .291 | .097 | | | | | | | | |
| .600 | .564 | .036 | .114 | .086 | 2 | .388 | .176 | | | | | | | | |
| .800 | .724 | .076 | .178 | .132 | 2 | .465 | .258 | | | | | | | | |
| .800 | .783 | .017 | .041 | .032 | 3 | .449 | .240 | .093 | | | | | | | |
| 1.000 | .866 | .134 | .246 | .180 | 2 | .529 | .337 | | | | | | | | |
| 1.000 | .964 | .036 | .069 | .052 | 3 | .508 | .310 | .146 | | | | | | | |
| 1.200 | .991 | .209 | .315 | .226 | 2 | .582 | .409 | | | | | | | | |
| 1.200 | 1.136 | .064 | .102 | .077 | 3 | .557 | .375 | .203 | | | | | | | |
| 1.200 | 1.183 | .017 | .028 | .021 | 4 | .549 | .363 | .192 | .079 | | | | | | |
| 1.400 | 1.100 | .300 | .383 | .270 | 2 | .626 | .473 | | | | | | | | |
| 1.400 | 1.296 | .104 | .140 | .105 | 3 | .599 | .434 | .263 | | | | | | | |
| 1.400 | 1.369 | .031 | .043 | .033 | 4 | .588 | .418 | .246 | .116 | | | | | | |
| 1.600 | 1.444 | .156 | .181 | .135 | 3 | .636 | .486 | .322 | | | | | | | |
| 1.600 | 1.548 | .052 | .062 | .048 | 4 | .622 | .467 | .300 | .159 | | | | | | |
| 1.800 | 1.580 | .220 | .225 | .165 | 3 | .667 | .534 | .380 | | | | | | | |
| 1.800 | 1.720 | .080 | .085 | .065 | 4 | .652 | .511 | .352 | .205 | | | | | | |
| 1.800 | 1.774 | .026 | .028 | .022 | 5 | .646 | .502 | .341 | .194 | .091 | | | | | |
| 2.000 | 1.704 | .296 | .271 | .196 | 3 | .695 | .576 | .433 | | | | | | | |
| 2.000 | 1.883 | .117 | .111 | .084 | 4 | .678 | .551 | .401 | .253 | | | | | | |
| 2.000 | 1.959 | .041 | .040 | .030 | 5 | .671 | .540 | .387 | .238 | .123 | | | | | |
| 2.200 | 1.815 | .385 | .317 | .227 | 3 | .719 | .613 | .483 | | | | | | | |
| 2.200 | 2.037 | .163 | .139 | .104 | 4 | .702 | .587 | .448 | .301 | | | | | | |
| 2.200 | 2.138 | .062 | .054 | .041 | 5 | .693 | .574 | .430 | .283 | .159 | | | | | |
| 2.400 | 1.915 | .485 | .363 | .257 | 3 | .740 | .646 | .528 | | | | | | | |
| 2.400 | 2.182 | .218 | .170 | .127 | 4 | .722 | .619 | .491 | .350 | | | | | | |
| 2.400 | 2.312 | .088 | .070 | .054 | 5 | .713 | .604 | .471 | .327 | .197 | | | | | |
| 2.600 | 2.004 | .596 | .407 | .285 | 3 | .759 | .676 | .569 | | | | | | | |
| 2.600 | 2.316 | .284 | .203 | .150 | 4 | .741 | .648 | .531 | .397 | | | | | | |
| 2.600 | 2.478 | .122 | .080 | .068 | 5 | .731 | .632 | .508 | .370 | .238 | | | | | |
| 2.600 | 2.553 | .047 | .035 | .027 | 6 | .726 | .624 | .497 | .358 | .226 | .123 | | | | |
| 2.800 | 2.441 | .359 | .236 | .173 | 4 | .758 | .674 | .567 | .442 | | | | | | |
| 2.800 | 2.638 | .162 | .110 | .083 | 5 | .747 | .657 | .543 | .412 | .279 | | | | | |
| 2.800 | 2.733 | .067 | .046 | .035 | 6 | .741 | .648 | .530 | .397 | .264 | .153 | | | | |
| 3.000 | 2.555 | .445 | .271 | .197 | 4 | .773 | .698 | .600 | .484 | | | | | | |
| 3.000 | 2.789 | .211 | .132 | .100 | 5 | .762 | .680 | .575 | .451 | .321 | | | | | |
| 3.000 | 2.909 | .091 | .058 | .045 | 6 | .755 | .670 | .561 | .434 | .303 | .186 | | | | |
| 3.200 | 2.660 | .540 | .306 | .220 | 4 | .787 | .719 | .631 | .523 | | | | | | |
| 3.200 | 2.932 | .268 | .157 | .117 | 5 | .775 | .701 | .604 | .489 | .363 | | | | | |

JEWETT EQUIVALENT QUEUE EXTENDED ERLANG B(EQEEB): 1 MINUTE MAXIMUM WAIT (IF NOT CONNECTED AFTER 1 MINUTE WAIT, 100% OVERFLOW)

| HOURS FIRST ATTEMPT VMR = 1.0 | HOURS CON-NECT | HOURS OVFL | DELAYED PORTION | DELAY LENGTH CALL FACTOR | LINE | HOURS CONNECTED TRAFFIC ON LINES 1 | 2 | 3 | 4 | 5 | 6 | 7 | 8 | 9 | 10 |
|---|---|---|---|---|---|---|---|---|---|---|---|---|---|---|---|
| 3.200 | 3.079 | .121 | .072 | .055 | 6 | .768 | .690 | .589 | .469 | .341 | .221 | | | | |
| 3.200 | 3.150 | .050 | .030 | .023 | 7 | .765 | .684 | .581 | .459 | .330 | .211 | .119 | | | |
| 3.400 | 2.756 | .644 | .341 | .243 | 4 | .800 | .738 | .658 | .559 | | | | | | |
| 3.400 | 3.067 | .333 | .183 | .135 | 5 | .788 | .720 | .632 | .524 | .404 | | | | | |
| 3.400 | 3.242 | .158 | .088 | .067 | 6 | .780 | .708 | .615 | .503 | .379 | .258 | | | | |
| 3.400 | 3.331 | .069 | .039 | .030 | 7 | .776 | .702 | .606 | .491 | .366 | .245 | .145 | | | |
| 3.600 | 2.843 | .757 | .376 | .266 | 4 | .811 | .756 | .683 | .593 | | | | | | |
| 3.600 | 3.193 | .407 | .210 | .154 | 5 | .799 | .737 | .657 | .557 | .443 | | | | | |
| 3.600 | 3.400 | .200 | .106 | .080 | 6 | .791 | .725 | .639 | .534 | .415 | .295 | | | | |
| 3.600 | 3.509 | .091 | .049 | .037 | 7 | .787 | .718 | .629 | .521 | .400 | .279 | .174 | | | |
| 3.800 | 2.922 | .878 | .411 | .287 | 4 | .822 | .771 | .706 | .623 | | | | | | |
| 3.800 | 3.311 | .489 | .237 | .174 | 5 | .810 | .753 | .679 | .588 | .480 | | | | | |
| 3.800 | 3.550 | .250 | .124 | .094 | 6 | .801 | .740 | .661 | .564 | .451 | .332 | | | | |
| 3.800 | 3.682 | .118 | .060 | .046 | 7 | .796 | .733 | .651 | .549 | .434 | .314 | .205 | | | |
| 3.800 | 3.748 | .052 | .026 | .020 | 8 | .794 | .729 | .645 | .542 | .425 | .305 | .197 | .113 | | |
| 4.000 | 3.420 | .580 | .266 | .193 | 5 | .819 | .768 | .700 | .616 | .516 | | | | | |
| 4.000 | 3.693 | .307 | .144 | .108 | 6 | .811 | .755 | .682 | .591 | .485 | .369 | | | | |
| 4.000 | 3.849 | .151 | .072 | .055 | 7 | .805 | .747 | .670 | .576 | .466 | .348 | .237 | | | |
| 4.000 | 3.931 | .069 | .033 | .025 | 8 | .803 | .742 | .664 | .567 | .455 | .337 | .226 | .136 | | |
| 4.200 | 3.521 | .679 | .294 | .212 | 5 | .828 | .781 | .720 | .642 | .549 | | | | | |
| 4.200 | 3.828 | .372 | .166 | .123 | 6 | .820 | .768 | .701 | .617 | .517 | .406 | | | | |
| 4.200 | 4.011 | .189 | .086 | .065 | 7 | .814 | .759 | .689 | .600 | .496 | .382 | .270 | | | |
| 4.200 | 4.111 | .089 | .041 | .031 | 8 | .811 | .755 | .682 | .591 | .484 | .369 | .257 | .162 | | |
| 4.400 | 3.614 | .786 | .323 | .231 | 5 | .837 | .793 | .737 | .666 | .580 | | | | | |
| 4.400 | 3.956 | .444 | .188 | .139 | 6 | .828 | .780 | .719 | .641 | .547 | .441 | | | | |
| 4.400 | 4.167 | .233 | .100 | .076 | 7 | .822 | .771 | .706 | .624 | .525 | .415 | .304 | | | |
| 4.400 | 4.286 | .114 | .050 | .038 | 8 | .818 | .766 | .698 | .613 | .512 | .400 | .289 | .189 | | |
| 4.600 | 3.700 | .900 | .352 | .250 | 5 | .845 | .805 | .753 | .688 | .609 | | | | | |
| 4.600 | 4.076 | .524 | .211 | .155 | 6 | .836 | .792 | .735 | .663 | .576 | .475 | | | | |
| 4.600 | 4.317 | .283 | .116 | .088 | 7 | .829 | .782 | .722 | .645 | .553 | .448 | .338 | | | |
| 4.600 | 4.458 | .142 | .059 | .045 | 8 | .826 | .777 | .714 | .634 | .539 | .431 | .320 | .218 | | |
| 4.600 | 4.533 | .067 | .028 | .022 | 9 | .823 | .774 | .709 | .628 | .531 | .422 | .311 | .209 | .127 | |
| 4.800 | 3.779 | 1.021 | .380 | .268 | 5 | .852 | .815 | .768 | .709 | .635 | | | | | |
| 4.800 | 4.189 | .611 | .235 | .172 | 6 | .843 | .802 | .750 | .684 | .603 | .508 | | | | |
| 4.800 | 4.461 | .339 | .133 | .100 | 7 | .836 | .793 | .737 | .665 | .579 | .479 | .372 | | | |
| 4.800 | 4.624 | .176 | .070 | .054 | 8 | .832 | .787 | .728 | .654 | .564 | .460 | .352 | .247 | | |
| 4.800 | 4.714 | .086 | .034 | .026 | 9 | .830 | .783 | .723 | .647 | .555 | .450 | .340 | .237 | .150 | |
| 5.000 | 3.851 | 1.149 | .408 | .286 | 5 | .858 | .825 | .781 | .727 | .659 | | | | | |
| 5.000 | 4.295 | .705 | .259 | .188 | 6 | .850 | .812 | .764 | .703 | .628 | .539 | | | | |
| 5.000 | 4.597 | .403 | .151 | .113 | 7 | .843 | .803 | .750 | .684 | .603 | .509 | .405 | | | |

JEWETT EQUIVALENT QUEUE EXTENDED ERLANG B(EQEEB): 1 MINUTE MAXIMUM WAIT (IF NOT CONNECTED AFTER 1 MINUTE WAIT, 100% OVERFLOW)

| HOURS FIRST ATTEMPT VMR = 1.0 | HOURS CON-NECT | HOURS OVFL | DELAYED PORTION | DELAY LENGTH CALL FACTOR | LINE | HOURS CONNECTED TRAFFIC ON LINES 1 | 2 | 3 | 4 | 5 | 6 | 7 | 8 | 9 | 10 |
|---|---|---|---|---|---|---|---|---|---|---|---|---|---|---|---|
| 5.000 | 4.785 | .215 | .082 | .062 | 8 | .839 | .796 | .741 | .672 | .587 | .489 | .383 | .278 | | |
| 5.000 | 4.892 | .108 | .041 | .032 | 9 | .836 | .792 | .736 | .664 | .577 | .477 | .370 | .265 | .174 | |
| 5.500 | 4.528 | .972 | .320 | .229 | 6 | .865 | .834 | .794 | .744 | .683 | .608 | | | | |
| 5.500 | 4.909 | .591 | .200 | .147 | 7 | .858 | .824 | .781 | .726 | .658 | .577 | .484 | | | |
| 5.500 | 5.163 | .337 | .116 | .087 | 8 | .853 | .817 | .771 | .713 | .641 | .555 | .458 | .356 | | |
| 5.500 | 5.319 | .181 | .063 | .048 | 9 | .850 | .813 | .765 | .704 | .629 | .540 | .441 | .338 | .240 | |
| 5.500 | 5.409 | .091 | .032 | .025 | 10 | .848 | .810 | .761 | .699 | .622 | .532 | .431 | .327 | .230 | .149 |
| 6.000 | 4.722 | 1.278 | .381 | .269 | 6 | .878 | .852 | .819 | .779 | .728 | .666 | | | | |
| 6.000 | 5.179 | .821 | .251 | .183 | 7 | .871 | .843 | .807 | .761 | .705 | .636 | .556 | | | |
| 6.000 | 5.504 | .496 | .155 | .116 | 8 | .866 | .835 | .796 | .747 | .687 | .613 | .527 | .432 | | |
| 6.000 | 5.718 | .282 | .090 | .068 | 9 | .862 | .830 | .789 | .738 | .673 | .596 | .507 | .410 | .312 | |
| 6.000 | 5.848 | .152 | .049 | .037 | 10 | .860 | .827 | .785 | .731 | .665 | .586 | .494 | .396 | .297 | .207 |
| 6.500 | 5.408 | 1.092 | .305 | .219 | 7 | .882 | .858 | .828 | .791 | .744 | .687 | .618 | | | |
| 6.500 | 5.806 | .694 | .198 | .147 | 8 | .877 | .851 | .818 | .777 | .726 | .663 | .589 | .504 | | |
| 6.500 | 6.084 | .416 | .121 | .091 | 9 | .873 | .846 | .811 | .767 | .712 | .646 | .567 | .479 | .384 | |
| 6.500 | 6.263 | .237 | .070 | .053 | 10 | .871 | .842 | .805 | .760 | .703 | .633 | .552 | .461 | .365 | .272 |
| 6.500 | 6.372 | .128 | .038 | .029 | 11 | .869 | .839 | .802 | .755 | .696 | .626 | .543 | .450 | .354 | .260 |
| | | | | | | .178 | | | | | | | | | |
| 7.000 | 5.602 | 1.398 | .359 | .254 | 7 | .892 | .872 | .847 | .815 | .776 | .729 | .671 | | | |
| 7.000 | 6.070 | .930 | .244 | .179 | 8 | .887 | .865 | .837 | .802 | .759 | .707 | .643 | .569 | | |
| 7.000 | 6.414 | .586 | .157 | .117 | 9 | .883 | .859 | .829 | .792 | .745 | .689 | .621 | .542 | .455 | |
| 7.000 | 6.650 | .350 | .095 | .072 | 10 | .880 | .855 | .823 | .784 | .735 | .675 | .604 | .522 | .432 | .340 |
| 7.000 | 6.801 | .199 | .055 | .042 | 11 | .878 | .852 | .819 | .779 | .728 | .666 | .593 | .509 | .417 | .324 |
| | | | | | | .238 | | | | | | | | | |
| 7.000 | 6.893 | .107 | .030 | .020 | 12 | .877 | .850 | .817 | .775 | .724 | .661 | .586 | .500 | .408 | .315 |
| | | | | | | .227 | .153 | | | | | | | | |
| 7.500 | 5.765 | 1.735 | .411 | .288 | 7 | .901 | .884 | .862 | .836 | .803 | .764 | .715 | | | |
| 7.500 | 6.297 | 1.203 | .292 | .211 | 8 | .896 | .877 | .853 | .824 | .788 | .743 | .690 | .626 | | |
| 7.500 | 6.708 | .792 | .196 | .145 | 9 | .892 | .871 | .845 | .813 | .774 | .726 | .607 | .599 | .521 | |
| 7.500 | 7.005 | .495 | .125 | .094 | 10 | .888 | .867 | .839 | .805 | .763 | .712 | .650 | .577 | .495 | .407 |
| 7.500 | 7.205 | .295 | .075 | .057 | 11 | .886 | .863 | .835 | .799 | .756 | .702 | .637 | .562 | .478 | .388 |
| | | | | | | .299 | | | | | | | | | |
| 7.500 | 7.333 | .167 | .043 | .033 | 12 | .884 | .861 | .832 | .796 | .750 | .695 | .629 | .552 | .466 | .376 |
| | | | | | | .286 | .205 | | | | | | | | |
| 8.000 | 6.491 | 1.509 | .340 | .242 | 8 | .904 | .887 | .867 | .842 | .812 | .774 | .729 | .675 | | |
| 8.000 | 6.967 | 1.033 | .238 | .174 | 9 | .900 | .882 | .859 | .832 | .799 | .758 | .708 | .649 | .580 | |
| 8.000 | 7.326 | .674 | .158 | .118 | 10 | .896 | .877 | .853 | .824 | .788 | .744 | .690 | .627 | .554 | .473 |
| 8.000 | 7.581 | .419 | .099 | .075 | 11 | .894 | .873 | .849 | .818 | .780 | .733 | .677 | .610 | .534 | .450 |
| | | | | | | .363 | | | | | | | | | |
| 8.000 | 7.751 | .249 | .060 | .046 | 12 | .892 | .871 | .845 | .813 | .774 | .726 | .667 | .599 | .520 | .435 |

JEWETT EQUIVALENT QUEUE EXTENDED ERLANG B(EQEEB): 1 MINUTE MAXIMUM WAIT (IF NOT CONNECTED AFTER 1 MINUTE WAIT, 100% OVERFLOW)

| HOURS FIRST ATTEMPT VMR = 1.0 | HOURS CON-NECT | HOURS OVFL | DELAYED PORTION | DELAY LENGTH CALL FACTOR | LINE | HOURS CONNECTED TRAFFIC ON LINES 1 | 2 | 3 | 4 | 5 | 6 | 7 | 8 | 9 | 10 |
|---|---|---|---|---|---|---|---|---|---|---|---|---|---|---|---|
| | | | | | | | | .347 | .262 | | | | | | |
| 8.000 | 7.859 | .141 | .034 | .026 | 13 | .890 | .869 | .843 | .811 | .770 | .721 | .661 | .591 | .511 | .425 |
| | | | | | | .337 | .252 | .177 | | | | | | | |
| 8.500 | 6.656 | 1.844 | .387 | .273 | 8 | .911 | .896 | .879 | .858 | .832 | .800 | .762 | .717 | | |
| 8.500 | 7.192 | 1.308 | .281 | .203 | 9 | .907 | .891 | .872 | .848 | .820 | .785 | .743 | .692 | .633 | |
| 8.500 | 7.614 | .886 | .194 | .143 | 10 | .903 | .886 | .866 | .840 | .809 | .772 | .726 | .671 | .607 | .534 |
| 8.500 | 7.927 | .573 | .127 | .096 | 11 | .900 | .883 | .861 | .834 | .801 | .761 | .712 | .654 | .586 | .510 |
| | | | | | | .427 | | | | | | | | | |
| 8.500 | 8.146 | .354 | .079 | .061 | 12 | .898 | .880 | .857 | .829 | .795 | .753 | .701 | .641 | .570 | .492 |
| | | | | | | .407 | .322 | | | | | | | | |
| 8.500 | 8.290 | .210 | .047 | .036 | 13 | .897 | .878 | .855 | .826 | .790 | .747 | .694 | .632 | .560 | .480 |
| | | | | | | .394 | .309 | .229 | | | | | | | |
| 8.500 | 8.381 | .119 | .027 | .021 | 14 | .896 | .877 | .853 | .824 | .788 | .743 | .690 | .626 | .553 | .472 |
| | | | | | | .386 | .300 | .221 | .153 | | | | | | |
| 9.000 | 7.386 | 1.614 | .324 | .232 | 9 | .913 | .899 | .883 | .863 | .838 | .809 | .773 | .730 | .679 | |
| 9.000 | 7.869 | 1.131 | .232 | .170 | 10 | .910 | .895 | .877 | .855 | .828 | .796 | .756 | .709 | .654 | .589 |
| 9.000 | 8.242 | .758 | .158 | .118 | 11 | .907 | .891 | .872 | .848 | .820 | .785 | .743 | .692 | .632 | .564 |
| | | | | | | .488 | | | | | | | | | |
| 9.000 | 8.513 | .487 | .103 | .078 | 12 | .904 | .888 | .868 | .843 | .813 | .776 | .732 | .678 | .616 | .545 |
| | | | | | | .466 | .384 | | | | | | | | |
| 9.000 | 8.700 | .300 | .064 | .049 | 13 | .903 | .886 | .865 | .840 | .808 | .770 | .724 | .669 | .604 | .531 |
| | | | | | | .450 | .367 | .285 | | | | | | | |
| 9.000 | 8.823 | .177 | .038 | .029 | 14 | .902 | .884 | .863 | .837 | .805 | .766 | .718 | .662 | .596 | .521 |
| | | | | | | .440 | .356 | .274 | .200 | | | | | | |
| 9.500 | 7.553 | 1.947 | .367 | .260 | 9 | .919 | .907 | .892 | .875 | .854 | .829 | .798 | .762 | .718 | |
| 9.500 | 8.092 | 1.408 | .271 | .197 | 10 | .915 | .902 | .887 | .868 | .844 | .817 | .783 | .743 | .695 | .639 |
| 9.500 | 8.524 | .976 | .191 | .142 | 11 | .912 | .898 | .882 | .861 | .836 | .806 | .770 | .726 | .674 | .614 |
| | | | | | | .545 | | | | | | | | | |
| 9.500 | 8.851 | .649 | .129 | .097 | 12 | .910 | .895 | .877 | .856 | .829 | .797 | .759 | .712 | .657 | .593 |
| | | | | | | .522 | .444 | | | | | | | | |
| 9.500 | 9.086 | .414 | .083 | .063 | 13 | .908 | .893 | .874 | .852 | .824 | .791 | .750 | .701 | .644 | .578 |
| | | | | | | .504 | .424 | .343 | | | | | | | |
| 9.500 | 9.246 | .254 | .051 | .039 | 14 | .907 | .891 | .872 | .849 | .821 | .786 | .744 | .694 | .635 | .567 |
| | | | | | | .491 | .410 | .329 | .251 | | | | | | |
| 9.500 | 9.351 | .149 | .030 | .023 | 15 | .906 | .890 | .871 | .847 | .818 | .783 | .740 | .689 | .628 | .559 |
| | | | | | | .483 | .401 | .320 | .242 | .173 | | | | | |
| 10.000 | 7.696 | 2.304 | .409 | .287 | 9 | .924 | .913 | .901 | .885 | .867 | .846 | .820 | .789 | .751 | |
| 10.000 | 8.287 | 1.713 | .311 | .223 | 10 | .921 | .909 | .895 | .879 | .859 | .835 | .806 | .771 | .730 | .682 |
| 10.000 | 8.776 | 1.224 | .226 | .166 | 11 | .918 | .905 | .890 | .872 | .851 | .825 | .793 | .755 | .711 | .658 |
| | | | | | | .597 | | | | | | | | | |

JEWETT EQUIVALENT QUEUE EXTENDED ERLANG B(EQEEB): 1 MINUTE MAXIMUM WAIT (IF NOT CONNECTED AFTER 1 MINUTE WAIT, 100% OVERFLOW)

| HOURS FIRST ATTEMPT VMR = 1.0 | HOURS CON-NECT | HOURS OVFL | DELAYED PORTION | DELAY LENGTH CALL FACTOR | LINE | HOURS CONNECTED TRAFFIC ON LINES 1 | 2 | 3 | 4 | 5 | 6 | 7 | 8 | 9 | 10 |
|---|---|---|---|---|---|---|---|---|---|---|---|---|---|---|---|
| 10.000 | 9.160 | .840 | .158 | .118 | 12 | .915 | .902 | .886 | .867 | .844 | .816 | .782 | .742 | .694 | .637 |
| | | | | | | .573 | .501 | | | | | | | | |
| 10.000 | 9.446 | .554 | .105 | .080 | 13 | .913 | .900 | .883 | .863 | .839 | .809 | .773 | .731 | .680 | .621 |
| | | | | | | .554 | .480 | .401 | | | | | | | |
| 10.000 | 9.648 | .352 | .067 | .051 | 14 | .912 | .898 | .881 | .860 | .835 | .804 | .767 | .722 | .670 | .609 |
| | | | | | | .539 | .464 | .384 | .305 | | | | | | |
| 10.000 | 9.785 | .215 | .041 | .032 | 15 | .911 | .896 | .879 | .858 | .832 | .800 | .762 | .717 | .663 | .600 |
| | | | | | | .529 | .453 | .373 | .293 | .220 | | | | | |
| 11.000 | 8.601 | 2.399 | .389 | .274 | 10 | .930 | .921 | .910 | .897 | .882 | .864 | .843 | .817 | .787 | .751 |
| 11.000 | 9.192 | 1.808 | .299 | .215 | 11 | .927 | .917 | .905 | .891 | .875 | .855 | .832 | .804 | .770 | .731 |
| | | | | | | .684 | | | | | | | | | |
| 11.000 | 9.686 | 1.314 | .221 | .162 | 12 | .924 | .914 | .901 | .886 | .869 | .847 | .822 | .791 | .755 | .712 |
| | | | | | | .661 | .604 | | | | | | | | |
| 11.000 | 10.080 | .920 | .157 | .117 | 13 | .922 | .911 | .898 | .882 | .863 | .840 | .813 | .780 | .741 | .695 |
| | | | | | | .641 | .580 | .513 | | | | | | | |
| 11.000 | 10.379 | .621 | .107 | .081 | 14 | .920 | .909 | .895 | .878 | .859 | .835 | .806 | .771 | .730 | .681 |
| | | | | | | .625 | .562 | .491 | .417 | | | | | | |
| 11.000 | 10.596 | .404 | .070 | .054 | 15 | .919 | .907 | .893 | .876 | .855 | .830 | .800 | .764 | .721 | .671 |
| | | | | | | .613 | .547 | .475 | .400 | .323 | | | | | |
| 11.000 | 10.746 | .254 | .044 | .034 | 16 | .918 | .906 | .891 | .874 | .853 | .827 | .796 | .759 | .715 | .664 |
| | | | | | | .604 | .537 | .464 | .387 | .311 | .239 | | | | |
| 11.000 | 10.846 | .154 | .027 | .021 | 17 | .918 | .905 | .891 | .873 | .851 | .825 | .794 | .756 | .711 | .659 |
| | | | | | | .598 | .530 | .456 | .379 | .303 | .231 | .168 | | | |
| 12.000 | 9.510 | 2.490 | .372 | .263 | 11 | .935 | .927 | .917 | .907 | .894 | .879 | .861 | .840 | .815 | .785 |
| | | | | | | .751 | | | | | | | | | |
| 12.000 | 10.101 | 1.899 | .288 | .208 | 12 | .932 | .921 | .914 | .902 | .888 | .872 | .852 | .829 | .802 | .769 |
| | | | | | | .731 | .687 | | | | | | | | |
| 12.000 | 10.600 | 1.400 | .216 | .159 | 13 | .930 | .921 | .910 | .898 | .883 | .865 | .844 | .819 | .789 | .754 |
| | | | | | | .713 | .665 | .610 | | | | | | | |
| 12.000 | 11.003 | .997 | .156 | .116 | 14 | .928 | .919 | .907 | .894 | .878 | .859 | .837 | .810 | .778 | .740 |
| | | | | | | .696 | .645 | .587 | .523 | | | | | | |
| 12.000 | 11.313 | .687 | .108 | .082 | 15 | .927 | .917 | .905 | .891 | .874 | .855 | .831 | .803 | .769 | .729 |
| | | | | | | .683 | .629 | .569 | .502 | .430 | | | | | |
| 12.000 | 11.543 | .457 | .073 | .056 | 16 | .926 | .915 | .903 | .889 | .872 | .851 | .826 | .797 | .762 | .721 |
| | | | | | | .672 | .617 | .554 | .486 | .413 | .340 | | | | |
| 12.000 | 11.706 | .294 | .047 | .036 | 17 | .925 | .914 | .902 | .887 | .869 | .848 | .823 | .793 | .757 | .714 |
| | | | | | | .665 | .608 | .544 | .474 | .401 | .327 | .256 | | | |
| 12.000 | 11.817 | .183 | .029 | .023 | 18 | .924 | .913 | .901 | .886 | .868 | .846 | .821 | .790 | .753 | .710 |
| | | | | | | .659 | .601 | .536 | .466 | .392 | .318 | .248 | .184 | | |
| 13.000 | 10.423 | 2.577 | .356 | .253 | 12 | .939 | .932 | .924 | .915 | .904 | .891 | .876 | .858 | .837 | .813 |

JEWETT EQUIVALENT QUEUE EXTENDED ERLANG B(EQEEB): 1 MINUTE MAXIMUM WAIT (IF NOT CONNECTED AFTER 1 MINUTE WAIT, 100% OVERFLOW)

| HOURS FIRST ATTEMPT VMR = 1.0 | HOURS CON-NECT | HOURS OVFL | DELAYED PORTION | DELAY LENGTH CALL FACTOR | LINE | HOURS CONNECTED TRAFFIC ON LINES 1 | 2 | 3 | 4 | 5 | 6 | 7 | 8 | 9 | 10 |
|---|---|---|---|---|---|---|---|---|---|---|---|---|---|---|---|
| | | | | | | .784 | .751 | | | | | | | | |
| 13.000 | 11.014 | 1.986 | .279 | .202 | 13 | .937 | .929 | .921 | .911 | .899 | .885 | .869 | .849 | .827 | .800 |
| | | | | | | .768 | .731 | .689 | | | | | | | |
| 13.000 | 11.517 | 1.483 | .211 | .156 | 14 | .935 | .927 | .918 | .907 | .894 | .880 | .862 | .841 | .816 | .787 |
| | | | | | | .753 | .713 | .667 | .615 | | | | | | |
| 13.000 | 11.928 | 1.072 | .155 | .116 | 15 | .933 | .925 | .915 | .904 | .891 | .875 | .856 | .834 | .808 | .776 |
| | | | | | | .740 | .697 | .648 | .593 | .531 | | | | | |
| 13.000 | 12.248 | .752 | .110 | .083 | 16 | .932 | .923 | .913 | .901 | .887 | .871 | .851 | .828 | .800 | .767 |
| | | | | | | .729 | .684 | .633 | .575 | .511 | .443 | | | | |
| 13.000 | 12.490 | .510 | .075 | .057 | 17 | .931 | .922 | .912 | .899 | .885 | .868 | .847 | .823 | .794 | .760 |
| | | | | | | .720 | .673 | .620 | .560 | .495 | .425 | .354 | | | |
| 13.000 | 12.664 | .336 | .050 | .038 | 18 | .930 | .921 | .910 | .898 | .883 | .865 | .844 | .819 | .790 | .754 |
| | | | | | | .713 | .665 | .611 | .550 | .483 | .413 | .341 | .272 | | |
| 13.000 | 12.786 | .214 | .032 | .024 | 19 | .930 | .920 | .910 | .897 | .882 | .864 | .842 | .817 | .786 | .751 |
| | | | | | | .709 | .660 | .604 | .542 | .475 | .404 | .332 | .263 | .200 | |
| 14.000 | 10.670 | 3.330 | .422 | .294 | 12 | .945 | .939 | .932 | .925 | .916 | .906 | .894 | .880 | .864 | .846 |
| | | | | | | .824 | .799 | | | | | | | | |
| 14.000 | 11.339 | 2.661 | .343 | .244 | 13 | .943 | .937 | .929 | .921 | .912 | .901 | .888 | .873 | .855 | .835 |
| | | | | | | .811 | .783 | .750 | | | | | | | |
| 14.000 | 11.931 | 2.069 | .270 | .196 | 14 | .941 | .934 | .927 | .918 | .908 | .896 | .882 | .866 | .847 | .824 |
| | | | | | | .798 | .767 | .732 | .691 | | | | | | |
| 14.000 | 12.437 | 1.563 | .207 | .153 | 15 | .939 | .932 | .924 | .915 | .904 | .891 | .877 | .859 | .838 | .814 |
| | | | | | | .786 | .753 | .714 | .670 | .620 | | | | | |
| 14.000 | 12.854 | 1.146 | .154 | .115 | 16 | .938 | .931 | .922 | .912 | .901 | .887 | .872 | .853 | .831 | .805 |
| | | | | | | .775 | .740 | .699 | .651 | .598 | .539 | | | | |
| 14.000 | 13.185 | .815 | .110 | .083 | 17 | .937 | .929 | .920 | .910 | .898 | .884 | .867 | .848 | .825 | .798 |
| | | | | | | .766 | .728 | .685 | .636 | .580 | .519 | .454 | | | |
| 14.000 | 13.437 | .563 | .077 | .059 | 18 | .936 | .928 | .919 | .908 | .896 | .881 | .864 | .844 | .820 | .792 |
| | | | | | | .758 | .719 | .675 | .623 | .566 | .503 | .437 | .368 | | |
| 14.000 | 13.622 | .378 | .052 | .040 | 19 | .935 | .927 | .918 | .907 | .894 | .879 | .862 | .841 | .816 | .787 |
| | | | | | | .753 | .713 | .666 | .614 | .555 | .491 | .424 | .355 | .287 | |
| 14.000 | 13.754 | .246 | .034 | .026 | 20 | .934 | .926 | .917 | .906 | .893 | .878 | .860 | .838 | .813 | .783 |
| | | | | | | .748 | .708 | .660 | .607 | .547 | .483 | .414 | .345 | .277 | .215 |
| 15.000 | 11.591 | 3.409 | .404 | .283 | 13 | .948 | .943 | .937 | .930 | .922 | .914 | .903 | .892 | .878 | .862 |
| | | | | | | .844 | .822 | .797 | | | | | | | |
| 15.000 | 12.258 | 2.742 | .330 | .236 | 14 | .946 | .941 | .934 | .927 | .919 | .909 | .898 | .885 | .870 | .853 |
| | | | | | | .833 | .809 | .782 | .750 | | | | | | |
| 15.000 | 12.850 | 2.150 | .263 | .191 | 15 | .944 | .939 | .932 | .924 | .915 | .905 | .893 | .879 | .863 | .844 |
| | | | | | | .822 | .797 | .767 | .732 | .692 | | | | | |
| 15.000 | 13.360 | 1.640 | .203 | .150 | 16 | .943 | .937 | .930 | .922 | .912 | .901 | .889 | .874 | .856 | .836 |

JEWETT EQUIVALENT QUEUE EXTENDED ERLANG B(EQEEB): 1 MINUTE MAXIMUM WAIT (IF NOT CONNECTED AFTER 1 MINUTE WAIT, 100% OVERFLOW)

| HOURS FIRST ATTEMPT VMR = 1.0 | HOURS CON-NECT | HOURS OVFL | DELAYED PORTION | DELAY LENGTH CALL FACTOR | LINE | HOURS CONNECTED TRAFFIC ON LINES 1 | 2 | 3 | 4 | 5 | 6 | 7 | 8 | 9 | 10 |
|---|---|---|---|---|---|---|---|---|---|---|---|---|---|---|---|
| | | | | | | .812 | .784 | .752 | .715 | .672 | .624 | | | | |
| 15.000 | 13.783 | 1.217 | .152 | .114 | 17 | .942 | .935 | .928 | .919 | .909 | .898 | .884 | .869 | .850 | .829 |
| | | | | | | .803 | .774 | .739 | .700 | .654 | .603 | .547 | | | |
| 15.000 | 14.122 | .878 | .111 | .084 | 18 | .941 | .934 | .926 | .917 | .907 | .895 | .881 | .864 | .845 | .822 |
| | | | | | | .796 | .764 | .728 | .686 | .639 | .585 | .527 | .464 | | |
| 15.000 | 14.384 | .616 | .078 | .060 | 19 | .940 | .933 | .925 | .916 | .905 | .893 | .878 | .861 | .841 | .817 |
| | | | | | | .789 | .757 | .719 | .676 | .626 | .571 | .511 | .447 | .380 | |
| 15.000 | 14.580 | .420 | .054 | .041 | 20 | .939 | .932 | .924 | .915 | .904 | .891 | .876 | .858 | .837 | .813 |
| | | | | | | .784 | .751 | .712 | .667 | .617 | .560 | .499 | .434 | .367 | .300 |
| 15.000 | 14.721 | .279 | .036 | .028 | 21 | .939 | .931 | .923 | .914 | .903 | .889 | .874 | .856 | .835 | .810 |
| | | | | | | .781 | .746 | .707 | .661 | .610 | .552 | .490 | .424 | .357 | .291 |
| | | | | | | .229 | | | | | | | | | |
| 16.000 | 12.515 | 3.485 | .389 | .274 | 14 | .951 | .946 | .941 | .935 | .928 | .920 | .911 | .901 | .889 | .876 |
| | | | | | | .860 | .842 | .820 | .796 | | | | | | |
| 16.000 | 13.179 | 2.821 | .319 | .229 | 15 | .949 | .944 | .939 | .932 | .925 | .916 | .907 | .896 | .883 | .868 |
| | | | | | | .851 | .831 | .808 | .781 | .750 | | | | | |
| 16.000 | 13.772 | 2.228 | .256 | .186 | 16 | .948 | .942 | .936 | .930 | .922 | .913 | .903 | .891 | .877 | .861 |
| | | | | | | .842 | .820 | .795 | .766 | .732 | .694 | | | | |
| 16.000 | 14.284 | 1.716 | .199 | .147 | 17 | .946 | .941 | .935 | .927 | .919 | .910 | .899 | .886 | .871 | .854 |
| | | | | | | .834 | .810 | .783 | .752 | .716 | .674 | .628 | | | |
| 16.000 | 14.713 | 1.287 | .151 | .113 | 18 | .945 | .939 | .933 | .925 | .917 | .907 | .895 | .882 | .866 | .848 |
| | | | | | | .826 | .801 | .772 | .739 | .700 | .657 | .607 | .553 | | |
| 16.000 | 15.060 | .940 | .111 | .084 | 19 | .944 | .938 | .931 | .924 | .915 | .904 | .892 | .878 | .862 | .842 |
| | | | | | | .820 | .794 | .763 | .728 | .687 | .642 | .590 | .534 | .473 | |
| 16.000 | 15.332 | .668 | .080 | .061 | 20 | .943 | .937 | .930 | .922 | .913 | .902 | .890 | .875 | .858 | .838 |
| | | | | | | .814 | .787 | .755 | .719 | .677 | .029 | .576 | .518 | .456 | .392 |
| 16.000 | 15.537 | .463 | .055 | .043 | 21 | .040 | .937 | .929 | .921 | .912 | .901 | .888 | .873 | .855 | .834 |
| | | | | | | .810 | .782 | .749 | .712 | .668 | .619 | .565 | .506 | .443 | .378 |
| | | | | | | .313 | | | | | | | | | |
| 16.000 | 15.688 | .312 | .038 | .029 | 22 | .942 | .936 | .929 | .920 | .911 | .899 | .886 | .871 | .853 | .832 |
| | | | | | | .807 | .778 | .745 | .706 | .662 | .612 | .557 | .496 | .433 | .368 |
| | | | | | | .303 | .242 | | | | | | | | |
| 17.000 | 13.441 | 3.559 | .375 | .265 | 15 | .953 | .949 | .944 | .939 | .933 | .926 | .918 | .909 | .899 | .887 |
| | | | | | | .873 | .858 | .840 | .819 | .795 | | | | | |
| 17.000 | 14.103 | 2.897 | .309 | .222 | 16 | .952 | .947 | .942 | .937 | .930 | .923 | .914 | .905 | .893 | .881 |
| | | | | | | .866 | .849 | .829 | .806 | .780 | .750 | | | | |
| 17.000 | 14.696 | 2.304 | .249 | .182 | 17 | .950 | .946 | .940 | .934 | .927 | .920 | .911 | .900 | .888 | .874 |
| | | | | | | .859 | .840 | .810 | .794 | .765 | .733 | .695 | | | |
| 17.000 | 15.211 | 1.789 | .196 | .145 | 18 | .949 | .944 | .939 | .932 | .925 | .017 | .907 | .896 | .883 | .869 |
| | | | | | | .852 | .832 | .809 | .782 | .751 | .716 | .676 | .631 | | |

JEWETT EQUIVALENT QUEUE EXTENDED ERLANG B(EQEEB): 1 MINUTE MAXIMUM WAIT (IF NOT CONNECTED AFTER 1 MINUTE WAIT, 100% OVERFLOW)

| HOURS FIRST ATTEMPT VMR = 1.0 | HOURS CON-NECT | HOURS OVFL | DELAYED PORTION | DELAY LENGTH CALL FACTOR | LINE | HOURS CONNECTED TRAFFIC ON LINES 1 | 2 | 3 | 4 | 5 | 6 | 7 | 8 | 9 | 10 |
|---|---|---|---|---|---|---|---|---|---|---|---|---|---|---|---|
| 17.000 | 15.645 | 1.355 | .150 | .112 | 19 | .948 | .943 | .937 | .931 | .923 | .914 | .904 | .893 | .879 | .863 |
| | | | | | | .845 | .824 | .800 | .771 | .739 | .701 | .659 | .611 | .559 | |
| 17.000 | 15.999 | 1.001 | .111 | .084 | 20 | .947 | .942 | .936 | .929 | .921 | .912 | .902 | .889 | .875 | .859 |
| | | | | | | .840 | .818 | .792 | .762 | .728 | .688 | .644 | .594 | .540 | .481 |
| 17.000 | 16.279 | .721 | .081 | .062 | 21 | .946 | .941 | .935 | .928 | .920 | .910 | .899 | .887 | .872 | .855 |
| | | | | | | .835 | .812 | .785 | .754 | .719 | .678 | .632 | .580 | .524 | .464 |
| | | | | | | .402 | | | | | | | | | |
| 17.000 | 16.494 | .506 | .057 | .044 | 22 | .946 | .940 | .934 | .927 | .918 | .909 | .898 | .885 | .870 | .852 |
| | | | | | | .832 | .808 | .780 | .748 | .711 | .669 | .622 | .569 | .512 | .451 |
| | | | | | | .388 | .325 | | | | | | | | |
| 17.000 | 16.653 | .347 | .039 | .030 | 23 | .945 | .940 | .933 | .926 | .918 | .908 | .896 | .883 | .868 | .850 |
| | | | | | | .829 | .804 | .776 | .743 | .706 | .663 | .614 | .561 | .503 | .441 |
| | | | | | | .378 | .315 | .254 | | | | | | | |
| 17.000 | 16.768 | .232 | .026 | .020 | 24 | .945 | .939 | .933 | .925 | .917 | .907 | .895 | .882 | .866 | .848 |
| | | | | | | .827 | .802 | .773 | .740 | .701 | .658 | .609 | .555 | .496 | .434 |
| | | | | | | .370 | .307 | .247 | .192 | | | | | | |
| 18.000 | 13.649 | 4.351 | .428 | .298 | 15 | .957 | .953 | .949 | .944 | .939 | .934 | .927 | .920 | .911 | .902 |
| | | | | | | .891 | .879 | .864 | .848 | .829 | | | | | |
| 18.000 | 14.369 | 3.631 | .362 | .257 | 16 | .955 | .952 | .947 | .942 | .937 | .931 | .924 | .916 | .907 | .897 |
| | | | | | | .885 | .871 | .856 | .838 | .817 | .794 | | | | |
| 18.000 | 15.029 | 2.971 | .300 | .216 | 17 | .954 | .950 | .945 | .940 | .935 | .928 | .921 | .912 | .902 | .891 |
| | | | | | | .878 | .864 | .847 | .827 | .805 | .779 | .750 | | | |
| 18.000 | 15.622 | 2.378 | .243 | .178 | 18 | .953 | .949 | .944 | .938 | .932 | .925 | .917 | .908 | .898 | .886 |
| | | | | | | .872 | .856 | .838 | .817 | .793 | .765 | .733 | .696 | | |
| 18.000 | 16.140 | 1.860 | .192 | .142 | 19 | .952 | .947 | .942 | .937 | .930 | .923 | .915 | .905 | .894 | .881 |
| | | | | | | .866 | .849 | .830 | .807 | .781 | .751 | .717 | .678 | .634 | |
| 18.000 | 16.579 | 1.421 | .148 | .111 | 20 | .951 | .946 | .941 | .935 | .928 | .921 | .912 | .902 | .890 | .877 |
| | | | | | | .861 | .843 | .822 | .798 | .770 | .739 | .702 | .661 | .615 | .564 |
| 18.000 | 16.940 | 1.060 | .112 | .084 | 21 | .950 | .945 | .940 | .934 | .927 | .919 | .910 | .899 | .887 | .873 |
| | | | | | | .856 | .837 | .816 | .790 | .761 | .728 | .689 | .646 | .598 | .546 |
| | | | | | | .489 | | | | | | | | | |
| 18.000 | 17.228 | .772 | .082 | .062 | 22 | .949 | .944 | .939 | .933 | .925 | .917 | .908 | .897 | .884 | .869 |
| | | | | | | .853 | .833 | .810 | .784 | .753 | .718 | .679 | .634 | .584 | .530 |
| | | | | | | .472 | .412 | | | | | | | | |
| 18.000 | 17.450 | .550 | .059 | .045 | 23 | .949 | .944 | .938 | .932 | .924 | .916 | .906 | .895 | .882 | .867 |
| | | | | | | .849 | .829 | .806 | .778 | .747 | .711 | .670 | .624 | .573 | .518 |
| | | | | | | .459 | .398 | .336 | | | | | | | |
| 18.000 | 17.618 | .382 | .041 | .031 | 24 | .948 | .943 | .938 | .931 | .924 | .915 | .905 | .893 | .880 | .865 |
| | | | | | | .847 | .826 | .802 | .774 | .742 | .705 | .663 | .617 | .565 | .508 |
| | | | | | | .449 | .387 | .325 | .266 | | | | | | |

JEWETT EQUIVALENT QUEUE EXTENDED ERLANG B(EQEEB): 1 MINUTE MAXIMUM WAIT (IF NOT CONNECTED AFTER 1 MINUTE WAIT, 100% OVERFLOW)

| HOURS FIRST ATTEMPT VMR = 1.0 | HOURS CON-NECT | HOURS OVFL | DELAYED PORTION | DELAY LENGTH CALL FACTOR | LINE | HOURS CONNECTED TRAFFIC ON LINES 1 | 2 | 3 | 4 | 5 | 6 | 7 | 8 | 9 | 10 |
|---|---|---|---|---|---|---|---|---|---|---|---|---|---|---|---|
| 18.000 | 17.741 | .259 | .028 | .021 | 25 | .948 | .943 | .937 | .930 | .923 | .914 | .904 | .892 | .879 | .863 |
| | | | | | | .845 | .824 | .799 | .771 | .738 | .701 | .658 | .611 | .558 | .501 |
| | | | | | | .441 | .379 | .318 | .258 | .204 | | | | | |
| 19.000 | 14.581 | 4.419 | .413 | .289 | 16 | .959 | .955 | .952 | .947 | .943 | .938 | .932 | .925 | .918 | .910 |
| | | | | | | .900 | .889 | .877 | .863 | .846 | .828 | | | | |
| 19.000 | 15.298 | 3.702 | .350 | .249 | 17 | .957 | .954 | .950 | .946 | .941 | .935 | .929 | .922 | .914 | .905 |
| | | | | | | .895 | .883 | .869 | .854 | .836 | .816 | .793 | | | |
| 19.000 | 15.957 | 3.043 | .292 | .210 | 18 | .956 | .953 | .948 | .944 | .939 | .933 | .926 | .919 | .910 | .900 |
| | | | | | | .889 | .876 | .862 | .845 | .826 | .804 | .778 | .749 | | |
| 19.000 | 16.550 | 2.450 | .238 | .174 | 19 | .955 | .951 | .947 | .942 | .937 | .930 | .923 | .915 | .906 | .896 |
| | | | | | | .884 | .870 | .854 | .836 | .815 | .792 | .764 | .733 | .698 | |
| 19.000 | 17.071 | 1.929 | .189 | .140 | 20 | .954 | .950 | .946 | .941 | .935 | .928 | .921 | .912 | .903 | .892 |
| | | | | | | .879 | .864 | .847 | .828 | .806 | .780 | .751 | .717 | .680 | .637 |
| 19.000 | 17.513 | 1.487 | .147 | .110 | 21 | .953 | .949 | .944 | .939 | .933 | .926 | .918 | .910 | .899 | .888 |
| | | | | | | .874 | .859 | .841 | .820 | .797 | .770 | .738 | .703 | .663 | .618 |
| | | | | | | .569 | | | | | | | | | |
| 19.000 | 17.881 | 1.119 | .112 | .084 | 22 | .952 | .948 | .943 | .938 | .932 | .925 | .916 | .907 | .897 | .884 |
| | | | | | | .870 | .854 | .835 | .814 | .789 | .760 | .727 | .690 | .648 | .602 |
| | | | | | | .551 | .496 | | | | | | | | |
| 19.000 | 18.176 | .824 | .083 | .063 | 23 | .952 | .947 | .943 | .937 | .930 | .923 | .915 | .905 | .894 | .882 |
| | | | | | | .867 | .850 | .831 | .808 | .782 | .752 | .718 | .680 | .636 | .588 |
| | | | | | | .536 | .479 | .421 | | | | | | | |
| 19.000 | 18.407 | .593 | .060 | .046 | 24 | .951 | .947 | .942 | .936 | .929 | .922 | .913 | .904 | .892 | .879 |
| | | | | | | .864 | .847 | .827 | .803 | .777 | .746 | .711 | .671 | .626 | .577 |
| | | | | | | .523 | .466 | .407 | .346 | | | | | | |
| 19.000 | 18.583 | .417 | .042 | .032 | 25 | .951 | .946 | .941 | .935 | .920 | .921 | .912 | .902 | .891 | .877 |
| | | | | | | .862 | .844 | .824 | .800 | .772 | .741 | .705 | .664 | .619 | .568 |
| | | | | | | .514 | .456 | .396 | .335 | .277 | | | | | |
| 19.000 | 18.714 | .286 | .029 | .022 | 26 | .951 | .946 | .941 | .935 | .928 | .920 | .912 | .901 | .890 | .876 |
| | | | | | | .860 | .842 | .821 | .797 | .769 | .737 | .700 | .659 | .613 | .562 |
| | | | | | | .506 | .448 | .388 | .327 | .269 | .215 | | | | |
| 20.000 | 15.515 | 4.485 | .399 | .280 | 17 | .960 | .957 | .954 | .950 | .946 | .941 | .936 | .930 | .924 | .916 |
| | | | | | | .908 | .898 | .887 | .875 | .861 | .845 | .826 | | | |
| 20.000 | 16.230 | 3.770 | .340 | .242 | 18 | .959 | .956 | .952 | .949 | .044 | .939 | .934 | .927 | .920 | .912 |
| | | | | | | .903 | .893 | .881 | .867 | .852 | .835 | .814 | .792 | | |
| 20.000 | 16.887 | 3.113 | .284 | .205 | 19 | .958 | .955 | .951 | .947 | .942 | .937 | .931 | .924 | .917 | .908 |
| | | | | | | .898 | .887 | .874 | .860 | .843 | .824 | .802 | .778 | .749 | |
| 20.000 | 17.481 | 2.519 | .232 | .170 | 20 | .957 | .954 | .950 | .945 | .940 | .935 | .929 | .922 | .913 | .904 |
| | | | | | | .894 | .882 | .868 | .853 | 835 | .814 | .791 | .764 | .733 | .699 |
| 20.000 | 18.003 | 1.997 | .186 | .138 | 21 | .956 | .953 | .949 | .944 | .939 | .933 | .926 | .919 | .910 | .901 |

JEWETT EQUIVALENT QUEUE EXTENDED ERLANG B(EQEEB): 1 MINUTE MAXIMUM WAIT (IF NOT CONNECTED AFTER 1 MINUTE WAIT, 100% OVERFLOW)

| HOURS FIRST ATTEMPT VMR = 1.0 | HOURS CON-NECT | HOURS OVFL | DELAYED PORTION | DELAY LENGTH CALL FACTOR | LINE | HOURS CONNECTED TRAFFIC ON LINES 1 | 2 | 3 | 4 | 5 | 6 | 7 | 8 | 9 | 10 |
|---|---|---|---|---|---|---|---|---|---|---|---|---|---|---|---|
| | | | | | | .890 | .877 | .862 | .846 | .826 | .804 | .779 | .750 | .718 | .681 |
| | | | | | | .640 | | | | | | | | | |
| 20.000 | 18.450 | 1.550 | .146 | .109 | 22 | .955 | .952 | .947 | .943 | .937 | .931 | .924 | .916 | .907 | .897 |
| | | | | | | .886 | .872 | .857 | .839 | .819 | .795 | .769 | .738 | .704 | .665 |
| | | | | | | .622 | .574 | | | | | | | | |
| 20.000 | 18.823 | 1.177 | .111 | .084 | 23 | .955 | .951 | .946 | .942 | .936 | .930 | .922 | .914 | .905 | .894 |
| | | | | | | .882 | .868 | .852 | .833 | .812 | .787 | .759 | .727 | .691 | .651 |
| | | | | | | .605 | .556 | .503 | | | | | | | |
| 20.000 | 19.125 | .875 | .083 | .063 | 24 | .954 | .950 | .946 | .941 | .935 | .928 | .921 | .913 | .903 | .892 |
| | | | | | | .879 | .865 | .848 | .828 | .806 | .781 | .751 | .718 | .681 | .638 |
| | | | | | | .592 | .541 | .486 | .429 | | | | | | |
| 20.000 | 19.364 | .636 | .061 | .047 | 25 | .954 | .950 | .945 | .940 | .934 | .927 | .920 | .911 | .901 | .890 |
| | | | | | | .877 | .862 | .844 | .824 | .802 | .775 | .745 | .711 | .672 | .628 |
| | | | | | | .580 | .528 | .473 | .415 | .356 | | | | | |
| 20.000 | 19.548 | .452 | .044 | .033 | 26 | .953 | .949 | .945 | .939 | .933 | .927 | .919 | .910 | .900 | .888 |
| | | | | | | .875 | .859 | .842 | .821 | .798 | .771 | .740 | .705 | .665 | .621 |
| | | | | | | .572 | .519 | .462 | .404 | .345 | .287 | | | | |
| 20.000 | 19.685 | .315 | .030 | .023 | 27 | .953 | .949 | .944 | .939 | .933 | .926 | .918 | .909 | .899 | .887 |
| | | | | | | .873 | .858 | .840 | .819 | .795 | .767 | .736 | .700 | .660 | .615 |
| | | | | | | .565 | .511 | .454 | .396 | .337 | .279 | .225 | | | |
| 22.000 | 17.387 | 4.613 | .375 | .265 | 19 | .963 | .961 | .958 | .955 | .951 | .947 | .943 | .938 | .933 | .927 |
| | | | | | | .920 | .913 | .904 | .895 | .884 | .871 | .857 | .842 | .824 | |
| 22.000 | 18.097 | 3.903 | .321 | .230 | 20 | .962 | .960 | .957 | .953 | .950 | .946 | .941 | .936 | .930 | .924 |
| | | | | | | .917 | .909 | .899 | .889 | .877 | .864 | .849 | .832 | .812 | .790 |
| 22.000 | 18.752 | 3.248 | .270 | .196 | 21 | .962 | .959 | .956 | .952 | .948 | .944 | .939 | .934 | .928 | .921 |
| | | | | | | .913 | .905 | .895 | .884 | .871 | .856 | .840 | .821 | .800 | .776 |
| | | | | | | .749 | | | | | | | | | |
| 22.000 | 19.346 | 2.654 | .223 | .164 | 22 | .961 | .958 | .955 | .951 | .947 | .942 | .937 | .932 | .925 | .918 |
| | | | | | | .910 | .901 | .890 | .878 | .865 | .849 | .832 | .811 | .789 | .763 |
| | | | | | | .734 | .701 | | | | | | | | |
| 22.000 | 19.872 | 2.128 | .180 | .134 | 23 | .960 | .957 | .954 | .950 | .945 | .941 | .935 | .929 | .923 | .915 |
| | | | | | | .907 | .897 | .886 | .873 | .859 | .842 | .823 | .802 | .778 | .750 |
| | | | | | | .719 | .684 | .645 | | | | | | | |
| 22.000 | 20.326 | 1.674 | .143 | .107 | 24 | .959 | .956 | .953 | .949 | .944 | .939 | .934 | .928 | .921 | .913 |
| | | | | | | .904 | .893 | .882 | .868 | .853 | .836 | .816 | .793 | .767 | .738 |
| | | | | | | .705 | .668 | .627 | .582 | | | | | | |
| 22.000 | 20.710 | 1.290 | .111 | .084 | 25 | .959 | .955 | .952 | .948 | .943 | .938 | .932 | .926 | .919 | .910 |
| | | | | | | .901 | .890 | .878 | .864 | .848 | .830 | .809 | .785 | .758 | .727 |
| | | | | | | .693 | .654 | .611 | .565 | .514 | | | | | |
| 22.000 | 21.025 | .975 | .084 | .064 | 26 | .958 | .955 | .951 | .947 | .942 | .937 | .931 | .924 | .917 | .908 |

JEWETT EQUIVALENT QUEUE EXTENDED ERLANG B(EQEEB): 1 MINUTE MAXIMUM WAIT (IF NOT CONNECTED AFTER 1 MINUTE WAIT, 100% OVERFLOW)

| HOURS FIRST ATTEMPT VMR = 1.0 | HOURS CON-NECT | HOURS OVFL | DELAYED PORTION | DELAY LENGTH CALL FACTOR | LINE | HOURS CONNECTED TRAFFIC ON LINES 1 | 2 | 3 | 4 | 5 | 6 | 7 | 8 | 9 | 10 |
|---|---|---|---|---|---|---|---|---|---|---|---|---|---|---|---|
| 22.000 | 21.277 | .723 | .063 | .048 | 27 | .958 | .954 | .951 | .946 | .942 | .936 | .930 | .923 | .916 | .907 |
| | | | | | | .897 | .885 | .872 | .857 | .840 | .820 | .798 | .772 | .743 | .710 |
| | | | | | | .673 | .632 | .587 | .538 | .485 | .430 | .374 | | | |
| 22.000 | 21.476 | .524 | .046 | .035 | 28 | .957 | .954 | .950 | .946 | .941 | .935 | .929 | .922 | .914 | .905 |
| | | | | | | .895 | .883 | .870 | .855 | .837 | .817 | .794 | .768 | .738 | .704 |
| | | | | | | .666 | .624 | .578 | .528 | .474 | .419 | .362 | .306 | | |
| 22.000 | 21.627 | .373 | .033 | .025 | 29 | .957 | .954 | .950 | .945 | .940 | .935 | .929 | .922 | .914 | .904 |
| | | | | | | .894 | .882 | .868 | .853 | .835 | .814 | .791 | .764 | .734 | .699 |
| | | | | | | .661 | .618 | .571 | .520 | .466 | .410 | .354 | .298 | .245 | |
| 24.000 | 18.513 | 5.487 | .407 | .285 | 20 | .967 | .965 | .962 | .960 | .957 | .954 | .950 | .947 | .943 | .938 |
| | | | | | | .933 | .927 | .921 | .914 | .906 | .897 | .887 | .875 | .863 | .848 |
| 24.000 | 19.265 | 4.735 | .355 | .252 | 21 | .966 | .964 | .961 | .959 | .956 | .952 | .949 | .945 | .940 | .936 |
| | | | | | | .930 | .924 | .917 | .910 | .901 | .891 | .881 | .868 | .854 | .839 |
| | | | | | | .821 | | | | | | | | | |
| 24.000 | 19.970 | 4.030 | .305 | .219 | 22 | .965 | .963 | .960 | .958 | .954 | .951 | .947 | .943 | .938 | .933 |
| | | | | | | .927 | .921 | .914 | .905 | .896 | .886 | .874 | .861 | .846 | .829 |
| | | | | | | .810 | .788 | | | | | | | | |
| 24.000 | 20.624 | 3.376 | .258 | .188 | 23 | .965 | .962 | .959 | .956 | .953 | .950 | .946 | .941 | .936 | .931 |
| | | | | | | .925 | .918 | .910 | .901 | .891 | .880 | .868 | .853 | .837 | .819 |
| | | | | | | .798 | .775 | .749 | | | | | | | |
| 24.000 | 21.217 | 2.783 | .215 | .158 | 24 | .964 | .961 | .958 | .955 | .952 | .948 | .944 | .939 | .934 | .928 |
| | | | | | | .922 | .915 | .906 | .897 | .887 | .875 | .861 | .846 | .829 | .809 |
| | | | | | | .787 | .762 | .734 | .702 | | | | | | |
| 24.000 | 21.746 | 2.254 | .175 | .130 | 25 | .963 | .961 | .958 | .954 | .951 | .947 | .943 | .938 | .932 | .926 |
| | | | | | | .919 | .912 | .903 | .893 | .882 | .869 | .855 | .839 | .821 | .800 |
| | | | | | | .778 | .750 | .720 | .686 | .649 | | | | | |
| 24.000 | 22.207 | 1.793 | .141 | .105 | 26 | .963 | .960 | .957 | .954 | .950 | .946 | .941 | .936 | .931 | .924 |
| | | | | | | .917 | .909 | .900 | .890 | .878 | .865 | .850 | .832 | .813 | .791 |
| | | | | | | .766 | .738 | .706 | .671 | .632 | .589 | | | | |
| 24.000 | 22.600 | 1.400 | .111 | .084 | 27 | .962 | .959 | .956 | .953 | .949 | .945 | .940 | .935 | .929 | .922 |
| | | | | | | .915 | .907 | .897 | .886 | .874 | .860 | .844 | .827 | .806 | .783 |
| | | | | | | .757 | .727 | .694 | .657 | .617 | .572 | .525 | | | |
| 24.000 | 22.926 | 1.074 | .085 | .065 | 28 | .962 | .959 | .956 | .952 | .948 | .944 | .939 | .934 | .928 | .921 |
| | | | | | | .913 | .905 | .895 | .884 | .871 | .856 | .840 | .821 | .800 | .776 |
| | | | | | | .749 | .718 | .684 | .646 | .604 | .558 | .509 | .457 | | |
| 24.000 | 23.192 | .808 | .064 | .049 | 29 | .961 | .958 | .955 | .952 | .948 | .943 | .938 | .933 | .927 | .920 |
| | | | | | | .912 | .903 | .893 | .881 | .868 | .853 | .836 | .817 | .795 | .770 |
| | | | | | | .742 | .710 | .675 | .636 | .593 | .546 | .496 | .443 | .389 | |

JEWETT EQUIVALENT QUEUE EXTENDED ERLANG B(EQEEB): 1 MINUTE MAXIMUM WAIT (IF NOT CONNECTED AFTER 1 MINUTE WAIT, 100% OVERFLOW)

| HOURS FIRST ATTEMPT VMR=1.0 | HOURS CON-NECT | HOURS OVFL | DELAYED PORTION | DELAY LENGTH CALL FACTOR | LINE | HOURS CONNECTED TRAFFIC ON LINES 1 | 2 | 3 | 4 | 5 | 6 | 7 | 8 | 9 | 10 |
|---|---|---|---|---|---|---|---|---|---|---|---|---|---|---|---|
| 24.000 | 23.403 | .597 | .048 | .037 | 30 | .961 | .958 | .955 | .951 | .947 | .943 | .938 | .932 | .926 | .918 |
| | | | | | | .910 | .901 | .891 | .879 | .866 | .851 | .833 | .813 | .791 | .765 |
| | | | | | | .737 | .704 | .668 | .628 | .583 | .536 | .485 | .432 | .378 | .324 |
| 24.000 | 23.567 | .433 | .035 | .027 | 31 | .961 | .958 | .954 | .951 | .947 | .942 | .937 | .931 | .925 | .918 |
| | | | | | | .909 | .900 | .890 | .878 | .864 | .848 | .831 | .811 | .788 | .762 |
| | | | | | | .732 | .699 | .662 | .621 | .576 | .528 | .477 | .423 | .369 | .315 |
| | | | | | | .263 | | | | | | | | | |
| 26.000 | 20.400 | 5.600 | .385 | .271 | 22 | .969 | .967 | .965 | .963 | .960 | .958 | .955 | .952 | .948 | .944 |
| | | | | | | .940 | .935 | .930 | .924 | .918 | .911 | .903 | .894 | .884 | .873 |
| | | | | | | .860 | .846 | | | | | | | | |
| 26.000 | 21.147 | 4.853 | .337 | .240 | 23 | .968 | .966 | .964 | .962 | .959 | .957 | .954 | .950 | .947 | .942 |
| | | | | | | .938 | .933 | .927 | .921 | .914 | .907 | .898 | .888 | .878 | .865 |
| | | | | | | .852 | .836 | .819 | | | | | | | |
| 26.000 | 21.848 | 4.152 | .291 | .210 | 24 | .968 | .966 | .963 | .961 | .958 | .955 | .952 | .949 | .945 | .940 |
| | | | | | | .936 | .930 | .924 | .918 | .911 | .902 | .893 | .883 | .871 | .858 |
| | | | | | | .843 | .827 | .808 | .787 | | | | | | |
| 26.000 | 22.500 | 3.500 | .247 | .181 | 25 | .967 | .965 | .963 | .960 | .957 | .954 | .951 | .947 | .943 | .939 |
| | | | | | | .934 | .928 | .922 | .915 | .907 | .898 | .888 | .877 | .865 | .851 |
| | | | | | | .835 | .817 | .797 | .774 | .749 | | | | | |
| 26.000 | 23.093 | 2.907 | .207 | .153 | 26 | .966 | .964 | .962 | .959 | .956 | .953 | .950 | .946 | .941 | .937 |
| | | | | | | .931 | .925 | .919 | .911 | .903 | .894 | .883 | .872 | .858 | .843 |
| | | | | | | .826 | .807 | .786 | .761 | .734 | .704 | | | | |
| 26.000 | 23.626 | 2.374 | .171 | .127 | 27 | .966 | .964 | .961 | .958 | .955 | .952 | .948 | .944 | .940 | .935 |
| | | | | | | .929 | .923 | .916 | .908 | .900 | .890 | .879 | .866 | .852 | .836 |
| | | | | | | .818 | .798 | .775 | .749 | .720 | .688 | .653 | | | |
| 26.000 | 24.092 | 1.908 | .138 | .104 | 28 | .965 | .963 | .960 | .958 | .954 | .951 | .947 | .943 | .938 | .933 |
| | | | | | | .927 | .921 | .914 | .906 | .896 | .886 | .875 | .861 | .846 | .830 |
| | | | | | | .811 | .789 | .765 | .738 | .707 | .674 | .636 | .596 | | |
| 26.000 | 24.493 | 1.507 | .110 | .083 | 29 | .965 | .962 | .960 | .957 | .954 | .950 | .946 | .942 | .937 | .932 |
| | | | | | | .926 | .919 | .912 | .903 | .894 | .883 | .871 | .857 | .841 | .824 |
| | | | | | | .804 | .781 | .756 | .727 | .696 | .660 | .622 | .579 | .534 | |
| 26.000 | 24.829 | 1.171 | .086 | .065 | 30 | .964 | .962 | .959 | .956 | .953 | .949 | .945 | .941 | .936 | .931 |
| | | | | | | .924 | .917 | .910 | .901 | .891 | .880 | .867 | .853 | .837 | .818 |
| | | | | | | .798 | .774 | .748 | .718 | .685 | .649 | .609 | .565 | .518 | .469 |
| 26.000 | 25.106 | .894 | .066 | .050 | 31 | .964 | .962 | .959 | .956 | .953 | .949 | .945 | .940 | .935 | .929 |
| | | | | | | .923 | .916 | .908 | .899 | .889 | .877 | .864 | .850 | .833 | .814 |
| | | | | | | .792 | .768 | .741 | .710 | .676 | .639 | .598 | .553 | .505 | .455 |
| | | | | | | .404 | | | | | | | | | |
| 26.000 | 25.329 | .671 | .050 | .038 | 32 | .964 | .961 | .959 | .955 | .952 | .948 | .944 | .940 | .934 | .929 |
| | | | | | | .922 | .915 | .907 | .897 | .887 | .875 | .862 | .847 | .830 | .810 |

JEWETT EQUIVALENT QUEUE EXTENDED ERLANG B(EQEEB): 1 MINUTE MAXIMUM WAIT (IF NOT CONNECTED AFTER 1 MINUTE WAIT, 100% OVERFLOW)

| HOURS FIRST ATTEMPT VMR = 1.0 | HOURS CON-NECT | HOURS OVFL | DELAYED PORTION | DELAY LENGTH CALL FACTOR | LINE | HOURS CONNECTED TRAFFIC ON LINES 1 | 2 | 3 | 4 | 5 | 6 | 7 | 8 | 9 | 10 |
|---|---|---|---|---|---|---|---|---|---|---|---|---|---|---|---|
| | | | | | | .788 | .763 | .735 | .704 | .669 | .631 | .588 | .543 | .495 | .444 |
| | | | | | | .392 | .340 | | | | | | | | |
| 26.000 | 25.506 | .494 | .037 | .028 | 33 | .964 | .961 | .958 | .955 | .952 | .948 | .944 | .939 | .934 | .928 |
| | | | | | | .921 | .914 | .906 | .896 | .886 | .874 | .860 | .844 | .827 | .807 |
| | | | | | | .785 | .759 | .731 | .699 | .663 | .624 | .581 | .535 | .486 | .435 |
| | | | | | | .383 | .330 | .279 | | | | | | | |
| 26.000 | 25.642 | .358 | .027 | .021 | 34 | .963 | .961 | .958 | .955 | .951 | .948 | .943 | .939 | .933 | .927 |
| | | | | | | .921 | .913 | .905 | .895 | .884 | .872 | .858 | .843 | .825 | .805 |
| | | | | | | .782 | .756 | .727 | .695 | .658 | .619 | .575 | .529 | .479 | .428 |
| | | | | | | .375 | .323 | .272 | .224 | | | | | | |
| 28.000 | 21.511 | 6.489 | .412 | .288 | 23 | .971 | .970 | .968 | .966 | .964 | .962 | .960 | .957 | .954 | .951 |
| | | | | | | .948 | .944 | .940 | .936 | .931 | .925 | .919 | .912 | .905 | .896 |
| | | | | | | .887 | .877 | .865 | | | | | | | |
| 28.000 | 22.290 | 5.710 | .366 | .259 | 24 | .971 | .969 | .967 | .965 | .963 | .961 | .959 | .956 | .953 | .950 |
| | | | | | | .946 | .942 | .938 | .933 | .928 | .922 | .915 | .908 | .900 | .891 |
| | | | | | | .881 | .870 | .857 | .843 | | | | | | |
| 28.000 | 23.033 | 4.967 | .321 | .230 | 25 | .970 | .969 | .967 | .965 | .962 | .960 | .957 | .955 | .951 | .948 |
| | | | | | | .944 | .940 | .936 | .930 | .925 | .919 | .912 | .904 | .895 | .886 |
| | | | | | | .875 | .863 | .849 | .834 | .817 | | | | | |
| 28.000 | 23.731 | 4.269 | .278 | .202 | 26 | .970 | .968 | .966 | .964 | .962 | .959 | .956 | .953 | .950 | .946 |
| | | | | | | .942 | .938 | .933 | .928 | .922 | .915 | .908 | .900 | .890 | .880 |
| | | | | | | .868 | .855 | .841 | .824 | .806 | .786 | | | | |
| 28.000 | 24.381 | 3.619 | .238 | .174 | 27 | .969 | .967 | .965 | .963 | .961 | .958 | .955 | .952 | .949 | .945 |
| | | | | | | .941 | .936 | .931 | .925 | .919 | .912 | .904 | .895 | .885 | .874 |
| | | | | | | .862 | .848 | .832 | .815 | .795 | .773 | .748 | | | |
| 28.000 | 24.975 | 3.025 | .201 | .148 | 28 | .969 | .967 | .965 | .962 | .960 | .957 | .954 | .951 | .947 | .943 |
| | | | | | | .939 | .934 | .929 | .923 | .916 | .909 | .900 | .891 | .880 | .869 |
| | | | | | | .856 | .841 | .824 | .805 | .784 | .761 | .734 | .705 | | |
| 28.000 | 25.509 | 2.491 | .166 | .124 | 29 | .968 | .966 | .964 | .962 | .959 | .956 | .953 | .950 | .946 | .942 |
| | | | | | | .937 | .932 | .927 | .920 | .913 | .905 | .897 | .887 | .876 | .863 |
| | | | | | | .849 | .834 | .816 | .796 | .774 | .749 | .721 | .690 | .656 | |
| 28.000 | 25.980 | 2.020 | .136 | .102 | 30 | .968 | .966 | .963 | .961 | .958 | .955 | .952 | .949 | .945 | .941 |
| | | | | | | .936 | .931 | .925 | .918 | .911 | .903 | .893 | .883 | .871 | .858 |
| | | | | | | .844 | .827 | .808 | .787 | .764 | .738 | .708 | .676 | .640 | .601 |
| 28.000 | 26.389 | 1.611 | .109 | .082 | 31 | .967 | .965 | .963 | .960 | .958 | .955 | .951 | .948 | .944 | .939 |
| | | | | | | .934 | .929 | .923 | .916 | .908 | .900 | .890 | .880 | .867 | .854 |
| | | | | | | .838 | .821 | .802 | .780 | .755 | .727 | .697 | .663 | .626 | .585 |
| | | | | | | .542 | | | | | | | | | |
| 28.000 | 26.734 | 1.266 | .086 | .065 | 32 | .967 | .965 | .962 | .960 | .957 | .954 | .951 | .947 | .943 | .938 |
| | | | | | | .933 | .928 | .921 | .914 | .906 | .898 | .888 | .876 | .864 | .850 |

JEWETT EQUIVALENT QUEUE EXTENDED ERLANG B(EQEEB): 1 MINUTE MAXIMUM WAIT (IF NOT CONNECTED AFTER 1 MINUTE WAIT, 100% OVERFLOW)

| HOURS FIRST ATTEMPT VMR = 1.0 | HOURS CON-NECT | HOURS OVFL | DELAYED PORTION | DELAY LENGTH CALL FACTOR | LINE | HOURS CONNECTED TRAFFIC ON LINES 1 | 2 | 3 | 4 | 5 | 6 | 7 | 8 | 9 | 10 |
|---|---|---|---|---|---|---|---|---|---|---|---|---|---|---|---|
| | | | | | | .834 | .816 | .795 | .773 | .747 | .718 | .687 | .652 | .613 | .571 |
| | | | | | | .527 | .480 | | | | | | | | |
| 28.000 | 27.022 | .978 | .067 | .051 | 33 | .967 | .964 | .962 | .959 | .957 | .954 | .950 | .946 | .942 | .937 |
| | | | | | | .932 | .926 | .920 | .913 | .905 | .896 | .885 | .874 | .861 | .846 |
| | | | | | | .830 | .811 | .790 | .767 | .740 | .711 | .678 | .642 | .602 | .559 |
| | | | | | | .514 | .466 | .416 | | | | | | | |
| 28.000 | 27.256 | .744 | .051 | .039 | 34 | .966 | .964 | .962 | .959 | .956 | .953 | .950 | .946 | .941 | .937 |
| | | | | | | .931 | .925 | .919 | .911 | .903 | .894 | .883 | .872 | .858 | .843 |
| | | | | | | .826 | .807 | .786 | .761 | .734 | .704 | .670 | .633 | .593 | .550 |
| | | | | | | .503 | .455 | .404 | .354 | | | | | | |
| 28.000 | 27.444 | .556 | .038 | .029 | 35 | .966 | .964 | .962 | .959 | .956 | .953 | .949 | .945 | .941 | .936 |
| | | | | | | .931 | .925 | .918 | .910 | .902 | .893 | .882 | .870 | .856 | .841 |
| | | | | | | .824 | .804 | .782 | .757 | .729 | .699 | .664 | .627 | .586 | .541 |
| | | | | | | .494 | .445 | .395 | .344 | .295 | | | | | |
| 28.000 | 27.591 | .409 | .028 | .022 | 36 | .966 | .964 | .961 | .959 | .956 | .952 | .949 | .945 | .940 | .936 |
| | | | | | | .930 | .924 | .917 | .910 | .901 | .891 | .881 | .868 | .855 | .839 |
| | | | | | | .821 | .801 | .779 | .754 | .726 | .694 | .659 | .621 | .580 | .535 |
| | | | | | | .488 | .438 | .387 | .337 | .287 | .240 | | | | |
| 30.000 | 23.409 | 6.591 | .392 | .276 | 25 | .973 | .972 | .970 | .969 | .967 | .965 | .963 | .960 | .958 | .955 |
| | | | | | | .953 | .949 | .946 | .942 | .938 | .933 | .928 | .923 | .917 | .910 |
| | | | | | | .902 | .894 | .885 | .874 | .863 | | | | | |
| 30.000 | 24.183 | 5.817 | .349 | .248 | 26 | .973 | .971 | .969 | .968 | .966 | .964 | .962 | .959 | .957 | .954 |
| | | | | | | .951 | .948 | .944 | .940 | .936 | .931 | .925 | .919 | .913 | .906 |
| | | | | | | .898 | .889 | .879 | .867 | .855 | .841 | | | | |
| 30.000 | 24.922 | 5.078 | .307 | .221 | 27 | .972 | .971 | .969 | .967 | .965 | .963 | .961 | .958 | .956 | .953 |
| | | | | | | .950 | .946 | .942 | .938 | .933 | .928 | .922 | .916 | .909 | .901 |
| | | | | | | .893 | .883 | .872 | .860 | .847 | .832 | .815 | | | |
| 30.000 | 25.618 | 4.382 | .267 | .194 | 28 | .972 | .970 | .968 | .966 | .964 | .962 | .960 | .957 | .954 | .951 |
| | | | | | | .948 | .944 | .940 | .936 | .931 | .925 | .919 | .913 | .905 | .897 |
| | | | | | | .888 | .877 | .866 | .853 | .839 | .822 | .804 | .784 | | |
| 30.000 | 26.266 | 3.734 | .230 | .168 | 29 | .971 | .969 | .968 | .966 | .964 | .961 | .959 | .956 | .953 | .950 |
| | | | | | | .947 | .943 | .938 | .934 | .928 | .923 | .916 | .909 | .901 | .892 |
| | | | | | | .883 | .872 | .859 | .846 | .830 | .813 | .794 | .772 | .748 | |
| 30.000 | 26.860 | 3.140 | .195 | .144 | 30 | .971 | .969 | .967 | .965 | .963 | .961 | .958 | .955 | .952 | .949 |
| | | | | | | .945 | .941 | .937 | .932 | .926 | .920 | .913 | .906 | .897 | .888 |
| | | | | | | .878 | .866 | .853 | .838 | .822 | .804 | .783 | .760 | .735 | .706 |
| 30.000 | 27.397 | 2.603 | .162 | .121 | 31 | .970 | .968 | .967 | .964 | .962 | .960 | .957 | .954 | .951 | .948 |
| | | | | | | .944 | .940 | .935 | .930 | .924 | .918 | .911 | .903 | .894 | .884 |
| | | | | | | .873 | .861 | .847 | .831 | .814 | .795 | .773 | .749 | .722 | .692 |
| | | | | | | .659 | | | | | | | | | |

JEWETT EQUIVALENT QUEUE EXTENDED ERLANG B(EQEEB): 1 MINUTE MAXIMUM WAIT (IF NOT CONNECTED AFTER 1 MINUTE WAIT, 100% OVERFLOW)

| HOURS FIRST ATTEMPT VMR = 1.0 | HOURS CON-NECT | HOURS OVFL | DELAYED PORTION | DELAY LENGTH CALL FACTOR | LINE | HOURS CONNECTED TRAFFIC ON LINES 1 | 2 | 3 | 4 | 5 | 6 | 7 | 8 | 9 | 10 |
|---|---|---|---|---|---|---|---|---|---|---|---|---|---|---|---|
| 30.000 | 27.872 | 2.128 | .134 | .100 | 32 | .970 | .968 | .966 | .964 | .962 | .959 | .956 | .953 | .950 | .946 |
| | | | | | | .943 | .938 | .933 | .928 | .922 | .915 | .908 | .900 | .890 | .880 |
| | | | | | | .869 | .856 | .841 | .825 | .806 | .786 | .763 | .738 | .709 | .678 |
| | | | | | | .644 | .606 | | | | | | | | |
| 30.000 | 28.287 | 1.713 | .108 | .082 | 33 | .969 | .968 | .966 | .963 | .961 | .958 | .956 | .953 | .949 | .945 |
| | | | | | | .941 | .937 | .932 | .926 | .920 | .913 | .906 | .897 | .887 | .877 |
| | | | | | | .865 | .851 | .836 | .819 | .800 | .778 | .754 | .728 | .698 | .665 |
| | | | | | | .630 | .591 | .549 | | | | | | | |
| 30.000 | 28.641 | 1.359 | .086 | .066 | 34 | .969 | .967 | .965 | .963 | .961 | .958 | .955 | .952 | .948 | .945 |
| | | | | | | .940 | .936 | .931 | .925 | .918 | .911 | .903 | .895 | .885 | .873 |
| | | | | | | .861 | .847 | .831 | .813 | .793 | .771 | .746 | .719 | .688 | .654 |
| | | | | | | .617 | .577 | .534 | .489 | | | | | | |
| 30.000 | 28.939 | 1.061 | .068 | .052 | 35 | .969 | .967 | .965 | .963 | .960 | .957 | .955 | .951 | .948 | .944 |
| | | | | | | .940 | .935 | .929 | .924 | .917 | .910 | .902 | .892 | .882 | .871 |
| | | | | | | .858 | .843 | .827 | .809 | .788 | .765 | .739 | .711 | .679 | .644 |
| | | | | | | .606 | .565 | .522 | .475 | .428 | | | | | |
| 30.000 | 29.183 | .817 | .052 | .040 | 36 | .969 | .967 | .965 | .962 | .960 | .957 | .954 | .951 | .947 | .943 |
| | | | | | | .939 | .934 | .929 | .922 | .916 | .908 | .900 | .891 | .880 | .868 |
| | | | | | | .855 | .840 | .823 | .805 | .783 | .760 | .733 | .704 | .672 | .636 |
| | | | | | | .597 | .555 | .511 | .464 | .416 | .367 | | | | |
| 30.000 | 29.381 | .619 | .040 | .031 | 37 | .968 | .966 | .964 | .962 | .960 | .957 | .954 | .950 | .947 | .943 |
| | | | | | | .938 | .933 | .928 | .922 | .915 | .907 | .899 | .889 | .878 | .866 |
| | | | | | | .853 | .838 | .820 | .801 | .780 | .755 | .729 | .699 | .665 | .629 |
| | | | | | | .590 | .547 | .502 | .455 | .406 | .357 | .309 | | | |
| 30.000 | 29.538 | .462 | .030 | .023 | 38 | .968 | .966 | .964 | .962 | .959 | .956 | .953 | .950 | .946 | .942 |
| | | | | | | .938 | .933 | .927 | .921 | .914 | .906 | .898 | .888 | .877 | .005 |
| | | | | | | .851 | .836 | .818 | .799 | .777 | .752 | .725 | .694 | .660 | .624 |
| | | | | | | .584 | .541 | .495 | .447 | .399 | .349 | .301 | .254 | | |

JEWETT EQUIVALENT QUEUE EXTENDED ERLANG B(EQEEB): 5 MINUTES MAXIMUM WAIT (IF NOT CONNECTED AFTER 5 MINUTES WAIT, 100% OVERFLOW)

| HOURS FIRST ATTEMPT VMR = 1.0 | HOURS CON-NECT | HOURS OVFL | DELAYED PORTION | DELAY LENGTH CALL FACTOR | LINE | HOURS CONNECTED TRAFFIC ON LINES 1 | 2 | 3 | 4 | 5 | 6 | 7 | 8 | 9 | 10 |
|---|---|---|---|---|---|---|---|---|---|---|---|---|---|---|---|
| .200 | .194 | .006 | .202 | .272 | 1 | .193 | | | | | | | | | |
| .200 | .200 | .000 | .017 | .024 | 2 | .169 | .031 | | | | | | | | |
| .400 | .370 | .030 | .400 | .519 | 1 | .370 | | | | | | | | | |
| .400 | .397 | .003 | .060 | .083 | 2 | .296 | .100 | | | | | | | | |
| .600 | .521 | .079 | .589 | .732 | 1 | .521 | | | | | | | | | |
| .600 | .589 | .011 | .122 | .166 | 2 | .400 | .189 | | | | | | | | |
| .600 | .598 | .002 | .021 | .029 | 3 | .379 | .168 | .051 | | | | | | | |
| .800 | .640 | .160 | .762 | .900 | 1 | .640 | | | | | | | | | |
| .800 | .775 | .025 | .198 | .266 | 2 | .489 | .286 | | | | | | | | |
| .800 | .795 | .005 | .043 | .059 | 3 | .454 | .245 | .097 | | | | | | | |
| 1.000 | .951 | .049 | .285 | .378 | 2 | .565 | .386 | | | | | | | | |
| 1.000 | .989 | .011 | .073 | .100 | 3 | .516 | .320 | .154 | | | | | | | |
| 1.000 | .998 | .002 | .016 | .023 | 4 | .503 | .304 | .141 | .049 | | | | | | |
| 1.200 | 1.115 | .085 | .381 | .496 | 2 | .633 | .482 | | | | | | | | |
| 1.200 | 1.180 | .020 | .111 | .151 | 3 | .569 | .391 | .219 | | | | | | | |
| 1.200 | 1.195 | .005 | .029 | .039 | 4 | .552 | .367 | .196 | .081 | | | | | | |
| 1.400 | 1.263 | .137 | .481 | .614 | 2 | .692 | .571 | | | | | | | | |
| 1.400 | 1.366 | .034 | .157 | .212 | 3 | .617 | .459 | .291 | | | | | | | |
| 1.400 | 1.391 | .009 | .045 | .062 | 4 | .593 | .424 | .253 | .121 | | | | | | |
| 1.600 | 1.392 | .208 | .584 | .726 | 2 | .742 | .650 | | | | | | | | |
| 1.600 | 1.546 | .054 | .210 | .282 | 3 | .659 | .522 | .365 | | | | | | | |
| 1.600 | 1.585 | .015 | .067 | .091 | 4 | .629 | .477 | .311 | .168 | | | | | | |
| 1.600 | 1.596 | .004 | .019 | .026 | 5 | .619 | .462 | .295 | .154 | .065 | | | | | |
| 1.800 | 1.500 | .300 | .683 | .827 | 2 | | | | | | | | | | |
| | | | | | .78 | .715 | | | | | | | | | |
| 1.800 | 1.718 | .082 | .270 | .359 | 3 | .698 | .581 | .439 | | | | | | | |
| 1.800 | 1.775 | .025 | .093 | .127 | 4 | .661 | .525 | .369 | .221 | | | | | | |
| 1.800 | 1.793 | .007 | .029 | .040 | 5 | .649 | .506 | .345 | .199 | .094 | | | | | |
| 2.000 | 1.587 | .413 | .777 | .914 | 2 | .819 | .768 | | | | | | | | |
| 2.000 | 1.880 | .120 | .336 | .441 | 3 | .733 | .635 | .512 | | | | | | | |
| 2.000 | 1.963 | .037 | .124 | .168 | 4 | .690 | .570 | .425 | .277 | | | | | | |
| 2.000 | 1.988 | .012 | .042 | .057 | 5 | .675 | .545 | .394 | .246 | .128 | | | | | |
| 2.200 | 1.656 | .544 | .863 | .985 | 2 | .847 | .809 | | | | | | | | |
| 2.200 | 2.030 | .170 | .406 | .526 | 3 | .764 | .684 | .581 | | | | | | | |
| 2.200 | 2.145 | .055 | .160 | .216 | 4 | .717 | .611 | .480 | .337 | | | | | | |
| 2.200 | 2.182 | .018 | .058 | .079 | 5 | .698 | .581 | .440 | .294 | .168 | | | | | |
| 2.200 | 2.194 | .006 | .019 | .026 | 6 | .691 | .571 | .426 | .279 | .155 | .073 | | | | |
| 2.400 | 2.166 | .234 | .480 | .612 | 3 | .793 | .728 | .644 | | | | | | | |
| 2.400 | 2.323 | .077 | .201 | .270 | 4 | .742 | .649 | .532 | .399 | | | | | | |
| 2.400 | 2.373 | .027 | .077 | .105 | 5 | .719 | .614 | .485 | .343 | .212 | | | | | |
| 2.400 | 2.391 | .009 | .027 | .037 | 6 | .711 | .601 | .466 | .322 | .193 | .098 | | | | |

JEWETT EQUIVALENT QUEUE EXTENDED ERLANG B(EQEEB): 5 MINUTES MAXIMUM WAIT (IF NOT CONNECTED AFTER 5 MINUTES WAIT, 100% OVERFLOW)

| HOURS FIRST ATTEMPT VMR = 1.0 | HOURS CON-NECT | HOURS OVFL | DELAYED PORTION | DELAY LENGTH CALL FACTOR | LINE | HOURS CONNECTED TRAFFIC ON LINES 1 | 2 | 3 | 4 | 5 | 6 | 7 | 8 | 9 | 10 |
|---|---|---|---|---|---|---|---|---|---|---|---|---|---|---|---|
| 2.600 | 2.286 | .314 | .555 | .696 | 3 | .819 | .767 | .700 | | | | | | | |
| 2.600 | 2.493 | .107 | .247 | .329 | 4 | .765 | .685 | .582 | .461 | | | | | | |
| 2.600 | 2.562 | .038 | .099 | .135 | 5 | .739 | .645 | .526 | .392 | .259 | | | | | |
| 2.600 | 2.586 | .014 | .037 | .051 | 6 | .729 | .629 | .504 | .365 | .233 | .128 | | | | |
| 2.800 | 2.390 | .410 | .630 | .774 | 3 | .842 | .801 | .748 | | | | | | | |
| 2.800 | 2.656 | .144 | .297 | .393 | 4 | .787 | .718 | .630 | .521 | | | | | | |
| 2.800 | 2.747 | .053 | .125 | .170 | 5 | .758 | .673 | .566 | .440 | .310 | | | | | |
| 2.800 | 2.780 | .020 | .049 | .067 | 6 | .745 | .654 | .539 | .407 | .274 | .162 | | | | |
| 2.800 | 2.793 | .007 | .018 | .025 | 7 | .740 | .646 | .528 | .393 | .261 | .151 | .075 | | | |
| 3.000 | 2.478 | .522 | .702 | .845 | 3 | .861 | .829 | .787 | | | | | | | |
| 3.000 | 2.809 | .191 | .351 | .460 | 4 | .807 | .749 | .673 | .580 | | | | | | |
| 3.000 | 2.928 | .072 | .154 | .209 | 5 | .775 | .700 | .604 | .488 | .362 | | | | | |
| 3.000 | 2.973 | .027 | .063 | .087 | 6 | .760 | .677 | .572 | .447 | .317 | .199 | | | | |
| 3.000 | 2.990 | .010 | .024 | .033 | 7 | .754 | .668 | .558 | .430 | .299 | .183 | .098 | | | |
| 3.200 | 2.550 | .650 | .770 | .907 | 3 | .878 | .852 | .820 | | | | | | | |
| 3.200 | 2.950 | .250 | .409 | .529 | 4 | .826 | .777 | | | | | | | | |
| | | | | | | .714 | .634 | | | | | | | | |
| 3.200 | 3.105 | .095 | .188 | .253 | 5 | .791 | .725 | .639 | .534 | .415 | | | | | |
| 3.200 | 3.163 | .037 | .080 | .109 | 6 | .774 | .699 | .603 | .487 | .360 | .240 | | | | |
| 3.200 | 3.186 | .014 | .032 | .044 | 7 | .767 | .688 | .586 | .466 | .337 | .218 | .124 | | | |
| 3.400 | 2.609 | .791 | .833 | .961 | 3 | .892 | .871 | .846 | | | | | | | |
| 3.400 | 3.080 | .320 | .468 | .599 | 4 | .843 | .802 | .750 | .684 | | | | | | |
| 3.400 | 3.275 | .125 | .224 | .300 | 5 | .806 | .748 | .673 | .579 | .469 | | | | | |
| 3.400 | 3.350 | .050 | .099 | .136 | 6 | .788 | .720 | .632 | .521 | .404 | .283 | | | | |
| 3.400 | 3.380 | .020 | .041 | .057 | 7 | .779 | .706 | .613 | .500 | .375 | .254 | .153 | | | |
| 3.400 | 3.392 | .008 | .016 | .022 | 8 | .775 | .701 | .604 | .489 | .363 | .242 | .143 | .074 | | |
| 3.600 | 3.195 | .405 | .529 | .667 | 4 | .859 | .825 | .783 | .728 | | | | | | |
| 3.600 | 3.439 | .161 | .265 | .352 | 5 | .821 | .770 | .704 | .621 | .522 | | | | | |
| 3.600 | 3.534 | .066 | .121 | .165 | 6 | .800 | .739 | .659 | .561 | .447 | .328 | | | | |
| 3.600 | 3.573 | .027 | .053 | .072 | 7 | .790 | .724 | .637 | .532 | .413 | .292 | .185 | | | |
| 3.600 | 3.589 | .011 | .021 | .029 | 8 | .786 | .717 | .628 | .519 | .397 | .277 | .172 | .094 | | |
| 3.800 | 3.297 | .503 | .590 | .733 | 4 | .873 | .846 | .811 | .767 | | | | | | |
| 3.800 | 3.595 | .205 | .308 | .407 | 5 | .835 | .791 | .734 | .661 | .574 | | | | | |
| 3.800 | 3.715 | .085 | .146 | .198 | 6 | .812 | .757 | .685 | .595 | .490 | .375 | | | | |
| 3.800 | 3.764 | .036 | .065 | .090 | 7 | .801 | .740 | .661 | .562 | .449 | .331 | .221 | | | |
| 3.800 | 3.785 | .015 | .028 | .038 | 8 | .796 | .732 | .649 | .547 | .431 | .311 | .202 | .117 | | |
| 4.000 | 3.385 | .615 | .650 | .794 | 4 | .886 | .863 | .835 | .800 | | | | | | |
| 4.000 | 3.741 | .259 | .355 | .465 | 5 | .848 | .810 | .761 | .699 | .623 | | | | | |
| 4.000 | 3.891 | .109 | .174 | .235 | 6 | .824 | .774 | .710 | .629 | .532 | .423 | | | | |
| 4.000 | 3.953 | .047 | .080 | .110 | 7 | .811 | .755 | .682 | .592 | .485 | .370 | .258 | | | |
| 4.000 | 3.980 | .020 | .035 | .048 | 8 | .805 | .746 | .669 | .574 | .463 | .346 | .235 | .143 | | |

JEWETT EQUIVALENT QUEUE EXTENDED ERLANG B(EQEEB): 5 MINUTES MAXIMUM WAIT (IF NOT CONNECTED AFTER 5 MINUTES WAIT, 100% OVERFLOW)

| HOURS FIRST ATTEMPT VMR = 1.0 | HOURS CON-NECT | HOURS OVFL | DELAYED PORTION | DELAY LENGTH CALL FACTOR | LINE | HOURS CONNECTED TRAFFIC ON LINES 1 | 2 | 3 | 4 | 5 | 6 | 7 | 8 | 9 | 10 |
|---|---|---|---|---|---|---|---|---|---|---|---|---|---|---|---|
| 4.200 | 3.459 | .741 | .707 | .850 | 4 | .897 | .879 | .856 | .827 | | | | | | |
| 4.200 | 3.877 | .323 | .404 | .524 | 5 | .861 | .828 | .786 | .734 | .668 | | | | | |
| 4.200 | 4.062 | .138 | .205 | .275 | 6 | .835 | .790 | .733 | .660 | .572 | .471 | | | | |
| 4.200 | 4.140 | .060 | .097 | .133 | 7 | .821 | .769 | .703 | .620 | .520 | .410 | .298 | | | |
| 4.200 | 4.174 | .026 | .044 | .060 | 8 | .813 | .759 | .688 | .599 | .494 | .380 | .268 | .171 | | |
| 4.200 | 4.189 | .011 | .019 | .026 | 9 | .810 | .754 | .681 | .590 | .483 | .367 | .255 | .160 | .090 | |
| 4.400 | 3.523 | .877 | .762 | .900 | 4 | .907 | .892 | .873 | .850 | | | | | | |
| 4.400 | 4.002 | .398 | .455 | .584 | 5 | .873 | .845 | .809 | .765 | .710 | | | | | |
| 4.400 | 4.227 | .173 | .238 | .318 | 6 | .845 | .806 | .755 | .691 | .612 | .519 | | | | |
| 4.400 | 4.323 | .077 | .116 | .158 | 7 | .830 | .783 | .722 | .646 | .554 | .449 | .339 | | | |
| 4.400 | 4.366 | .034 | .054 | .074 | 8 | .822 | .771 | .706 | .623 | .525 | .415 | .303 | .202 | | |
| 4.400 | 4.385 | .015 | .024 | .033 | 9 | .818 | .765 | .697 | .612 | .510 | .398 | .287 | .187 | .110 | |
| 4.600 | 3.576 | 1.024 | .812 | .944 | 4 | .916 | .903 | .888 | .869 | | | | | | |
| 4.600 | 4.114 | .486 | .507 | .643 | 5 | .884 | .860 | .830 | .793 | .747 | | | | | |
| 4.600 | 4.385 | .215 | .274 | .364 | 6 | .856 | .821 | .776 | .719 | .649 | .565 | | | | |
| 4.600 | 4.504 | .096 | .137 | .186 | 7 | .839 | .796 | .741 | .671 | .587 | .488 | .382 | | | |
| 4.600 | 4.556 | .044 | .066 | .090 | 8 | .830 | .783 | .722 | .646 | .553 | .449 | .339 | .235 | | |
| 4.600 | 4.581 | .019 | .030 | .041 | 9 | .825 | .776 | .713 | .633 | .537 | .429 | .318 | .216 | .133 | |
| 4.800 | 3.620 | 1.180 | .860 | .983 | 4 | .923 | .913 | .900 | .884 | | | | | | |
| 4.800 | 4.214 | .586 | .559 | .700 | 5 | .894 | .874 | .849 | .818 | .780 | | | | | |
| 4.800 | 4.536 | .264 | .313 | .413 | 6 | .865 | .835 | .795 | .746 | .684 | .610 | | | | |
| 4.800 | 4.680 | .120 | .161 | .218 | 7 | .847 | .808 | .759 | .696 | .618 | .527 | .425 | | | |
| 4.800 | 4.745 | .055 | .079 | .108 | 8 | .837 | .794 | .738 | .667 | .581 | .482 | .375 | .270 | | |
| 4.800 | 4.775 | .025 | .037 | .051 | 9 | .832 | .786 | .727 | .653 | .563 | .459 | .350 | .246 | .158 | |
| 4.800 | 4.789 | .011 | .016 | .022 | 10 | .830 | .783 | .722 | .646 | .553 | .448 | .339 | .235 | .148 | .085 |
| 5.000 | 4.302 | .698 | .611 | .754 | 5 | .903 | .886 | .865 | .840 | .809 | | | | | |
| 5.000 | 4.677 | .323 | .355 | .464 | 6 | .875 | .848 | .814 | .771 | .717 | .653 | | | | |
| 5.000 | 4.852 | .148 | .187 | .252 | 7 | .855 | .820 | .775 | .719 | .648 | .564 | .469 | | | |
| 5.000 | 4.931 | .069 | .094 | .128 | 8 | .844 | .804 | .753 | .688 | .608 | .514 | .411 | .307 | | |
| 5.000 | 4.968 | .032 | .045 | .062 | 9 | .839 | .796 | .741 | .672 | .587 | .489 | .382 | .278 | .185 | |
| 5.000 | 4.986 | .014 | .020 | .028 | 10 | .836 | .792 | .735 | .663 | .576 | .476 | .368 | .264 | .173 | .103 |
| 5.500 | 4.474 | 1.026 | .732 | .873 | 5 | .922 | .911 | .897 | .881 | .862 | | | | | |
| 5.500 | 4.987 | .513 | .465 | .595 | 6 | .896 | .877 | .854 | .825 | .789 | .745 | | | | |
| 5.500 | 5.257 | .243 | .262 | .349 | 7 | .875 | .848 | .814 | .771 | .718 | .653 | .577 | | | |
| 5.500 | 5.383 | .117 | .139 | .188 | 8 | .861 | .829 | .788 | .735 | .670 | .592 | .502 | .405 | | |
| 5.500 | 5.444 | .056 | .070 | .096 | 9 | .854 | .818 | .773 | .715 | .643 | .558 | .462 | .360 | .262 | |
| 5.500 | 5.473 | .027 | .034 | .047 | 10 | .850 | .813 | .764 | .704 | .629 | .540 | .441 | .337 | .240 | .157 |
| 5.500 | 5.488 | .012 | .016 | .022 | 11 | .848 | .810 | .760 | .698 | .621 | .531 | .430 | .326 | .229 | .148 |
| | | | | | | .087 | | | | | | | | | |
| 6.000 | 4.591 | 1.409 | .838 | .965 | 5 | .936 | .929 | .920 | .909 | .897 | | | | | |

JEWETT EQUIVALENT QUEUE EXTENDED ERLANG B(EQEEB): 5 MINUTES MAXIMUM WAIT (IF NOT CONNECTED AFTER 5 MINUTES WAIT, 100% OVERFLOW)

| HOURS FIRST ATTEMPT VMR = 1.0 | HOURS CON-NECT | HOURS OVFL | DELAYED PORTION | DELAY LENGTH CALL FACTOR | LINE | HOURS CONNECTED TRAFFIC ON LINES 1 | 2 | 3 | 4 | 5 | 6 | 7 | 8 | 9 | 10 |
|---|---|---|---|---|---|---|---|---|---|---|---|---|---|---|---|
| 6.000 | 5.228 | .772 | .579 | .721 | 6 | .915 | .902 | .886 | .867 | .843 | .815 | | | | |
| 6.000 | 5.617 | .383 | .352 | .461 | 7 | .893 | .873 | .848 | .817 | .779 | .732 | .675 | | | |
| 6.000 | 5.812 | .188 | .196 | .264 | 8 | .877 | .851 | .819 | .778 | .727 | .664 | .590 | .506 | | |
| 6.000 | 5.907 | .093 | .104 | .142 | 9 | .868 | .838 | .800 | .753 | .694 | .622 | .538 | .445 | .348 | |
| 6.000 | 5.954 | .046 | .053 | .073 | 10 | .863 | .831 | .790 | .739 | .675 | .598 | .509 | .412 | .314 | .223 |
| 6.000 | 5.978 | .022 | .026 | .036 | 11 | .860 | .827 | .784 | .731 | .665 | .585 | .494 | .395 | .297 | .207 |
| | | | | | | .133 | | | | | | | | | |
| 6.500 | 5.404 | 1.096 | .688 | .831 | 6 | .930 | .921 | .910 | .897 | .882 | .864 | | | | |
| 6.500 | 5.920 | .580 | .451 | .579 | 7 | .910 | .895 | .877 | .855 | .829 | .796 | .757 | | | |
| 6.500 | 6.207 | .293 | .266 | .354 | 8 | .892 | .872 | .847 | .815 | .776 | .729 | .671 | .604 | | |
| 6.500 | 6.352 | .148 | .148 | .201 | 9 | .881 | .857 | .826 | .787 | .739 | .681 | .611 | .530 | .441 | |
| 6.500 | 6.425 | .075 | .079 | .109 | 10 | .874 | .847 | .813 | .770 | .716 | .651 | .574 | .487 | .393 | .299 |
| 6.500 | 6.462 | .038 | .041 | .056 | 11 | .871 | .842 | .806 | .760 | .703 | .634 | .553 | .463 | .367 | .273 |
| | | | | | | .190 | | | | | | | | | |
| 6.500 | 6.482 | .018 | .020 | .028 | 12 | .869 | .839 | .802 | .755 | .696 | .625 | .542 | .450 | .353 | .260 |
| | | | | | | .178 | .113 | | | | | | | | |
| 7.000 | 5.529 | 1.471 | .786 | .921 | 6 | .941 | .935 | .927 | .919 | .909 | .897 | | | | |
| 7.000 | 6.159 | .841 | .554 | .694 | 7 | .924 | .913 | .901 | .886 | .868 | .846 | .820 | | | |
| 7.000 | 6.559 | .441 | .348 | .456 | 8 | .907 | .891 | .872 | .849 | .820 | .785 | .743 | .693 | | |
| 7.000 | 6.772 | .228 | .203 | .272 | 9 | .894 | .874 | .849 | .818 | .780 | .734 | .678 | .611 | .535 | |
| 7.000 | 6.882 | .118 | .113 | .154 | 10 | .885 | .863 | .834 | .798 | .754 | .699 | .634 | .558 | .473 | .383 |
| 7.000 | 6.939 | .061 | .061 | .084 | 11 | .881 | .856 | .825 | .786 | .738 | .679 | .608 | .527 | .438 | .345 |
| | | | | | | .257 | | | | | | | | | |
| 7.000 | 6.969 | .031 | .032 | .044 | 12 | .878 | .852 | .820 | .779 | .728 | .667 | .593 | .509 | .418 | .325 |
| | | | | | | .237 | .161 | | | | | | | | |
| 7.000 | 6.985 | .015 | .010 | .022 | 13 | .876 | .850 | .817 | .775 | .723 | .660 | .585 | .500 | .408 | .314 |
| | | | | | | .227 | .153 | .095 | | | | | | | |
| 7.500 | 5.617 | 1.883 | .870 | .991 | 6 | .950 | .945 | .940 | .934 | .927 | .919 | | | | |
| 7.500 | 6.339 | 1.161 | .652 | .796 | 7 | .936 | .928 | .920 | .909 | .897 | .883 | .866 | | | |
| 7.500 | 6.857 | .643 | .438 | .564 | 8 | .920 | .908 | .894 | .877 | .857 | .832 | .802 | .767 | | |
| 7.500 | 7.160 | .340 | .268 | .356 | 9 | .906 | .890 | .870 | .846 | .817 | .781 | .738 | .686 | .625 | |
| 7.500 | 7.320 | .180 | .155 | .210 | 10 | .896 | .877 | .853 | .824 | .787 | .743 | .690 | .626 | .553 | .471 |
| 7.500 | 7.404 | .096 | .087 | .119 | 11 | .890 | .869 | .842 | .009 | .769 | .719 | .658 | .588 | .508 | .421 |
| | | | | | | .332 | | | | | | | | | |
| 7.500 | 7.450 | .050 | .047 | .065 | 12 | .887 | .864 | .836 | .801 | .757 | .704 | .640 | .565 | .481 | .392 |
| | | | | | | .303 | .220 | | | | | | | | |
| 7.500 | 7.474 | .026 | .025 | .034 | 13 | .085 | .861 | .832 | .796 | .751 | .695 | .629 | .552 | .467 | .376 |
| | | | | | | .287 | .205 | .137 | | | | | | | |
| 8.000 | 6.470 | 1.530 | .743 | .883 | 7 | .946 | .940 | .934 | .926 | .918 | .908 | .897 | | | |
| 8.000 | 7.095 | .905 | .532 | .670 | 8 | .931 | .923 | .912 | .900 | .886 | .869 | .849 | .825 | | |

JEWETT EQUIVALENT QUEUE EXTENDED ERLANG B(EQEEB): 5 MINUTES MAXIMUM WAIT (IF NOT CONNECTED AFTER 5 MINUTES WAIT, 100% OVERFLOW)

| HOURS FIRST ATTEMPT VMR = 1.0 | HOURS CON-NECT | HOURS OVFL | DELAYED PORTION | DELAY LENGTH CALL FACTOR | LINE | HOURS CONNECTED TRAFFIC ON LINES 1 | 2 | 3 | 4 | 5 | 6 | 7 | 8 | 9 | 10 |
|---|---|---|---|---|---|---|---|---|---|---|---|---|---|---|---|
| 8.000 | 7.504 | .496 | .344 | .451 | 9 | .917 | .904 | .889 | .871 | .849 | .823 | .791 | .752 | .707 | |
| 8.000 | 7.733 | .267 | .207 | .278 | 10 | .906 | .890 | .871 | .847 | .818 | .783 | .740 | .689 | .629 | .560 |
| 8.000 | 7.855 | .145 | .120 | .163 | 11 | .899 | .881 | .858 | .831 | .797 | .755 | .705 | .645 | .575 | .497 |
| | | | | | | .413 | | | | | | | | | |
| 8.000 | 7.922 | .078 | .068 | .092 | 12 | .895 | .875 | .851 | .820 | .783 | .737 | .682 | .617 | .542 | .459 |
| | | | | | | .373 | .287 | | | | | | | | |
| 8.000 | 7.958 | .042 | .037 | .051 | 13 | .892 | .871 | .846 | .814 | .775 | .727 | .669 | .601 | .523 | .438 |
| | | | | | | .350 | .265 | .189 | | | | | | | |
| 8.000 | 7.978 | .022 | .019 | .027 | 14 | .891 | .869 | .843 | .811 | .770 | .721 | .661 | .591 | .512 | .425 |
| | | | | | | .337 | .252 | .178 | .117 | | | | | | |
| 8.500 | 6.565 | 1.935 | .823 | .953 | 7 | .953 | .949 | .944 | .939 | .933 | .926 | .919 | | | |
| 8.500 | 7.278 | 1.222 | .623 | .766 | 8 | .941 | .935 | .927 | .919 | .909 | .897 | .883 | .867 | | |
| 8.500 | 7.797 | .703 | .427 | .550 | 9 | .928 | .918 | .907 | .893 | .877 | .858 | .835 | .807 | .775 | |
| 8.500 | 8.113 | .387 | .269 | .357 | 10 | .916 | .903 | .887 | .869 | .846 | .819 | .785 | .746 | .699 | .643 |
| 8.500 | 8.287 | .213 | .161 | .218 | 11 | .908 | .892 | .873 | .850 | .822 | .788 | .747 | .698 | .639 | .572 |
| | | | | | | .497 | | | | | | | | | |
| 8.500 | 8.383 | .117 | .094 | .128 | 12 | .902 | .885 | .864 | .838 | .807 | .768 | .721 | .665 | .600 | .526 |
| | | | | | | .445 | .361 | | | | | | | | |
| 8.500 | 8.436 | .064 | .053 | .072 | 13 | .899 | .881 | .858 | .831 | .797 | .755 | .705 | .645 | .576 | .498 |
| | | | | | | .414 | .329 | .248 | | | | | | | |
| 8.500 | 8.465 | .035 | .029 | .040 | 14 | .897 | .878 | .855 | .826 | .791 | .748 | .695 | .633 | .561 | .481 |
| | | | | | | .396 | .311 | .231 | .161 | | | | | | |
| 8.500 | 8.482 | .018 | .015 | .021 | 15 | .896 | .877 | .853 | .824 | .788 | .743 | .690 | .626 | .553 | .472 |
| | | | | | | .386 | .301 | .221 | .153 | .099 | | | | | |
| 9.000 | 7.414 | 1.586 | .707 | .849 | 8 | .949 | .945 | .939 | .933 | .926 | .917 | .908 | .897 | | |
| 9.000 | 8.035 | .965 | .513 | .649 | 9 | .937 | .930 | .921 | .911 | .900 | .886 | .870 | .851 | .828 | |
| 9.000 | 8.451 | .549 | .339 | .445 | 10 | .925 | .915 | .903 | .888 | .871 | .850 | .825 | .795 | .760 | .718 |
| 9.000 | 8.694 | .306 | .210 | .282 | 11 | .916 | .903 | .887 | .869 | .846 | .819 | .786 | .746 | .699 | .643 |
| | | | | | | .580 | | | | | | | | | |
| 9.000 | 8.829 | .171 | .126 | .171 | 12 | .910 | .895 | .877 | .855 | .828 | .796 | .757 | .710 | .654 | .590 |
| | | | | | | .518 | .440 | | | | | | | | |
| 9.000 | 8.904 | .096 | .073 | .100 | 13 | .906 | .890 | .870 | .846 | .817 | .781 | .738 | .686 | .625 | .556 |
| | | | | | | .478 | .397 | .315 | | | | | | | |
| 9.000 | 8.947 | .053 | .042 | .057 | 14 | .903 | .886 | .866 | .841 | .810 | .772 | .726 | .671 | .607 | .535 |
| | | | | | | .455 | .372 | .290 | .214 | | | | | | |
| 9.000 | 8.971 | .029 | .023 | .031 | 15 | .902 | .885 | .863 | .837 | .805 | .766 | .719 | .663 | .597 | .522 |
| | | | | | | .441 | .357 | .275 | .201 | .138 | | | | | |
| 9.500 | 7.515 | 1.985 | .783 | .919 | 8 | .956 | .952 | .948 | .943 | .938 | .932 | .926 | .918 | | |
| 9.500 | 8.220 | 1.280 | .597 | .740 | 9 | .946 | .940 | .934 | .926 | .918 | .908 | .897 | .884 | .868 | |
| 9.500 | 8.741 | .759 | .416 | .538 | 10 | .934 | .926 | .917 | .905 | .892 | .877 | .859 | .837 | .812 | .782 |

JEWETT EQUIVALENT QUEUE EXTENDED ERLANG B(EQEEB): 5 MINUTES MAXIMUM WAIT (IF NOT CONNECTED AFTER 5 MINUTES WAIT, 100% OVERFLOW)

| HOURS FIRST ATTEMPT VMR = 1.0 | HOURS CON-NECT | HOURS OVFL | DELAYED PORTION | DELAY LENGTH CALL FACTOR | LINE | HOURS CONNECTED TRAFFIC ON LINES 1 | 2 | 3 | 4 | 5 | 6 | 7 | 8 | 9 | 10 |
|---|---|---|---|---|---|---|---|---|---|---|---|---|---|---|---|
| 9.500 | 9.068 | .432 | .269 | .357 | 11 | .924 | .913 | .901 | .886 | .868 | .846 | .820 | .789 | .753 | .709 |
| | | | | | | .658 | | | | | | | | | |
| 9.500 | 9.255 | .245 | .165 | .224 | 12 | .917 | .904 | .889 | .871 | .848 | .822 | .789 | .751 | .705 | .651 |
| | | | | | | .589 | .520 | | | | | | | | |
| 9.500 | 9.361 | .139 | .099 | .135 | 13 | .912 | .898 | .881 | .860 | .835 | .805 | .768 | .724 | .672 | .611 |
| | | | | | | .542 | .466 | .387 | | | | | | | |
| 9.500 | 9.421 | .079 | .058 | .079 | 14 | .909 | .894 | .876 | .854 | .827 | .794 | .754 | .706 | .650 | .585 |
| | | | | | | .512 | .433 | .352 | .274 | | | | | | |
| 9.500 | 9.456 | .044 | .033 | .045 | 15 | .907 | .892 | .873 | .850 | .821 | .787 | .746 | .696 | .637 | .569 |
| | | | | | | .494 | .414 | .332 | .254 | .184 | | | | | |
| 9.500 | 9.476 | .024 | .018 | .025 | 16 | .906 | .890 | .871 | .847 | .818 | .783 | .740 | .689 | .629 | .560 |
| | | | | | | .484 | .402 | .320 | .243 | .174 | .118 | | | | |
| 10.000 | 7.590 | 2.410 | .850 | .975 | 8 | .961 | .958 | .955 | .952 | .948 | .943 | .938 | .933 | | |
| 10.000 | 8.361 | 1.639 | .676 | .820 | 9 | .953 | .948 | .943 | .938 | .932 | .925 | .917 | .908 | .897 | |
| 10.000 | 8.978 | 1.022 | .496 | .630 | 10 | .942 | .936 | .929 | .920 | .911 | .899 | .886 | .871 | .853 | .832 |
| 10.000 | 9.401 | .599 | .335 | .440 | 11 | .932 | .923 | .913 | .901 | .887 | .871 | .851 | .827 | .800 | .767 |
| | | | | | | .728 | | | | | | | | | |
| 10.000 | 9.655 | .345 | .213 | .285 | 12 | .924 | .913 | .900 | .885 | .867 | .845 | .819 | .788 | .751 | .707 |
| | | | | | | .656 | .598 | | | | | | | | |
| 10.000 | 9.802 | .198 | .131 | .178 | 13 | .918 | .906 | .891 | .874 | .852 | .827 | .796 | .759 | .715 | .663 |
| | | | | | | .603 | .536 | .463 | | | | | | | |
| 10.000 | 9.886 | .114 | .078 | .107 | 14 | .915 | .901 | .885 | .866 | .843 | .814 | .780 | .739 | .690 | .633 |
| | | | | | | .568 | .495 | .418 | .340 | | | | | | |
| 10.000 | 9.935 | .065 | .046 | .063 | 15 | .912 | .898 | .882 | .861 | .836 | .806 | .770 | .726 | .674 | .614 |
| | | | | | | .545 | .470 | .391 | .312 | .237 | | | | | |
| 10.000 | 9.963 | .037 | .026 | .036 | 16 | .911 | .897 | .879 | .858 | .832 | .801 | .760 | .718 | .664 | .602 |
| | | | | | | .532 | .455 | .375 | .296 | .222 | .158 | | | | |
| 11.000 | 8.540 | 2.454 | .813 | .945 | 9 | .963 | .960 | .958 | .954 | .951 | .947 | .943 | .938 | .932 | |
| 11.000 | 9.310 | 1.690 | .649 | .793 | 10 | .955 | .952 | .947 | .942 | .937 | .931 | .924 | .916 | .907 | .897 |
| 11.000 | 9.924 | 1.076 | .481 | .613 | 11 | .946 | .941 | .935 | .928 | .919 | .910 | .899 | .886 | .871 | .854 |
| | | | | | | .834 | | | | | | | | | |
| 11.000 | 10.352 | .648 | .330 | .434 | 12 | .937 | .930 | .922 | .912 | .900 | .887 | .871 | .852 | .829 | .803 |
| | | | | | | .773 | .737 | | | | | | | | |
| 11.000 | 10.618 | .382 | .214 | .288 | 13 | .930 | .921 | .911 | .898 | .883 | .866 | .845 | .820 | .791 | .756 |
| | | | | | | .715 | .667 | .613 | | | | | | | |
| 11.000 | 10.774 | .226 | .135 | .183 | 14 | .925 | .915 | .903 | .888 | .871 | .850 | .826 | .796 | .761 | .719 |
| | | | | | | .670 | .615 | .552 | .483 | | | | | | |
| 11.000 | 10.867 | .133 | .083 | .113 | 15 | .922 | .911 | .898 | .882 | .863 | .840 | .812 | .779 | .740 | .693 |
| | | | | | | .640 | .578 | .510 | .437 | .362 | | | | | |
| 11.000 | 10.922 | .078 | .050 | .068 | 16 | .920 | .908 | .894 | .077 | .857 | .833 | .803 | .768 | .726 | .677 |

JEWETT EQUIVALENT QUEUE EXTENDED ERLANG B(EQEEB): 5 MINUTES MAXIMUM WAIT (IF NOT CONNECTED AFTER 5 MINUTES WAIT, 100% OVERFLOW)

| HOURS FIRST ATTEMPT VMR = 1.0 | HOURS CON- NECT | HOURS OVFL | DELAYED PORTION | DELAY LENGTH CALL FACTOR | LINE | HOURS CONNECTED TRAFFIC ON LINES 1 | 2 | 3 | 4 | 5 | 6 | 7 | 8 | 9 | 10 |
|---|---|---|---|---|---|---|---|---|---|---|---|---|---|---|---|
| | | | | | | .620 | .555 | .484 | .409 | .333 | .260 | | | | |
| 11.000 | 10.955 | .045 | .029 | .040 | 17 | .919 | .906 | .892 | .874 | .853 | .828 | .797 | .761 | .717 | .666 |
| | | | | | | .607 | .540 | .467 | .391 | .315 | .242 | .178 | | | |
| 11.000 | 10.975 | .025 | .017 | .023 | 18 | .918 | .905 | .891 | .873 | .851 | .825 | .794 | .756 | .712 | .659 |
| | | | | | | .599 | .531 | .457 | .381 | .304 | .232 | .169 | .117 | | |
| 12.000 | 9.504 | 2.496 | .781 | 917 | 10 | .965 | .962 | .960 | .957 | .954 | .950 | .946 | .942 | .937 | .932 |
| 12.000 | 10.261 | 1.739 | .626 | .770 | 11 | .958 | .954 | .951 | .946 | .942 | .936 | .930 | .923 | .916 | .907 |
| | | | | | | .897 | | | | | | | | | |
| 12.000 | 10.872 | 1.128 | .468 | .598 | 12 | .950 | .945 | .940 | .934 | .927 | .918 | .909 | .899 | .886 | .872 |
| | | | | | | .856 | .837 | | | | | | | | |
| 12.000 | 11.305 | .695 | .326 | .429 | 13 | .942 | .936 | .929 | .920 | .910 | .899 | .886 | .871 | .853 | .831 |
| | | | | | | .807 | .778 | .744 | | | | | | | |
| 12.000 | 11.580 | .420 | .215 | .289 | 14 | .936 | .928 | .919 | .909 | .896 | .882 | .865 | .845 | .821 | .793 |
| | | | | | | .760 | .722 | .677 | .627 | | | | | | |
| 12.000 | 11.747 | .253 | .138 | .187 | 15 | .931 | .923 | .912 | .900 | .886 | .869 | .849 | .825 | .796 | .763 |
| | | | | | | .723 | .678 | .625 | .566 | .501 | | | | | |
| 12.000 | 11.847 | .153 | .087 | .119 | 16 | .928 | .919 | .908 | .894 | .878 | .860 | .837 | .810 | .779 | .741 |
| | | | | | | .697 | .646 | .588 | .524 | .455 | .383 | | | | |
| 12.000 | 11.908 | .092 | .053 | .073 | 17 | .926 | .916 | .904 | .890 | .874 | .853 | .829 | .801 | .767 | .726 |
| | | | | | | .679 | .625 | .564 | .496 | .425 | .351 | .280 | | | |
| 12.000 | 11.946 | .054 | .032 | .044 | 18 | .925 | .915 | .902 | .888 | .870 | .849 | .824 | .794 | .759 | .717 |
| | | | | | | .668 | .611 | .548 | .479 | .406 | .332 | .261 | .197 | | |
| 12.000 | 11.968 | .032 | .019 | .026 | 19 | .924 | .914 | .901 | .886 | .868 | .847 | .821 | .790 | .754 | .711 |
| | | | | | | .660 | .603 | .538 | .468 | .394 | .320 | .250 | .186 | .132 | |
| 13.000 | 10.464 | 2.536 | .752 | .891 | 11 | .966 | .964 | .962 | .959 | .956 | .953 | .949 | .945 | .941 | .936 |
| | | | | | | .931 | | | | | | | | | |
| 13.000 | 11.213 | 1.787 | .605 | .748 | 12 | .960 | .957 | .954 | .950 | .945 | .941 | .935 | .929 | .923 | .915 |
| | | | | | | .907 | .897 | | | | | | | | |
| 13.000 | 11.822 | 1.178 | .456 | .584 | 13 | .953 | .949 | .944 | .939 | .933 | .926 | .918 | .909 | .898 | .886 |
| | | | | | | .873 | .857 | .839 | | | | | | | |
| 13.000 | 12.260 | .740 | .321 | .423 | 14 | .946 | .941 | .934 | .927 | .919 | .909 | .898 | .886 | .871 | .853 |
| | | | | | | .833 | .810 | .782 | .751 | | | | | | |
| 13.000 | 12.544 | .456 | .216 | .290 | 15 | .941 | .934 | .926 | .917 | .907 | .895 | .881 | .864 | .845 | .822 |
| | | | | | | .795 | .764 | .728 | .686 | .638 | | | | | |
| 13.000 | 12.719 | .281 | .141 | .191 | 16 | .936 | .929 | .920 | .910 | .898 | .884 | .867 | .847 | .824 | .797 |
| | | | | | | .765 | .727 | .684 | .634 | .579 | .517 | | | | |
| 13.000 | 12.827 | .173 | .090 | .123 | 17 | .934 | .925 | .916 | .905 | .891 | .876 | .857 | .835 | .809 | .779 |
| | | | | | | .743 | .700 | .652 | .597 | .536 | .470 | .401 | | | |
| 13.000 | 12.894 | .106 | .057 | .078 | 18 | .932 | .923 | .913 | .901 | .887 | .870 | .850 | .827 | .799 | .766 |
| | | | | | | .727 | .682 | .630 | .572 | .508 | .439 | .369 | .299 | | |

JEWETT EQUIVALENT QUEUE EXTENDED ERLANG B(EQEEB): 5 MINUTES MAXIMUM WAIT (IF NOT CONNECTED AFTER 5 MINUTES WAIT, 100% OVERFLOW)

| HOURS FIRST ATTEMPT VMR = 1.0 | HOURS CON-NECT | HOURS OVFL | DELAYED PORTION | DELAY LENGTH CALL FACTOR | LINE | HOURS CONNECTED TRAFFIC ON LINES 1 | 2 | 3 | 4 | 5 | 6 | 7 | 8 | 9 | 10 |
|---|---|---|---|---|---|---|---|---|---|---|---|---|---|---|---|
| 13.000 | 12.936 | .064 | .035 | .048 | 19 | .931 | .922 | .911 | .899 | .884 | .867 | .846 | .821 | .792 | .757 |
| | | | | | | .717 | .670 | .616 | .555 | .489 | .419 | .348 | .279 | .214 | |
| 13.000 | 12.962 | .038 | .021 | .029 | 20 | .930 | .921 | .910 | .897 | .882 | .864 | .843 | .818 | .787 | .752 |
| | | | | | | .710 | .662 | .606 | .545 | .477 | .407 | .335 | .266 | .203 | .148 |
| 14.000 | 10.589 | 3.411 | .856 | .980 | 11 | .972 | .971 | .969 | .967 | .965 | .963 | .961 | .959 | .956 | .953 |
| | | | | | | .950 | | | | | | | | | |
| 14.000 | 11.424 | 2.576 | .726 | .867 | 12 | .968 | .966 | .963 | .961 | .958 | .955 | .952 | .949 | .945 | .940 |
| | | | | | | .936 | .930 | | | | | | | | |
| 14.000 | 12.168 | 1.832 | .586 | .728 | 13 | .962 | .959 | .956 | .953 | .949 | .945 | .940 | .935 | .929 | .922 |
| | | | | | | .915 | .906 | .897 | | | | | | | |
| 14.000 | 12.775 | 1.225 | .445 | .571 | 14 | .956 | .952 | .948 | .943 | .938 | .932 | .925 | .917 | .908 | .898 |
| | | | | | | .887 | .873 | .858 | .841 | | | | | | |
| 14.000 | 13.216 | .784 | .317 | .418 | 15 | .950 | .945 | .939 | .933 | .926 | .918 | .908 | .898 | .885 | .871 |
| | | | | | | .854 | .835 | .812 | .786 | .756 | | | | | |
| 14.000 | 13.508 | .492 | .216 | .290 | 16 | .945 | .939 | .932 | .924 | .916 | .905 | .894 | .880 | .864 | .845 |
| | | | | | | .823 | .797 | .768 | .733 | .694 | .649 | | | | |
| 14.000 | 13.692 | .308 | .144 | .195 | 17 | .941 | .934 | .927 | .918 | .908 | .896 | .882 | .866 | .846 | .824 |
| | | | | | | .798 | .767 | .731 | .690 | .643 | .590 | .532 | | | |
| 14.000 | 13.807 | .193 | .093 | .128 | 18 | .938 | .931 | .923 | .913 | .902 | .889 | .873 | .855 | .834 | .808 |
| | | | | | | .779 | .744 | .704 | .658 | .606 | .548 | .485 | .419 | | |
| 14.000 | 13.880 | .120 | .060 | .082 | 19 | .937 | .929 | .920 | .910 | .898 | .884 | .867 | .848 | .825 | .797 |
| | | | | | | .765 | .728 | .685 | .635 | .580 | .518 | .453 | .385 | .317 | |
| 14.000 | 13.926 | .074 | .038 | .052 | 20 | .935 | .928 | .918 | .908 | .895 | .881 | .863 | .843 | .818 | .790 |
| | | | | | | .756 | .717 | .672 | .620 | .562 | .499 | .432 | .363 | .295 | .231 |
| 14.000 | 13.955 | .045 | .023 | .032 | 21 | .935 | .927 | .917 | .906 | .894 | .878 | .861 | .839 | .814 | .785 |
| | | | | | | .750 | .710 | .663 | .610 | .551 | .486 | .418 | .349 | .281 | .218 |
| | | | | | | .163 | | | | | | | | | |
| 15.000 | 11.555 | 3.445 | .827 | .956 | 12 | .973 | .972 | .970 | .969 | .967 | .965 | .963 | .961 | .958 | .956 |
| | | | | | | .953 | .949 | | | | | | | | |
| 15.000 | 12.385 | 2.615 | .702 | .845 | 13 | .969 | .967 | .965 | .963 | .960 | .958 | .955 | .951 | .948 | .944 |
| | | | | | | .940 | .935 | .930 | | | | | | | |
| 15.000 | 13.123 | 1.877 | .569 | .710 | 14 | .964 | .961 | .958 | .955 | .952 | .948 | .944 | .939 | .934 | .928 |
| | | | | | | .922 | .914 | .906 | .897 | | | | | | |
| 15.000 | 13.728 | 1.272 | .434 | .560 | 15 | .958 | .955 | .951 | .947 | .942 | .937 | .931 | .924 | .916 | .908 |
| | | | | | | .898 | .887 | .874 | .859 | .842 | | | | | |
| 15.000 | 14.173 | .827 | .313 | .413 | 16 | .953 | .948 | .944 | .938 | .932 | .925 | .917 | .908 | .897 | .885 |
| | | | | | | .871 | .855 | .836 | .815 | .790 | .761 | | | | |
| 15.000 | 14.472 | .528 | .217 | .290 | 17 | .948 | .943 | .937 | .931 | .923 | .914 | .904 | .892 | .879 | .863 |
| | | | | | | .845 | .824 | .799 | .771 | .738 | .701 | .658 | | | |
| 15.000 | 14.664 | .336 | .146 | .197 | 18 | .945 | .939 | .932 | .925 | .916 | .906 | .894 | .880 | .864 | .846 |

JEWETT EQUIVALENT QUEUE EXTENDED ERLANG B(EQEEB): 5 MINUTES MAXIMUM WAIT (IF NOT CONNECTED AFTER 5 MINUTES WAIT, 100% OVERFLOW)

| HOURS FIRST ATTEMPT VMR = 1.0 | HOURS CON-NECT | HOURS OVFL | DELAYED PORTION | DELAY LENGTH CALL FACTOR | LINE | HOURS CONNECTED TRAFFIC ON LINES 1 | 2 | 3 | 4 | 5 | 6 | 7 | 8 | 9 | 10 |
|---|---|---|---|---|---|---|---|---|---|---|---|---|---|---|---|
| | | | | | | .824 | .798 | .769 | .735 | .695 | .651 | .601 | .545 | | |
| 15.000 | 14.787 | .213 | .096 | .131 | 19 | .942 | .936 | .929 | .920 | .911 | .900 | .886 | .871 | .853 | .832 |
| | | | | | | .808 | .779 | .746 | .707 | .663 | .613 | .558 | .498 | .434 | |
| 15.000 | 14.865 | .135 | .063 | .086 | 20 | .941 | .934 | .926 | .917 | .907 | .895 | .881 | .865 | .845 | .823 |
| | | | | | | .796 | .765 | .729 | .687 | .640 | .587 | .528 | .465 | .400 | .334 |
| 15.000 | 14.915 | .085 | .040 | .055 | 21 | .940 | .933 | .925 | .916 | .905 | .892 | .877 | .860 | .840 | .816 |
| | | | | | | .788 | .755 | .717 | .674 | .624 | .568 | .508 | .443 | .377 | .311 |
| | | | | | | .247 | | | | | | | | | |
| 15.000 | 14.947 | .053 | .025 | .034 | 22 | .939 | .932 | .924 | .914 | .903 | .890 | .875 | .857 | .836 | .812 |
| | | | | | | .783 | .749 | .709 | .664 | .613 | .556 | .494 | .429 | .362 | .296 |
| | | | | | | .233 | .177 | | | | | | | | |
| 15.000 | 14.968 | .032 | .015 | .021 | 23 | .938 | .931 | .923 | .913 | .902 | .889 | .873 | .855 | .834 | .809 |
| | | | | | | .779 | .744 | .704 | .658 | .606 | .548 | .486 | .419 | .352 | .286 |
| | | | | | | .224 | .169 | .123 | | | | | | | |
| 16.000 | 12.523 | 3.477 | .801 | .934 | 13 | .974 | .973 | .971 | .970 | .968 | .966 | .964 | .962 | .960 | .958 |
| | | | | | | .955 | .952 | .949 | | | | | | | |
| 16.000 | 13.347 | 2.653 | .681 | .825 | 14 | .970 | .968 | .966 | .964 | .962 | .960 | .957 | .954 | .951 | .947 |
| | | | | | | .943 | .939 | .934 | .929 | | | | | | |
| 16.000 | 14.081 | 1.919 | .553 | .693 | 15 | .965 | .963 | .960 | .958 | .955 | .951 | .947 | .943 | .939 | .933 |
| | | | | | | .928 | .921 | .914 | .906 | .897 | | | | | |
| 16.000 | 14.684 | 1.316 | .425 | .549 | 16 | .960 | .957 | .954 | .950 | .946 | .941 | .936 | .930 | .923 | .916 |
| | | | | | | .907 | .898 | .887 | .874 | .860 | .844 | | | | |
| 16.000 | 15.132 | .868 | .309 | .408 | 17 | .955 | .951 | .947 | .942 | .937 | .931 | .924 | .916 | .907 | .896 |
| | | | | | | .885 | .871 | .855 | .837 | .817 | .793 | .766 | | | |
| 16.000 | 15.438 | .562 | .216 | .290 | 18 | .951 | .947 | .942 | .936 | .929 | .922 | .913 | .903 | .891 | .878 |
| | | | | | | .863 | .845 | .825 | .801 | .774 | .743 | .707 | .667 | | |
| 16.000 | 15.637 | .363 | .147 | .200 | 19 | .948 | .943 | .937 | .930 | .923 | .914 | .904 | .892 | .879 | .863 |
| | | | | | | .845 | .824 | .799 | .771 | .738 | .701 | .658 | .610 | .558 | |
| 16.000 | 15.766 | .234 | .099 | .135 | 20 | .946 | .940 | .934 | .927 | .918 | .909 | .897 | .884 | .869 | .852 |
| | | | | | | .831 | .807 | .779 | .747 | .710 | .668 | .620 | .568 | .510 | .449 |
| 16.000 | 15.850 | .150 | .065 | .089 | 21 | .944 | .938 | .932 | .924 | .915 | .905 | .893 | .879 | .862 | .843 |
| | | | | | | .821 | .795 | .765 | .730 | .690 | .644 | .593 | .537 | .477 | .413 |
| | | | | | | .349 | | | | | | | | | |
| 16.000 | 15.904 | .096 | .042 | .058 | 22 | .943 | .937 | .930 | .922 | .913 | .902 | .889 | .875 | .857 | .837 |
| | | | | | | .814 | .786 | .755 | .718 | .676 | .628 | .574 | .516 | .454 | .390 |
| | | | | | | .325 | .263 | | | | | | | | |
| 16.000 | 15.939 | .061 | .027 | .037 | 23 | .942 | .936 | .929 | .921 | .911 | .900 | .887 | .872 | .854 | .833 |
| | | | | | | .809 | .781 | .748 | .709 | .666 | .616 | .562 | .502 | .439 | .374 |
| | | | | | | .309 | .248 | .191 | | | | | | | |
| 16.000 | 15.962 | .038 | .017 | .023 | 24 | .942 | .936 | .928 | .920 | .910 | .899 | .886 | .870 | .852 | .831 |

JEWETT EQUIVALENT QUEUE EXTENDED ERLANG B(EQEEB): 5 MINUTES MAXIMUM WAIT (IF NOT CONNECTED AFTER 5 MINUTES WAIT, 100% OVERFLOW)

| HOURS FIRST ATTEMPT VMR = 1.0 | HOURS CON-NECT | HOURS OVFL | DELAYED PORTION | DELAY LENGTH CALL FACTOR | LINE | HOURS CONNECTED TRAFFIC ON LINES 1 | 2 | 3 | 4 | 5 | 6 | 7 | 8 | 9 | 10 |
|---|---|---|---|---|---|---|---|---|---|---|---|---|---|---|---|
| | | | | | | .806 | .777 | .743 | .704 | .659 | .609 | .553 | .493 | .429 | .363 |
| | | | | | | .299 | .238 | .183 | .135 | | | | | | |
| 17.000 | 13.491 | 3.509 | .777 | .913 | 14 | .975 | .974 | .972 | .971 | .969 | .968 | .966 | .964 | .962 | .960 |
| | | | | | | .957 | .954 | .952 | .948 | | | | | | |
| 17.000 | 14.310 | 2.690 | .661 | .805 | 15 | .971 | .969 | .968 | .966 | .964 | .961 | .959 | .956 | .953 | .950 |
| | | | | | | .947 | .943 | .939 | .934 | .929 | | | | | |
| 17.000 | 15.039 | 1.961 | .539 | .678 | 16 | .967 | .965 | .962 | .960 | .957 | .954 | .950 | .947 | .943 | .938 |
| | | | | | | .933 | .927 | .921 | .914 | .906 | .897 | | | | |
| 17.000 | 15.640 | 1.360 | .416 | .538 | 17 | .962 | .959 | .956 | .953 | .949 | .945 | .940 | .935 | .929 | .923 |
| | | | | | | .915 | .907 | .898 | .887 | .875 | .861 | .845 | | | |
| 17.000 | 16.092 | .908 | .306 | .404 | 18 | .958 | .954 | .950 | .946 | .941 | .936 | .930 | .923 | .915 | .906 |
| | | | | | | .896 | .884 | .871 | .856 | .839 | .819 | .796 | .770 | | |
| 17.000 | 16.403 | .597 | .216 | .290 | 19 | .954 | .950 | .945 | .940 | .934 | .928 | .920 | .912 | .902 | .891 |
| | | | | | | .878 | .863 | .846 | .826 | .803 | .777 | .747 | .713 | .675 | |
| 17.000 | 16.610 | .390 | .149 | .202 | 20 | .951 | .946 | .941 | .935 | .929 | .921 | .912 | .902 | .891 | .878 |
| | | | | | | .862 | .844 | .824 | .800 | .773 | .741 | .705 | .665 | .619 | .569 |
| 17.000 | 16.745 | .255 | .101 | .138 | 21 | .949 | .944 | .938 | .932 | .925 | .916 | .907 | .895 | .883 | .868 |
| | | | | | | .850 | .830 | .807 | .780 | .749 | .713 | .673 | .627 | .576 | .521 |
| | | | | | | .463 | | | | | | | | | |
| 17.000 | 16.834 | .166 | .068 | .092 | 22 | .947 | .942 | .936 | .929 | .922 | .913 | .902 | .890 | .876 | .860 |
| | | | | | | .841 | .820 | .794 | .765 | .731 | .692 | .648 | .599 | .546 | .487 |
| | | | | | | .426 | .364 | | | | | | | | |
| 17.000 | 16.893 | .107 | .044 | .061 | 23 | .946 | .941 | .935 | .928 | .920 | .910 | .899 | .887 | .872 | .855 |
| | | | | | | .835 | .812 | .785 | .754 | .718 | .678 | .631 | .580 | .524 | .464 |
| | | | | | | .402 | .339 | .277 | | | | | | | |
| 17.000 | 16.931 | .069 | .029 | .040 | 24 | .946 | .940 | .934 | .927 | .918 | .008 | .897 | .884 | .869 | .851 |
| | | | | | | .831 | 807 | .779 | .747 | .710 | .667 | .620 | .567 | .509 | .448 |
| | | | | | | .385 | .322 | .261 | .205 | | | | | | |
| 17.000 | 16.957 | .043 | .010 | .025 | 25 | .945 | .940 | .933 | .926 | .917 | .907 | .896 | .883 | .867 | .849 |
| | | | | | | .828 | .803 | .775 | .742 | .704 | .660 | .612 | .558 | .499 | .437 |
| | | | | | | .374 | .311 | .251 | .195 | .147 | | | | | |
| 18.000 | 13.588 | 4.412 | .859 | .982 | 14 | .978 | .977 | .976 | .975 | .974 | .973 | .972 | .971 | .969 | .968 |
| | | | | | | .966 | .964 | .962 | .960 | | | | | | |
| 18.000 | 14.460 | 3.540 | .755 | .894 | 15 | .975 | .974 | .973 | .972 | .970 | .969 | .967 | .005 | .963 | .961 |
| | | | | | | .959 | .957 | .954 | 951 | .948 | | | | | |
| 18.000 | 15.274 | 2.726 | .644 | .788 | 16 | .072 | .971 | .969 | .967 | .965 | .963 | 961 | .958 | .956 | .953 |
| | | | | | | .950 | .946 | .942 | .938 | .933 | .928 | | | | |
| 18.000 | 15.998 | 2.002 | .526 | .663 | 17 | .968 | .966 | .964 | .962 | .959 | .956 | .953 | .950 | .946 | .942 |
| | | | | | | .937 | .932 | .927 | .920 | .913 | .905 | .897 | | | |
| 18.000 | 16.598 | 1.402 | .408 | .529 | 18 | .964 | .961 | .959 | .955 | .952 | .948 | .944 | .940 | .934 | .929 |

JEWETT EQUIVALENT QUEUE EXTENDED ERLANG B(EQEEB): 5 MINUTES MAXIMUM WAIT (IF NOT CONNECTED AFTER 5 MINUTES WAIT, 100% OVERFLOW)

| HOURS FIRST ATTEMPT VMR = 1.0 | HOURS CON-NECT | HOURS OVFL | DELAYED PORTION | DELAY LENGTH CALL FACTOR | LINE | HOURS CONNECTED TRAFFIC ON LINES 1 | 2 | 3 | 4 | 5 | 6 | 7 | 8 | 9 | 10 |
|---|---|---|---|---|---|---|---|---|---|---|---|---|---|---|---|
| | | | | | | .922 | .915 | .907 | .897 | .887 | .875 | .862 | .847 | | |
| 18.000 | 17.052 | .948 | .302 | .399 | 19 | .960 | .957 | .953 | .949 | .945 | .940 | .935 | .929 | .922 | .914 |
| | | | | | | .906 | .896 | .884 | .871 | .857 | .840 | .821 | .799 | .774 | |
| 18.000 | 17.370 | .630 | .216 | .289 | 20 | .956 | .953 | .949 | .944 | .939 | .933 | .926 | .919 | .911 | .901 |
| | | | | | | .890 | .877 | .863 | .846 | .827 | .805 | .780 | .751 | .719 | .682 |
| 18.000 | 17.583 | .417 | .150 | .204 | 21 | .954 | .950 | .945 | .940 | .934 | .927 | .920 | .911 | .901 | .890 |
| | | | | | | .876 | .861 | .844 | .824 | .801 | .775 | .744 | .710 | .671 | .627 |
| | | | | | | .579 | | | | | | | | | |
| 18.000 | 17.724 | .276 | .103 | .140 | 22 | .952 | .947 | .942 | .937 | .930 | .923 | .914 | .905 | .894 | .881 |
| | | | | | | .866 | .849 | .829 | .807 | .780 | .750 | .716 | .677 | .633 | .585 |
| | | | | | | .532 | .475 | | | | | | | | |
| 18.000 | 17.818 | .182 | .070 | .095 | 23 | .950 | .946 | .940 | .934 | .927 | .920 | .911 | .900 | .888 | .874 |
| | | | | | | .858 | .840 | .818 | .794 | .765 | .732 | .695 | .653 | .605 | .553 |
| | | | | | | .497 | .438 | .378 | | | | | | | |
| 18.000 | 17.881 | .119 | .047 | .064 | 24 | .949 | .944 | .939 | .933 | .925 | .917 | .908 | .897 | .884 | .870 |
| | | | | | | .853 | .833 | .810 | .784 | .754 | .719 | .680 | .635 | .585 | .531 |
| | | | | | | .473 | .413 | .352 | .291 | | | | | | |
| 18.000 | 17.923 | .077 | .031 | .042 | 25 | .949 | .944 | .938 | .932 | .924 | .916 | .906 | .895 | .882 | .866 |
| | | | | | | .849 | .828 | .805 | .777 | .746 | .710 | .669 | .623 | .571 | .516 |
| | | | | | | .457 | .395 | .334 | .274 | .218 | | | | | |
| 18.000 | 17.951 | .049 | .020 | .027 | 26 | .948 | .943 | .937 | .931 | .923 | .915 | .905 | .893 | .880 | .864 |
| | | | | | | .846 | .825 | .801 | .773 | .740 | .703 | .661 | .614 | .562 | .505 |
| | | | | | | .445 | .384 | .322 | .263 | .208 | .159 | | | | |
| 19.000 | 14.561 | 4.439 | .835 | .963 | 15 | .979 | .978 | .977 | .976 | .975 | .974 | .973 | .971 | .970 | .969 |
| | | | | | | .967 | .965 | .964 | .962 | .960 | | | | | |
| 19.000 | 15.429 | 3.571 | .735 | .876 | 16 | .976 | .975 | .974 | .973 | .971 | .970 | .968 | .967 | .965 | .963 |
| | | | | | | .961 | .958 | .956 | .953 | .950 | .947 | | | | |
| 19.000 | 16.238 | 2.762 | .627 | .771 | 17 | .973 | .972 | .970 | .968 | .967 | .965 | .963 | .960 | .958 | .955 |
| | | | | | | .952 | .949 | .946 | .942 | .938 | .933 | .928 | | | |
| 19.000 | 16.959 | 2.041 | .513 | .650 | 18 | .969 | .967 | .966 | .963 | .961 | .958 | .956 | .953 | .949 | .945 |
| | | | | | | .941 | .937 | .932 | .926 | .920 | .913 | .905 | .897 | | |
| 19.000 | 17.557 | 1.443 | .401 | .520 | 19 | .965 | .963 | .961 | .958 | .955 | .951 | .948 | .943 | .939 | .934 |
| | | | | | | .928 | .922 | .914 | .906 | .897 | .887 | .876 | .863 | .848 | |
| 19.000 | 18.014 | .986 | .299 | .395 | 20 | .962 | .959 | .956 | .952 | .948 | .944 | .939 | .934 | .928 | .921 |
| | | | | | | .914 | .905 | .895 | .884 | .872 | .857 | .841 | .823 | .802 | .778 |
| 19.000 | 18.337 | .663 | .215 | .289 | 21 | .959 | .955 | .952 | .947 | .943 | .938 | .932 | .925 | .918 | .910 |
| | | | | | | .900 | .889 | .877 | .863 | .846 | .828 | .807 | .782 | .755 | .724 |
| | | | | | | .689 | | | | | | | | | |
| 19.000 | 18.556 | .444 | .151 | .205 | 22 | .956 | .952 | .948 | .944 | .938 | .932 | .926 | .918 | .910 | .900 |
| | | | | | | .888 | .876 | .861 | .844 | .824 | .802 | .776 | .747 | .714 | .677 |

JEWETT EQUIVALENT QUEUE EXTENDED ERLANG B(EQEEB): 5 MINUTES MAXIMUM WAIT (IF NOT CONNECTED AFTER 5 MINUTES WAIT, 100% OVERFLOW)

| HOURS FIRST ATTEMPT VMR = 1.0 | HOURS CON-NECT | HOURS OVFL | DELAYED PORTION | DELAY LENGTH CALL FACTOR | LINE | HOURS CONNECTED TRAFFIC ON LINES 1 | 2 | 3 | 4 | 5 | 6 | 7 | 8 | 9 | 10 |
|---|---|---|---|---|---|---|---|---|---|---|---|---|---|---|---|
| | | | | | | .635 | .589 | | | | | | | | |
| 19.000 | 18.703 | .297 | .105 | .143 | 23 | .954 | .950 | .946 | .941 | .935 | .928 | .921 | .913 | .903 | .892 |
| | | | | | | .879 | .865 | .848 | .829 | .807 | .781 | .752 | .719 | .681 | .639 |
| | | | | | | .592 | .541 | .487 | | | | | | | |
| 19.000 | 18.802 | .198 | .072 | .098 | 24 | .953 | .949 | .944 | .939 | .932 | .926 | .918 | .909 | .898 | .886 |
| | | | | | | .873 | .857 | .838 | .817 | .793 | .765 | .733 | .697 | .656 | .611 |
| | | | | | | .561 | .507 | .450 | .391 | | | | | | |
| 19.000 | 18.869 | .131 | .048 | .066 | 25 | .952 | .948 | .943 | .937 | .931 | .923 | .915 | .906 | .895 | .882 |
| | | | | | | .868 | .851 | .831 | .809 | .783 | .754 | .720 | .681 | .638 | .590 |
| | | | | | | .538 | .482 | .424 | .364 | .305 | | | | | |
| 19.000 | 18.914 | .086 | .032 | .044 | 26 | .951 | .947 | .942 | .936 | .929 | .922 | .913 | .903 | .892 | .879 |
| | | | | | | .864 | .847 | .826 | .803 | .776 | .745 | .710 | .670 | .626 | .576 |
| | | | | | | .522 | .465 | .405 | .345 | .286 | .231 | | | | |
| 19.000 | 18.944 | .056 | .021 | .029 | 27 | .951 | .946 | .941 | .935 | .929 | .921 | .912 | .902 | .890 | .877 |
| | | | | | | .862 | .844 | .823 | .799 | .771 | .740 | .703 | .662 | .617 | .566 |
| | | | | | | .511 | .453 | .393 | .333 | .274 | .219 | .170 | | | |
| 20.000 | 15.535 | 4.465 | .814 | .945 | 16 | .979 | .979 | .978 | .977 | .976 | .975 | .974 | .972 | .971 | .970 |
| | | | | | | .968 | .967 | .965 | .963 | .961 | .959 | | | | |
| 20.000 | 16.398 | 3.602 | .716 | .858 | 17 | .977 | .976 | .975 | .973 | .972 | .971 | .969 | .968 | .966 | .964 |
| | | | | | | .962 | .960 | .958 | .956 | .953 | .950 | .947 | | | |
| 20.000 | 17.203 | 2.797 | .612 | .755 | 18 | .974 | .973 | .971 | .970 | .968 | .966 | .964 | .962 | .960 | .957 |
| | | | | | | .955 | .952 | .949 | .945 | .941 | .937 | .932 | .927 | | |
| 20.000 | 17.920 | 2.080 | .502 | .637 | 19 | .970 | .969 | .967 | .965 | .963 | .960 | .958 | .955 | .952 | .948 |
| | | | | | | .945 | .941 | .936 | .931 | .926 | .919 | .913 | .905 | .897 | |
| 20.000 | 18.517 | 1.483 | .394 | .511 | 20 | .967 | .965 | .962 | .960 | .957 | .954 | .951 | .947 | .943 | .938 |
| | | | | | | .933 | .927 | .921 | .914 | .908 | .897 | .887 | .876 | .863 | .849 |
| 20.000 | 18.976 | 1.024 | .205 | .391 | 21 | .963 | .961 | .958 | .955 | .951 | .948 | .943 | .938 | .933 | .927 |
| | | | | | | .920 | .913 | .904 | .895 | .884 | .872 | .858 | .842 | .824 | .804 |
| | | | | | | .781 | | | | | | | | | |
| 20.000 | 19.304 | .696 | .215 | .288 | 22 | .961 | .957 | .954 | .950 | .946 | .942 | .937 | .931 | .924 | .917 |
| | | | | | | .909 | .899 | .888 | .876 | .862 | .847 | .829 | .808 | .785 | .758 |
| | | | | | | .728 | .695 | | | | | | | | |
| 20.000 | 19.530 | .470 | .152 | .206 | 23 | .958 | .955 | .951 | .947 | .942 | .937 | .931 | .924 | .917 | .908 |
| | | | | | | .899 | .887 | .875 | .860 | .844 | .825 | .803 | .778 | .750 | .710 |
| | | | | | | .682 | .642 | .597 | | | | | | | |
| 20.000 | 19.682 | .318 | .107 | .145 | 24 | .956 | .953 | .949 | .944 | .939 | .933 | .927 | .919 | .911 | .901 |
| | | | | | | .891 | .878 | .864 | .847 | .828 | .807 | .782 | .754 | .721 | .685 |
| | | | | | | .645 | .600 | .551 | .498 | | | | | | |
| 20.000 | 19.786 | .214 | .074 | .101 | 25 | .955 | .951 | .947 | .942 | .937 | .931 | .924 | .916 | .907 | .896 |
| | | | | | | .884 | .871 | .855 | .837 | .817 | .793 | .766 | .735 | .700 | .660 |

JEWETT EQUIVALENT QUEUE EXTENDED ERLANG B(EQEEB): 5 MINUTES MAXIMUM WAIT (IF NOT CONNECTED AFTER 5 MINUTES WAIT, 100% OVERFLOW)

| HOURS FIRST ATTEMPT VMR = 1.0 | HOURS CON-NECT | HOURS OVFL | DELAYED PORTION | DELAY LENGTH CALL FACTOR | LINE | HOURS CONNECTED TRAFFIC ON LINES 1 | 2 | 3 | 4 | 5 | 6 | 7 | 8 | 9 | 10 |
|---|---|---|---|---|---|---|---|---|---|---|---|---|---|---|---|
| | | | | | | .616 | .568 | .516 | .460 | .403 | | | | | |
| 20.000 | 19.857 | .143 | .050 | .069 | 26 | .954 | .950 | .946 | .941 | .935 | .929 | .921 | .913 | .903 | .893 |
| | | | | | | .880 | .866 | .849 | .830 | .808 | .783 | .754 | .721 | .683 | .642 |
| | | | | | | .595 | .545 | .490 | .434 | .375 | .317 | | | | |
| 20.000 | 19.905 | .095 | .034 | .047 | 27 | .954 | .950 | .945 | .940 | .934 | .927 | .920 | .911 | .901 | .890 |
| | | | | | | .877 | .862 | .844 | .824 | .802 | .775 | .745 | .711 | .672 | .628 |
| | | | | | | .580 | .528 | .473 | .415 | .356 | .298 | .243 | | | |
| 20.000 | 19.937 | .063 | .022 | .031 | 28 | .953 | .949 | .944 | .939 | .933 | .926 | .919 | .910 | .900 | .888 |
| | | | | | | .875 | .859 | .841 | .821 | .797 | .770 | .739 | .703 | .664 | .619 |
| | | | | | | .570 | .517 | .460 | .402 | .343 | .285 | .231 | .181 | | |
| 20.000 | 19.959 | .041 | .015 | .020 | 29 | .953 | .949 | .944 | .939 | .933 | .926 | .918 | .909 | .898 | .887 |
| | | | | | | .873 | .857 | .839 | .818 | .794 | .766 | .735 | .698 | .658 | .613 |
| | | | | | | .563 | .509 | .452 | .393 | .334 | .276 | .223 | .174 | .131 | |
| 22.000 | 16.587 | 5.413 | .860 | .983 | 17 | .982 | .982 | .981 | .980 | .980 | .979 | .978 | .977 | .976 | .975 |
| | | | | | | .974 | .973 | .972 | .971 | .970 | .968 | .967 | | | |
| 22.000 | 17.483 | 4.517 | .775 | .911 | 18 | .980 | .980 | .979 | .978 | .977 | .976 | .975 | .974 | .973 | .972 |
| | | | | | | .970 | .969 | .967 | .966 | .964 | .962 | .961 | .958 | | |
| 22.000 | 18.339 | 3.661 | .683 | .826 | 19 | .978 | .977 | .976 | .975 | .974 | .973 | .971 | .970 | .969 | .967 |
| | | | | | | .965 | .963 | .961 | .959 | .957 | .955 | .952 | .949 | .946 | |
| 22.000 | 19.136 | 2.864 | .584 | .726 | 20 | .975 | .974 | .973 | .972 | .970 | .969 | .967 | .965 | .963 | .961 |
| | | | | | | .959 | .956 | .954 | .951 | .947 | .944 | .940 | .936 | .932 | .927 |
| 22.000 | 19.846 | 2.154 | .481 | .614 | 21 | .972 | .971 | .969 | .968 | .966 | .964 | .962 | .959 | .957 | .954 |
| | | | | | | .951 | .947 | .944 | .940 | .935 | .930 | .925 | .919 | .912 | .905 |
| | | | | | | .897 | | | | | | | | | |
| 22.000 | 20.440 | 1.560 | .381 | .496 | 22 | .969 | .968 | .966 | .963 | .961 | .958 | .956 | .953 | .949 | .945 |
| | | | | | | .941 | .937 | .932 | .926 | .920 | .913 | .906 | .897 | .887 | .877 |
| | | | | | | .865 | .851 | | | | | | | | |
| 22.000 | 20.903 | 1.097 | .289 | .383 | 23 | .966 | .964 | .962 | .959 | .956 | .953 | .950 | .946 | .942 | .937 |
| | | | | | | .932 | .926 | .919 | .912 | .904 | .894 | .884 | .872 | .859 | .844 |
| | | | | | | .827 | .808 | .787 | | | | | | | |
| 22.000 | 21.240 | .760 | .213 | .286 | 24 | .964 | .961 | .959 | .956 | .952 | .948 | .944 | .940 | .935 | .929 |
| | | | | | | .922 | .915 | .907 | .898 | .887 | .876 | .862 | .847 | .830 | .811 |
| | | | | | | .789 | .764 | .737 | .705 | | | | | | |
| 22.000 | 21.477 | .523 | .154 | .208 | 25 | .962 | .959 | .956 | .953 | .949 | .945 | .940 | .935 | .929 | .922 |
| | | | | | | .914 | .906 | .896 | .886 | .873 | .859 | .843 | .825 | .805 | .781 |
| | | | | | | .755 | .725 | .692 | .654 | .613 | | | | | |
| 22.000 | 21.640 | .360 | .110 | .149 | 26 | .960 | .957 | .954 | .950 | .946 | .942 | .936 | .931 | .924 | .917 |
| | | | | | | .908 | .899 | .888 | .876 | .862 | .846 | .828 | .807 | .783 | .757 |
| | | | | | | .726 | .692 | .655 | .613 | .567 | .518 | | | | |
| 22.000 | 21.753 | .247 | .077 | .105 | 27 | .959 | .956 | .953 | .949 | .944 | .939 | .934 | .927 | .920 | .912 |

JEWETT EQUIVALENT QUEUE EXTENDED ERLANG B(EQEEB): 5 MINUTES MAXIMUM WAIT (IF NOT CONNECTED AFTER 5 MINUTES WAIT, 100% OVERFLOW)

| HOURS FIRST ATTEMPT VMR = 1.0 | HOURS CON-NECT | HOURS OVFL | DELAYED PORTION | DELAY LENGTH CALL FACTOR | LINE | HOURS CONNECTED TRAFFIC ON LINES 1 | 2 | 3 | 4 | 5 | 6 | 7 | 8 | 9 | 10 |
|---|---|---|---|---|---|---|---|---|---|---|---|---|---|---|---|
| | | | | | | .903 | .893 | .881 | .868 | .853 | .835 | .815 | .792 | .767 | .737 |
| | | | | | | .704 | .667 | .626 | .581 | .532 | .480 | .426 | | | |
| 22.000 | 21.831 | .169 | .054 | .074 | 28 | .958 | .955 | .951 | .947 | .943 | .938 | .932 | .925 | .918 | .909 |
| | | | | | | .900 | .889 | .876 | .862 | .846 | .827 | .806 | .781 | .754 | .723 |
| | | | | | | .687 | .648 | .604 | .557 | .506 | .452 | .396 | .341 | | |
| 22.000 | 21.885 | .115 | .037 | .051 | 29 | .958 | .954 | .951 | .946 | .942 | .936 | .930 | .923 | .916 | .907 |
| | | | | | | .897 | .886 | .873 | .858 | .841 | .821 | .799 | .773 | .744 | .712 |
| | | | | | | .675 | .634 | .589 | .539 | .487 | .432 | .376 | .320 | .266 | |
| 22.000 | 21.923 | .077 | .025 | .034 | 30 | .957 | .954 | .950 | .946 | .941 | .935 | .929 | .922 | .914 | .905 |
| | | | | | | .895 | .883 | .870 | .855 | .837 | .817 | .794 | .768 | .738 | .704 |
| | | | | | | .666 | .624 | .577 | .527 | .474 | .418 | .362 | .306 | .252 | .203 |
| 22.000 | 21.949 | .051 | .017 | .023 | 31 | .957 | .954 | .950 | .945 | .940 | .935 | .929 | .921 | .913 | .904 |
| | | | | | | .894 | .882 | .868 | .852 | .834 | .814 | .790 | .763 | .733 | .698 |
| | | | | | | .660 | .617 | .569 | .518 | .464 | .408 | .352 | .296 | .243 | .194 |
| | | | | | | .151 | | | | | | | | | |
| 24.000 | 18.543 | 5.457 | .822 | .952 | 19 | .983 | .982 | .982 | .981 | .980 | .980 | .979 | .978 | .977 | .977 |
| | | | | | | .976 | .975 | .974 | .973 | .971 | .970 | .969 | .968 | .966 | |
| 24.000 | 19.432 | 4.568 | .741 | .881 | 20 | .981 | .980 | .980 | .979 | .978 | .977 | .976 | .975 | .974 | .973 |
| | | | | | | .972 | .971 | .970 | .968 | .967 | .965 | .963 | .962 | .960 | .958 |
| 24.000 | 20.281 | 3.719 | .653 | .797 | 21 | .979 | .978 | .977 | .976 | .975 | .974 | .973 | .972 | .971 | .969 |
| | | | | | | .968 | .966 | .964 | .963 | .961 | .958 | .956 | .954 | .951 | .948 |
| | | | | | | .945 | | | | | | | | | |
| 24.000 | 21.070 | 2.930 | .560 | .700 | 22 | .977 | .976 | .975 | .973 | .972 | .971 | .969 | .968 | .966 | .964 |
| | | | | | | .962 | .960 | .958 | .955 | .953 | .950 | .947 | .943 | .939 | .935 |
| | | | | | | .931 | .926 | | | | | | | | |
| 24.000 | 21.774 | 2.226 | .463 | .593 | 23 | .974 | .973 | .071 | .970 | .968 | .967 | .965 | .963 | .961 | .958 |
| | | | | | | .956 | .953 | .950 | .946 | .943 | .939 | .934 | .929 | .924 | .918 |
| | | | | | | .912 | .904 | .897 | | | | | | | |
| 24.000 | 22.367 | 1.633 | .369 | .482 | 24 | .972 | .970 | .968 | .966 | .964 | .962 | .960 | .957 | .954 | .951 |
| | | | | | | .948 | .944 | .940 | .936 | .931 | .925 | .919 | .913 | .905 | .897 |
| | | | | | | .888 | .877 | .866 | .853 | | | | | | |
| 24.000 | 22.833 | 1.167 | .283 | .376 | 25 | .969 | .967 | .965 | .963 | .960 | .958 | .955 | .952 | .948 | .944 |
| | | | | | | .940 | .935 | .930 | .924 | .918 | .911 | .903 | .894 | .884 | .873 |
| | | | | | | .860 | .846 | .830 | .812 | .792 | | | | | |
| 24.000 | 23.178 | .822 | .212 | .284 | 26 | .967 | .965 | .962 | .960 | .957 | .954 | .950 | .947 | .943 | .938 |
| | | | | | | .933 | .927 | .921 | .914 | .906 | .897 | .887 | .875 | .863 | .848 |
| | | | | | | .832 | .814 | .793 | .770 | .744 | .715 | | | | |
| 24.000 | 23.425 | .575 | .155 | .210 | 27 | .965 | .963 | .960 | .957 | .954 | .951 | .947 | .942 | .938 | .932 |
| | | | | | | .926 | .920 | .912 | .904 | .895 | .884 | .872 | .859 | .843 | .826 |
| | | | | | | .806 | .784 | .759 | .731 | .700 | .665 | .627 | | | |

JEWETT EQUIVALENT QUEUE EXTENDED ERLANG B(EQEEB): 5 MINUTES MAXIMUM WAIT (IF NOT CONNECTED AFTER 5 MINUTES WAIT, 100% OVERFLOW)

| HOURS FIRST ATTEMPT VMR = 1.0 | HOURS CON-NECT | HOURS OVFL | DELAYED PORTION | DELAY LENGTH CALL FACTOR | LINE | HOURS CONNECTED TRAFFIC ON LINES 1 | 2 | 3 | 4 | 5 | 6 | 7 | 8 | 9 | 10 |
|---|---|---|---|---|---|---|---|---|---|---|---|---|---|---|---|
| 24.000 | 23.598 | .402 | .112 | .152 | 28 | .964 | .961 | .958 | .955 | .952 | .948 | .944 | .939 | .934 | .928 |
| | | | | | | .921 | .914 | .906 | .896 | .886 | .874 | .860 | .845 | .827 | .807 |
| | | | | | | .785 | .760 | .731 | .699 | .664 | .624 | .582 | .536 | | |
| 24.000 | 23.720 | .280 | .080 | .109 | 29 | .963 | .960 | .957 | .954 | .950 | .946 | .942 | .936 | .931 | .925 |
| | | | | | | .917 | .909 | .900 | .890 | .879 | .865 | .850 | .833 | .814 | .792 |
| | | | | | | .768 | .740 | .708 | .673 | .635 | .592 | .546 | .497 | .446 | |
| 24.000 | 23.805 | .195 | .057 | .078 | 30 | .962 | .959 | .956 | .953 | .949 | .945 | .940 | .934 | .929 | .922 |
| | | | | | | .914 | .906 | .896 | .885 | .873 | .859 | .843 | .825 | .804 | .781 |
| | | | | | | .754 | .724 | .691 | .654 | .613 | .568 | .520 | .469 | .416 | .362 |
| 24.000 | 23.865 | .135 | .040 | .054 | 31 | .961 | .958 | .955 | .952 | .948 | .943 | .939 | .933 | .927 | .920 |
| | | | | | | .912 | .903 | .893 | .882 | .869 | .854 | .838 | .818 | .797 | .772 |
| | | | | | | .744 | .713 | .678 | .639 | .596 | .550 | .500 | .448 | .394 | .340 |
| | | | | | | .288 | | | | | | | | | |
| 24.000 | 23.908 | .092 | .027 | .038 | 32 | .961 | .958 | .955 | .951 | .947 | .943 | .938 | .932 | .926 | .919 |
| | | | | | | .910 | .901 | .891 | .879 | .866 | .851 | .833 | .814 | .791 | .766 |
| | | | | | | .737 | .705 | .668 | .628 | .584 | .537 | .486 | .433 | .379 | .325 |
| | | | | | | .272 | .223 | | | | | | | | |
| 24.000 | 23.938 | .062 | .019 | .026 | 33 | .961 | .958 | .954 | .951 | .947 | .942 | .937 | .931 | .925 | .918 |
| | | | | | | .909 | .900 | .889 | .877 | .864 | .848 | .830 | .810 | .787 | .761 |
| | | | | | | .732 | .698 | .661 | .620 | .575 | .527 | .476 | .422 | .368 | .314 |
| | | | | | | .262 | .213 | .169 | | | | | | | |
| 26.000 | 19.587 | 6.413 | .862 | .984 | 20 | .985 | .985 | .984 | .984 | .983 | .983 | .982 | .981 | .981 | .980 |
| | | | | | | .979 | .979 | .978 | .977 | .976 | .976 | .975 | .974 | .973 | .972 |
| 26.000 | 20.499 | 5.501 | .789 | .923 | 21 | .984 | .983 | .983 | .982 | .981 | .981 | .980 | .979 | .979 | .978 |
| | | | | | | .977 | .976 | .975 | .974 | .973 | .972 | .971 | .970 | .968 | .967 |
| | | | | | | .966 | | | | | | | | | |
| 26.000 | 21.382 | 4.618 | .710 | .853 | 22 | .982 | .981 | .981 | .980 | .979 | .978 | .978 | .977 | .976 | .975 |
| | | | | | | .974 | .973 | .971 | .970 | .969 | .968 | .966 | .964 | .963 | .961 |
| | | | | | | .959 | .957 | | | | | | | | |
| 26.000 | 22.224 | 3.776 | .627 | .770 | 23 | .980 | .979 | .979 | .978 | .977 | .976 | .975 | .974 | .972 | .971 |
| | | | | | | .970 | .968 | .967 | .965 | .964 | .962 | .960 | .958 | .955 | .953 |
| | | | | | | .950 | .947 | .944 | | | | | | | |
| 26.000 | 23.007 | 2.993 | .538 | .677 | 24 | .978 | .977 | .976 | .975 | .974 | .973 | .971 | .970 | .968 | .967 |
| | | | | | | .965 | .963 | .961 | .959 | .957 | .954 | .952 | .949 | .946 | .942 |
| | | | | | | .938 | .934 | .930 | .925 | | | | | | |
| 26.000 | 23.706 | 2.294 | .447 | .574 | 25 | .976 | .975 | .973 | .972 | .971 | .969 | .967 | .966 | .964 | .962 |
| | | | | | | .960 | .957 | .955 | .952 | .949 | .945 | .942 | .938 | .933 | .929 |
| | | | | | | .923 | .918 | .911 | .904 | .896 | | | | | |
| 26.000 | 24.297 | 1.703 | .358 | .469 | 26 | .973 | .972 | .970 | .969 | .967 | .965 | .963 | .961 | .959 | .956 |
| | | | | | | .953 | .950 | .947 | .943 | .939 | .935 | .930 | .924 | .919 | .912 |

JEWETT EQUIVALENT QUEUE EXTENDED ERLANG B(EQEEB): 5 MINUTES MAXIMUM WAIT (IF NOT CONNECTED AFTER 5 MINUTES WAIT, 100% OVERFLOW)

| HOURS FIRST ATTEMPT VMR = 1.0 | HOURS CON-NECT | HOURS OVFL | DELAYED PORTION | DELAY LENGTH CALL FACTOR | LINE | HOURS CONNECTED TRAFFIC ON LINES 1 | 2 | 3 | 4 | 5 | 6 | 7 | 8 | 9 | 10 |
|---|---|---|---|---|---|---|---|---|---|---|---|---|---|---|---|
| | | | | | | .905 | .897 | .888 | .878 | .867 | .854 | | | | |
| 26.000 | 24.766 | 1.234 | .278 | .369 | 27 | .971 | .970 | .968 | .966 | .964 | .962 | .959 | .956 | .954 | .950 |
| | | | | | | .947 | .943 | .939 | .934 | .929 | .923 | .917 | .910 | .902 | .893 |
| | | | | | | .884 | .873 | .861 | .847 | .832 | .815 | .796 | | | |
| 26.000 | 25.118 | .882 | .210 | .282 | 28 | .969 | .967 | .965 | .963 | .961 | .958 | .955 | .952 | .949 | .945 |
| | | | | | | .941 | .936 | .931 | .926 | .919 | .912 | .904 | .896 | .886 | .875 |
| | | | | | | .863 | .849 | .833 | .816 | .797 | .775 | .750 | .723 | | |
| 26.000 | 25.374 | .626 | .156 | .211 | 29 | .968 | .966 | .963 | .961 | .958 | .955 | .952 | .949 | .945 | .941 |
| | | | | | | .936 | .930 | .925 | .918 | .911 | .902 | .893 | .883 | .871 | .858 |
| | | | | | | .843 | .827 | .808 | .787 | .764 | .737 | .708 | .675 | .639 | |
| 26.000 | 25.556 | .444 | .114 | .155 | 30 | .966 | .964 | .962 | .959 | .956 | .953 | .950 | .946 | .942 | .937 |
| | | | | | | .932 | .926 | .919 | .912 | .903 | .894 | .884 | .872 | .859 | .844 |
| | | | | | | .827 | .808 | .786 | .762 | .735 | .705 | .672 | .635 | .595 | .551 |
| 26.000 | 25.686 | .314 | .083 | .113 | 31 | .965 | .963 | .961 | .958 | .955 | .951 | .948 | .944 | .939 | .934 |
| | | | | | | .928 | .922 | .915 | .907 | .898 | .888 | .876 | .863 | .849 | .832 |
| | | | | | | .814 | .793 | .769 | .742 | .712 | .679 | .643 | .603 | .559 | .513 |
| | | | | | | .464 | | | | | | | | | |
| 26.000 | 25.779 | .221 | .059 | .081 | 32 | .965 | .962 | .960 | .957 | .954 | .950 | .946 | .942 | .937 | .932 |
| | | | | | | .926 | .919 | .911 | .903 | .893 | .883 | .870 | .856 | .841 | .823 |
| | | | | | | .803 | .781 | .755 | .726 | .695 | .659 | .620 | .578 | .532 | .484 |
| | | | | | | .433 | .382 | | | | | | | | |
| 26.000 | 25.845 | .155 | .042 | .058 | 33 | .964 | .962 | .959 | .956 | .953 | .949 | .945 | .941 | .936 | .930 |
| | | | | | | .924 | .917 | .909 | .900 | .890 | .879 | .866 | .851 | .835 | .816 |
| | | | | | | .795 | .771 | .745 | .714 | .681 | .644 | .603 | .559 | .512 | .462 |
| | | | | | | .411 | .359 | .307 | | | | | | | |
| 26.000 | 25.892 | .108 | .030 | .041 | 34 | .964 | .961 | .959 | .956 | .952 | .948 | .944 | .940 | .035 | .929 |
| | | | | | | .922 | .015 | .907 | .898 | .887 | .876 | .002 | .847 | .830 | .811 |
| | | | | | | .789 | .764 | .737 | .705 | .671 | .632 | .591 | .545 | .497 | .446 |
| | | | | | | .394 | .342 | .291 | .242 | | | | | | |
| 26.000 | 25.925 | .075 | .021 | .028 | 35 | .964 | .961 | .958 | .955 | .952 | .948 | .944 | .939 | .934 | .928 |
| | | | | | | .921 | .914 | .906 | .896 | .886 | .874 | .860 | .845 | .827 | .807 |
| | | | | | | .785 | .759 | .731 | .699 | .663 | .624 | .581 | .535 | .486 | .435 |
| | | | | | | .383 | .330 | .279 | .231 | .187 | | | | | |
| 28.000 | 21.548 | 6.452 | .829 | .057 | 22 | .985 | .985 | .985 | .984 | .984 | .983 | .983 | .982 | .982 | .981 |
| | | | | | | .980 | .980 | .979 | .978 | .978 | .977 | .976 | .975 | .974 | .973 |
| | | | | | | .972 | .971 | | | | | | | | |
| 28.000 | 22.455 | 5.545 | .758 | .897 | 23 | .984 | .984 | .983 | .983 | .982 | .981 | .981 | .980 | .979 | .979 |
| | | | | | | .978 | .977 | .976 | .975 | .975 | .974 | .973 | .971 | .970 | .969 |
| | | | | | | .968 | .966 | .965 | | | | | | | |
| 28.000 | 23.333 | 4.667 | .683 | .827 | 24 | .983 | .982 | .981 | .981 | .980 | .979 | .979 | .978 | .977 | .976 |

JEWETT EQUIVALENT QUEUE EXTENDED ERLANG B(EQEEB): 5 MINUTES MAXIMUM WAIT (IF NOT CONNECTED AFTER 5 MINUTES WAIT, 100% OVERFLOW)

| HOURS FIRST ATTEMPT VMR = 1.0 | HOURS CON-NECT | HOURS OVFL | DELAYED PORTION | DELAY LENGTH CALL FACTOR | LINE | HOURS CONNECTED TRAFFIC ON LINES 1 | 2 | 3 | 4 | 5 | 6 | 7 | 8 | 9 | 10 |
|---|---|---|---|---|---|---|---|---|---|---|---|---|---|---|---|
| | | | | | | .975 | .974 | .973 | .972 | .971 | .970 | .968 | .967 | .965 | .964 |
| | | | | | | .962 | .960 | .958 | .956 | | | | | | |
| 28.000 | 24.169 | 3.831 | .603 | .746 | 25 | .981 | .980 | .980 | .979 | .978 | .977 | .976 | .975 | .974 | .973 |
| | | | | | | .972 | .970 | .969 | .968 | .966 | .965 | .963 | .961 | .959 | .957 |
| | | | | | | .955 | .952 | .949 | .946 | .943 | | | | | |
| 28.000 | 24.946 | 3.054 | .519 | .656 | 26 | .979 | .978 | .977 | .976 | .975 | .974 | .973 | .972 | .971 | .969 |
| | | | | | | .968 | .966 | .964 | .962 | .960 | .958 | .956 | .954 | .951 | .948 |
| | | | | | | .945 | .941 | .938 | .934 | .929 | .925 | | | | |
| 28.000 | 25.640 | 2.360 | .433 | .557 | 27 | .977 | .976 | .975 | .974 | .973 | .971 | .970 | .968 | .967 | .965 |
| | | | | | | .963 | .961 | .959 | .956 | .954 | .951 | .948 | .944 | .941 | .937 |
| | | | | | | .933 | .928 | .923 | .917 | .911 | .904 | .896 | | | |
| 28.000 | 26.229 | 1.771 | .349 | .457 | 28 | .975 | .974 | .972 | .971 | .970 | .968 | .966 | .964 | .962 | .960 |
| | | | | | | .958 | .955 | .952 | .949 | .946 | .942 | .938 | .934 | .929 | .924 |
| | | | | | | .918 | .911 | .904 | .896 | .888 | .878 | .868 | .856 | | |
| 28.000 | 26.701 | 1.299 | .273 | .362 | 29 | .973 | .972 | .970 | .968 | .967 | .965 | .963 | .960 | .958 | .955 |
| | | | | | | .952 | .949 | .946 | .942 | .938 | .933 | .928 | .922 | .916 | .909 |
| | | | | | | .902 | .893 | .884 | .873 | .862 | .849 | .834 | .818 | .800 | |
| 28.000 | 27.060 | .940 | .208 | .279 | 30 | .971 | .970 | .968 | .966 | .964 | .962 | .959 | .957 | .954 | .951 |
| | | | | | | .947 | .944 | .939 | .935 | .930 | .924 | .918 | .911 | .903 | .895 |
| | | | | | | .885 | .875 | .863 | .850 | .835 | .818 | .800 | .779 | .756 | .730 |
| 28.000 | 27.324 | .676 | .156 | .211 | 31 | .970 | .968 | .966 | .964 | .962 | .959 | .957 | .954 | .951 | .947 |
| | | | | | | .943 | .939 | .934 | .929 | .923 | .916 | .909 | .901 | .892 | .882 |
| | | | | | | .871 | .858 | .844 | .828 | .810 | .790 | .767 | .742 | .715 | .684 |
| | | | | | | .650 | | | | | | | | | |
| 28.000 | 27.515 | .485 | .116 | .157 | 32 | .969 | .967 | .965 | .963 | .960 | .957 | .955 | .951 | .948 | .944 |
| | | | | | | .940 | .935 | .929 | .924 | .917 | .910 | .902 | .892 | .882 | .871 |
| | | | | | | .858 | .843 | .827 | .809 | .788 | .765 | .739 | .711 | .679 | .644 |
| | | | | | | .606 | .565 | | | | | | | | |
| 28.000 | 27.652 | .348 | .085 | .116 | 33 | .968 | .966 | .964 | .961 | .959 | .956 | .953 | .949 | .946 | .941 |
| | | | | | | .937 | .932 | .926 | .919 | .912 | .904 | .895 | .885 | .874 | .862 |
| | | | | | | .847 | .831 | .813 | .793 | .770 | .745 | .716 | .685 | .650 | .612 |
| | | | | | | .571 | .527 | .481 | | | | | | | |
| 28.000 | 27.752 | .248 | .062 | .084 | 34 | .967 | .965 | .963 | .960 | .958 | .955 | .952 | .948 | .944 | .940 |
| | | | | | | .935 | .929 | .923 | .916 | .909 | .900 | .891 | .880 | .868 | .854 |
| | | | | | | .839 | .822 | .802 | .780 | .756 | .729 | .698 | .664 | .627 | .587 |
| | | | | | | .544 | .498 | .449 | .400 | | | | | | |
| 28.000 | 27.823 | .177 | .044 | .061 | 35 | .967 | .965 | .962 | .960 | .957 | .954 | .951 | .947 | .943 | .938 |
| | | | | | | .933 | .927 | .921 | .914 | .906 | .897 | .887 | .876 | .863 | .849 |
| | | | | | | .833 | .814 | .794 | .771 | .745 | .716 | .684 | .649 | .610 | .568 |
| | | | | | | .523 | .475 | .426 | .376 | .326 | | | | | |

JEWETT EQUIVALENT QUEUE EXTENDED ERLANG B(EQEEB): 5 MINUTES MAXIMUM WAIT (IF NOT CONNECTED AFTER 5 MINUTES WAIT, 100% OVERFLOW)

| HOURS FIRST ATTEMPT VMR = 1.0 | HOURS CON-NECT | HOURS OVFL | DELAYED PORTION | DELAY LENGTH CALL FACTOR | LINE | HOURS CONNECTED TRAFFIC ON LINES 1 | 2 | 3 | 4 | 5 | 6 | 7 | 8 | 9 | 10 |
|---|---|---|---|---|---|---|---|---|---|---|---|---|---|---|---|
| 28.000 | 27.875 | .125 | .032 | .044 | 36 | .966 | .964 | .962 | .959 | .956 | .953 | .950 | .946 | .942 | .937 |
| | | | | | | .932 | .926 | .919 | .912 | .904 | .895 | .884 | .872 | .859 | .844 |
| | | | | | | .828 | .809 | .787 | .763 | .736 | .706 | .673 | .637 | .596 | .553 |
| | | | | | | .507 | .459 | .409 | .358 | .308 | .260 | | | | |
| 28.000 | 27.912 | .088 | .022 | .031 | 37 | .966 | .964 | .962 | .959 | .956 | .953 | .949 | .945 | .941 | .936 |
| | | | | | | .931 | .925 | .918 | .911 | .902 | .893 | .882 | .870 | .857 | .841 |
| | | | | | | .824 | .804 | .783 | .758 | .730 | .699 | .665 | .628 | .587 | .542 |
| | | | | | | .496 | .447 | .396 | .346 | .296 | .248 | .204 | | | |
| 28.000 | 27.939 | .061 | .016 | .021 | 38 | .966 | .964 | .961 | .959 | .956 | .952 | .949 | .945 | .940 | .936 |
| | | | | | | .930 | .924 | .917 | .910 | .901 | .891 | .881 | .868 | .854 | .839 |
| | | | | | | .821 | .801 | .779 | .754 | .726 | .694 | .659 | .621 | .579 | .535 |
| | | | | | | .487 | .438 | .387 | .337 | .287 | .240 | .196 | .156 | | |
| 30.000 | 22.587 | 7.413 | .863 | .985 | 23 | .987 | .987 | .986 | .986 | .986 | .985 | .985 | .984 | .984 | .983 |
| | | | | | | .983 | .982 | .982 | .981 | .981 | .980 | .980 | .979 | .978 | .978 |
| | | | | | | .977 | .976 | .975 | | | | | | | |
| 30.000 | 23.511 | 6.489 | .799 | .932 | 24 | .986 | .986 | .985 | .985 | .984 | .984 | .983 | .983 | .982 | .982 |
| | | | | | | .981 | .981 | .980 | .979 | .979 | .978 | .977 | .976 | .975 | .975 |
| | | | | | | .974 | .973 | .972 | .971 | | | | | | |
| 30.000 | 24.413 | 5.587 | .731 | .872 | 25 | .985 | .984 | .984 | .983 | .983 | .982 | .982 | .981 | .980 | .980 |
| | | | | | | .979 | .978 | .978 | .977 | .976 | .975 | .974 | .973 | .972 | .971 |
| | | | | | | .970 | .968 | .967 | .966 | .964 | | | | | |
| 30.000 | 25.285 | 4.715 | .659 | .803 | 26 | .983 | .983 | .982 | .982 | .981 | .980 | .980 | .979 | .978 | .977 |
| | | | | | | .976 | .976 | .975 | .974 | .973 | .971 | .970 | .969 | .968 | .966 |
| | | | | | | .965 | .963 | .961 | .959 | .957 | .955 | | | | |
| 30.000 | 26.116 | 3.884 | .582 | .724 | 27 | .982 | .981 | .980 | .980 | .979 | .978 | .977 | .976 | .975 | .974 |
| | | | | | | .973 | .972 | .971 | .970 | .060 | .967 | .965 | .964 | .962 | .960 |
| | | | | | | .050 | .956 | .954 | .951 | .949 | .946 | .943 | | | |
| 30.000 | 26.886 | 3.114 | .501 | .636 | 28 | .980 | .979 | .978 | .978 | .977 | .976 | .975 | .974 | .972 | .971 |
| | | | | | | .970 | .968 | .967 | .965 | .964 | .962 | .960 | .958 | .955 | .953 |
| | | | | | | .950 | .947 | .944 | .941 | .937 | .933 | .929 | .924 | | |
| 30.000 | 27.575 | 2.425 | .420 | .542 | 29 | .978 | .977 | .976 | .975 | .974 | .973 | .972 | .970 | .969 | .967 |
| | | | | | | .966 | .964 | .962 | .960 | .958 | .955 | .953 | .950 | .947 | .944 |
| | | | | | | .940 | .936 | .932 | .927 | .922 | .917 | .910 | .904 | .896 | |
| 30.000 | 28.163 | 1.837 | .340 | .447 | 30 | .976 | .975 | .974 | .973 | .972 | .970 | .969 | .967 | .065 | .963 |
| | | | | | | .961 | .959 | .957 | .954 | .051 | .948 | .945 | .941 | .937 | .933 |
| | | | | | | .928 | .923 | .917 | .911 | .904 | .896 | .888 | .879 | .868 | .857 |
| 30.000 | 28.638 | 1.362 | .268 | .356 | 31 | .975 | .973 | .972 | .971 | .969 | .967 | .966 | .964 | .962 | .959 |
| | | | | | | .957 | .954 | .951 | .948 | .944 | .941 | .937 | .932 | .927 | .921 |
| | | | | | | .915 | .909 | .901 | .893 | .884 | .874 | .862 | .850 | .836 | .821 |
| | | | | | | .804 | | | | | | | | | |

JEWETT EQUIVALENT QUEUE EXTENDED ERLANG B(EQEEB): 5 MINUTES MAXIMUM WAIT (IF NOT CONNECTED AFTER 5 MINUTES WAIT, 100% OVERFLOW)

| HOURS FIRST ATTEMPT VMR = 1.0 | HOURS CON-NECT | HOURS OVFL | DELAYED PORTION | DELAY LENGTH CALL FACTOR | LINE | HOURS CONNECTED TRAFFIC ON LINES 1 | 2 | 3 | 4 | 5 | 6 | 7 | 8 | 9 | 10 |
|---|---|---|---|---|---|---|---|---|---|---|---|---|---|---|---|
| 30.000 | 29.003 | .997 | .206 | .277 | 32 | .973 | .972 | .970 | .969 | .967 | .965 | .963 | .961 | .958 | .955 |
| | | | | | | .953 | .949 | .946 | .942 | .938 | .934 | .929 | .923 | .917 | .910 |
| | | | | | | .903 | .894 | .885 | .875 | .863 | .850 | .836 | .820 | .803 | .783 |
| | | | | | | .761 | .737 | | | | | | | | |
| 30.000 | 29.274 | .726 | .156 | .211 | 33 | .972 | .970 | .969 | .967 | .965 | .963 | .960 | .958 | .955 | .952 |
| | | | | | | .949 | .945 | .941 | .937 | .932 | .927 | .921 | .915 | .908 | .900 |
| | | | | | | .891 | .881 | .870 | .858 | .844 | .829 | .811 | .792 | .771 | .747 |
| | | | | | | .721 | .692 | .660 | | | | | | | |
| 30.000 | 29.473 | .527 | .117 | .159 | 34 | .971 | .969 | .967 | .965 | .963 | .961 | .959 | .956 | .953 | .950 |
| | | | | | | .946 | .942 | .938 | .933 | .928 | .922 | .915 | .908 | .900 | .891 |
| | | | | | | .881 | .870 | .857 | .843 | .827 | .809 | .790 | .768 | .743 | .716 |
| | | | | | | .686 | .653 | .617 | .578 | | | | | | |
| 30.000 | 29.618 | .382 | .087 | .118 | 35 | .970 | .968 | .966 | .964 | .962 | .960 | .957 | .954 | .951 | .947 |
| | | | | | | .944 | .939 | .935 | .929 | .924 | .917 | .910 | .902 | .893 | .884 |
| | | | | | | .872 | .860 | .846 | .831 | .813 | .793 | .772 | .747 | .720 | .690 |
| | | | | | | .657 | .621 | .582 | .540 | .496 | | | | | |
| 30.000 | 29.724 | .276 | .064 | .087 | 36 | .969 | .968 | .966 | .964 | .961 | .959 | .956 | .953 | .949 | .946 |
| | | | | | | .942 | .937 | .932 | .927 | .921 | .914 | .906 | .898 | .888 | .878 |
| | | | | | | .866 | .852 | .837 | .821 | .802 | .781 | .757 | .731 | .701 | .669 |
| | | | | | | .634 | .596 | .554 | .510 | .464 | .417 | | | | |
| 30.000 | 29.801 | .199 | .047 | .064 | 37 | .969 | .967 | .965 | .963 | .961 | .958 | .955 | .952 | .948 | .945 |
| | | | | | | .940 | .936 | .930 | .925 | .918 | .911 | .903 | .894 | .884 | .873 |
| | | | | | | .861 | .846 | .831 | .813 | .793 | .770 | .745 | .718 | .687 | .653 |
| | | | | | | .616 | .576 | .533 | .487 | .440 | .392 | .343 | | | |
| 30.000 | 29.858 | .142 | .034 | .046 | 38 | .969 | .967 | .965 | .962 | .960 | .957 | .954 | .951 | .948 | .944 |
| | | | | | | .939 | .934 | .929 | .923 | .916 | .909 | .901 | .892 | .881 | .870 |
| | | | | | | .857 | .842 | .825 | .807 | .786 | .763 | .737 | .708 | .676 | .640 |
| | | | | | | .602 | .561 | .517 | .470 | .422 | .373 | .325 | .277 | | |
| 30.000 | 29.899 | .101 | .024 | .033 | 39 | .968 | .966 | .964 | .962 | .960 | .957 | .954 | .950 | .947 | .943 |
| | | | | | | .938 | .933 | .928 | .922 | .915 | .907 | .899 | .890 | .879 | .867 |
| | | | | | | .853 | .838 | .821 | .802 | .781 | .757 | .730 | .700 | .667 | .631 |
| | | | | | | .592 | .550 | .505 | .457 | .409 | .360 | .311 | .264 | .220 | |
| 30.000 | 29.929 | .071 | .017 | .023 | 40 | .968 | .966 | .964 | .962 | .959 | .956 | .953 | .950 | .946 | .942 |
| | | | | | | .938 | .933 | .927 | .921 | .914 | .906 | .898 | .888 | .877 | .865 |
| | | | | | | .851 | .836 | .818 | .799 | .777 | .752 | .725 | .694 | .661 | .624 |
| | | | | | | .584 | .541 | .496 | .448 | .399 | .350 | .302 | .255 | .211 | .171 |

JEWETT EQUIVALENT QUEUE EXTENDED ERLANG B(EQEEB): 10 MINUTES MAXIMUM WAIT (IF NOT CONNECTED AFTER 10 MINUTES WAIT, 100% OVERFLOW)

| HOURS FIRST ATTEMPT VMR = 1.0 | HOURS CON-NECT | HOURS OVFL | DELAYED PORTION | DELAY LENGTH CALL FACTOR | LINE | HOURS CONNECTED TRAFFIC ON LINES 1 | 2 | 3 | 4 | 5 | 6 | 7 | 8 | 9 | 10 |
|---|---|---|---|---|---|---|---|---|---|---|---|---|---|---|---|
| .200 | .198 | .002 | .204 | .311 | 1 | .198 | | | | | | | | | |
| .200 | .200 | .000 | .017 | .026 | 2 | .169 | .031 | | | | | | | | |
| .400 | .390 | .010 | .407 | .612 | 1 | .390 | | | | | | | | | |
| .400 | .399 | .001 | .061 | .093 | 2 | .298 | .101 | | | | | | | | |
| .600 | .570 | .030 | .609 | .894 | 1 | .570 | | | | | | | | | |
| .600 | .597 | .003 | .124 | .189 | 2 | .404 | .192 | | | | | | | | |
| .600 | .599 | .001 | .021 | .033 | 3 | .380 | .169 | .051 | | | | | | | |
| .800 | .724 | .076 | .805 | 1.136 | 1 | .724 | | | | | | | | | |
| .800 | .792 | .008 | .202 | .308 | 2 | .497 | .296 | | | | | | | | |
| .800 | .799 | .001 | .043 | .067 | 3 | .455 | .246 | .098 | | | | | | | |
| 1.000 | .984 | .016 | .295 | .447 | 2 | .579 | .405 | | | | | | | | |
| 1.000 | .997 | .003 | .074 | .113 | 3 | .518 | .323 | .156 | | | | | | | |
| 1.000 | .999 | .001 | .016 | .025 | 4 | .504 | .305 | .141 | .049 | | | | | | |
| 1.200 | 1.171 | .029 | .400 | .601 | 2 | .655 | .516 | | | | | | | | |
| 1.200 | 1.194 | .006 | .113 | .173 | 3 | .573 | .396 | .225 | | | | | | | |
| 1.200 | 1.199 | .001 | .029 | .044 | 4 | .552 | .368 | .197 | .082 | | | | | | |
| 1.400 | 1.348 | .052 | .516 | .767 | 2 | .725 | .623 | | | | | | | | |
| 1.400 | 1.390 | .010 | .161 | .246 | 3 | .622 | .467 | .300 | | | | | | | |
| 1.400 | 1.397 | .003 | .046 | .071 | 4 | .594 | .426 | .255 | .123 | | | | | | |
| 1.600 | 1.511 | .089 | .640 | .935 | 2 | .789 | .722 | | | | | | | | |
| 1.600 | 1.583 | .017 | .218 | .332 | 3 | .668 | .535 | .381 | | | | | | | |
| 1.600 | 1.595 | .005 | .068 | .104 | 4 | .631 | .480 | .314 | .171 | | | | | | |
| 1.600 | 1.599 | .001 | .019 | .030 | 5 | .620 | .463 | .296 | .155 | .065 | | | | | |
| 1.800 | 1.649 | .151 | .766 | 1.092 | 2 | .844 | .804 | | | | | | | | |
| 1.800 | 1.773 | .027 | .284 | .431 | 3 | .710 | .599 | .464 | | | | | | | |
| 1.800 | 1.793 | .007 | .095 | .145 | 4 | .664 | .529 | .374 | .226 | | | | | | |
| 1.800 | 1.798 | .002 | .029 | .045 | 5 | .649 | .507 | .347 | .200 | .095 | | | | | |
| 2.000 | 1.754 | .246 | .886 | 1.221 | 2 | .888 | .866 | | | | | | | | |
| 2.000 | 1.959 | .041 | .359 | .541 | 3 | .749 | .661 | .548 | | | | | | | |
| 2.000 | 1.989 | .011 | .128 | .195 | 4 | .694 | .576 | .433 | .286 | | | | | | |
| 2.000 | 1.997 | .003 | .042 | .065 | 5 | .676 | .547 | .396 | .248 | .130 | | | | | |
| 2.200 | 2.137 | .063 | .442 | .663 | 3 | .787 | .719 | .631 | | | | | | | |
| 2.200 | 2.183 | .017 | .166 | .254 | 4 | .723 | .619 | .491 | .350 | | | | | | |
| 2.200 | 2.195 | .005 | .059 | .090 | 5 | .700 | .584 | .444 | .297 | .171 | | | | | |
| 2.200 | 2.198 | .002 | .019 | .030 | 6 | .691 | .571 | .427 | .280 | .156 | .073 | | | | |
| 2.400 | 2.306 | .094 | .534 | .792 | 3 | .823 | .773 | .709 | | | | | | | |
| 2.400 | 2.376 | .024 | .211 | .321 | 4 | .749 | .660 | .548 | .418 | | | | | | |
| 2.400 | 2.392 | .008 | .078 | .120 | 5 | .722 | .618 | .489 | .348 | .216 | | | | | |
| 2.400 | 2.397 | .003 | .027 | .042 | 6 | .711 | .602 | .468 | .324 | .194 | .099 | | | | |
| 2.600 | 2.459 | .141 | .632 | .925 | 3 | .857 | .823 | .779 | | | | | | | |

JEWETT EQUIVALENT QUEUE EXTENDED ERLANG B(EQEEB): 10 MINUTES MAXIMUM WAIT (IF NOT CONNECTED AFTER 10 MINUTES WAIT, 100% OVERFLOW)

| HOURS FIRST ATTEMPT VMR = 1.0 | HOURS CON-NECT | HOURS OVFL | DELAYED PORTION | DELAY LENGTH CALL FACTOR | LINE | HOURS CONNECTED TRAFFIC ON LINES 1 | 2 | 3 | 4 | 5 | 6 | 7 | 8 | 9 | 10 |
|---|---|---|---|---|---|---|---|---|---|---|---|---|---|---|---|
| 2.600 | 2.565 | .035 | .262 | .398 | 4 | .775 | .700 | .603 | .487 | | | | | | |
| 2.600 | 2.589 | .011 | .102 | .156 | 5 | .742 | .649 | .532 | .399 | .266 | | | | | |
| 2.600 | 2.596 | .004 | .037 | .058 | 6 | .729 | .630 | .506 | .367 | .235 | .130 | | | | |
| 2.800 | 2.589 | .211 | .733 | 1.052 | 3 | .887 | .865 | .837 | | | | | | | |
| 2.800 | 2.751 | .049 | .320 | .484 | 4 | .799 | .737 | .657 | .558 | | | | | | |
| 2.800 | 2.784 | .016 | .129 | .198 | 5 | .761 | .679 | .574 | .450 | .320 | | | | | |
| 2.800 | 2.794 | .006 | .050 | .077 | 6 | .746 | .656 | .541 | .410 | .277 | .164 | | | | |
| 2.800 | 2.798 | .002 | .018 | .028 | 7 | .740 | .647 | .529 | .394 | .262 | .151 | .075 | | | |
| 3.000 | 2.692 | .308 | .829 | 1.162 | 3 | .912 | .898 | .881 | | | | | | | |
| 3.000 | 2.931 | .069 | .385 | .580 | 4 | .823 | .773 | .708 | .627 | | | | | | |
| 3.000 | 2.978 | .022 | .161 | .246 | 5 | .779 | .707 | .614 | .501 | .377 | | | | | |
| 3.000 | 2.992 | .008 | .065 | .099 | 6 | .762 | .680 | .575 | .451 | .321 | .203 | | | | |
| 3.000 | 2.997 | .003 | .025 | .038 | 7 | .754 | .668 | .559 | .432 | .300 | .184 | .099 | | | |
| 3.200 | 2.767 | .433 | .916 | 1.249 | 3 | .931 | .923 | .913 | | | | | | | |
| 3.200 | 3.104 | .096 | .457 | .684 | 4 | .846 | .807 | .757 | .693 | | | | | | |
| 3.200 | 3.170 | .030 | .198 | .302 | 5 | .797 | .734 | .652 | .551 | .436 | | | | | |
| 3.200 | 3.189 | .011 | .082 | .126 | 6 | .776 | .702 | .607 | .492 | .366 | .246 | | | | |
| 3.200 | 3.196 | .004 | .033 | .050 | 7 | .768 | .689 | .588 | .468 | .339 | .220 | .125 | | | |
| 3.400 | 3.266 | .134 | .536 | .795 | 4 | .869 | .839 | .802 | .755 | | | | | | |
| 3.400 | 3.359 | .041 | .240 | .364 | 5 | .814 | .760 | .689 | .601 | .496 | | | | | |
| 3.400 | 3.385 | .015 | .103 | .157 | 6 | .790 | .723 | .637 | .531 | .412 | .291 | | | | |
| 3.400 | 3.394 | .006 | .042 | .065 | 7 | .780 | .708 | .615 | .502 | .378 | .257 | .155 | | | |
| 3.400 | 3.398 | .002 | .016 | .025 | 8 | .775 | .701 | .605 | .490 | .364 | .243 | .144 | .075 | | |
| 3.600 | 3.412 | .188 | .620 | .909 | 4 | .890 | .869 | .843 | .810 | | | | | | |
| 3.600 | 3.546 | .054 | .287 | .434 | 5 | .831 | .784 | .724 | .649 | .557 | | | | | |
| 3.600 | 3.580 | .020 | .126 | .193 | 6 | .803 | .743 | .666 | .570 | .458 | .340 | | | | |
| 3.600 | 3.592 | .008 | .054 | .082 | 7 | .791 | .725 | .640 | .535 | .416 | .296 | .189 | | | |
| 3.600 | 3.597 | .003 | .022 | .033 | 8 | .786 | .717 | .628 | .520 | .399 | .278 | .173 | .095 | | |
| 3.800 | 3.538 | .262 | .705 | 1.018 | 4 | .910 | .895 | .877 | .856 | | | | | | |
| 3.800 | 3.728 | .072 | .339 | .513 | 5 | .847 | .808 | .759 | .696 | .618 | | | | | |
| 3.800 | 3.774 | .026 | .153 | .234 | 6 | .816 | .763 | .693 | .607 | .504 | .391 | | | | |
| 3.800 | 3.789 | .011 | .067 | .103 | 7 | .802 | .742 | .663 | .566 | .454 | .336 | .225 | | | |
| 3.800 | 3.796 | .004 | .028 | .043 | 8 | .796 | .732 | .650 | .549 | .433 | .313 | .204 | .119 | | |
| 4.000 | 3.640 | .360 | .788 | 1.117 | 4 | .927 | .917 | .905 | .891 | | | | | | |
| 4.000 | 3.904 | .096 | .398 | .599 | 5 | .863 | .832 | .791 | .740 | .677 | | | | | |
| 4.000 | 3.965 | .035 | .184 | .281 | 6 | .829 | .781 | .720 | .643 | .549 | .444 | | | | |
| 4.000 | 3.986 | .014 | .083 | .127 | 7 | .813 | .757 | .686 | .597 | .491 | .377 | .265 | | | |
| 4.000 | 3.994 | .006 | .036 | .055 | 8 | .805 | .747 | .670 | .576 | .466 | .348 | .237 | .145 | | |
| 4.000 | 3.998 | .002 | .015 | .022 | 9 | .802 | .742 | .664 | .566 | .454 | .336 | .225 | .136 | .073 | |
| 4.200 | 3.718 | .482 | .864 | 1.199 | 4 | .940 | .934 | .926 | .917 | | | | | | |

## JEWETT EQUIVALENT QUEUE EXTENDED ERLANG B(EQEEB): 10 MINUTES MAXIMUM WAIT (IF NOT CONNECTED AFTER 10 MINUTES WAIT, 100% OVERFLOW)

| HOURS FIRST ATTEMPT VMR = 1.0 | HOURS CON-NECT | HOURS OVFL | DELAYED PORTION | DELAY LENGTH CALL FACTOR | LINE | HOURS CONNECTED TRAFFIC ON LINES 1 | 2 | 3 | 4 | 5 | 6 | 7 | 8 | 9 | 10 |
|---|---|---|---|---|---|---|---|---|---|---|---|---|---|---|---|
| 4.200 | 4.071 | .129 | .463 | .692 | 5 | .879 | .854 | .822 | .783 | .733 | | | | | |
| 4.200 | 4.155 | .045 | .219 | .333 | 6 | .841 | .799 | .745 | .678 | .595 | .498 | | | | |
| 4.200 | 4.182 | .018 | .101 | .154 | 7 | .823 | .772 | .707 | .626 | .528 | .419 | .307 | | | |
| 4.200 | 4.192 | .008 | .045 | .069 | 8 | .814 | .760 | .690 | .602 | .497 | .384 | .272 | .174 | | |
| 4.200 | 4.197 | .003 | .019 | .029 | 9 | .810 | .754 | .681 | .590 | .484 | .368 | .256 | .161 | .091 | |
| 4.400 | 3.775 | .625 | .932 | 1.264 | 4 | .951 | .946 | .941 | .935 | | | | | | |
| 4.400 | 4.228 | .172 | .533 | .790 | 5 | .895 | .876 | .851 | .821 | .785 | | | | | |
| 4.400 | 4.342 | .058 | .258 | .392 | 6 | .853 | .817 | .770 | .711 | .639 | .553 | | | | |
| 4.400 | 4.377 | .023 | .121 | .185 | 7 | .832 | .787 | .728 | .654 | .564 | .461 | .352 | | | |
| 4.400 | 4.390 | .010 | .055 | .085 | 8 | .823 | .773 | .708 | .626 | .528 | .419 | .308 | .206 | | |
| 4.400 | 4.396 | .004 | .024 | .037 | 9 | .818 | .766 | .698 | .613 | .512 | .400 | .288 | .189 | .111 | |
| 4.600 | 4.370 | .230 | .607 | .891 | 5 | .910 | .896 | .878 | .856 | .830 | | | | | |
| 4.600 | 4.525 | .075 | .302 | .457 | 6 | .865 | .834 | .794 | .744 | .682 | .607 | | | | |
| 4.600 | 4.570 | .030 | .144 | .221 | 7 | .842 | .801 | .748 | .681 | .598 | .503 | .398 | | | |
| 4.600 | 4.587 | .013 | .067 | .103 | 8 | .831 | .785 | .725 | .649 | .558 | .454 | .345 | .241 | | |
| 4.600 | 4.594 | .006 | .030 | .046 | 9 | .826 | .777 | .714 | .634 | .539 | .431 | .321 | .218 | .135 | |
| 4.800 | 4.493 | .307 | .682 | .990 | 5 | .924 | .914 | .901 | .886 | .868 | | | | | |
| 4.800 | 4.704 | .096 | .351 | .530 | 6 | .877 | .850 | .817 | .775 | .724 | .661 | | | | |
| 4.800 | 4.762 | .038 | .171 | .261 | 7 | .851 | .814 | .767 | .707 | .633 | .545 | .446 | | | |
| 4.800 | 4.783 | .017 | .081 | .125 | 8 | .839 | .796 | .741 | .672 | .587 | .489 | .382 | .278 | | |
| 4.800 | 4.793 | .007 | .037 | .058 | 9 | .833 | .787 | .729 | .655 | .565 | .462 | .353 | .249 | .160 | |
| 4.800 | 4.797 | .003 | .016 | .025 | 10 | .830 | .783 | .723 | .646 | .554 | .450 | .340 | .236 | .149 | .085 |
| 5.000 | 4.594 | .406 | .756 | 1.080 | 5 | .937 | .929 | .920 | .910 | .898 | | | | | |
| 5.000 | 4.876 | .124 | .405 | .608 | 6 | .889 | .867 | .839 | .806 | .764 | .712 | | | | |
| 5.000 | 4.952 | .048 | .200 | .305 | 7 | .860 | .827 | .785 | .732 | .666 | .587 | .495 | | | |
| 5.000 | 4.979 | .021 | .097 | .149 | 8 | .846 | .807 | .757 | .693 | .615 | .523 | .421 | .317 | | |
| 5.000 | 4.991 | .009 | .046 | .070 | 9 | .839 | .797 | .743 | .674 | .590 | .492 | .386 | .281 | .188 | |
| 5.000 | 4.996 | .004 | .021 | .032 | 10 | .836 | .792 | .736 | .664 | .577 | .477 | .370 | .265 | .174 | .104 |
| 5.500 | 4.758 | .742 | .915 | 1.248 | 5 | .959 | .956 | .952 | .948 | .943 | | | | | |
| 5.500 | 5.264 | .236 | .561 | .820 | 6 | .918 | .905 | .891 | .873 | .851 | .825 | | | | |
| 5.500 | 5.415 | .085 | .291 | .441 | 7 | .883 | .859 | .829 | .791 | .745 | .688 | .620 | | | |
| 5.500 | 5.464 | .036 | .147 | .224 | 8 | .864 | .833 | .794 | .743 | .681 | .606 | .519 | .423 | | |
| 5.500 | 5.483 | .017 | .073 | .111 | 9 | .855 | .820 | .775 | .718 | .648 | .564 | .468 | .367 | .269 | |
| 5.500 | 5.492 | .008 | .035 | .053 | 10 | .850 | .813 | .765 | .705 | .631 | .542 | .443 | .340 | .243 | .159 |
| 5.500 | 5.496 | .004 | .016 | .025 | 11 | .848 | .810 | .761 | .699 | .622 | .532 | .431 | .328 | .230 | .149 |
| | | | | | | .088 | | | | | | | | | |
| 6.000 | 5.552 | .448 | .730 | 1.048 | 6 | .944 | .938 | .931 | .923 | .914 | .903 | | | | |
| 6.000 | 5.850 | .150 | .408 | .613 | 7 | .906 | .890 | .870 | .846 | .817 | .782 | .739 | | | |
| 6.000 | 5.938 | .062 | .213 | .324 | 8 | .882 | .858 | .828 | .790 | .743 | .685 | .616 | .537 | | |
| 6.000 | 5.972 | .028 | .109 | .167 | 9 | .870 | .841 | .804 | .758 | .700 | .631 | .549 | .457 | .361 | |

JEWETT EQUIVALENT QUEUE EXTENDED ERLANG B(EQEEB): 10 MINUTES MAXIMUM WAIT (IF NOT CONNECTED AFTER 10 MINUTES WAIT, 100% OVERFLOW)

| HOURS FIRST ATTEMPT VMR = 1.0 | HOURS CON-NECT | HOURS OVFL | DELAYED PORTION | DELAY LENGTH CALL FACTOR | LINE | HOURS CONNECTED TRAFFIC ON LINES 1 | 2 | 3 | 4 | 5 | 6 | 7 | 8 | 9 | 10 |
|---|---|---|---|---|---|---|---|---|---|---|---|---|---|---|---|
| 6.000 | 5.986 | .014 | .055 | .084 | 10 | .864 | .832 | .792 | .741 | .678 | .602 | .514 | .418 | .319 | .228 |
| 6.000 | 5.994 | .006 | .027 | .041 | 11 | .860 | .827 | .785 | .732 | .666 | .587 | .496 | .398 | .299 | .209 |
| | | | | | | .134 | | | | | | | | | |
| 6.500 | 5.723 | .777 | .877 | 1.212 | 6 | .962 | .959 | .956 | .952 | .949 | .944 | | | | |
| 6.500 | 6.229 | .271 | .552 | .817 | 7 | .929 | .919 | .908 | .895 | .879 | .861 | .838 | | | |
| 6.500 | 6.396 | .104 | .299 | .453 | 8 | .900 | .882 | .860 | .833 | .800 | .759 | .710 | .651 | | |
| 6.500 | 6.453 | .047 | .158 | .241 | 9 | .884 | .861 | .831 | .795 | .749 | .694 | .627 | .549 | .463 | |
| 6.500 | 6.477 | .023 | .082 | .126 | 10 | .876 | .849 | .815 | .773 | .721 | .657 | .581 | .495 | .402 | .308 |
| 6.500 | 6.489 | .011 | .042 | .064 | 11 | .871 | .843 | .807 | .762 | .705 | .637 | .557 | .466 | .371 | .277 |
| | | | | | | .193 | | | | | | | | | |
| 6.500 | 6.495 | .005 | .021 | .032 | 12 | .869 | .840 | .802 | .755 | .697 | .626 | .544 | .451 | .355 | .262 |
| | | | | | | .179 | .114 | | | | | | | | |
| 7.000 | 6.514 | .486 | .707 | 1.021 | 7 | .949 | .944 | .939 | .932 | .925 | .917 | .907 | | | |
| 7.000 | 6.824 | .176 | .409 | .615 | 8 | .918 | .906 | .891 | .874 | .853 | .827 | .796 | .759 | | |
| 7.000 | 6.924 | .076 | .222 | .338 | 9 | .898 | .880 | .857 | .829 | .795 | .752 | .701 | .641 | .570 | |
| 7.000 | 6.963 | .037 | .119 | .182 | 10 | .887 | .865 | .837 | .803 | .760 | .708 | .645 | .571 | .488 | .399 |
| 7.000 | 6.982 | .018 | .063 | .096 | 11 | .882 | .857 | .827 | .788 | .741 | .682 | .613 | .533 | .444 | .352 |
| | | | | | | .263 | | | | | | | | | |
| 7.000 | 6.991 | .009 | .032 | .050 | 12 | .878 | .853 | .820 | .780 | .730 | .669 | .596 | .512 | .421 | .328 |
| | | | | | | .240 | .164 | | | | | | | | |
| 7.000 | 6.996 | .004 | .016 | .025 | 13 | .877 | .850 | .817 | .776 | .724 | .661 | .586 | .501 | .409 | .316 |
| | | | | | | .228 | .154 | .096 | | | | | | | |
| 7.500 | 6.689 | .811 | .845 | 1.180 | 7 | .964 | .962 | .959 | .956 | .953 | .949 | .945 | | | |
| 7.500 | 7.196 | .304 | .544 | .805 | 8 | .937 | .929 | .921 | .910 | .898 | .884 | .868 | .848 | | |
| 7.500 | 7.377 | .123 | .305 | .461 | 9 | .913 | .899 | .882 | .862 | .837 | .807 | .771 | .728 | .677 | |
| 7.500 | 7.442 | .058 | .167 | .255 | 10 | .899 | .881 | .858 | .831 | .797 | .755 | .705 | .645 | .575 | .497 |
| 7.500 | 7.471 | .029 | .091 | .139 | 11 | .891 | .871 | .845 | .813 | .773 | .724 | .665 | .596 | .518 | .432 |
| | | | | | | .344 | | | | | | | | | |
| 7.500 | 7.485 | .015 | .048 | .074 | 12 | .887 | .865 | .837 | .802 | .759 | .707 | .643 | .569 | .486 | .397 |
| | | | | | | .308 | .225 | | | | | | | | |
| 7.500 | 7.492 | .008 | .025 | .039 | 13 | .885 | .862 | .833 | .797 | .752 | .697 | .631 | .554 | .469 | .378 |
| | | | | | | .289 | .208 | .139 | | | | | | | |
| 8.000 | 6.785 | 1.215 | .953 | 1.282 | 7 | .974 | .972 | .971 | .969 | .968 | .966 | .964 | | | |
| 8.000 | 7.479 | .521 | .688 | .997 | 8 | .954 | .949 | .945 | .940 | .934 | .927 | .919 | .911 | | |
| 8.000 | 7.799 | .201 | .409 | .615 | 9 | .928 | .918 | .907 | .893 | .877 | .858 | .835 | .808 | .775 | |
| 8.000 | 7.909 | .091 | .229 | .349 | 10 | .911 | .896 | .879 | .857 | .831 | .800 | .761 | .715 | .661 | .598 |
| 8.000 | 7.955 | .045 | .127 | .195 | 11 | .901 | .883 | .862 | .835 | .803 | .763 | .715 | .657 | .590 | .514 |
| | | | | | | .432 | | | | | | | | | |
| 8.000 | 7.977 | .023 | .070 | .107 | 12 | .895 | .876 | .852 | .823 | .786 | .741 | .687 | .623 | .549 | .467 |
| | | | | | | .381 | .296 | | | | | | | | |

JEWETT EQUIVALENT QUEUE EXTENDED ERLANG B(EQEEB): 10 MINUTES MAXIMUM WAIT (IF NOT CONNECTED AFTER 10 MINUTES WAIT, 100% OVERFLOW)

| HOURS FIRST ATTEMPT VMR = 1.0 | HOURS CON-NECT | HOURS OVFL | DELAYED PORTION | DELAY LENGTH CALL FACTOR | LINE | HOURS CONNECTED TRAFFIC ON LINES 1 | 2 | 3 | 4 | 5 | 6 | 7 | 8 | 9 | 10 |
|---|---|---|---|---|---|---|---|---|---|---|---|---|---|---|---|
| 8.000 | 7.988 | .012 | .038 | .058 | 13 | .892 | .872 | .847 | .815 | .776 | .729 | .671 | .603 | .526 | .441 |
| | | | | | | .354 | .269 | .192 | | | | | | | |
| 8.000 | 7.994 | .006 | .020 | .030 | 14 | .891 | .870 | .844 | .811 | .771 | .722 | .662 | .593 | .513 | .427 |
| | | | | | | .339 | .254 | .179 | .118 | | | | | | |
| 8.500 | 7.657 | .843 | .818 | 1.151 | 8 | .966 | .964 | .962 | .959 | .956 | .953 | .950 | .946 | | |
| 8.500 | 8.165 | .335 | .536 | .794 | 9 | .943 | .937 | .930 | .922 | .912 | .901 | .889 | .874 | .857 | |
| 8.500 | 8.358 | .142 | .309 | .468 | 10 | .923 | .912 | .898 | .883 | .864 | .841 | .814 | .782 | .743 | .698 |
| 8.500 | 8.431 | .069 | .175 | .267 | 11 | .910 | .896 | .878 | .857 | .831 | .799 | .761 | .715 | .660 | .597 |
| | | | | | | .526 | | | | | | | | | |
| 8.500 | 8.464 | .036 | .098 | .150 | 12 | .904 | .887 | .866 | .841 | .811 | .773 | .728 | .673 | .610 | .538 |
| | | | | | | .458 | .375 | | | | | | | | |
| 8.500 | 8.481 | .019 | .054 | .084 | 13 | .900 | .882 | .860 | .832 | .799 | .758 | .708 | .649 | .581 | .503 |
| | | | | | | .420 | .335 | .254 | | | | | | | |
| 8.500 | 8.490 | .010 | .029 | .045 | 14 | .897 | .879 | .856 | .827 | .792 | .749 | .697 | .635 | .564 | .484 |
| | | | | | | .399 | .314 | .234 | .164 | | | | | | |
| 8.500 | 8.495 | .005 | .015 | .024 | 15 | .896 | .877 | .853 | .824 | .788 | .744 | .691 | .627 | .554 | .473 |
| | | | | | | .388 | .302 | .223 | .154 | .100 | | | | | |
| 9.000 | 7.760 | 1.240 | .922 | 1.255 | 8 | .975 | .974 | .972 | .971 | .969 | .968 | .966 | .964 | | |
| 9.000 | 8.445 | .555 | .671 | .975 | 9 | .957 | .954 | .950 | .945 | .940 | .935 | .929 | .921 | .913 | |
| 9.000 | 8.774 | .226 | .409 | .614 | 10 | .935 | .927 | .918 | .907 | .895 | .880 | .863 | .842 | .818 | .789 |
| 9.000 | 8.895 | .105 | .235 | .358 | 11 | .920 | .909 | .895 | .878 | .858 | .833 | .804 | .769 | .728 | .679 |
| | | | | | | .622 | | | | | | | | | |
| 9.000 | 8.946 | .054 | .135 | .206 | 12 | .912 | .897 | .880 | .859 | .834 | .803 | .766 | .721 | .668 | .607 |
| | | | | | | .537 | .461 | | | | | | | | |
| 9.000 | 8.971 | .029 | .076 | .117 | 13 | .907 | .891 | .872 | .848 | .820 | .785 | .742 | .692 | .632 | .564 |
| | | | | | | .488 | .407 | .325 | | | | | | | |
| 9.000 | 8.984 | .016 | .043 | .065 | 14 | .904 | .887 | .867 | .842 | .811 | .774 | .728 | .674 | .611 | .539 |
| | | | | | | .459 | .376 | .294 | .218 | | | | | | |
| 9.000 | 8.992 | .008 | .023 | .036 | 15 | .902 | .885 | .864 | .838 | .806 | .767 | .720 | .664 | .599 | .524 |
| | | | | | | .443 | .359 | .277 | .203 | .140 | | | | | |
| 9.500 | 8.627 | .873 | .794 | 1.124 | 9 | .968 | .966 | .964 | .962 | .959 | .957 | .953 | .950 | .946 | |
| 9.500 | 9.135 | .365 | .528 | .783 | 10 | .948 | .943 | .937 | .931 | .923 | .914 | .904 | .893 | .879 | .863 |
| 9.500 | 9.339 | .161 | .312 | .472 | 11 | .930 | .921 | .911 | .898 | .884 | .866 | .845 | .820 | .791 | .756 |
| | | | | | | .715 | | | | | | | | | |
| 9.500 | 9.420 | .080 | .181 | .276 | 12 | .920 | .908 | .894 | .877 | .856 | .832 | .802 | .766 | .724 | .674 |
| | | | | | | .617 | .551 | | | | | | | | |
| 9.500 | 9.457 | .043 | .105 | .160 | 13 | .913 | .900 | .883 | .863 | .839 | .810 | .774 | .732 | .681 | .622 |
| | | | | | | .555 | .481 | .403 | | | | | | | |
| 9.500 | 9.476 | .024 | .060 | .092 | 14 | .910 | .895 | .877 | .855 | .829 | .796 | .757 | .710 | .655 | .591 |
| | | | | | | .519 | .441 | .360 | .281 | | | | | | |

JEWETT EQUIVALENT QUEUE EXTENDED ERLANG B(EQEEB): 10 MINUTES MAXIMUM WAIT (IF NOT CONNECTED AFTER 10 MINUTES WAIT, 100% OVERFLOW)

| HOURS FIRST ATTEMPT VMR = 1.0 | HOURS CON- NECT | HOURS OVFL | DELAYED PORTION | DELAY LENGTH CALL FACTOR | LINE | HOURS CONNECTED TRAFFIC ON LINES 1 | 2 | 3 | 4 | 5 | 6 | 7 | 8 | 9 | 10 |
|---|---|---|---|---|---|---|---|---|---|---|---|---|---|---|---|
| 9.500 | 9.487 | .013 | .034 | .052 | 15 | .908 | .892 | .873 | .850 | .822 | .788 | .747 | .698 | .639 | .572 |
| | | | | | | .498 | .417 | .336 | .258 | .187 | | | | | |
| 9.500 | 9.493 | .007 | .018 | .028 | 16 | .906 | .890 | .871 | .848 | .819 | .784 | .741 | .690 | .630 | .562 |
| | | | | | | .485 | .404 | .322 | .245 | .176 | .119 | | | | |
| 10.000 | 8.735 | 1.265 | .895 | 1.229 | 9 | .976 | .975 | .973 | .972 | .971 | .969 | .968 | .966 | .964 | |
| 10.000 | 9.413 | .587 | .656 | .955 | 10 | .960 | .957 | .954 | .950 | .946 | .941 | .936 | .930 | .923 | .916 |
| 10.000 | 9.751 | .249 | .407 | .612 | 11 | .941 | .935 | .927 | .919 | .908 | .897 | .883 | .867 | .848 | .826 |
| | | | | | | .800 | | | | | | | | | |
| 10.000 | 9.880 | .120 | .240 | .365 | 12 | .928 | .918 | .907 | .894 | .878 | .859 | .836 | .809 | .777 | .739 |
| | | | | | | .694 | .643 | | | | | | | | |
| 10.000 | 9.937 | .063 | .141 | .215 | 13 | .920 | .908 | .895 | .878 | .858 | .833 | .804 | .769 | .727 | .678 |
| | | | | | | .622 | .557 | .487 | | | | | | | |
| 10.000 | 9.965 | .035 | .082 | .126 | 14 | .916 | .903 | .887 | .868 | .845 | .817 | .784 | .744 | .696 | .641 |
| | | | | | | .577 | .506 | .430 | .352 | | | | | | |
| 10.000 | 9.981 | .019 | .047 | .072 | 15 | .913 | .899 | .882 | .862 | .838 | .808 | .772 | .729 | .677 | .618 |
| | | | | | | .550 | .476 | .397 | .318 | .243 | | | | | |
| 10.000 | 9.989 | .011 | .027 | .041 | 16 | .911 | .897 | .880 | .859 | .833 | .802 | .764 | .719 | .666 | .604 |
| | | | | | | .534 | .458 | .378 | .299 | .225 | .161 | | | | |
| 10.000 | 9.994 | .006 | .015 | .022 | 17 | .910 | .896 | .878 | .857 | .830 | .799 | .760 | .714 | .659 | .596 |
| | | | | | | .525 | .447 | .367 | .288 | .215 | .152 | .101 | | | |
| 11.000 | 9.711 | 1.289 | .870 | 1.205 | 10 | .977 | .976 | .975 | .973 | .972 | .971 | .969 | .968 | .966 | .964 |
| 11.000 | 10.383 | .617 | .642 | .937 | 11 | .963 | .960 | .957 | .954 | .950 | .946 | .942 | .937 | .931 | .925 |
| | | | | | | .918 | | | | | | | | | |
| 11.000 | 10.727 | .273 | .406 | .610 | 12 | .946 | .941 | .934 | .927 | .919 | .909 | .898 | .886 | .871 | .853 |
| | | | | | | .833 | .810 | | | | | | | | |
| 11.000 | 10.866 | .134 | .244 | .370 | 13 | .934 | .926 | .917 | .906 | .893 | .878 | .860 | .838 | .813 | .783 |
| | | | | | | .748 | .707 | .660 | | | | | | | |
| 11.000 | 10.928 | .072 | .146 | .223 | 14 | .927 | .918 | .906 | .892 | .876 | .856 | .833 | .805 | .772 | .733 |
| | | | | | | .688 | .635 | .575 | .509 | | | | | | |
| 11.000 | 10.959 | .041 | .087 | .133 | 15 | .923 | .912 | .899 | .884 | .865 | .843 | .816 | .784 | .746 | .701 |
| | | | | | | .649 | .589 | .522 | .450 | .376 | | | | | |
| 11.000 | 10.977 | .023 | .051 | .079 | 16 | .920 | .909 | .895 | .878 | .858 | .834 | .805 | .771 | .729 | .681 |
| | | | | | | .624 | .560 | .490 | .415 | .339 | .266 | | | | |
| 11.000 | 10.987 | .013 | .030 | .046 | 17 | .919 | .907 | .892 | .875 | .854 | .829 | .799 | .762 | .719 | .668 |
| | | | | | | .609 | .543 | .471 | .395 | .318 | .246 | .181 | | | |
| 11.000 | 10.993 | .007 | .017 | .026 | 18 | .918 | .906 | .891 | .873 | .852 | .826 | .795 | .757 | .713 | .660 |
| | | | | | | .600 | .533 | .459 | .382 | .306 | .234 | .170 | .118 | | |
| 12.000 | 10.687 | 1.313 | .849 | 1.183 | 11 | .978 | .977 | .976 | .974 | .973 | .972 | .971 | .969 | .968 | .966 |
| | | | | | | .964 | | | | | | | | | |
| 12.000 | 11.354 | .646 | .629 | .921 | 12 | .965 | .963 | .960 | .957 | .954 | .950 | .947 | .942 | .938 | .932 |

JEWETT EQUIVALENT QUEUE EXTENDED ERLANG B(EQEEB): 10 MINUTES MAXIMUM WAIT (IF NOT CONNECTED AFTER 10 MINUTES WAIT, 100% OVERFLOW)

| HOURS FIRST ATTEMPT VMR = 1.0 | HOURS CON- NECT | HOURS OVFL | DELAYED PORTION | DELAY LENGTH CALL FACTOR | LINE | HOURS CONNECTED TRAFFIC ON LINES 1 | 2 | 3 | 4 | 5 | 6 | 7 | 8 | 9 | 10 |
|---|---|---|---|---|---|---|---|---|---|---|---|---|---|---|---|
| 12.000 | 11.705 | .295 | .404 | .607 | 13 | .950 | .946 | .940 | .934 | .927 | .919 | .910 | .900 | .888 | .874 |
| | | | | | | .858 | .839 | .818 | | | | | | | |
| 12.000 | 11.851 | .149 | .247 | .375 | 14 | .940 | .933 | .925 | .916 | .905 | .893 | .878 | .861 | .841 | .817 |
| | | | | | | .789 | .757 | .719 | .676 | | | | | | |
| 12.000 | 11.918 | .082 | .151 | .230 | 15 | .933 | .925 | .915 | .904 | .890 | .874 | .856 | .833 | .807 | .776 |
| | | | | | | .739 | .696 | .647 | .592 | .530 | | | | | |
| 12.000 | 11.953 | .047 | .092 | .140 | 16 | .929 | .920 | .909 | .896 | .881 | .863 | .841 | .815 | .784 | .748 |
| | | | | | | .706 | .656 | .600 | .537 | .469 | .398 | | | | |
| 12.000 | 11.973 | .027 | .055 | .085 | 17 | .927 | .917 | .905 | .891 | .875 | .855 | .831 | .803 | .770 | .730 |
| | | | | | | .684 | .630 | .570 | .503 | .432 | .359 | .288 | | | |
| 12.000 | 11.984 | .016 | .033 | .051 | 18 | .925 | .915 | .903 | .888 | .871 | .850 | .825 | .796 | .761 | .719 |
| | | | | | | .670 | .614 | .551 | .483 | .410 | .336 | .265 | .200 | | |
| 12.000 | 11.991 | .009 | .019 | .029 | 19 | .924 | .914 | .901 | .886 | .869 | .847 | .822 | .791 | .755 | .712 |
| | | | | | | .662 | .604 | .540 | .470 | .396 | .322 | .252 | .188 | .134 | |
| 13.000 | 11.665 | 1.335 | .829 | 1.162 | 12 | .978 | .977 | .976 | .975 | .974 | .973 | .972 | .970 | .969 | .967 |
| | | | | | | .966 | .964 | | | | | | | | |
| 13.000 | 12.327 | .673 | .618 | .906 | 13 | .967 | .965 | .962 | .960 | .957 | .954 | .951 | .947 | .943 | .938 |
| | | | | | | .933 | .928 | .921 | | | | | | | |
| 13.000 | 12.683 | .317 | .402 | .604 | 14 | .954 | .950 | .945 | .940 | .934 | .927 | .920 | .911 | .901 | .890 |
| | | | | | | .877 | .862 | .845 | .825 | | | | | | |
| 13.000 | 12.836 | .164 | .249 | .379 | 15 | .944 | .938 | .932 | .924 | .915 | .905 | .893 | .879 | .862 | .843 |
| | | | | | | .821 | .795 | .765 | .730 | .690 | | | | | |
| 13.000 | 12.908 | .092 | .155 | .236 | 16 | .938 | .931 | .923 | .913 | .902 | .889 | .873 | .855 | .834 | .809 |
| | | | | | | .779 | .744 | .704 | .658 | .606 | .548 | | | | |
| 13.000 | 12.947 | .053 | .096 | .147 | 17 | .935 | .927 | .917 | .906 | .894 | .879 | .861 | .839 | .814 | .785 |
| | | | | | | .750 | .710 | .663 | .610 | .551 | .487 | .410 | | | |
| 13.000 | 12.968 | .032 | .060 | .090 | 18 | .932 | .924 | .914 | .902 | .888 | .872 | .852 | .829 | .802 | .769 |
| | | | | | | .731 | .687 | .636 | .579 | .516 | .448 | .378 | .308 | | |
| 13.000 | 12.981 | .019 | .036 | .055 | 19 | .931 | .922 | .912 | .899 | .885 | .867 | .847 | .823 | .794 | .759 |
| | | | | | | .719 | .673 | .619 | .559 | .494 | .424 | .353 | .283 | .219 | |
| 13.000 | 12.989 | .011 | .021 | .033 | 20 | .930 | .921 | .910 | .898 | .883 | .865 | .844 | .818 | .788 | .753 |
| | | | | | | .711 | .663 | .608 | .547 | .480 | .409 | .338 | .269 | .205 | .150 |
| 14.000 | 11.792 | 2.208 | .965 | 1.292 | 12 | .985 | .985 | .984 | .984 | .983 | .983 | .982 | .982 | .981 | .981 |
| | | | | | | .980 | .979 | | | | | | | | |
| 14.000 | 12.642 | 1.358 | .811 | 1.143 | 13 | .979 | .978 | .977 | .976 | .975 | .974 | .973 | .972 | .970 | .969 |
| | | | | | | .967 | .966 | .964 | | | | | | | |
| 14.000 | 13.300 | .700 | .607 | .891 | 14 | .969 | .967 | .965 | .962 | .960 | .957 | .954 | .951 | .947 | .943 |
| | | | | | | .939 | .934 | .929 | .923 | | | | | | |
| 14.000 | 13.661 | .339 | .399 | .601 | 15 | .957 | .953 | .949 | .945 | .940 | .934 | .928 | .920 | .912 | .903 |

JEWETT EQUIVALENT QUEUE EXTENDED ERLANG B(EQEEB): 10 MINUTES MAXIMUM WAIT (IF NOT CONNECTED AFTER 10 MINUTES WAIT, 100% OVERFLOW)

| HOURS FIRST ATTEMPT VMR = 1.0 | HOURS CON-NECT | HOURS OVFL | DELAYED PORTION | DELAY LENGTH CALL FACTOR | LINE | HOURS CONNECTED TRAFFIC ON LINES 1 | 2 | 3 | 4 | 5 | 6 | 7 | 8 | 9 | 10 |
|---|---|---|---|---|---|---|---|---|---|---|---|---|---|---|---|
| | | | | | | .892 | .880 | .866 | .850 | .831 | | | | | |
| 14.000 | 13.822 | .178 | .252 | .382 | 16 | .948 | .943 | .937 | .931 | .923 | .914 | .904 | .893 | .879 | .864 |
| | | | | | | .846 | .825 | .800 | .772 | .740 | .702 | | | | |
| 14.000 | 13.899 | .101 | .158 | .242 | 17 | .943 | .936 | .929 | .921 | .912 | .901 | .888 | .873 | .855 | .834 |
| | | | | | | .810 | .782 | .749 | .712 | .668 | .619 | .565 | | | |
| 14.000 | 13.940 | .060 | .100 | .153 | 18 | .939 | .932 | .924 | .915 | .904 | .891 | .876 | .859 | .838 | .814 |
| | | | | | | .786 | .753 | .714 | .670 | .620 | .564 | .502 | .437 | | |
| 14.000 | 13.964 | .036 | .062 | .096 | 19 | .937 | .930 | .921 | .911 | .899 | .885 | .869 | .850 | .827 | .801 |
| | | | | | | .769 | .733 | .690 | .642 | .587 | .527 | .462 | .395 | .327 | |
| 14.000 | 13.978 | .022 | .039 | .059 | 20 | .936 | .928 | .919 | .908 | .896 | .881 | .864 | .844 | .820 | .792 |
| | | | | | | .758 | .720 | .675 | .624 | .567 | .504 | .437 | .368 | .301 | .237 |
| 14.000 | 13.987 | .013 | .023 | .036 | 21 | .935 | .927 | .918 | .907 | .894 | .879 | .861 | .840 | .815 | .786 |
| | | | | | | .751 | .711 | .665 | .612 | .553 | .489 | .421 | .352 | .284 | .221 |
| | | | | | | .165 | | | | | | | | | |
| 14.000 | 13.992 | .008 | .014 | .022 | 22 | .934 | .926 | .917 | .906 | .893 | .877 | .859 | .838 | .812 | .782 |
| | | | | | | .747 | .706 | .658 | .605 | .545 | .480 | .411 | .342 | .274 | .212 |
| | | | | | | .157 | .112 | | | | | | | | |
| 15.000 | 12.776 | 2.224 | .945 | 1.275 | 13 | .986 | .985 | .985 | .984 | .984 | .983 | .983 | .982 | .982 | .981 |
| | | | | | | .981 | .980 | .979 | | | | | | | |
| 15.000 | 13.621 | 1.379 | .795 | 1.124 | 14 | .980 | .979 | .978 | .977 | .976 | .975 | .974 | .973 | .972 | .970 |
| | | | | | | .969 | .967 | .966 | .964 | | | | | | |
| 15.000 | 14.274 | .726 | .597 | .878 | 15 | .970 | .968 | .966 | .964 | .962 | .960 | .957 | .954 | .951 | .948 |
| | | | | | | .944 | .940 | .935 | .930 | .924 | | | | | |
| 15.000 | 14.640 | .360 | .397 | .597 | 16 | .959 | .956 | .953 | .949 | .945 | .940 | .934 | .928 | .921 | .913 |
| | | | | | | .904 | .894 | .882 | .869 | .854 | .837 | | | | |
| 15.000 | 14.807 | .193 | .253 | .385 | 17 | .952 | .947 | .942 | .936 | .930 | .922 | .914 | .904 | .893 | .880 |
| | | | | | | .865 | .848 | .828 | .805 | .779 | .748 | .713 | | | |
| 15.000 | 14.889 | .111 | .161 | .246 | 18 | .946 | .941 | .935 | .928 | .920 | .910 | .899 | .887 | .872 | .855 |
| | | | | | | .835 | .812 | .785 | .754 | .719 | .678 | .632 | .580 | | |
| 15.000 | 14.933 | .067 | .103 | .158 | 19 | .943 | .937 | .930 | .922 | .913 | .902 | .889 | .875 | .858 | .838 |
| | | | | | | .814 | .787 | .755 | .718 | .676 | .628 | .575 | .517 | .455 | |
| 15.000 | 14.959 | .041 | .065 | .100 | 20 | .941 | .935 | .927 | .918 | .908 | .897 | .883 | .867 | .848 | .826 |
| | | | | | | .800 | .770 | .734 | .694 | .647 | .595 | .538 | .476 | .411 | .345 |
| 15.000 | 14.975 | .025 | .041 | .063 | 21 | .940 | .933 | .925 | .916 | .905 | .893 | .878 | .861 | .841 | .818 |
| | | | | | | .790 | .758 | .720 | .677 | .628 | .573 | .513 | .449 | .383 | .317 |
| | | | | | | .253 | | | | | | | | | |
| 15.000 | 14.985 | .015 | .026 | .039 | 22 | .939 | .932 | .924 | .914 | .903 | .891 | .876 | .858 | .837 | .813 |
| | | | | | | .784 | .750 | .711 | .667 | .616 | .559 | .498 | .432 | .365 | .299 |
| | | | | | | .237 | .180 | | | | | | | | |
| 15.000 | 14.991 | .009 | .016 | .024 | 23 | .938 | .931 | .923 | .913 | .902 | .889 | .874 | .856 | .834 | .809 |

JEWETT EQUIVALENT QUEUE EXTENDED ERLANG B(EQEEB): 10 MINUTES MAXIMUM WAIT (IF NOT CONNECTED AFTER 10 MINUTES WAIT, 100% OVERFLOW)

| HOURS FIRST ATTEMPT VMR = 1.0 | HOURS CON-NECT | HOURS OVFL | DELAYED PORTION | DELAY LENGTH CALL FACTOR | LINE | HOURS CONNECTED TRAFFIC ON LINES 1 | 2 | 3 | 4 | 5 | 6 | 7 | 8 | 9 | 10 |
|---|---|---|---|---|---|---|---|---|---|---|---|---|---|---|---|
| | | | | | | .780 | .745 | .705 | .660 | .608 | .550 | .487 | .421 | .354 | .288 |
| | | | | | | .226 | .171 | .124 | | | | | | | |
| 16.000 | 13.761 | 2.239 | .927 | 1.259 | 14 | .986 | .985 | .985 | .985 | .984 | .984 | .983 | .983 | .982 | .982 |
| | | | | | | .981 | .980 | .980 | .979 | | | | | | |
| 16.000 | 14.600 | 1.400 | .779 | 1.107 | 15 | .980 | .980 | .979 | .978 | .977 | .976 | .975 | .974 | .973 | .972 |
| | | | | | | .970 | .969 | .967 | .966 | .964 | | | | | |
| 16.000 | 15.250 | .750 | .588 | .865 | 16 | .971 | .970 | .968 | .966 | .964 | .962 | .960 | .957 | .954 | .951 |
| | | | | | | .948 | .944 | .940 | .936 | .931 | .925 | | | | |
| 16.000 | 15.619 | .381 | .395 | .594 | 17 | .962 | .959 | .956 | .952 | .949 | .944 | .940 | .934 | .928 | .922 |
| | | | | | | .914 | .905 | .896 | .885 | .872 | .858 | .842 | | | |
| 16.000 | 15.793 | .207 | .255 | .387 | 18 | .955 | .951 | .946 | .941 | .936 | .929 | .922 | .914 | .904 | .893 |
| | | | | | | .881 | .867 | .850 | .831 | .810 | .785 | .756 | .724 | | |
| 16.000 | 15.879 | .121 | .164 | .251 | 19 | .950 | .945 | .940 | .933 | .926 | .918 | .909 | .898 | .886 | .872 |
| | | | | | | .855 | .836 | .814 | .788 | .759 | .725 | .686 | .643 | .594 | |
| 16.000 | 15.926 | .074 | .106 | .163 | 20 | .947 | .941 | .935 | .928 | .920 | .911 | .900 | .888 | .873 | .857 |
| | | | | | | .837 | .814 | .788 | .757 | .722 | .682 | .637 | .586 | .530 | .471 |
| 16.000 | 15.954 | .046 | .068 | .105 | 21 | .945 | .939 | .932 | .925 | .916 | .906 | .894 | .881 | .865 | .846 |
| | | | | | | .825 | .799 | .770 | .736 | .697 | .652 | .603 | .548 | .488 | .426 |
| | | | | | | .362 | | | | | | | | | |
| 16.000 | 15.971 | .029 | .044 | .067 | 22 | .944 | .937 | .931 | .923 | .913 | .903 | .890 | .876 | .859 | .839 |
| | | | | | | .816 | .789 | .758 | .721 | .680 | .633 | .580 | .522 | .461 | .396 |
| | | | | | | .332 | .270 | | | | | | | | |
| 16.000 | 15.982 | .018 | .028 | .042 | 23 | .943 | .936 | .929 | .921 | .912 | .901 | .888 | .873 | .855 | .834 |
| | | | | | | .810 | .782 | .749 | .712 | .668 | .619 | .565 | .506 | .443 | .378 |
| | | | | | | .313 | .251 | .195 | | | | | | | |
| 16.000 | 15.989 | .011 | .017 | .026 | 24 | .942 | .936 | .929 | .920 | .910 | .800 | .000 | .870 | .852 | .831 |
| | | | | | | .806 | .778 | .744 | .705 | .661 | .611 | .555 | .495 | .431 | .366 |
| | | | | | | .301 | .240 | .185 | .137 | | | | | | |
| 17.000 | 14.746 | 2.254 | .910 | 1.243 | 15 | .986 | .986 | .985 | .985 | .985 | .984 | .984 | .983 | .983 | .982 |
| | | | | | | .982 | .981 | .980 | .980 | .979 | | | | | |
| 17.000 | 15.579 | 1.421 | .765 | 1.090 | 16 | .981 | .980 | .979 | .979 | .978 | .977 | .976 | .975 | .974 | .973 |
| | | | | | | .971 | .970 | .969 | .967 | .966 | .964 | | | | |
| 17.000 | 16.226 | .774 | .579 | .853 | 17 | .973 | .971 | .970 | .968 | .966 | .964 | .962 | .960 | .957 | .954 |
| | | | | | | .951 | .948 | .945 | .941 | .036 | .902 | .928 | | | |
| 17.000 | 10.599 | .401 | .393 | .591 | 18 | .964 | .961 | .959 | .955 | .952 | .948 | .944 | .940 | .934 | .929 |
| | | | | | | .922 | .915 | .907 | .897 | .887 | .875 | .862 | .847 | | |
| 17.000 | 16.779 | .221 | .256 | .389 | 19 | .957 | .954 | .950 | .945 | .940 | .935 | .929 | .921 | .913 | .904 |
| | | | | | | .894 | .882 | .868 | .852 | .834 | .814 | .790 | .763 | .733 | |
| 17.000 | 16.869 | .131 | .167 | .255 | 20 | .953 | .949 | .944 | .938 | .932 | .925 | .917 | .908 | .898 | .885 |
| | | | | | | .872 | .856 | .837 | .816 | .791 | .763 | .731 | .694 | .653 | .607 |

JEWETT EQUIVALENT QUEUE EXTENDED ERLANG B(EQEEB): 10 MINUTES MAXIMUM WAIT (IF NOT CONNECTED AFTER 10 MINUTES WAIT, 100% OVERFLOW)

| HOURS FIRST ATTEMPT VMR = 1.0 | HOURS CON-NECT | HOURS OVFL | DELAYED PORTION | DELAY LENGTH CALL FACTOR | LINE | HOURS CONNECTED TRAFFIC ON LINES 1 | 2 | 3 | 4 | 5 | 6 | 7 | 8 | 9 | 10 |
|---|---|---|---|---|---|---|---|---|---|---|---|---|---|---|---|
| 17.000 | 16.919 | .081 | .109 | .167 | 21 | .950 | .945 | .940 | .934 | .927 | .918 | .909 | .899 | .886 | .872 |
| | | | | | | .856 | .837 | .815 | .789 | .760 | .726 | .688 | .644 | .596 | .543 |
| | | | | | | .486 | | | | | | | | | |
| 17.000 | 16.949 | .051 | .071 | .109 | 22 | .948 | .943 | .937 | .930 | .923 | .914 | .904 | .892 | .879 | .863 |
| | | | | | | .845 | .824 | .799 | .770 | .738 | .700 | .657 | .610 | .557 | .500 |
| | | | | | | .440 | .378 | | | | | | | | |
| 17.000 | 16.968 | .032 | .046 | .071 | 23 | .947 | .941 | .935 | .928 | .920 | .911 | .900 | .888 | .873 | .857 |
| | | | | | | .837 | .814 | .788 | .757 | .722 | .682 | .637 | .586 | .531 | .471 |
| | | | | | | .409 | .346 | .285 | | | | | | | |
| 17.000 | 16.980 | .020 | .030 | .045 | 24 | .946 | .940 | .934 | .927 | .919 | .909 | .898 | .885 | .870 | .852 |
| | | | | | | .832 | .808 | .781 | .749 | .712 | .670 | .623 | .570 | .513 | .452 |
| | | | | | | .389 | .326 | .265 | .209 | | | | | | |
| 17.000 | 16.987 | .013 | .019 | .029 | 25 | .945 | .940 | .933 | .926 | .917 | .908 | .896 | .883 | .867 | .850 |
| | | | | | | .829 | .804 | .776 | .743 | .705 | .662 | .613 | .560 | .502 | .440 |
| | | | | | | .376 | .313 | .253 | .198 | .149 | | | | | |
| 18.000 | 15.730 | 2.270 | .893 | 1.228 | 16 | .986 | .986 | .986 | .985 | .985 | .984 | .984 | .984 | .983 | .983 |
| | | | | | | .982 | .981 | .981 | .980 | .980 | .979 | | | | |
| 18.000 | 16.559 | 1.441 | .752 | 1.075 | 17 | .981 | .981 | .980 | .979 | .978 | .978 | .977 | .976 | .975 | .974 |
| | | | | | | .973 | .971 | .970 | .969 | .967 | .966 | .964 | | | |
| 18.000 | 17.202 | .798 | .571 | .842 | 18 | .974 | .972 | .971 | .970 | .968 | .966 | .964 | .962 | .960 | .957 |
| | | | | | | .955 | .952 | .948 | .945 | .941 | .937 | .932 | .927 | | |
| 18.000 | 17.579 | .421 | .390 | .587 | 19 | .966 | .963 | .961 | .958 | .955 | .952 | .948 | .944 | .940 | .935 |
| | | | | | | .929 | .923 | .916 | .908 | .899 | .889 | .878 | .865 | .851 | |
| 18.000 | 17.765 | .235 | .257 | .390 | 20 | .959 | .956 | .953 | .949 | .945 | .940 | .934 | .928 | .921 | .913 |
| | | | | | | .904 | .894 | .883 | .869 | .854 | .837 | .818 | .795 | .770 | .741 |
| 18.000 | 17.859 | .141 | .169 | .258 | 21 | .955 | .952 | .947 | .942 | .937 | .931 | .924 | .916 | .907 | .897 |
| | | | | | | .885 | .872 | .856 | .838 | .818 | .794 | .767 | .737 | .702 | .663 |
| | | | | | | .619 | | | | | | | | | |
| 18.000 | 17.912 | .088 | .112 | .171 | 22 | .953 | .948 | .944 | .938 | .932 | .925 | .917 | .908 | .897 | .885 |
| | | | | | | .871 | .855 | .837 | .815 | .791 | .762 | .730 | .693 | .652 | .605 |
| | | | | | | .555 | .500 | | | | | | | | |
| 18.000 | 17.944 | .056 | .074 | .113 | 23 | .951 | .946 | .941 | .935 | .929 | .921 | .912 | .902 | .890 | .877 |
| | | | | | | .861 | .843 | .823 | .799 | .771 | .739 | .703 | .662 | .616 | .566 |
| | | | | | | .511 | .453 | .393 | | | | | | | |
| 18.000 | 17.964 | .036 | .048 | .074 | 24 | .950 | .945 | .939 | .933 | .926 | .918 | .909 | .898 | .886 | .871 |
| | | | | | | .855 | .835 | .813 | .787 | .758 | .723 | .685 | .641 | .592 | .539 |
| | | | | | | .481 | .421 | .360 | .300 | | | | | | |
| 18.000 | 17.977 | .023 | .031 | .048 | 25 | .949 | .944 | .938 | .932 | .925 | .916 | .906 | .895 | .882 | .867 |
| | | | | | | .850 | .830 | .806 | .779 | .748 | .712 | .672 | .626 | .575 | .520 |
| | | | | | | .461 | .400 | .339 | .279 | .223 | | | | | |

JEWETT EQUIVALENT QUEUE EXTENDED ERLANG B(EQEEB): 10 MINUTES MAXIMUM WAIT (IF NOT CONNECTED AFTER 10 MINUTES WAIT, 100% OVERFLOW)

| HOURS FIRST ATTEMPT VMR = 1.0 | HOURS CON-NECT | HOURS OVFL | DELAYED PORTION | DELAY LENGTH CALL FACTOR | LINE | HOURS CONNECTED TRAFFIC ON LINES 1 | 2 | 3 | 4 | 5 | 6 | 7 | 8 | 9 | 10 |
|---|---|---|---|---|---|---|---|---|---|---|---|---|---|---|---|
| 18.000 | 17.986 | .014 | .020 | .031 | 26 | .948 | .943 | .938 | .931 | .923 | .915 | .905 | .893 | .880 | .865 |
| | | | | | | .847 | .826 | .802 | .774 | .742 | .705 | .663 | .616 | .564 | .508 |
| | | | | | | .448 | .387 | .325 | .265 | .210 | .161 | | | | |
| 19.000 | 16.715 | 2.285 | .878 | 1.213 | 17 | .987 | .986 | .986 | .986 | .985 | .985 | .984 | .984 | .983 | .983 |
| | | | | | | .982 | .982 | .981 | .981 | .980 | .980 | .979 | | | |
| 19.000 | 17.539 | 1.461 | .739 | 1.060 | 18 | .982 | .981 | .981 | .980 | .979 | .978 | .977 | .977 | .976 | .975 |
| | | | | | | .974 | .972 | .971 | .970 | .969 | .967 | .966 | .964 | | |
| 19.000 | 18.179 | .821 | .563 | .832 | 19 | .975 | .974 | .972 | .971 | .969 | .968 | .966 | .964 | .962 | .960 |
| | | | | | | .957 | .955 | .952 | .949 | .945 | .942 | .938 | .933 | .928 | |
| 19.000 | 18.560 | .440 | .388 | .584 | 20 | .967 | .965 | .963 | .961 | .958 | .955 | .952 | .948 | .944 | .940 |
| | | | | | | .935 | .929 | .923 | .916 | .909 | .900 | .891 | .880 | .868 | .855 |
| 19.000 | 18.751 | .249 | .258 | .391 | 21 | .962 | .959 | .956 | .952 | .948 | .944 | .939 | .934 | .928 | .921 |
| | | | | | | .913 | .904 | .895 | .883 | .871 | .856 | .840 | .821 | .800 | .776 |
| | | | | | | .749 | | | | | | | | | |
| 19.000 | 18.849 | .151 | .171 | .261 | 22 | .958 | .954 | .950 | .946 | .941 | .936 | .930 | .923 | .915 | .906 |
| | | | | | | .896 | .885 | .872 | .857 | .839 | .820 | .797 | .771 | .742 | .709 |
| | | | | | | .671 | .630 | | | | | | | | |
| 19.000 | 18.905 | .095 | .114 | .175 | 23 | .955 | .951 | .947 | .942 | .937 | .931 | .924 | .916 | .906 | .896 |
| | | | | | | .884 | .870 | .855 | .837 | .816 | .792 | .765 | .733 | .698 | .658 |
| | | | | | | .614 | .566 | .513 | | | | | | | |
| 19.000 | 18.939 | .061 | .076 | .117 | 24 | .953 | .949 | .945 | .939 | .933 | .927 | .919 | .910 | .900 | .889 |
| | | | | | | .875 | .860 | .842 | .822 | .799 | .772 | .741 | .706 | .667 | .623 |
| | | | | | | .574 | .521 | .465 | .407 | | | | | | |
| 19.000 | 18.961 | .039 | .050 | .077 | 25 | .952 | .948 | .943 | .938 | .931 | .924 | .916 | .907 | .896 | .884 |
| | | | | | | .869 | .853 | .834 | .812 | .787 | .758 | .725 | .687 | .645 | .598 |
| | | | | | | .546 | .491 | .433 | .373 | .314 | | | | | |
| 19.000 | 18.975 | .025 | .033 | .051 | 26 | .952 | .947 | .942 | .936 | .930 | .922 | .914 | .904 | .893 | .880 |
| | | | | | | .865 | .848 | .828 | .805 | .778 | .748 | .713 | .674 | .629 | .580 |
| | | | | | | .527 | .470 | .411 | .351 | .292 | .236 | | | | |
| 19.000 | 18.984 | .016 | .022 | .033 | 27 | .951 | .946 | .941 | .935 | .929 | .921 | .912 | .902 | .891 | .878 |
| | | | | | | .862 | .844 | .824 | .800 | .773 | .741 | .705 | .664 | .619 | .569 |
| | | | | | | .514 | .456 | .396 | .336 | .277 | .222 | .173 | | | |
| 19.000 | 18.990 | .010 | .014 | .021 | 28 | .951 | .946 | .941 | .935 | .928 | .920 | .911 | .901 | .889 | .876 |
| | | | | | | .860 | .842 | .821 | .797 | .769 | .737 | .700 | .658 | .612 | .561 |
| | | | | | | .506 | .447 | .387 | .326 | .268 | .214 | .165 | .123 | | |
| 20.000 | 16.795 | 3.205 | .971 | 1.297 | 17 | .990 | .990 | .989 | .989 | .989 | .989 | .988 | .988 | .988 | .988 |
| | | | | | | .987 | .987 | .987 | .987 | .986 | .986 | .986 | | | |
| 20.000 | 17.701 | 2.299 | .864 | 1.199 | 18 | .987 | .987 | .986 | .986 | .985 | .985 | .985 | .984 | .984 | .983 |
| | | | | | | .983 | .982 | .982 | .981 | .981 | .980 | .970 | .979 | | |
| 20.000 | 18.520 | 1.480 | .727 | 1.045 | 19 | .982 | .902 | .981 | .980 | .980 | .979 | .978 | .977 | .976 | .975 |

JEWETT EQUIVALENT QUEUE EXTENDED ERLANG B(EQEEB): 10 MINUTES MAXIMUM WAIT (IF NOT CONNECTED AFTER 10 MINUTES WAIT, 100% OVERFLOW)

| HOURS FIRST ATTEMPT VMR = 1.0 | HOURS CON-NECT | HOURS OVFL | DELAYED PORTION | DELAY LENGTH CALL FACTOR | LINE | HOURS CONNECTED TRAFFIC ON LINES 1 | 2 | 3 | 4 | 5 | 6 | 7 | 8 | 9 | 10 |
|---|---|---|---|---|---|---|---|---|---|---|---|---|---|---|---|
| | | | | | | .974 | .973 | .972 | .971 | .970 | .969 | .967 | .966 | .964 | |
| 20.000 | 19.157 | .843 | .555 | .822 | 20 | .976 | .975 | .973 | .972 | .971 | .969 | .968 | .966 | .964 | .962 |
| | | | | | | .960 | .957 | .955 | .952 | .949 | .946 | .942 | .938 | .934 | .929 |
| 20.000 | 19.541 | .459 | .386 | .580 | 21 | .969 | .967 | .965 | .963 | .960 | .958 | .955 | .951 | .948 | .944 |
| | | | | | | .940 | .935 | .930 | .924 | .917 | .910 | .902 | .893 | .882 | .871 |
| | | | | | | .858 | | | | | | | | | |
| 20.000 | 19.737 | .263 | .258 | .392 | 22 | .963 | .961 | .958 | .955 | .951 | .948 | .943 | .939 | .933 | .927 |
| | | | | | | .921 | .913 | .905 | .895 | .884 | .872 | .858 | .843 | .825 | .805 |
| | | | | | | .782 | .756 | | | | | | | | |
| 20.000 | 19.839 | .161 | .173 | .264 | 23 | .960 | .957 | .953 | .949 | .945 | .940 | .935 | .929 | .922 | .914 |
| | | | | | | .906 | .896 | .885 | .872 | .857 | .840 | .821 | .800 | .775 | .747 |
| | | | | | | .715 | .680 | .640 | | | | | | | |
| 20.000 | 19.898 | .102 | .116 | .178 | 24 | .957 | .954 | .950 | .946 | .941 | .935 | .929 | .922 | .914 | .905 |
| | | | | | | .895 | .883 | .870 | .854 | .837 | .816 | .793 | .767 | .737 | .703 |
| | | | | | | .665 | .623 | .576 | .526 | | | | | | |
| 20.000 | 19.934 | .066 | .078 | .120 | 25 | .956 | .952 | .948 | .943 | .938 | .932 | .925 | .917 | .909 | .899 |
| | | | | | | .887 | .874 | .859 | .842 | .822 | .799 | .773 | .743 | .709 | .671 |
| | | | | | | .629 | .582 | .531 | .477 | .420 | | | | | |
| 20.000 | 19.957 | .043 | .053 | .081 | 26 | .955 | .951 | .946 | .941 | .936 | .929 | .922 | .914 | .905 | .894 |
| | | | | | | .882 | .867 | .851 | .833 | .811 | .786 | .758 | .726 | .689 | .648 |
| | | | | | | .603 | .553 | .500 | .443 | .385 | .327 | | | | |
| 20.000 | 19.972 | .028 | .035 | .054 | 27 | .954 | .950 | .945 | .940 | .934 | .928 | .920 | .912 | .902 | .891 |
| | | | | | | .878 | .863 | .846 | .826 | .804 | .777 | .748 | .714 | .675 | .633 |
| | | | | | | .585 | .534 | .478 | .421 | .362 | .304 | .249 | | | |
| 20.000 | 19.982 | .018 | .023 | .035 | 28 | .953 | .949 | .945 | .939 | .933 | .927 | .919 | .910 | .900 | .888 |
| | | | | | | .875 | .860 | .842 | .822 | .798 | .771 | .741 | .705 | .666 | .622 |
| | | | | | | .573 | .520 | .464 | .405 | .347 | .289 | .234 | .184 | | |
| 20.000 | 19.988 | .012 | .015 | .023 | 29 | .953 | .949 | .944 | .939 | .933 | .926 | .918 | .909 | .899 | .887 |
| | | | | | | .873 | .858 | .840 | .819 | .795 | .767 | .736 | .700 | .659 | .614 |
| | | | | | | .565 | .511 | .454 | .395 | .336 | .279 | .225 | .176 | .133 | |
| 22.000 | 18.773 | 3.227 | .942 | 1.273 | 19 | .990 | .990 | .990 | .989 | .989 | .989 | .989 | .989 | .988 | .988 |
| | | | | | | .988 | .988 | .987 | .987 | .987 | .986 | .986 | .986 | .985 | |
| 22.000 | 19.672 | 2.328 | .838 | 1.172 | 20 | .987 | .987 | .987 | .986 | .986 | .986 | .985 | .985 | .984 | .984 |
| | | | | | | .984 | .983 | .983 | .982 | .982 | .981 | .980 | .980 | .979 | .979 |
| 22.000 | 20.482 | 1.518 | .706 | 1.019 | 21 | .983 | .983 | .982 | .981 | .981 | .980 | .979 | .979 | .978 | .977 |
| | | | | | | .976 | .975 | .974 | .973 | .972 | .971 | .970 | .968 | .967 | .966 |
| | | | | | | .964 | | | | | | | | | |
| 22.000 | 21.115 | .885 | .542 | .803 | 22 | .977 | .976 | .975 | .974 | .973 | .972 | .970 | .969 | .967 | .966 |
| | | | | | | .964 | .962 | .960 | .957 | .955 | .952 | .949 | .946 | .943 | .939 |
| | | | | | | .935 | .931 | | | | | | | | |

JEWETT EQUIVALENT QUEUE EXTENDED ERLANG B(EQEEB): 10 MINUTES MAXIMUM WAIT (IF NOT CONNECTED AFTER 10 MINUTES WAIT, 100% OVERFLOW)

| HOURS FIRST ATTEMPT VMR = 1.0 | HOURS CON-NECT | HOURS OVFL | DELAYED PORTION | DELAY LENGTH CALL FACTOR | LINE | HOURS CONNECTED TRAFFIC ON LINES 1 | 2 | 3 | 4 | 5 | 6 | 7 | 8 | 9 | 10 |
|---|---|---|---|---|---|---|---|---|---|---|---|---|---|---|---|
| 22.000 | 21.504 | .496 | .381 | .574 | 23 | .971 | .970 | .968 | .966 | .964 | .962 | .960 | .957 | .954 | .951 |
| | | | | | | .948 | .944 | .940 | .935 | .930 | .925 | .919 | .912 | .904 | .896 |
| | | | | | | .887 | .876 | .864 | | | | | | | |
| 22.000 | 21.709 | .291 | .259 | .394 | 24 | .967 | .965 | .962 | .960 | .957 | .954 | .950 | .947 | .942 | .938 |
| | | | | | | .933 | .927 | .920 | .913 | .905 | .896 | .886 | .875 | .862 | .847 |
| | | | | | | .831 | .813 | .792 | .768 | | | | | | |
| 22.000 | 21.819 | .181 | .176 | .269 | 25 | .963 | .961 | .958 | .955 | .951 | .948 | .943 | .939 | .933 | .927 |
| | | | | | | .921 | .913 | .905 | .895 | .884 | .872 | .858 | .843 | .825 | .805 |
| | | | | | | .782 | .756 | .727 | .695 | .659 | | | | | |
| 22.000 | 21.884 | .116 | .120 | .184 | 26 | .961 | .958 | .955 | .952 | .948 | .943 | .938 | .933 | .927 | .920 |
| | | | | | | .912 | .903 | .893 | .882 | .869 | .854 | .837 | .818 | .796 | .771 |
| | | | | | | .743 | .712 | .677 | .638 | .595 | .548 | | | | |
| 22.000 | 21.923 | .077 | .082 | .126 | 27 | .960 | .957 | .953 | .949 | .945 | .940 | .935 | .929 | .922 | .914 |
| | | | | | | .906 | .896 | .884 | .872 | .857 | .840 | .821 | .799 | .775 | .747 |
| | | | | | | .715 | .679 | .640 | .596 | .549 | .498 | .445 | | | |
| 22.000 | 21.949 | .051 | .056 | .086 | 28 | .959 | .956 | .952 | .948 | .943 | .938 | .933 | .926 | .919 | .911 |
| | | | | | | .901 | .890 | .878 | .864 | .849 | .830 | .810 | .786 | .759 | .728 |
| | | | | | | .694 | .656 | .613 | .566 | .516 | .463 | .408 | .352 | | |
| 22.000 | 21.966 | .034 | .038 | .059 | 29 | .958 | .955 | .951 | .947 | .942 | .937 | .931 | .924 | .916 | .908 |
| | | | | | | .898 | .887 | .874 | .859 | .842 | .823 | .801 | .776 | .748 | .715 |
| | | | | | | .679 | .639 | .594 | .546 | .494 | .439 | .383 | .327 | .273 | |
| 22.000 | 21.977 | .023 | .026 | .039 | 30 | .958 | .954 | .950 | .946 | .941 | .936 | .930 | .923 | .915 | .906 |
| | | | | | | .896 | .884 | .871 | .856 | .838 | .818 | .795 | .769 | .740 | .706 |
| | | | | | | .669 | .627 | .581 | .531 | .478 | .422 | .366 | .310 | .256 | .207 |
| 22.000 | 21.985 | .015 | .017 | .026 | 31 | .957 | .954 | .950 | .945 | .941 | .935 | .929 | .922 | .914 | .905 |
| | | | | | | .894 | .882 | .869 | .853 | .835 | .815 | .791 | .765 | .734 | .700 |
| | | | | | | .661 | .618 | .572 | .521 | .407 | .411 | .354 | .299 | .245 | .196 |
| | | | | | | .153 | | | | | | | | | |
| 24.000 | 20.751 | 3.249 | .916 | 1.249 | 21 | .990 | .990 | .990 | .990 | .990 | .989 | .989 | .989 | .989 | .988 |
| | | | | | | .988 | .988 | .988 | .987 | .987 | .987 | .987 | .986 | .986 | .986 |
| | | | | | | .985 | | | | | | | | | |
| 24.000 | 21.643 | 2.357 | .815 | 1.147 | 22 | .988 | .987 | .987 | .987 | .987 | .986 | .986 | .985 | .985 | .985 |
| | | | | | | .984 | .984 | .983 | .083 | .982 | .982 | .981 | .981 | .980 | .980 |
| | | | | | | .979 | .978 | | | | | | | | |
| 24.000 | 22.446 | 1.554 | .686 | .994 | 23 | .984 | .083 | .903 | .982 | .982 | .981 | .981 | .980 | .979 | .978 |
| | | | | | | .978 | .977 | .976 | .975 | .974 | .973 | .972 | .971 | .970 | .968 |
| | | | | | | .967 | .966 | .964 | | | | | | | |
| 24.000 | 23.074 | .926 | .530 | .786 | 24 | .979 | .978 | .977 | .976 | .975 | .974 | .973 | .971 | .970 | .969 |
| | | | | | | .967 | .965 | .964 | .962 | .960 | .957 | .955 | .953 | .950 | .947 |
| | | | | | | .943 | .940 | .936 | .032 | | | | | | |

JEWETT EQUIVALENT QUEUE EXTENDED ERLANG B(EQEEB): 10 MINUTES MAXIMUM WAIT (IF NOT CONNECTED AFTER 10 MINUTES WAIT, 100% OVERFLOW)

| HOURS FIRST ATTEMPT VMR = 1.0 | HOURS CON-NECT | HOURS OVFL | DELAYED PORTION | DELAY LENGTH CALL FACTOR | LINE | HOURS CONNECTED TRAFFIC ON LINES 1 | 2 | 3 | 4 | 5 | 6 | 7 | 8 | 9 | 10 |
|---|---|---|---|---|---|---|---|---|---|---|---|---|---|---|---|
| 24.000 | 23.468 | .532 | .377 | .567 | 25 | .974 | .972 | .971 | .969 | .967 | .966 | .964 | .961 | .959 | .957 |
| | | | | | | .954 | .951 | .948 | .944 | .940 | .936 | .931 | .926 | .920 | .914 |
| | | | | | | .907 | .899 | .890 | .880 | .870 | | | | | |
| 24.000 | 23.682 | .318 | .260 | .394 | 26 | .969 | .968 | .966 | .963 | .961 | .959 | .956 | .953 | .949 | .946 |
| | | | | | | .942 | .937 | .932 | .927 | .920 | .914 | .906 | .897 | .888 | .877 |
| | | | | | | .865 | .852 | .837 | .820 | .801 | .779 | | | | |
| 24.000 | 23.799 | .201 | .179 | .273 | 27 | .967 | .964 | .962 | .959 | .957 | .953 | .950 | .946 | .942 | .937 |
| | | | | | | .932 | .926 | .920 | .912 | .904 | .895 | .885 | .873 | .860 | .845 |
| | | | | | | .828 | .810 | .788 | .764 | .738 | .708 | .675 | | | |
| 24.000 | 23.869 | .131 | .124 | .190 | 28 | .965 | .962 | .959 | .956 | .953 | .950 | .946 | .941 | .936 | .931 |
| | | | | | | .925 | .918 | .910 | .901 | .892 | .881 | .868 | .854 | .838 | .820 |
| | | | | | | .799 | .776 | .750 | .720 | .688 | .651 | .611 | .568 | | |
| 24.000 | 23.912 | .088 | .086 | .132 | 29 | .963 | .961 | .958 | .954 | .951 | .947 | .943 | .938 | .932 | .926 |
| | | | | | | .919 | .912 | .903 | .893 | .882 | .870 | .856 | .839 | .821 | .800 |
| | | | | | | .777 | .750 | .720 | .687 | .650 | .609 | .565 | .517 | .467 | |
| 24.000 | 23.941 | .059 | .060 | .092 | 30 | .962 | .959 | .956 | .953 | .949 | .945 | .941 | .935 | .930 | .923 |
| | | | | | | .916 | .907 | .898 | .888 | .875 | .862 | .846 | .829 | .809 | .786 |
| | | | | | | .760 | .731 | .699 | .662 | .622 | .579 | .531 | .481 | .429 | .376 |
| 24.000 | 23.960 | .040 | .041 | .063 | 31 | .962 | .959 | .956 | .952 | .948 | .944 | .939 | .934 | .928 | .921 |
| | | | | | | .913 | .904 | .894 | .883 | .871 | .856 | .840 | .821 | .800 | .775 |
| | | | | | | .748 | .717 | .683 | .645 | .602 | .557 | .507 | .456 | .402 | .349 |
| | | | | | | .296 | | | | | | | | | |
| 24.000 | 23.973 | .027 | .028 | .043 | 32 | .961 | .958 | .955 | .951 | .947 | .943 | .938 | .932 | .926 | .919 |
| | | | | | | .911 | .902 | .892 | .880 | .867 | .852 | .835 | .815 | .793 | .768 |
| | | | | | | .739 | .707 | .672 | .632 | .588 | .541 | .491 | .438 | .384 | .330 |
| | | | | | | .277 | .228 | | | | | | | | |
| 24.000 | 23.982 | .018 | .019 | .029 | 33 | .961 | .958 | .955 | .951 | .947 | .942 | .937 | .931 | .925 | .918 |
| | | | | | | .910 | .900 | .890 | .878 | .864 | .849 | .831 | .811 | .788 | .763 |
| | | | | | | .733 | .700 | .663 | .623 | .578 | .530 | .479 | .425 | .371 | .317 |
| | | | | | | .265 | .216 | .172 | | | | | | | |
| 26.000 | 21.797 | 4.203 | .973 | 1.299 | 22 | .992 | .992 | .992 | .992 | .992 | .992 | .991 | .991 | .991 | .991 |
| | | | | | | .991 | .991 | .991 | .990 | .990 | .990 | .990 | .990 | .990 | .989 |
| | | | | | | .989 | .989 | | | | | | | | |
| 26.000 | 22.729 | 3.271 | .893 | 1.227 | 23 | .991 | .990 | .990 | .990 | .990 | .990 | .989 | .989 | .989 | .989 |
| | | | | | | .989 | .988 | .988 | .988 | .988 | .987 | .987 | .987 | .986 | .986 |
| | | | | | | .986 | .985 | .985 | | | | | | | |
| 26.000 | 23.615 | 2.385 | .794 | 1.124 | 24 | .988 | .988 | .988 | .987 | .987 | .987 | .986 | .986 | .986 | .985 |
| | | | | | | .985 | .984 | .984 | .984 | .983 | .983 | .982 | .982 | .981 | .981 |
| | | | | | | .980 | .979 | .979 | .978 | | | | | | |
| 26.000 | 24.410 | 1.590 | .668 | .972 | 25 | .985 | .984 | .984 | .983 | .983 | .982 | .982 | .981 | .980 | .980 |

JEWETT EQUIVALENT QUEUE EXTENDED ERLANG B(EQEEB): 10 MINUTES MAXIMUM WAIT (IF NOT CONNECTED AFTER 10 MINUTES WAIT, 100% OVERFLOW)

| HOURS FIRST ATTEMPT VMR = 1.0 | HOURS CON-NECT | HOURS OVFL | DELAYED PORTION | DELAY LENGTH CALL FACTOR | LINE | HOURS CONNECTED TRAFFIC ON LINES 1 | 2 | 3 | 4 | 5 | 6 | 7 | 8 | 9 | 10 |
|---|---|---|---|---|---|---|---|---|---|---|---|---|---|---|---|
| | | | | | | .979 | .978 | .977 | .977 | .976 | .975 | .974 | .973 | .972 | .971 |
| | | | | | | .970 | .968 | .967 | .966 | .964 | | | | | |
| 26.000 | 25.035 | .965 | .518 | .770 | 26 | .980 | .979 | .979 | .978 | .977 | .976 | .975 | .974 | .972 | .971 |
| | | | | | | .970 | .968 | .967 | .965 | .964 | .962 | .960 | .958 | .955 | .953 |
| | | | | | | .950 | .947 | .944 | .941 | .937 | .933 | | | | |
| 26.000 | 25.433 | .567 | .372 | .561 | 27 | .975 | .974 | .973 | .972 | .970 | .969 | .967 | .965 | .963 | .961 |
| | | | | | | .959 | .956 | .954 | .951 | .948 | .944 | .940 | .936 | .932 | .927 |
| | | | | | | .921 | .915 | .909 | .901 | .893 | .884 | .874 | | | |
| 26.000 | 25.655 | .345 | .260 | .394 | 28 | .972 | .970 | .968 | .967 | .965 | .963 | .960 | .958 | .955 | .952 |
| | | | | | | .949 | .945 | .941 | .937 | .932 | .926 | .920 | .914 | .907 | .898 |
| | | | | | | .889 | .879 | .868 | .856 | .842 | .826 | .808 | .789 | | |
| 26.000 | 25.780 | .220 | .181 | .276 | 29 | .969 | .967 | .965 | .963 | .961 | .958 | .955 | .952 | .949 | .945 |
| | | | | | | .941 | .936 | .931 | .925 | .919 | .912 | .904 | .895 | .885 | .874 |
| | | | | | | .861 | .847 | .832 | .814 | .794 | .772 | .747 | .720 | .689 | |
| 26.000 | 25.854 | .146 | .127 | .194 | 30 | .967 | .965 | .963 | .960 | .958 | .955 | .951 | .948 | .944 | .939 |
| | | | | | | .935 | .929 | .923 | .916 | .909 | .900 | .891 | .880 | .868 | .854 |
| | | | | | | .839 | .821 | .802 | .780 | .755 | .728 | .697 | .664 | .627 | .586 |
| 26.000 | 25.901 | .099 | .089 | .137 | 31 | .966 | .964 | .961 | .959 | .956 | .952 | .949 | .945 | .940 | .936 |
| | | | | | | .930 | .924 | .917 | .910 | .901 | .891 | .881 | .868 | .855 | .839 |
| | | | | | | .821 | .801 | .779 | .754 | .726 | .694 | .659 | .621 | .579 | .535 |
| | | | | | | .487 | | | | | | | | | |
| 26.000 | 25.932 | .068 | .063 | .096 | 32 | .965 | .963 | .960 | .957 | .954 | .951 | .947 | .943 | .938 | .933 |
| | | | | | | .927 | .920 | .913 | .905 | .895 | .885 | .873 | .860 | .845 | .828 |
| | | | | | | .808 | .786 | .762 | .734 | .703 | .669 | .631 | .590 | .545 | .498 |
| | | | | | | .448 | .397 | | | | | | | | |
| 26.000 | 25.953 | .047 | .044 | .067 | 33 | .964 | .962 | .959 | .956 | .953 | .950 | .946 | .941 | .936 | .931 |
| | | | | | | .925 | .918 | .910 | .901 | .891 | .880 | .868 | .853 | .837 | .819 |
| | | | | | | .708 | .775 | .740 | .719 | .687 | .650 | .610 | .567 | .520 | .471 |
| | | | | | | .420 | .368 | .317 | | | | | | | |
| 26.000 | 25.968 | .032 | .031 | .047 | 34 | .964 | .962 | .959 | .956 | .952 | .949 | .945 | .940 | .935 | .929 |
| | | | | | | .923 | .916 | .908 | .899 | .888 | .877 | .864 | .849 | .832 | .813 |
| | | | | | | .791 | .767 | .739 | .709 | .674 | .637 | .595 | .550 | .502 | .452 |
| | | | | | | .400 | .348 | .297 | .248 | | | | | | |
| 26.000 | 25.978 | .022 | .021 | .032 | 35 | .964 | .961 | .958 | .955 | .952 | .948 | .914 | .939 | .934 | .928 |
| | | | | | | .922 | .914 | .906 | .097 | .886 | .874 | .861 | .845 | .828 | .808 |
| | | | | | | .786 | .761 | .733 | .701 | .666 | .627 | .584 | .538 | .490 | .439 |
| | | | | | | .386 | .334 | .280 | .235 | .190 | | | | | |
| 26.000 | 25.985 | .015 | .014 | .022 | 36 | .963 | .961 | .958 | .955 | .951 | .948 | .943 | .939 | .933 | .927 |
| | | | | | | .921 | .913 | .905 | .895 | .885 | .872 | .859 | .843 | .825 | .805 |
| | | | | | | .782 | .757 | .728 | .695 | .659 | .620 | .577 | .530 | .481 | .429 |

JEWETT EQUIVALENT QUEUE EXTENDED ERLANG B(EQEEB): 10 MINUTES MAXIMUM WAIT (IF NOT CONNECTED AFTER 10 MINUTES WAIT, 100% OVERFLOW)

| HOURS FIRST ATTEMPT VMR = 1.0 | HOURS CON- NECT | HOURS OVFL | DELAYED PORTION | DELAY LENGTH CALL FACTOR | LINE | HOURS CONNECTED TRAFFIC ON LINES 1 | 2 | 3 | 4 | 5 | 6 | 7 | 8 | 9 | 10 |
|---|---|---|---|---|---|---|---|---|---|---|---|---|---|---|---|
| | | | | | | .377 | .324 | .274 | .226 | .182 | .143 | | | | |
| 28.000 | 23.780 | 4.220 | .951 | 1.280 | 24 | .992 | .992 | .992 | .992 | .992 | .992 | .992 | .991 | .991 | .991 |
| | | | | | | .991 | .991 | .991 | .991 | .990 | .990 | .990 | .990 | .990 | .990 |
| | | | | | | .989 | .989 | .989 | .989 | | | | | | |
| 28.000 | 24.707 | 3.293 | .871 | 1.207 | 25 | .991 | .991 | .990 | .990 | .990 | .990 | .990 | .989 | .989 | .989 |
| | | | | | | .989 | .989 | .988 | .988 | .988 | .988 | .987 | .987 | .987 | .987 |
| | | | | | | .986 | .986 | .986 | .985 | .985 | | | | | |
| 28.000 | 25.588 | 2.412 | .774 | 1.101 | 26 | .989 | .988 | .988 | .988 | .987 | .987 | .987 | .986 | .986 | .986 |
| | | | | | | .985 | .985 | .985 | .984 | .984 | .983 | .983 | .983 | .982 | .982 |
| | | | | | | .981 | .980 | .980 | .979 | .979 | .978 | | | | |
| 28.000 | 26.376 | 1.624 | .652 | .951 | 27 | .985 | .985 | .984 | .984 | .984 | .983 | .982 | .982 | .981 | .981 |
| | | | | | | .980 | .979 | .979 | .978 | .977 | .976 | .976 | .975 | .974 | .973 |
| | | | | | | .972 | .971 | .969 | .968 | .967 | .966 | .964 | | | |
| 28.000 | 26.997 | 1.003 | .508 | .755 | 28 | .981 | .981 | .980 | .979 | .978 | .977 | .976 | .975 | .974 | .973 |
| | | | | | | .972 | .971 | .970 | .968 | .967 | .965 | .963 | .962 | .960 | .958 |
| | | | | | | .955 | .953 | .950 | .948 | .945 | .941 | .938 | .934 | | |
| 28.000 | 27.400 | .600 | .368 | .555 | 29 | .977 | .976 | .975 | .974 | .972 | .971 | .970 | .968 | .966 | .965 |
| | | | | | | .963 | .961 | .958 | .956 | .953 | .951 | .948 | .944 | .941 | .937 |
| | | | | | | .932 | .927 | .922 | .917 | .910 | .903 | .896 | .888 | .878 | |
| 28.000 | 27.629 | .371 | .260 | .394 | 30 | .974 | .972 | .971 | .969 | .968 | .966 | .964 | .962 | .959 | .957 |
| | | | | | | .954 | .951 | .948 | .944 | .940 | .936 | .931 | .926 | .921 | .914 |
| | | | | | | .907 | .900 | .891 | .882 | .871 | .859 | .846 | .832 | .815 | .797 |
| 28.000 | 27.760 | .240 | .183 | .279 | 31 | .971 | .970 | .968 | .966 | .964 | .962 | .959 | .957 | .954 | .951 |
| | | | | | | .947 | .944 | .939 | .935 | .930 | .924 | .918 | .911 | .903 | .895 |
| | | | | | | .885 | .875 | .863 | .850 | .835 | .818 | .800 | .779 | .756 | .730 |
| | | | | | | .702 | | | | | | | | | |
| 28.000 | 27.839 | .161 | .130 | .198 | 32 | .970 | .968 | .966 | .964 | .961 | .959 | .956 | .953 | .950 | .946 |
| | | | | | | .942 | .938 | .933 | .928 | .921 | .915 | .907 | .899 | .890 | .879 |
| | | | | | | .868 | .854 | .840 | .823 | .805 | .784 | .761 | .735 | .706 | .675 |
| | | | | | | .640 | .602 | | | | | | | | |
| 28.000 | 27.890 | .110 | .092 | .141 | 33 | .968 | .967 | .964 | .962 | .960 | .957 | .954 | .951 | .947 | .943 |
| | | | | | | .938 | .934 | .928 | .922 | .915 | .908 | .899 | .890 | .879 | .867 |
| | | | | | | .854 | .839 | .822 | .803 | .781 | .757 | .731 | .701 | .668 | .632 |
| | | | | | | .593 | .551 | .506 | | | | | | | |
| 28.000 | 27.923 | .077 | .066 | .101 | 34 | .968 | .966 | .963 | .961 | .958 | .955 | .952 | .949 | .945 | .940 |
| | | | | | | .936 | .930 | .925 | .918 | .911 | .902 | .893 | .883 | .871 | .858 |
| | | | | | | .843 | .827 | .808 | .787 | .763 | .737 | .708 | .675 | .639 | .600 |
| | | | | | | .558 | .513 | .465 | .416 | | | | | | |
| 28.000 | 27.947 | .053 | .047 | .071 | 35 | .967 | .965 | .963 | .960 | .957 | .954 | .951 | .947 | .943 | .939 |
| | | | | | | .934 | .928 | .922 | .915 | .907 | .898 | .889 | .878 | .865 | .851 |

JEWETT EQUIVALENT QUEUE EXTENDED ERLANG B(EQEEB): 10 MINUTES MAXIMUM WAIT (IF NOT CONNECTED AFTER 10 MINUTES WAIT, 100% OVERFLOW)

| HOURS FIRST ATTEMPT VMR = 1.0 | HOURS CON-NECT | HOURS OVFL | DELAYED PORTION | DELAY LENGTH CALL FACTOR | LINE | HOURS CONNECTED TRAFFIC ON LINES 1 | 2 | 3 | 4 | 5 | 6 | 7 | 8 | 9 | 10 |
|---|---|---|---|---|---|---|---|---|---|---|---|---|---|---|---|
| | | | | | | .835 | .818 | .798 | .775 | .750 | .722 | .690 | .656 | .618 | .576 |
| | | | | | | .532 | .485 | .436 | .387 | .337 | | | | | |
| 28.000 | 27.963 | .037 | .033 | .050 | 36 | .967 | .964 | .962 | .959 | .957 | .953 | .950 | .946 | .942 | .937 |
| | | | | | | .932 | .926 | .920 | .913 | .905 | .895 | .885 | .874 | .861 | .846 |
| | | | | | | .830 | .811 | .790 | .766 | .740 | .710 | .677 | .641 | .602 | .559 |
| | | | | | | .513 | .465 | .416 | .365 | .315 | .267 | | | | |
| 28.000 | 27.974 | .026 | .023 | .035 | 37 | .966 | .964 | .962 | .959 | .956 | .953 | .949 | .946 | .941 | .936 |
| | | | | | | .931 | .925 | .919 | .911 | .903 | .893 | .883 | .871 | .857 | .842 |
| | | | | | | .825 | .806 | .784 | .760 | .732 | .702 | .668 | .631 | .590 | .546 |
| | | | | | | .500 | .451 | .401 | .350 | .300 | .252 | .208 | | | |
| 28.000 | 27.982 | .018 | .016 | .024 | 38 | .966 | .964 | .961 | .959 | .956 | .953 | .949 | .945 | .941 | .936 |
| | | | | | | .930 | .924 | .917 | .910 | .901 | .892 | .881 | .869 | .855 | .840 |
| | | | | | | .822 | .802 | .780 | .755 | .727 | .696 | .661 | .623 | .582 | .537 |
| | | | | | | .490 | .441 | .390 | .339 | .290 | .242 | .198 | .158 | | |
| 30.000 | 25.763 | 4.237 | .930 | 1.262 | 26 | .992 | .992 | .992 | .992 | .992 | .992 | .992 | .992 | .992 | .991 |
| | | | | | | .991 | .991 | .991 | .991 | .991 | .990 | .990 | .990 | .990 | .990 |
| | | | | | | .990 | .989 | .989 | .989 | .989 | .989 | | | | |
| 30.000 | 26.686 | 3.314 | .852 | 1.186 | 27 | .991 | .991 | .991 | .990 | .990 | .990 | .990 | .990 | .990 | .989 |
| | | | | | | .989 | .989 | .989 | .988 | .988 | .988 | .988 | .988 | .987 | .987 |
| | | | | | | .987 | .986 | .986 | .986 | .985 | .985 | .985 | | | |
| 30.000 | 27.561 | 2.439 | .756 | 1.080 | 28 | .989 | .989 | .988 | .988 | .988 | .988 | .987 | .987 | .987 | .986 |
| | | | | | | .986 | .986 | .985 | .985 | .985 | .984 | .984 | .983 | .983 | .982 |
| | | | | | | .982 | .981 | .981 | .980 | .980 | .979 | .979 | .978 | | |
| 30.000 | 28.343 | 1.657 | .637 | .931 | 29 | .986 | .986 | .985 | .985 | .984 | .984 | .983 | .983 | .982 | .982 |
| | | | | | | .981 | .981 | .980 | .979 | .979 | .978 | .977 | .976 | .975 | .975 |
| | | | | | | .974 | .973 | .972 | .970 | .969 | .968 | .967 | .965 | .964 | |
| 30.000 | 28.961 | 1.039 | .498 | .742 | 30 | .982 | .982 | .981 | .980 | .979 | .979 | .978 | .977 | .976 | .975 |
| | | | | | | .974 | .973 | .972 | .971 | .969 | .968 | .967 | .965 | .963 | .962 |
| | | | | | | .960 | .958 | .956 | .953 | .951 | .948 | .945 | .942 | .939 | .935 |
| 30.000 | 29.367 | .633 | .364 | .549 | 31 | .978 | .977 | .977 | .975 | .974 | .973 | .972 | .971 | .969 | .968 |
| | | | | | | .966 | .964 | .962 | .960 | .958 | .956 | .953 | .951 | .948 | .944 |
| | | | | | | .941 | .937 | .933 | .928 | .923 | .918 | .912 | .905 | .898 | .891 |
| | | | | | | .882 | | | | | | | | | |
| 30.000 | 29.603 | .397 | .259 | .394 | 32 | .975 | .974 | .973 | .972 | .970 | .969 | .967 | .965 | .963 | .961 |
| | | | | | | .959 | .956 | .953 | .951 | .947 | .944 | .940 | .936 | .931 | .926 |
| | | | | | | .921 | .915 | .908 | .901 | .893 | .884 | .874 | .863 | .850 | .837 |
| | | | | | | .821 | .805 | | | | | | | | |
| 30.000 | 29.740 | .260 | .184 | .281 | 33 | .973 | .972 | .970 | .969 | .967 | .965 | .963 | .961 | .958 | .956 |
| | | | | | | .953 | .950 | .946 | .943 | .938 | .934 | .929 | .923 | .917 | .911 |
| | | | | | | .903 | .895 | .886 | .876 | .864 | .852 | .838 | .822 | .805 | .785 |

JEWETT EQUIVALENT QUEUE EXTENDED ERLANG B(EQEEB): 10 MINUTES MAXIMUM WAIT (IF NOT CONNECTED AFTER 10 MINUTES WAIT, 100% OVERFLOW)

| HOURS FIRST ATTEMPT VMR = 1.0 | HOURS CON-NECT | HOURS OVFL | DELAYED PORTION | DELAY LENGTH CALL FACTOR | LINE | HOURS CONNECTED TRAFFIC ON LINES 1 | 2 | 3 | 4 | 5 | 6 | 7 | 8 | 9 | 10 |
|---|---|---|---|---|---|---|---|---|---|---|---|---|---|---|---|
| | | | | | | .764 | .740 | .713 | | | | | | | |
| 30.000 | 29.824 | .176 | .132 | .202 | 34 | .972 | .970 | .968 | .967 | .965 | .962 | .960 | .958 | .955 | .952 |
| | | | | | | .948 | .945 | .941 | .936 | .932 | .926 | .920 | .914 | .906 | .898 |
| | | | | | | .889 | .879 | .868 | .855 | .841 | .825 | .808 | .788 | .766 | .742 |
| | | | | | | .715 | .685 | .652 | .617 | | | | | | |
| 30.000 | 29.878 | .122 | .095 | .145 | 35 | .971 | .969 | .967 | .965 | .963 | .961 | .958 | .955 | .952 | .949 |
| | | | | | | .945 | .941 | .937 | .932 | .926 | .920 | .914 | .906 | .898 | .888 |
| | | | | | | .878 | .866 | .853 | .839 | .822 | .804 | .784 | .761 | .735 | .707 |
| | | | | | | .676 | .642 | .605 | .565 | .523 | | | | | |
| 30.000 | 29.915 | .085 | .068 | .105 | 36 | .970 | .968 | .966 | .964 | .962 | .959 | .957 | .954 | .950 | .947 |
| | | | | | | .943 | .938 | .934 | .928 | .922 | .916 | .908 | .900 | .891 | .881 |
| | | | | | | .869 | .857 | .842 | .826 | .808 | .788 | .765 | .740 | .712 | .681 |
| | | | | | | .647 | .610 | .570 | .527 | .482 | .435 | | | | |
| 30.000 | 29.940 | .060 | .049 | .075 | 37 | .969 | .967 | .965 | .963 | .961 | .958 | .955 | .952 | .949 | .945 |
| | | | | | | .941 | .936 | .931 | .926 | .919 | .912 | .905 | .896 | .886 | .875 |
| | | | | | | .863 | .849 | .834 | .817 | .797 | .775 | .751 | .724 | .694 | .661 |
| | | | | | | .624 | .585 | .543 | .498 | .452 | .403 | .355 | | | |
| 30.000 | 29.958 | .042 | .035 | .054 | 38 | .969 | .967 | .965 | .963 | .960 | .958 | .955 | .951 | .948 | .944 |
| | | | | | | .940 | .935 | .930 | .924 | .917 | .910 | .902 | .893 | .882 | .871 |
| | | | | | | .858 | .844 | .828 | .809 | .789 | .766 | .740 | .712 | .680 | .646 |
| | | | | | | .608 | .567 | .523 | .477 | .430 | .381 | .332 | .285 | | |
| 30.000 | 29.970 | .030 | .025 | .038 | 39 | .969 | .967 | .964 | .962 | .960 | .957 | .954 | .951 | .947 | .943 |
| | | | | | | .939 | .934 | .928 | .922 | .916 | .908 | .900 | .890 | .880 | .868 |
| | | | | | | .855 | .840 | .823 | .804 | .783 | .759 | .732 | .703 | .670 | .635 |
| | | | | | | .596 | .554 | .509 | .462 | .414 | .365 | .316 | .269 | .225 | |
| 30.000 | 29.979 | .021 | .017 | .027 | 40 | .968 | .966 | .964 | .962 | .959 | .957 | .954 | .950 | .947 | .942 |
| | | | | | | .938 | .933 | .927 | .921 | .914 | .907 | .898 | .889 | .878 | .866 |
| | | | | | | .852 | .837 | .819 | .800 | .778 | .754 | .727 | .696 | .663 | .626 |
| | | | | | | .587 | .544 | .499 | .451 | .402 | .353 | .305 | .258 | .214 | .174 |

ERLANG C: INFINITE QUEUE - ALL CALLERS BLOCKED WILL WAIT AS LONG AS NECESSARY FOR CIRCUIT

| HOURS FIRST ATTEMPT VMR = 1.0 | HOURS CON-NECT | HOURS OVFL | DELAYED PORTION | DELAY LENGTH CALL FACTOR | LINE | HOURS CONNECTED TRAFFIC ON LINES 1 | 2 | 3 | 4 | 5 | 6 | 7 | 8 | 9 | 10 |
|---|---|---|---|---|---|---|---|---|---|---|---|---|---|---|---|
| .200 | .200 | .000 | .200 | 1.250 | 1 | .200 | | | | | | | | | |
| .200 | .200 | .000 | .018 | .556 | 2 | .182 | .018 | | | | | | | | |
| .200 | .200 | .000 | .001 | .357 | 3 | .181 | .018 | .001 | | | | | | | |
| .400 | .400 | .000 | .400 | 1.667 | 1 | .400 | | | | | | | | | |
| .400 | .400 | .000 | .067 | .625 | 2 | .333 | .067 | | | | | | | | |
| .400 | .400 | .000 | .008 | .385 | 3 | .330 | .062 | .008 | | | | | | | |
| .400 | .400 | .000 | .001 | .278 | 4 | .330 | .062 | .008 | .001 | .400 | 600 | | | | |
| .600 | .600 | .000 | .138 | .714 | 2 | .462 | .138 | | | | | | | | |
| .600 | .600 | .000 | .025 | .417 | 3 | .452 | .123 | .025 | | | | | | | |
| .600 | .600 | .000 | .003 | .294 | 4 | .451 | .122 | .023 | .003 | | | | | | |
| .800 | .800 | .000 | .800 | 5.000 | 1 | .800 | | | | | | | | | |
| .800 | .800 | .000 | .229 | .833 | 2 | .571 | .229 | | | | | | | | |
| .800 | .800 | .000 | .052 | .455 | 3 | .553 | .195 | .052 | | | | | | | |
| .800 | .800 | .000 | .010 | .313 | 4 | .551 | .192 | .048 | .010 | | | | | | |
| .800 | .800 | .000 | .001 | .238 | 5 | .551 | .191 | .047 | .009 | .001 | | | | | |
| 1.000 | 1.000 | .000 | .333 | 1.000 | 2 | .667 | .333 | | | | | | | | |
| 1.000 | 1.000 | .000 | .091 | .500 | 3 | .636 | .273 | .091 | | | | | | | |
| 1.000 | 1.000 | .000 | .020 | .333 | 4 | .633 | .265 | .082 | .020 | | | | | | |
| 1.000 | 1.000 | .000 | .004 | .250 | 5 | .632 | .264 | .080 | .019 | .004 | | | | | |
| 1.000 | 1.000 | .000 | .001 | .200 | 6 | .632 | .264 | .080 | .019 | .004 | .001 | | | | |
| 1.200 | 1.200 | .000 | .450 | 1.250 | 2 | .750 | .450 | | | | | | | | |
| 1.200 | 1.200 | .000 | .141 | .556 | 3 | .706 | .353 | .141 | | | | | | | |
| 1.200 | 1.200 | .000 | .037 | .357 | 4 | .700 | .340 | .123 | .037 | | | | | | |
| 1.200 | 1.200 | .000 | .008 | .263 | 5 | .699 | .338 | .121 | .034 | .008 | | | | | |
| 1.200 | 1.200 | .000 | .002 | .208 | 6 | .699 | .337 | .121 | .034 | .008 | .002 | | | | |
| 1.400 | 1.400 | .000 | .576 | 1.667 | 2 | .824 | .576 | | | | | | | | |
| 1.400 | 1.400 | .000 | .202 | .625 | 3 | .764 | .434 | .202 | | | | | | | |
| 1.400 | 1.400 | .000 | .060 | .385 | 4 | .755 | .412 | .172 | .060 | | | | | | |
| 1.400 | 1.400 | .000 | .015 | .278 | 5 | .754 | .409 | .167 | .055 | .015 | | | | | |
| 1.400 | 1.400 | .000 | .003 | .217 | 6 | .753 | .408 | .167 | .054 | .014 | .003 | | | | |
| 1.400 | 1.400 | .000 | .001 | .179 | 7 | .753 | .408 | .167 | .054 | .014 | .003 | .001 | | | |
| 1.600 | 1.600 | .000 | .711 | 2.500 | 2 | .889 | .711 | | | | | | | | |
| 1.600 | 1.600 | .000 | .274 | .714 | 3 | .813 | .513 | .274 | | | | | | | |
| 1.600 | 1.600 | .000 | .091 | .417 | 4 | .801 | .482 | .227 | .091 | | | | | | |
| 1.600 | 1.600 | .000 | .026 | .294 | 5 | .799 | .476 | .218 | .081 | .026 | | | | | |
| 1.600 | 1.600 | .000 | .006 | .227 | 6 | .798 | .475 | .217 | .079 | .024 | .006 | | | | |
| 1.600 | 1.600 | .000 | .001 | .185 | 7 | .798 | .475 | .217 | .079 | .024 | .006 | .001 | | | |
| 1.800 | 1.800 | .000 | .853 | 5.000 | 2 | .947 | .853 | | | | | | | | |
| 1.800 | 1.800 | .000 | .355 | .833 | 3 | .854 | .591 | .355 | | | | | | | |

ERLANG C: INFINITE QUEUE - ALL CALLERS BLOCKED WILL WAIT AS LONG AS NECESSARY FOR CIRCUIT

| HOURS FIRST ATTEMPT VMR = 1.0 | HOURS CON-NECT | HOURS OVFL | DELAYED PORTION | DELAY LENGTH CALL FACTOR | LINE | HOURS CONNECTED TRAFFIC ON LINES 1 | 2 | 3 | 4 | 5 | 6 | 7 | 8 | 9 | 10 |
|---|---|---|---|---|---|---|---|---|---|---|---|---|---|---|---|
| 1.800 | 1.800 | .000 | .129 | .455 | 4 | .838 | .547 | .286 | .129 | | | | | | |
| 1.800 | 1.800 | .000 | .040 | .313 | 5 | .835 | .539 | .272 | .112 | .040 | | | | | |
| 1.800 | 1.800 | .000 | .011 | .238 | 6 | .835 | .538 | .270 | .109 | .037 | .011 | | | | |
| 1.800 | 1.800 | .000 | .003 | .192 | 7 | .835 | .537 | .269 | .109 | .037 | .011 | .003 | | | |
| 1.800 | 1.800 | .000 | .001 | .161 | 8 | .835 | .537 | .269 | .109 | .036 | .010 | .003 | .001 | | |
| 2.000 | 2.000 | .000 | .444 | 1.000 | 3 | .889 | .667 | .444 | | | | | | | |
| 2.000 | 2.000 | .000 | .174 | .500 | 4 | .870 | .609 | .348 | .174 | | | | | | |
| 2.000 | 2.000 | .000 | .060 | .333 | 5 | .866 | .597 | .328 | .149 | .060 | | | | | |
| 2.000 | 2.000 | .000 | .018 | .250 | 6 | .865 | .595 | .324 | .144 | .054 | .018 | | | | |
| 2.000 | 2.000 | .000 | .005 | .200 | 7 | .865 | .594 | .324 | .143 | .053 | .017 | .005 | | | |
| 2.000 | 2.000 | .000 | .001 | .167 | 8 | .865 | .594 | .323 | .143 | .053 | .017 | .005 | .001 | | |
| 2.200 | 2.200 | .000 | .542 | 1.250 | 3 | .919 | .739 | .542 | | | | | | | |
| 2.200 | 2.200 | .000 | .227 | .556 | 4 | .895 | .665 | .412 | .227 | | | | | | |
| 2.200 | 2.200 | .000 | .084 | .357 | 5 | .891 | .650 | .385 | .191 | .084 | | | | | |
| 2.200 | 2.200 | .000 | .027 | .263 | 6 | .889 | .646 | .379 | .183 | .075 | .027 | | | | |
| 2.200 | 2.200 | .000 | .008 | .208 | 7 | .889 | .646 | .378 | .181 | .073 | .025 | .008 | | | |
| 2.200 | 2.200 | .000 | .002 | .172 | 8 | .889 | .645 | .377 | .181 | .073 | .025 | .008 | .002 | | |
| 2.400 | 2.400 | .000 | .647 | 1.667 | 3 | .944 | .809 | .647 | | | | | | | |
| 2.400 | 2.400 | .000 | .287 | .625 | 4 | .917 | .718 | .478 | .287 | | | | | | |
| 2.400 | 2.400 | .000 | .114 | .385 | 5 | .911 | .698 | .441 | .236 | .114 | | | | | |
| 2.400 | 2.400 | .000 | .040 | .278 | 6 | .910 | .693 | .433 | .225 | .100 | .040 | | | | |
| 2.400 | 2.400 | .000 | .013 | .217 | 7 | .909 | .692 | .431 | .222 | .097 | .037 | .013 | | | |
| 2.400 | 2.400 | .000 | .004 | .179 | 8 | .909 | .692 | .430 | .221 | .096 | .036 | .012 | .004 | | |
| 2.400 | 2.400 | .000 | .001 | .152 | 9 | .909 | .692 | .430 | .221 | .096 | .036 | .012 | .003 | .001 | |
| 2.600 | 2.600 | .000 | .759 | 2.500 | 3 | .965 | .876 | .759 | | | | | | | |
| 2.600 | 2.600 | .000 | .354 | .714 | 4 | .935 | .765 | .545 | .354 | | | | | | |
| 2.600 | 2.600 | .000 | .149 | .417 | 5 | .928 | .741 | .497 | .286 | .149 | | | | | |
| 2.600 | 2.600 | .000 | .056 | .294 | 6 | .926 | .735 | .485 | .269 | .129 | .056 | | | | |
| 2.600 | 2.600 | .000 | .019 | .227 | 7 | .926 | .733 | .482 | .265 | .124 | .051 | .019 | | | |
| 2.600 | 2.600 | .000 | .006 | .185 | 8 | .926 | .733 | .482 | .264 | .123 | .049 | .018 | .006 | | |
| 2.600 | 2.600 | .000 | .002 | .156 | 9 | .926 | .733 | .482 | .264 | .123 | .049 | .017 | .005 | .002 | |
| 2.800 | 2.800 | .000 | .877 | 5.000 | 3 | .984 | .939 | .877 | | | | | | | |
| 2.800 | 2.800 | .000 | .429 | .833 | 4 | .950 | .809 | .612 | .429 | | | | | | |
| 2.800 | 2.800 | .000 | .190 | .455 | 5 | .942 | .779 | .551 | .338 | .190 | | | | | |
| 2.800 | 2.800 | .000 | .075 | .313 | 6 | .940 | .771 | .536 | .316 | .162 | .075 | | | | |
| 2.800 | 2.800 | .000 | .027 | .238 | 7 | .939 | .770 | .532 | .310 | .155 | .068 | .027 | | | |
| 2.800 | 2.800 | .000 | .009 | .192 | 8 | .939 | .769 | .531 | .309 | .153 | .066 | .025 | .009 | | |
| 2.800 | 2.800 | .000 | .003 | .161 | 9 | .939 | .769 | .531 | .308 | .152 | .065 | .025 | .008 | .003 | |
| 2.800 | 2.800 | .000 | .001 | .139 | 10 | .939 | .769 | .531 | .308 | .152 | .065 | .024 | .008 | .002 | .001 |
| 3.000 | 3.000 | .000 | .509 | 1.000 | 4 | .962 | .849 | .679 | .509 | | | | | | |

ERLANG C: INFINITE QUEUE - ALL CALLERS BLOCKED WILL WAIT AS LONG AS NECESSARY FOR CIRCUIT

| HOURS FIRST ATTEMPT VMR = 1.0 | HOURS CON-NECT | HOURS OVFL | DELAYED PORTION | DELAY LENGTH CALL FACTOR | LINE | HOURS CONNECTED TRAFFIC ON LINES 1 | 2 | 3 | 4 | 5 | 6 | 7 | 8 | 9 | 10 |
|---|---|---|---|---|---|---|---|---|---|---|---|---|---|---|---|
| 3.000 | 3.000 | .000 | .236 | .500 | 5 | .953 | .813 | .603 | .394 | .236 | | | | | |
| 3.000 | 3.000 | .000 | .099 | .333 | 6 | .951 | .804 | .584 | .364 | .198 | .099 | | | | |
| 3.000 | 3.000 | .000 | .038 | .250 | 7 | .950 | .802 | .579 | .356 | .188 | .088 | .038 | | | |
| 3.000 | 3.000 | .000 | .013 | .200 | 8 | .950 | .801 | .577 | .353 | .186 | .085 | .035 | .013 | | |
| 3.000 | 3.000 | .000 | .004 | .167 | 9 | .950 | .801 | .577 | .353 | .185 | .084 | .034 | .012 | .004 | |
| 3.000 | 3.000 | .000 | .001 | .143 | 10 | .950 | .801 | .577 | .353 | .185 | .084 | .034 | .012 | .004 | .001 |
| 3.200 | 3.200 | .000 | .596 | 1.250 | 4 | .973 | .885 | .746 | .596 | | | | | | |
| 3.200 | 3.200 | .000 | .289 | .556 | 5 | .963 | .844 | .654 | .451 | .289 | | | | | |
| 3.200 | 3.200 | .000 | .127 | .357 | 6 | .960 | .833 | .629 | .412 | .238 | .127 | | | | |
| 3.200 | 3.200 | .000 | .051 | .263 | 7 | .960 | .830 | .623 | .401 | .224 | .111 | .051 | | | |
| 3.200 | 3.200 | .000 | .018 | .208 | 8 | .959 | .829 | .621 | .399 | .221 | .107 | .046 | .018 | | |
| 3.200 | 3.200 | .000 | .006 | .172 | 9 | .959 | .829 | .620 | .398 | .220 | .106 | .045 | .017 | .006 | |
| 3.200 | 3.200 | .000 | .002 | .147 | 10 | .959 | .829 | .620 | .398 | .219 | .105 | .045 | .017 | .006 | .002 |
| 3.200 | 3.200 | .000 | .001 | .128 | 11 | .959 | .829 | .620 | .397 | .219 | .105 | .045 | .017 | .006 | .002 |
| | | | | | | .001 | | | | | | | | | |
| 3.400 | 3.400 | .000 | .689 | 1.667 | 4 | .981 | .918 | .811 | .689 | | | | | | |
| 3.400 | 3.400 | .000 | .347 | .625 | 5 | .971 | .871 | .702 | .510 | .347 | | | | | |
| 3.400 | 3.400 | .000 | .160 | .385 | 6 | .968 | .858 | .672 | .461 | .282 | .160 | | | | |
| 3.400 | 3.400 | .000 | .067 | .278 | 7 | .967 | .855 | .664 | .447 | .263 | .138 | .067 | | | |
| 3.400 | 3.400 | .000 | .026 | .217 | 8 | .967 | .854 | .661 | .443 | .258 | .132 | .060 | .026 | | |
| 3.400 | 3.400 | .000 | .009 | .179 | 9 | .967 | .853 | .660 | .442 | .256 | .130 | .058 | .024 | .009 | |
| 3.400 | 3.400 | .000 | .003 | .152 | 10 | .967 | .853 | .660 | .442 | .256 | .130 | .058 | .023 | .008 | .003 |
| 3.400 | 3.400 | .000 | .001 | .132 | 11 | .967 | .853 | .660 | .442 | .256 | .129 | .058 | .023 | .008 | .003 |
| | | | | | | .001 | | | | | | | | | |
| 3.600 | 3.600 | .000 | .788 | 2.500 | 4 | .989 | .948 | .875 | .788 | | | | | | |
| 3.600 | 3.600 | .000 | .410 | .714 | 5 | .977 | .895 | .747 | .570 | .410 | | | | | |
| 3.600 | 3.600 | .000 | .197 | .417 | 6 | .974 | .880 | .712 | .510 | .328 | .197 | | | | |
| 3.600 | 3.600 | .000 | .086 | .294 | 7 | .973 | .876 | .701 | .492 | .303 | .168 | .086 | | | |
| 3.600 | 3.600 | .000 | .035 | .227 | 8 | .973 | .875 | .698 | .487 | .296 | .159 | .077 | .035 | | |
| 3.600 | 3.600 | .000 | .013 | .185 | 9 | .973 | .874 | .698 | .485 | .294 | .157 | .074 | .032 | .013 | |
| 3.600 | 3.600 | .000 | .004 | .156 | 10 | .973 | .874 | .697 | .485 | .294 | .156 | .074 | .031 | .012 | .004 |
| 3.600 | 3.600 | .000 | .001 | .135 | 11 | .973 | .874 | .697 | .485 | .294 | .156 | .073 | .031 | .012 | .004 |
| | | | | | | .001 | | | | | | | | | |
| 3.800 | 3.800 | .000 | .891 | 5.000 | 4 | .995 | .975 | .938 | .891 | | | | | | |
| 3.800 | 3.800 | .000 | .480 | .833 | 5 | .983 | .916 | .790 | .631 | .480 | | | | | |
| 3.800 | 3.800 | .000 | .238 | .455 | 6 | .979 | .900 | .749 | .558 | .376 | .238 | | | | |
| 3.800 | 3.800 | .000 | .109 | .313 | 7 | .978 | .895 | .736 | .536 | .345 | .201 | .109 | | | |
| 3.800 | 3.800 | .000 | .046 | .238 | 8 | .978 | .893 | .733 | .529 | .336 | .189 | .096 | .046 | | |
| 3.800 | 3.800 | .000 | .018 | .192 | 9 | .978 | .893 | .732 | .527 | .333 | .186 | .092 | .042 | .018 | |
| 3.800 | 3.800 | .000 | .006 | .161 | 10 | .978 | .893 | .731 | .527 | .332 | .185 | .091 | .041 | .016 | .006 |

ERLANG C: INFINITE QUEUE - ALL CALLERS BLOCKED WILL WAIT AS LONG AS NECESSARY FOR CIRCUIT

| HOURS FIRST ATTEMPT VMR = 1.0 | HOURS CON-NECT | HOURS OVFL | DELAYED PORTION | DELAY LENGTH CALL FACTOR | LINE | HOURS CONNECTED TRAFFIC ON LINES 1 | 2 | 3 | 4 | 5 | 6 | 7 | 8 | 9 | 10 |
|---|---|---|---|---|---|---|---|---|---|---|---|---|---|---|---|
| 3.800 | 3.800 | .000 | .002 | .139 | 11 | .978 | .893 | .731 | .527 | .332 | .185 | .091 | .040 | .016 | .006 |
| | | | | | | .002 | | | | | | | | | |
| 3.800 | 3.800 | .000 | .001 | .122 | 12 | .978 | .893 | .731 | .527 | .332 | .184 | .091 | .040 | .016 | .006 |
| | | | | | | .002 | .001 | | | | | | | | |
| 4.000 | 4.000 | .000 | .554 | 1.000 | 5 | .987 | .935 | .831 | .693 | .554 | | | | | |
| 4.000 | 4.000 | .000 | .285 | .500 | 6 | .983 | .917 | .783 | .605 | .427 | .285 | | | | |
| 4.000 | 4.000 | .000 | .135 | .333 | 7 | .982 | .911 | .768 | .578 | .388 | .236 | .135 | | | |
| 4.000 | 4.000 | .000 | .059 | .250 | 8 | .982 | .909 | .764 | .570 | .376 | .221 | .118 | .059 | | |
| 4.000 | 4.000 | .000 | .024 | .200 | 9 | .982 | .909 | .762 | .568 | .373 | .217 | .113 | .053 | .024 | |
| 4.000 | 4.000 | .000 | .009 | .167 | 10 | .982 | .908 | .762 | .567 | .372 | .215 | .111 | .052 | .022 | .009 |
| 4.000 | 4.000 | .000 | .003 | .143 | 11 | .982 | .908 | .762 | .567 | .371 | .215 | .111 | .051 | .022 | .008 |
| | | | | | | .003 | | | | | | | | | |
| 4.000 | 4.000 | .000 | .001 | .125 | 12 | .982 | .908 | .762 | .567 | .371 | .215 | .111 | .051 | .021 | .008 |
| | | | | | | .003 | .001 | | | | | | | | |
| 4.200 | 4.200 | .000 | .634 | 1.250 | 5 | .991 | .952 | .869 | .754 | .634 | | | | | |
| 4.200 | 4.200 | .000 | .336 | .556 | 6 | .987 | .931 | .815 | .651 | .480 | .336 | | | | |
| 4.200 | 4.200 | .000 | .165 | .357 | 7 | .986 | .925 | .798 | .619 | .432 | .275 | .165 | | | |
| 4.200 | 4.200 | .000 | .075 | .263 | 8 | .985 | .923 | .792 | .609 | .417 | .256 | .143 | .075 | | |
| 4.200 | 4.200 | .000 | .031 | .208 | 9 | .985 | .922 | .791 | .606 | .412 | .250 | .136 | .067 | .031 | |
| 4.200 | 4.200 | .000 | .012 | .172 | 10 | .985 | .922 | .790 | .605 | .411 | .248 | .133 | .065 | .029 | .012 |
| 4.200 | 4.200 | .000 | .004 | .147 | 11 | .985 | .922 | .790 | .605 | .410 | .247 | .133 | .064 | .028 | .011 |
| | | | | | | .004 | | | | | | | | | |
| 4.200 | 4.200 | .000 | .001 | .128 | 12 | .985 | .922 | .790 | .605 | .410 | .247 | .133 | .064 | .028 | .011 |
| | | | | | | .004 | .001 | | | | | | | | |
| 4.400 | 4.400 | .000 | .718 | 1.667 | 5 | .994 | .966 | .905 | .816 | .718 | | | | | |
| 4.400 | 4.400 | .000 | .392 | .625 | 6 | .990 | .944 | .844 | .696 | .534 | .392 | | | | |
| 4.400 | 4.400 | .000 | .199 | .385 | 7 | .988 | .937 | .824 | .659 | .477 | .316 | .199 | | | |
| 4.400 | 4.400 | .000 | .094 | .278 | 8 | .988 | .935 | .818 | .646 | .458 | .292 | .170 | .094 | | |
| 4.400 | 4.400 | .000 | .041 | .217 | 9 | .988 | .934 | .816 | .642 | .452 | .284 | .161 | .083 | .041 | |
| 4.400 | 4.400 | .000 | .016 | .179 | 10 | .988 | .934 | .815 | .641 | .450 | .281 | .158 | .080 | .037 | .016 |
| 4.400 | 4.400 | .000 | .006 | .152 | 11 | .988 | .934 | .815 | .641 | .449 | .280 | .157 | .079 | .036 | .015 |
| | | | | | | .006 | | | | | | | | | |
| 4.400 | 4.400 | .000 | .002 | .132 | 12 | .988 | .934 | .815 | .641 | .449 | .280 | .156 | .079 | .036 | .015 |
| | | | | | | .006 | .002 | | | | | | | | |
| 4.400 | 4.400 | .000 | .001 | .116 | 13 | .988 | .934 | .815 | .641 | .449 | .280 | .156 | .079 | .036 | .015 |
| | | | | | | .006 | .002 | .001 | | | | | | | |
| 4.600 | 4.600 | .000 | .808 | 2.500 | 5 | .996 | .979 | .939 | .878 | .808 | | | | | |
| 4.600 | 4.600 | .000 | .453 | .714 | 6 | .992 | .955 | .870 | .740 | .590 | .453 | | | | |
| 4.600 | 4.600 | .000 | .237 | .417 | 7 | .991 | .947 | .848 | .696 | .521 | .360 | .237 | | | |
| 4.600 | 4.600 | .000 | .115 | .294 | 8 | .990 | .945 | .841 | .681 | .498 | .329 | .200 | .115 | | |

ERLANG C: INFINITE QUEUE - ALL CALLERS BLOCKED WILL WAIT AS LONG AS NECESSARY FOR CIRCUIT

| HOURS FIRST ATTEMPT VMR = 1.0 | HOURS CON-NECT | HOURS OVFL | DELAYED PORTION | DELAY LENGTH CALL FACTOR | LINE | HOURS CONNECTED TRAFFIC ON LINES 1 | 2 | 3 | 4 | 5 | 6 | 7 | 8 | 9 | 10 |
|---|---|---|---|---|---|---|---|---|---|---|---|---|---|---|---|
| 4.600 | 4.600 | .000 | .052 | .227 | 9 | .990 | .944 | .839 | .677 | .490 | .319 | .188 | .102 | .052 | |
| 4.600 | 4.600 | .000 | .022 | .185 | 10 | .990 | .944 | .838 | .675 | .488 | .316 | .184 | .097 | .047 | .022 |
| 4.600 | 4.600 | .000 | .008 | .156 | 11 | .990 | .944 | .837 | .675 | .487 | .315 | .183 | .096 | .046 | .020 |
| | | | | | | .008 | | | | | | | | | |
| 4.600 | 4.600 | .000 | .003 | .135 | 12 | .990 | .944 | .837 | .674 | .487 | .314 | .182 | .095 | .045 | .020 |
| | | | | | | .008 | .003 | | | | | | | | |
| 4.600 | 4.600 | .000 | .001 | .119 | 13 | .990 | .944 | .837 | .674 | .487 | .314 | .182 | .095 | .045 | .020 |
| | | | | | | .008 | .003 | .001 | | | | | | | |
| 4.800 | 4.800 | .000 | .902 | 5.000 | 5 | .998 | .990 | .971 | .939 | .902 | | | | | |
| 4.800 | 4.800 | .000 | .518 | .833 | 6 | .994 | .965 | .894 | .782 | .647 | .518 | | | | |
| 4.800 | 4.800 | .000 | .278 | .455 | 7 | .992 | .956 | .870 | .731 | .565 | .406 | .278 | | | |
| 4.800 | 4.800 | .000 | .140 | .313 | 8 | .992 | .954 | .862 | .714 | .538 | .368 | .233 | .140 | | |
| 4.800 | 4.800 | .000 | .065 | .238 | 9 | .992 | .953 | .859 | .709 | .528 | .355 | .217 | .122 | .065 | |
| 4.800 | 4.800 | .000 | .028 | .192 | 10 | .992 | .952 | .858 | .707 | .525 | .351 | .212 | .116 | .059 | .028 |
| 4.800 | 4.800 | .000 | .011 | .161 | 11 | .992 | .952 | .858 | .706 | .524 | .350 | .210 | .114 | .057 | .026 |
| | | | | | | .011 | | | | | | | | | |
| 4.800 | 4.800 | .000 | .004 | .139 | 12 | .992 | .952 | .858 | .706 | .524 | .349 | .209 | .114 | .056 | .025 |
| | | | | | | .011 | .004 | | | | | | | | |
| 4.800 | 4.800 | .000 | .002 | .122 | 13 | .992 | .952 | .857 | .706 | .524 | .349 | .209 | .113 | .056 | .025 |
| | | | | | | .010 | .004 | .002 | | | | | | | |
| 5.000 | 5.000 | .000 | .588 | 1.000 | 6 | .995 | .973 | .917 | .823 | .705 | .588 | | | | |
| 5.000 | 5.000 | .000 | .324 | .500 | 7 | .994 | .964 | .889 | .765 | .609 | .454 | .324 | | | |
| 5.000 | 5.000 | .000 | .167 | .333 | 8 | .994 | .961 | .880 | .745 | .577 | .408 | .268 | .167 | | |
| 5.000 | 5.000 | .000 | .081 | .250 | 9 | .993 | .960 | .877 | .739 | .565 | .392 | .248 | .146 | .081 | |
| 5.000 | 5.000 | .000 | .036 | .200 | 10 | .993 | .960 | .078 | .736 | .561 | .387 | .241 | .137 | .072 | .036 |
| 5.000 | 5.000 | .000 | .015 | .167 | 11 | .993 | .960 | .876 | .735 | .560 | .385 | .239 | .135 | .069 | .033 |
| | | | | | | .015 | | | | | | | | | |
| 5.000 | 5.000 | .000 | .006 | .143 | 12 | .993 | .960 | .875 | .735 | .560 | .384 | .238 | .134 | .068 | .032 |
| | | | | | | .014 | .006 | | | | | | | | |
| 5.000 | 5.000 | .000 | .002 | .125 | 13 | .993 | .960 | .875 | .735 | .560 | .384 | .238 | .133 | .068 | .032 |
| | | | | | | .014 | .006 | .002 | | | | | | | |
| 5.000 | 5.000 | .000 | .001 | .111 | 14 | .993 | .960 | .875 | .735 | .560 | .384 | .238 | .133 | .068 | .032 |
| | | | | | | .014 | .005 | .002 | .001 | | | | | | |
| 5.500 | 5.500 | .000 | .781 | 2.000 | 6 | .998 | .989 | .963 | .916 | .852 | .781 | | | | |
| 5.500 | 5.500 | .000 | .456 | .667 | 7 | .997 | .979 | .930 | .840 | .717 | .581 | .456 | | | |
| 5.500 | 5.500 | .000 | .251 | .400 | 8 | .996 | .975 | .918 | .813 | .669 | .511 | .365 | .251 | | |
| 5.500 | 5.500 | .000 | .130 | .286 | 9 | .996 | 974 | .014 | .804 | .652 | .485 | .332 | .212 | .130 | |
| 5.500 | 5.500 | .000 | .063 | .222 | 10 | .996 | .974 | .912 | .800 | .646 | .476 | .320 | .198 | .114 | .063 |
| 5.500 | 5.500 | .000 | .028 | .182 | 11 | .996 | .974 | .912 | .799 | .644 | .473 | .316 | .193 | .109 | .057 |
| | | | | | | .028 | | | | | | | | | |

ERLANG C: INFINITE QUEUE - ALL CALLERS BLOCKED WILL WAIT AS LONG AS NECESSARY FOR CIRCUIT

| HOURS FIRST ATTEMPT VMR = 1.0 | HOURS CON-NECT | HOURS OVFL | DELAYED PORTION | DELAY LENGTH CALL FACTOR | LINE | HOURS CONNECTED TRAFFIC ON LINES 1 | 2 | 3 | 4 | 5 | 6 | 7 | 8 | 9 | 10 |
|---|---|---|---|---|---|---|---|---|---|---|---|---|---|---|---|
| 5.500 | 5.500 | .000 | .012 | .154 | 12 | .996 | .973 | .912 | .799 | .643 | .472 | .315 | .191 | .107 | .055 |
| | | | | | | .026 | .012 | | | | | | | | |
| 5.500 | 5.500 | .000 | .005 | .133 | 13 | .996 | .973 | .912 | .798 | .643 | .471 | .314 | .191 | .106 | .054 |
| | | | | | | .026 | .011 | .005 | | | | | | | |
| 5.500 | 5.500 | .000 | .002 | .118 | 14 | .996 | .973 | .912 | .798 | .643 | .471 | .314 | .191 | .106 | .054 |
| | | | | | | .025 | .011 | .005 | .002 | | | | | | |
| 5.500 | 5.500 | .000 | .001 | .105 | 15 | .996 | .973 | .912 | .798 | .642 | .471 | .314 | .191 | .106 | .054 |
| | | | | | | .025 | .011 | .004 | .002 | .001 | | | | | |
| 6.000 | 6.000 | .000 | .614 | 1.000 | 7 | .998 | .989 | .961 | .904 | .818 | .716 | .614 | | | |
| 6.000 | 6.000 | .000 | .357 | .500 | 8 | .998 | .985 | .946 | .869 | .754 | .615 | .476 | .357 | | |
| 6.000 | 6.000 | .000 | .196 | .333 | 9 | .998 | .984 | .941 | .857 | .729 | .577 | .425 | .294 | .196 | |
| 6.000 | 6.000 | .000 | .101 | .250 | 10 | .998 | .983 | .939 | .852 | .720 | .563 | .405 | .270 | .169 | .101 |
| 6.000 | 6.000 | .000 | .049 | .200 | 11 | .998 | .983 | .938 | .850 | .717 | .557 | .398 | .261 | .159 | .090 |
| | | | | | | .049 | | | | | | | | | |
| 6.000 | 6.000 | .000 | .022 | .167 | 12 | .998 | .983 | .938 | .849 | .716 | .555 | .395 | .258 | .155 | .086 |
| | | | | | | .045 | .022 | | | | | | | | |
| 6.000 | 6.000 | .000 | .010 | .143 | 13 | .998 | .983 | .938 | .849 | .715 | .555 | .394 | .257 | .153 | .085 |
| | | | | | | .043 | .021 | .010 | | | | | | | |
| 6.000 | 6.000 | .000 | .004 | .125 | 14 | .998 | .983 | .938 | .849 | .715 | .554 | .394 | .256 | .153 | .084 |
| | | | | | | .043 | .020 | .009 | .004 | | | | | | |
| 6.000 | 6.000 | .000 | .001 | .111 | 15 | .998 | .983 | .938 | .849 | .715 | .554 | .394 | .256 | .153 | .084 |
| | | | | | | .043 | .020 | .009 | .004 | .001 | | | | | |
| 6.000 | 6.000 | .000 | .001 | .100 | 16 | .998 | .983 | .938 | .849 | .715 | .554 | .394 | .256 | .153 | .084 |
| | | | | | | .043 | .020 | .009 | .004 | .001 | .001 | | | | |
| 6.500 | 6.500 | .000 | .795 | 2.000 | 7 | .999 | .996 | .983 | .957 | .913 | .857 | .795 | | | |
| 6.500 | 6.500 | .000 | .485 | .667 | 8 | .999 | .991 | .967 | .914 | .829 | .718 | .597 | .485 | | |
| 6.500 | 6.500 | .000 | .281 | .400 | 9 | .999 | .990 | .961 | .898 | .797 | .665 | .522 | .389 | .281 | |
| 6.500 | 6.500 | .000 | .154 | .286 | 10 | .999 | .989 | .958 | .892 | .784 | .644 | .492 | .351 | .236 | .154 |
| 6.500 | 6.500 | .000 | .079 | .222 | 11 | .999 | .989 | .958 | .890 | .779 | .636 | .481 | .336 | .219 | .134 |
| | | | | | | .079 | | | | | | | | | |
| 6.500 | 6.500 | .000 | .039 | .182 | 12 | .999 | .989 | .957 | .889 | .777 | .633 | .476 | .331 | .212 | .127 |
| | | | | | | .072 | .039 | | | | | | | | |
| 6.500 | 6.500 | .000 | .018 | .154 | 13 | .998 | .989 | .957 | .888 | .777 | .632 | .474 | .328 | .210 | .124 |
| | | | | | | .069 | .036 | .018 | | | | | | | |
| 6.500 | 6.500 | .000 | .008 | .133 | 14 | .998 | .989 | .957 | .888 | .776 | .631 | .474 | .328 | .209 | .123 |
| | | | | | | .067 | .034 | .017 | .008 | | | | | | |
| 6.500 | 6.500 | .000 | .003 | .118 | 15 | .998 | .989 | .957 | .888 | .776 | .631 | .474 | .327 | .209 | .123 |
| | | | | | | .067 | .034 | .016 | .007 | .003 | | | | | |
| 6.500 | 6.500 | .000 | .001 | .105 | 16 | .998 | .989 | .957 | .888 | .776 | .631 | .474 | .327 | .208 | .123 |
| | | | | | | .067 | .034 | .016 | .007 | .003 | .001 | | | | |

ERLANG C: INFINITE QUEUE - ALL CALLERS BLOCKED WILL WAIT AS LONG AS NECESSARY FOR CIRCUIT

| HOURS FIRST ATTEMPT VMR = 1.0 | HOURS CON-NECT | HOURS OVFL | DELAYED PORTION | DELAY LENGTH CALL FACTOR | LINE | HOURS CONNECTED TRAFFIC ON LINES 1 | 2 | 3 | 4 | 5 | 6 | 7 | 8 | 9 | 10 |
|---|---|---|---|---|---|---|---|---|---|---|---|---|---|---|---|
| 7.000 | 7.000 | .000 | .635 | 1.000 | 8 | .999 | .996 | .982 | .950 | .895 | .817 | .726 | .635 | | |
| 7.000 | 7.000 | .000 | .385 | .500 | 9 | .999 | .994 | .975 | .931 | .854 | .746 | .621 | .495 | .385 | |
| 7.000 | 7.000 | .000 | .222 | .333 | 10 | .999 | .993 | .972 | .923 | .838 | .718 | .579 | .439 | .317 | .222 |
| 7.000 | 7.000 | .000 | .121 | .250 | 11 | .999 | .993 | .971 | .920 | .831 | .707 | .562 | .416 | .289 | .190 |
| | | | | | | .121 | | | | | | | | | |
| 7.000 | 7.000 | .000 | .063 | .200 | 12 | .999 | .993 | .971 | .919 | .829 | .702 | .555 | .407 | .278 | .178 |
| | | | | | | .107 | .063 | | | | | | | | |
| 7.000 | 7.000 | .000 | .031 | .167 | 13 | .999 | .993 | .970 | .919 | .828 | .700 | .552 | .404 | .274 | .173 |
| | | | | | | .102 | .057 | .031 | | | | | | | |
| 7.000 | 7.000 | .000 | .014 | .143 | 14 | .999 | .993 | .970 | .918 | .827 | .700 | .551 | .402 | .272 | .171 |
| | | | | | | .100 | .055 | .028 | .014 | | | | | | |
| 7.000 | 7.000 | .000 | .006 | .125 | 15 | .999 | .993 | .970 | .918 | .827 | .699 | .551 | .402 | .271 | .170 |
| | | | | | | .099 | .054 | .027 | .013 | .006 | | | | | |
| 7.000 | 7.000 | .000 | .003 | .111 | 16 | .999 | .993 | .970 | .918 | .827 | .699 | .550 | .401 | .271 | .170 |
| | | | | | | .099 | .054 | .027 | .013 | .006 | .003 | | | | |
| 7.000 | 7.000 | .000 | .001 | .100 | 17 | .999 | .993 | .970 | .918 | .827 | .699 | .550 | .401 | .271 | .170 |
| | | | | | | .099 | .053 | .027 | .013 | .006 | .002 | .001 | | | |
| 7.500 | 7.500 | .000 | .807 | 2.000 | 8 | .999 | .998 | .993 | .978 | .951 | .911 | .861 | .807 | | |
| 7.500 | 7.500 | .000 | .509 | .667 | 9 | .999 | .997 | .985 | .956 | .902 | .821 | .720 | .611 | .509 | |
| 7.500 | 7.500 | .000 | .307 | .400 | 10 | .999 | .996 | .982 | .947 | .882 | .784 | .662 | .531 | .409 | .307 |
| 7.500 | 7.500 | .000 | .176 | .286 | 11 | .999 | .996 | .981 | .943 | .874 | .769 | .639 | .499 | .367 | .258 |
| | | | | | | .176 | | | | | | | | | |
| 7.500 | 7.500 | .000 | .096 | .222 | 12 | .999 | .995 | .980 | .942 | .870 | .763 | .629 | .485 | .350 | .238 |
| | | | | | | .153 | .096 | | | | | | | | |
| 7.500 | 7.500 | .000 | .050 | .182 | 13 | .999 | .995 | .980 | .941 | .869 | .760 | .625 | .479 | .343 | .229 |
| | | | | | | .144 | .086 | .050 | | | | | | | |
| 7.500 | 7.500 | .000 | .024 | .154 | 14 | .999 | .995 | .980 | .941 | .868 | .759 | .623 | .477 | .340 | .226 |
| | | | | | | .140 | .082 | .045 | .024 | | | | | | |
| 7.500 | 7.500 | .000 | .011 | .133 | 15 | .999 | .995 | .980 | .941 | .868 | .759 | .622 | .476 | .339 | .224 |
| | | | | | | .139 | .080 | .044 | .023 | .011 | | | | | |
| 7.500 | 7.500 | .000 | .005 | .118 | 16 | .999 | .995 | .980 | .941 | .868 | .759 | .622 | .476 | .338 | .224 |
| | | | | | | .138 | .080 | .043 | .022 | .011 | .005 | | | | |
| 7.500 | 7.500 | .000 | .002 | .105 | 17 | .999 | .995 | .980 | .941 | .868 | .759 | .622 | .475 | .338 | .224 |
| | | | | | | .138 | .079 | .043 | .022 | .010 | .005 | .002 | | | |
| 7.500 | 7.500 | .000 | .001 | .095 | 18 | .999 | .995 | .980 | .941 | .868 | .759 | .622 | .475 | .338 | .224 |
| | | | | | | .138 | .079 | .043 | .022 | .010 | .005 | .002 | .001 | | |
| 8.000 | 8.000 | .000 | .653 | 1.000 | 9 | .999 | .998 | .992 | .975 | .942 | .888 | .817 | .735 | .653 | |
| 8.000 | 8.000 | .000 | .409 | .500 | 10 | .999 | .998 | .989 | .965 | .918 | .842 | .742 | .627 | .511 | .409 |
| 8.000 | 8.000 | .000 | .245 | .333 | 11 | .999 | .997 | .987 | .961 | .908 | .823 | .710 | .581 | .452 | .337 |
| | | | | | | .245 | | | | | | | | | |

ERLANG C: INFINITE QUEUE - ALL CALLERS BLOCKED WILL WAIT AS LONG AS NECESSARY FOR CIRCUIT

| HOURS FIRST ATTEMPT VMR = 1.0 | HOURS CON-NECT | HOURS OVFL | DELAYED PORTION | DELAY LENGTH CALL FACTOR | LINE | HOURS CONNECTED TRAFFIC ON LINES 1 | 2 | 3 | 4 | 5 | 6 | 7 | 8 | 9 | 10 |
|---|---|---|---|---|---|---|---|---|---|---|---|---|---|---|---|
| 8.000 | 8.000 | .000 | .140 | .250 | 12 | .999 | .997 | .987 | .959 | .903 | .815 | .696 | .561 | .426 | .306 |
| | | | | | | .210 | .140 | | | | | | | | |
| 8.000 | 8.000 | .000 | .076 | .200 | 13 | .999 | .997 | .986 | .958 | .902 | .811 | .691 | .553 | .415 | .293 |
| | | | | | | .195 | .123 | .076 | | | | | | | |
| 8.000 | 8.000 | .000 | .039 | .167 | 14 | .999 | .997 | .986 | .958 | .901 | .810 | .688 | .549 | .411 | .287 |
| | | | | | | .188 | .117 | .069 | .039 | | | | | | |
| 8.000 | 8.000 | .000 | .019 | .143 | 15 | .999 | .997 | .986 | .958 | .901 | .809 | .687 | .548 | .409 | .285 |
| | | | | | | .186 | .114 | .066 | .036 | .019 | | | | | |
| 8.000 | 8.000 | .000 | .009 | .125 | 16 | .999 | .997 | .986 | .958 | .900 | .809 | .687 | .547 | .408 | .284 |
| | | | | | | .185 | .113 | .065 | .035 | .018 | .009 | | | | |
| 8.000 | 8.000 | .000 | .004 | .111 | 17 | .999 | .997 | .986 | .958 | .900 | .809 | .687 | .547 | .408 | .284 |
| | | | | | | .184 | .112 | .064 | .034 | .018 | .009 | .004 | | | |
| 8.000 | 8.000 | .000 | .002 | .100 | 18 | .999 | .997 | .986 | .958 | .900 | .809 | .687 | .547 | .408 | .283 |
| | | | | | | .184 | .112 | .064 | .034 | .017 | .008 | .004 | .002 | | |
| 8.000 | 8.000 | .000 | .001 | .091 | 19 | .999 | .997 | .986 | .958 | .900 | .809 | .687 | .547 | .407 | .283 |
| | | | | | | .184 | .112 | .064 | .034 | .017 | .008 | .004 | .002 | .001 | |
| 8.500 | 8.500 | .000 | .817 | 2.000 | 9 | .999 | .999 | .997 | .989 | .974 | .948 | .910 | .865 | .817 | |
| 8.500 | 8.500 | .000 | .530 | .667 | 10 | .999 | .999 | .993 | .978 | .946 | .892 | .816 | .722 | .623 | .530 |
| 8.500 | 8.500 | .000 | .330 | .400 | 11 | .999 | .998 | .992 | .974 | .935 | .869 | .775 | .661 | .541 | .427 |
| | | | | | | .330 | | | | | | | | | |
| 8.500 | 8.500 | .000 | .196 | .286 | 12 | .999 | .998 | .991 | .971 | .930 | .858 | .757 | .635 | .505 | .382 |
| | | | | | | .277 | .196 | | | | | | | | |
| 8.500 | 8.500 | .000 | .112 | .222 | 13 | .999 | .998 | .991 | .971 | .927 | .854 | .750 | .623 | .489 | .362 |
| | | | | | | .254 | .171 | .112 | | | | | | | |
| 8.500 | 8.500 | .000 | .060 | .182 | 14 | .999 | .998 | .991 | .970 | .926 | .852 | .746 | .618 | .482 | .353 |
| | | | | | | .244 | .159 | .100 | .060 | | | | | | |
| 8.500 | 8.500 | .000 | .031 | .154 | 15 | .999 | .998 | .991 | .970 | .926 | .851 | .745 | .616 | .479 | .350 |
| | | | | | | .240 | .155 | .094 | .055 | .031 | | | | | |
| 8.500 | 8.500 | .000 | .015 | .133 | 16 | .999 | .998 | .991 | .970 | .926 | .851 | .744 | .615 | .478 | .348 |
| | | | | | | .238 | .153 | .092 | .053 | .029 | .015 | | | | |
| 8.500 | 8.500 | .000 | .007 | .118 | 17 | .999 | .998 | .991 | .970 | .926 | .850 | .744 | .615 | .477 | .347 |
| | | | | | | .237 | .152 | .091 | .052 | .028 | .014 | .007 | | | |
| 8.500 | 8.500 | .000 | .003 | .105 | 18 | .999 | .998 | .991 | .970 | .926 | .850 | .744 | .614 | .477 | .347 |
| | | | | | | .237 | .152 | .091 | .052 | .028 | .014 | .007 | .003 | | |
| 8.500 | 8.500 | .000 | .001 | .095 | 19 | .999 | .998 | .991 | .970 | .926 | .850 | .744 | .614 | .477 | .347 |
| | | | | | | .237 | .151 | .091 | .051 | .028 | .014 | .007 | .003 | .001 | |
| 8.500 | 8.500 | .000 | .001 | .087 | 20 | .999 | .998 | .991 | .970 | .926 | .850 | .744 | .614 | .477 | .347 |
| | | | | | | .237 | .151 | .091 | .051 | .027 | .014 | .007 | .003 | .001 | .001 |
| 9.000 | 9.000 | .000 | .669 | 1.000 | 10 | .999 | .999 | .996 | .988 | .969 | .935 | .883 | .817 | .743 | .669 |
| 9.000 | 9.000 | .000 | .430 | .500 | 11 | .999 | .999 | .995 | .983 | .956 | .907 | .833 | .739 | .632 | .526 |

ERLANG C: INFINITE QUEUE - ALL CALLERS BLOCKED WILL WAIT AS LONG AS NECESSARY FOR CIRCUIT

| HOURS FIRST ATTEMPT VMR = 1.0 | HOURS CON-NECT | HOURS OVFL | DELAYED PORTION | DELAY LENGTH CALL FACTOR | LINE | HOURS CONNECTED TRAFFIC ON LINES 1 | 2 | 3 | 4 | 5 | 6 | 7 | 8 | 9 | 10 |
|---|---|---|---|---|---|---|---|---|---|---|---|---|---|---|---|
| | | | | | | .430 | | | | | | | | | |
| 9.000 | 9.000 | .000 | .266 | .333 | 12 | .999 | .999 | .994 | .981 | .950 | .894 | .811 | .704 | .584 | .463 |
| | | | | | | .355 | .266 | | | | | | | | |
| 9.000 | 9.000 | .000 | .158 | .250 | 13 | .999 | .999 | .994 | .980 | .947 | .889 | .801 | .688 | .562 | .435 |
| | | | | | | .321 | .228 | .158 | | | | | | | |
| 9.000 | 9.000 | .000 | .089 | .200 | 14 | .999 | .999 | .994 | .979 | .946 | .886 | .797 | .681 | .552 | .422 |
| | | | | | | .306 | .210 | .139 | .089 | | | | | | |
| 9.000 | 9.000 | .000 | .048 | .167 | 15 | .999 | .999 | .994 | .979 | .945 | .885 | .795 | .678 | .548 | .417 |
| | | | | | | .299 | .203 | .130 | .080 | .048 | | | | | |
| 9.000 | 9.000 | .000 | .025 | .143 | 16 | .999 | .999 | .994 | .979 | .945 | .885 | .794 | .677 | .546 | .414 |
| | | | | | | .296 | .199 | .127 | .077 | .044 | .025 | | | | |
| 9.000 | 9.000 | .000 | .012 | .125 | 17 | .999 | .999 | .994 | .979 | .945 | .884 | .793 | .676 | .545 | .413 |
| | | | | | | .295 | .198 | .125 | .075 | .043 | .023 | .012 | | | |
| 9.000 | 9.000 | .000 | .006 | .111 | 18 | .999 | .999 | .994 | .979 | .945 | .884 | .793 | .676 | .545 | .413 |
| | | | | | | .294 | .197 | .125 | .074 | .042 | .022 | .012 | .006 | | |
| 9.000 | 9.000 | .000 | .003 | .100 | 19 | .999 | .999 | .994 | .979 | .945 | .884 | .793 | .676 | .544 | .413 |
| | | | | | | .294 | .197 | .124 | .074 | .042 | .022 | .011 | .005 | .003 | |
| 9.000 | 9.000 | .000 | .001 | .091 | 20 | .999 | .999 | .994 | .979 | .945 | .884 | .793 | .676 | .544 | .413 |
| | | | | | | .294 | .197 | .124 | .074 | .042 | .022 | .011 | .005 | .002 | .001 |
| 9.500 | 9.500 | .000 | .826 | 2.000 | 10 | .999 | .999 | .999 | .995 | .987 | .970 | .945 | .910 | .869 | .826 |
| 9.500 | 9.500 | .000 | .548 | .667 | 11 | .999 | .999 | .997 | .990 | .972 | .938 | .884 | .812 | .726 | .634 |
| | | | | | | .548 | | | | | | | | | |
| 9.500 | 9.500 | .000 | .350 | .400 | 12 | .999 | .999 | .996 | .987 | .965 | .924 | .857 | .768 | .661 | .549 |
| | | | | | | .442 | .350 | | | | | | | | |
| 9.500 | 9.500 | .000 | .215 | .286 | 13 | .999 | .999 | .996 | .986 | .962 | .917 | .845 | .748 | .632 | .510 |
| | | | | | | .394 | .294 | .215 | | | | | | | |
| 9.500 | 9.500 | .000 | .127 | .222 | 14 | .999 | .999 | .996 | .986 | .961 | .914 | .840 | .739 | .619 | .493 |
| | | | | | | .372 | .269 | .187 | .127 | | | | | | |
| 9.500 | 9.500 | .000 | .071 | .182 | 15 | .999 | .999 | .996 | .985 | .960 | .913 | .837 | .735 | .613 | .485 |
| | | | | | | .363 | .257 | .174 | .113 | .071 | | | | | |
| 9.500 | 9.500 | .000 | .039 | .154 | 16 | .999 | .999 | .996 | .985 | .960 | .912 | .836 | .733 | .610 | .481 |
| | | | | | | .358 | .252 | .168 | .107 | .065 | .039 | | | | |
| 9.500 | 9.500 | .000 | .020 | .133 | 17 | .999 | .999 | .996 | .985 | .060 | .912 | .835 | .732 | .609 | .479 |
| | | | | | | .356 | .250 | .165 | .104 | .062 | .036 | .020 | | | |
| 9.500 | 9.500 | .000 | .010 | .118 | 18 | .999 | .999 | .996 | .985 | .960 | .912 | .835 | .732 | .609 | .479 |
| | | | | | | .355 | .249 | .164 | .103 | .061 | .034 | .019 | .010 | | |
| 9.500 | 9.500 | .000 | .005 | .105 | 19 | .999 | .999 | .996 | .985 | .960 | .912 | .835 | .731 | .608 | .478 |
| | | | | | | .355 | .248 | .164 | .102 | .060 | .034 | .018 | .009 | .005 | |
| 9.500 | 9.500 | .000 | .002 | .095 | 20 | .999 | .999 | .996 | .985 | .960 | .911 | .835 | .731 | .608 | .478 |
| | | | | | | .355 | .248 | .164 | .102 | .060 | .034 | .018 | .009 | .004 | .002 |

ERLANG C: INFINITE QUEUE - ALL CALLERS BLOCKED WILL WAIT AS LONG AS NECESSARY FOR CIRCUIT

| HOURS FIRST ATTEMPT VMR = 1.0 | HOURS CON-NECT | HOURS OVFL | DELAYED PORTION | DELAY LENGTH CALL FACTOR | LINE | HOURS CONNECTED TRAFFIC ON LINES 1 | 2 | 3 | 4 | 5 | 6 | 7 | 8 | 9 | 10 |
|---|---|---|---|---|---|---|---|---|---|---|---|---|---|---|---|
| 9.500 | 9.500 | .000 | .001 | .087 | 21 | .999 | .999 | .996 | .985 | .960 | .911 | .835 | .731 | .608 | .478 |
| | | | | | | .355 | .248 | .164 | .102 | .060 | .034 | .018 | .009 | .004 | .002 |
| | | | | | | .001 | | | | | | | | | |
| 10.000 | 10.000 | .000 | .682 | 1.000 | 11 | .999 | .999 | .998 | .994 | .984 | .963 | .929 | .880 | .819 | .750 |
| | | | | | | .682 | | | | | | | | | |
| 10.000 | 10.000 | .000 | .449 | .500 | 12 | .999 | .999 | .998 | .992 | .977 | .947 | .897 | .826 | .737 | .638 |
| | | | | | | .539 | .449 | | | | | | | | |
| 10.000 | 10.000 | .000 | .285 | .333 | 13 | .999 | .999 | .997 | .991 | .974 | .939 | .882 | .801 | .699 | .587 |
| | | | | | | .474 | .371 | .285 | | | | | | | |
| 10.000 | 10.000 | .000 | .174 | .250 | 14 | .999 | .999 | .997 | .990 | .972 | .936 | .876 | .790 | .682 | .563 |
| | | | | | | .443 | .334 | .244 | .174 | | | | | | |
| 10.000 | 10.000 | .000 | .102 | .200 | 15 | .999 | .999 | .997 | .990 | .971 | .934 | .872 | .784 | .674 | .551 |
| | | | | | | .429 | .317 | .224 | .153 | .102 | | | | | |
| 10.000 | 10.000 | .000 | .057 | .167 | 16 | .999 | .999 | .997 | .990 | .971 | .934 | .871 | .782 | .670 | .546 |
| | | | | | | .422 | .310 | .216 | .143 | .092 | .057 | | | | |
| 10.000 | 10.000 | .000 | .031 | .143 | 17 | .999 | .999 | .997 | .990 | .971 | .933 | .870 | .781 | .668 | .544 |
| | | | | | | .419 | .306 | .212 | .139 | .087 | .052 | .031 | | | |
| 10.000 | 10.000 | .000 | .016 | .125 | 18 | .999 | .999 | .997 | .990 | .971 | .933 | .870 | .780 | .668 | .543 |
| | | | | | | .418 | .304 | .210 | .137 | .085 | .050 | .029 | .016 | | |
| 10.000 | 10.000 | .000 | .008 | .111 | 19 | .999 | .999 | .997 | .990 | .971 | .933 | .870 | .780 | .667 | .542 |
| | | | | | | .417 | .304 | .209 | .136 | .084 | .049 | .028 | .015 | .008 | |
| 10.000 | 10.000 | .000 | .004 | .100 | 20 | .999 | .999 | .997 | .990 | .971 | .933 | .870 | .780 | .667 | .542 |
| | | | | | | .417 | .303 | .209 | .136 | .084 | .049 | .027 | .015 | .007 | .004 |
| 10.000 | 10.000 | .000 | .002 | .091 | 21 | .999 | .999 | .997 | .990 | .971 | .933 | .870 | .780 | .667 | .542 |
| | | | | | | .417 | .303 | .209 | .136 | .084 | .049 | .027 | .014 | .007 | .004 |
| | | | | | | .002 | | | | | | | | | |
| 10.000 | 10.000 | .000 | .001 | .083 | 22 | .999 | .999 | .997 | .990 | .971 | .933 | .870 | .780 | .667 | .542 |
| | | | | | | .417 | .303 | .208 | .136 | .083 | .049 | .027 | .014 | .007 | .003 |
| | | | | | | .002 | .001 | | | | | | | | |
| 11.000 | 11.000 | .000 | .694 | 1.000 | 12 | .999 | .999 | .999 | .997 | .992 | .980 | .958 | .924 | .877 | .820 |
| | | | | | | .757 | .694 | | | | | | | | |
| 11.000 | 11.000 | .000 | .466 | .500 | 13 | .999 | .999 | .999 | .996 | .988 | .971 | .939 | .889 | .820 | .736 |
| | | | | | | .644 | .551 | .466 | | | | | | | |
| 11.000 | 11.000 | .000 | .303 | .333 | 14 | .999 | .999 | .999 | .996 | .987 | .967 | .930 | .872 | .793 | .696 |
| | | | | | | .590 | .483 | .386 | .303 | | | | | | |
| 11.000 | 11.000 | .000 | .190 | .250 | 15 | .999 | .999 | .999 | .995 | .986 | .964 | .925 | .864 | .780 | .677 |
| | | | | | | .564 | .450 | .347 | .259 | .190 | | | | | |
| 11.000 | 11.000 | .000 | .115 | .200 | 16 | .999 | .999 | .999 | .995 | .985 | .963 | .923 | .860 | .774 | .668 |
| | | | | | | .551 | .435 | .328 | .238 | .167 | .115 | | | | |
| 11.000 | 11.000 | .000 | .066 | .167 | 17 | .999 | .999 | .999 | .995 | .985 | .963 | .922 | .858 | .771 | .663 |

ERLANG C: INFINITE QUEUE - ALL CALLERS BLOCKED WILL WAIT AS LONG AS NECESSARY FOR CIRCUIT

| HOURS FIRST ATTEMPT VMR = 1.0 | HOURS CON-NECT | HOURS OVFL | DELAYED PORTION | DELAY LENGTH CALL FACTOR | LINE | HOURS CONNECTED TRAFFIC ON LINES 1 | 2 | 3 | 4 | 5 | 6 | 7 | 8 | 9 | 10 |
|---|---|---|---|---|---|---|---|---|---|---|---|---|---|---|---|
| | | | | | | .545 | .427 | .319 | .227 | .156 | .103 | .066 | | | |
| 11.000 | 11.000 | .000 | .037 | .143 | 18 | .999 | .999 | .999 | .995 | .985 | .963 | .922 | .858 | .769 | .661 |
| | | | | | | .542 | .424 | .315 | .223 | .150 | .097 | .061 | .037 | | |
| 11.000 | 11.000 | .000 | .020 | .125 | 19 | .999 | .999 | .999 | .995 | .985 | .963 | .922 | .857 | .769 | .660 |
| | | | | | | .541 | .422 | .313 | .220 | .148 | .095 | .058 | .034 | .020 | |
| 11.000 | 11.000 | .000 | .010 | .111 | 20 | .999 | .999 | .999 | .995 | .985 | .963 | .921 | .857 | .768 | .660 |
| | | | | | | .541 | .421 | .312 | .219 | .147 | .093 | .057 | .033 | .019 | .010 |
| 11.000 | 11.000 | .000 | .005 | .100 | 21 | .999 | .999 | .999 | .995 | .985 | .962 | .921 | .857 | .768 | .660 |
| | | | | | | .540 | .421 | .312 | .219 | .146 | .093 | .056 | .033 | .018 | .010 |
| | | | | | | .005 | | | | | | | | | |
| 11.000 | 11.000 | .000 | .002 | .091 | 22 | .999 | .999 | .999 | .995 | .985 | .962 | .921 | .857 | .768 | .660 |
| | | | | | | .540 | .421 | .311 | .219 | .146 | .093 | .056 | .032 | .018 | .009 |
| | | | | | | .005 | .002 | | | | | | | | |
| 11.000 | 11.000 | .000 | .001 | .083 | 23 | .999 | .999 | .999 | .995 | .985 | .962 | .921 | .857 | .768 | .660 |
| | | | | | | .540 | .421 | .311 | .219 | .146 | .093 | .056 | .032 | .018 | .009 |
| | | | | | | .005 | .002 | .001 | | | | | | | |
| 12.000 | 12.000 | .000 | .704 | 1.000 | 13 | .999 | .999 | .999 | .999 | .996 | .990 | .976 | .954 | .920 | .876 |
| | | | | | | .822 | .763 | .704 | | | | | | | |
| 12.000 | 12.000 | .000 | .482 | .500 | 14 | .999 | .999 | .999 | .998 | .994 | .985 | .965 | .932 | .882 | .816 |
| | | | | | | .736 | .649 | .562 | .482 | | | | | | |
| 12.000 | 12.000 | .000 | .319 | .333 | 15 | .999 | .999 | .999 | .998 | .993 | .982 | .960 | .921 | .863 | .786 |
| | | | | | | .694 | .593 | .492 | .399 | .319 | | | | | |
| 12.000 | 12.000 | .000 | .205 | .250 | 16 | .999 | .999 | .999 | .998 | .993 | .981 | .957 | .916 | .854 | .772 |
| | | | | | | .673 | .565 | .457 | .358 | .273 | .205 | | | | |
| 12.000 | 12.000 | .000 | .127 | .200 | 17 | .999 | .999 | .999 | .998 | .993 | .980 | .955 | .913 | .849 | .764 |
| | | | | | | .663 | .551 | .440 | .338 | .250 | .179 | .127 | | | |
| 12.000 | 12.000 | .000 | .076 | .167 | 18 | .999 | .999 | .999 | .998 | .993 | .980 | .955 | .912 | .847 | .761 |
| | | | | | | .657 | .545 | .432 | .328 | .238 | .167 | .113 | .076 | | |
| 12.000 | 12.000 | .000 | .044 | .143 | 19 | .999 | .999 | .999 | .998 | .992 | .980 | .954 | .911 | .846 | .759 |
| | | | | | | .655 | .541 | .428 | .323 | .233 | .161 | .107 | .069 | .044 | |
| 12.000 | 12.000 | .000 | .024 | .125 | 20 | .999 | .999 | .999 | .998 | .992 | .980 | .954 | .911 | .845 | .758 |
| | | | | | | .654 | .540 | .426 | .320 | .230 | .158 | .104 | .066 | .040 | .024 |
| 12.000 | 12.000 | .000 | .013 | .111 | 21 | .999 | .999 | .900 | .998 | .992 | .980 | .954 | .911 | .845 | .758 |
| | | | | | | .653 | .539 | .425 | .319 | .229 | .157 | .102 | .064 | .039 | .023 |
| | | | | | | .013 | | | | | | | | | |
| 12.000 | 12.000 | .000 | .007 | .100 | 22 | .999 | .999 | .999 | .998 | .992 | .980 | .954 | .911 | .845 | .758 |
| | | | | | | .653 | .539 | .424 | .319 | .228 | .156 | .102 | .064 | .038 | .022 |
| | | | | | | .012 | .007 | | | | | | | | |
| 12.000 | 12.000 | .000 | .003 | .091 | 23 | .999 | .999 | .999 | .998 | .992 | .980 | .954 | .911 | .845 | .758 |
| | | | | | | .653 | .539 | .424 | .319 | .228 | .156 | .102 | .063 | .038 | .022 |

ERLANG C: INFINITE QUEUE - ALL CALLERS BLOCKED WILL WAIT AS LONG AS NECESSARY FOR CIRCUIT

| HOURS FIRST ATTEMPT VMR = 1.0 | HOURS CON-NECT | HOURS OVFL | DELAYED PORTION | DELAY LENGTH CALL FACTOR | LINE | HOURS CONNECTED TRAFFIC ON LINES 1 | 2 | 3 | 4 | 5 | 6 | 7 | 8 | 9 | 10 |
|---|---|---|---|---|---|---|---|---|---|---|---|---|---|---|---|
| | | | | | | .012 | .006 | .003 | | | | | | | |
| 12.000 | 12.000 | .000 | .002 | .083 | 24 | .999 | .999 | .999 | .998 | .992 | .980 | .954 | .911 | .845 | .758 |
| | | | | | | .653 | .538 | .424 | .319 | .228 | .156 | .101 | .063 | .038 | .021 |
| | | | | | | .012 | .006 | .003 | .002 | | | | | | |
| 12.000 | 12.000 | .000 | .001 | .077 | 25 | .999 | .999 | .999 | .998 | .992 | .980 | .954 | .910 | .845 | .758 |
| | | | | | | .653 | .538 | .424 | .318 | .228 | .156 | .101 | .063 | .037 | .021 |
| | | | | | | .012 | .006 | .003 | .002 | .001 | | | | | |
| 13.000 | 13.000 | .000 | .714 | 1.000 | 14 | .999 | .999 | .999 | .999 | .998 | .995 | .987 | .973 | .950 | .917 |
| | | | | | | .874 | .824 | .769 | .714 | | | | | | |
| 13.000 | 13.000 | .000 | .496 | .500 | 15 | .999 | .999 | .999 | .999 | .997 | .992 | .981 | .960 | .925 | .876 |
| | | | | | | .812 | .736 | .654 | .572 | .496 | | | | | |
| 13.000 | 13.000 | .000 | .334 | .333 | 16 | .999 | .999 | .999 | .999 | .997 | .991 | .977 | .953 | .913 | .855 |
| | | | | | | .781 | .692 | .596 | .500 | .411 | .334 | | | | |
| 13.000 | 13.000 | .000 | .219 | .250 | 17 | .999 | .999 | .999 | .999 | .997 | .990 | .976 | .949 | .907 | .845 |
| | | | | | | .765 | .670 | .567 | .464 | .369 | .286 | .219 | | | |
| 13.000 | 13.000 | .000 | .138 | .200 | 18 | .999 | .999 | .999 | .999 | .996 | .990 | .975 | .948 | .903 | .840 |
| | | | | | | .756 | .658 | .552 | .445 | .347 | .261 | .192 | .138 | | |
| 13.000 | 13.000 | .000 | .085 | .167 | 19 | .999 | .999 | .999 | .999 | .996 | .989 | .975 | .947 | .902 | .837 |
| | | | | | | .752 | .652 | .544 | .436 | .336 | .249 | .178 | .124 | .085 | |
| 13.000 | 13.000 | .000 | .050 | .143 | 20 | .999 | .999 | .999 | .999 | .996 | .989 | .974 | .946 | .901 | .835 |
| | | | | | | .750 | .650 | .540 | .431 | .330 | .242 | .171 | .116 | .077 | .050 |
| 13.000 | 13.000 | .000 | .029 | .125 | 21 | .999 | .999 | .999 | .999 | .996 | .989 | .974 | .946 | .901 | .835 |
| | | | | | | .749 | .648 | .539 | .429 | .327 | .239 | .168 | .113 | .073 | .046 |
| | | | | | | .029 | | | | | | | | | |
| 13.000 | 13.000 | .000 | .016 | .111 | 22 | .999 | .999 | .999 | .999 | .996 | .989 | .974 | .946 | .900 | .834 |
| | | | | | | .749 | .647 | .538 | .428 | .326 | .238 | .166 | .111 | .071 | .044 |
| | | | | | | .027 | .016 | | | | | | | | |
| 13.000 | 13.000 | .000 | .008 | .100 | 23 | .999 | .999 | .999 | .999 | .996 | .989 | .974 | .946 | .900 | .834 |
| | | | | | | .749 | .647 | .537 | .427 | .325 | .237 | .165 | .110 | .071 | .043 |
| | | | | | | .026 | .015 | .008 | | | | | | | |
| 13.000 | 13.000 | .000 | .004 | .091 | 24 | .999 | .999 | .999 | .999 | .996 | .989 | .974 | .946 | .900 | .834 |
| | | | | | | .748 | .647 | .537 | .427 | .325 | .237 | .165 | .110 | .070 | .043 |
| | | | | | | .025 | .014 | .008 | .004 | | | | | | |
| 13.000 | 13.000 | .000 | .002 | .083 | 25 | .999 | .999 | .999 | .999 | .996 | .989 | .974 | .946 | .900 | .834 |
| | | | | | | .748 | .647 | .537 | .427 | .325 | .237 | .165 | .110 | .070 | .043 |
| | | | | | | .025 | .014 | .008 | .004 | .002 | | | | | |
| 13.000 | 13.000 | .000 | .001 | .077 | 26 | .999 | .999 | .999 | .999 | .996 | .989 | .974 | .946 | .900 | .834 |
| | | | | | | .748 | .647 | .537 | .427 | .325 | .236 | .165 | .110 | .070 | .043 |
| | | | | | | .025 | .014 | .008 | .004 | .002 | .001 | | | | |
| 14.000 | 14.000 | .000 | .722 | 1.000 | 15 | .999 | .999 | .999 | .999 | .999 | .997 | .993 | .985 | .970 | .947 |

ERLANG C: INFINITE QUEUE - ALL CALLERS BLOCKED WILL WAIT AS LONG AS NECESSARY FOR CIRCUIT

| HOURS FIRST ATTEMPT VMR = 1.0 | HOURS CON-NECT | HOURS OVFL | DELAYED PORTION | DELAY LENGTH CALL FACTOR | LINE | HOURS CONNECTED TRAFFIC ON LINES 1 | 2 | 3 | 4 | 5 | 6 | 7 | 8 | 9 | 10 |
|---|---|---|---|---|---|---|---|---|---|---|---|---|---|---|---|
| | | | | | | .914 | .873 | .826 | .774 | .722 | | | | | |
| 14.000 | 14.000 | .000 | .508 | .500 | 16 | .999 | .999 | .999 | .999 | .999 | .996 | .990 | .977 | .954 | .920 |
| | | | | | | .871 | .809 | .737 | .659 | .581 | .508 | | | | |
| 14.000 | 14.000 | .000 | .348 | .333 | 17 | .999 | .999 | .999 | .999 | .998 | .995 | .988 | .973 | .946 | .906 |
| | | | | | | .849 | .776 | .691 | .600 | .508 | .423 | .348 | | | |
| 14.000 | 14.000 | .000 | .232 | .250 | 18 | .999 | .999 | .999 | .999 | .998 | .995 | .987 | .971 | .942 | .898 |
| | | | | | | .837 | .758 | .667 | .569 | .470 | .378 | .298 | .232 | | |
| 14.000 | 14.000 | .000 | .150 | .200 | 19 | .999 | .999 | .999 | .999 | .998 | .995 | .986 | .970 | .940 | .895 |
| | | | | | | .831 | .749 | .655 | .553 | .450 | .355 | .272 | .203 | .150 | |
| 14.000 | 14.000 | .000 | .094 | .167 | 20 | .999 | .999 | .999 | .999 | .998 | .995 | .986 | .969 | .939 | .893 |
| | | | | | | .828 | .745 | .648 | .544 | .440 | .343 | .258 | .188 | .134 | .094 |
| 14.000 | 14.000 | .000 | .057 | .143 | 21 | .999 | .999 | .999 | .999 | .998 | .995 | .986 | .969 | .939 | .892 |
| | | | | | | .826 | .742 | .645 | .540 | .435 | .337 | .251 | .180 | .125 | .085 |
| | | | | | | .057 | | | | | | | | | |
| 14.000 | 14.000 | .000 | .033 | .125 | 22 | .999 | .999 | .999 | .999 | .998 | .994 | .986 | .969 | .938 | .891 |
| | | | | | | .825 | .741 | .643 | .538 | .432 | .334 | .247 | .177 | .121 | .081 |
| | | | | | | .052 | .033 | | | | | | | | |
| 14.000 | 14.000 | .000 | .019 | .111 | 23 | .999 | .999 | .999 | .999 | .998 | .994 | .986 | .968 | .938 | .891 |
| | | | | | | .825 | .741 | .642 | .537 | .431 | .332 | .246 | .175 | .119 | .078 |
| | | | | | | .050 | .031 | .019 | | | | | | | |
| 14.000 | 14.000 | .000 | .010 | .100 | 24 | .999 | .999 | .999 | .999 | .998 | .994 | .986 | .968 | .938 | .891 |
| | | | | | | .824 | .740 | .642 | .536 | .430 | .331 | .245 | .174 | .118 | .077 |
| | | | | | | .049 | .030 | .018 | .010 | | | | | | |
| 14.000 | 14.000 | .000 | .005 | .091 | 25 | .999 | .999 | .999 | .999 | .998 | .994 | .986 | .968 | .938 | .891 |
| | | | | | | .824 | .740 | .642 | .536 | .430 | .331 | .244 | .173 | .118 | .077 |
| | | | | | | .048 | .029 | .017 | .010 | .005 | | | | | |
| 14.000 | 14.000 | .000 | .003 | .083 | 26 | .999 | .999 | .999 | .999 | .998 | .994 | .986 | .968 | .938 | .891 |
| | | | | | | .824 | .740 | .642 | .536 | .430 | .331 | .244 | .173 | .118 | .077 |
| | | | | | | .048 | .029 | .017 | .010 | .005 | .003 | | | | |
| 14.000 | 14.000 | .000 | .001 | .077 | 27 | .999 | .999 | .999 | .999 | .998 | .994 | .986 | .968 | .938 | .891 |
| | | | | | | .824 | .740 | .642 | .536 | .430 | .331 | .244 | .173 | .117 | .077 |
| | | | | | | .048 | .029 | .017 | .009 | .005 | .003 | .001 | | | |
| 14.000 | 14.000 | .000 | .001 | .071 | 28 | .999 | .999 | .999 | .999 | .998 | .994 | .986 | .968 | .938 | .891 |
| | | | | | | .824 | .740 | .642 | .536 | .430 | .331 | .244 | .173 | .117 | .077 |
| | | | | | | .048 | .029 | .017 | .009 | .005 | .003 | .001 | .001 | | |
| 15.000 | 15.000 | .000 | .730 | 1.000 | 16 | .999 | .999 | .999 | .999 | .999 | .999 | .996 | .991 | .982 | .967 |
| | | | | | | .944 | .912 | .873 | .827 | .779 | .730 | | | | |
| 15.000 | 15.000 | .000 | .520 | .500 | 17 | .999 | .999 | .999 | .999 | .999 | .998 | .994 | .987 | .973 | .950 |
| | | | | | | .914 | .867 | .807 | .738 | .664 | .590 | .520 | | | |
| 15.000 | 15.000 | .000 | .361 | .333 | 18 | .999 | .909 | .999 | .999 | .999 | .998 | .993 | .985 | .968 | .940 |

ERLANG C: INFINITE QUEUE - ALL CALLERS BLOCKED WILL WAIT AS LONG AS NECESSARY FOR CIRCUIT

| HOURS FIRST ATTEMPT VMR = 1.0 | HOURS CON-NECT | HOURS OVFL | DELAYED PORTION | DELAY LENGTH CALL FACTOR | LINE | HOURS CONNECTED TRAFFIC ON LINES 1 | 2 | 3 | 4 | 5 | 6 | 7 | 8 | 9 | 10 |
|---|---|---|---|---|---|---|---|---|---|---|---|---|---|---|---|
| | | | | | | .899 | .842 | .772 | .690 | .603 | .516 | .434 | .361 | | |
| 15.000 | 15.000 | .000 | .244 | .250 | 19 | .999 | .999 | .999 | .999 | .999 | .997 | .993 | .983 | .965 | .936 |
| | | | | | | .891 | .830 | .753 | .665 | .571 | .476 | .387 | .309 | .244 | |
| 15.000 | 15.000 | .000 | .160 | .200 | 20 | .999 | .999 | .999 | .999 | .999 | .997 | .993 | .983 | .964 | .933 |
| | | | | | | .886 | .823 | .743 | .652 | .553 | .455 | .363 | .282 | .214 | .160 |
| 15.000 | 15.000 | .000 | .102 | .167 | 21 | .999 | .999 | .999 | .999 | .999 | .997 | .993 | .982 | .963 | .932 |
| | | | | | | .884 | .819 | .738 | .644 | .544 | .444 | .350 | .267 | .198 | .143 |
| | | | | | | .102 | | | | | | | | | |
| 15.000 | 15.000 | .000 | .063 | .143 | 22 | .999 | .999 | .999 | .999 | .999 | .997 | .992 | .982 | .963 | .931 |
| | | | | | | .883 | .817 | .735 | .641 | .539 | .438 | .343 | .259 | .189 | .134 |
| | | | | | | .093 | .063 | | | | | | | | |
| 15.000 | 15.000 | .000 | .038 | .125 | 23 | .999 | .999 | .999 | .999 | .999 | .997 | .992 | .982 | .963 | .931 |
| | | | | | | .882 | .816 | .734 | .639 | .537 | .435 | .339 | .255 | .185 | .130 |
| | | | | | | .088 | .058 | .038 | | | | | | | |
| 15.000 | 15.000 | .000 | .022 | .111 | 24 | .999 | .999 | .999 | .999 | .999 | .997 | .992 | .982 | .963 | .930 |
| | | | | | | .882 | .816 | .733 | .638 | .536 | .433 | .338 | .253 | .183 | .127 |
| | | | | | | .085 | .056 | .035 | .022 | | | | | | |
| 15.000 | 15.000 | .000 | .012 | .100 | 25 | .999 | .999 | .999 | .999 | .999 | .997 | .992 | .982 | .963 | .930 |
| | | | | | | .882 | .815 | .733 | .637 | .535 | .433 | .337 | .252 | .182 | .126 |
| | | | | | | .084 | .054 | .034 | .021 | .012 | | | | | |
| 15.000 | 15.000 | .000 | .007 | .091 | 26 | .999 | .999 | .999 | .999 | .999 | .997 | .992 | .982 | .963 | .930 |
| | | | | | | .882 | .815 | .733 | .637 | .535 | .432 | .336 | .252 | .181 | .125 |
| | | | | | | .084 | .054 | .033 | .020 | .012 | .007 | | | | |
| 15.000 | 15.000 | .000 | .004 | .083 | 27 | .999 | .999 | .999 | .999 | .999 | .997 | .992 | .982 | .963 | .930 |
| | | | | | | .882 | .815 | .732 | .637 | .534 | .432 | .336 | .251 | .181 | .125 |
| | | | | | | .083 | .053 | .033 | .020 | .011 | .006 | .004 | | | |
| 15.000 | 15.000 | .000 | .002 | .077 | 28 | .999 | .999 | .999 | .999 | .999 | .997 | .992 | .982 | .963 | .930 |
| | | | | | | .882 | .815 | .732 | .637 | .534 | .432 | .336 | .251 | .181 | .125 |
| | | | | | | .083 | .053 | .033 | .020 | .011 | .006 | .003 | .002 | | |
| 15.000 | 15.000 | .000 | .001 | .071 | 29 | .999 | .999 | .999 | .999 | .999 | .997 | .992 | .982 | .963 | .930 |
| | | | | | | .882 | .815 | .732 | .637 | .534 | .432 | .336 | .251 | .181 | .125 |
| | | | | | | .083 | .053 | .033 | .020 | .011 | .006 | .003 | .002 | .001 | |
| 16.000 | 16.000 | .000 | .737 | 1.000 | 17 | .999 | .999 | .999 | .999 | .999 | .999 | .998 | .995 | .990 | .980 |
| | | | | | | .964 | .941 | .910 | .873 | .829 | .783 | .737 | | | |
| 16.000 | 16.000 | .000 | .531 | .500 | 18 | .999 | .999 | .999 | .999 | .999 | .999 | .997 | .993 | .984 | .969 |
| | | | | | | .945 | .910 | .863 | .805 | .739 | .668 | .598 | .531 | | |
| 16.000 | 16.000 | .000 | .374 | .333 | 19 | .999 | .999 | .999 | .999 | .999 | .999 | .997 | .992 | .981 | .963 |
| | | | | | | .935 | .893 | .837 | .768 | .690 | .606 | .522 | .444 | .374 | |
| 16.000 | 16.000 | .000 | .256 | .250 | 20 | .999 | .999 | .999 | .999 | .999 | .999 | .996 | .991 | .980 | .960 |
| | | | | | | .929 | .884 | .823 | .749 | .663 | .573 | .482 | .396 | .320 | .256 |

ERLANG C: INFINITE QUEUE - ALL CALLERS BLOCKED WILL WAIT AS LONG AS NECESSARY FOR CIRCUIT

| HOURS FIRST ATTEMPT VMR = 1.0 | HOURS CON-NECT | HOURS OVFL | DELAYED PORTION | DELAY LENGTH CALL FACTOR | LINE | HOURS CONNECTED TRAFFIC ON LINES 1 | 2 | 3 | 4 | 5 | 6 | 7 | 8 | 9 | 10 |
|---|---|---|---|---|---|---|---|---|---|---|---|---|---|---|---|
| 16.000 | 16.000 | .000 | .171 | .200 | 21 | .999 | .999 | .999 | .999 | .999 | .999 | .996 | .990 | .979 | .959 |
| | | | | | | .926 | .879 | .816 | .738 | .649 | .554 | .460 | .370 | .291 | .224 |
| | | | | | | .171 | | | | | | | | | |
| 16.000 | 16.000 | .000 | .111 | .167 | 22 | .999 | .999 | .999 | .999 | .999 | .999 | .996 | .990 | .979 | .958 |
| | | | | | | .924 | .876 | .811 | .732 | .641 | .544 | .448 | .356 | .275 | .207 |
| | | | | | | .153 | .111 | | | | | | | | |
| 16.000 | 16.000 | .000 | .070 | .143 | 23 | .999 | .999 | .999 | .999 | .999 | .999 | .996 | .990 | .978 | .957 |
| | | | | | | .924 | .875 | .809 | .729 | .637 | .539 | .441 | .349 | .267 | .198 |
| | | | | | | .143 | .101 | .070 | | | | | | | |
| 16.000 | 16.000 | .000 | .043 | .125 | 24 | .999 | .999 | .999 | .999 | .999 | .999 | .996 | .990 | .978 | .957 |
| | | | | | | .923 | .874 | .808 | .727 | .635 | .536 | .438 | .345 | .262 | .193 |
| | | | | | | .137 | .095 | .064 | .043 | | | | | | |
| 16.000 | 16.000 | .000 | .025 | .111 | 25 | .999 | .999 | .999 | .999 | .999 | .999 | .996 | .990 | .978 | .957 |
| | | | | | | .923 | .873 | .808 | .726 | .634 | .535 | .436 | .343 | .260 | .190 |
| | | | | | | .135 | .092 | .061 | .040 | .025 | | | | | |
| 16.000 | 16.000 | .000 | .015 | .100 | 26 | .999 | .999 | .999 | .999 | .999 | .999 | .996 | .990 | .978 | .957 |
| | | | | | | .923 | .873 | .807 | .726 | .633 | .534 | .435 | .342 | .259 | .189 |
| | | | | | | .133 | .091 | .060 | .038 | .024 | .015 | | | | |
| 16.000 | 16.000 | .000 | .008 | .091 | 27 | .999 | .999 | .999 | .999 | .999 | .999 | .996 | .990 | .978 | .957 |
| | | | | | | .923 | .873 | .807 | .726 | .633 | .534 | .434 | .341 | .258 | .188 |
| | | | | | | .133 | .090 | .059 | .037 | .023 | .014 | .008 | | | |
| 16.000 | 16.000 | .000 | .004 | .083 | 28 | .999 | .999 | .999 | .999 | .999 | .999 | .996 | .990 | .978 | .957 |
| | | | | | | .923 | .873 | .807 | .726 | .633 | .533 | .434 | .341 | .258 | .188 |
| | | | | | | .132 | .090 | .059 | .037 | .023 | .013 | .008 | .004 | | |
| 16.000 | 16.000 | .000 | .002 | .077 | 29 | .999 | .999 | .999 | .999 | .999 | .999 | .996 | .990 | .978 | .957 |
| | | | | | | .923 | .873 | .807 | .726 | .633 | .533 | .434 | .341 | .258 | .188 |
| | | | | | | .132 | .089 | .058 | .007 | .022 | .013 | .008 | .004 | .002 | |
| 16.000 | 16.000 | .000 | .001 | .071 | 30 | .999 | .999 | .999 | .999 | .999 | .999 | .996 | .990 | .978 | .957 |
| | | | | | | .923 | .873 | .807 | .726 | .633 | .533 | .434 | .341 | .258 | .188 |
| | | | | | | .132 | .089 | .058 | .037 | .022 | .013 | .008 | .004 | .002 | .001 |
| 16.000 | 16.000 | .000 | .001 | .067 | 31 | .999 | .999 | .999 | .999 | .999 | .999 | .996 | .990 | .978 | .957 |
| | | | | | | .923 | .873 | .807 | .725 | .632 | .533 | .434 | .341 | .258 | .188 |
| | | | | | | .132 | .089 | .058 | .037 | .022 | .013 | .007 | .004 | .002 | .001 |
| | | | | | | .001 | | | | | | | | | |
| 17.000 | 17.000 | .000 | .744 | 1.000 | 18 | .999 | .999 | .999 | .999 | .999 | .999 | .999 | .998 | .994 | .988 |
| | | | | | | .978 | .962 | .939 | .909 | .872 | .831 | .787 | .744 | | |
| 17.000 | 17.000 | .000 | .541 | .500 | 19 | .999 | .999 | .999 | .999 | .999 | .999 | .999 | .996 | .991 | .982 |
| | | | | | | .966 | .941 | .905 | .859 | .803 | .740 | .672 | .605 | .541 | |
| 17.000 | 17.000 | .000 | .385 | .333 | 20 | .999 | .999 | .999 | .999 | .999 | .999 | .998 | .995 | .989 | .978 |
| | | | | | | .959 | .929 | .887 | .832 | .705 | .690 | .609 | .529 | .453 | .385 |

ERLANG C: INFINITE QUEUE - ALL CALLERS BLOCKED WILL WAIT AS LONG AS NECESSARY FOR CIRCUIT

| HOURS FIRST ATTEMPT VMR = 1.0 | HOURS CON-NECT | HOURS OVFL | DELAYED PORTION | DELAY LENGTH CALL FACTOR | LINE | HOURS CONNECTED TRAFFIC ON LINES | | | | | | | | | |
|---|---|---|---|---|---|---|---|---|---|---|---|---|---|---|---|
| | | | | | | 1 | 2 | 3 | 4 | 5 | 6 | 7 | 8 | 9 | 10 |
| 17.000 | 17.000 | .000 | .267 | .250 | 21 | .999 | .999 | .999 | .999 | .999 | .999 | .998 | .995 | .989 | .976 |
| | | | | | | .955 | .923 | .877 | .817 | .745 | .662 | .575 | .487 | .404 | .330 |
| | | | | | | .267 | | | | | | | | | |
| 17.000 | 17.000 | .000 | .181 | .200 | 22 | .999 | .999 | .999 | .999 | .999 | .999 | .998 | .995 | .988 | .975 |
| | | | | | | .953 | .920 | .872 | .809 | .733 | .647 | .555 | .464 | .377 | .300 |
| | | | | | | .234 | .181 | | | | | | | | |
| 17.000 | 17.000 | .000 | .119 | .167 | 23 | .999 | .999 | .999 | .999 | .999 | .999 | .998 | .995 | .988 | .975 |
| | | | | | | .952 | .918 | .869 | .804 | .727 | .638 | .545 | .451 | .362 | .283 |
| | | | | | | .216 | .161 | .119 | | | | | | | |
| 17.000 | 17.000 | .000 | .077 | .143 | 24 | .999 | .999 | .999 | .999 | .999 | .999 | .998 | .995 | .988 | .974 |
| | | | | | | .952 | .917 | .867 | .802 | .723 | .634 | .539 | .444 | .354 | .274 |
| | | | | | | .206 | .151 | .108 | .077 | | | | | | |
| 17.000 | 17.000 | .000 | .048 | .125 | 25 | .999 | .999 | .999 | .999 | .999 | .999 | .998 | .995 | .987 | .974 |
| | | | | | | .951 | .916 | .866 | .801 | .721 | .631 | .536 | .440 | .350 | .269 |
| | | | | | | .201 | .145 | .102 | .070 | .048 | | | | | |
| 17.000 | 17.000 | .000 | .029 | .111 | 26 | .999 | .999 | .999 | .999 | .999 | .999 | .998 | .995 | .987 | .974 |
| | | | | | | .951 | .916 | .865 | .800 | .720 | .630 | .534 | .438 | .348 | .267 |
| | | | | | | .198 | .142 | .099 | .067 | .044 | .029 | | | | |
| 17.000 | 17.000 | .000 | .017 | .100 | 27 | .999 | .999 | .999 | .999 | .999 | .999 | .998 | .995 | .987 | .974 |
| | | | | | | .951 | .915 | .865 | .800 | .720 | .629 | .533 | .437 | .346 | .265 |
| | | | | | | .196 | .140 | .097 | .065 | .043 | .027 | .017 | | | |
| 17.000 | 17.000 | .000 | .010 | .091 | 28 | .999 | .999 | .999 | .999 | .999 | .999 | .998 | .995 | .987 | .974 |
| | | | | | | .951 | .915 | .865 | .799 | .719 | .629 | .533 | .437 | .346 | .264 |
| | | | | | | .195 | .139 | .096 | .064 | .042 | .026 | .016 | .010 | | |
| 17.000 | 17.000 | .000 | .005 | .083 | 29 | .999 | .999 | .999 | .999 | .999 | .999 | .998 | .995 | .987 | .974 |
| | | | | | | .951 | .915 | .865 | .799 | .719 | .629 | .532 | .436 | .345 | .264 |
| | | | | | | .195 | .139 | .096 | .064 | .041 | .026 | .016 | .009 | .005 | |
| 17.000 | 17.000 | .000 | .003 | .077 | 30 | .999 | .999 | .999 | .999 | .999 | .999 | .998 | .995 | .987 | .974 |
| | | | | | | .951 | .915 | .865 | .799 | .719 | .629 | .532 | .436 | .345 | .264 |
| | | | | | | .195 | .139 | .095 | .064 | .041 | .025 | .015 | .009 | .005 | .003 |
| 17.000 | 17.000 | .000 | .002 | .071 | 31 | .999 | .999 | .999 | .999 | .999 | .999 | .998 | .995 | .987 | .974 |
| | | | | | | .951 | .915 | .865 | .799 | .719 | .629 | .532 | .436 | .345 | .264 |
| | | | | | | .195 | .139 | .095 | .063 | .041 | .025 | .015 | .009 | .005 | .003 |
| | | | | | | .002 | | | | | | | | | |
| 17.000 | 17.000 | .000 | .001 | .067 | 32 | .999 | .999 | .999 | .999 | .999 | .999 | .998 | .995 | .987 | .974 |
| | | | | | | .951 | .915 | .865 | .799 | .719 | .629 | .532 | .436 | .345 | .264 |
| | | | | | | .195 | .139 | .095 | .063 | .041 | .025 | .015 | .009 | .005 | .003 |
| | | | | | | .001 | .001 | | | | | | | | |
| 18.000 | 18.000 | .000 | .750 | 1.000 | 19 | .999 | .999 | .999 | .999 | .999 | .999 | .999 | .999 | .997 | .993 |
| | | | | | | .986 | .976 | .959 | .937 | .907 | .872 | .833 | .791 | .750 | |

# ERLANG C: INFINITE QUEUE - ALL CALLERS BLOCKED WILL WAIT AS LONG AS NECESSARY FOR CIRCUIT

| HOURS FIRST ATTEMPT VMR=1.0 | HOURS CON-NECT | HOURS OVFL | DELAYED PORTION | DELAY LENGTH CALL FACTOR | LINE | HOURS CONNECTED TRAFFIC ON LINES 1 | 2 | 3 | 4 | 5 | 6 | 7 | 8 | 9 | 10 |
|---|---|---|---|---|---|---|---|---|---|---|---|---|---|---|---|
| 18.000 | 18.000 | .000 | .551 | .500 | 20 | .999 | .999 | .999 | .999 | .999 | .999 | .999 | .998 | .995 | .989 |
| | | | | | | .979 | .962 | .937 | .902 | .856 | .802 | .741 | .677 | .612 | .551 |
| 18.000 | 18.000 | .000 | .396 | .333 | 21 | .999 | .999 | .999 | .999 | .999 | .999 | .999 | .998 | .994 | .987 |
| | | | | | | .975 | .955 | .924 | .882 | .828 | .763 | .690 | .613 | .535 | .462 |
| | | | | | | .396 | | | | | | | | | |
| 18.000 | 18.000 | .000 | .278 | .250 | 22 | .999 | .999 | .999 | .999 | .999 | .999 | .999 | .997 | .994 | .986 |
| | | | | | | .973 | .950 | .917 | .871 | .812 | .741 | .661 | .577 | .492 | .412 |
| | | | | | | .340 | .278 | | | | | | | | |
| 18.000 | 18.000 | .000 | .191 | .200 | 23 | .999 | .999 | .999 | .999 | .999 | .999 | .999 | .997 | .993 | .985 |
| | | | | | | .971 | .948 | .913 | .865 | .803 | .729 | .645 | .556 | .468 | .384 |
| | | | | | | .308 | .244 | .191 | | | | | | | |
| 18.000 | 18.000 | .000 | .128 | .167 | 24 | .999 | .999 | .999 | .999 | .999 | .999 | .999 | .997 | .993 | .985 |
| | | | | | | .971 | .947 | .911 | .862 | .798 | .722 | .636 | .545 | .454 | .368 |
| | | | | | | .291 | .224 | .170 | .128 | | | | | | |
| 18.000 | 18.000 | .000 | .083 | .143 | 25 | .999 | .999 | .999 | .999 | .999 | .999 | .999 | .997 | .993 | .985 |
| | | | | | | .970 | .946 | .910 | .860 | .795 | .718 | .631 | .539 | .447 | .359 |
| | | | | | | .281 | .214 | .159 | .115 | .083 | | | | | |
| 18.000 | 18.000 | .000 | .053 | .125 | 26 | .999 | .999 | .999 | .999 | .999 | .999 | .999 | .997 | .993 | .985 |
| | | | | | | .970 | .946 | .909 | .859 | .794 | .716 | .628 | .535 | .443 | .355 |
| | | | | | | .276 | .208 | .152 | .109 | .076 | .053 | | | | |
| 18.000 | 18.000 | .000 | .033 | .111 | 27 | .999 | .999 | .999 | .999 | .999 | .999 | .999 | .997 | .993 | .985 |
| | | | | | | .970 | .945 | .909 | .858 | .793 | .715 | .627 | .533 | .440 | .352 |
| | | | | | | .273 | .204 | .149 | .105 | .072 | .049 | .033 | | | |
| 18.000 | 18.000 | .000 | .020 | .100 | 28 | .999 | .999 | .999 | .999 | .999 | .999 | .999 | .997 | .993 | .985 |
| | | | | | | .970 | .945 | .909 | .858 | .792 | .714 | .626 | .532 | .439 | .351 |
| | | | | | | .271 | .203 | .147 | .103 | .070 | .047 | .030 | .020 | | |
| 18.000 | 18.000 | .000 | .011 | .091 | 29 | .999 | .909 | .999 | .999 | .990 | .999 | .999 | .997 | .993 | .985 |
| | | | | | | .970 | .945 | .908 | .858 | .792 | .714 | .625 | .532 | .438 | .350 |
| | | | | | | .270 | .202 | .146 | .102 | .069 | .046 | .029 | .018 | .011 | |
| 18.000 | 18.000 | .000 | .007 | .083 | 30 | .999 | .999 | .999 | .999 | .999 | .999 | .999 | .997 | .993 | .985 |
| | | | | | | .970 | .945 | .908 | .857 | .792 | .714 | .625 | .532 | .438 | .349 |
| | | | | | | .270 | .201 | .145 | .102 | .069 | .045 | .029 | .018 | .011 | .007 |
| 18.000 | 18.000 | .000 | .004 | .077 | 31 | .999 | .999 | .999 | .999 | .999 | .999 | .999 | .997 | .993 | .985 |
| | | | | | | .970 | .945 | .908 | .857 | .792 | .713 | .625 | .531 | .438 | .349 |
| | | | | | | .269 | .201 | .145 | .101 | .069 | .045 | .029 | .018 | .011 | .006 |
| | | | | | | .004 | | | | | | | | | |
| 18.000 | 18.000 | .000 | .002 | .071 | 32 | .999 | .999 | .999 | .999 | .999 | .999 | .999 | .997 | .993 | .905 |
| | | | | | | .970 | .945 | .908 | .857 | .792 | .713 | .625 | .531 | .438 | .349 |
| | | | | | | .269 | .201 | .145 | .101 | .068 | .045 | .028 | .017 | .010 | .006 |
| | | | | | | .003 | .002 | | | | | | | | |

ERLANG C: INFINITE QUEUE - ALL CALLERS BLOCKED WILL WAIT AS LONG AS NECESSARY FOR CIRCUIT

| HOURS FIRST ATTEMPT VMR = 1.0 | HOURS CON-NECT | HOURS OVFL | DELAYED PORTION | DELAY LENGTH CALL FACTOR | LINE | HOURS CONNECTED TRAFFIC ON LINES 1 | 2 | 3 | 4 | 5 | 6 | 7 | 8 | 9 | 10 |
|---|---|---|---|---|---|---|---|---|---|---|---|---|---|---|---|
| 18.000 | 18.000 | .000 | .001 | .067 | 33 | .999 | .999 | .999 | .999 | .999 | .999 | .999 | .997 | .993 | .985 |
| | | | | | | .970 | .945 | .908 | .857 | .792 | .713 | .625 | .531 | .438 | .349 |
| | | | | | | .269 | .201 | .145 | .101 | .068 | .045 | .028 | .017 | .010 | .006 |
| | | | | | | .003 | .002 | .001 | | | | | | | |
| 18.000 | 18.000 | .000 | .001 | .063 | 34 | .999 | .999 | .999 | .999 | .999 | .999 | .999 | .997 | .993 | .985 |
| | | | | | | .970 | .945 | .908 | .857 | .792 | .713 | .625 | .531 | .438 | .349 |
| | | | | | | .269 | .201 | .145 | .101 | .068 | .045 | .028 | .017 | .010 | .006 |
| | | | | | | .003 | .002 | .001 | .001 | | | | | | |
| 19.000 | 19.000 | .000 | .755 | 1.000 | 20 | .999 | .999 | .999 | .999 | .999 | .999 | .999 | .999 | .998 | .996 |
| | | | | | | .992 | .985 | .974 | .957 | .935 | .906 | .873 | .835 | .795 | .755 |
| 19.000 | 19.000 | .000 | .560 | .500 | 21 | .999 | .999 | .999 | .999 | .999 | .999 | .999 | .999 | .997 | .994 |
| | | | | | | .988 | .976 | .959 | .933 | .898 | .854 | .801 | .743 | .681 | .619 |
| | | | | | | .560 | | | | | | | | | |
| 19.000 | 19.000 | .000 | .406 | .333 | 22 | .999 | .999 | .999 | .999 | .999 | .999 | .999 | .999 | .997 | .993 |
| | | | | | | .985 | .972 | .950 | .919 | .877 | .824 | .761 | .690 | .616 | .541 |
| | | | | | | .470 | .406 | | | | | | | | |
| 19.000 | 19.000 | .000 | .288 | .250 | 23 | .999 | .999 | .999 | .999 | .999 | .999 | .999 | .999 | .997 | .992 |
| | | | | | | .984 | .969 | .946 | .912 | .866 | .807 | .738 | .660 | .579 | .497 |
| | | | | | | .419 | .349 | .288 | | | | | | | |
| 19.000 | 19.000 | .000 | .200 | .200 | 24 | .999 | .999 | .999 | .999 | .999 | .999 | .999 | .999 | .996 | .992 |
| | | | | | | .983 | .967 | .943 | .907 | .859 | .798 | .725 | .644 | .558 | .472 |
| | | | | | | .390 | .316 | .253 | .200 | | | | | | |
| 19.000 | 19.000 | .000 | .136 | .167 | 25 | .999 | .999 | .999 | .999 | .999 | .999 | .999 | .999 | .996 | .991 |
| | | | | | | .982 | .966 | .941 | .905 | .855 | .792 | .717 | .634 | .546 | .457 |
| | | | | | | .374 | .298 | .232 | .178 | .136 | | | | | |
| 19.000 | 19.000 | .000 | .090 | .143 | 26 | .999 | .999 | .999 | .999 | .999 | .999 | .999 | .999 | .996 | .991 |
| | | | | | | .982 | .966 | .941 | .903 | .853 | .789 | .713 | .628 | .539 | .449 |
| | | | | | | .364 | .287 | .221 | .166 | .123 | .090 | | | | |
| 19.000 | 19.000 | .000 | .058 | .125 | 27 | .999 | .999 | .999 | .999 | .999 | .999 | .999 | .999 | .996 | .991 |
| | | | | | | .982 | .966 | .940 | .903 | .852 | .787 | .711 | .625 | .535 | .445 |
| | | | | | | .359 | .282 | .215 | .159 | .115 | .082 | .058 | | | |
| 19.000 | 19.000 | .000 | .036 | .111 | 28 | .999 | .999 | .999 | .999 | .999 | .999 | .999 | .998 | .996 | .991 |
| | | | | | | .982 | .966 | .940 | .902 | .851 | .786 | .709 | .624 | .533 | .442 |
| | | | | | | .356 | .278 | .211 | .155 | .111 | .078 | .054 | .036 | | |
| 19.000 | 19.000 | .000 | .022 | .100 | 29 | .999 | .999 | .999 | .999 | .999 | .999 | .999 | .998 | .996 | .991 |
| | | | | | | .982 | .965 | .940 | .902 | .851 | .786 | .709 | .623 | .532 | .441 |
| | | | | | | .355 | .276 | .209 | .153 | .109 | .076 | .051 | .034 | .022 | |
| 19.000 | 19.000 | .000 | .013 | .091 | 30 | .999 | .999 | .999 | .999 | .999 | .999 | .999 | .998 | .996 | .991 |
| | | | | | | .982 | .965 | .940 | .902 | .850 | .786 | .708 | .622 | .531 | .440 |
| | | | | | | .354 | .276 | .208 | .152 | .108 | .074 | .050 | .033 | .021 | .013 |

ERLANG C: INFINITE QUEUE - ALL CALLERS BLOCKED WILL WAIT AS LONG AS NECESSARY FOR CIRCUIT

| HOURS FIRST ATTEMPT VMR = 1.0 | HOURS CON-NECT | HOURS OVFL | DELAYED PORTION | DELAY LENGTH CALL FACTOR | LINE | HOURS CONNECTED TRAFFIC ON LINES 1 | 2 | 3 | 4 | 5 | 6 | 7 | 8 | 9 | 10 |
|---|---|---|---|---|---|---|---|---|---|---|---|---|---|---|---|
| 19.000 | 19.000 | .000 | .008 | .083 | 31 | .999 | .999 | .999 | .999 | .999 | .999 | .999 | .998 | .996 | .991 |
| | | | | | | .982 | .965 | .939 | .902 | .850 | .785 | .708 | .622 | .531 | .440 |
| | | | | | | .353 | .275 | .207 | .152 | .107 | .074 | .049 | .032 | .020 | .013 |
| | | | | | | .008 | | | | | | | | | |
| 19.000 | 19.000 | .000 | .004 | .077 | 32 | .999 | .999 | .999 | .999 | .999 | .999 | .999 | .998 | .996 | .991 |
| | | | | | | .982 | .965 | .939 | .902 | .850 | .785 | .708 | .622 | .531 | .440 |
| | | | | | | .353 | .275 | .207 | .151 | .107 | .073 | .049 | .032 | .020 | .012 |
| | | | | | | .007 | .004 | | | | | | | | |
| 19.000 | 19.000 | .000 | .002 | .071 | 33 | .999 | .999 | .999 | .999 | .999 | .999 | .999 | .998 | .996 | .991 |
| | | | | | | .982 | .965 | .939 | .902 | .850 | .785 | .708 | .622 | .531 | .439 |
| | | | | | | .353 | .275 | .207 | .151 | .107 | .073 | .049 | .031 | .020 | .012 |
| | | | | | | .007 | .004 | .002 | | | | | | | |
| 19.000 | 19.000 | .000 | .001 | .067 | 34 | .999 | .999 | .999 | .999 | .999 | .999 | .999 | .998 | .996 | .991 |
| | | | | | | .982 | .965 | .939 | .902 | .850 | .785 | .708 | .622 | .531 | .439 |
| | | | | | | .353 | .275 | .207 | .151 | .107 | .073 | .049 | .031 | .020 | .012 |
| | | | | | | .007 | .004 | .002 | .001 | | | | | | |
| 19.000 | 19.000 | .000 | .001 | .063 | 35 | .999 | .999 | .999 | .999 | .999 | .999 | .999 | .998 | .996 | .991 |
| | | | | | | .982 | .965 | .939 | .902 | .850 | .785 | .708 | .622 | .531 | .439 |
| | | | | | | .353 | .275 | .207 | .151 | .107 | .073 | .049 | .031 | .020 | .012 |
| | | | | | | .007 | .004 | .002 | .001 | .001 | | | | | |
| 20.000 | 20.000 | .000 | .761 | 1.000 | 21 | .999 | .999 | .999 | .999 | .999 | .999 | .999 | .999 | .999 | .998 |
| | | | | | | .995 | .991 | .983 | .972 | .955 | .933 | .905 | .873 | .837 | .799 |
| | | | | | | .761 | | | | | | | | | |
| 20.000 | 20.000 | .000 | .568 | .500 | 22 | .999 | .999 | .999 | .999 | .999 | .999 | .999 | .999 | .999 | .997 |
| | | | | | | .993 | .986 | .974 | .956 | .930 | .805 | .852 | .801 | .744 | .684 |
| | | | | | | .625 | .568 | | | | | | | | |
| 20.000 | 20.000 | .000 | .416 | .333 | 23 | .999 | .999 | .999 | .999 | .999 | .999 | .999 | .999 | .998 | .996 |
| | | | | | | .991 | .983 | .968 | .946 | .915 | .873 | .821 | .759 | .691 | .619 |
| | | | | | | .547 | .478 | .416 | | | | | | | |
| 20.000 | 20.000 | .000 | .298 | .250 | 24 | .999 | .999 | .999 | .999 | .999 | .999 | .999 | .999 | .998 | .996 |
| | | | | | | .990 | .981 | .965 | .941 | .907 | .860 | .803 | .735 | .660 | .581 |
| | | | | | | .502 | .426 | .358 | .298 | | | | | | |
| 20.000 | 20.000 | .000 | .209 | .200 | 25 | .999 | .990 | .999 | .999 | .999 | .999 | .999 | .999 | .998 | .995 |
| | | | | | | .990 | .980 | .963 | .938 | .902 | .853 | .793 | .721 | .642 | .559 |
| | | | | | | .476 | .090 | .324 | .261 | .209 | | | | | |
| 20.000 | 20.000 | .000 | .143 | .167 | 26 | .999 | .999 | .999 | .999 | .999 | .999 | .999 | .999 | .998 | .995 |
| | | | | | | .990 | .979 | .962 | .936 | .899 | .849 | .787 | .713 | .632 | .546 |
| | | | | | | .461 | .379 | .305 | .240 | .186 | .143 | | | | |
| 20.000 | 20.000 | .000 | .096 | .143 | 27 | .999 | .999 | .999 | .999 | .999 | .999 | .999 | .999 | .998 | .995 |
| | | | | | | .989 | .979 | .062 | .935 | .897 | .847 | .783 | .709 | .626 | .539 |

## ERLANG C: INFINITE QUEUE - ALL CALLERS BLOCKED WILL WAIT AS LONG AS NECESSARY FOR CIRCUIT

| HOURS FIRST ATTEMPT VMR = 1.0 | HOURS CON-NECT | HOURS OVFL | DELAYED PORTION | DELAY LENGTH CALL FACTOR | LINE | HOURS CONNECTED TRAFFIC ON LINES 1 | 2 | 3 | 4 | 5 | 6 | 7 | 8 | 9 | 10 |
|---|---|---|---|---|---|---|---|---|---|---|---|---|---|---|---|
| | | | | | | .452 | .369 | .294 | .228 | .173 | .130 | .096 | | | |
| 20.000 | 20.000 | .000 | .063 | .125 | 28 | .999 | .999 | .999 | .999 | .999 | .999 | .999 | .999 | .998 | .995 |
| | | | | | | .989 | .979 | .961 | .935 | .896 | .845 | .781 | .706 | .623 | .535 |
| | | | | | | .447 | .363 | .287 | .221 | .166 | .122 | .088 | .063 | | |
| 20.000 | 20.000 | .000 | .040 | .111 | 29 | .999 | .999 | .999 | .999 | .999 | .999 | .999 | .999 | .998 | .995 |
| | | | | | | .989 | .979 | .961 | .934 | .896 | .844 | .780 | .705 | .621 | .533 |
| | | | | | | .444 | .360 | .284 | .217 | .162 | .117 | .083 | .058 | .040 | |
| 20.000 | 20.000 | .000 | .025 | .100 | 30 | .999 | .999 | .999 | .999 | .999 | .999 | .999 | .999 | .998 | .995 |
| | | | | | | .989 | .979 | .961 | .934 | .895 | .844 | .780 | .704 | .620 | .531 |
| | | | | | | .443 | .358 | .282 | .215 | .159 | .115 | .081 | .056 | .037 | .025 |
| 20.000 | 20.000 | .000 | .015 | .091 | 31 | .999 | .999 | .999 | .999 | .999 | .999 | .999 | .999 | .998 | .995 |
| | | | | | | .989 | .979 | .961 | .934 | .895 | .844 | .779 | .703 | .619 | .531 |
| | | | | | | .442 | .357 | .281 | .214 | .158 | .114 | .079 | .054 | .036 | .023 |
| | | | | | | .015 | | | | | | | | | |
| 20.000 | 20.000 | .000 | .009 | .083 | 32 | .999 | .999 | .999 | .999 | .999 | .999 | .999 | .999 | .998 | .995 |
| | | | | | | .989 | .979 | .961 | .934 | .895 | .844 | .779 | .703 | .619 | .530 |
| | | | | | | .441 | .357 | .280 | .213 | .158 | .113 | .079 | .053 | .035 | .023 |
| | | | | | | .014 | .009 | | | | | | | | |
| 20.000 | 20.000 | .000 | .005 | .077 | 33 | .999 | .999 | .999 | .999 | .999 | .999 | .999 | .999 | .998 | .995 |
| | | | | | | .989 | .979 | .961 | .934 | .895 | .844 | .779 | .703 | .619 | .530 |
| | | | | | | .441 | .357 | .280 | .213 | .157 | .113 | .078 | .053 | .035 | .022 |
| | | | | | | .014 | .009 | .005 | | | | | | | |
| 20.000 | 20.000 | .000 | .003 | .071 | 34 | .999 | .999 | .999 | .999 | .999 | .999 | .999 | .999 | .998 | .995 |
| | | | | | | .989 | .979 | .961 | .934 | .895 | .844 | .779 | .703 | .619 | .530 |
| | | | | | | .441 | .356 | .280 | .213 | .157 | .112 | .078 | .053 | .035 | .022 |
| | | | | | | .014 | .008 | .005 | .003 | | | | | | |
| 20.000 | 20.000 | .000 | .002 | .067 | 35 | .999 | .999 | .999 | .999 | .999 | .999 | .999 | .999 | .998 | .995 |
| | | | | | | .989 | .979 | .961 | .934 | .895 | .844 | .779 | .703 | .619 | .530 |
| | | | | | | .441 | .356 | .279 | .213 | .157 | .112 | .078 | .053 | .034 | .022 |
| | | | | | | .014 | .008 | .005 | .003 | .002 | | | | | |
| 20.000 | 20.000 | .000 | .001 | .063 | 36 | .999 | .999 | .999 | .999 | .999 | .999 | .999 | .999 | .998 | .995 |
| | | | | | | .989 | .979 | .961 | .934 | .895 | .843 | .779 | .703 | .619 | .530 |
| | | | | | | .441 | .356 | .279 | .213 | .157 | .112 | .078 | .053 | .034 | .022 |
| | | | | | | .014 | .008 | .005 | .003 | .002 | .001 | | | | |
| 22.000 | 22.000 | .000 | .770 | 1.000 | 23 | .999 | .999 | .999 | .999 | .999 | .999 | .999 | .999 | .999 | .999 |
| | | | | | | .999 | .997 | .994 | .989 | .980 | .968 | .952 | .930 | .904 | .874 |
| | | | | | | .840 | .805 | .770 | | | | | | | |
| 22.000 | 22.000 | .000 | .583 | .500 | 24 | .999 | .999 | .999 | .999 | .999 | .999 | .999 | .999 | .999 | .999 |
| | | | | | | .998 | .995 | .990 | .982 | .969 | .950 | .923 | .889 | .848 | .800 |
| | | | | | | .747 | .692 | .636 | .583 | | | | | | |

ERLANG C: INFINITE QUEUE - ALL CALLERS BLOCKED WILL WAIT AS LONG AS NECESSARY FOR CIRCUIT

| HOURS FIRST ATTEMPT VMR = 1.0 | HOURS CON-NECT | HOURS OVFL | DELAYED PORTION | DELAY LENGTH CALL FACTOR | LINE | HOURS CONNECTED TRAFFIC ON LINES 1 | 2 | 3 | 4 | 5 | 6 | 7 | 8 | 9 | 10 |
|---|---|---|---|---|---|---|---|---|---|---|---|---|---|---|---|
| 22.000 | 22.000 | .000 | .434 | .333 | 25 | .999 | .999 | .999 | .999 | .999 | .999 | .999 | .999 | .999 | .999 |
| | | | | | | .997 | .994 | .988 | .978 | .962 | .939 | .907 | .866 | .815 | .756 |
| | | | | | | .692 | .625 | .557 | .493 | .434 | | | | | |
| 22.000 | 22.000 | .000 | .316 | .250 | 26 | .999 | .999 | .999 | .999 | .999 | .999 | .999 | .999 | .999 | .999 |
| | | | | | | .997 | .993 | .987 | .976 | .958 | .932 | .897 | .851 | .795 | .731 |
| | | | | | | .660 | .585 | .511 | .439 | .374 | .316 | | | | |
| 22.000 | 22.000 | .000 | .226 | .200 | 27 | .999 | .999 | .999 | .999 | .999 | .999 | .999 | .999 | .999 | .999 |
| | | | | | | .997 | .993 | .986 | .974 | .956 | .929 | .891 | .843 | .784 | .716 |
| | | | | | | .640 | .562 | .483 | .408 | .338 | .278 | .226 | | | |
| 22.000 | 22.000 | .000 | .159 | .167 | 28 | .999 | .999 | .999 | .999 | .999 | .999 | .999 | .999 | .999 | .999 |
| | | | | | | .997 | .993 | .986 | .973 | .954 | .926 | .888 | .838 | .777 | .707 |
| | | | | | | .629 | .548 | .466 | .389 | .317 | .255 | .202 | .159 | | |
| 22.000 | 22.000 | .000 | .109 | .143 | 29 | .999 | .999 | .999 | .999 | .999 | .999 | .999 | .999 | .999 | .999 |
| | | | | | | .997 | .993 | .985 | .973 | .953 | .925 | .886 | .835 | .773 | .701 |
| | | | | | | .622 | .540 | .457 | .378 | .305 | .241 | .187 | .143 | .109 | |
| 22.000 | 22.000 | .000 | .073 | .125 | 30 | .999 | .999 | .999 | .999 | .999 | .999 | .999 | .999 | .999 | .999 |
| | | | | | | .997 | .992 | .985 | .973 | .953 | .924 | .885 | .833 | .771 | .698 |
| | | | | | | .618 | .535 | .451 | .371 | .298 | .233 | .179 | .134 | .099 | .073 |
| 22.000 | 22.000 | .000 | .048 | .111 | 31 | .999 | .999 | .999 | .999 | .999 | .999 | .999 | .999 | .999 | .999 |
| | | | | | | .996 | .992 | .985 | .972 | .953 | .924 | .884 | .832 | .769 | .696 |
| | | | | | | .616 | .532 | .448 | .367 | .294 | .229 | .174 | .129 | .094 | .067 |
| | | | | | | .048 | | | | | | | | | |
| 22.000 | 22.000 | .000 | .031 | .100 | 32 | .999 | .999 | .999 | .999 | .999 | .999 | .999 | .999 | .999 | .999 |
| | | | | | | .996 | .992 | .985 | .972 | .953 | .923 | .883 | .832 | .768 | .695 |
| | | | | | | .615 | .530 | .446 | .365 | .291 | .226 | .171 | .126 | .091 | .064 |
| | | | | | | .045 | .031 | | | | | | | | |
| 22.000 | 22.000 | .000 | .019 | .091 | 33 | .999 | .999 | .999 | .999 | .999 | .999 | .999 | .999 | .999 | .998 |
| | | | | | | .996 | .992 | .985 | .972 | .952 | .923 | .883 | .831 | .768 | .695 |
| | | | | | | .614 | .529 | .445 | .364 | .290 | .225 | .169 | .125 | .089 | .062 |
| | | | | | | .043 | .029 | .019 | | | | | | | |
| 22.000 | 22.000 | .000 | .012 | .083 | 34 | .999 | .999 | .999 | .999 | .999 | .999 | .999 | .999 | .999 | .998 |
| | | | | | | .996 | .992 | .985 | .972 | .952 | .923 | .883 | .831 | .768 | .694 |
| | | | | | | .614 | .529 | .444 | .363 | .289 | .224 | .169 | .124 | .088 | .061 |
| | | | | | | .042 | .028 | .018 | .012 | | | | | | |
| 22.000 | 22.000 | .000 | .007 | .077 | 35 | .999 | .999 | .999 | .999 | .999 | .999 | .999 | .999 | .999 | .998 |
| | | | | | | .996 | .992 | .985 | .972 | .952 | .923 | .883 | .831 | .768 | .694 |
| | | | | | | .613 | .529 | .444 | .363 | .289 | .223 | .168 | .123 | .088 | .061 |
| | | | | | | .041 | .027 | .018 | .011 | .007 | | | | | |
| 22.000 | 22.000 | .000 | .004 | .071 | 36 | .999 | .999 | .999 | .999 | .999 | .999 | .999 | .999 | .999 | .998 |
| | | | | | | .996 | .992 | .985 | .972 | .952 | .923 | .883 | .831 | .768 | .694 |

ERLANG C: INFINITE QUEUE - ALL CALLERS BLOCKED WILL WAIT AS LONG AS NECESSARY FOR CIRCUIT

| HOURS FIRST ATTEMPT VMR=1.0 | HOURS CON-NECT | HOURS OVFL | DELAYED PORTION | DELAY LENGTH CALL FACTOR | LINE | HOURS CONNECTED TRAFFIC ON LINES 1 | 2 | 3 | 4 | 5 | 6 | 7 | 8 | 9 | 10 |
|---|---|---|---|---|---|---|---|---|---|---|---|---|---|---|---|
| | | | | | | .613 | .529 | .444 | .363 | .289 | .223 | .168 | .123 | .087 | .061 |
| | | | | | | .041 | .027 | .017 | .011 | .007 | .004 | | | | |
| 22.000 | 22.000 | .000 | .002 | .067 | 37 | .999 | .999 | .999 | .999 | .999 | .999 | .999 | .999 | .999 | .998 |
| | | | | | | .996 | .992 | .985 | .972 | .952 | .923 | .883 | .831 | .768 | .694 |
| | | | | | | .613 | .528 | .444 | .363 | .288 | .223 | .168 | .123 | .087 | .060 |
| | | | | | | .041 | .027 | .017 | .011 | .007 | .004 | .002 | | | |
| 22.000 | 22.000 | .000 | .001 | .063 | 38 | .999 | .999 | .999 | .999 | .999 | .999 | .999 | .999 | .999 | .998 |
| | | | | | | .996 | .992 | .985 | .972 | .952 | .923 | .883 | .831 | .768 | .694 |
| | | | | | | .613 | .528 | .444 | .363 | .288 | .223 | .168 | .123 | .087 | .060 |
| | | | | | | .041 | .027 | .017 | .011 | .006 | .004 | .002 | .001 | | |
| 22.000 | 22.000 | .000 | .001 | .059 | 39 | .999 | .999 | .999 | .999 | .999 | .999 | .999 | .999 | .999 | .998 |
| | | | | | | .996 | .992 | .985 | .972 | .952 | .923 | .883 | .831 | .768 | .694 |
| | | | | | | .613 | .528 | .444 | .363 | .288 | .223 | .168 | .123 | .087 | .060 |
| | | | | | | .041 | .027 | .017 | .011 | .006 | .004 | .002 | .001 | .001 | |
| 24.000 | 24.000 | .000 | .779 | 1.000 | 25 | .999 | .999 | .999 | .999 | .999 | .999 | .999 | .999 | .999 | .999 |
| | | | | | | .999 | .999 | .998 | .996 | .992 | .986 | .978 | .965 | .949 | .928 |
| | | | | | | .903 | .875 | .843 | .811 | .779 | | | | | |
| 24.000 | 24.000 | .000 | .597 | .500 | 26 | .999 | .999 | .999 | .999 | .999 | .999 | .999 | .999 | .999 | .999 |
| | | | | | | .999 | .998 | .997 | .993 | .987 | .978 | .964 | .944 | .918 | .885 |
| | | | | | | .845 | .800 | .750 | .698 | .646 | .597 | | | | |
| 24.000 | 24.000 | .000 | .450 | .333 | 27 | .999 | .999 | .999 | .999 | .999 | .999 | .999 | .999 | .999 | .999 |
| | | | | | | .999 | .998 | .996 | .992 | .985 | .973 | .956 | .932 | .900 | .859 |
| | | | | | | .810 | .755 | .694 | .630 | .567 | .506 | .450 | | | |
| 24.000 | 24.000 | .000 | .333 | .250 | 28 | .999 | .999 | .999 | .999 | .999 | .999 | .999 | .999 | .999 | .999 |
| | | | | | | .999 | .998 | .995 | .991 | .983 | .970 | .951 | .924 | .889 | .843 |
| | | | | | | .789 | .727 | .660 | .589 | .519 | .451 | .389 | .333 | | |
| 24.000 | 24.000 | .000 | .242 | .200 | 29 | .999 | .999 | .999 | .999 | .999 | .999 | .999 | .999 | .999 | .999 |
| | | | | | | .999 | .998 | .995 | .990 | .982 | .968 | .948 | .920 | .882 | .834 |
| | | | | | | .776 | .711 | .639 | .564 | .490 | .418 | .352 | .293 | .242 | |
| 24.000 | 24.000 | .000 | .173 | .167 | 30 | .999 | .999 | .999 | .999 | .999 | .999 | .999 | .999 | .999 | .999 |
| | | | | | | .999 | .998 | .995 | .990 | .981 | .967 | .946 | .917 | .878 | .828 |
| | | | | | | .769 | .701 | .627 | .549 | .472 | .398 | .329 | .268 | .216 | .173 |
| 24.000 | 24.000 | .000 | .121 | .143 | 31 | .999 | .999 | .999 | .999 | .999 | .999 | .999 | .999 | .999 | .999 |
| | | | | | | .999 | .998 | .995 | .990 | .981 | .967 | .945 | .915 | .875 | .825 |
| | | | | | | .764 | .695 | .619 | .540 | .461 | .386 | .316 | .254 | .200 | .156 |
| | | | | | | .121 | | | | | | | | | |
| 24.000 | 24.000 | .000 | .083 | .125 | 32 | .999 | .999 | .999 | .999 | .999 | .999 | .999 | .999 | .999 | .999 |
| | | | | | | .999 | .998 | .995 | .989 | .980 | .966 | .945 | .914 | .874 | .823 |
| | | | | | | .761 | .691 | .615 | .535 | .455 | .378 | .308 | .245 | .191 | .146 |
| | | | | | | .111 | .083 | | | | | | | | |

ERLANG C: INFINITE QUEUE - ALL CALLERS BLOCKED WILL WAIT AS LONG AS NECESSARY FOR CIRCUIT

| HOURS FIRST ATTEMPT VMR=1.0 | HOURS CON-NECT | HOURS OVFL | DELAYED PORTION | DELAY LENGTH CALL FACTOR | LINE | HOURS CONNECTED TRAFFIC ON LINES 1 | 2 | 3 | 4 | 5 | 6 | 7 | 8 | 9 | 10 |
|---|---|---|---|---|---|---|---|---|---|---|---|---|---|---|---|
| 24.000 | 24.000 | .000 | .056 | .111 | 33 | .999 | .999 | .999 | .999 | .999 | .999 | .999 | .999 | .999 | .999 |
| | | | | | | .999 | .997 | .995 | .989 | .980 | .966 | .944 | .914 | .873 | .821 |
| | | | | | | .760 | .689 | .612 | .532 | .451 | .374 | .303 | .239 | .185 | .140 |
| | | | | | | .104 | .077 | .056 | | | | | | | |
| 24.000 | 24.000 | .000 | .037 | .100 | 34 | .999 | .999 | .999 | .999 | .999 | .999 | .999 | .999 | .999 | .999 |
| | | | | | | .999 | .997 | .995 | .989 | .980 | .966 | .944 | .913 | .872 | .821 |
| | | | | | | .759 | .688 | .610 | .530 | .449 | .371 | .300 | .236 | .182 | .137 |
| | | | | | | .101 | .073 | .052 | .037 | | | | | | |
| 24.000 | 24.000 | .000 | .024 | .091 | 35 | .999 | .999 | .999 | .999 | .999 | .999 | .999 | .999 | .999 | .999 |
| | | | | | | .999 | .997 | .995 | .989 | .980 | .966 | .944 | .913 | .872 | .820 |
| | | | | | | .758 | .687 | .609 | .529 | .448 | .370 | .298 | .235 | .180 | .135 |
| | | | | | | .099 | .071 | .050 | .034 | .024 | | | | | |
| 24.000 | 24.000 | .000 | .015 | .083 | 36 | .999 | .999 | .999 | .999 | .999 | .999 | .999 | .999 | .999 | .999 |
| | | | | | | .999 | .997 | .995 | .989 | .980 | .966 | .944 | .913 | .872 | .820 |
| | | | | | | .758 | .687 | .609 | .528 | .447 | .369 | .297 | .234 | .179 | .134 |
| | | | | | | .097 | .069 | .048 | .033 | .022 | .015 | | | | |
| 24.000 | 24.000 | .000 | .009 | .077 | 37 | .999 | .999 | .999 | .999 | .999 | .999 | .999 | .999 | .999 | .999 |
| | | | | | | .999 | .997 | .995 | .989 | .980 | .966 | .944 | .913 | .872 | .820 |
| | | | | | | .758 | .686 | .609 | .528 | .447 | .369 | .297 | .233 | .178 | .133 |
| | | | | | | .097 | .069 | .048 | .032 | .021 | .014 | .009 | | | |
| 24.000 | 24.000 | .000 | .005 | .071 | 38 | .999 | .999 | .999 | .999 | .999 | .999 | .999 | .999 | .999 | .999 |
| | | | | | | .999 | .997 | .995 | .989 | .980 | .966 | .944 | .913 | .872 | .820 |
| | | | | | | .757 | .686 | .608 | .527 | .446 | .368 | .297 | .233 | .178 | .133 |
| | | | | | | .096 | .068 | .047 | .032 | .021 | .014 | .009 | .005 | | |
| 24.000 | 24.000 | .000 | .003 | .067 | 39 | .999 | .999 | .999 | .999 | .999 | .999 | .999 | .999 | .999 | .999 |
| | | | | | | .999 | .997 | .995 | .989 | .980 | .966 | .944 | .913 | .872 | .820 |
| | | | | | | .757 | .686 | .608 | .527 | .440 | .368 | .296 | .232 | .178 | .132 |
| | | | | | | .096 | .068 | .047 | .032 | .021 | .013 | .008 | .005 | .003 | |
| 24.000 | 24.000 | .000 | .002 | .063 | 40 | .999 | .999 | .999 | .999 | .999 | .999 | .999 | .999 | .999 | .999 |
| | | | | | | .999 | .997 | .995 | .989 | .980 | .966 | .944 | .913 | .872 | .820 |
| | | | | | | .757 | .686 | .608 | .527 | .446 | .368 | .296 | .232 | .178 | .132 |
| | | | | | | .096 | .068 | .047 | .032 | .021 | .013 | .008 | .005 | .003 | .002 |
| 24.000 | 24.000 | .000 | .001 | .059 | 41 | .999 | .999 | .999 | .999 | .999 | .999 | .999 | .999 | .999 | .999 |
| | | | | | | .999 | .997 | .995 | .989 | .980 | .966 | .944 | .913 | .872 | .820 |
| | | | | | | .757 | .686 | .608 | .527 | .446 | .368 | .296 | .232 | .178 | .132 |
| | | | | | | .096 | .068 | .047 | .031 | .021 | .013 | .008 | .005 | .003 | .002 |
| | | | | | | .001 | | | | | | | | | |
| 24.000 | 24.000 | .000 | .001 | .056 | 42 | .999 | .999 | .999 | .999 | .999 | .999 | .999 | .999 | .999 | .999 |
| | | | | | | .999 | .997 | .995 | .989 | .980 | .966 | .944 | .913 | .872 | .020 |
| | | | | | | .757 | .686 | .608 | .527 | .446 | .368 | .296 | .232 | .177 | .132 |

ERLANG C: INFINITE QUEUE - ALL CALLERS BLOCKED WILL WAIT AS LONG AS NECESSARY FOR CIRCUIT

| HOURS FIRST ATTEMPT VMR = 1.0 | HOURS CON-NECT | HOURS OVFL | DELAYED PORTION | DELAY LENGTH CALL FACTOR | LINE | HOURS CONNECTED TRAFFIC ON LINES 1 | 2 | 3 | 4 | 5 | 6 | 7 | 8 | 9 | 10 |
|---|---|---|---|---|---|---|---|---|---|---|---|---|---|---|---|
| | | | | | | .096 | .068 | .047 | .031 | .021 | .013 | .008 | .005 | .003 | .002 |
| | | | | | | .001 | .001 | | | | | | | | |
| 26.000 | 26.000 | .000 | .786 | 1.000 | 27 | .999 | .999 | .999 | .999 | .999 | .999 | .999 | .999 | .999 | .999 |
| | | | | | | .999 | .999 | .999 | .999 | .997 | .995 | .990 | .984 | .975 | .962 |
| | | | | | | .946 | .926 | .902 | .876 | .847 | .816 | .786 | | | |
| 26.000 | 26.000 | .000 | .609 | .500 | 28 | .999 | .999 | .999 | .999 | .999 | .999 | .999 | .999 | .999 | .999 |
| | | | | | | .999 | .999 | .999 | .998 | .995 | .991 | .985 | .974 | .960 | .940 |
| | | | | | | .914 | .881 | .843 | .800 | .753 | .704 | .656 | .609 | | |
| 26.000 | 26.000 | .000 | .465 | .333 | 29 | .999 | .999 | .999 | .999 | .999 | .999 | .999 | .999 | .999 | .999 |
| | | | | | | .999 | .999 | .999 | .997 | .994 | .989 | .981 | .968 | .950 | .926 |
| | | | | | | .893 | .854 | .807 | .753 | .696 | .636 | .576 | .518 | .465 | |
| 26.000 | 26.000 | .000 | .349 | .250 | 30 | .999 | .999 | .999 | .999 | .999 | .999 | .999 | .999 | .999 | .999 |
| | | | | | | .999 | .999 | .998 | .997 | .993 | .988 | .979 | .965 | .945 | .917 |
| | | | | | | .881 | .837 | .784 | .725 | .660 | .593 | .527 | .462 | .402 | .349 |
| 26.000 | 26.000 | .000 | .257 | .200 | 31 | .999 | .999 | .999 | .999 | .999 | .999 | .999 | .999 | .999 | .999 |
| | | | | | | .999 | .999 | .998 | .996 | .993 | .987 | .977 | .962 | .941 | .912 |
| | | | | | | .873 | .826 | .770 | .707 | .638 | .567 | .496 | .428 | .364 | .307 |
| | | | | | | .257 | | | | | | | | | |
| 26.000 | 26.000 | .000 | .187 | .167 | 32 | .999 | .999 | .999 | .999 | .999 | .999 | .999 | .999 | .999 | .999 |
| | | | | | | .999 | .999 | .998 | .996 | .993 | .987 | .976 | .961 | .939 | .908 |
| | | | | | | .869 | .820 | .762 | .696 | .625 | .551 | .477 | .406 | .340 | .281 |
| | | | | | | .230 | .187 | | | | | | | | |
| 26.000 | 26.000 | .000 | .133 | .143 | 33 | .999 | .999 | .999 | .999 | .999 | .999 | .999 | .999 | .999 | .999 |
| | | | | | | .999 | .999 | .998 | .996 | .993 | .986 | .976 | .960 | .937 | .906 |
| | | | | | | .866 | .816 | .756 | .689 | .617 | .541 | .466 | .393 | .325 | .265 |
| | | | | | | .213 | .169 | .133 | | | | | | | |
| 26.000 | 26.000 | .000 | .093 | .125 | 34 | .999 | .999 | .999 | .999 | .999 | .999 | .999 | .999 | .999 | .999 |
| | | | | | | .999 | .999 | .998 | .996 | .993 | .986 | .976 | .960 | .937 | .905 |
| | | | | | | .864 | .813 | .753 | .685 | .612 | .535 | .459 | .385 | .317 | .255 |
| | | | | | | .202 | .158 | .121 | .093 | | | | | | |
| 26.000 | 26.000 | .000 | .064 | .111 | 35 | .999 | .999 | .999 | .999 | .999 | .999 | .999 | .999 | .999 | .999 |
| | | | | | | .999 | .999 | .998 | .996 | .992 | .986 | .975 | .959 | .936 | .904 |
| | | | | | | .863 | .812 | .751 | .683 | .609 | .531 | .454 | .380 | .311 | .249 |
| | | | | | | .196 | .151 | .114 | .086 | .064 | | | | | |
| 26.000 | 26.000 | .000 | .043 | .100 | 36 | .999 | .999 | .999 | .999 | .999 | .999 | .999 | .999 | .999 | .999 |
| | | | | | | .999 | .999 | .998 | .996 | .992 | .986 | .975 | .959 | .936 | .904 |
| | | | | | | .862 | .811 | .750 | .681 | .607 | .529 | .452 | .377 | .308 | .246 |
| | | | | | | .192 | .147 | .110 | .081 | .059 | .043 | | | | |
| 26.000 | 26.000 | .000 | .028 | .091 | 37 | .999 | .999 | .999 | .999 | .999 | .999 | .999 | .999 | .999 | .999 |
| | | | | | | .999 | .999 | .998 | .996 | .992 | .986 | .975 | .959 | .936 | .904 |

ERLANG C: INFINITE QUEUE - ALL CALLERS BLOCKED WILL WAIT AS LONG AS NECESSARY FOR CIRCUIT

| HOURS FIRST ATTEMPT VMR = 1.0 | HOURS CON-NECT | HOURS OVFL | DELAYED PORTION | DELAY LENGTH CALL FACTOR | LINE | HOURS CONNECTED TRAFFIC ON LINES 1 | 2 | 3 | 4 | 5 | 6 | 7 | 8 | 9 | 10 |
|---|---|---|---|---|---|---|---|---|---|---|---|---|---|---|---|
| | | | | | | .862 | .810 | .749 | .680 | .606 | .528 | .450 | .375 | .306 | .244 |
| | | | | | | .190 | .144 | .108 | .079 | .056 | .040 | .028 | | | |
| 26.000 | 26.000 | .000 | .018 | .083 | 38 | .999 | .999 | .999 | .999 | .999 | .999 | .999 | .999 | .999 | .999 |
| | | | | | | .999 | .999 | .998 | .996 | .992 | .986 | .975 | .959 | .936 | .903 |
| | | | | | | .862 | .810 | .749 | .680 | .605 | .527 | .449 | .374 | .305 | .242 |
| | | | | | | .188 | .143 | .106 | .077 | .055 | .038 | .027 | .018 | | |
| 26.000 | 26.000 | .000 | .011 | .077 | 39 | .999 | .999 | .999 | .999 | .999 | .999 | .999 | .999 | .999 | .999 |
| | | | | | | .999 | .999 | .998 | .996 | .992 | .986 | .975 | .959 | .935 | .903 |
| | | | | | | .861 | .810 | .749 | .679 | .605 | .527 | .449 | .374 | .304 | .242 |
| | | | | | | .188 | .142 | .105 | .076 | .054 | .037 | .026 | .017 | .011 | |
| 26.000 | 26.000 | .000 | .007 | .071 | 40 | .999 | .999 | .999 | .999 | .999 | .999 | .999 | .999 | .999 | .999 |
| | | | | | | .999 | .999 | .998 | .996 | .992 | .986 | .975 | .959 | .935 | .903 |
| | | | | | | .861 | .810 | .748 | .679 | .604 | .526 | .448 | .373 | .304 | .241 |
| | | | | | | .187 | .142 | .105 | .076 | .053 | .037 | .025 | .017 | .011 | .007 |
| 26.000 | 26.000 | .000 | .004 | .067 | 41 | .999 | .999 | .999 | .999 | .999 | .999 | .999 | .999 | .999 | .999 |
| | | | | | | .999 | .999 | .998 | .996 | .992 | .986 | .975 | .959 | .935 | .903 |
| | | | | | | .861 | .810 | .748 | .679 | .604 | .526 | .448 | .373 | .304 | .241 |
| | | | | | | .187 | .141 | .105 | .075 | .053 | .037 | .025 | .016 | .011 | .007 |
| | | | | | | .004 | | | | | | | | | |
| 26.000 | 26.000 | .000 | .003 | .063 | 42 | .999 | .999 | .999 | .999 | .999 | .999 | .999 | .999 | .999 | .999 |
| | | | | | | .999 | .999 | .998 | .996 | .992 | .986 | .975 | .959 | .935 | .903 |
| | | | | | | .861 | .810 | .748 | .679 | .604 | .526 | .448 | .373 | .303 | .241 |
| | | | | | | .187 | .141 | .104 | .075 | .053 | .037 | .025 | .016 | .010 | .007 |
| | | | | | | .004 | .003 | | | | | | | | |
| 26.000 | 26.000 | .000 | .001 | .059 | 43 | .999 | .999 | .999 | .999 | .999 | .999 | .999 | .999 | .999 | .999 |
| | | | | | | .999 | .999 | .998 | .996 | .992 | .986 | .975 | .959 | .935 | .903 |
| | | | | | | .861 | .810 | .748 | .679 | .604 | .526 | .448 | .373 | .303 | .241 |
| | | | | | | .187 | .141 | .104 | .075 | .053 | .036 | .024 | .016 | .010 | .007 |
| | | | | | | .004 | .002 | .001 | | | | | | | |
| 26.000 | 26.000 | .000 | .001 | .056 | 44 | .999 | .999 | .999 | .999 | .999 | .999 | .999 | .999 | .999 | .999 |
| | | | | | | .999 | .999 | .998 | .996 | .992 | .986 | .975 | .959 | .935 | .903 |
| | | | | | | .861 | .810 | .748 | .679 | .604 | .526 | .448 | .373 | .303 | .241 |
| | | | | | | .187 | .141 | .104 | .075 | .053 | .036 | .024 | .016 | .010 | .006 |
| | | | | | | .004 | .002 | .001 | .001 | | | | | | |
| 28.000 | 28.000 | .000 | .793 | 1.000 | 29 | .999 | .999 | .999 | .999 | .999 | .999 | .999 | .999 | .999 | .999 |
| | | | | | | .999 | .999 | .999 | .999 | .999 | .998 | .996 | .993 | .989 | .982 |
| | | | | | | .973 | .960 | .944 | .925 | .902 | .877 | .849 | .821 | .793 | |
| 28.000 | 28.000 | .000 | .620 | .500 | 30 | .999 | .999 | .999 | .999 | .999 | .999 | .999 | .999 | .999 | .999 |
| | | | | | | .999 | .999 | .999 | .999 | .998 | .997 | .994 | .989 | .982 | .971 |
| | | | | | | .956 | .935 | .910 | .878 | .841 | .800 | .756 | .710 | .664 | .620 |

## ERLANG C: INFINITE QUEUE - ALL CALLERS BLOCKED WILL WAIT AS LONG AS NECESSARY FOR CIRCUIT

| HOURS FIRST ATTEMPT VMR = 1.0 | HOURS CON-NECT | HOURS OVFL | DELAYED PORTION | DELAY LENGTH CALL FACTOR | LINE | HOURS CONNECTED TRAFFIC ON LINES 1 | 2 | 3 | 4 | 5 | 6 | 7 | 8 | 9 | 10 |
|---|---|---|---|---|---|---|---|---|---|---|---|---|---|---|---|
| 28.000 | 28.000 | .000 | .478 | .333 | 31 | .999 | .999 | .999 | .999 | .999 | .999 | .999 | .999 | .999 | .999 |
| | | | | | | .999 | .999 | .999 | .999 | .998 | .996 | .992 | .986 | .977 | .964 |
| | | | | | | .945 | .920 | .888 | .849 | .803 | .753 | .698 | .641 | .584 | .529 |
| | | | | | | .478 | | | | | | | | | |
| 28.000 | 28.000 | .000 | .363 | .250 | 32 | .999 | .999 | .999 | .999 | .999 | .999 | .999 | .999 | .999 | .999 |
| | | | | | | .999 | .999 | .999 | .999 | .998 | .995 | .991 | .985 | .975 | .959 |
| | | | | | | .938 | .910 | .874 | .831 | .780 | .723 | .661 | .598 | .534 | .472 |
| | | | | | | .415 | .363 | | | | | | | | |
| 28.000 | 28.000 | .000 | .271 | .200 | 33 | .999 | .999 | .999 | .999 | .999 | .999 | .999 | .999 | .999 | .999 |
| | | | | | | .999 | .999 | .999 | .999 | .998 | .995 | .991 | .984 | .973 | .957 |
| | | | | | | .934 | .904 | .866 | .819 | .765 | .704 | .638 | .570 | .502 | .436 |
| | | | | | | .375 | .320 | .271 | | | | | | | |
| 28.000 | 28.000 | .000 | .200 | .167 | 34 | .999 | .999 | .999 | .999 | .999 | .999 | .999 | .999 | .999 | .999 |
| | | | | | | .999 | .999 | .999 | .999 | .997 | .995 | .990 | .983 | .972 | .955 |
| | | | | | | .932 | .900 | .860 | .812 | .755 | .692 | .624 | .553 | .482 | .414 |
| | | | | | | .350 | .293 | .242 | .200 | | | | | | |
| 28.000 | 28.000 | .000 | .144 | .143 | 35 | .999 | .999 | .999 | .999 | .999 | .999 | .999 | .999 | .999 | .999 |
| | | | | | | .999 | .999 | .999 | .999 | .997 | .995 | .990 | .983 | .971 | .954 |
| | | | | | | .930 | .898 | .857 | .807 | .749 | .685 | .615 | .542 | .470 | .400 |
| | | | | | | .335 | .276 | .224 | .180 | .144 | | | | | |
| 28.000 | 28.000 | .000 | .102 | .125 | 36 | .999 | .999 | .999 | .999 | .999 | .999 | .999 | .999 | .999 | .999 |
| | | | | | | .999 | .999 | .999 | .999 | .997 | .995 | .990 | .982 | .971 | .953 |
| | | | | | | .929 | .896 | .855 | .805 | .746 | .680 | .609 | .536 | .462 | .391 |
| | | | | | | .325 | .265 | .213 | .168 | .132 | .102 | | | | |
| 28.000 | 28.000 | .000 | .071 | .111 | 37 | .999 | .999 | .999 | .999 | .999 | .999 | .999 | .999 | .999 | .999 |
| | | | | | | .999 | .999 | .999 | .999 | .997 | .995 | .990 | .982 | .970 | .953 |
| | | | | | | .928 | .895 | .854 | .803 | .744 | .677 | .606 | .531 | .457 | .386 |
| | | | | | | .319 | .258 | .206 | .161 | .124 | .094 | .071 | | | |
| 28.000 | 28.000 | .000 | .049 | .100 | 38 | .999 | .999 | .999 | .999 | .999 | .999 | .999 | .999 | .999 | .999 |
| | | | | | | .999 | .999 | .999 | .999 | .997 | .995 | .990 | .982 | .970 | .953 |
| | | | | | | .928 | .895 | .853 | .802 | .742 | .675 | .603 | .529 | .454 | .382 |
| | | | | | | .315 | .254 | .201 | .156 | .119 | .089 | .066 | .049 | | |
| 28.000 | 28.000 | .000 | .033 | .091 | 39 | .999 | .999 | .999 | .999 | .999 | .999 | .999 | .999 | .999 | .999 |
| | | | | | | .999 | .999 | .999 | .999 | .997 | .995 | .990 | .982 | .970 | .952 |
| | | | | | | .928 | .895 | .852 | .801 | .741 | .674 | .602 | .527 | .453 | .380 |
| | | | | | | .313 | .252 | .199 | .154 | .116 | .086 | .063 | .046 | .033 | |
| 28.000 | 28.000 | .000 | .022 | .083 | 40 | .999 | .999 | .999 | .999 | .999 | .999 | .999 | .999 | .999 | .999 |
| | | | | | | .999 | .999 | .999 | .999 | .997 | .995 | .990 | .982 | .970 | .952 |
| | | | | | | .927 | .894 | .852 | .801 | .741 | .674 | .601 | .526 | .451 | .379 |
| | | | | | | .312 | .251 | .197 | .152 | .115 | .085 | .061 | .044 | .031 | .022 |

ERLANG C: INFINITE QUEUE - ALL CALLERS BLOCKED WILL WAIT AS LONG AS NECESSARY FOR CIRCUIT

| HOURS FIRST ATTEMPT VMR = 1.0 | HOURS CON-NECT | HOURS OVFL | DELAYED PORTION | DELAY LENGTH CALL FACTOR | LINE | HOURS CONNECTED TRAFFIC ON LINES 1 | 2 | 3 | 4 | 5 | 6 | 7 | 8 | 9 | 10 |
|---|---|---|---|---|---|---|---|---|---|---|---|---|---|---|---|
| 28.000 | 28.000 | .000 | .014 | .077 | 41 | .999 | .999 | .999 | .999 | .999 | .999 | .999 | .999 | .999 | .999 |
| | | | | | | .999 | .999 | .999 | .999 | .997 | .995 | .990 | .982 | .970 | .952 |
| | | | | | | .927 | .894 | .852 | .801 | .741 | .673 | .601 | .526 | .451 | .378 |
| | | | | | | .311 | .250 | .196 | .151 | .114 | .084 | .060 | .043 | .030 | .021 |
| | | | | | | .014 | | | | | | | | | |
| 28.000 | 28.000 | .000 | .009 | .071 | 42 | .999 | .999 | .999 | .999 | .999 | .999 | .999 | .999 | .999 | .999 |
| | | | | | | .999 | .999 | .999 | .999 | .997 | .995 | .990 | .982 | .970 | .952 |
| | | | | | | .927 | .894 | .852 | .800 | .740 | .673 | .601 | .526 | .450 | .378 |
| | | | | | | .310 | .249 | .196 | .150 | .113 | .083 | .060 | .042 | .029 | .020 |
| | | | | | | .013 | .009 | | | | | | | | |
| 28.000 | 28.000 | .000 | .006 | .067 | 43 | .999 | .999 | .999 | .999 | .999 | .999 | .999 | .999 | .999 | .999 |
| | | | | | | .999 | .999 | .999 | .999 | .997 | .995 | .990 | .982 | .970 | .952 |
| | | | | | | .927 | .894 | .852 | .800 | .740 | .673 | .601 | .525 | .450 | .378 |
| | | | | | | .310 | .249 | .195 | .150 | .113 | .083 | .059 | .042 | .029 | .019 |
| | | | | | | .013 | .009 | .006 | | | | | | | |
| 28.000 | 28.000 | .000 | .003 | .063 | 44 | .999 | .999 | .999 | .999 | .999 | .999 | .999 | .999 | .999 | .999 |
| | | | | | | .999 | .999 | .999 | .999 | .997 | .995 | .990 | .982 | .970 | .952 |
| | | | | | | .927 | .894 | .852 | .800 | .740 | .673 | .600 | .525 | .450 | .378 |
| | | | | | | .310 | .249 | .195 | .150 | .112 | .082 | .059 | .042 | .029 | .019 |
| | | | | | | .013 | .008 | .005 | .003 | | | | | | |
| 28.000 | 28.000 | .000 | .002 | .059 | 45 | .999 | .999 | .999 | .999 | .999 | .999 | .999 | .999 | .999 | .999 |
| | | | | | | .999 | .999 | .999 | .999 | .997 | .995 | .990 | .982 | .970 | .952 |
| | | | | | | .927 | .894 | .852 | .800 | .740 | .673 | .600 | .525 | .450 | .377 |
| | | | | | | .310 | .249 | .195 | .150 | .112 | .082 | .059 | .041 | .028 | .019 |
| | | | | | | .013 | .008 | .005 | .003 | .002 | | | | | |
| 28.000 | 28.000 | .000 | .001 | .056 | 46 | .999 | .999 | .999 | .999 | .999 | .900 | .999 | .999 | .999 | .999 |
| | | | | | | .999 | .000 | .999 | .999 | .997 | .995 | .990 | .982 | .970 | .952 |
| | | | | | | .927 | .894 | .852 | .800 | .740 | .673 | .600 | .525 | .450 | .377 |
| | | | | | | .310 | .249 | .195 | .150 | .112 | .082 | .059 | .041 | .028 | .019 |
| | | | | | | .013 | .008 | .005 | .003 | .002 | .001 | | | | |
| 28.000 | 28.000 | .000 | .001 | .053 | 47 | .999 | .999 | .999 | .999 | .999 | .999 | .999 | .999 | .999 | .999 |
| | | | | | | .999 | .999 | .999 | .999 | .997 | .995 | .990 | .982 | .970 | .952 |
| | | | | | | .927 | .894 | .852 | .800 | .740 | .673 | .600 | .525 | .450 | .377 |
| | | | | | | .310 | .249 | .195 | .150 | .112 | .082 | .059 | .041 | .028 | .019 |
| | | | | | | .013 | .008 | .005 | .003 | .002 | .001 | .001 | | | |
| 30.000 | 30.000 | .000 | .799 | 1.000 | 31 | .999 | .999 | .999 | .999 | .999 | .999 | .999 | .999 | .999 | .999 |
| | | | | | | .999 | .999 | .999 | .999 | .999 | .999 | .999 | .997 | .995 | .992 |
| | | | | | | .987 | .980 | .970 | .958 | .942 | .924 | .902 | .078 | .852 | .826 |
| | | | | | | .799 | | | | | | | | | |
| 30.000 | 30.000 | .000 | .630 | .500 | 32 | .999 | .999 | .999 | .999 | .999 | .999 | .999 | .999 | .999 | .999 |

ERLANG C: INFINITE QUEUE - ALL CALLERS BLOCKED WILL WAIT AS LONG AS NECESSARY FOR CIRCUIT

| HOURS FIRST ATTEMPT VMR=1.0 | HOURS CON-NECT | HOURS OVFL | DELAYED PORTION | DELAY LENGTH CALL FACTOR | LINE | HOURS CONNECTED TRAFFIC ON LINES 1 | 2 | 3 | 4 | 5 | 6 | 7 | 8 | 9 | 10 |
|---|---|---|---|---|---|---|---|---|---|---|---|---|---|---|---|
| | | | | | | .999 | .999 | .999 | .999 | .999 | .999 | .998 | .996 | .992 | .987 |
| | | | | | | .979 | .967 | .952 | .931 | .906 | .875 | .840 | .801 | .759 | .716 |
| | | | | | | .672 | .630 | | | | | | | | |
| 30.000 | 30.000 | .000 | .490 | .333 | 33 | .999 | .999 | .999 | .999 | .999 | .999 | .999 | .999 | .999 | .999 |
| | | | | | | .999 | .999 | .999 | .999 | .999 | .999 | .997 | .995 | .990 | .984 |
| | | | | | | .974 | .959 | .940 | .915 | .883 | .845 | .801 | .752 | .700 | .646 |
| | | | | | | .592 | .540 | .490 | | | | | | | |
| 30.000 | 30.000 | .000 | .376 | .250 | 34 | .999 | .999 | .999 | .999 | .999 | .999 | .999 | .999 | .999 | .999 |
| | | | | | | .999 | .999 | .999 | .999 | .999 | .998 | .997 | .994 | .989 | .982 |
| | | | | | | .970 | .954 | .933 | .904 | .868 | .825 | .776 | .721 | .662 | .601 |
| | | | | | | .541 | .482 | .427 | .376 | | | | | | |
| 30.000 | 30.000 | .000 | .285 | .200 | 35 | .999 | .999 | .999 | .999 | .999 | .999 | .999 | .999 | .999 | .999 |
| | | | | | | .999 | .999 | .999 | .999 | .999 | .998 | .997 | .993 | .988 | .980 |
| | | | | | | .968 | .951 | .928 | .897 | .859 | .813 | .760 | .701 | .638 | .573 |
| | | | | | | .508 | .445 | .386 | .332 | .285 | | | | | |
| 30.000 | 30.000 | .000 | .212 | .167 | 36 | .999 | .999 | .999 | .999 | .999 | .999 | .999 | .999 | .999 | .999 |
| | | | | | | .999 | .999 | .999 | .999 | .999 | .998 | .996 | .993 | .988 | .980 |
| | | | | | | .967 | .949 | .925 | .893 | .853 | .805 | .750 | .689 | .623 | .555 |
| | | | | | | .487 | .421 | .360 | .304 | .254 | .212 | | | | |
| 30.000 | 30.000 | .000 | .155 | .143 | 37 | .999 | .999 | .999 | .999 | .999 | .999 | .999 | .999 | .999 | .999 |
| | | | | | | .999 | .999 | .999 | .999 | .999 | .998 | .996 | .993 | .988 | .979 |
| | | | | | | .966 | .948 | .923 | .890 | .849 | .800 | .743 | .681 | .613 | .544 |
| | | | | | | .474 | .406 | .343 | .286 | .235 | .191 | .155 | | | |
| 30.000 | 30.000 | .000 | .112 | .125 | 38 | .999 | .999 | .999 | .999 | .999 | .999 | .999 | .999 | .999 | .999 |
| | | | | | | .999 | .999 | .999 | .999 | .999 | .998 | .996 | .993 | .987 | .979 |
| | | | | | | .966 | .947 | .921 | .888 | .847 | .797 | .739 | .675 | .607 | .536 |
| | | | | | | .465 | .397 | .333 | .274 | .223 | .179 | .142 | .112 | | |
| 30.000 | 30.000 | .000 | .079 | .111 | 39 | .999 | .999 | .999 | .999 | .999 | .999 | .999 | .999 | .999 | .999 |
| | | | | | | .999 | .999 | .999 | .999 | .999 | .998 | .996 | .993 | .987 | .978 |
| | | | | | | .965 | .946 | .921 | .887 | .845 | .795 | .737 | .672 | .603 | .532 |
| | | | | | | .460 | .391 | .326 | .267 | .215 | .170 | .133 | .103 | .079 | |
| 30.000 | 30.000 | .000 | .055 | .100 | 40 | .999 | .999 | .999 | .999 | .999 | .999 | .999 | .999 | .999 | .999 |
| | | | | | | .999 | .999 | .999 | .999 | .999 | .998 | .996 | .993 | .987 | .978 |
| | | | | | | .965 | .946 | .920 | .886 | .844 | .794 | .735 | .670 | .601 | .529 |
| | | | | | | .457 | .387 | .322 | .263 | .210 | .165 | .128 | .098 | .074 | .055 |
| 30.000 | 30.000 | .000 | .038 | .091 | 41 | .999 | .999 | .999 | .999 | .999 | .999 | .999 | .999 | .999 | .999 |
| | | | | | | .999 | .999 | .999 | .999 | .999 | .998 | .996 | .993 | .987 | .978 |
| | | | | | | .965 | .946 | .920 | .886 | .844 | .793 | .734 | .669 | .599 | .527 |
| | | | | | | .455 | .385 | .319 | .260 | .207 | .162 | .125 | .094 | .070 | .052 |
| | | | | | | .038 | | | | | | | | | |

ERLANG C: INFINITE QUEUE - ALL CALLERS BLOCKED WILL WAIT AS LONG AS NECESSARY FOR CIRCUIT

| HOURS FIRST ATTEMPT VMR = 1.0 | HOURS CON-NECT | HOURS OVFL | DELAYED PORTION | DELAY LENGTH CALL FACTOR | LINE | HOURS CONNECTED TRAFFIC ON LINES 1 | 2 | 3 | 4 | 5 | 6 | 7 | 8 | 9 | 10 |
|---|---|---|---|---|---|---|---|---|---|---|---|---|---|---|---|
| 30.000 | 30.000 | .000 | .025 | .083 | 42 | .999 | .999 | .999 | .999 | .999 | .999 | .999 | .999 | .999 | .999 |
| | | | | | | .999 | .999 | .999 | .999 | .999 | .998 | .996 | .993 | .987 | .978 |
| | | | | | | .965 | .946 | .920 | .886 | .843 | .792 | .734 | .668 | .598 | .526 |
| | | | | | | .454 | .383 | .318 | .258 | .205 | .160 | .123 | .092 | .068 | .049 |
| | | | | | | .036 | .025 | | | | | | | | |
| 30.000 | 30.000 | .000 | .017 | .077 | 43 | .999 | .999 | .999 | .999 | .999 | .999 | .999 | .999 | .999 | .999 |
| | | | | | | .999 | .999 | .999 | .999 | .999 | .998 | .996 | .993 | .987 | .978 |
| | | | | | | .965 | .946 | .920 | .886 | .843 | .792 | .733 | .668 | .598 | .525 |
| | | | | | | .453 | .383 | .317 | .257 | .204 | .159 | .121 | .091 | .067 | .048 |
| | | | | | | .034 | .024 | .017 | | | | | | | |
| 30.000 | 30.000 | .000 | .011 | .071 | 44 | .999 | .999 | .999 | .999 | .999 | .999 | .999 | .999 | .999 | .999 |
| | | | | | | .999 | .999 | .999 | .999 | .999 | .998 | .996 | .993 | .987 | .978 |
| | | | | | | .965 | .946 | .920 | .885 | .843 | .792 | .733 | .668 | .597 | .525 |
| | | | | | | .452 | .382 | .316 | .256 | .204 | .158 | .121 | .090 | .066 | .047 |
| | | | | | | .033 | .023 | .016 | .011 | | | | | | |
| 30.000 | 30.000 | .000 | .007 | .067 | 45 | .999 | .999 | .999 | .999 | .999 | .999 | .999 | .999 | .999 | .999 |
| | | | | | | .999 | .999 | .999 | .999 | .999 | .998 | .996 | .993 | .987 | .978 |
| | | | | | | .965 | .946 | .919 | .885 | .843 | .792 | .733 | .667 | .597 | .525 |
| | | | | | | .452 | .382 | .316 | .256 | .203 | .158 | .120 | .090 | .065 | .047 |
| | | | | | | .033 | .023 | .015 | .010 | .007 | | | | | |
| 30.000 | 30.000 | .000 | .004 | .063 | 46 | .999 | .999 | .999 | .999 | .999 | .999 | .999 | .999 | .999 | .999 |
| | | | | | | .999 | .999 | .999 | .999 | .999 | .998 | .996 | .993 | .987 | .978 |
| | | | | | | .965 | .946 | .919 | .885 | .843 | .792 | .733 | .667 | .597 | .524 |
| | | | | | | .452 | .382 | .316 | .256 | .203 | .158 | .120 | .089 | .065 | .047 |
| | | | | | | .033 | .022 | .015 | .010 | .007 | .004 | | | | |
| 30.000 | 30.000 | .000 | .003 | .059 | 47 | .999 | .999 | .999 | .999 | .999 | .999 | .999 | .999 | .999 | .999 |
| | | | | | | .999 | .999 | .000 | .999 | .999 | .998 | .996 | .993 | .987 | .978 |
| | | | | | | .965 | .946 | .919 | .885 | .843 | .792 | .733 | .667 | .597 | .524 |
| | | | | | | .452 | .381 | .316 | .256 | .203 | .158 | .120 | .089 | .065 | .046 |
| | | | | | | .033 | .022 | .015 | .010 | .006 | .004 | .003 | | | |
| 30.000 | 30.000 | .000 | .002 | .056 | 48 | .999 | .999 | .999 | .999 | .999 | .999 | .999 | .999 | .999 | .999 |
| | | | | | | .999 | .999 | .999 | .999 | .999 | .998 | .996 | .993 | .987 | .978 |
| | | | | | | .965 | .946 | .919 | .885 | .843 | .792 | .733 | .667 | .597 | .524 |
| | | | | | | .452 | .381 | .316 | .256 | .203 | .157 | .120 | .089 | .065 | .046 |
| | | | | | | .032 | .022 | .015 | .010 | .006 | .004 | .003 | .002 | | |
| 30.000 | 30.000 | .000 | .001 | .053 | 49 | .999 | .000 | .999 | .999 | .999 | .999 | .999 | .999 | .999 | .999 |
| | | | | | | .999 | .999 | .999 | .999 | .999 | .998 | .996 | .993 | .987 | .978 |
| | | | | | | .965 | .946 | .919 | .885 | .843 | .792 | .733 | .667 | .597 | .524 |
| | | | | | | .452 | .381 | .316 | .256 | .203 | .157 | .120 | .080 | .065 | .046 |
| | | | | | | .032 | .022 | .015 | .010 | .006 | .004 | .003 | .002 | .001 | |

ERLANG C: INFINITE QUEUE - ALL CALLERS BLOCKED WILL WAIT AS LONG AS NECESSARY FOR CIRCUIT

| HOURS FIRST ATTEMPT VMR = 1.0 | HOURS CON- NECT | HOURS OVFL | DELAYED PORTION | DELAY LENGTH CALL FACTOR | LINE | HOURS CONNECTED TRAFFIC ON LINES 1 | 2 | 3 | 4 | 5 | 6 | 7 | 8 | 9 | 10 |
|---|---|---|---|---|---|---|---|---|---|---|---|---|---|---|---|
| 30.000 | 30.000 | .000 | .001 | .050 | 50 | .999 | .999 | .999 | .999 | .999 | .999 | .999 | .999 | .999 | .999 |
| | | | | | | .999 | .999 | .999 | .999 | .999 | .998 | .996 | .993 | .987 | .978 |
| | | | | | | .965 | .946 | .919 | .885 | .843 | .792 | .733 | .667 | .597 | .524 |
| | | | | | | .452 | .381 | .315 | .256 | .203 | .157 | .120 | .089 | .065 | .046 |
| | | | | | | .032 | .022 | .015 | .010 | .006 | .004 | .002 | .002 | .001 | .001 |

# SECTION 4
# NON-RANDOM TRAFFIC

ERLANG-ENGSET: FOR SMOOTH ATTEMPTING TRAFFIC AND FEW PERSONS DOING THE CALLING (FEW MEANS RATIO OF CALLERS TO CIRCUITS IS LESS THAN 8 - GREATER THAN 8 ASSUME TRAFFIC RANDOM)

| HOURS FIRST ATTEMPT | FIRST ATTEMPT VMR | NUMBER SOURCES | HOURS CON-NECT | HOURS OVFL | PERCENT BLOCKED (GRADE) | LINE | HOURS CONNECTED TRAFFIC ON LINES 1 | 2 | 3 | 4 | 5 | 6 | 7 | 8 | 9 | 10 |
|---|---|---|---|---|---|---|---|---|---|---|---|---|---|---|---|---|
| .200 | .90 | 2 | .180 | .020 | 9.9% | 1 | .180 | | | | | | | | | |
| .200 | .95 | 4 | .173 | .027 | 13.6% | 1 | .173 | | | | | | | | | |
| .200 | .97 | 6 | .171 | .029 | 14.6% | 1 | .171 | | | | | | | | | |
| .200 | .98 | 8 | .170 | .030 | 15.2% | 1 | .170 | | | | | | | | | |
| .400 | .80 | 2 | .323 | .077 | 19.3% | 1 | .323 | | | | | | | | | |
| .400 | .90 | 4 | .302 | .098 | 24.5% | 1 | .302 | | | | | | | | | |
| .400 | .93 | 6 | .296 | .104 | 26.0% | 1 | .296 | | | | | | | | | |
| .400 | .95 | 8 | .293 | .107 | 26.6% | 1 | .293 | | | | | | | | | |
| .400 | .87 | 3 | .393 | .007 | 1.8% | 2 | .308 | .085 | | | | | | | | |
| .400 | .93 | 6 | .386 | .014 | 3.6% | 2 | .296 | .089 | | | | | | | | |
| .400 | .96 | 9 | .383 | .017 | 4.2% | 2 | .293 | .091 | | | | | | | | |
| .400 | .97 | 12 | .382 | .018 | 4.5% | 2 | .291 | .091 | | | | | | | | |
| .400 | .97 | 15 | .381 | .019 | 4.7% | 2 | .290 | .092 | | | | | | | | |
| .600 | .70 | 2 | .434 | .166 | 27.7% | 1 | .434 | | | | | | | | | |
| .600 | .85 | 4 | .400 | .200 | 33.3% | 1 | .400 | | | | | | | | | |
| .600 | .90 | 6 | .391 | .209 | 34.8% | 1 | .391 | | | | | | | | | |
| .600 | .80 | 3 | .576 | .024 | 3.9% | 2 | .410 | .166 | | | | | | | | |
| .600 | .90 | 6 | .556 | .044 | 7.3% | 2 | .391 | .166 | | | | | | | | |
| .600 | .93 | 9 | .550 | .050 | 8.3% | 2 | .385 | .165 | | | | | | | | |
| .600 | .95 | 12 | .548 | .052 | 8.7% | 2 | .383 | .165 | | | | | | | | |
| .600 | .96 | 15 | .546 | .054 | 9.0% | 2 | .381 | .165 | | | | | | | | |
| .800 | .73 | 3 | .745 | .055 | 6.9% | 2 | .489 | .250 | | | | | | | | |
| .800 | .87 | 6 | .708 | .092 | 11.5% | 2 | .404 | .243 | | | | | | | | |
| .800 | .91 | 9 | .698 | .102 | 12.8% | 2 | .457 | .240 | | | | | | | | |
| .800 | .03 | 12 | .693 | .107 | 13.4% | 2 | .454 | .239 | | | | | | | | |
| .800 | .95 | 15 | .690 | .110 | 13.8% | 2 | .452 | .238 | | | | | | | | |
| .800 | .80 | 4 | .794 | .006 | .8% | 3 | .476 | .249 | .069 | | | | | | | |
| .800 | .90 | 8 | .782 | .018 | 2.3% | 3 | .459 | .241 | .082 | | | | | | | |
| .800 | .93 | 12 | .777 | .023 | 2.8% | 3 | .454 | .239 | .085 | | | | | | | |
| .800 | .95 | 16 | .775 | .025 | 3.1% | 3 | .452 | .238 | .086 | | | | | | | |
| .800 | .96 | 20 | .774 | .026 | 3.2% | 3 | .450 | .237 | .087 | | | | | | | |
| .800 | .97 | 24 | .773 | .027 | 3.3% | 3 | .449 | .237 | .087 | | | | | | | |
| 1.000 | .67 | 3 | .800 | .104 | 10.4% | 2 | .551 | .346 | | | | | | | | |
| 1.000 | .83 | 6 | .840 | .160 | 16.0% | 2 | .523 | .317 | | | | | | | | |
| 1.000 | .89 | 9 | .825 | .175 | 17.5% | 2 | .515 | .311 | | | | | | | | |
| 1.000 | .92 | 12 | .818 | .102 | 18.2% | 2 | .511 | .308 | | | | | | | | |
| 1.000 | .93 | 15 | .815 | .185 | 18.5% | 2 | .509 | .306 | | | | | | | | |
| 1.000 | .75 | 4 | .985 | .015 | 1.5% | 3 | .536 | .329 | .120 | | | | | | | |
| 1.000 | .88 | 8 | .960 | .040 | 4.0% | 3 | .517 | .312 | .131 | | | | | | | |
| 1.000 | .92 | 12 | .952 | .048 | 4.8% | 3 | .511 | .308 | .134 | | | | | | | |

ERLANG-ENGSET: FOR SMOOTH ATTEMPTING TRAFFIC AND FEW PERSONS DOING THE CALLING (FEW MEANS RATIO OF CALLERS TO CIRCUITS IS LESS THAN 8 - GREATER THAN 8 ASSUME TRAFFIC RANDOM)

| HOURS FIRST ATTEMPT | FIRST ATTEMPT VMR | NUMBER SOURCES | HOURS CON-NECT | HOURS OVFL | PERCENT BLOCKED (GRADE) | LINE | HOURS CONNECTED TRAFFIC ON LINES 1 | 2 | 3 | 4 | 5 | 6 | 7 | 8 | 9 | 10 |
|---|---|---|---|---|---|---|---|---|---|---|---|---|---|---|---|---|
| 1.000 | .94 | 16 | .949 | .051 | 5.1% | 3 | .508 | .306 | .135 | | | | | | | |
| 1.000 | .95 | 20 | .946 | .054 | 5.4% | 3 | .506 | .304 | .135 | | | | | | | |
| 1.000 | .96 | 24 | .945 | .055 | 5.5% | 3 | .505 | .304 | .136 | | | | | | | |
| 1.200 | .60 | 3 | 1.028 | .172 | 14.3% | 2 | .600 | .428 | | | | | | | | |
| 1.200 | .80 | 6 | .954 | .246 | 20.5% | 2 | .570 | .384 | | | | | | | | |
| 1.200 | .87 | 9 | .935 | .265 | 22.1% | 2 | .561 | .374 | | | | | | | | |
| 1.200 | .90 | 12 | .927 | .273 | 22.8% | 2 | .557 | .370 | | | | | | | | |
| 1.200 | .92 | 15 | .922 | .278 | 23.2% | 2 | .555 | .367 | | | | | | | | |
| 1.200 | .70 | 4 | 1.168 | .032 | 2.6% | 3 | .584 | .402 | .182 | | | | | | | |
| 1.200 | .85 | 8 | 1.127 | .073 | 6.1% | 3 | .564 | .376 | .187 | | | | | | | |
| 1.200 | .90 | 12 | 1.114 | .086 | 7.1% | 3 | .557 | .370 | .188 | | | | | | | |
| 1.200 | .92 | 16 | 1.109 | .091 | 7.6% | 3 | .554 | .367 | .188 | | | | | | | |
| 1.200 | .94 | 20 | 1.105 | .095 | 7.9% | 3 | .552 | .365 | .188 | | | | | | | |
| 1.200 | .95 | 24 | 1.103 | .097 | 8.1% | 3 | .551 | .364 | .188 | | | | | | | |
| 1.400 | .53 | 3 | 1.142 | .258 | 18.5% | 2 | .640 | .501 | | | | | | | | |
| 1.400 | .77 | 6 | 1.051 | .349 | 24.9% | 2 | .609 | .442 | | | | | | | | |
| 1.400 | .84 | 9 | 1.030 | .370 | 26.4% | 2 | .600 | .430 | | | | | | | | |
| 1.400 | .88 | 12 | 1.020 | .380 | 27.1% | 2 | .596 | .425 | | | | | | | | |
| 1.400 | .91 | 15 | 1.015 | .385 | 27.5% | 2 | .593 | .422 | | | | | | | | |
| 1.400 | .65 | 4 | 1.343 | .057 | 4.1% | 3 | .624 | .466 | .252 | | | | | | | |
| 1.400 | .83 | 8 | 1.280 | .120 | 8.6% | 3 | .602 | .433 | .245 | | | | | | | |
| 1.400 | .88 | 12 | 1.263 | .137 | 9.8% | 3 | .596 | .425 | .243 | | | | | | | |
| 1.400 | .91 | 16 | 1.255 | .145 | 10.4% | 3 | .592 | .421 | .242 | | | | | | | |
| 1.400 | .93 | 20 | 1.250 | .150 | 10.7% | 3 | .591 | .419 | .241 | | | | | | | |
| 1.400 | .94 | 24 | 1.247 | .153 | 10.9% | 3 | .589 | .417 | .241 | | | | | | | |
| 1.400 | .72 | 5 | 1.391 | .009 | .6% | 4 | .615 | .451 | .249 | .077 | | | | | | |
| 1.400 | .86 | 10 | 1.368 | .032 | 2.3% | 4 | .598 | .428 | .243 | .099 | | | | | | |
| 1.400 | .91 | 15 | 1.360 | .040 | 2.8% | 4 | .593 | .422 | .242 | .104 | | | | | | |
| 1.400 | .93 | 20 | 1.356 | .044 | 3.1% | 4 | .591 | .419 | .241 | .106 | | | | | | |
| 1.400 | .94 | 25 | 1.354 | .046 | 3.3% | 4 | .589 | .417 | .241 | .107 | | | | | | |
| 1.400 | .95 | 30 | 1.352 | .048 | 3.4% | 4 | .588 | .416 | .240 | .107 | | | | | | |
| 1.600 | .47 | 3 | 1.238 | .362 | 22.6% | 2 | .674 | .564 | | | | | | | | |
| 1.600 | .73 | 6 | 1.136 | .464 | 29.0% | 2 | .642 | .494 | | | | | | | | |
| 1.600 | .82 | 9 | 1.112 | .488 | 30.5% | 2 | .632 | .479 | | | | | | | | |
| 1.600 | .87 | 12 | 1.101 | .499 | 31.2% | 2 | .628 | .473 | | | | | | | | |
| 1.600 | .89 | 15 | 1.095 | .505 | 31.6% | 2 | .625 | .469 | | | | | | | | |
| 1.600 | .60 | 4 | 1.505 | .095 | 6.0% | 3 | .657 | .522 | .326 | | | | | | | |
| 1.600 | .80 | 8 | 1.419 | .181 | 11.3% | 3 | .635 | .483 | .302 | | | | | | | |
| 1.600 | .87 | 12 | 1.397 | .203 | 12.7% | 3 | .628 | .473 | .297 | | | | | | | |
| 1.600 | .90 | 16 | 1.388 | .212 | 13.3% | 3 | .625 | .469 | .294 | | | | | | | |

ERLANG-ENGSET: FOR SMOOTH ATTEMPTING TRAFFIC AND FEW PERSONS DOING THE CALLING (FEW MEANS RATIO OF CALLERS TO CIRCUITS IS LESS THAN 8 - GREATER THAN 8 ASSUME TRAFFIC RANDOM)

| HOURS FIRST ATTEMPT | FIRST ATTEMPT VMR | NUMBER SOURCES | HOURS CONNECT | HOURS OVFL | PERCENT BLOCKED (GRADE) | LINE | HOURS CONNECTED TRAFFIC ON LINES 1 | 2 | 3 | 4 | 5 | 6 | 7 | 8 | 9 | 10 |
|---|---|---|---|---|---|---|---|---|---|---|---|---|---|---|---|---|
| 1.600 | .92 | 20 | 1.382 | .218 | 13.6% | 3 | .623 | .466 | .293 | | | | | | | |
| 1.600 | .93 | 24 | 1.378 | .222 | 13.9% | 3 | .622 | .464 | .292 | | | | | | | |
| 1.600 | .68 | 5 | 1.583 | .017 | 1.0% | 4 | .648 | .504 | .314 | .118 | | | | | | |
| 1.600 | .84 | 10 | 1.545 | .055 | 3.4% | 4 | .631 | .477 | .298 | .139 | | | | | | |
| 1.600 | .89 | 15 | 1.533 | .067 | 4.2% | 4 | .625 | .469 | .295 | .143 | | | | | | |
| 1.600 | .92 | 20 | 1.527 | .073 | 4.6% | 4 | .623 | .466 | .293 | .145 | | | | | | |
| 1.600 | .94 | 25 | 1.523 | .077 | 4.8% | 4 | .621 | .464 | .292 | .146 | | | | | | |
| 1.600 | .95 | 30 | 1.521 | .079 | 4.9% | 4 | .620 | .463 | .291 | .147 | | | | | | |
| 1.800 | .40 | 3 | 1.319 | .481 | 26.7% | 2 | .701 | .617 | | | | | | | | |
| 1.800 | .70 | 6 | 1.208 | .592 | 32.9% | 2 | .670 | .538 | | | | | | | | |
| 1.800 | .80 | 9 | 1.182 | .618 | 34.3% | 2 | .660 | .522 | | | | | | | | |
| 1.800 | .85 | 12 | 1.171 | .629 | 34.9% | 2 | .656 | .515 | | | | | | | | |
| 1.800 | .55 | 4 | 1.653 | .147 | 8.2% | 3 | .685 | .570 | .398 | | | | | | | |
| 1.800 | .77 | 8 | 1.544 | .256 | 14.2% | 3 | .662 | .526 | .356 | | | | | | | |
| 1.800 | .85 | 12 | 1.519 | .281 | 15.6% | 3 | .656 | .515 | .348 | | | | | | | |
| 1.800 | .89 | 16 | 1.507 | .293 | 16.3% | 3 | .652 | .510 | .344 | | | | | | | |
| 1.800 | .91 | 20 | 1.500 | .300 | 16.7% | 3 | .650 | .508 | .342 | | | | | | | |
| 1.800 | .93 | 24 | 1.496 | .304 | 16.9% | 3 | .649 | .506 | .341 | | | | | | | |
| 1.800 | .64 | 5 | 1.770 | .030 | 1.6% | 4 | .675 | .550 | .376 | .169 | | | | | | |
| 1.800 | .82 | 10 | 1.713 | .087 | 4.8% | 4 | .658 | .519 | .351 | .184 | | | | | | |
| 1.800 | .88 | 15 | 1.696 | .104 | 5.8% | 4 | .653 | .511 | .345 | .187 | | | | | | |
| 1.800 | .91 | 20 | 1.688 | .112 | 6.2% | 4 | .650 | .508 | .342 | .108 | | | | | | |
| 1.800 | .93 | 25 | 1.683 | .117 | 6.5% | 4 | .649 | .505 | .341 | .188 | | | | | | |
| 1.800 | .94 | 30 | 1.000 | .120 | 6.7% | 4 | .648 | .504 | .340 | .188 | | | | | | |
| 2.000 | .33 | 3 | 1.387 | .613 | 30.6% | 2 | .725 | .662 | | | | | | | | |
| 2.000 | .50 | 4 | 1.786 | .214 | 10.7% | 3 | .709 | .611 | .467 | | | | | | | |
| 2.000 | .75 | 8 | 1.657 | .343 | 17.2% | 3 | .686 | .564 | .407 | | | | | | | |
| 2.000 | .83 | 12 | 1.627 | .373 | 18.6% | 3 | .680 | .552 | .395 | | | | | | | |
| 2.000 | .88 | 16 | 1.614 | .386 | 19.3% | 3 | .676 | .547 | .391 | | | | | | | |
| 2.000 | .90 | 20 | 1.607 | .393 | 19.7% | 3 | .674 | .544 | .388 | | | | | | | |
| 2.000 | .92 | 24 | 1.602 | .398 | 19.9% | 3 | .673 | .542 | .386 | | | | | | | |
| 2.000 | .60 | 5 | 1.951 | .049 | 2.5% | 4 | .699 | .589 | .435 | .227 | | | | | | |
| 2.000 | .80 | 10 | 1.870 | .130 | 6.5% | 4 | .682 | .557 | .400 | .231 | | | | | | |
| 2.000 | .87 | 15 | 1.848 | .152 | 7.6% | 4 | .677 | .540 | .392 | .231 | | | | | | |
| 2.000 | .90 | 20 | 1.838 | .162 | 8.1% | 4 | .674 | .544 | .388 | .231 | | | | | | |
| 2.000 | .92 | 25 | 1.832 | .168 | 8.4% | 4 | .673 | .542 | .386 | .231 | | | | | | |
| 2.000 | .93 | 30 | 1.828 | .172 | 8.6% | 4 | .672 | .540 | .385 | .231 | | | | | | |
| 2.200 | .27 | 3 | 1.445 | .755 | 34.3% | 2 | .746 | .699 | | | | | | | | |
| 2.200 | .45 | 4 | 1.905 | .295 | 13.4% | 3 | .729 | .646 | .530 | | | | | | | |
| 2.200 | .73 | 8 | 1.757 | .443 | 20.1% | 3 | .707 | .597 | .453 | | | | | | | |

ERLANG-ENGSET: FOR SMOOTH ATTEMPTING TRAFFIC AND FEW PERSONS DOING THE CALLING (FEW MEANS RATIO OF CALLERS TO CIRCUITS IS LESS THAN 8 - GREATER THAN 8 ASSUME TRAFFIC RANDOM)

| HOURS FIRST ATTEMPT | FIRST ATTEMPT VMR | NUMBER SOURCES | HOURS CONNECT | HOURS OVFL | PERCENT BLOCKED (GRADE) | LINE | HOURS CONNECTED TRAFFIC ON LINES 1 | 2 | 3 | 4 | 5 | 6 | 7 | 8 | 9 | 10 |
|---|---|---|---|---|---|---|---|---|---|---|---|---|---|---|---|---|
| 2.200 | .82 | 12 | 1.725 | .475 | 21.6% | 3 | .700 | .585 | .439 | | | | | | | |
| 2.200 | .86 | 16 | 1.710 | .490 | 22.3% | 3 | .697 | .580 | .434 | | | | | | | |
| 2.200 | .89 | 20 | 1.702 | .498 | 22.6% | 3 | .695 | .577 | .430 | | | | | | | |
| 2.200 | .91 | 24 | 1.697 | .503 | 22.9% | 3 | .694 | .575 | .428 | | | | | | | |
| 2.200 | .56 | 5 | 2.122 | .078 | 3.5% | 4 | .720 | .624 | .488 | .290 | | | | | | |
| 2.200 | .78 | 10 | 2.016 | .184 | 8.4% | 4 | .703 | .590 | .444 | .279 | | | | | | |
| 2.200 | .85 | 15 | 1.989 | .211 | 9.6% | 4 | .698 | .581 | .435 | .276 | | | | | | |
| 2.200 | .89 | 20 | 1.977 | .223 | 10.1% | 4 | .695 | .577 | .430 | .275 | | | | | | |
| 2.200 | .91 | 25 | 1.970 | .230 | 10.5% | 4 | .694 | .574 | .428 | .274 | | | | | | |
| 2.200 | .93 | 30 | 1.965 | .235 | 10.7% | 4 | .693 | .573 | .426 | .274 | | | | | | |
| 2.200 | .63 | 6 | 2.186 | .014 | .7% | 5 | .714 | .611 | .470 | .286 | .104 | | | | | |
| 2.200 | .82 | 12 | 2.138 | .062 | 2.8% | 5 | .700 | .585 | .439 | .278 | .136 | | | | | |
| 2.200 | .88 | 18 | 2.122 | .078 | 3.5% | 5 | .696 | .578 | .432 | .275 | .141 | | | | | |
| 2.200 | .91 | 24 | 2.115 | .085 | 3.9% | 5 | .694 | .575 | .428 | .274 | .144 | | | | | |
| 2.200 | .93 | 30 | 2.110 | .090 | 4.1% | 5 | .693 | .573 | .426 | .274 | .145 | | | | | |
| 2.200 | .94 | 36 | 2.107 | .093 | 4.2% | 5 | .692 | .571 | .425 | .273 | .146 | | | | | |
| 2.400 | .40 | 4 | 2.008 | .392 | 16.3% | 3 | .747 | .676 | .585 | | | | | | | |
| 2.400 | .70 | 8 | 1.846 | .554 | 23.1% | 3 | .725 | .626 | .495 | | | | | | | |
| 2.400 | .80 | 12 | 1.812 | .588 | 24.5% | 3 | .719 | .614 | .479 | | | | | | | |
| 2.400 | .85 | 16 | 1.796 | .604 | 25.1% | 3 | .715 | .608 | .473 | | | | | | | |
| 2.400 | .88 | 20 | 1.788 | .612 | 25.5% | 3 | .713 | .605 | .469 | | | | | | | |
| 2.400 | .90 | 24 | 1.782 | .618 | 25.7% | 3 | .712 | .603 | .467 | | | | | | | |
| 2.400 | .52 | 5 | 2.284 | .116 | 4.8% | 4 | .738 | .653 | .537 | .356 | | | | | | |
| 2.400 | .76 | 10 | 2.151 | .249 | 10.4% | 4 | .721 | .619 | .485 | .326 | | | | | | |
| 2.400 | .84 | 15 | 2.120 | .280 | 11.7% | 4 | .716 | .609 | .474 | .320 | | | | | | |
| 2.400 | .88 | 20 | 2.105 | .295 | 12.3% | 4 | .713 | .605 | .469 | .318 | | | | | | |
| 2.400 | .90 | 25 | 2.097 | .303 | 12.6% | 4 | .712 | .603 | .466 | .316 | | | | | | |
| 2.400 | .92 | 30 | 2.092 | .308 | 12.8% | 4 | .711 | .601 | .465 | .315 | | | | | | |
| 2.400 | .60 | 6 | 2.376 | .024 | 1.0% | 5 | .732 | .640 | .515 | .343 | .145 | | | | | |
| 2.400 | .80 | 12 | 2.308 | .092 | 3.8% | 5 | .719 | .614 | .479 | .323 | .173 | | | | | |
| 2.400 | .87 | 18 | 2.288 | .112 | 4.7% | 5 | .714 | .607 | .471 | .319 | .178 | | | | | |
| 2.400 | .90 | 24 | 2.278 | .122 | 5.1% | 5 | .712 | .603 | .467 | .317 | .179 | | | | | |
| 2.400 | .92 | 30 | 2.272 | .128 | 5.3% | 5 | .711 | .601 | .465 | .315 | .180 | | | | | |
| 2.400 | .93 | 36 | 2.268 | .132 | 5.5% | 5 | .710 | .600 | .463 | .315 | .181 | | | | | |
| 2.600 | .35 | 4 | 2.099 | .501 | 19.3% | 3 | .763 | .703 | .634 | | | | | | | |
| 2.600 | .68 | 8 | 1.926 | .674 | 25.9% | 3 | .741 | .652 | .533 | | | | | | | |
| 2.600 | .78 | 12 | 1.890 | .710 | 27.3% | 3 | .735 | .639 | .515 | | | | | | | |
| 2.600 | .84 | 16 | 1.874 | .726 | 27.9% | 3 | .732 | .634 | .508 | | | | | | | |
| 2.600 | .87 | 20 | 1.865 | .735 | 28.3% | 3 | .730 | .631 | .504 | | | | | | | |
| 2.600 | .89 | 24 | 1.859 | .741 | 28.5% | 3 | .728 | .629 | .502 | | | | | | | |

ERLANG-ENGSET: FOR SMOOTH ATTEMPTING TRAFFIC AND FEW PERSONS DOING THE CALLING (FEW MEANS RATIO OF CALLERS TO CIRCUITS IS LESS THAN 8 - GREATER THAN 8 ASSUME TRAFFIC RANDOM)

| HOURS FIRST ATTEMPT | FIRST ATTEMPT VMR | NUMBER SOURCES | HOURS CON-NECT | HOURS OVFL | PERCENT BLOCKED (GRADE) | LINE | HOURS CONNECTED TRAFFIC ON LINES 1 | 2 | 3 | 4 | 5 | 6 | 7 | 8 | 9 | 10 |
|---|---|---|---|---|---|---|---|---|---|---|---|---|---|---|---|---|
| 2.600 | .48 | 5 | 2.433 | .167 | 6.4% | 4 | .754 | .679 | .579 | .421 | | | | | | |
| 2.600 | .74 | 10 | 2.275 | .325 | 12.5% | 4 | .737 | .644 | .522 | .371 | | | | | | |
| 2.600 | .83 | 15 | 2.239 | .361 | 13.9% | 4 | .732 | .635 | .510 | .363 | | | | | | |
| 2.600 | .87 | 20 | 2.223 | .377 | 14.5% | 4 | .730 | .631 | .504 | .359 | | | | | | |
| 2.600 | .90 | 25 | 2.214 | .386 | 14.8% | 4 | .728 | .628 | .501 | .357 | | | | | | |
| 2.600 | .91 | 30 | 2.208 | .392 | 15.1% | 4 | .727 | .626 | .499 | .355 | | | | | | |
| 2.600 | .57 | 6 | 2.562 | .038 | 1.5% | 5 | .748 | .666 | .555 | .398 | .194 | | | | | |
| 2.600 | .78 | 12 | 2.470 | .130 | 5.0% | 5 | .735 | .639 | .515 | .367 | .214 | | | | | |
| 2.600 | .86 | 18 | 2.444 | .156 | 6.0% | 5 | .730 | .632 | .506 | .360 | .216 | | | | | |
| 2.600 | .89 | 24 | 2.432 | .168 | 6.5% | 5 | .728 | .629 | .502 | .357 | .217 | | | | | |
| 2.600 | .91 | 30 | 2.425 | .175 | 6.7% | 5 | .727 | .626 | .499 | .355 | .217 | | | | | |
| 2.600 | .93 | 36 | 2.421 | .179 | 6.9% | 5 | .726 | .625 | .498 | .354 | .217 | | | | | |
| 2.800 | .30 | 4 | 2.178 | .622 | 22.2% | 3 | .777 | .725 | .676 | | | | | | | |
| 2.800 | .65 | 8 | 1.997 | .803 | 28.7% | 3 | .756 | .675 | .567 | | | | | | | |
| 2.800 | .77 | 12 | 1.960 | .840 | 30.0% | 3 | .749 | .662 | .548 | | | | | | | |
| 2.800 | .83 | 16 | 1.943 | .857 | 30.6% | 3 | .746 | .657 | .540 | | | | | | | |
| 2.800 | .86 | 20 | 1.934 | .866 | 30.9% | 3 | .744 | .653 | .536 | | | | | | | |
| 2.800 | .88 | 24 | 1.928 | .872 | 31.2% | 3 | .743 | .651 | .534 | | | | | | | |
| 2.800 | .44 | 5 | 2.570 | .230 | 8.2% | 4 | .768 | .702 | .616 | .483 | | | | | | |
| 2.800 | .72 | 10 | 2.388 | .412 | 14.7% | 4 | .752 | .667 | .555 | .414 | | | | | | |
| 2.800 | .81 | 15 | 2.349 | .451 | 16.1% | 4 | .747 | .658 | .542 | .402 | | | | | | |
| 2.800 | .86 | 20 | 2.331 | .469 | 16.7% | 4 | .744 | .653 | .536 | .398 | | | | | | |
| 2.800 | .89 | 25 | 2.322 | .478 | 17.1% | 4 | .743 | .651 | .533 | .395 | | | | | | |
| 2.800 | .91 | 30 | 2.315 | .485 | 17.3% | 4 | .742 | .649 | .531 | .393 | | | | | | |
| 2.800 | .53 | 6 | 2.741 | .059 | 2.1% | 5 | .762 | .689 | .591 | .451 | .248 | | | | | |
| 2.800 | .77 | 12 | 2.623 | .177 | 6.3% | 5 | .749 | .662 | .548 | .408 | .255 | | | | | |
| 2.800 | .84 | 18 | 2.592 | .208 | 7.4% | 5 | .745 | .655 | .538 | .399 | .255 | | | | | |
| 2.800 | .88 | 24 | 2.578 | .222 | 7.9% | 5 | .743 | .651 | .534 | .395 | .254 | | | | | |
| 2.800 | .91 | 30 | 2.569 | .231 | 8.2% | 5 | .742 | .649 | .531 | .393 | .254 | | | | | |
| 2.800 | .92 | 36 | 2.564 | .236 | 8.4% | 5 | .741 | .648 | .529 | .392 | .254 | | | | | |
| 3.000 | .25 | 4 | 2.246 | .754 | 25,1% | 3 | .709 | .745 | .712 | | | | | | | |
| 3.000 | .63 | 8 | 2.061 | .939 | 31.3% | 3 | .769 | .695 | .597 | | | | | | | |
| 3.000 | .75 | 12 | 2.023 | .977 | 32.6% | 3 | .762 | .682 | .578 | | | | | | | |
| 3.000 | .81 | 16 | 2.006 | .994 | 33.1% | 3 | .759 | .677 | .570 | | | | | | | |
| 3.000 | .85 | 20 | 1.996 | 1.004 | 33.5% | 3 | .757 | .674 | .565 | | | | | | | |
| 3.000 | .88 | 24 | 1.990 | 1.010 | 33.7% | 3 | .756 | .672 | .562 | | | | | | | |
| 3.000 | .40 | 5 | 2.694 | .306 | 10.2% | 4 | .780 | .722 | .650 | .541 | | | | | | |
| 3.000 | .70 | 10 | 2.491 | .509 | 17.0% | 4 | .765 | .687 | .585 | .454 | | | | | | |
| 3.000 | .80 | 15 | 2.449 | .551 | 18.4% | 4 | .760 | .678 | .571 | .440 | | | | | | |
| 3.000 | .85 | 20 | 2.430 | .570 | 19.0% | 4 | .757 | .674 | .565 | .434 | | | | | | |

ERLANG-ENGSET: FOR SMOOTH ATTEMPTING TRAFFIC AND FEW PERSONS DOING THE CALLING (FEW MEANS RATIO OF CALLERS TO CIRCUITS IS LESS THAN 8 - GREATER THAN 8 ASSUME TRAFFIC RANDOM)

| HOURS FIRST ATTEMPT | FIRST ATTEMPT VMR | NUMBER SOURCES | HOURS CON-NECT | HOURS OVFL | PERCENT BLOCKED (GRADE) | LINE | HOURS CONNECTED TRAFFIC ON LINES 1 | 2 | 3 | 4 | 5 | 6 | 7 | 8 | 9 | 10 |
|---|---|---|---|---|---|---|---|---|---|---|---|---|---|---|---|---|
| 3.000 | .88 | 25 | 2.420 | .580 | 19.3% | 4 | .756 | .671 | .562 | .431 | | | | | | |
| 3.000 | .90 | 30 | 2.413 | .587 | 19.6% | 4 | .755 | .670 | .560 | .429 | | | | | | |
| 3.000 | .50 | 6 | 2.913 | .087 | 2.9% | 5 | .775 | .709 | .622 | .499 | .307 | | | | | |
| 3.000 | .75 | 12 | 2.766 | .234 | 7.8% | 5 | .762 | .682 | .578 | .446 | .297 | | | | | |
| 3.000 | .83 | 18 | 2.730 | .270 | 9.0% | 5 | .758 | .675 | .567 | .436 | .293 | | | | | |
| 3.000 | .88 | 24 | 2.713 | .287 | 9.6% | 5 | .756 | .672 | .562 | .432 | .292 | | | | | |
| 3.000 | .90 | 30 | 2.704 | .296 | 9.9% | 5 | .755 | .670 | .560 | .429 | .291 | | | | | |
| 3.000 | .92 | 36 | 2.698 | .302 | 10.1% | 5 | .754 | .668 | .558 | .427 | .291 | | | | | |
| 3.000 | .57 | 7 | 2.982 | .018 | .6% | 6 | .771 | .701 | .607 | .479 | .304 | .119 | | | | |
| 3.000 | .79 | 14 | 2.909 | .091 | 3.0% | 6 | .760 | .679 | .573 | .442 | .295 | .160 | | | | |
| 3.000 | .86 | 21 | 2.886 | .114 | 3.8% | 6 | .757 | .673 | .565 | .433 | .293 | .166 | | | | |
| 3.000 | .89 | 28 | 2.875 | .125 | 4.2% | 6 | .755 | .670 | .560 | .430 | .291 | .168 | | | | |
| 3.000 | .91 | 35 | 2.869 | .131 | 4.4% | 6 | .754 | .668 | .558 | .428 | .291 | .170 | | | | |
| 3.000 | .93 | 42 | 2.864 | .136 | 4.5% | 6 | .753 | .667 | .557 | .426 | .290 | .170 | | | | |
| 3.200 | .20 | 4 | 2.306 | .894 | 27.9% | 3 | .800 | .763 | .743 | | | | | | | |
| 3.200 | .60 | 8 | 2.118 | 1.082 | 33.8% | 3 | .780 | .713 | .625 | | | | | | | |
| 3.200 | .36 | 5 | 2.804 | .396 | 12.4% | 4 | .792 | .740 | .678 | .594 | | | | | | |
| 3.200 | .68 | 10 | 2.584 | .616 | 19.2% | 4 | .776 | .706 | .612 | .491 | | | | | | |
| 3.200 | .79 | 15 | 2.540 | .660 | 20.6% | 4 | .771 | .696 | .598 | .475 | | | | | | |
| 3.200 | .84 | 20 | 2.521 | .679 | 21.2% | 4 | .769 | .692 | .592 | .468 | | | | | | |
| 3.200 | .87 | 25 | 2.510 | .690 | 21.6% | 4 | .768 | .689 | .588 | .465 | | | | | | |
| 3.200 | .89 | 30 | 2.503 | .697 | 21.8% | 4 | .767 | .688 | .586 | .462 | | | | | | |
| 3.200 | .47 | 6 | 3.075 | .125 | 3.9% | 5 | .787 | .727 | .651 | .543 | .368 | | | | | |
| 3.200 | .73 | 12 | 2.899 | .301 | 9.4% | 5 | .774 | .701 | .604 | .482 | .337 | | | | | |
| 3.200 | .82 | 18 | 2.859 | .341 | 10.7% | 5 | .770 | .693 | .594 | .470 | .332 | | | | | |
| 3.200 | .87 | 24 | 2.840 | .360 | 11.2% | 5 | .768 | .690 | .589 | .465 | .329 | | | | | |
| 3.200 | .89 | 30 | 2.830 | .370 | 11.6% | 5 | .767 | .688 | .586 | .462 | .327 | | | | | |
| 3.200 | .91 | 36 | 2.823 | .377 | 11.8% | 5 | .766 | .686 | .584 | .461 | .327 | | | | | |
| 3.200 | .54 | 7 | 3.172 | .028 | .9% | 6 | .783 | .719 | .635 | .520 | .356 | .159 | | | | |
| 3.200 | .77 | 14 | 3.075 | .125 | 3.9% | 6 | .772 | .697 | .600 | .477 | .335 | .195 | | | | |
| 3.200 | .85 | 21 | 3.047 | .153 | 4.8% | 6 | .769 | .691 | .591 | .467 | .330 | .199 | | | | |
| 3.200 | .89 | 28 | 3.034 | .166 | 5.2% | 6 | .767 | .688 | .587 | .463 | .328 | .201 | | | | |
| 3.200 | .91 | 35 | 3.026 | .174 | 5.4% | 6 | .766 | .687 | .584 | .461 | .327 | .201 | | | | |
| 3.200 | .92 | 42 | 3.021 | .179 | 5.6% | 6 | .765 | .686 | .583 | .459 | .326 | .202 | | | | |
| 3.400 | .15 | 4 | 2.358 | 1.042 | 30.7% | 3 | .810 | .779 | .769 | | | | | | | |
| 3.400 | .32 | 5 | 2.903 | .497 | 14.6% | 4 | .802 | .756 | .704 | .641 | | | | | | |
| 3.400 | .66 | 10 | 2.669 | .731 | 21.5% | 4 | .787 | .722 | .636 | .524 | | | | | | |
| 3.400 | .77 | 15 | 2.623 | .777 | 22.8% | 4 | .782 | .713 | .622 | .507 | | | | | | |
| 3.400 | .83 | 20 | 2.603 | .797 | 23.4% | 4 | .780 | .709 | .615 | .500 | | | | | | |
| 3.400 | .86 | 25 | 2.592 | .808 | 23.8% | 4 | .778 | .706 | .612 | .496 | | | | | | |

ERLANG-ENGSET: FOR SMOOTH ATTEMPTING TRAFFIC AND FEW PERSONS DOING THE CALLING (FEW MEANS RATIO OF CALLERS TO CIRCUITS IS LESS THAN 8 - GREATER THAN 8 ASSUME TRAFFIC RANDOM)

| HOURS FIRST ATTEMPT | FIRST ATTEMPT VMR | NUMBER SOURCES | HOURS CONNECT | HOURS OVFL | PERCENT BLOCKED (GRADE) | LINE | HOURS CONNECTED TRAFFIC ON LINES 1 | 2 | 3 | 4 | 5 | 6 | 7 | 8 | 9 | 10 |
|---|---|---|---|---|---|---|---|---|---|---|---|---|---|---|---|---|
| 3.400 | .89 | 30 | 2.585 | .815 | 24.0% | 4 | .777 | .704 | .610 | .493 | | | | | | |
| 3.400 | .43 | 6 | 3.228 | .172 | 5.1% | 5 | .797 | .743 | .676 | .582 | .430 | | | | | |
| 3.400 | .72 | 12 | 3.022 | .378 | 11.1% | 5 | .784 | .717 | .628 | .515 | .377 | | | | | |
| 3.400 | .81 | 18 | 2.978 | .422 | 12.4% | 5 | .780 | .710 | .618 | .502 | .368 | | | | | |
| 3.400 | .86 | 24 | 2.958 | .442 | 13.0% | 5 | .779 | .706 | .613 | .496 | .365 | | | | | |
| 3.400 | .89 | 30 | 2.947 | .361 | 13.3% | 5 | .777 | .704 | .610 | .493 | .363 | | | | | |
| 3.400 | .91 | 36 | 2.940 | .460 | 13.5% | 5 | .777 | .703 | .608 | .491 | .361 | | | | | |
| 3.400 | .51 | 7 | 3.357 | .043 | 1.3% | 6 | .793 | .735 | .659 | .556 | .407 | .206 | | | | |
| 3.400 | .76 | 14 | 3.233 | .167 | 4.9% | 6 | .783 | .714 | .624 | .509 | .373 | .231 | | | | |
| 3.400 | .84 | 21 | 3.200 | .200 | 5.9% | 6 | .779 | .708 | .615 | .499 | .366 | .233 | | | | |
| 3.400 | .88 | 28 | 3.184 | .216 | 6.3% | 6 | .778 | .705 | .610 | .494 | .363 | .234 | | | | |
| 3.400 | .90 | 35 | 3.175 | .225 | 6.6% | 6 | .777 | .703 | .608 | .492 | .361 | .234 | | | | |
| 3.400 | .92 | 42 | 3.169 | .231 | 6.8% | 6 | .776 | .702 | .606 | .490 | .360 | .234 | | | | |
| 3.600 | .10 | 4 | 2.404 | 1.196 | 33.2% | 3 | .819 | .793 | .792 | | | | | | | |
| 3.600 | .28 | 5 | 2.990 | .610 | 16.9% | 4 | .811 | .770 | .726 | .682 | | | | | | |
| 3.600 | .64 | 10 | 2.747 | .853 | 23.7% | 4 | .796 | .737 | .658 | .556 | | | | | | |
| 3.600 | .76 | 15 | 2.700 | .900 | 25.0% | 4 | .792 | .728 | .644 | .537 | | | | | | |
| 3.600 | .82 | 20 | 2.679 | .921 | 25.6% | 4 | .789 | .723 | .637 | .529 | | | | | | |
| 3.600 | .86 | 25 | 2.667 | .933 | 25.9% | 4 | .788 | .721 | .634 | .525 | | | | | | |
| 3.600 | .88 | 30 | 2.660 | .940 | 26.1% | 4 | .787 | .719 | .631 | .522 | | | | | | |
| 3.600 | .40 | 6 | 3.369 | .231 | 6.4% | 5 | .806 | .758 | .607 | .610 | .489 | | | | | |
| 3.600 | .70 | 12 | 3.137 | .463 | 12.9% | 5 | .794 | .732 | .650 | .546 | .415 | | | | | |
| 3.600 | .80 | 18 | 3.089 | .511 | 14.2% | 5 | .790 | .725 | .639 | .532 | .403 | | | | | |
| 3.600 | .85 | 24 | 3.068 | .532 | 14.8% | 5 | .788 | .721 | .634 | .525 | .399 | | | | | |
| 3.600 | .88 | 30 | 3.056 | .544 | 15.1% | 5 | .787 | .719 | .631 | .522 | .396 | | | | | |
| 3.600 | .90 | 36 | 3.048 | .552 | 15.3% | 5 | .786 | .718 | .629 | .520 | .395 | | | | | |
| 3.600 | .49 | 7 | 3.537 | .063 | 1.8% | 6 | .803 | .750 | .682 | .590 | .456 | .257 | | | | |
| 3.600 | .74 | 14 | 3.383 | .217 | 6.0% | 6 | .792 | .729 | .645 | .539 | .410 | .268 | | | | |
| 3.600 | .83 | 21 | 3.345 | .255 | 7.1% | 6 | .789 | .723 | .630 | .528 | .401 | .268 | | | | |
| 3.600 | .87 | 28 | 3.327 | .273 | 7.6% | 6 | .787 | .720 | .632 | .523 | .397 | .268 | | | | |
| 3.600 | .90 | 35 | 3.317 | .283 | 7.9% | 6 | .786 | .718 | .630 | .520 | .395 | .267 | | | | |
| 3.600 | .91 | 42 | 3.310 | .290 | 8.1% | 6 | .786 | .717 | .628 | .518 | .393 | .267 | | | | |
| 3.800 | .24 | 5 | 3.067 | .733 | 19.3% | 4 | .820 | .783 | .746 | .718 | | | | | | |
| 3.800 | .62 | 10 | 2.817 | .983 | 25.9% | 4 | .805 | .750 | .678 | .584 | | | | | | |
| 3.800 | .75 | 15 | 2.769 | 1.031 | 27.1% | 4 | .801 | .741 | .663 | .564 | | | | | | |
| 3.800 | .81 | 20 | 2.748 | 1.052 | 27.7% | 4 | .798 | .737 | .657 | .556 | | | | | | |
| 3.800 | .85 | 25 | 2.736 | 1.064 | 28.0% | 4 | .797 | .735 | .653 | .551 | | | | | | |
| 3.800 | .87 | 30 | 2.729 | 1.071 | 28.2% | 4 | .796 | .733 | .651 | .548 | | | | | | |
| 3.800 | .37 | 6 | 3.498 | .302 | 8.0% | 5 | .815 | .771 | .718 | .650 | .545 | | | | | |
| 3.800 | .68 | 12 | 3.242 | .558 | 14.7% | 5 | .803 | .745 | .670 | .573 | .451 | | | | | |

## ERLANG-ENGSET: FOR SMOOTH ATTEMPTING TRAFFIC AND FEW PERSONS DOING THE CALLING (FEW MEANS RATIO OF CALLERS TO CIRCUITS IS LESS THAN 8 - GREATER THAN 8 ASSUME TRAFFIC RANDOM)

| HOURS FIRST ATTEMPT | FIRST ATTEMPT VMR | NUMBER SOURCES | HOURS CON-NECT | HOURS OVFL | PERCENT BLOCKED (GRADE) | LINE | HOURS CONNECTED TRAFFIC ON LINES 1 | 2 | 3 | 4 | 5 | 6 | 7 | 8 | 9 | 10 |
|---|---|---|---|---|---|---|---|---|---|---|---|---|---|---|---|---|
| 3.800 | .79 | 18 | 3.192 | .608 | 16.0% | 5 | .799 | .738 | .659 | .558 | .437 | | | | | |
| 3.800 | .84 | 24 | 3.170 | .630 | 16.6% | 5 | .797 | .735 | .654 | .552 | .431 | | | | | |
| 3.800 | .87 | 30 | 3.157 | .643 | 16.9% | 5 | .796 | .733 | .651 | .548 | .428 | | | | | |
| 3.800 | .89 | 36 | 3.149 | .651 | 17.1% | 5 | .795 | .732 | .649 | .546 | .426 | | | | | |
| 3.800 | .46 | 7 | 3.710 | .090 | 2.4% | 6 | .811 | .763 | .701 | .620 | .502 | .313 | | | | |
| 3.800 | .73 | 14 | 3.525 | .275 | 7.2% | 6 | .801 | .743 | .665 | .566 | .444 | .306 | | | | |
| 3.800 | .82 | 21 | 3.481 | .319 | 8.4% | 6 | .798 | .737 | .656 | .555 | .434 | .302 | | | | |
| 3.800 | .86 | 28 | 3.462 | .338 | 8.9% | 6 | .796 | .734 | .652 | .549 | .429 | .301 | | | | |
| 3.800 | .89 | 35 | 3.450 | .350 | 9.2% | 6 | .795 | .732 | .650 | .547 | .426 | .300 | | | | |
| 3.800 | .91 | 42 | 3.443 | .357 | 9.4% | 6 | .795 | .731 | .648 | .545 | .425 | .300 | | | | |
| 3.800 | .52 | 8 | 3.780 | .020 | .5% | 7 | .809 | .757 | .691 | .603 | .481 | .312 | .127 | | | |
| 3.800 | .76 | 16 | 3.683 | .117 | 3.1% | 7 | .800 | .740 | .662 | .562 | .440 | .304 | .176 | | | |
| 3.800 | .84 | 24 | 3.653 | .147 | 3.9% | 7 | .797 | .735 | .654 | .552 | .431 | .302 | .182 | | | |
| 3.800 | .88 | 32 | 3.639 | .161 | 4.2% | 7 | .796 | .733 | .650 | .548 | .427 | .300 | .185 | | | |
| 3.800 | .90 | 40 | 3.631 | .169 | 4.5% | 7 | .795 | .731 | .648 | .545 | .425 | .300 | .186 | | | |
| 3.800 | .92 | 48 | 3.625 | .175 | 4.6% | 7 | .794 | .730 | .647 | .543 | .424 | .299 | .187 | | | |
| 3.800 | .93 | 56 | 3.621 | .179 | 4.7% | 7 | .794 | .730 | .646 | .542 | .423 | .299 | .187 | | | |
| 4.000 | .20 | 5 | 3.135 | .865 | 21.6% | 4 | .827 | .795 | .764 | .750 | | | | | | |
| 4.000 | .60 | 10 | 2.881 | 1.119 | 28.0% | 4 | .813 | .762 | .696 | .609 | | | | | | |
| 4.000 | .73 | 15 | 2.833 | 1.167 | 29.2% | 4 | .809 | .754 | .681 | .589 | | | | | | |
| 4.000 | .80 | 20 | 2.812 | 1.188 | 29.7% | 4 | .807 | .750 | .675 | .580 | | | | | | |
| 4.000 | .84 | 25 | 2.800 | 1.200 | 30.0% | 4 | .805 | .747 | .672 | .676 | | | | | | |
| 4.000 | .87 | 30 | 2.792 | 1.208 | 30.2% | 4 | .804 | .746 | .669 | .573 | | | | | | |
| 4.000 | .33 | 6 | 3.615 | .385 | 9.6% | 5 | .822 | .783 | .735 | .678 | .596 | | | | | |
| 4.000 | .67 | 12 | 3.340 | .660 | 16.5% | 5 | .811 | .758 | .688 | .598 | .484 | | | | | |
| 4.000 | .78 | 18 | 3.287 | .713 | 17.8% | 5 | .807 | .751 | .677 | .583 | .469 | | | | | |
| 4.000 | .83 | 24 | 3.264 | .736 | 18.4% | 5 | .805 | .748 | .672 | .577 | .462 | | | | | |
| 4.000 | .87 | 30 | 3.251 | .749 | 18.7% | 5 | .804 | .746 | .669 | .573 | .458 | | | | | |
| 4.000 | .89 | 36 | 3.242 | .758 | 18.9% | 5 | .804 | .744 | .667 | .571 | .456 | | | | | |
| 4.000 | .43 | 7 | 3.875 | .125 | 3.1% | 6 | .819 | .774 | .720 | .647 | .544 | .371 | | | | |
| 4.000 | .71 | 14 | 3.658 | .342 | 8.6% | 6 | .809 | .755 | [illegible] | .591 | .477 | .342 | | | | |
| 4.000 | .81 | 21 | 3.610 | .390 | 9.8% | 6 | .806 | .749 | .674 | .579 | .465 | .336 | | | | |
| 4.000 | .86 | 28 | 3.588 | .412 | 10.3% | 6 | .805 | .746 | .670 | .574 | .459 | .334 | | | | |
| 4.000 | .89 | 35 | 3.576 | .424 | 10.6% | 6 | .804 | .745 | .668 | .571 | .456 | .333 | | | | |
| 4.000 | .90 | 42 | 3.568 | .432 | 10.8% | 6 | .803 | .744 | .666 | .569 | .455 | .332 | | | | |
| 4.000 | .50 | 8 | 3.970 | .030 | .8% | 7 | .817 | .769 | .709 | .629 | .519 | .361 | .164 | | | |
| 4.000 | .75 | 16 | 3.847 | .153 | 3.8% | 7 | .808 | .753 | .680 | .587 | .472 | .340 | .208 | | | |
| 4.000 | .83 | 24 | 3.811 | .189 | 4.7% | 7 | .805 | .748 | .672 | .577 | .462 | .335 | .212 | | | |
| 4.000 | .88 | 32 | 3.795 | .205 | 5.1% | 7 | .804 | .745 | .669 | .572 | .457 | .333 | .214 | | | |
| 4.000 | .90 | 40 | 3.785 | .215 | 5.4% | 7 | .803 | .744 | .666 | .569 | .455 | .332 | .215 | | | |

ERLANG-ENGSET: FOR SMOOTH ATTEMPTING TRAFFIC AND FEW PERSONS DOING THE CALLING (FEW MEANS RATIO OF CALLERS TO CIRCUITS IS LESS THAN 8 - GREATER THAN 8 ASSUME TRAFFIC RANDOM)

| HOURS FIRST ATTEMPT | FIRST ATTEMPT VMR | NUMBER SOURCES | HOURS CONNECT | HOURS OVFL | PERCENT BLOCKED (GRADE) | LINE | HOURS CONNECTED TRAFFIC ON LINES 1 | 2 | 3 | 4 | 5 | 6 | 7 | 8 | 9 | 10 |
|---|---|---|---|---|---|---|---|---|---|---|---|---|---|---|---|---|
| 4.000 | .92 | 48 | 3.779 | .221 | 5.5% | 7 | .803 | .743 | .665 | .568 | .453 | .331 | .216 | | | |
| 4.000 | .93 | 56 | 3.774 | .226 | 5.6% | 7 | .802 | .742 | .664 | .567 | .452 | .331 | .216 | | | |
| 4.200 | .16 | 5 | 3.195 | 1.005 | 23.9% | 4 | .835 | .804 | .780 | .776 | | | | | | |
| 4.200 | .58 | 10 | 2.939 | 1.261 | 30.0% | 4 | .821 | .773 | .712 | .633 | | | | | | |
| 4.200 | .72 | 15 | 2.891 | 1.309 | 31.2% | 4 | .816 | .765 | .698 | .612 | | | | | | |
| 4.200 | .79 | 20 | 2.870 | 1.330 | 31.7% | 4 | .814 | .761 | .692 | .603 | | | | | | |
| 4.200 | .83 | 25 | 2.858 | 1.342 | 32.0% | 4 | .813 | .759 | .688 | .598 | | | | | | |
| 4.200 | .86 | 30 | 2.851 | 1.349 | 32.1% | 4 | .812 | .757 | .686 | .596 | | | | | | |
| 4.200 | .30 | 6 | 3.720 | .480 | 11.4% | 5 | .830 | .793 | .751 | .704 | .642 | | | | | |
| 4.200 | .65 | 12 | 3.429 | .771 | 18.4% | 5 | .818 | .769 | .705 | .622 | .515 | | | | | |
| 4.200 | .77 | 18 | 3.374 | .826 | 19.7% | 5 | .815 | .762 | .694 | .606 | .498 | | | | | |
| 4.200 | .82 | 24 | 3.351 | .849 | 20.2% | 5 | .813 | .759 | .689 | .599 | .491 | | | | | |
| 4.200 | .86 | 30 | 3.337 | .863 | 20.5% | 5 | .812 | .757 | .686 | .596 | .487 | | | | | |
| 4.200 | .88 | 36 | 3.329 | .871 | 20.7% | 5 | .811 | .756 | .684 | .593 | .484 | | | | | |
| 4.200 | .40 | 7 | 4.031 | .169 | 4.0% | 6 | .827 | .786 | .736 | .671 | .583 | .429 | | | | |
| 4.200 | .70 | 14 | 3.782 | .418 | 9.9% | 6 | .817 | .766 | .700 | .614 | .507 | .378 | | | | |
| 4.200 | .80 | 21 | 3.730 | .470 | 11.2% | 6 | .814 | .760 | .691 | .602 | .493 | .369 | | | | |
| 4.200 | .85 | 28 | 3.707 | .493 | 11.7% | 6 | .812 | .758 | .687 | .597 | .488 | .366 | | | | |
| 4.200 | .88 | 35 | 3.694 | .506 | 12.1% | 6 | .811 | .756 | .684 | .594 | .485 | .364 | | | | |
| 4.200 | .90 | 42 | 3.685 | .515 | 12.3% | 6 | .811 | .755 | .683 | .592 | .483 | .363 | | | | |
| 4.200 | .48 | 8 | 4.156 | .044 | 1.1% | 7 | .824 | .780 | .725 | .654 | .555 | .409 | 208 | | | |
| 4.200 | .74 | 16 | 4.003 | .197 | 4.7% | 7 | .816 | .764 | .696 | .610 | .501 | .374 | .241 | | | |
| 4.200 | .82 | 24 | 3.962 | .238 | 5.7% | 7 | .813 | .759 | .089 | .599 | .491 | .368 | .244 | | | |
| 4.200 | .87 | 32 | 3.943 | .257 | 6.1% | 7 | .812 | .757 | .685 | .595 | .486 | .365 | .245 | | | |
| 4.200 | .90 | 40 | 3.932 | .268 | 6.4% | 7 | .811 | .755 | .683 | .592 | .483 | .363 | .245 | | | |
| 4.200 | .91 | 48 | 3.925 | .275 | 6.5% | 7 | .810 | .754 | .682 | .590 | .482 | .362 | .245 | | | |
| 4.200 | .93 | 56 | 3.920 | .280 | 6.7% | 7 | .810 | .754 | .681 | .589 | .480 | .361 | .245 | | | |
| 4.400 | .56 | 10 | 2.993 | 1.407 | 32.0% | 4 | .828 | .783 | .728 | .654 | | | | | | |
| 4.400 | .71 | 15 | 2.945 | 1.455 | 33.1% | 4 | .823 | .775 | .713 | .633 | | | | | | |
| 4.400 | .78 | 20 | 2.924 | 1.476 | 33.6% | 4 | .821 | .771 | .707 | .624 | | | | | | |
| 4.400 | .82 | 25 | 2.912 | 1.488 | 33.8% | 4 | .820 | 760 | .703 | .619 | | | | | | |
| 4.400 | .85 | 30 | 2.904 | 1.496 | 34.0% | 4 | .819 | .768 | .701 | .616 | | | | | | |
| 4.400 | .27 | 6 | 3.814 | .586 | 13.3% | 5 | .836 | .803 | .766 | .726 | .083 | | | | | |
| 4.400 | .63 | 12 | 3.511 | .889 | 20.2% | 5 | .825 | .779 | .720 | .643 | .544 | | | | | |
| 4.400 | .76 | 18 | 3.456 | .944 | 21.5% | 5 | .822 | .773 | .709 | .627 | .525 | | | | | |
| 4.400 | .82 | 24 | 3.431 | .969 | 22.0% | 5 | .820 | .770 | .704 | .620 | .518 | | | | | |
| 4.400 | .85 | 30 | 3.418 | .982 | 22.3% | 5 | .819 | .768 | .701 | .616 | .513 | | | | | |
| 4.400 | .88 | 36 | 3.409 | .991 | 22.5% | 5 | .818 | .766 | .699 | .614 | .511 | | | | | |
| 4.400 | .37 | 7 | 4.176 | .224 | 5.1% | 6 | .833 | .796 | .750 | .694 | .617 | .486 | | | | |
| 4.400 | .69 | 14 | 3.898 | .502 | 11.4% | 6 | .824 | .776 | .715 | .636 | .535 | .412 | | | | |

## ERLANG-ENGSET: FOR SMOOTH ATTEMPTING TRAFFIC AND FEW PERSONS DOING THE CALLING (FEW MEANS RATIO OF CALLERS TO CIRCUITS IS LESS THAN 8 - GREATER THAN 8 ASSUME TRAFFIC RANDOM)

| HOURS FIRST ATTEMPT | FIRST ATTEMPT VMR | NUMBER SOURCES | HOURS CON-NECT | HOURS OVFL | PERCENT BLOCKED (GRADE) | LINE | HOURS CONNECTED TRAFFIC ON LINES 1 | 2 | 3 | 4 | 5 | 6 | 7 | 8 | 9 | 10 |
|---|---|---|---|---|---|---|---|---|---|---|---|---|---|---|---|---|
| 4.400 | .79 | 21 | 3.843 | .557 | 12.7% | 6 | .821 | .771 | .706 | .623 | .521 | .401 | | | | |
| 4.400 | .84 | 28 | 3.818 | .582 | 13.2% | 6 | .819 | .768 | .702 | .618 | .515 | .397 | | | | |
| 4.400 | .87 | 35 | 3.804 | .596 | 13.5% | 6 | .818 | .767 | .700 | .614 | .511 | .394 | | | | |
| 4.400 | .90 | 42 | 3.795 | .605 | 13.7% | 6 | .818 | .766 | .698 | .612 | .509 | .393 | | | | |
| 4.400 | .45 | 8 | 4.337 | .063 | 1.4% | 7 | .831 | .790 | .740 | .675 | .587 | .456 | .258 | | | |
| 4.400 | .73 | 16 | 4.151 | .249 | 5.6% | 7 | .823 | .774 | .711 | .631 | .529 | .408 | .275 | | | |
| 4.400 | .82 | 24 | 4.105 | .295 | 6.7% | 7 | .820 | .770 | .704 | .620 | .518 | .399 | .275 | | | |
| 4.400 | .86 | 32 | 4.085 | .315 | 7.2% | 7 | .819 | .767 | .700 | .615 | .512 | .395 | .275 | | | |
| 4.400 | .89 | 40 | 4.073 | .327 | 7.4% | 7 | .818 | .766 | .698 | .613 | .510 | .393 | .275 | | | |
| 4.400 | .91 | 48 | 4.065 | .335 | 7.6% | 7 | .817 | .765 | .697 | .611 | .508 | .392 | .275 | | | |
| 4.400 | .92 | 56 | 4.059 | .341 | 7.7% | 7 | .817 | .764 | .696 | .610 | .506 | .391 | .275 | | | |
| 4.600 | .54 | 10 | 3.042 | 1.558 | 33.9% | 4 | .834 | .793 | .741 | .674 | | | | | | |
| 4.600 | .69 | 15 | 2.994 | 1.606 | 34.9% | 4 | .830 | .785 | .727 | .653 | | | | | | |
| 4.600 | .23 | 6 | 3.898 | .702 | 15.3% | 5 | .843 | .811 | .779 | .746 | .719 | | | | | |
| 4.600 | .62 | 12 | 3.587 | 1.013 | 22.0% | 5 | .832 | .789 | .734 | .663 | .570 | | | | | |
| 4.600 | .74 | 18 | 3.530 | 1.070 | 23.3% | 5 | .828 | .782 | .723 | .646 | .551 | | | | | |
| 4.600 | .81 | 24 | 3.506 | 1.094 | 23.8% | 5 | .827 | .779 | .718 | .640 | .543 | | | | | |
| 4.600 | .85 | 30 | 3.492 | 1.108 | 24.1% | 5 | .826 | .777 | .715 | .636 | .538 | | | | | |
| 4.600 | .87 | 36 | 3.483 | 1.117 | 24.3% | 5 | .825 | .776 | .713 | .633 | .535 | | | | | |
| 4.600 | .34 | 7 | 4.310 | .290 | 6.3% | 6 | .840 | .805 | .763 | .714 | .649 | .540 | | | | |
| 4.600 | .67 | 14 | 4.006 | .594 | 12.9% | 6 | .830 | .786 | .729 | .655 | .561 | .445 | | | | |
| 4.600 | .78 | 21 | 3.948 | .652 | 14.2% | 6 | .827 | .781 | .720 | .642 | .546 | .432 | | | | |
| 4.600 | .84 | 28 | 3.922 | .678 | 14.7% | 6 | .826 | .778 | .716 | .637 | .539 | .426 | | | | |
| 4.600 | .87 | 35 | 3.908 | .692 | 15.0% | 6 | .825 | .776 | .714 | .634 | .536 | .424 | | | | |
| 4.600 | .89 | 42 | 3.899 | .701 | 15.2% | 6 | .824 | .775 | .712 | .632 | .534 | .422 | | | | |
| 4.600 | .42 | 8 | 4.511 | .089 | 1.9% | 7 | .837 | .800 | .753 | .694 | .617 | .500 | .310 | | | |
| 4.600 | .71 | 16 | 4.292 | .308 | 6.7% | 7 | .829 | .784 | .725 | .650 | .555 | .440 | .309 | | | |
| 4.600 | .81 | 24 | 4.242 | .358 | 7.8% | 7 | .827 | .779 | .718 | .640 | .543 | .429 | .307 | | | |
| 4.600 | .86 | 32 | 4.219 | .381 | 8.3% | 7 | .825 | .777 | .714 | .635 | .537 | .425 | .305 | | | |
| 4.600 | .89 | 40 | 4.206 | .394 | 8.6% | 7 | .824 | .776 | .712 | .632 | .534 | .422 | .305 | | | |
| 4.600 | .90 | 48 | 4.197 | .403 | 8.8% | 7 | .824 | .775 | .711 | .630 | .532 | .420 | .304 | | | |
| 4.600 | .92 | 56 | 4.191 | .409 | 8.9% | 7 | .824 | .774 | .710 | .629 | .531 | .420 | .304 | | | |
| 4.800 | .20 | 6 | 3.975 | .825 | 17.2% | 5 | .848 | .820 | .791 | .764 | .752 | | | | | |
| 4.800 | .60 | 12 | 3.657 | 1.143 | 23.8% | 5 | .838 | .797 | .746 | .680 | .596 | | | | | |
| 4.800 | .73 | 18 | 3.600 | 1.200 | 25.0% | 5 | .834 | .791 | .735 | .664 | .574 | | | | | |
| 4.800 | .80 | 24 | 3.575 | 1.225 | 25.5% | 5 | .833 | .788 | .731 | .657 | .566 | | | | | |
| 4.800 | .84 | 30 | 3.561 | 1.239 | 25.8% | 5 | .832 | .786 | .728 | .654 | .561 | | | | | |
| 4.800 | .87 | 36 | 3.552 | 1.248 | 26.0% | 5 | .831 | .785 | .726 | .651 | .559 | | | | | |
| 4.800 | .31 | 7 | 4.433 | .367 | 7.6% | 6 | .845 | .813 | .776 | .730 | .677 | .591 | | | | |
| 4.800 | .66 | 14 | 4.107 | .693 | 14.4% | 6 | .836 | .795 | .742 | .673 | .586 | .476 | | | | |

ERLANG-ENGSET: FOR SMOOTH ATTEMPTING TRAFFIC AND FEW PERSONS DOING THE CALLING (FEW MEANS RATIO OF CALLERS TO CIRCUITS IS LESS THAN 8 - GREATER THAN 8 ASSUME TRAFFIC RANDOM)

| HOURS FIRST ATTEMPT | FIRST ATTEMPT VMR | NUMBER SOURCES | HOURS CONNECT | HOURS OVFL | PERCENT BLOCKED (GRADE) | LINE | HOURS CONNECTED TRAFFIC ON LINES 1 | 2 | 3 | 4 | 5 | 6 | 7 | 8 | 9 | 10 |
|---|---|---|---|---|---|---|---|---|---|---|---|---|---|---|---|---|
| 4.800 | .77 | 21 | 4.046 | .754 | 15.7% | 6 | .833 | .789 | .733 | .660 | .570 | .461 | | | | |
| 4.800 | .83 | 28 | 4.020 | .780 | 16.3% | 6 | .832 | .787 | .729 | .655 | .563 | .455 | | | | |
| 4.800 | .86 | 35 | 4.005 | .795 | 16.6% | 6 | .831 | .785 | .726 | .651 | .559 | .451 | | | | |
| 4.800 | .89 | 42 | 3.995 | .805 | 16.8% | 6 | .830 | .784 | .725 | .649 | .557 | .449 | | | | |
| 4.800 | .40 | 8 | 4.679 | .121 | 2.5% | 7 | .843 | .808 | .765 | .713 | .643 | .541 | .366 | | | |
| 4.800 | .70 | 16 | 4.425 | .375 | 7.8% | 7 | .835 | .793 | .738 | .668 | .579 | .470 | .343 | | | |
| 4.800 | .80 | 24 | 4.370 | .430 | 9.0% | 7 | .833 | .788 | .731 | .657 | .566 | .457 | .338 | | | |
| 4.800 | .85 | 32 | 4.346 | .454 | 9.5% | 7 | .831 | .786 | .727 | .653 | .560 | .452 | .335 | | | |
| 4.800 | .88 | 40 | 4.332 | .468 | 9.8% | 7 | .831 | .785 | .725 | .650 | .557 | .450 | .334 | | | |
| 4.800 | .90 | 48 | 4.323 | .477 | 9.9% | 7 | .830 | .784 | .724 | .648 | .555 | .448 | .333 | | | |
| 4.800 | .91 | 56 | 4.316 | .484 | 10.1% | 7 | .830 | .783 | .723 | .647 | .554 | .447 | .333 | | | |
| 4.800 | .47 | 9 | 4.769 | .031 | .6% | 8 | .841 | .805 | .759 | .700 | .624 | .517 | .360 | .164 | | |
| 4.800 | .73 | 18 | 4.622 | .178 | 3.7% | 8 | .834 | .791 | .735 | .664 | .574 | .465 | .341 | .216 | | |
| 4.800 | .82 | 27 | 4.580 | .220 | 4.6% | 8 | .832 | .787 | .729 | .655 | .564 | .455 | .337 | .221 | | |
| 4.800 | .87 | 36 | 4.561 | .239 | 5.0% | 8 | .831 | .785 | .726 | .651 | .559 | .451 | .335 | .223 | | |
| 4.800 | .89 | 45 | 4.549 | .251 | 5.2% | 8 | .830 | .784 | .724 | .649 | .556 | .449 | .334 | .224 | | |
| 4.800 | .91 | 54 | 4.542 | .258 | 5.4% | 8 | .830 | .783 | .723 | .647 | .554 | .447 | .333 | .224 | | |
| 4.800 | .92 | 63 | 4.537 | .263 | 5.5% | 8 | .829 | .783 | .723 | .646 | .553 | .446 | .332 | .225 | | |
| 5.000 | .17 | 6 | 4.041 | .959 | 19.2% | 5 | .854 | .828 | .801 | .781 | .778 | | | | | |
| 5.000 | .58 | 12 | 3.721 | 1.279 | 25.6% | 5 | .843 | .805 | .758 | .696 | .618 | | | | | |
| 5.000 | .72 | 18 | 3.663 | 1.337 | 26.7% | 5 | .840 | .799 | .747 | .681 | .597 | | | | | |
| 5.000 | .79 | 24 | 3.639 | 1.361 | 27.2% | 5 | .838 | .796 | .742 | .674 | .588 | | | | | |
| 5.000 | .83 | 30 | 3.625 | 1.375 | 27.5% | 5 | .837 | .795 | .740 | .670 | .583 | | | | | |
| 5.000 | .86 | 36 | 3.616 | 1.384 | 27.7% | 5 | .837 | .794 | .738 | .668 | .580 | | | | | |
| 5.000 | .29 | 7 | 4.545 | .455 | 9.1% | 6 | .850 | .821 | .787 | .747 | .701 | .638 | | | | |
| 5.000 | .64 | 14 | 4.200 | .800 | 16.0% | 6 | .842 | .803 | .753 | .689 | .608 | .505 | | | | |
| 5.000 | .76 | 21 | 4.137 | .863 | 17.3% | 6 | .839 | .798 | .745 | .677 | .591 | .488 | | | | |
| 5.000 | .82 | 28 | 4.110 | .890 | 17.8% | 6 | .838 | .795 | .740 | .671 | .584 | .481 | | | | |
| 5.000 | .86 | 35 | 4.095 | .905 | 18.1% | 6 | .837 | .794 | .738 | .668 | .581 | .478 | | | | |
| 5.000 | .88 | 42 | 4.085 | .915 | 18.3% | 6 | .836 | .793 | .737 | .666 | .578 | .475 | | | | |
| 5.000 | .38 | 8 | 4.838 | .162 | 3.2% | 7 | .848 | .816 | .777 | .728 | .668 | .570 | .422 | | | |
| 5.000 | .69 | 16 | 4.551 | .449 | 9.0% | 7 | .841 | .801 | .750 | .685 | .601 | .499 | .376 | | | |
| 5.000 | .79 | 24 | 4.492 | .508 | 10.2% | 7 | .838 | .796 | .742 | .674 | .588 | .485 | .368 | | | |
| 5.000 | .84 | 32 | 4.466 | .534 | 10.7% | 7 | .837 | .794 | .739 | .669 | .582 | .479 | .365 | | | |
| 5.000 | .88 | 40 | 4.451 | .549 | 11.0% | 7 | .836 | .793 | .737 | .666 | .579 | .476 | .363 | | | |
| 5.000 | .90 | 48 | 4.441 | .559 | 11.2% | 7 | .836 | .792 | .736 | .665 | .577 | .474 | .362 | | | |
| 5.000 | .91 | 56 | 4.435 | .565 | 11.3% | 7 | .835 | .792 | .735 | .663 | .575 | .473 | .361 | | | |
| 5.000 | .44 | 9 | 4.956 | .044 | .9% | 8 | .847 | .812 | .770 | .717 | .648 | .551 | .406 | .206 | | |
| 5.000 | .72 | 18 | 4.777 | .223 | 4.5% | 8 | .840 | .799 | .747 | .681 | .597 | .493 | .373 | .247 | | |
| 5.000 | .81 | 27 | 4.730 | .270 | 5.4% | 8 | .838 | .795 | .741 | .672 | .585 | .482 | .366 | .250 | | |

ERLANG-ENGSET: FOR SMOOTH ATTEMPTING TRAFFIC AND FEW PERSONS DOING THE CALLING (FEW MEANS RATIO OF CALLERS TO CIRCUITS IS LESS THAN 8 - GREATER THAN 8 ASSUME TRAFFIC RANDOM)

| HOURS FIRST ATTEMPT | FIRST ATTEMPT VMR | NUMBER SOURCES | HOURS CON-NECT | HOURS OVFL | PERCENT BLOCKED (GRADE) | LINE | HOURS CONNECTED TRAFFIC ON LINES 1 | 2 | 3 | 4 | 5 | 6 | 7 | 8 | 9 | 10 |
|---|---|---|---|---|---|---|---|---|---|---|---|---|---|---|---|---|
| 5.000 | .86 | 36 | 4.708 | .292 | 5.8% | 8 | .837 | .794 | .738 | .668 | .580 | .477 | .364 | .251 | | |
| 5.000 | .89 | 45 | 4.696 | .304 | 6.1% | 8 | .836 | .793 | .736 | .665 | .578 | .475 | .362 | .251 | | |
| 5.000 | .91 | 54 | 4.688 | .312 | 6.2% | 8 | .836 | .792 | .735 | .664 | .576 | .473 | .361 | .252 | | |
| 5.000 | .92 | 63 | 4.682 | .318 | 6.4% | 8 | .835 | .791 | .735 | .663 | .574 | .472 | .361 | .252 | | |
| 5.500 | .54 | 12 | 3.861 | 1.639 | 29.8% | 5 | .856 | .823 | .783 | .732 | .667 | | | | | |
| 5.500 | .69 | 18 | 3.804 | 1.696 | 30.8% | 5 | .852 | .818 | .773 | .717 | .645 | | | | | |
| 5.500 | .77 | 24 | 3.780 | 1.720 | 31.3% | 5 | .851 | .815 | .769 | .710 | .636 | | | | | |
| 5.500 | .82 | 30 | 3.766 | 1.734 | 31.5% | 5 | .850 | .813 | .766 | .706 | .631 | | | | | |
| 5.500 | .85 | 36 | 3.758 | 1.742 | 31.7% | 5 | .849 | .812 | .764 | .704 | .628 | | | | | |
| 5.500 | .21 | 7 | 4.781 | .719 | 13.1% | 6 | .863 | .837 | .810 | .781 | .753 | .736 | | | | |
| 5.500 | .61 | 14 | 4.405 | 1.095 | 19.9% | 6 | .854 | .821 | .778 | .724 | .657 | .570 | | | | |
| 5.500 | .74 | 21 | 4.340 | 1.160 | 21.1% | 6 | .851 | .816 | .770 | .713 | .639 | .550 | | | | |
| 5.500 | .80 | 28 | 4.312 | 1.188 | 21.6% | 6 | .850 | .814 | .766 | .707 | .633 | .542 | | | | |
| 5.500 | .84 | 35 | 4.296 | 1.204 | 21.9% | 6 | .849 | .812 | .765 | .704 | .628 | .538 | | | | |
| 5.500 | .87 | 42 | 4.286 | 1.214 | 22.1% | 6 | .849 | .811 | .763 | .702 | .626 | .535 | | | | |
| 5.500 | .31 | 8 | 5.193 | .307 | 5.6% | 7 | .860 | .834 | .801 | .763 | .719 | .660 | .558 | | | |
| 5.500 | .66 | 16 | 4.832 | .668 | 12.1% | 7 | .853 | .819 | .775 | .720 | .650 | .562 | .453 | | | |
| 5.500 | .77 | 24 | 4.765 | .735 | 13.4% | 7 | .851 | .815 | .769 | .710 | .636 | .546 | .439 | | | |
| 5.500 | .83 | 32 | 4.736 | .764 | 13.9% | 7 | .850 | .813 | .765 | .705 | .630 | .539 | .434 | | | |
| 5.500 | .86 | 40 | 4.720 | .780 | 14.2% | 7 | .849 | .812 | .764 | .702 | .627 | .535 | .431 | | | |
| 5.500 | .89 | 48 | 4.709 | .791 | 14.4% | 7 | .848 | .811 | .762 | .701 | .625 | .533 | .429 | | | |
| 5.500 | .90 | 56 | 4.702 | .798 | 14.5% | 7 | .848 | .810 | .761 | .700 | .623 | .531 | .428 | | | |
| 5.500 | .39 | 9 | 5.402 | .098 | 1.8% | 8 | .859 | .829 | .795 | .751 | .698 | .625 | .515 | .329 | | |
| 5.500 | .69 | 18 | 5.134 | .366 | 6.7% | 8 | .852 | .818 | .773 | .717 | .645 | .556 | .448 | .326 | | |
| 5.500 | .80 | 27 | 5.075 | .425 | 7.7% | 8 | .850 | .814 | .767 | .708 | .633 | .543 | .437 | .323 | | |
| 5.500 | .85 | 36 | 5.048 | .452 | 8.2% | 8 | .849 | .812 | .764 | .704 | .628 | .537 | .432 | .321 | | |
| 5.500 | .88 | 45 | 5.033 | .467 | 8.5% | 8 | .849 | .811 | .763 | .701 | .625 | .534 | .430 | .320 | | |
| 5.500 | .90 | 54 | 5.023 | .477 | 8.7% | 8 | .848 | .810 | .762 | .700 | .623 | .532 | .428 | .320 | | |
| 5.500 | .91 | 63 | 5.016 | .484 | 8.8% | 8 | .848 | .810 | .761 | .699 | .622 | .531 | .427 | .319 | | |
| 6.000 | .50 | 12 | 3.976 | 2.024 | 33.7% | 5 | .866 | .838 | .804 | .761 | .707 | | | | | |
| 6.000 | .67 | 18 | 3.921 | 2.079 | 34.6% | 5 | .863 | .833 | .795 | .746 | .685 | | | | | |
| 6.000 | .57 | 14 | 4.575 | 1.425 | 23.7% | 6 | .865 | .836 | .799 | .755 | .697 | .624 | | | | |
| 6.000 | .71 | 21 | 4.510 | 1.490 | 24.8% | 6 | .862 | .831 | .792 | .742 | .680 | .602 | | | | |
| 6.000 | .79 | 28 | 4.482 | 1.518 | 25.3% | 6 | .861 | .829 | .788 | .737 | .673 | .593 | | | | |
| 6.000 | .83 | 35 | 4.466 | 1.534 | 25.6% | 6 | .860 | .828 | .787 | .734 | .669 | .589 | | | | |
| 6.000 | .86 | 42 | 4.456 | 1.544 | 25.7% | 6 | .860 | .827 | .785 | .732 | .666 | .586 | | | | |
| 6.000 | .25 | 8 | 5.484 | .516 | 8.6% | 7 | .871 | .847 | .821 | .791 | .757 | .724 | .673 | | | |
| 6.000 | .63 | 16 | 5.070 | .930 | 15.5% | 7 | .864 | .834 | .797 | .749 | .689 | .615 | .521 | | | |
| 6.000 | .75 | 24 | 4.999 | 1.001 | 16.7% | 7 | .862 | .830 | .790 | .740 | .676 | .597 | .504 | | | |
| 6.000 | .81 | 32 | 4.968 | 1.032 | 17.2% | 7 | .860 | .828 | .787 | .735 | .671 | .590 | .496 | | | |

ERLANG-ENGSET: FOR SMOOTH ATTEMPTING TRAFFIC AND FEW PERSONS DOING THE CALLING (FEW MEANS RATIO OF CALLERS TO CIRCUITS IS LESS THAN 8 - GREATER THAN 8 ASSUME TRAFFIC RANDOM)

| HOURS FIRST ATTEMPT | FIRST ATTEMPT VMR | NUMBER SOURCES | HOURS CON-NECT | HOURS OVFL | PERCENT BLOCKED (GRADE) | LINE | HOURS CONNECTED TRAFFIC ON LINES | | | | | | | | | |
|---|---|---|---|---|---|---|---|---|---|---|---|---|---|---|---|---|
| | | | | | | | 1 | 2 | 3 | 4 | 5 | 6 | 7 | 8 | 9 | 10 |
| 6.000 | .85 | 40 | 4.951 | 1.049 | 17.5% | 7 | .860 | .827 | .786 | .733 | .667 | .587 | .492 | | | |
| 6.000 | .88 | 48 | 4.940 | 1.060 | 17.7% | 7 | .859 | .826 | .784 | .731 | .665 | .584 | .490 | | | |
| 6.000 | .89 | 56 | 4.932 | 1.068 | 17.8% | 7 | .859 | .826 | .784 | .730 | .663 | .582 | .488 | | | |
| 6.000 | .33 | 9 | 5.803 | .197 | 3.3% | 8 | .869 | .844 | .814 | .779 | .738 | .685 | .608 | .465 | | |
| 6.000 | .67 | 18 | 5.446 | .554 | 9.2% | 8 | .863 | .833 | .795 | .746 | .685 | .608 | .515 | .402 | | |
| 6.000 | .78 | 27 | 5.377 | .623 | 10.4% | 8 | .861 | .829 | .789 | .738 | .674 | .594 | .500 | .392 | | |
| 6.000 | .83 | 36 | 5.347 | .653 | 10.9% | 8 | .860 | .828 | .786 | .734 | .668 | .588 | .494 | .389 | | |
| 6.000 | .87 | 45 | 5.330 | .670 | 11.2% | 8 | .859 | .827 | .785 | .732 | .665 | .585 | .491 | .387 | | |
| 6.000 | .89 | 54 | 5.319 | .681 | 11.4% | 8 | .859 | .826 | .784 | .730 | .664 | .583 | .489 | .385 | | |
| 6.000 | .90 | 63 | 5.311 | .689 | 11.5% | 8 | .859 | .825 | .783 | .729 | .662 | .581 | .487 | .384 | | |
| 6.000 | .40 | 10 | 5.943 | .057 | 1.0% | 9 | .868 | .842 | .810 | .772 | .724 | .662 | .577 | .444 | .243 | |
| 6.000 | .70 | 20 | 5.704 | .296 | 4.9% | 9 | .863 | .831 | .793 | .743 | .682 | .604 | .510 | .399 | .280 | |
| 6.000 | .80 | 30 | 5.645 | .355 | 5.9% | 9 | .861 | .829 | .787 | .736 | .671 | .592 | .497 | .391 | .281 | |
| 6.000 | .85 | 40 | 5.619 | .381 | 6.3% | 9 | .860 | .827 | .786 | .733 | .667 | .587 | .492 | .388 | .281 | |
| 6.000 | .88 | 50 | 5.604 | .396 | 6.6% | 9 | .859 | .826 | .784 | .731 | .664 | .583 | .489 | .386 | .281 | |
| 6.000 | .90 | 60 | 5.594 | .406 | 6.8% | 9 | .859 | .826 | .783 | .730 | .663 | .582 | .487 | .384 | .281 | |
| 6.000 | .91 | 70 | 5.588 | .412 | 6.9% | 9 | .859 | .825 | .783 | .729 | .661 | .580 | .486 | .384 | .281 | |
| 6.500 | .54 | 14 | 4.717 | 1.783 | 27.4% | 6 | .874 | .849 | .817 | .779 | .730 | .669 | | | | |
| 6.500 | .69 | 21 | 4.653 | 1.847 | 28.4% | 6 | .871 | .844 | .810 | .767 | .714 | .647 | | | | |
| 6.500 | .77 | 28 | 4.625 | 1.875 | 28.8% | 6 | .870 | .842 | .807 | .762 | .707 | .637 | | | | |
| 6.500 | .81 | 35 | 4.610 | 1.890 | 29.1% | 6 | .870 | .841 | .805 | .759 | .703 | .633 | | | | |
| 6.500 | .85 | 42 | 4.600 | 1.900 | 29.2% | 6 | .869 | .840 | .803 | .758 | .700 | .630 | | | | |
| 6.500 | .19 | 8 | 5.713 | .787 | 12.1% | 7 | .880 | .859 | .837 | .814 | .790 | .770 | .762 | | | |
| 6.500 | .59 | 16 | 5.271 | 1.229 | 18.9% | 7 | .873 | .847 | .815 | .774 | .723 | .660 | .579 | | | |
| 6.500 | .73 | 24 | 5.197 | 1.303 | 20.0% | 7 | .871 | .843 | .808 | .764 | .710 | .642 | .559 | | | |
| 6.500 | .80 | 32 | 5.166 | 1.334 | 20.5% | 7 | .870 | .841 | .805 | .761 | .704 | .634 | .551 | | | |
| 6.500 | .84 | 40 | 5.149 | 1.351 | 20.8% | 7 | .869 | .840 | .804 | .758 | .701 | .630 | .546 | | | |
| 6.500 | .86 | 48 | 5.138 | 1.362 | 21.0% | 7 | .869 | .840 | .803 | .757 | .699 | .628 | .543 | | | |
| 6.500 | .88 | 56 | 5.130 | 1.370 | 21.1% | 7 | .868 | .839 | .802 | .756 | .697 | .626 | .542 | | | |
| 6.500 | .28 | 9 | 6.147 | .353 | 5.4% | 8 | .878 | .857 | .831 | .803 | .769 | .733 | .682 | .594 | | |
| 6.500 | .64 | 18 | 5.716 | .784 | 12.1% | 8 | .872 | .846 | .812 | .771 | .719 | .653 | .572 | .472 | | |
| 6.500 | .76 | 27 | 5.640 | .860 | 13.2% | 8 | .870 | .842 | .807 | .763 | .707 | .638 | .555 | .457 | | |
| 6.500 | .82 | 36 | 5.608 | .892 | 13.7% | 8 | .870 | .841 | .805 | .759 | .702 | .632 | .548 | .451 | | |
| 6.500 | .86 | 45 | 5.590 | .910 | 14.0% | 8 | .869 | .840 | .803 | .757 | .699 | .629 | .544 | .448 | | |
| 6.500 | .88 | 54 | 5.578 | .922 | 14.2% | 8 | .869 | .839 | .802 | .756 | .698 | .626 | .542 | .446 | | |
| 6.500 | .90 | 63 | 5.570 | .930 | 14.3% | 8 | .868 | .839 | .802 | .755 | .696 | .625 | .540 | .445 | | |
| 6.500 | .35 | 10 | 6.379 | .121 | 1.9% | 9 | .877 | .854 | .828 | .795 | .756 | .709 | .646 | .546 | .369 | |
| 6.500 | .67 | 20 | 6.045 | .455 | 7.0% | 9 | .872 | .845 | .811 | .768 | .715 | .648 | .566 | .467 | .353 | |
| 6.500 | .78 | 30 | 5.975 | .525 | 8.1% | 9 | .870 | .842 | .806 | .761 | .705 | .635 | .553 | .455 | .348 | |
| 6.500 | .84 | 40 | 5.944 | .556 | 8.5% | 9 | .869 | .840 | .804 | .758 | .701 | .630 | .546 | .450 | .346 | |

ERLANG-ENGSET: FOR SMOOTH ATTEMPTING TRAFFIC AND FEW PERSONS DOING THE CALLING (FEW MEANS RATIO OF CALLERS TO CIRCUITS IS LESS THAN 8 - GREATER THAN 8 ASSUME TRAFFIC RANDOM)

| HOURS FIRST ATTEMPT | FIRST ATTEMPT VMR | NUMBER SOURCES | HOURS CON-NECT | HOURS OVFL | PERCENT BLOCKED (GRADE) | LINE | HOURS CONNECTED TRAFFIC ON LINES 1 | 2 | 3 | 4 | 5 | 6 | 7 | 8 | 9 | 10 |
|---|---|---|---|---|---|---|---|---|---|---|---|---|---|---|---|---|
| 6.500 | .87 | 50 | 5.927 | .573 | 8.8% | 9 | .869 | .840 | .803 | .756 | .698 | .627 | .543 | .447 | .344 | |
| 6.500 | .89 | 60 | 5.916 | .584 | 9.0% | 9 | .868 | .839 | .802 | .755 | .697 | .625 | .541 | .445 | .344 | |
| 6.500 | .91 | 70 | 5.908 | .592 | 9.1% | 9 | .868 | .839 | .801 | .754 | .696 | .624 | .539 | .444 | .343 | |
| 6.500 | .41 | 11 | 6.467 | .033 | .5% | 10 | .876 | .852 | .824 | .789 | .746 | .694 | .624 | .521 | .369 | .171 |
| 6.500 | .70 | 22 | 6.262 | .238 | 3.7% | 10 | .871 | .844 | .810 | .766 | .712 | .645 | .562 | .463 | .352 | .237 |
| 6.500 | .80 | 33 | 6.206 | .294 | 4.5% | 10 | .870 | .841 | .805 | .760 | .704 | .634 | .550 | .452 | .348 | .242 |
| 6.500 | .85 | 44 | 6.180 | .320 | 4.9% | 10 | .869 | .840 | .803 | .757 | .699 | .629 | .545 | .448 | .345 | .244 |
| 6.500 | .88 | 55 | 6.166 | .334 | 5.1% | 10 | .868 | .839 | .802 | .756 | .698 | .626 | .542 | .446 | .344 | .245 |
| 6.500 | .90 | 66 | 6.156 | .344 | 5.3% | 10 | .868 | .839 | .801 | .755 | .696 | .624 | .540 | .444 | .343 | .245 |
| 6.500 | .92 | 77 | 6.149 | .351 | 5.4% | 10 | .868 | .838 | .801 | .754 | .695 | .623 | .539 | .443 | .343 | .246 |
| 7.000 | .50 | 14 | 4.836 | 2.164 | 30.9% | 6 | .882 | .860 | .832 | .799 | .758 | .705 | | | | |
| 7.000 | .67 | 21 | 4.774 | 2.226 | 31.8% | 6 | .880 | .856 | .825 | .788 | .742 | .683 | | | | |
| 7.000 | .75 | 28 | 4.748 | 2.252 | 32.2% | 6 | .878 | .853 | .823 | .784 | .735 | .675 | | | | |
| 7.000 | .80 | 35 | 4.733 | 2.267 | 32.4% | 6 | .878 | .853 | .821 | .781 | .731 | .670 | | | | |
| 7.000 | .83 | 42 | 4.723 | 2.277 | 32.5% | 6 | .877 | .852 | .820 | .779 | .729 | .667 | | | | |
| 7.000 | .56 | 16 | 5.440 | 1.560 | 22.3% | 7 | .881 | .858 | .830 | .795 | .751 | .697 | .628 | | | |
| 7.000 | .71 | 24 | 5.367 | 1.633 | 23.3% | 7 | .879 | .855 | .824 | .786 | .738 | .679 | .607 | | | |
| 7.000 | .78 | 32 | 5.337 | 1.663 | 23.8% | 7 | .878 | .853 | .821 | .782 | .733 | .672 | .598 | | | |
| 7.000 | .83 | 40 | 5.319 | 1.681 | 24.0% | 7 | .877 | .852 | .820 | .780 | .729 | .668 | .593 | | | |
| 7.000 | .85 | 48 | 5.308 | 1.692 | 24.2% | 7 | .877 | .851 | .819 | .778 | .727 | .665 | .590 | | | |
| 7.000 | .88 | 56 | 5.300 | 1.700 | 24.3% | 7 | .877 | .851 | .818 | .777 | .726 | .663 | .588 | | | |
| 7.000 | .22 | 9 | 6.429 | .571 | 8.2% | 8 | .886 | .867 | .845 | .822 | .797 | .769 | .740 | .703 | | |
| 7.000 | .61 | 18 | 5.948 | 1.052 | 15.0% | 8 | .880 | .857 | .828 | .792 | .746 | .691 | .621 | .533 | | |
| 7.000 | .74 | 27 | 5.868 | 1.132 | 16.2% | 8 | .879 | .854 | .823 | .784 | .736 | .676 | .603 | .515 | | |
| 7.000 | .81 | 36 | 5.834 | 1.166 | 16.7% | 8 | .878 | .852 | .821 | .781 | .731 | .669 | .595 | .508 | | |
| 7.000 | .84 | 45 | 5.815 | 1.185 | 16.9% | 8 | .877 | .852 | .819 | .779 | .728 | .666 | .591 | .504 | | |
| 7.000 | .87 | 54 | 5.803 | 1.197 | 17.1% | 8 | .877 | .851 | .818 | .777 | .726 | .664 | .588 | .501 | | |
| 7.000 | .89 | 63 | 5.795 | 1.205 | 17.2% | 8 | .877 | .850 | .818 | .777 | .725 | .662 | .587 | .499 | | |
| 7.000 | .30 | 10 | 6.770 | .230 | 3.3% | 9 | .885 | .865 | .842 | .814 | .783 | .747 | .699 | .633 | .503 | |
| 7.000 | .65 | 20 | 6.345 | .655 | 9.4% | 9 | .880 | .856 | .826 | .789 | .743 | .686 | .615 | .527 | .423 | |
| 7.000 | .77 | 30 | 6.266 | .734 | 10.5% | 9 | .878 | .853 | .822 | .782 | .734 | .673 | .600 | .512 | .412 | |
| 7.000 | .83 | 40 | 6.233 | .767 | 11.0% | 9 | .877 | .852 | .820 | .780 | .729 | .668 | .593 | .505 | .408 | |
| 7.000 | .86 | 50 | 6.213 | .787 | 11.2% | 9 | .877 | .851 | .819 | .778 | .727 | .665 | .590 | .502 | .406 | |
| 7.000 | .88 | 60 | 6.201 | .799 | 11.4% | 9 | .877 | .851 | .818 | .777 | .726 | .663 | .587 | .500 | .404 | |
| 7.000 | .90 | 70 | 6.193 | .807 | 11.5% | 9 | .876 | .850 | .817 | .776 | .725 | .661 | .586 | .498 | .403 | |
| 7.000 | .36 | 11 | 6.929 | .071 | 1.0% | 10 | .884 | .863 | .839 | .808 | .775 | .731 | .676 | .599 | .476 | .277 |
| 7.000 | .68 | 22 | 6.629 | .371 | 5.3% | 10 | .880 | .855 | .825 | .787 | .740 | .683 | .609 | .523 | .420 | .307 |
| 7.000 | .79 | 33 | 6.560 | .440 | 6.3% | 10 | .878 | .853 | .821 | .782 | .732 | .671 | .597 | .509 | .411 | .306 |
| 7.000 | .84 | 44 | 6.529 | .471 | 6.7% | 10 | .877 | .852 | .819 | .779 | .728 | .666 | .591 | .504 | .407 | .306 |
| 7.000 | .87 | 55 | 6.512 | .488 | 7.0% | 10 | .877 | .851 | .818 | .777 | .726 | .664 | .588 | .501 | .405 | .305 |

ERLANG-ENGSET: FOR SMOOTH ATTEMPTING TRAFFIC AND FEW PERSONS DOING THE CALLING (FEW MEANS RATIO OF CALLERS TO CIRCUITS IS LESS THAN 8 - GREATER THAN 8 ASSUME TRAFFIC RANDOM)

| HOURS FIRST ATTEMPT | FIRST ATTEMPT VMR | NUMBER SOURCES | HOURS CON-NECT | HOURS OVFL | PERCENT BLOCKED (GRADE) | LINE | HOURS CONNECTED TRAFFIC ON LINES 1 | 2 | 3 | 4 | 5 | 6 | 7 | 8 | 9 | 10 | |
|---|---|---|---|---|---|---|---|---|---|---|---|---|---|---|---|---|---|
| 7.000 | .89 | 66 | 6.501 | .499 | 7.1% | 10 | .876 | .850 | .818 | .776 | .725 | .662 | .586 | .499 | .403 | .305 | |
| 7.000 | .91 | 77 | 6.493 | .507 | 7.2% | 10 | .876 | .850 | .817 | .776 | .724 | .661 | .585 | .498 | .402 | .305 | |
| 7.500 | .46 | 14 | 4.938 | 2.562 | 34.2% | 6 | .889 | .869 | .845 | .815 | .782 | .737 | | | | | |
| 7.500 | .64 | 21 | 4.878 | 2.622 | 35.0% | 6 | .887 | .865 | .839 | .806 | .765 | .716 | | | | | |
| 7.500 | .53 | 16 | 5.585 | 1.915 | 25.5% | 7 | .888 | .868 | .843 | .812 | .775 | .727 | .672 | | | | |
| 7.500 | .69 | 24 | 5.513 | 1.987 | 26.5% | 7 | .886 | .864 | .837 | .804 | .762 | .711 | .647 | | | | |
| 7.500 | .77 | 32 | 5.483 | 2.017 | 26.9% | 7 | .885 | .863 | .835 | .800 | .757 | .703 | .639 | | | | |
| 7.500 | .81 | 40 | 5.466 | 2.034 | 27.1% | 7 | .885 | .862 | .834 | .798 | .754 | .700 | .634 | | | | |
| 7.500 | .84 | 48 | 5.455 | 2.045 | 27.3% | 7 | .884 | .861 | .833 | .797 | .752 | .698 | .631 | | | | |
| 7.500 | .87 | 56 | 5.447 | 2.053 | 27.4% | 7 | .884 | .861 | .832 | .796 | .751 | .696 | .628 | | | | |
| 7.500 | .58 | 18 | 6.145 | 1.355 | 18.1% | 8 | .888 | .867 | .841 | .809 | .770 | .721 | .663 | .586 | | | |
| 7.500 | .72 | 27 | 6.064 | 1.436 | 19.1% | 8 | .886 | .864 | .836 | .802 | .760 | .708 | .644 | .565 | | | |
| 7.500 | .79 | 36 | 6.030 | 1.470 | 19.6% | 8 | .885 | .862 | .834 | .799 | .755 | .701 | .636 | .557 | | | |
| 7.500 | .83 | 45 | 6.011 | 1.489 | 19.8% | 8 | .884 | .861 | .833 | .797 | .753 | .698 | .632 | .553 | | | |
| 7.500 | .86 | 54 | 5.999 | 1.501 | 20.0% | 8 | .884 | .861 | .832 | .796 | .751 | .696 | .629 | .550 | | | |
| 7.500 | .88 | 63 | 5.991 | 1.509 | 20.1% | 8 | .884 | .861 | .831 | .795 | .750 | .694 | .627 | .548 | | | |
| 7.500 | .25 | 10 | 7.104 | .396 | 5.3% | 9 | .892 | .874 | .853 | .831 | .806 | .776 | .743 | .703 | .625 | | |
| 7.500 | .63 | 20 | 6.606 | .894 | 11.9% | 9 | .887 | .866 | .840 | .807 | .768 | .717 | .656 | .580 | .487 | | |
| 7.500 | .75 | 30 | 6.522 | .978 | 13.0% | 9 | .885 | .863 | .836 | .801 | .758 | .705 | .640 | .562 | .471 | | |
| 7.500 | .81 | 40 | 6.486 | 1.014 | 13.5% | 9 | .885 | .862 | .834 | .798 | .754 | .700 | .634 | .555 | .465 | | |
| 7.500 | .85 | 50 | 6.466 | 1.034 | 13.8% | 9 | .884 | .861 | .832 | .706 | .752 | .697 | .630 | .551 | .462 | | |
| 7.500 | .88 | 60 | 6.453 | 1.047 | 14.0% | 9 | .884 | .861 | .832 | .795 | .750 | .695 | .628 | .549 | .460 | | |
| 7.500 | .89 | 70 | 6.444 | 1.056 | 14.1% | 9 | .884 | .860 | .831 | .794 | .749 | .694 | .626 | .547 | .458 | | |
| 7.500 | .32 | 11 | 7.357 | .143 | 1.9% | 10 | .891 | .873 | .851 | .825 | .796 | .761 | .720 | .663 | .572 | .405 | |
| 7.500 | .66 | 22 | 6.957 | .543 | 7.2% | 10 | .887 | .865 | .838 | .805 | .764 | .714 | .651 | .575 | .482 | .376 | |
| 7.500 | .77 | 33 | 6.878 | .622 | 8.3% | 10 | .885 | .863 | .835 | .800 | .757 | .703 | .638 | .560 | .469 | .369 | |
| 7.500 | .83 | 44 | 6.843 | .657 | 8.8% | 10 | .884 | .862 | .833 | .797 | .753 | .699 | .632 | .553 | .464 | .366 | |
| 7.500 | .86 | 55 | 6.823 | .677 | 9.0% | 10 | .884 | .861 | .832 | .796 | .751 | .696 | .629 | .550 | .461 | .364 | |
| 7.500 | .89 | 66 | 6.811 | .689 | 9.2% | 10 | .884 | .860 | .831 | .795 | .750 | .694 | .627 | .548 | .459 | .364 | |
| 7.500 | .90 | 77 | 6.802 | .698 | 9.3% | 10 | .884 | .860 | .831 | .794 | .749 | .693 | .625 | .547 | .457 | .363 | |
| 7.500 | .30 | 12 | 7.459 | .041 | .5% | 11 | .890 | .872 | .849 | .821 | .790 | .752 | .705 | .637 | .546 | .401 | .198 |
| 7.500 | .69 | 24 | 7.199 | .301 | 4.0% | 11 | .886 | .864 | .837 | .804 | .762 | .711 | .647 | .571 | .478 | .374 | .264 |
| 7.500 | .79 | 36 | 7.133 | .367 | 4.9% | 11 | .885 | .862 | .834 | .799 | .755 | .701 | .636 | .557 | .467 | .368 | .267 |
| 7.500 | .84 | 48 | 7.103 | .397 | 5.3% | 11 | .884 | .861 | .833 | .797 | .752 | .698 | .631 | .552 | .462 | .365 | .268 |
| 7.500 | .88 | 60 | 7.086 | .414 | 5.5% | 11 | .884 | .861 | .832 | .795 | .750 | .695 | .628 | .549 | .460 | .364 | .269 |

# ERLANG-ENGSET: FOR SMOOTH ATTEMPTING TRAFFIC AND FEW PERSONS DOING THE CALLING (FEW MEANS RATIO OF CALLERS TO CIRCUITS IS LESS THAN 8 - GREATER THAN 8 ASSUME TRAFFIC RANDOM)

| HOURS FIRST ATTEMPT | FIRST ATTEMPT VMR | NUMBER SOURCES | HOURS CONNECT | HOURS OVFL | PERCENT BLOCKED (GRADE) | LINE | HOURS CONNECTED TRAFFIC ON LINES 1 | 2 | 3 | 4 | 5 | 6 | 7 | 8 | 9 | 10 |
|---|---|---|---|---|---|---|---|---|---|---|---|---|---|---|---|---|
| 7.500 | .90 | 72 | 7.075 | .425 | 5.7% | 11 | .884 | .860 | .831 | .794 | .749 | .693 | .626 | .547 | .458 | .363 |
| | | | | | | | .269 | | | | | | | | | |
| 7.500 | .91 | 84 | 7.067 | .433 | 5.8% | 11 | .883 | .860 | .831 | .794 | .748 | .692 | .625 | .546 | .457 | .363 |
| | | | | | | | .269 | | | | | | | | | |
| 8.000 | .50 | 16 | 5.707 | 2.293 | 28.7% | 7 | .895 | .876 | .854 | .828 | .795 | .753 | .707 | | | |
| 8.000 | .67 | 24 | 5.638 | 2.362 | 29.5% | 7 | .893 | .873 | .849 | .819 | .784 | .738 | .683 | | | |
| 8.000 | .75 | 32 | 5.609 | 2.391 | 29.9% | 7 | .892 | .872 | .847 | .816 | .778 | .731 | .674 | | | |
| 8.000 | .80 | 40 | 5.592 | 2.408 | 30.1% | 7 | .891 | .871 | .845 | .814 | .775 | .727 | .669 | | | |
| 8.000 | .83 | 48 | 5.582 | 2.418 | 30.2% | 7 | .891 | .870 | .844 | .812 | .773 | .724 | .666 | | | |
| 8.000 | .86 | 56 | 5.575 | 2.425 | 30.3% | 7 | .891 | .870 | .844 | .812 | .772 | .723 | .664 | | | |
| 8.000 | .56 | 18 | 6.314 | 1.686 | 21.1% | 8 | .894 | .875 | .852 | .825 | .790 | .749 | .696 | .633 | | |
| 8.000 | .70 | 27 | 6.235 | 1.765 | 22.1% | 8 | .892 | .872 | .848 | .818 | .781 | .735 | .679 | .610 | | |
| 8.000 | .78 | 36 | 6.200 | 1.800 | 22.5% | 8 | .891 | .871 | .846 | .815 | .776 | .729 | .671 | .601 | | |
| 8.000 | .82 | 45 | 6.182 | 1.818 | 22.7% | 8 | .891 | .870 | .845 | .813 | .774 | .726 | .667 | .597 | | |
| 8.000 | .85 | 54 | 6.170 | 1.830 | 22.9% | 8 | .891 | .870 | .844 | .812 | .772 | .723 | .664 | .594 | | |
| 8.000 | .87 | 63 | 6.162 | 1.838 | 23.0% | 8 | .890 | .869 | .843 | .811 | .771 | .722 | .663 | .592 | | |
| 8.000 | .20 | 10 | 7.378 | .622 | 7.8% | 9 | .898 | .883 | .864 | .845 | .824 | .803 | .779 | .754 | .730 | |
| 8.000 | .60 | 20 | 6.832 | 1.168 | 14.6% | 9 | .893 | .874 | .851 | .822 | .788 | .744 | .691 | .625 | .543 | |
| 8.000 | .73 | 30 | 6.745 | 1.255 | 15.7% | 9 | .892 | .872 | .847 | .816 | .779 | .733 | .675 | .607 | .524 | |
| 8.000 | .80 | 40 | 6.708 | 1.292 | 16.2% | 9 | .891 | .871 | .845 | .814 | .775 | .727 | .669 | .599 | .517 | |
| 8.000 | .84 | 50 | 6.688 | 1.312 | 16.4% | 9 | .891 | .870 | .844 | .812 | .773 | .724 | .665 | .595 | .513 | |
| 8.000 | .87 | 60 | 6.674 | 1.326 | 16.6% | 9 | .890 | .869 | .844 | .811 | .771 | .722 | .663 | .592 | .510 | |
| 8.000 | .89 | 70 | 6.665 | 1.335 | 16.7% | 9 | .890 | .869 | .843 | .811 | .771 | .721 | .661 | .590 | .509 | |
| 8.000 | .27 | 11 | 7.739 | .261 | 3.3% | 10 | .897 | .880 | .863 | .840 | .814 | .789 | .755 | .713 | .652 | .536 |
| 8.000 | .64 | 22 | 7.247 | .753 | 9.4% | 10 | .893 | .874 | .850 | .821 | .785 | .741 | .686 | .620 | .538 | .440 |
| 8.000 | .76 | 33 | 7.159 | .841 | 10.5% | 10 | .892 | .871 | .847 | .816 | .777 | .731 | .673 | .603 | .522 | .428 |
| 8.000 | .82 | 44 | 7.122 | .878 | 11.0% | 10 | .891 | .870 | .845 | .813 | .774 | .726 | .667 | .597 | .515 | .423 |
| 8.000 | .85 | 55 | 7.101 | .899 | 11.2% | 10 | .890 | .870 | .844 | .812 | .772 | .723 | .664 | .593 | .512 | .421 |
| 8.000 | .88 | 66 | 7.088 | .912 | 11.4% | 10 | .890 | .869 | .843 | .811 | .771 | .722 | .662 | .591 | .509 | .419 |
| 8.000 | .90 | 77 | 7.078 | .922 | 11.5% | 10 | .890 | .869 | .843 | .810 | .770 | .720 | .661 | .590 | .508 | .418 |
| 8.000 | .33 | 12 | 7.915 | .085 | 1.1% | 11 | .896 | .879 | .861 | .836 | .810 | .777 | .738 | .688 | .617 | .504 |
| | | | | | | | .309 | | | | | | | | | |
| 8.000 | .67 | 24 | 7.553 | .447 | 5.6% | 11 | .893 | .873 | .849 | .819 | .784 | .738 | .683 | .615 | .533 | .437 |
| | | | | | | | .330 | | | | | | | | | |
| 8.000 | .78 | 36 | 7.475 | .525 | 6.6% | 11 | .891 | .871 | .846 | .815 | .776 | .729 | .671 | .601 | .520 | .426 |
| | | | | | | | .328 | | | | | | | | | |
| 8.000 | .83 | 48 | 7.440 | .560 | 7.0% | 11 | .891 | .870 | .844 | .812 | .773 | .724 | .666 | .595 | .514 | .422 |
| | | | | | | | .326 | | | | | | | | | |
| 8.000 | .87 | 60 | 7.420 | .580 | 7.2% | 11 | .890 | .869 | .844 | .811 | .771 | .722 | .663 | .592 | .510 | .420 |
| | | | | | | | .326 | | | | | | | | | |

ERLANG-ENGSET: FOR SMOOTH ATTEMPTING TRAFFIC AND FEW PERSONS DOING THE CALLING (FEW MEANS RATIO OF CALLERS TO CIRCUITS IS LESS THAN 8 - GREATER THAN 8 ASSUME TRAFFIC RANDOM)

| HOURS FIRST ATTEMPT | FIRST ATTEMPT VMR | NUMBER SOURCES | HOURS CON-NECT | HOURS OVFL | PERCENT BLOCKED (GRADE) | LINE | HOURS CONNECTED TRAFFIC ON LINES 1 | 2 | 3 | 4 | 5 | 6 | 7 | 8 | 9 | 10 |
|---|---|---|---|---|---|---|---|---|---|---|---|---|---|---|---|---|
| 8.000 | .89 | 72 | 7.408 | .592 | 7.4% | 11 | .890 | .869 | .843 | .810 | .770 | .721 | .661 | .590 | .508 | .419 |
| | | | | | | | .325 | | | | | | | | | |
| 8.000 | .90 | 84 | 7.399 | .601 | 7.5% | 11 | .890 | .869 | .843 | .810 | .769 | .720 | .660 | .589 | .507 | .418 |
| | | | | | | | .325 | | | | | | | | | |
| 8.500 | .47 | 16 | 5.813 | 2.687 | 31.6% | 7 | .900 | .884 | .864 | .839 | .812 | .778 | .735 | | | |
| 8.500 | .65 | 24 | 5.746 | 2.754 | 32.4% | 7 | .898 | .881 | .859 | .832 | .801 | .762 | .713 | | | |
| 8.500 | .73 | 32 | 5.717 | 2.783 | 32.7% | 7 | .897 | .879 | .857 | .830 | .795 | .755 | .703 | | | |
| 8.500 | .79 | 40 | 5.702 | 2.798 | 32.9% | 7 | .897 | .878 | .856 | .828 | .793 | .751 | .700 | | | |
| 8.500 | .82 | 48 | 5.692 | 2.808 | 33.0% | 7 | .897 | .878 | .855 | .826 | .792 | .748 | .697 | | | |
| 8.500 | .85 | 56 | 5.685 | 2.815 | 33.1% | 7 | .896 | .877 | .854 | .826 | .790 | .747 | .694 | | | |
| 8.500 | .53 | 18 | 6.460 | 2.040 | 24.0% | 8 | .900 | .883 | .862 | .837 | .808 | .771 | .727 | .672 | | |
| 8.500 | .69 | 27 | 6.382 | 2.118 | 24.9% | 8 | .898 | .880 | .858 | .831 | .798 | .758 | .710 | .648 | | |
| 8.500 | .76 | 36 | 6.349 | 2.151 | 25.3% | 8 | .897 | .879 | .856 | .828 | .795 | .753 | .701 | .640 | | |
| 8.500 | .81 | 45 | 6.330 | 2.170 | 25.5% | 8 | .897 | .878 | .855 | .827 | .792 | .750 | .698 | .635 | | |
| 8.500 | .84 | 54 | 6.319 | 2.181 | 25.7% | 8 | .896 | .877 | .854 | .826 | .791 | .748 | .695 | .632 | | |
| 8.500 | .87 | 63 | 6.310 | 2.190 | 25.8% | 8 | .896 | .877 | .854 | .825 | .790 | .746 | .693 | .630 | | |
| 8.500 | .58 | 20 | 7.027 | 1.473 | 17.3% | 9 | .899 | .882 | .861 | .836 | .805 | .767 | .721 | .663 | .593 | |
| 8.500 | .72 | 30 | 6.939 | 1.561 | 18.4% | 9 | .898 | .880 | .857 | .830 | .797 | .756 | .706 | .645 | .571 | |
| 8.500 | .79 | 40 | 6.903 | 1.597 | 18.8% | 9 | .897 | .878 | .856 | .828 | .793 | .751 | .700 | .637 | .563 | |
| 8.500 | .83 | 50 | 6.882 | 1.618 | 19.0% | 9 | .896 | .878 | .854 | .826 | .791 | .748 | .696 | .632 | .559 | |
| 8.500 | .86 | 60 | 6.869 | 1.631 | 19.2% | 9 | .896 | .877 | .854 | .825 | .790 | .746 | .694 | .630 | .556 | |
| 8.500 | .88 | 70 | 6.860 | 1.640 | 19.3% | 9 | .896 | .877 | .854 | .825 | .789 | .745 | .692 | .628 | .554 | |
| 8.500 | .20 | 11 | 8.062 | .438 | 5.1% | 10 | .903 | .888 | .870 | .853 | .831 | .810 | .782 | .755 | .719 | .652 |
| 8.500 | .61 | 22 | 7.500 | 1.000 | 11.8% | 10 | .899 | .881 | .860 | .834 | .802 | .764 | .716 | .659 | .586 | .499 |
| 8.500 | .74 | 33 | 7.409 | 1.091 | 12.8% | 10 | .897 | .879 | .857 | .829 | .796 | .754 | .704 | .642 | .568 | .484 |
| 8.500 | .81 | 44 | 7.369 | 1.131 | 13.3% | 10 | .897 | .878 | .855 | .827 | .792 | .750 | .698 | .635 | .561 | .476 |
| 8.500 | .85 | 55 | 7.348 | 1.152 | 13.6% | 10 | .896 | .877 | .854 | .826 | .791 | .747 | .694 | .632 | .557 | .473 |
| 8.500 | .87 | 66 | 7.334 | 1.166 | 13.7% | 10 | .896 | .877 | .854 | .825 | .789 | .740 | .693 | .620 | .555 | .471 |
| 8.500 | .89 | 77 | 7.324 | 1.176 | 13.8% | 10 | .806 | .877 | .853 | .824 | .789 | .744 | .691 | .627 | .553 | .470 |
| 8.500 | .29 | 12 | 8.335 | .165 | 1.9% | 11 | .902 | .886 | .870 | .849 | .825 | .798 | .768 | .728 | .677 | .596 |
| | | | | | | | .435 | | | | | | | | | |
| 8.500 | .65 | 24 | 7.870 | .630 | 7.4% | 11 | .898 | .881 | .859 | .832 | .801 | .762 | .713 | .653 | .581 | .494 |
| | | | | | | | .395 | | | | | | | | | |
| 8.500 | .76 | 36 | 7.781 | .719 | 8.5% | 11 | .897 | .879 | .856 | .828 | .795 | .753 | .701 | .640 | .560 | .481 |
| | | | | | | | .386 | | | | | | | | | |
| 8.500 | .82 | 48 | 7.743 | .757 | 8.9% | 11 | .897 | .878 | .855 | .826 | .792 | .748 | .697 | .634 | .559 | .475 |
| | | | | | | | .383 | | | | | | | | | |
| 8.500 | .86 | 60 | 7.722 | .778 | 9.2% | 11 | .896 | .877 | .854 | .825 | .790 | .746 | .694 | .630 | .556 | .472 |
| | | | | | | | .381 | | | | | | | | | |
| 8.500 | .88 | 72 | 7.708 | .792 | 9.3% | 11 | .896 | .877 | .853 | .825 | .789 | .745 | .692 | .628 | .554 | .470 |

## ERLANG-ENGSET: FOR SMOOTH ATTEMPTING TRAFFIC AND FEW PERSONS DOING THE CALLING (FEW MEANS RATIO OF CALLERS TO CIRCUITS IS LESS THAN 8 - GREATER THAN 8 ASSUME TRAFFIC RANDOM)

| HOURS FIRST ATTEMPT | FIRST ATTEMPT VMR | NUMBER SOURCES | HOURS CON-NECT | HOURS OVFL | PERCENT BLOCKED (GRADE) | LINE | HOURS CONNECTED TRAFFIC ON LINES 1 | 2 | 3 | 4 | 5 | 6 | 7 | 8 | 9 | 10 |
|---|---|---|---|---|---|---|---|---|---|---|---|---|---|---|---|---|
| | | | | | | | .380 | | | | | | | | | |
| 8.500 | .90 | 84 | 7.699 | .801 | 9.4% | 11 | .896 | .877 | .853 | .824 | .788 | .744 | .690 | .627 | .552 | .468 |
| | | | | | | | .379 | | | | | | | | | |
| 8.500 | .35 | 13 | 8.450 | .050 | .6% | 12 | .902 | .886 | .866 | .847 | .822 | .791 | .753 | .714 | .652 | .567 |
| | | | | | | | .428 | .223 | | | | | | | | |
| 8.500 | .67 | 26 | 8.135 | .365 | 4.3% | 12 | .898 | .880 | .858 | .832 | .799 | .760 | .710 | .649 | .578 | .491 |
| | | | | | | | .392 | .287 | | | | | | | | |
| 8.500 | .78 | 39 | 8.059 | .441 | 5.2% | 12 | .897 | .878 | .856 | .828 | .793 | .751 | .700 | .637 | .564 | .479 |
| | | | | | | | .386 | .289 | | | | | | | | |
| 8.500 | .84 | 52 | 8.025 | .475 | 5.6% | 12 | .896 | .878 | .855 | .826 | .791 | .748 | .695 | .632 | .558 | .474 |
| | | | | | | | .382 | .289 | | | | | | | | |
| 8.500 | .87 | 65 | 8.005 | .495 | 5.8% | 12 | .896 | .877 | .854 | .825 | .789 | .746 | .693 | .629 | .555 | .471 |
| | | | | | | | .381 | .290 | | | | | | | | |
| 8.500 | .89 | 78 | 7.993 | .507 | 6.0% | 12 | .896 | .877 | .853 | .824 | .788 | .744 | .691 | .627 | .553 | .469 |
| | | | | | | | .380 | .290 | | | | | | | | |
| 8.500 | .91 | 91 | 7.984 | .516 | 6.1% | 12 | .896 | .877 | .853 | .824 | .788 | .743 | .690 | .626 | .551 | .468 |
| | | | | | | | .379 | .290 | | | | | | | | |
| 9.000 | .44 | 16 | 5.903 | 3.097 | 34.4% | 7 | .905 | .890 | .873 | .851 | .827 | .795 | .762 | | | |
| 9.000 | .50 | 18 | 6.584 | 2.416 | 26.8% | 8 | .905 | .889 | .871 | .849 | .823 | .791 | .752 | .704 | | |
| 9.000 | .67 | 27 | 6.509 | 2.491 | 27.7% | 8 | .903 | .887 | .867 | .844 | .814 | .778 | .735 | .681 | | |
| 9.000 | .75 | 36 | 6.477 | 2.523 | 28.0% | 8 | .902 | .886 | .865 | .840 | .810 | .773 | .728 | .673 | | |
| 9.000 | .80 | 45 | 6.460 | 2.540 | 28.2% | 8 | .902 | .885 | .864 | .839 | .808 | .769 | .724 | .668 | | |
| 9.000 | .83 | 54 | 6.449 | 2.551 | 28.3% | 8 | .902 | .884 | .864 | .838 | .806 | .768 | .721 | .665 | | |
| 9.000 | .86 | 63 | 6.441 | 2.559 | 28.4% | 8 | .901 | .884 | .863 | .837 | .806 | .767 | .719 | .663 | | |
| 9.000 | .55 | 20 | 7.196 | 1.804 | 20.0% | 9 | .904 | .889 | .869 | .848 | .820 | .787 | .746 | .697 | .636 | |
| 9.000 | .70 | 30 | 7.109 | 1.891 | 21.0% | 9 | .903 | .886 | .866 | .842 | .813 | .776 | .731 | .678 | .613 | |
| 9.000 | .77 | 40 | 7.073 | 1.927 | 21.4% | 9 | .902 | .885 | .864 | .840 | .809 | .771 | .726 | .670 | .605 | |
| 9.000 | .82 | 50 | 7.052 | 1.948 | 21.6% | 9 | .902 | .885 | .864 | .838 | .807 | .768 | .723 | .666 | .599 | |
| 9.000 | .85 | 60 | 7.039 | 1.961 | 21.8% | 9 | .901 | .884 | .863 | .837 | .806 | .767 | .720 | .663 | .596 | |
| 9.000 | .87 | 70 | 7.031 | 1.969 | 21.9% | 9 | .901 | .884 | .863 | .837 | .805 | .766 | .718 | .662 | .594 | |
| 9.000 | .18 | 11 | 8.331 | .669 | 7.4% | 10 | .908 | .894 | .881 | .861 | .846 | .827 | .806 | .786 | .771 | .751 |
| 9.000 | .59 | 22 | 7.722 | 1.278 | 14.2% | 10 | .904 | .888 | .869 | .846 | .818 | .784 | .742 | .691 | .629 | .552 |
| 9.000 | .73 | 33 | 7.628 | 1.372 | 15.2% | 10 | .902 | .886 | .866 | .841 | .811 | .774 | .729 | .675 | .610 | .532 |
| 9.000 | .80 | 44 | 7.588 | 1.412 | 15.7% | 10 | .902 | .885 | .864 | .839 | .808 | .770 | .724 | .668 | .602 | .525 |
| 9.000 | .84 | 55 | 7.566 | 1.434 | 15.9% | 10 | .902 | .884 | .863 | .838 | .807 | .768 | .721 | .665 | .598 | .521 |
| 9.000 | .86 | 66 | 7.552 | 1.448 | 16.1% | 10 | .901 | .884 | .863 | .837 | .805 | .767 | .719 | .662 | .595 | .518 |
| 9.000 | .88 | 77 | 7.542 | 1.458 | 16.2% | 10 | .901 | .884 | .863 | .836 | .805 | .765 | .718 | .661 | .594 | .516 |
| 9.000 | .25 | 12 | 8.709 | .291 | 3.2% | 11 | .907 | .894 | .878 | .860 | .837 | .819 | .791 | .761 | .726 | .673 |
| | | | | | | | .564 | | | | | | | | | |
| 9.000 | .63 | 24 | 8.151 | .849 | 9.4% | 11 | .904 | .887 | .868 | .844 | .816 | .781 | .740 | .687 | .624 | .546 |

ERLANG-ENGSET: FOR SMOOTH ATTEMPTING TRAFFIC AND FEW PERSONS DOING THE CALLING (FEW MEANS RATIO OF CALLERS TO CIRCUITS IS LESS THAN 8 - GREATER THAN 8 ASSUME TRAFFIC RANDOM)

| HOURS FIRST ATTEMPT | FIRST ATTEMPT VMR | NUMBER SOURCES | HOURS CON-NECT | HOURS OVFL | PERCENT BLOCKED (GRADE) | LINE | HOURS CONNECTED TRAFFIC ON LINES | | | | | | | | | |
|---|---|---|---|---|---|---|---|---|---|---|---|---|---|---|---|---|
| | | | | | | | 1 | 2 | 3 | 4 | 5 | 6 | 7 | 8 | 9 | 10 |
| | | | | | | | .455 | | | | | | | | | |
| 9.000 | .75 | 36 | 8.056 | .944 | 10.5% | 11 | .902 | .886 | .865 | .840 | .810 | .773 | .728 | .673 | .607 | .530 |
| | | | | | | | .442 | | | | | | | | | |
| 9.000 | .81 | 48 | 8.016 | .984 | 10.9% | 11 | .902 | .885 | .864 | .839 | .807 | .769 | .723 | .667 | .600 | .524 |
| | | | | | | | .437 | | | | | | | | | |
| 9.000 | .85 | 60 | 7.993 | 1.007 | 11.2% | 11 | .901 | .884 | .863 | .837 | .806 | .767 | .720 | .663 | .596 | .520 |
| | | | | | | | .434 | | | | | | | | | |
| 9.000 | .88 | 72 | 7.979 | 1.021 | 11.3% | 11 | .901 | .884 | .863 | .837 | .805 | .766 | .719 | .661 | .594 | .517 |
| | | | | | | | .432 | | | | | | | | | |
| 9.000 | .89 | 84 | 7.968 | 1.032 | 11.5% | 11 | .901 | .884 | .862 | .836 | .804 | .765 | .717 | .660 | .593 | .516 |
| | | | | | | | .431 | | | | | | | | | |
| 9.000 | .31 | 13 | 8.900 | .100 | 1.1% | 12 | .906 | .893 | .876 | .857 | .835 | .810 | .781 | .744 | .699 | .634 |
| | | | | | | | .527 | .337 | | | | | | | | |
| 9.000 | .65 | 26 | 8.477 | .523 | 5.8% | 12 | .903 | .887 | .867 | .844 | .814 | .780 | .736 | .683 | .620 | .542 |
| | | | | | | | .451 | .349 | | | | | | | | |
| 9.000 | .77 | 39 | 8.389 | .611 | 6.8% | 12 | .902 | .885 | .865 | .840 | .809 | .772 | .726 | .671 | .605 | .527 |
| | | | | | | | .441 | .346 | | | | | | | | |
| 9.000 | .83 | 52 | 8.351 | .649 | 7.2% | 12 | .902 | .885 | .864 | .838 | .807 | .768 | .722 | .666 | .599 | .522 |
| | | | | | | | .436 | .344 | | | | | | | | |
| 9.000 | .86 | 65 | 8.330 | .670 | 7.4% | 12 | .901 | .884 | .863 | .837 | .805 | .766 | .719 | .663 | .595 | .518 |
| | | | | | | | .433 | .344 | | | | | | | | |
| 9.000 | .88 | 78 | 8.316 | .684 | 7.6% | 12 | .901 | .884 | .863 | .836 | .804 | .765 | .718 | .661 | .593 | .516 |
| | | | | | | | .432 | .343 | | | | | | | | |
| 9.000 | .90 | 91 | 8.306 | .694 | 7.7% | 12 | .901 | .883 | .862 | .830 | .804 | .765 | .717 | .659 | .592 | .515 |
| | | | | | | | .401 | .342 | | | | | | | | |
| 9.500 | .47 | 18 | 6.694 | 2.806 | 29.5% | 8 | .909 | .895 | .879 | .859 | .836 | .808 | .774 | .733 | | |
| 9.500 | .65 | 27 | 6.621 | 2.879 | 30.3% | 8 | .908 | .893 | .875 | .854 | .827 | .796 | .758 | .711 | | |
| 9.500 | .74 | 36 | 6.590 | 2.910 | 30.6% | 8 | .907 | .892 | .874 | .851 | .824 | .791 | .750 | .702 | | |
| 9.500 | .79 | 45 | 6.574 | 2.926 | 30.8% | 8 | .907 | .891 | .872 | .850 | .822 | .788 | .747 | .697 | | |
| 9.500 | .82 | 54 | 6.562 | 2.938 | 30.9% | 8 | .906 | .891 | .872 | .849 | .821 | .786 | .745 | .694 | | |
| 9.500 | .85 | 63 | 6.555 | 2.945 | 31.0% | 8 | .900 | .890 | .871 | .848 | .819 | .785 | .743 | .692 | | |
| 9.500 | .52 | 20 | 7.342 | 2.158 | 22.7% | 9 | .909 | .895 | .878 | .857 | .832 | .806 | .769 | .724 | .673 | |
| 9.500 | .68 | 30 | 7.258 | 2.242 | 23.6% | 9 | .907 | .893 | .874 | .852 | .827 | .793 | .754 | .708 | .649 | |
| 9.500 | .76 | 40 | 7.222 | 2.278 | 24.0% | 9 | .907 | .891 | .873 | .851 | .823 | .789 | .748 | .700 | .640 | |
| 9.500 | .81 | 50 | 7.203 | 2.297 | 24.2% | 9 | .906 | .891 | .872 | .840 | .822 | .787 | .745 | .696 | .635 | |
| 9.500 | .84 | 60 | 7.100 | 2.310 | 24.3% | 9 | .906 | .890 | .871 | .848 | .820 | .785 | .743 | .602 | .033 | |
| 9.500 | .86 | 70 | 7.181 | 2.319 | 24.4% | 9 | .906 | .890 | .871 | .848 | .819 | .784 | .742 | .691 | .630 | |
| 9.500 | .57 | 22 | 7.915 | 1.585 | 16.7% | 10 | .909 | .894 | .877 | .856 | .831 | .801 | .763 | .721 | .665 | .598 |
| 9.500 | .71 | 33 | 7.821 | 1.679 | 17.7% | 10 | .907 | .892 | .874 | .852 | .825 | .792 | .752 | .705 | .645 | .577 |
| 9.500 | .78 | 44 | 7.781 | 1.719 | 18.1% | 10 | .907 | .891 | .873 | .850 | .822 | .789 | .747 | .698 | .638 | .567 |

ERLANG-ENGSET: FOR SMOOTH ATTEMPTING TRAFFIC AND FEW PERSONS DOING THE CALLING (FEW MEANS RATIO OF CALLERS TO CIRCUITS IS LESS THAN 8 - GREATER THAN 8 ASSUME TRAFFIC RANDOM)

| HOURS FIRST ATTEMPT | FIRST ATTEMPT VMR | NUMBER SOURCES | HOURS CON-NECT | HOURS OVFL | PERCENT BLOCKED (GRADE) | LINE | HOURS CONNECTED TRAFFIC ON LINES | | | | | | | | | |
|---|---|---|---|---|---|---|---|---|---|---|---|---|---|---|---|---|
| | | | | | | | 1 | 2 | 3 | 4 | 5 | 6 | 7 | 8 | 9 | 10 |
| 9.500 | .83 | 55 | 7.759 | 1.741 | 18.3% | 10 | .906 | .891 | .871 | .849 | .820 | .786 | .744 | .694 | .634 | .564 |
| 9.500 | .86 | 66 | 7.745 | 1.755 | 18.5% | 10 | .906 | .890 | .871 | .848 | .819 | .784 | .743 | .692 | .631 | .561 |
| 9.500 | .88 | 77 | 7.735 | 1.765 | 18.6% | 10 | .906 | .890 | .871 | .847 | .819 | .783 | .741 | .690 | .629 | .559 |
| 9.500 | .21 | 12 | 9.027 | .473 | 5.0% | 11 | .911 | .898 | .886 | .867 | .853 | .833 | .812 | .788 | .764 | .734 |
| | | | | | | | .678 | | | | | | | | | |
| 9.500 | .60 | 24 | 8.399 | 1.101 | 11.6% | 11 | .908 | .893 | .876 | .855 | .830 | .799 | .761 | .716 | .660 | .591 |
| | | | | | | | .510 | | | | | | | | | |
| 9.500 | .74 | 36 | 8.299 | 1.201 | 12.6% | 11 | .907 | .892 | .874 | .851 | .824 | .791 | .750 | .702 | .643 | .573 |
| | | | | | | | .493 | | | | | | | | | |
| 9.500 | .80 | 48 | 8.258 | 1.242 | 13.1% | 11 | .906 | .891 | .872 | .849 | .821 | .787 | .746 | .696 | .636 | .567 |
| | | | | | | | .486 | | | | | | | | | |
| 9.500 | .84 | 60 | 8.235 | 1.265 | 13.3% | 11 | .906 | .890 | .871 | .848 | .820 | .785 | .743 | .692 | .633 | .562 |
| | | | | | | | .483 | | | | | | | | | |
| 9.500 | .87 | 72 | 8.220 | 1.280 | 13.5% | 11 | .906 | .890 | .871 | .848 | .819 | .784 | .741 | .691 | .630 | .560 |
| | | | | | | | .481 | | | | | | | | | |
| 9.500 | .89 | 84 | 8.210 | 1.290 | 13.6% | 11 | .906 | .890 | .871 | .847 | .818 | .783 | .741 | .689 | .628 | .558 |
| | | | | | | | .479 | | | | | | | | | |
| 9.500 | .27 | 13 | 9.314 | .186 | 2.0% | 12 | .911 | .898 | .883 | .868 | .848 | .824 | .803 | .772 | .737 | .690 |
| | | | | | | | .616 | .464 | | | | | | | | |
| 9.500 | .63 | 26 | 8.784 | .716 | 7.5% | 12 | .908 | .893 | .876 | .854 | .829 | .797 | .759 | .712 | .655 | .587 |
| | | | | | | | .505 | .410 | | | | | | | | |
| 9.500 | .76 | 39 | 8.688 | .812 | 8.6% | 12 | .907 | .891 | .873 | .850 | .824 | .790 | .748 | .701 | .641 | .571 |
| | | | | | | | .490 | .401 | | | | | | | | |
| 9.500 | .82 | 52 | 8.646 | .854 | 9.0% | 12 | .906 | .891 | .872 | .848 | .821 | .787 | .744 | .695 | .635 | .565 |
| | | | | | | | .485 | .398 | | | | | | | | |
| 9.500 | .85 | 65 | 8.623 | .877 | 9.2% | 12 | .906 | .890 | .871 | .848 | .820 | .785 | .742 | .692 | .631 | .561 |
| | | | | | | | .482 | .395 | | | | | | | | |
| 9.500 | .88 | 78 | 8.608 | .892 | 9.4% | 12 | .906 | .890 | .871 | .847 | .818 | .783 | .741 | .690 | .629 | .559 |
| | | | | | | | .480 | .394 | | | | | | | | |
| 9.500 | .90 | 91 | 8.598 | .902 | 9.5% | 12 | .906 | .890 | .870 | .847 | .818 | .783 | .740 | .689 | .628 | .557 |
| | | | | | | | .479 | .393 | | | | | | | | |
| 9.500 | .32 | 14 | 9.442 | .058 | .6% | 13 | .911 | .898 | .882 | .865 | .844 | .819 | .794 | .760 | .719 | .665 |
| | | | | | | | .585 | .452 | .247 | | | | | | | |
| 9.500 | .66 | 28 | 9.069 | .431 | 4.5% | 13 | .907 | .893 | .875 | .853 | .827 | .795 | .757 | .710 | .652 | .583 |
| | | | | | | | .502 | .408 | .307 | | | | | | | |
| 9.500 | .77 | 42 | 8.984 | .516 | 5.4% | 13 | .907 | .891 | .873 | .850 | .822 | .789 | .748 | .698 | .639 | .568 |
| | | | | | | | .489 | .400 | .308 | | | | | | | |
| 9.500 | .83 | 56 | 8.946 | .554 | 5.8% | 13 | .906 | .891 | .872 | .848 | .821 | .786 | .744 | .694 | .633 | .563 |
| | | | | | | | .484 | .397 | .308 | | | | | | | |
| 9.500 | .86 | 70 | 8.925 | .575 | 6.1% | 13 | .906 | .890 | .871 | .848 | .819 | .784 | .742 | .691 | .630 | .560 |

ERLANG-ENGSET: FOR SMOOTH ATTEMPTING TRAFFIC AND FEW PERSONS DOING THE CALLING (FEW MEANS RATIO OF CALLERS TO CIRCUITS IS LESS THAN 8 - GREATER THAN 8 ASSUME TRAFFIC RANDOM)

| HOURS FIRST ATTEMPT | FIRST ATTEMPT VMR | NUMBER SOURCES | HOURS CON-NECT | HOURS OVFL | PERCENT BLOCKED (GRADE) | LINE | HOURS CONNECTED TRAFFIC ON LINES 1 | 2 | 3 | 4 | 5 | 6 | 7 | 8 | 9 | 10 |
|---|---|---|---|---|---|---|---|---|---|---|---|---|---|---|---|---|
| | | | | | | | .480 | .395 | .308 | | | | | | | |
| 9.500 | .89 | 84 | 8.911 | .589 | 6.2% | 13 | .906 | .890 | .871 | .847 | .818 | .783 | .741 | .689 | .628 | .558 |
| | | | | | | | .479 | .394 | .307 | | | | | | | |
| 9.500 | .90 | 98 | 8.901 | .599 | 6.3% | 13 | .906 | .890 | .870 | .847 | .818 | .782 | .740 | .688 | .627 | .557 |
| | | | | | | | .478 | .393 | .307 | | | | | | | |
| 10.000 | .44 | 18 | 6.789 | 3.211 | 32.1% | 8 | .913 | .901 | .885 | .869 | .847 | .823 | .792 | .758 | | |
| 10.000 | .63 | 27 | 6.719 | 3.281 | 32.8% | 8 | .912 | .899 | .882 | .863 | .839 | .812 | .777 | .736 | | |
| 10.000 | .72 | 36 | 6.689 | 3.311 | 33.1% | 8 | .911 | .897 | .881 | .860 | .836 | .806 | .771 | .727 | | |
| 10.000 | .78 | 45 | 6.674 | 3.326 | 33.3% | 8 | .911 | .897 | .880 | .859 | .834 | .803 | .768 | .723 | | |
| 10.000 | .81 | 54 | 6.663 | 3.337 | 33.4% | 8 | .911 | .896 | .879 | .858 | .833 | .802 | .765 | .720 | | |
| 10.000 | .84 | 63 | 6.656 | 3.344 | 33.4% | 8 | .910 | .896 | .879 | .857 | .832 | .801 | .763 | .718 | | |
| 10.000 | .50 | 20 | 7.470 | 2.530 | 25.3% | 9 | .913 | .900 | .884 | .867 | .845 | .819 | .788 | .750 | .704 | |
| 10.000 | .67 | 30 | 7.388 | 2.612 | 26.1% | 9 | .912 | .898 | .882 | .861 | .839 | .809 | .774 | .733 | .680 | |
| 10.000 | .75 | 40 | 7.353 | 2.647 | 26.5% | 9 | .911 | .897 | .880 | .860 | .835 | .805 | .769 | .725 | .672 | |
| 10.000 | .80 | 50 | 7.335 | 2.665 | 26.7% | 9 | .911 | .896 | .879 | .859 | .833 | .803 | .765 | .721 | .667 | |
| 10.000 | .83 | 60 | 7.322 | 2.678 | 26.8% | 9 | .910 | .896 | .879 | .858 | .832 | .801 | .764 | .718 | .664 | |
| 10.000 | .86 | 70 | 7.314 | 2.686 | 26.9% | 9 | .910 | .896 | .878 | .857 | .832 | .800 | .763 | .717 | .662 | |
| 10.000 | .55 | 22 | 8.084 | 1.916 | 19.2% | 10 | .913 | .900 | .884 | .865 | .843 | .816 | .784 | .745 | .697 | .638 |
| 10.000 | .70 | 33 | 7.990 | 2.010 | 20.1% | 10 | .911 | .897 | .881 | .861 | .836 | .808 | .772 | .730 | .677 | .616 |
| 10.000 | .77 | 44 | 7.951 | 2.049 | 20.5% | 10 | .911 | .897 | .879 | .860 | .834 | .804 | .767 | .723 | .670 | .607 |
| 10.000 | .82 | 55 | 7.930 | 2.070 | 20.7% | 10 | .911 | .896 | .879 | .858 | .833 | .802 | .764 | .720 | .665 | .602 |
| 10.000 | .85 | 66 | 7.916 | 2.084 | 20.8% | 10 | .910 | .896 | .878 | .858 | .832 | .801 | .782 | .718 | .663 | .600 |
| 10.000 | .87 | 77 | 7.907 | 2.093 | 20.9% | 10 | .910 | .896 | .878 | .857 | .831 | .799 | .761 | .716 | .661 | .597 |
| 10.000 | .17 | 12 | 9.286 | .714 | 7.1% | 11 | .915 | .904 | .892 | .876 | .864 | .846 | .827 | .813 | .792 | .786 |
| | | | | | | | .769 | | | | | | | | | |
| 10.000 | .58 | 24 | 8.617 | 1.383 | 13.8% | 11 | .912 | .899 | .883 | .863 | .841 | .814 | .781 | .739 | .693 | .631 |
| | | | | | | | .560 | | | | | | | | | |
| 10.000 | .72 | 36 | 8.516 | 1.484 | 14.8% | 11 | .911 | .897 | .881 | .860 | .836 | .806 | .771 | .727 | .675 | .612 |
| | | | | | | | .539 | | | | | | | | | |
| 10.000 | .79 | 48 | 8.474 | 1.526 | 15.3% | 11 | .911 | .897 | .879 | .858 | .834 | .803 | .767 | .721 | .667 | .605 |
| | | | | | | | .532 | | | | | | | | | |
| 10.000 | .83 | 60 | 8.450 | 1.550 | 15.5% | 11 | .910 | .896 | .879 | .858 | .832 | .801 | .764 | .718 | .664 | .600 |
| | | | | | | | .528 | | | | | | | | | |
| 10.000 | .86 | 72 | 8.435 | 1.565 | 15.7% | 11 | .910 | .896 | .878 | .857 | .831 | .800 | .762 | .716 | .662 | .598 |
| | | | | | | | .524 | | | | | | | | | |
| 10.000 | .88 | 84 | 8.425 | 1.575 | 15.8% | 11 | .910 | .896 | .878 | .857 | .831 | .800 | .761 | .715 | .660 | .596 |
| | | | | | | | .523 | | | | | | | | | |
| 10.000 | .23 | 13 | 9.681 | .319 | 3.2% | 12 | .915 | .903 | .891 | .873 | .860 | .840 | .818 | .797 | .768 | .736 |
| | | | | | | | .689 | .590 | | | | | | | | |
| 10.000 | .62 | 26 | 9.058 | .942 | 9.4% | 12 | .912 | .899 | .883 | .863 | .840 | .812 | .778 | .737 | .687 | .627 |

## ERLANG-ENGSET: FOR SMOOTH ATTEMPTING TRAFFIC AND FEW PERSONS DOING THE CALLING (FEW MEANS RATIO OF CALLERS TO CIRCUITS IS LESS THAN 8 - GREATER THAN 8 ASSUME TRAFFIC RANDOM)

| HOURS FIRST ATTEMPT | FIRST ATTEMPT VMR | NUMBER SOURCES | HOURS CON-NECT | HOURS OVFL | PERCENT BLOCKED (GRADE) | LINE | HOURS CONNECTED TRAFFIC ON LINES 1 | 2 | 3 | 4 | 5 | 6 | 7 | 8 | 9 | 10 |
|---|---|---|---|---|---|---|---|---|---|---|---|---|---|---|---|---|
| | | | | | | | .554 | .466 | | | | | | | | |
| 10.000 | .74 | 39 | 8.956 | 1.044 | 10.4% | 12 | .911 | .897 | .880 | .860 | .835 | .805 | .769 | .725 | .674 | .609 |
| | | | | | | | .538 | .453 | | | | | | | | |
| 10.000 | .81 | 52 | 8.913 | 1.087 | 10.9% | 12 | .911 | .896 | .879 | .858 | .833 | .802 | .765 | .720 | .666 | .603 |
| | | | | | | | .529 | .449 | | | | | | | | |
| 10.000 | .85 | 65 | 8.888 | 1.112 | 11.1% | 12 | .910 | .896 | .879 | .857 | .832 | .801 | .763 | .717 | .664 | .599 |
| | | | | | | | .526 | .445 | | | | | | | | |
| 10.000 | .87 | 78 | 8.873 | 1.127 | 11.3% | 12 | .910 | .896 | .878 | .857 | .831 | .799 | .762 | .715 | .661 | .597 |
| | | | | | | | .524 | .443 | | | | | | | | |
| 10.000 | .89 | 91 | 8.862 | 1.138 | 11.4% | 12 | .910 | .895 | .878 | .856 | .831 | .799 | .760 | .714 | .660 | .595 |
| | | | | | | | .522 | .442 | | | | | | | | |
| 10.000 | .29 | 14 | 9.886 | .114 | 1.1% | 13 | .914 | .903 | .888 | .874 | .855 | .834 | .812 | .782 | .752 | .709 |
| | | | | | | | .648 | .551 | .363 | | | | | | | |
| 10.000 | .64 | 28 | 9.402 | .598 | 6.0% | 13 | .912 | .898 | .882 | .862 | .840 | .810 | .776 | .735 | .684 | .623 |
| | | | | | | | .549 | .464 | .367 | | | | | | | |
| 10.000 | .76 | 42 | 9.305 | .695 | 6.9% | 13 | .911 | .897 | .880 | .860 | .834 | .805 | .768 | .724 | .672 | .607 |
| | | | | | | | .535 | .452 | .362 | | | | | | | |
| 10.000 | .82 | 56 | 9.264 | .736 | 7.4% | 13 | .910 | .896 | .879 | .858 | .832 | .802 | .764 | .719 | .665 | .601 |
| | | | | | | | .529 | .447 | .360 | | | | | | | |
| 10.000 | .86 | 70 | 9.240 | .760 | 7.6% | 13 | .910 | .896 | .878 | .857 | .832 | .800 | .763 | .717 | .662 | .598 |
| | | | | | | | .525 | .444 | .359 | | | | | | | |
| 10.000 | .88 | 84 | 9.225 | .775 | 7.7% | 13 | .910 | .896 | .878 | .857 | .831 | .800 | .761 | .715 | .660 | .596 |
| | | | | | | | .523 | .442 | .358 | | | | | | | |
| 10.000 | .90 | 98 | 9.215 | .785 | 7.9% | 13 | .910 | .895 | .878 | .856 | .830 | .798 | .760 | .714 | .659 | .594 |
| | | | | | | | .521 | .441 | .357 | | | | | | | |
| 11.000 | .45 | 20 | 7.679 | 3.321 | 30.2% | 9 | .920 | .909 | .897 | .882 | .864 | .842 | .820 | .789 | .756 | |
| 11.000 | .63 | 30 | 7.604 | 3.396 | 30.9% | 9 | .919 | .908 | .894 | .878 | .858 | .834 | .807 | .773 | .734 | |
| 11.000 | .73 | 40 | 7.572 | 3.428 | 31.2% | 9 | .918 | .907 | .893 | .876 | .855 | .831 | .802 | .767 | .725 | |
| 11.000 | .78 | 50 | 7.555 | 3.445 | 31.3% | 9 | .918 | .906 | .892 | .874 | .854 | .829 | .799 | .763 | .720 | |
| 11.000 | .82 | 60 | 7.544 | 3.456 | 31.4% | 9 | .918 | .906 | .891 | .873 | .853 | .828 | .797 | .761 | .717 | |
| 11.000 | .84 | 70 | 7.536 | 3.464 | 31.5% | 9 | .918 | .906 | .891 | .873 | .852 | .827 | .796 | .759 | .715 | |
| 11.000 | .50 | 22 | 8.360 | 2.640 | 24.0% | 10 | .920 | .909 | .896 | .881 | .862 | .841 | .814 | .785 | .748 | .704 |
| 11.000 | .67 | 33 | 8.272 | 2.728 | 24.8% | 10 | .919 | .907 | .893 | .876 | .857 | .833 | .806 | .770 | .729 | .680 |
| 11.000 | .75 | 44 | 8.236 | 2.764 | 25.1% | 10 | .918 | .906 | .892 | .875 | .854 | .831 | .800 | .764 | .722 | .672 |
| 11.000 | .80 | 55 | 8.215 | 2.785 | 25.3% | 10 | .918 | .906 | .891 | .874 | .853 | .828 | .798 | .762 | .717 | .667 |
| 11.000 | .83 | 66 | 8.203 | 2.797 | 25.4% | 10 | .918 | .905 | .891 | .874 | .852 | .827 | .796 | .760 | .716 | .663 |
| 11.000 | .86 | 77 | 8.194 | 2.806 | 25.5% | 10 | .918 | .905 | .891 | .873 | .852 | .826 | .795 | .758 | .714 | .662 |
| 11.000 | .54 | 24 | 8.975 | 2.025 | 18.4% | 11 | .920 | .909 | .895 | .880 | .861 | .839 | .813 | .781 | .743 | .696 |
| | | | | | | | .640 | | | | | | | | | |
| 11.000 | .69 | 36 | 8.877 | 2.123 | 19.3% | 11 | .919 | .907 | .892 | .876 | .856 | .833 | .802 | .769 | .728 | .676 |

ERLANG-ENGSET: FOR SMOOTH ATTEMPTING TRAFFIC AND FEW PERSONS DOING THE CALLING (FEW MEANS RATIO OF CALLERS TO CIRCUITS IS LESS THAN 8 - GREATER THAN 8 ASSUME TRAFFIC RANDOM)

| HOURS FIRST ATTEMPT | FIRST ATTEMPT VMR | NUMBER SOURCES | HOURS CON-NECT | HOURS OVFL | PERCENT BLOCKED (GRADE) | LINE | HOURS CONNECTED TRAFFIC ON LINES 1 | 2 | 3 | 4 | 5 | 6 | 7 | 8 | 9 | 10 |
|---|---|---|---|---|---|---|---|---|---|---|---|---|---|---|---|---|
| | | | | | | | .618 | | | | | | | | | |
| 11.000 | .77 | 48 | 8.836 | 2.164 | 19.7% | 11 | .918 | .906 | .892 | .874 | .855 | .829 | .799 | .763 | .721 | .669 |
| | | | | | | | .609 | | | | | | | | | |
| 11.000 | .82 | 60 | 8.813 | 2.187 | 19.9% | 11 | .918 | .906 | .891 | .873 | .853 | .828 | .797 | .761 | .717 | .665 |
| | | | | | | | .604 | | | | | | | | | |
| 11.000 | .85 | 72 | 8.799 | 2.201 | 20.0% | 11 | .918 | .905 | .891 | .873 | .852 | .826 | .795 | .759 | .715 | .662 |
| | | | | | | | .601 | | | | | | | | | |
| 11.000 | .87 | 84 | 8.788 | 2.212 | 20.1% | 11 | .918 | .905 | .890 | .873 | .852 | .826 | .795 | .757 | .713 | .661 |
| | | | | | | | .599 | | | | | | | | | |
| 11.000 | .15 | 13 | 10.247 | .753 | 6.8% | 12 | .922 | .912 | .903 | .889 | .876 | .862 | .846 | .834 | .816 | .803 |
| | | | | | | | .792 | .791 | | | | | | | | |
| 11.000 | .58 | 26 | 9.516 | 1.484 | 13.5% | 12 | .919 | .908 | .895 | .878 | .861 | .837 | .811 | .778 | .738 | .692 |
| | | | | | | | .634 | .565 | | | | | | | | |
| 11.000 | .72 | 39 | 9.408 | 1.592 | 14.5% | 12 | .918 | .907 | .892 | .876 | .855 | .831 | .803 | .767 | .725 | .676 |
| | | | | | | | .615 | .544 | | | | | | | | |
| 11.000 | .79 | 52 | 9.364 | 1.636 | 14.9% | 12 | .918 | .906 | .892 | .874 | .853 | .829 | .798 | .763 | .719 | .667 |
| | | | | | | | .608 | .537 | | | | | | | | |
| 11.000 | .83 | 65 | 9.339 | 1.661 | 15.1% | 12 | .918 | .906 | .891 | .874 | .853 | .827 | .796 | .760 | .716 | .664 |
| | | | | | | | .602 | .534 | | | | | | | | |
| 11.000 | .86 | 78 | 9.324 | 1.676 | 15.2% | 12 | .918 | .905 | .891 | .873 | .852 | .826 | .795 | .758 | .714 | .662 |
| | | | | | | | .600 | .530 | | | | | | | | |
| 11.000 | .88 | 91 | 9.312 | 1.688 | 15.3% | 12 | .917 | .905 | .890 | .872 | .851 | .825 | .794 | .757 | .712 | .660 |
| | | | | | | | .598 | .529 | | | | | | | | |
| 11.000 | .21 | 14 | 10.652 | .348 | 3.2% | 13 | .922 | .912 | .901 | .887 | .875 | .857 | .839 | .822 | .800 | .774 |
| | | | | | | | .748 | .703 | .614 | | | | | | | |
| 11.000 | .61 | 28 | 9.967 | 1.033 | 9.4% | 13 | .919 | .908 | .894 | .878 | .859 | .836 | .807 | .776 | .736 | .686 |
| | | | | | | | .631 | .560 | .476 | | | | | | | |
| 11.000 | .74 | 42 | 9.859 | 1.141 | 10.4% | 13 | .918 | .906 | .892 | .875 | .855 | .831 | .801 | .765 | .724 | .672 |
| | | | | | | | .613 | .542 | .464 | | | | | | | |
| 11.000 | .80 | 56 | 9.813 | 1.187 | 10.8% | 13 | .918 | .906 | .891 | .874 | .853 | .829 | .797 | .762 | .718 | .666 |
| | | | | | | | .604 | .537 | .450 | | | | | | | |
| 11.000 | .84 | 70 | 9.787 | 1.213 | 11.0% | 13 | .918 | .906 | .891 | .873 | .852 | .827 | .796 | .759 | .715 | .662 |
| | | | | | | | .602 | .532 | .455 | | | | | | | |
| 11.000 | .87 | 84 | 9.771 | 1.229 | 11.2% | 13 | .918 | .905 | .890 | .873 | .852 | .826 | .795 | .757 | .713 | .661 |
| | | | | | | | .599 | .530 | .453 | | | | | | | |
| 11.000 | .89 | 98 | 9.759 | 1.241 | 11.3% | 13 | .917 | .905 | .890 | .872 | .851 | .825 | .794 | .756 | .712 | .659 |
| | | | | | | | .598 | .528 | .451 | | | | | | | |
| 11.000 | .27 | 15 | 10.873 | .127 | 1.2% | 14 | .921 | .911 | .900 | .886 | .872 | .852 | .834 | .814 | .785 | .759 |
| | | | | | | | .718 | .661 | .570 | .388 | | | | | | |
| 11.000 | .63 | 30 | 10.326 | .674 | 6.1% | 14 | .919 | .908 | .894 | .878 | .858 | .834 | .807 | .773 | .734 | .684 |

**ERLANG-ENGSET: FOR SMOOTH ATTEMPTING TRAFFIC AND FEW PERSONS DOING THE CALLING (FEW MEANS RATIO OF CALLERS TO CIRCUITS IS LESS THAN 8 - GREATER THAN 8 ASSUME TRAFFIC RANDOM)**

| HOURS FIRST ATTEMPT | FIRST ATTEMPT VMR | NUMBER SOURCES | HOURS CON-NECT | HOURS OVFL | PERCENT BLOCKED (GRADE) | LINE | HOURS CONNECTED TRAFFIC ON LINES 1 | 2 | 3 | 4 | 5 | 6 | 7 | 8 | 9 | 10 |
|---|---|---|---|---|---|---|---|---|---|---|---|---|---|---|---|---|
| | | | | | | | .626 | .555 | .475 | .382 | | | | | | |
| 11.000 | .76 | 45 | 10.222 | .778 | 7.1% | 14 | .918 | .906 | .892 | .875 | .854 | .830 | .800 | .764 | .723 | .670 |
| | | | | | | | .611 | .540 | .463 | .376 | | | | | | |
| 11.000 | .82 | 60 | 10.178 | .822 | 7.5% | 14 | .918 | .906 | .891 | .873 | .853 | .828 | .797 | .761 | .717 | .665 |
| | | | | | | | .604 | .534 | .457 | .373 | | | | | | |
| 11.000 | .85 | 75 | 10.152 | .848 | 7.7% | 14 | .918 | .905 | .891 | .873 | .852 | .826 | .796 | .759 | .714 | .662 |
| | | | | | | | .600 | .531 | .454 | .372 | | | | | | |
| 11.000 | .88 | 90 | 10.136 | .864 | 7.9% | 14 | .917 | .905 | .890 | .873 | .851 | .825 | .794 | .757 | .713 | .660 |
| | | | | | | | .598 | .529 | .452 | .371 | | | | | | |
| 11.000 | .90 | 105 | 10.125 | .875 | 8.0% | 14 | .917 | .905 | .890 | .872 | .851 | .825 | .793 | .756 | .712 | .658 |
| | | | | | | | .597 | .527 | .451 | .370 | | | | | | |
| 12.000 | .40 | 20 | 7.843 | 4.157 | 34.6% | 9 | .927 | .917 | .907 | .894 | .878 | .864 | .843 | .820 | .793 | |
| 12.000 | .45 | 22 | 8.574 | 3.426 | 28.5% | 10 | .926 | .916 | .905 | .894 | .877 | .861 | .840 | .816 | .787 | .752 |
| 12.000 | .64 | 33 | 8.494 | 3.506 | 29.2% | 10 | .925 | .915 | .904 | .889 | .873 | .854 | .831 | .803 | .770 | .730 |
| 12.000 | .73 | 44 | 8.459 | 3.541 | 29.5% | 10 | .925 | .914 | .902 | .888 | .872 | .851 | .826 | .798 | .762 | .722 |
| 12.000 | .78 | 55 | 8.442 | 3.558 | 29.7% | 10 | .924 | .914 | .902 | .887 | .870 | .849 | .824 | .795 | .759 | .717 |
| 12.000 | .82 | 66 | 8.430 | 3.570 | 29.8% | 10 | .924 | .914 | .901 | .886 | .869 | .848 | .823 | .793 | .757 | .714 |
| 12.000 | .84 | 77 | 8.422 | 3.578 | 29.8% | 10 | .924 | .914 | .901 | .886 | .868 | .847 | .822 | .792 | .755 | .712 |
| 12.000 | .50 | 24 | 9.254 | 2.746 | 22.9% | 11 | .926 | .916 | .905 | .891 | .878 | .859 | .837 | .813 | .782 | .744 |
| | | | | | | | .702 | | | | | | | | | |
| 12.000 | .67 | 36 | 9.162 | 2.838 | 23.6% | 11 | .925 | .915 | .903 | .889 | .872 | .853 | .830 | .801 | .767 | .728 |
| | | | | | | | .680 | | | | | | | | | |
| 12.000 | .75 | 48 | 9.123 | 2.877 | 24.0% | 11 | .924 | .914 | .902 | .888 | .871 | .850 | .825 | .797 | .760 | .720 |
| | | | | | | | .671 | | | | | | | | | |
| 12.000 | .80 | 60 | 9.102 | 2.898 | 24.2% | 11 | .924 | .914 | .902 | .887 | .869 | .848 | .824 | .794 | .758 | .715 |
| | | | | | | | .667 | | | | | | | | | |
| 12.000 | .83 | 72 | 9.088 | 2.912 | 24.3% | 11 | .924 | .914 | .901 | .886 | .869 | .848 | .822 | .792 | .756 | .713 |
| | | | | | | | .663 | | | | | | | | | |
| 12.000 | .86 | 84 | 9.078 | 2.922 | 24.4% | 11 | .924 | .913 | .901 | .886 | .868 | .847 | .821 | .791 | .755 | .711 |
| | | | | | | | .661 | | | | | | | | | |
| 12.000 | .54 | 26 | 9.873 | 2.127 | 17.7% | 12 | .926 | .916 | .905 | .891 | .875 | .859 | .835 | .808 | .778 | .741 |
| | | | | | | | .698 | .642 | | | | | | | | |
| 12.000 | .69 | 39 | 9.769 | 2.231 | 18.6% | 12 | .925 | .915 | .902 | .889 | .871 | .852 | .828 | .800 | .765 | .725 |
| | | | | | | | .677 | .620 | | | | | | | | |
| 12.000 | .77 | 52 | 9.725 | 2.275 | 19.0% | 12 | .924 | .914 | .902 | .887 | .871 | .849 | .825 | .795 | .760 | .718 |
| | | | | | | | .670 | .610 | | | | | | | | |
| 12.000 | .82 | 65 | 9.701 | 2.299 | 19.2% | 12 | .924 | .914 | .901 | .886 | .869 | .848 | .823 | .793 | .758 | .715 |
| | | | | | | | .664 | .606 | | | | | | | | |
| 12.000 | .85 | 78 | 9.686 | 2.314 | 19.3% | 12 | .924 | .913 | .901 | .886 | .869 | .847 | .822 | .792 | .755 | .712 |
| | | | | | | | .662 | .603 | | | | | | | | |

ERLANG-ENGSET: FOR SMOOTH ATTEMPTING TRAFFIC AND FEW PERSONS DOING THE CALLING (FEW MEANS RATIO OF CALLERS TO CIRCUITS IS LESS THAN 8 - GREATER THAN 8 ASSUME TRAFFIC RANDOM)

| HOURS FIRST ATTEMPT | FIRST ATTEMPT VMR | NUMBER SOURCES | HOURS CON-NECT | HOURS OVFL | PERCENT BLOCKED (GRADE) | LINE | HOURS CONNECTED TRAFFIC ON LINES 1 | 2 | 3 | 4 | 5 | 6 | 7 | 8 | 9 | 10 |
|---|---|---|---|---|---|---|---|---|---|---|---|---|---|---|---|---|
| 12.000 | .87 | 91 | 9.676 | 2.324 | 19.4% | 12 | .924 | .913 | .901 | .886 | .868 | .847 | .821 | .790 | .754 | .711 |
| | | | | | | | .660 | .601 | | | | | | | | |
| 12.000 | .57 | 28 | 10.419 | 1.581 | 13.2% | 13 | .926 | .916 | .904 | .891 | .875 | .856 | .833 | .807 | .775 | .739 |
| | | | | | | | .691 | .636 | .570 | | | | | | | |
| 12.000 | .71 | 42 | 10.306 | 1.694 | 14.1% | 13 | .925 | .914 | .903 | .888 | .871 | .852 | .826 | .799 | .765 | .722 |
| | | | | | | | .675 | .617 | .550 | | | | | | | |
| 12.000 | .79 | 56 | 10.259 | 1.741 | 14.5% | 13 | .924 | .914 | .902 | .887 | .869 | .849 | .825 | .795 | .759 | .717 |
| | | | | | | | .667 | .608 | .543 | | | | | | | |
| 12.000 | .83 | 70 | 10.233 | 1.767 | 14.7% | 13 | .924 | .914 | .901 | .886 | .869 | .848 | .823 | .792 | .756 | .714 |
| | | | | | | | .663 | .604 | .539 | | | | | | | |
| 12.000 | .86 | 84 | 10.216 | 1.784 | 14.9% | 13 | .924 | .013 | .901 | .886 | .868 | .847 | .821 | .791 | .755 | .711 |
| | | | | | | | .661 | .603 | .535 | | | | | | | |
| 12.000 | .88 | 98 | 10.204 | 1.796 | 15.0% | 13 | .924 | .913 | .901 | .886 | .868 | .846 | .821 | .790 | .753 | .710 |
| | | | | | | | .659 | .601 | .533 | | | | | | | |
| 12.000 | .20 | 15 | 11.625 | .375 | 3.1% | 14 | .927 | .920 | .909 | .897 | .887 | .873 | .856 | .838 | .826 | .802 |
| | | | | | | | .781 | .755 | .719 | .633 | | | | | | |
| 12.000 | .60 | 30 | 10.881 | 1.119 | 9.3% | 14 | .925 | .916 | .904 | .889 | .875 | .855 | .832 | .806 | .773 | .734 |
| | | | | | | | .688 | .632 | .566 | .487 | | | | | | |
| 12.000 | .73 | 45 | 10.764 | 1.236 | 10.3% | 14 | .925 | .915 | .902 | .888 | .870 | .851 | .825 | .798 | .762 | .722 |
| | | | | | | | .672 | .615 | .547 | .472 | | | | | | |
| 12.000 | .80 | 60 | 10.716 | 1.284 | 10.7% | 14 | .924 | .914 | .902 | .887 | .869 | .848 | .824 | .794 | .758 | .715 |
| | | | | | | | .667 | .608 | .541 | .466 | | | | | | |
| 12.000 | .84 | 75 | 10.689 | 1.311 | 10.9% | 14 | .924 | .914 | .901 | .886 | .869 | .847 | .022 | .792 | .755 | 713 |
| | | | | | | | .663 | .603 | .537 | .403 | | | | | | |
| 12.000 | .87 | 90 | 10.072 | 1.328 | 11.1% | 14 | .924 | .913 | .901 | .886 | .868 | .847 | .821 | .791 | .754 | .711 |
| | | | | | | | .660 | .601 | .535 | .461 | | | | | | |
| 12.000 | .89 | 105 | 10.660 | 1.340 | 11.2% | 14 | .924 | .913 | .900 | .886 | .868 | .846 | .820 | .790 | .753 | .709 |
| | | | | | | | .659 | .599 | .533 | .460 | | | | | | |
| 12.000 | .25 | 16 | 11.860 | .140 | 1.2% | 15 | .927 | .918 | .909 | .896 | .884 | .869 | .854 | .836 | .815 | 788 |
| | | | | | | | .763 | .724 | .677 | .508 | .412 | | | | | |
| 12.000 | .03 | 32 | 11.251 | .749 | 6.2% | 15 | .925 | .915 | .904 | .889 | .874 | .854 | .831 | .803 | 772 | .732 |
| | | | | | | | .685 | .627 | .502 | .484 | .394 | | | | | |
| 12.000 | 75 | 40 | 11.141 | .859 | 7.2% | 15 | .924 | .914 | .902 | .888 | .871 | .850 | .825 | .797 | .760 | .720 |
| | | | | | | | .671 | .612 | .547 | .471 | .388 | | | | | |
| 12.000 | .81 | 64 | 11.093 | .907 | 7.6% | 15 | .924 | .914 | .901 | .886 | .869 | 848 | .820 | .793 | .758 | .715 |
| | | | | | | | .665 | .606 | .540 | .465 | .385 | | | | | |
| 12.000 | .85 | 80 | 11.066 | .934 | 7.8% | 15 | .924 | .913 | .901 | .800 | .868 | .847 | .821 | .792 | .755 | .712 |
| | | | | | | | .662 | .603 | .536 | .462 | .384 | | | | | |
| 12.000 | .88 | 96 | 11.049 | .951 | 7.9% | 15 | .924 | .913 | .901 | .886 | .868 | .846 | .821 | .790 | .754 | .710 |
| | | | | | | | .659 | .600 | .534 | .461 | .383 | | | | | |

## ERLANG-ENGSET: FOR SMOOTH ATTEMPTING TRAFFIC AND FEW PERSONS DOING THE CALLING (FEW MEANS RATIO OF CALLERS TO CIRCUITS IS LESS THAN 8 - GREATER THAN 8 ASSUME TRAFFIC RANDOM)

| HOURS FIRST ATTEMPT | FIRST ATTEMPT VMR | NUMBER SOURCES | HOURS CONNECT | HOURS OVFL | PERCENT BLOCKED (GRADE) | LINE | HOURS CONNECTED TRAFFIC ON LINES | | | | | | | | | |
|---|---|---|---|---|---|---|---|---|---|---|---|---|---|---|---|---|
| | | | | | | | 1 | 2 | 3 | 4 | 5 | 6 | 7 | 8 | 9 | 10 |
| 12.000 | .89 | 112 | 11.037 | .963 | 8.0% | 15 | .924 | .913 | .901 | .885 | .867 | .846 | .820 | .789 | .752 | .709 |
| | | | | | | | .658 | .599 | .532 | .459 | .382 | | | | | |
| 13.000 | .41 | 22 | 8.742 | 4.258 | 32.8% | 10 | .932 | .923 | .914 | .903 | .891 | .876 | .859 | .838 | .816 | .791 |
| 13.000 | .61 | 33 | 8.671 | 4.329 | 33.3% | 10 | .931 | .922 | .912 | .900 | .886 | .870 | .851 | .829 | .802 | .769 |
| 13.000 | .70 | 44 | 8.639 | 4.361 | 33.5% | 10 | .930 | .921 | .911 | .899 | .885 | .867 | .847 | .823 | .796 | .761 |
| 13.000 | .76 | 55 | 8.623 | 4.377 | 33.7% | 10 | .930 | .921 | .910 | .897 | .884 | .866 | .846 | .820 | .793 | .757 |
| 13.000 | .80 | 66 | 8.613 | 4.387 | 33.7% | 10 | .930 | .920 | .910 | .898 | .882 | .865 | .844 | .819 | .790 | .755 |
| 13.000 | .83 | 77 | 8.605 | 4.395 | 33.8% | 10 | .929 | .920 | .910 | .897 | .882 | .864 | .843 | .818 | .788 | .753 |
| 13.000 | .46 | 24 | 9.474 | 3.526 | 27.1% | 11 | .931 | .923 | .913 | .903 | .889 | .874 | .856 | .838 | .812 | .784 |
| | | | | | | | .750 | | | | | | | | | |
| 13.000 | .64 | 36 | 9.387 | 3.613 | 27.8% | 11 | .930 | .922 | .912 | .899 | .886 | .870 | .849 | .827 | .799 | .766 |
| | | | | | | | .728 | | | | | | | | | |
| 13.000 | .73 | 48 | 9.352 | 3.648 | 28.1% | 11 | .930 | .921 | .910 | .898 | .883 | .866 | .846 | .823 | .794 | .759 |
| | | | | | | | .721 | | | | | | | | | |
| 13.000 | .78 | 60 | 9.332 | 3.668 | 28.2% | 11 | .930 | .920 | .910 | .898 | .882 | .866 | .845 | .819 | .791 | .757 |
| | | | | | | | .714 | | | | | | | | | |
| 13.000 | .82 | 72 | 9.321 | 3.679 | 28.3% | 11 | .929 | .920 | .910 | .897 | .882 | .865 | .844 | .818 | .790 | .754 |
| | | | | | | | .712 | | | | | | | | | |
| 13.000 | .85 | 84 | 9.312 | 3.688 | 28.4% | 11 | .929 | .920 | .909 | .896 | .882 | .864 | .843 | .818 | .788 | .752 |
| | | | | | | | .710 | | | | | | | | | |
| 13.000 | .50 | 26 | 10.155 | 2.845 | 21.9% | 12 | .931 | .922 | .913 | .902 | .889 | .871 | .857 | .833 | .808 | .781 |
| | | | | | | | .743 | .704 | | | | | | | | |
| 13.000 | .67 | 39 | 10.055 | 2.945 | 22.7% | 12 | .930 | .921 | .911 | .899 | .884 | .869 | .848 | .826 | .797 | .765 |
| | | | | | | | .725 | .679 | | | | | | | | |
| 13.000 | .75 | 52 | 10.014 | 2.986 | 23.0% | 12 | .930 | .921 | .910 | .898 | .883 | .866 | .846 | .821 | .793 | .757 |
| | | | | | | | .719 | .669 | | | | | | | | |
| 13.000 | .80 | 65 | 9.993 | 3.007 | 23.1% | 12 | .930 | .921 | .910 | .897 | .882 | .865 | .844 | .819 | .790 | .755 |
| | | | | | | | .714 | .665 | | | | | | | | |
| 13.000 | .83 | 78 | 9.978 | 3.022 | 23.2% | 12 | .929 | .920 | .910 | .897 | .882 | .864 | .843 | .818 | .789 | .753 |
| | | | | | | | .711 | .662 | | | | | | | | |
| 13.000 | .86 | 91 | 9.968 | 3.032 | 23.3% | 12 | .929 | .920 | .909 | .897 | .881 | .864 | .842 | .817 | .788 | .751 |
| | | | | | | | .709 | .660 | | | | | | | | |
| 13.000 | .54 | 28 | 10.775 | 2.225 | 17.1% | 13 | .931 | .922 | .912 | .901 | .888 | .872 | .854 | .831 | .806 | .776 |
| | | | | | | | .741 | .697 | .643 | | | | | | | |
| 13.000 | .69 | 42 | 10.665 | 2.335 | 18.0% | 13 | .930 | .921 | .911 | .898 | .885 | .868 | .848 | .824 | .796 | .763 |
| | | | | | | | .724 | .675 | .622 | | | | | | | |
| 13.000 | .77 | 56 | 10.619 | 2.381 | 18.3% | 13 | .930 | .921 | .910 | .898 | .883 | .866 | .846 | .820 | .791 | .758 |
| | | | | | | | .717 | .669 | .612 | | | | | | | |
| 13.000 | .81 | 70 | 10.594 | 2.406 | 18.5% | 13 | .929 | .920 | .909 | .897 | .882 | .865 | .844 | .819 | .789 | .753 |
| | | | | | | | .713 | .663 | .608 | | | | | | | |

ERLANG-ENGSET: FOR SMOOTH ATTEMPTING TRAFFIC AND FEW PERSONS DOING THE CALLING (FEW MEANS RATIO OF CALLERS TO CIRCUITS IS LESS THAN 8 - GREATER THAN 8 ASSUME TRAFFIC RANDOM)

| HOURS FIRST ATTEMPT | FIRST ATTEMPT VMR | NUMBER SOURCES | HOURS CON-NECT | HOURS OVFL | PERCENT BLOCKED (GRADE) | LINE | HOURS CONNECTED TRAFFIC ON LINES 1 | 2 | 3 | 4 | 5 | 6 | 7 | 8 | 9 | 10 |
|---|---|---|---|---|---|---|---|---|---|---|---|---|---|---|---|---|
| 13.000 | .85 | 84 | 10.578 | 2.422 | 18.6% | 13 | .929 | .920 | .909 | .896 | .882 | .864 | .843 | .818 | .788 | .752 |
| | | | | | | | .710 | .661 | .605 | | | | | | | |
| 13.000 | .87 | 98 | 10.567 | 2.433 | 18.7% | 13 | .929 | .920 | .909 | .897 | .881 | .863 | .842 | .817 | .786 | .751 |
| | | | | | | | .709 | .659 | .603 | | | | | | | |
| 13.000 | .57 | 30 | 11.324 | 1.676 | 12.9% | 14 | .931 | .922 | .913 | .900 | .887 | .871 | .852 | .829 | .806 | .773 |
| | | | | | | | .734 | .691 | .641 | .573 | | | | | | |
| 13.000 | .71 | 45 | 11.206 | 1.794 | 13.8% | 14 | .930 | .921 | .911 | .898 | .885 | .866 | .847 | .825 | .794 | .762 |
| | | | | | | | .722 | .673 | .619 | .555 | | | | | | |
| 13.000 | .78 | 60 | 11.156 | 1.844 | 14.2% | 14 | .930 | .920 | .910 | .898 | .882 | .866 | .845 | .819 | .791 | .757 |
| | | | | | | | .714 | .667 | .611 | .546 | | | | | | |
| 13.000 | .83 | 75 | 11.130 | 1.870 | 14.4% | 14 | .929 | .920 | .910 | .897 | .882 | .864 | .844 | .818 | .789 | .753 |
| | | | | | | | .712 | .663 | .606 | .543 | | | | | | |
| 13.000 | .86 | 90 | 11.112 | 1.888 | 14.5% | 14 | .929 | .920 | .909 | .897 | .882 | .864 | .842 | .817 | .787 | .752 |
| | | | | | | | 709 | .661 | .604 | .540 | | | | | | |
| 13.000 | .88 | 105 | 11.100 | 1.900 | 14.6% | 14 | .929 | .920 | .909 | .896 | .881 | .863 | .842 | .816 | .786 | .750 |
| | | | | | | | .708 | .659 | .602 | .538 | | | | | | |
| 13.000 | .19 | 16 | 12.601 | .399 | 3.1% | 15 | .933 | .924 | .918 | .905 | .898 | .882 | .872 | .856 | .844 | .826 |
| | | | | | | | .806 | .787 | .766 | .730 | .654 | | | | | |
| 13.000 | .59 | 32 | 11.796 | 1.204 | 9.3% | 15 | .931 | .922 | .912 | .900 | .887 | .870 | .851 | .829 | .802 | .771 |
| | | | | | | | .733 | .689 | .634 | .569 | .496 | | | | | |
| 13.000 | .73 | 48 | 11.673 | 1.327 | 10.2% | 15 | .930 | .921 | .910 | .898 | .883 | .866 | .846 | .823 | .794 | .759 |
| | | | | | | | .721 | .671 | .618 | .553 | .479 | | | | | |
| 13.000 | .80 | 64 | 11.622 | 1.378 | 10.6% | 15 | .930 | .921 | .910 | .897 | .883 | .865 | .844 | .820 | .790 | .754 |
| | | | | | | | .715 | .666 | .610 | .545 | .474 | | | | | |
| 13.000 | .84 | 80 | 11.594 | 1.406 | 10.8% | 15 | .929 | .920 | .909 | .897 | .882 | .864 | .843 | .818 | .788 | .753 |
| | | | | | | | .711 | .662 | .605 | .541 | .470 | | | | | |
| 13.000 | .86 | 96 | 11.576 | 1.424 | 11.0% | 15 | .929 | .920 | .909 | .896 | .882 | .863 | .842 | .817 | .786 | .751 |
| | | | | | | | .709 | .660 | .603 | .539 | .468 | | | | | |
| 13.000 | .88 | 112 | 11.563 | 1.437 | 11.1% | 15 | .929 | .920 | .909 | .896 | .881 | .863 | .842 | .816 | .785 | .750 |
| | | | | | | | .707 | .658 | .601 | .537 | .467 | | | | | |
| 13.000 | .24 | 17 | 12.846 | .154 | 1.2% | 16 | .933 | .925 | .916 | .906 | .892 | .882 | .871 | .851 | .835 | .817 |
| | | | | | | | .793 | .768 | .735 | .687 | .603 | .432 | | | | |
| 13.000 | .62 | 34 | 12.178 | .822 | 6.3% | 16 | .931 | .922 | .912 | .900 | .885 | .870 | .850 | .829 | .801 | .769 |
| | | | | | | | .730 | .685 | .631 | .566 | .493 | .407 | | | | |
| 13.000 | .75 | 51 | 12.060 | .940 | 7.2% | 16 | .930 | .921 | .910 | .808 | .884 | .867 | .846 | .822 | .792 | .759 |
| | | | | | | | .718 | .671 | .615 | .550 | .479 | .399 | | | | |
| 13.000 | .81 | 68 | 12.010 | .990 | 7.6% | 16 | .930 | .920 | .910 | .897 | .883 | .865 | .844 | .819 | .789 | .755 |
| | | | | | | | .712 | .666 | .608 | .545 | .473 | .396 | | | | |
| 13.000 | .85 | 85 | 11.981 | 1.019 | 7.8% | 16 | .929 | .920 | .910 | .897 | .882 | .864 | .843 | .818 | .788 | .752 |
| | | | | | | | .710 | .661 | .605 | .541 | .470 | .394 | | | | |

ERLANG-ENGSET: FOR SMOOTH ATTEMPTING TRAFFIC AND FEW PERSONS DOING THE CALLING (FEW MEANS RATIO OF CALLERS TO CIRCUITS IS LESS THAN 8 - GREATER THAN 8 ASSUME TRAFFIC RANDOM)

| HOURS FIRST ATTEMPT | FIRST ATTEMPT VMR | NUMBER SOURCES | HOURS CON-NECT | HOURS OVFL | PERCENT BLOCKED (GRADE) | LINE | HOURS CONNECTED TRAFFIC ON LINES 1 | 2 | 3 | 4 | 5 | 6 | 7 | 8 | 9 | 10 |
|---|---|---|---|---|---|---|---|---|---|---|---|---|---|---|---|---|
| 13.000 | .87 | 102 | 11.963 | 1.037 | 8.0% | 16 | .929 | .920 | .909 | .897 | .881 | .863 | .842 | .816 | .786 | .750 |
| | | | | | | | .708 | .659 | .602 | .538 | .468 | .393 | | | | |
| 13.000 | .89 | 119 | 11.951 | 1.049 | 8.1% | 16 | .929 | .920 | .909 | .896 | .881 | .863 | .842 | .816 | .785 | .749 |
| | | | | | | | .707 | .658 | .601 | .537 | .467 | .392 | | | | |
| 14.000 | .42 | 24 | 9.647 | 4.353 | 31.1% | 11 | .936 | .929 | .920 | .911 | .900 | .888 | .873 | .856 | .834 | .815 |
| | | | | | | | .785 | | | | | | | | | |
| 14.000 | .61 | 36 | 9.570 | 4.430 | 31.6% | 11 | .935 | .927 | .919 | .908 | .897 | .881 | .867 | .847 | .825 | .799 |
| | | | | | | | .766 | | | | | | | | | |
| 14.000 | .71 | 48 | 9.538 | 4.462 | 31.9% | 11 | .935 | .927 | .918 | .907 | .895 | .881 | .862 | .844 | .820 | .791 |
| | | | | | | | .759 | | | | | | | | | |
| 14.000 | .77 | 60 | 9.521 | 4.479 | 32.0% | 11 | .934 | .927 | .917 | .907 | .894 | .879 | .862 | .840 | .818 | .787 |
| | | | | | | | .756 | | | | | | | | | |
| 14.000 | .81 | 72 | 9.509 | 4.491 | 32.1% | 11 | .934 | .926 | .917 | .906 | .894 | .879 | .861 | .840 | .815 | .786 |
| | | | | | | | .752 | | | | | | | | | |
| 14.000 | .83 | 84 | 9.500 | 4.500 | 32.1% | 11 | .934 | .926 | .917 | .906 | .893 | .878 | .860 | .839 | .814 | .785 |
| | | | | | | | .749 | | | | | | | | | |
| 14.000 | .46 | 26 | 10.375 | 3.625 | 25.9% | 12 | .936 | .928 | .920 | .911 | .899 | .886 | .871 | .853 | .834 | .808 |
| | | | | | | | .781 | .748 | | | | | | | | |
| 14.000 | .64 | 39 | 10.286 | 3.714 | 26.5% | 12 | .935 | .927 | .919 | .907 | .896 | .882 | .866 | .846 | .823 | .795 |
| | | | | | | | .764 | .727 | | | | | | | | |
| 14.000 | .73 | 52 | 10.248 | 3.752 | 26.8% | 12 | .35 | .926 | .918 | .907 | .894 | .880 | .862 | .844 | .819 | .790 |
| | | | | | | | .756 | .717 | | | | | | | | |
| 14.000 | .78 | 65 | 10.228 | 3.772 | 26.9% | 12 | .934 | .926 | .917 | .907 | .893 | .879 | .862 | .840 | .817 | .787 |
| | | | | | | | .753 | .713 | | | | | | | | |
| 14.000 | .82 | 78 | 10.215 | 3.785 | 27.0% | 12 | .934 | .926 | .917 | .905 | .894 | .878 | .860 | .840 | .815 | .785 |
| | | | | | | | .751 | .710 | | | | | | | | |
| 14.000 | .85 | 91 | 10.206 | 3.794 | 27.1% | 12 | .934 | .926 | .917 | .906 | .893 | .878 | .860 | .839 | .813 | .784 |
| | | | | | | | .749 | .708 | | | | | | | | |
| 14.000 | .50 | 28 | 11.056 | 2.944 | 21.0% | 13 | .936 | .928 | .920 | .908 | .899 | .887 | .870 | .852 | .831 | .807 |
| | | | | | | | .776 | .743 | .700 | | | | | | | |
| 14.000 | .67 | 42 | 10.953 | 3.047 | 21.8% | 13 | .935 | .927 | .918 | .907 | .896 | .882 | .864 | .845 | .822 | .793 |
| | | | | | | | .762 | .724 | .678 | | | | | | | |
| 14.000 | .75 | 56 | 10.910 | 3.090 | 22.1% | 13 | .934 | .926 | .918 | .906 | .895 | .878 | .862 | .843 | .817 | .789 |
| | | | | | | | .756 | .716 | .668 | | | | | | | |
| 14.000 | .80 | 70 | 10.888 | 3.112 | 22.2% | 13 | .934 | .926 | .917 | .906 | .893 | .879 | .861 | .840 | .815 | .786 |
| | | | | | | | .752 | .712 | .665 | | | | | | | |
| 14.000 | .83 | 84 | 10.872 | 3.128 | 22.3% | 13 | .934 | .926 | .917 | .906 | .893 | .878 | .860 | .839 | .814 | .785 |
| | | | | | | | .749 | .709 | .662 | | | | | | | |
| 14.000 | .86 | 98 | 10.862 | 3.138 | 22.4% | 13 | .934 | .926 | .916 | .906 | .893 | .877 | .859 | .838 | .813 | .784 |
| | | | | | | | .748 | .708 | .660 | | | | | | | |

ERLANG-ENGSET: FOR SMOOTH ATTEMPTING TRAFFIC AND FEW PERSONS DOING THE CALLING (FEW MEANS RATIO OF CALLERS TO CIRCUITS IS LESS THAN 8 - GREATER THAN 8 ASSUME TRAFFIC RANDOM)

| HOURS FIRST ATTEMPT | FIRST ATTEMPT VMR | NUMBER SOURCES | HOURS CONNECT | HOURS OVFL | PERCENT BLOCKED (GRADE) | LINE | HOURS CONNECTED TRAFFIC ON LINES 1 | 2 | 3 | 4 | 5 | 6 | 7 | 8 | 9 | 10 |
|---|---|---|---|---|---|---|---|---|---|---|---|---|---|---|---|---|
| 14.000 | .53 | 30 | 11.679 | 2.321 | 16.6% | 14 | .935 | .927 | .920 | .908 | .899 | .883 | .868 | .851 | .830 | .803 |
| | | | | | | | .775 | .736 | .697 | .645 | | | | | | |
| 14.000 | .69 | 45 | 11.564 | 2.436 | 17.4% | 14 | .935 | .927 | .918 | .907 | .895 | .881 | .864 | .844 | .821 | .794 |
| | | | | | | | .760 | .721 | .676 | .622 | | | | | | |
| 14.000 | .77 | 60 | 11.517 | 2.483 | 17.7% | 14 | .934 | .927 | .917 | .907 | .894 | .879 | .862 | .840 | .818 | .787 |
| | | | | | | | .756 | .714 | .669 | .614 | | | | | | |
| 14.000 | .81 | 75 | 11.490 | 2.510 | 17.9% | 14 | .934 | .926 | .917 | .906 | .893 | .878 | .861 | .840 | .815 | .785 |
| | | | | | | | .752 | .711 | .663 | .609 | | | | | | |
| 14.000 | .84 | 90 | 11.474 | 2.526 | 18.0% | 14 | .934 | .926 | .917 | .905 | .893 | .878 | .860 | .839 | .814 | .784 |
| | | | | | | | .749 | .708 | .661 | .606 | | | | | | |
| 14.000 | .87 | 105 | 11.462 | 2.538 | 18.1% | 14 | .934 | .926 | .916 | .906 | .892 | .877 | .859 | .838 | .813 | .783 |
| | | | | | | | .748 | .706 | .660 | .604 | | | | | | |
| 14.000 | .56 | 32 | 12.235 | 1.765 | 12.6% | 15 | .935 | .927 | .919 | .909 | .897 | .884 | .868 | .849 | .828 | .801 |
| | | | | | | | .769 | .737 | .693 | .640 | .579 | | | | | |
| 14.000 | .71 | 48 | 12.110 | 1.890 | 13.5% | 15 | .935 | .927 | .918 | .907 | .895 | .881 | .862 | .844 | .820 | .791 |
| | | | | | | | .759 | .719 | .675 | .621 | .558 | | | | | |
| 14.000 | .78 | 64 | 12.059 | 1.941 | 13.9% | 15 | .934 | .926 | .917 | .907 | .894 | .879 | .861 | .840 | .817 | .786 |
| | | | | | | | .754 | .714 | .666 | .614 | .551 | | | | | |
| 14.000 | .83 | 80 | 12.031 | 1.969 | 14.1% | 15 | .934 | .926 | .917 | .906 | .893 | .879 | .860 | .839 | .815 | .785 |
| | | | | | | | .750 | .710 | .663 | .608 | .546 | | | | | |
| 14.000 | .85 | 96 | 12.012 | 1.988 | 14.2% | 15 | .934 | .926 | .917 | .906 | .893 | .877 | .860 | .838 | .813 | .784 |
| | | | | | | | .748 | .708 | .660 | .605 | .544 | | | | | |
| 14.000 | .88 | 112 | 12.000 | 2.000 | 14.3% | 15 | .934 | .926 | .916 | .905 | .892 | .877 | .859 | .838 | .813 | .783 |
| | | | | | | | .747 | .707 | .658 | .604 | .542 | | | | | |
| 14.000 | .18 | 17 | 13.579 | .421 | 3.0% | 16 | .937 | .931 | .923 | .913 | .904 | .894 | .884 | .870 | .859 | .843 |
| | | | | | | | .825 | .812 | .791 | .771 | .748 | .674 | | | | |
| 14.000 | .59 | 34 | 12.713 | 1.287 | 9.2% | 16 | .935 | .928 | .919 | .908 | .897 | .883 | .867 | .848 | .827 | .800 |
| | | | | | | | .768 | .732 | .688 | .636 | .575 | .503 | | | | |
| 14.000 | .73 | 51 | 12.585 | 1.415 | 10.1% | 16 | .935 | .927 | .918 | .907 | .895 | .880 | .863 | .841 | .820 | .791 |
| | | | | | | | .756 | .719 | .671 | .618 | .558 | .486 | | | | |
| 14.000 | .79 | 68 | 12.530 | 1.470 | 10.5% | 16 | .934 | .926 | .917 | .907 | .893 | .879 | .862 | .840 | .816 | .787 |
| | | | | | | | 753 | .711 | .666 | .612 | .549 | .480 | | | | |
| 14.000 | .84 | 85 | 12.501 | 1.409 | 10.7% | 16 | .934 | .926 | .917 | .906 | .893 | .878 | .860 | .839 | .814 | .785 |
| | | | | | | | .750 | 700 | .661 | .608 | .545 | .477 | | | | |
| 14.000 | .86 | 102 | 12.482 | 1.518 | 10.0% | 16 | .934 | .926 | .916 | .905 | .893 | .878 | .859 | .838 | .813 | .783 |
| | | | | | | | .748 | .707 | .659 | .605 | .543 | .475 | | | | |
| 14.000 | .88 | 119 | 12.469 | 1.531 | 10.9% | 16 | .934 | .926 | .916 | .905 | .892 | .877 | .859 | .837 | .812 | .782 |
| | | | | | | | .747 | .706 | .658 | .603 | .541 | .474 | | | | |
| 14.000 | 22 | 18 | 13.833 | .167 | 1.2% | 17 | .936 | .930 | .923 | .913 | .904 | .891 | .880 | .868 | .853 | .837 |
| | | | | | | | .818 | .798 | .772 | .740 | .698 | .617 | .454 | | | |

ERLANG-ENGSET: FOR SMOOTH ATTEMPTING TRAFFIC AND FEW PERSONS DOING THE CALLING (FEW MEANS RATIO OF CALLERS TO CIRCUITS IS LESS THAN 8 - GREATER THAN 8 ASSUME TRAFFIC RANDOM)

| HOURS FIRST ATTEMPT | FIRST ATTEMPT VMR | NUMBER SOURCES | HOURS CON-NECT | HOURS OVFL | PERCENT BLOCKED (GRADE) | LINE | HOURS CONNECTED TRAFFIC ON LINES 1 | 2 | 3 | 4 | 5 | 6 | 7 | 8 | 9 | 10 |
|---|---|---|---|---|---|---|---|---|---|---|---|---|---|---|---|---|
| 14.000 | .61 | 36 | 13.106 | .894 | 6.4% | 17 | .935 | .927 | .919 | .908 | .897 | .881 | .867 | .847 | .825 | .799 |
| | | | | | | | .766 | .730 | .684 | .634 | .569 | .501 | .418 | | | |
| 14.000 | .74 | 54 | 12.981 | 1.019 | 7.3% | 17 | .934 | .927 | .918 | .907 | .893 | .880 | .863 | .842 | .818 | .790 |
| | | | | | | | .755 | .718 | .670 | .617 | .554 | .486 | .409 | | | |
| 14.000 | .81 | 72 | 12.928 | 1.072 | 7.7% | 17 | .934 | .926 | .917 | .906 | .894 | .879 | .861 | .840 | .815 | .786 |
| | | | | | | | .752 | .711 | .665 | .610 | .548 | .479 | .406 | | | |
| 14.000 | .84 | 90 | 12.899 | 1.101 | 7.9% | 17 | .934 | .926 | .917 | .905 | .893 | .878 | .860 | .839 | .814 | .784 |
| | | | | | | | .749 | .708 | .661 | .606 | .545 | .476 | .404 | | | |
| 14.000 | .87 | 108 | 12.879 | 1.121 | 8.0% | 17 | .934 | .926 | .916 | .905 | .892 | .877 | .859 | .838 | .812 | .783 |
| | | | | | | | .748 | .706 | .659 | .604 | .542 | .474 | .402 | | | |
| 14.000 | .89 | 126 | 12.866 | 1.134 | 8.1% | 17 | .934 | .926 | .916 | .905 | .892 | .877 | .859 | .837 | .812 | .782 |
| | | | | | | | .747 | .705 | .657 | .602 | .541 | .473 | .402 | | | |
| 15.000 | .38 | 24 | 9.791 | 5.209 | 34.7% | 11 | .940 | .934 | .925 | .918 | .910 | .896 | .885 | .875 | .856 | .835 |
| | | | | | | | .818 | | | | | | | | | |
| 15.000 | .42 | 26 | 10.557 | 4.443 | 29.6% | 12 | .940 | .933 | .925 | .918 | .908 | .895 | .884 | .872 | .853 | .833 |
| | | | | | | | .811 | .784 | | | | | | | | |
| 15.000 | .62 | 39 | 10.474 | 4.526 | 30.2% | 12 | .939 | .932 | .924 | .916 | .905 | .893 | .879 | .863 | .842 | .823 |
| | | | | | | | .795 | .763 | | | | | | | | |
| 15.000 | .71 | 52 | 10.440 | 4.560 | 30.4% | 12 | .939 | .931 | .924 | .915 | .903 | .892 | .877 | .860 | .838 | .817 |
| | | | | | | | .788 | .757 | | | | | | | | |
| 15.000 | .77 | 65 | 10.421 | 4.579 | 30.5% | 12 | .938 | .932 | .923 | .914 | .902 | .891 | .876 | .858 | .837 | .813 |
| | | | | | | | .785 | .752 | | | | | | | | |
| 15.000 | .81 | 78 | 10.409 | 4.591 | 30.6% | 12 | .938 | .931 | .923 | .914 | .902 | .890 | .875 | .857 | .836 | .812 |
| | | | | | | | .782 | .749 | | | | | | | | |
| 15.000 | .84 | 91 | 10.401 | 4.599 | 30.7% | 12 | .938 | .931 | .923 | .913 | .903 | .889 | .874 | .856 | .835 | .810 |
| | | | | | | | .782 | .747 | | | | | | | | |
| 15.000 | .46 | 28 | 11.285 | 3.715 | 24.8% | 13 | .940 | .932 | .926 | .918 | .907 | .896 | .884 | .868 | .850 | .830 |
| | | | | | | | .805 | .782 | .747 | | | | | | | |
| 15.000 | .64 | 42 | 11.189 | 3.811 | 25.4% | 13 | .939 | .932 | .924 | .915 | .904 | .893 | .879 | .862 | .841 | .819 |
| | | | | | | | .795 | .760 | .726 | | | | | | | |
| 15.000 | .73 | 56 | 11.149 | 3.851 | 25.7% | 13 | .938 | .932 | .924 | .913 | .904 | .891 | .875 | .859 | .838 | .815 |
| | | | | | | | .787 | .753 | .717 | | | | | | | |
| 15.000 | .79 | 70 | 11.128 | 3.872 | 25.8% | 13 | .938 | .931 | .923 | .913 | .903 | .890 | .875 | .858 | .837 | .813 |
| | | | | | | | .784 | .751 | .711 | | | | | | | |
| 15.000 | .82 | 84 | 11.113 | 3.887 | 25.9% | 13 | .938 | .931 | .923 | .913 | .903 | .889 | .874 | .856 | .836 | .811 |
| | | | | | | | .782 | .748 | .708 | | | | | | | |
| 15.000 | .85 | 98 | 11.103 | 3.897 | 26.0% | 13 | .938 | .931 | .923 | .913 | .902 | .889 | .874 | .856 | .835 | .810 |
| | | | | | | | .780 | .747 | .706 | | | | | | | |
| 15.000 | .50 | 30 | 11.964 | 3.036 | 20.2% | 14 | .939 | .933 | .925 | .917 | .907 | .895 | .881 | .868 | .849 | .828 |
| | | | | | | | .804 | .775 | .742 | .701 | | | | | | |

ERLANG-ENGSET: FOR SMOOTH ATTEMPTING TRAFFIC AND FEW PERSONS DOING THE CALLING (FEW MEANS RATIO OF CALLERS TO CIRCUITS IS LESS THAN 8 - GREATER THAN 8 ASSUME TRAFFIC RANDOM)

| HOURS FIRST ATTEMPT | FIRST ATTEMPT VMR | NUMBER SOURCES | HOURS CON-NECT | HOURS OVFL | PERCENT BLOCKED (GRADE) | LINE | HOURS CONNECTED TRAFFIC ON LINES 1 | 2 | 3 | 4 | 5 | 6 | 7 | 8 | 9 | 10 |
|---|---|---|---|---|---|---|---|---|---|---|---|---|---|---|---|---|
| 15.000 | .67 | 45 | 11.855 | 3.145 | 21.0% | 14 | .939 | .932 | .923 | .915 | .904 | .892 | .878 | .862 | .841 | .819 |
| | | | | | | | .792 | .758 | .723 | .677 | | | | | | |
| 15.000 | .75 | 60 | 11.810 | 3.190 | 21.3% | 14 | .938 | .932 | .923 | .914 | .902 | .891 | .876 | .859 | .838 | .814 |
| | | | | | | | .786 | .754 | .714 | .668 | | | | | | |
| 15.000 | .80 | 75 | 11.785 | 3.215 | 21.4% | 14 | .938 | .931 | .923 | .914 | .903 | .890 | .875 | .856 | .837 | .812 |
| | | | | | | | .783 | .748 | .711 | .663 | | | | | | |
| 15.000 | .83 | 90 | 11.770 | 3.230 | 21.5% | 14 | .938 | .931 | .923 | .913 | .902 | .889 | .874 | .856 | .835 | .810 |
| | | | | | | | .781 | .747 | .707 | .662 | | | | | | |
| 15.000 | .86 | 105 | 11.759 | 3.241 | 21.6% | 14 | .938 | .931 | .923 | .913 | .902 | .889 | .873 | .856 | .834 | .810 |
| | | | | | | | .780 | .746 | .705 | .660 | | | | | | |
| 15.000 | .53 | 32 | 12.585 | 2.415 | 16.1% | 15 | .939 | .933 | .925 | .916 | .907 | .895 | .881 | .865 | .847 | .827 |
| | | | | | | | .801 | .772 | .735 | .696 | .646 | | | | | |
| 15.000 | .69 | 48 | 12.468 | 2.532 | 16.9% | 15 | .939 | .932 | .924 | .914 | .905 | .891 | .878 | .860 | .840 | .818 |
| | | | | | | | .788 | .760 | .720 | .676 | .624 | | | | | |
| 15.000 | .77 | 64 | 12.418 | 2.582 | 17.2% | 15 | .938 | .932 | .923 | .914 | .903 | .891 | .875 | .858 | .837 | .815 |
| | | | | | | | .785 | .751 | .712 | .668 | .616 | | | | | |
| 15.000 | .81 | 80 | 12.390 | 2.610 | 17.4% | 15 | .938 | .931 | .923 | .914 | .902 | .889 | .875 | .857 | .836 | .811 |
| | | | | | | | .783 | .748 | .709 | .663 | .610 | | | | | |
| 15.000 | .84 | 96 | 12.372 | 2.628 | 17.5% | 15 | .938 | .931 | .922 | .914 | .902 | .889 | .874 | .856 | .835 | .809 |
| | | | | | | | .781 | .747 | .707 | .660 | .607 | | | | | |
| 15.000 | .87 | 112 | 12.361 | 2.639 | 17.6% | 15 | .938 | .931 | .923 | .913 | .902 | .889 | .874 | .855 | .834 | .809 |
| | | | | | | | .780 | .745 | .705 | .658 | .606 | | | | | |
| 15.000 | .56 | 34 | 13.147 | 1.853 | 12.4% | 16 | .939 | .933 | .924 | .917 | .906 | .894 | .880 | .865 | .846 | .825 |
| | | | | | | | .798 | .769 | .734 | .691 | .642 | .583 | | | | |
| 15.000 | .71 | 51 | 13.017 | 1.983 | 13.2% | 16 | .939 | .932 | .924 | .914 | .905 | .890 | .878 | .858 | .840 | .818 |
| | | | | | | | .789 | .756 | .718 | .675 | .621 | .561 | | | | |
| 15.000 | .78 | 68 | 12.963 | 2.037 | 13.6% | 16 | .938 | .931 | .923 | .913 | .903 | .890 | .875 | .858 | .837 | .813 |
| | | | | | | | .784 | .750 | .713 | .666 | .612 | .555 | | | | |
| 15.000 | .82 | 85 | 12.934 | 2.066 | 13.8% | 16 | .938 | .931 | .923 | .913 | .902 | .890 | .874 | .856 | .835 | .811 |
| | | | | | | | .782 | .748 | .708 | .662 | .609 | .549 | | | | |
| 15.000 | .85 | 102 | 12.915 | 2.085 | 13.9% | 16 | .938 | .931 | .923 | .913 | .902 | .889 | .874 | .856 | .834 | .809 |
| | | | | | | | .781 | .746 | .706 | .660 | .607 | .547 | | | | |
| 15.000 | .87 | 119 | 12.903 | 2.097 | 14.0% | 16 | .938 | .931 | .920 | .913 | .902 | .888 | .873 | .855 | .834 | .808 |
| | | | | | | | .780 | .745 | .704 | .658 | .605 | .546 | | | | |
| 15.000 | .17 | 18 | 14.553 | .447 | 3.0% | 17 | .941 | .934 | .927 | .922 | .911 | .904 | .893 | .882 | .868 | .858 |
| | | | | | | | .844 | .831 | .812 | .802 | .780 | .752 | .690 | | | |
| 15.000 | .58 | 36 | 13.630 | 1.370 | 9.1% | 17 | .939 | .932 | .925 | .916 | .906 | .893 | .880 | .864 | .845 | .823 |
| | | | | | | | .795 | .769 | .731 | .688 | .637 | .579 | .508 | | | |
| 15.000 | .72 | 54 | 13.498 | 1.502 | 10.0% | 17 | .939 | .931 | .924 | .914 | .904 | .891 | .876 | .859 | .840 | .816 |
| | | | | | | | .786 | .756 | .716 | .671 | .620 | .561 | .494 | | | |

# ERLANG-ENGSET: FOR SMOOTH ATTEMPTING TRAFFIC AND FEW PERSONS DOING THE CALLING (FEW MEANS RATIO OF CALLERS TO CIRCUITS IS LESS THAN 8 - GREATER THAN 8 ASSUME TRAFFIC RANDOM)

| HOURS FIRST ATTEMPT | FIRST ATTEMPT VMR | NUMBER SOURCES | HOURS CON-NECT | HOURS OVFL | PERCENT BLOCKED (GRADE) | LINE | HOURS CONNECTED TRAFFIC ON LINES 1 | 2 | 3 | 4 | 5 | 6 | 7 | 8 | 9 | 10 |
|---|---|---|---|---|---|---|---|---|---|---|---|---|---|---|---|---|
| 15.000 | .79 | 72 | 13.442 | 1.558 | 10.4% | 17 | .938 | .931 | .923 | .914 | .903 | .889 | .875 | .858 | .837 | .812 |
| | | | | | | | .783 | .750 | .711 | .665 | .611 | .554 | .487 | | | |
| 15.000 | .83 | 90 | 13.411 | 1.589 | 10.6% | 17 | .938 | .931 | .923 | .913 | .902 | .889 | .874 | .856 | .835 | .810 |
| | | | | | | | .781 | .747 | .707 | .662 | .609 | .548 | .484 | | | |
| 15.000 | .86 | 108 | 13.391 | 1.609 | 10.7% | 17 | .938 | .931 | .923 | .913 | .902 | .889 | .873 | .856 | .834 | .809 |
| | | | | | | | .780 | .746 | .705 | .659 | .605 | .547 | .481 | | | |
| 15.000 | .88 | 126 | 13.378 | 1.622 | 10.8% | 17 | .938 | .931 | .923 | .913 | .902 | .888 | .873 | .855 | .834 | .808 |
| | | | | | | | .779 | .744 | .704 | .657 | .604 | .545 | .479 | | | |
| 15.000 | .21 | 19 | 14.821 | .179 | 1.2% | 18 | .941 | .935 | .928 | .919 | .912 | .902 | .891 | .878 | .865 | .854 |
| | | | | | | | .840 | .818 | .799 | .777 | .750 | .707 | .630 | .474 | | |
| 15.000 | .61 | 38 | 14.035 | .965 | 6.4% | 18 | .939 | .932 | .924 | .916 | .906 | .893 | .880 | .863 | .844 | .821 |
| | | | | | | | .796 | .765 | .726 | .688 | .635 | .575 | .506 | .427 | | |
| 15.000 | .74 | 57 | 13.903 | 1.097 | 7.3% | 18 | .938 | .932 | .924 | .914 | .904 | .890 | .877 | .859 | .838 | .815 |
| | | | | | | | .786 | .755 | .716 | .669 | .618 | .558 | .492 | .417 | | |
| 15.000 | .80 | 76 | 13.848 | 1.152 | 7.7% | 18 | .938 | .931 | .923 | .913 | .903 | .890 | .874 | .858 | .836 | .812 |
| | | | | | | | .783 | .749 | .710 | .664 | .612 | .551 | .487 | .414 | | |
| 15.000 | .84 | 95 | 13.817 | 1.183 | 7.9% | 18 | .938 | .931 | .923 | .913 | .902 | .889 | .874 | .856 | .835 | .810 |
| | | | | | | | .781 | .747 | .707 | .661 | .608 | .548 | .482 | .413 | | |
| 15.000 | .87 | 114 | 13.797 | 1.203 | 8.0% | 18 | .938 | .931 | .923 | .913 | .902 | .889 | .873 | .855 | .834 | .808 |
| | | | | | | | .780 | .745 | .704 | .659 | .606 | .546 | .480 | .411 | | |
| 15.000 | .89 | 133 | 13.783 | 1.217 | 8.1% | 18 | .938 | .931 | .922 | .913 | .902 | .888 | .873 | .855 | .833 | .808 |
| | | | | | | | .779 | .744 | .703 | .657 | .604 | .544 | .479 | .410 | | |
| 16.000 | .38 | 26 | 10.703 | 5.297 | 33.1% | 12 | .943 | .937 | .931 | .924 | .914 | .907 | .894 | .884 | .868 | .855 |
| | | | | | | | .833 | .811 | | | | | | | | |
| 16.000 | .59 | 39 | 10.629 | 5.371 | 33.6% | 12 | .943 | .937 | .930 | .922 | .912 | .902 | .891 | .876 | .860 | .841 |
| | | | | | | | .820 | .795 | | | | | | | | |
| 16.000 | .69 | 52 | 10.598 | 5.402 | 33.8% | 12 | .942 | .936 | .928 | .921 | .912 | .900 | .888 | .873 | .857 | .837 |
| | | | | | | | .815 | .787 | | | | | | | | |
| 16.000 | .75 | 65 | 10.581 | 5.419 | 33.9% | 12 | .942 | .936 | .929 | .920 | .911 | .900 | .887 | .873 | .854 | .834 |
| | | | | | | | .811 | .784 | | | | | | | | |
| 16.000 | .79 | 78 | 10.570 | 5.430 | 33.9% | 12 | .942 | .936 | .928 | .920 | .910 | .899 | .887 | .871 | .853 | .834 |
| | | | | | | | .809 | .781 | | | | | | | | |
| 16.000 | .82 | 91 | 10.562 | 5.438 | 34.0% | 12 | .942 | .936 | .928 | .920 | .910 | .899 | .886 | .870 | .854 | .832 |
| | | | | | | | .807 | .779 | | | | | | | | |
| 16.000 | .43 | 28 | 11.465 | 4.535 | 28.3% | 13 | .943 | .938 | .931 | .923 | .914 | .907 | .894 | .879 | .870 | .850 |
| | | | | | | | .828 | .810 | .779 | | | | | | | |
| 16.000 | .62 | 42 | 11.381 | 4.619 | 28.9% | 13 | .943 | .936 | .929 | .922 | .913 | .902 | .890 | .876 | .860 | .839 |
| | | | | | | | .818 | .791 | .762 | | | | | | | |
| 16.000 | .71 | 56 | 11.344 | 4.656 | 29.1% | 13 | .942 | .936 | .929 | .921 | .912 | .900 | .889 | .873 | .856 | .835 |
| | | | | | | | .815 | .785 | .752 | | | | | | | |

ERLANG-ENGSET: FOR SMOOTH ATTEMPTING TRAFFIC AND FEW PERSONS DOING THE CALLING (FEW MEANS RATIO OF CALLERS TO CIRCUITS IS LESS THAN 8 - GREATER THAN 8 ASSUME TRAFFIC RANDOM)

| HOURS FIRST ATTEMPT | FIRST ATTEMPT VMR | NUMBER SOURCES | HOURS CON-NECT | HOURS OVFL | PERCENT BLOCKED (GRADE) | LINE | HOURS CONNECTED TRAFFIC ON LINES 1 | 2 | 3 | 4 | 5 | 6 | 7 | 8 | 9 | 10 |
|---|---|---|---|---|---|---|---|---|---|---|---|---|---|---|---|---|
| 16.000 | .77 | 70 | 11.325 | 4.675 | 29.2% | 13 | .942 | .936 | .929 | .920 | .910 | .900 | .887 | .872 | .854 | .834 |
| | | | | | | | .810 | .782 | .749 | | | | | | | |
| 16.000 | .81 | 84 | 11.312 | 4.688 | 29.3% | 13 | .942 | .936 | .928 | .920 | .910 | .899 | .886 | .871 | .853 | .832 |
| | | | | | | | .808 | .779 | .747 | | | | | | | |
| 16.000 | .84 | 98 | 11.303 | 4.697 | 29.4% | 13 | .942 | .935 | .928 | .920 | .910 | .899 | .886 | .870 | .853 | .831 |
| | | | | | | | .807 | .778 | .743 | | | | | | | |
| 16.000 | .47 | 30 | 12.192 | 3.808 | 23.8% | 14 | .943 | .937 | .931 | .923 | .914 | .905 | .892 | .882 | .865 | .845 |
| | | | | | | | .831 | .805 | .774 | .745 | | | | | | |
| 16.000 | .64 | 45 | 12.094 | 3.906 | 24.4% | 14 | .942 | .936 | .930 | .921 | .912 | .902 | .889 | .875 | .858 | .840 |
| | | | | | | | .817 | .789 | .759 | .723 | | | | | | |
| 16.000 | .73 | 60 | 12.051 | 3.949 | 24.7% | 14 | .942 | .936 | .928 | .921 | .911 | .900 | .888 | .872 | .856 | .836 |
| | | | | | | | .811 | .783 | .752 | .714 | | | | | | |
| 16.000 | .79 | 75 | 12.030 | 3.970 | 24.8% | 14 | .942 | .935 | .929 | .921 | .911 | .900 | .886 | .872 | .854 | .833 |
| | | | | | | | .809 | .780 | .749 | .710 | | | | | | |
| 16.000 | .82 | 90 | 12.015 | 3.985 | 24.9% | 14 | .942 | .935 | .928 | .920 | .910 | .899 | .886 | .871 | .853 | .832 |
| | | | | | | | .807 | .780 | .745 | .707 | | | | | | |
| 16.000 | .85 | 105 | 12.005 | 3.995 | 25.0% | 14 | .942 | .935 | .928 | .920 | .910 | .899 | .886 | .870 | .852 | .831 |
| | | | | | | | .806 | .777 | .745 | .705 | | | | | | |
| 16.000 | .50 | 32 | 12.870 | 3.130 | 19.6% | 15 | .943 | .937 | .930 | .922 | .915 | .904 | .893 | .879 | .864 | .844 |
| | | | | | | | .825 | .803 | .775 | .740 | .696 | | | | | |
| 16.000 | .67 | 48 | 12.759 | 3.241 | 20.3% | 15 | .942 | .936 | .929 | .921 | .912 | .902 | .890 | .874 | .857 | .839 |
| | | | | | | | .815 | .789 | .758 | .720 | .677 | | | | | |
| 16.000 | .75 | 64 | 12.713 | 3.287 | 20.5% | 15 | .942 | .936 | .929 | .920 | .911 | .901 | .887 | .872 | .855 | .803 |
| | | | | | | | .812 | .784 | .750 | .712 | .660 | | | | | |
| 16.000 | .80 | 80 | 12.688 | 3.312 | 20.7% | 15 | .942 | .936 | .929 | .920 | .910 | .899 | .886 | .871 | .854 | .833 |
| | | | | | | | .808 | .781 | .747 | .708 | .664 | | | | | |
| 16.000 | .83 | 96 | 12.672 | 3.328 | 20.8% | 15 | .942 | .935 | .928 | .920 | .910 | .899 | .886 | .870 | .853 | .832 |
| | | | | | | | .807 | .778 | .746 | .705 | .661 | | | | | |
| 16.000 | .86 | 112 | 12.660 | 3.340 | 20.9% | 15 | .942 | .935 | .928 | .920 | .910 | .898 | .885 | .870 | .852 | .831 |
| | | | | | | | .806 | .777 | .743 | .704 | .659 | | | | | |
| 16.000 | .53 | 34 | 13.497 | 2.503 | 15.6% | 16 | .943 | .937 | .929 | .924 | .913 | .903 | .892 | .878 | 863 | .044 |
| | | | | | | | .822 | .801 | .771 | .735 | .690 | .647 | | | | |
| 16.000 | .69 | 51 | 13.373 | 2.627 | 16.4% | 10 | .942 | .936 | .929 | .922 | .911 | .901 | .889 | .874 | .857 | .836 |
| | | | | | | | .815 | .787 | .756 | .718 | .676 | .624 | | | | |
| 16.000 | .76 | 68 | 13.322 | 2.678 | 16.7% | 16 | .942 | .936 | .929 | .920 | .910 | .900 | .887 | .072 | .854 | .834 |
| | | | | | | | .812 | .781 | .751 | .711 | .666 | .618 | | | | |
| 16.000 | .81 | 85 | 13.293 | 2.707 | 16.9% | 16 | .942 | .936 | .928 | .920 | .911 | .099 | .885 | .872 | .853 | .833 |
| | | | | | | | .808 | .779 | .745 | .708 | .663 | .611 | | | | |
| 16.000 | .84 | 102 | 13.278 | 2.724 | 17.0% | 16 | .942 | .935 | .928 | .920 | .910 | .899 | .885 | .871 | .852 | .831 |
| | | | | | | | .806 | .778 | .744 | .704 | .661 | .609 | | | | |

ERLANG-ENGSET: FOR SMOOTH ATTEMPTING TRAFFIC AND FEW PERSONS DOING THE CALLING (FEW MEANS RATIO OF CALLERS TO CIRCUITS IS LESS THAN 8 - GREATER THAN 8 ASSUME TRAFFIC RANDOM)

| HOURS FIRST ATTEMPT * | FIRST ATTEMPT VMR | NUMBER SOURCES | HOURS CON-NECT | HOURS OVFL | PERCENT BLOCKED (GRADE) | LINE | HOURS CONNECTED TRAFFIC ON LINES 1 | 2 | 3 | 4 | 5 | 6 | 7 | 8 | 9 | 10 |
|---|---|---|---|---|---|---|---|---|---|---|---|---|---|---|---|---|
| 16.000 | .87 | 119 | 13.263 | 2.737 | 17.1% | 16 | .942 | .935 | .928 | .919 | .910 | .899 | .885 | .870 | .852 | .830 |
| | | | | | | | .806 | .777 | .743 | .703 | .658 | .607 | | | | |
| 16.000 | .56 | 36 | 14.058 | 1.942 | 12.1% | 17 | .943 | .937 | .930 | .923 | .913 | .903 | .889 | .879 | .863 | .843 |
| | | | | | | | .822 | .797 | .767 | .732 | .692 | .640 | .586 | | | |
| 16.000 | .70 | 54 | 13.928 | 2.072 | 13.0% | 17 | .942 | .936 | .929 | .920 | .912 | .900 | .889 | .873 | .856 | .836 |
| | | | | | | | .814 | .785 | .755 | .717 | .674 | .622 | .566 | | | |
| 16.000 | .78 | 72 | 13.872 | 2.128 | 13.3% | 17 | .942 | .936 | .928 | .920 | .911 | .900 | .886 | .873 | .854 | .834 |
| | | | | | | | .809 | .781 | .749 | .711 | .666 | .615 | .557 | | | |
| 16.000 | .82 | 90 | 13.841 | 2.159 | 13.5% | 17 | .942 | .935 | .928 | .920 | .910 | .899 | .886 | .871 | .853 | .832 |
| | | | | | | | .807 | .780 | .745 | .707 | .662 | .611 | .553 | | | |
| 16.000 | .85 | 108 | 13.821 | 2.179 | 13.6% | 17 | .942 | .935 | .928 | .919 | .910 | .899 | .885 | .870 | .852 | .831 |
| | | | | | | | .806 | .777 | .744 | .704 | .660 | .608 | .550 | | | |
| 16.000 | .87 | 126 | 13.807 | 2.193 | 13.7% | 17 | .942 | .935 | .928 | .920 | .910 | .898 | .885 | .870 | .852 | .830 |
| | | | | | | | .806 | .776 | .742 | .703 | .658 | .606 | .548 | | | |
| 16.000 | .16 | 19 | 15.533 | .467 | 2.9% | 18 | .944 | .938 | .934 | .926 | .916 | .913 | .900 | .891 | .884 | .870 |
| | | | | | | | .857 | .846 | .833 | .820 | .803 | .788 | .765 | .705 | | |
| 16.000 | .58 | 38 | 14.552 | 1.448 | 9.0% | 18 | .943 | .937 | .930 | .922 | .913 | .903 | .890 | .878 | .861 | .842 |
| | | | | | | | .820 | .795 | .763 | .733 | .687 | .640 | .580 | .516 | | |
| 16.000 | .72 | 57 | 14.413 | 1.587 | 9.9% | 18 | .942 | .936 | .929 | .921 | .910 | .901 | .887 | .874 | .855 | .836 |
| | | | | | | | .813 | .786 | .753 | .715 | .672 | .622 | .563 | .498 | | |
| 16.000 | .79 | 76 | 14.356 | 1.644 | 10.3% | 18 | .942 | .936 | .929 | .920 | .911 | .899 | .887 | .871 | .854 | .833 |
| | | | | | | | .809 | .780 | .748 | .709 | .665 | .613 | .556 | .493 | | |
| 16.000 | .83 | 95 | 14.323 | 1.677 | 10.5% | 18 | .942 | .935 | .928 | .920 | .910 | .898 | .886 | .870 | .852 | .832 |
| | | | | | | | .807 | .778 | .744 | .707 | .661 | .610 | .552 | .488 | | |
| 16.000 | .86 | 114 | 14.303 | 1.697 | 10.6% | 18 | .942 | .935 | .928 | .920 | .910 | .899 | .885 | .870 | .852 | .830 |
| | | | | | | | .806 | .777 | .743 | .704 | .659 | .607 | .550 | .486 | | |
| 16.000 | .88 | 133 | 14.288 | 1.712 | 10.7% | 18 | .942 | .935 | .928 | .920 | .910 | .898 | .885 | .870 | .851 | .830 |
| | | | | | | | .805 | .776 | .742 | .703 | .657 | .606 | .548 | .485 | | |
| 16.000 | .20 | 20 | 15.807 | .193 | 1.2% | 19 | .944 | .938 | .932 | .925 | .917 | .913 | .899 | .891 | .876 | .867 |
| | | | | | | | .855 | .837 | .822 | .803 | .783 | .754 | .717 | .642 | .492 | |
| 16.000 | .60 | 40 | 14.965 | 1.035 | 6.5% | 19 | .943 | .936 | .930 | .921 | .913 | .903 | .890 | .876 | .860 | .840 |
| | | | | | | | .819 | .794 | .763 | .727 | .685 | .636 | .579 | .513 | .436 | |
| 16.000 | .73 | 60 | 14.827 | 1.173 | 7.3% | 19 | .942 | .936 | .928 | .921 | .911 | .900 | .888 | .872 | .856 | .836 |
| | | | | | | | .811 | .783 | .752 | .714 | .671 | .620 | .562 | .497 | .427 | |
| 16.000 | .80 | 80 | 14.769 | 1.231 | 7.7% | 19 | .942 | .936 | .929 | .920 | .910 | .899 | .886 | .871 | .854 | .833 |
| | | | | | | | .808 | .781 | .747 | .708 | .664 | .612 | .556 | .491 | .422 | |
| 16.000 | .84 | 100 | 14.737 | 1.263 | 7.9% | 19 | .942 | .935 | .928 | .920 | .910 | .899 | .886 | .870 | .852 | .831 |
| | | | | | | | .808 | .778 | .744 | .706 | .659 | .610 | .551 | .489 | .420 | |
| 16.000 | .87 | 120 | 14.716 | 1.284 | 8.0% | 19 | .942 | .935 | .928 | .920 | .910 | .898 | .885 | .869 | .852 | .830 |
| | | | | | | | .806 | .777 | .743 | .703 | .659 | .607 | .549 | .486 | .419 | |

ERLANG-ENGSET: FOR SMOOTH ATTEMPTING TRAFFIC AND FEW PERSONS DOING THE CALLING (FEW MEANS RATIO OF CALLERS TO CIRCUITS IS LESS THAN 8 - GREATER THAN 8 ASSUME TRAFFIC RANDOM)

| HOURS FIRST ATTEMPT | FIRST ATTEMPT VMR | NUMBER SOURCES | HOURS CON-NECT | HOURS OVFL | PERCENT BLOCKED (GRADE) | LINE | HOURS CONNECTED TRAFFIC ON LINES 1 | 2 | 3 | 4 | 5 | 6 | 7 | 8 | 9 | 10 |
|---|---|---|---|---|---|---|---|---|---|---|---|---|---|---|---|---|
| 16.000 | .89 | 140 | 14.702 | 1.298 | 8.1% | 19 | .942 | .935 | .928 | .919 | .910 | .898 | .885 | .870 | .851 | .830 |
| | | | | | | | .805 | .775 | .742 | .702 | .656 | .605 | .548 | .484 | .418 | |
| 17.000 | .39 | 28 | 11.620 | 5.380 | 31.6% | 13 | .946 | .941 | .935 | .928 | .922 | .911 | .906 | .893 | .878 | .868 |
| | | | | | | | .846 | .837 | .809 | | | | | | | |
| 17.000 | .60 | 42 | 11.540 | 5.460 | 32.1% | 13 | .946 | .940 | .934 | .927 | .919 | .911 | .900 | .887 | .873 | .857 |
| | | | | | | | .839 | .814 | .793 | | | | | | | |
| 17.000 | .70 | 56 | 11.507 | 5.493 | 32.3% | 13 | .945 | .940 | .934 | .926 | .918 | .910 | .898 | .884 | .870 | .855 |
| | | | | | | | .833 | .811 | .784 | | | | | | | |
| 17.000 | .76 | 70 | 11.490 | 5.510 | 32.4% | 13 | .945 | .940 | .933 | .926 | .918 | .907 | .897 | .883 | .869 | .852 |
| | | | | | | | .830 | .809 | .780 | | | | | | | |
| 17.000 | .80 | 84 | 11.478 | 5.522 | 32.5% | 13 | .945 | .940 | .933 | .926 | .917 | .907 | .896 | .884 | .868 | .850 |
| | | | | | | | .830 | .806 | .777 | | | | | | | |
| 17.000 | .83 | 98 | 11.470 | 5.530 | 32.5% | 13 | .945 | .939 | .933 | .925 | .917 | .907 | .896 | .883 | .868 | .849 |
| | | | | | | | .828 | .803 | .777 | | | | | | | |
| 17.000 | .43 | 30 | 12.382 | 4.618 | 27.2% | 14 | .946 | .941 | .934 | .929 | .921 | .913 | .900 | .894 | .878 | .865 |
| | | | | | | | .848 | .824 | .810 | .778 | | | | | | |
| 17.000 | .62 | 45 | 12.290 | 4.710 | 27.7% | 14 | .946 | .940 | .934 | .927 | .919 | .910 | .900 | .887 | .871 | .857 |
| | | | | | | | .837 | .815 | .789 | .759 | | | | | | |
| 17.000 | .72 | 60 | 12.252 | 4.748 | 27.9% | 14 | .945 | .940 | .933 | .926 | .918 | .909 | .898 | .883 | .870 | .853 |
| | | | | | | | .834 | .809 | .782 | .751 | | | | | | |
| 17.000 | .77 | 75 | 12.231 | 4.769 | 28.1% | 14 | .945 | .940 | .933 | .925 | .918 | .908 | .897 | .884 | .868 | .851 |
| | | | | | | | .832 | .806 | .780 | .746 | | | | | | |
| 17.000 | .81 | 90 | 12.218 | 4.782 | 28.1% | 14 | .945 | .939 | .933 | .925 | .917 | .907 | .896 | .882 | .809 | .849 |
| | | | | | | | .830 | .806 | .777 | .743 | | | | | | |
| 17.000 | .84 | 105 | 12.210 | 4.790 | 28.2% | 14 | .945 | .939 | .933 | .925 | .917 | .907 | .895 | .882 | .867 | .849 |
| | | | | | | | .828 | .804 | .775 | .742 | | | | | | |
| 17.000 | .47 | 32 | 13.106 | 3.894 | 22.9% | 15 | .946 | .941 | .935 | .928 | .920 | .912 | .900 | .891 | .879 | .862 |
| | | | | | | | .845 | .825 | .803 | .774 | .743 | | | | | |
| 17.000 | .65 | 48 | 13.002 | 3.998 | 23.5% | 15 | .946 | .940 | .934 | .927 | .919 | .909 | .898 | .885 | .873 | .855 |
| | | | | | | | .838 | .813 | .788 | .755 | .723 | | | | | |
| 17.000 | .73 | 64 | 12.957 | 4.043 | 23.8% | 15 | .945 | .940 | .933 | .926 | .918 | .907 | .898 | .885 | .869 | .853 |
| | | | | | | | .831 | .808 | .783 | .749 | .712 | | | | | |
| 17.000 | .79 | 80 | 12.933 | 4.067 | 23.9% | 15 | .945 | .939 | .933 | .925 | .917 | .908 | .896 | .884 | .868 | .851 |
| | | | | | | | .830 | .806 | .778 | .746 | .706 | | | | | |
| 17.000 | .82 | 96 | 12.920 | 4.080 | 24.0% | 15 | .945 | .939 | .933 | .926 | .917 | .907 | .895 | .883 | .867 | .849 |
| | | | | | | | .829 | .804 | .776 | .743 | .706 | | | | | |
| 17.000 | .85 | 112 | 12.909 | 4.091 | 24.1% | 15 | .945 | .939 | .933 | .925 | .917 | .907 | .096 | .882 | .867 | .849 |
| | | | | | | | .828 | .803 | .774 | .742 | .703 | | | | | |
| 17.000 | .50 | 34 | 13.783 | 3.217 | 18.9% | 16 | .946 | .941 | .934 | .928 | .920 | .911 | .901 | .890 | .877 | .859 |
| | | | | | | | .843 | .824 | .800 | .771 | .736 | .701 | | | | |

# ERLANG-ENGSET: FOR SMOOTH ATTEMPTING TRAFFIC AND FEW PERSONS DOING THE CALLING (FEW MEANS RATIO OF CALLERS TO CIRCUITS IS LESS THAN 8 - GREATER THAN 8 ASSUME TRAFFIC RANDOM)

| HOURS FIRST ATTEMPT | FIRST ATTEMPT VMR | NUMBER SOURCES | HOURS CON-NECT | HOURS OVFL | PERCENT BLOCKED (GRADE) | LINE | HOURS CONNECTED TRAFFIC ON LINES 1 | 2 | 3 | 4 | 5 | 6 | 7 | 8 | 9 | 10 |
|---|---|---|---|---|---|---|---|---|---|---|---|---|---|---|---|---|
| 17.000 | .67 | 51 | 13.668 | 3.332 | 19.6% | 16 | .945 | .940 | .934 | .927 | .919 | .908 | .899 | .885 | .872 | .855 |
| | | | | | | | .836 | .811 | .786 | .755 | .719 | .677 | | | | |
| 17.000 | .75 | 68 | 13.619 | 3.381 | 19.9% | 16 | .945 | .940 | .933 | .926 | .918 | .908 | .896 | .885 | .868 | .852 |
| | | | | | | | .832 | .808 | .780 | .747 | .712 | .668 | | | | |
| 17.000 | .80 | 85 | 13.593 | 3.407 | 20.0% | 16 | .945 | .939 | .933 | .926 | .917 | .908 | .896 | .882 | .869 | .849 |
| | | | | | | | .830 | .806 | .777 | .744 | .707 | .665 | | | | |
| 17.000 | .83 | 102 | 13.576 | 3.424 | 20.1% | 16 | .945 | .939 | .933 | .925 | .917 | .907 | .896 | .882 | .867 | .849 |
| | | | | | | | .828 | .804 | .775 | .743 | .704 | .660 | | | | |
| 17.000 | .86 | 119 | 13.564 | 3.436 | 20.2% | 16 | .945 | .939 | .933 | .925 | .916 | .907 | .895 | .882 | .867 | .848 |
| | | | | | | | .828 | .803 | .774 | .741 | .703 | .659 | | | | |
| 17.000 | .53 | 36 | 14.412 | 2.588 | 15.2% | 17 | .946 | .941 | .934 | .928 | .920 | .911 | .900 | .888 | .877 | .861 |
| | | | | | | | .842 | .820 | .796 | .769 | .735 | .695 | .648 | | | |
| 17.000 | .69 | 54 | 14.282 | 2.718 | 16.0% | 17 | .945 | .940 | .934 | .926 | .918 | .910 | .898 | .885 | .870 | .855 |
| | | | | | | | .833 | .812 | .784 | .754 | .718 | .674 | .627 | | | |
| 17.000 | .76 | 72 | 14.230 | 2.770 | 16.3% | 17 | .945 | .940 | .933 | .926 | .917 | .908. | | | | |
| | | | | | | | | | | | | | ..882 | .870 | .851 | |
| | | | | | | | .831 | .807 | .779 | .748 | .711 | .667 | .618 | | | |
| 17.000 | .81 | 90 | 14.200 | 2.800 | 16.5% | 17 | .945 | .939 | .933 | .925 | .917 | .907 | .896 | .882 | .869 | .849 |
| | | | | | | | .830 | .805 | .777 | .743 | .706 | .663 | .613 | | | |
| 17.000 | .84 | 108 | 14.182 | 2.818 | 16.6% | 17 | .945 | .939 | .933 | .925 | .917 | .907 | .895 | .882 | .867 | .849 |
| | | | | | | | .828 | .803 | .775 | .742 | .704 | .660 | .610 | | | |
| 17.000 | .87 | 126 | 14.168 | 2.832 | 16.7% | 17 | .945 | .939 | .933 | .925 | .916 | .907 | .895 | .882 | .866 | .849 |
| | | | | | | | .827 | .802 | .774 | .741 | .702 | .658 | .608 | | | |
| 17.000 | .55 | 38 | 14.978 | 2.022 | 11.9% | 18 | .946 | .940 | .935 | .927 | .919 | .911 | .900 | .889 | .875 | .859 |
| | | | | | | | .841 | .820 | .791 | .768 | .733 | .692 | .644 | .589 | | |
| 17.000 | .70 | 57 | 14.840 | 2.160 | 12.7% | 18 | .945 | .940 | .934 | .926 | .918 | .909 | .897 | .885 | .871 | .853 |
| | | | | | | | .833 | .811 | .784 | .752 | .715 | .673 | .625 | .569 | | |
| 17.000 | .78 | 76 | 14.782 | 2.218 | 13.0% | 18 | .945 | .939 | .933 | .925 | .918 | .907 | .897 | .884 | .868 | .850 |
| | | | | | | | .830 | .808 | .779 | .746 | .709 | .665 | .617 | .561 | | |
| 17.000 | .82 | 95 | 14.750 | 2.250 | 13.2% | 18 | .945 | .939 | .933 | .926 | .917 | .907 | .896 | .883 | .867 | .849 |
| | | | | | | | .829 | .804 | .775 | .745 | .706 | .661 | .611 | .557 | | |
| 17.000 | .85 | 114 | 14.730 | 2.270 | 13.4% | 18 | .945 | .939 | .933 | .925 | .917 | .907 | .896 | .882 | .866 | .849 |
| | | | | | | | .827 | .804 | .775 | .741 | .703 | .659 | .609 | .553 | | |
| 17.000 | .87 | 133 | 14.716 | 2.284 | 13.4% | 18 | .945 | .939 | .933 | .925 | .916 | .907 | .895 | .882 | .866 | .848 |
| | | | | | | | .826 | .802 | .773 | .740 | .702 | .657 | .607 | .551 | | |
| 17.000 | .15 | 20 | 16.511 | .489 | 2.9% | 19 | .947 | .941 | .937 | .930 | .924 | .920 | .907 | .901 | .890 | .882 |
| | | | | | | | .871 | .860 | .846 | .835 | .822 | .811 | .792 | .771 | .725 | |
| 17.000 | .58 | 40 | 15.475 | 1.525 | 9.0% | 19 | .946 | .940 | .934 | .927 | .919 | .910 | .900 | .885 | .876 | .856 |
| | | | | | | | .842 | .815 | .796 | .764 | .729 | .688 | .642 | .585 | .520 | |
| 17.000 | .72 | 60 | 15.331 | 1.669 | 9.8% | 19 | .945 | .940 | .933 | .926 | .918 | .909 | .898 | .883 | .870 | .853 |
| | | | | | | | .834 | .809 | .782 | .751 | .714 | .672 | .622 | .566 | .504 | |

ERLANG-ENGSET: FOR SMOOTH ATTEMPTING TRAFFIC AND FEW PERSONS DOING THE CALLING (FEW MEANS RATIO OF CALLERS TO CIRCUITS IS LESS THAN 8 - GREATER THAN 8 ASSUME TRAFFIC RANDOM)

| HOURS FIRST ATTEMPT | FIRST ATTEMPT VMR | NUMBER SOURCES | HOURS CON-NECT | HOURS OVFL | PERCENT BLOCKED (GRADE) | LINE | HOURS CONNECTED TRAFFIC ON LINES 1 | 2 | 3 | 4 | 5 | 6 | 7 | 8 | 9 | 10 |
|---|---|---|---|---|---|---|---|---|---|---|---|---|---|---|---|---|
| 17.000 | .79 | 80 | 15.271 | 1.729 | 10.2% | 19 | .945 | .939 | .933 | .925 | .917 | .908 | .896 | .884 | .868 | .851 |
| | | | | | | | .830 | .806 | .778 | .746 | .706 | .666 | .615 | .560 | .497 | |
| 17.000 | .83 | 100 | 15.237 | 1.763 | 10.4% | 19 | .945 | .939 | .933 | .925 | .917 | .907 | .896 | .883 | .867 | .849 |
| | | | | | | | .828 | .804 | .776 | .743 | .705 | .661 | .611 | .555 | .494 | |
| 17.000 | .86 | 120 | 15.217 | 1.783 | 10.5% | 19 | .945 | .939 | .933 | .925 | .916 | .907 | .895 | .882 | .866 | .849 |
| | | | | | | | .827 | .803 | .774 | .741 | .703 | .658 | .610 | .552 | .492 | |
| 17.000 | .88 | 140 | 15.202 | 1.798 | 10.6% | 19 | .945 | .939 | .933 | .925 | .916 | .907 | .895 | .881 | .866 | .848 |
| | | | | | | | .826 | .802 | .772 | .741 | .701 | .657 | .607 | .551 | .490 | |
| 17.000 | .19 | 21 | 16.797 | .203 | 1.2% | 20 | .947 | .942 | .936 | .932 | .923 | .916 | .908 | .896 | .891 | .879 |
| | | | | | | | .863 | .857 | .839 | .818 | .810 | .789 | .760 | .720 | .662 | .508 |
| 17.000 | .60 | 42 | 15.896 | 1.104 | 6.5% | 20 | .946 | .940 | .934 | .927 | .919 | .911 | .900 | .887 | .873 | .857 |
| | | | | | | | .839 | .814 | .793 | .762 | .727 | .686 | .636 | .584 | .518 | .443 |
| 17.000 | .73 | 63 | 15.753 | 1.247 | 7.3% | 20 | .945 | .940 | .934 | .926 | .918 | .908 | .897 | .884 | .868 | .852 |
| | | | | | | | .833 | .808 | .783 | .749 | .715 | .668 | .623 | .565 | .503 | .433 |
| 17.000 | .80 | 84 | 15.691 | 1.309 | 7.7% | 20 | .945 | .940 | .933 | .926 | .917 | .907 | .896 | .884 | .868 | .850 |
| | | | | | | | .830 | .806 | .777 | .745 | .706 | .665 | .614 | .557 | .498 | .428 |
| 17.000 | .84 | 105 | 15.659 | 1.341 | 7.9% | 20 | .945 | .939 | .933 | .925 | .917 | .907 | .895 | .882 | .867 | .849 |
| | | | | | | | .828 | .804 | .775 | .742 | .704 | .660 | .611 | .554 | .493 | .427 |
| 17.000 | .87 | 126 | 15.637 | 1.363 | 8.0% | 20 | .945 | .939 | .933 | .925 | .916 | .907 | .895 | .882 | .866 | .849 |
| | | | | | | | .827 | .802 | .774 | .741 | .702 | .658 | .608 | .552 | .491 | .425 |
| 17.000 | .88 | 147 | 15.622 | 1.378 | 8.1% | 20 | .945 | .939 | .932 | .925 | .917 | .906 | .895 | .881 | .866 | .848 |
| | | | | | | | .827 | .802 | .773 | .739 | .701 | .657 | .606 | .551 | .489 | .424 |
| 18.000 | .36 | 28 | 11.746 | 6.254 | 34.7% | 13 | .949 | .945 | .938 | .935 | .927 | .918 | .912 | .900 | .894 | .878 |
| | | | | | | | .867 | .850 | .833 | | | | | | | |
| 18.000 | .40 | 30 | 12.539 | 5.461 | 30.3% | 14 | .949 | .944 | .939 | .933 | .927 | .918 | .912 | .902 | .888 | .878 |
| | | | | | | | .867 | .847 | .827 | .810 | | | | | | |
| 18.000 | .60 | 45 | 12.454 | 5.546 | 30.8% | 14 | .948 | .943 | .938 | .932 | .924 | .917 | .907 | .896 | .885 | .869 |
| | | | | | | | .855 | .837 | .812 | .790 | | | | | | |
| 18.000 | .70 | 60 | 12.419 | 5.581 | 31.0% | 14 | .948 | .943 | .938 | .930 | .924 | .915 | .906 | .895 | .883 | .866 |
| | | | | | | | .851 | .830 | .800 | .781 | | | | | | |
| 18.000 | .76 | 75 | 12.401 | 5.599 | 31.1% | 14 | .948 | .943 | .937 | .931 | .923 | .914 | .906 | .893 | .881 | .866 |
| | | | | | | | .847 | .829 | .803 | .778 | | | | | | |
| 18.000 | .80 | 90 | 12.389 | 5.611 | 31.2% | 14 | .948 | .943 | .937 | .931 | .923 | .914 | .905 | .893 | .880 | .865 |
| | | | | | | | .847 | .826 | .803 | .775 | | | | | | |
| 18.000 | .83 | 105 | 12.381 | 5.019 | 31.2% | 14 | .948 | .943 | .937 | .931 | .923 | .914 | .904 | .803 | .878 | .865 |
| | | | | | | | .845 | .827 | .801 | .773 | | | | | | |
| 18.000 | .44 | 32 | 13.298 | 4.702 | 26.1% | 15 | .949 | .944 | .939 | .933 | .926 | .918 | .911 | .899 | .889 | .877 |
| | | | | | | | .861 | .843 | .828 | .799 | .781 | | | | | |
| 18.000 | .63 | 48 | 13.202 | 4.798 | 26.7% | 15 | .948 | .943 | .938 | .931 | .925 | .916 | .907 | .896 | .885 | .869 |
| | | | | | | | .853 | .835 | .813 | .786 | .757 | | | | | |

ERLANG-ENGSET: FOR SMOOTH ATTEMPTING TRAFFIC AND FEW PERSONS DOING THE CALLING (FEW MEANS RATIO OF CALLERS TO CIRCUITS IS LESS THAN 8 - GREATER THAN 8 ASSUME TRAFFIC RANDOM)

| HOURS FIRST ATTEMPT | FIRST ATTEMPT VMR | NUMBER SOURCES | HOURS CONNECT | HOURS OVFL | PERCENT BLOCKED (GRADE) | LINE | HOURS CONNECTED TRAFFIC ON LINES 1 | 2 | 3 | 4 | 5 | 6 | 7 | 8 | 9 | 10 |
|---|---|---|---|---|---|---|---|---|---|---|---|---|---|---|---|---|
| 18.000 | .72 | 64 | 13.162 | 4.838 | 26.9% | 15 | .948 | .943 | .938 | .931 | .923 | .916 | .905 | .895 | .880 | .867 |
| | | | | | | | .851 | .830 | .805 | .780 | .750 | | | | | |
| 18.000 | .77 | 80 | 13.140 | 4.860 | 27.0% | 15 | .948 | .943 | .937 | .931 | .923 | .915 | .904 | .894 | .881 | .864 |
| | | | | | | | .849 | .827 | .805 | .776 | .743 | | | | | |
| 18.000 | .81 | 96 | 13.127 | 4.873 | 27.1% | 15 | .948 | .943 | .937 | .930 | .923 | .914 | .905 | .893 | .880 | .864 |
| | | | | | | | .847 | .826 | .802 | .773 | .743 | | | | | |
| 18.000 | .84 | 112 | 13.118 | 4.882 | 27.1% | 15 | .948 | .943 | .937 | .930 | .923 | .914 | .904 | .892 | .879 | .864 |
| | | | | | | | .846 | .825 | .801 | .772 | .741 | | | | | |
| 18.000 | .47 | 34 | 14.022 | 3.978 | 22.1% | 16 | .949 | .944 | .938 | .933 | .924 | .920 | .909 | .899 | .889 | .872 |
| | | | | | | | .860 | .846 | .820 | .801 | .776 | .742 | | | | |
| 18.000 | .65 | 51 | 13.913 | 4.087 | 22.7% | 16 | .948 | .943 | .938 | .931 | .924 | .916 | .906 | .896 | .884 | .868 |
| | | | | | | | .851 | .835 | .810 | .785 | .755 | .720 | | | | |
| 18.000 | .74 | 68 | 13.867 | 4.133 | 23.0% | 16 | .948 | .943 | .937 | .931 | .923 | .915 | .906 | .894 | .881 | .867 |
| | | | | | | | .849 | .829 | .806 | .778 | .748 | .712 | | | | |
| 18.000 | .79 | 85 | 13.841 | 4.159 | 23.1% | 16 | .948 | .943 | .937 | .931 | .924 | .914 | .905 | .893 | .879 | .866 |
| | | | | | | | .847 | .826 | .805 | .776 | .742 | .706 | | | | |
| 18.000 | .82 | 102 | 13.827 | 4.173 | 23.2% | 16 | .948 | .943 | .937 | .931 | .923 | .914 | .904 | .892 | .880 | .864 |
| | | | | | | | .847 | .824 | .803 | .773 | .740 | .705 | | | | |
| 18.000 | .85 | 119 | 13.816 | 4.184 | 23.2% | 16 | .948 | .943 | .937 | .930 | .922 | .914 | .904 | .892 | .879 | .863 |
| | | | | | | | .845 | .825 | .800 | .772 | .739 | .702 | | | | |
| 18.000 | .50 | 36 | 14.699 | 3.301 | 18.3% | 17 | .949 | .944 | .938 | .933 | .926 | .918 | .908 | .899 | .888 | .873 |
| | | | | | | | .858 | .841 | .822 | .797 | .769 | .738 | .698 | | | |
| 18.000 | .67 | 54 | 14.579 | 3.421 | 19.0% | 17 | .948 | .943 | .938 | .932 | .923 | .916 | .906 | .897 | .882 | .869 |
| | | | | | | | .851 | .832 | .809 | .784 | .753 | .720 | .677 | | | |
| 18.000 | .75 | 72 | 14.527 | 3.473 | 19.3% | 17 | .948 | .943 | .937 | .931 | .923 | .915 | .905 | .894 | .880 | .866 |
| | | | | | | | .849 | .829 | .805 | .778 | .744 | .711 | .668 | | | |
| 18.000 | .80 | 90 | 14.501 | 3.499 | 19.4% | 17 | .948 | .943 | .937 | .931 | .923 | .914 | .905 | .893 | .880 | .865 |
| | | | | | | | .847 | .826 | .803 | .775 | .743 | .704 | .665 | | | |
| 18.000 | .83 | 108 | 14.483 | 3.517 | 19.5% | 17 | .948 | .943 | .937 | .930 | .923 | .914 | .904 | .893 | .878 | .864 |
| | | | | | | | .846 | .825 | .801 | .773 | .741 | .703 | .660 | | | |
| 18.000 | .86 | 126 | 14.471 | 3.529 | 19.6% | 17 | .948 | .943 | .937 | .930 | .923 | .914 | .904 | .892 | .878 | .863 |
| | | | | | | | .845 | .824 | .799 | .772 | .739 | .701 | .658 | | | |
| 18.000 | .53 | 38 | 15.329 | 2.671 | 14.8% | 18 | .949 | .944 | .938 | .932 | .926 | .917 | .909 | .896 | .889 | .872 |
| | | | | | | | .858 | .838 | .819 | .796 | .765 | .735 | .695 | .649 | | |
| 18.000 | .68 | 57 | 15.195 | 2.805 | 15.6% | 18 | .948 | .943 | .937 | .931 | .924 | .916 | .906 | .896 | .882 | .868 |
| | | | | | | | .851 | .831 | .808 | .784 | .750 | .719 | .673 | .628 | | |
| 18.000 | .76 | 76 | 15.139 | 2.861 | 15.9% | 18 | .948 | .943 | .937 | .930 | .924 | .914 | .906 | .893 | .881 | .866 |
| | | | | | | | .847 | .829 | .803 | .778 | .747 | .708 | .668 | .618 | | |
| 18.000 | .81 | 95 | 15.109 | 2.891 | 16.1% | 18 | .948 | .943 | .937 | .931 | .923 | .915 | .904 | .892 | .880 | .864 |
| | | | | | | | .846 | .827 | .802 | .774 | .741 | .705 | .663 | .614 | | |

ERLANG-ENGSET: FOR SMOOTH ATTEMPTING TRAFFIC AND FEW PERSONS DOING THE CALLING (FEW MEANS RATIO OF CALLERS TO CIRCUITS IS LESS THAN 8 - GREATER THAN 8 ASSUME TRAFFIC RANDOM)

| HOURS FIRST ATTEMPT | FIRST ATTEMPT VMR | NUMBER SOURCES | HOURS CON-NECT | HOURS OVFL | PERCENT BLOCKED (GRADE) | LINE | HOURS CONNECTED TRAFFIC ON LINES 1 | 2 | 3 | 4 | 5 | 6 | 7 | 8 | 9 | 10 |
|---|---|---|---|---|---|---|---|---|---|---|---|---|---|---|---|---|
| 18.000 | .84 | 114 | 15.089 | 2.911 | 16.2% | 18 | .948 | .943 | .937 | .930 | .922 | .914 | .904 | .892 | .879 | .864 |
| | | | | | | | .845 | .824 | .802 | .772 | .740 | .702 | .659 | .611 | | |
| 18.000 | .86 | 133 | 15.075 | 2.925 | 16.2% | 18 | .948 | .943 | .937 | .930 | .923 | .914 | .904 | .892 | .879 | .863 |
| | | | | | | | .845 | .823 | .800 | .772 | .738 | .700 | .659 | .608 | | |
| 18.000 | .55 | 40 | 15.896 | 2.104 | 11.7% | 19 | .948 | .944 | .938 | .932 | .925 | .917 | .909 | .895 | .887 | .873 |
| | | | | | | | .856 | .838 | .815 | .795 | .765 | .729 | .690 | .649 | .590 | |
| 18.000 | .70 | 60 | 15.753 | 2.247 | 12.5% | 19 | .948 | .943 | .938 | .930 | .924 | .915 | .906 | .895 | .883 | .866 |
| | | | | | | | .851 | .830 | .808 | .781 | .749 | .716 | .673 | .626 | .571 | |
| 18.000 | .77 | 80 | 15.695 | 2.305 | 12.8% | 19 | .948 | .943 | .937 | .931 | .923 | .915 | .904 | .894 | .881 | .864 |
| | | | | | | | .849 | .827 | .805 | .776 | .743 | .707 | .667 | .617 | .563 | |
| 18.000 | .82 | 100 | 15.662 | 2.338 | 13.0% | 19 | .948 | .943 | .937 | .930 | .923 | .914 | .904 | .893 | .879 | .864 |
| | | | | | | | .846 | .826 | .802 | .774 | .741 | .704 | .662 | .613 | .558 | |
| 18.000 | .85 | 120 | 15.641 | 2.359 | 13.1% | 19 | .948 | .943 | .937 | .930 | .923 | .914 | .904 | .892 | .879 | .863 |
| | | | | | | | .845 | .825 | .800 | .771 | .740 | .701 | .659 | .610 | .557 | |
| 18.000 | .87 | 140 | 15.626 | 2.374 | 13.2% | 19 | .948 | .943 | .937 | .930 | .922 | .914 | .904 | .892 | .879 | .862 |
| | | | | | | | .844 | .824 | .799 | .771 | .738 | .701 | .657 | .608 | .554 | |
| 18.000 | .14 | 21 | 17.489 | .511 | 2.8% | 20 | .950 | .944 | .942 | .934 | .927 | .923 | .916 | .909 | .896 | .893 |
| | | | | | | | .879 | .873 | .854 | .851 | .839 | .823 | .810 | .805 | .784 | .736 |
| 18.000 | .57 | 42 | 16.399 | 1.601 | 8.9% | 20 | .949 | .944 | .938 | .932 | .925 | .917 | .907 | .898 | .885 | .871 |
| | | | | | | | .856 | .837 | .815 | .789 | .764 | .729 | .688 | .640 | .590 | .525 |
| 18.000 | .71 | 63 | 16.250 | 1.750 | 9.7% | 20 | .948 | .943 | .937 | .931 | .924 | .914 | .906 | .895 | .882 | .867 |
| | | | | | | | .849 | .830 | .807 | .780 | .748 | .714 | .671 | .625 | .570 | .507 |
| 18.000 | .79 | 84 | 16.188 | 1.812 | 10.1% | 20 | .948 | .943 | .937 | .930 | .923 | .915 | .904 | .894 | .881 | .865 |
| | | | | | | | .040 | .829 | .803 | .775 | .744 | .705 | .666 | .616 | .560 | .503 |
| 10.000 | .83 | 105 | 16.154 | 1.846 | 10.3% | 20 | .948 | .943 | .937 | .931 | .923 | .914 | .904 | .893 | .878 | .865 |
| | | | | | | | .845 | .827 | .801 | .773 | .741 | .703 | .661 | .612 | .558 | .498 |
| 18.000 | .86 | 126 | 16.132 | 1.868 | 10.4% | 20 | .948 | .943 | .937 | .930 | .923 | .914 | .904 | .892 | .878 | .863 |
| | | | | | | | .845 | .824 | .799 | .772 | .739 | .701 | .658 | .609 | .556 | .496 |
| 18.000 | .88 | 147 | 16.117 | 1.883 | 10.5% | 20 | .948 | .943 | .937 | .930 | .923 | .914 | .903 | .892 | .879 | .862 |
| | | | | | | | .845 | .823 | .800 | .770 | .737 | .699 | .657 | .608 | .553 | .405 |
| 18.000 | .18 | 22 | 17.783 | .217 | 1.2% | 21 | .949 | .945 | .940 | .935 | .929 | .920 | .914 | 909 | .800 | .088 |
| | | | | | | | .877 | .870 | .853 | .836 | .829 | .810 | .791 | .767 | .729 | .672 |
| | | | | | | | .524 | | | | | | | | | |
| 18.000 | .59 | 44 | 16.827 | 1.173 | 6.5% | 21 | .948 | .944 | .938 | .932 | .925 | .916 | .908 | .896 | .886 | .871 |
| | | | | | | | .855 | .835 | .814 | 788 | .761 | .720 | .686 | .640 | .585 | .522 |
| | | | | | | | .450 | | | | | | | | | |
| 18.000 | .73 | 66 | 16.679 | 1.321 | 7.3% | 21 | .948 | .943 | .937 | .930 | .924 | .914 | .906 | .895 | .881 | .865 |
| | | | | | | | .849 | .830 | .807 | .779 | .747 | .713 | .670 | .622 | .568 | .506 |
| | | | | | | | .441 | | | | | | | | | |
| 18.000 | .80 | 88 | 16.616 | 1.384 | 7.7% | 21 | .948 | .943 | .937 | .931 | .922 | .914 | .905 | .894 | .879 | .866 |

**ERLANG-ENGSET: FOR SMOOTH ATTEMPTING TRAFFIC AND FEW PERSONS DOING THE CALLING (FEW MEANS RATIO OF CALLERS TO CIRCUITS IS LESS THAN 8 - GREATER THAN 8 ASSUME TRAFFIC RANDOM)**

| HOURS FIRST ATTEMPT | FIRST ATTEMPT VMR | NUMBER SOURCES | HOURS CON-NECT | HOURS OVFL | PERCENT BLOCKED (GRADE) | LINE | HOURS CONNECTED TRAFFIC ON LINES 1 | 2 | 3 | 4 | 5 | 6 | 7 | 8 | 9 | 10 |
|---|---|---|---|---|---|---|---|---|---|---|---|---|---|---|---|---|
| | | | | | | | .846 | .827 | .803 | .775 | .742 | .708 | .663 | .614 | .562 | .502 |
| | | | | | | | .436 | | | | | | | | | |
| 18.000 | .84 | 110 | 16.581 | 1.419 | 7.9% | 21 | .948 | .943 | .937 | .930 | .922 | .914 | .904 | .893 | .879 | .864 |
| | | | | | | | .846 | .825 | .801 | .773 | .740 | .703 | .660 | .612 | .557 | .498 |
| | | | | | | | .433 | | | | | | | | | |
| 18.000 | .86 | 132 | 16.558 | 1.442 | 8.0% | 21 | .948 | .943 | .937 | .930 | .922 | .914 | .904 | .892 | .878 | .863 |
| | | | | | | | .845 | .824 | .800 | .771 | .739 | .701 | .658 | .609 | .555 | .496 |
| | | | | | | | .431 | | | | | | | | | |
| 18.000 | .88 | 154 | 16.544 | 1.456 | 8.1% | 21 | .948 | .943 | .937 | .930 | .922 | .914 | .903 | .892 | .878 | .863 |
| | | | | | | | .844 | .823 | .799 | .769 | .738 | .700 | .656 | .608 | .553 | .494 |
| | | | | | | | .431 | | | | | | | | | |
| 19.000 | .37 | 30 | 12.666 | 6.334 | 33.3% | 14 | .952 | .948 | .943 | .937 | .930 | .927 | .917 | .906 | .900 | .892 |
| | | | | | | | .872 | .867 | .845 | .831 | | | | | | |
| 19.000 | .58 | 45 | 12.592 | 6.408 | 33.7% | 14 | .951 | .947 | .941 | .937 | .930 | .922 | .914 | .904 | .896 | .882 |
| | | | | | | | .869 | .853 | .834 | .812 | | | | | | |
| 19.000 | .68 | 60 | 12.561 | 6.439 | 33.9% | 14 | .951 | .946 | .941 | .936 | .929 | .922 | .913 | .904 | .891 | .881 |
| | | | | | | | .865 | .848 | .830 | .806 | | | | | | |
| 19.000 | .75 | 75 | 12.544 | 6.456 | 34.0% | 14 | .951 | .946 | .941 | .935 | .929 | .921 | .913 | .903 | .890 | .878 |
| | | | | | | | .864 | .845 | .825 | .804 | | | | | | |
| 19.000 | .79 | 90 | 12.533 | 6.467 | 34.0% | 14 | .951 | .946 | .941 | .934 | .928 | .921 | .912 | .902 | .891 | .877 |
| | | | | | | | .862 | .844 | .824 | .800 | | | | | | |
| 19.000 | .82 | 105 | 12.526 | 6.474 | 34.1% | 14 | .950 | .946 | .940 | .935 | .928 | .920 | .912 | .901 | .890 | .876 |
| | | | | | | | .861 | .845 | .822 | .800 | | | | | | |
| 19.000 | .41 | 32 | 13.455 | 5.545 | 29.2% | 15 | .952 | .947 | .943 | .937 | .931 | .924 | .916 | .909 | .898 | .888 |
| | | | | | | | .875 | .857 | .849 | .827 | .802 | | | | | |
| 19.000 | .60 | 48 | 13.371 | 5.629 | 29.6% | 15 | .951 | .946 | .942 | .935 | .930 | .923 | .913 | .906 | .893 | .880 |
| | | | | | | | .869 | .852 | .832 | .811 | .787 | | | | | |
| 19.000 | .70 | 64 | 13.334 | 5.666 | 29.8% | 15 | .951 | .946 | .941 | .935 | .929 | .921 | .912 | .905 | .892 | .879 |
| | | | | | | | .863 | .848 | .827 | .807 | .778 | | | | | |
| 19.000 | .76 | 80 | 13.314 | 5.686 | 29.9% | 15 | .951 | .946 | .941 | .935 | .928 | .921 | .912 | .903 | .892 | .877 |
| | | | | | | | .863 | .845 | .825 | .802 | .774 | | | | | |
| 19.000 | .80 | 96 | 13.302 | 5.698 | 30.0% | 15 | .951 | .946 | .941 | .934 | .929 | .920 | .912 | .901 | .891 | .877 |
| | | | | | | | .861 | .844 | .823 | .801 | .773 | | | | | |
| 19.000 | .83 | 112 | 13.293 | 5.707 | 30.0% | 15 | .950 | .946 | .940 | .935 | .928 | .920 | .912 | .901 | .890 | .876 |
| | | | | | | | .861 | .843 | .822 | .798 | .770 | | | | | |
| 19.000 | .44 | 34 | 14.216 | 4.784 | 25.2% | 16 | .951 | .947 | .942 | .936 | .930 | .925 | .918 | .904 | .898 | .889 |
| | | | | | | | .874 | .856 | .846 | .823 | .798 | .778 | | | | |
| 19.000 | .63 | 51 | 14.117 | 4.883 | 25.7% | 16 | .951 | .946 | .942 | .936 | .929 | .921 | .914 | .905 | .892 | .881 |
| | | | | | | | .867 | .851 | .831 | .810 | .784 | .755 | | | | |
| 19.000 | .72 | 68 | 14.074 | 4.926 | 25.9% | 16 | .951 | .946 | .941 | .936 | .929 | .920 | .913 | .903 | .891 | .880 |

ERLANG-ENGSET: FOR SMOOTH ATTEMPTING TRAFFIC AND FEW PERSONS DOING THE CALLING (FEW MEANS RATIO OF CALLERS TO CIRCUITS IS LESS THAN 8 - GREATER THAN 8 ASSUME TRAFFIC RANDOM)

| HOURS FIRST ATTEMPT | FIRST ATTEMPT VMR | NUMBER SOURCES | HOURS CON-NECT | HOURS OVFL | PERCENT BLOCKED (GRADE) | LINE | HOURS CONNECTED TRAFFIC ON LINES 1 | 2 | 3 | 4 | 5 | 6 | 7 | 8 | 9 | 10 |
|---|---|---|---|---|---|---|---|---|---|---|---|---|---|---|---|---|
| | | | | | | | .862 | .848 | .827 | .804 | .777 | .747 | | | | |
| 19.000 | .78 | 85 | 14.051 | 4.949 | 26.0% | 16 | .951 | .946 | .941 | .935 | .928 | .921 | .912 | .902 | .891 | .877 |
| | | | | | | | .862 | .846 | .824 | .800 | .775 | .742 | | | | |
| 19.000 | .81 | 102 | 14.039 | 4.961 | 26.1% | 16 | .951 | .946 | .941 | .935 | .928 | .919 | .912 | .901 | .890 | .877 |
| | | | | | | | .861 | .844 | .823 | .798 | .772 | .740 | | | | |
| 19.000 | .84 | 119 | 14.028 | 4.972 | 26.2% | 16 | .950 | .946 | .941 | .935 | .928 | .920 | .911 | .901 | .890 | .876 |
| | | | | | | | .860 | .843 | .822 | .798 | .770 | .738 | | | | |
| 19.000 | .47 | 36 | 14.935 | 4.065 | 21.4% | 17 | .951 | .947 | .941 | .937 | .931 | .924 | .915 | .907 | .897 | .883 |
| | | | | | | | .876 | .858 | .840 | .822 | .795 | .774 | .736 | | | |
| 19.000 | .65 | 54 | 14.825 | 4.175 | 22.0% | 17 | .951 | .946 | .941 | .935 | .929 | .922 | .913 | .905 | .893 | .880 |
| | | | | | | | .867 | .848 | .832 | .808 | .783 | .751 | .719 | | | |
| 19.000 | .74 | 72 | 14.779 | 4.221 | 22.2% | 17 | .951 | .946 | .941 | .935 | .929 | .921 | .912 | .902 | .892 | .878 |
| | | | | | | | .864 | .848 | .825 | .802 | .777 | .745 | .712 | | | |
| 19.000 | .79 | 90 | 14.753 | 4.247 | 22.4% | 17 | .951 | .946 | .941 | .934 | .928 | .921 | .912 | .902 | .891 | .877 |
| | | | | | | | .862 | .844 | .824 | .800 | .772 | .742 | .706 | | | |
| 19.000 | .82 | 108 | 14.737 | 4.263 | 22.4% | 17 | .950 | .946 | .941 | .935 | .928 | .920 | .912 | .901 | .890 | .876 |
| | | | | | | | .861 | .844 | .821 | .800 | .771 | .738 | .704 | | | |
| 19.000 | .85 | 126 | 14.726 | 4.274 | 22.5% | 17 | .950 | .946 | .940 | .935 | .927 | .920 | .911 | .901 | .890 | .875 |
| | | | | | | | .861 | .842 | .821 | .798 | .770 | .737 | .700 | | | |
| 19.000 | .50 | 38 | 15.615 | 3.385 | 17.8% | 18 | .951 | .947 | .942 | .935 | .932 | .923 | .916 | .906 | .894 | .888 |
| | | | | | | | .871 | .857 | .839 | .818 | .795 | .767 | .738 | .696 | | |
| 19.000 | .67 | 57 | 15.490 | 3.510 | 18.5% | 18 | .951 | .946 | .941 | .936 | .928 | .922 | .913 | .905 | .891 | .880 |
| | | | | | | | .866 | .850 | .829 | .808 | .780 | .752 | .717 | .675 | | |
| 19.000 | .75 | 76 | 15.440 | 3.560 | 18.7% | 18 | .951 | .946 | .941 | .935 | .929 | .921 | .913 | .902 | .890 | .878 |
| | | | | | | | .864 | .847 | .824 | .804 | .776 | .743 | .710 | .668 | | |
| 19.000 | .80 | 95 | 15.411 | 3.589 | 18.9% | 18 | .951 | .946 | .941 | .935 | .928 | .920 | .913 | .902 | .890 | .877 |
| | | | | | | | .862 | .844 | .824 | .800 | .771 | .742 | .703 | .665 | | |
| 19.000 | .83 | 114 | 15.392 | 3.608 | 19.0% | 18 | .950 | .946 | .941 | .935 | .928 | .920 | .911 | .901 | .890 | .876 |
| | | | | | | | .861 | .843 | .823 | .798 | .770 | .738 | .702 | .659 | | |
| 19.000 | .86 | 133 | 15.380 | 3.620 | 19.1% | 18 | .950 | .946 | .940 | .935 | .928 | .920 | .911 | .901 | .880 | .876 |
| | | | | | | | .860 | .842 | .021 | .797 | .769 | .737 | .700 | .658 | | |
| 19.000 | .52 | 40 | 16.247 | 2.753 | 14.5% | 19 | .951 | .947 | .942 | .937 | .929 | .924 | .915 | .904 | .898 | .883 |
| | | | | | | | .870 | .856 | .837 | .816 | .788 | .767 | .735 | .690 | .650 | |
| 19.000 | .68 | 60 | 16.108 | 2.892 | 15.2% | 19 | .051 | .946 | .941 | .936 | .929 | .922 | .913 | .904 | .891 | .881 |
| | | | | | | | .865 | .848 | .830 | .806 | 780 | .749 | .716 | .674 | .628 | |
| 19.000 | .76 | 80 | 16.050 | 2.950 | 15.5% | 19 | .951 | .946 | .941 | .935 | .928 | .921 | .912 | .903 | .892 | .877 |
| | | | | | | | .863 | .845 | .825 | .802 | .774 | .744 | .708 | .666 | .618 | |
| 19.000 | .81 | 100 | 16.020 | 2.980 | 15.7% | 19 | .951 | .946 | .940 | .935 | .928 | .921 | .912 | .901 | .890 | .877 |
| | | | | | | | .862 | .844 | .823 | .801 | .772 | .739 | .704 | .662 | .615 | |
| 19.000 | .84 | 120 | 16.000 | 3.000 | 15.8% | 19 | .950 | .946 | .940 | .935 | .928 | .920 | .911 | .901 | .889 | .876 |

ERLANG-ENGSET: FOR SMOOTH ATTEMPTING TRAFFIC AND FEW PERSONS DOING THE CALLING (FEW MEANS RATIO OF CALLERS TO CIRCUITS IS LESS THAN 8 - GREATER THAN 8 ASSUME TRAFFIC RANDOM)

| HOURS FIRST ATTEMPT | FIRST ATTEMPT VMR | NUMBER SOURCES | HOURS CON-NECT | HOURS OVFL | PERCENT BLOCKED (GRADE) | LINE | HOURS CONNECTED TRAFFIC ON LINES 1 | 2 | 3 | 4 | 5 | 6 | 7 | 8 | 9 | 10 |
|---|---|---|---|---|---|---|---|---|---|---|---|---|---|---|---|---|
| | | | | | | | .860 | .843 | .822 | .798 | .769 | .739 | .701 | .660 | .612 | |
| 19.000 | .86 | 140 | 15.986 | 3.014 | 15.9% | 19 | .950 | .946 | .940 | .934 | .928 | .920 | .911 | .901 | .889 | .875 |
| | | | | | | | .860 | .842 | .821 | .796 | .769 | .736 | .700 | .657 | .610 | |
| 19.000 | .55 | 42 | 16.819 | 2.181 | 11.5% | 20 | .951 | .947 | .942 | .935 | .931 | .922 | .916 | .906 | .894 | .883 |
| | | | | | | | .867 | .858 | .835 | .815 | .789 | .765 | .729 | .692 | .646 | .595 |
| 19.000 | .70 | 63 | 16.670 | 2.330 | 12.3% | 20 | .951 | .946 | .942 | .935 | .929 | .920 | .914 | .903 | .892 | .881 |
| | | | | | | | .864 | .849 | .825 | .806 | .779 | .750 | .712 | .675 | .625 | .574 |
| 19.000 | .77 | 84 | 16.610 | 2.390 | 12.6% | 20 | .950 | .946 | .941 | .935 | .928 | .921 | .912 | .902 | .891 | .876 |
| | | | | | | | .864 | .844 | .826 | .801 | .774 | .741 | .708 | .664 | .619 | .566 |
| 19.000 | .82 | 105 | 16.576 | 2.424 | 12.8% | 20 | .950 | .946 | .940 | .935 | .928 | .920 | .912 | .901 | .890 | .876 |
| | | | | | | | .861 | .845 | .822 | .800 | .771 | .739 | .704 | .661 | .614 | .561 |
| 19.000 | .85 | 126 | 16.553 | 2.447 | 12.9% | 20 | .950 | .946 | .940 | .935 | .927 | .920 | .911 | .901 | .890 | .875 |
| | | | | | | | .861 | .842 | .821 | .798 | .770 | .737 | .700 | .659 | .611 | .558 |
| 19.000 | .87 | 147 | 16.539 | 2.461 | 13.0% | 20 | .950 | .946 | .940 | .934 | .928 | .920 | .911 | .900 | .889 | .875 |
| | | | | | | | .860 | .842 | .820 | .796 | .768 | .735 | .700 | .657 | .609 | .557 |
| 19.000 | .14 | 22 | 18.472 | .528 | 2.8% | 21 | .952 | .947 | .945 | .939 | .931 | .929 | .923 | .912 | .907 | .901 |
| | | | | | | | .886 | .881 | .870 | .864 | .847 | .840 | .830 | .820 | .808 | .785 |
| | | | | | | | .755 | | | | | | | | | |
| 19.000 | .57 | 44 | 17.323 | 1.677 | 8.8% | 21 | .951 | .946 | .942 | .936 | .930 | .922 | .914 | .906 | .894 | .883 |
| | | | | | | | .869 | .854 | .833 | .815 | .789 | .758 | .730 | .686 | .647 | .589 |
| | | | | | | | .529 | | | | | | | | | |
| 19.000 | .71 | 66 | 17.171 | 1.829 | 9.6% | 21 | .951 | .946 | .941 | .935 | .928 | .922 | .913 | .904 | .890 | .879 |
| | | | | | | | .866 | .845 | .829 | .804 | .779 | .747 | .712 | .670 | .627 | .570 |
| | | | | | | | .514 | | | | | | | | | |
| 19.000 | .78 | 88 | 17.107 | 1.893 | 10.0% | 21 | .951 | .946 | .941 | .935 | .928 | .921 | .911 | .902 | .891 | .876 |
| | | | | | | | .864 | .843 | .825 | .800 | .774 | .740 | .707 | .663 | .618 | .564 |
| | | | | | | | .506 | | | | | | | | | |
| 19.000 | .83 | 110 | 17.072 | 1.928 | 10.1% | 21 | .950 | .946 | .941 | .935 | .927 | .921 | .912 | .901 | .890 | .877 |
| | | | | | | | .861 | .843 | .823 | .798 | .771 | .739 | .702 | .660 | .613 | .560 |
| | | | | | | | .502 | | | | | | | | | |
| 19.000 | .86 | 132 | 17.050 | 1.950 | 10.3% | 21 | .950 | .946 | .940 | .935 | .928 | .920 | .912 | .901 | .889 | .876 |
| | | | | | | | .860 | .843 | .821 | .797 | .769 | .737 | .700 | .658 | .611 | .557 |
| | | | | | | | .500 | | | | | | | | | |
| 19.000 | .88 | 154 | 17.034 | 1.966 | 10.3% | 21 | .950 | .946 | .941 | .934 | .927 | .920 | .910 | .901 | .888 | .876 |
| | | | | | | | .860 | .841 | .820 | .796 | .768 | .736 | .699 | .656 | .609 | .556 |
| | | | | | | | .499 | | | | | | | | | |
| 19.000 | .17 | 23 | 18.773 | .227 | 1.2% | 22 | .952 | .949 | .944 | .939 | .931 | .929 | .917 | .913 | .909 | .896 |
| | | | | | | | .885 | .880 | .866 | .853 | .842 | .828 | .812 | .793 | .774 | .737 |
| | | | | | | | .681 | .542 | | | | | | | | |
| 19.000 | .59 | 46 | 17.761 | 1.239 | 6.5% | 22 | .951 | .947 | .942 | .936 | .930 | .923 | .913 | .906 | .894 | .882 |

ERLANG-ENGSET: FOR SMOOTH ATTEMPTING TRAFFIC AND FEW PERSONS DOING THE CALLING (FEW MEANS RATIO OF CALLERS TO CIRCUITS IS LESS THAN 8 - GREATER THAN 8 ASSUME TRAFFIC RANDOM)

| HOURS FIRST ATTEMPT | FIRST ATTEMPT VMR | NUMBER SOURCES | HOURS CONNECT | HOURS OVFL | PERCENT BLOCKED (GRADE) | LINE | HOURS CONNECTED TRAFFIC ON LINES 1 | 2 | 3 | 4 | 5 | 6 | 7 | 8 | 9 | 10 |
|---|---|---|---|---|---|---|---|---|---|---|---|---|---|---|---|---|
| | | | | | | | .868 | .850 | .835 | .814 | .786 | .756 | .728 | .687 | .641 | .586 |
| | | | | | | | .527 | .460 | | | | | | | | |
| 19.000 | .72 | 69 | 17.606 | 1.394 | 7.3% | 22 | .951 | .946 | .941 | .935 | .929 | .921 | .912 | .902 | .892 | .879 |
| | | | | | | | .865 | .847 | .825 | .804 | .777 | .748 | .709 | .672 | .623 | .569 |
| | | | | | | | .513 | .446 | | | | | | | | |
| 19.000 | .79 | 92 | 17.541 | 1.459 | 7.7% | 22 | .951 | .946 | .941 | .934 | .929 | .921 | .912 | .901 | .891 | .877 |
| | | | | | | | .862 | .844 | .824 | .799 | .774 | .741 | .705 | .663 | .616 | .562 |
| | | | | | | | .505 | .442 | | | | | | | | |
| 19.000 | .83 | 115 | 17.506 | 1.494 | 7.9% | 22 | .950 | .946 | .941 | .935 | .928 | .920 | .911 | .901 | .890 | .876 |
| | | | | | | | .859 | .844 | .822 | .798 | .771 | .738 | .702 | .660 | .612 | .561 |
| | | | | | | | .501 | .440 | | | | | | | | |
| 19.000 | .86 | 138 | 17.482 | 1.518 | 8.0% | 22 | .950 | .946 | .940 | .934 | .927 | .920 | .911 | .901 | .889 | .876 |
| | | | | | | | .860 | .842 | .821 | .796 | .769 | .737 | .700 | .657 | .610 | .557 |
| | | | | | | | .500 | .437 | | | | | | | | |
| 19.000 | .88 | 161 | 17.467 | 1.533 | 8.1% | 22 | .950 | .946 | .940 | .934 | .927 | .920 | .910 | .901 | .889 | .876 |
| | | | | | | | .859 | .842 | .820 | .796 | .768 | .735 | .699 | .655 | .609 | .556 |
| | | | | | | | .498 | .437 | | | | | | | | |
| 20.000 | .38 | 32 | 13.589 | 6.411 | 32.1% | 15 | .954 | .950 | .946 | .940 | .936 | .929 | .924 | .915 | .907 | .893 |
| | | | | | | | .891 | .875 | .860 | .843 | .825 | | | | | |
| 20.000 | .58 | 48 | 13.512 | 6.488 | 32.4% | 15 | .953 | .949 | .945 | .940 | .934 | .928 | .920 | .910 | .905 | .892 |
| | | | | | | | .880 | .865 | .850 | .830 | .810 | | | | | |
| 20.000 | .69 | 64 | 13.478 | 6.522 | 32.6% | 15 | .953 | .949 | .945 | .939 | .932 | .927 | .919 | .911 | .901 | .888 |
| | | | | | | | .876 | .863 | .846 | .825 | .803 | | | | | |
| 20.000 | .75 | 80 | 13.461 | 6.539 | 32.7% | 15 | .953 | .949 | .944 | .939 | .933 | .926 | .918 | .910 | .899 | .889 |
| | | | | | | | .877 | .859 | .842 | .825 | .799 | | | | | |
| 20.000 | .79 | 96 | 13.451 | 6.549 | 32.7% | 15 | .953 | .949 | .944 | .938 | .933 | .926 | .917 | .909 | .900 | .888 |
| | | | | | | | .874 | .859 | .842 | .821 | .798 | | | | | |
| 20.000 | .82 | 112 | 13.443 | 6.557 | 32.8% | 15 | .953 | .949 | .943 | .939 | .932 | .926 | .918 | .909 | .899 | .887 |
| | | | | | | | .872 | .859 | .840 | .820 | .796 | | | | | |
| 20.000 | .41 | 34 | 14.378 | 5.622 | 28.1% | 16 | .954 | .949 | .946 | .941 | .935 | .930 | .920 | .916 | .908 | .897 |
| | | | | | | | .885 | .872 | .857 | .842 | .824 | .804 | | | | |
| 20.000 | .61 | 51 | 14.290 | 5.710 | 28.6% | 16 | .953 | .949 | .945 | .940 | .933 | .928 | .920 | .911 | .903 | .892 |
| | | | | | | | .879 | .865 | .846 | .831 | .810 | .785 | | | | |
| 20.000 | .71 | 68 | 14.249 | 5.751 | 28.8% | 16 | .953 | .949 | .945 | .939 | .933 | .926 | .919 | .910 | .900 | .889 |
| | | | | | | | .878 | .860 | .847 | .825 | .800 | .776 | | | | |
| 20.000 | .76 | 85 | 14.231 | 5.769 | 28.8% | 16 | .953 | .949 | .944 | .939 | .933 | .928 | .917 | .911 | .900 | .887 |
| | | | | | | | .874 | .860 | .843 | .823 | .799 | .774 | | | | |
| 20.000 | .80 | 102 | 14.216 | 5.784 | 28.9% | 16 | .953 | .948 | .944 | .939 | .933 | .925 | .918 | .910 | .899 | .888 |
| | | | | | | | .874 | .858 | .841 | .821 | .797 | .769 | | | | |
| 20.000 | .83 | 119 | 14.208 | 5.792 | 29.0% | 16 | .953 | .949 | .944 | .938 | .933 | .926 | .918 | .909 | .898 | .887 |

ERLANG-ENGSET: FOR SMOOTH ATTEMPTING TRAFFIC AND FEW PERSONS DOING THE CALLING (FEW MEANS RATIO OF CALLERS TO CIRCUITS IS LESS THAN 8 - GREATER THAN 8 ASSUME TRAFFIC RANDOM)

| HOURS FIRST ATTEMPT | FIRST ATTEMPT VMR | NUMBER SOURCES | HOURS CON-NECT | HOURS OVFL | PERCENT BLOCKED (GRADE) | LINE | HOURS CONNECTED TRAFFIC ON LINES 1 | 2 | 3 | 4 | 5 | 6 | 7 | 8 | 9 | 10 |
|---|---|---|---|---|---|---|---|---|---|---|---|---|---|---|---|---|
| | | | | | | | .874 | .858 | .840 | .819 | .795 | .768 | | | | |
| 20.000 | .44 | 36 | 15.138 | 4.862 | 24.3% | 17 | .954 | .949 | .946 | .941 | .935 | .929 | .919 | .917 | .905 | .896 |
| | | | | | | | .880 | .872 | .860 | .840 | .822 | .798 | .775 | | | |
| 20.000 | .63 | 54 | 15.033 | 4.967 | 24.8% | 17 | .953 | .949 | .945 | .940 | .934 | .927 | .920 | .911 | .902 | .891 |
| | | | | | | | .878 | .865 | .847 | .829 | .806 | .782 | .753 | | | |
| 20.000 | .72 | 72 | 14.990 | 5.010 | 25.0% | 17 | .953 | .949 | .944 | .939 | .933 | .927 | .917 | .910 | .900 | .890 |
| | | | | | | | .876 | .860 | .844 | .825 | .802 | .775 | .746 | | | |
| 20.000 | .78 | 90 | 14.965 | 5.035 | 25.2% | 17 | .953 | .949 | .944 | .939 | .933 | .925 | .918 | .910 | .899 | .888 |
| | | | | | | | .875 | .860 | .842 | .822 | .798 | .770 | .741 | | | |
| 20.000 | .81 | 108 | 14.951 | 5.049 | 25.2% | 17 | .953 | .949 | .944 | .939 | .933 | .926 | .917 | .910 | .899 | .887 |
| | | | | | | | .874 | .858 | .841 | .820 | .797 | .769 | .737 | | | |
| 20.000 | .84 | 126 | 14.941 | 5.059 | 25.3% | 17 | .953 | .949 | .944 | .938 | .933 | .925 | .918 | .908 | .899 | .886 |
| | | | | | | | .873 | .858 | .839 | .820 | .795 | .768 | .735 | | | |
| 20.000 | .47 | 38 | 15.858 | 4.142 | 20.7% | 18 | .954 | .950 | .945 | .941 | .935 | .929 | .921 | .914 | .905 | .894 |
| | | | | | | | .883 | .871 | .855 | .838 | .819 | .796 | .769 | .740 | | |
| 20.000 | .65 | 57 | 15.740 | 4.260 | 21.3% | 18 | .953 | .949 | .945 | .940 | .933 | .927 | .919 | .911 | .903 | .890 |
| | | | | | | | .877 | .864 | .848 | .827 | .805 | .780 | .753 | .717 | | |
| 20.000 | .74 | 76 | 15.691 | 4.309 | 21.5% | 18 | .953 | .949 | .944 | .938 | .933 | .926 | .919 | .910 | .900 | .889 |
| | | | | | | | .875 | .861 | .842 | .825 | .799 | .774 | .745 | .707 | | |
| 20.000 | .79 | 95 | 15.666 | 4.334 | 21.7% | 18 | .953 | .948 | .944 | .939 | .932 | .927 | .918 | .909 | .898 | .887 |
| | | | | | | | .874 | .859 | .842 | .821 | .797 | .772 | .739 | .705 | | |
| 20.000 | .82 | 114 | 15.649 | 4.351 | 21.8% | 18 | .953 | .949 | .944 | .939 | .932 | .926 | .917 | .909 | .899 | .887 |
| | | | | | | | .874 | .858 | .840 | .819 | .797 | .769 | .737 | .702 | | |
| 20.000 | .85 | 133 | 15.638 | 4.362 | 21.8% | 18 | .953 | .948 | .944 | .938 | .932 | .926 | .918 | .908 | .898 | .887 |
| | | | | | | | .873 | .857 | .840 | .819 | .794 | .766 | .737 | .699 | | |
| 20.000 | .50 | 40 | 16.533 | 3.467 | 17.3% | 19 | .954 | .949 | .946 | .940 | .935 | .928 | .922 | .911 | .906 | .894 |
| | | | | | | | .883 | .869 | .853 | .834 | .819 | .793 | .766 | .736 | .695 | |
| 20.000 | .67 | 60 | 16.406 | 3.594 | 18.0% | 19 | .953 | .949 | .944 | .940 | .934 | .926 | .920 | .911 | .900 | .890 |
| | | | | | | | .878 | .862 | .848 | .826 | .805 | .780 | .749 | .715 | .676 | |
| 20.000 | .75 | 80 | 16.353 | 3.647 | 18.2% | 19 | .953 | .949 | .944 | .939 | .933 | .926 | .918 | .910 | .899 | .889 |
| | | | | | | | .877 | .859 | .842 | .825 | .799 | .774 | .742 | .707 | .669 | |
| 20.000 | .80 | 100 | 16.324 | 3.676 | 18.4% | 19 | .953 | .949 | .944 | .939 | .933 | .925 | .919 | .909 | .899 | .888 |
| | | | | | | | .874 | .859 | .841 | .821 | .797 | .769 | .740 | .702 | .665 | |
| 20.000 | .83 | 120 | 16.305 | 3.695 | 18.5% | 19 | .953 | .949 | .944 | .938 | .933 | .926 | .918 | .909 | .899 | .887 |
| | | | | | | | .874 | .858 | .840 | .819 | .795 | .769 | .736 | .699 | .661 | |
| 20.000 | .86 | 140 | 16.292 | 3.708 | 18.5% | 19 | .953 | .948 | .944 | .938 | .933 | .925 | .917 | .909 | .898 | .886 |
| | | | | | | | .873 | .857 | .839 | .819 | .793 | .768 | .735 | .699 | .657 | |
| 20.000 | .52 | 42 | 17.168 | 2.832 | 14.2% | 20 | .954 | .950 | .945 | .940 | .935 | .928 | .921 | .912 | .905 | .893 |
| | | | | | | | .880 | .870 | .852 | .834 | .817 | .791 | .764 | .729 | .698 | .652 |
| 20.000 | .68 | 63 | 17.022 | 2.978 | 14.9% | 20 | .953 | .949 | .945 | .939 | .933 | .927 | .919 | .911 | .900 | .891 |

# ERLANG-ENGSET: FOR SMOOTH ATTEMPTING TRAFFIC AND FEW PERSONS DOING THE CALLING (FEW MEANS RATIO OF CALLERS TO CIRCUITS IS LESS THAN 8 - GREATER THAN 8 ASSUME TRAFFIC RANDOM)

| HOURS FIRST ATTEMPT | FIRST ATTEMPT VMR | NUMBER SOURCES | HOURS CON-NECT | HOURS OVFL | PERCENT BLOCKED (GRADE) | LINE | HOURS CONNECTED TRAFFIC ON LINES 1 | 2 | 3 | 4 | 5 | 6 | 7 | 8 | 9 | 10 |
|---|---|---|---|---|---|---|---|---|---|---|---|---|---|---|---|---|
| | | | | | | | .876 | .862 | .847 | .824 | .805 | .779 | .748 | .714 | .675 | .626 |
| 20.000 | .76 | 84 | 16.966 | 3.034 | 15.2% | 20 | .953 | .949 | .944 | .939 | .933 | .926 | .919 | .909 | .900 | .889 |
| | | | | | | | .874 | .860 | .842 | .822 | .801 | .770 | .744 | .706 | .665 | .621 |
| 20.000 | .81 | 105 | 16.934 | 3.066 | 15.3% | 20 | .953 | .948 | .944 | .939 | .933 | .926 | .917 | .909 | .899 | .888 |
| | | | | | | | .874 | .858 | .841 | .821 | .796 | .770 | .739 | .702 | .660 | .617 |
| 20.000 | .84 | 126 | 16.913 | 3.087 | 15.4% | 20 | .953 | .949 | .944 | .938 | .933 | .925 | .918 | .908 | .899 | .886 |
| | | | | | | | .873 | .858 | .839 | .820 | .795 | .768 | .735 | .701 | .658 | .614 |
| 20.000 | .86 | 147 | 16.899 | 3.101 | 15.5% | 20 | .953 | .948 | .944 | .938 | .932 | .925 | .917 | .908 | .899 | .886 |
| | | | | | | | .872 | .857 | .839 | .818 | .793 | .767 | .734 | .699 | .657 | .611 |
| 20.000 | .55 | 44 | 17.740 | 2.260 | 11.3% | 21 | .953 | .949 | .944 | .941 | .933 | .929 | .019 | .914 | .902 | .895 |
| | | | | | | | .881 | .868 | .851 | .833 | .814 | .790 | .757 | .733 | .689 | .649 |
| | | | | | | | .595 | | | | | | | | | |
| 20.000 | .70 | 66 | 17.588 | 2.412 | 12.1% | 21 | .953 | .949 | .944 | .940 | .933 | .927 | .918 | .911 | .902 | .888 |
| | | | | | | | .877 | .862 | .844 | .826 | .804 | .775 | .749 | .711 | .672 | .629 |
| | | | | | | | .575 | | | | | | | | | |
| 20.000 | .77 | 88 | 17.524 | 2.476 | 12.4% | 21 | .953 | .949 | .944 | .939 | .933 | .926 | .918 | .909 | .900 | .889 |
| | | | | | | | .873 | .861 | .841 | .823 | .796 | .772 | .743 | .704 | .664 | .620 |
| | | | | | | | .567 | | | | | | | | | |
| 20.000 | .82 | 110 | 17.492 | 2.508 | 12.5% | 21 | .953 | .949 | .944 | .939 | .932 | .926 | .918 | .909 | .899 | .887 |
| | | | | | | | .874 | .858 | .840 | .820 | .796 | .770 | .738 | .700 | .663 | .615 |
| | | | | | | | .563 | | | | | | | | | |
| 20.000 | .85 | 132 | 17.470 | 2.530 | 12.7% | 21 | .953 | .949 | .944 | .938 | .932 | .925 | .918 | .908 | .899 | .887 |
| | | | | | | | .873 | .858 | .839 | .819 | .795 | .767 | .735 | .700 | .657 | .613 |
| | | | | | | | .561 | | | | | | | | | |
| 20.000 | .87 | 154 | 17.454 | 2.546 | 12.7% | 21 | .953 | .948 | .944 | .938 | .932 | .925 | .917 | .908 | .898 | .886 |
| | | | | | | | .873 | .856 | .839 | .818 | .794 | .766 | .734 | .698 | .656 | .611 |
| | | | | | | | .559 | | | | | | | | | |
| 20.000 | .57 | 46 | 18.253 | 1.747 | 8.7% | 22 | .953 | .950 | .944 | .940 | .934 | .928 | .920 | .912 | .904 | .891 |
| | | | | | | | .881 | .864 | .853 | .833 | .812 | .787 | .759 | .726 | .689 | .645 |
| | | | | | | | .593 | .534 | | | | | | | | |
| 20.000 | .71 | 69 | 18.004 | 1.906 | 9.5% | 22 | .953 | .949 | .944 | .940 | .932 | .927 | .918 | .911 | .901 | .889 |
| | | | | | | | .876 | .860 | .844 | .825 | .803 | .776 | .744 | .712 | .671 | .625 |
| | | | | | | | .574 | .518 | | | | | | | | |
| 20.000 | .78 | 92 | 18.028 | 1.972 | 9.9% | 22 | .953 | .949 | .944 | .939 | .933 | .925 | .919 | .909 | .899 | .887 |
| | | | | | | | .874 | .861 | .841 | .821 | .799 | .771 | .739 | .705 | .664 | .617 |
| | | | | | | | .568 | .511 | | | | | | | | |
| 20.000 | .83 | 115 | 17.991 | 2.009 | 10.0% | 22 | .953 | .949 | .944 | .939 | .932 | .926 | .918 | .909 | .890 | .000 |
| | | | | | | | .075 | .857 | .841 | .820 | .796 | .768 | .737 | .701 | .661 | .613 |
| | | | | | | | .564 | .505 | | | | | | | | |
| 20.000 | .86 | 138 | 17.969 | 2.031 | 10.2% | 22 | .953 | .948 | .944 | .938 | .932 | .925 | .917 | .909 | .898 | .887 |

ERLANG-ENGSET: FOR SMOOTH ATTEMPTING TRAFFIC AND FEW PERSONS DOING THE CALLING (FEW MEANS RATIO OF CALLERS TO CIRCUITS IS LESS THAN 8 - GREATER THAN 8 ASSUME TRAFFIC RANDOM)

| HOURS FIRST ATTEMPT | FIRST ATTEMPT VMR | NUMBER SOURCES | HOURS CON-NECT | HOURS OVFL | PERCENT BLOCKED (GRADE) | LINE | HOURS CONNECTED TRAFFIC ON LINES 1 | 2 | 3 | 4 | 5 | 6 | 7 | 8 | 9 | 10 |
|---|---|---|---|---|---|---|---|---|---|---|---|---|---|---|---|---|
| | | | | | | | .873 | .858 | .839 | .818 | .795 | .767 | .736 | .699 | .657 | .613 |
| | | | | | | | .560 | .504 | | | | | | | | |
| 20.000 | .88 | 161 | 17.952 | 2.048 | 10.2% | 22 | .953 | .948 | .944 | .938 | .932 | .925 | .918 | .908 | .898 | .886 |
| | | | | | | | .873 | .856 | .838 | .818 | .794 | .765 | .734 | .697 | .656 | .609 |
| | | | | | | | .558 | .502 | | | | | | | | |
| 20.000 | .17 | 24 | 19.761 | .239 | 1.2% | 23 | .955 | .949 | .948 | .942 | .938 | .930 | .925 | .919 | .912 | .904 |
| | | | | | | | .899 | .885 | .877 | .862 | .859 | .843 | .831 | .815 | .800 | .770 |
| | | | | | | | .756 | .688 | .555 | | | | | | | |
| 20.000 | .58 | 48 | 18.695 | 1.305 | 6.5% | 23 | .953 | .949 | .945 | .940 | .934 | .928 | .920 | .910 | .905 | .892 |
| | | | | | | | .880 | .865 | .850 | .830 | .810 | .786 | .758 | .724 | .686 | .643 |
| | | | | | | | .590 | .529 | .467 | | | | | | | |
| 20.000 | .72 | 72 | 18.534 | 1.466 | 7.3% | 23 | .953 | .949 | .944 | .939 | .933 | .927 | .917 | .910 | .900 | .890 |
| | | | | | | | .876 | .860 | .844 | .825 | .802 | .775 | .746 | .708 | .672 | .624 |
| | | | | | | | .574 | .516 | .451 | | | | | | | |
| 20.000 | .79 | 96 | 18.468 | 1.532 | 7.7% | 23 | .953 | .949 | .944 | .938 | .933 | .926 | .917 | .909 | .900 | .888 |
| | | | | | | | .874 | .859 | .842 | .821 | .798 | .770 | .740 | .702 | .665 | .616 |
| | | | | | | | .567 | .508 | .449 | | | | | | | |
| 20.000 | .83 | 120 | 18.431 | 1.569 | 7.8% | 23 | .953 | .949 | .944 | .938 | .933 | .926 | .918 | .909 | .899 | .887 |
| | | | | | | | .874 | .858 | .840 | .819 | .795 | .769 | .736 | .699 | .661 | .614 |
| | | | | | | | .562 | .504 | .446 | | | | | | | |
| 20.000 | .86 | 144 | 18.407 | 1.593 | 8.0% | 23 | .953 | .949 | .944 | .938 | .932 | .925 | .918 | .908 | .898 | .886 |
| | | | | | | | .873 | .857 | .839 | .818 | .795 | .766 | .735 | .698 | .657 | .611 |
| | | | | | | | .559 | .504 | .443 | | | | | | | |
| 20.000 | .88 | 168 | 18.391 | 1.609 | 8.0% | 23 | .953 | .948 | .944 | .938 | .932 | .925 | .917 | .908 | .898 | .886 |
| | | | | | | | .873 | .856 | .839 | .817 | .793 | .766 | .733 | .697 | .656 | .609 |
| | | | | | | | .557 | .502 | .442 | | | | | | | |
| 22.000 | .35 | 34 | 14.637 | 7.363 | 33.5% | 16 | .958 | .954 | .951 | .948 | .944 | .937 | .933 | .926 | .919 | .911 |
| | | | | | | | .904 | .896 | .882 | .874 | .853 | .848 | | | | |
| 22.000 | .57 | 51 | 14.560 | 7.440 | 33.8% | 16 | .957 | .954 | .949 | .946 | .942 | .934 | .932 | .923 | .915 | .910 |
| | | | | | | | .898 | .888 | .876 | .861 | .846 | .828 | | | | |
| 22.000 | .68 | 68 | 14.528 | 7.472 | 34.0% | 16 | .957 | .954 | .950 | .945 | .941 | .936 | .930 | .922 | .915 | .906 |
| | | | | | | | .897 | .883 | .873 | .858 | .840 | .822 | | | | |
| 22.000 | .74 | 85 | 14.511 | 7.489 | 34.0% | 16 | .957 | .954 | .949 | .945 | .940 | .936 | .929 | .923 | .914 | .905 |
| | | | | | | | .895 | .884 | .869 | .857 | .838 | .818 | | | | |
| 22.000 | .78 | 102 | 14.502 | 7.498 | 34.1% | 16 | .957 | .953 | .950 | .945 | .940 | .935 | .929 | .922 | .913 | .904 |
| | | | | | | | .895 | .883 | .869 | .854 | .836 | .817 | | | | |
| 22.000 | .82 | 119 | 14.494 | 7.506 | 34.1% | 16 | .957 | .953 | .949 | .945 | .939 | .935 | .928 | .922 | .913 | .904 |
| | | | | | | | .894 | .882 | .869 | .853 | .835 | .815 | | | | |
| 22.000 | .39 | 36 | 15.448 | 6.552 | 29.8% | 17 | .958 | .955 | .951 | .947 | .942 | .936 | .932 | .928 | .917 | .914 |
| | | | | | | | .903 | .891 | .885 | .866 | .861 | .840 | .824 | | | |

ERLANG-ENGSET: FOR SMOOTH ATTEMPTING TRAFFIC AND FEW PERSONS DOING THE CALLING (FEW MEANS RATIO OF CALLERS TO CIRCUITS IS LESS THAN 8 - GREATER THAN 8 ASSUME TRAFFIC RANDOM)

| HOURS FIRST ATTEMPT | FIRST ATTEMPT VMR | NUMBER SOURCES | HOURS CON-NECT | HOURS OVFL | PERCENT BLOCKED (GRADE) | LINE | HOURS CONNECTED TRAFFIC ON LINES 1 | 2 | 3 | 4 | 5 | 6 | 7 | 8 | 9 | 10 |
|---|---|---|---|---|---|---|---|---|---|---|---|---|---|---|---|---|
| 22.000 | .59 | 54 | 15.360 | 6.640 | 30.2% | 17 | .957 | .954 | .950 | .946 | .942 | .935 | .929 | .925 | .915 | .908 |
| | | | | | | | .898 | .887 | .875 | .860 | .845 | .827 | .806 | | | |
| 22.000 | .69 | 72 | 15.322 | 6.678 | 30.4% | 17 | .957 | .954 | .950 | .945 | .941 | .935 | .929 | .923 | .915 | .906 |
| | | | | | | | .896 | .884 | .873 | .854 | .842 | .820 | .798 | | | |
| 22.000 | .76 | 90 | 15.304 | 6.696 | 30.4% | 17 | .957 | .953 | .950 | .945 | .941 | .935 | .929 | .920 | .915 | .905 |
| | | | | | | | .894 | .883 | .870 | .856 | .838 | .818 | .795 | | | |
| 22.000 | .80 | 108 | 15.291 | 6.709 | 30.5% | 17 | .957 | .953 | .949 | .945 | .941 | .935 | .928 | .922 | .913 | .905 |
| | | | | | | | .894 | .882 | .869 | .854 | .836 | .816 | .792 | | | |
| 22.000 | .83 | 126 | 15.283 | 6.717 | 30.5% | 17 | .957 | .953 | .950 | .945 | .940 | .935 | .929 | .921 | .913 | .904 |
| | | | | | | | .894 | .882 | .868 | .853 | .835 | .815 | .791 | | | |
| 22.000 | .42 | 38 | 16.228 | 5.772 | 26.2% | 18 | .958 | .955 | .951 | .946 | .942 | .937 | .932 | .925 | .919 | .907 |
| | | | | | | | .902 | .896 | .879 | .870 | .853 | .840 | .819 | .796 | | |
| 22.000 | .61 | 57 | 16.129 | 5.871 | 26.7% | 18 | .957 | .954 | .950 | .946 | .942 | .936 | .929 | .924 | .915 | .908 |
| | | | | | | | .899 | .807 | .873 | .861 | .844 | .823 | .804 | .781 | | |
| 22.000 | .71 | 76 | 16.090 | 5.910 | 26.9% | 18 | .957 | .953 | .950 | .945 | .941 | .935 | .929 | .923 | .913 | .906 |
| | | | | | | | .895 | .883 | .872 | .858 | .839 | .820 | .798 | .771 | | |
| 22.000 | .77 | 95 | 16.067 | 5.933 | 27.0% | 18 | .957 | .954 | .950 | .945 | .941 | .935 | .928 | .922 | .912 | .906 |
| | | | | | | | .895 | .883 | .868 | .856 | .835 | .819 | .794 | .767 | | |
| 22.000 | .81 | 114 | 16.054 | 5.946 | 27.0% | 18 | .957 | .953 | .949 | .945 | .940 | .935 | .928 | .920 | .914 | .904 |
| | | | | | | | .894 | .882 | .869 | .853 | .836 | .815 | .791 | .767 | | |
| 22.000 | .83 | 133 | 16.044 | 5.956 | 27.1% | 18 | .957 | .953 | .949 | .945 | .940 | .934 | .928 | .921 | .912 | .903 |
| | | | | | | | .893 | .883 | .868 | .852 | .835 | .815 | .790 | .764 | | |
| 22.000 | .45 | 40 | 16.982 | 5.018 | 22.8% | 19 | .958 | .954 | .950 | .947 | .942 | .937 | .930 | .923 | .921 | .007 |
| | | | | | | | .904 | .009 | .880 | .869 | .852 | .831 | .822 | .794 | .769 | |
| 22.000 | .03 | 60 | 16.871 | 5.129 | 23.3% | 19 | .957 | .954 | .950 | .946 | .941 | .936 | .929 | .924 | .914 | .908 |
| | | | | | | | .896 | .886 | .873 | .860 | .843 | .824 | .802 | .776 | .750 | |
| 22.000 | .73 | 80 | 16.824 | 5.176 | 23.5% | 19 | .957 | .954 | .949 | .946 | .941 | .935 | .930 | .922 | .914 | .906 |
| | | | | | | | .895 | .882 | .870 | .857 | .840 | .818 | .798 | .771 | .739 | |
| 22.000 | .78 | 100 | 16.801 | 5.199 | 23.6% | 19 | .957 | .954 | .950 | .945 | .939 | .936 | .929 | .922 | .913 | .905 |
| | | | | | | | .895 | .883 | .870 | .852 | .839 | .817 | .794 | .766 | .738 | |
| 22.000 | .82 | 120 | 16.785 | 5.215 | 23.7% | 19 | .957 | .953 | .950 | .945 | .940 | .935 | .928 | .922 | .913 | .904 |
| | | | | | | | .894 | .882 | .868 | .853 | .835 | .815 | .790 | .766 | .704 | |
| 22.000 | .84 | 140 | 16.774 | 5.226 | 23.8% | 10 | .957 | .953 | .949 | .945 | .940 | .934 | .928 | .921 | .913 | .904 |
| | | | | | | | .893 | .881 | .868 | .852 | .834 | .814 | .700 | .762 | .734 | |
| 22.000 | .48 | 42 | 17.698 | 4.302 | 19.6% | 20 | .958 | .954 | .951 | 946 | .042 | .937 | .931 | .926 | .917 | .909 |
| | | | | | | | .901 | .891 | .878 | .864 | .855 | .830 | .819 | .792 | .767 | .703 |
| 22.000 | .65 | 63 | 17.576 | 4.424 | 20.1% | 20 | .957 | .954 | .950 | .946 | .941 | .935 | .930 | .923 | .914 | .907 |
| | | | | | | | .896 | .886 | .874 | .858 | .842 | .823 | .802 | .775 | .748 | .715 |
| 22.000 | .74 | 84 | 17.525 | 4.475 | 20.3% | 20 | .957 | .953 | .950 | .945 | .940 | .935 | .928 | .922 | .914 | .906 |
| | | | | | | | .894 | .884 | .872 | .855 | .838 | .818 | .798 | .768 | .743 | .706 |

ERLANG-ENGSET: FOR SMOOTH ATTEMPTING TRAFFIC AND FEW PERSONS DOING THE CALLING (FEW MEANS RATIO OF CALLERS TO CIRCUITS IS LESS THAN 8 - GREATER THAN 8 ASSUME TRAFFIC RANDOM)

| HOURS FIRST ATTEMPT | FIRST ATTEMPT VMR | NUMBER SOURCES | HOURS CONNECT | HOURS OVFL | PERCENT BLOCKED (GRADE) | LINE | HOURS CONNECTED TRAFFIC ON LINES 1 | 2 | 3 | 4 | 5 | 6 | 7 | 8 | 9 | 10 |
|---|---|---|---|---|---|---|---|---|---|---|---|---|---|---|---|---|
| 22.000 | .79 | 105 | 17.496 | 4.504 | 20.5% | 20 | .957 | .953 | .949 | .945 | .940 | .934 | .929 | .921 | .914 | .904 |
| | | | | | | | .893 | .884 | .869 | .853 | .836 | .818 | .793 | .766 | .735 | .702 |
| 22.000 | .83 | 126 | 17.480 | 4.520 | 20.5% | 20 | .957 | .953 | .950 | .945 | .940 | .935 | .929 | .921 | .913 | .904 |
| | | | | | | | .894 | .882 | .868 | .853 | .835 | .815 | .791 | .763 | .735 | .699 |
| 22.000 | .85 | 147 | 17.468 | 4.532 | 20.6% | 20 | .957 | .953 | .949 | .945 | .940 | .934 | .928 | .921 | .913 | .904 |
| | | | | | | | .893 | .881 | .868 | .852 | .834 | .814 | .790 | .763 | .732 | .697 |
| 22.000 | .50 | 44 | 18.381 | 3.619 | 16.5% | 21 | .958 | .954 | .950 | .946 | .942 | .937 | .931 | .924 | .918 | .910 |
| | | | | | | | .899 | .891 | .878 | .865 | .848 | .833 | .813 | .788 | .765 | .729 |
| | | | | | | | .702 | | | | | | | | | |
| 22.000 | .67 | 66 | 18.242 | 3.758 | 17.1% | 21 | .957 | .954 | .950 | .946 | .941 | .935 | .930 | .923 | .915 | .907 |
| | | | | | | | .895 | .887 | .872 | .857 | .840 | .823 | .802 | .774 | .746 | .713 |
| | | | | | | | .676 | | | | | | | | | |
| 22.000 | .75 | 88 | 18.184 | 3.816 | 17.3% | 21 | .957 | .954 | .949 | .945 | .941 | .935 | .928 | .922 | .915 | .905 |
| | | | | | | | .895 | .882 | .872 | .855 | .838 | .818 | .794 | .771 | .739 | .705 |
| | | | | | | | .666 | | | | | | | | | |
| 22.000 | .80 | 110 | 18.155 | 3.845 | 17.5% | 21 | .957 | .953 | .949 | .945 | .940 | .934 | .928 | .922 | .914 | .904 |
| | | | | | | | .894 | .881 | .870 | .852 | .837 | .816 | .792 | .766 | .736 | .701 |
| | | | | | | | .663 | | | | | | | | | |
| 22.000 | .83 | 132 | 18.135 | 3.865 | 17.6% | 21 | .957 | .953 | .949 | .945 | .940 | .934 | .929 | .921 | .913 | .904 |
| | | | | | | | .894 | .881 | .868 | .853 | .835 | .813 | .791 | .764 | .733 | .699 |
| | | | | | | | .659 | | | | | | | | | |
| 22.000 | .86 | 154 | 18.122 | 3.878 | 17.6% | 21 | .957 | .953 | .949 | .945 | .940 | .935 | .928 | .921 | .913 | .904 |
| | | | | | | | .893 | .881 | .867 | .852 | .833 | .813 | .789 | .762 | .732 | .697 |
| | | | | | | | .658 | | | | | | | | | |
| 22.000 | .52 | 46 | 19.014 | 2.986 | 13.6% | 22 | .958 | .954 | .950 | .946 | .941 | .937 | .929 | .925 | .916 | .909 |
| | | | | | | | .900 | .889 | .877 | .862 | .851 | .830 | .811 | .787 | .761 | .730 |
| | | | | | | | .694 | .653 | | | | | | | | |
| 22.000 | .68 | 69 | 18.863 | 3.137 | 14.3% | 22 | .957 | .954 | .950 | .946 | .940 | .936 | .929 | .922 | .915 | .906 |
| | | | | | | | .897 | .886 | .870 | .857 | .842 | .821 | .800 | .774 | .743 | .713 |
| | | | | | | | .673 | .632 | | | | | | | | |
| 22.000 | .76 | 92 | 18.799 | 3.201 | 14.6% | 22 | .957 | .953 | .949 | .946 | .940 | .935 | .928 | .922 | .914 | .904 |
| | | | | | | | .895 | .883 | .868 | .856 | .836 | .819 | .792 | .769 | .740 | .703 |
| | | | | | | | .667 | .619 | | | | | | | | |
| 22.000 | .81 | 115 | 18.767 | 3.233 | 14.7% | 22 | .957 | .953 | .950 | .945 | .941 | .935 | .928 | .921 | .913 | .903 |
| | | | | | | | .895 | .881 | .868 | .854 | .834 | .817 | .792 | .765 | .733 | .701 |
| | | | | | | | .663 | .617 | | | | | | | | |
| 22.000 | .84 | 138 | 18.745 | 3.255 | 14.8% | 22 | .957 | .953 | .949 | .945 | .940 | .935 | .928 | .920 | .914 | .903 |
| | | | | | | | .893 | .881 | .869 | .852 | .836 | .814 | .790 | .764 | .732 | .699 |
| | | | | | | | .659 | .614 | | | | | | | | |
| 22.000 | .86 | 161 | 18.731 | 3.269 | 14.9% | 22 | .957 | .953 | .949 | .945 | .940 | .934 | .928 | .921 | .913 | .903 |

ERLANG-ENGSET: FOR SMOOTH ATTEMPTING TRAFFIC AND FEW PERSONS DOING THE CALLING (FEW MEANS RATIO OF CALLERS TO CIRCUITS IS LESS THAN 8 - GREATER THAN 8 ASSUME TRAFFIC RANDOM)

| HOURS FIRST ATTEMPT | FIRST ATTEMPT VMR | NUMBER SOURCES | HOURS CON-NECT | HOURS OVFL | PERCENT BLOCKED (GRADE) | LINE | HOURS CONNECTED TRAFFIC ON LINES 1 | 2 | 3 | 4 | 5 | 6 | 7 | 8 | 9 | 10 |
|---|---|---|---|---|---|---|---|---|---|---|---|---|---|---|---|---|
| | | | | | | | .893 | .881 | .868 | .851 | .834 | .813 | .789 | .762 | .731 | .696 |
| | | | | | | | .657 | .613 | | | | | | | | |
| 22.000 | .54 | 48 | 19.593 | 2.407 | 10.9% | 23 | .957 | .954 | .950 | .946 | .940 | .938 | .930 | .924 | .917 | .908 |
| | | | | | | | .898 | .889 | .876 | .863 | .847 | .828 | .811 | .785 | .757 | .727 |
| | | | | | | | .690 | .653 | .601 | | | | | | | |
| 22.000 | .69 | 72 | 19.432 | 2.568 | 11.7% | 23 | .957 | .954 | .950 | .945 | .941 | .935 | .929 | .923 | .915 | .906 |
| | | | | | | | .896 | .884 | .873 | .854 | .842 | .820 | .798 | .772 | .744 | .711 |
| | | | | | | | .674 | .627 | .581 | | | | | | | |
| 22.000 | .77 | 96 | 19.365 | 2.635 | 12.0% | 23 | .957 | .954 | .949 | .945 | .940 | .935 | .929 | .922 | .914 | .905 |
| | | | | | | | .893 | .885 | .869 | .854 | .837 | .819 | .792 | .769 | .739 | .703 |
| | | | | | | | .664 | .620 | .573 | | | | | | | |
| 22.000 | .82 | 120 | 19.329 | 2.671 | 12.1% | 23 | .957 | .953 | .950 | .945 | .940 | .935 | .928 | .922 | .913 | .904 |
| | | | | | | | .894 | .882 | .868 | .853 | .835 | .815 | .790 | .766 | .734 | .700 |
| | | | | | | | .660 | .617 | .568 | | | | | | | |
| 22.000 | .85 | 144 | 19.305 | 2.695 | 12.2% | 23 | .957 | .953 | .949 | .945 | .940 | .935 | .928 | .921 | .912 | .905 |
| | | | | | | | .893 | .881 | .869 | .851 | .835 | .814 | .790 | .763 | .732 | .698 |
| | | | | | | | .658 | .614 | .565 | | | | | | | |
| 22.000 | .87 | 168 | 19.290 | 2.710 | 12.3% | 23 | .957 | .953 | .949 | .945 | .940 | .934 | .928 | .921 | .913 | .903 |
| | | | | | | | .894 | .881 | .867 | .851 | .834 | .812 | .789 | .762 | .731 | .696 |
| | | | | | | | .656 | .612 | .563 | | | | | | | |
| 22.000 | .56 | 50 | 20.112 | 1.888 | 8.6% | 24 | .958 | .954 | .950 | .946 | .941 | .937 | .931 | .923 | .917 | .909 |
| | | | | | | | .899 | .887 | .877 | .861 | .843 | .832 | .807 | .784 | .755 | .724 |
| | | | | | | | .693 | .646 | .596 | .542 | | | | | | |
| 22.000 | .71 | 75 | 19.944 | 2.056 | 9.3% | 24 | .957 | .953 | .950 | .945 | .941 | .936 | .927 | .924 | .914 | .905 |
| | | | | | | | .897 | .884 | .870 | .857 | .841 | .819 | .797 | .773 | .742 | .711 |
| | | | | | | | .669 | .628 | .578 | .525 | | | | | | |
| 22.000 | .78 | 100 | 19.875 | 2.125 | 9.7% | 24 | .957 | .954 | .950 | .945 | .939 | .936 | .929 | .922 | .913 | .905 |
| | | | | | | | .895 | .883 | .870 | .852 | .839 | .817 | .794 | .766 | .738 | .701 |
| | | | | | | | .666 | .618 | .572 | .518 | | | | | | |
| 22.000 | .82 | 125 | 19.837 | 2.163 | 9.8% | 24 | .957 | .953 | .949 | .945 | .940 | .934 | .928 | .921 | .913 | .903 |
| | | | | | | | .895 | .881 | .869 | .852 | .836 | .814 | .792 | .764 | .734 | .699 |
| | | | | | | | .660 | .615 | .567 | .513 | | | | | | |
| 22.000 | .85 | 150 | 19.812 | 2.188 | 9.9% | 24 | .957 | .953 | .949 | .945 | .940 | .934 | .929 | .921 | .913 | .903 |
| | | | | | | | .893 | .881 | .868 | .851 | .835 | .813 | .790 | .763 | .732 | .697 |
| | | | | | | | .657 | .613 | .564 | .511 | | | | | | |
| 22.000 | .87 | 175 | 19.795 | 2.205 | 10.0% | 24 | .957 | .953 | .949 | .945 | .940 | .934 | .928 | .921 | .913 | .903 |
| | | | | | | | .893 | .881 | .867 | .851 | .833 | .812 | .789 | .761 | .730 | .695 |
| | | | | | | | .656 | .611 | .562 | .510 | | | | | | |
| 22.000 | .58 | 52 | 20.566 | 1.434 | 6.5% | 25 | .957 | .954 | .950 | .946 | .941 | .936 | .931 | .923 | .916 | .909 |
| | | | | | | | .898 | .887 | .875 | .862 | .846 | .827 | .805 | .785 | .755 | .723 |

ERLANG-ENGSET: FOR SMOOTH ATTEMPTING TRAFFIC AND FEW PERSONS DOING THE CALLING (FEW MEANS RATIO OF CALLERS TO CIRCUITS IS LESS THAN 8 - GREATER THAN 8 ASSUME TRAFFIC RANDOM)

| HOURS FIRST ATTEMPT | FIRST ATTEMPT VMR | NUMBER SOURCES | HOURS CON-NECT | HOURS OVFL | PERCENT BLOCKED (GRADE) | LINE | HOURS CONNECTED TRAFFIC ON LINES 1 | 2 | 3 | 4 | 5 | 6 | 7 | 8 | 9 | 10 |
|---|---|---|---|---|---|---|---|---|---|---|---|---|---|---|---|---|
| | | | | | | | .687 | .640 | .598 | .539 | .476 | | | | | |
| 22.000 | .72 | 78 | 20.396 | 1.604 | 7.3% | 25 | .957 | .953 | .950 | .945 | .941 | .935 | .930 | .923 | .914 | .905 |
| | | | | | | | .896 | .884 | .869 | .855 | .839 | .820 | .797 | .771 | .742 | .709 |
| | | | | | | | .669 | .629 | .576 | .522 | .464 | | | | | |
| 22.000 | .79 | 104 | 20.325 | 1.675 | 7.6% | 25 | .957 | .953 | .950 | .945 | .940 | .935 | .929 | .920 | .915 | .905 |
| | | | | | | | .894 | .883 | .869 | .854 | .836 | .816 | .793 | .765 | .737 | .703 |
| | | | | | | | .661 | .621 | .570 | .516 | .458 | | | | | |
| 22.000 | .83 | 130 | 20.286 | 1.714 | 7.8% | 25 | .957 | .953 | .949 | .945 | .940 | .935 | .928 | .921 | .913 | .904 |
| | | | | | | | .894 | .882 | .868 | .852 | .835 | .814 | .791 | .763 | .734 | .699 |
| | | | | | | | .659 | .614 | .567 | .512 | .455 | | | | | |
| 22.000 | .86 | 156 | 20.261 | 1.739 | 7.9% | 25 | .957 | .953 | .949 | .945 | .940 | .934 | .928 | .921 | .913 | .903 |
| | | | | | | | .893 | .881 | .868 | .852 | .834 | .812 | .790 | .763 | .731 | .696 |
| | | | | | | | .657 | .613 | .564 | .511 | .453 | | | | | |
| 22.000 | .88 | 182 | 20.244 | 1.756 | 8.0% | 25 | .957 | .953 | .949 | .945 | .940 | .934 | .928 | .921 | .913 | .903 |
| | | | | | | | .893 | .881 | .867 | .851 | .833 | .813 | .788 | .762 | .730 | .695 |
| | | | | | | | .656 | .610 | .563 | .509 | .452 | | | | | |
| 24.000 | .33 | 36 | 15.676 | 8.324 | 34.7% | 17 | .961 | .958 | .955 | .952 | .946 | .944 | .942 | .934 | .931 | .923 |
| | | | | | | | .917 | .910 | .898 | .893 | .880 | .867 | .864 | | | |
| 24.000 | .56 | 54 | 15.602 | 8.398 | 35.0% | 17 | .961 | .958 | .955 | .950 | .948 | .943 | .936 | .934 | .927 | .918 |
| | | | | | | | .915 | .903 | .895 | .885 | .872 | .858 | .843 | | | |
| 24.000 | .58 | 57 | 16.418 | 7.582 | 31.6% | 18 | .961 | .958 | .955 | .951 | .948 | .942 | .938 | .932 | .928 | .919 |
| | | | | | | | .910 | .907 | .893 | .884 | .872 | .857 | .840 | .824 | | |
| 24.000 | .68 | 76 | 16.381 | 7.619 | 31.7% | 18 | .961 | .957 | .954 | .951 | .947 | .941 | .937 | .932 | .926 | .919 |
| | | | | | | | .911 | .901 | .891 | .882 | .868 | .852 | .837 | .815 | | |
| 24.000 | .75 | 95 | 16.365 | 7.635 | 31.8% | 18 | .960 | .958 | .954 | .951 | .947 | .942 | .936 | .931 | .926 | .917 |
| | | | | | | | .910 | .902 | .890 | .879 | .866 | .848 | .835 | .815 | | |
| 24.000 | .79 | 114 | 16.353 | 7.647 | 31.9% | 18 | .960 | .957 | .954 | .950 | .947 | .942 | .937 | .931 | .924 | .917 |
| | | | | | | | .910 | .901 | .890 | .878 | .864 | .849 | .832 | .812 | | |
| 24.000 | .82 | 133 | 16.345 | 7.655 | 31.9% | 18 | .960 | .957 | .954 | .950 | .946 | .941 | .936 | .931 | .925 | .917 |
| | | | | | | | .909 | .900 | .889 | .877 | .863 | .848 | .831 | .810 | | |
| 24.000 | .60 | 60 | 17.211 | 6.789 | 28.3% | 19 | .961 | .958 | .955 | .951 | .947 | .942 | .939 | .932 | .925 | .921 |
| | | | | | | | .912 | .902 | .893 | .883 | .871 | .853 | .841 | .824 | .801 | |
| 24.000 | .70 | 80 | 17.171 | 6.829 | 28.5% | 19 | .961 | .957 | .954 | .950 | .947 | .941 | .938 | .932 | .925 | .917 |
| | | | | | | | .911 | .902 | .891 | .881 | .867 | .850 | .837 | .815 | .793 | |
| 24.000 | .76 | 100 | 17.149 | 6.851 | 28.5% | 19 | .960 | .957 | .954 | .951 | .946 | .942 | .936 | .931 | .925 | .919 |
| | | | | | | | .909 | .901 | .890 | .877 | .867 | .849 | .832 | .815 | .788 | |
| 24.000 | .80 | 120 | 17.137 | 6.863 | 28.6% | 19 | .960 | .957 | .954 | .950 | .946 | .942 | .937 | .931 | .923 | .918 |
| | | | | | | | .909 | .900 | .889 | .878 | .864 | .849 | .831 | .811 | .787 | |
| 24.000 | .83 | 140 | 17.129 | 6.871 | 28.6% | 19 | .960 | .957 | .954 | .950 | .946 | .941 | .937 | .931 | .924 | .916 |
| | | | | | | | .910 | .899 | .889 | .877 | .863 | .849 | .828 | .811 | .787 | |

ERLANG-ENGSET: FOR SMOOTH ATTEMPTING TRAFFIC AND FEW PERSONS DOING THE CALLING (FEW MEANS RATIO OF CALLERS TO CIRCUITS IS LESS THAN 8 - GREATER THAN 8 ASSUME TRAFFIC RANDOM)

| HOURS FIRST ATTEMPT | FIRST ATTEMPT VMR | NUMBER SOURCES | HOURS CON-NECT | HOURS OVFL | PERCENT BLOCKED (GRADE) | LINE | HOURS CONNECTED TRAFFIC ON LINES 1 | 2 | 3 | 4 | 5 | 6 | 7 | 8 | 9 | 10 |
|---|---|---|---|---|---|---|---|---|---|---|---|---|---|---|---|---|
| 24.000 | .43 | 42 | 18.086 | 5.914 | 24.6% | 20 | .961 | .958 | .956 | .952 | .948 | .942 | .940 | .934 | .929 | .922 |
| | | | | | | | .915 | .908 | .898 | .884 | .875 | .866 | .855 | .831 | .816 | .797 |
| 24.000 | .62 | 63 | 17.978 | 6.022 | 25.1% | 20 | .961 | .958 | .954 | .951 | .947 | .942 | .937 | .933 | .925 | .920 |
| | | | | | | | .911 | .902 | .896 | .880 | .870 | .857 | .839 | .819 | .799 | .776 |
| 24.000 | .71 | 84 | 17.933 | 6.067 | 25.3% | 20 | .960 | .957 | .954 | .951 | .946 | .943 | .935 | .933 | .924 | .918 |
| | | | | | | | .909 | .902 | .891 | .881 | .866 | .852 | .836 | .813 | .794 | .767 |
| 24.000 | .77 | 105 | 17.912 | 6.088 | 25.4% | 20 | .960 | .958 | .954 | .951 | .947 | .941 | .937 | .932 | .923 | .919 |
| | | | | | | | .909 | .900 | .888 | .880 | .864 | .849 | .835 | .810 | .791 | .764 |
| 24.000 | .81 | 126 | 17.897 | 6.103 | 25.4% | 20 | .960 | .957 | .954 | .951 | .946 | .942 | .937 | .930 | .924 | .917 |
| | | | | | | | .910 | .900 | .889 | .877 | .864 | .848 | .831 | .810 | .787 | .763 |
| 24.000 | .84 | 147 | 17.885 | 6.115 | 25.5% | 20 | .960 | .957 | .954 | .950 | .946 | .941 | .937 | .931 | .924 | .917 |
| | | | | | | | .909 | .899 | .889 | .876 | .863 | .848 | .829 | .809 | .786 | .759 |
| 24.000 | .64 | 66 | 18.714 | 5.286 | 22.0% | 21 | .960 | .958 | .955 | .951 | .947 | .943 | .937 | .931 | .926 | .919 |
| | | | | | | | .910 | .905 | .892 | .882 | .870 | .854 | .838 | .819 | .796 | .777 |
| | | | | | | | .744 | | | | | | | | | |
| 24.000 | .73 | 88 | 18.668 | 5.332 | 22.2% | 21 | .960 | .957 | .954 | .950 | .947 | .942 | .936 | .932 | .924 | .919 |
| | | | | | | | .909 | .902 | .890 | .879 | .867 | .850 | .833 | .817 | .790 | .768 |
| | | | | | | | .740 | | | | | | | | | |
| 24.000 | .78 | 110 | 18.641 | 5.359 | 22.3% | 21 | .960 | .957 | .954 | .950 | .947 | .941 | .938 | .931 | .924 | .918 |
| | | | | | | | .908 | .900 | .890 | .879 | .865 | .849 | .832 | .812 | .788 | .765 |
| | | | | | | | .734 | | | | | | | | | |
| 24.000 | .82 | 132 | 18.625 | 5.375 | 22.4% | 21 | .960 | .957 | .954 | .950 | .946 | .942 | .937 | .931 | .924 | .917 |
| | | | | | | | .909 | .900 | .888 | .878 | .862 | .850 | .830 | .810 | .787 | .761 |
| | | | | | | | .732 | | | | | | | | | |
| 24.000 | .84 | 154 | 18.613 | 5.387 | 22.4% | 21 | .960 | .957 | .954 | .950 | .946 | .942 | .936 | .930 | .024 | .917 |
| | | | | | | | .908 | .899 | .888 | .877 | .863 | .847 | .829 | .809 | .786 | .759 |
| | | | | | | | .730 | | | | | | | | | |
| 24.000 | .48 | 46 | 19.549 | 4.451 | 18.5% | 22 | .961 | .958 | .955 | .952 | .947 | .944 | .939 | .931 | .929 | .921 |
| | | | | | | | .911 | .908 | .897 | .887 | .873 | .864 | .848 | .830 | .810 | .790 |
| | | | | | | | .764 | .729 | | | | | | | | |
| 24.000 | .65 | 69 | 19.421 | 4.579 | 19.1% | 22 | .961 | .958 | .954 | .951 | .947 | .942 | .937 | .933 | .025 | .919 |
| | | | | | | | .910 | .902 | .893 | .882 | .867 | .855 | .836 | .818 | .798 | .773 |
| | | | | | | | .744 | .715 | | | | | | | | |
| 24.000 | .74 | 92 | 19.365 | 4.635 | 19.3% | 22 | .961 | .958 | .954 | .950 | .947 | .942 | .936 | .931 | .926 | .918 |
| | | | | | | | .908 | .903 | .890 | .877 | .868 | .850 | .833 | .813 | .794 | .763 |
| | | | | | | | .740 | .703 | | | | | | | | |
| 24.000 | .79 | 115 | 19.337 | 4.663 | 19.4% | 22 | .960 | .957 | .954 | .951 | .948 | .941 | .937 | .931 | .925 | .916 |
| | | | | | | | .911 | .899 | .890 | .878 | .864 | .849 | .831 | .811 | .700 | .763 |
| | | | | | | | .732 | .701 | | | | | | | | |
| 24.000 | .83 | 138 | 19.317 | 4.683 | 19.5% | 22 | .960 | .957 | .954 | .950 | .946 | .941 | .936 | .931 | .925 | .917 |

ERLANG-ENGSET: FOR SMOOTH ATTEMPTING TRAFFIC AND FEW PERSONS DOING THE CALLING (FEW MEANS RATIO OF CALLERS TO CIRCUITS IS LESS THAN 8 - GREATER THAN 8 ASSUME TRAFFIC RANDOM)

| HOURS FIRST ATTEMPT | FIRST ATTEMPT VMR | NUMBER SOURCES | HOURS CON-NECT | HOURS OVFL | PERCENT BLOCKED (GRADE) | LINE | HOURS CONNECTED TRAFFIC ON LINES 1 | 2 | 3 | 4 | 5 | 6 | 7 | 8 | 9 | 10 |
|---|---|---|---|---|---|---|---|---|---|---|---|---|---|---|---|---|
| | | | | | | | .909 | .899 | .889 | .877 | .863 | .848 | .831 | .809 | .786 | .762 |
| | | | | | | | .731 | .696 | | | | | | | | |
| 24.000 | .85 | 161 | 19.305 | 4.695 | 19.6% | 22 | .960 | .957 | .954 | .950 | .946 | .942 | .936 | .931 | .924 | .917 |
| | | | | | | | .908 | .899 | .888 | .877 | .863 | .847 | .829 | .808 | .786 | .758 |
| | | | | | | | .731 | .695 | | | | | | | | |
| 24.000 | .50 | 48 | 20.233 | 3.767 | 15.7% | 23 | .961 | .958 | .955 | .952 | .947 | .944 | .937 | .934 | .927 | .922 |
| | | | | | | | .913 | .905 | .896 | .885 | .874 | .861 | .843 | .831 | .809 | .787 |
| | | | | | | | .759 | .733 | .698 | | | | | | | |
| 24.000 | .67 | 72 | 20.084 | 3.916 | 16.3% | 23 | .961 | .958 | .955 | .950 | .946 | .943 | .937 | .932 | .926 | .918 |
| | | | | | | | .910 | .904 | .889 | .883 | .867 | .855 | .837 | .817 | .795 | .773 |
| | | | | | | | .743 | .711 | .673 | | | | | | | |
| 24.000 | .75 | 96 | 20.027 | 3.973 | 16.6% | 23 | .960 | .957 | .954 | .951 | .947 | .941 | .938 | .931 | .924 | .918 |
| | | | | | | | .910 | .900 | .891 | .879 | .863 | .853 | .831 | .815 | .791 | .765 |
| | | | | | | | .737 | .701 | .668 | | | | | | | |
| 24.000 | .80 | 120 | 19.993 | 4.007 | 16.7% | 23 | .960 | .957 | .954 | .950 | .946 | .942 | .937 | .931 | .923 | .918 |
| | | | | | | | .909 | .900 | .889 | .878 | .864 | .849 | .831 | .811 | .787 | .764 |
| | | | | | | | .733 | .699 | .660 | | | | | | | |
| 24.000 | .83 | 144 | 19.973 | 4.027 | 16.8% | 23 | .960 | .957 | .954 | .950 | .946 | .942 | .936 | .930 | .924 | .917 |
| | | | | | | | .909 | .900 | .889 | .877 | .863 | .847 | .830 | .808 | .787 | .760 |
| | | | | | | | .730 | .697 | .658 | | | | | | | |
| 24.000 | .86 | 168 | 19.959 | 4.041 | 16.8% | 23 | .960 | .957 | .954 | .950 | .946 | .942 | .936 | .931 | .924 | .917 |
| | | | | | | | .908 | .899 | .888 | .876 | .862 | .847 | .829 | .808 | .785 | .758 |
| | | | | | | | .729 | .695 | .657 | | | | | | | |
| 24.000 | .52 | 50 | 20.867 | 3.133 | 13.1% | 24 | .961 | .958 | .955 | .951 | .946 | .944 | .939 | .933 | .928 | .921 |
| | | | | | | | .914 | .904 | .896 | .885 | .873 | .860 | .840 | .831 | .810 | .784 |
| | | | | | | | .758 | .730 | .693 | .655 | | | | | | |
| 24.000 | .68 | 75 | 20.708 | 3.292 | 13.7% | 24 | .961 | .957 | .955 | .950 | .947 | .941 | .938 | .931 | .927 | .918 |
| | | | | | | | .912 | .900 | .893 | .881 | .867 | .852 | .836 | .818 | .793 | .771 |
| | | | | | | | .744 | .711 | .673 | .631 | | | | | | |
| 24.000 | .76 | 100 | 20.642 | 3.358 | 14.0% | 24 | .960 | .957 | .954 | .951 | .946 | .942 | .936 | .931 | .925 | .919 |
| | | | | | | | .909 | .901 | .890 | .877 | .867 | .849 | .832 | .815 | .788 | .765 |
| | | | | | | | .738 | .701 | .667 | .621 | | | | | | |
| 24.000 | .81 | 125 | 20.608 | 3.392 | 14.1% | 24 | .960 | .957 | .954 | .950 | .946 | .941 | .936 | .931 | .925 | .917 |
| | | | | | | | .909 | .900 | .889 | .875 | .866 | .846 | .833 | .810 | .788 | .761 |
| | | | | | | | .733 | .698 | .661 | .619 | | | | | | |
| 24.000 | .84 | 150 | 20.583 | 3.417 | 14.2% | 24 | .960 | .957 | .954 | .950 | .946 | .941 | .937 | .930 | .924 | .917 |
| | | | | | | | .909 | .899 | .888 | .877 | .863 | .848 | .829 | .808 | .788 | .760 |
| | | | | | | | .730 | .696 | .658 | .614 | | | | | | |
| 24.000 | .86 | 175 | 20.569 | 3.431 | 14.3% | 24 | .960 | .957 | .954 | .950 | .946 | .942 | .936 | .931 | .924 | .916 |
| | | | | | | | .908 | .899 | .888 | .876 | .863 | .845 | .829 | .809 | .785 | .758 |

ERLANG-ENGSET: FOR SMOOTH ATTEMPTING TRAFFIC AND FEW PERSONS DOING THE CALLING (FEW MEANS RATIO OF CALLERS TO CIRCUITS IS LESS THAN 8 - GREATER THAN 8 ASSUME TRAFFIC RANDOM)

| HOURS FIRST ATTEMPT | FIRST ATTEMPT VMR | NUMBER SOURCES | HOURS CONNECT | HOURS OVFL | PERCENT BLOCKED (GRADE) | LINE | HOURS CONNECTED TRAFFIC ON LINES 1 | 2 | 3 | 4 | 5 | 6 | 7 | 8 | 9 | 10 |
|---|---|---|---|---|---|---|---|---|---|---|---|---|---|---|---|---|
| | | | | | | | .728 | .695 | .656 | .613 | | | | | | |
| 24.000 | .54 | 52 | 21.450 | 2.550 | 10.6% | 25 | .961 | .958 | .955 | .951 | .947 | .944 | .939 | .933 | .926 | .919 |
| | | | | | | | .915 | .901 | .899 | .884 | .873 | .859 | .842 | .826 | .807 | .781 |
| | | | | | | | .758 | .727 | .692 | .647 | .607 | | | | | |
| 24.000 | .69 | 78 | 21.281 | 2.719 | 11.3% | 25 | .961 | .957 | .954 | .950 | .948 | .942 | .936 | .933 | .924 | .919 |
| | | | | | | | .911 | .901 | .893 | .878 | .867 | .854 | .835 | .816 | .794 | .772 |
| | | | | | | | .741 | .709 | .671 | .632 | .584 | | | | | |
| 24.000 | .77 | 104 | 21.212 | 2.788 | 11.6% | 25 | .960 | .957 | .954 | .950 | .947 | .942 | .936 | .932 | .925 | .918 |
| | | | | | | | .908 | .900 | .890 | .879 | .866 | .850 | .832 | .813 | .790 | .762 |
| | | | | | | | .735 | .703 | .665 | .622 | .575 | | | | | |
| 24.000 | .82 | 130 | 21.173 | 2.827 | 11.8% | 25 | .960 | .957 | .954 | .950 | .946 | .942 | .937 | .930 | .924 | .917 |
| | | | | | | | .909 | .900 | .889 | .877 | .864 | .848 | .830 | .810 | .786 | .763 |
| | | | | | | | .731 | .698 | .660 | .618 | .571 | | | | | |
| 24.000 | .85 | 156 | 21.149 | 2.851 | 11.9% | 25 | .960 | .957 | .954 | .950 | .946 | .942 | .936 | .930 | .924 | .917 |
| | | | | | | | .909 | .899 | .889 | .876 | .863 | .847 | .829 | .808 | .787 | .759 |
| | | | | | | | .730 | .695 | .658 | .615 | .568 | | | | | |
| 24.000 | .87 | 182 | 21.132 | 2.868 | 12.0% | 25 | .960 | .957 | .954 | .950 | .946 | .941 | .936 | .930 | .923 | .917 |
| | | | | | | | .908 | .899 | .888 | .876 | .862 | .847 | .829 | .808 | .784 | .758 |
| | | | | | | | .728 | .695 | .656 | .614 | .566 | | | | | |
| 24.000 | .56 | 54 | 21.976 | 2.024 | 8.4% | 26 | .961 | .958 | .955 | .950 | .948 | .943 | .936 | .934 | .927 | .918 |
| | | | | | | | .915 | .903 | .895 | .885 | .872 | .858 | .843 | .819 | .810 | .781 |
| | | | | | | | .754 | .726 | .688 | .646 | .603 | .548 | | | | |
| 24.000 | .70 | 81 | 21.798 | 2.202 | 9.2% | 26 | .961 | .957 | .954 | .950 | .947 | .941 | .937 | .931 | .927 | .917 |
| | | | | | | | .911 | .901 | .893 | .877 | .869 | .852 | .836 | .813 | .795 | .768 |
| | | | | | | | .741 | .709 | .670 | .629 | .582 | .530 | | | | |
| 24.000 | .78 | 108 | 21.727 | 2.273 | 9.5% | 26 | .960 | .958 | .954 | .950 | .947 | .942 | .936 | .931 | .925 | .918 |
| | | | | | | | .908 | .900 | .890 | .879 | .865 | .850 | .832 | .812 | .788 | .765 |
| | | | | | | | .734 | .702 | .664 | .619 | .575 | .525 | | | | |
| 24.000 | .82 | 135 | 21.685 | 2.315 | 9.6% | 26 | .960 | .957 | .954 | .950 | .946 | .942 | .937 | .931 | .923 | .918 |
| | | | | | | | .908 | .899 | .889 | .877 | .863 | .848 | .820 | .811 | .787 | .760 |
| | | | | | | | .730 | .698 | .659 | .618 | .570 | .518 | | | | |
| 24.000 | .85 | 162 | 21.661 | 2.339 | 9.7% | 26 | .960 | .957 | .954 | .950 | .946 | .941 | .936 | .931 | .924 | .917 |
| | | | | | | | .909 | .899 | .889 | .876 | .863 | .846 | .829 | .809 | .785 | .760 |
| | | | | | | | .729 | .696 | .655 | .616 | .567 | .518 | | | | |
| 24.000 | .87 | 189 | 21.644 | 2.356 | 9.8% | 26 | .960 | .957 | .954 | .950 | .946 | .941 | .936 | .930 | .924 | .917 |
| | | | | | | | .908 | .899 | .888 | .876 | .861 | .847 | .828 | .807 | .785 | .758 |
| | | | | | | | .728 | .694 | .655 | .613 | .566 | .515 | | | | |
| 24.000 | .57 | 56 | 22.440 | 1.560 | 6.5% | 27 | .961 | .958 | .954 | .951 | .947 | .944 | .938 | .932 | .928 | .920 |
| | | | | | | | .912 | .904 | .893 | .884 | .870 | .854 | .844 | .825 | .804 | .780 |
| | | | | | | | .748 | .727 | .686 | .645 | .596 | .547 | .486 | | | |

ERLANG-ENGSET: FOR SMOOTH ATTEMPTING TRAFFIC AND FEW PERSONS DOING THE CALLING (FEW MEANS RATIO OF CALLERS TO CIRCUITS IS LESS THAN 8 - GREATER THAN 8 ASSUME TRAFFIC RANDOM)

| HOURS FIRST ATTEMPT | FIRST ATTEMPT VMR | NUMBER SOURCES | HOURS CON-NECT | HOURS OVFL | PERCENT BLOCKED (GRADE) | LINE | HOURS CONNECTED TRAFFIC ON LINES 1 | 2 | 3 | 4 | 5 | 6 | 7 | 8 | 9 | 10 |
|---|---|---|---|---|---|---|---|---|---|---|---|---|---|---|---|---|
| 24.000 | .71 | 84 | 22.260 | 1.740 | 7.2% | 27 | .960 | .957 | .954 | .951 | .946 | .943 | .935 | .933 | .924 | .918 |
| | | | | | | | .909 | .902 | .891 | .881 | .866 | .852 | .836 | .813 | .794 | .767 |
| | | | | | | | .740 | .707 | .669 | .629 | .581 | .530 | .471 | | | |
| 24.000 | .79 | 112 | 22.185 | 1.815 | 7.6% | 27 | .960 | .957 | .954 | .951 | .946 | .943 | .937 | .931 | .925 | .916 |
| | | | | | | | .909 | .900 | .891 | .878 | .865 | .849 | .832 | .812 | .788 | .764 |
| | | | | | | | .733 | .701 | .661 | .622 | .572 | .523 | .466 | | | |
| 24.000 | .83 | 140 | 22.144 | 1.856 | 7.7% | 27 | .960 | .957 | .954 | .950 | .946 | .941 | .937 | .931 | .924 | .916 |
| | | | | | | | .910 | .899 | .889 | .877 | .863 | .849 | .828 | .811 | .787 | .760 |
| | | | | | | | .730 | .698 | .659 | .617 | .570 | .518 | .464 | | | |
| 24.000 | .86 | 168 | 22.119 | 1.881 | 7.8% | 27 | .960 | .957 | .954 | .950 | .946 | .942 | .936 | .931 | .924 | .917 |
| | | | | | | | .908 | .899 | .888 | .876 | .862 | .847 | .829 | .808 | .785 | .758 |
| | | | | | | | .729 | .695 | .657 | .614 | .567 | .517 | .462 | | | |
| 24.000 | .88 | 196 | 22.101 | 1.899 | 7.9% | 27 | .960 | .957 | .954 | .950 | .946 | .941 | .936 | .930 | .924 | .916 |
| | | | | | | | .909 | .899 | .888 | .875 | .862 | .847 | .828 | .807 | .785 | .758 |
| | | | | | | | .728 | .693 | .655 | .612 | .566 | .515 | .460 | | | |
| 26.000 | .57 | 60 | 17.462 | 8.538 | 32.8% | 19 | .964 | .961 | .958 | .955 | .952 | .949 | .944 | .941 | .935 | .929 |
| | | | | | | | .923 | .914 | .912 | .899 | .891 | .880 | .865 | .855 | .834 | |
| 26.000 | .67 | 80 | 17.434 | 8.566 | 32.9% | 19 | .963 | .961 | .958 | .954 | .952 | .948 | .944 | .939 | .934 | .928 |
| | | | | | | | .922 | .914 | .909 | .896 | .889 | .876 | .862 | .849 | .834 | |
| 26.000 | .74 | 100 | 17.413 | 8.587 | 33.0% | 19 | .963 | .961 | .958 | .954 | .951 | .948 | .943 | .939 | .933 | .928 |
| | | | | | | | .920 | .916 | .906 | .895 | .886 | .875 | .862 | .848 | .826 | |
| 26.000 | .78 | 120 | 17.405 | 8.595 | 33.1% | 19 | .963 | .961 | .958 | .955 | .951 | .947 | .943 | .939 | .934 | .926 |
| | | | | | | | .922 | .913 | .906 | .896 | .885 | .873 | .860 | .845 | .827 | |
| 26.000 | .81 | 140 | 17.397 | 8.603 | 33.1% | 19 | .963 | .961 | .958 | .955 | .951 | .947 | .942 | .939 | .933 | .927 |
| | | | | | | | .921 | .913 | .905 | .896 | .885 | .873 | .859 | .844 | .826 | |
| 26.000 | .38 | 42 | 18.371 | 7.629 | 29.3% | 20 | .964 | .962 | .958 | .957 | .952 | .950 | .945 | .941 | .936 | .932 |
| | | | | | | | .925 | .916 | .912 | .909 | .895 | .889 | .877 | .863 | .847 | .841 |
| 26.000 | .59 | 63 | 18.278 | 7.722 | 29.7% | 20 | .964 | .961 | .958 | .955 | .953 | .948 | .944 | .939 | .936 | .929 |
| | | | | | | | .924 | .915 | .908 | .901 | .889 | .879 | .865 | .856 | .837 | .818 |
| 26.000 | .69 | 84 | 18.242 | 7.758 | 29.8% | 20 | .963 | .961 | .958 | .955 | .952 | .947 | .945 | .938 | .934 | .928 |
| | | | | | | | .924 | .913 | .907 | .897 | .889 | .873 | .865 | .848 | .832 | .814 |
| 26.000 | .75 | 105 | 18.220 | 7.780 | 29.9% | 20 | .963 | .961 | .958 | .955 | .952 | .947 | .944 | .937 | .935 | .928 |
| | | | | | | | .920 | .916 | .906 | .894 | .888 | .873 | .860 | .848 | .828 | .808 |
| 26.000 | .79 | 126 | 18.209 | 7.791 | 30.0% | 20 | .963 | .961 | .958 | .955 | .951 | .947 | .943 | .939 | .933 | .926 |
| | | | | | | | .922 | .913 | .905 | .896 | .885 | .873 | .859 | .845 | .827 | .807 |
| 26.000 | .82 | 147 | 18.201 | 7.799 | 30.0% | 20 | .963 | .960 | .958 | .955 | .951 | .947 | .943 | .939 | .933 | .927 |
| | | | | | | | .921 | .912 | .906 | .895 | .884 | .874 | .858 | .845 | .824 | .807 |
| 26.000 | .41 | 44 | 19.172 | 6.828 | 26.3% | 21 | .964 | .961 | .959 | .956 | .951 | .951 | .945 | .941 | .935 | .929 |
| | | | | | | | .928 | .916 | .910 | .904 | .898 | .889 | .871 | .864 | .851 | .831 |
| | | | | | | | .817 | | | | | | | | | |

ERLANG-ENGSET: FOR SMOOTH ATTEMPTING TRAFFIC AND FEW PERSONS DOING THE CALLING (FEW MEANS RATIO OF CALLERS TO CIRCUITS IS LESS THAN 8 - GREATER THAN 8 ASSUME TRAFFIC RANDOM)

| HOURS FIRST ATTEMPT | FIRST ATTEMPT VMR | NUMBER SOURCES | HOURS CONNECT | HOURS OVFL | PERCENT BLOCKED (GRADE) | LINE | HOURS CONNECTED TRAFFIC ON LINES 1 | 2 | 3 | 4 | 5 | 6 | 7 | 8 | 9 | 10 |
|---|---|---|---|---|---|---|---|---|---|---|---|---|---|---|---|---|
| 26.000 | .61 | 66 | 19.068 | 6.932 | 26.7% | 21 | .964 | .961 | .958 | .955 | .952 | .948 | .945 | .939 | .935 | .928 |
| | | | | | | | .923 | .917 | .907 | .900 | .890 | .878 | .865 | .850 | .838 | .820 |
| | | | | | | | .796 | | | | | | | | | |
| 26.000 | .70 | 88 | 19.026 | 6.974 | 26.8% | 21 | .963 | .961 | .958 | .955 | .952 | .948 | .943 | .940 | .934 | .927 |
| | | | | | | | .923 | .914 | .905 | .898 | .889 | .876 | .860 | .847 | .834 | .811 |
| | | | | | | | .789 | | | | | | | | | |
| 26.000 | .76 | 110 | 19.005 | 6.995 | 26.9% | 21 | .963 | .961 | .958 | .955 | .952 | .948 | .943 | .939 | .934 | .926 |
| | | | | | | | .921 | .914 | .907 | .895 | .888 | .873 | .860 | .845 | .828 | .807 |
| | | | | | | | .789 | | | | | | | | | |
| 26.000 | .80 | 132 | 18.990 | 7.010 | 27.0% | 21 | .963 | .960 | .958 | .955 | .951 | .948 | .943 | .937 | .934 | .927 |
| | | | | | | | .921 | .914 | .905 | .895 | .886 | .873 | .859 | .844 | .826 | .807 |
| | | | | | | | .783 | | | | | | | | | |
| 26.000 | .83 | 154 | 18.981 | 7.019 | 27.0% | 21 | .963 | .960 | .958 | .955 | .951 | .947 | .943 | .938 | .933 | .927 |
| | | | | | | | .921 | .913 | .905 | .895 | .885 | .872 | .859 | .843 | .826 | .805 |
| | | | | | | | .783 | | | | | | | | | |
| 26.000 | .43 | 46 | 19.946 | 6.054 | 23.3% | 22 | .964 | .961 | .959 | .956 | .952 | .948 | .946 | .941 | .935 | .932 |
| | | | | | | | .925 | .918 | .912 | .903 | .895 | .884 | .873 | .862 | .847 | .831 |
| | | | | | | | .805 | .797 | | | | | | | | |
| 26.000 | .62 | 69 | 19.834 | 6.166 | 23.7% | 22 | .964 | .961 | .958 | .955 | .952 | .949 | .943 | .939 | .935 | .929 |
| | | | | | | | .920 | .916 | .910 | .898 | .890 | .877 | .866 | .849 | .836 | .817 |
| | | | | | | | .795 | .774 | | | | | | | | |
| 26.000 | .72 | 92 | 19.787 | 6.213 | 23.9% | 22 | .963 | .961 | .958 | .955 | .952 | .947 | .944 | .939 | .932 | .928 |
| | | | | | | | .923 | .915 | .905 | .897 | .888 | .875 | .862 | .844 | .832 | .812 |
| | | | | | | | .790 | .764 | | | | | | | | |
| 26.000 | .77 | 115 | 19.760 | 6.240 | 24.0% | 22 | .963 | .961 | .958 | .955 | .951 | .947 | .943 | .938 | .934 | .928 |
| | | | | | | | .920 | .913 | .906 | .897 | .886 | .871 | .863 | .842 | .830 | .809 |
| | | | | | | | .786 | .759 | | | | | | | | |
| 26.000 | .81 | 138 | 19.746 | 6.254 | 24.1% | 22 | .963 | .961 | .958 | .954 | .952 | .947 | .943 | .939 | .933 | .927 |
| | | | | | | | .921 | .913 | .904 | .897 | .884 | .872 | .861 | .842 | .828 | .806 |
| | | | | | | | .784 | .758 | | | | | | | | |
| 26.000 | .84 | 161 | 19.733 | 6.267 | 24.1% | 22 | .963 | .960 | .958 | .955 | .951 | .947 | .943 | .938 | .933 | .927 |
| | | | | | | | .920 | .913 | .904 | .895 | .884 | .872 | .858 | .843 | .825 | .805 |
| | | | | | | | .782 | .755 | | | | | | | | |
| 26.000 | .46 | 48 | 20.695 | 5.305 | 20.4% | 23 | .964 | .961 | .959 | .956 | .952 | .949 | .946 | .940 | .937 | .930 |
| | | | | | | | .925 | .918 | .912 | .900 | .896 | .884 | .872 | .860 | .845 | .826 |
| | | | | | | | .813 | .789 | .763 | | | | | | | |
| 26.000 | .64 | 72 | 20.567 | 5.433 | 20.9% | 23 | .963 | .961 | .958 | .955 | .952 | .949 | .944 | .940 | .933 | .928 |
| | | | | | | | .923 | .917 | .908 | .897 | .888 | .880 | .865 | .850 | .833 | .817 |
| | | | | | | | .794 | .770 | .744 | | | | | | | |
| 26.000 | .73 | 96 | 20.514 | 5.486 | 21.1% | 23 | .963 | .961 | .957 | .955 | .952 | .948 | .944 | .939 | .932 | .930 |

ERLANG-ENGSET: FOR SMOOTH ATTEMPTING TRAFFIC AND FEW PERSONS DOING THE CALLING (FEW MEANS RATIO OF CALLERS TO CIRCUITS IS LESS THAN 8 - GREATER THAN 8 ASSUME TRAFFIC RANDOM)

| HOURS FIRST ATTEMPT | FIRST ATTEMPT VMR | NUMBER SOURCES | HOURS CONNECT | HOURS OVFL | PERCENT BLOCKED (GRADE) | LINE | HOURS CONNECTED TRAFFIC ON LINES 1 | 2 | 3 | 4 | 5 | 6 | 7 | 8 | 9 | 10 |
|---|---|---|---|---|---|---|---|---|---|---|---|---|---|---|---|---|
| | | | | | | | .919 | .916 | .903 | .899 | .886 | .876 | .859 | .849 | .828 | .812 |
| | | | | | | | .789 | .763 | .733 | | | | | | | |
| 26.000 | .78 | 120 | 20.490 | 5.510 | 21.2% | 23 | .963 | .961 | .958 | .955 | .951 | .947 | .943 | .939 | .934 | .926 |
| | | | | | | | .922 | .913 | .906 | .896 | .885 | .873 | .860 | .845 | .827 | .808 |
| | | | | | | | .785 | .761 | .732 | | | | | | | |
| 26.000 | .82 | 144 | 20.471 | 5.529 | 21.3% | 23 | .963 | .961 | .958 | .955 | .951 | .947 | .944 | .938 | .933 | .927 |
| | | | | | | | .921 | .913 | .905 | .895 | .885 | .873 | .859 | .844 | .826 | .806 |
| | | | | | | | .783 | .758 | .728 | | | | | | | |
| 26.000 | .85 | 168 | 20.459 | 5.541 | 21.3% | 23 | .963 | .961 | .958 | .955 | .951 | .947 | .943 | .938 | .933 | .927 |
| | | | | | | | .920 | .913 | .905 | .894 | .884 | .872 | .858 | .843 | .825 | .804 |
| | | | | | | | .782 | .755 | .728 | | | | | | | |
| 26.000 | .48 | 50 | 21.408 | 4.592 | 17.7% | 24 | .964 | .961 | .959 | .956 | .951 | .950 | .943 | .942 | .934 | .933 |
| | | | | | | | .924 | .918 | .911 | .900 | .897 | .883 | .869 | .861 | .842 | .825 |
| | | | | | | | .811 | .786 | .762 | .729 | | | | | | |
| 26.000 | .65 | 75 | 21.271 | 4.729 | 18.2% | 24 | .963 | .961 | .958 | .955 | .951 | .949 | .944 | .939 | .935 | .927 |
| | | | | | | | .924 | .915 | .907 | .899 | .889 | .875 | .864 | .852 | .832 | .814 |
| | | | | | | | .794 | .772 | .740 | .712 | | | | | | |
| 26.000 | .74 | 100 | 21.215 | 4.785 | 18.4% | 24 | .963 | .961 | .958 | .954 | .951 | .948 | .943 | .939 | .933 | .928 |
| | | | | | | | .920 | .916 | .906 | .895 | .886 | .875 | .862 | .848 | .826 | .811 |
| | | | | | | | .788 | .765 | .735 | .703 | | | | | | |
| 26.000 | .79 | 125 | 21.181 | 4.819 | 18.5% | 24 | .963 | .961 | .957 | .955 | .951 | .948 | .943 | .938 | .932 | .927 |
| | | | | | | | .921 | .914 | .905 | .896 | .885 | .871 | .862 | .842 | .829 | .808 |
| | | | | | | | .784 | .758 | .732 | .697 | | | | | | |
| 26.000 | .83 | 150 | 21.162 | 4.838 | 18.6% | 24 | .963 | .961 | .958 | .955 | .951 | .947 | .944 | .938 | .933 | .927 |
| | | | | | | | .920 | .913 | .905 | .895 | .885 | .872 | .859 | .843 | .826 | .805 |
| | | | | | | | .781 | .759 | .728 | .694 | | | | | | |
| 26.000 | .85 | 175 | 21.150 | 4.850 | 18.7% | 24 | .963 | .961 | .958 | .955 | .951 | .947 | .943 | .938 | .933 | .927 |
| | | | | | | | .920 | .913 | .904 | .893 | .886 | .870 | .859 | .843 | .824 | .804 |
| | | | | | | | .782 | .756 | .725 | .695 | | | | | | |
| 26.000 | .50 | 52 | 22.087 | 3.913 | 15.1% | 25 | .964 | .961 | .958 | .956 | .951 | .950 | .944 | .941 | .936 | .930 |
| | | | | | | | .924 | .917 | .911 | .901 | .888 | .887 | .868 | .861 | .841 | .826 |
| | | | | | | | .804 | .786 | .753 | .736 | .693 | | | | | |
| 26.000 | .67 | 78 | 21.935 | 4.065 | 15.6% | 25 | .963 | .961 | .958 | .956 | .951 | .947 | .945 | .939 | .933 | .930 |
| | | | | | | | .922 | .914 | .909 | .897 | .887 | .878 | .864 | .848 | .833 | .812 |
| | | | | | | | .794 | .769 | .740 | .709 | .675 | | | | | |
| 26.000 | .75 | 104 | 21.872 | 4.128 | 15.9% | 25 | .963 | .961 | .958 | .955 | .951 | .948 | .943 | .939 | .934 | .927 |
| | | | | | | | .921 | .915 | .905 | .896 | .887 | .875 | .859 | .848 | .826 | .811 |
| | | | | | | | .788 | .761 | .735 | .704 | .663 | | | | | |
| 26.000 | .80 | 130 | 21.840 | 4.160 | 16.0% | 25 | .963 | .961 | .958 | .955 | .951 | .947 | .943 | .939 | .933 | .927 |
| | | | | | | | .921 | .912 | .906 | .895 | .885 | .873 | .859 | .844 | .828 | .805 |

ERLANG-ENGSET: FOR SMOOTH ATTEMPTING TRAFFIC AND FEW PERSONS DOING THE CALLING (FEW MEANS RATIO OF CALLERS TO CIRCUITS IS LESS THAN 8 - GREATER THAN 8 ASSUME TRAFFIC RANDOM)

| HOURS FIRST ATTEMPT | FIRST ATTEMPT VMR | NUMBER SOURCES | HOURS CON-NECT | HOURS OVFL | PERCENT BLOCKED (GRADE) | LINE | HOURS CONNECTED TRAFFIC ON LINES 1 | 2 | 3 | 4 | 5 | 6 | 7 | 8 | 9 | 10 |
|---|---|---|---|---|---|---|---|---|---|---|---|---|---|---|---|---|
| | | | | | | | .786 | .759 | .728 | .699 | .662 | | | | | |
| 26.000 | .83 | 156 | 21.818 | 4.182 | 16.1% | 25 | .963 | .961 | .958 | .955 | .951 | .948 | .943 | .938 | .933 | .927 |
| | | | | | | | .920 | .913 | .905 | .895 | .885 | .872 | .858 | .843 | .825 | .805 |
| | | | | | | | .781 | .759 | .728 | .694 | .659 | | | | | |
| 26.000 | .86 | 182 | 21.803 | 4.197 | 16.1% | 25 | .963 | .961 | .958 | .955 | .951 | .947 | .943 | .939 | .933 | .927 |
| | | | | | | | .920 | .913 | .904 | .895 | .884 | .871 | .858 | .842 | .825 | .804 |
| | | | | | | | .781 | .755 | .726 | .693 | .657 | | | | | |
| 26.000 | .52 | 54 | 22.725 | 3.275 | 12.6% | 26 | .964 | .961 | .958 | .955 | .952 | .948 | .945 | .939 | .936 | .929 |
| | | | | | | | .925 | .914 | .913 | .896 | .895 | .883 | .870 | .856 | .841 | .824 |
| | | | | | | | .800 | .784 | .758 | .727 | .694 | .655 | | | | |
| 26.000 | .68 | 81 | 22.558 | 3.442 | 13.2% | 26 | .963 | .961 | .958 | .954 | .952 | .948 | .945 | .939 | .933 | .930 |
| | | | | | | | .921 | .916 | .905 | .898 | .890 | .874 | .863 | .850 | .832 | .813 |
| | | | | | | | .792 | .766 | .743 | .708 | .672 | .632 | | | | |
| 26.000 | .76 | 108 | 22.491 | 3.509 | 13.5% | 26 | .963 | .961 | .958 | .954 | .951 | .948 | .943 | .939 | .934 | .926 |
| | | | | | | | .921 | .914 | .907 | .895 | .885 | .874 | .861 | .846 | .828 | .808 |
| | | | | | | | .790 | .762 | .732 | .701 | .665 | .624 | | | | |
| 26.000 | .81 | 135 | 22.452 | 3.548 | 13.6% | 26 | .963 | .961 | .958 | .955 | .951 | .948 | .942 | .938 | .933 | .928 |
| | | | | | | | .921 | .913 | .905 | .895 | .885 | .873 | .859 | .844 | .827 | .806 |
| | | | | | | | .784 | .757 | .731 | .696 | .661 | .618 | | | | |
| 26.000 | .84 | 162 | 22.431 | 3.569 | 13.7% | 26 | .963 | .961 | .958 | .954 | .951 | .947 | .943 | .939 | .933 | .927 |
| | | | | | | | .920 | .913 | .905 | .895 | .884 | .872 | .858 | .843 | .825 | .805 |
| | | | | | | | .781 | .758 | .727 | .694 | .658 | .617 | | | | |
| 26.000 | .86 | 189 | 22.415 | 3.585 | 13.8% | 26 | .963 | .961 | .958 | .955 | .951 | .947 | .943 | .938 | .933 | .926 |
| | | | | | | | .920 | .912 | .904 | .895 | .883 | .872 | .858 | .841 | .825 | .804 |
| | | | | | | | .781 | .753 | .727 | .693 | .656 | .615 | | | | |
| 26.000 | .54 | 56 | 23.312 | 2.688 | 10.3% | 27 | .964 | .961 | .959 | .955 | .952 | .949 | .943 | .941 | .935 | .929 |
| | | | | | | | .926 | .915 | .911 | .901 | .889 | .882 | .870 | .854 | .840 | .823 |
| | | | | | | | .803 | .781 | .756 | .725 | .691 | .652 | .607 | | | |
| 26.000 | .69 | 84 | 23.136 | 2.864 | 11.0% | 27 | .963 | .961 | .958 | .955 | .952 | .947 | .945 | .938 | .934 | .928 |
| | | | | | | | .924 | .913 | .907 | .897 | .889 | .873 | .865 | .840 | .832 | .814 |
| | | | | | | | .788 | .760 | .739 | .705 | .674 | .632 | .586 | | | |
| 26.000 | .77 | 112 | 23.063 | 2.937 | 11.3% | 27 | .963 | .961 | .958 | .955 | .952 | .947 | .943 | .939 | .934 | .027 |
| | | | | | | | .021 | .915 | .906 | .894 | .887 | .873 | .860 | .845 | .827 | .811 |
| | | | | | | | .783 | .761 | .735 | .700 | .664 | .624 | .578 | | | |
| 26.000 | .81 | 140 | 23.022 | 2.978 | 11.5% | 27 | .963 | .961 | .958 | .955 | .951 | .947 | .942 | .939 | .933 | .927 |
| | | | | | | | .921 | .913 | .905 | .896 | .885 | .873 | .059 | .844 | .826 | .806 |
| | | | | | | | .782 | .760 | .729 | .695 | .661 | .619 | .573 | | | |
| 26.000 | .85 | 168 | 22.998 | 3.002 | 11.5% | 27 | .963 | .961 | .958 | .955 | .951 | .947 | .943 | .938 | .933 | .927 |
| | | | | | | | .920 | .913 | .905 | .894 | .884 | .872 | .858 | .843 | .825 | .804 |
| | | | | | | | .782 | .755 | .728 | .694 | .657 | .617 | .571 | | | |

ERLANG-ENGSET: FOR SMOOTH ATTEMPTING TRAFFIC AND FEW PERSONS DOING THE CALLING (FEW MEANS RATIO OF CALLERS TO CIRCUITS IS LESS THAN 8 - GREATER THAN 8 ASSUME TRAFFIC RANDOM)

| HOURS FIRST ATTEMPT | FIRST ATTEMPT VMR | NUMBER SOURCES | HOURS CON-NECT | HOURS OVFL | PERCENT BLOCKED (GRADE) | LINE | HOURS CONNECTED TRAFFIC ON LINES 1 | 2 | 3 | 4 | 5 | 6 | 7 | 8 | 9 | 10 |
|---|---|---|---|---|---|---|---|---|---|---|---|---|---|---|---|---|
| 26.000 | .87 | 196 | 22.980 | 3.020 | 11.6% | 27 | .963 | .961 | .958 | .955 | .951 | .947 | .943 | .937 | .934 | .926 |
| | | | | | | | .921 | .912 | .903 | .895 | .883 | .872 | .858 | .841 | .824 | .804 |
| | | | | | | | .780 | .755 | .725 | .693 | .655 | .614 | .569 | | | |
| 26.000 | .55 | 58 | 23.844 | 2.156 | 8.3% | 28 | .964 | .961 | .958 | .956 | .952 | .947 | .946 | .939 | .936 | .929 |
| | | | | | | | .923 | .917 | .908 | .901 | .892 | .880 | .868 | .855 | .839 | .821 |
| | | | | | | | .801 | .780 | .747 | .729 | .688 | .649 | .602 | .555 | | |
| 26.000 | .70 | 87 | 23.660 | 2.340 | 9.0% | 28 | .963 | .961 | .958 | .954 | .952 | .948 | .945 | .938 | .934 | .930 |
| | | | | | | | .920 | .914 | .908 | .895 | .890 | .873 | .865 | .848 | .831 | .813 |
| | | | | | | | .790 | .764 | .738 | .706 | .673 | .627 | .588 | .534 | | |
| 26.000 | .78 | 116 | 23.583 | 2.417 | 9.3% | 28 | .963 | .961 | .958 | .955 | .951 | .948 | .944 | .937 | .935 | .927 |
| | | | | | | | .921 | .914 | .904 | .898 | .885 | .874 | .861 | .845 | .827 | .807 |
| | | | | | | | .787 | .761 | .730 | .700 | .664 | .620 | .579 | .528 | | |
| 26.000 | .82 | 145 | 23.543 | 2.457 | 9.5% | 28 | .963 | .961 | .958 | .955 | .951 | .947 | .942 | .939 | .933 | .927 |
| | | | | | | | .921 | .913 | .905 | .895 | .885 | .873 | .859 | .843 | .826 | .805 |
| | | | | | | | .781 | .759 | .729 | .694 | .661 | .619 | .573 | .525 | | |
| 26.000 | .85 | 174 | 23.516 | 2.484 | 9.6% | 28 | .963 | .961 | .958 | .955 | .951 | .947 | .943 | .938 | .932 | .928 |
| | | | | | | | .920 | .913 | .904 | .894 | .885 | .871 | .859 | .841 | .826 | .804 |
| | | | | | | | .782 | .756 | .726 | .693 | .657 | .615 | .572 | .522 | | |
| 26.000 | .87 | 203 | 23.498 | 2.502 | 9.6% | 28 | .963 | .961 | .958 | .955 | .951 | .947 | .943 | .938 | .933 | .927 |
| | | | | | | | .920 | .913 | .904 | .894 | .884 | .872 | .858 | .841 | .824 | .803 |
| | | | | | | | .780 | .755 | .725 | .692 | .655 | .613 | .571 | .520 | | |
| 26.000 | .57 | 60 | 24.314 | 1.686 | 6.5% | 29 | .964 | .961 | .958 | .955 | .952 | .949 | .944 | .941 | .935 | .929 |
| | | | | | | | .923 | .914 | .912 | .899 | .891 | .880 | .865 | .855 | .834 | .821 |
| | | | | | | | .804 | .777 | .750 | .721 | .688 | .645 | .604 | .548 | .493 | |
| 26.000 | .71 | 90 | 24.129 | 1.871 | 7.2% | 29 | .963 | .961 | .958 | .955 | .952 | .947 | .943 | .940 | .934 | .928 |
| | | | | | | | .920 | .917 | .905 | .897 | .889 | .875 | .859 | .849 | .833 | .810 |
| | | | | | | | .791 | .765 | .734 | .707 | .670 | .628 | .586 | .535 | .479 | |
| 26.000 | .78 | 120 | 24.052 | 1.948 | 7.5% | 29 | .963 | .961 | .958 | .955 | .951 | .947 | .943 | .939 | .934 | .926 |
| | | | | | | | .922 | .913 | .906 | .896 | .885 | .873 | .860 | .845 | .827 | .808 |
| | | | | | | | .785 | .761 | .732 | .696 | .665 | .621 | .576 | .529 | .475 | |
| 26.000 | .83 | 150 | 24.007 | 1.993 | 7.7% | 29 | .963 | .961 | .958 | .955 | .951 | .947 | .944 | .938 | .933 | .927 |
| | | | | | | | .920 | .913 | .905 | .895 | .885 | .872 | .859 | .843 | .826 | .805 |
| | | | | | | | .781 | .759 | .728 | .694 | .660 | .617 | .574 | .525 | .470 | |
| 26.000 | .86 | 180 | 23.981 | 2.019 | 7.8% | 29 | .963 | .960 | .958 | .954 | .951 | .947 | .943 | .938 | .933 | .927 |
| | | | | | | | .920 | .913 | .904 | .895 | .884 | .872 | .857 | .843 | .824 | .804 |
| | | | | | | | .782 | .755 | .726 | .693 | .655 | .617 | .571 | .522 | .470 | |
| 26.000 | .88 | 210 | 23.962 | 2.038 | 7.8% | 29 | .963 | .961 | .958 | .955 | .951 | .947 | .942 | .938 | .932 | .927 |
| | | | | | | | .920 | .912 | .904 | .895 | .883 | .872 | .856 | .843 | .823 | .804 |
| | | | | | | | .780 | .754 | .725 | .692 | .655 | .614 | .569 | .519 | .469 | |
| 28.000 | .56 | 63 | 18.504 | 9.496 | 33.9% | 20 | .966 | .964 | .962 | .959 | .957 | .952 | .950 | .946 | .941 | .937 |

ERLANG-ENGSET: FOR SMOOTH ATTEMPTING TRAFFIC AND FEW PERSONS DOING THE CALLING (FEW MEANS RATIO OF CALLERS TO CIRCUITS IS LESS THAN 8 - GREATER THAN 8 ASSUME TRAFFIC RANDOM)

| HOURS FIRST ATTEMPT | FIRST ATTEMPT VMR | NUMBER SOURCES | HOURS CON-NECT | HOURS OVFL | PERCENT BLOCKED (GRADE) | LINE | HOURS CONNECTED TRAFFIC ON LINES 1 | 2 | 3 | 4 | 5 | 6 | 7 | 8 | 9 | 10 |
|---|---|---|---|---|---|---|---|---|---|---|---|---|---|---|---|---|
| | | | | | | | .932 | .928 | .917 | .916 | .906 | .897 | .887 | .876 | .865 | .846 |
| 28.000 | .67 | 84 | 18.475 | 9.525 | 34.0% | 20 | .966 | .964 | .961 | .959 | .956 | .953 | .949 | .945 | .940 | .938 |
| | | | | | | | .930 | .927 | .917 | .912 | .904 | .894 | .883 | .874 | .860 | .843 |
| 28.000 | .73 | 105 | 18.456 | 9.544 | 34.1% | 20 | .966 | .964 | .962 | .959 | .955 | .953 | .949 | .944 | .941 | .937 |
| | | | | | | | .930 | .925 | .917 | .910 | .903 | .891 | .885 | .871 | .858 | .839 |
| 28.000 | .78 | 126 | 18.450 | 9.550 | 34.1% | 20 | .966 | .964 | .961 | .959 | .955 | .952 | .949 | .944 | .940 | .936 |
| | | | | | | | .931 | .924 | .916 | .909 | .901 | .893 | .882 | .870 | .856 | .841 |
| 28.000 | .81 | 147 | 18.442 | 9.558 | 34.1% | 20 | .966 | .963 | .961 | .958 | .956 | .952 | .949 | .945 | .940 | .934 |
| | | | | | | | .931 | .924 | .917 | .910 | .901 | .892 | .880 | .869 | .855 | .839 |
| 28.000 | .58 | 66 | 19.335 | 8.665 | 30.9% | 21 | .966 | .964 | .961 | .959 | .956 | .953 | .950 | .945 | .939 | .938 |
| | | | | | | | .932 | .927 | .920 | .913 | .904 | .896 | .887 | .874 | .865 | .851 |
| | | | | | | | .832 | | | | | | | | | |
| 28.000 | .68 | 88 | 19.299 | 8.701 | 31.1% | 21 | .966 | .964 | .961 | .958 | .956 | .952 | .948 | .946 | .941 | .936 |
| | | | | | | | .930 | .924 | .919 | .911 | .902 | .895 | .884 | .871 | .859 | .846 |
| | | | | | | | .826 | | | | | | | | | |
| 28.000 | .75 | 110 | 19.283 | 8.717 | 31.1% | 21 | .966 | .964 | .961 | .958 | .955 | .953 | .948 | .945 | .942 | .935 |
| | | | | | | | .930 | .926 | .917 | .910 | .903 | .890 | .885 | .869 | .856 | .841 |
| | | | | | | | .828 | | | | | | | | | |
| 28.000 | .79 | 132 | 19.270 | 8.730 | 31.2% | 21 | .966 | .963 | .961 | .959 | .956 | .951 | .950 | .945 | .940 | .936 |
| | | | | | | | .930 | .922 | .919 | .909 | .901 | .892 | .881 | .869 | .856 | .840 |
| | | | | | | | .823 | | | | | | | | | |
| 28.000 | .82 | 154 | 19.262 | 8.738 | 31.2% | 21 | .966 | .964 | .961 | .959 | .955 | .952 | .949 | .944 | .940 | .936 |
| | | | | | | | .930 | .924 | .915 | .911 | .900 | .893 | .879 | .867 | .857 | .839 |
| | | | | | | | .822 | | | | | | | | | |
| 28.000 | .59 | 69 | 20.145 | 7.855 | 28.1% | 22 | .966 | .964 | .962 | .959 | .956 | .953 | .949 | .944 | .943 | .937 |
| | | | | | | | .932 | .926 | .917 | .914 | .907 | .895 | .888 | .871 | .863 | .853 |
| | | | | | | | .833 | .815 | | | | | | | | |
| 28.000 | .70 | 92 | 20.104 | 7.896 | 28.2% | 22 | .966 | .964 | .961 | .959 | .955 | .953 | .949 | .945 | .939 | .936 |
| | | | | | | | .933 | .923 | .918 | .913 | .900 | .896 | .884 | .870 | .859 | .845 |
| | | | | | | | .825 | .810 | | | | | | | | |
| 28.000 | .70 | 115 | 20.084 | 7.910 | 28.3% | 22 | .966 | .964 | .961 | .958 | .956 | .952 | .947 | .946 | .940 | .936 |
| | | | | | | | .929 | .924 | .918 | .912 | .900 | .891 | .886 | .070 | .857 | .842 |
| | | | | | | | .824 | .804 | | | | | | | | |
| 28.000 | .80 | 138 | 20.071 | 7.929 | 28.3% | 22 | .966 | .964 | .961 | .959 | .955 | .052 | .949 | .945 | .940 | .934 |
| | | | | | | | .929 | .924 | .918 | .911 | .901 | .892 | .881 | .869 | .855 | .840 |
| | | | | | | | .823 | .803 | | | | | | | | |
| 20.000 | .83 | 161 | 20.061 | 7.939 | 28.4% | 22 | .966 | .964 | .961 | .958 | .956 | .952 | .948 | .945 | .040 | .935 |
| | | | | | | | .930 | .924 | .917 | .910 | .900 | .891 | .881 | .868 | .854 | .839 |
| | | | | | | | .821 | .802 | | | | | | | | |
| 28.000 | .61 | 72 | 20.931 | 7.069 | 25.2% | 23 | .966 | .964 | .962 | .959 | .955 | .953 | .950 | .946 | .941 | .937 |

ERLANG-ENGSET: FOR SMOOTH ATTEMPTING TRAFFIC AND FEW PERSONS DOING THE CALLING (FEW MEANS RATIO OF CALLERS TO CIRCUITS IS LESS THAN 8 - GREATER THAN 8 ASSUME TRAFFIC RANDOM)

| HOURS FIRST ATTEMPT | FIRST ATTEMPT VMR | NUMBER SOURCES | HOURS CON-NECT | HOURS OVFL | PERCENT BLOCKED (GRADE) | LINE | HOURS CONNECTED TRAFFIC ON LINES 1 | 2 | 3 | 4 | 5 | 6 | 7 | 8 | 9 | 10 |
|---|---|---|---|---|---|---|---|---|---|---|---|---|---|---|---|---|
| | | | | | | | .931 | .925 | .922 | .912 | .903 | .896 | .887 | .872 | .863 | .850 |
| | | | | | | | .831 | .815 | .793 | | | | | | | |
| 28.000 | .71 | 96 | 20.885 | 7.115 | 25.4% | 23 | .966 | .964 | .962 | .958 | .957 | .951 | .949 | .946 | .941 | .935 |
| | | | | | | | .931 | .924 | .919 | .910 | .904 | .892 | .884 | .869 | .861 | .843 |
| | | | | | | | .827 | .807 | .785 | | | | | | | |
| 28.000 | .77 | 120 | 20.861 | 7.139 | 25.5% | 23 | .966 | .964 | .961 | .958 | .955 | .953 | .949 | .944 | .940 | .937 |
| | | | | | | | .928 | .926 | .917 | .910 | .902 | .890 | .884 | .868 | .858 | .839 |
| | | | | | | | .826 | .801 | .784 | | | | | | | |
| 28.000 | .81 | 144 | 20.848 | 7.152 | 25.5% | 23 | .966 | .964 | .961 | .958 | .955 | .953 | .949 | .945 | .940 | .935 |
| | | | | | | | .930 | .924 | .916 | .912 | .900 | .891 | .880 | .868 | .858 | .839 |
| | | | | | | | .824 | .802 | .781 | | | | | | | |
| 28.000 | .83 | 168 | 20.838 | 7.162 | 25.6% | 23 | .966 | .963 | .961 | .958 | .955 | .952 | .949 | .945 | .940 | .935 |
| | | | | | | | .930 | .923 | .917 | .909 | .901 | .891 | .880 | .868 | .854 | .839 |
| | | | | | | | .821 | .801 | .779 | | | | | | | |
| 28.000 | .63 | 75 | 21.694 | 6.306 | 22.5% | 24 | .966 | .964 | .962 | .958 | .957 | .953 | .950 | .945 | .941 | .937 |
| | | | | | | | .931 | .926 | .921 | .912 | .902 | .897 | .885 | .874 | .859 | .848 |
| | | | | | | | .831 | .814 | .792 | .771 | | | | | | |
| 28.000 | .72 | 100 | 21.644 | 6.356 | 22.7% | 24 | .966 | .964 | .961 | .958 | .956 | .952 | .949 | .944 | .941 | .935 |
| | | | | | | | .931 | .923 | .918 | .912 | .902 | .894 | .881 | .870 | .858 | .844 |
| | | | | | | | .827 | .807 | .785 | .763 | | | | | | |
| 28.000 | .78 | 125 | 21.615 | 6.385 | 22.8% | 24 | .966 | .963 | .961 | .959 | .956 | .951 | .950 | .945 | .940 | .935 |
| | | | | | | | .929 | .924 | .918 | .911 | .902 | .893 | .879 | .872 | .854 | .843 |
| | | | | | | | .821 | .806 | .783 | .756 | | | | | | |
| 28.000 | .81 | 150 | 21.599 | 6.401 | 22.9% | 24 | .966 | .964 | .961 | .958 | .955 | .951 | .949 | .944 | .940 | .935 |
| | | | | | | | .930 | .924 | .917 | .910 | .901 | .891 | .881 | .869 | .855 | .839 |
| | | | | | | | .822 | .802 | .780 | .753 | | | | | | |
| 28.000 | .84 | 175 | 21.588 | 6.412 | 22.9% | 24 | .966 | .964 | .961 | .958 | .955 | .952 | .948 | .945 | .940 | .935 |
| | | | | | | | .929 | .924 | .917 | .909 | .900 | .892 | .880 | .867 | .854 | .839 |
| | | | | | | | .821 | .801 | .779 | .753 | | | | | | |
| 28.000 | .64 | 78 | 22.426 | 5.574 | 19.9% | 25 | .966 | .964 | .961 | .959 | .955 | .953 | .949 | .947 | .941 | .937 |
| | | | | | | | .930 | .927 | .916 | .912 | .907 | .893 | .886 | .875 | .862 | .846 |
| | | | | | | | .829 | .812 | .791 | .767 | .742 | | | | | |
| 28.000 | .73 | 104 | 22.371 | 5.629 | 20.1% | 25 | .966 | .963 | .961 | .959 | .956 | .951 | .950 | .945 | .940 | .936 |
| | | | | | | | .931 | .924 | .918 | .909 | .902 | .894 | .880 | .874 | .857 | .842 |
| | | | | | | | .825 | .806 | .787 | .761 | .731 | | | | | |
| 28.000 | .78 | 130 | 22.343 | 5.657 | 20.2% | 25 | .966 | .964 | .961 | .958 | .955 | .952 | .949 | .943 | .942 | .935 |
| | | | | | | | .930 | .924 | .917 | .908 | .904 | .891 | .883 | .868 | .855 | .843 |
| | | | | | | | .821 | .806 | .782 | .756 | .729 | | | | | |
| 28.000 | .82 | 156 | 22.323 | 5.677 | 20.3% | 25 | .966 | .964 | .961 | .958 | .955 | .951 | .949 | .944 | .940 | .935 |
| | | | | | | | .930 | .924 | .917 | .909 | .901 | .891 | .881 | .868 | .855 | .840 |

ERLANG-ENGSET: FOR SMOOTH ATTEMPTING TRAFFIC AND FEW PERSONS DOING THE CALLING (FEW MEANS RATIO OF CALLERS TO CIRCUITS IS LESS THAN 8 - GREATER THAN 8 ASSUME TRAFFIC RANDOM)

| HOURS FIRST ATTEMPT | FIRST ATTEMPT VMR | NUMBER SOURCES | HOURS CON-NECT | HOURS OVFL | PERCENT BLOCKED (GRADE) | LINE | HOURS CONNECTED TRAFFIC ON LINES 1 | 2 | 3 | 4 | 5 | 6 | 7 | 8 | 9 | 10 |
|---|---|---|---|---|---|---|---|---|---|---|---|---|---|---|---|---|
| | | | | | | | .821 | .802 | .779 | .755 | .726 | | | | | |
| 28.000 | .85 | 182 | 22.310 | 5.690 | 20.3% | 25 | .966 | .963 | .961 | .958 | .956 | .952 | .948 | .945 | .940 | .935 |
| | | | | | | | .930 | .922 | .918 | .908 | .902 | .890 | .879 | .869 | .854 | .839 |
| | | | | | | | .821 | .801 | .778 | .753 | .724 | | | | | |
| 28.000 | .48 | 54 | 23.275 | 4.725 | 16.9% | 26 | .966 | .964 | .962 | .959 | .955 | .955 | .950 | .946 | .943 | .938 |
| | | | | | | | .933 | .927 | .922 | .914 | .908 | .899 | .886 | .885 | .867 | .854 |
| | | | | | | | .841 | .818 | .808 | .786 | .759 | .732 | | | | |
| 28.000 | .65 | 81 | 23.126 | 4.874 | 17.4% | 26 | .966 | .964 | .961 | .959 | .955 | .953 | .948 | .946 | .940 | .936 |
| | | | | | | | .933 | .925 | .918 | .911 | .903 | .897 | .883 | .875 | .858 | .845 |
| | | | | | | | .830 | .813 | .792 | .764 | .739 | .711 | | | | |
| 28.000 | .74 | 108 | 23.066 | 4.934 | 17.6% | 26 | .966 | .964 | .961 | .958 | .956 | .952 | .948 | .945 | .942 | .935 |
| | | | | | | | .931 | .923 | .920 | .910 | .900 | .895 | .882 | .871 | .858 | .843 |
| | | | | | | | .825 | .806 | .783 | .761 | .731 | .701 | | | | |
| 28.000 | .79 | 135 | 23.035 | 4.965 | 17.7% | 26 | .966 | .964 | .961 | .958 | .955 | .953 | .949 | .945 | .940 | .935 |
| | | | | | | | .929 | .926 | .917 | .910 | .901 | .892 | .881 | .869 | .856 | .840 |
| | | | | | | | .823 | .803 | .779 | .759 | .728 | .698 | | | | |
| 28.000 | .83 | 162 | 23.014 | 4.986 | 17.8% | 26 | .966 | .964 | .961 | .958 | .956 | .952 | .948 | .945 | .941 | .936 |
| | | | | | | | .930 | .924 | .917 | .909 | .901 | .891 | .880 | .869 | .855 | .839 |
| | | | | | | | .821 | .802 | .778 | .754 | .726 | .694 | | | | |
| 28.000 | .85 | 189 | 22.999 | 5.001 | 17.9% | 26 | .966 | .963 | .961 | .958 | .956 | .952 | .949 | .944 | .940 | .935 |
| | | | | | | | .929 | .924 | .916 | .909 | .900 | .891 | .880 | .868 | .854 | .838 |
| | | | | | | | .820 | .801 | .779 | .752 | .724 | .692 | | | | |
| 28.000 | .67 | 84 | 23.791 | 4.209 | 15.0% | 27 | .966 | .964 | .961 | .959 | .956 | .953 | .949 | .945 | .940 | .938 |
| | | | | | | | .930 | .927 | .917 | .912 | .904 | .894 | .883 | .874 | .860 | .843 |
| | | | | | | | .832 | .810 | .789 | .764 | .742 | .708 | .672 | | | |
| 28.000 | .75 | 112 | 23.727 | 4.273 | 15.3% | 27 | .966 | .964 | .962 | .958 | .956 | .953 | .948 | .945 | .941 | .935 |
| | | | | | | | .930 | .925 | .916 | .910 | .902 | .894 | .880 | .873 | .855 | .840 |
| | | | | | | | .828 | .803 | .785 | .758 | .733 | .698 | .669 | | | |
| 28.000 | .80 | 140 | 23.691 | 4.309 | 15.4% | 27 | .966 | .964 | .961 | .958 | .956 | .951 | .950 | .945 | .940 | .936 |
| | | | | | | | .930 | .924 | .917 | .908 | .903 | .891 | .880 | .868 | .857 | .839 |
| | | | | | | | .821 | .803 | .781 | .756 | .725 | .697 | .663 | | | |
| 28.000 | .83 | 168 | 23.667 | 4.333 | 15.5% | 27 | .966 | .963 | .961 | .958 | .955 | .952 | .949 | .945 | .940 | .935 |
| | | | | | | | .930 | .923 | .917 | .909 | .901 | .891 | .880 | .868 | .854 | .839 |
| | | | | | | | .821 | .801 | .779 | .752 | .727 | .693 | .656 | | | |
| 28.000 | .86 | 196 | 23.652 | 4.348 | 15.5% | 27 | .966 | .964 | .961 | .958 | .956 | .952 | .949 | .944 | .940 | .935 |
| | | | | | | | .930 | .923 | .917 | .909 | .900 | .890 | .880 | .868 | .853 | .838 |
| | | | | | | | .820 | .800 | .777 | .751 | .724 | .692 | .656 | | | |
| 28.000 | .52 | 58 | 24.590 | 3.410 | 12.2% | 28 | .966 | .964 | .961 | .959 | .957 | .953 | .949 | .947 | .942 | .937 |
| | | | | | | | .933 | .927 | .921 | .914 | .905 | .899 | .884 | .881 | .868 | .853 |
| | | | | | | | .839 | .815 | .802 | .785 | .753 | .727 | .693 | .656 | | |

ERLANG-ENGSET: FOR SMOOTH ATTEMPTING TRAFFIC AND FEW PERSONS DOING THE CALLING (FEW MEANS RATIO OF CALLERS TO CIRCUITS IS LESS THAN 8 - GREATER THAN 8 ASSUME TRAFFIC RANDOM)

| HOURS FIRST ATTEMPT | FIRST ATTEMPT VMR | NUMBER SOURCES | HOURS CON-NECT | HOURS OVFL | PERCENT BLOCKED (GRADE) | LINE | HOURS CONNECTED TRAFFIC ON LINES 1 | 2 | 3 | 4 | 5 | 6 | 7 | 8 | 9 | 10 |
|---|---|---|---|---|---|---|---|---|---|---|---|---|---|---|---|---|
| 28.000 | .68 | 87 | 24.416 | 3.584 | 12.8% | 28 | .966 | .964 | .961 | .959 | .956 | .952 | .949 | .946 | .939 | .936 |
| | | | | | | | .931 | .927 | .917 | .912 | .903 | .893 | .885 | .869 | .861 | .844 |
| | | | | | | | .828 | .810 | .788 | .767 | .737 | .707 | .675 | .633 | | |
| 28.000 | .76 | 116 | 24.343 | 3.657 | 13.1% | 28 | .966 | .963 | .962 | .958 | .956 | .952 | .949 | .945 | .939 | .935 |
| | | | | | | | .931 | .926 | .916 | .910 | .902 | .894 | .879 | .873 | .854 | .844 |
| | | | | | | | .822 | .807 | .784 | .757 | .732 | .702 | .663 | .624 | | |
| 28.000 | .81 | 145 | 24.306 | 3.694 | 13.2% | 28 | .966 | .964 | .961 | .958 | .955 | .951 | .949 | .944 | .940 | .936 |
| | | | | | | | .930 | .924 | .917 | .910 | .899 | .894 | .879 | .871 | .853 | .842 |
| | | | | | | | .822 | .803 | .780 | .754 | .729 | .696 | .660 | .620 | | |
| 28.000 | .84 | 174 | 24.282 | 3.718 | 13.3% | 28 | .966 | .964 | .961 | .958 | .955 | .952 | .949 | .944 | .940 | .935 |
| | | | | | | | .930 | .924 | .917 | .908 | .902 | .890 | .882 | .866 | .856 | .838 |
| | | | | | | | .821 | .800 | .779 | .754 | .723 | .694 | .657 | .618 | | |
| 28.000 | .86 | 203 | 24.265 | 3.735 | 13.3% | 28 | .966 | .963 | .961 | .958 | .955 | .952 | .948 | .945 | .939 | .936 |
| | | | | | | | .929 | .923 | .916 | .909 | .900 | .891 | .880 | .867 | .853 | .838 |
| | | | | | | | .821 | .800 | .778 | .752 | .723 | .691 | .656 | .616 | | |
| 28.000 | .53 | 60 | 25.177 | 2.823 | 10.1% | 29 | .966 | .963 | .962 | .960 | .956 | .952 | .951 | .946 | .942 | .937 |
| | | | | | | | .933 | .923 | .923 | .915 | .902 | .901 | .889 | .875 | .868 | .851 |
| | | | | | | | .834 | .821 | .800 | .780 | .747 | .730 | .690 | .653 | .608 | |
| 28.000 | .69 | 90 | 24.995 | 3.005 | 10.7% | 29 | .966 | .964 | .961 | .958 | .956 | .951 | .951 | .944 | .942 | .936 |
| | | | | | | | .930 | .926 | .916 | .914 | .901 | .894 | .885 | .872 | .857 | .847 |
| | | | | | | | .827 | .808 | .790 | .761 | .739 | .705 | .673 | .632 | .588 | |
| 28.000 | .77 | 120 | 24.918 | 3.082 | 11.0% | 29 | .966 | .964 | .961 | .958 | .955 | .953 | .949 | .944 | .940 | .937 |
| | | | | | | | .928 | .926 | .917 | .910 | .902 | .890 | .884 | .868 | .858 | .839 |
| | | | | | | | .826 | .801 | .784 | .761 | .730 | .697 | .663 | .625 | .581 | |
| 28.000 | .81 | 150 | 24.879 | 3.121 | 11.1% | 29 | .966 | .964 | .961 | .958 | .955 | .951 | .949 | .944 | .940 | .935 |
| | | | | | | | .930 | .924 | .917 | .910 | .901 | .891 | .881 | .869 | .855 | .839 |
| | | | | | | | .822 | .802 | .780 | .753 | .728 | .695 | .660 | .620 | .577 | |
| 28.000 | .84 | 180 | 24.852 | 3.148 | 11.2% | 29 | .966 | .963 | .961 | .958 | .955 | .952 | .948 | .945 | .940 | .935 |
| | | | | | | | .930 | .924 | .917 | .910 | .900 | .891 | .880 | .868 | .854 | .838 |
| | | | | | | | .819 | .802 | .779 | .753 | .725 | .691 | .659 | .618 | .574 | |
| 28.000 | .87 | 210 | 24.834 | 3.166 | 11.3% | 29 | .966 | .963 | .961 | .958 | .955 | .952 | .948 | .944 | .940 | .935 |
| | | | | | | | .929 | .923 | .916 | .909 | .900 | .890 | .880 | .868 | .853 | .838 |
| | | | | | | | .820 | .800 | .778 | .751 | .723 | .691 | .656 | .615 | .572 | |
| 28.000 | .55 | 62 | 25.714 | 2.286 | 8.2% | 30 | .966 | .964 | .962 | .958 | .956 | .954 | .950 | .943 | .943 | .938 |
| | | | | | | | .933 | .926 | .921 | .912 | .903 | .901 | .887 | .876 | .866 | .850 |
| | | | | | | | .836 | .815 | .799 | .780 | .751 | .722 | .690 | .647 | .611 | .554 |
| 28.000 | .70 | 93 | 25.524 | 2.476 | 8.8% | 30 | .966 | .964 | .961 | .959 | .956 | .952 | .949 | .945 | .942 | .934 |
| | | | | | | | .933 | .923 | .918 | .910 | .903 | .896 | .884 | .870 | .858 | .844 |
| | | | | | | | .828 | .809 | .787 | .761 | .739 | .705 | .666 | .636 | .586 | .540 |
| 28.000 | .77 | 124 | 25.447 | 2.553 | 9.1% | 30 | .966 | .964 | .961 | .959 | .955 | .952 | .949 | .946 | .939 | .937 |

ERLANG-ENGSET: FOR SMOOTH ATTEMPTING TRAFFIC AND FEW PERSONS DOING THE CALLING (FEW MEANS RATIO OF CALLERS TO CIRCUITS IS LESS THAN 8 - GREATER THAN 8 ASSUME TRAFFIC RANDOM)

| HOURS FIRST ATTEMPT | FIRST ATTEMPT VMR | NUMBER SOURCES | HOURS CON-NECT | HOURS OVFL | PERCENT BLOCKED (GRADE) | LINE | HOURS CONNECTED TRAFFIC ON LINES 1 | 2 | 3 | 4 | 5 | 6 | 7 | 8 | 9 | 10 |
|---|---|---|---|---|---|---|---|---|---|---|---|---|---|---|---|---|
| | | | | | | | .930 | .924 | .917 | .908 | .904 | .891 | .881 | .870 | .856 | .841 |
| | | | | | | | .823 | .807 | .783 | .758 | .728 | .699 | .665 | .621 | .583 | .533 |
| 28.000 | .82 | 155 | 25.402 | 2.598 | 9.3% | 30 | .966 | .963 | .961 | .958 | .955 | .953 | .949 | .945 | .940 | .935 |
| | | | | | | | .930 | .924 | .917 | .909 | .901 | .891 | .881 | .868 | .855 | .839 |
| | | | | | | | .822 | .802 | .779 | .753 | .728 | .695 | .659 | .620 | .576 | .528 |
| 28.000 | .85 | 186 | 25.375 | 2.625 | 9.4% | 30 | .966 | .963 | .961 | .958 | .956 | .952 | .948 | .944 | .940 | .935 |
| | | | | | | | .930 | .924 | .916 | .910 | .900 | .891 | .880 | .868 | .854 | .838 |
| | | | | | | | .821 | .801 | .778 | .753 | .724 | .693 | .656 | .617 | .574 | .527 |
| 28.000 | .87 | 217 | 25.356 | 2.644 | 9.4% | 30 | .966 | .963 | .961 | .958 | .955 | .952 | .948 | .944 | .940 | .935 |
| | | | | | | | .929 | .924 | .916 | .909 | .900 | .890 | .879 | .867 | .853 | .837 |
| | | | | | | | .821 | .799 | .777 | .751 | .723 | .691 | .654 | .616 | .572 | .525 |
| 28.000 | .56 | 64 | 26.195 | 1.805 | 6.4% | 31 | .966 | .963 | .962 | .959 | .956 | .953 | .949 | .946 | .942 | .937 |
| | | | | | | | .931 | .926 | .922 | .912 | .906 | .896 | .888 | .872 | .865 | .855 |
| | | | | | | | .835 | .817 | .794 | .773 | .748 | .727 | .682 | .654 | .605 | .557 |
| | | | | | | | .497 | | | | | | | | | |
| 28.000 | .71 | 96 | 26.001 | 1.999 | 7.1% | 31 | .966 | .964 | .962 | .958 | .957 | .951 | .949 | .946 | .941 | .935 |
| | | | | | | | .931 | .924 | .919 | .910 | .904 | .892 | .884 | .869 | .861 | .843 |
| | | | | | | | .827 | .807 | .785 | .763 | .737 | .703 | .668 | .631 | .588 | .540 |
| | | | | | | | .486 | | | | | | | | | |
| 28.000 | .78 | 128 | 25.920 | 2.080 | 7.4% | 31 | .966 | .963 | .961 | .959 | .956 | .952 | .948 | .945 | .941 | .936 |
| | | | | | | | .929 | .923 | .917 | .911 | .903 | .892 | .882 | .867 | .859 | .837 |
| | | | | | | | .826 | .805 | .782 | .755 | .730 | .699 | .662 | .623 | .580 | .532 |
| | | | | | | | .481 | | | | | | | | | |
| 28.000 | .83 | 160 | 25.876 | 2.124 | 7.6% | 31 | .966 | .963 | .961 | .959 | .955 | .952 | .949 | .945 | .940 | .936 |
| | | | | | | | .930 | .923 | .917 | .908 | .902 | .890 | .882 | .867 | .856 | .836 |
| | | | | | | | .824 | .802 | .779 | .754 | .726 | .694 | .658 | .620 | .575 | .530 |
| | | | | | | | .478 | | | | | | | | | |
| 28.000 | .85 | 192 | 25.847 | 2.153 | 7.7% | 31 | .966 | .964 | .961 | .958 | .955 | .952 | .948 | .945 | .940 | .934 |
| | | | | | | | .930 | .923 | .917 | .908 | .900 | .892 | .879 | .869 | .852 | .839 |
| | | | | | | | .821 | .800 | .778 | .753 | .724 | .691 | .657 | .617 | .573 | .526 |
| | | | | | | | .477 | | | | | | | | | |
| 28.000 | .88 | 224 | 25.827 | 2.173 | 7.8% | 31 | .966 | .963 | .961 | .958 | .955 | .952 | .948 | .944 | .940 | .935 |
| | | | | | | | .930 | .923 | .916 | .909 | .900 | .890 | .879 | .867 | .853 | .838 |
| | | | | | | | .819 | .800 | .776 | .751 | .722 | .691 | .655 | .615 | .571 | .525 |
| | | | | | | | .475 | | | | | | | | | |
| 30.000 | .55 | 66 | 19.539 | 10.461 | 34.9% | 21 | .968 | .966 | .965 | .962 | .959 | .958 | .953 | .951 | .947 | .944 |
| | | | | | | | .939 | .934 | .930 | .920 | .919 | .912 | .904 | .892 | .887 | .873 |
| | | | | | | | .855 | | | | | | | | | |
| 30.000 | .66 | 88 | 19.513 | 10.487 | 35.0% | 21 | .968 | .966 | .964 | .962 | .959 | .957 | .953 | .950 | .948 | .941 |
| | | | | | | | .939 | .934 | .928 | .919 | .918 | .907 | .902 | .892 | .881 | .868 |

ERLANG-ENGSET: FOR SMOOTH ATTEMPTING TRAFFIC AND FEW PERSONS DOING THE CALLING (FEW MEANS RATIO OF CALLERS TO CIRCUITS IS LESS THAN 8 - GREATER THAN 8 ASSUME TRAFFIC RANDOM)

| HOURS FIRST ATTEMPT | FIRST ATTEMPT VMR | NUMBER SOURCES | HOURS CON-NECT | HOURS OVFL | PERCENT BLOCKED (GRADE) | LINE | HOURS CONNECTED TRAFFIC ON LINES 1 | 2 | 3 | 4 | 5 | 6 | 7 | 8 | 9 | 10 |
|---|---|---|---|---|---|---|---|---|---|---|---|---|---|---|---|---|
| | | | | | | | .857 | | | | | | | | | |
| 30.000 | .57 | 69 | 20.385 | 9.615 | 32.1% | 22 | .968 | .966 | .964 | .962 | .958 | .958 | .955 | .950 | .947 | .941 |
| | | | | | | | .941 | .935 | .930 | .919 | .917 | .913 | .902 | .894 | .882 | .874 |
| | | | | | | | .862 | .846 | | | | | | | | |
| 30.000 | .67 | 92 | 20.351 | 9.649 | 32.2% | 22 | .968 | .966 | .964 | .961 | .960 | .957 | .953 | .949 | .947 | .943 |
| | | | | | | | .938 | .932 | .929 | .922 | .915 | .906 | .901 | .890 | .882 | .868 |
| | | | | | | | .857 | .840 | | | | | | | | |
| 30.000 | .74 | 115 | 20.331 | 9.669 | 32.2% | 22 | .968 | .966 | .964 | .962 | .958 | .957 | .953 | .949 | .946 | .943 |
| | | | | | | | .939 | .932 | .928 | .919 | .916 | .907 | .899 | .887 | .881 | .866 |
| | | | | | | | .853 | .838 | | | | | | | | |
| 30.000 | .78 | 138 | 20.322 | 9.678 | 32.3% | 22 | .968 | .966 | .964 | .961 | .959 | .957 | .953 | .949 | .947 | .942 |
| | | | | | | | .937 | .933 | .927 | .919 | .916 | .906 | .898 | .888 | .878 | .866 |
| | | | | | | | .852 | .837 | | | | | | | | |
| 30.000 | .81 | 161 | 20.315 | 9.685 | 32.3% | 22 | .968 | .966 | .964 | .962 | .958 | .957 | .953 | .950 | .946 | .942 |
| | | | | | | | .937 | .932 | .927 | .920 | .914 | .904 | .899 | .886 | .879 | .863 |
| | | | | | | | .853 | .835 | | | | | | | | |
| 30.000 | .58 | 72 | 21.209 | 8.791 | 29.3% | 23 | .968 | .966 | .964 | .962 | .959 | .957 | .954 | .950 | .947 | .944 |
| | | | | | | | .938 | .935 | .929 | .921 | .919 | .910 | .902 | .893 | .880 | .873 |
| | | | | | | | .863 | .845 | .830 | | | | | | | |
| 30.000 | .69 | 96 | 21.171 | 8.829 | 29.4% | 23 | .968 | .966 | .964 | .961 | .959 | .956 | .952 | .950 | .948 | .941 |
| | | | | | | | .938 | .932 | .929 | .922 | .914 | .908 | .902 | .887 | .883 | .869 |
| | | | | | | | .853 | .843 | .823 | | | | | | | |
| 30.000 | .75 | 120 | 21.152 | 8.848 | 29.5% | 23 | .968 | .966 | .964 | .962 | .960 | .956 | .953 | .950 | .945 | .942 |
| | | | | | | | .939 | .931 | .929 | .921 | .914 | .905 | .901 | .888 | .878 | .867 |
| | | | | | | | .853 | .838 | .821 | | | | | | | |
| 30.000 | .79 | 144 | 21.140 | 8.860 | 29.5% | 23 | .968 | .966 | .964 | .962 | .958 | .956 | .953 | .950 | .946 | .942 |
| | | | | | | | .936 | .934 | .927 | .921 | .914 | .907 | .898 | .888 | .877 | .865 |
| | | | | | | | .851 | .837 | .819 | | | | | | | |
| 30.000 | .82 | 168 | 21.130 | 8.870 | 29.6% | 23 | .968 | .966 | .964 | .961 | .959 | .956 | .953 | .950 | .946 | .942 |
| | | | | | | | .938 | .932 | .927 | .920 | .914 | .906 | .897 | .887 | .876 | .865 |
| | | | | | | | .851 | .835 | .818 | | | | | | | |
| 30.000 | .60 | 75 | 22.015 | 7.985 | 26.6% | 24 | .968 | .966 | .965 | .961 | .960 | .956 | .954 | .951 | .946 | .943 |
| | | | | | | | .940 | .930 | .929 | .927 | .915 | .909 | .903 | .893 | .882 | .872 |
| | | | | | | | .860 | .844 | .828 | .813 | | | | | | |
| 30.000 | .70 | 100 | 21.970 | 8.030 | 26.8% | 24 | .968 | .966 | .963 | .963 | .958 | .956 | .954 | .948 | .948 | .943 |
| | | | | | | | .937 | .934 | .927 | .920 | .915 | .909 | .899 | .891 | .878 | .868 |
| | | | | | | | .856 | .842 | .821 | .806 | | | | | | |
| 30.000 | .76 | 125 | 21.950 | 8.050 | 26.8% | 24 | .968 | .966 | .964 | .962 | .959 | .956 | .954 | .949 | .948 | .942 |
| | | | | | | | .938 | .931 | .929 | .921 | .915 | .907 | .896 | .891 | .877 | .865 |
| | | | | | | | .856 | .836 | .819 | .804 | | | | | | |

ERLANG-ENGSET: FOR SMOOTH ATTEMPTING TRAFFIC AND FEW PERSONS DOING THE CALLING (FEW MEANS RATIO OF CALLERS TO CIRCUITS IS LESS THAN 8 - GREATER THAN 8 ASSUME TRAFFIC RANDOM)

| HOURS FIRST ATTEMPT | FIRST ATTEMPT VMR | NUMBER SOURCES | HOURS CONNECT | HOURS OVFL | PERCENT BLOCKED (GRADE) | LINE | HOURS CONNECTED TRAFFIC ON LINES 1 | 2 | 3 | 4 | 5 | 6 | 7 | 8 | 9 | 10 |
|---|---|---|---|---|---|---|---|---|---|---|---|---|---|---|---|---|
| 30.000 | .80 | 150 | 21.933 | 8.067 | 26.9% | 24 | .968 | .966 | .964 | .961 | .959 | .955 | .954 | .950 | .946 | .942 |
| | | | | | | | .938 | .932 | .927 | .919 | .916 | .905 | .897 | .887 | .877 | .868 |
| | | | | | | | .851 | .835 | .821 | .796 | | | | | | |
| 30.000 | .83 | 175 | 21.927 | 8.073 | 26.9% | 24 | .968 | .966 | .964 | .961 | .960 | .956 | .953 | .950 | .946 | .942 |
| | | | | | | | .937 | .932 | .926 | .920 | .914 | .905 | .898 | .887 | .876 | .864 |
| | | | | | | | .851 | .834 | .818 | .799 | | | | | | |
| 30.000 | .62 | 78 | 22.797 | 7.203 | 24.0% | 25 | .968 | .966 | .964 | .962 | .959 | .957 | .953 | .950 | .947 | .942 |
| | | | | | | | .941 | .932 | .928 | .923 | .918 | .909 | .899 | .894 | .882 | .871 |
| | | | | | | | .856 | .845 | .828 | .812 | .789 | | | | | |
| 30.000 | .71 | 104 | 22.748 | 7.252 | 24.2% | 25 | .968 | .966 | .964 | .962 | .959 | .955 | .955 | .949 | .946 | .943 |
| | | | | | | | .938 | .934 | .927 | .923 | .914 | .908 | .897 | .889 | .880 | .869 |
| | | | | | | | .852 | .843 | .821 | .807 | .780 | | | | | |
| 30.000 | .77 | 130 | 22.724 | 7.276 | 24.3% | 25 | .968 | .966 | .964 | .962 | .959 | .956 | .952 | .949 | .946 | .943 |
| | | | | | | | .936 | .935 | .927 | .921 | .914 | .907 | .896 | .891 | .876 | .865 |
| | | | | | | | .855 | .836 | .822 | .798 | .781 | | | | | |
| 30.000 | .81 | 156 | 22.711 | 7.289 | 24.3% | 25 | .968 | .966 | .964 | .962 | .959 | .955 | .954 | .949 | .946 | .942 |
| | | | | | | | .937 | .932 | .927 | .920 | .913 | .906 | .898 | .886 | .880 | .862 |
| | | | | | | | .853 | .835 | .819 | .799 | .777 | | | | | |
| 30.000 | .84 | 182 | 22.699 | 7.301 | 24.3% | 25 | .968 | .966 | .964 | .961 | .959 | .956 | .953 | .950 | .946 | .942 |
| | | | | | | | .937 | .932 | .927 | .921 | .913 | .906 | .897 | .887 | .876 | .864 |
| | | | | | | | .850 | .835 | .818 | .798 | .775 | | | | | |
| 30.000 | .63 | 81 | 23.554 | 6.446 | 21.5% | 26 | .968 | .967 | .964 | .962 | .958 | .957 | .955 | .949 | .948 | .942 |
| | | | | | | | .938 | .934 | .929 | .921 | .915 | .912 | .898 | .892 | .883 | .871 |
| | | | | | | | .850 | .843 | .822 | .812 | .792 | .782 | | | | |
| 30.000 | .72 | 108 | 23.503 | 6.497 | 21.7% | 26 | .968 | .966 | .965 | .961 | .960 | .956 | .953 | .950 | .947 | .942 |
| | | | | | | | .936 | .935 | .928 | .920 | .914 | .908 | .897 | .893 | .879 | .864 |
| | | | | | | | .856 | .841 | .820 | .805 | .783 | .757 | | | | |
| 30.000 | .78 | 135 | 23.475 | 6.525 | 21.8% | 26 | .968 | .966 | .964 | .961 | .959 | .956 | .954 | .950 | .945 | .944 |
| | | | | | | | .937 | .933 | .927 | .921 | .912 | .908 | .897 | .888 | .878 | .866 |
| | | | | | | | .852 | .837 | .819 | .803 | .775 | .755 | | | | |
| 30.000 | .81 | 162 | 23.460 | 6.540 | 21.8% | 26 | .968 | .966 | .964 | .961 | .960 | .955 | .954 | .949 | .946 | .942 |
| | | | | | | | .938 | .932 | .927 | .921 | .913 | .906 | .897 | .888 | .877 | .865 |
| | | | | | | | .851 | .836 | .818 | .798 | .778 | .753 | | | | |
| 30.000 | .84 | 189 | 23.448 | 6.552 | 21.8% | 26 | .968 | .966 | .964 | .961 | .959 | .956 | .953 | .950 | .946 | .942 |
| | | | | | | | .936 | .933 | .926 | .921 | .913 | .905 | .898 | .887 | .875 | .865 |
| | | | | | | | .849 | .836 | .818 | .797 | .776 | .750 | | | | |
| 30.000 | .64 | 84 | 24.288 | 5.712 | 19.0% | 27 | .968 | .966 | .964 | .961 | .960 | .958 | .954 | .951 | .945 | .944 |
| | | | | | | | .940 | .932 | .930 | .921 | .915 | .911 | .899 | .890 | .880 | .871 |
| | | | | | | | .857 | .845 | .827 | .810 | .786 | .765 | .740 | | | |
| 30.000 | .73 | 112 | 24.230 | 5.770 | 19.2% | 27 | .968 | .966 | .964 | .962 | .959 | .957 | .952 | .950 | .947 | .941 |

ERLANG-ENGSET: FOR SMOOTH ATTEMPTING TRAFFIC AND FEW PERSONS DOING THE CALLING (FEW MEANS RATIO OF CALLERS TO CIRCUITS IS LESS THAN 8 - GREATER THAN 8 ASSUME TRAFFIC RANDOM)

| HOURS FIRST ATTEMPT | FIRST ATTEMPT VMR | NUMBER SOURCES | HOURS CON-NECT | HOURS OVFL | PERCENT BLOCKED (GRADE) | LINE | HOURS CONNECTED TRAFFIC ON LINES 1 | 2 | 3 | 4 | 5 | 6 | 7 | 8 | 9 | 10 |
|---|---|---|---|---|---|---|---|---|---|---|---|---|---|---|---|---|
| | | | | | | | .938 | .934 | .927 | .922 | .912 | .910 | .899 | .889 | .876 | .870 |
| | | | | | | | .852 | .837 | .825 | .800 | .783 | .757 | .732 | | | |
| 30.000 | .79 | 140 | 24.200 | 5.800 | 19.3% | 27 | .968 | .966 | .964 | .962 | .958 | .957 | .953 | .950 | .946 | .942 |
| | | | | | | | .936 | .934 | .927 | .921 | .914 | .906 | .898 | .888 | .878 | .862 |
| | | | | | | | .855 | .837 | .819 | .800 | .778 | .755 | .726 | | | |
| 30.000 | .82 | 168 | 24.181 | 5.819 | 19.4% | 27 | .968 | .966 | .964 | .961 | .959 | .956 | .953 | .950 | .946 | .942 |
| | | | | | | | .938 | .932 | .927 | .920 | .914 | .906 | .897 | .887 | .876 | .865 |
| | | | | | | | .851 | .835 | .818 | .799 | .776 | .749 | .726 | | | |
| 30.000 | .85 | 196 | 24.167 | 5.833 | 19.4% | 27 | .968 | .966 | .964 | .961 | .959 | .956 | .953 | .949 | .946 | .942 |
| | | | | | | | .937 | .932 | .926 | .920 | .914 | .905 | .897 | .887 | .876 | .864 |
| | | | | | | | .850 | .834 | .817 | .796 | .777 | .749 | .722 | | | |
| 30.000 | .66 | 87 | 24.987 | 5.013 | 16.7% | 28 | .968 | .966 | .964 | .962 | .959 | .957 | .953 | .950 | .948 | .942 |
| | | | | | | | .940 | .932 | .926 | .923 | .916 | .909 | .900 | .894 | .879 | .870 |
| | | | | | | | .856 | .843 | .828 | .806 | .786 | .765 | .740 | .707 | | |
| 30.000 | .74 | 116 | 24.922 | 5.078 | 16.9% | 28 | .968 | .966 | .964 | .961 | .959 | .957 | .953 | .950 | .947 | .941 |
| | | | | | | | .938 | .934 | .926 | .921 | .916 | .905 | .901 | .889 | .879 | .867 |
| | | | | | | | .854 | .835 | .822 | .803 | .785 | .754 | .734 | .695 | | |
| 30.000 | .79 | 145 | 24.892 | 5.108 | 17.0% | 28 | .968 | .966 | .964 | .962 | .959 | .955 | .954 | .950 | .946 | .942 |
| | | | | | | | .937 | .933 | .925 | .922 | .913 | .906 | .898 | .888 | .877 | .866 |
| | | | | | | | .852 | .836 | .819 | .798 | .781 | .754 | .725 | .696 | | |
| 30.000 | .83 | 174 | 24.870 | 5.130 | 17.1% | 28 | .968 | .966 | .964 | .961 | .959 | .956 | .953 | .949 | .946 | .942 |
| | | | | | | | .938 | .932 | .927 | .920 | .914 | .906 | .895 | .889 | .877 | .865 |
| | | | | | | | .850 | .835 | .817 | .798 | .776 | .750 | .725 | .692 | | |
| 30.000 | .85 | 203 | 24.855 | 5.145 | 17.2% | 28 | .968 | .966 | .964 | .961 | .959 | .956 | .953 | .950 | .946 | .942 |
| | | | | | | | .937 | .932 | .926 | .920 | .914 | .905 | .897 | .886 | .876 | .863 |
| | | | | | | | .850 | .834 | .817 | .796 | .775 | .749 | .722 | .691 | | |
| 30.000 | .67 | 90 | 25.653 | 4.347 | 14.5% | 29 | .968 | .966 | .964 | .962 | .958 | .956 | .955 | .949 | .947 | .943 |
| | | | | | | | .938 | .933 | .927 | .923 | .913 | .908 | .902 | .889 | .881 | .871 |
| | | | | | | | .857 | .840 | .828 | .806 | .785 | .764 | .739 | .706 | .674 | |
| 30.000 | .75 | 120 | 25.581 | 4.419 | 14.7% | 29 | .968 | .966 | .964 | .962 | .960 | .956 | .953 | .950 | .945 | .942 |
| | | | | | | | .939 | .931 | .929 | .921 | .914 | .905 | .901 | .888 | .878 | .867 |
| | | | | | | | .853 | .838 | .821 | .801 | .778 | .758 | .732 | .699 | .661 | |
| 30.000 | .80 | 150 | 25.548 | 4.452 | 14.8% | 29 | .968 | .966 | .964 | .961 | .959 | .955 | .954 | .950 | .946 | .942 |
| | | | | | | | .938 | .932 | .927 | .919 | .916 | .905 | .897 | .887 | .877 | .868 |
| | | | | | | | .851 | .835 | .821 | .796 | .780 | .753 | .723 | .696 | .662 | |
| 30.000 | .83 | 180 | 25.524 | 4.476 | 14.9% | 29 | .968 | .966 | .964 | .962 | .959 | .956 | .953 | .950 | .945 | .942 |
| | | | | | | | .936 | .934 | .926 | .919 | .915 | .905 | .897 | .887 | .878 | .863 |
| | | | | | | | .849 | .837 | .817 | .798 | .775 | .751 | .722 | .693 | .658 | |
| 30.000 | .86 | 210 | 25.508 | 4.492 | 15.0% | 29 | .968 | .966 | .964 | .961 | .959 | .956 | .952 | .949 | .947 | .942 |
| | | | | | | | .937 | .932 | .926 | .920 | .913 | .905 | .897 | .887 | .876 | .864 |

ERLANG-ENGSET: FOR SMOOTH ATTEMPTING TRAFFIC AND FEW PERSONS DOING THE CALLING (FEW MEANS RATIO OF CALLERS TO CIRCUITS IS LESS THAN 8 - GREATER THAN 8 ASSUME TRAFFIC RANDOM)

| HOURS FIRST ATTEMPT | FIRST ATTEMPT VMR | NUMBER SOURCES | HOURS CON-NECT | HOURS OVFL | PERCENT BLOCKED (GRADE) | LINE | HOURS CONNECTED TRAFFIC ON LINES 1 | 2 | 3 | 4 | 5 | 6 | 7 | 8 | 9 | 10 |
|---|---|---|---|---|---|---|---|---|---|---|---|---|---|---|---|---|
| | | | | | | | .849 | .833 | .817 | .797 | .774 | .750 | .721 | .690 | .656 | |
| 30.000 | .68 | 93 | 26.278 | 3.722 | 12.4% | 30 | .968 | .966 | .964 | .962 | .960 | .955 | .953 | .952 | .945 | .943 |
| | | | | | | | .938 | .935 | .926 | .922 | .918 | .906 | .900 | .893 | .877 | .871 |
| | | | | | | | .856 | .842 | .823 | .808 | .787 | .762 | .736 | .703 | .672 | .636 |
| 30.000 | .76 | 124 | 26.204 | 3.796 | 12.7% | 30 | .968 | .966 | .963 | .962 | .959 | .956 | .952 | .949 | .947 | .943 |
| | | | | | | | .936 | .932 | .928 | .923 | .912 | .909 | .898 | .889 | .878 | .866 |
| | | | | | | | .852 | .838 | .821 | .801 | .778 | .758 | .728 | .699 | .661 | .629 |
| 30.000 | .81 | 155 | 26.163 | 3.837 | 12.8% | 30 | .968 | .966 | .964 | .961 | .959 | .956 | .953 | .950 | .947 | .942 |
| | | | | | | | .937 | .932 | .927 | .921 | .912 | .907 | .896 | .887 | .879 | .864 |
| | | | | | | | .850 | .838 | .819 | .800 | .777 | .752 | .722 | .696 | .658 | .623 |
| 30.000 | .84 | 186 | 26.139 | 3.861 | 12.9% | 30 | .968 | .966 | .964 | .961 | .959 | .955 | .954 | .949 | .946 | .942 |
| | | | | | | | .938 | .932 | .926 | .920 | .913 | .906 | .897 | .888 | .876 | .864 |
| | | | | | | | .851 | .834 | .817 | .796 | .777 | .750 | .723 | .691 | .657 | .619 |
| 30.000 | .86 | 217 | 26.122 | 3.878 | 12.9% | 30 | .968 | .966 | .964 | .961 | .959 | .956 | .953 | .950 | .945 | .942 |
| | | | | | | | .937 | .932 | .926 | .919 | .913 | .906 | .897 | .886 | .877 | .863 |
| | | | | | | | .849 | .834 | .816 | .797 | .773 | .749 | .722 | .690 | .655 | .618 |
| 30.000 | .69 | 96 | 26.859 | 3.141 | 10.5% | 31 | .968 | .966 | .964 | .961 | .959 | .956 | .952 | .950 | .948 | .941 |
| | | | | | | | .938 | .932 | .929 | .922 | .914 | .908 | .902 | .887 | .883 | .869 |
| | | | | | | | .853 | .843 | .823 | .804 | .787 | .761 | .736 | .703 | .672 | .633 |
| | | | | | | | .591 | | | | | | | | | |
| 30.000 | .77 | 128 | 26.779 | 3.221 | 10.7% | 31 | .968 | .966 | .964 | .961 | .960 | .956 | .953 | .949 | .948 | .942 |
| | | | | | | | .938 | .933 | .926 | .920 | .914 | .907 | .900 | .889 | .875 | .869 |
| | | | | | | | .850 | .840 | .817 | .803 | .779 | .758 | .728 | .695 | .663 | .626 |
| | | | | | | | .584 | | | | | | | | | |
| 30.000 | .81 | 160 | 26.738 | 3.262 | 10.9% | 31 | .008 | .966 | .964 | .961 | .959 | .956 | .953 | .950 | .946 | .942 |
| | | | | | | | .937 | .932 | .927 | .919 | .915 | .905 | .897 | .887 | .879 | .864 |
| | | | | | | | .850 | .838 | .816 | .801 | .777 | .753 | .722 | .696 | .658 | .622 |
| | | | | | | | .578 | | | | | | | | | |
| 30.000 | .84 | 192 | 26.712 | 3.288 | 11.0% | 31 | .968 | .966 | .964 | .962 | .959 | .956 | .953 | .950 | .946 | .942 |
| | | | | | | | .937 | .932 | .927 | .920 | .913 | .905 | .897 | .887 | .875 | .865 |
| | | | | | | | .849 | .835 | .816 | .798 | .775 | .749 | .724 | .691 | .656 | .619 |
| | | | | | | | .577 | | | | | | | | | |
| 30.000 | .87 | 224 | 26.693 | 3.307 | 11.0% | 31 | .968 | .966 | .964 | .962 | .959 | .050 | .953 | .949 | .946 | .942 |
| | | | | | | | .937 | .932 | .926 | .921 | .913 | .905 | .896 | .886 | .876 | .862 |
| | | | | | | | .850 | .834 | .815 | .796 | .774 | .750 | .721 | .689 | .655 | .617 |
| | | | | | | | .575 | | | | | | | | | |
| 30.000 | .55 | 66 | 27.594 | 2.406 | 8.0% | 32 | .968 | .966 | .965 | .962 | .959 | .958 | .050 | .951 | .947 | .944 |
| | | | | | | | .939 | .934 | .930 | .920 | .919 | .912 | .904 | .892 | .887 | .873 |
| | | | | | | | .855 | .848 | .838 | .816 | .796 | .767 | .757 | .714 | .694 | .653 |
| | | | | | | | .608 | .564 | | | | | | | | |

ERLANG-ENGSET: FOR SMOOTH ATTEMPTING TRAFFIC AND FEW PERSONS DOING THE CALLING (FEW MEANS RATIO OF CALLERS TO CIRCUITS IS LESS THAN 8 - GREATER THAN 8 ASSUME TRAFFIC RANDOM)

| HOURS FIRST ATTEMPT | FIRST ATTEMPT VMR | NUMBER SOURCES | HOURS CON-NECT | HOURS OVFL | PERCENT BLOCKED (GRADE) | LINE | HOURS CONNECTED TRAFFIC ON LINES 1 | 2 | 3 | 4 | 5 | 6 | 7 | 8 | 9 | 10 |
|---|---|---|---|---|---|---|---|---|---|---|---|---|---|---|---|---|
| 30.000 | .70 | 99 | 27.394 | 2.606 | 8.7% | 32 | .968 | .966 | .964 | .961 | .959 | .956 | .954 | .950 | .947 | .942 |
| | | | | | | | .937 | .933 | .927 | .922 | .914 | .909 | .898 | .891 | .882 | .868 |
| | | | | | | | .856 | .837 | .825 | .806 | .784 | .759 | .734 | .704 | .670 | .631 |
| | | | | | | | .593 | .543 | | | | | | | | |
| 30.000 | .77 | 132 | 27.313 | 2.687 | 9.0% | 32 | .968 | .966 | .964 | .962 | .958 | .956 | .953 | .951 | .946 | .941 |
| | | | | | | | .937 | .933 | .928 | .921 | .912 | .909 | .897 | .888 | .878 | .866 |
| | | | | | | | .853 | .837 | .819 | .799 | .782 | .756 | .726 | .698 | .664 | .623 |
| | | | | | | | .582 | .540 | | | | | | | | |
| 30.000 | .82 | 165 | 27.268 | 2.732 | 9.1% | 32 | .968 | .966 | .964 | .961 | .959 | .956 | .953 | .950 | .946 | .941 |
| | | | | | | | .937 | .931 | .928 | .920 | .913 | .906 | .898 | .888 | .876 | .865 |
| | | | | | | | .851 | .835 | .816 | .801 | .776 | .752 | .723 | .694 | .659 | .621 |
| | | | | | | | .579 | .534 | | | | | | | | |
| 30.000 | .85 | 198 | 27.239 | 2.761 | 9.2% | 32 | .968 | .966 | .964 | .961 | .959 | .956 | .952 | .950 | .945 | .941 |
| | | | | | | | .938 | .932 | .927 | .921 | .913 | .905 | .897 | .888 | .876 | .864 |
| | | | | | | | .849 | .833 | .819 | .797 | .774 | .750 | .722 | .690 | .657 | .618 |
| | | | | | | | .575 | .532 | | | | | | | | |
| 30.000 | .87 | 231 | 27.219 | 2.781 | 9.3% | 32 | .968 | .966 | .964 | .962 | .959 | .956 | .953 | .949 | .946 | .942 |
| | | | | | | | .937 | .932 | .926 | .920 | .913 | .905 | .896 | .887 | .875 | .863 |
| | | | | | | | .849 | .834 | .816 | .795 | .774 | .748 | .721 | .688 | .656 | .616 |
| | | | | | | | .575 | .530 | | | | | | | | |
| 30.000 | .56 | 68 | 28.082 | 1.918 | 6.4% | 33 | .968 | .966 | .964 | .961 | .960 | .957 | .954 | .951 | .946 | .944 |
| | | | | | | | .939 | .935 | .926 | .926 | .918 | .910 | .899 | .896 | .885 | .872 |
| | | | | | | | .856 | .848 | .837 | .816 | .795 | .772 | .749 | .719 | .686 | .650 |
| | | | | | | | .603 | .564 | .510 | | | | | | | |
| 30.000 | .71 | 102 | 27.877 | 2.123 | 7.1% | 33 | .968 | .966 | .964 | .961 | .959 | .957 | .952 | .952 | .945 | .944 |
| | | | | | | | .936 | .935 | .925 | .924 | .912 | .910 | .900 | .888 | .879 | .868 |
| | | | | | | | .856 | .842 | .821 | .806 | .784 | .758 | .733 | .704 | .670 | .631 |
| | | | | | | | .590 | .544 | .493 | | | | | | | |
| 30.000 | .78 | 136 | 27.794 | 2.206 | 7.4% | 33 | .968 | .966 | .964 | .962 | .958 | .957 | .953 | .950 | .946 | .940 |
| | | | | | | | .939 | .932 | .927 | .921 | .914 | .906 | .895 | .891 | .876 | .868 |
| | | | | | | | .850 | .839 | .817 | .803 | .779 | .753 | .729 | .695 | .663 | .621 |
| | | | | | | | .585 | .535 | .490 | | | | | | | |
| 30.000 | .82 | 170 | 27.747 | 2.253 | 7.5% | 33 | .968 | .966 | .964 | .961 | .959 | .956 | .953 | .949 | .947 | .941 |
| | | | | | | | .937 | .933 | .927 | .920 | .914 | .905 | .897 | .887 | .876 | .865 |
| | | | | | | | .851 | .835 | .818 | .797 | .777 | .752 | .723 | .691 | .660 | .620 |
| | | | | | | | .579 | .532 | .485 | | | | | | | |
| 30.000 | .85 | 204 | 27.718 | 2.282 | 7.6% | 33 | .968 | .966 | .964 | .961 | .959 | .956 | .953 | .950 | .946 | .941 |
| | | | | | | | .938 | .932 | .926 | .921 | .913 | .905 | .897 | .887 | .876 | .864 |
| | | | | | | | .849 | .834 | .816 | .798 | .774 | .748 | .723 | .691 | .655 | .616 |
| | | | | | | | .578 | .531 | .483 | | | | | | | |

ERLANG-ENGSET: FOR SMOOTH ATTEMPTING TRAFFIC AND FEW PERSONS DOING THE CALLING (FEW MEANS RATIO OF CALLERS TO CIRCUITS IS LESS THAN 8 - GREATER THAN 8 ASSUME TRAFFIC RANDOM)

| HOURS FIRST ATTEMPT | FIRST ATTEMPT VMR | NUMBER SOURCES | HOURS CON-NECT | HOURS OVFL | PERCENT BLOCKED (GRADE) | LINE | HOURS CONNECTED TRAFFIC ON LINES 1 | 2 | 3 | 4 | 5 | 6 | 7 | 8 | 9 | 10 |
|---|---|---|---|---|---|---|---|---|---|---|---|---|---|---|---|---|
| 30.000 | .87 | 238 | 27.697 | 2.303 | 7.7% | 33 | .968 | .966 | .964 | .961 | .959 | .956 | .953 | .949 | .946 | .941 |
| | | | | | | | .937 | .931 | .926 | .920 | .912 | .905 | .896 | .887 | .875 | .864 |
| | | | | | | | .849 | .833 | .815 | .796 | .773 | .749 | .720 | .688 | .655 | .616 |
| | | | | | | | .575 | .529 | .481 | | | | | | | |

*EQUIVALENT RANDOM THEORY: VMR(VARIANCE TO MEAN RATIO)– 1.5 - NO RETRIES (PEAKED TRAFFIC - ROUTE ADVANCEMENT OR DAY-TO-DAY VARIATION)

| HOURS FIRST ATTEMPT VMR=1.5 | HOURS CON-NECT | HOURS RECALL /RETRY | HOURS OVFL | CON-NECT VMR | OVFL VMR | PERCENT BLOCKED (GRADE) | LINE | HOURS CONNECTED TRAFFIC ON LINES 1 | 2 | 3 | 4 | 5 | 6 | 7 | 8 | 9 | 10 |
|---|---|---|---|---|---|---|---|---|---|---|---|---|---|---|---|---|---|
| .200 | .169 | .000 | .031 | .95 | 1.45 | 15.3% | 2 | .116 | .053 | | | | | | | | |
| .200 | .190 | .000 | .010 | .98 | 1.39 | 5.0% | 3 | .116 | .053 | .021 | | | | | | | |
| .200 | .197 | .000 | .003 | .99 | 1.35 | 1.4% | 4 | .116 | .053 | .021 | .007 | | | | | | |
| .400 | .322 | .000 | .078 | .89 | 1.49 | 19.6% | 2 | .211 | .111 | | | | | | | | |
| .400 | .372 | .000 | .028 | .95 | 1.44 | 7.1% | 3 | .211 | .111 | .050 | | | | | | | |
| .400 | .391 | .000 | .009 | .98 | 1.39 | 2.3% | 4 | .211 | .111 | .050 | .019 | | | | | | |
| .400 | .397 | .000 | .003 | .99 | 1.34 | .6% | 5 | .211 | .111 | .050 | .019 | .006 | | | | | |
| .600 | .451 | .000 | .149 | .83 | 1.56 | 24.9% | 2 | .282 | .169 | | | | | | | | |
| .600 | .538 | .000 | .062 | .91 | 1.52 | 10.3% | 3 | .282 | .169 | .087 | | | | | | | |
| .600 | .577 | .000 | .023 | .96 | 1.47 | 3.8% | 4 | .282 | .169 | .087 | .039 | | | | | | |
| .600 | .593 | .000 | .007 | .98 | 1.41 | 1.2% | 5 | .282 | .169 | .087 | .039 | .015 | | | | | |
| .800 | .565 | .000 | .235 | .78 | 1.61 | 29.4% | 2 | .342 | .223 | | | | | | | | |
| .800 | .692 | .000 | .108 | .87 | 1.59 | 13.5% | 3 | .342 | .223 | .127 | | | | | | | |
| .800 | .756 | .000 | .044 | .94 | 1.54 | 5.5% | 4 | .342 | .223 | .127 | .064 | | | | | | |
| .800 | .784 | .000 | .016 | .97 | 1.48 | 2.0% | 5 | .342 | .223 | .127 | .064 | .028 | | | | | |
| .800 | .795 | .000 | .005 | .99 | 1.43 | .7% | 6 | .342 | .223 | .127 | .064 | .028 | .011 | | | | |
| 1.000 | .666 | .000 | .334 | .73 | 1.65 | 33.4% | 2 | .393 | .272 | | | | | | | | |
| 1.000 | .834 | .000 | .166 | .83 | 1.64 | 16.6% | 3 | .393 | .272 | .168 | | | | | | | |
| 1.000 | .926 | .000 | .074 | .91 | 1.60 | 7.4% | 4 | .393 | .272 | .168 | .092 | | | | | | |
| 1.000 | .970 | .000 | .030 | .96 | 1.54 | 3.0% | 5 | .393 | .272 | .168 | .092 | .044 | | | | | |
| 1.000 | .989 | .000 | .011 | .98 | 1.48 | 1.1% | 6 | .393 | .272 | .168 | .092 | .044 | .019 | | | | |
| 1.200 | .965 | .000 | .235 | .79 | 1.68 | 19.6% | 3 | .438 | .318 | .209 | | | | | | | |
| 1.200 | 1.088 | .000 | .112 | .88 | 1.65 | 9.4% | 4 | .438 | .318 | .209 | .123 | | | | | | |
| 1.200 | 1.151 | .000 | .049 | .94 | 1.60 | 4.0% | 5 | .438 | .318 | .209 | .123 | .064 | | | | | |
| 1.200 | 1.181 | .000 | .019 | .97 | 1.54 | 1.6% | 6 | .438 | .318 | .209 | .123 | .064 | .030 | | | | |
| 1.200 | 1.193 | .000 | .007 | .99 | 1.48 | .6% | 7 | .438 | .318 | .209 | .123 | .064 | .030 | .012 | | | |
| 1.400 | 1.086 | .000 | .314 | .75 | 1.72 | 22.5% | 3 | .477 | .360 | .249 | | | | | | | |
| 1.400 | 1.240 | .000 | .160 | .85 | 1.70 | 11.4% | 4 | .477 | .360 | .249 | .155 | | | | | | |
| 1.400 | 1.326 | .000 | .074 | .91 | 1.65 | 5.3% | 5 | .477 | .360 | .249 | .155 | .086 | | | | | |
| 1.400 | 1.369 | .000 | .031 | .96 | 1.59 | 2.2% | 6 | .477 | .360 | .249 | .155 | .086 | .043 | | | | |
| 1.400 | 1.388 | .000 | .012 | .98 | 1.53 | .8% | 7 | .477 | .360 | .249 | .155 | .086 | .043 | .019 | | | |
| 1.600 | 1.197 | .000 | .403 | .71 | 1.75 | 25.2% | 3 | .511 | .399 | .287 | | | | | | | |
| 1.600 | 1.384 | .000 | .216 | .81 | 1.74 | 13.5% | 4 | .511 | .399 | .287 | .188 | | | | | | |
| 1.600 | 1.495 | .000 | .105 | .89 | 1.70 | 6.6% | 5 | .511 | .399 | .287 | .188 | .111 | | | | | |
| 1.600 | 1.553 | .000 | .047 | .94 | 1.64 | 2.9% | 6 | .511 | .399 | .287 | .188 | .111 | .058 | | | | |
| 1.600 | 1.581 | .000 | .019 | .97 | 1.58 | 1.2% | 7 | .511 | .399 | .287 | .188 | .111 | .058 | .028 | | | |
| 1.800 | 1.299 | .000 | .501 | .68 | 1.77 | 27.8% | 3 | .541 | .434 | .323 | | | | | | | |
| 1.800 | 1.520 | .000 | .280 | .78 | 1.77 | 15.6% | 4 | .541 | .434 | .323 | .221 | | | | | | |
| 1.800 | 1.656 | .000 | .144 | .86 | 1.74 | 8.0% | 5 | .541 | .434 | .323 | .221 | .137 | | | | | |
| 1.800 | 1.733 | .000 | .067 | .92 | 1.69 | 3.7% | 6 | .541 | .434 | .323 | .221 | .137 | .076 | | | | |

*SIMILAR TO A.T.&T. NEAL-WILKINSON TABLES(SEE SECTION 1)

*EQUIVALENT RANDOM THEORY: VMR(VARIANCE TO MEAN RATIO)– 1.5 - NO RETRIES (PEAKED TRAFFIC - ROUTE ADVANCEMENT OR DAY-TO-DAY VARIATION)

| HOURS FIRST ATTEMPT VMR=1.5 | HOURS CON-NECT | HOURS RECALL /RETRY | HOURS OVFL | CON-NECT VMR | OVFL VMR | PERCENT BLOCKED (GRADE) | LINE | HOURS CONNECTED TRAFFIC ON LINES 1 | 2 | 3 | 4 | 5 | 6 | 7 | 8 | 9 | 10 |
|---|---|---|---|---|---|---|---|---|---|---|---|---|---|---|---|---|---|
| 1.800 | 1.771 | .000 | .029 | .96 | 1.62 | 1.6% | 7 | .541 | .434 | .323 | .221 | .137 | .076 | .038 | | | |
| 1.800 | 1.789 | .000 | .011 | .98 | 1.56 | .6% | 8 | .541 | .434 | .323 | .221 | .137 | .076 | .038 | .017 | | |
| 2.000 | 1.393 | .000 | .607 | .64 | 1.79 | 30.4% | 3 | .568 | .466 | .358 | | | | | | | |
| 2.000 | 1.647 | .000 | .353 | .75 | 1.80 | 17.7% | 4 | .568 | .466 | .358 | .254 | | | | | | |
| 2.000 | 1.811 | .000 | .189 | .84 | 1.78 | 9.5% | 5 | .568 | .466 | .358 | .254 | .164 | | | | | |
| 2.000 | 1.907 | .000 | .093 | .90 | 1.73 | 4.7% | 6 | .568 | .466 | .358 | .254 | .164 | .096 | | | | |
| 2.000 | 1.958 | .000 | .042 | .95 | 1.67 | 2.1% | 7 | .568 | .466 | .358 | .254 | .164 | .096 | .051 | | | |
| 2.000 | 1.983 | .000 | .017 | .98 | 1.61 | .9% | 8 | .568 | .466 | .358 | .254 | .164 | .096 | .051 | .025 | | |
| 2.200 | 1.480 | .000 | .720 | .61 | 1.80 | 32.7% | 3 | .593 | .496 | .391 | | | | | | | |
| 2.200 | 1.765 | .000 | .435 | .71 | 1.83 | 19.8% | 4 | .593 | .496 | .391 | .286 | | | | | | |
| 2.200 | 1.958 | .000 | .242 | .81 | 1.81 | 11.0% | 5 | .593 | .496 | .391 | .286 | .192 | | | | | |
| 2.200 | 2.076 | .000 | .124 | .88 | 1.77 | 5.7% | 6 | .593 | .496 | .391 | .286 | .192 | .118 | | | | |
| 2.200 | 2.141 | .000 | .059 | .93 | 1.71 | 2.7% | 7 | .593 | .496 | .391 | .286 | .192 | .118 | .066 | | | |
| 2.200 | 2.174 | .000 | .026 | .97 | 1.65 | 1.2% | 8 | .593 | .496 | .391 | .286 | .192 | .118 | .066 | .033 | | |
| 2.400 | 1.877 | .000 | .523 | .68 | 1.85 | 21.8% | 4 | .615 | .523 | .421 | .317 | | | | | | |
| 2.400 | 2.098 | .000 | .302 | .78 | 1.84 | 12.6% | 5 | .615 | .523 | .421 | .317 | .221 | | | | | |
| 2.400 | 2.238 | .000 | .162 | .86 | 1.81 | 6.7% | 6 | .615 | .523 | .421 | .317 | .221 | .141 | | | | |
| 2.400 | 2.320 | .000 | .080 | .92 | 1.76 | 3.3% | 7 | .615 | .523 | .421 | .317 | .221 | .141 | .082 | | | |
| 2.400 | 2.364 | .000 | .036 | .96 | 1.69 | 1.5% | 8 | .615 | .523 | .421 | .317 | .221 | .141 | .082 | .043 | | |
| 2.400 | 2.385 | .000 | .015 | .98 | 1.63 | .6% | 9 | .615 | .523 | .421 | .317 | .221 | .141 | .082 | .043 | .021 | |
| 2.600 | 1.981 | .000 | .619 | .65 | 1.86 | 23.8% | 4 | .635 | .548 | .450 | .348 | | | | | | |
| 2.600 | 2.230 | .000 | .370 | .75 | 1.87 | 14.2% | 5 | .635 | .548 | .450 | .348 | .250 | | | | | |
| 2.600 | 2.395 | .000 | .205 | .84 | 1.85 | 7.9% | 6 | .635 | .548 | .450 | .348 | .250 | .165 | | | | |
| 2.600 | 2,495 | .000 | .105 | .90 | 1.80 | 4.0% | 7 | .635 | .548 | .450 | .348 | .250 | .165 | .100 | | | |
| 2.600 | 2.550 | .000 | .050 | .94 | 1.74 | 1.9% | 8 | .635 | .548 | .450 | .348 | .250 | .165 | .100 | .055 | | |
| 2.600 | 2.578 | .000 | .022 | .97 | 1.67 | .8% | 9 | .635 | .548 | .450 | .348 | .250 | .165 | .100 | .055 | .028 | |
| 2.800 | 2.078 | .000 | .722 | .62 | 1.88 | 25.8% | 4 | .653 | .571 | .477 | .377 | | | | | | |
| 2.800 | 2.356 | .000 | .444 | .72 | 1.90 | 15.9% | 5 | .653 | .571 | .477 | .377 | .278 | | | | | |
| 2.800 | 2.546 | .000 | .254 | .81 | 1.88 | 9.1% | 6 | .653 | .571 | .477 | .377 | .278 | .190 | | | | |
| 2.800 | 2.665 | .000 | .135 | .88 | 1.84 | 4.8% | 7 | .653 | .571 | .477 | .377 | .278 | .190 | .119 | | | |
| 2.800 | 2.733 | .000 | .067 | .93 | 1.78 | 2.4% | 8 | .653 | .571 | .477 | .377 | .278 | .190 | .119 | .069 | | |
| 2.800 | 2.770 | .000 | .030 | .96 | 1.71 | 1.1% | 9 | .653 | .571 | .477 | .377 | .278 | .190 | .119 | .069 | .036 | |
| 3.000 | 2.169 | .000 | .831 | .60 | 1.89 | 27.7% | 4 | .670 | .592 | .502 | .404 | | | | | | |
| 3.000 | 2.475 | .000 | .525 | .69 | 1.92 | 17.5% | 5 | .670 | .592 | .502 | .404 | .306 | | | | | |
| 3.000 | 2.690 | .000 | .310 | .78 | 1.91 | 10.0% | 6 | .670 | .592 | .502 | .404 | .306 | .215 | | | | |
| 3.000 | 2.830 | .000 | .170 | .86 | 1.87 | 5.7% | 7 | .670 | .592 | .502 | .404 | .306 | .215 | .140 | | | |
| 3.000 | 2.913 | .000 | .087 | .92 | 1.82 | 2.9% | 8 | .670 | .592 | .502 | .404 | .306 | .215 | .140 | .083 | | |
| 3.000 | 2.959 | .000 | .041 | .95 | 1.75 | 1.4% | 9 | .670 | .592 | .502 | .404 | .306 | .215 | .140 | .083 | .046 | |
| 3.000 | 2.982 | .000 | .018 | .98 | 1.69 | .6% | 10 | .670 | .592 | .502 | .404 | .306 | .215 | .140 | .083 | .046 | .023 |
| 3.200 | 2.254 | .000 | .946 | .57 | 1.89 | 29.6% | 4 | .686 | .612 | .526 | .431 | | | | | | |

*SIMILAR TO A.T.&T. NEAL-WILKINSON TABLES(SEE SECTION 1)

*EQUIVALENT RANDOM THEORY: VMR(VARIANCE TO MEAN RATIO)– 1.5 - NO RETRIES (PEAKED TRAFFIC - ROUTE ADVANCEMENT OR DAY-TO-DAY VARIATION)

| HOURS FIRST ATTEMPT VMR = 1.5 | HOURS CON-NECT | HOURS RECALL /RETRY | HOURS OVFL | CON-NECT VMR | OVFL VMR | PERCENT BLOCKED (GRADE) | LINE | HOURS CONNECTED TRAFFIC ON LINES 1 | 2 | 3 | 4 | 5 | 6 | 7 | 8 | 9 | 10 |
|---|---|---|---|---|---|---|---|---|---|---|---|---|---|---|---|---|---|
| 3.200 | 2.587 | .000 | .613 | .67 | 1.93 | 19.2% | 5 | .686 | .612 | .526 | .431 | .333 | | | | | |
| 3.200 | 2.828 | .000 | .372 | .76 | 1.94 | 11.6% | 6 | .686 | .612 | .526 | .431 | .333 | .241 | | | | |
| 3.200 | 2.989 | .000 | .211 | .84 | 1.91 | 6.6% | 7 | .686 | .612 | .526 | .431 | .333 | .241 | .161 | | | |
| 3.200 | 3.089 | .000 | .111 | .90 | 1.86 | 3.5% | 8 | .686 | .612 | .526 | .431 | .333 | .241 | .161 | .100 | | |
| 3.200 | 3.145 | .000 | .055 | .94 | 1.79 | 1.7% | 9 | .686 | .612 | .526 | .431 | .333 | .241 | .161 | .100 | .057 | |
| 3.200 | 3.175 | .000 | .025 | .97 | 1.73 | .8% | 10 | .686 | .612 | .526 | .431 | .333 | .241 | .161 | .100 | .057 | .030 |
| 3.400 | 2.333 | .000 | 1.067 | .54 | 1.90 | 31.4% | 4 | .700 | .630 | .548 | .456 | | | | | | |
| 3.400 | 2.693 | .000 | .707 | .64 | 1.95 | 20.8% | 5 | .700 | .630 | .548 | .456 | .360 | | | | | |
| 3.400 | 2.959 | .000 | .441 | .73 | 1.96 | 13.0% | 6 | .700 | .630 | .548 | .456 | .360 | .266 | | | | |
| 3.400 | 3.143 | .000 | .257 | .82 | 1.94 | 7.6% | 7 | .700 | .630 | .548 | .456 | .360 | .266 | .184 | | | |
| 3.400 | 3.260 | .000 | .140 | .88 | 1.89 | 4.1% | 8 | .700 | .630 | .548 | .456 | .360 | .266 | .184 | .117 | | |
| 3.400 | 3.329 | .000 | .071 | .93 | 1.83 | 2.1% | 9 | .700 | .630 | .548 | .456 | .360 | .266 | .184 | .117 | .069 | |
| 3.400 | 3.366 | .000 | .034 | .96 | 1.76 | 1.0% | 10 | .700 | .630 | .548 | .456 | .360 | .266 | .184 | .117 | .069 | .037 |
| 3.600 | 2.367 | .000 | 1.233 | .53 | 1.98 | 34.3% | 4 | .700 | .635 | .559 | .474 | | | | | | |
| 3.600 | 2.751 | .000 | .849 | .62 | 2.03 | 23.6% | 5 | .700 | .635 | .559 | .474 | .384 | | | | | |
| 3.600 | 3.046 | .000 | .554 | .71 | 2.05 | 15.4% | 6 | .700 | .635 | .559 | .474 | .384 | .295 | | | | |
| 3.600 | 3.259 | .000 | .341 | .79 | 2.04 | 9.5% | 7 | .700 | .635 | .559 | .474 | .384 | .295 | .213 | | | |
| 3.600 | 3.402 | .000 | .198 | .86 | 2.00 | 5.5% | 8 | .700 | .635 | .559 | .474 | .384 | .295 | .213 | .143 | | |
| 3.600 | 3.492 | .000 | .108 | .91 | 1.94 | 3.0% | 9 | .700 | .635 | .559 | .474 | .384 | .295 | .213 | .143 | .090 | |
| 3.600 | 3.545 | .000 | .055 | .95 | 1.87 | 1.5% | 10 | .700 | .635 | .559 | .474 | .384 | .295 | .213 | .143 | .090 | .053 |
| 3.600 | 3.573 | .000 | .027 | .97 | 1.81 | .7% | 11 | .700 | .635 | .559 | .474 | .384 | .295 | .213 | .143 | .090 | .053 |
| | | | | | | | | .029 | | | | | | | | | |
| 3.800 | 2.844 | .000 | .956 | .59 | 2.04 | 25.1% | 5 | .712 | .650 | .578 | .496 | .408 | | | | | |
| 3.800 | 3.163 | .000 | .637 | .68 | 2.07 | 16.8% | 6 | .712 | .650 | .578 | .496 | .408 | .319 | | | | |
| 3.800 | 3.399 | .000 | .401 | .76 | 2.06 | 10.6% | 7 | .712 | .650 | .578 | .496 | .408 | .319 | .235 | | | |
| 3.800 | 3.562 | .000 | .238 | .84 | 2.03 | 6.3% | 8 | .712 | .650 | .578 | .496 | .408 | .319 | .235 | .163 | | |
| 3.800 | 3.667 | .000 | .133 | .89 | 1.98 | 3.5% | 9 | .712 | .650 | .578 | .496 | .408 | .319 | .235 | .163 | .105 | |
| 3.800 | 3.730 | .000 | .070 | .94 | 1.91 | 1.8% | 10 | .712 | .650 | .578 | .496 | .408 | .319 | .235 | .163 | .105 | .063 |
| 3.800 | 3.765 | .000 | .035 | .96 | 1.84 | .9% | 11 | .712 | .650 | .578 | .496 | .408 | .319 | .235 | .163 | .105 | .063 |
| | | | | | | | | .035 | | | | | | | | | |
| 4.000 | 2.932 | .000 | 1.068 | .57 | 2.05 | 26.7% | 5 | .724 | .665 | .596 | .517 | .431 | | | | | |
| 4.000 | 3.275 | .000 | .725 | .66 | 2.08 | 18.1% | 6 | .724 | .665 | .596 | .517 | .431 | .343 | | | | |
| 4.000 | 3.534 | .000 | .466 | .74 | 2.09 | 11.7% | 7 | .724 | .665 | .596 | .517 | .431 | .343 | .258 | | | |
| 4.000 | 3.716 | .000 | .284 | .82 | 2.06 | 7.1% | 8 | .724 | .665 | .596 | .517 | .431 | .343 | .258 | .183 | | |
| 4.000 | 3.837 | .000 | .163 | .88 | 2.01 | 4.1% | 9 | .724 | .665 | .596 | .517 | .431 | .343 | .258 | .183 | .121 | |
| 4.000 | 3.912 | .000 | .088 | .93 | 1.95 | 2.2% | 10 | .724 | .665 | .596 | .517 | .431 | .343 | .258 | .183 | .121 | .075 |
| 4.000 | 3.955 | .000 | .045 | .96 | 1.88 | 1.1% | 11 | .724 | .665 | .596 | .517 | .431 | .343 | .258 | .183 | .121 | .075 |
| | | | | | | | | .043 | | | | | | | | | |
| 4.000 | 3.979 | .000 | .021 | .98 | 1.81 | .5% | 12 | .724 | .665 | .596 | .517 | .431 | .343 | .258 | .183 | .121 | .075 |
| | | | | | | | | .043 | .023 | | | | | | | | |

*SIMILAR TO A.T.&T. NEAL-WILKINSON TABLES(SEE SECTION 1)

*EQUIVALENT RANDOM THEORY: VMR(VARIANCE TO MEAN RATIO)– 1.5 - NO RETRIES (PEAKED TRAFFIC - ROUTE ADVANCEMENT OR DAY-TO-DAY VARIATION)

| HOURS FIRST ATTEMPT VMR=1.5 | HOURS CON-NECT | HOURS RECALL /RETRY | HOURS OVFL | CON-NECT VMR | OVFL VMR | PERCENT BLOCKED (GRADE) | LINE | HOURS CONNECTED TRAFFIC ON LINES 1 | 2 | 3 | 4 | 5 | 6 | 7 | 8 | 9 | 10 |
|---|---|---|---|---|---|---|---|---|---|---|---|---|---|---|---|---|---|
| 4.200 | 3.016 | .000 | 1.184 | .55 | 2.05 | 28.2% | 5 | .735 | .679 | .612 | .537 | .453 | | | | | |
| 4.200 | 3.382 | .000 | .818 | .63 | 2.09 | 19.5% | 6 | .735 | .679 | .612 | .537 | .453 | .366 | | | | |
| 4.200 | 3.663 | .000 | .537 | .72 | 2.11 | 12.8% | 7 | .735 | .679 | .612 | .537 | .453 | .366 | .281 | | | |
| 4.200 | 3.866 | .000 | .334 | .80 | 2.09 | 8.0% | 8 | .735 | .679 | .612 | .537 | .453 | .366 | .281 | .203 | | |
| 4.200 | 4.004 | .000 | .196 | .86 | 2.05 | 4.7% | 9 | .735 | .679 | .612 | .537 | .453 | .366 | .281 | .203 | .138 | |
| 4.200 | 4.091 | .000 | .109 | .91 | 1.99 | 2.6% | 10 | .735 | .679 | .612 | .537 | .453 | .366 | .281 | .203 | .138 | .087 |
| 4.200 | 4.143 | .000 | .057 | .95 | 1.92 | 1.4% | 11 | .735 | .679 | .612 | .537 | .453 | .366 | .281 | .203 | .138 | .087 |
| | | | | | | | | .052 | | | | | | | | | |
| 4.200 | 4.172 | .000 | .028 | .97 | 1.85 | .7% | 12 | .735 | .679 | .612 | .537 | .453 | .366 | .281 | .203 | .138 | .087 |
| | | | | | | | | .052 | .029 | | | | | | | | |
| 4.400 | 3.094 | .000 | 1.306 | .53 | 2.05 | 29.7% | 5 | .745 | .691 | .628 | .555 | .475 | | | | | |
| 4.400 | 3.483 | .000 | .917 | .61 | 2.10 | 20.8% | 6 | .745 | .691 | .628 | .555 | .475 | .389 | | | | |
| 4.400 | 3.787 | .000 | .613 | .70 | 2.12 | 13.9% | 7 | .745 | .691 | .628 | .555 | .475 | .389 | .304 | | | |
| 4.400 | 4.011 | .000 | .389 | .78 | 2.11 | 8.8% | 8 | .745 | .691 | .628 | .555 | .475 | .389 | .304 | .224 | | |
| 4.400 | 4.166 | .000 | .234 | .84 | 2.08 | 5.3% | 9 | .745 | .691 | .628 | .555 | .475 | .389 | .304 | .224 | .156 | |
| 4.400 | 4.267 | .000 | .133 | .90 | 2.02 | 3.0% | 10 | .745 | .691 | .628 | .555 | .475 | .389 | .304 | .224 | .156 | .101 |
| 4.400 | 4.329 | .000 | .071 | .94 | 1.95 | 1.6% | 11 | .745 | .691 | .628 | .555 | .475 | .389 | .304 | .224 | .156 | .101 |
| | | | | | | | | .062 | | | | | | | | | |
| 4.400 | 4.364 | .000 | .036 | .96 | 1.88 | .8% | 12 | .745 | .691 | .628 | .555 | .475 | .389 | .304 | .224 | .156 | .101 |
| | | | | | | | | .062 | .035 | | | | | | | | |
| 4.600 | 3.168 | .000 | 1.432 | .51 | 2.05 | 31.1% | 5 | .754 | .703 | .643 | .573 | .495 | | | | | |
| 4.600 | 3.579 | .000 | 1.021 | .59 | 2.11 | 22.2% | 6 | .754 | .703 | .643 | .573 | .495 | .411 | | | | |
| 4.600 | 3.905 | .000 | .695 | .67 | 2.14 | 15.1% | 7 | .754 | .703 | .643 | .573 | .495 | .411 | .326 | | | |
| 4.600 | 4.150 | .000 | .450 | .75 | 2.14 | 9.8% | 8 | .754 | .703 | .643 | .573 | .495 | .411 | .326 | .245 | | |
| 4.600 | 4.324 | .000 | .276 | .83 | 2.11 | 6.0% | 9 | .754 | .703 | .643 | .573 | .495 | .411 | .326 | .245 | .174 | |
| 4.600 | 4.440 | .000 | .160 | .88 | 2.05 | 3.5% | 10 | .754 | .703 | .643 | .573 | .495 | .411 | .326 | .245 | .174 | .116 |
| 4.600 | 4.512 | .000 | .088 | .93 | 1.99 | 1.9% | 11 | .754 | .703 | .643 | .573 | .495 | .411 | .326 | .245 | .174 | .116 |
| | | | | | | | | .072 | | | | | | | | | |
| 4.600 | 4.554 | .000 | .046 | .96 | 1.92 | 1.0% | 12 | .754 | .703 | .643 | .573 | .495 | .411 | .326 | .245 | .174 | .116 |
| | | | | | | | | .072 | .042 | | | | | | | | |
| 4.800 | 3.239 | .000 | 1.561 | .49 | 2.05 | 32.5% | 5 | .763 | .715 | .657 | .590 | .514 | | | | | |
| 4.800 | 3.671 | .000 | 1.129 | .57 | 2.12 | 23.5% | 6 | .763 | .715 | .657 | .590 | .514 | .432 | | | | |
| 4.800 | 4.019 | .000 | .781 | .65 | 2.15 | 16.3% | 7 | .763 | .715 | .657 | .590 | .514 | .432 | .348 | | | |
| 4.800 | 4.285 | .000 | .515 | .73 | 2.16 | 10.7% | 8 | .763 | .715 | .657 | .590 | .514 | .432 | .348 | .266 | | |
| 4.800 | 4.478 | .000 | .322 | .81 | 2.13 | 6.7% | 9 | .763 | .715 | .657 | .590 | .514 | .432 | .348 | .266 | .193 | |
| 4.800 | 4.609 | .000 | .191 | .87 | 2.09 | 4.0% | 10 | .763 | .715 | .657 | .590 | .514 | .432 | .348 | .266 | .193 | .131 |
| 4.800 | 4.693 | .000 | .107 | .92 | 2.02 | 2.2% | 11 | .763 | .715 | .657 | .590 | .514 | .432 | .348 | .266 | .193 | .131 |
| | | | | | | | | .084 | | | | | | | | | |
| 4.800 | 4.743 | .000 | .057 | .95 | 1.96 | 1.2% | 12 | .763 | .715 | .657 | .590 | .514 | .432 | .348 | .266 | .193 | .131 |
| | | | | | | | | .084 | .050 | | | | | | | | |

*SIMILAR TO A.T.&T. NEAL-WILKINSON TABLES(SEE SECTION 1)

*EQUIVALENT RANDOM THEORY: VMR(VARIANCE TO MEAN RATIO)– 1.5 - NO RETRIES (PEAKED TRAFFIC - ROUTE ADVANCEMENT OR DAY-TO-DAY VARIATION)

| HOURS FIRST ATTEMPT VMR=1.5 | HOURS CON-NECT | HOURS RECALL /RETRY | HOURS OVFL | CON-NECT VMR | OVFL VMR | PERCENT BLOCKED (GRADE) | LINE | HOURS CONNECTED TRAFFIC ON LINES 1 | 2 | 3 | 4 | 5 | 6 | 7 | 8 | 9 | 10 |
|---|---|---|---|---|---|---|---|---|---|---|---|---|---|---|---|---|---|
| 4.800 | 4.771 | .000 | .029 | .97 | 1.88 | .6% | 13 | .763 | .715 | .657 | .590 | .514 | .432 | .348 | .266 | .193 | .131 |
| | | | | | | | | .084 | .050 | .028 | | | | | | | |
| 5.000 | 3.305 | .000 | 1.695 | .47 | 2.05 | 33.9% | 5 | .771 | .725 | .670 | .606 | .533 | | | | | |
| 5.000 | 3.758 | .000 | 1.242 | .55 | 2.12 | 24.8% | 6 | .771 | .725 | .670 | .606 | .533 | .453 | | | | |
| 5.000 | 4.127 | .000 | .873 | .63 | 2.16 | 17.5% | 7 | .771 | .725 | .670 | .606 | .533 | .453 | .369 | | | |
| 5.000 | 4.415 | .000 | .585 | .71 | 2.18 | 11.7% | 8 | .771 | .725 | .670 | .606 | .533 | .453 | .369 | .287 | | |
| 5.000 | 4.627 | .000 | .373 | .79 | 2.16 | 7.5% | 9 | .771 | .725 | .670 | .606 | .533 | .453 | .369 | .287 | .212 | |
| 5.000 | 4.774 | .000 | .226 | .85 | 2.12 | 4.5% | 10 | .771 | .725 | .670 | .606 | .533 | .453 | .369 | .287 | .212 | .147 |
| 5.000 | 4.870 | .000 | .130 | .90 | 2.06 | 2.6% | 11 | .771 | .725 | .670 | .606 | .533 | .453 | .369 | .287 | .212 | .147 |
| | | | | | | | | .096 | | | | | | | | | |
| 5.000 | 4.930 | .000 | .070 | .94 | 1.99 | 1.4% | 12 | .771 | .725 | .670 | .606 | .533 | .453 | .369 | .287 | .212 | .147 |
| | | | | | | | | .096 | .059 | | | | | | | | |
| 5.000 | 4.964 | .000 | .036 | .97 | 1.92 | .7% | 13 | .771 | .725 | .670 | .606 | .533 | .453 | .369 | .287 | .212 | .147 |
| | | | | | | | | .096 | .059 | .034 | | | | | | | |
| 5.500 | 3.956 | .000 | 1.544 | .50 | 2.12 | 28.1% | 6 | .790 | .749 | .700 | .642 | .575 | .501 | | | | |
| 5.500 | 4.377 | .000 | 1.123 | .58 | 2.18 | 20.4% | 7 | .790 | .749 | .700 | .642 | .575 | .501 | .421 | | | |
| 5.500 | 4.717 | .000 | .783 | .66 | 2.21 | 14.2% | 8 | .790 | .749 | .700 | .642 | .575 | .501 | .421 | .340 | | |
| 5.500 | 4.978 | .000 | .522 | .74 | 2.21 | 9.5% | 9 | .790 | .749 | .700 | .642 | .575 | .501 | .421 | .340 | .261 | |
| 5.500 | 5.169 | .000 | .331 | .81 | 2.19 | 6.0% | 10 | .790 | .749 | .700 | .642 | .575 | .501 | .421 | .340 | .261 | .191 |
| 5.500 | 5.300 | .000 | .200 | .87 | 2.14 | 3.6% | 11 | .790 | .749 | .700 | .642 | .575 | .501 | .421 | .340 | .261 | .191 |
| | | | | | | | | .131 | | | | | | | | | |
| 5.500 | 5.386 | .000 | .114 | .91 | 2.08 | 2.1% | 12 | .790 | .749 | .700 | .642 | .575 | .501 | .421 | .340 | .261 | .191 |
| | | | | | | | | .131 | .085 | | | | | | | | |
| 5.500 | 5.438 | .000 | .062 | .95 | 2.01 | 1.1% | 13 | .790 | .749 | .700 | .642 | .575 | .501 | .421 | .340 | .261 | .191 |
| | | | | | | | | .131 | .085 | .052 | | | | | | | |
| 5.500 | 5.468 | .000 | .032 | .97 | 1.94 | .6% | 14 | .790 | .749 | .700 | .642 | .575 | .501 | .421 | .340 | .261 | .191 |
| | | | | | | | | .131 | .085 | .052 | .030 | | | | | | |
| 6.000 | 4.087 | .000 | 1.913 | .46 | 2.18 | 31.9% | 6 | .797 | .761 | .717 | .666 | .606 | .540 | | | | |
| 6.000 | 4.554 | .000 | 1.446 | .53 | 2.25 | 24.1% | 7 | .797 | .761 | .717 | .666 | .606 | .540 | .467 | | | |
| 6.000 | 4.945 | .000 | 1.055 | .61 | 2.30 | 17.6% | 8 | .797 | .761 | .717 | .666 | .606 | .540 | .467 | .391 | | |
| 6.000 | 5.259 | .000 | .741 | .69 | 2.32 | 12.3% | 9 | .797 | .761 | .717 | .666 | .606 | .540 | .467 | .391 | .314 | |
| 6.000 | 5.501 | .000 | .499 | .76 | 2.31 | 8.3% | 10 | .797 | .761 | .717 | .666 | .606 | .540 | .467 | .391 | .314 | .242 |
| 6.000 | 5.679 | .000 | .321 | .82 | 2.28 | 5.4% | 11 | .797 | .761 | .717 | .666 | .606 | .540 | .467 | .391 | .314 | .242 |
| | | | | | | | | .178 | | | | | | | | | |
| 6.000 | 5.802 | .000 | .198 | .88 | 2.22 | 3.3% | 12 | .797 | .761 | .717 | .666 | .606 | .540 | .467 | .391 | .314 | .242 |
| | | | | | | | | .178 | .124 | | | | | | | | |
| 6.000 | 5.884 | .000 | .116 | .92 | 2.16 | 1.9% | 13 | .797 | .761 | .717 | .666 | .606 | .540 | .467 | .391 | .314 | .242 |
| | | | | | | | | .178 | .124 | .082 | | | | | | | |
| 6.000 | 5.935 | .000 | .065 | .95 | 2.09 | 1.1% | 14 | .797 | .761 | .717 | .666 | .606 | .540 | .467 | .391 | .314 | .242 |
| | | | | | | | | .178 | .124 | .082 | .051 | | | | | | |

*SIMILAR TO A.T.&T. NEAL-WILKINSON TABLES(SEE SECTION 1)

*EQUIVALENT RANDOM THEORY: VMR(VARIANCE TO MEAN RATIO)– 1.5 - NO RETRIES (PEAKED TRAFFIC - ROUTE ADVANCEMENT OR DAY-TO-DAY VARIATION)

| HOURS FIRST ATTEMPT VMR = 1.5 | HOURS CON-NECT | HOURS RECALL /RETRY | HOURS OVFL | CON-NECT VMR | OVFL VMR | PERCENT BLOCKED (GRADE) | LINE | HOURS CONNECTED TRAFFIC ON LINES 1 | 2 | 3 | 4 | 5 | 6 | 7 | 8 | 9 | 10 |
|---|---|---|---|---|---|---|---|---|---|---|---|---|---|---|---|---|---|
| 6.000 | 5.965 | .000 | .035 | .97 | 2.01 | .6% | 15 | .797 | .761 | .717 | .666 | .606 | .540 | .467 | .391 | .314 | .242 |
| | | | | | | | | .178 | .124 | .082 | .051 | .030 | | | | | |
| 6.500 | 4.241 | .000 | 2.259 | .42 | 2.17 | 34.8% | 6 | .812 | .779 | .740 | .693 | .639 | .578 | | | | |
| 6.500 | 4.751 | .000 | 1.749 | .49 | 2.25 | 26.9% | 7 | .812 | .779 | .740 | .693 | .639 | .578 | .510 | | | |
| 6.500 | 5.188 | .000 | 1.312 | .56 | 2.31 | 20.2% | 8 | .812 | .779 | .740 | .693 | .639 | .578 | .510 | .437 | | |
| 6.500 | 5.550 | .000 | .950 | .64 | 2.35 | 14.6% | 9 | .812 | .779 | .740 | .693 | .639 | .578 | .510 | .437 | .362 | |
| 6.500 | 5.838 | .000 | .662 | .71 | 2.36 | 10.2% | 10 | .812 | .779 | .740 | .693 | .639 | .578 | .510 | .437 | .362 | .288 |
| 6.500 | 6.057 | .000 | .443 | .78 | 2.34 | 6.8% | 11 | .812 | .779 | .740 | .693 | .639 | .578 | .510 | .437 | .362 | .288 |
| | | | | | | | | .219 | | | | | | | | | |
| 6.500 | 6.216 | .000 | .284 | .84 | 2.30 | 4.4% | 12 | .812 | .779 | .740 | .693 | .639 | .578 | .510 | .437 | .362 | .288 |
| | | | | | | | | .219 | .159 | | | | | | | | |
| 6.500 | 6.326 | .000 | .174 | .89 | 2.24 | 2.7% | 13 | .812 | .779 | .740 | .693 | .639 | .578 | .510 | .437 | .362 | .288 |
| | | | | | | | | .219 | .159 | .110 | | | | | | | |
| 6.500 | 6.398 | .000 | .102 | .93 | 2.17 | 1.6% | 14 | .812 | .779 | .740 | .693 | .639 | .578 | .510 | .437 | .362 | .288 |
| | | | | | | | | .219 | .159 | .110 | .072 | | | | | | |
| 6.500 | 6.443 | .000 | .057 | .96 | 2.10 | .9% | 15 | .812 | .779 | .740 | .693 | .639 | .578 | .510 | .437 | .362 | .288 |
| | | | | | | | | .219 | .159 | .110 | .072 | .045 | | | | | |
| 7.000 | 4.927 | .000 | 2.073 | .45 | 2.24 | 29.6% | 7 | .825 | .795 | .759 | .718 | .669 | .612 | .549 | | | |
| 7.000 | 5.407 | .000 | 1.593 | .52 | 2.32 | 22.8% | 8 | .825 | .795 | .759 | .718 | .669 | .612 | .549 | .480 | | |
| 7.000 | 5.814 | .000 | 1.186 | .59 | 2.37 | 16.9% | 9 | .825 | .795 | .759 | .718 | .669 | .612 | .549 | .480 | .407 | |
| 7.000 | 6.148 | .000 | .852 | .67 | 2.39 | 12.2% | 10 | .825 | .795 | .759 | .718 | .669 | .612 | .549 | .480 | .407 | .334 |
| 7.000 | 6.410 | .000 | .590 | .74 | 2.39 | 8.4% | 11 | .825 | .795 | .759 | .718 | .669 | .612 | .549 | .480 | .407 | .334 |
| | | | | | | | | .263 | | | | | | | | | |
| 7.000 | 6.608 | .000 | .392 | .80 | 2.36 | 5.6% | 12 | .825 | .795 | .759 | .718 | .669 | .612 | .549 | .480 | .407 | .334 |
| | | | | | | | | .263 | .198 | | | | | | | | |
| 7.000 | 6.751 | .000 | .249 | .86 | 2.31 | 3.6% | 13 | .825 | .795 | .759 | .718 | .669 | .612 | .549 | .480 | .407 | .334 |
| | | | | | | | | .263 | .198 | .142 | | | | | | | |
| 7.000 | 6.848 | .000 | .152 | .90 | 2.25 | 2.2% | 14 | .825 | .795 | .759 | .718 | .669 | .612 | .549 | .480 | .407 | .334 |
| | | | | | | | | .263 | .198 | .142 | .097 | | | | | | |
| 7.000 | 6.911 | .000 | .089 | .94 | 2.18 | 1.3% | 15 | .825 | .795 | .759 | .718 | .669 | .612 | .549 | .480 | .407 | .334 |
| | | | | | | | | .263 | .198 | .142 | .097 | .063 | | | | | |
| 7.000 | 6.950 | .000 | .050 | .96 | 2.10 | .7% | 16 | .825 | .795 | .759 | .718 | .669 | .612 | .549 | .480 | .407 | .334 |
| | | | | | | | | .263 | .198 | .142 | .097 | .063 | .039 | | | | |
| 7.500 | 5.084 | .000 | 2.416 | .42 | 2.23 | 32.2% | 7 | .836 | .809 | .777 | .739 | .695 | .643 | .585 | | | |
| 7.500 | 5.604 | .000 | 1.896 | .48 | 2.31 | 25.3% | 8 | .836 | .809 | .777 | .739 | .695 | .643 | .585 | .520 | | |
| 7.500 | 6.054 | .000 | 1.446 | .55 | 2.38 | 19.3% | 9 | .836 | .809 | .777 | .739 | .695 | .643 | .585 | .520 | .450 | |
| 7.500 | 6.432 | .000 | 1.068 | .62 | 2.42 | 14.2% | 10 | .836 | .809 | .777 | .739 | .695 | .643 | .585 | .520 | .450 | .378 |
| 7.500 | 6.739 | .000 | .761 | .69 | 2.43 | 10.2% | 11 | .836 | .809 | .777 | .739 | .695 | .643 | .585 | .520 | .450 | .378 |
| | | | | | | | | .306 | | | | | | | | | |
| 7.500 | 6.977 | .000 | .523 | .76 | 2.42 | 7.0% | 12 | .836 | .809 | .777 | .739 | .695 | .643 | .585 | .520 | .450 | .378 |

*SIMILAR TO A.T.&T. NEAL-WILKINSON TABLES(SEE SECTION 1)

*EQUIVALENT RANDOM THEORY: VMR(VARIANCE TO MEAN RATIO)– 1.5 - NO RETRIES (PEAKED TRAFFIC - ROUTE ADVANCEMENT OR DAY-TO-DAY VARIATION)

| HOURS FIRST ATTEMPT VMR=1.5 | HOURS CON- NECT | HOURS RECALL /RETRY | HOURS OVFL | CON- NECT VMR | OVFL VMR | PERCENT BLOCKED (GRADE) | LINE | HOURS CONNECTED TRAFFIC ON LINES 1 | 2 | 3 | 4 | 5 | 6 | 7 | 8 | 9 | 10 |
|---|---|---|---|---|---|---|---|---|---|---|---|---|---|---|---|---|---|
| | | | | | | | | .306 | .239 | | | | | | | | |
| 7.500 | 7.155 | .000 | .345 | .82 | 2.38 | 4.6% | 13 | .836 | .809 | .777 | .739 | .695 | .643 | .585 | .520 | .450 | .378 |
| | | | | | | | | .306 | .239 | .178 | | | | | | | |
| 7.500 | 7.281 | .000 | .219 | .87 | 2.33 | 2.9% | 14 | .836 | .809 | .777 | .739 | .695 | .643 | .585 | .520 | .450 | .378 |
| | | | | | | | | .306 | .239 | .178 | .127 | | | | | | |
| 7.500 | 7.367 | .000 | .133 | .91 | 2.26 | 1.8% | 15 | .836 | .809 | .777 | .739 | .695 | .643 | .585 | .520 | .450 | .378 |
| | | | | | | | | .306 | .239 | .178 | .127 | .086 | | | | | |
| 7.500 | 7.423 | .000 | .077 | .94 | 2.19 | 1.0% | 16 | .836 | .809 | .777 | .739 | .695 | .643 | .585 | .520 | .450 | .378 |
| | | | | | | | | .306 | .239 | .178 | .127 | .086 | .055 | | | | |
| 7.500 | 7.457 | .000 | .043 | .97 | 2.11 | .6% | 17 | .836 | .809 | .777 | .739 | .695 | .643 | .585 | .520 | .450 | .378 |
| | | | | | | | | .306 | .239 | .178 | .127 | .086 | .055 | .034 | | | |
| 8.000 | 5.735 | .000 | 2.265 | .45 | 2.36 | 28.3% | 8 | .840 | .815 | .786 | .752 | .711 | .665 | .612 | .553 | | |
| 8.000 | 6.224 | .000 | 1.776 | .51 | 2.44 | 22.2% | 9 | .840 | .815 | .786 | .752 | .711 | .665 | .612 | .553 | .489 | |
| 8.000 | 6.646 | .000 | 1.354 | .58 | 2.49 | 16.9% | 10 | .840 | .815 | .786 | .752 | .711 | .665 | .612 | .553 | .489 | .421 |
| 8.000 | 6.998 | .000 | 1.002 | .65 | 2.52 | 12.5% | 11 | .840 | .815 | .786 | .752 | .711 | .665 | .612 | .553 | .489 | .421 |
| | | | | | | | | .352 | | | | | | | | | |
| 8.000 | 7.283 | .000 | .717 | .72 | 2.52 | 9.0% | 12 | .840 | .815 | .786 | .752 | .711 | .665 | .612 | .553 | .489 | .421 |
| | | | | | | | | .352 | .285 | | | | | | | | |
| 8.000 | 7.504 | .000 | .496 | .78 | 2.50 | 6.2% | 13 | .840 | .815 | .786 | .752 | .711 | .665 | .612 | .553 | .489 | .421 |
| | | | | | | | | .352 | .285 | .222 | | | | | | | |
| 8.000 | 7.670 | .000 | .330 | .83 | 2.46 | 4.1% | 14 | .840 | .815 | .786 | .752 | .711 | .665 | .612 | .553 | .489 | .421 |
| | | | | | | | | .352 | .285 | .222 | .166 | | | | | | |
| 8.000 | 7.788 | .000 | .212 | .88 | 2.40 | 2.6% | 15 | .840 | .815 | .786 | .752 | .711 | .665 | .612 | .553 | .489 | .421 |
| | | | | | | | | .352 | .285 | .222 | .166 | .118 | | | | | |
| 8.000 | 7.869 | .000 | .131 | .92 | 2.33 | 1.6% | 16 | .840 | .815 | .786 | .752 | .711 | .665 | .612 | .553 | .489 | .421 |
| | | | | | | | | .352 | .285 | .222 | .166 | .118 | .081 | | | | |
| 8.000 | 7.922 | .000 | .078 | .95 | 2.25 | 1.0% | 17 | .840 | .815 | .786 | .752 | .711 | .665 | .612 | .553 | .489 | .421 |
| | | | | | | | | .352 | .285 | .222 | .166 | .118 | .081 | .053 | | | |
| 8.000 | 7.956 | .000 | .044 | .97 | 2.18 | .6% | 18 | .840 | .815 | .786 | .752 | .711 | .665 | .612 | .553 | .489 | .421 |
| | | | | | | | | .352 | .285 | .222 | .166 | .118 | .081 | .053 | .033 | | |
| 8.500 | 5.894 | .000 | 2.606 | .41 | 2.35 | 30.7% | 8 | .850 | .827 | .800 | .769 | .732 | .690 | .641 | .586 | | |
| 8.500 | 6.420 | .000 | 2.080 | .47 | 2.43 | 24.5% | 9 | .850 | .827 | .800 | .769 | .732 | .690 | .641 | .586 | .526 | |
| 8.500 | 6.881 | .000 | 1.619 | .54 | 2.50 | 19.0% | 10 | .850 | .827 | .800 | .769 | .732 | .690 | .641 | .586 | .526 | .461 |
| 8.500 | 7.275 | .000 | 1.225 | .61 | 2.54 | 14.4% | 11 | .850 | .827 | .800 | .769 | .732 | .690 | .641 | .586 | .526 | .461 |
| | | | | | | | | .393 | | | | | | | | | |
| 8.500 | 7.600 | .000 | .900 | .67 | 2.56 | 10.6% | 12 | .850 | .827 | .800 | .769 | .732 | .690 | .641 | .586 | .526 | .461 |
| | | | | | | | | .393 | .326 | | | | | | | | |
| 8.500 | 7.861 | .000 | .639 | .74 | 2.55 | 7.5% | 13 | .850 | .827 | .800 | .769 | .732 | .690 | .641 | .586 | .526 | .461 |
| | | | | | | | | .393 | .326 | .260 | | | | | | | |
| 8.500 | 8.061 | .000 | .439 | .80 | 2.52 | 5.2% | 14 | .850 | .827 | .800 | .769 | .732 | .690 | .641 | .586 | .526 | .461 |

*SIMILAR TO A.T.&T. NEAL-WILKINSON TABLES(SEE SECTION 1)

*EQUIVALENT RANDOM THEORY: VMR(VARIANCE TO MEAN RATIO)– 1.5 - NO RETRIES (PEAKED TRAFFIC - ROUTE ADVANCEMENT OR DAY-TO-DAY VARIATION)

| HOURS FIRST ATTEMPT VMR = 1.5 | HOURS CON-NECT | HOURS RECALL /RETRY | HOURS OVFL | CON-NECT VMR | OVFL VMR | PERCENT BLOCKED (GRADE) | LINE | HOURS CONNECTED TRAFFIC ON LINES 1 | 2 | 3 | 4 | 5 | 6 | 7 | 8 | 9 | 10 |
|---|---|---|---|---|---|---|---|---|---|---|---|---|---|---|---|---|---|
| | | | | | | | | .393 | .326 | .260 | .201 | | | | | | |
| 8.500 | 8.210 | .000 | .290 | .85 | 2.47 | 3.4% | 15 | .850 | .827 | .800 | .769 | .732 | .690 | .641 | .586 | .526 | .461 |
| | | | | | | | | .393 | .326 | .260 | .201 | .148 | | | | | |
| 8.500 | 8.315 | .000 | .185 | .89 | 2.41 | 2.2% | 16 | .850 | .827 | .800 | .769 | .732 | .690 | .641 | .586 | .526 | .461 |
| | | | | | | | | .393 | .326 | .260 | .201 | .148 | .105 | | | | |
| 8.500 | 8.386 | .000 | .114 | .93 | 2.33 | 1.3% | 17 | .850 | .827 | .800 | .769 | .732 | .690 | .641 | .586 | .526 | .461 |
| | | | | | | | | .393 | .326 | .260 | .201 | .148 | .105 | .071 | | | |
| 8.500 | 8.433 | .000 | .067 | .95 | 2.26 | .8% | 18 | .850 | .827 | .800 | .769 | .732 | .690 | .641 | .586 | .526 | .461 |
| | | | | | | | | .393 | .326 | .260 | .201 | .148 | .105 | .071 | .046 | | |
| 9.000 | 6.038 | .000 | 2.962 | .38 | 2.33 | 32.9% | 8 | .858 | .837 | .813 | .784 | .751 | .712 | .667 | .616 | | |
| 9.000 | 6.597 | .000 | 2.403 | .44 | 2.42 | 26.7% | 9 | .858 | .837 | .813 | .784 | .751 | .712 | .667 | .616 | .560 | |
| 9.000 | 7.096 | .000 | 1.904 | .50 | 2.49 | 21.2% | 10 | .858 | .837 | .813 | .784 | .751 | .712 | .667 | .616 | .560 | .498 |
| 9.000 | 7.529 | .000 | 1.471 | .57 | 2.55 | 16.3% | 11 | .858 | .837 | .813 | .784 | .751 | .712 | .667 | .616 | .560 | .498 |
| | | | | | | | | .433 | | | | | | | | | |
| 9.000 | 7.895 | .000 | 1.105 | .63 | 2.58 | 12.3% | 12 | .858 | .837 | .813 | .784 | .751 | .712 | .667 | .616 | .560 | .498 |
| | | | | | | | | .433 | .366 | | | | | | | | |
| 9.000 | 8.194 | .000 | .806 | .70 | 2.59 | 9.0% | 13 | .858 | .837 | .813 | .784 | .751 | .712 | .667 | .616 | .560 | .498 |
| | | | | | | | | .433 | .366 | .300 | | | | | | | |
| 9.000 | 8.432 | .000 | .568 | .76 | 2.57 | 6.3% | 14 | .858 | .837 | .813 | .784 | .751 | .712 | .667 | .616 | .560 | .498 |
| | | | | | | | | .433 | .366 | .300 | .237 | | | | | | |
| 9.000 | 8.613 | .000 | .387 | .82 | 2.53 | 4.3% | 15 | .858 | .837 | .813 | .784 | .751 | .712 | .667 | .616 | .560 | .498 |
| | | | | | | | | .433 | .366 | .300 | .237 | .181 | | | | | |
| 9.000 | 8.745 | .000 | .255 | .87 | 2.48 | 2.8% | 16 | .858 | .837 | .813 | .784 | .751 | .712 | .667 | .616 | .560 | .498 |
| | | | | | | | | .433 | .366 | .300 | .237 | .181 | .133 | | | | |
| 9.000 | 8.839 | .000 | .161 | .91 | 2.41 | 1.8% | 17 | .858 | .837 | .813 | .784 | .751 | .712 | .667 | .616 | .560 | .498 |
| | | | | | | | | .433 | .366 | .300 | .237 | .181 | .133 | .093 | | | |
| 9.000 | 8.901 | .000 | .099 | .94 | 2.34 | 1.1% | 18 | .858 | .837 | .813 | .784 | .751 | .712 | .667 | .616 | .560 | .498 |
| | | | | | | | | .433 | .366 | .300 | .237 | .181 | .133 | .093 | .063 | | |
| 9.000 | 8.942 | .000 | .058 | .96 | 2.26 | .6% | 19 | .858 | .837 | .813 | .784 | .751 | .712 | .667 | .616 | .560 | .498 |
| | | | | | | | | .433 | .366 | .300 | .237 | .181 | .133 | .093 | .063 | .041 | |
| 9.500 | 6.758 | .000 | 2.742 | .41 | 2.40 | 28.9% | 9 | .866 | .847 | .824 | .798 | .767 | .732 | .690 | .644 | .591 | |
| 9.500 | 7.291 | .000 | 2.209 | .47 | 2.49 | 23.3% | 10 | .866 | .847 | .824 | .798 | .767 | .732 | .690 | .644 | .591 | .533 |
| 9.500 | 7.761 | .000 | 1.739 | .53 | 2.55 | 18.3% | 11 | .866 | .847 | .824 | .798 | .767 | .732 | .690 | .644 | .591 | .533 |
| | | | | | | | | .470 | | | | | | | | | |
| 9.500 | 8.167 | .000 | 1.333 | .59 | 2.60 | 14.0% | 12 | .866 | .847 | .824 | .798 | .767 | .732 | .690 | .644 | .591 | .533 |
| | | | | | | | | .470 | .405 | | | | | | | | |
| 9.500 | 8.506 | .000 | .994 | .66 | 2.62 | 10.5% | 13 | .866 | .847 | .824 | .798 | .767 | .732 | .690 | .644 | .591 | .533 |
| | | | | | | | | .470 | .405 | .339 | | | | | | | |
| 9.500 | 8.781 | .000 | .719 | .72 | 2.62 | 7.6% | 14 | .866 | .847 | .824 | .798 | .767 | .732 | .690 | .644 | .591 | .533 |
| | | | | | | | | .470 | .405 | .339 | .275 | | | | | | |

*SIMILAR TO A.T.&T. NEAL-WILKINSON TABLES(SEE SECTION 1)

*EQUIVALENT RANDOM THEORY: VMR(VARIANCE TO MEAN RATIO)– 1.5 - NO RETRIES (PEAKED TRAFFIC - ROUTE ADVANCEMENT OR DAY-TO-DAY VARIATION)

| HOURS FIRST ATTEMPT VMR = 1.5 | HOURS CON-NECT | HOURS RECALL /RETRY | HOURS OVFL | CON-NECT VMR | OVFL VMR | PERCENT BLOCKED (GRADE) | LINE | HOURS CONNECTED TRAFFIC ON LINES 1 | 2 | 3 | 4 | 5 | 6 | 7 | 8 | 9 | 10 |
|---|---|---|---|---|---|---|---|---|---|---|---|---|---|---|---|---|---|
| 9.500 | 8.997 | .000 | .503 | .78 | 2.59 | 5.3% | 15 | .866 | .847 | .824 | .798 | .767 | .732 | .690 | .644 | .591 | .533 |
| | | | | | | | | .470 | .405 | .339 | .275 | .216 | | | | | |
| 9.500 | 9.159 | .000 | .341 | .84 | 2.54 | 3.6% | 16 | .866 | .847 | .824 | .798 | .767 | .732 | .690 | .644 | .591 | .533 |
| | | | | | | | | .470 | .405 | .339 | .275 | .216 | .163 | | | | |
| 9.500 | 9.277 | .000 | .223 | .88 | 2.48 | 2.3% | 17 | .866 | .847 | .824 | .798 | .767 | .732 | .690 | .644 | .591 | .533 |
| | | | | | | | | .470 | .405 | .339 | .275 | .216 | .163 | .118 | | | |
| 9.500 | 9.360 | .000 | .140 | .92 | 2.41 | 1.5% | 18 | .866 | .847 | .824 | .798 | .767 | .732 | .690 | .644 | .591 | .533 |
| | | | | | | | | .470 | .405 | .339 | .275 | .216 | .163 | .118 | .082 | | |
| 9.500 | 9.415 | .000 | .085 | .95 | 2.34 | .9% | 19 | .866 | .847 | .824 | .798 | .767 | .732 | .690 | .644 | .591 | .533 |
| | | | | | | | | .470 | .405 | .339 | .275 | .216 | .163 | .118 | .082 | .055 | |
| 9.500 | 9.450 | .000 | .050 | .96 | 2.26 | .5% | 20 | .866 | .847 | .824 | .798 | .767 | .732 | .690 | .644 | .591 | .533 |
| | | | | | | | | .470 | .405 | .339 | .275 | .216 | .163 | .118 | .082 | .055 | .035 |
| 10.000 | 6.861 | .000 | 3.139 | .38 | 2.44 | 31.4% | 9 | .868 | .850 | .829 | .805 | .777 | .744 | .707 | .664 | .616 | |
| 10.000 | 7.424 | .000 | 2.576 | .44 | 2.53 | 25.8% | 10 | .868 | .850 | .829 | .805 | .777 | .744 | .707 | .664 | .616 | .563 |
| 10.000 | 7.928 | .000 | 2.072 | .50 | 2.60 | 20.7% | 11 | .868 | .850 | .829 | .805 | .777 | .744 | .707 | .664 | .616 | .563 |
| | | | | | | | | .505 | | | | | | | | | |
| 10.000 | 8.371 | .000 | 1.629 | .56 | 2.66 | 16.3% | 12 | .868 | .850 | .829 | .805 | .777 | .744 | .707 | .664 | .616 | .563 |
| | | | | | | | | .505 | .443 | | | | | | | | |
| 10.000 | 8.751 | .000 | 1.249 | .62 | 2.69 | 12.5% | 13 | .868 | .850 | .829 | .805 | .777 | .744 | .707 | .664 | .616 | .563 |
| | | | | | | | | .505 | .443 | .380 | | | | | | | |
| 10.000 | 9.068 | .000 | .932 | .68 | 2.70 | 9.3% | 14 | .868 | .850 | .829 | .805 | .777 | .744 | .707 | .664 | .616 | .563 |
| | | | | | | | | .505 | .443 | .380 | .317 | | | | | | |
| 10.000 | 9.324 | .000 | .676 | .74 | 2.69 | 6.8% | 15 | .868 | .850 | .829 | .805 | .777 | .744 | .707 | .664 | .616 | .563 |
| | | | | | | | | .505 | .443 | .380 | .317 | .256 | | | | | |
| 10.000 | 9.525 | .000 | .475 | .80 | 2.66 | 4.8% | 16 | .868 | .850 | .829 | .805 | .777 | .744 | .707 | .664 | .616 | .563 |
| | | | | | | | | .505 | .443 | .380 | .317 | .256 | .201 | | | | |
| 10.000 | 9.676 | .000 | .324 | .85 | 2.61 | 3.2% | 17 | .868 | .850 | .829 | .805 | .777 | .744 | .707 | .664 | .616 | .563 |
| | | | | | | | | .505 | .443 | .380 | .317 | .256 | .201 | .152 | | | |
| 10.000 | 9.787 | .000 | .213 | .89 | 2.55 | 2.1% | 18 | .868 | .850 | .829 | .805 | .777 | .744 | .707 | .664 | .616 | .563 |
| | | | | | | | | .505 | .443 | .380 | .317 | .256 | .201 | .152 | .110 | | |
| 10.000 | 9.864 | .000 | .136 | .92 | 2.47 | 1.4% | 19 | .868 | .850 | .829 | .805 | .777 | .744 | .707 | .664 | .616 | .563 |
| | | | | | | | | .505 | .443 | .380 | .317 | .256 | .201 | .152 | .110 | .077 | |
| 10.000 | 9.916 | .000 | .084 | .95 | 2.39 | .8% | 20 | .868 | .850 | .829 | .805 | .777 | .744 | .707 | .664 | .616 | .563 |
| | | | | | | | | .505 | .443 | .380 | .317 | .256 | .201 | .152 | .110 | .077 | .052 |
| 10.000 | 9.950 | .000 | .050 | .97 | 2.32 | .5% | 21 | .868 | .850 | .829 | .805 | .777 | .744 | .707 | .664 | .616 | .563 |
| | | | | | | | | .505 | .443 | .380 | .317 | .256 | .201 | .152 | .110 | .077 | .052 |
| | | | | | | | | .034 | | | | | | | | | |
| 11.000 | 7.734 | .000 | 3.266 | .38 | 2.49 | 29.7% | 10 | .881 | .865 | .848 | .827 | .803 | .775 | .744 | .707 | .665 | .619 |
| 11.000 | 8.301 | .000 | 2.699 | .43 | 2.58 | 24.5% | 11 | .881 | .865 | .848 | .827 | .803 | .775 | .744 | .707 | .665 | .619 |
| | | | | | | | | .567 | | | | | | | | | |

*SIMILAR TO A.T.&T. NEAL-WILKINSON TABLES(SEE SECTION 1)

*EQUIVALENT RANDOM THEORY: VMR(VARIANCE TO MEAN RATIO)– 1.5 - NO RETRIES (PEAKED TRAFFIC - ROUTE ADVANCEMENT OR DAY-TO-DAY VARIATION)

| HOURS FIRST ATTEMPT VMR = 1.5 | HOURS CON-NECT | HOURS RECALL /RETRY | HOURS OVFL | CON-NECT VMR | OVFL VMR | PERCENT BLOCKED (GRADE) | LINE | HOURS CONNECTED TRAFFIC ON LINES 1 | 2 | 3 | 4 | 5 | 6 | 7 | 8 | 9 | 10 |
|---|---|---|---|---|---|---|---|---|---|---|---|---|---|---|---|---|---|
| 11.000 | 8.812 | .000 | 2.188 | .49 | 2.65 | 19.9% | 12 | .881 | .865 | .848 | .827 | .803 | .775 | .744 | .707 | .665 | .619 |
| | | | | | | | | .567 | .511 | | | | | | | | |
| 11.000 | 9.264 | .000 | 1.736 | .55 | 2.71 | 15.8% | 13 | .881 | .865 | .848 | .827 | .803 | .775 | .744 | .707 | .665 | .619 |
| | | | | | | | | .567 | .511 | .452 | | | | | | | |
| 11.000 | 9.655 | .000 | 1.345 | .61 | 2.75 | 12.2% | 14 | .881 | .865 | .848 | .827 | .803 | .775 | .744 | .707 | .665 | .619 |
| | | | | | | | | .567 | .511 | .452 | .390 | | | | | | |
| 11.000 | 9.983 | .000 | 1.017 | .67 | 2.76 | 9.2% | 15 | .881 | .865 | .848 | .827 | .803 | .775 | .744 | .707 | .665 | .619 |
| | | | | | | | | .567 | .511 | .452 | .390 | .329 | | | | | |
| 11.000 | 10.252 | .000 | .748 | .73 | 2.76 | 6.8% | 16 | .881 | .865 | .848 | .827 | .803 | .775 | .744 | .707 | .665 | .619 |
| | | | | | | | | .567 | .511 | .452 | .390 | .329 | .269 | | | | |
| 11.000 | 10.466 | .000 | .534 | .79 | 2.73 | 4.9% | 17 | .881 | .865 | .848 | .827 | .803 | .775 | .744 | .707 | .665 | .619 |
| | | | | | | | | .567 | .511 | .452 | .390 | .329 | .269 | .213 | | | |
| 11.000 | 10.629 | .000 | .371 | .84 | 2.68 | 3.4% | 18 | .881 | .865 | .848 | .827 | .803 | .775 | .744 | .707 | .665 | .619 |
| | | | | | | | | .567 | .511 | .452 | .390 | .329 | .269 | .213 | .164 | | |
| 11.000 | 10.751 | .000 | .249 | .88 | 2.62 | 2.3% | 19 | .881 | .865 | .848 | .827 | .803 | .775 | .744 | .707 | .665 | .619 |
| | | | | | | | | .567 | .511 | .452 | .390 | .329 | .269 | .213 | .164 | .121 | |
| 11.000 | 10.838 | .000 | .162 | .91 | 2.55 | 1.5% | 20 | .881 | .865 | .848 | .827 | .803 | .775 | .744 | .707 | .665 | .619 |
| | | | | | | | | .567 | .511 | .452 | .390 | .329 | .269 | .213 | .164 | .121 | .087 |
| 11.000 | 10.898 | .000 | .102 | .94 | 2.47 | .9% | 21 | .881 | .865 | .848 | .827 | .803 | .775 | .744 | .707 | .665 | .619 |
| | | | | | | | | .567 | .511 | .452 | .390 | .329 | .269 | .213 | .164 | .121 | .087 |
| | | | | | | | | .060 | | | | | | | | | |
| 11.000 | 10.937 | .000 | .063 | .96 | 2.39 | .6% | 22 | .881 | .865 | .848 | .827 | .803 | .775 | .744 | .707 | .665 | .619 |
| | | | | | | | | .567 | .511 | .452 | .390 | .329 | .269 | .213 | .164 | .121 | .087 |
| | | | | | | | | .060 | .040 | | | | | | | | |
| 12.000 | 7.953 | .000 | 4.047 | .34 | 2.50 | 33.7% | 10 | .888 | .874 | .859 | .841 | .820 | .797 | .770 | .738 | .703 | .663 |
| 12.000 | 8.571 | .000 | 3.420 | .00 | 2.59 | 28.6% | 11 | .888 | .874 | .859 | .841 | .820 | .797 | .770 | .738 | .703 | .663 |
| | | | | | | | | .618 | | | | | | | | | |
| 12.000 | 9.141 | .000 | 2.859 | .43 | 2.68 | 23.8% | 12 | .888 | .874 | .859 | .841 | .820 | .797 | .770 | .738 | .703 | .663 |
| | | | | | | | | .618 | .569 | | | | | | | | |
| 12.000 | 9.657 | .000 | 2.343 | .48 | 2.76 | 19.5% | 13 | .888 | .874 | .859 | .841 | .820 | .797 | .770 | .738 | .703 | .663 |
| | | | | | | | | .618 | .569 | .516 | | | | | | | |
| 12.000 | 10.116 | .000 | 1.884 | .54 | 2.81 | 15.7% | 14 | .888 | .874 | .860 | .041 | .820 | .797 | .770 | .738 | .703 | .663 |
| | | | | | | | | .618 | .569 | .516 | .460 | | | | | | |
| 12.000 | 10.517 | .000 | 1.483 | .60 | 2.85 | 12.4% | 15 | .888 | .874 | .859 | .841 | .820 | .707 | .770 | .738 | .703 | .663 |
| | | | | | | | | .618 | .569 | .516 | .460 | .401 | | | | | |
| 12.000 | 10.859 | .000 | 1.141 | .66 | 2.87 | 9.5% | 16 | .888 | .874 | .859 | .841 | .820 | .797 | .770 | .738 | .703 | .663 |
| | | | | | | | | .618 | .569 | .516 | .460 | .401 | .342 | | | | |
| 12.000 | 11.144 | .000 | .856 | .72 | 2.87 | 7.1% | 17 | .888 | .074 | .859 | .841 | .820 | .797 | .770 | .738 | .703 | .663 |
| | | | | | | | | .618 | .569 | .516 | .460 | .401 | .342 | .281 | | | |
| 12.000 | 11.374 | .000 | .626 | .77 | 2.84 | 5.2% | 18 | .888 | .874 | .859 | .841 | .820 | .797 | .770 | .738 | .703 | .663 |

*SIMILAR TO A.T.&T. NEAL-WILKINSON TABLES(SEE SECTION 1)

*EQUIVALENT RANDOM THEORY: VMR(VARIANCE TO MEAN RATIO)– 1.5 - NO RETRIES (PEAKED TRAFFIC - ROUTE ADVANCEMENT OR DAY-TO-DAY VARIATION)

| HOURS FIRST ATTEMPT VMR = 1.5 | HOURS CON-NECT | HOURS RECALL /RETRY | HOURS OVFL | CON-NECT VMR | OVFL VMR | PERCENT BLOCKED (GRADE) | LINE | HOURS CONNECTED TRAFFIC ON LINES 1 | 2 | 3 | 4 | 5 | 6 | 7 | 8 | 9 | 10 |
|---|---|---|---|---|---|---|---|---|---|---|---|---|---|---|---|---|---|
| | | | | | | | | .618 | .569 | .516 | .460 | .401 | .342 | .284 | .230 | | |
| 12.000 | 11.554 | .000 | .446 | .82 | 2.80 | 3.7% | 19 | .888 | .874 | .859 | .841 | .820 | .797 | .770 | .738 | .703 | .663 |
| | | | | | | | | .618 | .569 | .516 | .460 | .401 | .342 | .284 | .230 | .181 | |
| 12.000 | 11.692 | .000 | .308 | .86 | 2.74 | 2.6% | 20 | .888 | .874 | .859 | .841 | .820 | .797 | .770 | .738 | .703 | .663 |
| | | | | | | | | .618 | .569 | .516 | .460 | .401 | .342 | .284 | .230 | .181 | .137 |
| 12.000 | 11.793 | .000 | .207 | .90 | 2.67 | 1.7% | 21 | .888 | .874 | .859 | .841 | .820 | .797 | .770 | .738 | .703 | .663 |
| | | | | | | | | .618 | .569 | .516 | .460 | .401 | .342 | .284 | .230 | .181 | .137 |
| | | | | | | | | .101 | | | | | | | | | |
| 12.000 | 11.865 | .000 | .135 | .93 | 2.60 | 1.1% | 22 | .888 | .874 | .859 | .841 | .820 | .797 | .770 | .738 | .703 | .663 |
| | | | | | | | | .618 | .569 | .516 | .460 | .401 | .342 | .284 | .230 | .181 | .137 |
| | | | | | | | | .101 | .072 | | | | | | | | |
| 12.000 | 11.914 | .000 | .086 | .95 | 2.52 | .7% | 23 | .888 | .874 | .859 | .841 | .820 | .797 | .770 | .738 | .703 | .663 |
| | | | | | | | | .618 | .569 | .516 | .460 | .401 | .342 | .284 | .230 | .181 | .137 |
| | | | | | | | | .101 | .072 | .050 | | | | | | | |
| 13.000 | 8.836 | .000 | 4.164 | .34 | 2.54 | 32.0% | 11 | .897 | .885 | .872 | .857 | .839 | .818 | .795 | .768 | .738 | .703 |
| | | | | | | | | .664 | | | | | | | | | |
| 13.000 | 9.457 | .000 | 3.543 | .38 | 2.64 | 27.3% | 12 | .897 | .885 | .872 | .857 | .839 | .818 | .795 | .768 | .738 | .703 |
| | | | | | | | | .664 | .621 | | | | | | | | |
| 13.000 | 10.030 | .000 | 2.970 | .43 | 2.73 | 22.8% | 13 | .897 | .885 | .872 | .857 | .839 | .818 | .795 | .768 | .738 | .703 |
| | | | | | | | | .664 | .621 | .573 | | | | | | | |
| 13.000 | 10.551 | .000 | 2.449 | .48 | 2.80 | 18.8% | 14 | .897 | .885 | .872 | .857 | .839 | .818 | .795 | .768 | .738 | .703 |
| | | | | | | | | .664 | .621 | .573 | .521 | | | | | | |
| 13.000 | 11.017 | .000 | 1.983 | .53 | 2.86 | 15.3% | 15 | .897 | .885 | .872 | .857 | .839 | .818 | .795 | .768 | .738 | .703 |
| | | | | | | | | .664 | .621 | .573 | .521 | .466 | | | | | |
| 13.000 | 11.427 | .000 | 1.573 | .59 | 2.90 | 12.1% | 16 | .897 | .885 | .872 | .857 | .839 | .818 | .795 | .768 | .738 | .703 |
| | | | | | | | | .664 | .621 | .573 | .521 | .466 | .409 | | | | |
| 13.000 | 11.778 | .000 | 1.222 | .65 | 2.92 | 9.4% | 17 | .897 | .885 | .872 | .857 | .839 | .818 | .795 | .768 | .738 | .703 |
| | | | | | | | | .664 | .621 | .573 | .521 | .466 | .409 | .352 | | | |
| 13.000 | 12.073 | .000 | .927 | .71 | 2.92 | 7.1% | 18 | .897 | .885 | .872 | .857 | .839 | .818 | .795 | .768 | .738 | .703 |
| | | | | | | | | .664 | .621 | .573 | .521 | .466 | .409 | .352 | .295 | | |
| 13.000 | 12.314 | .000 | .686 | .76 | 2.90 | 5.3% | 19 | .897 | .885 | .872 | .857 | .839 | .818 | .795 | .768 | .738 | .703 |
| | | | | | | | | .664 | .621 | .573 | .521 | .466 | .409 | .352 | .295 | .241 | |
| 13.000 | 12.505 | .000 | .495 | .81 | 2.86 | 3.8% | 20 | .897 | .885 | .872 | .857 | .839 | .818 | .795 | .768 | .738 | .703 |
| | | | | | | | | .664 | .621 | .573 | .521 | .466 | .409 | .352 | .295 | .241 | .191 |
| 13.000 | 12.652 | .000 | .348 | .85 | 2.81 | 2.7% | 21 | .897 | .885 | .872 | .857 | .839 | .818 | .795 | .768 | .738 | .703 |
| | | | | | | | | .664 | .621 | .573 | .521 | .466 | .409 | .352 | .295 | .241 | .191 |
| | | | | | | | | .147 | | | | | | | | | |
| 13.000 | 12.763 | .000 | .237 | .89 | 2.74 | 1.8% | 22 | .897 | .885 | .872 | .857 | .839 | .818 | .795 | .768 | .738 | .703 |
| | | | | | | | | .664 | .621 | .573 | .521 | .466 | .409 | .352 | .295 | .241 | .191 |
| | | | | | | | | .147 | .110 | | | | | | | | |

*SIMILAR TO A.T.&T. NEAL-WILKINSON TABLES(SEE SECTION 1)

*EQUIVALENT RANDOM THEORY: VMR(VARIANCE TO MEAN RATIO)– 1.5 - NO RETRIES (PEAKED TRAFFIC - ROUTE ADVANCEMENT OR DAY-TO-DAY VARIATION)

| HOURS FIRST ATTEMPT VMR=1.5 | HOURS CON-NECT | HOURS RECALL /RETRY | HOURS OVFL | CON-NECT VMR | OVFL VMR | PERCENT BLOCKED (GRADE) | LINE | HOURS CONNECTED TRAFFIC ON LINES 1 | 2 | 3 | 4 | 5 | 6 | 7 | 8 | 9 | 10 |
|---|---|---|---|---|---|---|---|---|---|---|---|---|---|---|---|---|---|
| 13.000 | 12.842 | .000 | .158 | .92 | 2.67 | 1.2% | 23 | .897 | .885 | .872 | .857 | .839 | .818 | .795 | .768 | .738 | .703 |
| | | | | | | | | .664 | .621 | .573 | .521 | .466 | .409 | .352 | .295 | .241 | .191 |
| | | | | | | | | .147 | .110 | .080 | | | | | | | |
| 13.000 | 12.898 | .000 | .102 | .94 | 2.59 | .8% | 24 | .897 | .885 | .872 | .857 | .839 | .818 | .795 | .768 | .738 | .703 |
| | | | | | | | | .664 | .621 | .573 | .521 | .466 | .409 | .352 | .295 | .241 | .191 |
| | | | | | | | | .147 | .110 | .080 | .056 | | | | | | |
| 14.000 | 9.685 | .000 | 4.315 | .34 | 2.64 | 30.8% | 12 | .902 | .892 | .880 | .866 | .851 | .833 | .813 | .790 | .763 | .733 |
| | | | | | | | | .700 | .662 | | | | | | | | |
| 14.000 | 10.305 | .000 | 3.695 | .38 | 2.74 | 26.4% | 13 | .902 | .892 | .880 | .866 | .851 | .833 | .813 | .790 | .763 | .733 |
| | | | | | | | | .700 | .662 | .620 | | | | | | | |
| 14.000 | 10.880 | .000 | 3.120 | .43 | 2.82 | 22.3% | 14 | .902 | .892 | .880 | .866 | .851 | .833 | .813 | .790 | .763 | .733 |
| | | | | | | | | .700 | .662 | .620 | .574 | | | | | | |
| 14.000 | 11.405 | .000 | 2.595 | .47 | 2.90 | 18.5% | 15 | .902 | .892 | .880 | .866 | .851 | .833 | .813 | .790 | .763 | .733 |
| | | | | | | | | .700 | .662 | .620 | .574 | .525 | | | | | |
| 14.000 | 11.877 | .000 | 2.123 | .53 | 2.96 | 15.2% | 16 | .902 | .892 | .880 | .866 | .851 | .833 | .813 | .790 | .763 | .733 |
| | | | | | | | | .700 | .662 | .620 | .574 | .525 | .473 | | | | |
| 14.000 | 12.295 | .000 | 1.705 | .58 | 3.00 | 12.2% | 17 | .902 | .892 | .880 | .866 | .851 | .833 | .813 | .790 | .763 | .733 |
| | | | | | | | | .700 | .662 | .620 | .574 | .525 | .473 | .418 | | | |
| 14.000 | 12.658 | .000 | 1.342 | .64 | 3.02 | 9.6% | 18 | .902 | .892 | .880 | .866 | .851 | .833 | .813 | .790 | .763 | .733 |
| | | | | | | | | .700 | .662 | .620 | .574 | .525 | .473 | .418 | .362 | | |
| 14.000 | 12.965 | .000 | 1.035 | .69 | 3.02 | 7.4% | 19 | .902 | .892 | .880 | .866 | .851 | .833 | .813 | .790 | .763 | .733 |
| | | | | | | | | .700 | .662 | .620 | .574 | .525 | .473 | .418 | .362 | .308 | |
| 14.000 | 13.220 | .000 | .780 | .75 | 3.01 | 5.6% | 20 | .902 | .892 | .880 | .866 | .851 | .833 | .813 | .790 | .763 | .733 |
| | | | | | | | | .700 | .662 | .620 | .574 | .525 | .473 | .418 | .362 | .308 | .255 |
| 14.000 | 13.426 | .000 | .574 | .79 | 2.97 | 4.1% | 21 | .902 | .892 | .880 | .866 | .851 | .833 | .813 | .790 | .763 | .733 |
| | | | | | | | | .700 | .662 | .620 | .574 | .525 | .473 | .418 | .362 | .308 | .255 |
| | | | | | | | | .206 | | | | | | | | | |
| 14.000 | 13.588 | .000 | .412 | .84 | 2.92 | 2.9% | 22 | .902 | .892 | .880 | .866 | .851 | .833 | .813 | .790 | .763 | .733 |
| | | | | | | | | .700 | .662 | .620 | .574 | .525 | .473 | .418 | .362 | .308 | .255 |
| | | | | | | | | .206 | .162 | | | | | | | | |
| 14.000 | 13.712 | .000 | .288 | .88 | 2.86 | 2.1% | 23 | .902 | .892 | .880 | .866 | .851 | .833 | .813 | .790 | .763 | .733 |
| | | | | | | | | .700 | .662 | .620 | .574 | .525 | .473 | .418 | .362 | .308 | .255 |
| | | | | | | | | .206 | .162 | .124 | | | | | | | |
| 14.000 | 13.803 | .000 | .197 | .91 | 2.78 | 1.4% | 24 | .902 | .892 | .880 | .866 | .851 | .833 | .813 | .790 | .763 | .733 |
| | | | | | | | | .700 | .662 | .620 | .574 | .525 | .473 | .418 | .362 | .308 | .255 |
| | | | | | | | | .206 | .162 | .124 | .092 | | | | | | |
| 14.000 | 13.869 | .000 | .131 | .93 | 2.70 | .9% | 25 | .902 | .892 | .880 | .866 | .851 | .833 | .813 | .790 | .763 | .733 |
| | | | | | | | | .700 | .662 | .620 | .574 | .525 | .473 | .418 | .362 | .308 | .255 |
| | | | | | | | | .206 | .162 | .124 | .092 | .066 | | | | | |
| 14.000 | 13.915 | .000 | .085 | .95 | 2.62 | .6% | 26 | .902 | .892 | .880 | .866 | .851 | .833 | .813 | .790 | .763 | .733 |

*SIMILAR TO A.T.&T. NEAL-WILKINSON TABLES(SEE SECTION 1)

*EQUIVALENT RANDOM THEORY: VMR(VARIANCE TO MEAN RATIO)– 1.5 - NO RETRIES (PEAKED TRAFFIC - ROUTE ADVANCEMENT OR DAY-TO-DAY VARIATION)

| HOURS FIRST ATTEMPT VMR = 1.5 | HOURS CON-NECT | HOURS RECALL /RETRY | HOURS OVFL | CON-NECT VMR | OVFL VMR | PERCENT BLOCKED (GRADE) | LINE | HOURS CONNECTED TRAFFIC ON LINES 1 | 2 | 3 | 4 | 5 | 6 | 7 | 8 | 9 | 10 |
|---|---|---|---|---|---|---|---|---|---|---|---|---|---|---|---|---|---|
| | | | | | | | | .700 | .662 | .620 | .574 | .525 | .473 | .418 | .362 | .308 | .255 |
| | | | | | | | | .206 | .162 | .124 | .092 | .066 | .046 | | | | |
| 15.000 | 9.915 | .000 | 5.085 | .30 | 2.59 | 33.9% | 12 | .910 | .900 | .890 | .878 | .864 | .849 | .831 | .811 | .788 | .762 |
| | | | | | | | | .733 | .700 | | | | | | | | |
| 15.000 | 10.578 | .000 | 4.422 | .34 | 2.69 | 29.5% | 13 | .910 | .900 | .890 | .878 | .864 | .849 | .831 | .811 | .788 | .762 |
| | | | | | | | | .733 | .700 | .663 | | | | | | | |
| 15.000 | 11.200 | .000 | 3.800 | .38 | 2.78 | 25.3% | 14 | .910 | .900 | .890 | .878 | .864 | .849 | .831 | .811 | .788 | .762 |
| | | | | | | | | .733 | .700 | .663 | .622 | | | | | | |
| 15.000 | 11.778 | .000 | 3.222 | .42 | 2.87 | 21.5% | 15 | .910 | .900 | .890 | .878 | .864 | .849 | .831 | .811 | .788 | .762 |
| | | | | | | | | .733 | .700 | .663 | .622 | .577 | | | | | |
| 15.000 | 12.307 | .000 | 2.693 | .47 | 2.94 | 18.0% | 16 | .910 | .900 | .890 | .878 | .864 | .849 | .831 | .811 | .788 | .762 |
| | | | | | | | | .733 | .700 | .663 | .622 | .577 | .529 | | | | |
| 15.000 | 12.785 | .000 | 2.215 | .52 | 3.00 | 14.8% | 17 | .910 | .900 | .890 | .878 | .864 | .849 | .831 | .811 | .788 | .762 |
| | | | | | | | | .733 | .700 | .663 | .622 | .577 | .529 | .478 | | | |
| 15.000 | 13.209 | .000 | 1.791 | .58 | 3.05 | 11.9% | 18 | .910 | .900 | .890 | .878 | .864 | .849 | .831 | .811 | .788 | .762 |
| | | | | | | | | .733 | .700 | .663 | .622 | .577 | .529 | .478 | .425 | | |
| 15.000 | 13.580 | .000 | 1.420 | .63 | 3.07 | 9.5% | 19 | .910 | .900 | .890 | .878 | .864 | .849 | .831 | .811 | .788 | .762 |
| | | | | | | | | .733 | .700 | .663 | .622 | .577 | .529 | .478 | .425 | .370 | |
| 15.000 | 13.896 | .000 | 1.104 | .68 | 3.08 | 7.4% | 20 | .910 | .900 | .890 | .878 | .864 | .849 | .831 | .811 | .788 | .762 |
| | | | | | | | | .733 | .700 | .663 | .622 | .577 | .529 | .478 | .425 | .370 | .316 |
| 15.000 | 14.160 | .000 | .840 | .74 | 3.06 | 5.6% | 21 | .910 | .900 | .890 | .878 | .864 | .849 | .831 | .811 | .788 | .762 |
| | | | | | | | | .733 | .700 | .663 | .622 | .577 | .529 | .478 | .425 | .370 | .316 |
| | | | | | | | | .264 | | | | | | | | | |
| 15.000 | 14.375 | .000 | .625 | .78 | 3.03 | 4.2% | 22 | .910 | .900 | .890 | .878 | .864 | .849 | .831 | .811 | .788 | .762 |
| | | | | | | | | .733 | .700 | .663 | .622 | .577 | .529 | .478 | .425 | .370 | .316 |
| | | | | | | | | .264 | .215 | | | | | | | | |
| 15.000 | 14.546 | .000 | .454 | .83 | 2.98 | 3.0% | 23 | .910 | .900 | .890 | .878 | .864 | .849 | .831 | .811 | .788 | .762 |
| | | | | | | | | .733 | .700 | .663 | .622 | .577 | .529 | .478 | .425 | .370 | .316 |
| | | | | | | | | .264 | .215 | .171 | | | | | | | |
| 15.000 | 14.678 | .000 | .322 | .87 | 2.92 | 2.1% | 24 | .910 | .900 | .890 | .878 | .864 | .849 | .831 | .811 | .788 | .762 |
| | | | | | | | | .733 | .700 | .663 | .622 | .577 | .529 | .478 | .425 | .370 | .316 |
| | | | | | | | | .264 | .215 | .171 | .132 | | | | | | |
| 15.000 | 14.778 | .000 | .222 | .90 | 2.85 | 1.5% | 25 | .910 | .900 | .890 | .878 | .864 | .849 | .831 | .811 | .788 | .762 |
| | | | | | | | | .733 | .700 | .663 | .622 | .577 | .529 | .478 | .425 | .370 | .316 |
| | | | | | | | | .264 | .215 | .171 | .132 | .099 | | | | | |
| 15.000 | 14.850 | .000 | .150 | .93 | 2.77 | 1.0% | 26 | .910 | .900 | .890 | .878 | .864 | .849 | .831 | .811 | .788 | .762 |
| | | | | | | | | .733 | .700 | .663 | .622 | .577 | .529 | .478 | .425 | .370 | .316 |
| | | | | | | | | .264 | .215 | .171 | .132 | .099 | .072 | | | | |
| 15.000 | 14.901 | .000 | .099 | .95 | 2.69 | .7% | 27 | .910 | .900 | .890 | .878 | .864 | .849 | .831 | .811 | .788 | .762 |
| | | | | | | | | .733 | .700 | .663 | .622 | .577 | .529 | .478 | .425 | .370 | .316 |

*SIMILAR TO A.T.&T. NEAL-WILKINSON TABLES(SEE SECTION 1)

*EQUIVALENT RANDOM THEORY: VMR(VARIANCE TO MEAN RATIO)– 1.5 - NO RETRIES (PEAKED TRAFFIC - ROUTE ADVANCEMENT OR DAY-TO-DAY VARIATION)

| HOURS FIRST ATTEMPT VMR = 1.5 | HOURS CON- NECT | HOURS RECALL /RETRY | HOURS OVFL | CON- NECT VMR | OVFL VMR | PERCENT BLOCKED (GRADE) | LINE | HOURS CONNECTED TRAFFIC ON LINES 1 | 2 | 3 | 4 | 5 | 6 | 7 | 8 | 9 | 10 |
|---|---|---|---|---|---|---|---|---|---|---|---|---|---|---|---|---|---|
| | | | | | | | | .264 | .215 | .171 | .132 | .099 | .072 | .051 | | | |
| 16.000 | 10.775 | .000 | 5.225 | .30 | 2.68 | 32.7% | 13 | .914 | .905 | .896 | .885 | .873 | .859 | .844 | .826 | .806 | .783 |
| | | | | | | | | .758 | .729 | .697 | | | | | | | |
| 16.000 | 11.436 | .000 | 4.564 | .34 | 2.78 | 28.5% | 14 | .914 | .905 | .896 | .885 | .873 | .859 | .844 | .826 | .806 | .783 |
| | | | | | | | | .758 | .729 | .697 | .661 | | | | | | |
| 16.000 | 12.057 | .000 | 3.943 | .38 | 2.87 | 24.6% | 15 | .914 | .905 | .896 | .885 | .873 | .859 | .844 | .826 | .806 | .783 |
| | | | | | | | | .758 | .729 | .697 | .661 | .622 | | | | | |
| 16.000 | 12.636 | .000 | 3.364 | .42 | 2.96 | 21.0% | 16 | .914 | .905 | .896 | .885 | .873 | .859 | .844 | .826 | .806 | .783 |
| | | | | | | | | .758 | .729 | .697 | .661 | .622 | .579 | | | | |
| 16.000 | 13.168 | .000 | 2.832 | .47 | 3.03 | 17.7% | 17 | .914 | .905 | .896 | .885 | .873 | .859 | .844 | .826 | .806 | .783 |
| | | | | | | | | .758 | .729 | .697 | .661 | .622 | .579 | .532 | | | |
| 16.000 | 13.651 | .000 | 2.349 | .52 | 3.09 | 14.7% | 18 | .914 | .905 | .896 | .885 | .873 | .859 | .844 | .826 | .806 | .783 |
| | | | | | | | | .758 | .729 | .697 | .661 | .622 | .579 | .532 | .483 | | |
| 16.000 | 14.083 | .000 | 1.917 | .57 | 3.14 | 12.0% | 19 | .914 | .905 | .896 | .885 | .873 | .859 | .844 | .826 | .806 | .783 |
| | | | | | | | | .758 | .729 | .697 | .661 | .622 | .579 | .532 | .483 | .432 | |
| 16.000 | 14.463 | .000 | 1.537 | .62 | 3.16 | 9.6% | 20 | .914 | .905 | .896 | .885 | .873 | .859 | .844 | .826 | .806 | .783 |
| | | | | | | | | .758 | .729 | .697 | .661 | .622 | .579 | .532 | .483 | .432 | .379 |
| 16.000 | 14.790 | .000 | 1.210 | .67 | 3.17 | 7.6% | 21 | .914 | .905 | .896 | .885 | .873 | .859 | .844 | .826 | .806 | .783 |
| | | | | | | | | .758 | .729 | .697 | .661 | .622 | .579 | .532 | .483 | .432 | .379 |
| | | | | | | | | .327 | | | | | | | | | |
| 16.000 | 15.066 | .000 | .934 | .72 | 3.16 | 5.8% | 22 | .914 | .905 | .896 | .885 | .873 | .859 | .844 | .826 | .806 | .783 |
| | | | | | | | | .758 | .729 | .697 | .661 | .622 | .579 | .532 | .483 | .432 | .379 |
| | | | | | | | | .327 | .276 | | | | | | | | |
| 16.000 | 15.294 | .000 | .706 | .77 | 3.13 | 4.4% | 23 | .914 | .905 | .896 | .885 | .873 | .859 | .844 | .826 | .806 | .783 |
| | | | | | | | | .758 | .729 | .697 | .661 | .622 | .579 | .532 | .483 | .432 | .379 |
| | | | | | | | | .327 | .276 | .228 | | | | | | | |
| 16.000 | 15.478 | .000 | .522 | .82 | 3.09 | 3.3% | 24 | .914 | .905 | .896 | .885 | .873 | .859 | .844 | .826 | .806 | .783 |
| | | | | | | | | .758 | .729 | .697 | .661 | .622 | .579 | .532 | .483 | .432 | .379 |
| | | | | | | | | .327 | .276 | .228 | .184 | | | | | | |
| 16.000 | 15.623 | .000 | .377 | .86 | 3.03 | 2.4% | 25 | .914 | .905 | .896 | .885 | .873 | .859 | .844 | .826 | .806 | .783 |
| | | | | | | | | .758 | .729 | .697 | .661 | .622 | .579 | .532 | .483 | .432 | .379 |
| | | | | | | | | .327 | .276 | .228 | .184 | .145 | | | | | |
| 16.000 | 15.733 | .000 | .267 | .89 | 2.96 | 1.7% | 26 | .914 | .905 | .896 | .885 | .873 | .859 | .844 | .826 | .806 | .783 |
| | | | | | | | | .758 | .729 | .697 | .661 | .622 | .579 | .532 | .483 | .432 | .379 |
| | | | | | | | | .327 | .276 | .228 | .184 | .145 | .111 | | | | |
| 16.000 | 15.816 | .000 | .184 | .92 | 2.88 | 1.1% | 27 | .914 | .905 | .896 | .885 | .873 | .859 | .844 | .826 | .806 | .783 |
| | | | | | | | | .758 | .729 | .697 | .661 | .622 | .579 | .532 | .483 | .432 | .379 |
| | | | | | | | | .327 | .276 | .228 | .184 | .145 | .111 | .083 | | | |
| 16.000 | 15.876 | .000 | .124 | .94 | 2.80 | .8% | 28 | .914 | .905 | .896 | .885 | .873 | .859 | .844 | .826 | .806 | .783 |
| | | | | | | | | .758 | .729 | .697 | .661 | .622 | .579 | .532 | .483 | .432 | .379 |

*SIMILAR TO A.T.&T. NEAL-WILKINSON TABLES(SEE SECTION 1)

## *EQUIVALENT RANDOM THEORY: VMR(VARIANCE TO MEAN RATIO)– 1.5 - NO RETRIES (PEAKED TRAFFIC - ROUTE ADVANCEMENT OR DAY-TO-DAY VARIATION)

| HOURS FIRST ATTEMPT VMR = 1.5 | HOURS CON-NECT | HOURS RECALL /RETRY | HOURS OVFL | CON-NECT VMR | OVFL VMR | PERCENT BLOCKED (GRADE) | LINE | HOURS CONNECTED TRAFFIC ON LINES 1 | 2 | 3 | 4 | 5 | 6 | 7 | 8 | 9 | 10 |
|---|---|---|---|---|---|---|---|---|---|---|---|---|---|---|---|---|---|
| | | | | | | | | .327 | .276 | .228 | .184 | .145 | .111 | .083 | .060 | | |
| 16.000 | 15.918 | .000 | .082 | .96 | 2.72 | .5% | 29 | .914 | .905 | .896 | .885 | .873 | .859 | .844 | .826 | .806 | .783 |
| | | | | | | | | .758 | .729 | .697 | .661 | .622 | .579 | .532 | .483 | .432 | .379 |
| | | | | | | | | .327 | .276 | .228 | .184 | .145 | .111 | .083 | .060 | .042 | |
| 17.000 | 11.675 | .000 | 5.325 | .30 | 2.72 | 31.3% | 14 | .919 | .912 | .903 | .894 | .883 | .871 | .857 | .842 | .824 | .804 |
| | | | | | | | | .782 | .757 | .729 | .697 | | | | | | |
| 17.000 | 12.337 | .000 | 4.663 | .34 | 2.82 | 27.4% | 15 | .919 | .912 | .903 | .894 | .883 | .871 | .857 | .842 | .824 | .804 |
| | | | | | | | | .782 | .757 | .729 | .697 | .662 | | | | | |
| 17.000 | 12.960 | .000 | 4.040 | .38 | 2.91 | 23.8% | 16 | .919 | .912 | .903 | .894 | .883 | .871 | .857 | .842 | .824 | .804 |
| | | | | | | | | .782 | .757 | .729 | .697 | .662 | .623 | | | | |
| 17.000 | 13.541 | .000 | 3.459 | .42 | 3.00 | 20.3% | 17 | .919 | .912 | .903 | .894 | .883 | .871 | .857 | .842 | .824 | .804 |
| | | | | | | | | .782 | .757 | .729 | .697 | .662 | .623 | .581 | | | |
| 17.000 | 14.077 | .000 | 2.923 | .46 | 3.07 | 17.2% | 18 | .919 | .912 | .903 | .894 | .883 | .871 | .857 | .842 | .824 | .804 |
| | | | | | | | | .782 | .757 | .729 | .697 | .662 | .623 | .581 | .536 | | |
| 17.000 | 14.564 | .000 | 2.436 | .51 | 3.13 | 14.3% | 19 | .919 | .912 | .903 | .894 | .883 | .871 | .857 | .842 | .824 | .804 |
| | | | | | | | | .782 | .757 | .729 | .697 | .662 | .623 | .581 | .536 | .488 | |
| 17.000 | 15.002 | .000 | 1.998 | .56 | 3.18 | 11.8% | 20 | .919 | .912 | .903 | .894 | .883 | .871 | .857 | .842 | .824 | .804 |
| | | | | | | | | .782 | .757 | .729 | .697 | .662 | .623 | .581 | .536 | .488 | .437 |
| 17.000 | 15.388 | .000 | 1.612 | .61 | 3.21 | 9.5% | 21 | .919 | .912 | .903 | .894 | .883 | .871 | .857 | .842 | .824 | .804 |
| | | | | | | | | .782 | .757 | .903 | | | | .386 | | | |
| 17.000 | 15.722 | .000 | 1.278 | .67 | 3.22 | 7.5% | 22 | .919 | .912 | .903 | .894 | .883 | .871 | .857 | .842 | .824 | .804 |
| | | | | | | | | .782 | .757 | .729 | .697 | .662 | .623 | .581 | .536 | .488 | .437 |
| | | | | | | | | .386 | .334 | | | | | | | | |
| 17.000 | 16.006 | .000 | .994 | .72 | 3.21 | 5.8% | 23 | .919 | .912 | .903 | .894 | .883 | .871 | .857 | .842 | .824 | .804 |
| | | | | | | | | .782 | .757 | .729 | .697 | .662 | .623 | .581 | .536 | .488 | .437 |
| | | | | | | | | .386 | .334 | .284 | | | | | | | |
| 17.000 | 16.243 | .000 | .757 | .76 | 3.18 | 4.5% | 24 | .919 | .912 | .903 | .894 | .883 | .871 | .857 | .842 | .824 | .804 |
| | | | | | | | | .782 | .757 | .729 | .697 | .662 | .623 | .581 | .536 | .488 | .437 |
| | | | | | | | | .386 | .334 | .284 | .236 | | | | | | |
| 17.000 | 16.435 | .000 | .565 | .81 | 3.14 | 3.3% | 25 | .919 | .912 | .903 | .894 | .883 | .871 | .857 | .842 | .824 | .804 |
| | | | | | | | | .782 | .757 | .729 | .697 | .662 | .623 | .581 | .536 | .488 | .437 |
| | | | | | | | | .386 | .334 | .284 | .236 | .192 | | | | | |
| 17.000 | 16.587 | .000 | .413 | .85 | 3.09 | 2.4% | 26 | .919 | .912 | .903 | .894 | .883 | .871 | .857 | .842 | .824 | .804 |
| | | | | | | | | .782 | .757 | .729 | .697 | .662 | .623 | .581 | .536 | .488 | .437 |
| | | | | | | | | .306 | .334 | .284 | .236 | .192 | .152 | | | | |
| 17.000 | 16.705 | .000 | .295 | .88 | 3.02 | 1.7% | 27 | .919 | .912 | .903 | .894 | .883 | .871 | .857 | .842 | .824 | .804 |
| | | | | | | | | .782 | .757 | .729 | .697 | .662 | .623 | .581 | .536 | .488 | .437 |
| | | | | | | | | .386 | .334 | .284 | .236 | .192 | .152 | .118 | | | |
| 17.000 | 16.794 | .000 | .206 | .91 | 2.94 | 1.2% | 28 | .919 | .912 | .903 | .894 | .883 | .871 | .857 | .842 | .824 | .804 |

*SIMILAR TO A.T.&T. NEAL-WILKINSON TABLES(SEE SECTION 1)

*EQUIVALENT RANDOM THEORY: VMR(VARIANCE TO MEAN RATIO)– 1.5 - NO RETRIES (PEAKED TRAFFIC - ROUTE ADVANCEMENT OR DAY-TO-DAY VARIATION)

| HOURS FIRST ATTEMPT VMR = 1.5 | HOURS CON-NECT | HOURS RECALL /RETRY | HOURS OVFL | CON-NECT VMR | OVFL VMR | PERCENT BLOCKED (GRADE) | LINE | HOURS CONNECTED TRAFFIC ON LINES 1 | 2 | 3 | 4 | 5 | 6 | 7 | 8 | 9 | 10 |
|---|---|---|---|---|---|---|---|---|---|---|---|---|---|---|---|---|---|
| | | | | | | | | .782 | .757 | .729 | .697 | .662 | .623 | .581 | .536 | .488 | .437 |
| | | | | | | | | .386 | .334 | .284 | .236 | .192 | .152 | .118 | .089 | | |
| 17.000 | 16.860 | .000 | .140 | .93 | 2.86 | .8% | 29 | .919 | .912 | .903 | .894 | .883 | .871 | .857 | .842 | .824 | .804 |
| | | | | | | | | .782 | .757 | .729 | .697 | .662 | .623 | .581 | .536 | .488 | .437 |
| | | | | | | | | .386 | .334 | .284 | .236 | .192 | .152 | .118 | .089 | .065 | |
| 17.000 | 16.906 | .000 | .094 | .95 | 2.78 | .6% | 30 | .919 | .912 | .903 | .894 | .883 | .871 | .857 | .842 | .824 | .804 |
| | | | | | | | | .782 | .757 | .729 | .697 | .662 | .623 | .581 | .536 | .488 | .437 |
| | | | | | | | | .386 | .334 | .284 | .236 | .192 | .152 | .118 | .089 | .065 | .047 |
| 18.000 | 11.847 | .000 | 6.153 | .27 | 2.71 | 34.2% | 14 | .922 | .916 | .908 | .899 | .890 | .879 | .867 | .853 | .837 | .820 |
| | | | | | | | | .800 | .778 | .753 | .725 | | | | | | |
| 18.000 | 12.541 | .000 | 5.459 | .31 | 2.81 | 30.3% | 15 | .922 | .916 | .908 | .899 | .890 | .879 | .867 | .853 | .837 | .820 |
| | | | | | | | | .800 | .778 | .753 | .725 | .694 | | | | | |
| 18.000 | 13.202 | .000 | 4.798 | .34 | 2.91 | 26.7% | 16 | .922 | .916 | .908 | .899 | .890 | .879 | .867 | .853 | .837 | .820 |
| | | | | | | | | .800 | .778 | .753 | .725 | .694 | .660 | | | | |
| 18.000 | 13.825 | .000 | 4.175 | .38 | 3.00 | 23.2% | 17 | .922 | .916 | .908 | .899 | .890 | .879 | .867 | .853 | .837 | .820 |
| | | | | | | | | .800 | .778 | .753 | .725 | .694 | .660 | .623 | | | |
| 18.000 | 14.407 | .000 | 3.593 | .42 | 3.08 | 20.0% | 18 | .922 | .916 | .908 | .899 | .890 | .879 | .867 | .853 | .837 | .820 |
| | | | | | | | | .800 | .778 | .753 | .725 | .694 | .660 | .623 | .582 | | |
| 18.000 | 14.945 | .000 | 3.055 | .46 | 3.16 | 17.0% | 19 | .922 | .916 | .908 | .899 | .890 | .879 | .867 | .853 | .837 | .820 |
| | | | | | | | | .800 | .778 | .753 | .725 | .694 | .660 | .623 | .582 | .538 | |
| 18.000 | 15.437 | .000 | 2.563 | .51 | 3.22 | 14.2% | 20 | .922 | .916 | .908 | .899 | .890 | .879 | .867 | .853 | .837 | .820 |
| | | | | | | | | .800 | .778 | .753 | .725 | .694 | .660 | .623 | .582 | .538 | .492 |
| 18.000 | 15.880 | .000 | 2.120 | .56 | 3.26 | 11.8% | 21 | .922 | .916 | .908 | .899 | .890 | .879 | .867 | .853 | .837 | .820 |
| | | | | | | | | .800 | .778 | .753 | .725 | .694 | .660 | .623 | .582 | .538 | .492 |
| | | | | | | | | .443 | | | | | | | | | |
| 18.000 | 16.274 | .000 | 1.726 | .61 | 3.29 | 9.0% | 22 | .922 | .916 | .908 | .899 | .890 | .879 | .867 | .853 | .837 | .820 |
| | | | | | | | | .800 | .778 | .753 | .725 | .694 | .660 | .623 | .582 | .538 | .492 |
| | | | | | | | | .443 | .394 | | | | | | | | |
| 18.000 | 16.618 | .000 | 1.382 | .66 | 3.31 | 7.7% | 23 | .922 | .916 | .908 | .899 | .890 | .879 | .867 | .853 | .837 | .820 |
| | | | | | | | | .800 | .778 | .753 | .725 | .694 | .660 | .623 | .582 | .538 | .492 |
| | | | | | | | | .443 | .394 | .344 | | | | | | | |
| 18.000 | 16.912 | .000 | 1.088 | .71 | 3.30 | 6.0% | 24 | .922 | .916 | .908 | .899 | .890 | .879 | .867 | .853 | .837 | .820 |
| | | | | | | | | .800 | .778 | .753 | .725 | .694 | .660 | .623 | .582 | .538 | .492 |
| | | | | | | | | .443 | .394 | .344 | .295 | | | | | | |
| 18.000 | 17.160 | .000 | .840 | .75 | 3.28 | 4.7% | 25 | .922 | .916 | .908 | .899 | .890 | .879 | .867 | .853 | .837 | .820 |
| | | | | | | | | .800 | .778 | .753 | .725 | .694 | .660 | .623 | .582 | .538 | .492 |
| | | | | | | | | .443 | .394 | .344 | .295 | .248 | | | | | |
| 18.000 | 17.364 | .000 | .636 | .80 | 3.24 | 3.5% | 26 | .922 | .916 | .908 | .899 | .890 | .879 | .867 | .853 | .837 | .820 |
| | | | | | | | | .800 | .778 | .753 | .725 | .694 | .660 | .623 | .582 | .538 | .492 |
| | | | | | | | | .443 | .394 | .344 | .295 | .248 | .204 | | | | |

*SIMILAR TO A.T.&T. NEAL-WILKINSON TABLES(SEE SECTION 1)

*EQUIVALENT RANDOM THEORY: VMR(VARIANCE TO MEAN RATIO)– 1.5 - NO RETRIES (PEAKED TRAFFIC - ROUTE ADVANCEMENT OR DAY-TO-DAY VARIATION)

| HOURS FIRST ATTEMPT VMR = 1.5 | HOURS CON-NECT | HOURS RECALL /RETRY | HOURS OVFL | CON-NECT VMR | OVFL VMR | PERCENT BLOCKED (GRADE) | LINE | HOURS CONNECTED TRAFFIC ON LINES 1 | 2 | 3 | 4 | 5 | 6 | 7 | 8 | 9 | 10 |
|---|---|---|---|---|---|---|---|---|---|---|---|---|---|---|---|---|---|
| 18.000 | 17.528 | .000 | .472 | .84 | 3.19 | 2.6% | 27 | .922 | .916 | .908 | .899 | .890 | .879 | .867 | .853 | .837 | .820 |
| | | | | | | | | .800 | .778 | .753 | .725 | .694 | .660 | .623 | .582 | .538 | .492 |
| | | | | | | | | .443 | .394 | .344 | .295 | .248 | .204 | .164 | | | |
| 18.000 | 17.657 | .000 | .343 | .87 | 3.12 | 1.9% | 28 | .922 | .916 | .908 | .899 | .890 | .879 | .867 | .853 | .837 | .820 |
| | | | | | | | | .800 | .778 | .753 | .725 | .694 | .660 | .623 | .582 | .538 | .492 |
| | | | | | | | | .443 | .394 | .344 | .295 | .248 | .204 | .164 | .129 | | |
| 18.000 | 17.756 | .000 | .244 | .90 | 3.05 | 1.4% | 29 | .922 | .916 | .908 | .899 | .890 | .879 | .867 | .853 | .837 | .820 |
| | | | | | | | | .800 | .778 | .753 | .725 | .694 | .660 | .623 | .582 | .538 | .492 |
| | | | | | | | | .443 | .394 | .344 | .295 | .248 | .204 | .164 | .129 | .099 | |
| 18.000 | 17.830 | .000 | .170 | .93 | 2.97 | .9% | 30 | .922 | .916 | .908 | .899 | .890 | .879 | .867 | .853 | .837 | .820 |
| | | | | | | | | .800 | .778 | .753 | .725 | .694 | .660 | .623 | .582 | .538 | .492 |
| | | | | | | | | .443 | .394 | .344 | .295 | .248 | .204 | .164 | .129 | .099 | .074 |
| 18.000 | 17.884 | .000 | .116 | .95 | 2.88 | .6% | 31 | .922 | .916 | .908 | .899 | .890 | .879 | .867 | .853 | .837 | .820 |
| | | | | | | | | .800 | .778 | .753 | .725 | .694 | .660 | .623 | .582 | .538 | .492 |
| | | | | | | | | .443 | .394 | .344 | .295 | .248 | .204 | .164 | .129 | .099 | .074 |
| | | | | | | | | .054 | | | | | | | | | |
| 19.000 | 12.754 | .000 | 6.246 | .28 | 2.75 | 32.9% | 15 | .927 | .921 | .914 | .906 | .898 | .888 | .877 | .865 | .851 | .836 |
| | | | | | | | | .818 | .799 | .777 | .752 | .725 | | | | | |
| 19.000 | 13.448 | .000 | 5.552 | .31 | 2.85 | 29.2% | 16 | .927 | .921 | .914 | .906 | .898 | .888 | .877 | .865 | .851 | .836 |
| | | | | | | | | .818 | .799 | .777 | .752 | .725 | .694 | | | | |
| 19.000 | 14.109 | .000 | 4.891 | .34 | 2.95 | 25.7% | 17 | .927 | .921 | .914 | .906 | .898 | .888 | .877 | .865 | .851 | .836 |
| | | | | | | | | .818 | .799 | .777 | .752 | .725 | .694 | .661 | | | |
| 19.000 | 14.733 | .000 | 4.267 | .38 | 3.04 | 22.5% | 18 | .927 | .921 | .914 | .906 | .898 | .888 | .877 | .865 | .851 | .836 |
| | | | | | | | | .818 | .799 | .777 | .752 | .725 | .694 | .661 | .624 | | |
| 19.000 | 15.317 | .000 | 3.683 | .42 | 3.12 | 19.4% | 19 | .927 | .921 | .914 | .906 | .898 | .888 | .877 | .865 | .851 | .836 |
| | | | | | | | | .818 | .799 | .777 | .752 | .725 | .694 | .661 | .624 | .584 | |
| 19.000 | 15.859 | .000 | 3.141 | .46 | 3.20 | 16.5% | 20 | .927 | .921 | .914 | .906 | .898 | .888 | .877 | .865 | .851 | .836 |
| | | | | | | | | .818 | .799 | .777 | .752 | .725 | .694 | .661 | .624 | .584 | .541 |
| 19.000 | 16.354 | .000 | 2.646 | .50 | 3.26 | 13.9% | 21 | .927 | .921 | .914 | .906 | .898 | .888 | .877 | .865 | .851 | .836 |
| | | | | | | | | .818 | .799 | .777 | .752 | .725 | .694 | .661 | .624 | .584 | .541 |
| | | | | | | | | .496 | | | | | | | | | |
| 19.000 | 16.802 | .000 | 2.198 | .55 | 3.31 | 11.6% | 22 | .927 | .921 | .914 | .906 | .898 | .888 | .877 | .865 | .851 | .836 |
| | | | | | | | | .818 | .799 | .777 | .752 | .725 | .694 | .661 | .624 | .584 | .541 |
| | | | | | | | | .496 | .448 | | | | | | | | |
| 19.000 | 17.202 | .000 | 1.798 | .60 | 3.34 | 9.5% | 23 | .927 | .921 | .914 | .906 | .898 | .888 | .877 | .865 | .851 | .836 |
| | | | | | | | | .818 | .799 | .777 | .752 | .725 | .694 | .661 | .624 | .584 | .541 |
| | | | | | | | | .496 | .448 | .399 | | | | | | | |
| 19.000 | 17.552 | .000 | 1.448 | .65 | 3.35 | 7.6% | 24 | .927 | .921 | .914 | .906 | .898 | .888 | .877 | .865 | .851 | .836 |
| | | | | | | | | .818 | .799 | .777 | .752 | .725 | .694 | .661 | .624 | .584 | .541 |
| | | | | | | | | .496 | .448 | .399 | .350 | | | | | | |

*SIMILAR TO A.T.&T. NEAL-WILKINSON TABLES(SEE SECTION 1)

*EQUIVALENT RANDOM THEORY: VMR(VARIANCE TO MEAN RATIO)– 1.5 - NO RETRIES (PEAKED TRAFFIC - ROUTE ADVANCEMENT OR DAY-TO-DAY VARIATION)

| HOURS FIRST ATTEMPT VMR = 1.5 | HOURS CON-NECT | HOURS RECALL /RETRY | HOURS OVFL | CON-NECT VMR | OVFL VMR | PERCENT BLOCKED (GRADE) | LINE | HOURS CONNECTED TRAFFIC ON LINES 1 | 2 | 3 | 4 | 5 | 6 | 7 | 8 | 9 | 10 |
|---|---|---|---|---|---|---|---|---|---|---|---|---|---|---|---|---|---|
| 19.000 | 17.854 | .000 | 1.146 | .70 | 3.35 | 6.0% | 25 | .927 | .921 | .914 | .906 | .898 | .888 | .877 | .865 | .851 | .836 |
| | | | | | | | | .818 | .799 | .777 | .752 | .725 | .694 | .661 | .624 | .584 | .541 |
| | | | | | | | | .496 | .448 | .399 | .350 | .302 | | | | | |
| 19.000 | 18.109 | .000 | .891 | .74 | 3.33 | 4.7% | 26 | .927 | .921 | .914 | .906 | .898 | .888 | .877 | .865 | .851 | .836 |
| | | | | | | | | .818 | .799 | .777 | .752 | .725 | .694 | .661 | .624 | .584 | .541 |
| | | | | | | | | .496 | .448 | .399 | .350 | .302 | .255 | | | | |
| 19.000 | 18.320 | .000 | .680 | .79 | 3.29 | 3.6% | 27 | .927 | .921 | .914 | .906 | .898 | .888 | .877 | .865 | .851 | .836 |
| | | | | | | | | .818 | .799 | .777 | .752 | .725 | .694 | .661 | .624 | .584 | .541 |
| | | | | | | | | .496 | .448 | .399 | .350 | .302 | .255 | .211 | | | |
| 19.000 | 18.491 | .000 | .509 | .83 | 3.24 | 2.7% | 28 | .927 | .921 | .914 | .906 | .898 | .888 | .877 | .865 | .851 | .836 |
| | | | | | | | | .818 | .799 | .777 | .752 | .725 | .694 | .661 | .624 | .584 | .541 |
| | | | | | | | | .496 | .448 | .399 | .350 | .302 | .255 | .211 | .171 | | |
| 19.000 | 18.627 | .000 | .373 | .86 | 3.18 | 2.0% | 29 | .927 | .921 | .914 | .906 | .898 | .888 | .877 | .865 | .851 | .836 |
| | | | | | | | | .818 | .799 | .777 | .752 | .725 | .694 | .661 | .624 | .584 | .541 |
| | | | | | | | | .496 | .448 | .399 | .350 | .302 | .255 | .211 | .171 | .136 | |
| 19.000 | 18.732 | .000 | .268 | .89 | 3.10 | 1.4% | 30 | .927 | .921 | .914 | .906 | .898 | .888 | .877 | .865 | .851 | .836 |
| | | | | | | | | .818 | .799 | .777 | .752 | .725 | .694 | .661 | .624 | .584 | .541 |
| | | | | | | | | .496 | .448 | .399 | .350 | .302 | .255 | .211 | .171 | .136 | .105 |
| 19.000 | 18.811 | .000 | .189 | .92 | 3.03 | 1.0% | 31 | .927 | .921 | .914 | .906 | .898 | .888 | .877 | .865 | .851 | .836 |
| | | | | | | | | .818 | .799 | .777 | .752 | .725 | .694 | .661 | .624 | .584 | .541 |
| | | | | | | | | .496 | .448 | .399 | .350 | .302 | .255 | .211 | .171 | .136 | .105 |
| | | | | | | | | .079 | | | | | | | | | |
| 19.000 | 18.870 | .000 | .130 | .94 | 2.94 | .7% | 32 | .927 | .921 | .914 | .906 | .898 | .888 | .877 | .865 | .851 | .836 |
| | | | | | | | | .818 | .799 | .777 | .752 | .725 | .694 | .661 | .624 | .584 | .541 |
| | | | | | | | | .496 | .448 | .399 | .350 | .302 | .255 | .211 | .171 | .136 | .105 |
| | | | | | | | | .079 | .059 | | | | | | | | |
| 20.000 | 13.629 | .000 | 6.371 | .28 | 2.83 | 31.9% | 16 | .930 | .924 | .918 | .911 | .903 | .894 | .884 | .873 | .861 | .847 |
| | | | | | | | | .832 | .814 | .795 | .773 | .749 | .722 | | | | |
| 20.000 | 14.321 | .000 | 5.679 | .31 | 2.93 | 28.4% | 17 | .930 | .924 | .918 | .911 | .903 | .894 | .884 | .873 | .861 | .847 |
| | | | | | | | | .832 | .814 | .795 | .773 | .749 | .722 | .692 | | | |
| 20.000 | 14.980 | .000 | 5.020 | .34 | 3.03 | 25.1% | 18 | .930 | .924 | .918 | .911 | .903 | .894 | .884 | .873 | .861 | .847 |
| | | | | | | | | .832 | .814 | .705 | .773 | .749 | .722 | .692 | .660 | | |
| 20.000 | 15.604 | .000 | 4.306 | .38 | 3.12 | 22.0% | 19 | .930 | .924 | .918 | .911 | .903 | .894 | .884 | .873 | .861 | .847 |
| | | | | | | | | .832 | .814 | .795 | .773 | .749 | .722 | .092 | .660 | .624 | |
| 20.000 | 16.180 | .000 | 3.811 | .42 | 3.20 | 19.1% | 20 | .930 | .924 | .918 | .911 | .903 | .894 | .884 | .873 | .861 | .847 |
| | | | | | | | | .832 | .814 | .795 | .773 | .749 | .722 | .692 | .660 | .624 | .585 |
| 20.000 | 16.733 | .000 | 3.267 | .46 | 3.28 | 16.3% | 21 | .930 | .924 | .918 | .911 | .903 | .894 | .884 | .873 | .861 | .847 |
| | | | | | | | | .832 | .814 | .795 | .773 | .749 | .722 | .692 | .660 | .624 | .585 |
| | | | | | | | | .543 | | | | | | | | | |
| 20.000 | 17.232 | .000 | 2.768 | .50 | 3.34 | 13.8% | 22 | .930 | .924 | .918 | .911 | .903 | .894 | .884 | .873 | .861 | .847 |

*SIMILAR TO A.T.&T. NEAL-WILKINSON TABLES(SEE SECTION 1)

*EQUIVALENT RANDOM THEORY: VMR(VARIANCE TO MEAN RATIO)– 1.5 - NO RETRIES (PEAKED TRAFFIC - ROUTE ADVANCEMENT OR DAY-TO-DAY VARIATION)

| HOURS FIRST ATTEMPT VMR=1.5 | HOURS CON-NECT | HOURS RECALL /RETRY | HOURS OVFL | CON-NECT VMR | OVFL VMR | PERCENT BLOCKED (GRADE) | LINE | HOURS CONNECTED TRAFFIC ON LINES 1 | 2 | 3 | 4 | 5 | 6 | 7 | 8 | 9 | 10 |
|---|---|---|---|---|---|---|---|---|---|---|---|---|---|---|---|---|---|
| | | | | | | | | .832 | .814 | .795 | .773 | .749 | .722 | .692 | .660 | .624 | .585 |
| | | | | | | | | .543 | .499 | | | | | | | | |
| 20.000 | 17.685 | .000 | 2.315 | .55 | 3.39 | 11.6% | 23 | .930 | .924 | .918 | .911 | .903 | .894 | .884 | .873 | .861 | .847 |
| | | | | | | | | .832 | .814 | .795 | .773 | .749 | .722 | .692 | .660 | .624 | .585 |
| | | | | | | | | .543 | .499 | .453 | | | | | | | |
| 20.000 | 18.091 | .000 | 1.909 | .59 | 3.42 | 9.5% | 24 | .930 | .924 | .918 | .911 | .903 | .894 | .884 | .873 | .861 | .847 |
| | | | | | | | | .832 | .814 | .795 | .773 | .749 | .722 | .692 | .660 | .624 | .585 |
| | | | | | | | | .543 | .499 | .453 | .406 | | | | | | |
| 20.000 | 18.449 | .000 | 1.551 | .64 | 3.44 | 7.8% | 25 | .930 | .924 | .918 | .911 | .903 | .894 | .884 | .873 | .861 | .847 |
| | | | | | | | | .832 | .814 | .795 | .773 | .749 | .722 | .692 | .660 | .624 | .585 |
| | | | | | | | | .543 | .499 | .453 | .406 | .358 | | | | | |
| 20.000 | 18.760 | .000 | 1.240 | .69 | 3.44 | 6.2% | 26 | .930 | .924 | .918 | .911 | .903 | .894 | .884 | .873 | .861 | .847 |
| | | | | | | | | .832 | .814 | .795 | .773 | .749 | .722 | .692 | .660 | .624 | .585 |
| | | | | | | | | .543 | .499 | .453 | .406 | .358 | .311 | | | | |
| 20.000 | 19.025 | .000 | .975 | .73 | 3.42 | 4.9% | 27 | .930 | .924 | .918 | .911 | .903 | .894 | .884 | .873 | .861 | .847 |
| | | | | | | | | .832 | .814 | .795 | .773 | .749 | .722 | .692 | .660 | .624 | .585 |
| | | | | | | | | .543 | .499 | .453 | .406 | .358 | .311 | .265 | | | |
| 20.000 | 19.247 | .000 | .753 | .78 | 3.38 | 3.8% | 28 | .930 | .924 | .918 | .911 | .903 | .894 | .884 | .873 | .861 | .847 |
| | | | | | | | | .832 | .814 | .795 | .773 | .749 | .722 | .692 | .660 | .624 | .585 |
| | | | | | | | | .543 | .499 | .453 | .406 | .358 | .311 | .265 | .222 | | |
| 20.000 | 19.429 | .000 | .571 | .82 | 3.34 | 2.9% | 29 | .930 | .924 | .918 | .911 | .903 | .894 | .884 | .873 | .861 | .847 |
| | | | | | | | | .832 | .814 | .795 | .773 | .749 | .722 | .692 | .660 | .624 | .585 |
| | | | | | | | | .543 | .499 | .453 | .406 | .358 | .311 | .265 | .222 | .182 | |
| 20.000 | 19.575 | .000 | .425 | .85 | 3.28 | 2.1% | 30 | .930 | .924 | .918 | .911 | .903 | .894 | .884 | .873 | .861 | .847 |
| | | | | | | | | .832 | .814 | .795 | .773 | .749 | .722 | .692 | .660 | .624 | .585 |
| | | | | | | | | .543 | .499 | .453 | .406 | .358 | .311 | .265 | .222 | .182 | .146 |
| 20.000 | 19.690 | .000 | .310 | .89 | 3.21 | 1.5% | 31 | .930 | .924 | .918 | .911 | .903 | .894 | .884 | .873 | .861 | .847 |
| | | | | | | | | .832 | .814 | .795 | .773 | .749 | .722 | .692 | .660 | .624 | .585 |
| | | | | | | | | .543 | .499 | .453 | .406 | .358 | .311 | .265 | .222 | .182 | .146 |
| | | | | | | | | .115 | | | | | | | | | |
| 20.000 | 19.778 | .000 | .222 | .91 | 3.13 | 1.1% | 32 | .930 | .924 | .918 | .911 | .903 | .894 | .884 | .873 | .861 | .847 |
| | | | | | | | | .832 | .814 | .795 | .773 | .749 | .722 | .692 | .660 | .624 | .585 |
| | | | | | | | | .543 | .499 | .453 | .406 | .358 | .311 | .265 | .222 | .182 | .146 |
| | | | | | | | | .115 | .088 | | | | | | | | |
| 20.000 | 19.844 | .000 | .156 | .93 | 3.04 | .8% | 33 | .930 | .924 | .918 | .911 | .903 | .894 | .884 | .873 | .861 | .847 |
| | | | | | | | | .832 | .814 | .795 | .773 | .749 | .722 | .692 | .660 | .624 | .585 |
| | | | | | | | | .543 | .499 | .453 | .406 | .358 | .311 | .265 | .222 | .182 | .146 |
| | | | | | | | | .115 | .088 | .066 | | | | | | | |
| 20.000 | 19.893 | .000 | .107 | .95 | 2.96 | .5% | 34 | .930 | .924 | .918 | .911 | .903 | .894 | .884 | .873 | .861 | .847 |
| | | | | | | | | .832 | .814 | .795 | .773 | .749 | .722 | .692 | .660 | .624 | .585 |

*SIMILAR TO A.T.&T. NEAL-WILKINSON TABLES(SEE SECTION 1)

*EQUIVALENT RANDOM THEORY: VMR(VARIANCE TO MEAN RATIO)– 1.5 - NO RETRIES (PEAKED TRAFFIC - ROUTE ADVANCEMENT OR DAY-TO-DAY VARIATION)

| HOURS FIRST ATTEMPT VMR = 1.5 | HOURS CON-NECT | HOURS RECALL /RETRY | HOURS OVFL | CON-NECT VMR | OVFL VMR | PERCENT BLOCKED (GRADE) | LINE | HOURS CONNECTED TRAFFIC ON LINES 1 | 2 | 3 | 4 | 5 | 6 | 7 | 8 | 9 | 10 |
|---|---|---|---|---|---|---|---|---|---|---|---|---|---|---|---|---|---|
| | | | | | | | | .543 | .499 | .453 | .406 | .358 | .311 | .265 | .222 | .182 | .146 |
| | | | | | | | | .115 | .088 | .066 | .049 | | | | | | |
| 22.000 | 14.702 | .000 | 7.298 | .25 | 2.85 | 33.2% | 17 | .936 | .931 | .926 | .920 | .913 | .906 | .898 | .889 | .879 | .868 |
| | | | | | | | | .856 | .842 | .826 | .809 | .790 | .769 | .745 | | | |
| 22.000 | 15.421 | .000 | 6.579 | .28 | 2.95 | 29.9% | 18 | .986 | .931 | .926 | .920 | .913 | .906 | .898 | .889 | .879 | .868 |
| | | | | | | | | .856 | .842 | .826 | .809 | .790 | .769 | .745 | .719 | | |
| 22.000 | 16.111 | .000 | 5.889 | .31 | 3.05 | 26.8% | 19 | .936 | .931 | .926 | .920 | .913 | .906 | .898 | .889 | .879 | .868 |
| | | | | | | | | .856 | .842 | .826 | .809 | .790 | .769 | .745 | .719 | .690 | |
| 22.000 | 16.770 | .000 | 5.230 | .34 | 3.14 | 23.8% | 20 | .936 | .931 | .926 | .920 | .913 | .906 | .898 | .889 | .879 | .868 |
| | | | | | | | | .856 | .842 | .826 | .809 | .790 | .769 | .745 | .719 | .690 | .659 |
| 22.000 | 17.394 | .000 | 4.606 | .38 | 3.23 | 20.9% | 21 | .936 | .931 | .926 | .920 | .913 | .906 | .898 | .889 | .879 | .868 |
| | | | | | | | | .856 | .842 | .826 | .809 | .790 | .769 | .745 | .719 | .690 | .659 |
| | | | | | | | | .625 | | | | | | | | | |
| 22.000 | 17.982 | .000 | 4.018 | .41 | 3.32 | 18.3% | 22 | .936 | .931 | .926 | .920 | .913 | .906 | .898 | .889 | .879 | .868 |
| | | | | | | | | .856 | .842 | .826 | .809 | .790 | .769 | .745 | .719 | .690 | .659 |
| | | | | | | | | .625 | .587 | | | | | | | | |
| 22.000 | 18.530 | .000 | 3.470 | .45 | 3.39 | 15.8% | 23 | .936 | .931 | .926 | .920 | .913 | .906 | .898 | .889 | .879 | .868 |
| | | | | | | | | .856 | .842 | .826 | .809 | .790 | .769 | .745 | .719 | .690 | .659 |
| | | | | | | | | .625 | .587 | .548 | | | | | | | |
| 22.000 | 19.035 | .000 | 2.965 | .49 | 3.45 | 13.5% | 24 | .936 | .931 | .926 | .920 | .913 | .906 | .898 | .889 | .879 | .868 |
| | | | | | | | | .856 | .842 | .826 | .809 | .790 | .769 | .745 | .719 | .690 | .659 |
| | | | | | | | | .625 | .587 | .548 | .506 | | | | | | |
| 22.000 | 19.497 | .000 | 2.503 | .54 | 3.50 | 11.4% | 25 | .936 | .931 | .926 | .920 | .913 | .906 | .898 | .889 | .879 | .868 |
| | | | | | | | | .856 | .842 | .826 | .809 | .790 | .769 | .745 | .719 | .690 | .659 |
| | | | | | | | | .625 | .587 | .548 | .506 | .462 | | | | | |
| 22.000 | 19.914 | .000 | 2.086 | .58 | 3.54 | 9.5% | 26 | .936 | .931 | .926 | .920 | .913 | .906 | .898 | .889 | .879 | .868 |
| | | | | | | | | .856 | .842 | .826 | .809 | .790 | .769 | .745 | .719 | .690 | .659 |
| | | | | | | | | .625 | .587 | .548 | .506 | .462 | .417 | | | | |
| 22.000 | 20.285 | .000 | 1.715 | .63 | 3.56 | 7.8% | 27 | .936 | .931 | .926 | .920 | .913 | .906 | .898 | .889 | .879 | .868 |
| | | | | | | | | .856 | .842 | .826 | .809 | .790 | .769 | .745 | .719 | .690 | .659 |
| | | | | | | | | .625 | .587 | .548 | .506 | .462 | .417 | .371 | | | |
| 22.000 | 20.610 | .000 | 1.390 | .67 | 3.56 | 6.3% | 28 | .936 | .931 | .926 | .920 | .913 | .906 | .898 | .889 | .879 | .868 |
| | | | | | | | | .856 | .842 | .826 | .809 | .790 | .769 | .745 | .719 | .690 | .659 |
| | | | | | | | | .625 | .587 | .548 | .506 | .462 | .417 | .371 | .325 | | |
| 22.000 | 20.891 | .000 | 1.109 | .72 | 3.55 | 5.0% | 29 | .936 | .931 | .926 | .920 | .913 | .906 | .898 | .889 | .879 | .868 |
| | | | | | | | | .856 | .842 | .826 | .809 | .790 | .769 | .745 | .719 | .690 | .659 |
| | | | | | | | | .625 | .587 | .548 | .506 | .462 | .417 | .371 | .325 | .281 | |
| 22.000 | 21.129 | .000 | .871 | .76 | 3.52 | 4.0% | 30 | .936 | .931 | .926 | .920 | .913 | .906 | .898 | .889 | .879 | .868 |
| | | | | | | | | .856 | .842 | .826 | .809 | .790 | .769 | .745 | .719 | .690 | .659 |
| | | | | | | | | .625 | .587 | .548 | .506 | .462 | .417 | .371 | .325 | .281 | .238 |

*SIMILAR TO A.T.&T. NEAL-WILKINSON TABLES(SEE SECTION 1)

*EQUIVALENT RANDOM THEORY: VMR(VARIANCE TO MEAN RATIO)– 1.5 - NO RETRIES (PEAKED TRAFFIC - ROUTE ADVANCEMENT OR DAY-TO-DAY VARIATION)

| HOURS FIRST ATTEMPT VMR=1.5 | HOURS CON-NECT | HOURS RECALL /RETRY | HOURS OVFL | CON-NECT VMR | OVFL VMR | PERCENT BLOCKED (GRADE) | LINE | HOURS CONNECTED TRAFFIC ON LINES 1 | 2 | 3 | 4 | 5 | 6 | 7 | 8 | 9 | 10 |
|---|---|---|---|---|---|---|---|---|---|---|---|---|---|---|---|---|---|
| 22.000 | 21.327 | .000 | .673 | .80 | 3.48 | 3.1% | 31 | .936 | .931 | .926 | .920 | .913 | .906 | .898 | .889 | .879 | .868 |
| | | | | | | | | .856 | .842 | .826 | .809 | .790 | .769 | .745 | .719 | .690 | .659 |
| | | | | | | | | .625 | .587 | .548 | .506 | .462 | .417 | .371 | .325 | .281 | .238 |
| | | | | | | | | .198 | | | | | | | | | |
| 22.000 | 21.489 | .000 | .511 | .84 | 3.42 | 2.3% | 32 | .936 | .931 | .926 | .920 | .913 | .906 | .898 | .889 | .879 | .868 |
| | | | | | | | | .856 | .842 | .826 | .809 | .790 | .769 | .745 | .719 | .690 | .659 |
| | | | | | | | | .625 | .587 | .548 | .506 | .462 | .417 | .371 | .325 | .281 | .238 |
| | | | | | | | | .198 | .162 | | | | | | | | |
| 22.000 | 21.619 | .000 | .381 | .87 | 3.35 | 1.7% | 33 | .936 | .931 | .926 | .920 | .913 | .906 | .898 | .889 | .879 | .868 |
| | | | | | | | | .856 | .842 | .826 | .809 | .790 | .769 | .745 | .719 | .690 | .659 |
| | | | | | | | | .625 | .587 | .548 | .506 | .462 | .417 | .371 | .325 | .281 | .238 |
| | | | | | | | | .198 | .162 | .130 | | | | | | | |
| 22.000 | 21.721 | .000 | .279 | .90 | 3.28 | 1.3% | 34 | .936 | .931 | .926 | .920 | .913 | .906 | .898 | .889 | .879 | .868 |
| | | | | | | | | .856 | .842 | .826 | .809 | .790 | .769 | .745 | .719 | .690 | .659 |
| | | | | | | | | .625 | .587 | .548 | .506 | .462 | .417 | .371 | .325 | .281 | .238 |
| | | | | | | | | .198 | .162 | .130 | .102 | | | | | | |
| 22.000 | 21.800 | .000 | .200 | .92 | 3.20 | .9% | 35 | .936 | .931 | .926 | .920 | .913 | .906 | .898 | .889 | .879 | .868 |
| | | | | | | | | .856 | .842 | .826 | .809 | .790 | .769 | .745 | .719 | .690 | .659 |
| | | | | | | | | .625 | .587 | .548 | .506 | .462 | .417 | .371 | .325 | .281 | .238 |
| | | | | | | | | .198 | .162 | .130 | .102 | .078 | | | | | |
| 22.000 | 21.859 | .000 | .141 | .94 | 3.11 | .6% | 36 | .936 | .931 | .926 | .920 | .913 | .906 | .898 | .889 | .879 | .868 |
| | | | | | | | | .856 | .842 | .826 | .809 | .790 | .769 | .745 | .719 | .690 | .659 |
| | | | | | | | | .625 | .587 | .548 | .506 | .462 | .417 | .371 | .325 | .281 | .238 |
| | | | | | | | | .198 | .162 | .130 | .102 | .078 | .059 | | | | |
| 24.000 | 15.764 | .000 | 8.236 | .24 | 2.87 | 34.3% | 18 | .941 | .937 | .932 | .927 | .922 | .916 | .909 | .902 | .893 | .884 |
| | | | | | | | | .874 | .863 | .851 | .837 | .822 | .805 | .786 | .765 | | |
| 24.000 | 16.506 | .000 | 7.494 | .26 | 2.97 | 31.2% | 19 | .941 | .937 | .932 | .927 | .922 | .916 | .909 | .902 | .893 | .884 |
| | | | | | | | | .874 | .863 | .851 | .837 | .822 | .805 | .786 | .765 | .742 | |
| 24.000 | 17.222 | .000 | 6.778 | .28 | 3.07 | 28.2% | 20 | .941 | .937 | .932 | .927 | .922 | .916 | .909 | .902 | .893 | .884 |
| | | | | | | | | .874 | .863 | .851 | .837 | .822 | .805 | .786 | .765 | .742 | .716 |
| 24.000 | 17.911 | .000 | 6.089 | .31 | 3.16 | 25.4% | 21 | .941 | .937 | .932 | .927 | .922 | .916 | .909 | .902 | .893 | .884 |
| | | | | | | | | .874 | .863 | .851 | .837 | .822 | .805 | .786 | .765 | .742 | .716 |
| | | | | | | | | .689 | | | | | | | | | |
| 24.000 | 18.569 | .000 | 5.431 | .34 | 3.26 | 22.6% | 22 | .941 | .937 | .932 | .927 | .922 | .916 | .909 | .902 | .893 | .884 |
| | | | | | | | | .874 | .863 | .851 | .837 | .822 | .805 | .786 | .765 | .742 | .716 |
| | | | | | | | | .689 | .658 | | | | | | | | |
| 24.000 | 19.194 | .000 | 4.806 | .37 | 3.35 | 20.0% | 23 | .941 | .937 | .932 | .927 | .922 | .916 | .909 | .902 | .893 | .884 |
| | | | | | | | | .874 | .863 | .851 | .837 | .822 | .805 | .786 | .765 | .742 | .716 |
| | | | | | | | | .689 | .658 | .625 | | | | | | | |
| 24.000 | 19.784 | .000 | 4.216 | .41 | 3.43 | 17.6% | 24 | .941 | .937 | .932 | .927 | .922 | .916 | .909 | .902 | .893 | .884 |

*SIMILAR TO A.T.&T. NEAL-WILKINSON TABLES(SEE SECTION 1)

*EQUIVALENT RANDOM THEORY: VMR(VARIANCE TO MEAN RATIO)– 1.5 - NO RETRIES (PEAKED TRAFFIC - ROUTE ADVANCEMENT OR DAY-TO-DAY VARIATION)

| HOURS FIRST ATTEMPT VMR=1.5 | HOURS CON-NECT | HOURS RECALL /RETRY | HOURS OVFL | CON-NECT VMR | OVFL VMR | PERCENT BLOCKED (GRADE) | LINE | HOURS CONNECTED TRAFFIC ON LINES 1 | 2 | 3 | 4 | 5 | 6 | 7 | 8 | 9 | 10 |
|---|---|---|---|---|---|---|---|---|---|---|---|---|---|---|---|---|---|
| | | | | | | | | .874 | .863 | .851 | .837 | .822 | .805 | .786 | .765 | .742 | .716 |
| | | | | | | | | .689 | .658 | .625 | .590 | | | | | | |
| 24.000 | 20.335 | .000 | 3.665 | .45 | 3.50 | 15.3% | 25 | .941 | .937 | .932 | .927 | .922 | .916 | .909 | .902 | .893 | .884 |
| | | | | | | | | .874 | .863 | .851 | .837 | .822 | .805 | .786 | .765 | .742 | .716 |
| | | | | | | | | .689 | .658 | .625 | .590 | .552 | | | | | |
| 24.000 | 20.847 | .000 | 3.153 | .49 | 3.56 | 13.1% | 26 | .941 | .937 | .932 | .927 | .922 | .916 | .909 | .902 | .893 | .884 |
| | | | | | | | | .874 | .863 | .851 | .837 | .822 | .805 | .786 | .765 | .742 | .716 |
| | | | | | | | | .689 | .658 | .625 | .590 | .552 | .511 | | | | |
| 24.000 | 21.316 | .000 | 2.684 | .53 | 3.62 | 11.2% | 27 | .941 | .937 | .932 | .927 | .922 | .916 | .909 | .902 | .893 | .884 |
| | | | | | | | | .874 | .863 | .851 | .837 | .822 | .805 | .786 | .765 | .742 | .716 |
| | | | | | | | | .689 | .658 | .625 | .590 | .552 | .511 | .469 | | | |
| 24.000 | 21.742 | .000 | 2.258 | .57 | 3.65 | 9.4% | 28 | .941 | .937 | .932 | .927 | .922 | .916 | .909 | .902 | .893 | .884 |
| | | | | | | | | .874 | .863 | .851 | .837 | .822 | .805 | .786 | .765 | .742 | .716 |
| | | | | | | | | .689 | .658 | .625 | .590 | .552 | .511 | .469 | .426 | | |
| 24.000 | 22.124 | .000 | 1.876 | .62 | 3.68 | 7.8% | 29 | .941 | .937 | .932 | .927 | .922 | .916 | .909 | .902 | .893 | .884 |
| | | | | | | | | .874 | .863 | .851 | .837 | .822 | .805 | .786 | .765 | .742 | .716 |
| | | | | | | | | .689 | .658 | .625 | .590 | .552 | .511 | .469 | .426 | .382 | |
| 24.000 | 22.462 | .000 | 1.538 | .66 | 3.68 | 6.4% | 30 | .941 | .937 | .932 | .927 | .922 | .916 | .909 | .902 | .893 | .884 |
| | | | | | | | | .874 | .863 | .851 | .837 | .822 | .805 | .786 | .765 | .742 | .716 |
| | | | | | | | | .689 | .658 | .625 | .590 | .552 | .511 | .469 | .426 | .382 | .338 |
| 24.000 | 22.757 | .000 | 1.243 | .71 | 3.67 | 5.2% | 31 | .941 | .937 | .932 | .927 | .922 | .916 | .909 | .902 | .893 | .884 |
| | | | | | | | | .874 | .863 | .851 | .837 | .822 | .805 | .786 | .765 | .742 | .716 |
| | | | | | | | | .689 | .658 | .625 | .590 | .552 | .511 | .469 | .426 | .382 | .338 |
| | | | | | | | | .295 | | | | | | | | | |
| 24.000 | 23.009 | .000 | .991 | .75 | 3.65 | 4.1% | 32 | .941 | .937 | .932 | .927 | .922 | .916 | .909 | .902 | .893 | .884 |
| | | | | | | | | .874 | .863 | .851 | .837 | .822 | .805 | .786 | .765 | .742 | .716 |
| | | | | | | | | .689 | .658 | .625 | .590 | .552 | .511 | .469 | .426 | .382 | .338 |
| | | | | | | | | .295 | .253 | | | | | | | | |
| 24.000 | 23.223 | .000 | .777 | .79 | 3.61 | 3.2% | 33 | .941 | .937 | .932 | .927 | .922 | .916 | .909 | .902 | .893 | .884 |
| | | | | | | | | .874 | .863 | .851 | .837 | .822 | .805 | .786 | .765 | .742 | .716 |
| | | | | | | | | .689 | .658 | .625 | .590 | .552 | .511 | .469 | .426 | .382 | .338 |
| | | | | | | | | .295 | .253 | .213 | | | | | | | |
| 24.000 | 23.400 | .000 | .600 | .82 | 3.56 | 2.5% | 34 | .941 | .937 | .932 | .927 | .922 | .916 | .909 | .902 | .893 | .884 |
| | | | | | | | | .874 | .863 | .851 | .837 | .822 | .805 | .786 | .765 | .742 | .716 |
| | | | | | | | | .689 | .658 | .625 | .590 | .552 | .511 | .469 | .426 | .382 | .338 |
| | | | | | | | | .295 | .253 | .213 | .177 | | | | | | |
| 24.000 | 23.544 | .000 | .456 | .86 | 3.50 | 1.9% | 35 | .941 | .937 | .932 | .927 | .922 | .916 | .909 | .902 | .893 | .884 |
| | | | | | | | | .874 | .863 | .851 | .837 | .822 | .805 | .786 | .765 | .742 | .716 |
| | | | | | | | | .689 | .658 | .625 | .590 | .552 | .511 | .469 | .426 | .382 | .338 |
| | | | | | | | | .295 | .253 | .213 | .177 | .144 | | | | | |

*SIMILAR TO A.T.&T. NEAL-WILKINSON TABLES(SEE SECTION 1)

*EQUIVALENT RANDOM THEORY: VMR(VARIANCE TO MEAN RATIO)– 1.5 - NO RETRIES (PEAKED TRAFFIC - ROUTE ADVANCEMENT OR DAY-TO-DAY VARIATION)

| HOURS FIRST ATTEMPT VMR = 1.5 | HOURS CON-NECT | HOURS RECALL /RETRY | HOURS OVFL | CON-NECT VMR | OVFL VMR | PERCENT BLOCKED (GRADE) | LINE | HOURS CONNECTED TRAFFIC ON LINES 1 | 2 | 3 | 4 | 5 | 6 | 7 | 8 | 9 | 10 |
|---|---|---|---|---|---|---|---|---|---|---|---|---|---|---|---|---|---|
| 24.000 | 23.660 | .000 | .340 | .88 | 3.42 | 1.4% | 36 | .941 | .937 | .932 | .927 | .922 | .916 | .909 | .902 | .893 | .884 |
| | | | | | | | | .874 | .863 | .851 | .837 | .822 | .805 | .786 | .765 | .742 | .716 |
| | | | | | | | | .689 | .658 | .625 | .590 | .552 | .511 | .469 | .426 | .382 | .338 |
| | | | | | | | | .295 | .253 | .213 | .177 | .144 | .115 | | | | |
| 24.000 | 23.750 | .000 | .250 | .91 | 3.34 | 1.0% | 37 | .941 | .937 | .932 | .927 | .922 | .916 | .909 | .902 | .893 | .884 |
| | | | | | | | | .874 | .863 | .851 | .837 | .822 | .805 | .786 | .765 | .742 | .716 |
| | | | | | | | | .689 | .658 | .625 | .590 | .552 | .511 | .469 | .426 | .382 | .338 |
| | | | | | | | | .295 | .253 | .213 | .177 | .144 | .115 | .090 | | | |
| 24.000 | 23.820 | .000 | .180 | .93 | 3.26 | .8% | 38 | .941 | .937 | .932 | .927 | .922 | .916 | .909 | .902 | .893 | .884 |
| | | | | | | | | .874 | .863 | .851 | .837 | .822 | .805 | .786 | .765 | .742 | .716 |
| | | | | | | | | .689 | .658 | .625 | .590 | .552 | .511 | .469 | .426 | .382 | .338 |
| | | | | | | | | .295 | .253 | .213 | .177 | .144 | .115 | .090 | .070 | | |
| 24.000 | 23.872 | .000 | .128 | .95 | 3.17 | .5% | 39 | .941 | .937 | .932 | .927 | .922 | .916 | .909 | .902 | .893 | .884 |
| | | | | | | | | .874 | .863 | .851 | .837 | .822 | .805 | .786 | .765 | .742 | .716 |
| | | | | | | | | .689 | .658 | .625 | .590 | .552 | .511 | .469 | .426 | .382 | .338 |
| | | | | | | | | .295 | .253 | .213 | .177 | .144 | .115 | .090 | .070 | .052 | |
| 26.000 | 17.578 | .000 | 8.422 | .24 | 2.98 | 32.4% | 20 | .945 | .942 | .938 | .933 | .929 | .924 | .918 | .912 | .905 | .897 |
| | | | | | | | | .889 | .880 | .870 | .859 | .846 | .832 | .817 | .800 | .782 | .761 |
| 26.000 | 18.317 | .000 | 7.683 | .26 | 3.08 | 29.5% | 21 | .945 | .942 | .938 | .933 | .929 | .924 | .918 | .912 | .905 | .897 |
| | | | | | | | | .889 | .880 | .870 | .859 | .846 | .832 | .817 | .800 | .782 | .761 |
| | | | | | | | | .739 | | | | | | | | | |
| 26.000 | 19.031 | .000 | 6.969 | .29 | 3.17 | 26.8% | 22 | .945 | .942 | .938 | .933 | .929 | .924 | .918 | .912 | .905 | .897 |
| | | | | | | | | .889 | .880 | .870 | .859 | .846 | .832 | .817 | .800 | .782 | .761 |
| | | | | | | | | .739 | .714 | | | | | | | | |
| 26.000 | 19.718 | .000 | 6.282 | .31 | 3.27 | 24.2% | 23 | .945 | .942 | .938 | .933 | .929 | .924 | .918 | .912 | .905 | .897 |
| | | | | | | | | .889 | .880 | .870 | .859 | .846 | .832 | .817 | .800 | .782 | .761 |
| | | | | | | | | .739 | .714 | .687 | | | | | | | |
| 26.000 | 20.376 | .000 | 5.624 | .34 | 3.36 | 21.6% | 24 | .945 | .942 | .938 | .933 | .929 | .924 | .918 | .912 | .905 | .897 |
| | | | | | | | | .889 | .880 | .870 | .859 | .846 | .832 | .817 | .800 | .782 | .761 |
| | | | | | | | | .739 | .714 | .687 | .658 | | | | | | |
| 26.000 | 21.002 | .000 | 4.998 | .37 | 3.45 | 19.2% | 25 | .945 | .942 | .938 | .933 | .929 | .924 | .918 | .912 | .905 | .897 |
| | | | | | | | | .889 | .880 | .870 | .859 | .846 | .832 | .817 | .800 | .782 | .761 |
| | | | | | | | | .739 | .714 | .687 | .658 | .626 | | | | | |
| 26.000 | 21.593 | .000 | 4.407 | .41 | 3.53 | 16.9% | 26 | .945 | .942 | .938 | .933 | .929 | .924 | .918 | .912 | .905 | .897 |
| | | | | | | | | .000 | .880 | .870 | .859 | .846 | .832 | .817 | .800 | .782 | .761 |
| | | | | | | | | .739 | .714 | .687 | .658 | .626 | .591 | | | | |
| 26.000 | 22.148 | .000 | 3.852 | .45 | 3.61 | 14.8% | 27 | .945 | .942 | .938 | .933 | .929 | .924 | .918 | .912 | .905 | .897 |
| | | | | | | | | .889 | .880 | .870 | .859 | .846 | .832 | .817 | .800 | .782 | .761 |
| | | | | | | | | .739 | .714 | .687 | .658 | .626 | .591 | .555 | | | |
| 26.000 | 22.664 | .000 | 3.336 | .48 | 3.67 | 12.8% | 28 | .945 | .942 | .938 | .933 | .929 | .924 | .918 | .912 | .905 | .897 |

*SIMILAR TO A.T.&T. NEAL-WILKINSON TABLES(SEE SECTION 1)

*EQUIVALENT RANDOM THEORY: VMR(VARIANCE TO MEAN RATIO)– 1.5 - NO RETRIES (PEAKED TRAFFIC - ROUTE ADVANCEMENT OR DAY-TO-DAY VARIATION)

| HOURS FIRST ATTEMPT VMR = 1.5 | HOURS CON-NECT | HOURS RECALL /RETRY | HOURS OVFL | CON-NECT VMR | OVFL VMR | PERCENT BLOCKED (GRADE) | LINE | HOURS CONNECTED TRAFFIC ON LINES 1 | 2 | 3 | 4 | 5 | 6 | 7 | 8 | 9 | 10 |
|---|---|---|---|---|---|---|---|---|---|---|---|---|---|---|---|---|---|
| | | | | | | | | .889 | .880 | .870 | .859 | .846 | .832 | .817 | .800 | .782 | .761 |
| | | | | | | | | .739 | .714 | .687 | .658 | .626 | .591 | .555 | .516 | | |
| 26.000 | 23.140 | .000 | 2.860 | .52 | 3.72 | 11.0% | 29 | .945 | .942 | .938 | .933 | .929 | .924 | .918 | .912 | .905 | .897 |
| | | | | | | | | .889 | .880 | .870 | .859 | .846 | .832 | .817 | .800 | .782 | .761 |
| | | | | | | | | .739 | .714 | .687 | .658 | .626 | .591 | .555 | .516 | .476 | |
| 26.000 | 23.575 | .000 | 2.425 | .57 | 3.76 | 9.3% | 30 | .945 | .942 | .938 | .933 | .929 | .924 | .918 | .912 | .905 | .897 |
| | | | | | | | | .889 | .880 | .870 | .859 | .846 | .832 | .817 | .800 | .782 | .761 |
| | | | | | | | | .739 | .714 | .687 | .658 | .626 | .591 | .555 | .516 | .476 | .434 |
| 26.000 | 23.967 | .000 | 2.033 | .61 | 3.79 | 7.8% | 31 | .945 | .942 | .938 | .933 | .929 | .924 | .918 | .912 | .905 | .897 |
| | | | | | | | | .889 | .880 | .870 | .859 | .846 | .832 | .817 | .800 | .782 | .761 |
| | | | | | | | | .739 | .714 | .687 | .658 | .626 | .591 | .555 | .516 | .476 | .434 |
| | | | | | | | | .392 | | | | | | | | | |
| 26.000 | 24.316 | .000 | 1.684 | .65 | 3.80 | 6.5% | 32 | .945 | .942 | .938 | .933 | .929 | .924 | .918 | .912 | .905 | .897 |
| | | | | | | | | .889 | .880 | .870 | .859 | .846 | .832 | .817 | .800 | .782 | .761 |
| | | | | | | | | .739 | .714 | .687 | .658 | .626 | .591 | .555 | .516 | .476 | .434 |
| | | | | | | | | .392 | .349 | | | | | | | | |
| 26.000 | 24.623 | .000 | 1.377 | .69 | 3.79 | 5.3% | 33 | .945 | .942 | .938 | .933 | .929 | .924 | .918 | .912 | .905 | .897 |
| | | | | | | | | .889 | .880 | .870 | .859 | .846 | .832 | .817 | .800 | .782 | .761 |
| | | | | | | | | .739 | .714 | .687 | .658 | .626 | .591 | .555 | .516 | .476 | .434 |
| | | | | | | | | .392 | .349 | .307 | | | | | | | |
| 26.000 | 24.890 | .000 | 1.110 | .73 | 3.77 | 4.3% | 34 | .945 | .942 | .938 | .933 | .929 | .924 | .918 | .912 | .905 | .897 |
| | | | | | | | | .889 | .880 | .870 | .859 | .846 | .832 | .817 | .800 | .782 | .761 |
| | | | | | | | | .739 | .714 | .687 | .658 | .626 | .591 | .555 | .516 | .476 | .434 |
| | | | | | | | | .392 | .349 | .307 | .266 | | | | | | |
| 26.000 | 25.117 | .000 | .883 | .77 | 3.74 | 3.4% | 35 | .945 | .942 | .938 | .933 | .929 | .924 | .918 | .912 | .905 | .897 |
| | | | | | | | | .889 | .880 | .870 | .859 | .846 | .832 | .817 | .800 | .782 | .761 |
| | | | | | | | | .739 | .714 | .687 | .658 | .626 | .591 | .555 | .516 | .476 | .434 |
| | | | | | | | | .392 | .349 | .307 | .266 | .227 | | | | | |
| 26.000 | 25.308 | .000 | .692 | .81 | 3.69 | 2.7% | 36 | .945 | .942 | .938 | .933 | .929 | .924 | .918 | .912 | .905 | .897 |
| | | | | | | | | .889 | .880 | .870 | .859 | .846 | .832 | .817 | .800 | .782 | .761 |
| | | | | | | | | .739 | .714 | .687 | .658 | .626 | .591 | .555 | .516 | .476 | .434 |
| | | | | | | | | .392 | .349 | .307 | .266 | .227 | .191 | | | | |
| 26.000 | 25.466 | .000 | .534 | .84 | 3.63 | 2.1% | 37 | .945 | .942 | .938 | .933 | .929 | .924 | .918 | .912 | .905 | .897 |
| | | | | | | | | .889 | .880 | .870 | .859 | .846 | .832 | .817 | .800 | .782 | .761 |
| | | | | | | | | .739 | .714 | .687 | .658 | .626 | .591 | .555 | .516 | .476 | .434 |
| | | | | | | | | .392 | .349 | .307 | .266 | .227 | .191 | .158 | | | |
| 26.000 | 25.594 | .000 | .406 | .87 | 3.56 | 1.6% | 38 | .945 | .942 | .938 | .933 | .929 | .924 | .918 | .912 | .905 | .897 |
| | | | | | | | | .889 | .880 | .870 | .859 | .846 | .832 | .817 | .800 | .782 | .761 |
| | | | | | | | | .739 | .714 | .687 | .658 | .626 | .591 | .555 | .516 | .476 | .434 |
| | | | | | | | | .392 | .349 | .307 | .266 | .227 | .191 | .158 | .128 | | |

*SIMILAR TO A.T.&T. NEAL-WILKINSON TABLES(SEE SECTION 1)

*EQUIVALENT RANDOM THEORY: VMR(VARIANCE TO MEAN RATIO)– 1.5 - NO RETRIES (PEAKED TRAFFIC - ROUTE ADVANCEMENT OR DAY-TO-DAY VARIATION)

| HOURS FIRST ATTEMPT VMR=1.5 | HOURS CON-NECT | HOURS RECALL /RETRY | HOURS OVFL | CON-NECT VMR | OVFL VMR | PERCENT BLOCKED (GRADE) | LINE | HOURS CONNECTED TRAFFIC ON LINES 1 | 2 | 3 | 4 | 5 | 6 | 7 | 8 | 9 | 10 |
|---|---|---|---|---|---|---|---|---|---|---|---|---|---|---|---|---|---|
| 26.000 | 25.697 | .000 | .303 | .90 | 3.48 | 1.2% | 39 | .945 | .942 | .938 | .933 | .929 | .924 | .918 | .912 | .905 | .897 |
| | | | | | | | | .889 | .880 | .870 | .859 | .846 | .832 | .817 | .800 | .782 | .761 |
| | | | | | | | | .739 | .714 | .687 | .658 | .626 | .591 | .555 | .516 | .476 | .434 |
| | | | | | | | | .392 | .349 | .307 | .266 | .227 | .191 | .158 | .128 | .102 | |
| 26.000 | 25.777 | .000 | .223 | .92 | 3.40 | .9% | 40 | .945 | .942 | .938 | .933 | .929 | .924 | .918 | .912 | .905 | .897 |
| | | | | | | | | .889 | .880 | .870 | .859 | .846 | .832 | .817 | .800 | .782 | .761 |
| | | | | | | | | .739 | .714 | .687 | .658 | .626 | .591 | .555 | .516 | .476 | .434 |
| | | | | | | | | .392 | .349 | .307 | .266 | .227 | .191 | .158 | .128 | .102 | .080 |
| 26.000 | 25.838 | .000 | .162 | .94 | 3.32 | .6% | 41 | .945 | .942 | .938 | .933 | .929 | .924 | .918 | .912 | .905 | .897 |
| | | | | | | | | .889 | .880 | .870 | .859 | .846 | .832 | .817 | .800 | .782 | .761 |
| | | | | | | | | .739 | .714 | .687 | .658 | .626 | .591 | .555 | .516 | .476 | .434 |
| | | | | | | | | .392 | .349 | .307 | .266 | .227 | .191 | .158 | .128 | .102 | .080 |
| | | | | | | | | .062 | | | | | | | | | |
| 28.000 | 18.641 | .000 | 9.359 | .22 | 2.99 | 33.4% | 21 | .949 | .946 | .942 | .939 | .935 | .930 | .925 | .920 | .914 | .908 |
| | | | | | | | | .901 | .893 | .885 | .876 | .866 | .854 | .842 | .828 | .813 | .797 |
| | | | | | | | | .778 | | | | | | | | | |
| 28.000 | 19.400 | .000 | 8.600 | .24 | 3.08 | 30.7% | 22 | .949 | .946 | .942 | .939 | .935 | .930 | .925 | .920 | .914 | .908 |
| | | | | | | | | .901 | .893 | .885 | .876 | .866 | .854 | .842 | .828 | .813 | .797 |
| | | | | | | | | .778 | .758 | | | | | | | | |
| 28.000 | 20.136 | .000 | 7.864 | .26 | 3.18 | 28.1% | 23 | .949 | .946 | .942 | .939 | .935 | .930 | .925 | .920 | .914 | .908 |
| | | | | | | | | .901 | .893 | .885 | .876 | .866 | .854 | .842 | .828 | .813 | .797 |
| | | | | | | | | .778 | .758 | .736 | | | | | | | |
| 28.000 | 20.848 | .000 | 7.152 | .29 | 3.28 | 25.5% | 24 | .949 | .946 | .942 | .939 | .935 | .930 | .925 | .920 | .914 | .908 |
| | | | | | | | | .901 | .893 | .885 | .876 | .866 | .854 | .842 | .828 | .813 | .797 |
| | | | | | | | | .778 | .758 | .736 | .712 | | | | | | |
| 28.000 | 21.533 | .000 | 6.467 | .31 | 3.38 | 23.1% | 25 | .949 | .946 | .942 | .939 | .935 | .930 | .925 | .920 | .914 | .908 |
| | | | | | | | | .901 | .893 | .885 | .876 | .866 | .854 | .842 | .828 | .813 | .797 |
| | | | | | | | | .778 | .758 | .736 | .712 | .686 | | | | | |
| 28.000 | 22.190 | .000 | 5.810 | .34 | 3.47 | 20.7% | 26 | .949 | .946 | .942 | .939 | .935 | .930 | .925 | .920 | .914 | .908 |
| | | | | | | | | .901 | .893 | .885 | .876 | .866 | .854 | .842 | .828 | .813 | .797 |
| | | | | | | | | .778 | .758 | .736 | .712 | .686 | .657 | | | | |
| 28.000 | 22.816 | .000 | 5.184 | .37 | 3.56 | 18.5% | 27 | .949 | .946 | .942 | .939 | .935 | .930 | .925 | .920 | .914 | .908 |
| | | | | | | | | .901 | .893 | .885 | .876 | .866 | .854 | .842 | .828 | .813 | .797 |
| | | | | | | | | .778 | .758 | .736 | .712 | .686 | .657 | .626 | | | |
| 28.000 | 23.410 | .000 | 4.500 | .41 | 3.64 | 16.4% | 28 | .949 | .946 | .942 | .939 | .935 | .930 | .925 | .920 | .914 | .908 |
| | | | | | | | | .901 | .803 | .885 | .876 | .866 | .854 | .842 | .828 | .813 | .797 |
| | | | | | | | | .778 | .758 | .736 | .712 | .686 | .657 | .626 | .593 | | |
| 28.000 | 23.967 | .000 | 4.033 | .44 | 3.71 | 14.4% | 29 | .949 | .946 | .942 | .939 | .935 | .930 | .925 | .920 | .914 | .908 |
| | | | | | | | | .901 | .893 | .885 | .876 | .866 | .854 | .842 | .828 | .813 | .797 |
| | | | | | | | | .778 | .758 | .736 | .712 | .686 | .657 | .626 | .593 | .558 | |

*SIMILAR TO A.T.&T. NEAL-WILKINSON TABLES(SEE SECTION 1)

*EQUIVALENT RANDOM THEORY: VMR(VARIANCE TO MEAN RATIO)– 1.5 - NO RETRIES (PEAKED TRAFFIC - ROUTE ADVANCEMENT OR DAY-TO-DAY VARIATION)

| HOURS FIRST ATTEMPT VMR=1.5 | HOURS CON-NECT | HOURS RECALL /RETRY | HOURS OVFL | CON-NECT VMR | OVFL VMR | PERCENT BLOCKED (GRADE) | LINE | HOURS CONNECTED TRAFFIC ON LINES 1 | 2 | 3 | 4 | 5 | 6 | 7 | 8 | 9 | 10 |
|---|---|---|---|---|---|---|---|---|---|---|---|---|---|---|---|---|---|
| 28.000 | 24.488 | .000 | 3.512 | .48 | 3.77 | 12.5% | 30 | .949 | .946 | .942 | .939 | .935 | .930 | .925 | .920 | .914 | .908 |
| | | | | | | | | .901 | .893 | .885 | .876 | .866 | .854 | .842 | .828 | .813 | .797 |
| | | | | | | | | .778 | .758 | .736 | .712 | .686 | .657 | .626 | .593 | .558 | .521 |
| 28.000 | 24.970 | .000 | 3.030 | .52 | 3.83 | 10.8% | 31 | .949 | .946 | .942 | .939 | .935 | .930 | .925 | .920 | .914 | .908 |
| | | | | | | | | .901 | .893 | .885 | .876 | .866 | .854 | .842 | .828 | .813 | .797 |
| | | | | | | | | .778 | .758 | .736 | .712 | .686 | .657 | .626 | .593 | .558 | .521 |
| | | | | | | | | .482 | | | | | | | | | |
| 28.000 | 25.412 | .000 | 2.588 | .56 | 3.87 | 9.2% | 32 | .949 | .946 | .942 | .939 | .935 | .930 | .925 | .920 | .914 | .908 |
| | | | | | | | | .901 | .893 | .885 | .876 | .866 | .854 | .842 | .828 | .813 | .797 |
| | | | | | | | | .778 | .758 | .736 | .712 | .686 | .657 | .626 | .593 | .558 | .521 |
| | | | | | | | | .482 | .442 | | | | | | | | |
| 28.000 | 25.813 | .000 | 2.187 | .60 | 3.89 | 7.8% | 33 | .949 | .946 | .942 | .939 | .935 | .930 | .925 | .920 | .914 | .908 |
| | | | | | | | | .901 | .893 | .885 | .876 | .866 | .854 | .842 | .828 | .813 | .797 |
| | | | | | | | | .778 | .758 | .736 | .712 | .686 | .657 | .626 | .593 | .558 | .521 |
| | | | | | | | | .482 | .442 | .401 | | | | | | | |
| 28.000 | 26.173 | .000 | 1.827 | .64 | 3.91 | 6.5% | 34 | .949 | .946 | .942 | .939 | .935 | .930 | .925 | .920 | .914 | .908 |
| | | | | | | | | .901 | .893 | .885 | .876 | .866 | .854 | .842 | .828 | .813 | .797 |
| | | | | | | | | .778 | .758 | .736 | .712 | .686 | .657 | .626 | .593 | .558 | .521 |
| | | | | | | | | .482 | .442 | .401 | .360 | | | | | | |
| 28.000 | 26.491 | .000 | 1.509 | .68 | 3.91 | 5.4% | 35 | .949 | .946 | .942 | .939 | .935 | .930 | .925 | .920 | .914 | .908 |
| | | | | | | | | .901 | .893 | .885 | .876 | .866 | .854 | .842 | .828 | .813 | .797 |
| | | | | | | | | .778 | .758 | .736 | .712 | .686 | .657 | .626 | .593 | .558 | .521 |
| | | | | | | | | .482 | .442 | .401 | .360 | .319 | | | | | |
| 28.000 | 26.770 | .000 | 1.230 | .72 | 3.89 | 4.4% | 36 | .949 | .946 | .942 | .939 | .935 | .930 | .925 | .920 | .914 | .908 |
| | | | | | | | | .901 | .893 | .885 | .876 | .866 | .854 | .842 | .828 | .813 | .797 |
| | | | | | | | | .778 | .758 | .736 | .712 | .686 | .657 | .626 | .593 | .558 | .521 |
| | | | | | | | | .482 | .442 | .401 | .360 | .319 | .279 | | | | |
| 28.000 | 27.010 | .000 | .990 | .76 | 3.86 | 3.5% | 37 | .949 | .946 | .942 | .939 | .935 | .930 | .925 | .920 | .914 | .908 |
| | | | | | | | | .901 | .893 | .885 | .876 | .866 | .854 | .842 | .828 | .813 | .797 |
| | | | | | | | | .778 | .758 | .736 | .712 | .686 | .657 | .626 | .593 | .558 | .521 |
| | | | | | | | | .482 | .442 | .401 | .360 | .319 | .279 | .240 | | | |
| 28.000 | 27.214 | .000 | .786 | .80 | 3.81 | 2.8% | 38 | .949 | .946 | .942 | .939 | .935 | .930 | .925 | .920 | .914 | .908 |
| | | | | | | | | .901 | .893 | .885 | .876 | .866 | .854 | .842 | .828 | .813 | .797 |
| | | | | | | | | .778 | .758 | .736 | .712 | .686 | .657 | .626 | .593 | .558 | .521 |
| | | | | | | | | .482 | .442 | .401 | .360 | .319 | .279 | .240 | .204 | | |
| 28.000 | 27.385 | .000 | .615 | .83 | 3.76 | 2.2% | 39 | .949 | .946 | .942 | .939 | .935 | .930 | .925 | .920 | .914 | .908 |
| | | | | | | | | .901 | .893 | .885 | .876 | .866 | .854 | .842 | .828 | .813 | .797 |
| | | | | | | | | .778 | .758 | .736 | .712 | .686 | .657 | .626 | .593 | .558 | .521 |
| | | | | | | | | .482 | .442 | .401 | .360 | .319 | .279 | .240 | .204 | .171 | |
| 28.000 | 27.526 | .000 | .474 | .86 | 3.69 | 1.7% | 40 | .949 | .946 | .942 | .939 | .935 | .930 | .925 | .920 | .914 | .908 |

*SIMILAR TO A.T.&T. NEAL-WILKINSON TABLES(SEE SECTION 1)

*EQUIVALENT RANDOM THEORY: VMR(VARIANCE TO MEAN RATIO)– 1.5 - NO RETRIES (PEAKED TRAFFIC - ROUTE ADVANCEMENT OR DAY-TO-DAY VARIATION)

| HOURS FIRST ATTEMPT VMR = 1.5 | HOURS CON-NECT | HOURS RECALL /RETRY | HOURS OVFL | CON-NECT VMR | OVFL VMR | PERCENT BLOCKED (GRADE) | LINE | HOURS CONNECTED TRAFFIC ON LINES 1 | 2 | 3 | 4 | 5 | 6 | 7 | 8 | 9 | 10 |
|---|---|---|---|---|---|---|---|---|---|---|---|---|---|---|---|---|---|
| | | | | | | | | .901 | .893 | .885 | .876 | .866 | .854 | .842 | .828 | .813 | .797 |
| | | | | | | | | .778 | .758 | .736 | .712 | .686 | .657 | .626 | .593 | .558 | .521 |
| | | | | | | | | .482 | .442 | .401 | .360 | .319 | .279 | .240 | .204 | .171 | .141 |
| 28.000 | 27.640 | .000 | .360 | .89 | 3.62 | 1.3% | 41 | .949 | .946 | .942 | .939 | .935 | .930 | .925 | .920 | .914 | .908 |
| | | | | | | | | .901 | .893 | .885 | .876 | .866 | .854 | .842 | .828 | .813 | .797 |
| | | | | | | | | .778 | .758 | .736 | .712 | .686 | .657 | .626 | .593 | .558 | .521 |
| | | | | | | | | .482 | .442 | .401 | .360 | .319 | .279 | .240 | .204 | .171 | .141 |
| | | | | | | | | .114 | | | | | | | | | |
| 28.000 | 27.730 | .000 | .270 | .91 | 3.54 | 1.0% | 42 | .949 | .946 | .942 | .939 | .935 | .930 | .925 | .920 | .914 | .908 |
| | | | | | | | | .901 | .893 | .885 | .876 | .866 | .854 | .842 | .828 | .813 | .797 |
| | | | | | | | | .778 | .758 | .736 | .712 | .686 | .657 | .626 | .593 | .558 | .521 |
| | | | | | | | | .482 | .442 | .401 | .360 | .319 | .279 | .240 | .204 | .171 | .141 |
| | | | | | | | | .114 | .091 | | | | | | | | |
| 28.000 | 27.801 | .000 | .199 | .93 | 3.45 | .7% | 43 | .949 | .946 | .942 | .939 | .935 | .930 | .925 | .920 | .914 | .908 |
| | | | | | | | | .901 | .893 | .885 | .876 | .866 | .854 | .842 | .828 | .813 | .797 |
| | | | | | | | | .778 | .758 | .736 | .712 | .686 | .657 | .626 | .593 | .558 | .521 |
| | | | | | | | | .482 | .442 | .401 | .360 | .319 | .279 | .240 | .204 | .171 | .141 |
| | | | | | | | | .114 | .091 | .071 | | | | | | | |
| 28.000 | 27.856 | .000 | .144 | .95 | 3.37 | .5% | 44 | .949 | .946 | .942 | .939 | .935 | .930 | .925 | .920 | .914 | .908 |
| | | | | | | | | .901 | .893 | .885 | .876 | .866 | .854 | .842 | .828 | .813 | .797 |
| | | | | | | | | .778 | .758 | .736 | .712 | .686 | .657 | .626 | .593 | .558 | .521 |
| | | | | | | | | .482 | .442 | .401 | .360 | .319 | .279 | .240 | .204 | .171 | .141 |
| | | | | | | | | .114 | .091 | .071 | .054 | | | | | | |
| 30.000 | 19.697 | .000 | 10.303 | .21 | 2.99 | 34.3% | 22 | .952 | .949 | .946 | .943 | .940 | .936 | .932 | .927 | .922 | .917 |
| | | | | | | | | .911 | .904 | .897 | .890 | .881 | .872 | .862 | .850 | .838 | .824 |
| | | | | | | | | .809 | .793 | | | | | | | | |
| 30.000 | 20.472 | .000 | 9.528 | .23 | 3.09 | 31.8% | 23 | .952 | .949 | .946 | .943 | .940 | .936 | .932 | .927 | .922 | .917 |
| | | | | | | | | .911 | .904 | .897 | .890 | .881 | .872 | .862 | .850 | .838 | .824 |
| | | | | | | | | .809 | .793 | .775 | | | | | | | |
| 30.000 | 21.227 | .000 | 8.773 | .24 | 3.19 | 29.2% | 24 | .952 | .949 | .946 | .943 | .940 | .936 | .932 | .927 | .922 | .917 |
| | | | | | | | | .911 | .904 | .897 | .890 | .881 | .872 | .862 | .850 | .838 | .824 |
| | | | | | | | | .809 | .793 | .775 | .755 | | | | | | |
| 30.000 | 21.960 | .000 | 8.040 | .27 | 3.29 | 26.8% | 25 | .952 | .949 | .946 | .943 | .940 | .936 | .932 | .927 | .922 | .917 |
| | | | | | | | | .911 | .904 | .897 | .890 | .881 | .872 | .862 | .850 | .838 | .824 |
| | | | | | | | | .809 | .793 | .775 | .755 | .733 | | | | | |
| 30.000 | 22.670 | .000 | 7.330 | .29 | 3.38 | 24.4% | 26 | .952 | .949 | .946 | .943 | .940 | .936 | .932 | .927 | .922 | .917 |
| | | | | | | | | .911 | .904 | .897 | .890 | .881 | .872 | .862 | .850 | .838 | .824 |
| | | | | | | | | .809 | .793 | .775 | .755 | .733 | .710 | | | | |
| 30.000 | 23.354 | .000 | 6.646 | .32 | 3.48 | 22.2% | 27 | .952 | .949 | .946 | .943 | .940 | .936 | .932 | .927 | .922 | .917 |
| | | | | | | | | .911 | .904 | .897 | .890 | .881 | .872 | .862 | .850 | .838 | .824 |

*SIMILAR TO A.T.&T. NEAL-WILKINSON TABLES(SEE SECTION 1)

*EQUIVALENT RANDOM THEORY: VMR(VARIANCE TO MEAN RATIO)– 1.5 - NO RETRIES (PEAKED TRAFFIC - ROUTE ADVANCEMENT OR DAY-TO-DAY VARIATION)

| HOURS FIRST ATTEMPT VMR = 1.5 | HOURS CON-NECT | HOURS RECALL /RETRY | HOURS OVFL | CON-NECT VMR | OVFL VMR | PERCENT BLOCKED (GRADE) | LINE | HOURS CONNECTED TRAFFIC ON LINES 1 | 2 | 3 | 4 | 5 | 6 | 7 | 8 | 9 | 10 |
|---|---|---|---|---|---|---|---|---|---|---|---|---|---|---|---|---|---|
| | | | | | | | | .809 | .793 | .775 | .755 | .733 | .710 | .684 | | | |
| 30.000 | 24.011 | .000 | 5.989 | .34 | 3.57 | 20.0% | 28 | .952 | .949 | .946 | .943 | .940 | .936 | .932 | .927 | .922 | .917 |
| | | | | | | | | .911 | .904 | .897 | .890 | .881 | .872 | .862 | .850 | .838 | .824 |
| | | | | | | | | .809 | .793 | .775 | .755 | .733 | .710 | .684 | .657 | | |
| 30.000 | 24.638 | .000 | 5.362 | .37 | 3.66 | 17.9% | 29 | .952 | .949 | .946 | .943 | .940 | .936 | .932 | .927 | .922 | .917 |
| | | | | | | | | .911 | .904 | .897 | .890 | .881 | .872 | .862 | .850 | .838 | .824 |
| | | | | | | | | .809 | .793 | .775 | .755 | .733 | .710 | .684 | .657 | .627 | |
| 30.000 | 25.232 | .000 | 4.768 | .41 | 3.74 | 15.9% | 30 | .952 | .949 | .946 | .943 | .940 | .936 | .932 | .927 | .922 | .917 |
| | | | | | | | | .911 | .904 | .897 | .890 | .881 | .872 | .862 | .850 | .838 | .824 |
| | | | | | | | | .809 | .793 | .775 | .755 | .733 | .710 | .684 | .657 | .627 | .595 |
| 30.000 | 25.793 | .000 | 4.207 | .44 | 3.81 | 14.0% | 31 | .952 | .949 | .946 | .943 | .940 | .936 | .932 | .927 | .922 | .917 |
| | | | | | | | | .911 | .904 | .897 | .890 | .881 | .872 | .862 | .850 | .838 | .824 |
| | | | | | | | | .809 | .793 | .775 | .755 | .733 | .710 | .684 | .657 | .627 | .595 |
| | | | | | | | | .561 | | | | | | | | | |
| 30.000 | 26.318 | .000 | 3.682 | .48 | 3.87 | 12.3% | 32 | .952 | .949 | .946 | .943 | .940 | .936 | .932 | .927 | .922 | .917 |
| | | | | | | | | .911 | .904 | .897 | .890 | .881 | .872 | .862 | .850 | .838 | .824 |
| | | | | | | | | .809 | .793 | .775 | .755 | .733 | .710 | .684 | .657 | .627 | .595 |
| | | | | | | | | .561 | .525 | | | | | | | | |
| 30.000 | 26.805 | .000 | 3.195 | .51 | 3.93 | 10.6% | 33 | .952 | .949 | .946 | .943 | .940 | .936 | .932 | .927 | .922 | .917 |
| | | | | | | | | .911 | .904 | .897 | .890 | .881 | .872 | .862 | .850 | .838 | .824 |
| | | | | | | | | .809 | .793 | .775 | .755 | .733 | .710 | .684 | .657 | .627 | .595 |
| | | | | | | | | .561 | .525 | .487 | | | | | | | |
| 30.000 | 27.254 | .000 | 2.746 | .55 | 3.97 | 9.2% | 34 | .952 | .949 | .946 | .943 | .940 | .936 | .932 | .927 | .922 | .917 |
| | | | | | | | | .911 | .904 | .897 | .890 | .881 | .872 | .862 | .850 | .838 | .824 |
| | | | | | | | | .809 | .793 | .775 | .755 | .733 | .710 | .684 | .657 | .627 | .595 |
| | | | | | | | | .561 | .525 | .487 | .449 | | | | | | |
| 30.000 | 27.663 | .000 | 2.337 | .59 | 4.00 | 7.0% | 35 | .952 | .949 | .946 | .943 | .940 | .936 | .932 | .927 | .922 | .917 |
| | | | | | | | | .911 | .904 | .897 | .890 | .881 | .872 | .862 | .850 | .838 | .824 |
| | | | | | | | | .809 | .793 | .775 | .755 | .733 | .710 | .684 | .657 | .627 | .595 |
| | | | | | | | | .561 | .525 | .487 | .449 | .409 | | | | | |
| 30.000 | 28.032 | .000 | 1.968 | .63 | 4.01 | 6.6% | 36 | .952 | .949 | .946 | .943 | .940 | .936 | .932 | .927 | .922 | .917 |
| | | | | | | | | .911 | .904 | .897 | .890 | .881 | .872 | .862 | .850 | .838 | .824 |
| | | | | | | | | .809 | .793 | .775 | .755 | .733 | .710 | .684 | .657 | .627 | .595 |
| | | | | | | | | .561 | .525 | .487 | .449 | .409 | .369 | | | | |
| 30.000 | 28.361 | .000 | 1.639 | .67 | 4.01 | 5.5% | 37 | .952 | .949 | .946 | .943 | .940 | .936 | .932 | .927 | .922 | .917 |
| | | | | | | | | .911 | .904 | .897 | .890 | .881 | .872 | .862 | .850 | .838 | .824 |
| | | | | | | | | .809 | .793 | .775 | .755 | .733 | .710 | .684 | .657 | .627 | .595 |
| | | | | | | | | .561 | .525 | .487 | .449 | .409 | .369 | .329 | | | |
| 30.000 | 28.651 | .000 | 1.349 | .71 | 4.00 | 4.5% | 38 | .952 | .949 | .946 | .943 | .940 | .936 | .932 | .927 | .922 | .917 |
| | | | | | | | | .911 | .904 | .897 | .890 | .881 | .872 | .862 | .850 | .838 | .824 |

*SIMILAR TO A.T.&T. NEAL-WILKINSON TABLES(SEE SECTION 1)

*EQUIVALENT RANDOM THEORY: VMR(VARIANCE TO MEAN RATIO)– 1.5 - NO RETRIES (PEAKED TRAFFIC - ROUTE ADVANCEMENT OR DAY-TO-DAY VARIATION)

| HOURS FIRST ATTEMPT VMR = 1.5 | HOURS CON-NECT | HOURS RECALL /RETRY | HOURS OVFL | CON-NECT VMR | OVFL VMR | PERCENT BLOCKED (GRADE) | LINE | HOURS CONNECTED TRAFFIC ON LINES 1 | 2 | 3 | 4 | 5 | 6 | 7 | 8 | 9 | 10 |
|---|---|---|---|---|---|---|---|---|---|---|---|---|---|---|---|---|---|
| | | | | | | | | .809 | .793 | .775 | .755 | .733 | .710 | .684 | .657 | .627 | .595 |
| | | | | | | | | .561 | .525 | .487 | .449 | .409 | .369 | .329 | .290 | | |
| 30.000 | 28.903 | .000 | 1.097 | .75 | 3.97 | 3.7% | 39 | .952 | .949 | .946 | .943 | .940 | .936 | .932 | .927 | .922 | .917 |
| | | | | | | | | .911 | .904 | .897 | .890 | .881 | .872 | .862 | .850 | .838 | .824 |
| | | | | | | | | .809 | .793 | .775 | .755 | .733 | .710 | .684 | .657 | .627 | .595 |
| | | | | | | | | .561 | .525 | .487 | .449 | .409 | .369 | .329 | .290 | .252 | |
| 30.000 | 29.119 | .000 | .881 | .78 | 3.93 | 2.9% | 40 | .952 | .949 | .946 | .943 | .940 | .936 | .932 | .927 | .922 | .917 |
| | | | | | | | | .911 | .904 | .897 | .890 | .881 | .872 | .862 | .850 | .838 | .824 |
| | | | | | | | | .809 | .793 | .775 | .755 | .733 | .710 | .684 | .657 | .627 | .595 |
| | | | | | | | | .561 | .525 | .487 | .449 | .409 | .369 | .329 | .290 | .252 | .216 |
| 30.000 | 29.302 | .000 | .698 | .82 | 3.88 | 2.3% | 41 | .952 | .949 | .946 | .943 | .940 | .936 | .932 | .927 | .922 | .917 |
| | | | | | | | | .911 | .904 | .897 | .890 | .881 | .872 | .862 | .850 | .838 | .824 |
| | | | | | | | | .809 | .793 | .775 | .755 | .733 | .710 | .684 | .657 | .627 | .595 |
| | | | | | | | | .561 | .525 | .487 | .449 | .409 | .369 | .329 | .290 | .252 | .216 |
| | | | | | | | | .183 | | | | | | | | | |
| 30.000 | 29.454 | .000 | .546 | .85 | 3.82 | 1.8% | 42 | .952 | .949 | .946 | .943 | .940 | .936 | .932 | .927 | .922 | .917 |
| | | | | | | | | .911 | .904 | .897 | .890 | .881 | .872 | .862 | .850 | .838 | .824 |
| | | | | | | | | .809 | .793 | .775 | .755 | .733 | .710 | .684 | .657 | .627 | .595 |
| | | | | | | | | .561 | .525 | .487 | .449 | .409 | .369 | .329 | .290 | .252 | .216 |
| | | | | | | | | .183 | .152 | | | | | | | | |
| 30.000 | 29.579 | .000 | .421 | .88 | 3.75 | 1.4% | 43 | .952 | .949 | .946 | .943 | .940 | .936 | .932 | .927 | .922 | .917 |
| | | | | | | | | .911 | .904 | .897 | .890 | .881 | .872 | .862 | .850 | .838 | .824 |
| | | | | | | | | .809 | .793 | .775 | .755 | .733 | .710 | .684 | .657 | .627 | .595 |
| | | | | | | | | .561 | .525 | .487 | .449 | .409 | .369 | .329 | .290 | .252 | .216 |
| | | | | | | | | .183 | .152 | .125 | | | | | | | |
| 30.000 | 29.680 | .000 | .320 | .90 | 3.67 | 1.1% | 44 | .952 | .949 | .946 | .943 | .940 | .936 | .932 | .927 | .922 | .917 |
| | | | | | | | | .911 | .904 | .897 | .890 | .881 | .872 | .862 | .850 | .838 | .824 |
| | | | | | | | | .809 | .793 | .775 | .755 | .733 | .710 | .684 | .657 | .627 | .595 |
| | | | | | | | | .561 | .525 | .487 | .449 | .409 | .369 | .329 | .290 | .252 | .216 |
| | | | | | | | | .183 | .152 | .125 | .101 | | | | | | |
| 30.000 | 29.760 | .000 | .240 | .92 | 3.59 | .8% | 45 | .952 | .949 | .946 | .943 | .940 | .936 | .932 | .927 | .922 | .917 |
| | | | | | | | | .911 | .904 | .897 | .890 | .881 | .872 | .862 | .850 | .838 | .824 |
| | | | | | | | | .809 | .793 | .775 | .755 | .733 | .710 | .684 | .657 | .627 | .595 |
| | | | | | | | | .561 | .525 | .487 | .449 | .409 | .369 | .329 | .290 | .252 | .216 |
| | | | | | | | | .183 | .152 | .125 | .101 | .080 | | | | | |
| 30.000 | 29.823 | .000 | .177 | .94 | 3.50 | .6% | 46 | .952 | .949 | .946 | .943 | .940 | .936 | .932 | .927 | .922 | .917 |
| | | | | | | | | .911 | .904 | .897 | .890 | .881 | .872 | .862 | .850 | .838 | .824 |
| | | | | | | | | .809 | .793 | .775 | .755 | .733 | .710 | .684 | .657 | .627 | .595 |
| | | | | | | | | .561 | .525 | .487 | .449 | .409 | .369 | .329 | .290 | .252 | .216 |
| | | | | | | | | .183 | .152 | .125 | .101 | .080 | .063 | | | | |

*SIMILAR TO A.T.&T. NEAL-WILKINSON TABLES(SEE SECTION 1)

*EQUIVALENT RANDOM THEORY: VMR(VARIANCE TO MEAN RATIO)– 2.0 - NO RETRIES (PEAKED TRAFFIC - ROUTE ADVANCEMENT OR DAY-TO-DAY VARIATION)

| HOURS FIRST ATTEMPT VMR = 2.0 | HOURS CON-NECT | HOURS RECALL /RETRY | HOURS OVFL | CON-NECT VMR | OVFL VMR | PERCENT BLOCKED (GRADE) | LINE | HOURS CONNECTED TRAFFIC ON LINES 1 | 2 | 3 | 4 | 5 | 6 | 7 | 8 | 9 | 10 |
|---|---|---|---|---|---|---|---|---|---|---|---|---|---|---|---|---|---|
| .200 | .146 | .000 | .054 | .95 | 1.83 | 26.8% | 2 | .093 | .053 | | | | | | | | |
| .200 | .175 | .000 | .025 | .97 | 1.76 | 12.6% | 3 | .093 | .053 | .028 | | | | | | | |
| .200 | .189 | .000 | .011 | .99 | 1.70 | 5.5% | 4 | .093 | .053 | .028 | .014 | | | | | | |
| .200 | .195 | .000 | .005 | .99 | 1.64 | 2.3% | 5 | .093 | .053 | .028 | .014 | .006 | | | | | |
| .200 | .198 | .000 | .0021 | .00 | 1.59 | .9% | 6 | .093 | .053 | .028 | .014 | .006 | .003 | | | | |
| .400 | .272 | .000 | .128 | .89 | 1.92 | 32.0% | 2 | .166 | .106 | | | | | | | | |
| .400 | .334 | .000 | .066 | .94 | 1.86 | 16.4% | 3 | .166 | .106 | .062 | | | | | | | |
| .400 | .368 | .000 | .032 | .97 | 1.79 | 7.9% | 4 | .166 | .106 | .062 | .034 | | | | | | |
| .400 | .386 | .000 | .014 | .98 | 1.73 | 3.6% | 5 | .166 | .106 | .062 | .034 | .017 | | | | | |
| .400 | .394 | .000 | .006 | .99 | 1.67 | 1.5% | 6 | .166 | .106 | .062 | .034 | .017 | .008 | | | | |
| .400 | .398 | .000 | .0021 | .00 | 1.61 | .6% | 7 | .166 | .106 | .062 | .034 | .017 | .008 | .004 | | | |
| .600 | .485 | .000 | .115 | .90 | 1.91 | 19.1% | 3 | .231 | .156 | .098 | | | | | | | |
| .600 | .542 | .000 | .058 | .94 | 1.84 | 9.7% | 4 | .231 | .156 | .098 | .057 | | | | | | |
| .600 | .573 | .000 | .027 | .97 | 1.77 | 4.6% | 5 | .231 | .156 | .098 | .057 | .031 | | | | | |
| .600 | .588 | .000 | .012 | .98 | 1.71 | 2.0% | 6 | .231 | .156 | .098 | .057 | .031 | .015 | | | | |
| .600 | .595 | .000 | .005 | .99 | 1.65 | .9% | 7 | .231 | .156 | .098 | .057 | .031 | .015 | .007 | | | |
| .800 | .620 | .000 | .180 | .87 | 1.98 | 22.5% | 3 | .283 | .202 | .135 | | | | | | | |
| .800 | .703 | .000 | .097 | .92 | 1.92 | 12.1% | 4 | .283 | .202 | .135 | .083 | | | | | | |
| .800 | .751 | .000 | .049 | .95 | 1.86 | 6.1% | 5 | .283 | .202 | .135 | .083 | .048 | | | | | |
| .800 | .777 | .000 | .023 | .97 | 1.79 | 2.9% | 6 | .283 | .202 | .135 | .083 | .048 | .026 | | | | |
| .800 | .790 | .000 | .010 | .99 | 1.72 | 1.3% | 7 | .283 | .202 | .135 | .083 | .048 | .026 | .013 | | | |
| .800 | .796 | .000 | .004 | .99 | 1.67 | .6% | 8 | .283 | .202 | .135 | .083 | .048 | .026 | .013 | .006 | | |
| 1.000 | .756 | .000 | .244 | .83 | 1.98 | 24.4% | 3 | .337 | .248 | .171 | | | | | | | |
| 1.000 | .866 | .000 | .134 | .80 | 1.93 | 13.4% | 4 | .337 | .248 | .171 | .110 | | | | | | |
| 1.000 | .931 | .000 | .069 | .93 | 1.87 | 6.9% | 5 | .337 | .248 | .171 | .110 | .065 | | | | | |
| 1.000 | .966 | .000 | .034 | .96 | 1.80 | 3.4% | 6 | .337 | .248 | .171 | .110 | .065 | .036 | | | | |
| 1.000 | .985 | .000 | .015 | .98 | 1.74 | 1.5% | 7 | .337 | .248 | .171 | .110 | .065 | .036 | .018 | | | |
| 1.000 | .993 | .000 | .007 | .99 | 1.67 | .7% | 8 | .337 | .248 | .171 | .110 | .065 | .036 | .018 | .009 | | |
| 1.200 | .873 | .000 | .327 | .79 | 2.03 | 27.2% | 3 | .378 | .288 | .207 | | | | | | | |
| 1.200 | 1.012 | .000 | .188 | .86 | 1.99 | 15.7% | 4 | .378 | .288 | .207 | .139 | | | | | | |
| 1.200 | 1.098 | .000 | .102 | .91 | 1.93 | 8.5% | 5 | .378 | .288 | .207 | .139 | .086 | | | | | |
| 1.200 | 1.148 | .000 | .052 | .95 | 1.86 | 4.3% | 6 | .378 | .288 | .207 | .139 | .086 | .050 | | | | |
| 1.200 | 1.175 | .000 | .025 | .97 | 1.80 | 2.1% | 7 | .378 | .288 | .207 | .139 | .086 | .050 | .027 | | | |
| 1.200 | 1.189 | .000 | .011 | .99 | 1.73 | .9% | 8 | .378 | .288 | .207 | .139 | .086 | .050 | .027 | .014 | | |
| 1.400 | .082 | .000 | .418 | .76 | 2.07 | 29.9% | 3 | .414 | .326 | .242 | | | | | | | |
| 1.400 | 1.150 | .000 | .250 | .83 | 2.04 | 17.9% | 4 | .414 | .326 | .242 | .168 | | | | | | |
| 1.400 | 1.259 | .000 | .141 | .89 | 1.99 | 10.1% | 5 | .414 | .326 | .242 | .168 | .109 | | | | | |
| 1.400 | 1.325 | .000 | .075 | .93 | 1.92 | 5.3% | 6 | .414 | .326 | .242 | .168 | .109 | .066 | | | | |
| 1.400 | 1.363 | .000 | .037 | .96 | 1.85 | 2.7% | 7 | .414 | .326 | .242 | .168 | .109 | .066 | .037 | | | |
| 1.400 | 1.382 | .000 | .018 | .98 | 1.79 | 1.3% | 8 | .414 | .326 | .242 | .168 | .109 | .066 | .037 | .020 | | |

*SIMILAR TO A.T.&T. NEAL-WILKINSON TABLES(SEE SECTION 1)

*EQUIVALENT RANDOM THEORY: VMR(VARIANCE TO MEAN RATIO)– 2.0 - NO RETRIES (PEAKED TRAFFIC - ROUTE ADVANCEMENT OR DAY-TO-DAY VARIATION)

| HOURS FIRST ATTEMPT VMR = 2.0 | HOURS CON-NECT | HOURS RECALL /RETRY | HOURS OVFL | CON-NECT VMR | OVFL VMR | PERCENT BLOCKED (GRADE) | LINE | HOURS CONNECTED TRAFFIC ON LINES 1 | 2 | 3 | 4 | 5 | 6 | 7 | 8 | 9 | 10 |
|---|---|---|---|---|---|---|---|---|---|---|---|---|---|---|---|---|---|
| 1.400 | 1.392 | .000 | .008 | .99 | 1.72 | .6% | 9 | .414 | .326 | .242 | .168 | .109 | .066 | .037 | .020 | .010 | |
| 1.600 | 1.065 | .000 | .535 | .73 | 2.16 | 33.4% | 3 | .438 | .354 | .272 | | | | | | | |
| 1.600 | 1.263 | .000 | .337 | .80 | 2.14 | 21.0% | 4 | .438 | .354 | .272 | .198 | | | | | | |
| 1.600 | 1.399 | .000 | .201 | .86 | 2.10 | 12.6% | 5 | .438 | .354 | .272 | .198 | .136 | | | | | |
| 1.600 | 1.487 | .000 | .113 | .91 | 2.04 | 7.1% | 6 | .438 | .354 | .272 | .198 | .136 | .087 | | | | |
| 1.600 | 1.539 | .000 | .061 | .95 | 1.97 | 3.8% | 7 | .438 | .354 | .272 | .198 | .136 | .087 | .053 | | | |
| 1.600 | 1.569 | .000 | .031 | .97 | 1.89 | 1.9% | 8 | .438 | .354 | .272 | .198 | .136 | .087 | .053 | .030 | | |
| 1.600 | 1.585 | .000 | .015 | .98 | 1.83 | .9% | 9 | .438 | .354 | .272 | .198 | .136 | .087 | .053 | .030 | .016 | |
| 1.800 | 1.384 | .000 | .416 | .77 | 2.18 | 23.1% | 4 | .468 | .386 | .304 | .227 | | | | | | |
| 1.800 | 1.545 | .000 | .255 | .84 | 2.14 | 14.2% | 5 | .468 | .386 | .304 | .227 | .160 | | | | | |
| 1.800 | 1.651 | .000 | .149 | .89 | 2.08 | 8.3% | 6 | .468 | .386 | .304 | .227 | .160 | .107 | | | | |
| 1.800 | 1.718 | .000 | .082 | .93 | 2.02 | 4.6% | 7 | .468 | .386 | .304 | .227 | .160 | .107 | .067 | | | |
| 1.800 | 1.757 | .000 | .043 | .96 | 1.95 | 2.4% | 8 | .468 | .386 | .304 | .227 | .160 | .107 | .067 | .039 | | |
| 1.800 | 1.779 | .000 | .021 | .98 | 1.88 | 1.2% | 9 | .468 | .386 | .304 | .227 | .160 | .107 | .067 | .039 | .022 | |
| 1.800 | 1.790 | .000 | .010 | .99 | 1.81 | .6% | 10 | .468 | .386 | .304 | .227 | .160 | .107 | .067 | .039 | .022 | .011 |
| 2.000 | 1.498 | .000 | .502 | .74 | 2.21 | 25.1% | 4 | .495 | .415 | .333 | .255 | | | | | | |
| 2.000 | 1.684 | .000 | .316 | .81 | 2.18 | 15.8% | 5 | .495 | .415 | .333 | .255 | .185 | | | | | |
| 2.000 | 1.810 | .000 | .190 | .87 | 2.13 | 9.5% | 6 | .495 | .415 | .333 | .255 | .185 | .127 | | | | |
| 2.000 | 1.892 | .000 | .108 | .92 | 2.07 | 5.4% | 7 | .495 | .415 | .333 | .255 | .185 | .127 | .082 | | | |
| 2.000 | 1.942 | .000 | .058 | .95 | 2.00 | 2.9% | 8 | .495 | .415 | .333 | .255 | .185 | .127 | .082 | .050 | | |
| 2.000 | 1.970 | .000 | .030 | .97 | 1.92 | 1.5% | 9 | .495 | .415 | .333 | .255 | .185 | .127 | .082 | .050 | .028 | |
| 2.000 | 1.985 | .000 | .015 | .98 | 1.86 | .7% | 10 | .495 | .415 | .333 | .255 | .185 | .127 | .082 | .050 | .028 | .015 |
| 2.200 | 1.606 | .000 | .594 | .72 | 2.23 | 27.0% | 4 | .520 | .442 | .362 | .283 | | | | | | |
| 2.200 | 1.816 | .000 | .384 | .79 | 2.22 | 17.4% | 5 | .520 | .442 | .362 | .283 | .210 | | | | | |
| 2.200 | 1.964 | .000 | .236 | .85 | 2.17 | 10.7% | 6 | .520 | .442 | .362 | .283 | .210 | .148 | | | | |
| 2.200 | 2.062 | .000 | .138 | .90 | 2.11 | 6.3% | 7 | .520 | .442 | .362 | .283 | .210 | .148 | .098 | | | |
| 2.200 | 2.123 | .000 | .077 | .94 | 2.04 | 3.5% | 8 | .520 | .442 | .362 | .283 | .210 | .148 | .098 | .061 | | |
| 2.200 | 2.159 | .000 | .041 | .96 | 1.97 | 1.8% | 9 | .520 | .442 | .362 | .283 | .210 | .148 | .098 | .061 | .036 | |
| 2.200 | 2.180 | .000 | .020 | .98 | 1.90 | .9% | 10 | .520 | .442 | .362 | .283 | .210 | .148 | .098 | .061 | .036 | .020 |
| 2.400 | 1.708 | .000 | .692 | .69 | 2.26 | 28.8% | 4 | .543 | .467 | .388 | .309 | | | | | | |
| 2.400 | 1.943 | .000 | .457 | .77 | 2.25 | 19.1% | 5 | .543 | .467 | .388 | .309 | .235 | | | | | |
| 2.400 | 2.112 | .000 | .288 | .83 | 2.21 | 12.0% | 6 | .543 | .467 | .388 | .309 | .235 | .169 | | | | |
| 2.400 | 2.227 | .000 | .173 | .88 | 2.16 | 7.2% | 7 | .543 | .467 | .388 | .309 | .235 | .169 | .115 | | | |
| 2.400 | 2.301 | .000 | .099 | .93 | 2.09 | 4.1% | 8 | .543 | .467 | .388 | .309 | .235 | .169 | .115 | .074 | | |
| 2.400 | 2.346 | .000 | .054 | .95 | 2.02 | 2.2% | 9 | .543 | .467 | .388 | .309 | .235 | .169 | .115 | .074 | .045 | |
| 2.400 | 2.372 | .000 | .028 | .97 | 1.95 | 1.2% | 10 | .543 | .467 | .388 | .309 | .235 | .169 | .115 | .074 | .045 | .026 |
| 2.400 | 2.386 | .000 | .014 | .99 | 1.88 | .6% | 11 | .543 | .467 | .388 | .309 | .235 | .169 | .115 | .074 | .045 | .026 |
| | | | | | | | | .014 | | | | | | | | | |
| 2.600 | 1.804 | .000 | .796 | .66 | 2.27 | 30.6% | 4 | .564 | .491 | .414 | .335 | | | | | | |
| 2.600 | 2.063 | .000 | .537 | .74 | 2.27 | 20.7% | 5 | .564 | .491 | .414 | .335 | .259 | | | | | |

*SIMILAR TO A.T.&T. NEAL-WILKINSON TABLES(SEE SECTION 1)

## *EQUIVALENT RANDOM THEORY: VMR(VARIANCE TO MEAN RATIO)– 2.0 - NO RETRIES (PEAKED TRAFFIC - ROUTE ADVANCEMENT OR DAY-TO-DAY VARIATION)

| HOURS FIRST ATTEMPT VMR=2.0 | HOURS CONNECT | HOURS RECALL /RETRY | HOURS OVFL | CONNECT VMR | OVFL VMR | PERCENT BLOCKED (GRADE) | LINE | HOURS CONNECTED TRAFFIC ON LINES 1 | 2 | 3 | 4 | 5 | 6 | 7 | 8 | 9 | 10 |
|---|---|---|---|---|---|---|---|---|---|---|---|---|---|---|---|---|---|
| 2.600 | 2.254 | .000 | .346 | .81 | 2.24 | 13.3% | 6 | .564 | .491 | .414 | .335 | .259 | .191 | | | | |
| 2.600 | 2.388 | .000 | .212 | .87 | 2.20 | 8.2% | 7 | .564 | .491 | .414 | .335 | .259 | .191 | .133 | | | |
| 2.600 | 2.476 | .000 | .124 | .91 | 2.13 | 4.8% | 8 | .564 | .491 | .414 | .335 | .259 | .191 | .133 | .088 | | |
| 2.600 | 2.531 | .000 | .069 | .94 | 2.06 | 2.7% | 9 | .564 | .491 | .414 | .335 | .259 | .191 | .133 | .088 | .055 | |
| 2.600 | 2.563 | .000 | .037 | .97 | 1.99 | 1.4% | 10 | .564 | .491 | .414 | .335 | .259 | .191 | .133 | .088 | .055 | .032 |
| 2.600 | 2.581 | .000 | .019 | .98 | 1.92 | .7% | 11 | .564 | .491 | .414 | .335 | .259 | .191 | .133 | .088 | .055 | .032 |
| | | | | | | | | .018 | | | | | | | | | |
| 2.800 | 1.894 | .000 | .906 | .64 | 2.29 | 32.4% | 4 | .583 | .513 | .438 | .360 | | | | | | |
| 2.800 | 2.178 | .000 | .622 | .72 | 2.30 | 22.2% | 5 | .583 | .513 | .438 | .360 | .284 | | | | | |
| 2.800 | 2.391 | .000 | .409 | .79 | 2.28 | 14.6% | 6 | .583 | .513 | .438 | .360 | .284 | .213 | | | | |
| 2.800 | 2.543 | .000 | .257 | .85 | 2.23 | 9.2% | 7 | .583 | .513 | .438 | .360 | .284 | .213 | .152 | | | |
| 2.800 | 2.646 | .000 | .154 | .90 | 2.17 | 5.5% | 8 | .583 | .513 | .438 | .360 | .284 | .213 | .152 | .103 | | |
| 2.800 | 2.712 | .000 | .088 | .93 | 2.10 | 3.1% | 9 | .583 | .513 | .438 | .360 | .284 | .213 | .152 | .103 | .066 | |
| 2.800 | 2.752 | .000 | .048 | .96 | 2.03 | 1.7% | 10 | .583 | .513 | .438 | .360 | .284 | .213 | .152 | .103 | .066 | .040 |
| 2.800 | 2.775 | .000 | .025 | .98 | 1.96 | .9% | 11 | .583 | .513 | .438 | .360 | .284 | .213 | .152 | .103 | .066 | .040 |
| | | | | | | | | .023 | | | | | | | | | |
| 3.000 | 1.980 | .000 | 1.020 | .62 | 2.30 | 34.0% | 4 | .601 | .534 | .461 | .384 | | | | | | |
| 3.000 | 2.287 | .000 | .713 | .69 | 2.32 | 23.8% | 5 | .601 | .534 | .461 | .384 | .307 | | | | | |
| 3.000 | 2.522 | .000 | .478 | .76 | 2.30 | 15.9% | 6 | .601 | .534 | .461 | .384 | .307 | .235 | | | | |
| 3.000 | 2.694 | .000 | .306 | .83 | 2.27 | 10.2% | 7 | .601 | .534 | .461 | .384 | .307 | .235 | .172 | | | |
| 3.000 | 2.813 | .000 | .187 | .88 | 2.21 | 6.2% | 8 | .601 | .534 | .461 | .384 | .307 | .235 | .172 | .119 | | |
| 3.000 | 2.891 | .000 | .109 | .92 | 2.15 | 3.6% | 9 | .601 | .534 | .461 | .384 | .307 | .235 | .172 | .119 | .078 | |
| 3.000 | 2.939 | .000 | .061 | .95 | 2.07 | 2.0% | 10 | .601 | .534 | .461 | .384 | .307 | .235 | .172 | .119 | .078 | .048 |
| 3.000 | 2.968 | .000 | .032 | .97 | 2.00 | 1.1% | 11 | .601 | .534 | .461 | .384 | .307 | .235 | .172 | .119 | .078 | .048 |
| | | | | | | | | .028 | | | | | | | | | |
| 3.000 | 2.984 | .000 | .016 | .98 | 1.93 | .5% | 12 | .601 | .534 | .461 | .384 | .307 | .235 | .172 | .119 | .078 | .048 |
| | | | | | | | | .028 | .016 | | | | | | | | |
| 3.200 | 2.365 | .000 | .835 | .67 | 2.39 | 26.1% | 5 | .610 | .546 | .477 | .403 | .330 | | | | | |
| 3.200 | 2.624 | .000 | .576 | .74 | 2.39 | 18.0% | 6 | .610 | .546 | .477 | .403 | .330 | .259 | | | | |
| 3.200 | 2.819 | .000 | .381 | .81 | 2.36 | 11.9% | 7 | .610 | .546 | .477 | .403 | .330 | .259 | .194 | | | |
| 3.200 | 2.958 | .000 | .242 | .86 | 2.31 | 7.6% | 8 | .610 | .546 | .477 | .403 | .330 | .259 | .194 | .139 | | |
| 3.200 | 3.053 | .000 | .147 | .91 | 2.24 | 4.6% | 9 | .610 | .546 | .477 | .403 | .330 | .259 | .194 | .139 | .095 | |
| 3.200 | 3.114 | .000 | .086 | .94 | 2.17 | 2.7% | 10 | .610 | .546 | .477 | .403 | .330 | .259 | .194 | .139 | .095 | .062 |
| 3.200 | 3.152 | .000 | .048 | .96 | 2.10 | 1.5% | 11 | .610 | .546 | .477 | .403 | .330 | .259 | .194 | .139 | .095 | .062 |
| | | | | | | | | .038 | | | | | | | | | |
| 3.200 | 3.175 | .000 | .025 | .98 | 2.02 | .8% | 12 | .610 | .546 | .477 | .403 | .330 | .259 | .194 | .139 | .095 | .062 |
| | | | | | | | | .038 | .022 | | | | | | | | |
| 3.400 | 2.463 | .000 | .937 | .65 | 2.40 | 27.6% | 5 | .625 | .564 | .497 | .425 | .352 | | | | | |
| 3.400 | 2.744 | .000 | .656 | .72 | 2.41 | 19.3% | 6 | .625 | .564 | .497 | .425 | .352 | .280 | | | | |
| 3.400 | 2.958 | .000 | .442 | .79 | 2.39 | 13.0% | 7 | .625 | .564 | .497 | .425 | .352 | .280 | .214 | | | |

*SIMILAR TO A.T.&T. NEAL-WILKINSON TABLES(SEE SECTION 1)

*EQUIVALENT RANDOM THEORY: VMR(VARIANCE TO MEAN RATIO)– 2.0 - NO RETRIES (PEAKED TRAFFIC - ROUTE ADVANCEMENT OR DAY-TO-DAY VARIATION)

| HOURS FIRST ATTEMPT VMR = 2.0 | HOURS CON-NECT | HOURS RECALL /RETRY | HOURS OVFL | CON-NECT VMR | OVFL VMR | PERCENT BLOCKED (GRADE) | LINE | HOURS CONNECTED TRAFFIC ON LINES 1 | 2 | 3 | 4 | 5 | 6 | 7 | 8 | 9 | 10 |
|---|---|---|---|---|---|---|---|---|---|---|---|---|---|---|---|---|---|
| 3.400 | 3.114 | .000 | .286 | .84 | 2.34 | 8.4% | 8 | .625 | .564 | .497 | .425 | .352 | .280 | .214 | .156 | | |
| 3.400 | 3.223 | .000 | .177 | .89 | 2.28 | 5.2% | 9 | .625 | .564 | .497 | .425 | .352 | .280 | .214 | .156 | .109 | |
| 3.400 | 3.295 | .000 | .105 | .93 | 2.21 | 3.1% | 10 | .625 | .564 | .497 | .425 | .352 | .280 | .214 | .156 | .109 | .072 |
| 3.400 | 3.340 | .000 | .060 | .95 | 2.14 | 1.8% | 11 | .625 | .564 | .497 | .425 | .352 | .280 | .214 | .156 | .109 | .072 |
| | | | | | | | | .045 | | | | | | | | | |
| 3.400 | 3.367 | .000 | .033 | .97 | 2.06 | 1.0% | 12 | .625 | .564 | .497 | .425 | .352 | .280 | .214 | .156 | .109 | .072 |
| | | | | | | | | .045 | .027 | | | | | | | | |
| 3.400 | 3.383 | .000 | .017 | .98 | 1.99 | .5% | 13 | .625 | .564 | .497 | .425 | .352 | .280 | .214 | .156 | .109 | .072 |
| | | | | | | | | .045 | .027 | .016 | | | | | | | |
| 3.600 | 2.556 | .000 | 1.044 | .63 | 2.42 | 29.0% | 5 | .640 | .581 | .516 | .446 | .373 | | | | | |
| 3.600 | 2.858 | .000 | .742 | .70 | 2.43 | 20.6% | 6 | .640 | .581 | .516 | .446 | .373 | .302 | | | | |
| 3.600 | 3.092 | .000 | .508 | .77 | 2.41 | 14.1% | 7 | .640 | .581 | .516 | .446 | .373 | .302 | .234 | | | |
| 3.600 | 3.266 | .000 | .334 | .83 | 2.37 | 9.3% | 8 | .640 | .581 | .516 | .446 | .373 | .302 | .234 | .174 | | |
| 3.600 | 3.389 | .000 | .211 | .88 | 2.32 | 5.8% | 9 | .640 | .581 | .516 | .446 | .373 | .302 | .234 | .174 | .123 | |
| 3.600 | 3.473 | .000 | .127 | .92 | 2.25 | 3.5% | 10 | .640 | .581 | .516 | .446 | .373 | .302 | .234 | .174 | .123 | .083 |
| 3.600 | 3.526 | .000 | .074 | .95 | 2.18 | 2.1% | 11 | .640 | .581 | .516 | .446 | .373 | .302 | .234 | .174 | .123 | .083 |
| | | | | | | | | .054 | | | | | | | | | |
| 3.600 | 3.559 | .000 | .041 | .97 | 2.10 | 1.1% | 12 | .640 | .581 | .516 | .446 | .373 | .302 | .234 | .174 | .123 | .083 |
| | | | | | | | | .054 | .033 | | | | | | | | |
| 3.600 | 3.578 | .000 | .022 | .98 | 2.03 | .6% | 13 | .640 | .581 | .516 | .446 | .373 | .302 | .234 | .174 | .123 | .083 |
| | | | | | | | | .054 | .033 | .019 | | | | | | | |
| 3.800 | 2.645 | .000 | 1.155 | .61 | 2.42 | 30.4% | 5 | .654 | .597 | .534 | .466 | .394 | | | | | |
| 3.800 | 2.968 | .000 | .832 | .68 | 2.44 | 21.9% | 6 | .654 | .597 | .534 | .466 | .394 | .323 | | | | |
| 3.800 | 3.222 | .000 | .578 | .75 | 2.44 | 15.2% | 7 | .654 | .597 | .534 | .466 | .394 | .323 | .254 | | | |
| 3.800 | 3.414 | .000 | .386 | .81 | 2.40 | 10.2% | 8 | .654 | .597 | .534 | .466 | .394 | .323 | .254 | .192 | | |
| 3.800 | 3.552 | .000 | .248 | .86 | 2.35 | 6.5% | 9 | .654 | .597 | .534 | .466 | .394 | .323 | .254 | .192 | .138 | |
| 3.800 | 3.647 | .000 | .153 | .90 | 2.29 | 4.0% | 10 | .654 | .597 | .534 | .466 | .394 | .323 | .254 | .192 | .138 | .095 |
| 3.800 | 3.710 | .000 | .090 | .94 | 2.22 | 2.4% | 11 | .654 | .597 | .534 | .466 | .394 | .323 | .254 | .192 | .138 | .095 |
| | | | | | | | | .062 | | | | | | | | | |
| 3.800 | 3.749 | .000 | .051 | .96 | 2.14 | 1.3% | 12 | .654 | .597 | .534 | .466 | .394 | .323 | .254 | .192 | .138 | .095 |
| | | | | | | | | .062 | .039 | | | | | | | | |
| 3.800 | 3.772 | .000 | .028 | .98 | 2.07 | .7% | 13 | .654 | .597 | .534 | .466 | .394 | .323 | .254 | .192 | .138 | .095 |
| | | | | | | | | .062 | .039 | .023 | | | | | | | |
| 4.000 | 2.730 | .000 | 1.270 | .59 | 2.43 | 31.8% | 5 | .667 | .612 | .551 | .485 | .415 | | | | | |
| 4.000 | 3.073 | .000 | .927 | .66 | 2.46 | 23.2% | 6 | .667 | .612 | .551 | .485 | .415 | .343 | | | | |
| 4.000 | 3.347 | .000 | .653 | .73 | 2.46 | 16.3% | 7 | .667 | .612 | .551 | .485 | .415 | .343 | .274 | | | |
| 4.000 | 3.557 | .000 | .443 | .79 | 2.43 | 11.1% | 8 | .667 | .612 | .551 | .485 | .415 | .343 | .274 | .210 | | |
| 4.000 | 3.711 | .000 | .289 | .85 | 2.39 | 7.2% | 9 | .667 | .612 | .551 | .485 | .415 | .343 | .274 | .210 | .154 | |
| 4.000 | 3.819 | .000 | .181 | .89 | 2.33 | 4.5% | 10 | .667 | .612 | .551 | .485 | .415 | .343 | .274 | .210 | .154 | .108 |
| 4.000 | 3.891 | .000 | .109 | .93 | 2.26 | 2.7% | 11 | .667 | .612 | .551 | .485 | .415 | .343 | .274 | .210 | .154 | .108 |

*SIMILAR TO A.T.&T. NEAL-WILKINSON TABLES(SEE SECTION 1)

*EQUIVALENT RANDOM THEORY: VMR(VARIANCE TO MEAN RATIO)– 2.0 - NO RETRIES (PEAKED TRAFFIC - ROUTE ADVANCEMENT OR DAY-TO-DAY VARIATION)

| HOURS FIRST ATTEMPT VMR = 2.0 | HOURS CON-NECT | HOURS RECALL /RETRY | HOURS OVFL | CON-NECT VMR | OVFL VMR | PERCENT BLOCKED (GRADE) | LINE | HOURS CONNECTED TRAFFIC ON LINES 1 | 2 | 3 | 4 | 5 | 6 | 7 | 8 | 9 | 10 |
|---|---|---|---|---|---|---|---|---|---|---|---|---|---|---|---|---|---|
| | | | | | | | | .072 | | | | | | | | | |
| 4.000 | 3.937 | .000 | .063 | .95 | 2.18 | 1.6% | 12 | .667 | .612 | .551 | .485 | .415 | .343 | .274 | .210 | .154 | .108 |
| | | | | | | | | .072 | .046 | | | | | | | | |
| 4.000 | 3.965 | .000 | .035 | .97 | 2.10 | .9% | 13 | .667 | .612 | .551 | .485 | .415 | .343 | .274 | .210 | .154 | .108 |
| | | | | | | | | .072 | .046 | .028 | | | | | | | |
| 4.200 | 2.810 | .000 | 1.390 | .57 | 2.43 | 33.1% | 5 | .679 | .626 | .568 | .503 | .434 | | | | | |
| 4.200 | 3.173 | .000 | 1.027 | .64 | 2.47 | 24.4% | 6 | .679 | .626 | .568 | .503 | .434 | .363 | | | | |
| 4.200 | 3.467 | .000 | .733 | .71 | 2.48 | 17.4% | 7 | .679 | .626 | .568 | .503 | .434 | .363 | .294 | | | |
| 4.200 | 3.696 | .000 | .504 | .77 | 2.46 | 12.0% | 8 | .679 | .626 | .568 | .503 | .434 | .363 | .294 | .228 | | |
| 4.200 | 3.866 | .000 | .334 | .83 | 2.42 | 8.0% | 9 | .679 | .626 | .568 | .503 | .434 | .363 | .294 | .228 | .170 | |
| 4.200 | 3.987 | .000 | .213 | .88 | 2.36 | 5.1% | 10 | .679 | .626 | .568 | .503 | .434 | .363 | .294 | .228 | .170 | .121 |
| 4.200 | 4.070 | .000 | .130 | .92 | 2.29 | 3.1% | 11 | .679 | .626 | .568 | .503 | .434 | .363 | .294 | .228 | .170 | .121 |
| | | | | | | | | .083 | | | | | | | | | |
| 4.200 | 4.124 | .000 | .076 | .95 | 2.22 | 1.8% | 12 | .679 | .626 | .568 | .503 | .434 | .363 | .294 | .228 | .170 | .121 |
| | | | | | | | | .083 | .054 | | | | | | | | |
| 4.200 | 4.157 | .000 | .043 | .97 | 2.14 | 1.0% | 13 | .679 | .626 | .568 | .503 | .434 | .363 | .294 | .228 | .170 | .121 |
| | | | | | | | | .083 | .054 | .033 | | | | | | | |
| 4.200 | 4.177 | .000 | .023 | .98 | 2.07 | .6% | 14 | .679 | .626 | .568 | .503 | .434 | .363 | .294 | .228 | .170 | .121 |
| | | | | | | | | .083 | .054 | .033 | .020 | | | | | | |
| 4.400 | 3.241 | .000 | 1.159 | .62 | 2.53 | 26.3% | 6 | .683 | .633 | .577 | .516 | .450 | .382 | | | | |
| 4.400 | 3.555 | .000 | .845 | .69 | 2.55 | 19.2% | 7 | .683 | .633 | .577 | .516 | .450 | .382 | .314 | | | |
| 4.400 | 3.804 | .000 | .596 | .75 | 2.54 | 13.5% | 8 | .683 | .633 | .577 | .516 | .450 | .382 | .314 | .249 | | |
| 4.400 | 3.994 | .000 | .406 | .81 | 2.50 | 9.2% | 9 | .683 | .633 | .577 | .516 | .450 | .382 | .314 | .249 | .190 | |
| 4.400 | 4.134 | .000 | .266 | .86 | 2.45 | 6.1% | 10 | .683 | .633 | .577 | .516 | .450 | .382 | .314 | .249 | .190 | .140 |
| 4.400 | 4.202 | .000 | .168 | .90 | 2.38 | 3.8% | 11 | .683 | .633 | .577 | .516 | .450 | .382 | .314 | .249 | .190 | .140 |
| | | | | | | | | .098 | | | | | | | | | |
| 4.400 | 4.298 | .000 | .102 | .93 | 2.31 | 2.3% | 12 | .683 | .633 | .577 | .516 | .450 | .382 | .314 | .249 | .190 | .140 |
| | | | | | | | | .098 | .066 | | | | | | | | |
| 4.400 | 4.340 | .000 | .060 | .96 | 2.23 | 1.4% | 13 | .683 | .633 | .577 | .516 | .450 | .382 | .314 | .249 | .190 | .140 |
| | | | | | | | | .098 | .066 | .042 | | | | | | | |
| 4.400 | 4.366 | .000 | .034 | .97 | 2.16 | .8% | 14 | .683 | .633 | .577 | .516 | .450 | .382 | .314 | .249 | .190 | .140 |
| | | | | | | | | .098 | .066 | .042 | .026 | | | | | | |
| 4.600 | 3.332 | .000 | 1.268 | .60 | 2.54 | 27.6% | 6 | .694 | .646 | .592 | .532 | .468 | .401 | | | | |
| 4.600 | 3.665 | .000 | .935 | .67 | 2.56 | 20.3% | 7 | .694 | .646 | .592 | .532 | .468 | .401 | .333 | | | |
| 4.600 | 3.933 | .000 | .667 | .73 | 2.56 | 14.5% | 8 | .694 | .646 | .592 | .532 | .468 | .401 | .333 | .267 | | |
| 4.600 | 4.140 | .000 | .460 | .79 | 2.53 | 10.0% | 9 | .694 | .646 | .592 | .532 | .468 | .401 | .333 | .267 | .207 | |
| 4.600 | 4.294 | .000 | .306 | .85 | 2.48 | 6.7% | 10 | .694 | .646 | .592 | .532 | .468 | .401 | .333 | .267 | .207 | .154 |
| 4.600 | 4.404 | .000 | .196 | .89 | 2.42 | 4.3% | 11 | .694 | .646 | .592 | .532 | .468 | .401 | .333 | .267 | .207 | .154 |
| | | | | | | | | .110 | | | | | | | | | |
| 4.600 | 4.479 | .000 | .121 | .92 | 2.35 | 2.6% | 12 | .694 | .646 | .592 | .532 | .468 | .401 | .333 | .267 | .207 | .154 |

*SIMILAR TO A.T.&T. NEAL-WILKINSON TABLES(SEE SECTION 1)

*EQUIVALENT RANDOM THEORY: VMR(VARIANCE TO MEAN RATIO)– 2.0 - NO RETRIES (PEAKED TRAFFIC - ROUTE ADVANCEMENT OR DAY-TO-DAY VARIATION)

| HOURS FIRST ATTEMPT VMR = 2.0 | HOURS CON-NECT | HOURS RECALL /RETRY | HOURS OVFL | CON-NECT VMR | OVFL VMR | PERCENT BLOCKED (GRADE) | LINE | HOURS CONNECTED TRAFFIC ON LINES 1 | 2 | 3 | 4 | 5 | 6 | 7 | 8 | 9 | 10 |
|---|---|---|---|---|---|---|---|---|---|---|---|---|---|---|---|---|---|
| | | | | | | | | .110 | .075 | | | | | | | | |
| 4.600 | 4.528 | .000 | .072 | .95 | 2.27 | 1.6% | 13 | .694 | .646 | .592 | .532 | .468 | .401 | .333 | .267 | .207 | .154 |
| | | | | | | | | .110 | .075 | .049 | | | | | | | |
| 4.600 | 4.559 | .000 | .041 | .97 | 2.19 | .9% | 14 | .694 | .646 | .592 | .532 | .468 | .401 | .333 | .267 | .207 | .154 |
| | | | | | | | | .110 | .075 | .049 | .031 | | | | | | |
| 4.800 | 3.420 | .000 | 1.380 | .58 | 2.55 | 28.8% | 6 | .704 | .658 | .606 | .548 | .485 | .419 | | | | |
| 4.800 | 3.771 | .000 | 1.029 | .65 | 2.57 | 21.4% | 7 | .704 | .658 | .606 | .548 | .485 | .419 | .352 | | | |
| 4.800 | 4.057 | .000 | .743 | .71 | 2.57 | 15.5% | 8 | .704 | .658 | .606 | .548 | .485 | .419 | .352 | .286 | | |
| 4.800 | 4.281 | .000 | .519 | .78 | 2.55 | 10.8% | 9 | .704 | .658 | .606 | .548 | .485 | .419 | .352 | .286 | .224 | |
| 4.800 | 4.450 | .000 | .350 | .83 | 2.51 | 7.3% | 10 | .704 | .658 | .606 | .548 | .485 | .419 | .352 | .286 | .224 | .169 |
| 4.800 | 4.572 | .000 | .228 | .88 | 2.45 | 4.7% | 11 | .704 | .658 | .606 | .548 | .485 | .419 | .352 | .286 | .224 | .169 |
| | | | | | | | | .122 | | | | | | | | | |
| 4.800 | 4.657 | .000 | .143 | .92 | 2.38 | 3.0% | 12 | .704 | .658 | .606 | .548 | .485 | .419 | .352 | .286 | .224 | .169 |
| | | | | | | | | .122 | .085 | | | | | | | | |
| 4.800 | 4.714 | .000 | .086 | .94 | 2.31 | 1.8% | 13 | .704 | .658 | .606 | .548 | .485 | .419 | .352 | .286 | .224 | .169 |
| | | | | | | | | .122 | .085 | .057 | | | | | | | |
| 4.800 | 4.750 | .000 | .050 | .96 | 2.23 | 1.0% | 14 | .704 | .658 | .606 | .548 | .485 | .419 | .352 | .286 | .224 | .169 |
| | | | | | | | | .122 | .085 | .057 | .036 | | | | | | |
| 4.800 | 4.772 | .000 | .028 | .98 | 2.15 | .6% | 15 | .704 | .658 | .606 | .548 | .485 | .419 | .352 | .286 | .224 | .169 |
| | | | | | | | | .122 | .085 | .057 | .036 | .022 | | | | | |
| 5.000 | 3.503 | .000 | 1.497 | .56 | 2.55 | 29.9% | 6 | .714 | .669 | .619 | .563 | .502 | .437 | | | | |
| 5.000 | 3.873 | .000 | 1.127 | .63 | 2.58 | 22.5% | 7 | .714 | .669 | .619 | .563 | .502 | .437 | .370 | | | |
| 5.000 | 4.177 | .000 | .823 | .70 | 2.59 | 16.5% | 8 | .714 | .669 | .619 | .563 | .502 | .437 | .370 | .304 | | |
| 5.000 | 4.418 | .000 | .582 | .76 | 2.58 | 11.6% | 9 | .714 | .669 | .619 | .563 | .502 | .437 | .370 | .304 | .241 | |
| 5.000 | 4.602 | .000 | .398 | .82 | 2.54 | 8.0% | 10 | .714 | .669 | .619 | .563 | .502 | .437 | .370 | .304 | .241 | .184 |
| 5.000 | 4.738 | .000 | .262 | .86 | 2.48 | 5.2% | 11 | .714 | .669 | .619 | .563 | .502 | .437 | .370 | .304 | .241 | .184 |
| | | | | | | | | .135 | | | | | | | | | |
| 5.000 | 4.833 | .000 | .167 | .90 | 2.42 | 3.3% | 12 | .714 | .669 | .619 | .563 | .502 | .437 | .370 | .304 | .241 | .184 |
| | | | | | | | | .135 | .095 | | | | | | | | |
| 5.000 | 4.898 | .000 | .102 | .94 | 2.34 | 2.0% | 13 | .714 | .669 | .619 | .563 | .502 | .437 | .370 | .304 | .241 | .184 |
| | | | | | | | | .135 | .095 | .065 | | | | | | | |
| 5.000 | 4.940 | .000 | .060 | .96 | 2.27 | 1.2% | 14 | .714 | .669 | .619 | .563 | .502 | .437 | .370 | .304 | .241 | .184 |
| | | | | | | | | .135 | .095 | .065 | .042 | | | | | | |
| 5.000 | 4.966 | .000 | .034 | .97 | 2.19 | .7% | 15 | .714 | .669 | .619 | .563 | .502 | .437 | .370 | .304 | .241 | .184 |
| | | | | | | | | .135 | .095 | .065 | .042 | .026 | | | | | |
| 5.500 | 3.668 | .000 | 1.832 | .52 | 2.61 | 33.3% | 6 | .730 | .689 | .644 | .593 | .536 | .476 | | | | |
| 5.500 | 4.081 | .000 | 1.419 | .59 | 2.65 | 25.8% | 7 | .730 | .689 | .644 | .593 | .536 | .476 | .413 | | | |
| 5.500 | 4.429 | .000 | 1.071 | .65 | 2.68 | 19.5% | 8 | .730 | .689 | .644 | .593 | .536 | .476 | .413 | .349 | | |
| 5.500 | 4.715 | .000 | .785 | .71 | 2.68 | 14.3% | 9 | .730 | .689 | .644 | .593 | .536 | .476 | .413 | .349 | .286 | |
| 5.500 | 4.942 | .000 | .558 | .77 | 2.65 | 10.1% | 10 | .730 | .689 | .644 | .593 | .536 | .476 | .413 | .349 | .286 | .227 |

*SIMILAR TO A.T.&T. NEAL-WILKINSON TABLES(SEE SECTION 1)

*EQUIVALENT RANDOM THEORY: VMR(VARIANCE TO MEAN RATIO)– 2.0 - NO RETRIES (PEAKED TRAFFIC - ROUTE ADVANCEMENT OR DAY-TO-DAY VARIATION)

| HOURS FIRST ATTEMPT VMR = 2.0 | HOURS CON-NECT | HOURS RECALL /RETRY | HOURS OVFL | CON-NECT VMR | OVFL VMR | PERCENT BLOCKED (GRADE) | LINE | HOURS CONNECTED TRAFFIC ON LINES 1 | 2 | 3 | 4 | 5 | 6 | 7 | 8 | 9 | 10 |
|---|---|---|---|---|---|---|---|---|---|---|---|---|---|---|---|---|---|
| 5.500 | 5.116 | .000 | .384 | .83 | 2.61 | 7.0% | 11 | .730 | .689 | .644 | .593 | .536 | .476 | .413 | .349 | .286 | .227 |
| | | | | | | | | .174 | | | | | | | | | |
| 5.500 | 5.244 | .000 | .256 | .87 | 2.55 | 4.7% | 12 | .730 | .689 | .644 | .593 | .536 | .476 | .413 | .349 | .286 | .227 |
| | | | | | | | | .174 | .128 | | | | | | | | |
| 5.500 | 5.335 | .000 | .165 | .91 | 2.48 | 3.0% | 13 | .730 | .689 | .644 | .593 | .536 | .476 | .413 | .349 | .286 | .227 |
| | | | | | | | | .174 | .128 | .091 | | | | | | | |
| 5.500 | 5.397 | .000 | .103 | .94 | 2.41 | 1.9% | 14 | .730 | .689 | .644 | .593 | .536 | .476 | .413 | .349 | .286 | .227 |
| | | | | | | | | .174 | .128 | .091 | .062 | | | | | | |
| 5.500 | 5.438 | .000 | .062 | .96 | 2.33 | 1.1% | 15 | .730 | .689 | .644 | .593 | .536 | .476 | .413 | .349 | .286 | .227 |
| | | | | | | | | .174 | .128 | .091 | .062 | .041 | | | | | |
| 5.500 | 5.464 | .000 | .036 | .97 | 2.25 | .7% | 16 | .730 | .689 | .644 | .593 | .536 | .476 | .413 | .349 | .286 | .227 |
| | | | | | | | | .174 | .128 | .091 | .062 | .041 | .026 | | | | |
| 6.000 | 4.295 | .000 | 1.705 | .55 | 2.66 | 28.4% | 7 | .749 | .713 | .671 | .624 | .571 | .514 | .453 | | | |
| 6.000 | 4.686 | .000 | 1.314 | .61 | 2.70 | 21.9% | 8 | .749 | .713 | .671 | .624 | .571 | .514 | .453 | .391 | | |
| 6.000 | 5.013 | .000 | .987 | .67 | 2.71 | 16.4% | 9 | .749 | .713 | .671 | .624 | .571 | .514 | .453 | .391 | .328 | |
| 6.000 | 5.280 | .000 | .720 | .73 | 2.71 | 12.0% | 10 | .749 | .713 | .671 | .624 | .571 | .514 | .453 | .391 | .328 | .267 |
| 6.000 | 5.490 | .000 | .510 | .79 | 2.68 | 8.5% | 11 | .749 | .713 | .671 | .624 | .571 | .514 | .453 | .391 | .328 | .267 |
| | | | | | | | | .210 | | | | | | | | | |
| 6.000 | 5.650 | .000 | .350 | .84 | 2.63 | 5.8% | 12 | .749 | .713 | .671 | .624 | .571 | .514 | .453 | .391 | .328 | .267 |
| | | | | | | | | .210 | .160 | | | | | | | | |
| 6.000 | 5.768 | .000 | .232 | .88 | 2.57 | 3.9% | 13 | .749 | .713 | .671 | .624 | .571 | .514 | .453 | .391 | .328 | .267 |
| | | | | | | | | .210 | .160 | .117 | | | | | | | |
| 6.000 | 5.851 | .000 | .149 | .92 | 2.49 | 2.5% | 14 | .749 | .713 | .671 | .624 | .571 | .514 | .453 | .391 | .328 | .267 |
| | | | | | | | | .210 | .160 | .117 | .083 | | | | | | |
| 6.000 | 5.907 | .000 | .093 | .94 | 2.42 | 1.5% | 15 | .749 | .713 | .671 | .024 | .571 | .514 | .453 | .391 | .328 | .207 |
| | | | | | | | | .210 | .160 | .117 | .083 | .056 | | | | | |
| 6.000 | 5.944 | .000 | .056 | .96 | 2.34 | .9% | 16 | .749 | .713 | .671 | .624 | .571 | .514 | .453 | .391 | .328 | .267 |
| | | | | | | | | .210 | .160 | .117 | .083 | .056 | .037 | | | | |
| 6.000 | 5.967 | .000 | .033 | .98 | 2.26 | .5% | 17 | .749 | .713 | .671 | .624 | .571 | .514 | .453 | .391 | .328 | .267 |
| | | | | | | | | .210 | .160 | .117 | .083 | .056 | .037 | .023 | | | |
| 6.500 | 4.458 | .000 | 2.042 | .51 | 2.71 | 31.4% | 7 | .761 | .728 | .690 | .647 | .599 | .546 | .489 | | | |
| 6.500 | 4.888 | .000 | 1.612 | .57 | 2.76 | 24.8% | 8 | .761 | .728 | .690 | .647 | .599 | .546 | .489 | .430 | | |
| 6.500 | 5.257 | .000 | 1.243 | .63 | 2.79 | 19.1% | 9 | .761 | .728 | .690 | .647 | .599 | .546 | .489 | .430 | .009 | |
| 6.500 | 5.565 | .000 | .935 | .69 | 2.80 | 14.4% | 10 | .761 | .728 | .690 | .647 | .599 | .546 | .489 | .430 | .369 | .308 |
| 6.500 | 5.816 | .000 | .684 | .75 | 2.78 | 10.5% | 11 | .761 | .728 | .690 | .647 | .599 | .546 | .489 | .430 | .369 | .308 |
| | | | | | | | | .251 | | | | | | | | | |
| 6.500 | 6.013 | .000 | .487 | .80 | 2.75 | 7.5% | 12 | .761 | .728 | .690 | .647 | .599 | .546 | .489 | .430 | .369 | .308 |
| | | | | | | | | .251 | .197 | | | | | | | | |
| 6.500 | 6.164 | .000 | .336 | .85 | 2.69 | 5.2% | 13 | .761 | .728 | .690 | .647 | .599 | .546 | .489 | .430 | .369 | .308 |
| | | | | | | | | .251 | .197 | .151 | | | | | | | |

*SIMILAR TO A.T.&T. NEAL-WILKINSON TABLES(SEE SECTION 1)

*EQUIVALENT RANDOM THEORY: VMR(VARIANCE TO MEAN RATIO)– 2.0 - NO RETRIES (PEAKED TRAFFIC - ROUTE ADVANCEMENT OR DAY-TO-DAY VARIATION)

| HOURS FIRST ATTEMPT VMR = 2.0 | HOURS CON-NECT | HOURS RECALL /RETRY | HOURS OVFL | CON-NECT VMR | OVFL VMR | PERCENT BLOCKED (GRADE) | LINE | HOURS CONNECTED TRAFFIC ON LINES 1 | 2 | 3 | 4 | 5 | 6 | 7 | 8 | 9 | 10 |
|---|---|---|---|---|---|---|---|---|---|---|---|---|---|---|---|---|---|
| 6.500 | 6.275 | .000 | .225 | .89 | 2.63 | 3.5% | 14 | .761 | .728 | .690 | .647 | .599 | .546 | .489 | .430 | .369 | .308 |
| | | | | | | | | .251 | .197 | .151 | .111 | | | | | | |
| 6.500 | 6.354 | .000 | .146 | .92 | 2.55 | 2.3% | 15 | .761 | .728 | .690 | .647 | .599 | .546 | .489 | .430 | .369 | .308 |
| | | | | | | | | .251 | .197 | .151 | .111 | .079 | | | | | |
| 6.500 | 6.408 | .000 | .092 | .95 | 2.48 | 1.4% | 16 | .761 | .728 | .690 | .647 | .599 | .546 | .489 | .430 | .369 | .308 |
| | | | | | | | | .251 | .197 | .151 | .111 | .079 | .054 | | | | |
| 6.500 | 6.444 | .000 | .056 | .96 | 2.40 | .9% | 17 | .761 | .728 | .690 | .647 | .599 | .546 | .489 | .430 | .369 | .308 |
| | | | | | | | | .251 | .197 | .151 | .111 | .079 | .054 | .036 | | | |
| 6.500 | 6.467 | .000 | .033 | .98 | 2.32 | .5% | 18 | .761 | .728 | .690 | .647 | .599 | .546 | .489 | .430 | .369 | .308 |
| | | | | | | | | .251 | .197 | .151 | .111 | .079 | .054 | .036 | .023 | | |
| | | | | | | | | | | | | | | | | | |
| 7.000 | 4.634 | .000 | 2.366 | .48 | 2.70 | 33.8% | 7 | .777 | .746 | .711 | .671 | .627 | .578 | .524 | | | |
| 7.000 | 5.101 | .000 | 1.899 | .53 | 2.76 | 27.1% | 8 | .777 | .746 | .711 | .671 | .627 | .578 | .524 | .467 | | |
| 7.000 | 5.508 | .000 | 1.492 | .59 | 2.80 | 21.3% | 9 | .777 | .746 | .711 | .671 | .627 | .578 | .524 | .467 | .407 | |
| 7.000 | 5.856 | .000 | 1.144 | .65 | 2.83 | 16.3% | 10 | .777 | .746 | .711 | .671 | .627 | .578 | .524 | .467 | .407 | .347 |
| 7.000 | 6.144 | .000 | .856 | .71 | 2.82 | 12.2% | 11 | .777 | .746 | .711 | .671 | .627 | .578 | .524 | .467 | .407 | .347 |
| | | | | | | | | .288 | | | | | | | | | |
| 7.000 | 6.377 | .000 | .623 | .77 | 2.80 | 8.9% | 12 | .777 | .746 | .711 | .671 | .627 | .578 | .524 | .467 | .407 | .347 |
| | | | | | | | | .288 | .233 | | | | | | | | |
| 7.000 | 6.559 | .000 | .441 | .82 | 2.76 | 6.3% | 13 | .777 | .746 | .711 | .671 | .627 | .578 | .524 | .467 | .407 | .347 |
| | | | | | | | | .288 | .233 | .182 | | | | | | | |
| 7.000 | 6.696 | .000 | .304 | .86 | 2.70 | 4.3% | 14 | .777 | .746 | .711 | .671 | .627 | .578 | .524 | .467 | .407 | .347 |
| | | | | | | | | .288 | .233 | .182 | .138 | | | | | | |
| 7.000 | 6.797 | .000 | .203 | .90 | 2.63 | 2.9% | 15 | .777 | .746 | .711 | .671 | .627 | .578 | .524 | .467 | .407 | .347 |
| | | | | | | | | .288 | .233 | .182 | .138 | .101 | | | | | |
| 7.000 | 6.869 | .000 | .131 | .93 | 2.56 | 1.9% | 16 | .777 | .746 | .711 | .671 | .627 | .578 | .524 | .467 | .407 | .347 |
| | | | | | | | | .288 | .233 | .182 | .138 | .101 | .071 | | | | |
| 7.000 | 6.918 | .000 | .082 | .95 | 2.48 | 1.2% | 17 | .777 | .746 | .711 | .671 | .627 | .578 | .524 | .467 | .407 | .347 |
| | | | | | | | | .288 | .233 | .182 | .138 | .101 | .071 | .049 | | | |
| 7.000 | 6.950 | .000 | .050 | .97 | 2.40 | .7% | 18 | .777 | .746 | .711 | .671 | .627 | .578 | .524 | .467 | .407 | .347 |
| | | | | | | | | .288 | .233 | .182 | .138 | .101 | .071 | .049 | .032 | | |
| | | | | | | | | | | | | | | | | | |
| 7.500 | 5.295 | .000 | 2.205 | .50 | 2.76 | 29.4% | 8 | .791 | .763 | .731 | .694 | .653 | .607 | .556 | .502 | | |
| 7.500 | 5.739 | .000 | 1.761 | .56 | 2.81 | 23.5% | 9 | .791 | .763 | .731 | .694 | .653 | .607 | .556 | .502 | .444 | |
| 7.500 | 6.124 | .000 | 1.376 | .61 | 2.85 | 18.3% | 10 | .791 | .763 | .731 | .694 | .653 | .607 | .556 | .502 | .444 | .385 |
| 7.500 | 6.450 | .000 | 1.050 | .67 | 2.86 | 14.0% | 11 | .701 | .763 | .731 | .694 | .653 | .607 | .556 | .502 | .444 | .385 |
| | | | | | | | | .326 | | | | | | | | | |
| 7.500 | 6.719 | .000 | .781 | .73 | 2.85 | 10.4% | 12 | .791 | .763 | .731 | .694 | .653 | .607 | .556 | .502 | .444 | .385 |
| | | | | | | | | .326 | .269 | | | | | | | | |
| 7.500 | 6.934 | .000 | .566 | .78 | 2.82 | 7.5% | 13 | .791 | .763 | .731 | .694 | .653 | .607 | .556 | .502 | .444 | .385 |
| | | | | | | | | .326 | .269 | .215 | | | | | | | |
| 7.500 | 7.101 | .000 | .399 | .83 | 2.77 | 5.3% | 14 | .791 | .763 | .731 | .694 | .653 | .607 | .556 | .502 | .444 | .385 |

*SIMILAR TO A.T.&T. NEAL-WILKINSON TABLES(SEE SECTION 1)

*EQUIVALENT RANDOM THEORY: VMR(VARIANCE TO MEAN RATIO)– 2.0 - NO RETRIES (PEAKED TRAFFIC - ROUTE ADVANCEMENT OR DAY-TO-DAY VARIATION)

| HOURS FIRST ATTEMPT VMR = 2.0 | HOURS CON-NECT | HOURS RECALL /RETRY | HOURS OVFL | CON-NECT VMR | OVFL VMR | PERCENT BLOCKED (GRADE) | LINE | HOURS CONNECTED TRAFFIC ON LINES 1 | 2 | 3 | 4 | 5 | 6 | 7 | 8 | 9 | 10 |
|---|---|---|---|---|---|---|---|---|---|---|---|---|---|---|---|---|---|
| | | | | | | | | .326 | .269 | .215 | .167 | | | | | | |
| 7.500 | 7.227 | .000 | .273 | .87 | 2.71 | 3.6% | 15 | .791 | .763 | .731 | .694 | .653 | .607 | .556 | .502 | .444 | .385 |
| | | | | | | | | .326 | .269 | .215 | .167 | .126 | | | | | |
| 7.500 | 7.319 | .000 | .181 | .91 | 2.64 | 2.4% | 16 | .791 | .763 | .731 | .694 | .653 | .607 | .556 | .502 | .444 | .385 |
| | | | | | | | | .326 | .269 | .215 | .167 | .126 | .091 | | | | |
| 7.500 | 7.383 | .000 | .117 | .94 | 2.56 | 1.6% | 17 | .791 | .763 | .731 | .694 | .653 | .607 | .556 | .502 | .444 | .385 |
| | | | | | | | | .326 | .269 | .215 | .167 | .126 | .091 | .064 | | | |
| 7.500 | 7.427 | .000 | .073 | .96 | 2.48 | 1.0% | 18 | .791 | .763 | .731 | .694 | .653 | .607 | .556 | .502 | .444 | .385 |
| | | | | | | | | .326 | .269 | .215 | .167 | .126 | .091 | .064 | .044 | | |
| 7.500 | 7.455 | .000 | .045 | .97 | 2.40 | .6% | 19 | .791 | .763 | .731 | .694 | .653 | .607 | .556 | .502 | .444 | .385 |
| | | | | | | | | .326 | .269 | .215 | .167 | .126 | .091 | .064 | .044 | .029 | |
| 8.000 | 5.441 | .000 | 2.559 | .47 | 2.80 | 32.0% | 8 | .799 | .773 | .743 | .710 | .672 | .629 | .583 | .532 | | |
| 8.000 | 5.918 | .000 | 2.082 | .52 | 2.86 | 26.0% | 9 | .799 | .773 | .743 | .710 | .672 | .629 | .583 | .532 | .478 | |
| 8.000 | 6.340 | .000 | 1.660 | .58 | 2.91 | 20.8% | 10 | .799 | .773 | .743 | .710 | .672 | .629 | .583 | .532 | .478 | .421 |
| 8.000 | 6.703 | .000 | 1.297 | .64 | 2.93 | 16.2% | 11 | .799 | .773 | .743 | .710 | .672 | .629 | .583 | .532 | .478 | .421 |
| | | | | | | | | .364 | | | | | | | | | |
| 8.000 | 7.010 | .000 | .990 | .69 | 2.93 | 12.4% | 12 | .799 | .773 | .743 | .710 | .672 | .629 | .583 | .532 | .478 | .421 |
| | | | | | | | | .364 | .307 | | | | | | | | |
| 8.000 | 7.262 | .000 | .738 | .75 | 2.92 | 9.2% | 13 | .799 | .773 | .743 | .710 | .672 | .629 | .583 | .532 | .478 | .421 |
| | | | | | | | | .364 | .307 | .252 | | | | | | | |
| 8.000 | 7.464 | .000 | .536 | .80 | 2.88 | 6.7% | 14 | .799 | .773 | .743 | .710 | .672 | .629 | .583 | .532 | .478 | .421 |
| | | | | | | | | .364 | .307 | .252 | .202 | | | | | | |
| 8.000 | 7.621 | .000 | .379 | .84 | 2.83 | 4.7% | 15 | .799 | .773 | .743 | .710 | .672 | .629 | .583 | .532 | .478 | .421 |
| | | | | | | | | .364 | .307 | .252 | .202 | .157 | | | | | |
| 8.000 | 7.739 | .000 | .261 | .88 | 2.77 | 3.3% | 16 | .799 | .773 | .743 | .710 | .672 | .629 | .583 | .532 | .478 | .421 |
| | | | | | | | | .364 | .307 | .252 | .202 | .157 | .118 | | | | |
| 8.000 | 7.825 | .000 | .175 | .91 | 2.69 | 2.2% | 17 | .799 | .773 | .743 | .710 | .672 | .629 | .583 | .532 | .478 | .421 |
| | | | | | | | | .364 | .307 | .252 | .202 | .157 | .118 | .086 | | | |
| 8.000 | 7.886 | .000 | .114 | .94 | 2.61 | 1.4% | 18 | .799 | .773 | .743 | .710 | .672 | .629 | .583 | .532 | .478 | .421 |
| | | | | | | | | .364 | .307 | .252 | .202 | .157 | .118 | .086 | .061 | | |
| 8.000 | 7.928 | .000 | .072 | .96 | 2.53 | .9% | 19 | .799 | .773 | .743 | .710 | .672 | .629 | .583 | .532 | .478 | .421 |
| | | | | | | | | .364 | .307 | .252 | .202 | .157 | .118 | .086 | .061 | .042 | |
| 8.000 | 7.956 | .000 | .044 | .97 | 2.45 | .6% | 20 | .799 | .773 | .743 | .710 | .672 | .629 | .583 | .532 | .478 | .421 |
| | | | | | | | | .364 | .307 | .252 | .202 | .157 | .118 | .086 | .061 | .042 | .028 |
| 8.500 | 5.603 | .000 | 2.897 | .44 | 2.78 | 34.1% | 8 | .811 | .787 | .759 | .728 | .693 | .653 | .610 | .562 | | |
| 8.500 | 6.113 | .000 | 2.387 | .49 | 2.85 | 28.1% | 9 | .811 | .787 | .759 | .728 | .693 | .653 | .610 | .562 | .510 | |
| 8.500 | 6.568 | .000 | 1.932 | .54 | 2.91 | 22.7% | 10 | .811 | .787 | .759 | .728 | .693 | .653 | .610 | .562 | .510 | .455 |
| 8.500 | 6.067 | .000 | 1.533 | .60 | 2.95 | 18.0% | 11 | .811 | .787 | .759 | .728 | .693 | .653 | .610 | .562 | .510 | .455 |
| | | | | | | | | .399 | | | | | | | | | |
| 8.500 | 7.309 | .000 | 1.191 | .66 | 2.96 | 14.0% | 12 | .811 | .787 | .759 | .728 | .693 | .653 | .610 | .562 | .510 | .455 |

*SIMILAR TO A.T.&T. NEAL-WILKINSON TABLES(SEE SECTION 1)

*EQUIVALENT RANDOM THEORY: VMR(VARIANCE TO MEAN RATIO)– 2.0 - NO RETRIES (PEAKED TRAFFIC - ROUTE ADVANCEMENT OR DAY-TO-DAY VARIATION)

| HOURS FIRST ATTEMPT VMR = 2.0 | HOURS CON-NECT | HOURS RECALL /RETRY | HOURS OVFL | CON-NECT VMR | OVFL VMR | PERCENT BLOCKED (GRADE) | LINE | HOURS CONNECTED TRAFFIC ON LINES 1 | 2 | 3 | 4 | 5 | 6 | 7 | 8 | 9 | 10 |
|---|---|---|---|---|---|---|---|---|---|---|---|---|---|---|---|---|---|
| | | | | | | | | .399 | .342 | | | | | | | | |
| 8.500 | 7.596 | .000 | .904 | .71 | 2.96 | 10.6% | 13 | .811 | .787 | .759 | .728 | .693 | .653 | .610 | .562 | .510 | .455 |
| | | | | | | | | .399 | .342 | .287 | | | | | | | |
| 8.500 | 7.830 | .000 | .670 | .77 | 2.94 | 7.9% | 14 | .811 | .787 | .759 | .728 | .693 | .653 | .610 | .562 | .510 | .455 |
| | | | | | | | | .399 | .342 | .287 | .234 | | | | | | |
| 8.500 | 8.016 | .000 | .484 | .81 | 2.89 | 5.7% | 15 | .811 | .787 | .759 | .728 | .693 | .653 | .610 | .562 | .510 | .455 |
| | | | | | | | | .399 | .342 | .287 | .234 | .186 | | | | | |
| 8.500 | 8.159 | .000 | .341 | .86 | 2.84 | 4.0% | 16 | .811 | .787 | .759 | .728 | .693 | .653 | .610 | .562 | .510 | .455 |
| | | | | | | | | .399 | .342 | .287 | .234 | .186 | .143 | | | | |
| 8.500 | 8.266 | .000 | .234 | .89 | 2.77 | 2.7% | 17 | .811 | .787 | .759 | .728 | .693 | .653 | .610 | .562 | .510 | .455 |
| | | | | | | | | .399 | .342 | .287 | .234 | .186 | .143 | .107 | | | |
| 8.500 | 8.344 | .000 | .156 | .92 | 2.69 | 1.8% | 18 | .811 | .787 | .759 | .728 | .693 | .653 | .610 | .562 | .510 | .455 |
| | | | | | | | | .399 | .342 | .287 | .234 | .186 | .143 | .107 | .078 | | |
| 8.500 | 8.399 | .000 | .101 | .95 | 2.61 | 1.2% | 19 | .811 | .787 | .759 | .728 | .693 | .653 | .610 | .562 | .510 | .455 |
| | | | | | | | | .399 | .342 | .287 | .234 | .186 | .143 | .107 | .078 | .055 | |
| 8.500 | 8.436 | .000 | .064 | .96 | 2.53 | .8% | 20 | .811 | .787 | .759 | .728 | .693 | .653 | .610 | .562 | .510 | .455 |
| | | | | | | | | .399 | .342 | .287 | .234 | .186 | .143 | .107 | .078 | .055 | .037 |
| 9.000 | 6.259 | .000 | 2.741 | .46 | 2.89 | 30.5% | 9 | .817 | .795 | .770 | .741 | .708 | .672 | .631 | .587 | .538 | |
| 9.000 | 6.745 | .000 | 2.255 | .51 | 2.95 | 25.1% | 10 | .817 | .795 | .770 | .741 | .708 | .672 | .631 | .587 | .538 | .487 |
| 9.000 | 7.178 | .000 | 1.822 | .57 | 3.00 | 20.2% | 11 | .817 | .795 | .770 | .741 | .708 | .672 | .631 | .587 | .538 | .487 |
| | | | | | | | | .433 | | | | | | | | | |
| 9.000 | 7.556 | .000 | 1.444 | .62 | 3.03 | 16.0% | 12 | .817 | .795 | .770 | .741 | .708 | .672 | .631 | .587 | .538 | .487 |
| | | | | | | | | .433 | .378 | | | | | | | | |
| 9.000 | 7.879 | .000 | 1.121 | .68 | 3.04 | 12.5% | 13 | .817 | .795 | .770 | .741 | .708 | .672 | .631 | .587 | .538 | .487 |
| | | | | | | | | .433 | .378 | .323 | | | | | | | |
| 9.000 | 8.148 | .000 | .852 | .73 | 3.03 | 9.5% | 14 | .817 | .795 | .770 | .741 | .708 | .672 | .631 | .587 | .538 | .487 |
| | | | | | | | | .433 | .378 | .323 | .270 | .220 | | | | | |
| 9.000 | 8.368 | .000 | .632 | .78 | 3.00 | 7.0% | 15 | .817 | .795 | .770 | .741 | .708 | .672 | .631 | .587 | .538 | .487 |
| | | | | | | | | .433 | .378 | .323 | .270 | .220 | | | | | |
| 9.000 | 8.542 | .000 | .458 | .83 | 2.95 | 5.1% | 16 | .817 | .795 | .770 | .741 | .708 | .672 | .631 | .587 | .538 | .487 |
| | | | | | | | | .433 | .378 | .323 | .270 | .220 | .174 | | | | |
| 9.000 | 8.677 | .000 | .323 | .87 | 2.89 | 3.6% | 17 | .817 | .795 | .770 | .741 | .708 | .672 | .631 | .587 | .538 | .487 |
| | | | | | | | | .433 | .378 | .323 | .270 | .220 | .174 | .134 | | | |
| 9.000 | 8.777 | .000 | .223 | .90 | 2.82 | 2.5% | 18 | .817 | .795 | .770 | .741 | .708 | .672 | .631 | .587 | .538 | .487 |
| | | | | | | | | .433 | .378 | .323 | .270 | .220 | .174 | .134 | .101 | | |
| 9.000 | 8.850 | .000 | .150 | .93 | 2.74 | 1.7% | 19 | .817 | .795 | .770 | .741 | .708 | .672 | .631 | .587 | .538 | .487 |
| | | | | | | | | .433 | .378 | .323 | .270 | .220 | .174 | .134 | .101 | .073 | |
| 9.000 | 8.902 | .000 | .098 | .95 | 2.66 | 1.1% | 20 | .817 | .795 | .770 | .741 | .708 | .672 | .631 | .587 | .538 | .487 |
| | | | | | | | | .433 | .378 | .323 | .270 | .220 | .174 | .134 | .101 | .073 | .052 |
| 9.000 | 8.938 | .000 | .062 | .96 | 2.58 | .7% | 21 | .817 | .795 | .770 | .741 | .708 | .672 | .631 | .587 | .538 | .487 |

*SIMILAR TO A.T.&T. NEAL-WILKINSON TABLES(SEE SECTION 1)

*EQUIVALENT RANDOM THEORY: VMR(VARIANCE TO MEAN RATIO)– 2.0 - NO RETRIES (PEAKED TRAFFIC - ROUTE ADVANCEMENT OR DAY-TO-DAY VARIATION)

| HOURS FIRST ATTEMPT VMR = 2.0 | HOURS CON-NECT | HOURS RECALL /RETRY | HOURS OVFL | CON-NECT VMR | OVFL VMR | PERCENT BLOCKED (GRADE) | LINE | HOURS CONNECTED TRAFFIC ON LINES 1 | 2 | 3 | 4 | 5 | 6 | 7 | 8 | 9 | 10 |
|---|---|---|---|---|---|---|---|---|---|---|---|---|---|---|---|---|---|
| | | | | | | | | .433 | .378 | .323 | .270 | .220 | .174 | .134 | .101 | .073 | .052 |
| | | | | | | | | .035 | | | | | | | | | |
| 9.500 | 6.422 | .000 | 3.078 | .43 | 2.87 | 32.4% | 9 | .827 | .806 | .783 | .756 | .726 | .692 | .654 | .612 | .566 | |
| 9.500 | 6.940 | .000 | 2.560 | .48 | 2.95 | 27.0% | 10 | .827 | .806 | .783 | .756 | .726 | .692 | .654 | .612 | .566 | .517 |
| 9.500 | 7.404 | .000 | 2.096 | .54 | 3.00 | 22.1% | 11 | .827 | .806 | .783 | .756 | .726 | .692 | .654 | .612 | .566 | .517 |
| | | | | | | | | .465 | | | | | | | | | |
| 9.500 | 7.816 | .000 | 1.684 | .59 | 3.04 | 17.7% | 12 | .827 | .806 | .783 | .756 | .726 | .692 | .654 | .612 | .566 | .517 |
| | | | | | | | | .465 | .411 | | | | | | | | |
| 9.500 | 8.172 | .000 | 1.328 | .64 | 3.06 | 14.0% | 13 | .827 | .806 | .783 | .756 | .726 | .692 | .654 | .612 | .566 | .517 |
| | | | | | | | | .465 | .411 | .356 | | | | | | | |
| 9.500 | 8.474 | .000 | 1.026 | .70 | 3.06 | 10.8% | 14 | .827 | .806 | .783 | .756 | .726 | .692 | .654 | .612 | .566 | .517 |
| | | | | | | | | .465 | .411 | .356 | .303 | | | | | | |
| 9.500 | 8.725 | .000 | .775 | .75 | 3.04 | 8.2% | 15 | .827 | .806 | .783 | .756 | .726 | .692 | .654 | .612 | .566 | .517 |
| | | | | | | | | .465 | .411 | .356 | .303 | .251 | | | | | |
| 9.500 | 8.928 | .000 | .572 | .80 | 3.01 | 6.0% | 16 | .827 | .806 | .783 | .756 | .726 | .692 | .654 | .612 | .566 | .517 |
| | | | | | | | | .465 | .411 | .356 | .303 | .251 | .203 | | | | |
| 9.500 | 9.088 | .000 | .412 | .84 | 2.96 | 4.3% | 17 | .827 | .806 | .783 | .756 | .726 | .692 | .654 | .612 | .566 | .517 |
| | | | | | | | | .465 | .411 | .356 | .303 | .251 | .203 | .160 | | | |
| 9.500 | 9.210 | .000 | .290 | .88 | 2.89 | 3.0% | 18 | .827 | .806 | .783 | .756 | .726 | .692 | .654 | .612 | .566 | .517 |
| | | | | | | | | .465 | .411 | .356 | .303 | .251 | .203 | .160 | .122 | | |
| 9.500 | 9.301 | .000 | .199 | .91 | 2.82 | 2.1% | 19 | .827 | .806 | .783 | .756 | .726 | .692 | .654 | .612 | .566 | .517 |
| | | | | | | | | .465 | .411 | .356 | .303 | .251 | .203 | .160 | .122 | .091 | |
| 9.500 | 9.367 | .000 | .133 | .93 | 2.74 | 1.4% | 20 | .827 | .806 | .783 | .756 | .726 | .692 | .654 | .612 | .566 | .517 |
| | | | | | | | | .465 | .411 | .356 | .303 | .251 | .203 | .160 | .122 | .091 | .066 |
| 9.500 | 9.414 | .000 | .086 | .95 | 2.65 | .9% | 21 | .827 | .806 | .783 | .756 | .726 | .692 | .654 | .612 | .566 | .517 |
| | | | | | | | | .465 | .411 | .356 | .303 | .251 | .203 | .160 | .122 | .091 | .066 |
| | | | | | | | | .046 | | | | | | | | | |
| 9.500 | 9.445 | .000 | .055 | .97 | 2.57 | .6% | 22 | .827 | .806 | .783 | .756 | .726 | .692 | .654 | .612 | .566 | .517 |
| | | | | | | | | .465 | .411 | .356 | .303 | .251 | .203 | .160 | .122 | .091 | .066 |
| | | | | | | | | .046 | .032 | | | | | | | | |
| 10.000 | 6.542 | .000 | 3.458 | .41 | 2.90 | 34.6% | 9 | .833 | .813 | .791 | .766 | .738 | .707 | .672 | .633 | .500 | |
| 10.000 | 7.086 | .000 | 2.914 | .46 | 2.98 | 29.1% | 10 | .833 | .813 | .791 | .766 | .738 | .707 | .672 | .633 | .590 | .544 |
| 10.000 | 7.580 | .000 | 2.420 | .51 | 3.04 | 24.2% | 11 | .833 | .813 | .791 | .766 | .738 | .707 | .672 | .600 | .590 | .544 |
| | | | | | | | | .494 | | | | | | | | | |
| 10.000 | 8.023 | .000 | 1.977 | .56 | 3.09 | 19.8% | 12 | .833 | .813 | .791 | .766 | .738 | .707 | .672 | .633 | .590 | .544 |
| | | | | | | | | .494 | .443 | | | | | | | | |
| 10.000 | 8.413 | .000 | 1.587 | .61 | 3.12 | 15.9% | 13 | .833 | .813 | .701 | .700 | .738 | .707 | .672 | .633 | .590 | .544 |
| | | | | | | | | .494 | .443 | .390 | | | | | | | |
| 10.000 | 8.750 | .000 | 1.250 | .66 | 3.14 | 12.5% | 14 | .833 | .813 | .791 | .766 | .738 | .707 | .672 | .633 | .590 | .544 |
| | | | | | | | | .494 | .443 | .390 | .337 | | | | | | |

*SIMILAR TO A.T.&T. NEAL-WILKINSON TABLES(SEE SECTION 1)

*EQUIVALENT RANDOM THEORY: VMR(VARIANCE TO MEAN RATIO)– 2.0 - NO RETRIES (PEAKED TRAFFIC - ROUTE ADVANCEMENT OR DAY-TO-DAY VARIATION)

| HOURS FIRST ATTEMPT VMR = 2.0 | HOURS CON-NECT | HOURS RECALL /RETRY | HOURS OVFL | CON-NECT VMR | OVFL VMR | PERCENT BLOCKED (GRADE) | LINE | HOURS CONNECTED TRAFFIC ON LINES 1 | 2 | 3 | 4 | 5 | 6 | 7 | 8 | 9 | 10 |
|---|---|---|---|---|---|---|---|---|---|---|---|---|---|---|---|---|---|
| 10.000 | 9.035 | .000 | .965 | .72 | 3.13 | 9.6% | 15 | .833 | .813 | .791 | .766 | .738 | .707 | .672 | .633 | .590 | .544 |
| | | | | | | | | .494 | .443 | .390 | .337 | .285 | | | | | |
| 10.000 | 9.271 | .000 | .729 | .76 | 3.10 | 7.3% | 16 | .833 | .813 | .791 | .766 | .738 | .707 | .672 | .633 | .590 | .544 |
| | | | | | | | | .494 | .443 | .390 | .337 | .285 | .236 | | | | |
| 10.000 | 9.461 | .000 | .539 | .81 | 3.06 | 5.4% | 17 | .833 | .813 | .791 | .766 | .738 | .707 | .672 | .633 | .590 | .544 |
| | | | | | | | | .494 | .443 | .390 | .337 | .285 | .236 | .190 | | | |
| 10.000 | 9.611 | .000 | .389 | .85 | 3.01 | 3.9% | 18 | .833 | .813 | .791 | .766 | .738 | .707 | .672 | .633 | .590 | .544 |
| | | | | | | | | .494 | .443 | .390 | .337 | .285 | .236 | .190 | .150 | | |
| 10.000 | 9.725 | .000 | .275 | .89 | 2.94 | 2.7% | 19 | .833 | .813 | .791 | .766 | .738 | .707 | .672 | .633 | .590 | .544 |
| | | | | | | | | .494 | .443 | .390 | .337 | .285 | .236 | .190 | .150 | .115 | |
| 10.000 | 9.811 | .000 | .189 | .91 | 2.86 | 1.9% | 20 | .833 | .813 | .791 | .766 | .738 | .707 | .672 | .633 | .590 | .544 |
| | | | | | | | | .494 | .443 | .390 | .337 | .285 | .236 | .190 | .150 | .115 | .085 |
| 10.000 | 9.873 | .000 | .127 | .94 | 2.78 | 1.3% | 21 | .833 | .813 | .791 | .766 | .738 | .707 | .672 | .633 | .590 | .544 |
| | | | | | | | | .494 | .443 | .390 | .337 | .285 | .236 | .190 | .150 | .115 | .085 |
| | | | | | | | | .062 | | | | | | | | | |
| 10.000 | 9.917 | .000 | .083 | .96 | 2.70 | .8% | 22 | .833 | .813 | .791 | .766 | .738 | .707 | .672 | .633 | .590 | .544 |
| | | | | | | | | .494 | .443 | .390 | .337 | .285 | .236 | .190 | .150 | .115 | .085 |
| | | | | | | | | .062 | .044 | | | | | | | | |
| 10.000 | 9.947 | .000 | .053 | .97 | 2.62 | .5% | 23 | .833 | .813 | .791 | .766 | .738 | .707 | .672 | .633 | .590 | .544 |
| | | | | | | | | .494 | .443 | .390 | .337 | .285 | .236 | .190 | .150 | .115 | .085 |
| | | | | | | | | .062 | .044 | .030 | | | | | | | |
| 11.000 | 7.373 | .000 | 3.627 | .41 | 2.99 | 33.0% | 10 | .846 | .829 | .809 | .788 | .763 | .736 | .705 | .671 | .634 | .593 |
| 11.000 | 7.922 | .000 | 3.078 | .45 | 3.06 | 28.0% | 11 | .846 | .829 | .809 | .788 | .763 | .736 | .705 | .671 | .634 | .593 |
| | | | | | | | | .548 | | | | | | | | | |
| 11.000 | 8.423 | .000 | 2.577 | .50 | 3.13 | 23.4% | 12 | .846 | .829 | .809 | .788 | .763 | .736 | .705 | .671 | .634 | .593 |
| | | | | | | | | .548 | .501 | | | | | | | | |
| 11.000 | 8.874 | .000 | 2.126 | .55 | 3.18 | 19.3% | 13 | .846 | .829 | .809 | .788 | .763 | .736 | .705 | .671 | .634 | .593 |
| | | | | | | | | .548 | .501 | .452 | | | | | | | |
| 11.000 | 9.275 | .000 | 1.725 | .60 | 3.21 | 15.7% | 14 | .846 | .829 | .809 | .788 | .763 | .736 | .705 | .671 | .634 | .593 |
| | | | | | | | | .548 | .501 | .452 | .401 | | | | | | |
| 11.000 | 9.624 | .000 | 1.376 | .65 | 3.23 | 12.5% | 15 | .846 | .829 | .809 | .788 | .763 | .736 | .705 | .671 | .634 | .593 |
| | | | | | | | | .548 | .501 | .452 | .401 | .349 | | | | | |
| 11.000 | 9.923 | .000 | 1.077 | .70 | 3.23 | 9.8% | 16 | .846 | .829 | .809 | .788 | .763 | .736 | .705 | .671 | .634 | .593 |
| | | | | | | | | .548 | .501 | .452 | .401 | .349 | .299 | | | | |
| 11.000 | 10.173 | .000 | .827 | .75 | 3.21 | 7.5% | 17 | .846 | .829 | .809 | .788 | .763 | .736 | .705 | .671 | .634 | .593 |
| | | | | | | | | .548 | .501 | .452 | .401 | .349 | .299 | .250 | | | |
| 11.000 | 10.378 | .000 | .622 | .79 | 3.17 | 5.7% | 18 | .846 | .829 | .809 | .788 | .763 | .736 | .705 | .671 | .634 | .593 |
| | | | | | | | | .548 | .501 | .452 | .401 | .349 | .299 | .250 | .205 | | |
| 11.000 | 10.542 | .000 | .458 | .84 | 3.12 | 4.2% | 19 | .846 | .829 | .809 | .788 | .763 | .736 | .705 | .671 | .634 | .593 |
| | | | | | | | | .548 | .501 | .452 | .401 | .349 | .299 | .250 | .205 | .164 | |

*SIMILAR TO A.T.&T. NEAL-WILKINSON TABLES(SEE SECTION 1)

*EQUIVALENT RANDOM THEORY: VMR(VARIANCE TO MEAN RATIO)– 2.0 - NO RETRIES (PEAKED TRAFFIC - ROUTE ADVANCEMENT OR DAY-TO-DAY VARIATION)

| HOURS FIRST ATTEMPT VMR = 2.0 | HOURS CON-NECT | HOURS RECALL /RETRY | HOURS OVFL | CON-NECT VMR | OVFL VMR | PERCENT BLOCKED (GRADE) | LINE | HOURS CONNECTED TRAFFIC ON LINES 1 | 2 | 3 | 4 | 5 | 6 | 7 | 8 | 9 | 10 |
|---|---|---|---|---|---|---|---|---|---|---|---|---|---|---|---|---|---|
| 11.000 | 10.670 | .000 | .330 | .87 | 3.05 | 3.0% | 20 | .846 | .829 | .809 | .788 | .763 | .736 | .705 | .671 | .634 | .593 |
| | | | | | | | | .548 | .501 | .452 | .401 | .349 | .299 | .250 | .205 | .164 | .128 |
| 11.000 | 10.768 | .000 | .232 | .90 | 2.98 | 2.1% | 21 | .846 | .829 | .809 | .788 | .763 | .736 | .705 | .671 | .634 | .593 |
| | | | | | | | | .548 | .501 | .452 | .401 | .349 | .299 | .250 | .205 | .164 | .128 |
| | | | | | | | | .098 | | | | | | | | | |
| 11.000 | 10.840 | .000 | .160 | .93 | 2.90 | 1.5% | 22 | .846 | .829 | .809 | .788 | .763 | .736 | .705 | .671 | .634 | .593 |
| | | | | | | | | .548 | .501 | .452 | .401 | .349 | .299 | .250 | .205 | .164 | .128 |
| | | | | | | | | .098 | .072 | | | | | | | | |
| 11.000 | 10.893 | .000 | .107 | .95 | 2.81 | 1.0% | 23 | .846 | .829 | .809 | .788 | .763 | .736 | .705 | .671 | .634 | .593 |
| | | | | | | | | .548 | .501 | .452 | .401 | .349 | .299 | .250 | .205 | .164 | .128 |
| | | | | | | | | .098 | .072 | .052 | | | | | | | |
| 11.000 | 10.929 | .000 | .071 | .96 | 2.73 | .6% | 24 | .846 | .829 | .809 | .788 | .763 | .736 | .705 | .671 | .634 | .593 |
| | | | | | | | | .548 | .501 | .452 | .401 | .349 | .299 | .250 | .205 | .164 | .128 |
| | | | | | | | | .098 | .072 | .052 | .037 | | | | | | |
| 12.000 | 8.212 | .000 | 3.788 | .40 | 3.07 | 31.6% | 11 | .857 | .842 | .825 | .806 | .784 | .760 | .734 | .704 | .671 | .635 |
| | | | | | | | | .595 | | | | | | | | | |
| 12.000 | 8.764 | .000 | 3.236 | .45 | 3.15 | 27.0% | 12 | .857 | .842 | .825 | .806 | .784 | .760 | .734 | .704 | .671 | .635 |
| | | | | | | | | .595 | .552 | | | | | | | | |
| 12.000 | 9.271 | .000 | 2.729 | .49 | 3.21 | 22.7% | 13 | .857 | .842 | .825 | .806 | .784 | .760 | .734 | .704 | .671 | .635 |
| | | | | | | | | .595 | .552 | .507 | | | | | | | |
| 12.000 | 9.730 | .000 | 2.270 | .54 | 3.26 | 18.9% | 14 | .857 | .842 | .825 | .806 | .784 | .760 | .734 | .704 | .671 | .635 |
| | | | | | | | | .595 | .552 | .507 | .459 | | | | | | |
| 12.000 | 10.140 | .000 | 1.860 | .59 | 3.30 | 15.5% | 15 | .857 | .842 | .825 | .806 | .784 | .760 | .734 | .704 | .671 | .635 |
| | | | | | | | | .595 | .552 | .507 | .459 | .410 | | | | | |
| 12.000 | 10.501 | .000 | 1.400 | .64 | 3.32 | 12.5% | 16 | .857 | .842 | .825 | .806 | .784 | .760 | .734 | .704 | .671 | .635 |
| | | | | | | | | .595 | .552 | .507 | .459 | .410 | .360 | | | | |
| 12.000 | 10.812 | .000 | 1.188 | .69 | 3.32 | 9.9% | 17 | .857 | .842 | .825 | .806 | .784 | .760 | .734 | .704 | .671 | .635 |
| | | | | | | | | .595 | .552 | .507 | .459 | .410 | .360 | .311 | | | |
| 12.000 | 11.076 | .000 | .924 | .74 | 3.30 | 7.7% | 18 | .857 | .842 | .825 | .806 | .784 | .760 | .734 | .704 | .671 | .635 |
| | | | | | | | | .595 | .552 | .507 | .459 | .410 | .360 | .311 | .264 | | |
| 12.000 | 11.294 | .000 | .706 | .78 | 3.27 | 5.9% | 19 | .857 | .842 | .825 | .806 | .784 | .760 | .734 | .704 | .671 | .635 |
| | | | | | | | | .595 | .552 | .507 | .459 | .410 | .360 | .311 | .264 | .219 | |
| 12.000 | 11.472 | .000 | .528 | .82 | 3.22 | 4.4% | 20 | .857 | .842 | .825 | .806 | .784 | .760 | .734 | .704 | .671 | .635 |
| | | | | | | | | .595 | .552 | .507 | .459 | .410 | .360 | .311 | .264 | .219 | .178 |
| 12.000 | 11.613 | .000 | .387 | .86 | 3.16 | 3.2% | 21 | .857 | .842 | .825 | .806 | .784 | .760 | .734 | .704 | .671 | .635 |
| | | | | | | | | .595 | .552 | .507 | .459 | .410 | .360 | .311 | .264 | .219 | .178 |
| | | | | | | | | .141 | | | | | | | | | |
| 12.000 | 11.722 | .000 | .278 | .89 | 3.00 | 2.3% | 22 | .857 | .842 | .825 | .806 | .784 | .760 | .734 | .704 | .671 | .635 |
| | | | | | | | | .595 | .552 | .507 | .459 | .410 | .360 | .311 | .264 | .219 | .178 |
| | | | | | | | | .141 | .109 | | | | | | | | |

*SIMILAR TO A.T.&T. NEAL-WILKINSON TABLES(SEE SECTION 1)

*EQUIVALENT RANDOM THEORY: VMR(VARIANCE TO MEAN RATIO)– 2.0 - NO RETRIES (PEAKED TRAFFIC - ROUTE ADVANCEMENT OR DAY-TO-DAY VARIATION)

| HOURS FIRST ATTEMPT VMR = 2.0 | HOURS CON- NECT | HOURS RECALL /RETRY | HOURS OVFL | CON- NECT VMR | OVFL VMR | PERCENT BLOCKED (GRADE) | LINE | HOURS CONNECTED TRAFFIC ON LINES 1 | 2 | 3 | 4 | 5 | 6 | 7 | 8 | 9 | 10 |
|---|---|---|---|---|---|---|---|---|---|---|---|---|---|---|---|---|---|
| 12.000 | 11.805 | .000 | .195 | .92 | 3.01 | 1.6% | 23 | .857 | .842 | .825 | .806 | .784 | .760 | .734 | .704 | .671 | .635 |
| | | | | | | | | .595 | .552 | .507 | .459 | .410 | .360 | .311 | .264 | .219 | .178 |
| | | | | | | | | .141 | .109 | .083 | | | | | | | |
| 12.000 | 11.866 | .000 | .134 | .94 | 2.93 | 1.1% | 24 | .857 | .842 | .825 | .806 | .784 | .760 | .734 | .704 | .671 | .635 |
| | | | | | | | | .595 | .552 | .507 | .459 | .410 | .360 | .311 | .264 | .219 | .178 |
| | | | | | | | | .141 | .109 | .083 | .061 | | | | | | |
| 12.000 | 11.909 | .000 | .091 | .96 | 2.84 | .8% | 25 | .857 | .842 | .825 | .806 | .784 | .760 | .734 | .704 | .671 | .635 |
| | | | | | | | | .595 | .552 | .507 | .459 | .410 | .360 | .311 | .264 | .219 | .178 |
| | | | | | | | | .141 | .109 | .083 | .061 | .044 | | | | | |
| 13.000 | 8.460 | .000 | 4.540 | .36 | 3.06 | 34.9% | 11 | .866 | .853 | .838 | .821 | .802 | .781 | .758 | .732 | .703 | .671 |
| | | | | | | | | .635 | | | | | | | | | |
| 13.000 | 9.057 | .000 | 3.943 | .40 | 3.15 | 30.3% | 12 | .866 | .853 | .838 | .821 | .802 | .781 | .758 | .732 | .703 | .671 |
| | | | | | | | | .635 | .597 | | | | | | | | |
| 13.000 | 9.613 | .000 | 3.387 | .44 | 3.23 | 26.1% | 13 | .866 | .853 | .838 | .821 | .802 | .781 | .758 | .732 | .703 | .671 |
| | | | | | | | | .635 | .597 | .556 | | | | | | | |
| 13.000 | 10.125 | .000 | 2.875 | .49 | 3.29 | 22.1% | 14 | .866 | .853 | .838 | .821 | .802 | .781 | .758 | .732 | .703 | .671 |
| | | | | | | | | .635 | .597 | .556 | .512 | | | | | | |
| 13.000 | 10.591 | .000 | 2.409 | .53 | 3.35 | 18.5% | 15 | .866 | .853 | .838 | .821 | .802 | .781 | .758 | .732 | .703 | .671 |
| | | | | | | | | .635 | .597 | .556 | .512 | .466 | | | | | |
| 13.000 | 11.010 | .000 | 1.990 | .58 | 3.38 | 15.3% | 16 | .866 | .853 | .838 | .821 | .802 | .781 | .758 | .732 | .703 | .671 |
| | | | | | | | | .635 | .597 | .556 | .512 | .466 | .419 | | | | |
| 13.000 | 11.380 | .000 | 1.620 | .63 | 3.40 | 12.5% | 17 | .866 | .853 | .838 | .821 | .802 | .781 | .758 | .732 | .703 | .671 |
| | | | | | | | | .635 | .597 | .556 | .512 | .466 | .419 | .370 | | | |
| 13.000 | 11.702 | .000 | 1.298 | .68 | 3.41 | 10.0% | 18 | .866 | .853 | .838 | .821 | .802 | .781 | .758 | .732 | .703 | .671 |
| | | | | | | | | .635 | .597 | .556 | .512 | .466 | .419 | .370 | .322 | | |
| 13.000 | 11.978 | .000 | 1.022 | .72 | 3.40 | 7.9% | 19 | .866 | .853 | .838 | .821 | .802 | .781 | .758 | .732 | .703 | .671 |
| | | | | | | | | .635 | .597 | .556 | .512 | .466 | .419 | .370 | .322 | .276 | |
| 13.000 | 12.209 | .000 | .791 | .77 | 3.37 | 6.1% | 20 | .866 | .853 | .838 | .821 | .802 | .781 | .758 | .732 | .703 | .671 |
| | | | | | | | | .635 | .597 | .556 | .512 | .466 | .419 | .370 | .322 | .276 | .231 |
| 13.000 | 12.400 | .000 | .600 | .81 | 3.32 | 4.6% | 21 | .866 | .853 | .838 | .821 | .802 | .781 | .758 | .732 | .703 | .671 |
| | | | | | | | | .635 | .597 | .556 | .512 | .466 | .419 | .370 | .322 | .276 | .231 |
| | | | | | | | | .190 | | | | | | | | | |
| 13.000 | 12.553 | .000 | .447 | .85 | 3.26 | 3.4% | 22 | .866 | .853 | .838 | .821 | .802 | .781 | .758 | .732 | .703 | .671 |
| | | | | | | | | .635 | .597 | .556 | .512 | .466 | .419 | .370 | .322 | .276 | .231 |
| | | | | | | | | .190 | .153 | | | | | | | | |
| 13.000 | 12.673 | .000 | .327 | .88 | 3.19 | 2.5% | 23 | .866 | .853 | .838 | .821 | .802 | .781 | .758 | .732 | .703 | .671 |
| | | | | | | | | .635 | .597 | .556 | .512 | .466 | .419 | .370 | .322 | .276 | .231 |
| | | | | | | | | .190 | .153 | .121 | | | | | | | |
| 13.000 | 12.766 | .000 | .234 | .91 | 3.12 | 1.8% | 24 | .866 | .853 | .838 | .821 | .802 | .781 | .758 | .732 | .703 | .671 |
| | | | | | | | | .635 | .597 | .556 | .512 | .466 | .419 | .370 | .322 | .276 | .231 |

*SIMILAR TO A.T.&T. NEAL-WILKINSON TABLES(SEE SECTION 1)

# *EQUIVALENT RANDOM THEORY: VMR(VARIANCE TO MEAN RATIO)– 2.0 - NO RETRIES (PEAKED TRAFFIC - ROUTE ADVANCEMENT OR DAY-TO-DAY VARIATION)

| HOURS FIRST ATTEMPT VMR = 2.0 | HOURS CON-NECT | HOURS RECALL /RETRY | HOURS OVFL | CON-NECT VMR | OVFL VMR | PERCENT BLOCKED (GRADE) | LINE | HOURS CONNECTED TRAFFIC ON LINES 1 | 2 | 3 | 4 | 5 | 6 | 7 | 8 | 9 | 10 |
|---|---|---|---|---|---|---|---|---|---|---|---|---|---|---|---|---|---|
| | | | | | | | | .190 | .153 | .121 | .093 | | | | | | |
| 13.000 | 12.836 | .000 | .164 | .93 | 3.04 | 1.3% | 25 | .866 | .853 | .838 | .821 | .802 | .781 | .758 | .732 | .703 | .671 |
| | | | | | | | | .635 | .597 | .556 | .512 | .466 | .419 | .370 | .322 | .276 | .231 |
| | | | | | | | | .190 | .153 | .121 | .093 | .070 | | | | | |
| 13.000 | 12.887 | .000 | .113 | .95 | 2.95 | .9% | 26 | .866 | .853 | .838 | .821 | .802 | .781 | .758 | .732 | .703 | .671 |
| | | | | | | | | .635 | .597 | .556 | .512 | .466 | .419 | .370 | .322 | .276 | .231 |
| | | | | | | | | .190 | .153 | .121 | .093 | .070 | .051 | | | | |
| 13.000 | 12.924 | .000 | .076 | .96 | 2.87 | .6% | 27 | .866 | .853 | .838 | .821 | .802 | .781 | .758 | .732 | .703 | .671 |
| | | | | | | | | .635 | .597 | .556 | .512 | .466 | .419 | .370 | .322 | .276 | .231 |
| | | | | | | | | .190 | .153 | .121 | .093 | .070 | .051 | .037 | | | |
| 14.000 | 9.309 | .000 | 4.691 | .36 | 3.14 | 33.5% | 12 | .875 | .863 | .849 | .834 | .818 | .799 | .778 | .755 | .730 | .701 |
| | | | | | | | | .670 | .636 | | | | | | | | |
| 14.000 | 9.908 | .000 | 4.092 | .40 | 3.23 | 29.2% | 13 | .875 | .863 | .849 | .834 | .818 | .799 | .778 | .755 | .730 | .701 |
| | | | | | | | | .670 | .636 | .599 | | | | | | | |
| 14.000 | 10.467 | .000 | 3.533 | .44 | 3.30 | 25.2% | 14 | .875 | .863 | .849 | .834 | .818 | .799 | .778 | .755 | .730 | .701 |
| | | | | | | | | .670 | .636 | .599 | .559 | | | | | | |
| 14.000 | 10.984 | .000 | 3.016 | .48 | 3.37 | 21.5% | 15 | .875 | .863 | .849 | .834 | .818 | .799 | .778 | .755 | .730 | .701 |
| | | | | | | | | .670 | .636 | .599 | .559 | .517 | | | | | |
| 14.000 | 11.456 | .000 | 2.544 | .53 | 3.42 | 18.2% | 16 | .875 | .863 | .849 | .834 | .818 | .799 | .778 | .755 | .730 | .701 |
| | | | | | | | | .670 | .636 | .599 | .559 | .517 | .472 | | | | |
| 14.000 | 11.882 | .000 | 2.118 | .57 | 3.46 | 15.1% | 17 | .875 | .863 | .849 | .834 | .818 | .799 | .778 | .755 | .730 | .701 |
| | | | | | | | | .670 | .636 | .599 | .559 | .517 | .472 | .426 | | | |
| 14.000 | 12.262 | .000 | 1.738 | .62 | 3.49 | 12.4% | 18 | .875 | .863 | .849 | .834 | .818 | .799 | .778 | .755 | .730 | .701 |
| | | | | | | | | .670 | .636 | .599 | .559 | .517 | .472 | .426 | .379 | | |
| 14.000 | 12.594 | .000 | 1.406 | .67 | 3.49 | 10.0% | 19 | .875 | .863 | .849 | .834 | .818 | .799 | .778 | .755 | .730 | .701 |
| | | | | | | | | .670 | .636 | .599 | .559 | .517 | .472 | .426 | .379 | .333 | |
| 14.000 | 12.881 | .000 | 1.119 | .71 | 3.48 | 8.0% | 20 | .875 | .863 | .849 | .834 | .818 | .799 | .778 | .755 | .730 | .701 |
| | | | | | | | | .670 | .636 | .599 | .559 | .517 | .472 | .426 | .379 | .333 | .287 |
| 14.000 | 13.124 | .000 | .876 | .76 | 3.46 | 6.3% | 21 | .875 | .863 | .849 | .834 | .818 | .799 | .778 | .755 | .730 | .701 |
| | | | | | | | | .670 | .636 | .599 | .559 | .517 | .472 | .426 | .379 | .333 | .287 |
| | | | | | | | | .243 | | | | | | | | | |
| 14.000 | 13.326 | .000 | .674 | .80 | 3.42 | 4.8% | 22 | .875 | .863 | .849 | .834 | .818 | .799 | .778 | .755 | .730 | .701 |
| | | | | | | | | .670 | .636 | .599 | .559 | .517 | .472 | .426 | .379 | .333 | .287 |
| | | | | | | | | .243 | .202 | | | | | | | | |
| 14.000 | 13.491 | .000 | .509 | .84 | 3.36 | 3.6% | 23 | .875 | .863 | .849 | .834 | .818 | .799 | .778 | .755 | .730 | .701 |
| | | | | | | | | .670 | .636 | .599 | .559 | .517 | .472 | .426 | .379 | .333 | .287 |
| | | | | | | | | .243 | .202 | .165 | | | | | | | |
| 14.000 | 13.623 | .000 | .377 | .87 | 3.30 | 2.7% | 24 | .875 | .863 | .849 | .834 | .818 | .799 | .778 | .755 | .730 | .701 |
| | | | | | | | | .670 | .636 | .599 | .559 | .517 | .472 | .426 | .379 | .333 | .287 |
| | | | | | | | | .243 | .202 | .165 | .131 | | | | | | |

*SIMILAR TO A.T.&T. NEAL-WILKINSON TABLES(SEE SECTION 1)

*EQUIVALENT RANDOM THEORY: VMR(VARIANCE TO MEAN RATIO)– 2.0 - NO RETRIES (PEAKED TRAFFIC - ROUTE ADVANCEMENT OR DAY-TO-DAY VARIATION)

| HOURS FIRST ATTEMPT VMR=2.0 | HOURS CON-NECT | HOURS RECALL /RETRY | HOURS OVFL | CON-NECT VMR | OVFL VMR | PERCENT BLOCKED (GRADE) | LINE | HOURS CONNECTED TRAFFIC ON LINES 1 | 2 | 3 | 4 | 5 | 6 | 7 | 8 | 9 | 10 |
|---|---|---|---|---|---|---|---|---|---|---|---|---|---|---|---|---|---|
| 14.000 | 13.725 | .000 | .275 | .90 | 3.22 | 2.0% | 25 | .875 | .863 | .849 | .834 | .818 | .799 | .778 | .755 | .730 | .701 |
| | | | | | | | | .670 | .636 | .599 | .559 | .517 | .472 | .426 | .379 | .333 | .287 |
| | | | | | | | | .243 | .202 | .165 | .131 | .103 | | | | | |
| 14.000 | 13.804 | .000 | .196 | .92 | 3.14 | 1.4% | 26 | .875 | .863 | .849 | .834 | .818 | .799 | .778 | .755 | .730 | .701 |
| | | | | | | | | .670 | .636 | .599 | .559 | .517 | .472 | .426 | .379 | .333 | .287 |
| | | | | | | | | .243 | .202 | .165 | .131 | .103 | .079 | | | | |
| 14.000 | 13.863 | .000 | .137 | .94 | 3.06 | 1.0% | 27 | .875 | .863 | .849 | .834 | .818 | .799 | .778 | .755 | .730 | .701 |
| | | | | | | | | .670 | .636 | .599 | .559 | .517 | .472 | .426 | .379 | .333 | .287 |
| | | | | | | | | .243 | .202 | .165 | .131 | .103 | .079 | .059 | | | |
| 14.000 | 13.906 | .000 | .094 | .96 | 2.97 | .7% | 28 | .875 | .863 | .849 | .834 | .818 | .799 | .778 | .755 | .730 | .701 |
| | | | | | | | | .670 | .636 | .599 | .559 | .517 | .472 | .426 | .379 | .333 | .287 |
| | | | | | | | | .243 | .202 | .165 | .131 | .103 | .079 | .059 | .043 | | |
| 15.000 | 10.164 | .000 | 4.836 | .36 | 3.22 | 32.2% | 13 | .882 | .871 | .859 | .846 | .831 | .815 | .796 | .776 | .753 | .728 |
| | | | | | | | | .700 | .670 | .636 | | | | | | | |
| 15.000 | 10.764 | .000 | 4.236 | .40 | 3.30 | 28.2% | 14 | .882 | .871 | .859 | .846 | .831 | .815 | .796 | .776 | .753 | .728 |
| | | | | | | | | .700 | .670 | .636 | .600 | | | | | | |
| 15.000 | 11.326 | .000 | 3.674 | .44 | 3.38 | 24.5% | 15 | .882 | .871 | .859 | .846 | .831 | .815 | .796 | .776 | .753 | .728 |
| | | | | | | | | .700 | .670 | .636 | .600 | .562 | | | | | |
| 15.000 | 11.847 | .000 | 3.153 | .48 | 3.45 | 21.0% | 16 | .882 | .871 | .859 | .846 | .831 | .815 | .796 | .776 | .753 | .728 |
| | | | | | | | | .700 | .670 | .636 | .600 | .562 | .521 | | | | |
| 15.000 | 12.325 | .000 | 2.675 | .52 | 3.50 | 17.8% | 17 | .882 | .871 | .859 | .846 | .831 | .815 | .796 | .776 | .753 | .728 |
| | | | | | | | | .700 | .670 | .636 | .600 | .562 | .521 | .478 | | | |
| 15.000 | 12.758 | .000 | 2.242 | .57 | 3.54 | 14.9% | 18 | .882 | .871 | .859 | .846 | .831 | .815 | .796 | .776 | .753 | .728 |
| | | | | | | | | .700 | .670 | .636 | .600 | .562 | .521 | .478 | .433 | | |
| 15.000 | 13.146 | .000 | 1.854 | .61 | 3.57 | 12.4% | 19 | .882 | .871 | .859 | .846 | .831 | .815 | .796 | .776 | .753 | .728 |
| | | | | | | | | .700 | .670 | .636 | .600 | .562 | .521 | .478 | .433 | .388 | |
| 15.000 | 13.488 | .000 | 1.512 | .66 | 3.58 | 10.1% | 20 | .882 | .871 | .859 | .846 | .831 | .815 | .796 | .776 | .753 | .728 |
| | | | | | | | | .700 | .670 | .636 | .600 | .562 | .521 | .478 | .433 | .388 | .342 |
| 15.000 | 13.785 | .000 | 1.215 | .70 | 3.57 | 8.1% | 21 | .882 | .871 | .859 | .846 | .831 | .815 | .796 | .776 | .753 | .728 |
| | | | | | | | | .700 | .670 | .636 | .600 | .562 | .521 | .478 | .433 | .388 | .342 |
| | | | | | | | | .297 | | | | | | | | | |
| 15.000 | 14.039 | .000 | .961 | .75 | 3.55 | 6.4% | 22 | .882 | .871 | .859 | .846 | .831 | .815 | .796 | .776 | .753 | .728 |
| | | | | | | | | .700 | .670 | .636 | .600 | .562 | .521 | .478 | .433 | .388 | .342 |
| | | | | | | | | .297 | .254 | | | | | | | | |
| 15.000 | 14.252 | .000 | .748 | .79 | 3.51 | 5.0% | 23 | .882 | .871 | .859 | .846 | .831 | .815 | .796 | .776 | .753 | .728 |
| | | | | | | | | .700 | .670 | .636 | .600 | .562 | .521 | .478 | .433 | .388 | .342 |
| | | | | | | | | .297 | .254 | .213 | | | | | | | |
| 15.000 | 14.428 | .000 | .572 | .82 | 3.46 | 3.8% | 24 | .882 | .871 | .859 | .846 | .831 | .815 | .796 | .776 | .753 | .728 |
| | | | | | | | | .700 | .670 | .636 | .600 | .562 | .521 | .478 | .433 | .388 | .342 |
| | | | | | | | | .297 | .254 | .213 | .176 | | | | | | |

*SIMILAR TO A.T.&T. NEAL-WILKINSON TABLES(SEE SECTION 1)

*EQUIVALENT RANDOM THEORY: VMR(VARIANCE TO MEAN RATIO)– 2.0 - NO RETRIES (PEAKED TRAFFIC - ROUTE ADVANCEMENT OR DAY-TO-DAY VARIATION)

| HOURS FIRST ATTEMPT VMR = 2.0 | HOURS CON-NECT | HOURS RECALL /RETRY | HOURS OVFL | CON-NECT VMR | OVFL VMR | PERCENT BLOCKED (GRADE) | LINE | HOURS CONNECTED TRAFFIC ON LINES 1 | 2 | 3 | 4 | 5 | 6 | 7 | 8 | 9 | 10 |
|---|---|---|---|---|---|---|---|---|---|---|---|---|---|---|---|---|---|
| 15.000 | 14.570 | .000 | .430 | .86 | 3.39 | 2.9% | 25 | .882 | .871 | .859 | .846 | .831 | .815 | .796 | .776 | .753 | .728 |
| | | | | | | | | .700 | .670 | .636 | .600 | .562 | .521 | .478 | .433 | .388 | .342 |
| | | | | | | | | .297 | .254 | .213 | .176 | .142 | | | | | |
| 15.000 | 14.682 | .000 | .318 | .89 | 3.32 | 2.1% | 26 | .882 | .871 | .859 | .846 | .831 | .815 | .796 | .776 | .753 | .728 |
| | | | | | | | | .700 | .670 | .636 | .600 | .562 | .521 | .478 | .433 | .388 | .342 |
| | | | | | | | | .297 | .254 | .213 | .176 | .142 | .113 | | | | |
| 15.000 | 14.770 | .000 | .230 | .91 | 3.24 | 1.5% | 27 | .882 | .871 | .859 | .846 | .831 | .815 | .796 | .776 | .753 | .728 |
| | | | | | | | | .700 | .670 | .636 | .600 | .562 | .521 | .478 | .433 | .388 | .342 |
| | | | | | | | | .297 | .254 | .213 | .176 | .142 | .113 | .087 | | | |
| 15.000 | 14.836 | .000 | .164 | .93 | 3.16 | 1.1% | 28 | .882 | .871 | .859 | .846 | .831 | .815 | .796 | .776 | .753 | .728 |
| | | | | | | | | .700 | .670 | .636 | .600 | .562 | .521 | .478 | .433 | .388 | .342 |
| | | | | | | | | .297 | .254 | .213 | .176 | .142 | .113 | .087 | .066 | | |
| 15.000 | 14.885 | .000 | .115 | .95 | 3.07 | .8% | 29 | .882 | .871 | .859 | .846 | .831 | .815 | .796 | .776 | .753 | .728 |
| | | | | | | | | .700 | .670 | .636 | .600 | .562 | .521 | .478 | .433 | .388 | .342 |
| | | | | | | | | .297 | .254 | .213 | .176 | .142 | .113 | .087 | .066 | .049 | |
| 15.000 | 14.921 | .000 | .079 | .96 | 2.99 | .5% | 30 | .882 | .871 | .859 | .846 | .831 | .815 | .796 | .776 | .753 | .728 |
| | | | | | | | | .700 | .670 | .636 | .600 | .562 | .521 | .478 | .433 | .388 | .342 |
| | | | | | | | | .297 | .254 | .213 | .176 | .142 | .113 | .087 | .066 | .049 | .036 |
| 16.000 | 11.023 | .000 | 4.977 | .36 | 3.29 | 31.1% | 14 | .889 | .879 | .868 | .856 | .843 | .828 | .812 | .793 | .773 | .751 |
| | | | | | | | | .726 | .699 | .669 | .637 | | | | | | |
| 16.000 | 11.625 | .000 | 4.375 | .40 | 3.38 | 27.3% | 15 | .889 | .879 | .868 | .856 | .843 | .828 | .812 | .793 | .773 | .751 |
| | | | | | | | | .726 | .699 | .669 | .637 | .602 | | | | | |
| 16.000 | 12.190 | .000 | 3.810 | .44 | 3.45 | 23.8% | 16 | .889 | .879 | .868 | .856 | .843 | .828 | .812 | .793 | .773 | .751 |
| | | | | | | | | .726 | .699 | .669 | .637 | .602 | .564 | | | | |
| 16.000 | 12.714 | .000 | 3.286 | .48 | 3.52 | 20.5% | 17 | .889 | .879 | .868 | .856 | .843 | .828 | .812 | .793 | .773 | .751 |
| | | | | | | | | .726 | .699 | .669 | .637 | .602 | .564 | .525 | | | |
| 16.000 | 13.197 | .000 | 2.803 | .52 | 3.58 | 17.5% | 18 | .889 | .879 | .868 | .856 | .843 | .828 | .812 | .793 | .773 | .751 |
| | | | | | | | | .726 | .699 | .669 | .637 | .602 | .564 | .525 | .483 | | |
| 16.000 | 13.636 | .000 | 2.364 | .56 | 3.62 | 14.8% | 19 | .889 | .879 | .868 | .856 | .843 | .828 | .812 | .793 | .773 | .751 |
| | | | | | | | | .726 | .699 | .669 | .637 | .602 | .564 | .525 | .483 | .439 | |
| 16.000 | 14.032 | .000 | 1.968 | .60 | 3.64 | 12.3% | 20 | .889 | .879 | .868 | .856 | .843 | .828 | .812 | .793 | .773 | .751 |
| | | | | | | | | .726 | .699 | .669 | .637 | .602 | .564 | .525 | .483 | .439 | .395 |
| 16.000 | 14.382 | .000 | 1.618 | .65 | 3.66 | 10.1% | 21 | .889 | .879 | .868 | .856 | .843 | .828 | .812 | .793 | .773 | .751 |
| | | | | | | | | .726 | .699 | .669 | .637 | .602 | .564 | .525 | .483 | .439 | .395 |
| | | | | | | | | .351 | | | | | | | | | |
| 16.000 | 14.689 | .000 | 1.311 | .69 | 3.65 | 8.2% | 22 | .889 | .879 | .868 | .856 | .843 | .828 | .812 | .793 | .773 | .751 |
| | | | | | | | | .726 | .699 | .669 | .637 | .602 | .564 | .525 | .483 | .439 | .395 |
| | | | | | | | | .351 | .307 | | | | | | | | |
| 16.000 | 14.953 | .000 | 1.047 | .74 | 3.63 | 6.5% | 23 | .889 | .879 | .868 | .856 | .843 | .828 | .812 | .793 | .773 | .751 |
| | | | | | | | | .726 | .699 | .669 | .637 | .602 | .564 | .525 | .483 | .439 | .395 |

*SIMILAR TO A.T.&T. NEAL-WILKINSON TABLES(SEE SECTION 1)

*EQUIVALENT RANDOM THEORY: VMR(VARIANCE TO MEAN RATIO)– 2.0 - NO RETRIES (PEAKED TRAFFIC - ROUTE ADVANCEMENT OR DAY-TO-DAY VARIATION)

| HOURS FIRST ATTEMPT VMR = 2.0 | HOURS CON-NECT | HOURS RECALL /RETRY | HOURS OVFL | CON-NECT VMR | OVFL VMR | PERCENT BLOCKED (GRADE) | LINE | HOURS CONNECTED TRAFFIC ON LINES 1 | 2 | 3 | 4 | 5 | 6 | 7 | 8 | 9 | 10 |
|---|---|---|---|---|---|---|---|---|---|---|---|---|---|---|---|---|---|
| | | | | | | | | .351 | .307 | .264 | | | | | | | |
| 16.000 | 15.177 | .000 | .823 | .78 | 3.60 | 5.1% | 24 | .889 | .879 | .868 | .856 | .843 | .828 | .812 | .793 | .773 | .751 |
| | | | | | | | | .726 | .699 | .669 | .637 | .602 | .564 | .525 | .483 | .439 | .395 |
| | | | | | | | | .351 | .307 | .264 | .224 | | | | | | |
| 16.000 | 15.363 | .000 | .637 | .81 | 3.55 | 4.0% | 25 | .889 | .879 | .868 | .856 | .843 | .828 | .812 | .793 | .773 | .751 |
| | | | | | | | | .726 | .699 | .669 | .637 | .602 | .564 | .525 | .483 | .439 | .395 |
| | | | | | | | | .351 | .307 | .264 | .224 | .186 | | | | | |
| 16.000 | 15.515 | .000 | .485 | .85 | 3.49 | 3.0% | 26 | .889 | .879 | .868 | .856 | .843 | .828 | .812 | .793 | .773 | .751 |
| | | | | | | | | .726 | .699 | .669 | .637 | .602 | .564 | .525 | .483 | .439 | .395 |
| | | | | | | | | .351 | .307 | .264 | .224 | .186 | .152 | | | | |
| 16.000 | 15.637 | .000 | .363 | .88 | 3.42 | 2.3% | 27 | .889 | .879 | .868 | .856 | .843 | .828 | .812 | .793 | .773 | .751 |
| | | | | | | | | .726 | .699 | .669 | .637 | .602 | .564 | .525 | .483 | .439 | .395 |
| | | | | | | | | .351 | .307 | .264 | .224 | .186 | .152 | .122 | | | |
| 16.000 | 15.733 | .000 | .267 | .90 | 3.34 | 1.7% | 28 | .889 | .879 | .868 | .856 | .843 | .828 | .812 | .793 | .773 | .751 |
| | | | | | | | | .726 | .699 | .669 | .637 | .602 | .564 | .525 | .483 | .439 | .395 |
| | | | | | | | | .351 | .307 | .264 | .224 | .186 | .152 | .122 | .096 | | |
| 16.000 | 15.807 | .000 | .193 | .93 | 3.26 | 1.2% | 29 | .889 | .879 | .868 | .856 | .843 | .828 | .812 | .793 | .773 | .751 |
| | | | | | | | | .726 | .699 | .669 | .637 | .602 | .564 | .525 | .483 | .439 | .395 |
| | | | | | | | | .351 | .307 | .264 | .224 | .186 | .152 | .122 | .096 | .074 | |
| 16.000 | 15.863 | .000 | .137 | .94 | 3.17 | .9% | 30 | .889 | .879 | .868 | .856 | .843 | .828 | .812 | .793 | .773 | .751 |
| | | | | | | | | .726 | .699 | .669 | .637 | .602 | .564 | .525 | .483 | .439 | .395 |
| | | | | | | | | .351 | .307 | .264 | .224 | .186 | .152 | .122 | .096 | .074 | .056 |
| 16.000 | 15.904 | .000 | .096 | .96 | 3.09 | .6% | 31 | .889 | .879 | .868 | .856 | .843 | .828 | .812 | .793 | .773 | .751 |
| | | | | | | | | .726 | .699 | .669 | .637 | .602 | .564 | .525 | .483 | .439 | .395 |
| | | | | | | | | .351 | .307 | .264 | .224 | .186 | .152 | .122 | .096 | .074 | .056 |
| | | | | | | | | .041 | | | | | | | | | |
| 17.000 | 11.250 | .000 | 5.750 | .33 | 3.27 | 33.8% | 14 | .894 | .886 | .876 | .865 | .853 | .840 | .825 | .809 | .791 | .771 |
| | | | | | | | | .749 | .725 | .698 | .669 | | | | | | |
| 17.000 | 11.887 | .000 | 5.113 | .36 | 3.36 | 30.1% | 15 | .894 | .886 | .876 | .865 | .853 | .840 | .825 | .809 | .791 | .771 |
| | | | | | | | | .749 | .725 | .698 | .669 | .637 | | | | | |
| 17.000 | 12.490 | .000 | 4.510 | .40 | 3.45 | 26.5% | 16 | .894 | .886 | .876 | .865 | .853 | .840 | .825 | .809 | .791 | .771 |
| | | | | | | | | .749 | .725 | .698 | .669 | .637 | .603 | | | | |
| 17.000 | 13.057 | .000 | 3.943 | .43 | 3.52 | 23.2% | 17 | .894 | .886 | .876 | .865 | .853 | .840 | .825 | .809 | .791 | .771 |
| | | | | | | | | .749 | .725 | .698 | .669 | .637 | .603 | .567 | | | |
| 17.000 | 13.585 | .000 | 3.415 | .47 | 3.59 | 20.1% | 18 | .894 | .886 | .876 | .865 | .853 | .840 | .825 | .809 | .791 | .771 |
| | | | | | | | | .749 | .725 | .698 | .669 | .637 | .603 | .567 | .528 | | |
| 17.000 | 14.073 | .000 | 2.927 | .51 | 3.65 | 17.2% | 19 | .894 | .886 | .876 | .865 | .853 | .840 | .825 | .809 | .791 | .771 |
| | | | | | | | | .749 | .725 | .698 | .669 | .637 | .603 | .567 | .528 | .487 | |
| 17.000 | 14.518 | .000 | 2.482 | .55 | 3.69 | 14.6% | 20 | .894 | .886 | .876 | .865 | .853 | .840 | .825 | .809 | .791 | .771 |
| | | | | | | | | .749 | .725 | .698 | .669 | .637 | .603 | .567 | .528 | .487 | .445 |

*SIMILAR TO A.T.&T. NEAL-WILKINSON TABLES(SEE SECTION 1)

*EQUIVALENT RANDOM THEORY: VMR(VARIANCE TO MEAN RATIO)– 2.0 - NO RETRIES (PEAKED TRAFFIC - ROUTE ADVANCEMENT OR DAY-TO-DAY VARIATION)

| HOURS FIRST ATTEMPT VMR=2.0 | HOURS CON-NECT | HOURS RECALL /RETRY | HOURS OVFL | CON-NECT VMR | OVFL VMR | PERCENT BLOCKED (GRADE) | LINE | HOURS CONNECTED TRAFFIC ON LINES 1 | 2 | 3 | 4 | 5 | 6 | 7 | 8 | 9 | 10 |
|---|---|---|---|---|---|---|---|---|---|---|---|---|---|---|---|---|---|
| 17.000 | 14.920 | .000 | 2.080 | .60 | 3.72 | 12.2% | 21 | .894 | .886 | .876 | .865 | .853 | .840 | .825 | .809 | .791 | .771 |
| | | | | | | | | .749 | .725 | .698 | .669 | .637 | .603 | .567 | .528 | .487 | .445 |
| | | | | | | | | .402 | | | | | | | | | |
| 17.000 | 15.279 | .000 | 1.721 | .64 | 3.73 | 10.1% | 22 | .894 | .886 | .876 | .865 | .853 | .840 | .825 | .809 | .791 | .771 |
| | | | | | | | | .749 | .725 | .698 | .669 | .637 | .603 | .567 | .528 | .487 | .445 |
| | | | | | | | | .402 | .359 | | | | | | | | |
| 17.000 | 15.594 | .000 | 1.406 | .68 | 3.73 | 8.3% | 23 | .894 | .886 | .876 | .865 | .853 | .840 | .825 | .809 | .791 | .771 |
| | | | | | | | | .749 | .725 | .698 | .669 | .637 | .603 | .567 | .528 | .487 | .445 |
| | | | | | | | | .402 | .359 | .316 | | | | | | | |
| 17.000 | 15.868 | .000 | 1.132 | .73 | 3.72 | 6.7% | 24 | .894 | .886 | .876 | .865 | .853 | .840 | .825 | .809 | .791 | .771 |
| | | | | | | | | .749 | .725 | .698 | .669 | .637 | .603 | .567 | .528 | .487 | .445 |
| | | | | | | | | .402 | .359 | .316 | .274 | | | | | | |
| 17.000 | 16.101 | .000 | .899 | .77 | 3.68 | 5.3% | 25 | .894 | .886 | .876 | .865 | .853 | .840 | .825 | .809 | .791 | .771 |
| | | | | | | | | .749 | .725 | .698 | .669 | .637 | .603 | .567 | .528 | .487 | .445 |
| | | | | | | | | .402 | .359 | .316 | .274 | .234 | | | | | |
| 17.000 | 16.297 | .000 | .703 | .80 | 3.64 | 4.1% | 26 | .894 | .886 | .876 | .865 | .853 | .840 | .825 | .809 | .791 | .771 |
| | | | | | | | | .749 | .725 | .698 | .669 | .637 | .603 | .567 | .528 | .487 | .445 |
| | | | | | | | | .402 | .359 | .316 | .274 | .234 | .196 | | | | |
| 17.000 | 16.459 | .000 | .541 | .84 | 3.58 | 3.2% | 27 | .894 | .886 | .876 | .865 | .853 | .840 | .825 | .809 | .791 | .771 |
| | | | | | | | | .749 | .725 | .698 | .669 | .637 | .603 | .567 | .528 | .487 | .445 |
| | | | | | | | | .402 | .359 | .316 | .274 | .234 | .196 | .162 | | | |
| 17.000 | 16.591 | .000 | .409 | .87 | 3.51 | 2.4% | 28 | .894 | .886 | .876 | .865 | .853 | .840 | .825 | .809 | .791 | .771 |
| | | | | | | | | .749 | .725 | .698 | .669 | .637 | .603 | .567 | .528 | .487 | .445 |
| | | | | | | | | .402 | .359 | .316 | .274 | .234 | .196 | .162 | .131 | | |
| 17.000 | 16.695 | .000 | .305 | .90 | 3.44 | 1.8% | 29 | .894 | .886 | .876 | .865 | .853 | .840 | .825 | .809 | .791 | .771 |
| | | | | | | | | .749 | .725 | .690 | .669 | .637 | .603 | .567 | .528 | .407 | .445 |
| | | | | | | | | .402 | .359 | .316 | .274 | .234 | .196 | .162 | .131 | .104 | |
| 17.000 | 16.776 | .000 | .224 | .92 | 3.35 | 1.3% | 30 | .894 | .886 | .876 | .865 | .853 | .840 | .825 | .809 | .791 | .771 |
| | | | | | | | | .749 | .725 | .698 | .669 | .637 | .603 | .567 | .528 | .487 | .445 |
| | | | | | | | | .402 | .359 | .316 | .274 | .234 | .196 | .162 | .131 | .104 | .082 |
| 17.000 | 16.839 | .000 | .161 | .94 | 3.27 | .9% | 31 | .894 | .886 | .876 | .865 | .853 | .840 | .825 | .809 | .791 | .771 |
| | | | | | | | | .749 | .725 | .698 | .669 | .637 | .603 | .567 | .528 | .487 | .445 |
| | | | | | | | | .402 | .359 | .316 | .274 | .234 | .190 | .162 | .131 | .104 | .082 |
| | | | | | | | | .062 | | | | | | | | | |
| 17.000 | 16.886 | .000 | .114 | .95 | 3.18 | .7% | 32 | .894 | .886 | .876 | .865 | .853 | .840 | .825 | .809 | .791 | .771 |
| | | | | | | | | .749 | .725 | .698 | .669 | .637 | .603 | .567 | .528 | .487 | .445 |
| | | | | | | | | .402 | .359 | .316 | .274 | .234 | .196 | .162 | .131 | .104 | .082 |
| | | | | | | | | .062 | .047 | | | | | | | | |
| 18.000 | 12.118 | .000 | 5.882 | .33 | 3.34 | 32.7% | 15 | .900 | .892 | .883 | .873 | .862 | .850 | .837 | .822 | .806 | .708 |
| | | | | | | | | .769 | .747 | .723 | .697 | .668 | | | | | |

*SIMILAR TO A.T.&T. NEAL-WILKINSON TABLES(SEE SECTION 1)

## *EQUIVALENT RANDOM THEORY: VMR(VARIANCE TO MEAN RATIO)– 2.0 - NO RETRIES (PEAKED TRAFFIC - ROUTE ADVANCEMENT OR DAY-TO-DAY VARIATION)

| HOURS FIRST ATTEMPT VMR=2.0 | HOURS CON-NECT | HOURS RECALL /RETRY | HOURS OVFL | CON-NECT VMR | OVFL VMR | PERCENT BLOCKED (GRADE) | LINE | HOURS CONNECTED TRAFFIC ON LINES 1 | 2 | 3 | 4 | 5 | 6 | 7 | 8 | 9 | 10 |
|---|---|---|---|---|---|---|---|---|---|---|---|---|---|---|---|---|---|
| 18.000 | 12.755 | .000 | 5.245 | .36 | 3.43 | 29.1% | 16 | .900 | .892 | .883 | .873 | .862 | .850 | .837 | .822 | .806 | .788 |
| | | | | | | | | .769 | .747 | .723 | .697 | .668 | .638 | | | | |
| 18.000 | 13.360 | .000 | 4.640 | .40 | 3.52 | 25.8% | 17 | .900 | .892 | .883 | .873 | .862 | .850 | .837 | .822 | .806 | .788 |
| | | | | | | | | .769 | .747 | .723 | .697 | .668 | .638 | .604 | | | |
| 18.000 | 13.928 | .000 | 4.072 | .43 | 3.59 | 22.6% | 18 | .900 | .892 | .883 | .873 | .862 | .850 | .837 | .822 | .806 | .788 |
| | | | | | | | | .769 | .747 | .723 | .697 | .668 | .638 | .604 | .569 | | |
| 18.000 | 14.460 | .000 | 3.540 | .47 | 3.66 | 19.7% | 19 | .900 | .892 | .883 | .873 | .862 | .850 | .837 | .822 | .806 | .788 |
| | | | | | | | | .769 | .747 | .723 | .697 | .668 | .638 | .604 | .569 | .531 | |
| 18.000 | 14.951 | .000 | 3.049 | .51 | 3.72 | 16.9% | 20 | .900 | .892 | .883 | .873 | .862 | .850 | .837 | .822 | .806 | .788 |
| | | | | | | | | .769 | .747 | .723 | .697 | .668 | .638 | .604 | .569 | .531 | .492 |
| 18.000 | 15.402 | .000 | 2.598 | .55 | 3.76 | 14.4% | 21 | .900 | .892 | .883 | .873 | .862 | .850 | .837 | .822 | .806 | .788 |
| | | | | | | | | .769 | .747 | .723 | .697 | .668 | .638 | .604 | .569 | .531 | .492 |
| | | | | | | | | .451 | | | | | | | | | |
| 18.000 | 15.810 | .000 | 2.190 | .59 | 3.79 | 12.2% | 22 | .900 | .892 | .883 | .873 | .862 | .850 | .837 | .822 | .806 | .788 |
| | | | | | | | | .769 | .747 | .723 | .697 | .668 | .638 | .604 | .569 | .531 | .492 |
| | | | | | | | | .451 | .408 | | | | | | | | |
| 18.000 | 16.176 | .000 | 1.824 | .63 | 3.81 | 10.1% | 23 | .900 | .892 | .883 | .873 | .862 | .850 | .837 | .822 | .806 | .788 |
| | | | | | | | | .769 | .747 | .723 | .697 | .668 | .638 | .604 | .569 | .531 | .492 |
| | | | | | | | | .451 | .408 | .366 | | | | | | | |
| 18.000 | 16.500 | .000 | 1.500 | .68 | 3.81 | 8.3% | 24 | .900 | .892 | .883 | .873 | .862 | .850 | .837 | .822 | .806 | .788 |
| | | | | | | | | .769 | .747 | .723 | .697 | .668 | .638 | .604 | .569 | .531 | .492 |
| | | | | | | | | .451 | .408 | .366 | .324 | | | | | | |
| 18.000 | 16.782 | .000 | 1.218 | .72 | 3.80 | 6.8% | 25 | .900 | .892 | .883 | .873 | .862 | .850 | .837 | .822 | .806 | .788 |
| | | | | | | | | .769 | .747 | .723 | .697 | .668 | .638 | .604 | .569 | .531 | .492 |
| | | | | | | | | .451 | .408 | .366 | .324 | .283 | | | | | |
| 18.000 | 17.025 | .000 | .975 | .76 | 3.77 | 5.4% | 26 | .900 | .892 | .883 | .873 | .862 | .850 | .837 | .822 | .806 | .788 |
| | | | | | | | | .769 | .747 | .723 | .697 | .668 | .638 | .604 | .569 | .531 | .492 |
| | | | | | | | | .451 | .408 | .366 | .324 | .283 | .243 | | | | |
| 18.000 | 17.231 | .000 | .769 | .79 | 3.72 | 4.3% | 27 | .900 | .892 | .883 | .873 | .862 | .850 | .837 | .822 | .806 | .788 |
| | | | | | | | | .769 | .747 | .723 | .697 | .668 | .638 | .604 | .569 | .531 | .492 |
| | | | | | | | | .451 | .408 | .366 | .324 | .283 | .243 | .206 | | | |
| 18.000 | 17.402 | .000 | .598 | .83 | 3.67 | 3.3% | 28 | .900 | .892 | .883 | .873 | .862 | .850 | .837 | .822 | .806 | .788 |
| | | | | | | | | .769 | .747 | .723 | .697 | .668 | .638 | .604 | .569 | .531 | .492 |
| | | | | | | | | .451 | .408 | .366 | .324 | .283 | .243 | .206 | .171 | | |
| 18.000 | 17.542 | .000 | .458 | .86 | 3.60 | 2.5% | 29 | .900 | .892 | .883 | .873 | .862 | .850 | .837 | .822 | .806 | .788 |
| | | | | | | | | .769 | .747 | .723 | .697 | .668 | .638 | .604 | .569 | .531 | .492 |
| | | | | | | | | .451 | .408 | .366 | .324 | .283 | .243 | .206 | .171 | .140 | |
| 18.000 | 17.655 | .000 | .345 | .89 | 3.53 | 1.9% | 30 | .900 | .892 | .883 | .873 | .862 | .850 | .837 | .822 | .806 | .788 |
| | | | | | | | | .769 | .747 | .723 | .697 | .668 | .638 | .604 | .569 | .531 | .492 |
| | | | | | | | | .451 | .408 | .366 | .324 | .283 | .243 | .206 | .171 | .140 | .113 |

*SIMILAR TO A.T.&T. NEAL-WILKINSON TABLES(SEE SECTION 1)

*EQUIVALENT RANDOM THEORY: VMR(VARIANCE TO MEAN RATIO)– 2.0 - NO RETRIES (PEAKED TRAFFIC - ROUTE ADVANCEMENT OR DAY-TO-DAY VARIATION)

| HOURS FIRST ATTEMPT VMR = 2.0 | HOURS CON- NECT | HOURS RECALL /RETRY | HOURS OVFL | CON- NECT VMR | OVFL VMR | PERCENT BLOCKED (GRADE) | LINE | HOURS CONNECTED TRAFFIC ON LINES 1 | 2 | 3 | 4 | 5 | 6 | 7 | 8 | 9 | 10 |
|---|---|---|---|---|---|---|---|---|---|---|---|---|---|---|---|---|---|
| 18.000 | 17.744 | .000 | .256 | .91 | 3.45 | 1.4% | 31 | .900 | .892 | .883 | .873 | .862 | .850 | .837 | .822 | .806 | .788 |
| | | | | | | | | .769 | .747 | .723 | .697 | .668 | .638 | .604 | .569 | .531 | .492 |
| | | | | | | | | .451 | .408 | .366 | .324 | .283 | .243 | .206 | .171 | .140 | .113 |
| | | | | | | | | .089 | | | | | | | | | |
| 18.000 | 17.813 | .000 | .187 | .93 | 3.36 | 1.0% | 32 | .900 | .892 | .883 | .873 | .862 | .850 | .837 | .822 | .806 | .788 |
| | | | | | | | | .769 | .747 | .723 | .697 | .668 | .638 | .604 | .569 | .531 | .492 |
| | | | | | | | | .451 | .408 | .366 | .324 | .283 | .243 | .206 | .171 | .140 | .113 |
| | | | | | | | | .089 | .069 | | | | | | | | |
| 18.000 | 17.866 | .000 | .134 | .95 | 3.28 | .7% | 33 | .900 | .892 | .883 | .873 | .862 | .850 | .837 | .822 | .806 | .788 |
| | | | | | | | | .769 | .747 | .723 | .697 | .668 | .638 | .604 | .569 | .531 | .492 |
| | | | | | | | | .451 | .408 | .366 | .324 | .283 | .243 | .206 | .171 | .140 | .113 |
| | | | | | | | | .089 | .069 | .053 | | | | | | | |
| 18.000 | 17.905 | .000 | .095 | .96 | 3.19 | .5% | 34 | .900 | .892 | .883 | .873 | .862 | .850 | .837 | .822 | .806 | .788 |
| | | | | | | | | .769 | .747 | .723 | .697 | .668 | .638 | .604 | .569 | .531 | .492 |
| | | | | | | | | .451 | .408 | .366 | .324 | .283 | .243 | .206 | .171 | .140 | .113 |
| | | | | | | | | .089 | .069 | .053 | .039 | | | | | | |
| 19.000 | 12.989 | .000 | 6.011 | .33 | 3.41 | 31.6% | 16 | .904 | .897 | .889 | .880 | .870 | .859 | .848 | .834 | .820 | .804 |
| | | | | | | | | .786 | .766 | .745 | .722 | .696 | .668 | | | | |
| 19.000 | 13.627 | .000 | 5.373 | .36 | 3.50 | 28.3% | 17 | .904 | .897 | .889 | .880 | .870 | .859 | .848 | .834 | .820 | .804 |
| | | | | | | | | .786 | .766 | .745 | .722 | .696 | .668 | .638 | | | |
| 19.000 | 14.232 | .000 | 4.768 | .39 | 3.59 | 25.1% | 18 | .904 | .897 | .889 | .880 | .870 | .859 | .848 | .834 | .820 | .804 |
| | | | | | | | | .786 | .766 | .745 | .722 | .696 | .668 | .638 | .605 | | |
| 19.000 | 14.803 | .000 | 4.197 | .43 | 3.66 | 22.1% | 19 | .904 | .897 | .889 | .880 | .870 | .859 | .848 | .834 | .820 | .804 |
| | | | | | | | | .786 | .766 | .745 | .722 | .696 | .668 | .638 | .605 | .571 | |
| 19.000 | 15.337 | .000 | 3.663 | .47 | 3.73 | 19.3% | 20 | .904 | .897 | .889 | .880 | .870 | .859 | .848 | .834 | .820 | .804 |
| | | | | | | | | .786 | .766 | .745 | .722 | .696 | .668 | .638 | .605 | .571 | .534 |
| 19.000 | 15.832 | .000 | 3.168 | .50 | 3.79 | 16.7% | 21 | .904 | .897 | .889 | .880 | .870 | .859 | .848 | .834 | .820 | .804 |
| | | | | | | | | .786 | .766 | .745 | .722 | .696 | .668 | .638 | .605 | .571 | .534 |
| | | | | | | | | .495 | | | | | | | | | |
| 19.000 | 16.288 | .000 | 2.712 | .54 | 3.83 | 14.3% | 22 | .904 | .897 | .889 | .880 | .870 | .859 | .848 | .834 | .820 | .804 |
| | | | | | | | | .786 | .766 | .745 | .722 | .696 | .668 | .638 | .605 | .571 | .534 |
| | | | | | | | | .495 | .455 | | | | | | | | |
| 19.000 | 16.702 | .000 | 2.298 | .59 | 3.86 | 12.1% | 23 | .904 | .897 | .889 | .880 | .870 | .859 | .848 | .834 | .820 | .804 |
| | | | | | | | | .786 | .766 | .745 | .722 | .696 | .668 | .638 | .605 | .571 | .534 |
| | | | | | | | | .495 | .455 | .414 | | | | | | | |
| 19.000 | 17.075 | .000 | 1.925 | .63 | 3.88 | 10.1% | 24 | .904 | .897 | .889 | .880 | .870 | .859 | .848 | .834 | .820 | .804 |
| | | | | | | | | .786 | .766 | .745 | .722 | .696 | .668 | .638 | .605 | .571 | .534 |
| | | | | | | | | .495 | .455 | .414 | .373 | | | | | | |
| 19.000 | 17.407 | .000 | 1.593 | .67 | 3.88 | 8.4% | 25 | .904 | .897 | .889 | .880 | .870 | .859 | .848 | .834 | .820 | .804 |
| | | | | | | | | .786 | .766 | .745 | .722 | .696 | .668 | .638 | .605 | .571 | .534 |

*SIMILAR TO A.T.&T. NEAL-WILKINSON TABLES(SEE SECTION 1)

*EQUIVALENT RANDOM THEORY: VMR(VARIANCE TO MEAN RATIO)– 2.0 - NO RETRIES (PEAKED TRAFFIC - ROUTE ADVANCEMENT OR DAY-TO-DAY VARIATION)

| HOURS FIRST ATTEMPT VMR = 2.0 | HOURS CON-NECT | HOURS RECALL /RETRY | HOURS OVFL | CON-NECT VMR | OVFL VMR | PERCENT BLOCKED (GRADE) | LINE | HOURS CONNECTED TRAFFIC ON LINES 1 | 2 | 3 | 4 | 5 | 6 | 7 | 8 | 9 | 10 |
|---|---|---|---|---|---|---|---|---|---|---|---|---|---|---|---|---|---|
| | | | | | | | | .495 | .455 | .414 | .373 | .331 | | | | | |
| 19.000 | 17.698 | .000 | 1.302 | .71 | 3.87 | 6.9% | 26 | .904 | .897 | .889 | .880 | .870 | .859 | .848 | .834 | .820 | .804 |
| | | | | | | | | .786 | .766 | .745 | .722 | .696 | .668 | .638 | .605 | .571 | .534 |
| | | | | | | | | .495 | .455 | .414 | .373 | .331 | .291 | | | | |
| 19.000 | 17.949 | .000 | 1.051 | .75 | 3.85 | 5.5% | 27 | .904 | .897 | .889 | .880 | .870 | .859 | .848 | .834 | .820 | .804 |
| | | | | | | | | .786 | .766 | .745 | .722 | .696 | .668 | .638 | .605 | .571 | .534 |
| | | | | | | | | .495 | .455 | .414 | .373 | .331 | .291 | .252 | | | |
| 19.000 | 18.164 | .000 | .836 | .79 | 3.81 | 4.4% | 28 | .904 | .897 | .889 | .880 | .870 | .859 | .848 | .834 | .820 | .804 |
| | | | | | | | | .786 | .766 | .745 | .722 | .696 | .668 | .638 | .605 | .571 | .534 |
| | | | | | | | | .495 | .455 | .414 | .373 | .331 | .291 | .252 | .215 | | |
| 19.000 | 18.344 | .000 | .656 | .82 | 3.75 | 3.5% | 29 | .904 | .897 | .889 | .880 | .870 | .859 | .848 | .834 | .820 | .804 |
| | | | | | | | | .786 | .766 | .745 | .722 | .696 | .668 | .638 | .605 | .571 | .534 |
| | | | | | | | | .495 | .455 | .414 | .373 | .331 | .291 | .252 | .215 | .180 | |
| 19.000 | 18.493 | .000 | .507 | .85 | 3.69 | 2.7% | 30 | .904 | .897 | .889 | .880 | .870 | .859 | .848 | .834 | .820 | .804 |
| | | | | | | | | .786 | .766 | .745 | .722 | .696 | .668 | .638 | .605 | .571 | .534 |
| | | | | | | | | .495 | .455 | .414 | .373 | .331 | .291 | .252 | .215 | .180 | .149 |
| 19.000 | 18.613 | .000 | .387 | .88 | 3.62 | 2.0% | 31 | .904 | .897 | .889 | .880 | .870 | .859 | .848 | .834 | .820 | .804 |
| | | | | | | | | .786 | .766 | .745 | .722 | .696 | .668 | .638 | .605 | .571 | .534 |
| | | | | | | | | .495 | .455 | .414 | .373 | .331 | .291 | .252 | .215 | .180 | .149 |
| | | | | | | | | .121 | | | | | | | | | |
| 19.000 | 18.710 | .000 | .290 | .90 | 3.54 | 1.5% | 32 | .904 | .897 | .889 | .880 | .870 | .859 | .848 | .834 | .820 | .804 |
| | | | | | | | | .786 | .766 | .745 | .722 | .696 | .668 | .638 | .605 | .571 | .534 |
| | | | | | | | | .495 | .455 | .414 | .373 | .331 | .291 | .252 | .215 | .180 | .149 |
| | | | | | | | | .121 | .096 | | | | | | | | |
| 19.000 | 18.786 | .000 | .214 | .92 | 3.46 | 1.1% | 33 | .904 | .897 | .889 | .880 | .870 | .859 | .848 | .834 | .820 | .804 |
| | | | | | | | | .786 | .766 | .745 | .722 | .696 | .668 | .638 | .605 | .571 | .534 |
| | | | | | | | | .495 | .455 | .414 | .373 | .331 | .291 | .252 | .215 | .180 | .149 |
| | | | | | | | | .121 | .096 | .076 | | | | | | | |
| 19.000 | 18.844 | .000 | .156 | .94 | 3.37 | .8% | 34 | .904 | .897 | .889 | .880 | .870 | .859 | .848 | .834 | .820 | .804 |
| | | | | | | | | .786 | .766 | .745 | .722 | .696 | .668 | .638 | .605 | .571 | .534 |
| | | | | | | | | .495 | .455 | .414 | .373 | .331 | .291 | .252 | .215 | .180 | .149 |
| | | | | | | | | .121 | .096 | .076 | .058 | | | | | | |
| 19.000 | 18.888 | .000 | .112 | .96 | 3.28 | .6% | 35 | .904 | .897 | .889 | .880 | .870 | .859 | .848 | .834 | .820 | .804 |
| | | | | | | | | .786 | .766 | .745 | .722 | .696 | .668 | .638 | .605 | .571 | .534 |
| | | | | | | | | .495 | .455 | .414 | .373 | .331 | .291 | .252 | .215 | .180 | .149 |
| | | | | | | | | .121 | .098 | .070 | .058 | .041 | | | | | |
| 20.000 | 13.196 | .000 | 6.804 | .31 | 3.39 | 34.0% | 16 | .909 | .902 | .895 | .887 | .878 | .868 | .857 | .845 | .832 | .817 |
| | | | | | | | | .801 | .784 | .764 | .743 | .720 | .695 | | | | |
| 20.000 | 13.004 | .000 | 6.136 | .33 | 3.48 | 30.7% | 17 | .909 | .902 | .895 | .887 | .878 | .868 | .857 | .845 | .832 | .817 |
| | | | | | | | | .801 | .784 | .764 | .743 | .720 | .695 | .668 | | | |

*SIMILAR TO A.T.&T. NEAL-WILKINSON TABLES(SEE SECTION 1)

*EQUIVALENT RANDOM THEORY: VMR(VARIANCE TO MEAN RATIO)– 2.0 - NO RETRIES (PEAKED TRAFFIC - ROUTE ADVANCEMENT OR DAY-TO-DAY VARIATION)

| HOURS FIRST ATTEMPT VMR=2.0 | HOURS CON-NECT | HOURS RECALL /RETRY | HOURS OVFL | CON-NECT VMR | OVFL VMR | PERCENT BLOCKED (GRADE) | LINE | HOURS CONNECTED TRAFFIC ON LINES 1 | 2 | 3 | 4 | 5 | 6 | 7 | 8 | 9 | 10 |
|---|---|---|---|---|---|---|---|---|---|---|---|---|---|---|---|---|---|
| 20.000 | 14.502 | .000 | 5.498 | .36 | 3.57 | 27.5% | 18 | .909 | .902 | .895 | .887 | .878 | .868 | .857 | .845 | .832 | .817 |
| | | | | | | | | .801 | .784 | .764 | .743 | .720 | .695 | .668 | .638 | | |
| 20.000 | 15.108 | .000 | 4.892 | .39 | 3.65 | 24.5% | 19 | .909 | .902 | .895 | .887 | .878 | .868 | .857 | .845 | .832 | .817 |
| | | | | | | | | .801 | .784 | .764 | .743 | .720 | .695 | .668 | .638 | .606 | |
| 20.000 | 15.681 | .000 | 4.319 | .43 | 3.73 | 21.6% | 20 | .909 | .902 | .895 | .887 | .878 | .868 | .857 | .845 | .832 | .817 |
| | | | | | | | | .801 | .784 | .764 | .743 | .720 | .695 | .668 | .638 | .606 | .573 |
| 20.000 | 16.217 | .000 | 3.783 | .46 | 3.80 | 18.9% | 21 | .909 | .902 | .895 | .887 | .878 | .868 | .857 | .845 | .832 | .817 |
| | | | | | | | | .801 | .784 | .764 | .743 | .720 | .695 | .668 | .638 | .606 | .573 |
| | | | | | | | | .537 | | | | | | | | | |
| 20.000 | 16.716 | .000 | 3.284 | .50 | 3.85 | 16.4% | 22 | .909 | .902 | .895 | .887 | .878 | .868 | .857 | .845 | .832 | .817 |
| | | | | | | | | .801 | .784 | .764 | .743 | .720 | .695 | .668 | .638 | .606 | .573 |
| | | | | | | | | .537 | .499 | | | | | | | | |
| 20.000 | 17.176 | .000 | 2.824 | .54 | 3.90 | 14.1% | 23 | .909 | .902 | .895 | .887 | .878 | .868 | .857 | .845 | .832 | .817 |
| | | | | | | | | .801 | .784 | .764 | .743 | .720 | .695 | .668 | .638 | .606 | .573 |
| | | | | | | | | .537 | .499 | .460 | | | | | | | |
| 20.000 | 17.596 | .000 | 2.404 | .58 | 3.93 | 12.0% | 24 | .909 | .902 | .895 | .887 | .878 | .868 | .857 | .845 | .832 | .817 |
| | | | | | | | | .801 | .784 | .764 | .743 | .720 | .695 | .668 | .638 | .606 | .573 |
| | | | | | | | | .537 | .499 | .460 | .420 | | | | | | |
| 20.000 | 17.976 | .000 | 2.024 | .62 | 3.95 | 10.1% | 25 | .909 | .902 | .895 | .887 | .878 | .868 | .857 | .845 | .832 | .817 |
| | | | | | | | | .801 | .784 | .764 | .743 | .720 | .695 | .668 | .638 | .606 | .573 |
| | | | | | | | | .537 | .499 | .460 | .420 | .379 | | | | | |
| 20.000 | 18.314 | .000 | 1.686 | .66 | 3.96 | 8.4% | 26 | .909 | .902 | .895 | .887 | .878 | .868 | .857 | .845 | .832 | .817 |
| | | | | | | | | .801 | .784 | .764 | .743 | .720 | .695 | .668 | .638 | .606 | .573 |
| | | | | | | | | .537 | .499 | .460 | .420 | .379 | .339 | | | | |
| 20.000 | 18.613 | .000 | 1.387 | .70 | 3.95 | 6.9% | 27 | .909 | .902 | .895 | .887 | .878 | .868 | .857 | .845 | .832 | .817 |
| | | | | | | | | .801 | .784 | .764 | .743 | .720 | .695 | .668 | .638 | .606 | .573 |
| | | | | | | | | .537 | .499 | .460 | .420 | .379 | .339 | .299 | | | |
| 20.000 | 18.873 | .000 | 1.127 | .74 | 3.92 | 5.6% | 28 | .909 | .902 | .895 | .887 | .878 | .868 | .857 | .845 | .832 | .817 |
| | | | | | | | | .801 | .784 | .764 | .743 | .720 | .695 | .668 | .638 | .606 | .573 |
| | | | | | | | | .537 | .499 | .460 | .420 | .379 | .339 | .299 | .260 | | |
| 20.000 | 19.096 | .000 | .904 | .78 | 3.89 | 4.5% | 29 | .909 | .902 | .895 | .887 | .878 | .868 | .857 | .845 | .832 | .817 |
| | | | | | | | | .801 | .784 | .764 | .743 | .720 | .695 | .668 | .638 | .606 | .573 |
| | | | | | | | | .537 | .499 | .460 | .420 | .379 | .339 | .299 | .260 | .223 | |
| 20.000 | 19.285 | .000 | .715 | .81 | 3.84 | 3.6% | 30 | .909 | .902 | .895 | .887 | .878 | .868 | .857 | .845 | .832 | .817 |
| | | | | | | | | .801 | .784 | .764 | .743 | .720 | .695 | .668 | .638 | .606 | .573 |
| | | | | | | | | .537 | .499 | .460 | .420 | .379 | .339 | .299 | .260 | .223 | .189 |
| 20.000 | 19.442 | .000 | .558 | .84 | 3.78 | 2.8% | 31 | .909 | .902 | .895 | .887 | .878 | .868 | .857 | .845 | .832 | .817 |
| | | | | | | | | .801 | .784 | .764 | .743 | .720 | .695 | .668 | .638 | .606 | .573 |
| | | | | | | | | .537 | .499 | .460 | .420 | .379 | .339 | .299 | .260 | .223 | .189 |
| | | | | | | | | .157 | | | | | | | | | |

*SIMILAR TO A.T.&T. NEAL-WILKINSON TABLES(SEE SECTION 1)

*EQUIVALENT RANDOM THEORY: VMR(VARIANCE TO MEAN RATIO)– 2.0 - NO RETRIES (PEAKED TRAFFIC - ROUTE ADVANCEMENT OR DAY-TO-DAY VARIATION)

| HOURS FIRST ATTEMPT VMR=2.0 | HOURS CON-NECT | HOURS RECALL /RETRY | HOURS OVFL | CON-NECT VMR | OVFL VMR | PERCENT BLOCKED (GRADE) | LINE | HOURS CONNECTED TRAFFIC ON LINES 1 | 2 | 3 | 4 | 5 | 6 | 7 | 8 | 9 | 10 |
|---|---|---|---|---|---|---|---|---|---|---|---|---|---|---|---|---|---|
| 20.000 | 19.571 | .000 | .429 | .87 | 3.71 | 2.1% | 32 | .909 | .902 | .895 | .887 | .878 | .868 | .857 | .845 | .832 | .817 |
| | | | | | | | | .801 | .784 | .764 | .743 | .720 | .695 | .668 | .638 | .606 | .573 |
| | | | | | | | | .537 | .499 | .460 | .420 | .379 | .339 | .299 | .260 | .223 | .189 |
| | | | | | | | | .157 | .129 | | | | | | | | |
| 20.000 | 19.674 | .000 | .326 | .90 | 3.63 | 1.6% | 33 | .909 | .902 | .895 | .887 | .878 | .868 | .857 | .845 | .832 | .817 |
| | | | | | | | | .801 | .784 | .764 | .743 | .720 | .695 | .668 | .638 | .606 | .573 |
| | | | | | | | | .537 | .499 | .460 | .420 | .379 | .339 | .299 | .260 | .223 | .189 |
| | | | | | | | | .157 | .129 | .104 | | | | | | | |
| 20.000 | 19.757 | .000 | .243 | .92 | 3.55 | 1.2% | 34 | .909 | .902 | .895 | .887 | .878 | .868 | .857 | .845 | .832 | .817 |
| | | | | | | | | .801 | .784 | .764 | .743 | .720 | .695 | .668 | .638 | .606 | .573 |
| | | | | | | | | .537 | .499 | .460 | .420 | .379 | .339 | .299 | .260 | .223 | .189 |
| | | | | | | | | .157 | .129 | .104 | .082 | | | | | | |
| 20.000 | 19.821 | .000 | .179 | .94 | 3.46 | .9% | 35 | .909 | .902 | .895 | .887 | .878 | .868 | .857 | .845 | .832 | .817 |
| | | | | | | | | .801 | .784 | .764 | .743 | .720 | .695 | .668 | .638 | .606 | .573 |
| | | | | | | | | .537 | .499 | .460 | .420 | .379 | .339 | .299 | .260 | .223 | .189 |
| | | | | | | | | .157 | .129 | .104 | .082 | .064 | | | | | |
| 20.000 | 19.870 | .000 | .130 | .95 | 3.37 | .6% | 36 | .909 | .902 | .895 | .887 | .878 | .868 | .857 | .845 | .832 | .817 |
| | | | | | | | | .801 | .784 | .764 | .743 | .720 | .695 | .668 | .638 | .606 | .573 |
| | | | | | | | | .537 | .499 | .460 | .420 | .379 | .339 | .299 | .260 | .223 | .189 |
| | | | | | | | | .157 | .129 | .104 | .082 | .064 | .049 | | | | |
| 22.000 | 14.955 | .000 | 7.045 | .31 | 3.52 | 32.0% | 18 | .916 | .911 | .904 | .898 | .890 | .882 | .873 | .863 | .852 | .840 |
| | | | | | | | | .827 | .813 | .797 | .780 | .761 | .740 | .718 | .693 | | |
| 22.000 | 15.622 | .000 | 6.378 | .33 | 3.61 | 29.0% | 19 | .916 | .911 | .904 | .898 | .890 | .882 | .873 | .863 | .852 | .840 |
| | | | | | | | | .827 | .813 | .797 | .780 | .761 | .740 | .718 | .693 | .667 | |
| 22.000 | 16.261 | .000 | 5.739 | .36 | 3.70 | 26.1% | 20 | .916 | .911 | .904 | .898 | .890 | .882 | .873 | .863 | .852 | .840 |
| | | | | | | | | .827 | .813 | .797 | .780 | .761 | .740 | .718 | .693 | .667 | .638 |
| 22.000 | 16.869 | .000 | 5.131 | .39 | 3.78 | 23.3% | 21 | .916 | .911 | .904 | .898 | .890 | .882 | .873 | .863 | .852 | .840 |
| | | | | | | | | .827 | .813 | .797 | .780 | .761 | .740 | .718 | .693 | .667 | .638 |
| | | | | | | | | .608 | | | | | | | | | |
| 22.000 | 17.444 | .000 | 4.556 | .42 | 3.86 | 20.7% | 22 | .916 | .911 | .904 | .898 | .890 | .882 | .873 | .863 | .852 | .840 |
| | | | | | | | | .827 | .813 | .797 | .780 | .761 | .740 | .718 | .693 | .667 | .638 |
| | | | | | | | | .608 | .576 | | | | | | | | |
| 22.000 | 17.986 | .000 | 4.014 | .46 | 3.93 | 18.2% | 23 | .916 | .911 | .904 | .898 | .890 | .882 | .873 | .863 | .852 | .840 |
| | | | | | | | | .827 | .813 | .797 | .780 | .761 | .740 | .718 | .693 | .667 | .638 |
| | | | | | | | | .608 | .576 | .541 | | | | | | | |
| 22.000 | 18.491 | .000 | 3.509 | .49 | 3.98 | 15.9% | 24 | .916 | .911 | .904 | .898 | .890 | .882 | .873 | .863 | .852 | .840 |
| | | | | | | | | .827 | .813 | .797 | .780 | .761 | .740 | .718 | .693 | .667 | .638 |
| | | | | | | | | .608 | .576 | .541 | .506 | | | | | | |
| 22.000 | 18.960 | .000 | 3.040 | .53 | 4.03 | 13.8% | 25 | .916 | .911 | .904 | .898 | .890 | .882 | .873 | .863 | .852 | .840 |
| | | | | | | | | .827 | .813 | .797 | .780 | .761 | .740 | .718 | .693 | .667 | .638 |

*SIMILAR TO A.T.&T. NEAL-WILKINSON TABLES(SEE SECTION 1)

*EQUIVALENT RANDOM THEORY: VMR(VARIANCE TO MEAN RATIO)– 2.0 - NO RETRIES (PEAKED TRAFFIC - ROUTE ADVANCEMENT OR DAY-TO-DAY VARIATION)

| HOURS FIRST ATTEMPT VMR=2.0 | HOURS CON-NECT | HOURS RECALL /RETRY | HOURS OVFL | CON-NECT VMR | OVFL VMR | PERCENT BLOCKED (GRADE) | LINE | HOURS CONNECTED TRAFFIC ON LINES 1 | 2 | 3 | 4 | 5 | 6 | 7 | 8 | 9 | 10 |
|---|---|---|---|---|---|---|---|---|---|---|---|---|---|---|---|---|---|
| | | | | | | | | .608 | .576 | .541 | .506 | .468 | | | | | |
| 22.000 | 19.390 | .000 | 2.610 | .57 | 4.07 | 11.9% | 26 | .916 | .911 | .904 | .898 | .890 | .882 | .873 | .863 | .852 | .840 |
| | | | | | | | | .827 | .813 | .797 | .780 | .761 | .740 | .718 | .693 | .667 | .638 |
| | | | | | | | | .608 | .576 | .541 | .506 | .468 | .430 | | | | |
| 22.000 | 19.780 | .000 | 2.220 | .61 | 4.09 | 10.1% | 27 | .916 | .911 | .904 | .898 | .890 | .882 | .873 | .863 | .852 | .840 |
| | | | | | | | | .827 | .813 | .797 | .780 | .761 | .740 | .718 | .693 | .667 | .638 |
| | | | | | | | | .608 | .576 | .541 | .506 | .468 | .430 | .391 | | | |
| 22.000 | 20.132 | .000 | 1.868 | .65 | 4.10 | 8.5% | 28 | .916 | .911 | .904 | .898 | .890 | .882 | .873 | .863 | .852 | .840 |
| | | | | | | | | .827 | .813 | .797 | .780 | .761 | .740 | .718 | .693 | .667 | .638 |
| | | | | | | | | .608 | .576 | .541 | .506 | .468 | .430 | .391 | .352 | | |
| 22.000 | 20.445 | .000 | 1.555 | .69 | 4.09 | 7.1% | 29 | .916 | .911 | .904 | .898 | .890 | .882 | .873 | .863 | .852 | .840 |
| | | | | | | | | .827 | .813 | .797 | .780 | .761 | .740 | .718 | .693 | .667 | .638 |
| | | | | | | | | .608 | .576 | .541 | .506 | .468 | .430 | .391 | .352 | .313 | |
| 22.000 | 20.721 | .000 | 1.279 | .72 | 4.07 | 5.8% | 30 | .916 | .911 | .904 | .898 | .890 | .882 | .873 | .863 | .852 | .840 |
| | | | | | | | | .827 | .813 | .797 | .780 | .761 | .740 | .718 | .693 | .667 | .638 |
| | | | | | | | | .608 | .576 | .541 | .506 | .468 | .430 | .391 | .352 | .313 | .275 |
| 22.000 | 20.960 | .000 | 1.040 | .76 | 4.04 | 4.7% | 31 | .916 | .911 | .904 | .898 | .890 | .882 | .873 | .863 | .852 | .840 |
| | | | | | | | | .827 | .813 | .797 | .780 | .761 | .740 | .718 | .693 | .667 | .638 |
| | | | | | | | | .608 | .576 | .541 | .506 | .468 | .430 | .391 | .352 | .313 | .275 |
| | | | | | | | | .239 | | | | | | | | | |
| 22.000 | 21.164 | .000 | .836 | .80 | 4.00 | 3.8% | 32 | .916 | .911 | .904 | .898 | .890 | .882 | .873 | .863 | .852 | .840 |
| | | | | | | | | .827 | .813 | .797 | .780 | .761 | .740 | .718 | .693 | .667 | .638 |
| | | | | | | | | .608 | .576 | .541 | .506 | .468 | .430 | .391 | .352 | .313 | .275 |
| | | | | | | | | .239 | .205 | | | | | | | | |
| 22.000 | 21.337 | .000 | .663 | .83 | 3.94 | 3.0% | 33 | .916 | .911 | .904 | .898 | .890 | .882 | .873 | .863 | .852 | .840 |
| | | | | | | | | .827 | .813 | .797 | .780 | .761 | .740 | .718 | .693 | .667 | .638 |
| | | | | | | | | .608 | .576 | .541 | .506 | .468 | .430 | .391 | .352 | .313 | .275 |
| | | | | | | | | .239 | .205 | .173 | | | | | | | |
| 22.000 | 21.481 | .000 | .519 | .86 | 3.87 | 2.4% | 34 | .916 | .911 | .904 | .898 | .890 | .882 | .873 | .863 | .852 | .840 |
| | | | | | | | | .827 | .813 | .797 | .780 | .761 | .740 | .718 | .693 | .667 | .638 |
| | | | | | | | | .608 | .576 | .541 | .506 | .468 | .430 | .391 | .352 | .313 | .275 |
| | | | | | | | | .239 | .205 | .173 | .144 | | | | | | |
| 22.000 | 21.599 | .000 | .401 | .88 | 3.80 | 1.8% | 35 | .916 | .911 | .904 | .898 | .890 | .882 | .873 | .863 | .852 | .840 |
| | | | | | | | | .827 | .813 | .797 | .780 | .761 | .740 | .718 | .693 | .667 | .638 |
| | | | | | | | | .608 | .576 | .541 | .506 | .468 | .430 | .391 | .352 | .313 | .275 |
| | | | | | | | | .239 | .205 | .173 | .144 | .118 | | | | | |
| 22.000 | 21.695 | .000 | .305 | .90 | 3.72 | 1.4% | 36 | .916 | .911 | .904 | .898 | .890 | .882 | .873 | .863 | .852 | .840 |
| | | | | | | | | .827 | .813 | .797 | .780 | .761 | .740 | .718 | .693 | .667 | .638 |
| | | | | | | | | .608 | .576 | .541 | .506 | .468 | .430 | .391 | .352 | .313 | .275 |
| | | | | | | | | .239 | .205 | .173 | .144 | .118 | .095 | | | | |

*SIMILAR TO A.T.&T. NEAL-WILKINSON TABLES(SEE SECTION 1)

*EQUIVALENT RANDOM THEORY: VMR(VARIANCE TO MEAN RATIO)– 2.0 - NO RETRIES (PEAKED TRAFFIC - ROUTE ADVANCEMENT OR DAY-TO-DAY VARIATION)

| HOURS FIRST ATTEMPT VMR=2.0 | HOURS CON- NECT | HOURS RECALL /RETRY | HOURS OVFL | CON- NECT VMR | OVFL VMR | PERCENT BLOCKED (GRADE) | LINE | HOURS CONNECTED TRAFFIC ON LINES 1 | 2 | 3 | 4 | 5 | 6 | 7 | 8 | 9 | 10 |
|---|---|---|---|---|---|---|---|---|---|---|---|---|---|---|---|---|---|
| 22.000 | 21.770 | .000 | .230 | .92 | 3.64 | 1.0% | 37 | .916 | .911 | .904 | .898 | .890 | .882 | .873 | .863 | .852 | .840 |
| | | | | | | | | .827 | .813 | .797 | .780 | .761 | .740 | .718 | .693 | .667 | .638 |
| | | | | | | | | .608 | .576 | .541 | .506 | .468 | .430 | .391 | .352 | .313 | .275 |
| | | | | | | | | .239 | .205 | .173 | .144 | .118 | .095 | .076 | | | |
| 22.000 | 21.830 | .000 | .170 | .94 | 3.55 | .8% | 38 | .916 | .911 | .904 | .898 | .890 | .882 | .873 | .863 | .852 | .840 |
| | | | | | | | | .827 | .813 | .797 | .780 | .761 | .740 | .718 | .693 | .667 | .638 |
| | | | | | | | | .608 | .576 | .541 | .506 | .468 | .430 | .391 | .352 | .313 | .275 |
| | | | | | | | | .239 | .205 | .173 | .144 | .118 | .095 | .076 | .059 | | |
| 22.000 | 21.876 | .000 | .124 | .95 | 3.46 | .6% | 39 | .916 | .911 | .904 | .898 | .890 | .882 | .873 | .863 | .852 | .840 |
| | | | | | | | | .827 | .813 | .797 | .780 | .761 | .740 | .718 | .693 | .667 | .638 |
| | | | | | | | | .608 | .576 | .541 | .506 | .468 | .430 | .391 | .352 | .313 | .275 |
| | | | | | | | | .239 | .205 | .173 | .144 | .118 | .095 | .076 | .059 | .046 | |
| 24.000 | 16.034 | .000 | 7.966 | .28 | 3.55 | 33.2% | 19 | .923 | .918 | .913 | .907 | .900 | .893 | .886 | .877 | .868 | .858 |
| | | | | | | | | .847 | .835 | .822 | .808 | .793 | .776 | .757 | .737 | .715 | |
| 24.000 | 16.726 | .000 | 7.274 | .31 | 3.65 | 30.3% | 20 | .923 | .918 | .913 | .907 | .900 | .893 | .886 | .877 | .868 | .858 |
| | | | | | | | | .847 | .835 | .822 | .808 | .793 | .776 | .757 | .737 | .715 | .692 |
| 24.000 | 17.392 | .000 | 6.608 | .33 | 3.74 | 27.5% | 21 | .923 | .918 | .913 | .907 | .900 | .893 | .886 | .877 | .868 | .858 |
| | | | | | | | | .847 | .835 | .822 | .808 | .793 | .776 | .757 | .737 | .715 | .692 |
| | | | | | | | | .666 | | | | | | | | | |
| 24.000 | 18.030 | .000 | 5.970 | .36 | 3.82 | 24.9% | 22 | .923 | .918 | .913 | .907 | .900 | .893 | .886 | .877 | .868 | .858 |
| | | | | | | | | .847 | .835 | .822 | .808 | .793 | .776 | .757 | .737 | .715 | .692 |
| | | | | | | | | .666 | .639 | | | | | | | | |
| 24.000 | 18.640 | .000 | 5.360 | .39 | 3.91 | 22.3% | 23 | .923 | .918 | .913 | .907 | .900 | .893 | .886 | .877 | .868 | .858 |
| | | | | | | | | .847 | .835 | .822 | .808 | .793 | .776 | .757 | .737 | .715 | .692 |
| | | | | | | | | .666 | .639 | .610 | | | | | | | |
| 24.000 | 19.218 | .000 | 4.782 | .42 | 3.98 | 19.9% | 24 | .923 | .918 | .913 | .907 | .900 | .893 | .886 | .877 | .868 | .858 |
| | | | | | | | | .847 | .835 | .822 | .808 | .793 | .776 | .757 | .737 | .715 | .692 |
| | | | | | | | | .666 | .639 | .610 | .578 | | | | | | |
| 24.000 | 19.764 | .000 | 4.236 | .45 | 4.05 | 17.7% | 25 | .923 | .918 | .913 | .907 | .900 | .893 | .886 | .877 | .868 | .858 |
| | | | | | | | | .847 | .835 | .822 | .808 | .793 | .776 | .757 | .737 | .715 | .692 |
| | | | | | | | | .666 | .639 | .610 | .578 | .546 | | | | | |
| 24.000 | 20.275 | .000 | 3.725 | .49 | 4.11 | 15.5% | 26 | .923 | .918 | .913 | .907 | .900 | .893 | .886 | .877 | .868 | .858 |
| | | | | | | | | .847 | .835 | .822 | .808 | .793 | .776 | .757 | .737 | .715 | .692 |
| | | | | | | | | .666 | .639 | .610 | .578 | .546 | .511 | | | | |
| 24.000 | 20.751 | .000 | 3.249 | .52 | 4.16 | 13.5% | 27 | .923 | .918 | .913 | .907 | .900 | .893 | .886 | .877 | .868 | .858 |
| | | | | | | | | .847 | .835 | .822 | .808 | .793 | .776 | .757 | .737 | .715 | .692 |
| | | | | | | | | .666 | .639 | .610 | .578 | .546 | .511 | .475 | | | |
| 24.000 | 21.189 | .000 | 2.811 | .56 | 4.20 | 11.7% | 28 | .923 | .918 | .913 | .907 | .900 | .893 | .886 | .877 | .868 | .858 |
| | | | | | | | | .847 | .835 | .822 | .808 | .793 | .776 | .757 | .737 | .715 | .692 |
| | | | | | | | | .666 | .639 | .610 | .578 | .546 | .511 | .475 | .409 | | |

*SIMILAR TO A.T.&T. NEAL-WILKINSON TABLES(SEE SECTION 1)

*EQUIVALENT RANDOM THEORY: VMR(VARIANCE TO MEAN RATIO)– 2.0 - NO RETRIES (PEAKED TRAFFIC - ROUTE ADVANCEMENT OR DAY-TO-DAY VARIATION)

| HOURS FIRST ATTEMPT VMR=2.0 | HOURS CON-NECT | HOURS RECALL /RETRY | HOURS OVFL | CON-NECT VMR | OVFL VMR | PERCENT BLOCKED (GRADE) | LINE | HOURS CONNECTED TRAFFIC ON LINES 1 | 2 | 3 | 4 | 5 | 6 | 7 | 8 | 9 | 10 |
|---|---|---|---|---|---|---|---|---|---|---|---|---|---|---|---|---|---|
| 24.000 | 21.590 | .000 | 2.410 | .60 | 4.22 | 10.0% | 29 | .923 | .918 | .913 | .907 | .900 | .893 | .886 | .877 | .868 | .858 |
| | | | | | | | | .847 | .835 | .822 | .808 | .793 | .776 | .757 | .737 | .715 | .692 |
| | | | | | | | | .666 | .639 | .610 | .578 | .546 | .511 | .475 | .439 | .401 | |
| 24.000 | 21.954 | .000 | 2.046 | .64 | 4.23 | 8.5% | 30 | .923 | .918 | .913 | .907 | .900 | .893 | .886 | .877 | .868 | .858 |
| | | | | | | | | .847 | .835 | .822 | .808 | .793 | .776 | .757 | .737 | .715 | .692 |
| | | | | | | | | .666 | .639 | .610 | .578 | .546 | .511 | .475 | .439 | .401 | .363 |
| 24.000 | 22.280 | .000 | 1.720 | .67 | 4.23 | 7.2% | 31 | .923 | .918 | .913 | .907 | .900 | .893 | .886 | .877 | .868 | .858 |
| | | | | | | | | .847 | .835 | .822 | .808 | .793 | .776 | .757 | .737 | .715 | .692 |
| | | | | | | | | .666 | .639 | .610 | .578 | .546 | .511 | .475 | .439 | .401 | .363 |
| | | | | | | | | .326 | | | | | | | | | |
| 24.000 | 22.569 | .000 | 1.431 | .71 | 4.22 | 6.0% | 32 | .923 | .918 | .913 | .907 | .900 | .893 | .886 | .877 | .868 | .858 |
| | | | | | | | | .847 | .835 | .822 | .808 | .793 | .776 | .757 | .737 | .715 | .692 |
| | | | | | | | | .666 | .639 | .610 | .578 | .546 | .511 | .475 | .439 | .401 | .363 |
| | | | | | | | | .326 | .289 | | | | | | | | |
| 24.000 | 22.822 | .000 | 1.178 | .75 | 4.19 | 4.9% | 33 | .923 | .918 | .913 | .907 | .900 | .893 | .886 | .877 | .868 | .858 |
| | | | | | | | | .847 | .835 | .822 | .808 | .793 | .776 | .757 | .737 | .715 | .692 |
| | | | | | | | | .666 | .639 | .610 | .578 | .546 | .511 | .475 | .439 | .401 | .363 |
| | | | | | | | | .326 | .289 | .253 | | | | | | | |
| 24.000 | 23.042 | .000 | .958 | .78 | 4.15 | 4.0% | 34 | .923 | .918 | .913 | .907 | .900 | .893 | .886 | .877 | .868 | .858 |
| | | | | | | | | .847 | .835 | .822 | .808 | .793 | .776 | .757 | .737 | .715 | .692 |
| | | | | | | | | .666 | .639 | .610 | .578 | .546 | .511 | .475 | .439 | .401 | .363 |
| | | | | | | | | .326 | .289 | .253 | .219 | | | | | | |
| 24.000 | 23.229 | .000 | .771 | .81 | 4.10 | 3.2% | 35 | .923 | .918 | .913 | .907 | .900 | .893 | .886 | .877 | .868 | .858 |
| | | | | | | | | .847 | .835 | .822 | .808 | .793 | .776 | .757 | .737 | .715 | .692 |
| | | | | | | | | .666 | .639 | .610 | .578 | .546 | .511 | .475 | .439 | .401 | .363 |
| | | | | | | | | .326 | .289 | .253 | .219 | .188 | | | | | |
| 24.000 | 23.388 | .000 | .612 | .84 | 4.03 | 2.6% | 36 | .923 | .918 | .913 | .907 | .900 | .893 | .886 | .877 | .868 | .858 |
| | | | | | | | | .847 | .835 | .822 | .808 | .793 | .776 | .757 | .737 | .715 | .692 |
| | | | | | | | | .666 | .639 | .610 | .578 | .546 | .511 | .475 | .439 | .401 | .363 |
| | | | | | | | | .326 | .289 | .253 | .219 | .188 | .158 | | | | |
| 24.000 | 23.519 | .000 | .481 | .87 | 3.96 | 2.0% | 37 | .923 | .918 | .913 | .907 | .900 | .893 | .886 | .877 | .868 | .858 |
| | | | | | | | | .847 | .835 | .822 | .808 | .793 | .776 | .757 | .737 | .715 | .692 |
| | | | | | | | | .666 | .639 | .610 | .578 | .546 | .511 | .475 | .439 | .401 | .363 |
| | | | | | | | | .326 | .289 | .253 | .219 | .188 | .158 | .132 | | | |
| 24.000 | 23.628 | .000 | .372 | .89 | 3.89 | 1.6% | 38 | .923 | .918 | .913 | .907 | .900 | .893 | .886 | .877 | .868 | .858 |
| | | | | | | | | .847 | .835 | .822 | .808 | .793 | .776 | .757 | .737 | .715 | .692 |
| | | | | | | | | .666 | .639 | .610 | .578 | .546 | .511 | .475 | .439 | .401 | .363 |
| | | | | | | | | .326 | .289 | .253 | .219 | .188 | .158 | .132 | .108 | | |
| 24.000 | 23.715 | .000 | .285 | .91 | 3.80 | 1.2% | 39 | .923 | .918 | .913 | .907 | .900 | .893 | .886 | .877 | .868 | .858 |
| | | | | | | | | .847 | .835 | .822 | .808 | .793 | .776 | .757 | .737 | .715 | .692 |

*SIMILAR TO A.T.&T. NEAL-WILKINSON TABLES(SEE SECTION 1)

*EQUIVALENT RANDOM THEORY: VMR(VARIANCE TO MEAN RATIO)– 2.0 - NO RETRIES (PEAKED TRAFFIC - ROUTE ADVANCEMENT OR DAY-TO-DAY VARIATION)

| HOURS FIRST ATTEMPT VMR = 2.0 | HOURS CON-NECT | HOURS RECALL /RETRY | HOURS OVFL | CON-NECT VMR | OVFL VMR | PERCENT BLOCKED (GRADE) | LINE | HOURS CONNECTED TRAFFIC ON LINES 1 | 2 | 3 | 4 | 5 | 6 | 7 | 8 | 9 | 10 |
|---|---|---|---|---|---|---|---|---|---|---|---|---|---|---|---|---|---|
| | | | | | | | | .666 | .639 | .610 | .578 | .546 | .511 | .475 | .439 | .401 | .363 |
| | | | | | | | | .326 | .289 | .253 | .219 | .188 | .158 | .132 | .108 | .087 | |
| 24.000 | 23.785 | .000 | .215 | .93 | 3.72 | .9% | 40 | .923 | .918 | .913 | .907 | .900 | .893 | .886 | .877 | .868 | .858 |
| | | | | | | | | .847 | .835 | .822 | .808 | .793 | .776 | .757 | .737 | .715 | .692 |
| | | | | | | | | .666 | .639 | .610 | .578 | .546 | .511 | .475 | .439 | .401 | .363 |
| | | | | | | | | .326 | .289 | .253 | .219 | .188 | .158 | .132 | .108 | .087 | .070 |
| 24.000 | 23.839 | .000 | .161 | .95 | 3.63 | .7% | 41 | .923 | .918 | .913 | .907 | .900 | .893 | .886 | .877 | .868 | .858 |
| | | | | | | | | .847 | .835 | .822 | .808 | .793 | .776 | .757 | .737 | .715 | .692 |
| | | | | | | | | .666 | .639 | .610 | .578 | .546 | .511 | .475 | .439 | .401 | .363 |
| | | | | | | | | .326 | .289 | .253 | .219 | .188 | .158 | .132 | .108 | .087 | .070 |
| | | | | | | | | .055 | | | | | | | | | |
| 26.000 | 17.102 | .000 | 8.898 | .27 | 3.58 | 34.2% | 20 | .928 | .924 | .919 | .914 | .909 | .903 | .896 | .889 | .881 | .873 |
| | | | | | | | | .864 | .854 | .843 | .831 | .818 | .804 | .789 | .772 | .754 | .734 |
| 26.000 | 17.815 | .000 | 8.185 | .29 | 3.67 | 31.5% | 21 | .928 | .924 | .919 | .914 | .909 | .903 | .896 | .889 | .881 | .873 |
| | | | | | | | | .864 | .854 | .843 | .831 | .818 | .804 | .789 | .772 | .754 | .734 |
| | | | | | | | | .713 | | | | | | | | | |
| 26.000 | 18.505 | .000 | 7.495 | .31 | 3.77 | 28.8% | 22 | .928 | .924 | .919 | .914 | .909 | .903 | .896 | .889 | .881 | .873 |
| | | | | | | | | .864 | .854 | .843 | .831 | .818 | .804 | .789 | .772 | .754 | .734 |
| | | | | | | | | .713 | .690 | | | | | | | | |
| 26.000 | 19.170 | .000 | 6.830 | .33 | 3.86 | 26.3% | 23 | .928 | .924 | .919 | .914 | .909 | .903 | .896 | .889 | .881 | .873 |
| | | | | | | | | .864 | .854 | .843 | .831 | .818 | .804 | .789 | .772 | .754 | .734 |
| | | | | | | | | .713 | .690 | .665 | | | | | | | |
| 26.000 | 19.809 | .000 | 6.191 | .36 | 3.95 | 23.8% | 24 | .928 | .924 | .919 | .914 | .909 | .903 | .896 | .889 | .881 | .873 |
| | | | | | | | | .864 | .854 | .843 | .831 | .818 | .804 | .789 | .772 | .754 | .734 |
| | | | | | | | | .713 | .690 | .665 | .639 | | | | | | |
| 26.000 | 20.420 | .000 | 5.580 | .39 | 4.03 | 21.5% | 25 | .928 | .924 | .919 | .914 | .909 | .903 | .896 | .889 | .881 | .873 |
| | | | | | | | | .864 | .854 | .843 | .831 | .818 | .804 | .789 | .772 | .754 | .734 |
| | | | | | | | | .713 | .690 | .665 | .639 | .611 | | | | | |
| 26.000 | 21.001 | .000 | 4.999 | .42 | 4.10 | 19.2% | 26 | .928 | .924 | .919 | .914 | .909 | .903 | .896 | .889 | .881 | .873 |
| | | | | | | | | .864 | .854 | .843 | .831 | .818 | .804 | .789 | .772 | .754 | .734 |
| | | | | | | | | .713 | .690 | .665 | .639 | .611 | .581 | | | | |
| 26.000 | 21.550 | .000 | 4.450 | .45 | 4.17 | 17.1% | 27 | .928 | .924 | .919 | .914 | .909 | .903 | .896 | .889 | .881 | .873 |
| | | | | | | | | .864 | .854 | .843 | .831 | .818 | .804 | .789 | .772 | .754 | .734 |
| | | | | | | | | .713 | .690 | .665 | .639 | .611 | .581 | .549 | | | |
| 26.000 | 22.066 | .000 | 3.934 | .48 | 4.23 | 15.1% | 28 | .928 | .924 | .919 | .914 | .909 | .903 | .896 | .889 | .881 | .873 |
| | | | | | | | | .864 | .854 | .843 | .831 | .818 | .804 | .789 | .772 | .754 | .734 |
| | | | | | | | | .713 | .690 | .665 | .639 | .611 | .581 | .549 | .516 | | |
| 26.000 | 22.548 | .000 | 3.452 | .52 | 4.28 | 13.3% | 29 | .928 | .924 | .919 | .914 | .909 | .903 | .896 | .889 | .881 | .873 |
| | | | | | | | | .864 | .854 | .843 | .831 | .818 | .804 | .789 | .772 | .754 | .734 |
| | | | | | | | | .713 | .690 | .665 | .639 | .611 | .581 | .549 | .516 | .482 | |

*SIMILAR TO A.T.&T. NEAL-WILKINSON TABLES(SEE SECTION 1)

*EQUIVALENT RANDOM THEORY: VMR(VARIANCE TO MEAN RATIO)– 2.0 - NO RETRIES (PEAKED TRAFFIC - ROUTE ADVANCEMENT OR DAY-TO-DAY VARIATION)

| HOURS FIRST ATTEMPT VMR = 2.0 | HOURS CON-NECT | HOURS RECALL /RETRY | HOURS OVFL | CON-NECT VMR | OVFL VMR | PERCENT BLOCKED (GRADE) | LINE | HOURS CONNECTED TRAFFIC ON LINES 1 | 2 | 3 | 4 | 5 | 6 | 7 | 8 | 9 | 10 |
|---|---|---|---|---|---|---|---|---|---|---|---|---|---|---|---|---|---|
| 26.000 | 22.995 | .000 | 3.005 | .55 | 4.32 | 11.6% | 30 | .928 | .924 | .919 | .914 | .909 | .903 | .896 | .889 | .881 | .873 |
| | | | | | | | | .864 | .854 | .843 | .831 | .818 | .804 | .789 | .772 | .754 | .734 |
| | | | | | | | | .713 | .690 | .665 | .639 | .611 | .581 | .549 | .516 | .482 | .446 |
| 26.000 | 23.405 | .000 | 2.595 | .59 | 4.35 | 10.0% | 31 | .928 | .924 | .919 | .914 | .909 | .903 | .896 | .889 | .881 | .873 |
| | | | | | | | | .864 | .854 | .843 | .831 | .818 | .804 | .789 | .772 | .754 | .734 |
| | | | | | | | | .713 | .690 | .665 | .639 | .611 | .581 | .549 | .516 | .482 | .446 |
| | | | | | | | | .410 | | | | | | | | | |
| 26.000 | 23.779 | .000 | 2.221 | .63 | 4.36 | 8.5% | 32 | .928 | .924 | .919 | .914 | .909 | .903 | .896 | .889 | .881 | .873 |
| | | | | | | | | .864 | .854 | .843 | .831 | .818 | .804 | .789 | .772 | .754 | .734 |
| | | | | | | | | .713 | .690 | .665 | .639 | .611 | .581 | .549 | .516 | .482 | .446 |
| | | | | | | | | .410 | .374 | | | | | | | | |
| 26.000 | 24.116 | .000 | 1.884 | .66 | 4.36 | 7.2% | 33 | .928 | .924 | .919 | .914 | .909 | .903 | .896 | .889 | .881 | .873 |
| | | | | | | | | .864 | .854 | .843 | .831 | .818 | .804 | .789 | .772 | .754 | .734 |
| | | | | | | | | .713 | .690 | .665 | .639 | .611 | .581 | .549 | .516 | .482 | .446 |
| | | | | | | | | .410 | .374 | .337 | | | | | | | |
| 26.000 | 24.418 | .000 | 1.582 | .70 | 4.35 | 6.1% | 34 | .928 | .924 | .919 | .914 | .909 | .903 | .896 | .889 | .881 | .873 |
| | | | | | | | | .864 | .854 | .843 | .831 | .818 | .804 | .789 | .772 | .754 | .734 |
| | | | | | | | | .713 | .690 | .665 | .639 | .611 | .581 | .549 | .516 | .482 | .446 |
| | | | | | | | | .410 | .374 | .337 | | | | | | | |
| 26.000 | 24.684 | .000 | 1.316 | .73 | 4.33 | 5.1% | 35 | .928 | .924 | .919 | .914 | .909 | .903 | .896 | .889 | .881 | .873 |
| | | | | | | | | .864 | .854 | .843 | .831 | .818 | .804 | .789 | .772 | .754 | .734 |
| | | | | | | | | .713 | .690 | .665 | .639 | .611 | .581 | .549 | .516 | .482 | .446 |
| | | | | | | | | .410 | .374 | .337 | .302 | .267 | | | | | |
| 26.000 | 24.917 | .000 | 1.083 | .77 | 4.29 | 4.2% | 36 | .928 | .924 | .919 | .914 | .909 | .903 | .896 | .889 | .881 | .873 |
| | | | | | | | | .864 | .854 | .843 | .831 | .818 | .804 | .789 | .772 | .754 | .734 |
| | | | | | | | | .713 | .690 | .665 | .639 | .611 | .581 | .549 | .516 | .482 | .446 |
| | | | | | | | | .410 | .374 | .337 | .302 | .267 | .233 | | | | |
| 26.000 | 25.119 | .000 | .881 | .80 | 4.24 | 3.4% | 37 | .928 | .924 | .919 | .914 | .909 | .903 | .896 | .889 | .881 | .873 |
| | | | | | | | | .864 | .854 | .843 | .831 | .818 | .804 | .789 | .772 | .754 | .734 |
| | | | | | | | | .713 | .690 | .665 | .639 | .611 | .581 | .549 | .516 | .482 | .446 |
| | | | | | | | | .410 | .374 | .337 | .302 | .267 | .233 | .201 | | | |
| 26.000 | 25.291 | .000 | .709 | .83 | 4.19 | 2.7% | 38 | .928 | .924 | .919 | .914 | .909 | .903 | .896 | .889 | .881 | .873 |
| | | | | | | | | .864 | .854 | .843 | .831 | .818 | .804 | .789 | .772 | .754 | .734 |
| | | | | | | | | .713 | .690 | .665 | .639 | .611 | .581 | .549 | .516 | .482 | .446 |
| | | | | | | | | .410 | .374 | .337 | .302 | .267 | .233 | .201 | .172 | | |
| 26.000 | 25.436 | .000 | .564 | .86 | 4.12 | 2.2% | 39 | .928 | .924 | .919 | .914 | .909 | .903 | .896 | .889 | .881 | .873 |
| | | | | | | | | .864 | .854 | .843 | .831 | .818 | .804 | .789 | .772 | .754 | .734 |
| | | | | | | | | .713 | .690 | .665 | .639 | .611 | .581 | .549 | .516 | .482 | .446 |
| | | | | | | | | .410 | .374 | .337 | .302 | .267 | .233 | .201 | .172 | .145 | |
| 26.000 | 25.550 | .000 | .444 | .88 | 4.04 | 1.7% | 40 | .928 | .924 | .919 | .914 | .909 | .903 | .896 | .889 | .881 | .873 |

*SIMILAR TO A.T.&T. NEAL-WILKINSON TABLES(SEE SECTION 1)

*EQUIVALENT RANDOM THEORY: VMR(VARIANCE TO MEAN RATIO)– 2.0 - NO RETRIES (PEAKED TRAFFIC - ROUTE ADVANCEMENT OR DAY-TO-DAY VARIATION)

| HOURS FIRST ATTEMPT VMR=2.0 | HOURS CON-NECT | HOURS RECALL /RETRY | HOURS OVFL | CON-NECT VMR | OVFL VMR | PERCENT BLOCKED (GRADE) | LINE | HOURS CONNECTED TRAFFIC ON LINES 1 | 2 | 3 | 4 | 5 | 6 | 7 | 8 | 9 | 10 |
|---|---|---|---|---|---|---|---|---|---|---|---|---|---|---|---|---|---|
| | | | | | | | | .864 | .854 | .843 | .831 | .818 | .804 | .789 | .772 | .754 | .734 |
| | | | | | | | | .713 | .690 | .665 | .639 | .611 | .581 | .549 | .516 | .482 | .446 |
| | | | | | | | | .410 | .374 | .337 | .302 | .267 | .233 | .201 | .172 | .145 | .120 |
| 26.000 | 25.655 | .000 | .345 | .90 | 3.96 | 1.3% | 41 | .928 | .924 | .919 | .914 | .909 | .903 | .896 | .889 | .881 | .873 |
| | | | | | | | | .864 | .854 | .843 | .831 | .818 | .804 | .789 | .772 | .754 | .734 |
| | | | | | | | | .713 | .690 | .665 | .639 | .611 | .581 | .549 | .516 | .482 | .446 |
| | | | | | | | | .410 | .374 | .337 | .302 | .267 | .233 | .201 | .172 | .145 | .120 |
| | | | | | | | | .099 | | | | | | | | | |
| 26.000 | 25.735 | .000 | .265 | .92 | 3.88 | 1.0% | 42 | .928 | .924 | .919 | .914 | .909 | .903 | .896 | .889 | .881 | .873 |
| | | | | | | | | .864 | .854 | .843 | .831 | .818 | .804 | .789 | .772 | .754 | .734 |
| | | | | | | | | .713 | .690 | .665 | .639 | .611 | .581 | .549 | .516 | .482 | .446 |
| | | | | | | | | .410 | .374 | .337 | .302 | .267 | .233 | .201 | .172 | .145 | .120 |
| | | | | | | | | .099 | .080 | | | | | | | | |
| 26.000 | 25.799 | .000 | .201 | .94 | 3.79 | .8% | 43 | .928 | .924 | .919 | .914 | .909 | .903 | .896 | .889 | .881 | .873 |
| | | | | | | | | .864 | .854 | .843 | .831 | .818 | .804 | .789 | .772 | .754 | .734 |
| | | | | | | | | .713 | .690 | .665 | .639 | .611 | .581 | .549 | .516 | .482 | .446 |
| | | | | | | | | .410 | .374 | .337 | .302 | .267 | .233 | .201 | .172 | .145 | .120 |
| | | | | | | | | .099 | .080 | .064 | | | | | | | |
| 26.000 | 25.849 | .000 | .151 | .95 | 3.70 | .6% | 44 | .928 | .924 | .919 | .914 | .909 | .903 | .896 | .889 | .881 | .873 |
| | | | | | | | | .864 | .854 | .843 | .831 | .818 | .804 | .789 | .772 | .754 | .734 |
| | | | | | | | | .713 | .690 | .665 | .639 | .611 | .581 | .549 | .516 | .482 | .446 |
| | | | | | | | | .410 | .374 | .337 | .302 | .267 | .233 | .201 | .172 | .145 | .120 |
| | | | | | | | | .099 | .080 | .064 | .050 | | | | | | |
| 28.000 | 18.893 | .000 | 9.107 | .27 | 3.70 | 32.5% | 22 | .933 | .929 | .925 | .921 | .916 | .911 | .905 | .899 | .893 | .885 |
| | | | | | | | | .878 | .869 | .860 | .850 | .839 | .827 | .814 | .801 | .785 | .769 |
| | | | | | | | | .751 | .732 | | | | | | | | |
| 28.000 | 19.604 | .000 | 8.396 | .29 | 3.79 | 30.0% | 23 | .933 | .929 | .925 | .921 | .916 | .911 | .905 | .899 | .893 | .885 |
| | | | | | | | | .878 | .869 | .860 | .850 | .839 | .827 | .814 | .801 | .785 | .769 |
| | | | | | | | | .751 | .732 | .711 | | | | | | | |
| 28.000 | 20.293 | .000 | 7.707 | .31 | 3.89 | 27.5% | 24 | .933 | .929 | .925 | .921 | .916 | .911 | .905 | .899 | .893 | .885 |
| | | | | | | | | .878 | .869 | .860 | .850 | .839 | .827 | .814 | .801 | .785 | .769 |
| | | | | | | | | .751 | .732 | .711 | .689 | | | | | | |
| 28.000 | 20.958 | .000 | 7.042 | .34 | 3.98 | 25.2% | 25 | .933 | .929 | .925 | .921 | .916 | .911 | .905 | .899 | .893 | .885 |
| | | | | | | | | .878 | .869 | .860 | .850 | .839 | .827 | .814 | .801 | .785 | .769 |
| | | | | | | | | .751 | .732 | .711 | .689 | .665 | | | | | |
| 28.000 | 21.597 | .000 | 6.403 | .36 | 4.06 | 22.9% | 26 | .933 | .929 | .925 | .921 | .916 | .911 | .905 | .899 | .893 | .885 |
| | | | | | | | | .878 | .869 | .860 | .850 | .839 | .827 | .814 | .801 | .785 | .769 |
| | | | | | | | | .751 | .732 | .711 | .689 | .665 | .639 | | | | |
| 28.000 | 22.209 | .000 | 5.791 | .39 | 4.14 | 20.7% | 27 | .933 | .929 | .925 | .921 | .916 | .911 | .905 | .899 | .893 | .885 |
| | | | | | | | | .878 | .869 | .860 | .850 | .839 | .827 | .814 | .801 | .785 | .769 |

*SIMILAR TO A.T.&T. NEAL-WILKINSON TABLES(SEE SECTION 1)

*EQUIVALENT RANDOM THEORY: VMR(VARIANCE TO MEAN RATIO)– 2.0 - NO RETRIES (PEAKED TRAFFIC - ROUTE ADVANCEMENT OR DAY-TO-DAY VARIATION)

| HOURS FIRST ATTEMPT VMR=2.0 | HOURS CON-NECT | HOURS RECALL /RETRY | HOURS OVFL | CON-NECT VMR | OVFL VMR | PERCENT BLOCKED (GRADE) | LINE | HOURS CONNECTED TRAFFIC ON LINES 1 | 2 | 3 | 4 | 5 | 6 | 7 | 8 | 9 | 10 |
|---|---|---|---|---|---|---|---|---|---|---|---|---|---|---|---|---|---|
| | | | | | | | | .751 | .732 | .711 | .689 | .665 | .639 | .612 | | | |
| 28.000 | 22.792 | .000 | 5.208 | .42 | 4.22 | 18.6% | 28 | .933 | .929 | .925 | .921 | .916 | .911 | .905 | .899 | .893 | .885 |
| | | | | | | | | .878 | .869 | .860 | .850 | .839 | .827 | .814 | .801 | .785 | .769 |
| | | | | | | | | .751 | .732 | .711 | .689 | .665 | .639 | .612 | .583 | | |
| 28.000 | 23.344 | .000 | 4.656 | .45 | 4.29 | 16.6% | 29 | .933 | .929 | .925 | .921 | .916 | .911 | .905 | .899 | .893 | .885 |
| | | | | | | | | .878 | .869 | .860 | .850 | .839 | .827 | .814 | .801 | .785 | .769 |
| | | | | | | | | .751 | .732 | .711 | .689 | .665 | .639 | .612 | .583 | .552 | |
| 28.000 | 23.865 | .000 | 4.135 | .48 | 4.35 | 14.8% | 30 | .933 | .929 | .925 | .921 | .916 | .911 | .905 | .899 | .893 | .885 |
| | | | | | | | | .878 | .869 | .860 | .850 | .839 | .827 | .814 | .801 | .785 | .769 |
| | | | | | | | | .751 | .732 | .711 | .689 | .665 | .639 | .612 | .583 | .552 | .521 |
| 28.000 | 24.352 | .000 | 3.648 | .51 | 4.40 | 13.0% | 31 | .933 | .929 | .925 | .921 | .916 | .911 | .905 | .899 | .893 | .885 |
| | | | | | | | | .878 | .869 | .860 | .850 | .839 | .827 | .814 | .801 | .785 | .769 |
| | | | | | | | | .751 | .732 | .711 | .689 | .665 | .639 | .612 | .583 | .552 | .521 |
| | | | | | | | | .488 | | | | | | | | | |
| 28.000 | 24.806 | .000 | 3.194 | .55 | 4.44 | 11.4% | 32 | .933 | .929 | .925 | .921 | .916 | .911 | .905 | .899 | .893 | .885 |
| | | | | | | | | .878 | .869 | .860 | .850 | .839 | .827 | .814 | .801 | .785 | .769 |
| | | | | | | | | .751 | .732 | .711 | .689 | .665 | .639 | .612 | .583 | .552 | .521 |
| | | | | | | | | .488 | .453 | | | | | | | | |
| 28.000 | 25.224 | .000 | 2.776 | .58 | 4.47 | 9.9% | 33 | .933 | .929 | .925 | .921 | .916 | .911 | .905 | .899 | .893 | .885 |
| | | | | | | | | .878 | .869 | .860 | .850 | .839 | .827 | .814 | .801 | .785 | .769 |
| | | | | | | | | .751 | .732 | .711 | .689 | .665 | .639 | .612 | .583 | .552 | .521 |
| | | | | | | | | .488 | .453 | .419 | | | | | | | |
| 28.000 | 25.607 | .000 | 2.393 | .62 | 4.49 | 8.5% | 34 | .933 | .929 | .925 | .921 | .916 | .911 | .905 | .899 | .893 | .885 |
| | | | | | | | | .878 | .869 | .860 | .850 | .839 | .827 | .814 | .801 | .785 | .769 |
| | | | | | | | | .751 | .732 | .711 | .689 | .665 | .639 | .612 | .583 | .552 | .521 |
| | | | | | | | | .488 | .453 | .419 | .383 | | | | | | |
| 28.000 | 25.955 | .000 | 2.045 | .65 | 4.49 | 7.3% | 35 | .933 | .929 | .925 | .921 | .010 | .911 | .905 | .899 | .893 | .885 |
| | | | | | | | | .878 | .869 | .860 | .850 | .839 | .827 | .814 | .801 | .785 | .769 |
| | | | | | | | | .751 | .732 | .711 | .689 | .665 | .639 | .612 | .583 | .552 | .521 |
| | | | | | | | | .488 | .453 | .419 | .383 | .348 | | | | | |
| 28.000 | 26.268 | .000 | 1.732 | .69 | 4.48 | 6.2% | 36 | .933 | .929 | .925 | .921 | .916 | .911 | .905 | .899 | .893 | .885 |
| | | | | | | | | .878 | .869 | .860 | .850 | .839 | .827 | .814 | .801 | .785 | .769 |
| | | | | | | | | .751 | .732 | .711 | .689 | .665 | .639 | .612 | .583 | .552 | .521 |
| | | | | | | | | .488 | .453 | .419 | .383 | .348 | .313 | | | | |
| 28.000 | 26.547 | .000 | 1.453 | .72 | 4.46 | 5.2% | 37 | .933 | .929 | .925 | .921 | .916 | .911 | .905 | .899 | .893 | .885 |
| | | | | | | | | .878 | .869 | .860 | .850 | .039 | .827 | .814 | .801 | .785 | .769 |
| | | | | | | | | .751 | .732 | .711 | .689 | .665 | .639 | .612 | .583 | .552 | .521 |
| | | | | | | | | .488 | .453 | .419 | .383 | .348 | .313 | .279 | | | |
| 28.000 | 26.792 | .000 | 1.208 | 75 | 4.43 | 4.3% | 38 | .933 | .929 | .925 | .921 | .916 | .911 | .905 | .809 | .893 | .885 |
| | | | | | | | | .878 | .869 | .860 | .850 | .839 | .827 | .814 | .801 | .785 | .769 |

*SIMILAR TO A.T.&T. NEAL-WILKINSON TABLES(SEE SECTION 1)

# *EQUIVALENT RANDOM THEORY: VMR(VARIANCE TO MEAN RATIO)– 2.0 - NO RETRIES (PEAKED TRAFFIC - ROUTE ADVANCEMENT OR DAY-TO-DAY VARIATION)

| HOURS FIRST ATTEMPT VMR=2.0 | HOURS CON-NECT | HOURS RECALL /RETRY | HOURS OVFL | CON-NECT VMR | OVFL VMR | PERCENT BLOCKED (GRADE) | LINE | HOURS CONNECTED TRAFFIC ON LINES 1 | 2 | 3 | 4 | 5 | 6 | 7 | 8 | 9 | 10 |
|---|---|---|---|---|---|---|---|---|---|---|---|---|---|---|---|---|---|
| | | | | | | | | .751 | .732 | .711 | .689 | .665 | .639 | .612 | .583 | .552 | .521 |
| | | | | | | | | .488 | .453 | .419 | .383 | .348 | .313 | .279 | .246 | | |
| 28.000 | 27.007 | .000 | .993 | .79 | 4.39 | 3.5% | 39 | .933 | .929 | .925 | .921 | .916 | .911 | .905 | .899 | .893 | .885 |
| | | | | | | | | .878 | .869 | .860 | .850 | .839 | .827 | .814 | .801 | .785 | .769 |
| | | | | | | | | .751 | .732 | .711 | .689 | .665 | .639 | .612 | .583 | .552 | .521 |
| | | | | | | | | .488 | .453 | .419 | .383 | .348 | .313 | .279 | .246 | .214 | |
| 28.000 | 27.191 | .000 | .809 | .82 | 4.33 | 2.9% | 40 | .933 | .929 | .925 | .921 | .916 | .911 | .905 | .899 | .893 | .885 |
| | | | | | | | | .878 | .869 | .860 | .850 | .839 | .827 | .814 | .801 | .785 | .769 |
| | | | | | | | | .751 | .732 | .711 | .689 | .665 | .639 | .612 | .583 | .552 | .521 |
| | | | | | | | | .488 | .453 | .419 | .383 | .348 | .313 | .279 | .246 | .214 | .185 |
| 28.000 | 27.349 | .000 | .651 | .84 | 4.27 | 2.3% | 41 | .933 | .929 | .925 | .921 | .916 | .911 | .905 | .899 | .893 | .885 |
| | | | | | | | | .878 | .869 | .860 | .850 | .839 | .827 | .814 | .801 | .785 | .769 |
| | | | | | | | | .751 | .732 | .711 | .689 | .665 | .639 | .612 | .583 | .552 | .521 |
| | | | | | | | | .488 | .453 | .419 | .383 | .348 | .313 | .279 | .246 | .214 | .185 |
| | | | | | | | | .157 | | | | | | | | | |
| 28.000 | 27.481 | .000 | .519 | .87 | 4.20 | 1.9% | 42 | .933 | .929 | .925 | .921 | .916 | .911 | .905 | .899 | .893 | .885 |
| | | | | | | | | .878 | .869 | .860 | .850 | .839 | .827 | .814 | .801 | .785 | .769 |
| | | | | | | | | .751 | .732 | .711 | .689 | .665 | .639 | .612 | .583 | .552 | .521 |
| | | | | | | | | .488 | .453 | .419 | .383 | .348 | .313 | .279 | .246 | .214 | .185 |
| | | | | | | | | .157 | .132 | | | | | | | | |
| 28.000 | 27.591 | .000 | .409 | .89 | 4.12 | 1.5% | 43 | .933 | .929 | .925 | .921 | .916 | .911 | .905 | .899 | .893 | .885 |
| | | | | | | | | .878 | .869 | .860 | .850 | .839 | .827 | .814 | .801 | .785 | .769 |
| | | | | | | | | .751 | .732 | .711 | .689 | .665 | .639 | .612 | .583 | .552 | .521 |
| | | | | | | | | .488 | .453 | .419 | .383 | .348 | .313 | .279 | .246 | .214 | .185 |
| | | | | | | | | .157 | .132 | .110 | | | | | | | |
| 28.000 | 27.681 | .000 | .319 | .91 | 4.03 | 1.1% | 44 | .933 | .929 | .925 | .921 | .916 | .911 | .905 | .899 | .893 | .885 |
| | | | | | | | | .878 | .869 | .860 | .850 | .839 | .827 | .814 | .801 | .785 | .769 |
| | | | | | | | | .751 | .732 | .711 | .689 | .665 | .639 | .612 | .583 | .552 | .521 |
| | | | | | | | | .488 | .453 | .419 | .383 | .348 | .313 | .279 | .246 | .214 | .185 |
| | | | | | | | | .157 | .132 | .110 | .090 | | | | | | |
| 28.000 | 27.755 | .000 | .245 | .93 | 3.95 | .9% | 45 | .933 | .929 | .925 | .921 | .916 | .911 | .905 | .899 | .893 | .885 |
| | | | | | | | | .878 | .869 | .860 | .850 | .839 | .827 | .814 | .801 | .785 | .769 |
| | | | | | | | | .751 | .732 | .711 | .689 | .665 | .639 | .612 | .583 | .552 | .521 |
| | | | | | | | | .488 | .453 | .419 | .383 | .348 | .313 | .279 | .246 | .214 | .185 |
| | | | | | | | | .157 | .132 | .110 | .090 | .073 | | | | | |
| 28.000 | 27.813 | .000 | .187 | .94 | 3.86 | .7% | 46 | .933 | .929 | .925 | .921 | .916 | .911 | .905 | .899 | .893 | .885 |
| | | | | | | | | .878 | .869 | .860 | .850 | .839 | .827 | .814 | .801 | .785 | .769 |
| | | | | | | | | .751 | .732 | .711 | .689 | .665 | .639 | .612 | .583 | .552 | .521 |
| | | | | | | | | .488 | .453 | .419 | .383 | .348 | .313 | .279 | .246 | .214 | .185 |
| | | | | | | | | .157 | .132 | .110 | .090 | .073 | .059 | | | | |

*SIMILAR TO A.T.&T. NEAL-WILKINSON TABLES(SEE SECTION 1)

*EQUIVALENT RANDOM THEORY: VMR(VARIANCE TO MEAN RATIO)– 2.0 - NO RETRIES (PEAKED TRAFFIC - ROUTE ADVANCEMENT OR DAY-TO-DAY VARIATION)

| HOURS FIRST ATTEMPT VMR = 2.0 | HOURS CON-NECT | HOURS RECALL /RETRY | HOURS OVFL | CON-NECT VMR | OVFL VMR | PERCENT BLOCKED (GRADE) | LINE | HOURS CONNECTED TRAFFIC ON LINES 1 | 2 | 3 | 4 | 5 | 6 | 7 | 8 | 9 | 10 |
|---|---|---|---|---|---|---|---|---|---|---|---|---|---|---|---|---|---|
| 28.000 | 27.859 | .000 | .141 | .95 | 3.77 | .5% | 47 | .933 | .929 | .925 | .921 | .916 | .911 | .905 | .899 | .893 | .885 |
| | | | | | | | | .878 | .869 | .860 | .850 | .839 | .827 | .814 | .801 | .785 | .769 |
| | | | | | | | | .751 | .732 | .711 | .689 | .665 | .639 | .612 | .583 | .552 | .521 |
| | | | | | | | | .488 | .453 | .419 | .383 | .348 | .313 | .279 | .246 | .214 | .185 |
| | | | | | | | | .157 | .132 | .110 | .090 | .073 | .059 | .046 | | | |
| 30.000 | 19.962 | .000 | 10.038 | .25 | 3.72 | 33.5% | 23 | .937 | .934 | .930 | .926 | .922 | .918 | .913 | .908 | .902 | .896 |
| | | | | | | | | .889 | .882 | .874 | .865 | .856 | .846 | .835 | .824 | .811 | .797 |
| | | | | | | | | .782 | .766 | .748 | | | | | | | |
| 30.000 | 20.691 | .000 | 9.309 | .27 | 3.81 | 31.0% | 24 | .937 | .934 | .930 | .926 | .922 | .918 | .913 | .908 | .902 | .896 |
| | | | | | | | | .889 | .882 | .874 | .865 | .856 | .846 | .835 | .824 | .811 | .797 |
| | | | | | | | | .782 | .766 | .748 | .730 | | | | | | |
| 30.000 | 21.400 | .000 | 8.600 | .29 | 3.91 | 28.7% | 25 | .937 | .934 | .930 | .926 | .922 | .918 | .913 | .908 | .902 | .896 |
| | | | | | | | | .889 | .882 | .874 | .865 | .856 | .846 | .835 | .824 | .811 | .797 |
| | | | | | | | | .782 | .766 | .748 | .730 | .709 | | | | | |
| 30.000 | 22.088 | .000 | 7.912 | .31 | 4.00 | 26.4% | 26 | .937 | .934 | .930 | .926 | .922 | .918 | .913 | .908 | .902 | .896 |
| | | | | | | | | .889 | .882 | .874 | .865 | .856 | .846 | .835 | .824 | .811 | .797 |
| | | | | | | | | .782 | .766 | .748 | .730 | .709 | .688 | | | | |
| 30.000 | 22.752 | .000 | 7.248 | .34 | 4.09 | 24.2% | 27 | .937 | .934 | .930 | .926 | .922 | .918 | .913 | .908 | .902 | .896 |
| | | | | | | | | .889 | .882 | .874 | .865 | .856 | .846 | .835 | .824 | .811 | .797 |
| | | | | | | | | .782 | .766 | .748 | .730 | .709 | .688 | .664 | | | |
| 30.000 | 23.391 | .000 | 6.609 | .36 | 4.18 | 22.0% | 28 | .937 | .934 | .930 | .926 | .922 | .918 | .913 | .908 | .902 | .896 |
| | | | | | | | | .889 | .882 | .874 | .865 | .856 | .846 | .835 | .824 | .811 | .797 |
| | | | | | | | | .782 | .766 | .748 | .730 | .709 | .688 | .664 | .639 | | |
| 30.000 | 24.004 | .000 | 5.996 | .39 | 4.26 | 20.0% | 29 | .937 | .934 | .930 | .926 | .922 | .918 | .913 | .908 | .902 | .896 |
| | | | | | | | | .889 | .882 | .874 | .865 | .856 | .846 | .835 | .824 | .811 | .797 |
| | | | | | | | | .782 | .766 | .748 | .730 | .709 | .688 | .664 | .639 | .613 | |
| 30.000 | 24.589 | .000 | 5.411 | .42 | 4.33 | 18.0% | 30 | .937 | .934 | .930 | .926 | .922 | .918 | .913 | .908 | .902 | .896 |
| | | | | | | | | .889 | .882 | .874 | .865 | .856 | .846 | .835 | .824 | .811 | .797 |
| | | | | | | | | .782 | .766 | .748 | .730 | .709 | .688 | .664 | .639 | .613 | .585 |
| 30.000 | 25.145 | .000 | 4.855 | .44 | 4.40 | 16.2% | 31 | .937 | .934 | .930 | .926 | .922 | .918 | .913 | .908 | .902 | .896 |
| | | | | | | | | .889 | .882 | .874 | .865 | .856 | .846 | .835 | .824 | .811 | .797 |
| | | | | | | | | .782 | .766 | .748 | .730 | .709 | .688 | .664 | .639 | .613 | .585 |
| | | | | | | | | .555 | | | | | | | | | |
| 30.000 | 25.669 | .000 | 4.331 | .48 | 4.46 | 14.4% | 32 | .937 | .934 | .930 | .926 | .922 | .918 | .913 | .908 | .902 | .896 |
| | | | | | | | | .889 | .882 | .874 | .865 | .856 | .846 | .835 | .824 | .811 | .797 |
| | | | | | | | | .782 | .766 | .748 | .730 | .709 | .688 | .664 | .639 | .613 | .585 |
| | | | | | | | | .555 | .525 | | | | | | | | |
| 30.000 | 26.162 | .000 | 3.838 | .51 | 4.51 | 12.8% | 33 | .937 | .934 | .930 | .926 | .922 | .918 | .913 | .908 | .902 | .896 |
| | | | | | | | | .889 | .882 | .874 | .865 | .856 | .846 | .835 | .824 | .811 | .797 |
| | | | | | | | | .782 | .766 | .748 | .730 | .709 | .688 | .664 | .639 | .613 | .585 |

*SIMILAR TO A.T.&T. NEAL-WILKINSON TABLES(SEE SECTION 1)

*EQUIVALENT RANDOM THEORY: VMR(VARIANCE TO MEAN RATIO)– 2.0 - NO RETRIES (PEAKED TRAFFIC - ROUTE ADVANCEMENT OR DAY-TO-DAY VARIATION)

| HOURS FIRST ATTEMPT VMR = 2.0 | HOURS CON- NECT | HOURS RECALL /RETRY | HOURS OVFL | CON- NECT VMR | OVFL VMR | PERCENT BLOCKED (GRADE) | LINE | HOURS CONNECTED TRAFFIC ON LINES 1 | 2 | 3 | 4 | 5 | 6 | 7 | 8 | 9 | 10 |
|---|---|---|---|---|---|---|---|---|---|---|---|---|---|---|---|---|---|
| | | | | | | | | .555 | .525 | .493 | | | | | | | |
| 30.000 | 26.622 | .000 | 3.378 | .54 | 4.55 | 11.3% | 34 | .937 | .934 | .930 | .926 | .922 | .918 | .913 | .908 | .902 | .896 |
| | | | | | | | | .889 | .882 | .874 | .865 | .856 | .846 | .835 | .824 | .811 | .797 |
| | | | | | | | | .782 | .766 | .748 | .730 | .709 | .688 | .664 | .639 | .613 | .585 |
| | | | | | | | | .555 | .525 | .493 | .460 | | | | | | |
| 30.000 | 27.047 | .000 | 2.953 | .57 | 4.58 | 9.8% | 35 | .937 | .934 | .930 | .926 | .922 | .918 | .913 | .908 | .902 | .896 |
| | | | | | | | | .889 | .882 | .874 | .865 | .856 | .846 | .835 | .824 | .811 | .797 |
| | | | | | | | | .782 | .766 | .748 | .730 | .709 | .688 | .664 | .639 | .613 | .585 |
| | | | | | | | | .555 | .525 | .493 | .460 | .426 | | | | | |
| 30.000 | 27.439 | .000 | 2.561 | .61 | 4.60 | 8.5% | 36 | .937 | .934 | .930 | .926 | .922 | .918 | .913 | .908 | .902 | .896 |
| | | | | | | | | .889 | .882 | .874 | .865 | .856 | .846 | .835 | .824 | .811 | .797 |
| | | | | | | | | .782 | .766 | .748 | .730 | .709 | .688 | .664 | .639 | .613 | .585 |
| | | | | | | | | .555 | .525 | .493 | .460 | .426 | .392 | | | | |
| 30.000 | 27.797 | .000 | 2.203 | .64 | 4.61 | 7.3% | 37 | .937 | .934 | .930 | .926 | .922 | .918 | .913 | .908 | .902 | .896 |
| | | | | | | | | .889 | .882 | .874 | .865 | .856 | .846 | .835 | .824 | .811 | .797 |
| | | | | | | | | .782 | .766 | .748 | .730 | .709 | .688 | .664 | .639 | .613 | .585 |
| | | | | | | | | .555 | .525 | .493 | .460 | .426 | .392 | .357 | | | |
| 30.000 | 28.120 | .000 | 1.880 | .68 | 4.61 | 6.3% | 38 | .937 | .934 | .930 | .926 | .922 | .918 | .913 | .908 | .902 | .896 |
| | | | | | | | | .889 | .882 | .874 | .865 | .856 | .846 | .835 | .824 | .811 | .797 |
| | | | | | | | | .782 | .766 | .748 | .730 | .709 | .688 | .664 | .639 | .613 | .585 |
| | | | | | | | | .555 | .525 | .493 | .460 | .426 | .392 | .357 | .323 | | |
| 30.000 | 28.410 | .000 | 1.590 | .71 | 4.59 | 5.3% | 39 | .937 | .934 | .930 | .926 | .922 | .918 | .913 | .908 | .902 | .896 |
| | | | | | | | | .889 | .882 | .874 | .865 | .856 | .846 | .835 | .824 | .811 | .797 |
| | | | | | | | | .782 | .766 | .748 | .730 | .709 | .688 | .664 | .639 | .613 | .585 |
| | | | | | | | | .555 | .525 | .493 | .460 | .426 | .392 | .357 | .323 | .290 | |
| 30.000 | 28.667 | .000 | 1.333 | .74 | 4.56 | 4.4% | 40 | .937 | .934 | .930 | .926 | .922 | .918 | .913 | .908 | .902 | .896 |
| | | | | | | | | .889 | .882 | .874 | .865 | .856 | .846 | .835 | .824 | .811 | .797 |
| | | | | | | | | .782 | .766 | .748 | .730 | .709 | .688 | .664 | .639 | .613 | .585 |
| | | | | | | | | .555 | .525 | .493 | .460 | .426 | .392 | .357 | .323 | .290 | .257 |
| 30.000 | 28.893 | .000 | 1.107 | .77 | 4.52 | 3.7% | 41 | .937 | .934 | .930 | .926 | .922 | .918 | .913 | .908 | .902 | .896 |
| | | | | | | | | .889 | .882 | .874 | .865 | .856 | .846 | .835 | .824 | .811 | .797 |
| | | | | | | | | .782 | .766 | .748 | .730 | .709 | .688 | .664 | .639 | .613 | .585 |
| | | | | | | | | .555 | .525 | .493 | .460 | .426 | .392 | .357 | .323 | .290 | .257 |
| | | | | | | | | .226 | | | | | | | | | |
| 30.000 | 29.090 | .000 | .910 | .80 | 4.47 | 3.0% | 42 | .937 | .934 | .930 | .926 | .922 | .918 | .913 | .908 | .902 | .896 |
| | | | | | | | | .889 | .882 | .874 | .865 | .856 | .846 | .835 | .824 | .811 | .797 |
| | | | | | | | | .782 | .766 | .748 | .730 | .709 | .688 | .664 | .639 | .613 | .585 |
| | | | | | | | | .555 | .525 | .493 | .460 | .426 | .392 | .357 | .323 | .290 | .257 |
| | | | | | | | | .226 | .197 | | | | | | | | |
| 30.000 | 29.259 | .000 | .741 | .83 | 4.41 | 2.5% | 43 | .937 | .934 | .930 | .926 | .922 | .918 | .913 | .908 | .902 | .896 |

*SIMILAR TO A.T.&T. NEAL-WILKINSON TABLES(SEE SECTION 1)

*EQUIVALENT RANDOM THEORY: VMR(VARIANCE TO MEAN RATIO)– 2.0 - NO RETRIES (PEAKED TRAFFIC - ROUTE ADVANCEMENT OR DAY-TO-DAY VARIATION)

| HOURS FIRST ATTEMPT VMR=2.0 | HOURS CON-NECT | HOURS RECALL /RETRY | HOURS OVFL | CON-NECT VMR | OVFL VMR | PERCENT BLOCKED (GRADE) | LINE | HOURS CONNECTED TRAFFIC ON LINES 1 | 2 | 3 | 4 | 5 | 6 | 7 | 8 | 9 | 10 |
|---|---|---|---|---|---|---|---|---|---|---|---|---|---|---|---|---|---|
| | | | | | | | | .889 | .882 | .874 | .865 | .856 | .846 | .835 | .824 | .811 | .797 |
| | | | | | | | | .782 | .766 | .748 | .730 | .709 | .688 | .664 | .639 | .613 | .585 |
| | | | | | | | | .555 | .525 | .493 | .460 | .426 | .392 | .357 | .323 | .290 | .257 |
| | | | | | | | | .226 | .197 | .169 | | | | | | | |
| 30.000 | 29.403 | .000 | .597 | .86 | 4.34 | 2.0% | 44 | .937 | .934 | .930 | .926 | .922 | .918 | .913 | .908 | .902 | .896 |
| | | | | | | | | .889 | .882 | .874 | .865 | .856 | .846 | .835 | .824 | .811 | .797 |
| | | | | | | | | .782 | .766 | .748 | .730 | .709 | .688 | .664 | .639 | .613 | .585 |
| | | | | | | | | .555 | .525 | .493 | .460 | .426 | .392 | .357 | .323 | .290 | .257 |
| | | | | | | | | .226 | .197 | .169 | .144 | | | | | | |
| 30.000 | 29.524 | .000 | .476 | .88 | 4.27 | 1.6% | 45 | .937 | .934 | .930 | .926 | .922 | .918 | .913 | .908 | .902 | .896 |
| | | | | | | | | .889 | .882 | .874 | .865 | .856 | .846 | .835 | .824 | .811 | .797 |
| | | | | | | | | .782 | .766 | .748 | .730 | .709 | .688 | .664 | .639 | .613 | .585 |
| | | | | | | | | .555 | .525 | .493 | .460 | .426 | .392 | .357 | .323 | .290 | .257 |
| | | | | | | | | .226 | .197 | .169 | .144 | .121 | | | | | |
| 30.000 | 29.624 | .000 | .376 | .90 | 4.18 | 1.3% | 46 | .937 | .934 | .930 | .926 | .922 | .918 | .913 | .908 | .902 | .896 |
| | | | | | | | | .889 | .882 | .874 | .865 | .856 | .846 | .835 | .824 | .811 | .797 |
| | | | | | | | | .782 | .766 | .748 | .730 | .709 | .688 | .664 | .639 | .613 | .585 |
| | | | | | | | | .555 | .525 | .493 | .460 | .426 | .392 | .357 | .323 | .290 | .257 |
| | | | | | | | | .226 | .197 | .169 | .144 | .121 | .100 | | | | |
| 30.000 | 29.707 | .000 | .293 | .92 | 4.10 | 1.0% | 47 | .937 | .934 | .930 | .926 | .922 | .918 | .913 | .908 | .902 | .896 |
| | | | | | | | | .889 | .882 | .874 | .865 | .856 | .846 | .835 | .824 | .811 | .797 |
| | | | | | | | | .782 | .766 | .748 | .730 | .709 | .688 | .664 | .639 | .613 | .585 |
| | | | | | | | | .555 | .525 | .493 | .460 | .426 | .392 | .357 | .323 | .290 | .257 |
| | | | | | | | | .226 | .197 | .169 | .144 | .121 | .100 | .082 | | | |
| 30.000 | 29.773 | .000 | .227 | .93 | 4.01 | .8% | 48 | .937 | .934 | .930 | .926 | .922 | .918 | .913 | .908 | .902 | .896 |
| | | | | | | | | .889 | .882 | .874 | .865 | .856 | .846 | .835 | .824 | .811 | .797 |
| | | | | | | | | .782 | .766 | .748 | .700 | .709 | .688 | .664 | .639 | .613 | .585 |
| | | | | | | | | .555 | .525 | .493 | .460 | .426 | .392 | .357 | .323 | .290 | .257 |
| | | | | | | | | .226 | .197 | .169 | .144 | .121 | .100 | .082 | .067 | | |
| 30.000 | 29.827 | .000 | .173 | .95 | 3.92 | .6% | 49 | .937 | .934 | .930 | .926 | .922 | .918 | .913 | .908 | .902 | .896 |
| | | | | | | | | .889 | .882 | .874 | .865 | .856 | .846 | .835 | .824 | .811 | .797 |
| | | | | | | | | .782 | .766 | .748 | .730 | .709 | .688 | .664 | .639 | .613 | .585 |
| | | | | | | | | .555 | .525 | .493 | .460 | .426 | .392 | .357 | .323 | .290 | .257 |
| | | | | | | | | .226 | .197 | .169 | .144 | .121 | .100 | .082 | .067 | .053 | |

*SIMILAR TO A.T.&T. NEAL-WILKINSON TABLES(SEE SECTION 1)

*EQUIVALENT RANDOM THEORY: VMR(VARIANCE TO MEAN RATIO)– 3.0 - NO RETRIES (PEAKED TRAFFIC - ROUTE ADVANCEMENT OR DAY-TO-DAY VARIATION)

| HOURS FIRST ATTEMPT VMR=3.0 | HOURS CON-NECT | HOURS RECALL /RETRY | HOURS OVFL | CON-NECT VMR | OVFL VMR | PERCENT BLOCKED (GRADE) | LINE | HOURS CONNECTED TRAFFIC ON LINES 1 | 2 | 3 | 4 | 5 | 6 | 7 | 8 | 9 | 10 |
|---|---|---|---|---|---|---|---|---|---|---|---|---|---|---|---|---|---|
| .200 | .152 | .000 | .048 | .97 | 2.36 | 24.2% | 3 | .071 | .048 | .032 | | | | | | | |
| .200 | .172 | .000 | .028 | .98 | 2.29 | 14.2% | 4 | .071 | .048 | .032 | .020 | | | | | | |
| .200 | .184 | .000 | .016 | .99 | 2.21 | 8.0% | 5 | .071 | .048 | .032 | .020 | .012 | | | | | |
| .200 | .191 | .000 | .009 | .99 | 2.15 | 4.4% | 6 | .071 | .048 | .032 | .020 | .012 | .007 | | | | |
| .200 | .195 | .000 | .005 | 1.00 | 2.08 | 2.4 | 7 | .071 | .048 | .032 | .020 | .012 | .007 | .004 | | | |
| .200 | .198 | .000 | .002 | 1.00 | 2.03 | 1.2% | 8 | .071 | .048 | .032 | .020 | .012 | .007 | .004 | .002 | | |
| .200 | .199 | .000 | .001 | 1.00 | 1.97 | .6% | 9 | .071 | .048 | .032 | .020 | .012 | .007 | .004 | .002 | .001 | |
| .400 | .285 | .000 | .115 | .94 | 2.53 | 28.9% | 3 | .128 | .092 | .065 | | | | | | | |
| .400 | .328 | .000 | .072 | .96 | 2.45 | 17.9% | 4 | .128 | .092 | .065 | .044 | | | | | | |
| .400 | .357 | .000 | .043 | .97 | 2.37 | 10.8% | 5 | .128 | .092 | .065 | .044 | .028 | | | | | |
| .400 | .375 | .000 | .025 | .98 | 2.29 | 6.3% | 6 | .128 | .092 | .065 | .044 | .028 | .018 | | | | |
| .400 | .386 | .000 | .014 | .99 | 2.22 | 3.6% | 7 | .128 | .092 | .065 | .044 | .028 | .018 | .011 | | | |
| .400 | .392 | .000 | .008 | .99 | 2.15 | 2.0% | 8 | .128 | .092 | .065 | .044 | .028 | .018 | .011 | .006 | | |
| .400 | .396 | .000 | .004 | 1.00 | 2.09 | 1.1% | 9 | .128 | .092 | .065 | .044 | .028 | .018 | .011 | .006 | .004 | |
| .400 | .398 | .000 | .002 | 1.00 | 2.03 | .6% | 10 | .128 | .092 | .065 | .044 | .028 | .018 | .011 | .006 | .004 | .002 |
| .600 | .404 | .000 | .196 | .90 | 2.66 | 32.7% | 3 | .175 | .132 | .097 | | | | | | | |
| .600 | .473 | .000 | .127 | .93 | 2.58 | 21.2% | 4 | .175 | .132 | .097 | .069 | | | | | | |
| .600 | .520 | .000 | .080 | .95 | 2.50 | 13.4% | 5 | .175 | .132 | .097 | .069 | .047 | | | | | |
| .600 | .551 | .000 | .049 | .97 | 2.42 | 8.2% | 6 | .175 | .132 | .097 | .069 | .047 | .031 | | | | |
| .600 | .571 | .000 | .029 | .98 | 2.34 | 4.9% | 7 | .175 | .132 | .097 | .069 | .047 | .031 | .020 | | | |
| .600 | .583 | .000 | .017 | .99 | 2.27 | 2.8% | 8 | .175 | .132 | .097 | .069 | .047 | .031 | .020 | .012 | | |
| .600 | .590 | .000 | .010 | .99 | 2.20 | 1.6% | 9 | .175 | .132 | .097 | .069 | .047 | .031 | .020 | .012 | .007 | |
| .600 | .595 | .000 | .005 | 1.00 | 2.14 | .9% | 10 | .175 | .132 | .097 | .069 | .047 | .031 | .020 | .012 | .007 | .004 |
| .800 | .608 | .000 | .192 | .91 | 2.69 | 24.0% | 4 | .217 | .169 | .128 | .094 | | | | | | |
| .800 | .675 | .000 | .125 | .93 | 2.61 | 15.7% | 5 | .217 | .169 | .128 | .094 | .067 | | | | | |
| .800 | .721 | .000 | .079 | .95 | 2.53 | 9.9% | 6 | .217 | .169 | .128 | .094 | .067 | .046 | | | | |
| .800 | .751 | .000 | .049 | .97 | 2.45 | 6.1% | 7 | .217 | .169 | .128 | .094 | .067 | .046 | .030 | | | |
| .800 | .771 | .000 | .029 | .98 | 2.37 | 3.7% | 8 | .217 | .169 | .128 | .094 | .067 | .046 | .030 | .020 | | |
| .800 | .783 | .000 | .017 | .99 | 2.29 | 2.1% | 9 | .217 | .169 | .128 | .094 | .067 | .046 | .030 | .020 | .012 | |
| .800 | .790 | .000 | .010 | .99 | 2.23 | 1.2% | 10 | .217 | .169 | .128 | .094 | .067 | .046 | .030 | .020 | .012 | .007 |
| .800 | .795 | .000 | .005 | 1.00 | 2.16 | .7% | 11 | .217 | .169 | .128 | .094 | .067 | .046 | .030 | .020 | .012 | .007 |
| | | | | | | | | .004 | | | | | | | | | |
| 1.000 | .735 | .000 | .265 | .88 | 2.77 | 26.5% | 4 | .254 | .203 | .158 | .119 | | | | | | |
| 1.000 | .822 | .000 | .178 | .91 | 2.70 | 17.8% | 5 | .254 | .203 | .158 | .119 | .087 | | | | | |
| 1.000 | .884 | .000 | .116 | .94 | 2.62 | 11.6% | 6 | .254 | .203 | .158 | .119 | .087 | .062 | | | | |
| 1.000 | .927 | .000 | .073 | .96 | 2.54 | 7.3% | 7 | .254 | .203 | .158 | .119 | .087 | .062 | .042 | | | |
| 1.000 | .955 | .000 | .045 | .07 | 2.46 | 4.5% | 8 | .254 | .203 | .158 | .119 | .087 | .062 | .042 | .028 | | |
| 1.000 | .973 | .000 | .027 | .98 | 2.38 | 2.7% | 9 | .254 | .203 | .158 | .119 | .087 | .062 | .042 | .028 | .018 | |
| 1.000 | .984 | .000 | .016 | .99 | 2.30 | 1.6% | 10 | .254 | .203 | .158 | .119 | .087 | .062 | .042 | .028 | .018 | .011 |
| 1.000 | .991 | .000 | .009 | .99 | 2.24 | .9% | 11 | .254 | .203 | .158 | .119 | .087 | .062 | .042 | .028 | .018 | .011 |

*SIMILAR TO A.T.&T. NEAL-WILKINSON TABLES(SEE SECTION 1)

*EQUIVALENT RANDOM THEORY: VMR(VARIANCE TO MEAN RATIO)– 3.0 - NO RETRIES (PEAKED TRAFFIC - ROUTE ADVANCEMENT OR DAY-TO-DAY VARIATION)

| HOURS FIRST ATTEMPT VMR=3.0 | HOURS CON-NECT | HOURS RECALL /RETRY | HOURS OVFL | CON-NECT VMR | OVFL VMR | PERCENT BLOCKED (GRADE) | LINE | HOURS CONNECTED TRAFFIC ON LINES 1 | 2 | 3 | 4 | 5 | 6 | 7 | 8 | 9 | 10 |
|---|---|---|---|---|---|---|---|---|---|---|---|---|---|---|---|---|---|
| | | | | | | | | .007 | | | | | | | | | |
| 1.000 | .995 | .000 | .005 | 1.00 | 2.17 | .5% | 12 | .254 | .203 | .158 | .119 | .087 | .062 | .042 | .028 | .018 | .011 |
| | | | | | | | | .007 | .004 | | | | | | | | |
| 1.200 | .855 | .000 | .345 | .86 | 2.84 | 28.7% | 4 | .288 | .235 | .187 | .145 | | | | | | |
| 1.200 | .964 | .000 | .236 | .89 | 2.77 | 19.7% | 5 | .288 | .235 | .187 | .145 | .108 | | | | | |
| 1.200 | 1.042 | .000 | .158 | .92 | 2.70 | 13.2% | 6 | .288 | .235 | .187 | .145 | .108 | .079 | | | | |
| 1.200 | 1.097 | .000 | .103 | .94 | 2.62 | 8.5% | 7 | .288 | .235 | .187 | .145 | .108 | .079 | .055 | | | |
| 1.200 | 1.135 | .000 | .065 | .96 | 2.53 | 5.4% | 8 | .288 | .235 | .187 | .145 | .108 | .079 | .055 | .038 | | |
| 1.200 | 1.160 | .000 | .040 | .98 | 2.45 | 3.3% | 9 | .288 | .235 | .187 | .145 | .108 | .079 | .055 | .038 | .025 | |
| 1.200 | 1.176 | .000 | .024 | .98 | 2.38 | 2.0% | 10 | .288 | .235 | .187 | .145 | .108 | .079 | .055 | .038 | .025 | .016 |
| 1.200 | 1.186 | .000 | .014 | .99 | 2.30 | 1.2% | 11 | .288 | .235 | .187 | .145 | .108 | .079 | .055 | .038 | .025 | .016 |
| | | | | | | | | .010 | | | | | | | | | |
| 1.200 | 1.192 | .000 | .008 | .99 | 2.24 | .7% | 12 | .288 | .235 | .187 | .145 | .108 | .079 | .055 | .038 | .025 | .016 |
| | | | | | | | | .010 | .006 | | | | | | | | |
| 1.400 | .969 | .000 | .431 | .83 | 2.90 | 30.8% | 4 | .319 | .265 | .215 | .169 | | | | | | |
| 1.400 | 1.098 | .000 | .302 | .87 | 2.84 | 21.5% | 5 | .319 | .265 | .215 | .169 | .129 | | | | | |
| 1.400 | 1.194 | .000 | .206 | .90 | 2.77 | 14.7% | 6 | .319 | .265 | .215 | .169 | .129 | .096 | | | | |
| 1.400 | 1.264 | .000 | .136 | .93 | 2.69 | 9.7% | 7 | .319 | .265 | .215 | .169 | .129 | .096 | .069 | | | |
| 1.400 | 1.312 | .000 | .088 | .95 | 2.61 | 6.3% | 8 | .319 | .265 | .215 | .169 | .129 | .096 | .069 | .048 | | |
| 1.400 | 1.345 | .000 | .055 | .97 | 2.52 | 4.0% | 9 | .319 | .265 | .215 | .169 | .129 | .096 | .069 | .048 | .033 | |
| 1.400 | 1.366 | .000 | .034 | .98 | 2.45 | 2.4% | 10 | .319 | .265 | .215 | .169 | .129 | .096 | .069 | .048 | .033 | .021 |
| 1.400 | 1.380 | .000 | .020 | .99 | 2.37 | 1.4% | 11 | .319 | .265 | .215 | .169 | .129 | .096 | .069 | .048 | .033 | .021 |
| | | | | | | | | .014 | | | | | | | | | |
| 1.400 | 1.388 | .000 | .012 | .99 | 2.30 | .8% | 12 | .319 | .265 | .215 | .169 | .129 | .096 | .069 | .048 | .033 | .021 |
| | | | | | | | | .014 | .008 | | | | | | | | |
| 1.600 | 1.077 | .000 | .523 | .81 | 2.95 | 32.7% | 4 | .348 | .294 | .241 | .193 | | | | | | |
| 1.600 | 1.227 | .000 | .373 | .85 | 2.89 | 23.3% | 5 | .348 | .294 | .241 | .193 | .151 | | | | | |
| 1.600 | 1.341 | .000 | .259 | .89 | 2.83 | 16.2% | 6 | .348 | .294 | .241 | .193 | .151 | .114 | | | | |
| 1.600 | 1.425 | .000 | .175 | .92 | 2.75 | 10.9% | 7 | .348 | .294 | .241 | .193 | .151 | .114 | .084 | | | |
| 1.600 | 1.485 | .000 | .115 | .94 | 2.67 | 7.2% | 8 | .348 | .294 | .241 | .193 | .151 | .114 | .084 | .060 | | |
| 1.600 | 1.526 | .000 | .074 | .96 | 2.59 | 4.6% | 9 | .348 | .294 | .241 | .193 | .151 | .114 | .084 | .060 | .041 | |
| 1.600 | 1.554 | .000 | .046 | .97 | 2.51 | 2.9% | 10 | .348 | .294 | .241 | .193 | .151 | .114 | .084 | .060 | .041 | .028 |
| 1.600 | 1.572 | .000 | .028 | .98 | 2.43 | 1.8% | 11 | .348 | .294 | .241 | .193 | .151 | .114 | .084 | .060 | .041 | .028 |
| | | | | | | | | .018 | | | | | | | | | |
| 1.600 | 1.583 | .000 | .017 | .99 | 2.36 | 1.0% | 12 | .348 | .294 | .241 | .193 | .151 | .114 | .084 | .060 | .041 | .028 |
| | | | | | | | | .018 | .011 | | | | | | | | |
| 1.600 | 1.590 | .000 | .010 | .99 | 2.29 | .6% | 13 | .348 | .294 | .241 | .193 | .151 | .114 | .084 | .060 | .041 | .028 |
| | | | | | | | | .018 | .011 | .007 | | | | | | | |
| 1.800 | 1.179 | .000 | .621 | .78 | 2.99 | 34.5% | 4 | .375 | .320 | .267 | .217 | | | | | | |
| 1.800 | 1.351 | .000 | .449 | .83 | 2.94 | 25.0% | 5 | .375 | .320 | .267 | .217 | .172 | | | | | |

*SIMILAR TO A.T.&T. NEAL-WILKINSON TABLES(SEE SECTION 1)

*EQUIVALENT RANDOM THEORY: VMR(VARIANCE TO MEAN RATIO)– 3.0 - NO RETRIES (PEAKED TRAFFIC - ROUTE ADVANCEMENT OR DAY-TO-DAY VARIATION)

| HOURS FIRST ATTEMPT VMR = 3.0 | HOURS CON-NECT | HOURS RECALL /RETRY | HOURS OVFL | CON-NECT VMR | OVFL VMR | PERCENT BLOCKED (GRADE) | LINE | HOURS CONNECTED TRAFFIC ON LINES 1 | 2 | 3 | 4 | 5 | 6 | 7 | 8 | 9 | 10 |
|---|---|---|---|---|---|---|---|---|---|---|---|---|---|---|---|---|---|
| 1.800 | 1.483 | .000 | .317 | .87 | 2.88 | 17.6% | 6 | .375 | .320 | .267 | .217 | .172 | .132 | | | | |
| 1.800 | 1.582 | .000 | .218 | .90 | 2.81 | 12.1% | 7 | .375 | .320 | .267 | .217 | .172 | .132 | .099 | | | |
| 1.800 | 1.654 | .000 | .146 | .93 | 2.73 | 8.1% | 8 | .375 | .320 | .267 | .217 | .172 | .132 | .099 | .072 | | |
| 1.800 | 1.705 | .000 | .095 | .95 | 2.65 | 5.3% | 9 | .375 | .320 | .267 | .217 | .172 | .132 | .099 | .072 | .051 | |
| 1.800 | 1.739 | .000 | .061 | .97 | 2.57 | 3.4% | 10 | .375 | .320 | .267 | .217 | .172 | .132 | .099 | .072 | .051 | .035 |
| 1.800 | 1.762 | .000 | .038 | .98 | 2.49 | 2.1% | 11 | .375 | .320 | .267 | .217 | .172 | .132 | .099 | .072 | .051 | .035 |
| | | | | | | | | .023 | | | | | | | | | |
| 1.800 | 1.777 | .000 | .023 | .99 | 2.41 | 1.3% | 12 | .375 | .320 | .267 | .217 | .172 | .132 | .099 | .072 | .051 | .035 |
| | | | | | | | | .023 | .015 | | | | | | | | |
| 1.800 | 1.787 | .000 | .013 | .99 | 2.34 | .7% | 13 | .375 | .320 | .267 | .217 | .172 | .132 | .099 | .072 | .051 | .035 |
| | | | | | | | | .023 | .015 | .009 | | | | | | | |
| 2.000 | 1.469 | .000 | .531 | .81 | 2.99 | 26.6% | 5 | .400 | .345 | .291 | .240 | .193 | | | | | |
| 2.000 | 1.620 | .000 | .380 | .85 | 2.93 | 19.0% | 6 | .400 | .345 | .291 | .240 | .193 | .151 | | | | |
| 2.000 | 1.735 | .000 | .265 | .89 | 2.87 | 13.3% | 7 | .400 | .345 | .291 | .240 | .193 | .151 | .115 | | | |
| 2.000 | 1.819 | .000 | .181 | .92 | 2.79 | 9.0% | 8 | .400 | .345 | .291 | .240 | .193 | .151 | .115 | .085 | | |
| 2.000 | 1.880 | .000 | .120 | .94 | 2.71 | 6.0% | 9 | .400 | .345 | .291 | .240 | .193 | .151 | .115 | .085 | .061 | |
| 2.000 | 1.923 | .000 | .077 | .96 | 2.63 | 3.9% | 10 | .400 | .345 | .291 | .240 | .193 | .151 | .115 | .085 | .061 | .042 |
| 2.000 | 1.951 | ,000 | .049 | .97 | 2.55 | 2.4% | 11 | .400 | .345 | .291 | .240 | .193 | .151 | .115 | .085 | .061 | .042 |
| | | | | | | | | .029 | | | | | | | | | |
| 2.000 | 1.970 | .000 | .030 | .98 | 2.47 | 1.5% | 12 | .400 | .345 | .291 | .240 | .193 | .151 | .115 | .085 | .061 | .042 |
| | | | | | | | | .029 | .019 | | | | | | | | |
| 2.000 | 1.982 | .000 | .018 | .99 | 2.39 | .9% | 13 | .400 | .345 | .291 | .240 | .193 | .151 | .115 | .085 | .061 | .042 |
| | | | | | | | | .029 | .019 | .012 | | | | | | | |
| 2.000 | 1.989 | .000 | .011 | .99 | 2.32 | .5% | 14 | .400 | .345 | .291 | .240 | .193 | .151 | .115 | .085 | .061 | .042 |
| | | | | | | | | .029 | .019 | .012 | .007 | | | | | | |
| 2.200 | 1.582 | .000 | .618 | .79 | 3.03 | 28.1% | 5 | .423 | .369 | .315 | .262 | .214 | | | | | |
| 2.200 | 1.752 | .000 | .448 | .83 | 2.98 | 20.4% | 6 | .423 | .369 | .315 | .262 | .214 | .170 | | | | |
| 2.200 | 1.883 | .000 | .317 | .87 | 2.92 | 14.4% | 7 | .423 | .369 | .315 | .262 | .214 | .170 | .131 | | | |
| 2.200 | 1.981 | .000 | .219 | .90 | 2.84 | 10.0% | 8 | .423 | .369 | .315 | .262 | .214 | .170 | .131 | .098 | | |
| 2.200 | 2.052 | .000 | .148 | .93 | 2.77 | 6.7% | 9 | .423 | .369 | .315 | .262 | .214 | .170 | .131 | .098 | .072 | |
| 2.200 | 2.103 | .000 | .097 | .95 | 2.68 | 4.4% | 10 | .423 | .369 | .315 | .262 | .214 | .170 | .131 | .098 | .072 | .051 |
| 2.200 | 2.138 | .000 | .062 | .97 | 2.60 | 2.8% | 11 | .423 | .369 | .315 | .262 | .214 | .170 | .131 | .098 | .072 | .051 |
| | | | | | | | | .035 | | | | | | | | | |
| 2.200 | 2.161 | .000 | .039 | .98 | 2.52 | 1.8% | 12 | .423 | .369 | .315 | .262 | .214 | .170 | .131 | .098 | .072 | .051 |
| | | | | | | | | .035 | .023 | | | | | | | | |
| 2.200 | 2.176 | .000 | .024 | .98 | 2.44 | 1.1% | 13 | .423 | .369 | .315 | .262 | .214 | .170 | .131 | .098 | .072 | .051 |
| | | | | | | | | .035 | .023 | .015 | | | | | | | |
| 2.200 | 2.186 | .000 | .014 | .99 | 2.37 | .6% | 14 | .423 | .369 | .315 | .262 | .214 | .170 | .131 | .098 | .072 | .051 |
| | | | | | | | | .035 | .023 | .015 | .010 | | | | | | |
| 2.400 | 1.691 | .000 | .709 | .77 | 3.06 | 29.6% | 5 | .445 | .391 | .337 | .284 | .234 | | | | | |

*SIMILAR TO A.T.&T. NEAL-WILKINSON TABLES(SEE SECTION 1)

*EQUIVALENT RANDOM THEORY: VMR(VARIANCE TO MEAN RATIO)– 3.0 - NO RETRIES (PEAKED TRAFFIC - ROUTE ADVANCEMENT OR DAY-TO-DAY VARIATION)

| HOURS FIRST ATTEMPT VMR = 3.0 | HOURS CON- NECT | HOURS RECALL /RETRY | HOURS OVFL | CON- NECT VMR | OVFL VMR | PERCENT BLOCKED (GRADE) | LINE | HOURS CONNECTED TRAFFIC ON LINES 1 | 2 | 3 | 4 | 5 | 6 | 7 | 8 | 9 | 10 |
|---|---|---|---|---|---|---|---|---|---|---|---|---|---|---|---|---|---|
| 2.400 | 1.879 | .000 | .521 | .81 | 3.02 | 21.7% | 6 | .445 | .391 | .337 | .284 | .234 | .188 | | | | |
| 2.400 | 2.026 | .000 | .374 | .85 | 2.96 | 15.6% | 7 | .445 | .391 | .337 | .284 | .234 | .188 | .147 | | | |
| 2.400 | 2.138 | .000 | .262 | .89 | 2.89 | 10.9% | 8 | .445 | .391 | .337 | .284 | .234 | .188 | .147 | .112 | | |
| 2.400 | 2.221 | .000 | .179 | .92 | 2.82 | 7.5% | 9 | .445 | .391 | .337 | .284 | .234 | .188 | .147 | .112 | .083 | |
| 2.400 | 2.281 | .000 | .119 | .94 | 2.74 | 5.0% | 10 | .445 | .391 | .337 | .284 | .234 | .188 | .147 | .112 | .083 | .060 |
| 2.400 | 2.323 | .000 | .077 | .96 | 2.65 | 3.2% | 11 | .445 | .391 | .337 | .284 | .234 | .188 | .147 | .112 | .083 | .060 |
| | | | | | | | | .042 | | | | | | | | | |
| 2.400 | 2.351 | .000 | .049 | .97 | 2.57 | 2.0% | 12 | .445 | .391 | .337 | .284 | .234 | .188 | .147 | .112 | .083 | .060 |
| | | | | | | | | .042 | .028 | | | | | | | | |
| 2.400 | 2.370 | .000 | .030 | .98 | 2.49 | 1.3% | 13 | .445 | .391 | .337 | .284 | .234 | .188 | .147 | .112 | .083 | .060 |
| | | | | | | | | .042 | .028 | .019 | | | | | | | |
| 2.400 | 2.382 | .000 | .018 | .99 | 2.42 | .8% | 14 | .445 | .391 | .337 | .284 | .234 | .188 | .147 | .112 | .083 | .060 |
| | | | | | | | | .042 | .028 | .019 | .012 | | | | | | |
| 2.600 | 1.795 | .000 | .805 | .75 | 3.09 | 31.0% | 5 | .465 | .412 | .358 | .305 | .254 | | | | | |
| 2.600 | 2.001 | .000 | .599 | .79 | 3.05 | 23.0% | 6 | .465 | .412 | .358 | .305 | .254 | .207 | | | | |
| 2.600 | 2.165 | .000 | .435 | .84 | 3.00 | 16.7% | 7 | .465 | .412 | .358 | .305 | .254 | .207 | .164 | | | |
| 2.600 | 2.292 | .000 | .308 | .87 | 2.94 | 11.9% | 8 | .465 | .412 | .358 | .305 | .254 | .207 | .164 | .126 | | |
| 2.600 | 2.387 | .000 | .213 | .91 | 2.87 | 8.2% | 9 | .465 | .412 | .358 | .305 | .254 | .207 | .164 | .126 | .095 | |
| 2.600 | 2.456 | .000 | .144 | .93 | 2.79 | 5.5% | 10 | .465 | .412 | .358 | .305 | .254 | .207 | .164 | .126 | .095 | .069 |
| 2.600 | 2.505 | .000 | .095 | .95 | 2.70 | 3.7% | 11 | .465 | .412 | .358 | .305 | .254 | .207 | .164 | .126 | .095 | .069 |
| | | | | | | | | .049 | | | | | | | | | |
| 2.600 | 2.539 | .000 | .061 | .97 | 2.62 | 2.3% | 12 | .465 | .412 | .358 | .305 | .254 | .207 | .164 | .126 | .095 | .069 |
| | | | | | | | | .049 | .034 | | | | | | | | |
| 2.600 | 2.562 | .000 | .038 | .98 | 2.54 | 1.5% | 13 | .465 | .412 | .358 | .305 | .254 | .207 | .164 | .126 | .095 | .069 |
| | | | | | | | | .049 | .034 | .023 | | | | | | | |
| 2.600 | 2.577 | .000 | .023 | .99 | 2.46 | .0% | 14 | .465 | .412 | .358 | .305 | .254 | .207 | .164 | .126 | .095 | .069 |
| | | | | | | | | .049 | .034 | .023 | .015 | | | | | | |
| 2.600 | 2.586 | .000 | .014 | .99 | 2.39 | .5% | 15 | .465 | .412 | .358 | .305 | .254 | .207 | .164 | .126 | .095 | .069 |
| | | | | | | | | .049 | .034 | .023 | .015 | .009 | | | | | |
| 2.800 | 1.882 | .000 | .918 | .73 | 3.16 | 32.8% | 5 | .480 | .429 | .376 | .324 | .273 | | | | | |
| 2.800 | 2.108 | .000 | .692 | .77 | 3.13 | 24.7% | 6 | .480 | .429 | .376 | .324 | .273 | .225 | | | | |
| 2.800 | 2.280 | .000 | .511 | .82 | 3.08 | 18.3% | 7 | .480 | .429 | .376 | .324 | .273 | .225 | .181 | | | |
| 2.800 | 2.431 | .000 | .369 | .86 | 3.02 | 13.2% | 8 | .480 | .429 | .376 | .324 | .273 | .225 | .181 | .142 | | |
| 2.800 | 2.540 | .000 | .260 | .80 | 2.05 | 9.3% | 9 | .480 | .429 | .376 | .324 | .273 | .225 | .181 | .142 | .109 | |
| 2.800 | 2.621 | .000 | .179 | .92 | 2.87 | 6.4% | 10 | .480 | .429 | .376 | .324 | .273 | .225 | .181 | .142 | .109 | .081 |
| 2.800 | 2.680 | .000 | .120 | .94 | 2.79 | 4.3% | 11 | .480 | .429 | .376 | .324 | .273 | .225 | .181 | .142 | .109 | .081 |
| | | | | | | | | .059 | | | | | | | | | |
| 2.800 | 2.721 | .000 | .079 | .96 | 2.71 | 2.8% | 12 | .480 | .429 | .376 | .324 | .273 | .225 | .181 | .142 | .109 | .081 |
| | | | | | | | | .059 | .041 | | | | | | | | |
| 2.800 | 2.749 | .000 | .051 | .97 | 2.63 | 1.8% | 13 | .480 | .429 | .376 | .324 | .273 | .225 | .181 | .142 | .109 | .081 |

*SIMILAR TO A.T.&T. NEAL-WILKINSON TABLES(SEE SECTION 1)

*EQUIVALENT RANDOM THEORY: VMR(VARIANCE TO MEAN RATIO)– 3.0 - NO RETRIES (PEAKED TRAFFIC - ROUTE ADVANCEMENT OR DAY-TO-DAY VARIATION)

| HOURS FIRST ATTEMPT VMR=3.0 | HOURS CON-NECT | HOURS RECALL /RETRY | HOURS OVFL | CON-NECT VMR | OVFL VMR | PERCENT BLOCKED (GRADE) | LINE | HOURS CONNECTED TRAFFIC ON LINES 1 | 2 | 3 | 4 | 5 | 6 | 7 | 8 | 9 | 10 |
|---|---|---|---|---|---|---|---|---|---|---|---|---|---|---|---|---|---|
| | | | | | | | | .059 | .041 | .028 | | | | | | | |
| 2.800 | 2.768 | .000 | .032 | .98 | 2.55 | 1.1% | 14 | .480 | .429 | .376 | .324 | .273 | .225 | .181 | .142 | .109 | .081 |
| | | | | | | | | .059 | .041 | .028 | .019 | | | | | | |
| 2.800 | 2.781 | .000 | .019 | .99 | 2.47 | .7% | 15 | .480 | .429 | .376 | .324 | .273 | .225 | .181 | .142 | .109 | .081 |
| | | | | | | | | .059 | .041 | .028 | .019 | .012 | | | | | |
| 3.000 | 1.977 | .000 | 1.023 | .71 | 3.18 | 34.1% | 5 | .498 | .448 | .396 | .344 | .292 | | | | | |
| 3.000 | 2.221 | .000 | .779 | .76 | 3.16 | 26.0% | 6 | .498 | .448 | .396 | .344 | .292 | .243 | | | | |
| 3.000 | 2.419 | .000 | .581 | .80 | 3.12 | 19.4% | 7 | .498 | .448 | .396 | .344 | .292 | .243 | .198 | | | |
| 3.000 | 2.576 | .000 | .424 | .84 | 3.06 | 14.1% | 8 | .498 | .448 | .396 | .344 | .292 | .243 | .198 | .157 | | |
| 3.000 | 2.698 | .000 | .302 | .88 | 3.00 | 10.1% | 9 | .498 | .448 | .396 | .344 | .292 | .243 | .198 | .157 | .122 | |
| 3.000 | 2.789 | .000 | .211 | .91 | 2.92 | 7.0% | 10 | .498 | .448 | .396 | .344 | .292 | .243 | .198 | .157 | .122 | .092 |
| 3.000 | 2.857 | .000 | .143 | .93 | 2.84 | 4.8% | 11 | .498 | .448 | .396 | .344 | .292 | .243 | .198 | .157 | .122 | .092 |
| | | | | | | | | .067 | | | | | | | | | |
| 3.000 | 2.905 | .000 | .095 | .95 | 2.76 | 3.2% | 12 | .498 | .448 | .396 | .344 | .292 | .243 | .198 | .157 | .122 | .092 |
| | | | | | | | | .067 | .048 | | | | | | | | |
| 3.000 | 2.938 | .000 | .062 | .97 | 2.67 | 2.1% | 13 | .498 | .448 | .396 | .344 | .292 | .243 | .198 | .157 | .122 | .092 |
| | | | | | | | | .067 | .048 | .033 | | | | | | | |
| 3.000 | 2.961 | .000 | .039 | .98 | 2.59 | 1.3% | 14 | .498 | .448 | .396 | .344 | .292 | .243 | .198 | .157 | .122 | .092 |
| | | | | | | | | .067 | .048 | .033 | .023 | | | | | | |
| 3.000 | 2.976 | .000 | .024 | .99 | 2.51 | .8% | 15 | .498 | .448 | .396 | .344 | .292 | .243 | .198 | .157 | .122 | .092 |
| | | | | | | | | .067 | .048 | .033 | .023 | .015 | | | | | |
| 3.200 | 2.330 | .000 | .870 | .74 | 3.18 | 27.2% | 6 | .515 | .466 | .414 | .362 | .311 | .261 | | | | |
| 3.200 | 2.545 | .000 | .655 | .79 | 3.15 | 20.5% | 7 | .515 | .466 | .414 | .362 | .311 | .261 | .215 | | | |
| 3.200 | 2.717 | .000 | .483 | .83 | 3.10 | 15.1% | 8 | .515 | .466 | .414 | .362 | .311 | .261 | .215 | .172 | | |
| 3.200 | 2.852 | .000 | .348 | .87 | 3.04 | 10.9% | 9 | .515 | .466 | .414 | .362 | .311 | .261 | .215 | .172 | .135 | |
| 3.200 | 2.955 | .000 | .245 | .90 | 2.96 | 7.7% | 10 | .515 | .466 | .414 | .362 | .311 | .261 | .215 | .172 | .135 | .103 |
| 3.200 | 3.031 | .000 | .169 | .92 | 2.89 | 5.3% | 11 | .515 | .466 | .414 | .362 | .311 | .261 | .215 | .172 | .135 | .103 |
| | | | | | | | | .076 | | | | | | | | | |
| 3.200 | 3.086 | .000 | .114 | .94 | 2.80 | 3.5% | 12 | .515 | .466 | .414 | .362 | .311 | .261 | .215 | .172 | .135 | .103 |
| | | | | | | | | .076 | .055 | | | | | | | | |
| 3.200 | 3.125 | .000 | .075 | .96 | 2.72 | 2.3% | 13 | .515 | .466 | .414 | .362 | .311 | .261 | .215 | .172 | .135 | .103 |
| | | | | | | | | .076 | .055 | .039 | | | | | | | |
| 3.200 | 3.152 | .000 | .048 | .97 | 2.64 | 1.5% | 14 | .515 | .466 | .414 | .362 | .311 | .261 | .215 | .172 | .135 | .103 |
| | | | | | | | | .076 | .055 | .039 | .027 | | | | | | |
| 3.200 | 3.170 | .000 | .030 | .98 | 2.56 | .9% | 15 | .515 | .466 | .414 | .362 | .311 | .261 | .215 | .172 | .135 | .103 |
| | | | | | | | | .076 | .055 | .039 | .027 | .018 | | | | | |
| 3.200 | 3.182 | .000 | .018 | .99 | 2.48 | .6% | 16 | .515 | .466 | .414 | .362 | .311 | .261 | .215 | .172 | .135 | .103 |
| | | | | | | | | .076 | .055 | .039 | .027 | .018 | .012 | | | | |
| 3.400 | 2.435 | .000 | .965 | .72 | 3.20 | 28.4% | 6 | .531 | .483 | .432 | .381 | .329 | .279 | | | | |
| 3.400 | 2.666 | .000 | .734 | .77 | 3.18 | 21.6% | 7 | .531 | .483 | .432 | .381 | .329 | .279 | .231 | | | |

*SIMILAR TO A.T.&T. NEAL-WILKINSON TABLES(SEE SECTION 1)

*EQUIVALENT RANDOM THEORY: VMR(VARIANCE TO MEAN RATIO)– 3.0 - NO RETRIES (PEAKED TRAFFIC - ROUTE ADVANCEMENT OR DAY-TO-DAY VARIATION)

| HOURS FIRST ATTEMPT VMR=3.0 | HOURS CON-NECT | HOURS RECALL /RETRY | HOURS OVFL | CON-NECT VMR | OVFL VMR | PERCENT BLOCKED (GRADE) | LINE | HOURS CONNECTED TRAFFIC ON LINES 1 | 2 | 3 | 4 | 5 | 6 | 7 | 8 | 9 | 10 |
|---|---|---|---|---|---|---|---|---|---|---|---|---|---|---|---|---|---|
| 3.400 | 2.854 | .000 | .546 | .81 | 3.13 | 16.1% | 8 | .531 | .483 | .432 | .381 | .329 | .279 | .231 | .188 | | |
| 3.400 | 3.003 | .000 | .397 | .85 | 3.07 | 11.7% | 9 | .531 | .483 | .432 | .381 | .329 | .279 | .231 | .188 | .148 | |
| 3.400 | 3.117 | .000 | .283 | .89 | 3.01 | 8.3% | 10 | .531 | .483 | .432 | .381 | .329 | .279 | .231 | .188 | .148 | .114 |
| 3.400 | 3.203 | .000 | .197 | .91 | 2.93 | 5.8% | 11 | .531 | .483 | .432 | .381 | .329 | .279 | .231 | .188 | .148 | .114 |
| | | | | | | | | .086 | | | | | | | | | |
| 3.400 | 3.266 | .000 | .134 | .94 | 2.85 | 3.9% | 12 | .531 | .483 | .432 | .381 | .329 | .279 | .231 | .188 | .148 | .114 |
| | | | | | | | | .086 | .063 | | | | | | | | |
| 3.400 | 3.311 | .000 | .089 | .96 | 2.76 | 2.6% | 13 | .531 | .483 | .432 | .381 | .329 | .279 | .231 | .188 | .148 | .114 |
| | | | | | | | | .086 | .063 | .045 | | | | | | | |
| 3.400 | 3.342 | .000 | .058 | .97 | 2.68 | 1.7% | 14 | .531 | .483 | .432 | .381 | .329 | .279 | .231 | .188 | .148 | .114 |
| | | | | | | | | .086 | .063 | .045 | .031 | | | | | | |
| 3.400 | 3.363 | .000 | .037 | .98 | 2.60 | 1.1% | 15 | .531 | .483 | .432 | .381 | .329 | .279 | .231 | .188 | .148 | .114 |
| | | | | | | | | .086 | .063 | .045 | .031 | .021 | | | | | |
| 3.400 | 3.377 | .000 | .023 | .99 | 2.52 | .7% | 16 | .531 | .483 | .432 | .381 | .329 | .279 | .231 | .188 | .148 | .114 |
| | | | | | | | | .086 | .063 | .045 | .031 | .021 | .014 | | | | |
| 3.600 | 2.523 | .000 | 1.077 | .70 | 3.26 | 29.9% | 6 | .543 | .496 | .447 | .396 | .346 | .296 | | | | |
| 3.600 | 2.771 | .000 | .829 | .75 | 3.24 | 23.0% | 7 | .543 | .496 | .447 | .396 | .346 | .296 | .248 | | | |
| 3.600 | 2.974 | .000 | .626 | .80 | 3.20 | 17.4% | 8 | .543 | .496 | .447 | .396 | .346 | .296 | .248 | .204 | | |
| 3.600 | 3.138 | .000 | .462 | .84 | 3.15 | 12.8% | 9 | .543 | .496 | .447 | .396 | .346 | .296 | .248 | .204 | .163 | |
| 3.600 | 3.266 | .000 | .334 | .87 | 3.08 | 9.3% | 10 | .543 | .496 | .447 | .396 | .346 | .296 | .248 | .204 | .163 | .128 |
| 3.600 | 3.364 | .000 | .236 | .90 | 3.01 | 6.6% | 11 | .543 | .496 | .447 | .396 | .346 | .296 | .248 | .204 | .163 | .128 |
| | | | | | | | | .098 | | | | | | | | | |
| 3.600 | 3.436 | .000 | .164 | .93 | 2.93 | 4.5% | 12 | .543 | .496 | .447 | .396 | .346 | .296 | .248 | .204 | .163 | .128 |
| | | | | | | | | .098 | .073 | | | | | | | | |
| 3.600 | 3.489 | .000 | .111 | .95 | 2.85 | 3.1% | 13 | .543 | .496 | .447 | .396 | .346 | .296 | .248 | .204 | .163 | .128 |
| | | | | | | | | .098 | .073 | .053 | | | | | | | |
| 3.600 | 3.527 | .000 | .073 | .96 | 2.76 | 2.0% | 14 | .543 | .496 | .447 | .396 | .346 | .296 | .248 | .204 | .163 | .128 |
| | | | | | | | | .098 | .073 | .053 | .037 | | | | | | |
| 3.600 | 3.553 | .000 | .047 | .97 | 2.68 | 1.3% | 15 | .543 | .496 | .447 | .396 | .346 | .296 | .248 | .204 | .163 | .128 |
| | | | | | | | | .098 | .073 | .053 | .037 | .026 | | | | | |
| 3.600 | 3.570 | .000 | .030 | .98 | 2.60 | .8% | 16 | .543 | .496 | .447 | .396 | .346 | .296 | .248 | .204 | .163 | .128 |
| | | | | | | | | .098 | .073 | .053 | .037 | .026 | .017 | | | | |
| 3.600 | 3.581 | .000 | .019 | .99 | 2.53 | .5% | 17 | .543 | .496 | .447 | .396 | .346 | .296 | .248 | .204 | .163 | .128 |
| | | | | | | | | .098 | .073 | .053 | .037 | .026 | .017 | .011 | | | |
| 3.800 | 2.620 | .000 | 1.180 | .69 | 3.28 | 31.1% | 6 | .557 | .511 | .463 | .413 | .363 | .313 | | | | |
| 3.800 | 2.884 | .000 | .910 | .74 | 3.27 | 24.1% | 7 | .557 | .511 | .463 | .413 | .363 | .313 | .265 | | | |
| 3.800 | 3.103 | .000 | .697 | .78 | 3.23 | 18.3% | 8 | .557 | .511 | .463 | .413 | .363 | .313 | .265 | .219 | | |
| 3.800 | 3.281 | .000 | .519 | .82 | 3.18 | 13.7% | 9 | .557 | .511 | .463 | .413 | .363 | .313 | .265 | .219 | .177 | |
| 3.800 | 3.421 | .000 | .379 | .86 | 3.12 | 10.0% | 10 | .557 | .511 | .463 | .413 | .363 | .313 | .265 | .219 | .177 | .140 |
| 3.800 | 3.529 | .000 | .271 | .89 | 3.05 | 7.1% | 11 | .557 | .511 | .463 | .413 | .363 | .313 | .265 | .219 | .177 | .140 |

*SIMILAR TO A.T.&T. NEAL-WILKINSON TABLES(SEE SECTION 1)

*EQUIVALENT RANDOM THEORY: VMR(VARIANCE TO MEAN RATIO)– 3.0 - NO RETRIES (PEAKED TRAFFIC - ROUTE ADVANCEMENT OR DAY-TO-DAY VARIATION)

| HOURS FIRST ATTEMPT VMR=3.0 | HOURS CON- NECT | HOURS RECALL /RETRY | HOURS OVFL | CON- NECT VMR | OVFL VMR | PERCENT BLOCKED (GRADE) | LINE | HOURS CONNECTED TRAFFIC ON LINES 1 | 2 | 3 | 4 | 5 | 6 | 7 | 8 | 9 | 10 |
|---|---|---|---|---|---|---|---|---|---|---|---|---|---|---|---|---|---|
| | | | | | | | | .108 | | | | | | | | | |
| 3.800 | 3.611 | .000 | .189 | .92 | 2.97 | 5.0% | 12 | .557 | .511 | .463 | .413 | .363 | .313 | .265 | .219 | .177 | .140 |
| | | | | | | | | .108 | .082 | | | | | | | | |
| 3.800 | 3.671 | .000 | .129 | .94 | 2.89 | 3.4% | 13 | .557 | .511 | .463 | .413 | .363 | .313 | .265 | .219 | .177 | .140 |
| | | | | | | | | .108 | .082 | .060 | | | | | | | |
| 3.800 | 3.713 | .000 | .087 | .96 | 2.81 | 2.3% | 14 | .557 | .511 | .463 | .413 | .363 | .313 | .265 | .219 | .177 | .140 |
| | | | | | | | | .108 | .082 | .060 | .043 | | | | | | |
| 3.800 | 3.743 | .000 | .057 | .97 | 2.72 | 1.5% | 15 | .557 | .511 | .463 | .413 | .363 | .313 | .265 | .219 | .177 | .140 |
| | | | | | | | | .108 | .082 | .060 | .043 | .030 | | | | | |
| 3.800 | 3.764 | .000 | .036 | .98 | 2.64 | 1.0% | 16 | .557 | .511 | .463 | .413 | .363 | .313 | .265 | .219 | .177 | .140 |
| | | | | | | | | .108 | .082 | .060 | .043 | .030 | .020 | | | | |
| 3.800 | 3.777 | .000 | .023 | .99 | 2.56 | .6% | 17 | .557 | .511 | .463 | .413 | .363 | .313 | .265 | .219 | .177 | .140 |
| | | | | | | | | .108 | .082 | .060 | .043 | .030 | .020 | .014 | | | |
| 4.000 | 2.714 | .000 | 1.286 | .67 | 3.30 | 32.2% | 6 | .571 | .526 | .479 | .430 | .379 | .329 | | | | |
| 4.000 | 2.994 | .000 | 1.006 | .72 | 3.29 | 25.1% | 7 | .571 | .526 | .479 | .430 | .379 | .329 | .281 | | | |
| 4.000 | 3.229 | .000 | .771 | .77 | 3.26 | 19.3% | 8 | .571 | .526 | .479 | .430 | .379 | .329 | .281 | .234 | | |
| 4.000 | 3.420 | .000 | .580 | .81 | 3.21 | 14.5% | 9 | .571 | .526 | .479 | .430 | .379 | .329 | .281 | .234 | .191 | |
| 4.000 | 3.573 | .000 | .427 | .85 | 3.16 | 10.7% | 10 | .571 | .526 | .479 | .430 | .379 | .329 | .281 | .234 | .191 | .153 |
| 4.000 | 3.692 | .000 | .308 | .88 | 3.09 | 7.7% | 11 | .571 | .526 | .479 | .430 | .379 | .329 | .281 | .234 | .191 | .153 |
| | | | | | | | | .119 | | | | | | | | | |
| 4.000 | 3.783 | .000 | .217 | .91 | 3.01 | 5.4% | 12 | .571 | .526 | .479 | .430 | .379 | .329 | .281 | .234 | .191 | .153 |
| | | | | | | | | .119 | .091 | | | | | | | | |
| 4.000 | 3.850 | .000 | .150 | .93 | 2.93 | 3.8% | 13 | .571 | .526 | .479 | .430 | .379 | .329 | .281 | .234 | .191 | .153 |
| | | | | | | | | .119 | .091 | .067 | | | | | | | |
| 4.000 | 3.899 | .000 | .101 | .95 | 2.85 | 2.5% | 14 | .571 | .526 | .479 | .430 | .379 | .329 | .281 | .234 | .191 | .153 |
| | | | | | | | | .119 | .091 | .067 | .049 | | | | | | |
| 4.000 | 3.933 | .000 | .067 | .97 | 2.76 | 1.7% | 15 | .571 | .526 | .479 | .430 | .379 | .329 | .281 | .234 | .191 | .153 |
| | | | | | | | | .119 | .091 | .067 | .049 | .034 | | | | | |
| 4.000 | 3.957 | .000 | .043 | .98 | 2.68 | 1.1% | 16 | .571 | .526 | .479 | .430 | .379 | .329 | .281 | .234 | .191 | .153 |
| | | | | | | | | .119 | .091 | .067 | .049 | .034 | .024 | | | | |
| 4.000 | 3.973 | .000 | .027 | .98 | 2.60 | .7% | 17 | .571 | .526 | .479 | .430 | .379 | .329 | .281 | .234 | .191 | .153 |
| | | | | | | | | .119 | .091 | .067 | .049 | .034 | .024 | .016 | | | |
| 4.200 | 2.789 | .000 | 1.411 | .66 | 3.35 | 33.6% | 6 | .580 | .537 | .491 | .443 | .394 | .345 | | | | |
| 4.200 | 3.086 | .000 | 1.114 | .70 | 3.34 | 26.5% | 7 | .580 | .537 | .491 | .443 | .394 | .345 | .296 | | | |
| 4.200 | 3.336 | .000 | .864 | .75 | 3.32 | 20.6% | 8 | .580 | .537 | .491 | .443 | .394 | .345 | .296 | .250 | | |
| 4.200 | 3.542 | .000 | .658 | .79 | 3.28 | 15.7% | 9 | .580 | .537 | .491 | .443 | .394 | .345 | .296 | .250 | .207 | |
| 4.200 | 3.709 | .000 | .491 | .83 | 3.23 | 11.7% | 10 | .580 | .537 | .491 | .443 | .394 | .345 | .296 | .250 | .207 | .167 |
| 4.200 | 3.841 | .000 | .359 | .87 | 3.17 | 8.5% | 11 | .580 | .537 | .491 | .443 | .394 | .345 | .296 | .250 | .207 | .167 |
| | | | | | | | | .132 | | | | | | | | | |
| 4.200 | 3.943 | .000 | .257 | .90 | 3.09 | 6.1% | 12 | .580 | .537 | .491 | .443 | .394 | .345 | .296 | .250 | .207 | .167 |

*SIMILAR TO A.T.&T. NEAL-WILKINSON TABLES(SEE SECTION 1)

*EQUIVALENT RANDOM THEORY: VMR(VARIANCE TO MEAN RATIO)– 3.0 - NO RETRIES (PEAKED TRAFFIC - ROUTE ADVANCEMENT OR DAY-TO-DAY VARIATION)

| HOURS FIRST ATTEMPT VMR = 3.0 | HOURS CON- NECT | HOURS RECALL /RETRY | HOURS OVFL | CON- NECT VMR | OVFL VMR | PERCENT BLOCKED (GRADE) | LINE | HOURS CONNECTED TRAFFIC ON LINES 1 | 2 | 3 | 4 | 5 | 6 | 7 | 8 | 9 | 10 |
|---|---|---|---|---|---|---|---|---|---|---|---|---|---|---|---|---|---|
| | | | | | | | | .132 | .102 | | | | | | | | |
| 4.200 | 4.020 | .000 | .180 | .92 | 3.01 | 4.3% | 13 | .580 | .537 | .491 | .443 | .394 | .345 | .296 | .250 | .207 | .167 |
| | | | | | | | | .132 | .102 | .077 | | | | | | | |
| 4.200 | 4.076 | .000 | .124 | .94 | 2.93 | 2.9% | 14 | .580 | .537 | .491 | .443 | .394 | .345 | .296 | .250 | .207 | .167 |
| | | | | | | | | .132 | .102 | .077 | .056 | | | | | | |
| 4.200 | 4.117 | .000 | .083 | .96 | 2.84 | 2.0% | 15 | .580 | .537 | .491 | .443 | .394 | .345 | .296 | .250 | .207 | .167 |
| | | | | | | | | .132 | .102 | .077 | .056 | .041 | | | | | |
| 4.200 | 4.145 | .000 | .055 | .97 | 2.76 | 1.3% | 16 | .580 | .537 | .491 | .443 | .394 | .345 | .296 | .250 | .207 | .167 |
| | | | | | | | | .132 | .102 | .077 | .056 | .041 | .028 | | | | |
| 4.200 | 4.165 | .000 | .035 | .98 | 2.68 | .8% | 17 | .580 | .537 | .491 | .443 | .394 | .345 | .296 | .250 | .207 | .167 |
| | | | | | | | | .132 | .102 | .077 | .056 | .041 | .028 | .019 | | | |
| 4.200 | 4.178 | .000 | .022 | .99 | 2.60 | .5% | 18 | .580 | .537 | .491 | .443 | .394 | .345 | .296 | .250 | .207 | .167 |
| | | | | | | | | .132 | .102 | .077 | .056 | .041 | .028 | .019 | .013 | | |
| 4.400 | 2.876 | .000 | 1.524 | .64 | 3.36 | 34.6% | 6 | .592 | .550 | .505 | .458 | .410 | .361 | | | | |
| 4.400 | 3.188 | .000 | 1.212 | .69 | 3.36 | 27.5% | 7 | .592 | .550 | .505 | .458 | .410 | .361 | .312 | | | |
| 4.400 | 3.453 | .000 | .947 | .73 | 3.34 | 21.5% | 8 | .592 | .550 | .505 | .458 | .410 | .361 | .312 | .265 | | |
| 4.400 | 3.674 | .000 | .726 | .78 | 3.31 | 16.5% | 9 | .592 | .550 | .505 | .458 | .410 | .361 | .312 | .265 | .221 | |
| 4.400 | 3.854 | .000 | .546 | .82 | 3.26 | 12.4% | 10 | .592 | .550 | .505 | .458 | .410 | .361 | .312 | .265 | .221 | .180 |
| 4.400 | 3.997 | .000 | .403 | .86 | 3.20 | 9.2% | 11 | .592 | .550 | .505 | .458 | .410 | .361 | .312 | .265 | .221 | .180 |
| | | | | | | | | .143 | | | | | | | | | |
| 4.400 | 4.109 | .000 | .291 | .89 | 3.13 | 6.6% | 12 | .592 | .550 | .505 | .458 | .410 | .361 | .312 | .265 | .221 | .180 |
| | | | | | | | | .143 | .112 | | | | | | | | |
| 4.400 | 4.194 | .000 | .206 | .91 | 3.05 | 4.7% | 13 | .592 | .550 | .505 | .458 | .410 | .361 | .312 | .265 | .221 | .180 |
| | | | | | | | | .143 | .112 | .085 | | | | | | | |
| 4.400 | 4.257 | .000 | .143 | .94 | 2.97 | 3.2% | 14 | .592 | .550 | .505 | .458 | .410 | .361 | .312 | .265 | .221 | .180 |
| | | | | | | | | .143 | .112 | .085 | .063 | | | | | | |
| 4.400 | 4.303 | .000 | .097 | .95 | 2.88 | 2.2% | 15 | .592 | .550 | .505 | .458 | .410 | .361 | .312 | .265 | .221 | .180 |
| | | | | | | | | .143 | .112 | .085 | .063 | .046 | | | | | |
| 4.400 | 4.336 | .000 | .064 | .97 | 2.80 | 1.5% | 16 | .592 | .550 | .505 | .458 | .410 | .361 | .312 | .265 | .221 | .180 |
| | | | | | | | | .143 | .112 | .085 | .063 | .046 | .032 | | | | |
| 4.400 | 4.358 | .000 | .042 | .98 | 2.72 | 1.0% | 17 | .592 | .550 | .505 | .458 | .410 | .361 | .312 | .265 | .221 | .180 |
| | | | | | | | | .143 | .112 | .085 | .063 | .046 | .032 | .022 | | | |
| 4.400 | 4.373 | .000 | .027 | .98 | 2.64 | .6% | 18 | .592 | .550 | .505 | .458 | .410 | .361 | .312 | .265 | .221 | .180 |
| | | | | | | | | .143 | .112 | .085 | .063 | .046 | .032 | .022 | .015 | | |
| 4.600 | 3.287 | .000 | 1.313 | .67 | 3.37 | 28.5% | 7 | .604 | .563 | .519 | .473 | .425 | .376 | .327 | | | |
| 4.600 | 3.568 | .000 | 1.032 | .72 | 3.36 | 22.4% | 8 | .604 | .563 | .519 | .473 | .425 | .376 | .327 | .280 | | |
| 4.600 | 3.802 | .000 | .798 | .77 | 3.33 | 17.3% | 9 | .604 | .563 | .519 | .473 | .425 | .376 | .327 | .280 | .235 | |
| 4.600 | 3.995 | .000 | .605 | .81 | 3.29 | 13.1% | 10 | .604 | .563 | .519 | .473 | .425 | .376 | .327 | .280 | .235 | .193 |
| 4.600 | 4.151 | .000 | .449 | .84 | 3.23 | 9.8% | 11 | .604 | .563 | .519 | .473 | .425 | .376 | .327 | .280 | .235 | .193 |
| | | | | | | | | .155 | | | | | | | | | |

*SIMILAR TO A.T.&T. NEAL-WILKINSON TABLES(SEE SECTION 1)

# *EQUIVALENT RANDOM THEORY: VMR(VARIANCE TO MEAN RATIO)– 3.0 - NO RETRIES (PEAKED TRAFFIC - ROUTE ADVANCEMENT OR DAY-TO-DAY VARIATION)

| HOURS FIRST ATTEMPT VMR=3.0 | HOURS CON-NECT | HOURS RECALL /RETRY | HOURS OVFL | CON-NECT VMR | OVFL VMR | PERCENT BLOCKED (GRADE) | LINE | HOURS CONNECTED TRAFFIC ON LINES 1 | 2 | 3 | 4 | 5 | 6 | 7 | 8 | 9 | 10 |
|---|---|---|---|---|---|---|---|---|---|---|---|---|---|---|---|---|---|
| 4.600 | 4.273 | .000 | .327 | .88 | 3.17 | 7.1% | 12 | .604 | .563 | .519 | .473 | .425 | .376 | .327 | .280 | .235 | .193 |
| | | | | | | | | .155 | .122 | | | | | | | | |
| 4.600 | 4.366 | .000 | .234 | .91 | 3.09 | 5.1% | 13 | .604 | .563 | .519 | .473 | .425 | .376 | .327 | .280 | .235 | .193 |
| | | | | | | | | .155 | .122 | .094 | | | | | | | |
| 4.600 | 4.437 | .000 | .163 | .93 | 3.01 | 3.6% | 14 | .604 | .563 | .519 | .473 | .425 | .376 | .327 | .280 | .235 | .193 |
| | | | | | | | | .155 | .122 | .094 | .070 | | | | | | |
| 4.600 | 4.488 | .000 | .112 | .95 | 2.92 | 2.4% | 15 | .604 | .563 | .519 | .473 | .425 | .376 | .327 | .280 | .235 | .193 |
| | | | | | | | | .155 | .122 | .094 | .070 | .051 | | | | | |
| 4.600 | 4.525 | .000 | .075 | .96 | 2.84 | 1.6% | 16 | .604 | .563 | .519 | .473 | .425 | .376 | .327 | .280 | .235 | .193 |
| | | | | | | | | .155 | .122 | .094 | .070 | .051 | .037 | | | | |
| 4.600 | 4.551 | .000 | .049 | .97 | 2.76 | 1.1% | 17 | .604 | .563 | .519 | .473 | .425 | .376 | .327 | .280 | .235 | .193 |
| | | | | | | | | .155 | .122 | .094 | .070 | .051 | .037 | .026 | | | |
| 4.600 | 4.568 | .000 | .032 | .98 | 2.68 | .7% | 18 | .604 | .563 | .519 | .473 | .425 | .376 | .327 | .280 | .235 | .193 |
| | | | | | | | | .155 | .122 | .094 | .070 | .051 | .037 | .026 | .018 | | |
| 4.800 | 3.367 | .000 | 1.433 | .66 | 3.42 | 29.8% | 7 | .612 | .572 | .529 | .484 | .438 | .390 | .342 | | | |
| 4.800 | 3.662 | .000 | 1.138 | .71 | 3.42 | 23.7% | 8 | .612 | .572 | .529 | .484 | .438 | .390 | .342 | .295 | | |
| 4.800 | 3.912 | .000 | .888 | .75 | 3.40 | 18.5% | 9 | .612 | .572 | .529 | .484 | .438 | .390 | .342 | .295 | .250 | |
| 4.800 | 4.119 | .000 | .681 | .79 | 3.36 | 14.2% | 10 | .612 | .572 | .529 | .484 | .438 | .390 | .342 | .295 | .250 | .207 |
| 4.800 | 4.288 | .000 | .512 | .83 | 3.30 | 10.7% | 11 | .612 | .572 | .529 | .484 | .438 | .390 | .342 | .295 | .250 | .207 |
| | | | | | | | | .169 | | | | | | | | | |
| 4.800 | 4.422 | .000 | .378 | .87 | 3.24 | 7.9% | 12 | .612 | .572 | .529 | .484 | .438 | .390 | .342 | .295 | .250 | .207 |
| | | | | | | | | .169 | .134 | | | | | | | | |
| 4.800 | 4.527 | .000 | .273 | .90 | 3.17 | 5.7% | 13 | .612 | .572 | .529 | .484 | .438 | .390 | .342 | .295 | .250 | .207 |
| | | | | | | | | .169 | .134 | .104 | | | | | | | |
| 4.800 | 4.606 | .000 | .194 | .92 | 3.09 | 4.0% | 14 | .612 | .572 | .529 | .484 | .438 | .390 | .342 | .295 | .250 | .207 |
| | | | | | | | | .169 | .134 | .104 | .079 | | | | | | |
| 4.800 | 4.665 | .000 | .135 | .94 | 3.00 | 2.8% | 15 | .612 | .572 | .529 | .484 | .438 | .390 | .342 | .295 | .250 | .207 |
| | | | | | | | | .169 | .134 | .104 | .079 | .059 | | | | | |
| 4.800 | 4.708 | .000 | .092 | .96 | 2.92 | 1.9% | 16 | .612 | .572 | .529 | .484 | .438 | .390 | .342 | .295 | .250 | .207 |
| | | | | | | | | .169 | .134 | .104 | .079 | .059 | .043 | | | | |
| 4.800 | 4.739 | .000 | .061 | .97 | 2.83 | 1.3% | 17 | .612 | .572 | .529 | .484 | .438 | .390 | .342 | .295 | .250 | .207 |
| | | | | | | | | .169 | .134 | .104 | .079 | .059 | .043 | .031 | | | |
| 4.800 | 4.760 | .000 | .040 | .98 | 2.75 | .8% | 18 | .612 | .572 | .529 | .484 | .438 | .390 | .342 | .295 | .250 | .207 |
| | | | | | | | | .169 | .134 | .104 | .079 | .059 | .043 | .031 | .021 | | |
| 4.800 | 4.774 | .000 | .026 | .99 | 2.67 | .5% | 19 | .612 | .572 | .529 | .484 | .438 | .390 | .342 | .295 | .250 | .207 |
| | | | | | | | | .169 | .134 | .104 | .079 | .059 | .043 | .031 | .021 | .014 | |
| 5.000 | 3.460 | .000 | 1.540 | .64 | 3.44 | 30.8% | 7 | .623 | .584 | .542 | .498 | .452 | .405 | .357 | | | |
| 5.000 | 3.770 | .000 | 1.230 | .69 | 3.43 | 24.6% | 8 | .623 | .584 | .542 | .498 | .452 | .405 | .357 | .310 | | |
| 5.000 | 4.033 | .000 | .967 | .74 | 3.42 | 19.3% | 9 | .623 | .584 | .542 | .498 | .452 | .405 | .357 | .310 | .264 | |
| 5.000 | 4.254 | .000 | .746 | .78 | 3.38 | 14.9% | 10 | .623 | .584 | .542 | .498 | .452 | .405 | .357 | .310 | .264 | .221 |

*SIMILAR TO A.T.&T. NEAL-WILKINSON TABLES(SEE SECTION 1)

*EQUIVALENT RANDOM THEORY: VMR(VARIANCE TO MEAN RATIO)– 3.0 - NO RETRIES (PEAKED TRAFFIC - ROUTE ADVANCEMENT OR DAY-TO-DAY VARIATION)

| HOURS FIRST ATTEMPT VMR = 3.0 | HOURS CON-NECT | HOURS RECALL /RETRY | HOURS OVFL | CON-NECT VMR | OVFL VMR | PERCENT BLOCKED (GRADE) | LINE | HOURS CONNECTED TRAFFIC ON LINES 1 | 2 | 3 | 4 | 5 | 6 | 7 | 8 | 9 | 10 |
|---|---|---|---|---|---|---|---|---|---|---|---|---|---|---|---|---|---|
| 5.000 | 4.435 | .000 | .565 | .82 | 3.33 | 11.3% | 11 | .623 | .584 | .542 | .498 | .452 | .405 | .357 | .310 | .264 | .221 |
| | | | | | | | | .181 | | | | | | | | | |
| 5.000 | 4.580 | .000 | .420 | .85 | 3.27 | 8.4% | 12 | .623 | .584 | .542 | .498 | .452 | .405 | .357 | .310 | .264 | .221 |
| | | | | | | | | .181 | .145 | | | | | | | | |
| 5.000 | 4.693 | .000 | .307 | .89 | 3.20 | 6.1% | 13 | .623 | .584 | .542 | .498 | .452 | .405 | .357 | .310 | .264 | .221 |
| | | | | | | | | .181 | .145 | .114 | | | | | | | |
| 5.000 | 4.781 | .000 | .219 | .91 | 3.12 | 4.4% | 14 | .623 | .584 | .542 | .498 | .452 | .405 | .357 | .310 | .264 | .221 |
| | | | | | | | | .181 | .145 | .114 | .087 | | | | | | |
| 5.000 | 4.846 | .000 | .154 | .93 | 3.04 | 3.1% | 15 | .623 | .584 | .542 | .498 | .452 | .405 | .357 | .310 | .264 | .221 |
| | | | | | | | | .181 | .145 | .114 | .087 | .066 | | | | | |
| 5.000 | 4.894 | .000 | .106 | .95 | 2.96 | 2.1% | 16 | .623 | .584 | .542 | .498 | .452 | .405 | .357 | .310 | .264 | .221 |
| | | | | | | | | .181 | .145 | .114 | .087 | .066 | .048 | | | | |
| 5.000 | 4.929 | .000 | .071 | .97 | 2.87 | 1.4% | 17 | .623 | .584 | .542 | .498 | .452 | .405 | .357 | .310 | .264 | .221 |
| | | | | | | | | .181 | .145 | .114 | .087 | .066 | .048 | .034 | | | |
| 5.000 | 4.953 | .000 | .047 | .98 | 2.79 | .9% | 18 | .623 | .584 | .542 | .498 | .452 | .405 | .357 | .310 | .264 | .221 |
| | | | | | | | | .181 | .145 | .114 | .087 | .066 | .048 | .034 | .024 | | |
| 5.000 | 4.970 | .000 | .030 | .98 | 2.71 | .6% | 19 | .623 | .584 | .542 | .498 | .452 | .405 | .357 | .310 | .264 | .221 |
| | | | | | | | | .181 | .145 | .114 | .087 | .066 | .048 | .034 | .024 | .017 | |
| 5.500 | 3.662 | .000 | 1.838 | .61 | 3.49 | 33.4% | 7 | .644 | .608 | .569 | .527 | .484 | .438 | .392 | | | |
| 5.500 | 4.007 | .000 | 1.493 | .66 | 3.50 | 27.1% | 8 | .644 | .608 | .569 | .527 | .484 | .438 | .392 | .345 | | |
| 5.500 | 4.306 | .000 | 1.194 | .70 | 3.50 | 21.7% | 9 | .644 | .608 | .569 | .527 | .484 | .438 | .392 | .345 | .299 | |
| 5.500 | 4.560 | .000 | .940 | .75 | 3.48 | 17.1% | 10 | .644 | .608 | .569 | .527 | .484 | .438 | .392 | .345 | .299 | .255 |
| 5.500 | 4.773 | .000 | .727 | .79 | 3.44 | 13.2% | 11 | .644 | .608 | .569 | .527 | .484 | .438 | .392 | .345 | .299 | .255 |
| | | | | | | | | .213 | | | | | | | | | |
| 5.500 | 4.947 | .000 | .553 | .83 | 3.39 | 10.0% | 12 | .644 | .608 | .560 | .527 | .484 | .438 | .392 | .345 | .299 | .255 |
| | | | | | | | | .213 | .175 | | | | | | | | |
| 5.500 | 5.088 | .000 | .412 | .86 | 3.32 | 7.5% | 13 | .644 | .608 | .569 | .527 | .484 | .438 | .392 | .345 | .299 | .255 |
| | | | | | | | | .213 | .175 | .140 | | | | | | | |
| 5.500 | 5.198 | .000 | .302 | .89 | 3.25 | 5.5% | 14 | .644 | .608 | .569 | .527 | .484 | .438 | .392 | .345 | .299 | .255 |
| | | | | | | | | .213 | .175 | .140 | .110 | | | | | | |
| 5.500 | 5.283 | .000 | .217 | .92 | 3.17 | 3.9% | 15 | .644 | .608 | .569 | .527 | .484 | .438 | .392 | .345 | .299 | .255 |
| | | | | | | | | .213 | .175 | .140 | .110 | .085 | | | | | |
| 5.500 | 5.347 | .000 | .153 | .94 | 3.09 | 2.8% | 16 | .644 | .608 | .569 | .527 | .484 | .438 | .392 | .345 | .299 | .255 |
| | | | | | | | | .213 | .175 | .140 | .110 | .085 | .064 | | | | |
| 5.500 | 5.394 | .000 | .106 | .95 | 3.00 | 1.9% | 17 | .644 | .608 | .569 | .527 | .484 | .438 | .392 | .345 | .299 | .255 |
| | | | | | | | | .213 | .175 | .140 | .110 | .085 | .064 | .047 | | | |
| 5.500 | 5.428 | .000 | .072 | .97 | 2.92 | 1.3% | 18 | .644 | .608 | .569 | .527 | .484 | .438 | .392 | .345 | .299 | .255 |
| | | | | | | | | .213 | .175 | .140 | .110 | .085 | .064 | .047 | .034 | | |
| 5.500 | 5.452 | .000 | .048 | .98 | 2.83 | .9% | 19 | .644 | .608 | .569 | .527 | .484 | .438 | .392 | .345 | .299 | .255 |
| | | | | | | | | .213 | .175 | .140 | .110 | .085 | .064 | .047 | .034 | .024 | |

*SIMILAR TO A.T.&T. NEAL-WILKINSON TABLES(SEE SECTION 1)

*EQUIVALENT RANDOM THEORY: VMR(VARIANCE TO MEAN RATIO)– 3.0 - NO RETRIES (PEAKED TRAFFIC - ROUTE ADVANCEMENT OR DAY-TO-DAY VARIATION)

| HOURS FIRST ATTEMPT VMR = 3.0 | HOURS CON-NECT | HOURS RECALL /RETRY | HOURS OVFL | CON-NECT VMR | OVFL VMR | PERCENT BLOCKED (GRADE) | LINE | HOURS CONNECTED TRAFFIC ON LINES 1 | 2 | 3 | 4 | 5 | 6 | 7 | 8 | 9 | 10 |
|---|---|---|---|---|---|---|---|---|---|---|---|---|---|---|---|---|---|
| 5.500 | 5.469 | .000 | .031 | .98 | 2.75 | .6% | 20 | .644 | .608 | .569 | .527 | .484 | .438 | .392 | .345 | .299 | .255 |
| | | | | | | | | .213 | .175 | .140 | .110 | .085 | .064 | .047 | .034 | .024 | .017 |
| 6.000 | 4.224 | .000 | 1.776 | .62 | 3.56 | 29.6% | 8 | .664 | .630 | .593 | .554 | .512 | .469 | .424 | .378 | | |
| 6.000 | 4.557 | .000 | 1.443 | .67 | 3.57 | 24.1% | 9 | .664 | .630 | .593 | .554 | .512 | .469 | .424 | .378 | .333 | |
| 6.000 | 4.845 | .000 | 1.155 | .71 | 3.56 | 19.3% | 10 | .664 | .630 | .593 | .554 | .512 | .469 | .424 | .378 | .333 | .288 |
| 6.000 | 5.090 | .000 | .910 | .75 | 3.53 | 15.2% | 11 | .664 | .630 | .593 | .554 | .512 | .469 | .424 | .378 | .333 | .288 |
| | | | | | | | | .245 | | | | | | | | | |
| 6.000 | 5.295 | .000 | .705 | .80 | 3.49 | 11.8% | 12 | .664 | .630 | .593 | .554 | .512 | .469 | .424 | .378 | .333 | .288 |
| | | | | | | | | .245 | .205 | | | | | | | | |
| 6.000 | 5.463 | .000 | .537 | .83 | 3.44 | 9.0% | 13 | .664 | .630 | .593 | .554 | .512 | .469 | .424 | .378 | .333 | .288 |
| | | | | | | | | .245 | .205 | .168 | | | | | | | |
| 6.000 | 5.598 | .000 | .402 | .86 | 3.37 | 6.7% | 14 | .664 | .630 | .593 | .554 | .512 | .469 | .424 | .378 | .333 | .288 |
| | | | | | | | | .245 | .205 | .168 | .135 | | | | | | |
| 6.000 | 5.705 | .000 | .295 | .89 | 3.30 | 4.9% | 15 | .664 | .630 | .593 | .554 | .512 | .469 | .424 | .378 | .333 | .288 |
| | | | | | | | | .245 | .205 | .168 | .135 | .107 | | | | | |
| 6.000 | 5.787 | .000 | .213 | .92 | 3.22 | 3.6% | 16 | .664 | .630 | .593 | .554 | .512 | .469 | .424 | .378 | .333 | .288 |
| | | | | | | | | .245 | .205 | .168 | .135 | .107 | .082 | | | | |
| 6.000 | 5.849 | .000 | .151 | .94 | 3.13 | 2.5% | 17 | .664 | .630 | .593 | .554 | .512 | .469 | .424 | .378 | .333 | .288 |
| | | | | | | | | .245 | .205 | .168 | .135 | .107 | .082 | .062 | | | |
| 6.000 | 5.895 | .000 | .105 | .95 | 3.05 | 1.8% | 18 | .664 | .630 | .593 | .554 | .512 | .469 | .424 | .378 | .333 | .288 |
| | | | | | | | | .245 | .205 | .168 | .135 | .107 | .082 | .062 | .046 | | |
| 6.000 | 5.928 | .000 | .072 | .97 | 2.96 | 1.2% | 19 | .664 | .630 | .593 | .554 | .512 | .469 | .424 | .378 | .333 | .288 |
| | | | | | | | | .245 | .205 | .168 | .135 | .107 | .082 | .062 | .046 | .033 | |
| 6.000 | 5.952 | .000 | .048 | .98 | 2.88 | .8% | 20 | .664 | .630 | .593 | .554 | .512 | .469 | .424 | .378 | .333 | .288 |
| | | | | | | | | .245 | .205 | .168 | .135 | .107 | .082 | .062 | .046 | .033 | .024 |
| 6.000 | 5.968 | .000 | .032 | .98 | 2.80 | .5% | 21 | .664 | .630 | .593 | .554 | .512 | .469 | .424 | .378 | .333 | .288 |
| | | | | | | | | .245 | .205 | .168 | .135 | .107 | .082 | .062 | .046 | .033 | .024 |
| | | | | | | | | .016 | | | | | | | | | |
| 6.500 | 4.424 | .000 | 2.076 | .59 | 3.61 | 31.9% | 8 | .681 | .649 | .615 | .578 | .539 | .497 | .454 | .410 | | |
| 6.500 | 4.789 | .000 | 1.711 | .64 | 3.63 | 26.3% | 9 | .681 | .649 | .615 | .578 | .539 | .497 | .454 | .410 | .365 | |
| 6.500 | 5.109 | .000 | 1.391 | .68 | 3.63 | 21.4% | 10 | .681 | .649 | .615 | .578 | .539 | .497 | .454 | .410 | .365 | .320 |
| 6.500 | 5.386 | .000 | 1.114 | .72 | 3.62 | 17.1% | 11 | .681 | .649 | .615 | .578 | .539 | .497 | .454 | .410 | .365 | .320 |
| | | | | | | | | .277 | | | | | | | | | |
| 6.500 | 5.622 | .000 | .878 | .76 | 3.59 | 13.5% | 12 | .681 | .649 | .615 | .578 | .539 | .497 | .454 | .410 | .365 | .320 |
| | | | | | | | | .277 | .235 | | | | | | | | |
| 6.500 | 5.819 | .000 | .681 | .80 | 3.54 | 10.5% | 13 | .681 | .649 | .615 | .578 | .539 | .497 | .454 | .410 | .365 | .320 |
| | | | | | | | | .277 | .235 | .197 | | | | | | | |
| 6.500 | 5.980 | .000 | .520 | .84 | 3.48 | 8.0% | 14 | .681 | .649 | .615 | .578 | .539 | .497 | .454 | .410 | .365 | .320 |
| | | | | | | | | .277 | .235 | .197 | .161 | | | | | | |
| 6.500 | 6.110 | .000 | .390 | .87 | 3.42 | 6.0% | 15 | .681 | .649 | .615 | .578 | .539 | .497 | .454 | .410 | .365 | .320 |

*SIMILAR TO A.T.&T. NEAL-WILKINSON TABLES(SEE SECTION 1)

# *EQUIVALENT RANDOM THEORY: VMR(VARIANCE TO MEAN RATIO)– 3.0 - NO RETRIES (PEAKED TRAFFIC - ROUTE ADVANCEMENT OR DAY-TO-DAY VARIATION)

| HOURS FIRST ATTEMPT VMR=3.0 | HOURS CON-NECT | HOURS RECALL /RETRY | HOURS OVFL | CON-NECT VMR | OVFL VMR | PERCENT BLOCKED (GRADE) | LINE | HOURS CONNECTED TRAFFIC ON LINES 1 | 2 | 3 | 4 | 5 | 6 | 7 | 8 | 9 | 10 |
|---|---|---|---|---|---|---|---|---|---|---|---|---|---|---|---|---|---|
| | | | | | | | | .277 | .235 | .197 | .161 | .130 | | | | | |
| 6.500 | 6.212 | .000 | .288 | .90 | 3.34 | 4.4% | 16 | .681 | .649 | .615 | .578 | .539 | .497 | .454 | .410 | .365 | .320 |
| | | | | | | | | .277 | .235 | .197 | .161 | .130 | .103 | | | | |
| 6.500 | 6.292 | .000 | .208 | .92 | 3.26 | 3.2% | 17 | .681 | .649 | .615 | .578 | .539 | .497 | .454 | .410 | .365 | .320 |
| | | | | | | | | .277 | .235 | .197 | .161 | .130 | .103 | .079 | | | |
| 6.500 | 6.352 | .000 | .148 | .94 | 3.17 | 2.3% | 18 | .681 | .649 | .615 | .578 | .539 | .497 | .454 | .410 | .365 | .320 |
| | | | | | | | | .277 | .235 | 197 | .161 | .130 | .103 | .079 | .060 | | |
| 6.500 | 6.396 | .000 | .104 | .96 | 3.09 | 1.6% | 19 | .681 | .649 | .615 | .578 | .539 | .497 | .454 | .410 | .365 | .320 |
| | | | | | | | | .277 | .235 | .197 | .161 | .130 | .103 | .079 | .060 | .045 | |
| 6.500 | 6.429 | .000 | .071 | .97 | 3.00 | 1.1% | 20 | .681 | .649 | .615 | .578 | .539 | .497 | .454 | .410 | .365 | .320 |
| | | | | | | | | .277 | .235 | .197 | .161 | .130 | .103 | .079 | .060 | .045 | .032 |
| 6.500 | 6.452 | .000 | .048 | .98 | 2.92 | .7% | 21 | .681 | .649 | .615 | .578 | .539 | .497 | .454 | .410 | .365 | .320 |
| | | | | | | | | .277 | .235 | .197 | .161 | .130 | .103 | .079 | .060 | .045 | .032 |
| | | | | | | | | .023 | | | | | | | | | |
| 7.000 | 4.607 | .000 | 2.393 | .56 | 3.65 | 34.2% | 8 | .697 | .667 | .635 | .600 | .563 | .524 | .482 | .439 | | |
| 7.000 | 5.003 | .000 | 1.997 | .60 | 3.68 | 28.5% | 9 | .697 | .667 | .635 | .600 | .563 | .524 | .482 | .439 | .396 | |
| 7.000 | 5.355 | .000 | 1.645 | .65 | 3.69 | 23.5% | 10 | .697 | .667 | .635 | .600 | .563 | .524 | .482 | .439 | .396 | .352 |
| 7.000 | 5.663 | .000 | 1.337 | .69 | 3.69 | 19.1% | 11 | .697 | .667 | .635 | .600 | .563 | .524 | .482 | .439 | .396 | .352 |
| | | | | | | | | .308 | | | | | | | | | |
| 7.000 | 5.929 | .000 | 1.071 | .73 | 3.67 | 15.3% | 12 | .697 | .667 | .635 | .600 | .563 | .524 | .482 | .439 | .396 | .352 |
| | | | | | | | | .308 | .266 | | | | | | | | |
| 7.000 | 6.155 | .000 | .845 | .77 | 3.64 | 12.1% | 13 | .697 | .667 | .635 | .600 | .563 | .524 | .482 | .439 | .396 | .352 |
| | | | | | | | | .308 | .266 | .226 | | | | | | | |
| 7.000 | 6.344 | .000 | .656 | .81 | 3.59 | 9.4% | 14 | .697 | .667 | .635 | .600 | .563 | .524 | .482 | .439 | .396 | .352 |
| | | | | | | | | .308 | .266 | .226 | .189 | | | | | | |
| 7.000 | 6.400 | .000 | .502 | .85 | 3.53 | 7.2% | 15 | .697 | .667 | .635 | .600 | .563 | .524 | .482 | .439 | .396 | .352 |
| | | | | | | | | .308 | .266 | .226 | .189 | .155 | | | | | |
| 7.000 | 6.623 | .000 | .377 | .88 | 3.46 | 5.4% | 16 | .697 | .667 | .635 | .600 | .563 | .524 | .482 | .439 | .396 | .352 |
| | | | | | | | | .308 | .266 | .226 | .189 | .155 | .125 | | | | |
| 7.000 | 6.721 | .000 | .279 | .90 | 3.38 | 4.0% | 17 | .697 | .667 | .635 | .600 | .563 | .524 | .482 | .439 | .396 | .352 |
| | | | | | | | | .308 | .266 | .226 | .189 | .155 | .125 | .098 | | | |
| 7.000 | 6.797 | .000 | .203 | .92 | 3.30 | 2.9% | 18 | .697 | .667 | .635 | .600 | .563 | .524 | .482 | .439 | .396 | .352 |
| | | | | | | | | .308 | .266 | .226 | .189 | .155 | .125 | .098 | .076 | | |
| 7.000 | 6.855 | .000 | .145 | .94 | 3.21 | 2.1% | 19 | .697 | .667 | .635 | .600 | .563 | .524 | .482 | .439 | .396 | .352 |
| | | | | | | | | .308 | .266 | .226 | .189 | .155 | .125 | .098 | .076 | .058 | |
| 7.000 | 6.898 | .000 | .102 | .96 | 3.13 | 1.5% | 20 | .697 | .667 | .635 | .600 | .563 | .524 | .482 | .439 | .396 | .352 |
| | | | | | | | | .308 | .266 | .226 | .189 | .155 | .125 | .098 | .076 | .058 | .043 |
| 7.000 | 6.930 | .000 | .070 | .97 | 3.04 | 1.0% | 21 | .097 | .667 | .635 | .600 | .563 | .524 | .482 | .439 | .396 | .352 |
| | | | | | | | | .308 | .266 | .226 | .189 | 155 | .125 | .008 | .070 | .058 | .043 |
| | | | | | | | | .031 | | | | | | | | | |

*SIMILAR TO A.T.&T. NEAL-WILKINSON TABLES(SEE SECTION 1)

*EQUIVALENT RANDOM THEORY: VMR(VARIANCE TO MEAN RATIO)– 3.0 - NO RETRIES (PEAKED TRAFFIC - ROUTE ADVANCEMENT OR DAY-TO-DAY VARIATION)

| HOURS FIRST ATTEMPT VMR=3.0 | HOURS CON-NECT | HOURS RECALL /RETRY | HOURS OVFL | CON-NECT VMR | OVFL VMR | PERCENT BLOCKED (GRADE) | LINE | HOURS CONNECTED TRAFFIC ON LINES 1 | 2 | 3 | 4 | 5 | 6 | 7 | 8 | 9 | 10 |
|---|---|---|---|---|---|---|---|---|---|---|---|---|---|---|---|---|---|
| 7.000 | 6.952 | .000 | .048 | .98 | 2.96 | .7% | 22 | .697 | .667 | .635 | .600 | .563 | .524 | .482 | .439 | .396 | .352 |
| | | | | | | | | .308 | .266 | .226 | .189 | .155 | .125 | .098 | .076 | .058 | .043 |
| | | | | | | | | .031 | .023 | | | | | | | | |
| 7.500 | 5.201 | .000 | 2.299 | .58 | 3.73 | 30.7% | 9 | .711 | .683 | .653 | .620 | .585 | .548 | .508 | .467 | .425 | |
| 7.500 | 5.583 | .000 | 1.917 | .62 | 3.75 | 25.6% | 10 | .711 | .683 | .653 | .620 | .585 | .548 | .508 | .467 | .425 | .382 |
| 7.500 | 5.921 | .000 | 1.579 | .66 | 3.75 | 21.1% | 11 | .711 | .683 | .653 | .620 | .585 | .548 | .508 | .467 | .425 | .382 |
| | | | | | | | | .338 | | | | | | | | | |
| 7.500 | 6.217 | .000 | 1.283 | .70 | 3.74 | 17.1% | 12 | .711 | .683 | .653 | .620 | .585 | .548 | .508 | .467 | .425 | .382 |
| | | | | | | | | .338 | .296 | | | | | | | | |
| 7.500 | 6.472 | .000 | 1.028 | .74 | 3.72 | 13.7% | 13 | .711 | .683 | .653 | .620 | .585 | .548 | .508 | .467 | .425 | .382 |
| | | | | | | | | .338 | .296 | .255 | | | | | | | |
| 7.500 | 6.689 | .000 | .811 | .78 | 3.68 | 10.8% | 14 | .711 | .683 | .653 | .620 | .585 | .548 | .508 | .467 | .425 | .382 |
| | | | | | | | | .338 | .296 | .255 | .216 | | | | | | |
| 7.500 | 6.869 | .000 | .631 | .82 | 3.63 | 8.4% | 15 | .711 | .683 | .653 | .620 | .585 | .548 | .508 | .467 | .425 | .382 |
| | | | | | | | | .338 | .296 | .255 | .216 | .181 | | | | | |
| 7.500 | 7.017 | .000 | .483 | .85 | 3.57 | 6.4% | 16 | .711 | .683 | .653 | .620 | .585 | .548 | .508 | .467 | .425 | .382 |
| | | | | | | | | .338 | .296 | .255 | .216 | .181 | .148 | | | | |
| 7.500 | 7.136 | .000 | .364 | .88 | 3.50 | 4.8% | 17 | .711 | .683 | .653 | .620 | .585 | .548 | .508 | .467 | .425 | .382 |
| | | | | | | | | .338 | .296 | .255 | .216 | .181 | .148 | .119 | | | |
| 7.500 | 7.231 | .000 | .269 | .91 | 3.42 | 3.6% | 18 | .711 | .683 | .653 | .620 | .585 | .548 | .508 | .467 | .425 | .382 |
| | | | | | | | | .338 | .296 | .255 | .216 | .181 | .148 | .119 | .094 | | |
| 7.500 | 7.304 | .000 | .196 | .93 | 3.33 | 2.6% | 19 | .711 | .683 | .653 | .620 | .585 | .548 | .508 | .467 | .425 | .382 |
| | | | | | | | | .338 | .296 | .255 | .216 | .181 | .148 | .119 | .094 | .073 | |
| 7.500 | 7.359 | .000 | .141 | .94 | 3.25 | 1.9% | 20 | .711 | .683 | .653 | .620 | .585 | .548 | .508 | .467 | .425 | .382 |
| | | | | | | | | .338 | .296 | .255 | .216 | .181 | .148 | .119 | .094 | .073 | .056 |
| 7.500 | 7.401 | .000 | .099 | .96 | 3.16 | 1.3% | 21 | .711 | .683 | .653 | .620 | .585 | .548 | .508 | .467 | .425 | .382 |
| | | | | | | | | .338 | .296 | .255 | .216 | .181 | .148 | .119 | .094 | .073 | .056 |
| | | | | | | | | .042 | | | | | | | | | |
| 7.500 | 7.431 | .000 | .069 | .97 | 3.08 | .9% | 22 | .711 | .683 | .653 | .620 | .585 | .548 | .508 | .467 | .425 | .382 |
| | | | | | | | | .338 | .296 | .255 | .216 | .181 | .148 | .119 | .094 | .073 | .056 |
| | | | | | | | | .042 | .030 | | | | | | | | |
| 7.500 | 7.453 | .000 | .047 | .98 | 2.99 | .6% | 23 | .711 | .683 | .653 | .620 | .585 | .548 | .508 | .467 | .425 | .382 |
| | | | | | | | | .338 | .296 | .255 | .216 | .181 | .148 | .119 | .094 | .073 | .056 |
| | | | | | | | | .042 | .030 | .022 | | | | | | | |
| 8.000 | 5.384 | .000 | 2.616 | .55 | 3.76 | 32.7% | 9 | .724 | .698 | .670 | .639 | .606 | .570 | .533 | .493 | .452 | |
| 8.000 | 5.794 | .000 | 2.206 | .59 | 3.79 | 27.6% | 10 | .724 | .698 | .670 | .639 | .606 | .570 | .533 | .493 | .452 | .410 |
| 8.000 | 6.162 | .000 | 1.838 | .63 | 3.81 | 23.0% | 11 | .724 | .698 | .670 | .639 | .606 | .570 | .533 | .493 | .452 | .410 |
| | | | | | | | | .368 | | | | | | | | | |
| 8.000 | 6.487 | .000 | 1.513 | .67 | 3.81 | 18.9% | 12 | .724 | .698 | .670 | .639 | .606 | .570 | .533 | .493 | .452 | .410 |
| | | | | | | | | .368 | .325 | | | | | | | | |

*SIMILAR TO A.T.&T. NEAL-WILKINSON TABLES(SEE SECTION 1)

*EQUIVALENT RANDOM THEORY: VMR(VARIANCE TO MEAN RATIO)– 3.0 - NO RETRIES (PEAKED TRAFFIC - ROUTE ADVANCEMENT OR DAY-TO-DAY VARIATION)

| HOURS FIRST ATTEMPT VMR=3.0 | HOURS CON-NECT | HOURS RECALL /RETRY | HOURS OVFL | CON-NECT VMR | OVFL VMR | PERCENT BLOCKED (GRADE) | LINE | HOURS CONNECTED TRAFFIC ON LINES 1 | 2 | 3 | 4 | 5 | 6 | 7 | 8 | 9 | 10 |
|---|---|---|---|---|---|---|---|---|---|---|---|---|---|---|---|---|---|
| 8.000 | 6.771 | .000 | 1.229 | .72 | 3.80 | 15.4% | 13 | .724 | .698 | .670 | .639 | .606 | .570 | .533 | .493 | .452 | .410 |
| | | | | | | | | .368 | .325 | .284 | | | | | | | |
| 8.000 | 7.016 | .000 | .984 | .76 | 3.77 | 12.3% | 14 | .724 | .698 | .670 | .639 | .606 | .570 | .533 | .493 | .452 | .410 |
| | | | | | | | | .368 | .325 | .284 | .244 | | | | | | |
| 8.000 | 7.223 | .000 | .777 | .79 | 3.73 | 9.7% | 15 | .724 | .698 | .670 | .639 | .606 | .570 | .533 | .493 | .452 | .410 |
| | | | | | | | | .368 | .325 | .284 | .244 | .207 | | | | | |
| 8.000 | 7.395 | .000 | .605 | .83 | 3.67 | 7.6% | 16 | .724 | .698 | .670 | .639 | .606 | .570 | .533 | .493 | .452 | .410 |
| | | | | | | | | .368 | .325 | .284 | .244 | .207 | .173 | | | | |
| 8.000 | 7.537 | .000 | .463 | .86 | 3.61 | 5.8% | 17 | .724 | .698 | .670 | .639 | .606 | .570 | .533 | .493 | .452 | .410 |
| | | | | | | | | .368 | .325 | .284 | .244 | .207 | .173 | .141 | | | |
| 8.000 | 7.651 | .000 | .349 | .89 | 3.53 | 4.4% | 18 | .724 | .698 | .670 | .639 | .606 | .570 | .533 | .493 | .452 | .410 |
| | | | | | | | | .368 | .325 | .284 | .244 | .207 | .173 | .141 | .114 | | |
| 8.000 | /.741 | .000 | .259 | .91 | 3.45 | 3.2% | 19 | .724 | .698 | .670 | .639 | .606 | .570 | .533 | .493 | .452 | .410 |
| | | | | | | | | .368 | .325 | .284 | .244 | .207 | .173 | .141 | .114 | .090 | |
| 8.000 | 7.810 | .000 | .190 | .93 | 3.37 | 2.4% | 20 | .724 | .698 | .670 | .639 | .606 | .570 | .533 | .493 | .452 | .410 |
| | | | | | | | | .368 | .325 | .284 | .244 | .207 | .173 | .141 | .114 | .090 | .070 |
| 8.000 | 7.864 | .000 | .136 | .95 | 3.28 | 1.7% | 21 | .724 | .698 | .670 | .639 | .606 | .570 | .533 | .493 | .452 | .410 |
| | | | | | | | | .368 | .325 | .284 | .244 | .207 | .173 | .141 | .114 | .090 | .070 |
| | | | | | | | | .053 | | | | | | | | | |
| 8.000 | 7.904 | .000 | .096 | .96 | 3.19 | 1.2% | 22 | .724 | .698 | .670 | .639 | .606 | .570 | .533 | .493 | .452 | .410 |
| | | | | | | | | .368 | .325 | .284 | .244 | .207 | .173 | .141 | .114 | .090 | .070 |
| | | | | | | | | .053 | .040 | | | | | | | | |
| 8.000 | 7.933 | .000 | .067 | .97 | 3.11 | .8% | 23 | .724 | .698 | .670 | .639 | .606 | .570 | .533 | .493 | .452 | .410 |
| | | | | | | | | .368 | .325 | .284 | .244 | .207 | .173 | .141 | .114 | .090 | .070 |
| | | | | | | | | .053 | .040 | .029 | | | | | | | |
| 8.000 | 7.954 | .000 | 046 | .98 | 3.02 | .0% | 24 | .724 | .698 | .670 | .639 | 606 | .570 | .633 | .400 | .452 | .410 |
| | | | | | | | | .368 | .325 | .284 | .244 | .207 | .173 | .141 | .114 | .090 | .070 |
| | | | | | | | | .053 | .040 | .029 | .021 | | | | | | |
| 8.500 | 5.553 | .000 | 2.947 | .52 | 3.79 | 34.7% | 9 | .736 | .711 | .685 | .656 | .624 | .591 | .555 | .517 | .478 | |
| 8.500 | 5.990 | .000 | 2.510 | .56 | 3.83 | 29.5% | 10 | .736 | .711 | .685 | .656 | .624 | .591 | .555 | .517 | .478 | .437 |
| 8.500 | 6.386 | .000 | 2.114 | .60 | 3.86 | 24.9% | 11 | .736 | .711 | .685 | .656 | .624 | .591 | .555 | .517 | .478 | .437 |
| | | | | | | | | .396 | | | | | | | | | |
| 8.500 | 6.740 | .000 | 1.760 | .65 | 3.87 | 20.7% | 12 | .736 | .711 | .685 | .656 | .624 | .591 | .555 | .517 | .478 | .437 |
| | | | | | | | | .396 | .354 | | | | | | | | |
| 8.500 | 7.053 | .000 | 1.447 | .69 | 3.87 | 17.0% | 13 | .736 | .711 | .685 | .656 | .624 | .591 | .555 | .517 | .478 | .437 |
| | | | | | | | | .396 | .354 | .313 | | | | | | | |
| 8.500 | 7.325 | .000 | 1.175 | .73 | 3.85 | 13.8% | 14 | .736 | .711 | .685 | .656 | .624 | .591 | .555 | .517 | .478 | .437 |
| | | | | | | | | .398 | .354 | .313 | .272 | | | | | | |
| 8.500 | 7.559 | .000 | .941 | .77 | 3.81 | 11.1% | 15 | .736 | .711 | .685 | .656 | .624 | 591 | .556 | .517 | .470 | .437 |
| | | | | | | | | .396 | .354 | .313 | .272 | .234 | | | | | |

*SIMILAR TO A.T.&T. NEAL-WILKINSON TABLES(SEE SECTION 1)

# *EQUIVALENT RANDOM THEORY: VMR(VARIANCE TO MEAN RATIO)– 3.0 - NO RETRIES (PEAKED TRAFFIC - ROUTE ADVANCEMENT OR DAY-TO-DAY VARIATION)

| HOURS FIRST ATTEMPT VMR=3.0 | HOURS CON-NECT | HOURS RECALL /RETRY | HOURS OVFL | CON-NECT VMR | OVFL VMR | PERCENT BLOCKED (GRADE) | LINE | HOURS CONNECTED TRAFFIC ON LINES 1 | 2 | 3 | 4 | 5 | 6 | 7 | 8 | 9 | 10 |
|---|---|---|---|---|---|---|---|---|---|---|---|---|---|---|---|---|---|
| 8.500 | 7.757 | .000 | .743 | .80 | 3.77 | 8.7% | 16 | .736 | .711 | .685 | .656 | .624 | .591 | .555 | .517 | .478 | .437 |
| | | | | | | | | .396 | .354 | .313 | .272 | .234 | .198 | | | | |
| 8.500 | 7.922 | .000 | .578 | .84 | 3.71 | 6.8% | 17 | .736 | .711 | .685 | .656 | .624 | .591 | .555 | .517 | .478 | .437 |
| | | | | | | | | .396 | .354 | .313 | .272 | .234 | .198 | .165 | | | |
| 8.500 | 8.056 | .000 | .444 | .87 | 3.64 | 5.2% | 18 | .736 | .711 | .685 | .656 | .624 | .591 | .555 | .517 | .478 | .437 |
| | | | | | | | | .396 | .354 | .313 | .272 | .234 | .198 | .165 | .135 | | |
| 8.500 | 8.165 | .000 | .335 | .89 | 3.57 | 3.9% | 19 | .736 | .711 | .685 | .656 | .624 | .591 | .555 | .517 | .478 | .437 |
| | | | | | | | | .396 | .354 | .313 | .272 | .234 | .198 | .165 | .135 | .109 | |
| 8.500 | 8.251 | .000 | .249 | .91 | 3.49 | 2.9% | 20 | .736 | .711 | .685 | .656 | .624 | .591 | .555 | .517 | .478 | .437 |
| | | | | | | | | .396 | .354 | .313 | .272 | .234 | .198 | .165 | .135 | .109 | .086 |
| 8.500 | 8.318 | .000 | .182 | .93 | 3.40 | 2.1% | 21 | .736 | .711 | .685 | .656 | .624 | .591 | .555 | .517 | .478 | .437 |
| | | | | | | | | .396 | .354 | .313 | .272 | .234 | .198 | .165 | .135 | .109 | .086 |
| | | | | | | | | .067 | | | | | | | | | |
| 8.500 | 8.368 | .000 | .132 | .95 | 3.31 | 1.5% | 22 | .736 | .711 | .685 | .656 | .624 | .591 | .555 | .517 | .478 | .437 |
| | | | | | | | | .396 | .354 | .313 | .272 | .234 | .198 | .165 | .135 | .109 | .086 |
| | | | | | | | | .067 | .051 | | | | | | | | |
| 8.500 | 8.407 | .000 | .093 | .96 | 3.23 | 1.1% | 23 | .736 | .711 | .685 | .656 | .624 | .591 | .555 | .517 | .478 | .437 |
| | | | | | | | | .396 | .354 | .313 | .272 | .234 | .198 | .165 | .135 | .109 | .086 |
| | | | | | | | | .067 | .051 | .038 | | | | | | | |
| 8.500 | 8.435 | .000 | .065 | .97 | 3.14 | .8% | 24 | .736 | .711 | .685 | .656 | .624 | .591 | .555 | .517 | .478 | .437 |
| | | | | | | | | .396 | .354 | .313 | .272 | .234 | .198 | .165 | .135 | .109 | .086 |
| | | | | | | | | .067 | .051 | .038 | .028 | | | | | | |
| 8.500 | 8.455 | .000 | .045 | .98 | 3.06 | .5% | 25 | .736 | .711 | .685 | .656 | .624 | .591 | .555 | .517 | .478 | .437 |
| | | | | | | | | .396 | .354 | .313 | .272 | .234 | .198 | .165 | .135 | .109 | .086 |
| | | | | | | | | .067 | .051 | .038 | .028 | .020 | | | | | |
| 9.000 | 6.173 | .000 | 2.827 | .54 | 3.87 | 31.4% | 10 | .747 | .724 | .699 | .671 | .642 | .610 | .576 | .540 | .502 | .463 |
| 9.000 | 6.595 | .000 | 2.405 | .58 | 3.90 | 26.7% | 11 | .747 | .724 | .699 | .671 | .642 | .610 | .576 | .540 | .502 | .463 |
| | | | | | | | | .422 | | | | | | | | | |
| 9.000 | 6.976 | .000 | 2.024 | .62 | 3.92 | 22.5% | 12 | .747 | .724 | .699 | .671 | .642 | .610 | .576 | .540 | .502 | .463 |
| | | | | | | | | .422 | .381 | | | | | | | | |
| 9.000 | 7.317 | .000 | 1.683 | .66 | 3.93 | 18.7% | 13 | .747 | .724 | .699 | .671 | .642 | .610 | .576 | .540 | .502 | .463 |
| | | | | | | | | .422 | .381 | .340 | | | | | | | |
| 9.000 | 7.617 | .000 | 1.383 | .70 | 3.92 | 15.4% | 14 | .747 | .724 | .699 | .671 | .642 | .610 | .576 | .540 | .502 | .463 |
| | | | | | | | | .422 | .381 | .340 | .300 | | | | | | |
| 9.000 | 7.878 | .000 | 1.122 | .74 | 3.89 | 12.5% | 15 | .747 | .724 | .699 | .671 | .642 | .610 | .576 | .540 | .502 | .463 |
| | | | | | | | | .422 | .381 | .340 | .300 | .261 | | | | | |
| 9.000 | 8.102 | .000 | .898 | .78 | 3.86 | 10.0% | 16 | .747 | .724 | .699 | .671 | .642 | .610 | .576 | .540 | .502 | .463 |
| | | | | | | | | .422 | .381 | .340 | .300 | .261 | .224 | | | | |
| 9.000 | 8.290 | .000 | .710 | .81 | 3.81 | 7.9% | 17 | .747 | .724 | .699 | .671 | .642 | .610 | .576 | .540 | .502 | .463 |
| | | | | | | | | .422 | .381 | .340 | .300 | .261 | .224 | .189 | | | |

*SIMILAR TO A.T.&T. NEAL-WILKINSON TABLES(SEE SECTION 1)

*EQUIVALENT RANDOM THEORY: VMR(VARIANCE TO MEAN RATIO)– 3.0 - NO RETRIES (PEAKED TRAFFIC - ROUTE ADVANCEMENT OR DAY-TO-DAY VARIATION)

| HOURS FIRST ATTEMPT VMR=3.0 | HOURS CON-NECT | HOURS RECALL /RETRY | HOURS OVFL | CON-NECT VMR | OVFL VMR | PERCENT BLOCKED (GRADE) | LINE | HOURS CONNECTED TRAFFIC ON LINES 1 | 2 | 3 | 4 | 5 | 6 | 7 | 8 | 9 | 10 |
|---|---|---|---|---|---|---|---|---|---|---|---|---|---|---|---|---|---|
| 9.000 | 8.448 | .000 | .552 | .84 | 3.75 | 6.1% | 18 | .747 | .724 | .699 | .671 | .642 | .610 | .576 | .540 | .502 | .463 |
| | | | | | | | | .422 | .381 | .340 | .300 | .261 | .224 | .189 | .157 | | |
| 9.000 | 8.576 | .000 | .424 | .87 | 3.68 | 4.7% | 19 | .747 | .724 | .699 | .671 | .642 | .610 | .576 | .540 | .502 | .463 |
| | | | | | | | | .422 | .381 | .340 | .300 | .261 | .224 | .189 | .157 | .128 | |
| 9.000 | 8.679 | .000 | .321 | .90 | 3.60 | 3.6% | 20 | .747 | .724 | .699 | .671 | .642 | .610 | .576 | .540 | .502 | .463 |
| | | | | | | | | .422 | .381 | .340 | .300 | .261 | .224 | .189 | .157 | .128 | .103 |
| 9.000 | 8.761 | .000 | .239 | .92 | 3.52 | 2.7% | 21 | .747 | .724 | .699 | .671 | .642 | .610 | .576 | .540 | .502 | .463 |
| | | | | | | | | .422 | .381 | .340 | .300 | .261 | .224 | .189 | .157 | .128 | .103 |
| | | | | | | | | .082 | | | | | | | | | |
| 9.000 | 8.825 | .000 | .175 | .94 | 3.43 | 1.9% | 22 | .747 | .724 | .699 | .671 | .642 | .610 | .576 | .540 | .502 | .463 |
| | | | | | | | | .422 | .381 | .340 | .300 | .261 | .224 | .189 | .157 | .128 | .103 |
| | | | | | | | | .082 | .064 | | | | | | | | |
| 9.000 | 8.873 | .000 | .127 | .95 | 3.34 | 1.4% | 23 | .747 | .724 | .699 | .671 | .642 | .610 | .576 | .540 | .502 | .463 |
| | | | | | | | | .422 | .381 | .340 | .300 | .261 | .224 | .189 | .157 | .128 | .103 |
| | | | | | | | | .082 | .064 | .049 | | | | | | | |
| 9.000 | 8.910 | .000 | .090 | .96 | 3.26 | 1.0% | 24 | .747 | .724 | .699 | .671 | .642 | .610 | .576 | .540 | .502 | .463 |
| | | | | | | | | .422 | .381 | .340 | .300 | .261 | .224 | .189 | .157 | .128 | .103 |
| | | | | | | | | .082 | .064 | .049 | .036 | | | | | | |
| 9.000 | 8.937 | .000 | .063 | .97 | 3.17 | .7% | 25 | .747 | .724 | .699 | .671 | .642 | .610 | .576 | .540 | .502 | .463 |
| | | | | | | | | .422 | .381 | .340 | .300 | .261 | .224 | .189 | .157 | .128 | .103 |
| | | | | | | | | .082 | .064 | .049 | .036 | .027 | | | | | |
| 9.500 | 6.342 | .000 | 3.158 | .51 | 3.89 | 33.2% | 10 | .757 | .735 | .711 | .686 | .658 | .628 | .595 | .561 | .525 | .487 |
| 9.500 | 6.790 | .000 | 2.710 | .55 | 3.94 | 28.5% | 11 | .757 | .735 | .711 | .686 | .658 | .628 | .595 | .561 | .525 | .487 |
| | | | | | | | | .448 | | | | | | | | | |
| 9.500 | 7.198 | .000 | 2.302 | .59 | 3.96 | 24.2% | 12 | .757 | .735 | .711 | .686 | .658 | .628 | .595 | .561 | .525 | .487 |
| | | | | | | | | .448 | .400 | | | | | | | | |
| 9.500 | 7.565 | .000 | 1.935 | .63 | 3.98 | 20.4% | 13 | .757 | .735 | .711 | .686 | .658 | .628 | .595 | .561 | .525 | .487 |
| | | | | | | | | .448 | .408 | .367 | | | | | | | |
| 9.500 | 7.892 | .000 | 1.608 | .67 | 3.98 | 16.9% | 14 | .757 | .735 | .711 | .686 | .658 | .628 | .595 | .561 | .525 | .487 |
| | | | | | | | | .448 | .408 | .367 | .327 | | | | | | |
| 9.500 | 8.180 | .000 | 1.320 | .71 | 3.97 | 13.9% | 15 | .757 | .735 | .711 | .686 | .658 | .628 | .595 | .561 | .525 | .487 |
| | | | | | | | | .448 | .408 | .367 | .327 | .288 | | | | | |
| 9.500 | 8.430 | .000 | 1.070 | .75 | 3.94 | 11.3% | 16 | .757 | .735 | .711 | .686 | .658 | .628 | .595 | .561 | .525 | .487 |
| | | | | | | | | .448 | .408 | .367 | .327 | .288 | .250 | | | | |
| 9.500 | 8.643 | .000 | .857 | .79 | 3.90 | 9.0% | 17 | .757 | .735 | .711 | .686 | .658 | .628 | .595 | .561 | .525 | .487 |
| | | | | | | | | .448 | .408 | .367 | .327 | .288 | .250 | .214 | | | |
| 9.500 | 8.824 | .000 | .676 | .82 | 3.84 | 7.1% | 18 | .757 | .735 | .711 | .686 | .658 | .628 | .595 | .561 | .525 | .487 |
| | | | | | | | | .448 | .408 | .367 | .327 | .288 | .250 | .214 | .180 | | |
| 9.500 | 8.973 | .000 | .527 | .85 | 3.78 | 5.5% | 19 | .757 | .735 | .711 | .686 | .658 | .628 | .595 | .561 | .525 | .487 |
| | | | | | | | | .448 | .400 | .367 | .327 | .288 | .250 | .214 | .180 | .150 | |

*SIMILAR TO A.T.&T. NEAL-WILKINSON TABLES(SEE SECTION 1)

*EQUIVALENT RANDOM THEORY: VMR(VARIANCE TO MEAN RATIO)– 3.0 - NO RETRIES (PEAKED TRAFFIC - ROUTE ADVANCEMENT OR DAY-TO-DAY VARIATION)

| HOURS FIRST ATTEMPT VMR=3.0 | HOURS CON-NECT | HOURS RECALL /RETRY | HOURS OVFL | CON-NECT VMR | OVFL VMR | PERCENT BLOCKED (GRADE) | LINE | HOURS CONNECTED TRAFFIC ON LINES 1 | 2 | 3 | 4 | 5 | 6 | 7 | 8 | 9 | 10 |
|---|---|---|---|---|---|---|---|---|---|---|---|---|---|---|---|---|---|
| 9.500 | 9.095 | .000 | .405 | .88 | 3.71 | 4.3% | 20 | .757 | .735 | .711 | .686 | .658 | .628 | .595 | .561 | .525 | .487 |
| | | | | | | | | .448 | .408 | .367 | .327 | .288 | .250 | .214 | .180 | .150 | .122 |
| 9.500 | 9.194 | .000 | .306 | .90 | 3.63 | 3.2% | 21 | .757 | .735 | .711 | .686 | .658 | .628 | .595 | .561 | .525 | .487 |
| | | | | | | | | .448 | .408 | .367 | .327 | .288 | .250 | .214 | .180 | .150 | .122 |
| | | | | | | | | .098 | | | | | | | | | |
| 9.500 | 9.272 | .000 | .228 | .92 | 3.55 | 2.4% | 22 | .757 | .735 | .711 | .686 | .658 | .628 | .595 | .561 | .525 | .487 |
| | | | | | | | | .448 | .408 | .367 | .327 | .288 | .250 | .214 | .180 | .150 | .122 |
| | | | | | | | | .098 | .078 | | | | | | | | |
| 9.500 | 9.332 | .000 | .168 | .94 | 3.46 | 1.8% | 23 | .757 | .735 | .711 | .686 | .658 | .628 | .595 | .561 | .525 | .487 |
| | | | | | | | | .448 | .408 | .367 | .327 | .288 | .250 | .214 | .180 | .150 | .122 |
| | | | | | | | | .098 | .078 | .061 | | | | | | | |
| 9.500 | 9.378 | .000 | .122 | .95 | 3.37 | 1.3% | 24 | .757 | .735 | .711 | .686 | .658 | .628 | .595 | .561 | .525 | .487 |
| | | | | | | | | .448 | .408 | .367 | .327 | .288 | .250 | .214 | .180 | .150 | .122 |
| | | | | | | | | .098 | .078 | .061 | .046 | | | | | | |
| 9.500 | 9.413 | .000 | .087 | .97 | 3.28 | .9% | 25 | .757 | .735 | .711 | .686 | .658 | .628 | .595 | .561 | .525 | .487 |
| | | | | | | | | .448 | .408 | .367 | .327 | .288 | .250 | .214 | .180 | .150 | .122 |
| | | | | | | | | .098 | .078 | .061 | .046 | .035 | | | | | |
| 9.500 | 9.439 | .000 | .061 | .97 | 3.20 | .6% | 26 | .757 | .735 | .711 | .686 | .658 | .628 | .595 | .561 | .525 | .487 |
| | | | | | | | | .448 | .408 | .367 | .327 | .288 | .250 | .214 | .180 | .150 | .122 |
| | | | | | | | | .098 | .078 | .061 | .046 | .035 | .026 | | | | |
| 10.000 | 6.519 | .000 | 3.481 | .49 | 3.88 | 34.8% | 10 | .768 | .748 | .725 | .701 | .675 | .646 | .615 | .582 | .548 | .511 |
| 10.000 | 6.992 | .000 | 3.008 | .53 | 3.93 | 30.1% | 11 | .768 | .748 | .725 | .701 | .675 | .646 | .615 | .582 | .548 | .511 |
| | | | | | | | | .473 | | | | | | | | | |
| 10.000 | 7.425 | .000 | 2.575 | .57 | 3.97 | 25.7% | 12 | .768 | .748 | .725 | .701 | .675 | .646 | .615 | .582 | .548 | .511 |
| | | | | | | | | .473 | .433 | | | | | | | | |
| 10.000 | 7.819 | .000 | 2.181 | .61 | 3.99 | 21.8% | 13 | .768 | .748 | .725 | .701 | .675 | .646 | .615 | .582 | .548 | .511 |
| | | | | | | | | .473 | .433 | .393 | | | | | | | |
| 10.000 | 8.172 | .000 | 1.828 | .65 | 4.00 | 18.3% | 14 | .768 | .748 | .725 | .701 | .675 | .646 | .615 | .582 | .548 | .511 |
| | | | | | | | | .473 | .433 | .393 | .353 | | | | | | |
| 10.000 | 8.486 | .000 | 1.514 | .69 | 4.00 | 15.1% | 15 | .768 | .748 | .725 | .701 | .675 | .646 | .615 | .582 | .548 | .511 |
| | | | | | | | | .473 | .433 | .393 | .353 | .314 | | | | | |
| 10.000 | 8.760 | .000 | 1.240 | .73 | 3.98 | 12.4% | 16 | .768 | .748 | .725 | .701 | .675 | .646 | .615 | .582 | .548 | .511 |
| | | | | | | | | .473 | .433 | .393 | .353 | .314 | .275 | | | | |
| 10.000 | 8.998 | .000 | 1.002 | .76 | 3.95 | 10.0% | 17 | .768 | .748 | .725 | .701 | .675 | .646 | .615 | .582 | .548 | .511 |
| | | | | | | | | .473 | .433 | .393 | .353 | .314 | .275 | .238 | | | |
| 10.000 | 9.200 | .000 | .800 | .80 | 3.90 | 8.0% | 18 | .768 | .748 | .725 | .701 | .675 | .646 | .615 | .582 | .548 | .511 |
| | | | | | | | | .473 | .433 | .393 | .353 | .314 | .275 | .238 | .202 | | |
| 10.000 | 9.370 | .000 | .630 | .83 | 3.85 | 6.3% | 19 | .768 | .748 | .725 | .701 | .675 | .646 | .615 | .582 | .548 | .511 |
| | | | | | | | | .473 | .433 | .393 | .353 | .314 | .275 | .238 | .202 | .170 | |
| 10.000 | 9.511 | .000 | .489 | .86 | 3.78 | 4.9% | 20 | .768 | .748 | .725 | .701 | .675 | .646 | .615 | .582 | .548 | .511 |

*SIMILAR TO A.T.&T. NEAL-WILKINSON TABLES(SEE SECTION 1)

*EQUIVALENT RANDOM THEORY: VMR(VARIANCE TO MEAN RATIO)– 3.0 - NO RETRIES (PEAKED TRAFFIC - ROUTE ADVANCEMENT OR DAY-TO-DAY VARIATION)

| HOURS FIRST ATTEMPT VMR=3.0 | HOURS CON-NECT | HOURS RECALL /RETRY | HOURS OVFL | CON-NECT VMR | OVFL VMR | PERCENT BLOCKED (GRADE) | LINE | HOURS CONNECTED TRAFFIC ON LINES 1 | 2 | 3 | 4 | 5 | 6 | 7 | 8 | 9 | 10 |
|---|---|---|---|---|---|---|---|---|---|---|---|---|---|---|---|---|---|
| | | | | | | | | .473 | .433 | .393 | .353 | .314 | .275 | .238 | .202 | .170 | .141 |
| 10.000 | 9.626 | .000 | .374 | .89 | 3.70 | 3.7% | 21 | .768 | .748 | .725 | .701 | .675 | .646 | .615 | .582 | .548 | .511 |
| | | | | | | | | .473 | .433 | .393 | .353 | .314 | .275 | .238 | .202 | .170 | .141 |
| | | | | | | | | .115 | | | | | | | | | |
| 10.000 | 9.717 | .000 | .283 | .91 | 3.62 | 2.8% | 22 | .768 | .748 | .725 | .701 | .675 | .646 | .615 | .582 | .548 | .511 |
| | | | | | | | | .473 | .433 | .393 | .353 | .314 | .275 | .238 | .202 | .170 | .141 |
| | | | | | | | | .115 | .092 | | | | | | | | |
| 10.000 | 9.790 | .000 | .210 | .93 | 3.54 | 2.1% | 23 | .768 | .748 | .725 | .701 | .675 | .646 | .615 | .582 | .548 | .511 |
| | | | | | | | | .473 | .433 | .393 | .353 | .314 | .275 | .238 | .202 | .170 | .141 |
| | | | | | | | | .115 | .092 | .072 | | | | | | | |
| 10.000 | 9.846 | .000 | .154 | .94 | 3.45 | 1.5% | 24 | .768 | .748 | .725 | .701 | .675 | .646 | .615 | .582 | .548 | .511 |
| | | | | | | | | .473 | .433 | .393 | .353 | .314 | .275 | .238 | .202 | .170 | .141 |
| | | | | | | | | .115 | .092 | .072 | .056 | | | | | | |
| 10.000 | 9.889 | .000 | .111 | .96 | 3.36 | 1.1% | 25 | .768 | .748 | .725 | .701 | .675 | .646 | .615 | .582 | .548 | .511 |
| | | | | | | | | .473 | .433 | .393 | .353 | .314 | .275 | .238 | .202 | .170 | .141 |
| | | | | | | | | .115 | .092 | .072 | .056 | .043 | | | | | |
| 10.000 | 9.921 | .000 | .079 | .97 | 3.28 | .8% | 26 | .768 | .748 | .725 | .701 | .675 | .646 | .615 | .582 | .548 | .511 |
| | | | | | | | | .473 | .433 | .393 | .353 | .314 | .275 | .238 | .202 | .170 | .141 |
| | | | | | | | | .115 | .092 | .072 | .056 | .043 | .032 | | | | |
| 10.000 | 9.944 | .000 | .056 | .98 | 3.19 | .6% | 27 | .768 | .748 | .725 | .701 | .675 | .646 | .615 | .582 | .548 | .511 |
| | | | | | | | | .473 | .433 | .393 | .353 | .314 | .275 | .238 | .202 | .170 | .141 |
| | | | | | | | | .115 | .092 | .072 | .056 | .043 | .032 | .024 | | | |
| 11.000 | 7.320 | .000 | 3.680 | .48 | 3.98 | 33.5% | 11 | .785 | .766 | .746 | .725 | .701 | .675 | .648 | .618 | .586 | .553 |
| | | | | | | | | .517 | | | | | | | | | |
| 11.000 | 7.801 | .000 | 3.199 | .52 | 4.03 | 29.1% | 12 | .785 | .766 | .746 | .725 | .701 | .675 | .648 | .618 | .586 | .553 |
| | | | | | | | | .517 | .481 | | | | | | | | |
| 11.000 | 8.243 | .000 | 2.757 | .56 | 4.07 | 25.1% | 13 | .785 | .766 | .746 | .725 | .701 | .675 | .648 | .618 | .586 | .553 |
| | | | | | | | | .517 | .481 | .443 | | | | | | | |
| 11.000 | 8.647 | .000 | 2.353 | .60 | 4.09 | 21.4% | 14 | .785 | .766 | .746 | .725 | .701 | .675 | .648 | .618 | .586 | .553 |
| | | | | | | | | .517 | .481 | .443 | .404 | | | | | | |
| 11.000 | 9.012 | .000 | 1.988 | .64 | 4.11 | 18.1% | 15 | .785 | .766 | .746 | .725 | .701 | .675 | .648 | .618 | .586 | .553 |
| | | | | | | | | .517 | .481 | .443 | .404 | .365 | | | | | |
| 11.000 | 9.338 | .000 | 1.662 | .67 | 4.11 | 15.1% | 16 | .785 | .766 | .746 | .725 | .701 | .675 | .648 | .618 | .586 | .553 |
| | | | | | | | | .517 | .481 | .443 | .404 | .365 | .326 | | | | |
| 11.000 | 9.626 | .000 | 1.374 | .71 | 4.09 | 12.5% | 17 | .785 | .766 | .746 | .725 | .701 | .675 | .648 | .618 | .586 | .553 |
| | | | | | | | | .517 | .481 | .443 | .404 | .365 | .326 | .288 | | | |
| 11.000 | 9.877 | .000 | 1.123 | .75 | 4.07 | 10.2% | 18 | .785 | .766 | .746 | .725 | .701 | .675 | .648 | .618 | .586 | .553 |
| | | | | | | | | .517 | .481 | .443 | .404 | .365 | .326 | .288 | .251 | | |
| 11.000 | 10.004 | .000 | .906 | .78 | 4.02 | 8.2% | 19 | .785 | .766 | .746 | .725 | .701 | .675 | .648 | .618 | .586 | .553 |
| | | | | | | | | .517 | .481 | .443 | .404 | .365 | .326 | .288 | .251 | .216 | |

*SIMILAR TO A.T.&T. NEAL-WILKINSON TABLES(SEE SECTION 1)

*EQUIVALENT RANDOM THEORY: VMR(VARIANCE TO MEAN RATIO)– 3.0 - NO RETRIES (PEAKED TRAFFIC - ROUTE ADVANCEMENT OR DAY-TO-DAY VARIATION)

| HOURS FIRST ATTEMPT VMR=3.0 | HOURS CON-NECT | HOURS RECALL /RETRY | HOURS OVFL | CON-NECT VMR | OVFL VMR | PERCENT BLOCKED (GRADE) | LINE | HOURS CONNECTED TRAFFIC ON LINES 1 | 2 | 3 | 4 | 5 | 6 | 7 | 8 | 9 | 10 |
|---|---|---|---|---|---|---|---|---|---|---|---|---|---|---|---|---|---|
| 11.000 | 10.278 | .000 | .722 | .82 | 3.97 | 6.6% | 20 | .785 | .766 | .746 | .725 | .701 | .675 | .648 | .618 | .586 | .553 |
| | | | | | | | | .517 | .481 | .443 | .404 | .365 | .326 | .288 | .251 | .216 | .184 |
| 11.000 | 10.432 | .000 | .568 | .85 | 3.91 | 5.2% | 21 | .785 | .766 | .746 | .725 | .701 | .675 | .648 | .618 | .586 | .553 |
| | | | | | | | | .517 | .481 | .443 | .404 | .365 | .326 | .288 | .251 | .216 | .184 |
| | | | | | | | | .154 | | | | | | | | | |
| 11.000 | 10.558 | .000 | .442 | .87 | 3.84 | 4.0% | 22 | .785 | .766 | .746 | .725 | .701 | .675 | .648 | .618 | .586 | .553 |
| | | | | | | | | .517 | .481 | .443 | .404 | .365 | .326 | .288 | .251 | .216 | .184 |
| | | | | | | | | .154 | .127 | | | | | | | | |
| 11.000 | 10.662 | .000 | .338 | .90 | 3.76 | 3.1% | 23 | .785 | .766 | .746 | .725 | .701 | .675 | .648 | .618 | .586 | .553 |
| | | | | | | | | .517 | .481 | .443 | .404 | .365 | .326 | .288 | .251 | .216 | .184 |
| | | | | | | | | .154 | .127 | .103 | | | | | | | |
| 11.000 | 10.744 | .000 | .256 | .92 | 3.68 | 2.3% | 24 | .785 | .766 | .746 | .725 | .701 | .675 | .648 | .618 | .586 | .553 |
| | | | | | | | | .517 | .481 | .443 | .404 | .365 | .326 | .288 | .251 | .216 | .184 |
| | | | | | | | | .154 | .127 | .103 | .083 | | | | | | |
| 11.000 | 10.809 | .000 | .191 | .93 | 3.59 | 1.7% | 25 | .785 | .766 | .746 | .725 | .701 | .675 | .648 | .618 | .586 | .553 |
| | | | | | | | | .517 | .481 | .443 | .404 | .365 | .326 | .288 | .251 | .216 | .184 |
| | | | | | | | | .154 | .127 | .103 | .083 | .065 | | | | | |
| 11.000 | 10.860 | .000 | .140 | .95 | 3.50 | 1.3% | 26 | .785 | .766 | .746 | .725 | .701 | .675 | .648 | .618 | .586 | .553 |
| | | | | | | | | .517 | .481 | .443 | .404 | .365 | .326 | .288 | .251 | .216 | .184 |
| | | | | | | | | .154 | .127 | .103 | .083 | .065 | .050 | | | | |
| 11.000 | 10.898 | .000 | .102 | .96 | 3.41 | .9% | 27 | .785 | .766 | .746 | .725 | .701 | .675 | .648 | .618 | .586 | .553 |
| | | | | | | | | .517 | .481 | .443 | .404 | .365 | .326 | .288 | .251 | .216 | .184 |
| | | | | | | | | .154 | .127 | .103 | .083 | .065 | .050 | .039 | | | |
| 11.000 | 10.927 | .000 | .073 | .97 | 3.32 | .7% | 28 | .785 | .766 | .746 | .725 | .701 | .675 | .648 | .618 | .586 | .553 |
| | | | | | | | | .517 | .481 | .443 | .404 | .365 | .326 | .288 | .251 | .216 | .184 |
| | | | | | | | | .154 | .127 | .103 | .083 | .065 | .050 | .039 | .029 | | |
| 12.000 | 8.130 | .000 | 3.870 | .48 | 4.07 | 32.3% | 12 | .799 | .783 | .765 | .745 | .724 | .701 | .676 | .649 | .620 | .589 |
| | | | | | | | | .557 | .523 | | | | | | | | |
| 12.000 | 8.617 | .000 | 3.383 | .51 | 4.12 | 28.2% | 13 | .799 | .783 | .765 | .745 | .724 | .701 | .676 | .649 | .620 | .589 |
| | | | | | | | | .557 | .523 | .487 | | | | | | | |
| 12.000 | 9.068 | .000 | 2.932 | .55 | 4.16 | 24.4% | 14 | .799 | .783 | .765 | .745 | .724 | .701 | .676 | .649 | .620 | .589 |
| | | | | | | | | .557 | .523 | .487 | .451 | | | | | | |
| 12.000 | 9.481 | .000 | 2.519 | .59 | 4.19 | 21.0% | 15 | .799 | .783 | .765 | .745 | .724 | .701 | .676 | .649 | .620 | .589 |
| | | | | | | | | .557 | .523 | .487 | .451 | .413 | | | | | |
| 12.000 | 9.856 | .000 | 2.144 | .62 | 4.21 | 17.9% | 16 | .799 | .783 | .765 | .745 | .724 | .701 | .676 | .649 | .620 | .589 |
| | | | | | | | | .557 | .523 | .487 | .451 | .413 | .375 | | | | |
| 12.000 | 10.194 | .000 | 1.806 | .66 | 4.21 | 15.1% | 17 | .799 | .783 | .765 | .745 | .724 | .701 | .676 | .649 | .620 | .589 |
| | | | | | | | | .557 | .523 | .487 | .451 | .413 | .375 | .337 | | | |
| 12.000 | 10.494 | .000 | 1.506 | .70 | 4.20 | 12.6% | 18 | .799 | .783 | .765 | .745 | .724 | .701 | .676 | .649 | .620 | .589 |
| | | | | | | | | .557 | .523 | .487 | .451 | .413 | .375 | .337 | .300 | | |

*SIMILAR TO A.T.&T. NEAL-WILKINSON TABLES(SEE SECTION 1)

*EQUIVALENT RANDOM THEORY: VMR(VARIANCE TO MEAN RATIO)– 3.0 - NO RETRIES (PEAKED TRAFFIC - ROUTE ADVANCEMENT OR DAY-TO-DAY VARIATION)

| HOURS FIRST ATTEMPT VMR = 3.0 | HOURS CON-NECT | HOURS RECALL /RETRY | HOURS OVFL | CON-NECT VMR | OVFL VMR | PERCENT BLOCKED (GRADE) | LINE | HOURS CONNECTED TRAFFIC ON LINES 1 | 2 | 3 | 4 | 5 | 6 | 7 | 8 | 9 | 10 |
|---|---|---|---|---|---|---|---|---|---|---|---|---|---|---|---|---|---|
| 12.000 | 10.758 | .000 | 1.242 | .74 | 4.18 | 10.4% | 19 | .799 | .783 | .765 | .745 | .724 | .701 | .676 | .649 | .620 | .589 |
| | | | | | | | | .557 | .523 | .487 | .451 | .413 | .375 | .337 | .300 | .264 | |
| 12.000 | 10.987 | .000 | 1.013 | .77 | 4.14 | 8.4% | 20 | .799 | .783 | .765 | .745 | .724 | .701 | .676 | .649 | .620 | .589 |
| | | | | | | | | .557 | .523 | .487 | .451 | .413 | .375 | .337 | .300 | .264 | .229 |
| 12.000 | 11.184 | .000 | .816 | .80 | 4.09 | 6.8% | 21 | .799 | .783 | .765 | .745 | .724 | .701 | .676 | .649 | .620 | .589 |
| | | | | | | | | .557 | .523 | .487 | .451 | .413 | .375 | .337 | .300 | .264 | .229 |
| | | | | | | | | .197 | | | | | | | | | |
| 12.000 | 11.350 | .000 | .650 | .83 | 4.03 | 5.4% | 22 | .799 | .783 | .765 | .745 | .724 | .701 | .676 | .649 | .620 | .589 |
| | | | | | | | | .557 | .523 | .487 | .451 | .413 | .375 | .337 | .300 | .264 | .229 |
| | | | | | | | | .197 | .166 | | | | | | | | |
| 12.000 | 11.489 | .000 | .511 | .86 | 3.96 | 4.3% | 23 | .799 | .783 | .765 | .745 | .724 | .701 | .676 | .649 | .620 | .589 |
| | | | | | | | | .557 | .523 | .487 | .451 | .413 | .375 | .337 | .300 | .264 | .229 |
| | | | | | | | | .197 | .166 | .139 | | | | | | | |
| 12.000 | 11.603 | .000 | .397 | .89 | 3.89 | 3.3% | 24 | .799 | .783 | .765 | .745 | .724 | .701 | .676 | .649 | .620 | .589 |
| | | | | | | | | .557 | .523 | .487 | .451 | .413 | .375 | .337 | .300 | .264 | .229 |
| | | | | | | | | .197 | .166 | .139 | .114 | | | | | | |
| 12.000 | 11.696 | .000 | .304 | .91 | 3.81 | 2.5% | 25 | .799 | .783 | .765 | .745 | .724 | .701 | .676 | .649 | .620 | .589 |
| | | | | | | | | .557 | .523 | .487 | .451 | .413 | .375 | .337 | .300 | .264 | .229 |
| | | | | | | | | .197 | .166 | .139 | .114 | .093 | | | | | |
| 12.000 | 11.770 | .000 | .230 | .93 | 3.72 | 1.9% | 26 | .799 | .783 | .765 | .745 | .724 | .701 | .676 | .649 | .620 | .589 |
| | | | | | | | | .557 | .523 | .487 | .451 | .413 | .375 | .337 | .300 | .264 | .229 |
| | | | | | | | | .197 | .166 | .139 | .114 | .093 | .074 | | | | |
| 12.000 | 11.828 | .000 | .172 | .94 | 3.63 | 1.4% | 27 | .799 | .783 | .765 | .745 | .724 | .701 | .676 | .649 | .620 | .589 |
| | | | | | | | | .557 | .523 | .487 | .451 | .413 | .375 | .337 | .300 | .264 | .229 |
| | | | | | | | | .197 | .166 | .139 | .114 | .093 | .074 | .058 | | | |
| 12.000 | 11.873 | .000 | .127 | .95 | 3.54 | 1.1% | 28 | .799 | .783 | .765 | .745 | .724 | .701 | .676 | .649 | .620 | .580 |
| | | | | | | | | .557 | .523 | .487 | .451 | .413 | .375 | .337 | .300 | .264 | .229 |
| | | | | | | | | .197 | .166 | .139 | .114 | .093 | .074 | .058 | .045 | | |
| 12.000 | 11.908 | .000 | .092 | .97 | 3.45 | .8% | 29 | .799 | .783 | .765 | .745 | .724 | .701 | .676 | .649 | .620 | .589 |
| | | | | | | | | .557 | .523 | .487 | .451 | .413 | .375 | .337 | .300 | .264 | .229 |
| | | | | | | | | .197 | .166 | .139 | .114 | .093 | .074 | .058 | .045 | .035 | |
| 12.000 | 11.934 | .000 | .066 | .97 | 3.37 | .6% | 30 | .799 | .783 | .765 | .745 | .724 | .701 | .676 | .649 | .620 | .589 |
| | | | | | | | | .557 | .523 | .487 | .451 | .413 | .375 | .337 | .300 | .264 | .229 |
| | | | | | | | | .197 | .166 | .139 | .114 | .093 | .074 | .058 | .045 | .035 | .026 |
| 13.000 | 8.947 | .000 | 4.053 | .47 | 4.16 | 31.2% | 13 | .812 | .797 | .780 | .763 | .744 | .723 | .700 | .676 | .650 | .622 |
| | | | | | | | | .592 | .561 | .528 | | | | | | | |
| 13.000 | 9.440 | .000 | 3.560 | .51 | 4.21 | 27.4% | 14 | .812 | .797 | .780 | .763 | .744 | .723 | .700 | .676 | .650 | .622 |
| | | | | | | | | .592 | .561 | .528 | .493 | | | | | | |
| 13.000 | 0.890 | .000 | 3.102 | .54 | 4.26 | 23.9% | 15 | .812 | .797 | .780 | .763 | .744 | .723 | .700 | .676 | .650 | .622 |
| | | | | | | | | .592 | .561 | .528 | .493 | .458 | | | | | |

*SIMILAR TO A.T.&T. NEAL-WILKINSON TABLES(SEE SECTION 1)

*EQUIVALENT RANDOM THEORY: VMR(VARIANCE TO MEAN RATIO)– 3.0 - NO RETRIES (PEAKED TRAFFIC - ROUTE ADVANCEMENT OR DAY-TO-DAY VARIATION)

| HOURS FIRST ATTEMPT VMR=3.0 | HOURS CON-NECT | HOURS RECALL /RETRY | HOURS OVFL | CON-NECT VMR | OVFL VMR | PERCENT BLOCKED (GRADE) | LINE | HOURS CONNECTED TRAFFIC ON LINES 1 | 2 | 3 | 4 | 5 | 6 | 7 | 8 | 9 | 10 |
|---|---|---|---|---|---|---|---|---|---|---|---|---|---|---|---|---|---|
| 13.000 | 10.320 | .000 | 2.680 | .58 | 4.29 | 20.6% | 16 | .812 | .797 | .780 | .763 | .744 | .723 | .700 | .676 | .650 | .622 |
| | | | | | | | | .592 | .561 | .528 | .493 | .458 | .422 | | | | |
| 13.000 | 10.705 | .000 | 2.295 | .62 | 4.31 | 17.7% | 17 | .812 | .797 | .780 | .763 | .744 | .723 | .700 | .676 | .650 | .622 |
| | | | | | | | | .592 | .561 | .528 | .493 | .458 | .422 | .385 | | | |
| 13.000 | 11.052 | .000 | 1.948 | .65 | 4.31 | 15.0% | 18 | .812 | .797 | .780 | .763 | .744 | .723 | .700 | .676 | .650 | .622 |
| | | | | | | | | .592 | .561 | .528 | .493 | .458 | .422 | .385 | .348 | | |
| 13.000 | 11.363 | .000 | 1.637 | .69 | 4.31 | 12.6% | 19 | .812 | .797 | .780 | .763 | .744 | .723 | .700 | .676 | .650 | .622 |
| | | | | | | | | .592 | .561 | .528 | .493 | .458 | .422 | .385 | .348 | .311 | |
| 13.000 | 11.639 | .000 | 1.361 | .72 | 4.29 | 10.5% | 20 | .812 | .797 | .780 | .763 | .744 | .723 | .700 | .676 | .650 | .622 |
| | | | | | | | | .592 | .561 | .528 | .493 | .458 | .422 | .385 | .348 | .311 | .275 |
| 13.000 | 11.880 | .000 | 1.120 | .76 | 4.25 | 8.6% | 21 | .812 | .797 | .780 | .763 | .744 | .723 | .700 | .676 | .650 | .622 |
| | | | | | | | | .592 | .561 | .528 | .493 | .458 | .422 | .385 | .348 | .311 | .275 |
| | | | | | | | | .241 | | | | | | | | | |
| 13.000 | 12.088 | .000 | .912 | .79 | 4.21 | 7.0% | 22 | .812 | .797 | .780 | .763 | .744 | .723 | .700 | .676 | .650 | .622 |
| | | | | | | | | .592 | .561 | .528 | .493 | .458 | .422 | .385 | .348 | .311 | .275 |
| | | | | | | | | .241 | .209 | | | | | | | | |
| 13.000 | 12.266 | .000 | .734 | .82 | 4.15 | 5.6% | 23 | .812 | .797 | .780 | .763 | .744 | .723 | .700 | .676 | .650 | .622 |
| | | | | | | | | .592 | .561 | .528 | .493 | .458 | .422 | .385 | .348 | .311 | .275 |
| | | | | | | | | .241 | .209 | .178 | | | | | | | |
| 13.000 | 12.417 | .000 | .583 | .85 | 4.09 | 4.5% | 24 | .812 | .797 | .780 | .763 | .744 | .723 | .700 | .676 | .650 | .622 |
| | | | | | | | | .592 | .561 | .528 | .493 | .458 | .422 | .385 | .348 | .311 | .275 |
| | | | | | | | | .241 | .209 | .178 | .150 | | | | | | |
| 13.000 | 12.542 | .000 | .458 | .88 | 4.01 | 3.5% | 25 | .812 | .797 | .780 | .763 | .744 | .723 | .700 | .676 | .650 | .622 |
| | | | | | | | | .592 | .561 | .528 | .493 | .458 | .422 | .385 | .348 | .311 | .275 |
| | | | | | | | | .241 | .209 | .178 | .150 | .125 | | | | | |
| 13.000 | 12.644 | .000 | .356 | .90 | 3.93 | 2.7% | 26 | .812 | .797 | .780 | .763 | .744 | .723 | .700 | .676 | .650 | .622 |
| | | | | | | | | .592 | .561 | .528 | .493 | .458 | .422 | .385 | .348 | .311 | .275 |
| | | | | | | | | .241 | .209 | .178 | .150 | .125 | .103 | | | | |
| 13.000 | 12.727 | .000 | .273 | .92 | 3.85 | 2.1% | 27 | .812 | .797 | .780 | .763 | .744 | .723 | .700 | .676 | .650 | .622 |
| | | | | | | | | .592 | .561 | .528 | .493 | .458 | .422 | .385 | .348 | .311 | .275 |
| | | | | | | | | .241 | .209 | .178 | .150 | .125 | .103 | .083 | | | |
| 13.000 | 12.793 | .000 | .207 | .93 | 3.76 | 1.6% | 28 | .812 | .797 | .780 | .763 | .744 | .723 | .700 | .676 | .650 | .622 |
| | | | | | | | | .592 | .561 | .528 | .493 | .458 | .422 | .385 | .348 | .311 | .275 |
| | | | | | | | | .241 | .209 | .178 | .150 | .125 | .103 | .083 | .066 | | |
| 13.000 | 12.845 | .000 | .155 | .95 | 3.67 | 1.2% | 29 | .812 | .797 | .780 | .763 | .744 | .723 | .700 | .676 | .650 | .622 |
| | | | | | | | | .592 | .561 | .528 | .493 | .458 | .422 | .385 | .348 | .311 | .275 |
| | | | | | | | | .241 | .209 | .178 | .150 | .125 | .103 | .083 | .066 | .052 | |
| 13.000 | 12.886 | .000 | .114 | .96 | 3.58 | .9% | 30 | .812 | .797 | .780 | .763 | .744 | .723 | .700 | .676 | .650 | .622 |
| | | | | | | | | .592 | .561 | .528 | .493 | .458 | .422 | .385 | .348 | .311 | .275 |
| | | | | | | | | .241 | .209 | .178 | .150 | .125 | .103 | .083 | .066 | .052 | .040 |

*SIMILAR TO A.T.&T. NEAL-WILKINSON TABLES(SEE SECTION 1)

*EQUIVALENT RANDOM THEORY: VMR(VARIANCE TO MEAN RATIO)– 3.0 - NO RETRIES (PEAKED TRAFFIC - ROUTE ADVANCEMENT OR DAY-TO-DAY VARIATION)

| HOURS FIRST ATTEMPT VMR=3.0 | HOURS CON-NECT | HOURS RECALL /RETRY | HOURS OVFL | CON-NECT VMR | OVFL VMR | PERCENT BLOCKED (GRADE) | LINE | HOURS CONNECTED TRAFFIC ON LINES 1 | 2 | 3 | 4 | 5 | 6 | 7 | 8 | 9 | 10 |
|---|---|---|---|---|---|---|---|---|---|---|---|---|---|---|---|---|---|
| 13.000 | 12.917 | .000 | .083 | .97 | 3.49 | .6% | 31 | .812 | .797 | .780 | .763 | .744 | .723 | .700 | .676 | .650 | .622 |
| | | | | | | | | .592 | .561 | .528 | .493 | .458 | .422 | .385 | .348 | .311 | .275 |
| | | | | | | | | .241 | .209 | .178 | .150 | .125 | .103 | .083 | .066 | .052 | .040 |
| | | | | | | | | .031 | | | | | | | | | |
| 14.000 | 9.239 | .000 | 4.761 | .44 | 4.18 | 34.0% | 13 | .823 | .809 | .794 | .778 | .761 | .742 | .722 | .700 | .676 | .651 |
| | | | | | | | | .624 | .595 | .564 | | | | | | | |
| 14.000 | 9.771 | .000 | 4.229 | .47 | 4.25 | 30.2% | 14 | .823 | .809 | .794 | .778 | .761 | .742 | .722 | .700 | .676 | .651 |
| | | | | | | | | .624 | .595 | .564 | .532 | | | | | | |
| 14.000 | 10.270 | .000 | 3.730 | .50 | 4.30 | 26.6% | 15 | .823 | .809 | .794 | .778 | .761 | .742 | .722 | .700 | .676 | .651 |
| | | | | | | | | .624 | .595 | .564 | .532 | .499 | | | | | |
| 14.000 | 10.734 | .000 | 3.266 | .54 | 4.35 | 23.3% | 16 | .823 | .809 | .794 | .778 | .761 | .742 | .722 | .700 | .676 | .651 |
| | | | | | | | | .624 | .595 | .564 | .532 | .499 | .464 | | | | |
| 14.000 | 11.163 | .000 | 2.837 | .57 | 4.38 | 20.3% | 17 | .823 | .809 | .794 | .778 | .761 | .742 | .722 | .700 | .676 | .651 |
| | | | | | | | | .624 | .595 | .564 | .532 | .499 | .464 | .429 | | | |
| 14.000 | 11.556 | .000 | 2.444 | .61 | 4.40 | 17.5% | 18 | .823 | .809 | .794 | .778 | .761 | .742 | .722 | .700 | .676 | .651 |
| | | | | | | | | .624 | .595 | .564 | .532 | .499 | .464 | .429 | .393 | | |
| 14.000 | 11.913 | .000 | 2.087 | .64 | 4.41 | 14.9% | 19 | .823 | .809 | .794 | .778 | .761 | .742 | .722 | .700 | .676 | .651 |
| | | | | | | | | .624 | .595 | .564 | .532 | .499 | .464 | .429 | .393 | .357 | |
| 14.000 | 12.235 | .000 | 1.765 | .68 | 4.41 | 12.6% | 20 | .823 | .809 | .794 | .778 | .761 | .742 | .722 | .700 | .676 | .651 |
| | | | | | | | | .624 | .595 | .564 | .532 | .499 | .464 | .429 | .393 | .357 | .321 |
| 14.000 | 12.521 | .000 | 1.479 | .71 | 4.39 | 10.6% | 21 | .823 | .809 | .794 | .778 | .761 | .742 | .722 | .700 | .676 | .651 |
| | | | | | | | | .624 | .595 | .564 | .532 | .499 | .464 | .429 | .393 | .357 | .321 |
| | | | | | | | | .286 | | | | | | | | | |
| 14.000 | 12.773 | .000 | 1.227 | .75 | 4.36 | 8.8% | 22 | .823 | .809 | .794 | .778 | .761 | .742 | .722 | .700 | .676 | .651 |
| | | | | | | | | .624 | .595 | .564 | .532 | .499 | .464 | .429 | .393 | .357 | .321 |
| | | | | | | | | .286 | .252 | | | | | | | | |
| 14.000 | 12.992 | .000 | 1.008 | .78 | 4.32 | 7.2% | 23 | .823 | .809 | .794 | .778 | .761 | .742 | .722 | .700 | .676 | .651 |
| | | | | | | | | .624 | .595 | .564 | .532 | .499 | .464 | .429 | .393 | .357 | .321 |
| | | | | | | | | .286 | .252 | .220 | | | | | | | |
| 14.000 | 13.182 | .000 | .818 | .81 | 4.27 | 5.8% | 24 | .823 | .809 | .794 | .778 | .761 | .742 | .722 | .700 | .676 | .651 |
| | | | | | | | | .624 | .595 | .564 | .532 | .499 | .464 | .429 | .393 | .357 | .321 |
| | | | | | | | | .286 | .252 | .220 | .189 | | | | | | |
| 14.000 | 13.343 | .000 | .657 | .84 | 4.20 | 4.7% | 25 | .823 | .809 | .794 | .778 | .761 | .742 | .722 | .700 | .676 | .651 |
| | | | | | | | | .624 | .595 | .564 | .532 | .499 | .464 | .429 | .393 | .357 | .321 |
| | | | | | | | | .286 | .252 | .220 | .189 | .161 | | | | | |
| 14.000 | 13.478 | .000 | .522 | .86 | 4.13 | 3.7% | 26 | .823 | .809 | .794 | .778 | .761 | .742 | .722 | .700 | .676 | .651 |
| | | | | | | | | .624 | .595 | .564 | .532 | .499 | .464 | .429 | .393 | .357 | .321 |
| | | | | | | | | .286 | .252 | .220 | .189 | .161 | .135 | | | | |
| 14.000 | 13.590 | .000 | .410 | .89 | 4.06 | 2.9% | 27 | .823 | .809 | .794 | .778 | .761 | .742 | .722 | .700 | .676 | .651 |
| | | | | | | | | .624 | .595 | .564 | .532 | .499 | .464 | .429 | .393 | .357 | .321 |

*SIMILAR TO A.T.&T. NEAL-WILKINSON TABLES(SEE SECTION 1)

*EQUIVALENT RANDOM THEORY: VMR(VARIANCE TO MEAN RATIO)– 3.0 - NO RETRIES (PEAKED TRAFFIC - ROUTE ADVANCEMENT OR DAY-TO-DAY VARIATION)

| HOURS FIRST ATTEMPT VMR=3.0 | HOURS CON-NECT | HOURS RECALL /RETRY | HOURS OVFL | CON-NECT VMR | OVFL VMR | PERCENT BLOCKED (GRADE) | LINE | HOURS CONNECTED TRAFFIC ON LINES 1 | 2 | 3 | 4 | 5 | 6 | 7 | 8 | 9 | 10 |
|---|---|---|---|---|---|---|---|---|---|---|---|---|---|---|---|---|---|
| | | | | | | | | .286 | .252 | .220 | .189 | .161 | .135 | .112 | | | |
| 14.000 | 13.682 | .000 | .318 | .91 | 3.97 | 2.3% | 28 | .823 | .809 | .794 | .778 | .761 | .742 | .722 | .700 | .676 | .651 |
| | | | | | | | | .624 | .595 | .564 | .532 | .499 | .464 | .429 | .393 | .357 | .321 |
| | | | | | | | | .286 | .252 | .220 | .189 | .161 | .135 | .112 | .092 | | |
| 14.000 | 13.756 | .000 | .244 | .93 | 3.89 | 1.7% | 29 | .823 | .809 | .794 | .778 | .761 | .742 | .722 | .700 | .676 | .651 |
| | | | | | | | | .624 | .595 | .564 | .532 | .499 | .464 | .429 | .393 | .357 | .321 |
| | | | | | | | | .286 | .252 | .220 | .189 | .161 | .135 | .112 | .092 | .074 | |
| 14.000 | 13.815 | .000 | .185 | .94 | 3.80 | 1.3% | 30 | .823 | .809 | .794 | .778 | .761 | .742 | .722 | .700 | .676 | .651 |
| | | | | | | | | .624 | .595 | .564 | .532 | .499 | .464 | .429 | .393 | .357 | .321 |
| | | | | | | | | .286 | .252 | .220 | .189 | .161 | .135 | .112 | .092 | .074 | .059 |
| 14.000 | 13.862 | .000 | .138 | .95 | 3.71 | 1.0% | 31 | .823 | .809 | .794 | .778 | .761 | .742 | .722 | .700 | .676 | .651 |
| | | | | | | | | .624 | .595 | .564 | .532 | .499 | .464 | .429 | .393 | .357 | .321 |
| | | | | | | | | .286 | .252 | .220 | .189 | .161 | .135 | .112 | .092 | .074 | .059 |
| | | | | | | | | .046 | | | | | | | | | |
| 14.000 | 13.898 | .000 | .102 | .96 | 3.62 | .7% | 32 | .823 | .809 | .794 | .778 | .761 | .742 | .722 | .700 | .676 | .651 |
| | | | | | | | | .624 | .595 | .564 | .532 | .499 | .464 | .429 | .393 | .357 | .321 |
| | | | | | | | | .286 | .252 | .220 | .189 | .161 | .135 | .112 | .092 | .074 | .059 |
| | | | | | | | | .046 | .036 | | | | | | | | |
| 14.000 | 13.925 | .000 | .075 | .97 | 3.53 | .5% | 33 | .823 | .809 | .794 | .778 | .761 | .742 | .722 | .700 | .676 | .651 |
| | | | | | | | | .624 | .595 | .564 | .532 | .499 | .464 | .429 | .393 | .357 | .321 |
| | | | | | | | | .286 | .252 | .220 | .189 | .161 | .135 | .112 | .092 | .074 | .059 |
| | | | | | | | | .046 | .036 | .028 | | | | | | | |
| 15.000 | 10.065 | .000 | 4.935 | .43 | 4.27 | 32.9% | 14 | .833 | .820 | .807 | .792 | .776 | .759 | .741 | .721 | .699 | .676 |
| | | | | | | | | .651 | .625 | .597 | .567 | | | | | | |
| 15.000 | 10.601 | .000 | 4.399 | .46 | 4.33 | 29.3% | 15 | .833 | .820 | .807 | .792 | .776 | .759 | .741 | .721 | .699 | .676 |
| | | | | | | | | .651 | .625 | .597 | .567 | .536 | | | | | |
| 15.000 | 11.105 | .000 | 3.895 | .50 | 4.39 | 26.0% | 16 | .833 | .820 | .807 | .792 | .776 | .759 | .741 | .721 | .699 | .676 |
| | | | | | | | | .651 | .625 | .597 | .567 | .536 | .504 | | | | |
| 15.000 | 11.575 | .000 | 3.425 | .53 | 4.43 | 22.8% | 17 | .833 | .820 | .807 | .792 | .776 | .759 | .741 | .721 | .699 | .676 |
| | | | | | | | | .651 | .625 | .597 | .567 | .536 | .504 | .470 | | | |
| 15.000 | 12.011 | .000 | 2.989 | .56 | 4.47 | 19.9% | 18 | .833 | .820 | .807 | .792 | .776 | .759 | .741 | .721 | .699 | .676 |
| | | | | | | | | .651 | .625 | .597 | .567 | .536 | .504 | .470 | .436 | | |
| 15.000 | 12.412 | .000 | 2.588 | .60 | 4.49 | 17.3% | 19 | .833 | .820 | .807 | .792 | .776 | .759 | .741 | .721 | .699 | .676 |
| | | | | | | | | .651 | .625 | .597 | .567 | .536 | .504 | .470 | .436 | .401 | |
| 15.000 | 12.777 | .000 | 2.223 | .63 | 4.50 | 14.8% | 20 | .833 | .820 | .807 | .792 | .776 | .759 | .741 | .721 | .699 | .676 |
| | | | | | | | | .651 | .625 | .597 | .567 | .536 | .504 | .470 | .436 | .401 | .366 |
| 15.000 | 13.108 | .000 | 1.892 | .67 | 4.50 | 12.6% | 21 | .833 | .820 | .807 | .792 | .776 | .759 | .741 | .721 | .699 | .676 |
| | | | | | | | | .651 | .625 | .597 | .567 | .536 | .504 | .470 | .436 | .401 | .366 |
| | | | | | | | | .330 | | | | | | | | | |
| 15.000 | 13.404 | .000 | 1.596 | .70 | 4.49 | 10.6% | 22 | .833 | .820 | .807 | .792 | .776 | .759 | .741 | .721 | .699 | .676 |

*SIMILAR TO A.T.&T. NEAL-WILKINSON TABLES(SEE SECTION 1)

# *EQUIVALENT RANDOM THEORY: VMR(VARIANCE TO MEAN RATIO)– 3.0 - NO RETRIES (PEAKED TRAFFIC - ROUTE ADVANCEMENT OR DAY-TO-DAY VARIATION)

| HOURS FIRST ATTEMPT VMR=3.0 | HOURS CON-NECT | HOURS RECALL /RETRY | HOURS OVFL | CON-NECT VMR | OVFL VMR | PERCENT BLOCKED (GRADE) | LINE | HOURS CONNECTED TRAFFIC ON LINES 1 | 2 | 3 | 4 | 5 | 6 | 7 | 8 | 9 | 10 |
|---|---|---|---|---|---|---|---|---|---|---|---|---|---|---|---|---|---|
| | | | | | | | | .651 | .625 | .597 | .567 | .536 | .504 | .470 | .436 | .401 | .366 |
| | | | | | | | | .330 | .296 | | | | | | | | |
| 15.000 | 13.666 | .000 | 1.334 | .74 | 4.46 | 8.9% | 23 | .833 | .820 | .807 | .792 | .776 | .759 | .741 | .721 | .699 | .676 |
| | | | | | | | | .651 | .625 | .597 | .567 | .536 | .504 | .470 | .436 | .401 | .366 |
| | | | | | | | | .330 | .296 | .262 | | | | | | | |
| 15.000 | 13.896 | .000 | 1.104 | .77 | 4.42 | 7.4% | 24 | .833 | .820 | .807 | .792 | .776 | .759 | .741 | .721 | .699 | .676 |
| | | | | | | | | .651 | .625 | .597 | .567 | .536 | .504 | .470 | .436 | .401 | .366 |
| | | | | | | | | .330 | .296 | .262 | .230 | | | | | | |
| 15.000 | 14.096 | .000 | .904 | .80 | 4.38 | 6.0% | 25 | .833 | .820 | .807 | .792 | .776 | .759 | .741 | .721 | .699 | .676 |
| | | | | | | | | .651 | .625 | .597 | .567 | .536 | .504 | .470 | .436 | .401 | .366 |
| | | | | | | | | .330 | .296 | .262 | .230 | .200 | | | | | |
| 15.000 | 14.267 | .000 | .733 | .83 | 4.32 | 4.9% | 26 | .833 | .820 | .807 | .792 | .776 | .759 | .741 | .721 | .699 | .676 |
| | | | | | | | | .651 | .625 | .597 | .567 | .536 | .504 | .470 | .436 | .401 | .366 |
| | | | | | | | | .330 | .296 | .262 | .230 | .200 | .171 | | | | |
| 15.000 | 14.412 | .000 | .588 | .85 | 4.25 | 3.9% | 27 | .833 | .820 | .807 | .792 | .776 | .759 | .741 | .721 | .699 | .676 |
| | | | | | | | | .651 | .625 | .597 | .567 | .536 | .504 | .470 | .436 | .401 | .366 |
| | | | | | | | | .330 | .296 | .262 | .230 | .200 | .171 | .145 | | | |
| 15.000 | 14.534 | .000 | .466 | .88 | 4.17 | 3.1% | 28 | .833 | .820 | .807 | .792 | .776 | .759 | .741 | .721 | .699 | .676 |
| | | | | | | | | .651 | .625 | .597 | .567 | .536 | .504 | .470 | .436 | .401 | .366 |
| | | | | | | | | .330 | .296 | .262 | .230 | .200 | .171 | .145 | .122 | | |
| 15.000 | 14.634 | .000 | .366 | .90 | 4.09 | 2.4% | 29 | .833 | .820 | .807 | .792 | .776 | .759 | .741 | .721 | .699 | .676 |
| | | | | | | | | .651 | .625 | .597 | .567 | .536 | .504 | .470 | .436 | .401 | .366 |
| | | | | | | | | .330 | .296 | .262 | .230 | .200 | .171 | .145 | .122 | .100 | |
| 15.000 | 14.716 | .000 | .284 | .92 | 4.01 | 1.9% | 30 | .833 | .820 | .807 | .792 | .776 | .759 | .741 | .721 | .699 | .676 |
| | | | | | | | | .651 | .625 | .597 | .567 | .536 | .504 | .470 | .436 | .401 | .366 |
| | | | | | | | | .330 | .296 | .262 | .230 | .200 | .171 | .145 | .122 | .100 | .082 |
| 15.000 | 14.782 | .000 | .218 | .93 | 3.92 | 1.5% | 31 | .833 | .820 | .807 | .792 | .776 | .759 | .741 | .721 | .699 | .676 |
| | | | | | | | | .651 | .625 | .597 | .567 | .536 | .504 | .470 | .436 | .401 | .366 |
| | | | | | | | | .330 | .296 | .262 | .230 | .200 | .171 | .145 | .122 | .100 | .082 |
| | | | | | | | | .066 | | | | | | | | | |
| 15.000 | 14.835 | .000 | .165 | .95 | 3.83 | 1.1% | 32 | .833 | .820 | .807 | .792 | .776 | .759 | .741 | .721 | .699 | .676 |
| | | | | | | | | .651 | .625 | .597 | .567 | .536 | .504 | .470 | .436 | .401 | .366 |
| | | | | | | | | .330 | .296 | .262 | .230 | .200 | .171 | .145 | .122 | .100 | .082 |
| | | | | | | | | .066 | .053 | | | | | | | | |
| 15.000 | 14.876 | .000 | .124 | .96 | 3.74 | .8% | 33 | .833 | .820 | .807 | .792 | .776 | .759 | .741 | .721 | .699 | .676 |
| | | | | | | | | .651 | .625 | .597 | .567 | .536 | .504 | .470 | .436 | .401 | .366 |
| | | | | | | | | .330 | .296 | .262 | .230 | .200 | .171 | 145 | .122 | .100 | .082 |
| | | | | | | | | .066 | .053 | .041 | | | | | | | |
| 15.000 | 14.908 | .000 | .092 | .97 | 3.65 | .6% | 34 | .833 | .820 | .807 | .792 | .776 | .759 | .741 | .721 | .699 | .676 |
| | | | | | | | | .651 | .625 | .597 | .567 | .536 | .504 | .470 | .436 | .401 | .366 |

*SIMILAR TO A.T.&T. NEAL-WILKINSON TABLES(SEE SECTION 1)

*EQUIVALENT RANDOM THEORY: VMR(VARIANCE TO MEAN RATIO)– 3.0 - NO RETRIES (PEAKED TRAFFIC - ROUTE ADVANCEMENT OR DAY-TO-DAY VARIATION)

| HOURS FIRST ATTEMPT VMR=3.0 | HOURS CON-NECT | HOURS RECALL /RETRY | HOURS OVFL | CON-NECT VMR | OVFL VMR | PERCENT BLOCKED (GRADE) | LINE | HOURS CONNECTED TRAFFIC ON LINES 1 | 2 | 3 | 4 | 5 | 6 | 7 | 8 | 9 | 10 |
|---|---|---|---|---|---|---|---|---|---|---|---|---|---|---|---|---|---|
| | | | | | | | | .330 | .296 | .262 | .230 | .200 | .171 | .145 | .122 | .100 | .082 |
| | | | | | | | | .066 | .053 | .041 | .032 | | | | | | |
| 16.000 | 10.897 | .000 | 5.103 | .43 | 4.35 | 31.9% | 15 | .841 | .830 | .818 | .805 | .790 | .775 | .758 | .739 | .720 | .699 |
| | | | | | | | | .676 | .652 | .626 | .599 | .570 | | | | | |
| 16.000 | 11.437 | .000 | 4.563 | .46 | 4.42 | 28.5% | 16 | .841 | .830 | .818 | .805 | .790 | .775 | .758 | .739 | .720 | .699 |
| | | | | | | | | .676 | .652 | .626 | .599 | .570 | .540 | | | | |
| 16.000 | 11.945 | .000 | 4.055 | .49 | 4.47 | 25.3% | 17 | .841 | .830 | .818 | .805 | .790 | .775 | .758 | .739 | .720 | .699 |
| | | | | | | | | .676 | .652 | .626 | .599 | .570 | .540 | .508 | | | |
| 16.000 | 12.420 | .000 | 3.580 | .52 | 4.52 | 22.4% | 18 | .841 | .830 | .818 | .805 | .790 | .775 | .758 | .739 | .720 | .699 |
| | | | | | | | | .676 | .652 | .626 | .599 | .570 | .540 | .508 | .476 | | |
| 16.000 | 12.862 | .000 | 3.138 | .56 | 4.56 | 19.6% | 19 | .841 | .830 | .818 | .805 | .790 | .775 | .758 | .739 | .720 | .699 |
| | | | | | | | | .676 | .652 | .626 | .599 | .570 | .540 | .508 | .476 | .442 | |
| 16.000 | 13.270 | .000 | 2.730 | .59 | 4.58 | 17.1% | 20 | .841 | .830 | .818 | .805 | .790 | .775 | .758 | .739 | .720 | .699 |
| | | | | | | | | .676 | .652 | .626 | .599 | .570 | .540 | .508 | .476 | .442 | .408 |
| 16.000 | 13.643 | .000 | 2.357 | .63 | 4.59 | 14.7% | 21 | .841 | .830 | .818 | .805 | .790 | .775 | .758 | .739 | .720 | .699 |
| | | | | | | | | .676 | .652 | .626 | .599 | .570 | .540 | .508 | .476 | .442 | .408 |
| | | | | | | | | .373 | | | | | | | | | |
| 16.000 | 13.982 | .000 | 2.018 | .66 | 4.60 | 12.6% | 22 | .841 | .830 | .818 | .805 | .790 | .775 | .758 | .739 | .720 | .699 |
| | | | | | | | | .676 | .652 | .626 | .599 | .570 | .540 | .508 | .476 | .442 | .408 |
| | | | | | | | | .373 | .339 | | | | | | | | |
| 16.000 | 14.288 | .000 | 1.712 | .69 | 4.58 | 10.7% | 23 | .841 | .830 | .818 | .805 | .790 | .775 | .758 | .739 | .720 | .699 |
| | | | | | | | | .676 | .652 | .626 | .599 | .570 | .540 | .508 | .476 | .442 | .408 |
| | | | | | | | | .373 | .339 | .305 | | | | | | | |
| 16.000 | 14.559 | .000 | 1.441 | .73 | 4.56 | 9.0% | 24 | .841 | .830 | .818 | .805 | .790 | .775 | .758 | .739 | .720 | .699 |
| | | | | | | | | .676 | .652 | .626 | .599 | .570 | .540 | .508 | .476 | .442 | .408 |
| | | | | | | | | .373 | .339 | .305 | .272 | | | | | | |
| 16.000 | 14.799 | .000 | 1.201 | .76 | 4.53 | 7.5% | 25 | .841 | .830 | .818 | .805 | .790 | .775 | .758 | .739 | .720 | .699 |
| | | | | | | | | .676 | .652 | .626 | .599 | .570 | .540 | .508 | .476 | .442 | .408 |
| | | | | | | | | .373 | .339 | .305 | .272 | .240 | | | | | |
| 16.000 | 15.009 | .000 | .991 | .79 | 4.48 | 6.2% | 26 | .841 | .830 | .818 | .805 | .790 | .775 | .758 | .739 | .720 | .699 |
| | | | | | | | | .676 | .652 | .626 | .599 | .570 | .540 | .508 | .476 | .442 | .408 |
| | | | | | | | | .373 | .339 | .305 | .272 | .240 | .210 | | | | |
| 16.000 | 15.190 | .000 | .810 | .82 | 4.42 | 5.1% | 27 | .841 | .830 | .818 | .805 | .790 | .775 | .758 | .739 | .720 | .699 |
| | | | | | | | | .676 | .652 | .626 | .599 | .570 | .540 | .508 | .476 | .442 | .408 |
| | | | | | | | | .373 | .339 | .305 | .272 | .240 | .210 | .181 | | | |
| 16.000 | 15.345 | .000 | .655 | .85 | 4.36 | 4.1% | 28 | .841 | .830 | .818 | .805 | .790 | .775 | .758 | .739 | .720 | .699 |
| | | | | | | | | .676 | .652 | .626 | .599 | .570 | .540 | .508 | .476 | .442 | .408 |
| | | | | | | | | .373 | .339 | .305 | .272 | .240 | .210 | .181 | .155 | | |
| 16.000 | 15.476 | .000 | .524 | .87 | 4.29 | 3.3% | 29 | .841 | .830 | .818 | .805 | .790 | .775 | .758 | .739 | .720 | .699 |
| | | | | | | | | .676 | .652 | .626 | .599 | .570 | .540 | .508 | .476 | .442 | .408 |

*SIMILAR TO A.T.&T. NEAL-WILKINSON TABLES(SEE SECTION 1)

*EQUIVALENT RANDOM THEORY: VMR(VARIANCE TO MEAN RATIO)– 3.0 - NO RETRIES (PEAKED TRAFFIC - ROUTE ADVANCEMENT OR DAY-TO-DAY VARIATION)

| HOURS FIRST ATTEMPT VMR=3.0 | HOURS CON-NECT | HOURS RECALL /RETRY | HOURS OVFL | CON-NECT VMR | OVFL VMR | PERCENT BLOCKED (GRADE) | LINE | HOURS CONNECTED TRAFFIC ON LINES 1 | 2 | 3 | 4 | 5 | 6 | 7 | 8 | 9 | 10 |
|---|---|---|---|---|---|---|---|---|---|---|---|---|---|---|---|---|---|
| | | | | | | | | .373 | .339 | .305 | .272 | .240 | .210 | .181 | .155 | .131 | |
| 16.000 | 15.585 | .000 | .415 | .89 | 4.21 | 2.6% | 30 | .841 | .830 | .818 | .805 | .790 | .775 | .758 | .739 | .720 | .699 |
| | | | | | | | | .676 | .652 | .626 | .599 | .570 | .540 | .508 | .476 | .442 | .408 |
| | | | | | | | | .373 | .339 | .305 | .272 | .240 | .210 | .181 | .155 | .131 | .109 |
| 16.000 | 15.674 | .000 | .326 | .91 | 4.12 | 2.0% | 31 | .841 | .830 | .818 | .805 | .790 | .775 | .758 | .739 | .720 | .699 |
| | | | | | | | | .676 | .652 | .626 | .599 | .570 | .540 | .508 | .476 | .442 | .408 |
| | | | | | | | | .373 | .339 | .305 | .272 | .240 | .210 | .181 | .155 | .131 | .109 |
| | | | | | | | | .090 | | | | | | | | | |
| 16.000 | 15.748 | .000 | .252 | .93 | 4.04 | 1.6% | 32 | .841 | .830 | .818 | .805 | .790 | .775 | .758 | .739 | .720 | .699 |
| | | | | | | | | .676 | .652 | .626 | .599 | .570 | .540 | .508 | .476 | .442 | .408 |
| | | | | | | | | .373 | .339 | .305 | .272 | .240 | .210 | .181 | .155 | .131 | .109 |
| | | | | | | | | .090 | .073 | | | | | | | | |
| 16.000 | 15.806 | .000 | .194 | .94 | 3.95 | 1.2% | 33 | .841 | .830 | .818 | .805 | .790 | .775 | .758 | .739 | .720 | .699 |
| | | | | | | | | .676 | .652 | .626 | .599 | .570 | .540 | .508 | .476 | .442 | .408 |
| | | | | | | | | .373 | .339 | .305 | .272 | .240 | .210 | .181 | .155 | .131 | .109 |
| | | | | | | | | .090 | .073 | .059 | | | | | | | |
| 16.000 | 15.853 | .000 | .147 | .95 | 3.86 | .9% | 34 | .841 | .830 | .818 | .805 | .790 | .775 | .758 | .739 | .720 | .699 |
| | | | | | | | | .676 | .652 | .626 | .599 | .570 | .540 | .508 | .476 | .442 | .408 |
| | | | | | | | | .373 | .339 | .305 | .272 | .240 | .210 | .181 | .155 | .131 | .109 |
| | | | | | | | | .090 | .073 | .059 | .047 | | | | | | |
| 16.000 | 15.890 | .000 | .110 | .96 | 3.76 | .7% | 35 | .841 | .830 | .818 | .805 | .790 | .775 | .758 | .739 | .720 | .699 |
| | | | | | | | | .676 | .652 | .626 | .599 | .570 | .540 | .508 | .476 | .442 | .408 |
| | | | | | | | | .373 | .339 | .305 | .272 | .240 | .210 | .181 | .155 | .131 | .109 |
| | | | | | | | | .090 | .073 | .059 | .047 | .037 | | | | | |
| 16.000 | 15.918 | .000 | .082 | .97 | 3.68 | .5% | 36 | .841 | .830 | .818 | .805 | .790 | .775 | .758 | .739 | .720 | .699 |
| | | | | | | | | .676 | .652 | .626 | .599 | .570 | .540 | .508 | .476 | .442 | .408 |
| | | | | | | | | .073 | .339 | .305 | .272 | .240 | .210 | .181 | .155 | .131 | .109 |
| | | | | | | | | .090 | .073 | .059 | .047 | .037 | .028 | | | | |
| 17.000 | 11.162 | .000 | 5.838 | .40 | 4.36 | 34.3% | 15 | .849 | .839 | .828 | .815 | .802 | .788 | .773 | .756 | .738 | .719 |
| | | | | | | | | .698 | .676 | .652 | .627 | .600 | | | | | |
| 17.000 | 11.734 | .000 | 5.266 | .43 | 4.43 | 31.0% | 16 | .849 | .839 | .828 | .815 | .802 | .788 | .773 | .756 | .738 | .719 |
| | | | | | | | | .698 | .676 | .652 | .627 | .600 | .572 | | | | |
| 17.000 | 12.277 | .000 | 4.723 | .46 | 4.50 | 27.8% | 17 | .049 | .839 | .828 | .815 | .802 | .788 | .773 | .756 | .738 | .719 |
| | | | | | | | | .698 | .676 | .652 | .627 | .600 | .572 | .543 | | | |
| 17.000 | 12.789 | .000 | 4.211 | .49 | 4.56 | 24.8% | 18 | .849 | .839 | .828 | .815 | .802 | .788 | .773 | .756 | .738 | .719 |
| | | | | | | | | .698 | .676 | .652 | .627 | .600 | .572 | .543 | .512 | | |
| 17.000 | 13.270 | .000 | 3.730 | .52 | 4.60 | 21.9% | 19 | .849 | .839 | .828 | .815 | .802 | .788 | .773 | .756 | .738 | .719 |
| | | | | | | | | .698 | .676 | .652 | .627 | .000 | .572 | .543 | .512 | .480 | |
| 17.000 | 13.717 | .000 | 3.283 | .55 | 4.64 | 19.3% | 20 | .849 | .839 | .828 | .815 | .802 | .788 | .773 | .756 | .738 | .719 |
| | | | | | | | | .698 | .676 | .652 | .627 | .600 | .572 | .543 | .512 | .480 | .448 |

*SIMILAR TO A.T.&T. NEAL-WILKINSON TABLES(SEE SECTION 1)

*EQUIVALENT RANDOM THEORY: VMR(VARIANCE TO MEAN RATIO)– 3.0 - NO RETRIES (PEAKED TRAFFIC - ROUTE ADVANCEMENT OR DAY-TO-DAY VARIATION)

| HOURS FIRST ATTEMPT VMR=3.0 | HOURS CON-NECT | HOURS RECALL /RETRY | HOURS OVFL | CON-NECT VMR | OVFL VMR | PERCENT BLOCKED (GRADE) | LINE | HOURS CONNECTED TRAFFIC ON LINES 1 | 2 | 3 | 4 | 5 | 6 | 7 | 8 | 9 | 10 |
|---|---|---|---|---|---|---|---|---|---|---|---|---|---|---|---|---|---|
| 17.000 | 14.131 | .000 | 2.869 | .59 | 4.67 | 16.9% | 21 | .849 | .839 | .828 | .815 | .802 | .788 | .773 | .756 | .738 | .719 |
| | | | | | | | | .698 | .676 | .652 | .627 | .600 | .572 | .543 | .512 | .480 | .448 |
| | | | | | | | | .414 | | | | | | | | | |
| 17.000 | 14.512 | .000 | 2.488 | .62 | 4.68 | 14.6% | 22 | .849 | .839 | .828 | .815 | .802 | .788 | .773 | .756 | .738 | .719 |
| | | | | | | | | .698 | .676 | .652 | .627 | .600 | .572 | .543 | .512 | .480 | .448 |
| | | | | | | | | .414 | .381 | | | | | | | | |
| 17.000 | 14.859 | .000 | 2.141 | .65 | 4.69 | 12.6% | 23 | .849 | .839 | .828 | .815 | .802 | .788 | .773 | .756 | .738 | .719 |
| | | | | | | | | .698 | .676 | .652 | .627 | .600 | .572 | .543 | .512 | .480 | .448 |
| | | | | | | | | .414 | .381 | .347 | | | | | | | |
| 17.000 | 15.173 | .000 | 1.827 | .69 | 4.68 | 10.7% | 24 | .849 | .839 | .828 | .815 | .802 | .788 | .773 | .756 | .738 | .719 |
| | | | | | | | | .698 | .676 | .652 | .627 | .600 | .572 | .543 | .512 | .480 | .448 |
| | | | | | | | | .414 | .381 | .347 | .314 | | | | | | |
| 17.000 | 15.454 | .000 | 1.546 | .72 | 4.66 | 9.1% | 25 | .849 | .839 | .828 | .815 | .802 | .788 | .773 | .756 | .738 | .719 |
| | | | | | | | | .698 | .676 | .652 | .627 | .600 | .572 | .543 | .512 | .480 | .448 |
| | | | | | | | | .414 | .381 | .347 | .314 | .281 | | | | | |
| 17.000 | 15.703 | .000 | 1.297 | .75 | 4.63 | 7.6% | 26 | .849 | .839 | .828 | .815 | .802 | .788 | .773 | .756 | .738 | .719 |
| | | | | | | | | .698 | .676 | .652 | .627 | .600 | .572 | .543 | .512 | .480 | .448 |
| | | | | | | | | .414 | .381 | .347 | .314 | .281 | .249 | | | | |
| 17.000 | 15.922 | .000 | 1.078 | .78 | 4.58 | 6.3% | 27 | .849 | .839 | .828 | .815 | .802 | .788 | .773 | .756 | .738 | .719 |
| | | | | | | | | .698 | .676 | .652 | .627 | .600 | .572 | .543 | .512 | .480 | .448 |
| | | | | | | | | .414 | .381 | .347 | .314 | .281 | .249 | .219 | | | |
| 17.000 | 16.112 | .000 | .888 | .81 | 4.53 | 5.2% | 28 | .849 | .839 | .828 | .815 | .802 | .788 | .773 | .756 | .738 | .719 |
| | | | | | | | | .698 | .676 | .652 | .627 | .600 | .572 | .543 | .512 | .480 | .448 |
| | | | | | | | | .414 | .381 | .347 | .314 | .281 | .249 | .219 | .190 | | |
| 17.000 | 16.276 | .000 | .724 | .84 | 4.47 | 4.3% | 29 | .849 | .839 | .828 | .815 | .802 | .788 | .773 | .756 | .738 | .719 |
| | | | | | | | | .698 | .676 | .652 | .627 | .600 | .572 | .543 | .512 | .480 | .448 |
| | | | | | | | | .414 | .381 | .347 | .314 | .281 | .249 | .219 | .190 | .164 | |
| 17.000 | 16.416 | .000 | .584 | .86 | 4.40 | 3.4% | 30 | .849 | .839 | .828 | .815 | .802 | .788 | .773 | .756 | .738 | .719 |
| | | | | | | | | .698 | .676 | .652 | .627 | .600 | .572 | .543 | .512 | .480 | .448 |
| | | | | | | | | .414 | .381 | .347 | .314 | .281 | .249 | .219 | .190 | .164 | .139 |
| 17.000 | 16.533 | .000 | .467 | .88 | 4.32 | 2.7% | 31 | .849 | .839 | .828 | .815 | .802 | .788 | .773 | .756 | .738 | .719 |
| | | | | | | | | .698 | .676 | .652 | .627 | .600 | .572 | .543 | .512 | .480 | .448 |
| | | | | | | | | .414 | .381 | .347 | .314 | .281 | .249 | .219 | .190 | .164 | .139 |
| | | | | | | | | .117 | | | | | | | | | |
| 17.000 | 16.630 | .000 | .370 | .90 | 4.24 | 2.2% | 32 | .849 | .839 | .828 | .815 | .802 | .788 | .773 | .756 | .738 | .719 |
| | | | | | | | | .698 | .676 | .652 | .627 | .600 | .572 | .543 | .512 | .480 | .448 |
| | | | | | | | | .414 | .381 | .347 | .314 | .281 | .249 | .219 | .190 | .164 | .139 |
| | | | | | | | | .117 | .098 | | | | | | | | |
| 17.000 | 16.711 | .000 | .289 | .92 | 4.15 | 1.7% | 33 | .849 | .839 | .828 | .815 | .802 | .788 | .773 | .756 | .738 | .719 |
| | | | | | | | | .698 | .676 | .652 | .627 | .600 | .572 | .543 | .512 | .480 | .448 |

*SIMILAR TO A.T.&T. NEAL-WILKINSON TABLES(SEE SECTION 1)

*EQUIVALENT RANDOM THEORY: VMR(VARIANCE TO MEAN RATIO)– 3.0 - NO RETRIES (PEAKED TRAFFIC - ROUTE ADVANCEMENT OR DAY-TO-DAY VARIATION)

| HOURS FIRST ATTEMPT VMR=3.0 | HOURS CON-NECT | HOURS RECALL /RETRY | HOURS OVFL | CON-NECT VMR | OVFL VMR | PERCENT BLOCKED (GRADE) | LINE | HOURS CONNECTED TRAFFIC ON LINES 1 | 2 | 3 | 4 | 5 | 6 | 7 | 8 | 9 | 10 |
|---|---|---|---|---|---|---|---|---|---|---|---|---|---|---|---|---|---|
| | | | | | | | | .414 | .381 | .347 | .314 | .281 | .249 | .219 | .190 | .164 | .139 |
| | | | | | | | | .117 | .098 | .080 | | | | | | | |
| 17.000 | 16.776 | .000 | .224 | .93 | 4.06 | 1.3% | 34 | .849 | .839 | .828 | .815 | .802 | .788 | .773 | .756 | .738 | .719 |
| | | | | | | | | .698 | .676 | .652 | .627 | .600 | .572 | .543 | .512 | .480 | .448 |
| | | | | | | | | .414 | .381 | .347 | .314 | .281 | .249 | .219 | .190 | .164 | .139 |
| | | | | | | | | .117 | .098 | .080 | .065 | | | | | | |
| 17.000 | 16.828 | .000 | .172 | .95 | 3.97 | 1.0% | 35 | .849 | .839 | .828 | .815 | .802 | .788 | .773 | .756 | .738 | .719 |
| | | | | | | | | .698 | .676 | .652 | .627 | .600 | .572 | .543 | .512 | .480 | .448 |
| | | | | | | | | .414 | .381 | .347 | .314 | .281 | .249 | .219 | .190 | .164 | .139 |
| | | | | | | | | .117 | .098 | .080 | .065 | .052 | | | | | |
| 17.000 | 16.870 | .000 | .130 | .96 | 3.88 | .8% | 36 | .849 | .839 | .828 | .815 | .802 | .788 | .773 | .756 | .738 | .719 |
| | | | | | | | | .698 | .676 | .652 | .627 | .600 | .572 | .543 | .512 | .480 | .448 |
| | | | | | | | | .414 | .381 | .347 | .314 | .281 | .249 | .219 | .190 | .164 | .139 |
| | | | | | | | | .117 | .098 | .080 | .065 | .052 | .041 | | | | |
| 17.000 | 16.902 | .000 | .098 | .97 | 3.79 | .6% | 37 | .849 | .839 | .828 | .815 | .802 | .788 | .773 | .756 | .738 | .719 |
| | | | | | | | | .698 | .676 | .652 | .627 | .600 | .572 | .543 | .512 | .480 | .448 |
| | | | | | | | | .414 | .381 | .347 | .314 | .281 | .249 | .219 | .190 | .164 | .139 |
| | | | | | | | | .117 | .098 | .080 | .065 | .052 | .041 | .033 | | | |
| 18.000 | 12.002 | .000 | 5.998 | .40 | 4.44 | 33.3% | 16 | .856 | .847 | .837 | .825 | .813 | .800 | .786 | .771 | .755 | .737 |
| | | | | | | | | .718 | .698 | .676 | .653 | .628 | .602 | | | | |
| 18.000 | 12.576 | .000 | 5.424 | .43 | 4.51 | 30.1% | 17 | .856 | .847 | .837 | .825 | .813 | .800 | .786 | .771 | .755 | .737 |
| | | | | | | | | .718 | .698 | .676 | .653 | .628 | .602 | .575 | | | |
| 18.000 | 13.122 | .000 | 4.878 | .45 | 4.58 | 27.1% | 18 | .856 | .847 | .837 | .825 | .813 | .800 | .786 | .771 | .755 | .737 |
| | | | | | | | | .718 | .698 | .676 | .653 | .628 | .602 | .575 | .546 | | |
| 18.000 | 13.638 | .000 | 4.362 | .48 | 4.64 | 24.2% | 19 | .856 | .847 | .837 | .825 | .813 | .800 | .786 | .771 | .755 | .737 |
| | | | | | | | | .718 | .698 | .676 | .653 | .628 | .602 | .575 | .546 | .516 | |
| 18.000 | 14.123 | .000 | 3.877 | .52 | 4.68 | 21.5% | 20 | .856 | .847 | .837 | .825 | .813 | .800 | .786 | .771 | .755 | .737 |
| | | | | | | | | .718 | .698 | .676 | .653 | .628 | .602 | .575 | .546 | .516 | .485 |
| 18.000 | 14.575 | .000 | 3.425 | .55 | 4.72 | 19.0% | 21 | .856 | .847 | .837 | .825 | .813 | .800 | .786 | .771 | .755 | .737 |
| | | | | | | | | .718 | .698 | .676 | .653 | .628 | .602 | .575 | .546 | .516 | .485 |
| | | | | | | | | .453 | | | | | | | | | |
| 18.000 | 14.996 | .000 | 3.004 | .58 | 4.75 | 16.7% | 22 | .856 | .847 | .837 | .825 | .813 | .800 | .786 | .771 | .755 | .737 |
| | | | | | | | | .718 | .698 | .676 | .653 | .628 | .602 | .575 | .546 | .516 | .485 |
| | | | | | | | | .453 | .420 | | | | | | | | |
| 18.000 | 15.383 | .000 | 2.617 | .61 | 4.77 | 14.5% | 23 | .856 | .847 | .837 | .825 | .813 | .800 | .786 | .771 | .755 | .737 |
| | | | | | | | | .718 | .698 | .676 | .653 | .628 | .602 | .575 | .546 | .516 | .485 |
| | | | | | | | | .453 | .420 | .387 | | | | | | | |
| 18.000 | 15.737 | .000 | 2.263 | .65 | 4.77 | 12.6% | 24 | .856 | .847 | .837 | .825 | .813 | .800 | .786 | .771 | .755 | .737 |
| | | | | | | | | .718 | .698 | .676 | .653 | .628 | .602 | .575 | .546 | .516 | .485 |
| | | | | | | | | .453 | .420 | .387 | .354 | | | | | | |

*SIMILAR TO A.T.&T. NEAL-WILKINSON TABLES(SEE SECTION 1)

*EQUIVALENT RANDOM THEORY: VMR(VARIANCE TO MEAN RATIO)– 3.0 - NO RETRIES (PEAKED TRAFFIC - ROUTE ADVANCEMENT OR DAY-TO-DAY VARIATION)

| HOURS FIRST ATTEMPT VMR=3.0 | HOURS CON- NECT | HOURS RECALL /RETRY | HOURS OVFL | CON- NECT VMR | OVFL VMR | PERCENT BLOCKED (GRADE) | LINE | HOURS CONNECTED TRAFFIC ON LINES 1 | 2 | 3 | 4 | 5 | 6 | 7 | 8 | 9 | 10 |
|---|---|---|---|---|---|---|---|---|---|---|---|---|---|---|---|---|---|
| 18.000 | 16.059 | .000 | 1.941 | .68 | 4.77 | 10.8% | 25 | .856 | .847 | .837 | .825 | .813 | .800 | .786 | .771 | .755 | .737 |
| | | | | | | | | .718 | .698 | .676 | .653 | .628 | .602 | .575 | .546 | .516 | .485 |
| | | | | | | | | .453 | .420 | .387 | .354 | .322 | | | | | |
| 18.000 | 16.348 | .000 | 1.652 | .71 | 4.75 | 9.2% | 26 | .856 | .847 | .837 | .825 | .813 | .800 | .786 | .771 | .755 | .737 |
| | | | | | | | | .718 | .698 | .676 | .653 | .628 | .602 | .575 | .546 | .516 | .485 |
| | | | | | | | | .453 | .420 | .387 | .354 | .322 | .289 | | | | |
| 18.000 | 16.606 | .000 | 1.394 | .74 | 4.72 | 7.7% | 27 | .856 | .847 | .837 | .825 | .813 | .800 | .786 | .771 | .755 | .737 |
| | | | | | | | | .718 | .698 | .676 | .653 | .628 | .602 | .575 | .546 | .516 | .485 |
| | | | | | | | | .453 | .420 | .387 | .354 | .322 | .289 | .258 | | | |
| 18.000 | 16.834 | .000 | 1.166 | .77 | 4.68 | 6.5% | 28 | .856 | .847 | .837 | .825 | .813 | .800 | .786 | .771 | .755 | .737 |
| | | | | | | | | .718 | .698 | .676 | .653 | .628 | .602 | .575 | .546 | .516 | .485 |
| | | | | | | | | .453 | .420 | .387 | .354 | .322 | .289 | .258 | .228 | | |
| 18.000 | 17.033 | .000 | .967 | .80 | 4.63 | 5.4% | 29 | .856 | .847 | .837 | .825 | .813 | .800 | .786 | .771 | .755 | .737 |
| | | | | | | | | .718 | .698 | .676 | .653 | .628 | .602 | .575 | .546 | .516 | .485 |
| | | | | | | | | .453 | .420 | .387 | .354 | .322 | .289 | .258 | .228 | .199 | |
| 18.000 | 17.206 | .000 | .794 | .83 | 4.57 | 4.4% | 30 | .856 | .847 | .837 | .825 | .813 | .800 | .786 | .771 | .755 | .737 |
| | | | | | | | | .718 | .698 | .676 | .653 | .628 | .602 | .575 | .546 | .516 | .485 |
| | | | | | | | | .453 | .420 | .387 | .354 | .322 | .289 | .258 | .228 | .199 | .173 |
| 18.000 | 17.354 | .000 | .646 | .85 | 4.50 | 3.6% | 31 | .856 | .847 | .837 | .825 | .813 | .800 | .786 | .771 | .755 | .737 |
| | | | | | | | | .718 | .698 | .676 | .653 | .628 | .602 | .575 | .546 | .516 | .485 |
| | | | | | | | | .453 | .420 | .387 | .354 | .322 | .289 | .258 | .228 | .199 | .173 |
| | | | | | | | | .148 | | | | | | | | | |
| 18.000 | 17.479 | .000 | .521 | .87 | 4.43 | 2.9% | 32 | .856 | .847 | .837 | .825 | .813 | .800 | .786 | .771 | .755 | .737 |
| | | | | | | | | .718 | .698 | .676 | .653 | .628 | .602 | .575 | .546 | .516 | .485 |
| | | | | | | | | .453 | .420 | .387 | .354 | .322 | .289 | .258 | .228 | .199 | .173 |
| | | | | | | | | .148 | .125 | | | | | | | | |
| 18.000 | 17.585 | .000 | .415 | .89 | 4.35 | 2.3% | 33 | .856 | .847 | .837 | .825 | .813 | .800 | .786 | .771 | .755 | .737 |
| | | | | | | | | .718 | .698 | .676 | .653 | .628 | .602 | .575 | .546 | .516 | .485 |
| | | | | | | | | .453 | .420 | .387 | .354 | .322 | .289 | .258 | .228 | .199 | .173 |
| | | | | | | | | .148 | .125 | .105 | | | | | | | |
| 18.000 | 17.672 | .000 | .328 | .91 | 4.26 | 1.8% | 34 | .856 | .847 | .837 | .825 | .813 | .800 | .786 | .771 | .755 | .737 |
| | | | | | | | | .718 | .698 | .676 | .653 | .628 | .602 | .575 | .546 | .516 | .485 |
| | | | | | | | | .453 | .420 | .387 | .354 | .322 | .289 | .258 | .228 | .199 | .173 |
| | | | | | | | | .148 | .125 | .105 | .087 | | | | | | |
| 18.000 | 17.743 | .000 | .257 | .93 | 4.17 | 1.4% | 35 | .856 | .847 | .837 | .825 | .813 | .800 | .786 | .771 | .755 | .737 |
| | | | | | | | | .718 | .698 | .676 | .653 | .628 | .602 | .575 | .546 | .516 | .485 |
| | | | | | | | | .453 | .420 | .387 | .354 | .322 | .289 | .258 | .228 | .199 | .173 |
| | | | | | | | | .148 | .125 | .105 | .087 | .071 | | | | | |
| 18.000 | 17.801 | .000 | .199 | .94 | 4.08 | 1.1% | 36 | .856 | .847 | .837 | .825 | .813 | .800 | .786 | .771 | .755 | .737 |
| | | | | | | | | .718 | .698 | .676 | .653 | .628 | .602 | .575 | .546 | .516 | .485 |

*SIMILAR TO A.T.&T. NEAL-WILKINSON TABLES(SEE SECTION 1)

*EQUIVALENT RANDOM THEORY: VMR(VARIANCE TO MEAN RATIO)– 3.0 - NO RETRIES (PEAKED TRAFFIC - ROUTE ADVANCEMENT OR DAY-TO-DAY VARIATION)

| HOURS FIRST ATTEMPT VMR=3.0 | HOURS CON-NECT | HOURS RECALL /RETRY | HOURS OVFL | CON-NECT VMR | OVFL VMR | PERCENT BLOCKED (GRADE) | LINE | HOURS CONNECTED TRAFFIC ON LINES 1 | 2 | 3 | 4 | 5 | 6 | 7 | 8 | 9 | 10 |
|---|---|---|---|---|---|---|---|---|---|---|---|---|---|---|---|---|---|
| | | | | | | | | .453 | .420 | .387 | .354 | .322 | .289 | .258 | .228 | .199 | .173 |
| | | | | | | | | .148 | .125 | .105 | .087 | .071 | .058 | | | | |
| 18.000 | 17.848 | .000 | .152 | .95 | 3.99 | .8% | 37 | .856 | .847 | .837 | .825 | .813 | .800 | .786 | .771 | .755 | .737 |
| | | | | | | | | .718 | .698 | .676 | .653 | .628 | .602 | .575 | .546 | .516 | .485 |
| | | | | | | | | .453 | .420 | .387 | .354 | .322 | .289 | .258 | .228 | .199 | .173 |
| | | | | | | | | .148 | .125 | .105 | .087 | .071 | .058 | .046 | | | |
| 18.000 | 17.884 | .000 | .116 | .96 | 3.90 | .6% | 38 | .856 | .847 | .837 | .825 | .813 | .800 | .786 | .771 | .755 | .737 |
| | | | | | | | | .710 | .898 | .676 | .653 | .628 | .602 | .575 | .546 | .516 | .485 |
| | | | | | | | | .453 | .420 | .387 | .354 | .322 | .289 | .258 | .228 | .199 | .173 |
| | | | | | | | | .148 | .125 | .105 | .087 | .071 | .058 | .046 | .037 | | |
| 19.000 | 12.846 | .000 | 6.154 | .40 | 4.52 | 32.4% | 17 | .863 | .854 | .845 | .834 | .823 | .811 | .798 | .784 | .769 | .753 |
| | | | | | | | | .736 | .717 | .697 | .676 | .653 | .629 | .603 | | | |
| 19.000 | 13.422 | .000 | 5.578 | .42 | 4.59 | 29.4% | 18 | .863 | .854 | .845 | .834 | .823 | .811 | .798 | .784 | .769 | .753 |
| | | | | | | | | .736 | .717 | .697 | .676 | .653 | .629 | .603 | .577 | | |
| 19.000 | 13.971 | .000 | 5.029 | .45 | 4.66 | 26.5% | 19 | .863 | .854 | .845 | .834 | .823 | .811 | .798 | .784 | .769 | .753 |
| | | | | | | | | .736 | .717 | .697 | .676 | .653 | .629 | .603 | .577 | .548 | |
| 19.000 | 14.490 | .000 | 4.510 | .48 | 4.71 | 23.7% | 20 | .863 | .854 | .845 | .834 | .823 | .811 | .798 | .784 | .769 | .753 |
| | | | | | | | | .736 | .717 | .697 | .676 | .653 | .629 | .603 | .577 | .548 | .519 |
| 19.000 | 14.979 | .000 | 4.021 | .51 | 4.76 | 21.2% | 21 | .863 | .854 | .845 | .834 | .823 | .811 | .798 | .784 | .769 | .753 |
| | | | | | | | | .736 | .717 | .697 | .676 | .653 | .629 | .603 | .577 | .548 | .519 |
| | | | | | | | | .489 | | | | | | | | | |
| 19.000 | 15.437 | .000 | 3.563 | .54 | 4.80 | 18.8% | 22 | .863 | .854 | .845 | .834 | .823 | .811 | .798 | .784 | .769 | .753 |
| | | | | | | | | .736 | .717 | .697 | .676 | .653 | .629 | .603 | .577 | .548 | .519 |
| | | | | | | | | .489 | .458 | | | | | | | | |
| 19.000 | 15.863 | .000 | 3.137 | .57 | 4.83 | 16.5% | 23 | .863 | .854 | .845 | .834 | .823 | .811 | .790 | .784 | .769 | .753 |
| | | | | | | | | .736 | .717 | .097 | .078 | .653 | .629 | .603 | .577 | .548 | .519 |
| | | | | | | | | .489 | .458 | .426 | | | | | | | |
| 19.000 | 16.256 | .000 | 2.744 | .61 | 4.85 | 14.4% | 24 | .863 | .854 | .845 | .834 | .823 | .811 | .798 | .784 | .769 | .753 |
| | | | | | | | | .736 | .717 | .697 | .676 | .653 | .629 | .603 | .577 | .548 | .519 |
| | | | | | | | | .489 | .458 | .426 | .394 | | | | | | |
| 19.000 | 16.617 | .000 | 2.383 | .64 | 4.86 | 12.5% | 25 | .863 | .854 | .845 | .834 | .823 | .811 | .798 | .784 | .769 | .753 |
| | | | | | | | | .736 | .717 | .697 | .676 | .653 | .629 | .603 | .577 | .548 | .519 |
| | | | | | | | | .489 | .458 | .420 | .394 | .361 | | | | | |
| 10.000 | 10.946 | .000 | 2.054 | .67 | 4.86 | 10.8% | 26 | .863 | .854 | .845 | .834 | .823 | .811 | .798 | .784 | .769 | .753 |
| | | | | | | | | .736 | .717 | .697 | .676 | .653 | .629 | .603 | .577 | .548 | .519 |
| | | | | | | | | .489 | .458 | .426 | .394 | .361 | .329 | | | | |
| 19.000 | 17.243 | .000 | 1.757 | .70 | 4.84 | 9.2% | 27 | .863 | .854 | .845 | .834 | .823 | .811 | .798 | .784 | .769 | .753 |
| | | | | | | | | .736 | .717 | .697 | .676 | .653 | .629 | .603 | .577 | .548 | .519 |
| | | | | | | | | .489 | .458 | .426 | .394 | .361 | .329 | .297 | | | |
| 19.000 | 17.510 | .000 | 1.490 | .73 | 4.82 | 7.8% | 28 | .863 | .854 | .845 | .834 | .823 | .811 | .798 | .784 | .769 | .753 |

*SIMILAR TO A.T.&T. NEAL-WILKINSON TABLES(SEE SECTION 1)

*EQUIVALENT RANDOM THEORY: VMR(VARIANCE TO MEAN RATIO)– 3.0 - NO RETRIES (PEAKED TRAFFIC - ROUTE ADVANCEMENT OR DAY-TO-DAY VARIATION)

| HOURS FIRST ATTEMPT VMR=3.0 | HOURS CON-NECT | HOURS RECALL /RETRY | HOURS OVFL | CON-NECT VMR | OVFL VMR | PERCENT BLOCKED (GRADE) | LINE | HOURS CONNECTED TRAFFIC ON LINES 1 | 2 | 3 | 4 | 5 | 6 | 7 | 8 | 9 | 10 |
|---|---|---|---|---|---|---|---|---|---|---|---|---|---|---|---|---|---|
| | | | | | | | | .736 | .717 | .697 | .676 | .653 | .629 | .603 | .577 | .548 | .519 |
| | | | | | | | | .489 | .458 | .426 | .394 | .361 | .329 | .297 | .266 | | |
| 19.000 | 17.746 | .000 | 1.254 | .76 | 4.78 | 6.6% | 29 | .863 | .854 | .845 | .834 | .823 | .811 | .798 | .784 | .769 | .753 |
| | | | | | | | | .736 | .717 | .697 | .676 | .653 | .629 | .603 | .577 | .548 | .519 |
| | | | | | | | | .489 | .458 | .426 | .394 | .361 | .329 | .297 | .266 | .236 | |
| 19.000 | 17.954 | .000 | 1.046 | .79 | 4.73 | 5.5% | 30 | .863 | .854 | .845 | .834 | .823 | .811 | .798 | .784 | .769 | .753 |
| | | | | | | | | .736 | .717 | .697 | .676 | .653 | .629 | .603 | .577 | .548 | .519 |
| | | | | | | | | .489 | .458 | .426 | .394 | .361 | .329 | .297 | .266 | .236 | .208 |
| 19.000 | 18.135 | .000 | .865 | .82 | 4.67 | 4.6% | 31 | .863 | .854 | .845 | .834 | .823 | .811 | .798 | .784 | .769 | .753 |
| | | | | | | | | .736 | .717 | .697 | .676 | .653 | .629 | .603 | .577 | .548 | .519 |
| | | | | | | | | .489 | .458 | .426 | .394 | .361 | .329 | .297 | .266 | .236 | .208 |
| | | | | | | | | .181 | | | | | | | | | |
| 19.000 | 18.291 | .000 | .709 | .84 | 4.61 | 3.7% | 32 | .863 | .854 | .845 | .834 | .823 | .811 | .798 | .784 | .769 | .753 |
| | | | | | | | | .736 | .717 | .697 | .676 | .653 | .629 | .603 | .577 | .548 | .519 |
| | | | | | | | | .489 | .458 | .426 | .394 | .361 | .329 | .297 | .266 | .236 | .208 |
| | | | | | | | | .181 | .156 | | | | | | | | |
| 19.000 | 18.425 | .000 | .575 | .87 | 4.53 | 3.0% | 33 | .863 | .854 | .845 | .834 | .823 | .811 | .798 | .784 | .769 | .753 |
| | | | | | | | | .736 | .717 | .697 | .676 | .653 | .629 | .603 | .577 | .548 | .519 |
| | | | | | | | | .489 | .458 | .426 | .394 | .361 | .329 | .297 | .266 | .236 | .208 |
| | | | | | | | | .181 | .156 | .133 | | | | | | | |
| 19.000 | 18.537 | .000 | .463 | .89 | 4.45 | 2.4% | 34 | .863 | .854 | .845 | .834 | .823 | .811 | .798 | .784 | .769 | .753 |
| | | | | | | | | .736 | .717 | .697 | .676 | .653 | .629 | .603 | .577 | .548 | .519 |
| | | | | | | | | .489 | .458 | .426 | .394 | .361 | .329 | .297 | .266 | .236 | .208 |
| | | | | | | | | .181 | .156 | .133 | .113 | | | | | | |
| 19.000 | 18.631 | .000 | .369 | .91 | 4.37 | 1.9% | 35 | .863 | .854 | .845 | .834 | .823 | .811 | .798 | .784 | .769 | .753 |
| | | | | | | | | .736 | .717 | .697 | .676 | .653 | .629 | .603 | .577 | .548 | .519 |
| | | | | | | | | .489 | .458 | .426 | .394 | .361 | .329 | .297 | .266 | .236 | .208 |
| | | | | | | | | .181 | .156 | .133 | .113 | .094 | | | | | |
| 19.000 | 18.709 | .000 | .291 | .92 | 4.28 | 1.5% | 36 | .863 | .854 | .845 | .834 | .823 | .811 | .798 | .784 | .769 | .753 |
| | | | | | | | | .736 | .717 | .697 | .676 | .653 | .629 | .603 | .577 | .548 | .519 |
| | | | | | | | | .489 | .458 | .426 | .394 | .361 | .329 | .297 | .266 | .236 | .208 |
| | | | | | | | | .181 | .156 | .133 | .113 | .094 | .078 | | | | |
| 19.000 | 18.773 | .000 | .227 | .94 | 4.19 | 1.2% | 37 | .863 | .854 | .845 | .834 | .823 | .811 | .798 | .784 | .769 | .753 |
| | | | | | | | | .736 | .717 | .697 | .676 | .653 | .629 | .603 | .577 | .548 | .519 |
| | | | | | | | | .489 | .458 | .426 | .394 | .361 | .329 | .297 | .266 | .236 | .208 |
| | | | | | | | | .181 | .156 | .133 | .113 | .094 | .078 | .064 | | | |
| 19.000 | 18.824 | .000 | .176 | .95 | 4.10 | .9% | 38 | .863 | .854 | .845 | .834 | .823 | .811 | .798 | .784 | .769 | .753 |
| | | | | | | | | .736 | .717 | .697 | .676 | .653 | .629 | .603 | .577 | .548 | .519 |
| | | | | | | | | .489 | .458 | .426 | .394 | .361 | .329 | .297 | .266 | .236 | .208 |
| | | | | | | | | .181 | .156 | .133 | .113 | .094 | .078 | .064 | .051 | | |

*SIMILAR TO A.T.&T. NEAL-WILKINSON TABLES(SEE SECTION 1)

*EQUIVALENT RANDOM THEORY: VMR(VARIANCE TO MEAN RATIO)– 3.0 - NO RETRIES (PEAKED TRAFFIC - ROUTE ADVANCEMENT OR DAY-TO-DAY VARIATION)

| HOURS FIRST ATTEMPT VMR=3.0 | HOURS CON-NECT | HOURS RECALL /RETRY | HOURS OVFL | CON-NECT VMR | OVFL VMR | PERCENT BLOCKED (GRADE) | LINE | HOURS CONNECTED TRAFFIC ON LINES 1 | 2 | 3 | 4 | 5 | 6 | 7 | 8 | 9 | 10 |
|---|---|---|---|---|---|---|---|---|---|---|---|---|---|---|---|---|---|
| 19.000 | 18.865 | .000 | .135 | .96 | 4.01 | .7% | 39 | .863 | .854 | .845 | .834 | .823 | .811 | .798 | .784 | .769 | .753 |
| | | | | | | | | .736 | .717 | .697 | .676 | .653 | .629 | .603 | .577 | .548 | .519 |
| | | | | | | | | .489 | .458 | .426 | .394 | .361 | .329 | .297 | .266 | .236 | .208 |
| | | | | | | | | .181 | .156 | .133 | .113 | .094 | .078 | .064 | .051 | .041 | |
| 19.000 | 18.898 | .000 | .102 | .97 | 3.92 | .5% | 40 | .863 | .854 | .845 | .834 | .823 | .811 | .798 | .784 | .769 | .753 |
| | | | | | | | | .736 | .717 | .697 | .676 | .653 | .629 | .603 | .577 | .548 | .519 |
| | | | | | | | | .489 | .458 | .426 | .394 | .361 | .329 | .297 | .266 | .236 | .208 |
| | | | | | | | | .181 | .156 | .133 | .113 | .094 | .078 | .064 | .051 | .041 | .033 |
| 20.000 | 13.090 | .000 | 6.910 | .37 | 4.52 | 34.6% | 17 | .869 | .861 | .852 | .842 | .832 | .821 | .809 | .796 | .782 | .768 |
| | | | | | | | | .752 | .734 | .716 | .696 | .675 | .653 | .630 | | | |
| 20.000 | 13.694 | .000 | 6.306 | .40 | 4.60 | 31.5% | 18 | .869 | .861 | .852 | .842 | .832 | .821 | .809 | .796 | .782 | .768 |
| | | | | | | | | .752 | .734 | .716 | .696 | .675 | .653 | .630 | .605 | | |
| 20.000 | 14.273 | .000 | 5.727 | .42 | 4.67 | 28.6% | 19 | .869 | .861 | .852 | .842 | .832 | .821 | .809 | .796 | .782 | .768 |
| | | | | | | | | .752 | .734 | .716 | .696 | .675 | .653 | .630 | .605 | .578 | |
| 20.000 | 14.824 | .000 | 5.176 | .45 | 4.73 | 25.9% | 20 | .869 | .861 | .852 | .842 | .832 | .821 | .809 | .796 | .782 | .768 |
| | | | | | | | | .752 | .734 | .716 | .696 | .675 | .653 | .630 | .605 | .578 | .551 |
| 20.000 | 15.346 | .000 | 4.654 | .48 | 4.79 | 23.3% | 21 | .869 | .861 | .852 | .842 | .832 | .821 | .809 | .796 | .782 | .768 |
| | | | | | | | | .752 | .734 | .716 | .696 | .675 | .653 | .630 | .605 | .578 | .551 |
| | | | | | | | | .522 | | | | | | | | | |
| 20.000 | 15.839 | .000 | 4.161 | .51 | 4.84 | 20.8% | 22 | .869 | .861 | .852 | .842 | .832 | .821 | .809 | .796 | .782 | .768 |
| | | | | | | | | .752 | .734 | .716 | .696 | .675 | .653 | .630 | .605 | .578 | .551 |
| | | | | | | | | .522 | .493 | | | | | | | | |
| 20.000 | 16.301 | .000 | 3.699 | .54 | 4.88 | 18.5% | 23 | .869 | .861 | .852 | .842 | .832 | .821 | .809 | .796 | .782 | .768 |
| | | | | | | | | .752 | .734 | .716 | .696 | .675 | .653 | .630 | .605 | .578 | .551 |
| | | | | | | | | .522 | .493 | .462 | | | | | | | |
| 20.000 | 16.732 | .000 | 3.268 | .57 | 4.91 | 16.3% | 24 | .869 | .861 | .852 | .842 | .832 | .821 | .809 | .796 | .782 | .768 |
| | | | | | | | | .752 | .734 | .716 | .696 | .675 | .653 | .630 | .605 | .578 | .551 |
| | | | | | | | | .522 | .493 | .462 | .431 | | | | | | |
| 20.000 | 17.131 | .000 | 2.869 | .60 | 4.93 | 14.3% | 25 | .869 | .861 | .852 | .842 | .832 | .821 | .809 | .796 | .782 | .768 |
| | | | | | | | | .752 | .734 | .716 | .696 | .675 | .653 | .630 | .605 | .578 | .551 |
| | | | | | | | | .522 | .493 | .462 | .431 | .399 | | | | | |
| 20.000 | 17.499 | .000 | 2.501 | .63 | 4.94 | 12.5% | 26 | .869 | .861 | .852 | .842 | .832 | .821 | .809 | .796 | .782 | .768 |
| | | | | | | | | .752 | .734 | .716 | .696 | .675 | .653 | .630 | .605 | .578 | .551 |
| | | | | | | | | .522 | .493 | .462 | .431 | .399 | .368 | | | | |
| 20.000 | 17.835 | .000 | 2.165 | .66 | 4.94 | 10.8% | 27 | .869 | .861 | .852 | .842 | .832 | .821 | .809 | .796 | .782 | .768 |
| | | | | | | | | .752 | .734 | .716 | .696 | .675 | .653 | .630 | .605 | .578 | .551 |
| | | | | | | | | .522 | .493 | .462 | .431 | .399 | .368 | .336 | | | |
| 20.000 | 18.139 | .000 | 1.861 | .70 | 4.93 | 9.3% | 28 | .869 | .861 | .852 | .842 | .832 | .821 | .809 | .796 | .782 | .768 |
| | | | | | | | | .752 | .734 | .716 | .696 | .675 | .653 | .630 | .605 | .578 | .551 |
| | | | | | | | | .522 | .493 | .462 | .431 | .399 | .368 | .336 | .305 | | |

*SIMILAR TO A.T.&T. NEAL-WILKINSON TABLES(SEE SECTION 1)

*EQUIVALENT RANDOM THEORY: VMR(VARIANCE TO MEAN RATIO)– 3.0 - NO RETRIES (PEAKED TRAFFIC - ROUTE ADVANCEMENT OR DAY-TO-DAY VARIATION)

| HOURS FIRST ATTEMPT VMR = 3.0 | HOURS CON-NECT | HOURS RECALL /RETRY | HOURS OVFL | CON-NECT VMR | OVFL VMR | PERCENT BLOCKED (GRADE) | LINE | HOURS CONNECTED TRAFFIC ON LINES 1 | 2 | 3 | 4 | 5 | 6 | 7 | 8 | 9 | 10 |
|---|---|---|---|---|---|---|---|---|---|---|---|---|---|---|---|---|---|
| 20.000 | 18.413 | .000 | 1.587 | .73 | 4.91 | 7.9% | 29 | .869 | .861 | .852 | .842 | .832 | .821 | .809 | .796 | .782 | .768 |
| | | | | | | | | .752 | .734 | .716 | .696 | .675 | .653 | .630 | .605 | .578 | .551 |
| | | | | | | | | .522 | .493 | .462 | .431 | .399 | .368 | .336 | .305 | .274 | |
| 20.000 | 18.658 | .000 | 1.342 | .76 | 4.87 | 6.7% | 30 | .869 | .861 | .852 | .842 | .832 | .821 | .809 | .796 | .782 | .768 |
| | | | | | | | | .752 | .734 | .716 | .696 | .675 | .653 | .630 | .605 | .578 | .551 |
| | | | | | | | | .522 | .493 | .462 | .431 | .399 | .368 | .336 | .305 | .274 | .244 |
| 20.000 | 18.874 | .000 | 1.126 | .78 | 4.83 | 5.6% | 31 | .869 | .861 | .852 | .842 | .832 | .821 | .809 | .796 | .782 | .768 |
| | | | | | | | | .752 | .734 | .716 | .696 | .675 | .653 | .630 | .605 | .578 | .551 |
| | | | | | | | | .522 | .493 | .462 | .431 | .399 | .368 | .336 | .305 | .274 | .244 |
| | | | | | | | | .216 | | | | | | | | | |
| 20.000 | 19.063 | .000 | .937 | .81 | 4.77 | 4.7% | 32 | .869 | .861 | .852 | .842 | .832 | .821 | .809 | .796 | .782 | .768 |
| | | | | | | | | .752 | .734 | .716 | .696 | .675 | .653 | .630 | .605 | .578 | .551 |
| | | | | | | | | .522 | .493 | .462 | .431 | .399 | .368 | .336 | .305 | .274 | .244 |
| | | | | | | | | .216 | .189 | | | | | | | | |
| 20.000 | 19.227 | .000 | .773 | .84 | 4.71 | 3.9% | 33 | .869 | .861 | .852 | .842 | .832 | .821 | .809 | .796 | .782 | .768 |
| | | | | | | | | .752 | .734 | .716 | .696 | .675 | .653 | .630 | .605 | .578 | .551 |
| | | | | | | | | .522 | .493 | .462 | .431 | .399 | .368 | .336 | .305 | .274 | .244 |
| | | | | | | | | .216 | .189 | .164 | | | | | | | |
| 20.000 | 19.368 | .000 | .632 | .86 | 4.64 | 3.2% | 34 | .869 | .861 | .852 | .842 | .832 | .821 | .809 | .796 | .782 | .768 |
| | | | | | | | | .752 | .734 | .716 | .696 | .675 | .653 | .630 | .605 | .578 | .551 |
| | | | | | | | | .522 | .493 | .462 | .431 | .399 | .368 | .336 | .305 | .274 | .244 |
| | | | | | | | | .216 | .189 | .164 | .141 | | | | | | |
| 20.000 | 19.488 | .000 | .512 | .88 | 4.56 | 2.6% | 35 | .869 | .861 | .852 | .842 | .832 | .821 | .809 | .796 | .782 | .768 |
| | | | | | | | | .752 | .734 | .716 | .696 | .675 | .653 | .630 | .605 | .578 | .551 |
| | | | | | | | | .522 | .493 | .462 | .431 | .399 | .368 | .336 | .305 | .274 | .244 |
| | | | | | | | | .216 | .189 | .164 | .141 | .120 | | | | | |
| 20.000 | 19.589 | .000 | .411 | .90 | 4.48 | 2.1% | 36 | .869 | .861 | .852 | .842 | .832 | .821 | .809 | .796 | .782 | .768 |
| | | | | | | | | .752 | .734 | .716 | .696 | .675 | .653 | .630 | .605 | .578 | .551 |
| | | | | | | | | .522 | .493 | .462 | .431 | .399 | .368 | .336 | .305 | .274 | .244 |
| | | | | | | | | .216 | .189 | .164 | .141 | .120 | .101 | | | | |
| 20.000 | 19.673 | .000 | .327 | .92 | 4.39 | 1.6% | 37 | .869 | .861 | .852 | .842 | .832 | .821 | .809 | .796 | .782 | .768 |
| | | | | | | | | .752 | .734 | .716 | .696 | .675 | .653 | .630 | .605 | .578 | .551 |
| | | | | | | | | .522 | .493 | .462 | .431 | .399 | .368 | .336 | .305 | .274 | .244 |
| | | | | | | | | .216 | .189 | .164 | .141 | .120 | .101 | .084 | | | |
| 20.000 | 19.742 | .000 | .258 | .93 | 4.30 | 1.3% | 38 | .869 | .861 | .852 | .842 | .832 | .821 | .809 | .796 | .782 | .768 |
| | | | | | | | | .752 | .734 | .716 | .696 | .675 | .653 | .630 | .605 | .578 | .551 |
| | | | | | | | | .522 | .493 | .462 | .431 | .399 | .368 | .336 | .305 | .274 | .244 |
| | | | | | | | | .216 | .189 | .164 | .141 | .120 | .101 | .084 | .069 | | |
| 20.000 | 19.799 | .000 | .201 | .94 | 4.21 | 1.0% | 39 | .869 | .861 | .852 | .842 | .832 | .821 | .809 | .796 | .782 | .768 |
| | | | | | | | | .752 | .734 | .716 | .696 | .675 | .653 | .630 | .605 | .578 | .551 |

*SIMILAR TO A.T.&T. NEAL-WILKINSON TABLES(SEE SECTION 1)

*EQUIVALENT RANDOM THEORY: VMR(VARIANCE TO MEAN RATIO)– 3.0 - NO RETRIES (PEAKED TRAFFIC - ROUTE ADVANCEMENT OR DAY-TO-DAY VARIATION)

| HOURS FIRST ATTEMPT VMR=3.0 | HOURS CON-NECT | HOURS RECALL /RETRY | HOURS OVFL | CON-NECT VMR | OVFL VMR | PERCENT BLOCKED (GRADE) | LINE | HOURS CONNECTED TRAFFIC ON LINES 1 | 2 | 3 | 4 | 5 | 6 | 7 | 8 | 9 | 10 |
|---|---|---|---|---|---|---|---|---|---|---|---|---|---|---|---|---|---|
| | | | | | | | | .522 | .493 | .462 | .431 | .399 | .368 | .336 | .305 | .274 | .244 |
| | | | | | | | | .216 | .189 | .164 | .141 | .120 | .101 | .084 | .069 | .057 | |
| 20.000 | 19.845 | .000 | .155 | .95 | 4.12 | .8% | 40 | .869 | .861 | .852 | .842 | .832 | .821 | .809 | .796 | .782 | .768 |
| | | | | | | | | .752 | .734 | .716 | .696 | .675 | .653 | .630 | .605 | .578 | .551 |
| | | | | | | | | .522 | .493 | .462 | .431 | .399 | .368 | .336 | .305 | .274 | .244 |
| | | | | | | | | .216 | .189 | .164 | .141 | .120 | .101 | .084 | .069 | .057 | .046 |
| 20.000 | 19.881 | .000 | .119 | .96 | 4.03 | .6% | 41 | .869 | .861 | .852 | .842 | .832 | .821 | .809 | .796 | .782 | .768 |
| | | | | | | | | .752 | .734 | .716 | .696 | .675 | .653 | .630 | .605 | .578 | .551 |
| | | | | | | | | .522 | .493 | .462 | .431 | .399 | .368 | .336 | .305 | .274 | .244 |
| | | | | | | | | .216 | .189 | .164 | .141 | .120 | .101 | .084 | .069 | .057 | .046 |
| | | | | | | | | .036 | | | | | | | | | |
| 22.000 | 14.796 | .000 | 7.204 | .37 | 4.67 | 32.7% | 19 | .879 | .872 | .865 | .857 | .848 | .838 | .828 | .817 | .805 | .793 |
| | | | | | | | | .779 | .764 | .749 | .732 | .714 | .695 | .675 | .654 | .631 | |
| 22.000 | 15.402 | .000 | 6.598 | .39 | 4.75 | 30.0% | 20 | .879 | .872 | .865 | .857 | .848 | .838 | .828 | .817 | .805 | .793 |
| | | | | | | | | .779 | .764 | .749 | .732 | .714 | .695 | .675 | .654 | .631 | .607 |
| 22.000 | 15.984 | .000 | 6.016 | .42 | 4.82 | 27.3% | 21 | .879 | .872 | .865 | .857 | .848 | .838 | .828 | .817 | .805 | .793 |
| | | | | | | | | .779 | .764 | .749 | .732 | .714 | .695 | .675 | .654 | .631 | .607 |
| | | | | | | | | .582 | | | | | | | | | |
| 22.000 | 16.539 | .000 | 5.461 | .44 | 4.88 | 24.8% | 22 | .879 | .872 | .865 | .857 | .848 | .838 | .828 | .817 | .805 | .793 |
| | | | | | | | | .779 | .764 | .749 | .732 | .714 | .695 | .675 | .654 | .631 | .607 |
| | | | | | | | | .582 | .555 | | | | | | | | |
| 22.000 | 17.067 | .000 | 4.933 | .47 | 4.94 | 22.4% | 23 | .879 | .872 | .865 | .857 | .848 | .838 | .828 | .817 | .805 | .793 |
| | | | | | | | | .779 | .764 | .749 | .732 | .714 | .695 | .675 | .654 | .631 | .607 |
| | | | | | | | | .582 | .555 | .528 | | | | | | | |
| 22.000 | 17.567 | .000 | 4.433 | .50 | 4.99 | 20.2% | 24 | .879 | .872 | .865 | .857 | .848 | .838 | .828 | .817 | .805 | .793 |
| | | | | | | | | .779 | .764 | .749 | .732 | .714 | .695 | .675 | .654 | .631 | .607 |
| | | | | | | | | .582 | .555 | .528 | .500 | | | | | | |
| 22.000 | 18.037 | .000 | 3.963 | .53 | 5.04 | 18.0% | 25 | .879 | .872 | .865 | .857 | .848 | .838 | .828 | .817 | .805 | .793 |
| | | | | | | | | .779 | .764 | .749 | .732 | .714 | .695 | .675 | .654 | .631 | .607 |
| | | | | | | | | .582 | .555 | .528 | .500 | .470 | | | | | |
| 22.000 | 18.477 | .000 | 3.523 | .56 | 5.07 | 16.0% | 26 | .879 | .872 | .865 | .857 | .848 | .838 | .828 | .817 | .805 | .793 |
| | | | | | | | | .779 | .764 | .749 | .732 | .714 | .695 | .675 | .654 | .631 | .607 |
| | | | | | | | | .582 | .555 | .528 | .500 | .470 | .440 | | | | |
| 22.000 | 18.887 | .000 | 3.113 | .59 | 5.09 | 14.1% | 27 | .879 | .872 | .865 | .857 | .848 | .838 | .828 | .817 | .805 | .793 |
| | | | | | | | | .779 | .764 | .749 | .732 | .714 | .695 | .675 | .654 | .631 | .607 |
| | | | | | | | | .582 | .555 | .528 | .500 | .470 | .440 | .410 | | | |
| 22.000 | 19.267 | .000 | 2.733 | .62 | 5.11 | 12.4% | 28 | .879 | .872 | .865 | .857 | .848 | .838 | .828 | .817 | .805 | .793 |
| | | | | | | | | .779 | .764 | .749 | .732 | .714 | .695 | .675 | .654 | .631 | .607 |
| | | | | | | | | .582 | .555 | .528 | .500 | .470 | .440 | .410 | .379 | | |
| 22.000 | 19.615 | .000 | 2.385 | .65 | 5.11 | 10.8% | 29 | .879 | .872 | .865 | .857 | .848 | .838 | .828 | .817 | .805 | .793 |

*SIMILAR TO A.T.&T. NEAL-WILKINSON TABLES(SEE SECTION 1)

*EQUIVALENT RANDOM THEORY: VMR(VARIANCE TO MEAN RATIO)– 3.0 - NO RETRIES (PEAKED TRAFFIC - ROUTE ADVANCEMENT OR DAY-TO-DAY VARIATION)

| HOURS FIRST ATTEMPT VMR=3.0 | HOURS CON-NECT | HOURS RECALL /RETRY | HOURS OVFL | CON-NECT VMR | OVFL VMR | PERCENT BLOCKED (GRADE) | LINE | HOURS CONNECTED TRAFFIC ON LINES 1 | 2 | 3 | 4 | 5 | 6 | 7 | 8 | 9 | 10 |
|---|---|---|---|---|---|---|---|---|---|---|---|---|---|---|---|---|---|
| | | | | | | | | .779 | .764 | .749 | .732 | .714 | .695 | .675 | .654 | .631 | .607 |
| | | | | | | | | .582 | .555 | .528 | .500 | .470 | .440 | .410 | .379 | .349 | |
| 22.000 | 19.934 | .000 | 2.066 | .68 | 5.10 | 9.4% | 30 | .879 | .872 | .865 | .857 | .848 | .838 | .828 | .817 | .805 | .793 |
| | | | | | | | | .779 | .764 | .749 | .732 | .714 | .695 | .675 | .654 | .631 | .607 |
| | | | | | | | | .582 | .555 | .528 | .500 | .470 | .440 | .410 | .379 | .349 | .318 |
| 22.000 | 20.222 | .000 | 1.778 | .71 | 5.08 | 8.1% | 31 | .879 | .872 | .865 | .857 | .848 | .838 | .828 | .817 | .805 | .793 |
| | | | | | | | | .779 | .764 | .749 | .732 | .714 | .695 | .675 | .654 | .631 | .607 |
| | | | | | | | | .582 | .555 | .528 | .500 | .470 | .440 | .410 | .379 | .349 | .318 |
| | | | | | | | | .288 | | | | | | | | | |
| 22.000 | 20.481 | .000 | 1.519 | .74 | 5.05 | 6.9% | 32 | .879 | .872 | .865 | .857 | .848 | .838 | .828 | .817 | .805 | .793 |
| | | | | | | | | .779 | .764 | .749 | .732 | .714 | .695 | .675 | .654 | .631 | .607 |
| | | | | | | | | .582 | .555 | .528 | .500 | .470 | .440 | .410 | .379 | .349 | .318 |
| | | | | | | | | .288 | .259 | | | | | | | | |
| 22.000 | 20.712 | .000 | 1.288 | .77 | 5.01 | 5.9% | 33 | .879 | .872 | .865 | .857 | .848 | .838 | .828 | .817 | .805 | .793 |
| | | | | | | | | .779 | .764 | .749 | .732 | .714 | .695 | .675 | .654 | .631 | .607 |
| | | | | | | | | .582 | .555 | .528 | .500 | .470 | .440 | .410 | .379 | .349 | .318 |
| | | | | | | | | .288 | .259 | .231 | | | | | | | |
| 22.000 | 20.917 | .000 | 1.083 | .80 | 4.96 | 4.9% | 34 | .879 | .872 | .865 | .857 | .848 | .838 | .828 | .817 | .805 | .793 |
| | | | | | | | | .779 | .764 | .749 | .732 | .714 | .695 | .675 | .654 | .631 | .607 |
| | | | | | | | | .582 | .555 | .528 | .500 | .470 | .440 | .410 | .379 | .349 | .318 |
| | | | | | | | | .288 | .259 | .231 | .204 | | | | | | |
| 22.000 | 21.096 | .000 | .904 | .82 | 4.90 | 4.1% | 35 | .879 | .872 | .865 | .857 | .848 | .838 | .828 | .817 | .805 | .793 |
| | | | | | | | | .779 | .764 | .749 | .732 | .714 | .695 | .675 | .654 | .631 | .607 |
| | | | | | | | | .582 | .555 | .528 | .500 | .470 | .440 | .410 | .379 | .349 | .318 |
| | | | | | | | | .288 | .259 | .231 | .204 | .179 | | | | | |
| 22.000 | 21.252 | .000 | .748 | .84 | 4.83 | 3.4% | 36 | .879 | .872 | .865 | .857 | .848 | .838 | .828 | .817 | .805 | .793 |
| | | | | | | | | .779 | .764 | .749 | .732 | .714 | .695 | .675 | .654 | .631 | .607 |
| | | | | | | | | .582 | .555 | .528 | .500 | .470 | .440 | .410 | .379 | .349 | .318 |
| | | | | | | | | .288 | .259 | .231 | .204 | .179 | .156 | | | | |
| 22.000 | 21.386 | .000 | .614 | .87 | 4.76 | 2.8% | 37 | .879 | .872 | .865 | .857 | .848 | .838 | .828 | .817 | .805 | .793 |
| | | | | | | | | .779 | .764 | .749 | .732 | .714 | .695 | .675 | .654 | .631 | .607 |
| | | | | | | | | .582 | .555 | .528 | .500 | .470 | .440 | .410 | .379 | .349 | .318 |
| | | | | | | | | .288 | .259 | .231 | .204 | .179 | .156 | .134 | | | |
| 22.000 | 21.500 | .000 | .500 | .89 | 4.68 | 2.3% | 38 | .879 | .872 | .865 | .857 | .848 | .838 | .828 | .817 | .805 | .793 |
| | | | | | | | | .779 | .764 | .749 | .732 | .714 | .695 | .675 | .654 | .631 | .607 |
| | | | | | | | | .582 | .555 | .528 | .500 | .470 | .440 | .410 | .379 | .349 | .318 |
| | | | | | | | | .288 | .259 | .231 | .204 | .179 | .156 | .134 | .114 | | |
| 22.000 | 21.597 | .000 | .403 | .90 | 4.60 | 1.8% | 39 | .879 | .872 | .865 | .857 | .848 | .838 | .828 | .817 | .805 | .793 |
| | | | | | | | | .779 | .764 | .749 | .732 | .714 | .695 | .675 | .654 | .631 | .607 |
| | | | | | | | | .582 | .555 | .528 | .500 | .470 | .440 | .410 | .379 | .349 | .318 |

*SIMILAR TO A.T.&T. NEAL-WILKINSON TABLES(SEE SECTION 1)

*EQUIVALENT RANDOM THEORY: VMR(VARIANCE TO MEAN RATIO)– 3.0 - NO RETRIES (PEAKED TRAFFIC - ROUTE ADVANCEMENT OR DAY-TO-DAY VARIATION)

| HOURS FIRST ATTEMPT VMR = 3.0 | HOURS CON-NECT | HOURS RECALL /RETRY | HOURS OVFL | CON-NECT VMR | OVFL VMR | PERCENT BLOCKED (GRADE) | LINE | HOURS CONNECTED TRAFFIC ON LINES 1 | 2 | 3 | 4 | 5 | 6 | 7 | 8 | 9 | 10 |
|---|---|---|---|---|---|---|---|---|---|---|---|---|---|---|---|---|---|
| | | | | | | | | .288 | .259 | .231 | .204 | .179 | .156 | .134 | .114 | .096 | |
| 22.000 | 21.677 | .000 | .323 | .92 | 4.51 | 1.5% | 40 | .879 | .872 | .865 | .857 | .848 | .838 | .828 | .817 | .805 | .793 |
| | | | | | | | | .779 | .764 | .749 | .732 | .714 | .695 | .675 | .654 | .631 | .607 |
| | | | | | | | | .582 | .555 | .528 | .500 | .470 | .440 | .410 | .379 | .349 | .318 |
| | | | | | | | | .288 | .259 | .231 | .204 | .179 | .156 | .134 | .114 | .096 | .081 |
| 22.000 | 21.744 | .000 | .256 | .93 | 4.42 | 1.2% | 41 | .879 | .872 | .865 | .857 | .848 | .838 | .828 | .817 | .805 | .793 |
| | | | | | | | | .779 | .764 | .749 | .732 | .714 | .695 | .675 | .654 | .631 | .607 |
| | | | | | | | | .582 | .555 | .528 | .500 | .470 | .440 | .410 | .379 | .349 | .318 |
| | | | | | | | | .288 | .259 | .231 | .204 | .179 | .156 | .134 | .114 | .096 | .081 |
| | | | | | | | | .067 | | | | | | | | | |
| 22.000 | 21.799 | .000 | .201 | .95 | 4.33 | .9% | 42 | .879 | .872 | .865 | .857 | .848 | .838 | .828 | .817 | .805 | .793 |
| | | | | | | | | .779 | .764 | .749 | .732 | .714 | .695 | .675 | .654 | .631 | .607 |
| | | | | | | | | .582 | .555 | .528 | .500 | .470 | .440 | .410 | .379 | .349 | .318 |
| | | | | | | | | .288 | .259 | .231 | .204 | .179 | .156 | .134 | .114 | .096 | .081 |
| | | | | | | | | .067 | .055 | | | | | | | | |
| 22.000 | 21.843 | .000 | .157 | .96 | 4.23 | .7% | 43 | .879 | .872 | .865 | .857 | .848 | .838 | .828 | .817 | .805 | .793 |
| | | | | | | | | .779 | .764 | .749 | .732 | .714 | .695 | .675 | .654 | .631 | .607 |
| | | | | | | | | .582 | .555 | .528 | .500 | .470 | .440 | .410 | .379 | .349 | .318 |
| | | | | | | | | .288 | .259 | .231 | .204 | .179 | .156 | .134 | .114 | .096 | .081 |
| | | | | | | | | .067 | .055 | .044 | | | | | | | |
| 22.000 | 21.879 | .000 | .121 | .96 | 4.14 | .6% | 44 | .879 | .872 | .865 | .857 | .848 | .838 | .828 | .817 | .805 | .793 |
| | | | | | | | | .779 | .764 | .749 | .732 | .714 | .695 | .675 | .654 | .631 | .607 |
| | | | | | | | | .582 | .555 | .528 | .500 | .470 | .440 | .410 | .379 | .349 | .318 |
| | | | | | | | | .288 | .259 | .231 | .204 | .179 | .156 | .134 | .114 | .096 | .081 |
| | | | | | | | | .067 | .055 | .044 | .036 | | | | | | |

| HOURS FIRST ATTEMPT VMR = 3.0 | HOURS CON-NECT | HOURS RECALL /RETRY | HOURS OVFL | CON-NECT VMR | OVFL VMR | PERCENT BLOCKED (GRADE) | LINE | 1 | 2 | 3 | 4 | 5 | 6 | 7 | 8 | 9 | 10 |
|---|---|---|---|---|---|---|---|---|---|---|---|---|---|---|---|---|---|
| 24.000 | 15.883 | .000 | 0.117 | .35 | 4.73 | 33.8% | 20 | .888 | .882 | .876 | .868 | .861 | .853 | .844 | .834 | .824 | .813 |
| | | | | | | | | .802 | .789 | .776 | .762 | .746 | .730 | .713 | .694 | .675 | .654 |
| 24.000 | 16.515 | .000 | 7.485 | .37 | 4.81 | 31.2% | 21 | .888 | .882 | .876 | .868 | .861 | .853 | .844 | .834 | .824 | .813 |
| | | | | | | | | .802 | .789 | .776 | .762 | .746 | .730 | .713 | .694 | .675 | .654 |
| | | | | | | | | .632 | | | | | | | | | |
| 24.000 | 17.124 | .000 | 6.876 | .39 | 4.89 | 28.7% | 22 | .888 | .882 | .876 | .868 | .861 | .853 | .844 | .834 | .824 | .813 |
| | | | | | | | | .802 | .789 | .776 | .762 | .746 | .730 | .713 | .694 | .675 | .654 |
| | | | | | | | | .632 | .609 | | | | | | | | |
| 24.000 | 17.708 | .000 | 6.292 | .42 | 4.96 | 26.2% | 23 | .888 | .882 | .876 | .868 | .861 | .853 | .844 | .834 | .824 | .813 |
| | | | | | | | | .802 | .789 | .776 | .762 | .746 | .730 | .713 | .694 | .675 | .654 |
| | | | | | | | | .632 | .609 | .584 | | | | | | | |
| 24.000 | 18.267 | .000 | 5.733 | .44 | 5.03 | 23.9% | 24 | .888 | .882 | .876 | .868 | .861 | .853 | .844 | .834 | .824 | .813 |
| | | | | | | | | .802 | .789 | .776 | .762 | .746 | .730 | .713 | .694 | .675 | .654 |
| | | | | | | | | .632 | .609 | .584 | .559 | | | | | | |
| 24.000 | 18.800 | .000 | 5.200 | .47 | 5.09 | 21.7% | 25 | .888 | .882 | .876 | .868 | .861 | .853 | .844 | .834 | .824 | .813 |

*SIMILAR TO A.T.&T. NEAL-WILKINSON TABLES(SEE SECTION 1)

*EQUIVALENT RANDOM THEORY: VMR(VARIANCE TO MEAN RATIO)– 3.0 - NO RETRIES (PEAKED TRAFFIC - ROUTE ADVANCEMENT OR DAY-TO-DAY VARIATION)

| HOURS FIRST ATTEMPT VMR=3.0 | HOURS CON-NECT | HOURS RECALL /RETRY | HOURS OVFL | CON-NECT VMR | OVFL VMR | PERCENT BLOCKED (GRADE) | LINE | HOURS CONNECTED TRAFFIC ON LINES 1 | 2 | 3 | 4 | 5 | 6 | 7 | 8 | 9 | 10 |
|---|---|---|---|---|---|---|---|---|---|---|---|---|---|---|---|---|---|
| | | | | | | | | .802 | .789 | .776 | .762 | .746 | .730 | .713 | .694 | .675 | .654 |
| | | | | | | | | .632 | .609 | .584 | .559 | .533 | | | | | |
| 24.000 | 19.305 | .000 | 4.695 | .49 | 5.14 | 19.6% | 26 | .888 | .882 | .876 | .868 | .861 | .853 | .844 | .834 | .824 | .813 |
| | | | | | | | | .802 | .789 | .776 | .762 | .746 | .730 | .713 | .694 | .675 | .654 |
| | | | | | | | | .632 | .609 | .584 | .559 | .533 | .505 | | | | |
| 24.000 | 19.783 | .000 | 4.217 | .52 | 5.18 | 17.6% | 27 | .888 | .882 | .876 | .868 | .861 | .853 | .844 | .834 | .824 | .813 |
| | | | | | | | | .802 | .789 | .776 | .762 | .746 | .730 | .713 | .694 | .675 | .654 |
| | | | | | | | | .632 | .609 | .584 | .559 | .533 | .505 | .477 | | | |
| 24.000 | 20.231 | .000 | 3.769 | .55 | 5.22 | 15.7% | 28 | .888 | .882 | .876 | .868 | .861 | .853 | .844 | .834 | .824 | .813 |
| | | | | | | | | .802 | .789 | .776 | .762 | .746 | .730 | .713 | .694 | .675 | .654 |
| | | | | | | | | .632 | .609 | .584 | .559 | .533 | .505 | .477 | .449 | | |
| 24.000 | 20.651 | .000 | 3.349 | .58 | 5.25 | 14.0% | 29 | .888 | .882 | .876 | .868 | .861 | .853 | .844 | .834 | .824 | .813 |
| | | | | | | | | .802 | .789 | .776 | .762 | .746 | .730 | .713 | .694 | .675 | .654 |
| | | | | | | | | .632 | .609 | .584 | .559 | .533 | .505 | .477 | .449 | .419 | |
| 24.000 | 21.040 | .000 | 2.960 | .61 | 5.26 | 12.3% | 30 | .888 | .882 | .876 | .868 | .861 | .853 | .844 | .834 | .824 | .813 |
| | | | | | | | | .802 | .789 | .776 | .762 | .746 | .730 | .713 | .694 | .675 | .654 |
| | | | | | | | | .632 | .609 | .584 | .559 | .533 | .505 | .477 | .449 | .419 | .390 |
| 24.000 | 21.400 | .000 | 2.600 | .64 | 5.27 | 10.8% | 31 | .888 | .882 | .876 | .868 | .861 | .853 | .844 | .834 | .824 | .813 |
| | | | | | | | | .802 | .789 | .776 | .762 | .746 | .730 | .713 | .694 | .675 | .654 |
| | | | | | | | | .632 | .609 | .584 | .559 | .533 | .505 | .477 | .449 | .419 | .390 |
| | | | | | | | | .360 | | | | | | | | | |
| 24.000 | 21.731 | .000 | 2.269 | .67 | 5.26 | 9.5% | 32 | .888 | .882 | .876 | .868 | .861 | .853 | .844 | .834 | .824 | .813 |
| | | | | | | | | .802 | .789 | .776 | .762 | .746 | .730 | .713 | .694 | .675 | .654 |
| | | | | | | | | .632 | .609 | .584 | .559 | .533 | .505 | .477 | .449 | .419 | .390 |
| | | | | | | | | .360 | .330 | | | | | | | | |
| 24.000 | 22.032 | .000 | 1.968 | .70 | 5.25 | 8.2% | 33 | .888 | .882 | .876 | .868 | .861 | .853 | .844 | .834 | .824 | .813 |
| | | | | | | | | .802 | .789 | .776 | .762 | .746 | .730 | .713 | .694 | .675 | .654 |
| | | | | | | | | .632 | .609 | .584 | .559 | .533 | .505 | .477 | .449 | .419 | .390 |
| | | | | | | | | .360 | .330 | .301 | | | | | | | |
| 24.000 | 22.305 | .000 | 1.695 | .73 | 5.22 | 7.1% | 34 | .888 | .882 | .876 | .868 | .861 | .853 | .844 | .834 | .824 | .813 |
| | | | | | | | | .802 | .789 | .776 | .762 | .746 | .730 | .713 | .694 | .675 | .654 |
| | | | | | | | | .632 | .609 | .584 | .559 | .533 | .505 | .477 | .449 | .419 | .390 |
| | | | | | | | | .360 | .330 | .301 | .273 | | | | | | |
| 24.000 | 22.550 | .000 | 1.450 | .75 | 5.19 | 6.0% | 35 | .888 | .882 | .876 | .868 | .861 | .853 | .844 | .834 | .824 | .813 |
| | | | | | | | | .802 | .789 | .776 | .762 | .746 | .730 | .713 | .694 | .675 | .654 |
| | | | | | | | | .632 | .609 | .584 | .559 | .533 | .505 | .477 | .449 | .419 | .390 |
| | | | | | | | | .360 | .330 | .301 | .273 | .245 | | | | | |
| 24.000 | 22.769 | .000 | 1.231 | .78 | 5.14 | 5.1% | 36 | .888 | .882 | .876 | .868 | .861 | .853 | .844 | .834 | .824 | .813 |
| | | | | | | | | .802 | .789 | .776 | .762 | .746 | .730 | .713 | .694 | .675 | .654 |
| | | | | | | | | .632 | .609 | .584 | .559 | .533 | .505 | .477 | .449 | .419 | .390 |

*SIMILAR TO A.T.&T. NEAL-WILKINSON TABLES(SEE SECTION 1)

*EQUIVALENT RANDOM THEORY: VMR(VARIANCE TO MEAN RATIO)– 3.0 - NO RETRIES (PEAKED TRAFFIC - ROUTE ADVANCEMENT OR DAY-TO-DAY VARIATION)

| HOURS FIRST ATTEMPT VMR=3.0 | HOURS CON-NECT | HOURS RECALL /RETRY | HOURS OVFL | CON-NECT VMR | OVFL VMR | PERCENT BLOCKED (GRADE) | LINE | HOURS CONNECTED TRAFFIC ON LINES 1 | 2 | 3 | 4 | 5 | 6 | 7 | 8 | 9 | 10 |
|---|---|---|---|---|---|---|---|---|---|---|---|---|---|---|---|---|---|
| | | | | | | | | .360 | .330 | .301 | .273 | .245 | .219 | | | | |
| 24.000 | 22.962 | .000 | 1.038 | .81 | 5.08 | 4.3% | 37 | .888 | .882 | .876 | .868 | .861 | .853 | .844 | .834 | .824 | .813 |
| | | | | | | | | .802 | .789 | .776 | .762 | .746 | .730 | .713 | .694 | .675 | .654 |
| | | | | | | | | .632 | .609 | .584 | .559 | .533 | .505 | .477 | .449 | .419 | .390 |
| | | | | | | | | .360 | .330 | .301 | .273 | .245 | .219 | .193 | | | |
| 24.000 | 23.131 | .000 | .869 | .83 | 5.02 | 3.6% | 38 | .888 | .882 | .876 | .868 | .861 | .853 | .844 | .834 | .824 | .813 |
| | | | | | | | | .802 | .789 | .776 | .762 | .746 | .730 | .713 | .694 | .675 | .654 |
| | | | | | | | | .632 | .609 | .584 | .559 | .533 | .505 | .477 | .449 | .419 | .390 |
| | | | | | | | | .360 | .330 | .301 | .273 | .245 | .219 | .193 | .170 | | |
| 24.000 | 23.279 | .000 | .721 | .85 | 4.95 | 3.0% | 39 | .888 | .882 | .876 | .868 | .861 | .853 | .844 | .834 | .824 | .813 |
| | | | | | | | | .802 | .789 | .776 | .762 | .746 | .730 | .713 | .694 | .675 | .654 |
| | | | | | | | | .632 | .609 | .584 | .559 | .533 | .505 | .477 | .449 | .419 | .390 |
| | | | | | | | | .360 | .330 | .301 | .273 | .245 | .219 | .193 | .170 | .147 | |
| 24.000 | 23.406 | .000 | .594 | .87 | 4.88 | 2.5% | 40 | .888 | .882 | .876 | .868 | .861 | .853 | .844 | .834 | .824 | .813 |
| | | | | | | | | .802 | .789 | .776 | .762 | .746 | .730 | .713 | .694 | .675 | .654 |
| | | | | | | | | .632 | .609 | .584 | .559 | .533 | .505 | .477 | .449 | .419 | .390 |
| | | | | | | | | .360 | .330 | .301 | .273 | .245 | .219 | .193 | .170 | .147 | .127 |
| 24.000 | 23.515 | .000 | .485 | .89 | 4.79 | 2.0% | 41 | .888 | .882 | .876 | .868 | .861 | .853 | .844 | .834 | .824 | .813 |
| | | | | | | | | .802 | .789 | .776 | .762 | .746 | .730 | .713 | .694 | .675 | .654 |
| | | | | | | | | .632 | .609 | .584 | .559 | .533 | .505 | .477 | .449 | .419 | .390 |
| | | | | | | | | .360 | .330 | .301 | .273 | .245 | .219 | .193 | .170 | .147 | .127 |
| | | | | | | | | .109 | | | | | | | | | |
| 24.000 | 23.607 | .000 | .393 | .91 | 4.71 | 1.6% | 42 | .888 | .882 | .876 | .868 | .861 | .853 | .844 | .834 | .824 | .813 |
| | | | | | | | | .802 | .789 | .776 | .762 | .746 | .730 | .713 | .694 | .675 | .654 |
| | | | | | | | | .632 | .609 | .584 | .559 | .533 | .505 | .477 | .449 | .419 | .390 |
| | | | | | | | | .360 | .330 | .301 | .273 | .245 | .219 | .193 | .170 | .147 | .127 |
| | | | | | | | | .109 | .092 | | | | | | | | |
| 24.000 | 23.684 | .000 | .316 | .92 | 4.62 | 1.3% | 43 | .888 | .882 | .876 | .868 | .861 | .853 | .844 | .834 | .824 | .813 |
| | | | | | | | | .802 | .789 | .776 | .762 | .746 | .730 | .713 | .694 | .675 | .654 |
| | | | | | | | | .632 | .609 | .584 | .559 | .533 | .505 | .477 | .449 | .419 | .390 |
| | | | | | | | | .360 | .330 | .301 | .273 | .245 | .219 | .193 | .170 | .147 | .127 |
| | | | | | | | | .109 | .092 | .077 | | | | | | | |
| 24.000 | 23.748 | .000 | .252 | .94 | 4.53 | 1.1% | 44 | .888 | .882 | .876 | .868 | .861 | .853 | .844 | .834 | .824 | .813 |
| | | | | | | | | .802 | .789 | .776 | .762 | .746 | .730 | .713 | .694 | .675 | .654 |
| | | | | | | | | .632 | .609 | .584 | .559 | .533 | .505 | .477 | .449 | .419 | .390 |
| | | | | | | | | .360 | .330 | .301 | .273 | .245 | .219 | .193 | .170 | .147 | .127 |
| | | | | | | | | .109 | .092 | .077 | .064 | | | | | | |
| 24.000 | 23.801 | .000 | .199 | .95 | 4.43 | .8% | 45 | .888 | .882 | .876 | .868 | .861 | .853 | .844 | .834 | .824 | .813 |
| | | | | | | | | .802 | .789 | .770 | .762 | .746 | .730 | .713 | .694 | .675 | .654 |
| | | | | | | | | .632 | .609 | .584 | .559 | .533 | .505 | .477 | .449 | .419 | .390 |

*SIMILAR TO A.T.&T. NEAL-WILKINSON TABLES(SEE SECTION 1)

*EQUIVALENT RANDOM THEORY: VMR(VARIANCE TO MEAN RATIO)– 3.0 - NO RETRIES (PEAKED TRAFFIC - ROUTE ADVANCEMENT OR DAY-TO-DAY VARIATION)

| HOURS FIRST ATTEMPT VMR=3.0 | HOURS CON-NECT | HOURS RECALL /RETRY | HOURS OVFL | CON-NECT VMR | OVFL VMR | PERCENT BLOCKED (GRADE) | LINE | HOURS CONNECTED TRAFFIC ON LINES 1 | 2 | 3 | 4 | 5 | 6 | 7 | 8 | 9 | 10 |
|---|---|---|---|---|---|---|---|---|---|---|---|---|---|---|---|---|---|
| | | | | | | | | .360 | .330 | .301 | .273 | .245 | .219 | .193 | .170 | .147 | .127 |
| | | | | | | | | .109 | .092 | .077 | .064 | .053 | | | | | |
| 24.000 | 23.844 | .000 | .156 | .96 | 4.34 | .7% | 46 | .888 | .882 | .876 | .868 | .861 | .853 | .844 | .834 | .824 | .813 |
| | | | | | | | | .802 | .789 | .776 | .762 | .746 | .730 | .713 | .694 | .675 | .654 |
| | | | | | | | | .632 | .609 | .584 | .559 | .533 | .505 | .477 | .449 | .419 | .390 |
| | | | | | | | | .360 | .330 | .301 | .273 | .245 | .219 | .193 | .170 | .147 | .127 |
| | | | | | | | | .109 | .092 | .077 | .064 | .053 | .043 | | | | |
| 24.000 | 23.878 | .000 | .122 | .97 | 4.25 | .5% | 47 | .888 | .882 | .876 | .868 | .861 | .853 | .844 | .834 | .824 | .813 |
| | | | | | | | | .802 | .789 | .776 | .762 | .746 | .730 | .713 | .694 | .675 | .654 |
| | | | | | | | | .632 | .609 | .584 | .559 | .533 | .505 | .477 | .449 | .419 | .390 |
| | | | | | | | | .360 | .330 | .301 | .273 | .245 | .219 | .193 | .170 | .147 | .127 |
| | | | | | | | | .109 | .092 | .077 | .064 | .053 | .043 | .035 | | | |
| 26.000 | 16.960 | .000 | 9.040 | .33 | 4.79 | 34.8% | 21 | .896 | .891 | .885 | .879 | .872 | .865 | .857 | .849 | .840 | .831 |
| | | | | | | | | .821 | .810 | .798 | .786 | .773 | .759 | .744 | .728 | .711 | .693 |
| | | | | | | | | .674 | | | | | | | | | |
| 26.000 | 17.614 | .000 | 8.386 | .35 | 4.87 | 32.3% | 22 | .896 | .891 | .885 | .879 | .872 | .865 | .857 | .849 | .840 | .831 |
| | | | | | | | | .821 | .810 | .798 | .786 | .773 | .759 | .744 | .728 | .711 | .693 |
| | | | | | | | | .674 | .654 | | | | | | | | |
| 26.000 | 18.246 | .000 | 7.754 | .37 | 4.95 | 29.8% | 23 | .896 | .891 | .885 | .879 | .872 | .865 | .857 | .849 | .840 | .831 |
| | | | | | | | | .821 | .810 | .798 | .786 | .773 | .759 | .744 | .728 | .711 | .693 |
| | | | | | | | | .674 | .654 | .633 | | | | | | | |
| 26.000 | 18.856 | .000 | 7.144 | .39 | 5.03 | 27.5% | 24 | .896 | .891 | .885 | .879 | .872 | .865 | .857 | .849 | .840 | .831 |
| | | | | | | | | .821 | .810 | .798 | .786 | .773 | .759 | .744 | .728 | .711 | .693 |
| | | | | | | | | .674 | .654 | .633 | .610 | | | | | | |
| 26.000 | 19.443 | .000 | 6.557 | .41 | 5.10 | 25.2% | 25 | .896 | .891 | .885 | .879 | .872 | .865 | .857 | .849 | .840 | .831 |
| | | | | | | | | .821 | .810 | .798 | .786 | .773 | .759 | .744 | .728 | .711 | .693 |
| | | | | | | | | .674 | .654 | .633 | .610 | .587 | | | | | |
| 26.000 | 20.006 | .000 | 5.994 | .44 | 5.17 | 23.1% | 26 | .896 | .891 | .885 | .879 | .872 | .865 | .857 | .849 | .840 | .831 |
| | | | | | | | | .821 | .810 | .798 | .786 | .773 | .759 | .744 | .728 | .711 | .693 |
| | | | | | | | | .674 | .654 | .633 | .610 | .587 | .562 | | | | |
| 26.000 | 20.543 | .000 | 5.457 | .46 | 5.23 | 21.0% | 27 | .896 | .891 | .885 | .879 | .872 | .865 | .857 | .849 | .840 | .831 |
| | | | | | | | | .821 | .810 | .798 | .786 | .773 | .759 | .744 | .728 | .711 | .693 |
| | | | | | | | | .674 | .654 | .633 | .610 | .587 | .562 | .537 | | | |
| 26.000 | 21.053 | .000 | 4.947 | .49 | 5.28 | 19.0% | 28 | .896 | .891 | .885 | .879 | .872 | .865 | .857 | .849 | .840 | .831 |
| | | | | | | | | .821 | .810 | .798 | .786 | .773 | .759 | .744 | .728 | .711 | .693 |
| | | | | | | | | .674 | .654 | .633 | .610 | .587 | .562 | .537 | .511 | | |
| 26.000 | 21.537 | .000 | 4.463 | .52 | 5.33 | 17.2% | 29 | .896 | .891 | .885 | .879 | .872 | .865 | .857 | .849 | .840 | .831 |
| | | | | | | | | .821 | .810 | .798 | .786 | .773 | .759 | .744 | .728 | .711 | .693 |
| | | | | | | | | .674 | .654 | .633 | .610 | .587 | .562 | .537 | .511 | .484 | |
| 26.000 | 21.993 | .000 | 4.007 | .54 | 5.36 | 15.4% | 30 | .896 | .891 | .885 | .879 | .872 | .865 | .857 | .849 | .840 | .831 |

*SIMILAR TO A.T.&T. NEAL-WILKINSON TABLES(SEE SECTION 1)

*EQUIVALENT RANDOM THEORY: VMR(VARIANCE TO MEAN RATIO)– 3.0 - NO RETRIES (PEAKED TRAFFIC - ROUTE ADVANCEMENT OR DAY-TO-DAY VARIATION)

| HOURS FIRST ATTEMPT VMR = 3.0 | HOURS CON-NECT | HOURS RECALL /RETRY | HOURS OVFL | CON-NECT VMR | OVFL VMR | PERCENT BLOCKED (GRADE) | LINE | HOURS CONNECTED TRAFFIC ON LINES 1 | 2 | 3 | 4 | 5 | 6 | 7 | 8 | 9 | 10 |
|---|---|---|---|---|---|---|---|---|---|---|---|---|---|---|---|---|---|
| | | | | | | | | .821 | .810 | .798 | .786 | .773 | .759 | .744 | .728 | .711 | .693 |
| | | | | | | | | .674 | .654 | .633 | .610 | .587 | .562 | .537 | .511 | .484 | .456 |
| 26.000 | 22.421 | .000 | 3.579 | .57 | 5.39 | 13.8% | 31 | .896 | .891 | .885 | .879 | .872 | .865 | .857 | .849 | .840 | .831 |
| | | | | | | | | .821 | .810 | .798 | .786 | .773 | .759 | .744 | .728 | .711 | .693 |
| | | | | | | | | .674 | .654 | .633 | .610 | .587 | .562 | .537 | .511 | .484 | .456 |
| | | | | | | | | .428 | | | | | | | | | |
| 26.000 | 22.820 | .000 | 3.180 | .60 | 5.41 | 12.2% | 32 | .896 | .891 | .885 | .879 | .872 | .865 | .857 | .849 | .840 | .831 |
| | | | | | | | | .821 | .810 | .798 | .786 | .773 | .759 | .744 | .728 | .711 | .693 |
| | | | | | | | | .674 | .654 | .633 | .610 | .587 | .562 | .537 | .511 | .484 | .456 |
| | | | | | | | | .428 | .399 | | | | | | | | |
| 26.000 | 23.190 | .000 | 2.810 | .63 | 5.42 | 10.8% | 33 | .896 | .891 | .885 | .879 | .872 | .865 | .857 | .849 | .840 | .831 |
| | | | | | | | | .821 | .810 | .798 | .786 | .773 | .759 | .744 | .728 | .711 | .693 |
| | | | | | | | | .674 | .654 | .633 | .610 | .587 | .562 | .537 | .511 | .484 | .456 |
| | | | | | | | | .428 | .399 | .370 | | | | | | | |
| 26.000 | 23.531 | .000 | 2.469 | .66 | 5.42 | 9.5% | 34 | .896 | .891 | .885 | .879 | .872 | .865 | .857 | .849 | .840 | .831 |
| | | | | | | | | .821 | .810 | .798 | .786 | .773 | .759 | .744 | .728 | .711 | .693 |
| | | | | | | | | .674 | .654 | .633 | .610 | .587 | .562 | .537 | .511 | .484 | .456 |
| | | | | | | | | .428 | .399 | .370 | .341 | | | | | | |
| 26.000 | 23.844 | .000 | 2.156 | .69 | 5.41 | 8.3% | 35 | .896 | .891 | .885 | .879 | .872 | .865 | .857 | .849 | .840 | .831 |
| | | | | | | | | .821 | .810 | .798 | .786 | .773 | .759 | .744 | .728 | .711 | .693 |
| | | | | | | | | .674 | .654 | .633 | .610 | .587 | .562 | .537 | .511 | .484 | .456 |
| | | | | | | | | .428 | .399 | .370 | .341 | .313 | | | | | |
| 26.000 | 24.129 | .000 | 1.871 | .71 | 5.38 | 7.2% | 36 | .896 | .891 | .885 | .879 | .872 | .865 | .857 | .849 | .840 | .831 |
| | | | | | | | | .821 | .810 | .798 | .786 | .773 | .759 | .744 | .728 | .711 | .693 |
| | | | | | | | | .674 | .654 | .633 | .610 | .587 | .562 | .537 | .511 | .484 | .456 |
| | | | | | | | | .428 | .399 | .370 | .341 | .313 | .285 | | | | |
| 26.000 | 24.387 | .000 | 1.613 | .74 | 5.35 | 6.2% | 37 | .896 | .891 | .885 | .879 | .872 | .865 | .857 | .849 | .840 | .831 |
| | | | | | | | | .821 | .810 | .798 | .786 | .773 | .759 | .744 | .728 | .711 | .693 |
| | | | | | | | | .674 | .654 | .633 | .610 | .587 | .562 | .537 | .511 | .484 | .456 |
| | | | | | | | | .428 | .399 | .370 | .341 | .313 | .285 | .258 | | | |
| 26.000 | 24.619 | .000 | 1.381 | .77 | 5.31 | 5.3% | 38 | .896 | .891 | .885 | .879 | .872 | .865 | .857 | .849 | .840 | .831 |
| | | | | | | | | .821 | .810 | .798 | .786 | .773 | .759 | .744 | .728 | .711 | .693 |
| | | | | | | | | .674 | .654 | .633 | .610 | .587 | .562 | .537 | .511 | .484 | .456 |
| | | | | | | | | .428 | .399 | .370 | .341 | .313 | .285 | .258 | .232 | | |
| 26.000 | 24.825 | .000 | 1.175 | .79 | 5.26 | 4.5% | 39 | .896 | .891 | .885 | .879 | .872 | .865 | .857 | .849 | .840 | .831 |
| | | | | | | | | .821 | .810 | .798 | .786 | .773 | .759 | .744 | .728 | .711 | .693 |
| | | | | | | | | .674 | .654 | .633 | .610 | .587 | .562 | .537 | .511 | .484 | .456 |
| | | | | | | | | .428 | .399 | .370 | .341 | .313 | .285 | .258 | .232 | .206 | |
| 26.000 | 25.008 | .000 | .992 | .82 | 5.20 | 3.8% | 40 | .896 | .891 | .885 | .879 | .872 | .865 | .857 | .849 | .840 | .831 |
| | | | | | | | | .821 | .810 | .798 | .786 | .773 | .759 | .744 | .728 | .711 | .693 |

*SIMILAR TO A.T.&T. NEAL-WILKINSON TABLES(SEE SECTION 1)

*EQUIVALENT RANDOM THEORY: VMR(VARIANCE TO MEAN RATIO)– 3.0 - NO RETRIES (PEAKED TRAFFIC - ROUTE ADVANCEMENT OR DAY-TO-DAY VARIATION)

| HOURS FIRST ATTEMPT VMR=3.0 | HOURS CON-NECT | HOURS RECALL /RETRY | HOURS OVFL | CON-NECT VMR | OVFL VMR | PERCENT BLOCKED (GRADE) | LINE | HOURS CONNECTED TRAFFIC ON LINES 1 | 2 | 3 | 4 | 5 | 6 | 7 | 8 | 9 | 10 |
|---|---|---|---|---|---|---|---|---|---|---|---|---|---|---|---|---|---|
| | | | | | | | | .674 | .654 | .633 | .610 | .587 | .562 | .537 | .511 | .484 | .456 |
| | | | | | | | | .428 | .399 | .370 | .341 | .313 | .285 | .258 | .232 | .206 | .183 |
| 26.000 | 25.168 | .000 | .832 | .84 | 5.14 | 3.2% | 41 | .896 | .891 | .885 | .879 | .872 | .865 | .857 | .849 | .840 | .831 |
| | | | | | | | | .821 | .810 | .798 | .786 | .773 | .759 | .744 | .728 | .711 | .693 |
| | | | | | | | | .674 | .654 | .633 | .610 | .587 | .562 | .537 | .511 | .484 | .456 |
| | | | | | | | | .428 | .399 | .370 | .341 | .313 | .285 | .258 | .232 | .206 | .183 |
| | | | | | | | | .160 | | | | | | | | | |
| 26.000 | 25.307 | .000 | .693 | .86 | 5.06 | 2.7% | 42 | .896 | .891 | .885 | .879 | .872 | .865 | .857 | .849 | .840 | .831 |
| | | | | | | | | .821 | .810 | .798 | .786 | .773 | .759 | .744 | .728 | .711 | .693 |
| | | | | | | | | .674 | .654 | .633 | .610 | .587 | .562 | .537 | .511 | .484 | .456 |
| | | | | | | | | .428 | .399 | .370 | .341 | .313 | .285 | .258 | .232 | .206 | .183 |
| | | | | | | | | .160 | .139 | | | | | | | | |
| 26.000 | 25.428 | .000 | .572 | .88 | 4.98 | 2.2% | 43 | .896 | .891 | .885 | .879 | .872 | .865 | .857 | .849 | .840 | .831 |
| | | | | | | | | .821 | .810 | .798 | .786 | .773 | .759 | .744 | .728 | .711 | .693 |
| | | | | | | | | .674 | .654 | .633 | .610 | .587 | .562 | .537 | .511 | .484 | .456 |
| | | | | | | | | .428 | .399 | .370 | .341 | .313 | .285 | .258 | .232 | .206 | .183 |
| | | | | | | | | .160 | .139 | .120 | | | | | | | |
| 26.000 | 25.531 | .000 | .469 | .90 | 4.90 | 1.8% | 44 | .896 | .891 | .885 | .879 | .872 | .865 | .857 | .849 | .840 | .831 |
| | | | | | | | | .821 | .810 | .798 | .786 | .773 | .759 | .744 | .728 | .711 | .693 |
| | | | | | | | | .674 | .654 | .633 | .610 | .587 | .562 | .537 | .511 | .484 | .456 |
| | | | | | | | | .428 | .399 | .370 | .341 | .313 | .285 | .258 | .232 | .206 | .183 |
| | | | | | | | | .160 | .139 | .120 | .103 | | | | | | |
| 26.000 | 25.618 | .000 | .382 | .91 | 4.81 | 1.5% | 45 | .896 | .891 | .885 | .879 | .872 | .865 | .857 | .849 | .840 | .831 |
| | | | | | | | | .821 | .810 | .798 | .786 | .773 | .759 | .744 | .728 | .711 | .693 |
| | | | | | | | | .674 | .654 | .633 | .610 | .587 | .562 | .537 | .511 | .484 | .456 |
| | | | | | | | | .428 | .399 | .370 | .341 | .313 | .285 | .258 | .232 | .206 | .183 |
| | | | | | | | | .160 | .139 | .120 | .103 | .087 | | | | | |
| 26.000 | 25.692 | .000 | .308 | .93 | 4.72 | 1.2% | 46 | .896 | .891 | .885 | .879 | .872 | .865 | .857 | .849 | .840 | .831 |
| | | | | | | | | .821 | .810 | .798 | .786 | .773 | .759 | .744 | .728 | .711 | .693 |
| | | | | | | | | .674 | .654 | .633 | .610 | .587 | .562 | .537 | .511 | .484 | .456 |
| | | | | | | | | .428 | .399 | .370 | .341 | .313 | .285 | .258 | .232 | .206 | .183 |
| | | | | | | | | .160 | .139 | .120 | .103 | .087 | .074 | | | | |
| 26.000 | 25.753 | .000 | .247 | .94 | 4.63 | 1.0% | 47 | .896 | .891 | .885 | .879 | .872 | .865 | .857 | .849 | .840 | .831 |
| | | | | | | | | .821 | .810 | .798 | .786 | .773 | .759 | .744 | .728 | .711 | .693 |
| | | | | | | | | .674 | .654 | .633 | .610 | .587 | .562 | .537 | .511 | .484 | .456 |
| | | | | | | | | .428 | .399 | .370 | .341 | .313 | .285 | .258 | .232 | .206 | .183 |
| | | | | | | | | .160 | .139 | .120 | .103 | .087 | .074 | .061 | | | |
| 26.000 | 25.804 | .000 | .196 | .95 | 4.54 | .8% | 48 | .896 | .891 | .885 | .879 | .872 | .865 | .857 | .849 | .840 | .831 |
| | | | | | | | | .821 | .810 | .798 | .786 | .773 | .759 | .744 | .728 | .711 | .693 |
| | | | | | | | | .674 | .654 | .633 | .610 | .587 | .562 | .537 | .511 | .484 | .456 |

*SIMILAR TO A.T.&T. NEAL-WILKINSON TABLES(SEE SECTION 1)

*EQUIVALENT RANDOM THEORY: VMR(VARIANCE TO MEAN RATIO)– 3.0 - NO RETRIES (PEAKED TRAFFIC - ROUTE ADVANCEMENT OR DAY-TO-DAY VARIATION)

| HOURS FIRST ATTEMPT VMR = 3.0 | HOURS CON-NECT | HOURS RECALL /RETRY | HOURS OVFL | CON-NECT VMR | OVFL VMR | PERCENT BLOCKED (GRADE) | LINE | HOURS CONNECTED TRAFFIC ON LINES 1 | 2 | 3 | 4 | 5 | 6 | 7 | 8 | 9 | 10 |
|---|---|---|---|---|---|---|---|---|---|---|---|---|---|---|---|---|---|
| | | | | | | | | .428 | .399 | .370 | .341 | .313 | .285 | .258 | .232 | .206 | .183 |
| | | | | | | | | .160 | .139 | .120 | .103 | .087 | .074 | .061 | .051 | | |
| 26.000 | 25.845 | .000 | .155 | .96 | 4.44 | .6% | 49 | .896 | .891 | .885 | .879 | .872 | .865 | .857 | .849 | .840 | .831 |
| | | | | | | | | .821 | .810 | .798 | .786 | .773 | .759 | .744 | .728 | .711 | .693 |
| | | | | | | | | .674 | .654 | .633 | .610 | .587 | .562 | .537 | .511 | .484 | .456 |
| | | | | | | | | .428 | .399 | .370 | .341 | .313 | .285 | .258 | .232 | .206 | .183 |
| | | | | | | | | .160 | .139 | .120 | .103 | .087 | .074 | .061 | .051 | .041 | |
| 28.000 | 18.700 | .000 | 9.300 | .33 | 4.92 | 33.2% | 23 | .903 | .898 | .893 | .887 | .881 | .875 | .868 | .861 | .853 | .845 |
| | | | | | | | | .837 | .827 | .817 | .807 | .795 | .783 | .770 | .756 | .742 | .726 |
| | | | | | | | | .710 | .692 | .673 | | | | | | | |
| 28.000 | 19.354 | .000 | 8.646 | .35 | 5.01 | 30.9% | 24 | .903 | .898 | .893 | .887 | .881 | .875 | .868 | .861 | .853 | .845 |
| | | | | | | | | .837 | .827 | .817 | .807 | .795 | .783 | .770 | .756 | .742 | .726 |
| | | | | | | | | .710 | .692 | .673 | .654 | | | | | | |
| 28.000 | 19.988 | .000 | 8.012 | .37 | 5.09 | 28.6% | 25 | .903 | .898 | .893 | .887 | .881 | .875 | .868 | .861 | .853 | .845 |
| | | | | | | | | .837 | .827 | .817 | .807 | .795 | .783 | .770 | .756 | .742 | .726 |
| | | | | | | | | .710 | .692 | .673 | .654 | .633 | | | | | |
| 28.000 | 20.599 | .000 | 7.401 | .39 | 5.16 | 26.4% | 26 | .903 | .898 | .893 | .887 | .881 | .875 | .868 | .861 | .853 | .845 |
| | | | | | | | | .837 | .827 | .817 | .807 | .795 | .783 | .770 | .756 | .742 | .726 |
| | | | | | | | | .710 | .692 | .673 | .654 | .633 | .612 | | | | |
| 28.000 | 21.188 | .000 | 6.812 | .41 | 5.24 | 24.3% | 27 | .903 | .898 | .893 | .887 | .881 | .875 | .868 | .861 | .853 | .845 |
| | | | | | | | | .837 | .827 | .817 | .807 | .795 | .783 | .770 | .756 | .742 | .726 |
| | | | | | | | | .710 | .692 | .673 | .654 | .633 | .612 | .589 | | | |
| 28.000 | 21.754 | .000 | 6.246 | .43 | 5.30 | 22.3% | 28 | .903 | .898 | .893 | .887 | .881 | .875 | .868 | .861 | .853 | .845 |
| | | | | | | | | .837 | .827 | .817 | .807 | .795 | .783 | .770 | .756 | .742 | .726 |
| | | | | | | | | .710 | .602 | .670 | .054 | .633 | .612 | .589 | .565 | | |
| 28.000 | 22.294 | 000 | 5.706 | .46 | 5.00 | 20.4% | 29 | .903 | .898 | .893 | .887 | .881 | .875 | .868 | .861 | .853 | .845 |
| | | | | | | | | .837 | .827 | .817 | .807 | .795 | .783 | .770 | .756 | .742 | .726 |
| | | | | | | | | .710 | .692 | .673 | .654 | .633 | .612 | .589 | .565 | .541 | |
| 28.000 | 22.810 | .000 | 5.190 | .48 | 5.42 | 18.5% | 30 | .903 | .898 | .893 | .887 | .881 | .875 | .868 | .861 | .853 | .845 |
| | | | | | | | | .837 | .827 | .817 | .807 | .795 | .783 | .770 | .756 | .742 | .726 |
| | | | | | | | | .710 | .692 | .673 | .654 | .633 | .612 | .589 | .565 | .541 | .515 |
| 28.000 | 23.299 | .000 | 4.701 | .51 | 5.46 | 16.8% | 31 | .903 | .898 | .893 | .887 | .881 | .875 | .868 | .861 | .853 | .845 |
| | | | | | | | | .837 | .827 | .817 | .807 | .795 | .783 | .770 | .756 | .742 | .726 |
| | | | | | | | | .710 | .692 | .673 | .654 | .633 | .612 | .589 | .565 | .541 | .515 |
| | | | | | | | | .489 | | | | | | | | | |
| 28.000 | 23.762 | .000 | 4.238 | .54 | 5.50 | 15.1% | 32 | .903 | .898 | .893 | .887 | .881 | .875 | .868 | .861 | .853 | .845 |
| | | | | | | | | .837 | .827 | .817 | .807 | .795 | .783 | .770 | .756 | .742 | .726 |
| | | | | | | | | .710 | .692 | .673 | .654 | .033 | .812 | .589 | .565 | .541 | .515 |
| | | | | | | | | .489 | .462 | | | | | | | | |
| 28.000 | 24.197 | .000 | 3.803 | .56 | 5.50 | 13.6% | 33 | .903 | .898 | .893 | .887 | .881 | .875 | .868 | .861 | .853 | .845 |

*SIMILAR TO A.T.&T. NEAL-WILKINSON TABLES(SEE SECTION 1)

*EQUIVALENT RANDOM THEORY: VMR(VARIANCE TO MEAN RATIO)– 3.0 - NO RETRIES (PEAKED TRAFFIC - ROUTE ADVANCEMENT OR DAY-TO-DAY VARIATION)

| HOURS FIRST ATTEMPT VMR=3.0 | HOURS CON-NECT | HOURS RECALL /RETRY | HOURS OVFL | CON-NECT VMR | OVFL VMR | PERCENT BLOCKED (GRADE) | LINE | HOURS CONNECTED TRAFFIC ON LINES 1 | 2 | 3 | 4 | 5 | 6 | 7 | 8 | 9 | 10 |
|---|---|---|---|---|---|---|---|---|---|---|---|---|---|---|---|---|---|
| | | | | | | | | .837 | .827 | .817 | .807 | .795 | .783 | .770 | .756 | .742 | .726 |
| | | | | | | | | .710 | .692 | .673 | .654 | .633 | .612 | .589 | .565 | .541 | .515 |
| | | | | | | | | .489 | .462 | .435 | | | | | | | |
| 28.000 | 24.604 | .000 | 3.396 | .59 | 5.55 | 12.1% | 34 | .903 | .898 | .893 | .887 | .881 | .875 | .868 | .861 | .853 | .845 |
| | | | | | | | | .837 | .827 | .817 | .807 | .795 | .783 | .770 | .756 | .742 | .726 |
| | | | | | | | | .710 | .692 | .673 | .654 | .633 | .612 | .589 | .565 | .541 | .515 |
| | | | | | | | | .489 | .462 | .435 | .407 | | | | | | |
| 28.000 | 24.983 | .000 | 3.017 | .62 | 5.57 | 10.8% | 35 | .903 | .898 | .893 | .887 | .881 | .875 | .868 | .861 | .853 | .845 |
| | | | | | | | | .837 | .827 | .817 | .807 | .795 | .783 | .770 | .756 | .742 | .726 |
| | | | | | | | | .710 | .692 | .673 | .654 | .633 | .612 | .589 | .565 | .541 | .515 |
| | | | | | | | | .489 | .462 | .435 | .407 | .379 | | | | | |
| 28.000 | 25.335 | .000 | 2.665 | .65 | 5.57 | 9.5% | 36 | .903 | .898 | .893 | .887 | .881 | .875 | .868 | .861 | .853 | .845 |
| | | | | | | | | .837 | .827 | .817 | .807 | .795 | .783 | .770 | .756 | .742 | .726 |
| | | | | | | | | .710 | .692 | .673 | .654 | .633 | .612 | .589 | .565 | .541 | .515 |
| | | | | | | | | .489 | .462 | .435 | .407 | .379 | .352 | | | | |
| 28.000 | 25.659 | .000 | 2.341 | .68 | 5.56 | 8.4% | 37 | .903 | .898 | .893 | .887 | .881 | .875 | .868 | .861 | .853 | .845 |
| | | | | | | | | .837 | .827 | .817 | .807 | .795 | .783 | .770 | .756 | .742 | .726 |
| | | | | | | | | .710 | .692 | .673 | .654 | .633 | .612 | .589 | .565 | .541 | .515 |
| | | | | | | | | .489 | .462 | .435 | .407 | .379 | .352 | .324 | | | |
| 28.000 | 25.955 | .000 | 2.045 | .70 | 5.54 | 7.3% | 38 | .903 | .898 | .893 | .887 | .881 | .875 | .868 | .861 | .853 | .845 |
| | | | | | | | | .837 | .827 | .817 | .807 | .795 | .783 | .770 | .756 | .742 | .726 |
| | | | | | | | | .710 | .692 | .673 | .654 | .633 | .612 | .589 | .565 | .541 | .515 |
| | | | | | | | | .489 | .462 | .435 | .407 | .379 | .352 | .324 | .296 | | |
| 28.000 | 26.225 | .000 | 1.775 | .73 | 5.51 | 6.3% | 39 | .903 | .898 | .893 | .887 | .881 | .875 | .868 | .861 | .853 | .845 |
| | | | | | | | | .837 | .827 | .817 | .807 | .795 | .783 | .770 | .756 | .742 | .726 |
| | | | | | | | | .710 | .692 | .673 | .654 | .633 | .612 | .589 | .565 | .541 | .515 |
| | | | | | | | | .489 | .462 | .435 | .407 | .379 | .352 | .324 | .296 | .270 | |
| 28.000 | 26.468 | .000 | 1.532 | .76 | 5.47 | 5.5% | 40 | .903 | .898 | .893 | .887 | .881 | .875 | .868 | .861 | .853 | .845 |
| | | | | | | | | .837 | .827 | .817 | .807 | .795 | .783 | .770 | .756 | .742 | .726 |
| | | | | | | | | .710 | .692 | .673 | .654 | .633 | .612 | .589 | .565 | .541 | .515 |
| | | | | | | | | .489 | .462 | .435 | .407 | .379 | .352 | .324 | .296 | .270 | .244 |
| 28.000 | 26.687 | .000 | 1.313 | .78 | 5.43 | 4.7% | 41 | .903 | .898 | .893 | .887 | .881 | .875 | .868 | .861 | .853 | .845 |
| | | | | | | | | .837 | .827 | .817 | .807 | .795 | .783 | .770 | .756 | .742 | .726 |
| | | | | | | | | .710 | .692 | .673 | .654 | .633 | .612 | .589 | .565 | .541 | .515 |
| | | | | | | | | .489 | .462 | .435 | .407 | .379 | .352 | .324 | .296 | .270 | .244 |
| | | | | | | | | .219 | | | | | | | | | |
| 28.000 | 26.882 | .000 | 1.118 | .81 | 5.37 | 4.0% | 42 | .903 | .898 | .893 | .887 | .881 | .875 | .868 | .861 | .853 | .845 |
| | | | | | | | | .837 | .827 | .817 | .807 | .795 | .783 | .770 | .756 | .742 | .726 |
| | | | | | | | | .710 | .692 | .673 | .654 | .633 | .612 | .589 | .565 | .541 | .515 |
| | | | | | | | | .489 | .462 | .435 | .407 | .379 | .352 | .324 | .296 | .270 | .244 |

*SIMILAR TO A.T.&T. NEAL-WILKINSON TABLES(SEE SECTION 1)

*EQUIVALENT RANDOM THEORY: VMR(VARIANCE TO MEAN RATIO)– 3.0 - NO RETRIES (PEAKED TRAFFIC - ROUTE ADVANCEMENT OR DAY-TO-DAY VARIATION)

| HOURS FIRST ATTEMPT VMR = 3.0 | HOURS CON-NECT | HOURS RECALL /RETRY | HOURS OVFL | CON-NECT VMR | OVFL VMR | PERCENT BLOCKED (GRADE) | LINE | HOURS CONNECTED TRAFFIC ON LINES 1 | 2 | 3 | 4 | 5 | 6 | 7 | 8 | 9 | 10 |
|---|---|---|---|---|---|---|---|---|---|---|---|---|---|---|---|---|---|
| | | | | | | | | .219 | .195 | | | | | | | | |
| 28.000 | 27.054 | .000 | .946 | .83 | 5.31 | 3.4% | 43 | .903 | .898 | .893 | .887 | .881 | .875 | .868 | .861 | .853 | .845 |
| | | | | | | | | .837 | .827 | .817 | .807 | .795 | .783 | .770 | .756 | .742 | .726 |
| | | | | | | | | .710 | .692 | .673 | .654 | .633 | .612 | .589 | .565 | .541 | .515 |
| | | | | | | | | .489 | .462 | .435 | .407 | .379 | .352 | .324 | .296 | .270 | .244 |
| | | | | | | | | .219 | .195 | .172 | | | | | | | |
| 28.000 | 27.205 | .000 | .795 | .85 | 5.24 | 2.8% | 44 | .903 | .898 | .893 | .887 | .881 | .875 | .868 | .861 | .853 | .845 |
| | | | | | | | | .837 | .827 | .817 | .807 | .795 | .783 | .770 | .756 | .742 | .726 |
| | | | | | | | | .710 | .692 | .673 | .654 | .633 | .612 | .589 | .565 | .541 | .515 |
| | | | | | | | | .489 | .462 | .435 | .407 | .379 | .352 | .324 | .296 | .270 | .244 |
| | | | | | | | | .219 | .195 | .172 | .151 | | | | | | |
| 28.000 | 27.337 | .000 | .663 | .87 | 5.16 | 2.4% | 45 | .903 | .898 | .893 | .887 | .881 | .875 | .868 | .861 | .853 | .845 |
| | | | | | | | | .837 | .827 | .817 | .807 | .795 | .783 | .770 | .756 | .742 | .726 |
| | | | | | | | | .710 | .692 | .673 | .654 | .633 | .612 | .589 | .565 | .541 | .515 |
| | | | | | | | | .489 | .462 | .435 | .407 | .379 | .352 | .324 | .296 | .270 | .244 |
| | | | | | | | | .219 | .195 | .172 | .151 | .132 | | | | | |
| 28.000 | 27.451 | .000 | .549 | .89 | 5.08 | 2.0% | 46 | .903 | .898 | .893 | .887 | .881 | .875 | .868 | .861 | .853 | .845 |
| | | | | | | | | .837 | .827 | .817 | .807 | .795 | .783 | .770 | .756 | .742 | .726 |
| | | | | | | | | .710 | .692 | .673 | .654 | .633 | .612 | .589 | .565 | .541 | .515 |
| | | | | | | | | .489 | .462 | .435 | .407 | .379 | .352 | .324 | .296 | .270 | .244 |
| | | | | | | | | .219 | .195 | .172 | .151 | .132 | .114 | | | | |
| 28.000 | 27.548 | .000 | .452 | .90 | 5.00 | 1.6% | 47 | .903 | .898 | .893 | .887 | .881 | .875 | .868 | .861 | .853 | .845 |
| | | | | | | | | .837 | .827 | .817 | .807 | .795 | .783 | .770 | .756 | .742 | .726 |
| | | | | | | | | .710 | .692 | .673 | .654 | .633 | .612 | .589 | .565 | .541 | .515 |
| | | | | | | | | .489 | .462 | .435 | .407 | .379 | .352 | .324 | .296 | .270 | .244 |
| | | | | | | | | .219 | .105 | .172 | .151 | .132 | .114 | .098 | | | |
| 20.000 | 27.631 | .000 | .369 | .92 | 1.01 | 1.3% | 48 | .903 | .898 | .893 | .887 | .881 | .875 | .868 | .861 | .853 | .845 |
| | | | | | | | | .837 | .827 | .817 | .807 | .795 | .783 | .770 | .756 | .742 | .726 |
| | | | | | | | | .710 | .692 | .673 | .654 | .633 | .612 | .589 | .565 | .541 | .515 |
| | | | | | | | | .489 | .462 | .435 | .407 | .379 | .352 | .324 | .296 | .270 | .244 |
| | | | | | | | | .219 | .195 | .172 | .151 | .132 | .114 | .098 | .083 | | |
| 28.000 | 27.701 | .000 | .299 | .93 | 4.82 | 1.1% | 49 | .903 | .898 | .893 | .887 | .881 | .875 | .868 | .861 | .853 | .845 |
| | | | | | | | | .837 | .827 | .817 | .807 | .795 | .783 | .770 | .756 | .742 | .720 |
| | | | | | | | | .710 | .692 | .673 | .654 | .633 | .612 | .589 | .565 | .541 | .515 |
| | | | | | | | | .489 | .462 | .435 | .407 | .379 | .352 | .324 | .206 | .270 | .244 |
| | | | | | | | | .219 | .195 | .172 | .151 | .132 | .114 | .098 | .083 | .070 | |
| 28.000 | 27.759 | .000 | .241 | .94 | 4.72 | .9% | 50 | .903 | .898 | .893 | .887 | .881 | .875 | .868 | .861 | .853 | .845 |
| | | | | | | | | .837 | .827 | .817 | .807 | .795 | .783 | .770 | .756 | .742 | .726 |
| | | | | | | | | .710 | .692 | .673 | .654 | .633 | .612 | .589 | .565 | .541 | .515 |
| | | | | | | | | .489 | .462 | .435 | .407 | .379 | .352 | .324 | .296 | .270 | .244 |

*SIMILAR TO A.T.&T. NEAL-WILKINSON TABLES(SEE SECTION 1)

*EQUIVALENT RANDOM THEORY: VMR(VARIANCE TO MEAN RATIO)– 3.0 - NO RETRIES (PEAKED TRAFFIC - ROUTE ADVANCEMENT OR DAY-TO-DAY VARIATION)

| HOURS FIRST ATTEMPT VMR=3.0 | HOURS CON-NECT | HOURS RECALL /RETRY | HOURS OVFL | CON-NECT VMR | OVFL VMR | PERCENT BLOCKED (GRADE) | LINE | HOURS CONNECTED TRAFFIC ON LINES | | | | | | | | | |
|---|---|---|---|---|---|---|---|---|---|---|---|---|---|---|---|---|---|
| | | | | | | | | 1 | 2 | 3 | 4 | 5 | 6 | 7 | 8 | 9 | 10 |
| | | | | | | | | .219 | .195 | .172 | .151 | .132 | .114 | .098 | .083 | .070 | .058 |
| 28.000 | 27.808 | .000 | .192 | .95 | 4.63 | .7% | 51 | .903 | .898 | .893 | .887 | .881 | .875 | .868 | .861 | .853 | .845 |
| | | | | | | | | .837 | .827 | .817 | .807 | .795 | .783 | .770 | .756 | .742 | .726 |
| | | | | | | | | .710 | .692 | .673 | .654 | .633 | .612 | .589 | .565 | .541 | .515 |
| | | | | | | | | .489 | .462 | .435 | .407 | .379 | .352 | .324 | .296 | .270 | .244 |
| | | | | | | | | .219 | .195 | .172 | .151 | .132 | .114 | .098 | .083 | .070 | .058 |
| | | | | | | | | .048 | | | | | | | | | |
| 28.000 | 27.848 | .000 | .152 | .96 | 4.54 | .5% | 52 | .903 | .898 | .893 | .887 | .881 | .875 | .868 | .861 | .853 | .845 |
| | | | | | | | | .837 | .827 | .817 | .807 | .795 | .783 | .770 | .756 | .742 | .726 |
| | | | | | | | | .710 | .692 | .673 | .654 | .633 | .612 | .589 | .565 | .541 | .515 |
| | | | | | | | | .489 | .462 | .435 | .407 | .379 | .352 | .324 | .296 | .270 | .244 |
| | | | | | | | | .219 | .195 | .172 | .151 | .132 | .114 | .098 | .083 | .070 | .058 |
| | | | | | | | | .048 | .040 | | | | | | | | |
| 30.000 | 19.778 | .000 | 10.222 | .31 | 4.97 | 34.1% | 24 | .909 | .904 | .900 | .895 | .890 | .884 | .878 | .872 | .865 | .858 |
| | | | | | | | | .850 | .842 | .833 | .824 | .814 | .803 | .792 | .780 | .768 | .754 |
| | | | | | | | | .740 | .724 | .708 | .691 | | | | | | |
| 30.000 | 20.450 | .000 | 9.550 | .33 | 5.05 | 31.8% | 25 | .909 | .904 | .900 | .895 | .890 | .884 | .878 | .872 | .865 | .858 |
| | | | | | | | | .850 | .842 | .833 | .824 | .814 | .803 | .792 | .780 | .768 | .754 |
| | | | | | | | | .740 | .724 | .708 | .691 | .673 | | | | | |
| 30.000 | 21.104 | .000 | 8.896 | .35 | 5.14 | 29.7% | 26 | .909 | .904 | .900 | .895 | .890 | .884 | .878 | .872 | .865 | .858 |
| | | | | | | | | .850 | .842 | .833 | .824 | .814 | .803 | .792 | .780 | .768 | .754 |
| | | | | | | | | .740 | .724 | .708 | .691 | .673 | .654 | | | | |
| 30.000 | 21.738 | .000 | 8.262 | .37 | 5.22 | 27.5% | 27 | .909 | .904 | .900 | .895 | .890 | .884 | .878 | .872 | .865 | .858 |
| | | | | | | | | .850 | .842 | .833 | .824 | .814 | .803 | .792 | .780 | .768 | .754 |
| | | | | | | | | .740 | .724 | .708 | .691 | .673 | .654 | .634 | | | |
| 30.000 | 22.351 | .000 | 7.649 | .39 | 5.30 | 25.5% | 28 | .909 | .904 | .900 | .895 | .890 | .884 | .878 | .872 | .865 | .858 |
| | | | | | | | | .850 | .842 | .833 | .824 | .814 | .803 | .792 | .780 | .768 | .754 |
| | | | | | | | | .740 | .724 | .708 | .691 | .673 | .654 | .634 | .613 | | |
| 30.000 | 22.942 | .000 | 7.058 | .41 | 5.37 | 23.5% | 29 | .909 | .904 | .900 | .895 | .890 | .884 | .878 | .872 | .865 | .858 |
| | | | | | | | | .850 | .842 | .833 | .824 | .814 | .803 | .792 | .780 | .768 | .754 |
| | | | | | | | | .740 | .724 | .708 | .691 | .673 | .654 | .634 | .613 | .591 | |
| 30.000 | 23.510 | .000 | 6.490 | .43 | 5.43 | 21.6% | 30 | .909 | .904 | .900 | .895 | .890 | .884 | .878 | .872 | .865 | .858 |
| | | | | | | | | .850 | .842 | .833 | .824 | .814 | .803 | .792 | .780 | .768 | .754 |
| | | | | | | | | .740 | .724 | .708 | .691 | .673 | .654 | .634 | .613 | .591 | .568 |
| 30.000 | 24.054 | .000 | 5.946 | .46 | 5.50 | 19.8% | 31 | .909 | .904 | .900 | .895 | .890 | .884 | .878 | .872 | .865 | .858 |
| | | | | | | | | .850 | .842 | .833 | .824 | .814 | .803 | .792 | .780 | .768 | .754 |
| | | | | | | | | .740 | .724 | .708 | .691 | .673 | .654 | .634 | .613 | .591 | .568 |
| | | | | | | | | .544 | | | | | | | | | |
| 30.000 | 24.574 | .000 | 5.426 | .48 | 5.55 | 18.1% | 32 | .909 | .904 | .900 | .895 | .890 | .884 | .878 | .872 | .865 | .858 |
| | | | | | | | | .850 | .842 | .833 | .824 | .814 | .803 | .792 | .780 | .768 | .754 |

*SIMILAR TO A.T.&T. NEAL-WILKINSON TABLES(SEE SECTION 1)

*EQUIVALENT RANDOM THEORY: VMR(VARIANCE TO MEAN RATIO)– 3.0 - NO RETRIES (PEAKED TRAFFIC - ROUTE ADVANCEMENT OR DAY-TO-DAY VARIATION)

| HOURS FIRST ATTEMPT VMR=3.0 | HOURS CONNECT | HOURS RECALL /RETRY | HOURS OVFL | CONNECT VMR | OVFL VMR | PERCENT BLOCKED (GRADE) | LINE | HOURS CONNECTED TRAFFIC ON LINES 1 | 2 | 3 | 4 | 5 | 6 | 7 | 8 | 9 | 10 |
|---|---|---|---|---|---|---|---|---|---|---|---|---|---|---|---|---|---|
| | | | | | | | | .740 | .724 | .708 | .691 | .673 | .654 | .634 | .613 | .591 | .568 |
| | | | | | | | | .544 | .520 | | | | | | | | |
| 30.000 | 25.068 | .000 | 4.932 | .51 | 5.60 | 16.4% | 33 | .909 | .904 | .900 | .895 | .890 | .884 | .878 | .872 | .865 | .858 |
| | | | | | | | | .850 | .842 | .833 | .824 | .814 | .803 | .792 | .780 | .768 | .754 |
| | | | | | | | | .740 | .724 | .708 | .691 | .673 | .654 | .634 | .613 | .591 | .568 |
| | | | | | | | | .544 | .520 | .494 | | | | | | | |
| 30.000 | 25.537 | .000 | 4.463 | .53 | 5.64 | 14.9% | 34 | .909 | .904 | .900 | .895 | .890 | .884 | .878 | .872 | .865 | .858 |
| | | | | | | | | .850 | .842 | .833 | .824 | .814 | .803 | .792 | .780 | .768 | .754 |
| | | | | | | | | .740 | .724 | .708 | .691 | .673 | .654 | .634 | .613 | .591 | .568 |
| | | | | | | | | .544 | .520 | .494 | .468 | | | | | | |
| 30.000 | 25.978 | .000 | 4.022 | .56 | 5.67 | 13.4% | 35 | .909 | .904 | .900 | .895 | .890 | .884 | .878 | .872 | .865 | .858 |
| | | | | | | | | .850 | .842 | .833 | .824 | .814 | .803 | .792 | .780 | .768 | .754 |
| | | | | | | | | .740 | .724 | .708 | .691 | .673 | .654 | .634 | .613 | .591 | .568 |
| | | | | | | | | .544 | .520 | .494 | .468 | .442 | | | | | |
| 30.000 | 26.393 | .000 | 3.607 | .59 | 5.69 | 12.0% | 36 | .909 | .904 | .900 | .895 | .890 | .884 | .878 | .872 | .865 | .858 |
| | | | | | | | | .850 | .842 | .833 | .824 | .814 | .803 | .792 | .780 | .768 | .754 |
| | | | | | | | | .740 | .724 | .708 | .691 | .673 | .654 | .634 | .613 | .591 | .568 |
| | | | | | | | | .544 | .520 | .494 | .468 | .442 | .415 | | | | |
| 30.000 | 26.781 | .000 | 3.219 | .61 | 5.71 | 10.7% | 37 | .909 | .904 | .900 | .895 | .890 | .884 | .878 | .872 | .865 | .858 |
| | | | | | | | | .850 | .842 | .833 | .824 | .814 | .803 | .792 | .780 | .768 | .754 |
| | | | | | | | | .740 | .724 | .708 | .691 | .673 | .654 | .634 | .613 | .591 | .568 |
| | | | | | | | | .544 | .520 | .494 | .468 | .442 | .415 | .388 | | | |
| 30.000 | 27.142 | .000 | 2.858 | .64 | 5.71 | 9.5% | 38 | .909 | .904 | .900 | .895 | .890 | .884 | .878 | .872 | .865 | .858 |
| | | | | | | | | .850 | .842 | .833 | .824 | .814 | .803 | .792 | .780 | .768 | .754 |
| | | | | | | | | .740 | .724 | .708 | .691 | .673 | .654 | .634 | .613 | .591 | .568 |
| | | | | | | | | .544 | .520 | .494 | .468 | .442 | .415 | .388 | .361 | | |
| 30.000 | 27.475 | .000 | 2.525 | .67 | 5.71 | 8.4% | 39 | .909 | .904 | .900 | .895 | .890 | .884 | .878 | .872 | .865 | .858 |
| | | | | | | | | .850 | .842 | .833 | .824 | .814 | .803 | .792 | .780 | .768 | .754 |
| | | | | | | | | .740 | .724 | .708 | .691 | .673 | .654 | .634 | .613 | .591 | .568 |
| | | | | | | | | .544 | .520 | .494 | .468 | .442 | .415 | .388 | .361 | .334 | |
| 30.000 | 27.782 | .000 | 2.218 | .69 | 5.69 | 7.4% | 40 | .909 | .904 | .900 | .895 | .890 | .884 | .878 | .872 | .865 | .858 |
| | | | | | | | | .850 | .842 | .833 | .824 | .814 | .803 | .792 | .780 | .768 | .754 |
| | | | | | | | | .740 | .724 | .708 | .691 | .673 | .654 | .634 | .613 | .591 | .568 |
| | | | | | | | | .544 | .520 | .494 | .468 | .442 | .415 | .388 | .361 | .334 | .307 |
| 30.000 | 28.063 | .000 | 1.937 | .72 | 5.67 | 6.5% | 41 | .909 | .904 | .900 | .895 | .890 | .884 | .878 | .872 | .865 | .858 |
| | | | | | | | | .850 | .842 | .833 | .824 | .814 | .803 | .792 | .780 | .768 | .754 |
| | | | | | | | | .740 | .724 | .708 | .691 | .673 | .654 | .634 | .613 | .591 | .568 |
| | | | | | | | | .544 | .520 | .494 | .468 | .442 | .415 | .388 | .361 | .334 | .307 |
| | | | | | | | | .280 | | | | | | | | | |
| 30.000 | 28.317 | .000 | 1.683 | .75 | 5.63 | 5.6% | 42 | .909 | .904 | .900 | .895 | .890 | .884 | .878 | .872 | .865 | .858 |

*SIMILAR TO A.T.&T. NEAL-WILKINSON TABLES(SEE SECTION 1)

*EQUIVALENT RANDOM THEORY: VMR(VARIANCE TO MEAN RATIO)– 3.0 - NO RETRIES (PEAKED TRAFFIC - ROUTE ADVANCEMENT OR DAY-TO-DAY VARIATION)

| HOURS FIRST ATTEMPT VMR=3.0 | HOURS CON-NECT | HOURS RECALL /RETRY | HOURS OVFL | CON-NECT VMR | OVFL VMR | PERCENT BLOCKED (GRADE) | LINE | HOURS CONNECTED TRAFFIC ON LINES 1 | 2 | 3 | 4 | 5 | 6 | 7 | 8 | 9 | 10 |
|---|---|---|---|---|---|---|---|---|---|---|---|---|---|---|---|---|---|
| | | | | | | | | .850 | .842 | .833 | .824 | .814 | .803 | .792 | .780 | .768 | .754 |
| | | | | | | | | .740 | .724 | .708 | .691 | .673 | .654 | .634 | .613 | .591 | .568 |
| | | | | | | | | .544 | .520 | .494 | .468 | .442 | .415 | .388 | .361 | .334 | .307 |
| | | | | | | | | .280 | .255 | | | | | | | | |
| 30.000 | 28.547 | .000 | 1.453 | .77 | 5.59 | 4.8% | 43 | .909 | .904 | .900 | .895 | .890 | .884 | .878 | .872 | .865 | .858 |
| | | | | | | | | .850 | .842 | .833 | .824 | .814 | .803 | .792 | .780 | .768 | .754 |
| | | | | | | | | .740 | .724 | .708 | .691 | .673 | .654 | .634 | .613 | .591 | .568 |
| | | | | | | | | .544 | .520 | .494 | .468 | .442 | .415 | .388 | .361 | .334 | .307 |
| | | | | | | | | .280 | .255 | .230 | | | | | | | |
| 30.000 | 28.754 | .000 | 1.246 | .79 | 5.54 | 4.2% | 44 | .909 | .904 | .900 | .895 | .890 | .884 | .878 | .872 | .865 | .858 |
| | | | | | | | | .850 | .842 | .833 | .824 | .814 | .803 | .792 | .780 | .768 | .754 |
| | | | | | | | | .740 | .724 | .708 | .691 | .673 | .654 | .634 | .613 | .591 | .568 |
| | | | | | | | | .544 | .520 | .494 | .468 | .442 | .415 | .388 | .361 | .334 | .307 |
| | | | | | | | | .280 | .255 | .230 | .206 | | | | | | |
| 30.000 | 28.937 | .000 | 1.063 | .82 | 5.48 | 3.5% | 45 | .909 | .904 | .900 | .895 | .890 | .884 | .878 | .872 | .865 | .858 |
| | | | | | | | | .850 | .842 | .833 | .824 | .814 | .803 | .792 | .780 | .768 | .754 |
| | | | | | | | | .740 | .724 | .708 | .691 | .673 | .654 | .634 | .613 | .591 | .568 |
| | | | | | | | | .544 | .520 | .494 | .468 | .442 | .415 | .388 | .361 | .334 | .307 |
| | | | | | | | | .280 | .255 | .230 | .206 | .184 | | | | | |
| 30.000 | 29.100 | .000 | .900 | .84 | 5.41 | 3.0% | 46 | .909 | .904 | .900 | .895 | .890 | .884 | .878 | .872 | .865 | .858 |
| | | | | | | | | .850 | .842 | .833 | .824 | .814 | .803 | .792 | .780 | .768 | .754 |
| | | | | | | | | .740 | .724 | .708 | .691 | .673 | .654 | .634 | .613 | .591 | .568 |
| | | | | | | | | .544 | .520 | .494 | .468 | .442 | .415 | .388 | .361 | .334 | .307 |
| | | | | | | | | .280 | .255 | .230 | .206 | .184 | .162 | | | | |
| 30.000 | 29.242 | .000 | .758 | .86 | 5.34 | 2.5% | 47 | .909 | .904 | .900 | .895 | .890 | .884 | .878 | .872 | .865 | .858 |
| | | | | | | | | .850 | .842 | .833 | .824 | .814 | .803 | .792 | .780 | .768 | .754 |
| | | | | | | | | .740 | .724 | .708 | .691 | .673 | .654 | .634 | .613 | .591 | .568 |
| | | | | | | | | .544 | .520 | .494 | .468 | .442 | .415 | .388 | .361 | .334 | .307 |
| | | | | | | | | .280 | .255 | .230 | .206 | .184 | .162 | .143 | | | |
| 30.000 | 29.367 | .000 | .633 | .88 | 5.26 | 2.1% | 48 | .909 | .904 | .900 | .895 | .890 | .884 | .878 | .872 | .865 | .858 |
| | | | | | | | | .850 | .842 | .833 | .824 | .814 | .803 | .792 | .780 | .768 | .754 |
| | | | | | | | | .740 | .724 | .708 | .691 | .673 | .654 | .634 | .613 | .591 | .568 |
| | | | | | | | | .544 | .520 | .494 | .468 | .442 | .415 | .388 | .361 | .334 | .307 |
| | | | | | | | | .280 | .255 | .230 | .206 | .184 | .162 | .143 | .124 | | |
| 30.000 | 29.474 | .000 | .526 | .89 | 5.18 | 1.8% | 49 | .909 | .904 | .900 | .895 | .890 | .884 | .878 | .872 | .865 | .858 |
| | | | | | | | | .850 | .842 | .833 | .824 | .814 | .803 | .792 | .780 | .768 | .754 |
| | | | | | | | | .740 | .724 | .708 | .691 | .673 | .654 | .634 | .613 | .591 | .568 |
| | | | | | | | | .544 | .520 | .494 | .468 | .442 | .415 | .388 | .361 | .334 | .307 |
| | | | | | | | | .280 | .255 | .230 | .206 | .184 | .162 | .143 | .124 | .107 | |
| 30.000 | 29.566 | .000 | .434 | .91 | 5.09 | 1.4% | 50 | .909 | .904 | .900 | .895 | .890 | .884 | .878 | .872 | .865 | .858 |

*SIMILAR TO A.T.&T. NEAL-WILKINSON TABLES(SEE SECTION 1)

*EQUIVALENT RANDOM THEORY: VMR(VARIANCE TO MEAN RATIO)– 3.0 - NO RETRIES (PEAKED TRAFFIC - ROUTE ADVANCEMENT OR DAY-TO-DAY VARIATION)

| HOURS FIRST ATTEMPT VMR=3.0 | HOURS CON-NECT | HOURS RECALL /RETRY | HOURS OVFL | CON-NECT VMR | OVFL VMR | PERCENT BLOCKED (GRADE) | LINE | HOURS CONNECTED TRAFFIC ON LINES 1 | 2 | 3 | 4 | 5 | 6 | 7 | 8 | 9 | 10 |
|---|---|---|---|---|---|---|---|---|---|---|---|---|---|---|---|---|---|
| | | | | | | | | .850 | .842 | .833 | .824 | .814 | .803 | .792 | .780 | .768 | .754 |
| | | | | | | | | .740 | .724 | .708 | .691 | .673 | .654 | .634 | .613 | .591 | .568 |
| | | | | | | | | .544 | .520 | .494 | .468 | .442 | .415 | .388 | .361 | .334 | .307 |
| | | | | | | | | .280 | .255 | .230 | .206 | .184 | .162 | .143 | .124 | .107 | .092 |
| 30.000 | 29.645 | .000 | .355 | .92 | 5.00 | 1.2% | 51 | .909 | .904 | .900 | .895 | .890 | .884 | .878 | .872 | .865 | .858 |
| | | | | | | | | .850 | .842 | .833 | .824 | .814 | .803 | .792 | .780 | .768 | .754 |
| | | | | | | | | .740 | .724 | .708 | .691 | .673 | .654 | .634 | .613 | .591 | .568 |
| | | | | | | | | .544 | .520 | .494 | .468 | .442 | .415 | .388 | .361 | .334 | .307 |
| | | | | | | | | .280 | .255 | .230 | .206 | .184 | .162 | .143 | .124 | .107 | .092 |
| | | | | | | | | .079 | | | | | | | | | |
| 30.000 | 29.711 | .000 | .289 | .93 | 4.91 | 1.0% | 52 | .909 | .904 | .900 | .895 | .890 | .884 | .878 | .872 | .865 | .858 |
| | | | | | | | | .850 | .842 | .833 | .824 | .814 | .803 | .792 | .780 | .768 | .754 |
| | | | | | | | | .740 | .724 | .708 | .691 | .673 | .654 | .634 | .613 | .591 | .568 |
| | | | | | | | | .544 | .520 | .494 | .468 | .442 | .415 | .388 | .361 | .334 | .307 |
| | | | | | | | | .280 | .255 | .230 | .206 | .184 | .162 | .143 | .124 | .107 | .092 |
| | | | | | | | | .079 | .066 | | | | | | | | |
| 30.000 | 29.767 | .000 | .233 | .94 | 4.81 | .8% | 53 | .909 | .904 | .900 | .895 | .890 | .884 | .878 | .872 | .865 | .858 |
| | | | | | | | | .850 | .842 | .833 | .824 | .814 | .803 | .792 | .780 | .768 | .754 |
| | | | | | | | | .740 | .724 | .708 | .691 | .673 | .654 | .634 | .613 | .591 | .568 |
| | | | | | | | | .544 | .520 | .494 | .468 | .442 | .415 | .388 | .361 | .334 | .307 |
| | | | | | | | | .280 | .255 | .230 | .206 | .184 | .162 | .143 | .124 | .107 | .092 |
| | | | | | | | | .079 | .066 | .056 | | | | | | | |
| 30.000 | 29.813 | .000 | .187 | .95 | 4.72 | .6% | 54 | .909 | .904 | .900 | .895 | .890 | .884 | .878 | .872 | .865 | .858 |
| | | | | | | | | .850 | .842 | .833 | .824 | .814 | .803 | .792 | .780 | .768 | .754 |
| | | | | | | | | .740 | .724 | .708 | .691 | .673 | .654 | .634 | .613 | .591 | .568 |
| | | | | | | | | .544 | .520 | .494 | .468 | .442 | .415 | .388 | .361 | .334 | .307 |
| | | | | | | | | .280 | .255 | .230 | .200 | .184 | .162 | .143 | .124 | .107 | .092 |
| | | | | | | | | .079 | .066 | .056 | .046 | | | | | | |

*SIMILAR TO A.T.&T. NEAL-WILKINSON TABLES(SEE SECTION 1)

# *EQUIVALENT RANDOM THEORY: VMR(VARIANCE TO MEAN RATIO)– 4.0 - NO RETRIES (PEAKED TRAFFIC - ROUTE ADVANCEMENT OR DAY-TO-DAY VARIATION)

| HOURS FIRST ATTEMPT VMR=4.0 | HOURS CON-NECT | HOURS RECALL /RETRY | HOURS OVFL | CON-NECT VMR | OVFL VMR | PERCENT BLOCKED (GRADE) | LINE | HOURS CONNECTED TRAFFIC ON LINES 1 | 2 | 3 | 4 | 5 | 6 | 7 | 8 | 9 | 10 |
|---|---|---|---|---|---|---|---|---|---|---|---|---|---|---|---|---|---|
| .200 | .132 | .000 | .068 | .97 | 2.99 | 33.9% | 3 | .058 | .043 | .031 | | | | | | | |
| .200 | .154 | .000 | .046 | .98 | 2.91 | 22.8% | 4 | .058 | .043 | .031 | .022 | | | | | | |
| .200 | .170 | .000 | .030 | .98 | 2.82 | 15.1% | 5 | .058 | .043 | .031 | .022 | .015 | | | | | |
| .200 | .180 | .000 | .020 | .99 | 2.75 | 9.8% | 6 | .058 | .043 | .031 | .022 | .015 | .011 | | | | |
| .200 | .187 | .000 | .013 | .99 | 2.67 | 6.3% | 7 | .058 | .043 | .031 | .022 | .015 | .011 | .007 | | | |
| .200 | .192 | .000 | .008 | 1.00 | 2.61 | 3.9% | 8 | .058 | .043 | .031 | .022 | .015 | .011 | .007 | .005 | | |
| .200 | .195 | .000 | .005 | 1.00 | 2.54 | 2.4% | 9 | .058 | .043 | .031 | .022 | .015 | .011 | .007 | .005 | .003 | |
| .200 | .197 | .000 | .003 | 1.00 | 2.48 | 1.5% | 10 | .058 | .043 | .031 | .022 | .015 | .011 | .007 | .005 | .003 | .002 |
| .200 | .198 | .000 | .002 | 1.00 | 2.43 | .9% | 11 | .058 | .043 | .031 | .022 | .015 | .011 | .007 | .005 | .003 | .002 |
| | | | | | | | | .001 | | | | | | | | | |
| .200 | .199 | .000 | .001 | 1.00 | 2.37 | .5% | 12 | .058 | .043 | .031 | .022 | .015 | .011 | .007 | .005 | .003 | .002 |
| | | | | | | | | .001 | .001 | | | | | | | | |
| .400 | .292 | .000 | .108 | .95 | 3.12 | 27.1% | 4 | .104 | .081 | .061 | .046 | | | | | | |
| .400 | .325 | .000 | .075 | .97 | 3.04 | 18.7% | 5 | .104 | .081 | .061 | .046 | .034 | | | | | |
| .400 | .349 | .000 | .051 | .98 | 2.95 | 12.7% | 6 | .104 | .081 | .061 | .046 | .034 | .024 | | | | |
| .400 | .366 | .000 | .034 | .98 | 2.87 | 8.5% | 7 | .104 | .081 | .061 | .046 | .034 | .024 | .017 | | | |
| .400 | .378 | .000 | .022 | .99 | 2.79 | 5.5% | 8 | .104 | .081 | .061 | .046 | .034 | .024 | .017 | .012 | | |
| .400 | .386 | .000 | .014 | .99 | 2.72 | 3.6% | 9 | .104 | .081 | .061 | .046 | .034 | .024 | .017 | .012 | .008 | |
| .400 | .391 | .000 | .009 | .99 | 2.65 | 2.3% | 10 | .104 | .081 | .061 | .046 | .034 | .024 | .017 | .012 | .008 | .005 |
| .400 | .394 | .000 | .006 | 1.00 | 2.58 | 1.4% | 11 | .104 | .081 | .061 | .046 | .034 | .024 | .017 | .012 | .008 | .005 |
| | | | | | | | | .003 | | | | | | | | | |
| .400 | .397 | .000 | .003 | 1.00 | 2.52 | .9% | 12 | .104 | .081 | .061 | .046 | .034 | .024 | .017 | .012 | .008 | .005 |
| | | | | | | | | .003 | .002 | | | | | | | | |
| .400 | .398 | .000 | .002 | 1.00 | 2.46 | .5% | 13 | .104 | .081 | .061 | .046 | .034 | .024 | .017 | .012 | .008 | .005 |
| | | | | | | | | .003 | .002 | .001 | | | | | | | |
| .600 | .420 | .000 | .180 | .93 | 3.26 | 30.1% | 4 | .144 | .115 | .090 | .070 | | | | | | |
| .600 | .472 | .000 | .128 | .95 | 3.17 | 21.3% | 5 | .144 | .115 | .090 | .070 | .052 | | | | | |
| .600 | .511 | .000 | .089 | .96 | 3.09 | 14.9% | 6 | .144 | .115 | .090 | .070 | .052 | .039 | | | | |
| .600 | .539 | .000 | .061 | .97 | 3.00 | 10.2% | 7 | .144 | .115 | .090 | .070 | .052 | .039 | .028 | | | |
| .600 | .559 | .000 | .041 | .98 | 2.92 | 6.9% | 8 | .144 | .115 | .090 | .070 | .052 | .039 | .028 | .020 | | |
| .600 | .573 | .000 | .027 | .99 | 2.84 | 4.5% | 9 | .144 | .115 | .090 | .070 | .052 | .039 | .028 | .020 | .014 | |
| .600 | .582 | .000 | .018 | .99 | 2.76 | 3.0% | 10 | .144 | .115 | .090 | .070 | .052 | .039 | .028 | .020 | .014 | .010 |
| .600 | .589 | .000 | .011 | .99 | 2.69 | 1.9% | 11 | .144 | .115 | .090 | .070 | .052 | .039 | .028 | .020 | .014 | .010 |
| | | | | | | | | .006 | | | | | | | | | |
| .600 | .593 | .000 | .007 | 1.00 | 2.62 | 1.2% | 12 | .144 | .115 | .090 | .070 | .052 | .039 | .028 | .020 | .014 | .010 |
| | | | | | | | | .006 | .004 | | | | | | | | |
| .600 | .596 | .000 | .004 | 1.00 | 2.56 | .7% | 13 | .144 | .115 | .090 | .070 | .052 | .039 | .028 | .020 | .014 | .010 |
| | | | | | | | | .006 | .004 | .003 | | | | | | | |
| .800 | .538 | .000 | .262 | .91 | 3.38 | 32.8% | 4 | .180 | .147 | .118 | .093 | | | | | | |
| .800 | .610 | .000 | .190 | .93 | 3.30 | 23.8% | 5 | .180 | .147 | .118 | .093 | .072 | | | | | |

*SIMILAR TO A.T.&T. NEAL-WILKINSON TABLES(SEE SECTION 1)

*EQUIVALENT RANDOM THEORY: VMR(VARIANCE TO MEAN RATIO)– 4.0 - NO RETRIES (PEAKED TRAFFIC - ROUTE ADVANCEMENT OR DAY-TO-DAY VARIATION)

| HOURS FIRST ATTEMPT VMR=4.0 | HOURS CON-NECT | HOURS RECALL /RETRY | HOURS OVFL | CON-NECT VMR | OVFL VMR | PERCENT BLOCKED (GRADE) | LINE | HOURS CONNECTED TRAFFIC ON LINES 1 | 2 | 3 | 4 | 5 | 6 | 7 | 8 | 9 | 10 |
|---|---|---|---|---|---|---|---|---|---|---|---|---|---|---|---|---|---|
| .800 | .664 | .000 | .136 | .95 | 3.21 | 17.0% | 6 | .180 | .147 | .118 | .093 | .072 | .054 | | | | |
| .800 | .705 | .000 | .095 | .96 | 3.13 | 11.9% | 7 | .180 | .147 | .118 | .093 | .072 | .054 | .041 | | | |
| .800 | .734 | .000 | .066 | .97 | 3.04 | 8.2% | 8 | .180 | .147 | .118 | .093 | .072 | .054 | .041 | .030 | | |
| .800 | .755 | .000 | .045 | .98 | 2.96 | 5.6% | 9 | .180 | .147 | .118 | .093 | .072 | .054 | .041 | .030 | .021 | |
| .800 | .770 | .000 | .030 | .99 | 2.88 | 3.7% | 10 | .180 | .147 | .118 | .093 | .072 | .054 | .041 | .030 | .021 | .015 |
| .800 | .781 | .000 | .019 | .99 | 2.80 | 2.4% | 11 | .180 | .147 | .118 | .093 | .072 | .054 | .041 | .030 | .021 | .015 |
| | | | | | | | | .010 | | | | | | | | | |
| .800 | .787 | .000 | .013 | .99 | 2.73 | 1.6% | 12 | .180 | .147 | .118 | .093 | .072 | .054 | .041 | .030 | .021 | .015 |
| | | | | | | | | .010 | .007 | | | | | | | | |
| .800 | .792 | .000 | .008 | 1.00 | 2.66 | 1.0% | 13 | .180 | .147 | .118 | .093 | .072 | .054 | .041 | .030 | .021 | .015 |
| | | | | | | | | .010 | .007 | .005 | | | | | | | |
| .800 | .795 | .000 | .005 | 1.00 | 2.59 | .6% | 14 | .180 | .147 | .118 | .093 | .072 | .054 | .041 | .030 | .021 | .015 |
| | | | | | | | | .010 | .007 | .005 | .003 | | | | | | |
| 1.000 | .741 | .000 | .259 | .91 | 3.41 | 25.9% | 5 | .212 | .177 | .144 | .116 | .091 | | | | | |
| 1.000 | .811 | .000 | .189 | .93 | 3.32 | 18.9% | 6 | .212 | .177 | .144 | .116 | .091 | .071 | | | | |
| 1.000 | .865 | .000 | .135 | .95 | 3.23 | 13.5% | 7 | .212 | .177 | .144 | .116 | .091 | .071 | .054 | | | |
| 1.000 | .905 | .000 | .095 | .96 | 3.15 | 9.5% | 8 | .212 | .177 | .144 | .116 | .091 | .071 | .054 | .040 | | |
| 1.000 | .934 | .000 | .066 | .97 | 3.06 | 6.6% | 9 | .212 | .177 | .144 | .116 | .091 | .071 | .054 | .040 | .029 | |
| 1.000 | .955 | .000 | .045 | .98 | 2.98 | 4.5% | 10 | .212 | .177 | .144 | .116 | .091 | .071 | .054 | .040 | .029 | .021 |
| 1.000 | .970 | .000 | .030 | .99 | 2.90 | 3.0% | 11 | .212 | .177 | .144 | .116 | .091 | .071 | .054 | .040 | .029 | .021 |
| | | | | | | | | .015 | | | | | | | | | |
| 1.000 | .980 | .000 | .020 | .99 | 2.82 | 2.0% | 12 | .212 | .177 | .144 | .116 | .091 | .071 | .054 | .040 | .029 | .021 |
| | | | | | | | | .015 | .010 | | | | | | | | |
| 1.000 | .987 | .000 | .013 | .99 | 2.75 | 1.3% | 13 | .212 | .177 | .144 | .116 | .091 | .071 | .054 | .040 | .029 | .021 |
| | | | | | | | | .015 | .010 | .007 | | | | | | | |
| 1.000 | .992 | .000 | .008 | 1.00 | 2.68 | .8% | 14 | .212 | .177 | .144 | .116 | .091 | .071 | .054 | .040 | .029 | .021 |
| | | | | | | | | .015 | .010 | .007 | .005 | | | | | | |
| 1.000 | .995 | .000 | .005 | 1.00 | 2.61 | .5% | 15 | .212 | .177 | .144 | .116 | .091 | .071 | .054 | .040 | .029 | .021 |
| | | | | | | | | .015 | .010 | .007 | .005 | .003 | | | | | |
| 1.200 | .865 | .000 | .335 | .89 | 3.49 | 27.9% | 5 | .242 | .204 | .170 | .138 | .111 | | | | | |
| 1.200 | .953 | .000 | .247 | .91 | 3.41 | 20.6% | 6 | .242 | .204 | .170 | .138 | .111 | .087 | | | | |
| 1.200 | 1.020 | .000 | .180 | .93 | 3.33 | 15.0% | 7 | .242 | .204 | .170 | .138 | .111 | .087 | .068 | | | |
| 1.200 | 1.072 | .000 | .128 | .95 | 3.24 | 10.7% | 8 | .242 | .204 | .170 | .138 | .111 | .087 | .068 | .051 | | |
| 1.200 | 1.110 | .000 | .090 | .96 | 3.15 | 7.6% | 9 | .242 | .204 | .170 | .138 | .111 | .087 | .068 | .051 | .038 | |
| 1.200 | 1.138 | .000 | .062 | .97 | 3.07 | 5.2% | 10 | .242 | .204 | .170 | .138 | .111 | .087 | .068 | .051 | .038 | .028 |
| 1.200 | 1.157 | .000 | .043 | .98 | 2.98 | 3.5% | 11 | .242 | .204 | .170 | .138 | .111 | .087 | .068 | .051 | .038 | .028 |
| | | | | | | | | .020 | | | | | | | | | |
| 1.200 | 1.172 | .000 | .028 | .99 | 2.90 | 2.4% | 12 | .242 | .204 | .170 | .138 | .111 | .087 | .068 | .051 | .038 | .028 |
| | | | | | | | | .020 | .014 | | | | | | | | |
| 1.200 | 1.181 | .000 | .019 | .99 | 2.83 | 1.6% | 13 | .242 | .204 | .170 | .138 | .111 | .087 | .068 | .051 | .038 | .028 |

*SIMILAR TO A.T.&T. NEAL-WILKINSON TABLES(SEE SECTION 1)

*EQUIVALENT RANDOM THEORY: VMR(VARIANCE TO MEAN RATIO)– 4.0 - NO RETRIES (PEAKED TRAFFIC - ROUTE ADVANCEMENT OR DAY-TO-DAY VARIATION)

| HOURS FIRST ATTEMPT VMR = 4.0 | HOURS CON-NECT | HOURS RECALL /RETRY | HOURS OVFL | CON-NECT VMR | OVFL VMR | PERCENT BLOCKED (GRADE) | LINE | HOURS CONNECTED TRAFFIC ON LINES 1 | 2 | 3 | 4 | 5 | 6 | 7 | 8 | 9 | 10 |
|---|---|---|---|---|---|---|---|---|---|---|---|---|---|---|---|---|---|
| | | | | | | | | .020 | .014 | .010 | | | | | | | |
| 1.200 | 1.188 | .000 | .012 | .99 | 2.75 | 1.0% | 14 | .242 | .204 | .170 | .138 | .111 | .087 | .068 | .051 | .038 | .028 |
| | | | | | | | | .020 | .014 | .010 | .007 | | | | | | |
| 1.200 | 1.192 | .000 | .008 | 1.00 | 2.69 | .6% | 15 | .242 | .204 | .170 | .138 | .111 | .087 | .068 | .051 | .038 | .028 |
| | | | | | | | | .020 | .014 | .010 | .007 | .004 | | | | | |
| 1.400 | .981 | .000 | .419 | .87 | 3.60 | 30.0% | 5 | .268 | .229 | .193 | .160 | .130 | | | | | |
| 1.400 | 1.085 | .000 | .315 | .90 | 3.52 | 22.5% | 6 | .268 | .229 | .193 | .160 | .130 | .104 | | | | |
| 1.400 | 1.167 | .000 | .233 | .92 | 3.44 | 16.6% | 7 | .268 | .229 | .193 | .160 | .130 | .104 | .082 | | | |
| 1.400 | 1.231 | .000 | .169 | .94 | 3.35 | 12.1% | 8 | .268 | .229 | .193 | .160 | .130 | .104 | .082 | .063 | | |
| 1.400 | 1.279 | .000 | .121 | .95 | 3.27 | 8.7% | 9 | .268 | .229 | .193 | .160 | .130 | .104 | .082 | .063 | .048 | |
| 1.400 | 1.314 | .000 | .086 | .96 | 3.18 | 6.1% | 10 | .268 | .229 | .193 | .160 | .130 | .104 | .082 | .063 | .048 | .036 |
| 1.400 | 1.341 | .000 | .059 | .97 | 3.09 | 4.2% | 11 | .268 | .229 | .193 | .160 | .130 | .104 | .082 | .063 | .048 | .036 |
| | | | | | | | | .026 | | | | | | | | | |
| 1.400 | 1.359 | .000 | .041 | .98 | 3.01 | 2.9% | 12 | .268 | .229 | .193 | .160 | .130 | .104 | .082 | .063 | .048 | .036 |
| | | | | | | | | .026 | .019 | | | | | | | | |
| 1.400 | 1.373 | .000 | .027 | .99 | 2.93 | 1.9% | 13 | .268 | .229 | .193 | .160 | .130 | .104 | .082 | .063 | .048 | .036 |
| | | | | | | | | .026 | .019 | .013 | | | | | | | |
| 1.400 | 1.382 | .000 | .018 | .99 | 2.85 | 1.3% | 14 | .268 | .229 | .193 | .160 | .130 | .104 | .082 | .063 | .048 | .036 |
| | | | | | | | | .026 | .019 | .013 | .009 | | | | | | |
| 1.400 | 1.388 | .000 | .012 | .99 | 2.78 | .8% | 15 | .268 | .229 | .193 | .160 | .130 | .104 | .082 | .063 | .048 | .036 |
| | | | | | | | | .026 | .019 | .013 | .009 | .006 | | | | | |
| 1.400 | 1.392 | .000 | .008 | 1.00 | 2.71 | .5% | 16 | .268 | .229 | .193 | .160 | .130 | .104 | .082 | .063 | .048 | .036 |
| | | | | | | | | .026 | .019 | .013 | .009 | .006 | .004 | | | | |
| 1.600 | 1.094 | .000 | .506 | .85 | 3.66 | 31.6% | 5 | .293 | .254 | .216 | .181 | .150 | | | | | |
| 1.600 | 1.215 | .000 | .385 | .88 | 3.59 | 24.0% | 6 | .293 | .254 | .216 | .181 | .150 | .121 | | | | |
| 1.600 | 1.312 | .000 | .288 | .90 | 3.51 | 18.0% | 7 | .293 | .254 | .216 | .181 | .150 | .121 | .097 | | | |
| 1.600 | 1.388 | .000 | .212 | .92 | 3.43 | 13.3% | 8 | .293 | .254 | .216 | .181 | .150 | .121 | .097 | .076 | | |
| 1.600 | 1.446 | .000 | .154 | .94 | 3.34 | 9.6% | 9 | .293 | .254 | .216 | .181 | .150 | .121 | .097 | .076 | .058 | |
| 1.600 | 1.490 | .000 | .110 | .96 | 3.26 | 6.9% | 10 | .293 | .254 | .216 | .181 | .150 | .121 | .097 | .076 | .058 | .044 |
| 1.600 | 1.523 | .000 | .077 | .97 | 3.17 | 4.8% | 11 | .293 | .254 | .216 | .181 | .150 | .121 | .097 | .076 | .058 | .044 |
| | | | | | | | | .033 | | | | | | | | | |
| 1.600 | 1.547 | .000 | .053 | .98 | 3.08 | 3.3% | 12 | .293 | .254 | .216 | .181 | .150 | .121 | .097 | .076 | .058 | .044 |
| | | | | | | | | .033 | .024 | | | | | | | | |
| 1.600 | 1.564 | .000 | .036 | .98 | 3.00 | 2.3% | 13 | .293 | .254 | .216 | .181 | .150 | .121 | .097 | .076 | .058 | .044 |
| | | | | | | | | .033 | .024 | .017 | | | | | | | |
| 1.600 | 1.576 | .000 | .024 | .99 | 2.92 | 1.5% | 14 | .293 | .254 | .216 | .181 | .150 | .121 | .097 | .076 | .058 | .044 |
| | | | | | | | | .033 | .024 | .017 | .012 | | | | | | |
| 1.600 | 1.584 | .000 | .016 | .99 | 2.85 | 1.0% | 15 | .293 | .254 | .216 | .181 | .150 | .121 | .097 | .076 | .058 | .044 |
| | | | | | | | | .033 | .024 | .017 | .012 | .008 | | | | | |
| 1.600 | 1.590 | .000 | .010 | .99 | 2.78 | .7% | 16 | .293 | .254 | .216 | .181 | .150 | .121 | .097 | .076 | .058 | .044 |

*SIMILAR TO A.T.&T. NEAL-WILKINSON TABLES(SEE SECTION 1)

*EQUIVALENT RANDOM THEORY: VMR(VARIANCE TO MEAN RATIO)– 4.0 - NO RETRIES (PEAKED TRAFFIC - ROUTE ADVANCEMENT OR DAY-TO-DAY VARIATION)

| HOURS FIRST ATTEMPT VMR = 4.0 | HOURS CON-NECT | HOURS RECALL /RETRY | HOURS OVFL | CON-NECT VMR | OVFL VMR | PERCENT BLOCKED (GRADE) | LINE | HOURS CONNECTED TRAFFIC ON LINES 1 | 2 | 3 | 4 | 5 | 6 | 7 | 8 | 9 | 10 |
|---|---|---|---|---|---|---|---|---|---|---|---|---|---|---|---|---|---|
| | | | | | | | | .033 | .024 | .017 | .012 | .008 | .006 | | | | |
| 1.800 | 1.203 | .000 | .597 | .83 | 3.72 | 33.2% | 5 | .317 | .277 | .238 | .202 | .168 | | | | | |
| 1.800 | 1.341 | .000 | .459 | .86 | 3.66 | 25.5% | 6 | .317 | .277 | .238 | .202 | .168 | .138 | | | | |
| 1.800 | 1.453 | .000 | .347 | .89 | 3.58 | 19.3% | 7 | .317 | .277 | .238 | .202 | .168 | .138 | .112 | | | |
| 1.800 | 1.541 | .000 | .259 | .91 | 3.50 | 14.4% | 8 | .317 | .277 | .238 | .202 | .168 | .138 | .112 | .088 | | |
| 1.800 | 1.610 | .000 | .190 | .93 | 3.41 | 10.6% | 9 | .317 | .277 | .238 | .202 | .168 | .138 | .112 | .088 | .069 | |
| 1.800 | 1.663 | .000 | .137 | .95 | 3.33 | 7.6% | 10 | .317 | .277 | .238 | .202 | .168 | .138 | .112 | .088 | .069 | .053 |
| 1.800 | 1.702 | .000 | .098 | .96 | 3.24 | 5.4% | 11 | .317 | .277 | .238 | .202 | .168 | .138 | .112 | .088 | .069 | .053 |
| | | | | | | | | .040 | | | | | | | | | |
| 1.800 | 1.732 | .000 | .068 | .97 | 3.15 | 3.8% | 12 | .317 | .277 | .238 | .202 | .168 | .138 | .112 | .088 | .069 | .053 |
| | | | | | | | | .040 | .029 | | | | | | | | |
| 1.800 | 1.753 | .000 | .047 | .98 | 3.07 | 2.6% | 13 | .317 | .277 | .238 | .202 | .168 | .138 | .112 | .088 | .069 | .053 |
| | | | | | | | | .040 | .029 | .021 | | | | | | | |
| 1.800 | 1.768 | .000 | .032 | .98 | 2.99 | 1.8% | 14 | .317 | .277 | .238 | .202 | .168 | .138 | .112 | .088 | .069 | .053 |
| | | | | | | | | .040 | .029 | .021 | .015 | | | | | | |
| 1.800 | 1.779 | .000 | .021 | .99 | 2.91 | 1.2% | 15 | .317 | .277 | .238 | .202 | .168 | .138 | .112 | .088 | .069 | .053 |
| | | | | | | | | .040 | .029 | .021 | .015 | .011 | | | | | |
| 1.800 | 1.786 | .000 | .014 | .99 | 2.83 | .8% | 16 | .317 | .277 | .238 | .202 | .168 | .138 | .112 | .088 | .069 | .053 |
| | | | | | | | | .040 | .029 | .021 | .015 | .011 | .007 | | | | |
| 1.800 | 1.791 | .000 | .009 | 1.00 | 2.76 | .5% | 17 | .317 | .277 | .238 | .202 | .168 | .138 | .112 | .088 | .069 | .053 |
| | | | | | | | | .040 | .029 | .021 | .015 | .011 | .007 | .005 | | | |
| 2.000 | 1.302 | .000 | .698 | .81 | 3.80 | 34.9% | 5 | .338 | .298 | .258 | .221 | .187 | | | | | |
| 2.000 | 1.457 | .000 | .543 | .84 | 3.74 | 27.1% | 6 | .338 | .298 | .258 | .221 | .187 | .155 | | | | |
| 2.000 | 1.584 | .000 | .416 | .87 | 3.67 | 20.8% | 7 | .338 | .298 | .258 | .221 | .187 | .155 | 127 | | | |
| 2.000 | 1.686 | .000 | .314 | .90 | 3.59 | 15.7% | 8 | .338 | .298 | .258 | .221 | .187 | .155 | .127 | .102 | | |
| 2.000 | 1.766 | .000 | .234 | .92 | 3.51 | 11.7% | 9 | .338 | .298 | .258 | .221 | .187 | .155 | .127 | .102 | .080 | |
| 2.000 | 1.829 | .000 | .171 | .94 | 3.42 | 8.6% | 10 | .338 | .298 | .258 | .221 | .187 | .155 | .127 | .102 | .080 | .062 |
| 2.000 | 1.876 | .000 | .124 | .95 | 3.34 | 6.2% | 11 | .338 | .298 | .258 | .221 | .187 | .155 | .127 | .102 | .080 | .062 |
| | | | | | | | | .048 | | | | | | | | | |
| 2.000 | 1.912 | .000 | .088 | .96 | 3.25 | 4.4% | 12 | .338 | .298 | .258 | .221 | .187 | .155 | .127 | .102 | .080 | .062 |
| | | | | | | | | .048 | .036 | | | | | | | | |
| 2.000 | 1.938 | .000 | .062 | .97 | 3.16 | 3.1% | 13 | .338 | .298 | .258 | .221 | .187 | .155 | .127 | .102 | .080 | .062 |
| | | | | | | | | .048 | .036 | .026 | | | | | | | |
| 2.000 | 1.958 | .000 | .042 | .98 | 3.08 | 2.1% | 14 | .338 | .298 | .258 | .221 | .187 | .155 | .127 | .102 | .080 | .062 |
| | | | | | | | | .048 | .036 | .026 | .019 | | | | | | |
| 2.000 | 1.971 | .000 | .029 | .99 | 3.00 | 1.4% | 15 | .338 | .298 | .258 | .221 | .187 | .155 | .127 | .102 | .080 | .062 |
| | | | | | | | | .048 | .036 | .026 | .019 | .014 | | | | | |
| 2.000 | 1.981 | .000 | .019 | .99 | 2.92 | 1.0% | 16 | .338 | .298 | .258 | .221 | .187 | .155 | .127 | .102 | .080 | .062 |
| | | | | | | | | .048 | .036 | .026 | .019 | .014 | .010 | | | | |
| 2.000 | 1.987 | .000 | .013 | .99 | 2.85 | .6% | 17 | .338 | .298 | .258 | .221 | .187 | .155 | .127 | .102 | .080 | .062 |

*SIMILAR TO A.T.&T. NEAL-WILKINSON TABLES(SEE SECTION 1)

*EQUIVALENT RANDOM THEORY: VMR(VARIANCE TO MEAN RATIO)– 4.0 - NO RETRIES (PEAKED TRAFFIC - ROUTE ADVANCEMENT OR DAY-TO-DAY VARIATION)

| HOURS FIRST ATTEMPT VMR=4.0 | HOURS CON-NECT | HOURS RECALL /RETRY | HOURS OVFL | CON-NECT VMR | OVFL VMR | PERCENT BLOCKED (GRADE) | LINE | HOURS CONNECTED TRAFFIC ON LINES 1 | 2 | 3 | 4 | 5 | 6 | 7 | 8 | 9 | 10 |
|---|---|---|---|---|---|---|---|---|---|---|---|---|---|---|---|---|---|
| | | | | | | | | .048 | .036 | .026 | .019 | .014 | .010 | .007 | | | |
| 2.200 | 1.574 | .000 | .626 | .83 | 3.79 | 28.5% | 6 | .359 | .318 | .279 | .241 | .205 | .172 | | | | |
| 2.200 | 1.716 | .000 | .484 | .86 | 3.73 | 22.0% | 7 | .359 | .318 | .279 | .241 | .205 | .172 | .142 | | | |
| 2.200 | 1.831 | .000 | .369 | .88 | 3.65 | 16.8% | 8 | .359 | .318 | .279 | .241 | .205 | .172 | .142 | .115 | | |
| 2.200 | 1.922 | .000 | .278 | .91 | 3.57 | 12.6% | 9 | .359 | .318 | .279 | .241 | .205 | .172 | .142 | .115 | .092 | |
| 2.200 | 1.995 | .000 | .205 | .93 | 3.49 | 9.3% | 10 | .359 | .318 | .279 | .241 | .205 | .172 | .142 | .115 | .092 | .072 |
| 2.200 | 2.050 | .000 | .150 | .94 | 3.40 | 6.8% | 11 | .359 | .318 | .279 | .241 | .205 | .172 | .142 | .115 | .092 | .072 |
| | | | | | | | | .056 | | | | | | | | | |
| 2.200 | 2.092 | .000 | .108 | .96 | 3.31 | 4.9% | 12 | .359 | .318 | .279 | .241 | .205 | .172 | .142 | .115 | .092 | .072 |
| | | | | | | | | .056 | .042 | | | | | | | | |
| 2.200 | 2.124 | .000 | .076 | .97 | 3.22 | 3.5% | 13 | .359 | .318 | .279 | .241 | .205 | .172 | .142 | .115 | .092 | .072 |
| | | | | | | | | .056 | .042 | .031 | | | | | | | |
| 2.200 | 2.147 | .000 | .053 | .98 | 3.14 | 2.4% | 14 | .359 | .318 | .279 | .241 | .205 | .172 | .142 | .115 | .092 | .072 |
| | | | | | | | | .056 | .042 | .031 | .023 | | | | | | |
| 2.200 | 2.164 | .000 | .036 | .98 | 3.06 | 1.7% | 15 | .359 | .318 | .279 | .241 | .205 | .172 | .142 | .115 | .092 | .072 |
| | | | | | | | | .056 | .042 | .031 | .023 | .017 | | | | | |
| 2.200 | 2.175 | .000 | .025 | .99 | 2.98 | 1.1% | 16 | .359 | .318 | .279 | .241 | .205 | .172 | .142 | .115 | .092 | .072 |
| | | | | | | | | .056 | .042 | .031 | .023 | .017 | .012 | | | | |
| 2.200 | 2.184 | .000 | .016 | .99 | 2.90 | .7% | 17 | .359 | .318 | .279 | .241 | .205 | .172 | .142 | .115 | .092 | .072 |
| | | | | | | | | .056 | .042 | .031 | .023 | .017 | .012 | .008 | | | |
| 2.400 | 1.686 | .000 | .714 | .81 | 3.84 | 29.7% | 6 | .379 | .338 | .298 | .260 | .223 | .188 | | | | |
| 2.400 | 1.843 | .000 | .557 | .84 | 3.78 | 23.2% | 7 | .379 | .338 | .298 | .260 | .223 | .188 | .157 | | | |
| 2.400 | 1.972 | .000 | .428 | .87 | 3.71 | 17.8% | 8 | .379 | .338 | .298 | .260 | .223 | .188 | .157 | .129 | | |
| 2.400 | 2.075 | .000 | .325 | .90 | 3.63 | 13.5% | 9 | .379 | .338 | .298 | .260 | .223 | .188 | .157 | .129 | .104 | |
| 2.400 | 2.157 | .000 | .243 | .92 | 3.55 | 10.1% | 10 | .379 | .338 | .298 | .260 | .223 | .188 | .157 | .129 | .104 | .082 |
| 2.400 | 2.221 | .000 | .179 | .94 | 3.46 | 7.4% | 11 | .379 | .338 | .298 | .260 | .223 | .188 | .157 | .129 | .104 | .082 |
| | | | | | | | | .064 | | | | | | | | | |
| 2.400 | 2.271 | .000 | .129 | .95 | 3.37 | 5.4% | 12 | .379 | .338 | .298 | .260 | .223 | .188 | .157 | .129 | .104 | .082 |
| | | | | | | | | .064 | .049 | | | | | | | | |
| 2.400 | 2.308 | .000 | .092 | .96 | 3.28 | 3.9% | 13 | .379 | .338 | .298 | .260 | .223 | .188 | .157 | .129 | .104 | .082 |
| | | | | | | | | .064 | .049 | .037 | | | | | | | |
| 2.400 | 2.335 | .000 | .065 | .97 | 3.20 | 2.7% | 14 | .379 | .338 | .298 | .260 | .223 | .188 | .157 | .129 | .104 | .082 |
| | | | | | | | | .064 | .049 | .037 | .027 | | | | | | |
| 2.400 | 2.355 | .000 | .045 | .98 | 3.11 | 1.9% | 15 | .379 | .338 | .298 | .260 | .223 | .188 | .157 | .129 | .104 | .082 |
| | | | | | | | | .064 | .049 | .037 | .027 | .020 | | | | | |
| 2.400 | 2.369 | .000 | .031 | .99 | 3.03 | 1.3% | 16 | .379 | .338 | .298 | .260 | .223 | .188 | .157 | .129 | .104 | .082 |
| | | | | | | | | .064 | .049 | .037 | .027 | .020 | .014 | | | | |
| 2.400 | 2.379 | .000 | .021 | .99 | 2.95 | .9% | 17 | .379 | .338 | .298 | .260 | .223 | .188 | .157 | .129 | .104 | .082 |
| | | | | | | | | .064 | .049 | .037 | .027 | .020 | .014 | .010 | | | |
| 2.400 | 2.386 | .000 | .014 | .99 | 2.88 | .6% | 18 | .379 | .338 | .298 | .260 | .223 | .188 | .157 | .129 | .104 | .082 |

*SIMILAR TO A.T.&T. NEAL-WILKINSON TABLES(SEE SECTION 1)

*EQUIVALENT RANDOM THEORY: VMR(VARIANCE TO MEAN RATIO)– 4.0 - NO RETRIES (PEAKED TRAFFIC - ROUTE ADVANCEMENT OR DAY-TO-DAY VARIATION)

| HOURS FIRST ATTEMPT VMR=4.0 | HOURS CON-NECT | HOURS RECALL /RETRY | HOURS OVFL | CON-NECT VMR | OVFL VMR | PERCENT BLOCKED (GRADE) | LINE | HOURS CONNECTED TRAFFIC ON LINES 1 | 2 | 3 | 4 | 5 | 6 | 7 | 8 | 9 | 10 |
|---|---|---|---|---|---|---|---|---|---|---|---|---|---|---|---|---|---|
| | | | | | | | | .064 | .049 | .037 | .027 | .020 | .014 | .010 | .007 | | |
| 2.600 | 1.788 | .000 | .812 | .80 | 3.91 | 31.2% | 6 | .396 | .356 | .316 | .277 | .240 | .205 | | | | |
| 2.600 | 1.960 | .000 | .640 | .83 | 3.85 | 24.6% | 7 | .396 | .356 | .316 | .277 | .240 | .205 | .172 | | | |
| 2.600 | 2.103 | .000 | .497 | .86 | 3.79 | 19.1% | 8 | .396 | .356 | .316 | .277 | .240 | .205 | .172 | .142 | | |
| 2.600 | 2.219 | .000 | .381 | .88 | 3.71 | 14.7% | 9 | .396 | .356 | .316 | .277 | .240 | .205 | .172 | .142 | .116 | |
| 2.600 | 2.312 | .000 | .288 | .91 | 3.63 | 11.1% | 10 | .396 | .356 | .316 | .277 | .240 | .205 | .172 | .142 | .116 | .093 |
| 2.600 | 2.386 | .000 | .214 | .93 | 3.55 | 8.2% | 11 | .396 | .356 | .316 | .277 | .240 | .205 | .172 | .142 | .116 | .093 |
| | | | | | | | | .073 | | | | | | | | | |
| 2.600 | 2.443 | .000 | .157 | .94 | 3.46 | 6.1% | 12 | .396 | .356 | .316 | .277 | .240 | .205 | .172 | .142 | .116 | .093 |
| | | | | | | | | .073 | .057 | | | | | | | | |
| 2.600 | 2.486 | .000 | .114 | .96 | 3.37 | 4.4% | 13 | .396 | .356 | .316 | .277 | .240 | .205 | .172 | .142 | .116 | .093 |
| | | | | | | | | .073 | .057 | .044 | | | | | | | |
| 2.600 | 2.519 | .000 | .081 | .97 | 3.28 | 3.1% | 14 | .396 | .356 | .316 | .277 | .240 | .205 | .172 | .142 | .116 | .093 |
| | | | | | | | | .073 | .057 | .044 | .033 | | | | | | |
| 2.600 | 2.543 | .000 | .057 | .98 | 3.20 | 2.2% | 15 | .396 | .356 | .316 | .277 | .240 | .205 | .172 | .142 | .116 | .093 |
| | | | | | | | | .073 | .057 | .044 | .033 | .024 | | | | | |
| 2.600 | 2.560 | .000 | .040 | .98 | 3.11 | 1.5% | 16 | .396 | .356 | .316 | .277 | .240 | .205 | .172 | .142 | .116 | .093 |
| | | | | | | | | .073 | .057 | .044 | .033 | .024 | .018 | | | | |
| 2.600 | 2.573 | .000 | .027 | .99 | 3.03 | 1.0% | 17 | .396 | .356 | .316 | .277 | .240 | .205 | .172 | .142 | .116 | .093 |
| | | | | | | | | .073 | .057 | .044 | .033 | .024 | .018 | .013 | | | |
| 2.600 | 2.582 | .000 | .018 | .99 | 2.96 | .7% | 18 | .396 | .356 | .316 | .277 | .240 | .205 | .172 | .142 | .116 | .093 |
| | | | | | | | | .073 | .057 | .044 | .033 | .024 | .018 | .013 | .009 | | |
| 2.800 | 1.893 | .000 | .907 | .78 | 3.94 | 32.4% | 6 | .414 | .374 | .334 | .295 | .257 | .221 | | | | |
| 2.800 | 2.080 | .000 | .720 | .81 | 3.89 | 25.7% | 7 | .414 | .374 | .334 | .295 | .257 | .221 | .187 | | | |
| 2.800 | 2.236 | .000 | .561 | .84 | 3.83 | 20.1% | 8 | .414 | .374 | .334 | .295 | .257 | .221 | .187 | .156 | | |
| 2.800 | 2.364 | .000 | .436 | .87 | 3.76 | 15.6% | 9 | .414 | .374 | .334 | .295 | .257 | .221 | .187 | .156 | .128 | |
| 2.800 | 2.468 | .000 | .332 | .90 | 3.68 | 11.8% | 10 | .414 | .374 | .334 | .295 | .257 | .221 | .187 | .156 | .128 | .104 |
| 2.800 | 2.551 | .000 | .249 | .92 | 3.60 | 8.9% | 11 | .414 | .374 | .334 | .295 | .257 | .221 | .187 | .156 | .128 | .104 |
| | | | | | | | | .083 | | | | | | | | | |
| 2.800 | 2.616 | .000 | .184 | .94 | 3.51 | 6.6% | 12 | .414 | .374 | .334 | .295 | .257 | .221 | .187 | .156 | .128 | .104 |
| | | | | | | | | .083 | .065 | | | | | | | | |
| 2.800 | 2.665 | .000 | .135 | .95 | 3.43 | 4.8% | 13 | .414 | .374 | .334 | .295 | .257 | .221 | .187 | .156 | .128 | .104 |
| | | | | | | | | .083 | .065 | .050 | | | | | | | |
| 2.800 | 2.703 | .000 | .097 | .96 | 3.34 | 3.5% | 14 | .414 | .374 | .334 | .295 | .257 | .221 | .187 | .156 | .128 | .104 |
| | | | | | | | | .083 | .065 | .050 | .038 | | | | | | |
| 2.800 | 2.731 | .000 | .069 | .97 | 3.25 | 2.4% | 15 | .414 | .374 | .334 | .295 | .257 | .221 | .187 | .156 | .128 | .104 |
| | | | | | | | | .083 | .065 | .050 | .038 | .028 | | | | | |
| 2.800 | 2.752 | .000 | .048 | .98 | 3.16 | 1.7% | 16 | .414 | .374 | .334 | .295 | .257 | .221 | .187 | .156 | .128 | .104 |
| | | | | | | | | .083 | .065 | .050 | .038 | .028 | .021 | | | | |
| 2.800 | 2.767 | .000 | .033 | .99 | 3.08 | 1.2% | 17 | .414 | .374 | .334 | .295 | .257 | .221 | .187 | .156 | .128 | .104 |

*SIMILAR TO A.T.&T. NEAL-WILKINSON TABLES(SEE SECTION 1)

*EQUIVALENT RANDOM THEORY: VMR(VARIANCE TO MEAN RATIO)– 4.0 - NO RETRIES (PEAKED TRAFFIC - ROUTE ADVANCEMENT OR DAY-TO-DAY VARIATION)

| HOURS FIRST ATTEMPT VMR=4.0 | HOURS CON-NECT | HOURS RECALL /RETRY | HOURS OVFL | CON-NECT VMR | OVFL VMR | PERCENT BLOCKED (GRADE) | LINE | HOURS CONNECTED TRAFFIC ON LINES 1 | 2 | 3 | 4 | 5 | 6 | 7 | 8 | 9 | 10 |
|---|---|---|---|---|---|---|---|---|---|---|---|---|---|---|---|---|---|
| | | | | | | | | .083 | .065 | .050 | .038 | .028 | .021 | .015 | | | |
| 2.800 | 2.778 | .000 | .022 | .99 | 3.00 | .8% | 18 | .414 | .374 | .334 | .295 | .257 | .221 | .187 | .156 | .128 | .104 |
| | | | | | | | | .083 | .065 | .050 | .038 | .028 | .021 | .015 | .011 | | |
| 2.800 | 2.785 | .000 | .015 | .99 | 2.93 | .5% | 19 | .414 | .374 | .334 | .295 | .257 | .221 | .187 | .156 | .128 | .104 |
| | | | | | | | | .083 | .065 | .050 | .038 | .028 | .021 | .015 | .011 | .007 | |
| 3.000 | 1.987 | .000 | 1.013 | .76 | 4.01 | 33.8% | 6 | .429 | .389 | .350 | .311 | .273 | .236 | | | | |
| 3.000 | 2.189 | .000 | .811 | .80 | 3.96 | 27.0% | 7 | .429 | .389 | .350 | .311 | .273 | .236 | .202 | | | |
| 3.000 | 2.359 | .000 | .641 | .83 | 3.90 | 21.4% | 8 | .429 | .389 | .350 | .311 | .273 | .236 | .202 | .170 | | |
| 3.000 | 2.500 | .000 | .500 | .86 | 3.84 | 16.7% | 9 | .429 | .389 | .350 | .311 | .273 | .236 | .202 | .170 | .141 | |
| 3.000 | 2.615 | .000 | .385 | .88 | 3.76 | 12.8% | 10 | .429 | .389 | .350 | .311 | .273 | .236 | .202 | .170 | .141 | .115 |
| 3.000 | 2.708 | .000 | .292 | .91 | 3.68 | 9.7% | 11 | .429 | .389 | .350 | .311 | .273 | .236 | .202 | .170 | .141 | .115 |
| | | | | | | | | .093 | | | | | | | | | |
| 3.000 | 2.782 | .000 | .218 | .93 | 3.60 | 7.3% | 12 | .429 | .389 | .350 | .311 | .273 | .236 | .202 | .170 | .141 | .115 |
| | | | | | | | | .093 | .073 | | | | | | | | |
| 3.000 | 2.839 | .000 | .161 | .94 | 3.51 | 5.4% | 13 | .429 | .389 | .350 | .311 | .273 | .236 | .202 | .170 | .141 | .115 |
| | | | | | | | | .093 | .073 | .057 | | | | | | | |
| 3.000 | 2.883 | .000 | .117 | .96 | 3.42 | 3.9% | 14 | .429 | .389 | .350 | .311 | .273 | .236 | .202 | .170 | .141 | .115 |
| | | | | | | | | .093 | .073 | .057 | .044 | | | | | | |
| 3.000 | 2.916 | .000 | .084 | .97 | 3.33 | 2.8% | 15 | .429 | .389 | .350 | .311 | .273 | .236 | .202 | .170 | .141 | .115 |
| | | | | | | | | .093 | .073 | .057 | .044 | .033 | | | | | |
| 3.000 | 2.941 | .000 | .059 | .98 | 3.24 | 2.0% | 16 | .429 | .389 | .350 | .311 | .273 | .236 | .202 | .170 | .141 | .115 |
| | | | | | | | | .093 | .073 | .057 | .044 | .033 | .025 | | | | |
| 3.000 | 2.959 | .000 | .041 | .98 | 3.16 | 1.4% | 17 | .429 | .389 | .350 | .311 | .273 | .236 | .202 | .170 | .141 | .115 |
| | | | | | | | | .093 | .073 | .057 | .044 | .033 | .025 | .018 | | | |
| 3.000 | 2.971 | .000 | .029 | .99 | 3.08 | 1.0% | 18 | .429 | .389 | .350 | .311 | .273 | .236 | .202 | .170 | .141 | .115 |
| | | | | | | | | .093 | .073 | .057 | .044 | .033 | .025 | .018 | .013 | | |
| 3.000 | 2.981 | .000 | .019 | .99 | 3.00 | .6% | 19 | .429 | .389 | .350 | .311 | .273 | .236 | .202 | .170 | .141 | .115 |
| | | | | | | | | .093 | .073 | .057 | .044 | .033 | .025 | .018 | .013 | .009 | |
| 3.200 | 2.085 | .000 | 1.115 | .75 | 4.04 | 34.9% | 6 | .445 | .406 | .366 | .327 | .289 | .252 | | | | |
| 3.200 | 2.301 | .000 | .899 | .78 | 3.99 | 28.1% | 7 | .445 | .406 | .366 | .327 | .289 | .252 | .217 | | | |
| 3.200 | 2.485 | .000 | .715 | .82 | 3.94 | 22.3% | 8 | .445 | .406 | .366 | .327 | .289 | .252 | .217 | .184 | | |
| 3.200 | 2.638 | .000 | .562 | .85 | 3.88 | 17.5% | 9 | .445 | .406 | .366 | .327 | .289 | .252 | .217 | .184 | .154 | |
| 3.200 | 2.765 | .000 | .435 | .87 | 3.81 | 13.6% | 10 | .445 | .406 | .366 | .327 | .289 | .252 | .217 | .184 | .154 | .126 |
| 3.200 | 2.867 | .000 | .333 | .90 | 3.73 | 10.4% | 11 | .445 | .406 | .366 | .327 | .289 | .252 | .217 | .184 | .154 | .126 |
| | | | | | | | | .103 | | | | | | | | | |
| 3.200 | 2.949 | .000 | .251 | .92 | 3.65 | 7.8% | 12 | .445 | .406 | .366 | .327 | .289 | .252 | .217 | .184 | .154 | .126 |
| | | | | | | | | .103 | .082 | | | | | | | | |
| 3.200 | 3.014 | .000 | .186 | .94 | 3.56 | 5.8% | 13 | .445 | .406 | .366 | .327 | .289 | .252 | .217 | .184 | .154 | .126 |
| | | | | | | | | .103 | .082 | .064 | | | | | | | |
| 3.200 | 3.063 | .000 | .137 | .95 | 3.47 | 4.3% | 14 | .445 | .406 | .366 | .327 | .289 | .252 | .217 | .184 | .154 | .126 |

*SIMILAR TO A.T.&T. NEAL-WILKINSON TABLES(SEE SECTION 1)

*EQUIVALENT RANDOM THEORY: VMR(VARIANCE TO MEAN RATIO)– 4.0 - NO RETRIES (PEAKED TRAFFIC - ROUTE ADVANCEMENT OR DAY-TO-DAY VARIATION)

| HOURS FIRST ATTEMPT VMR=4.0 | HOURS CON-NECT | HOURS RECALL /RETRY | HOURS OVFL | CON-NECT VMR | OVFL VMR | PERCENT BLOCKED (GRADE) | LINE | HOURS CONNECTED TRAFFIC ON LINES 1 | 2 | 3 | 4 | 5 | 6 | 7 | 8 | 9 | 10 |
|---|---|---|---|---|---|---|---|---|---|---|---|---|---|---|---|---|---|
| | | | | | | | | .103 | .082 | .064 | .050 | | | | | | |
| 3.200 | 3.101 | .000 | .099 | .96 | 3.38 | 3.1% | 15 | .445 | .406 | .366 | .327 | .289 | .252 | .217 | .184 | .154 | .126 |
| | | | | | | | | .103 | .082 | .064 | .050 | .038 | | | | | |
| 3.200 | 3.130 | .000 | .070 | .97 | 3.30 | 2.2% | 16 | .445 | .406 | .366 | .327 | .289 | .252 | .217 | .184 | .154 | .126 |
| | | | | | | | | .103 | .082 | .064 | .050 | .038 | .028 | | | | |
| 3.200 | 3.151 | .000 | .049 | .98 | 3.21 | 1.5% | 17 | .445 | .406 | .366 | .327 | .289 | .252 | .217 | .184 | .154 | .126 |
| | | | | | | | | .103 | .082 | .064 | .050 | .038 | .028 | .021 | | | |
| 3.200 | 3.166 | .000 | .034 | .98 | 3.13 | 1.1% | 18 | .445 | .406 | .366 | .327 | .289 | .252 | .217 | .184 | .154 | .126 |
| | | | | | | | | .103 | .082 | .064 | .050 | .038 | .028 | .021 | .015 | | |
| 3.200 | 3.177 | .000 | .023 | .99 | 3.05 | .7% | 19 | .445 | .406 | .366 | .327 | .289 | .252 | .217 | .184 | .154 | .126 |
| | | | | | | | | .103 | .082 | .064 | .050 | .038 | .028 | .021 | .015 | .011 | |
| 3.400 | 2.410 | .000 | .990 | .77 | 4.03 | 29.1% | 7 | .460 | .422 | .382 | .343 | .305 | .267 | .231 | | | |
| 3.400 | 2.608 | .000 | .792 | .80 | 3.98 | 23.3% | 8 | .460 | .422 | .382 | .343 | .305 | .267 | .231 | .197 | | |
| 3.400 | 2.774 | .000 | .626 | .83 | 3.92 | 18.4% | 9 | .460 | .422 | .382 | .343 | .305 | .267 | .231 | .197 | .166 | |
| 3.400 | 2.911 | .000 | .489 | .86 | 3.85 | 14.4% | 10 | .460 | .422 | .382 | .343 | .305 | .267 | .231 | .197 | .166 | .138 |
| 3.400 | 3.024 | .000 | .376 | .89 | 3.78 | 11.1% | 11 | .460 | .422 | .382 | .343 | .305 | .267 | .231 | .197 | .166 | .138 |
| | | | | | | | | .113 | | | | | | | | | |
| 3.400 | 3.115 | .000 | .285 | .91 | 3.69 | 8.4% | 12 | .460 | .422 | .382 | .343 | .305 | .267 | .231 | .197 | .166 | .138 |
| | | | | | | | | .113 | .091 | | | | | | | | |
| 3.400 | 3.186 | .000 | .214 | .93 | 3.61 | 6.3% | 13 | .460 | .422 | .382 | .343 | .305 | .267 | .231 | .197 | .166 | .138 |
| | | | | | | | | .113 | .091 | .072 | | | | | | | |
| 3.400 | 3.242 | .000 | .158 | .94 | 3.52 | 4.6% | 14 | .460 | .422 | .382 | .343 | .305 | .267 | .231 | .197 | .166 | .138 |
| | | | | | | | | .113 | .091 | .072 | .056 | | | | | | |
| 3.400 | 3.285 | .000 | .115 | .96 | 3.43 | 3.4% | 15 | .460 | .422 | .382 | .343 | .305 | .267 | .231 | .197 | .166 | .138 |
| | | | | | | | | .113 | .091 | .072 | .056 | .043 | | | | | |
| 3.400 | 3.317 | .000 | .083 | .97 | 3.34 | 2.4% | 16 | .460 | .422 | .382 | .343 | .305 | .267 | .231 | .197 | .166 | .138 |
| | | | | | | | | .113 | .091 | .072 | .056 | .043 | .032 | | | | |
| 3.400 | 3.342 | .000 | .058 | .98 | 3.26 | 1.7% | 17 | .460 | .422 | .382 | .343 | .305 | .267 | .231 | .197 | .166 | .138 |
| | | | | | | | | .113 | .091 | .072 | .056 | .043 | .032 | .024 | | | |
| 3.400 | 3.359 | .000 | .041 | .98 | 3.17 | 1.2% | 18 | .460 | .422 | .382 | .343 | .305 | .267 | .231 | .197 | .166 | .138 |
| | | | | | | | | .113 | .091 | .072 | .056 | .043 | .032 | .024 | .018 | | |
| 3.400 | 3.372 | .000 | .028 | .99 | 3.09 | .8% | 19 | .460 | .422 | .382 | .343 | .305 | .267 | .231 | .197 | .166 | .138 |
| | | | | | | | | .113 | .091 | .072 | .056 | .043 | .032 | .024 | .018 | .013 | |
| 3.400 | 3.381 | .000 | .019 | .99 | 3.02 | .6% | 20 | .460 | .422 | .382 | .343 | .305 | .267 | .231 | .197 | .166 | .138 |
| | | | | | | | | .113 | .091 | .072 | .056 | .043 | .032 | .024 | .018 | .013 | .009 |
| 3.600 | 2.508 | .000 | 1.092 | .75 | 4.08 | 30.0% | 7 | .473 | .435 | .396 | .358 | .319 | .282 | .245 | | | |
| 3.600 | 2.719 | .000 | .881 | .79 | 4.04 | 24.5% | 8 | .473 | .435 | .396 | .358 | .319 | .282 | .245 | .211 | | |
| 3.600 | 2.898 | .000 | .702 | .02 | 3.99 | 19.5% | 9 | .473 | .435 | .396 | .358 | .319 | .282 | .245 | .211 | .179 | |
| 3.600 | 3.047 | .000 | .553 | .85 | 3.92 | 15.3% | 10 | .473 | .435 | .396 | .358 | .319 | .282 | .245 | .211 | .179 | .150 |
| 3.600 | 3.171 | .000 | .429 | .88 | 3.85 | 11.9% | 11 | .473 | .435 | .396 | .358 | .319 | .282 | .245 | .211 | .179 | .150 |

*SIMILAR TO A.T.&T. NEAL-WILKINSON TABLES(SEE SECTION 1)

*EQUIVALENT RANDOM THEORY: VMR(VARIANCE TO MEAN RATIO)– 4.0 - NO RETRIES (PEAKED TRAFFIC - ROUTE ADVANCEMENT OR DAY-TO-DAY VARIATION)

| HOURS FIRST ATTEMPT VMR=4.0 | HOURS CON-NECT | HOURS RECALL /RETRY | HOURS OVFL | CON-NECT VMR | OVFL VMR | PERCENT BLOCKED (GRADE) | LINE | HOURS CONNECTED TRAFFIC ON LINES 1 | 2 | 3 | 4 | 5 | 6 | 7 | 8 | 9 | 10 |
|---|---|---|---|---|---|---|---|---|---|---|---|---|---|---|---|---|---|
| | | | | | | | | .123 | | | | | | | | | |
| 3.600 | 3.271 | .000 | .329 | .90 | 3.77 | 9.1% | 12 | .473 | .435 | .396 | .358 | .319 | .282 | .245 | .211 | .179 | .150 |
| | | | | | | | | .123 | .100 | | | | | | | | |
| 3.600 | 3.351 | .000 | .249 | .92 | 3.69 | 6.9% | 13 | .473 | .435 | .396 | .358 | .319 | .282 | .245 | .211 | .179 | .150 |
| | | | | | | | | .123 | .100 | .080 | | | | | | | |
| 3.600 | 3.415 | .000 | .185 | .94 | 3.60 | 5.2% | 14 | .473 | .435 | .396 | .358 | .319 | .282 | .245 | .211 | .179 | .150 |
| | | | | | | | | .123 | .100 | .080 | .063 | | | | | | |
| 3.600 | 3.464 | .000 | .136 | .95 | 3.51 | 3.8% | 15 | .473 | .435 | .396 | .358 | .319 | .282 | .245 | .211 | .179 | .150 |
| | | | | | | | | .123 | .100 | .080 | .063 | .049 | | | | | |
| 3.600 | 3.501 | .000 | .099 | .96 | 3.42 | 2.8% | 16 | .473 | .435 | .396 | .358 | .319 | .282 | .245 | .211 | .179 | .150 |
| | | | | | | | | .123 | .100 | .080 | .063 | .049 | .037 | | | | |
| 3.600 | 3.529 | .000 | .071 | .97 | 3.33 | 2.0% | 17 | .473 | .435 | .396 | .358 | .319 | .282 | .245 | .211 | .179 | .150 |
| | | | | | | | | .123 | .100 | .080 | .063 | .049 | .037 | .028 | | | |
| 3.600 | 3.550 | .000 | .050 | .98 | 3.25 | 1.4% | 18 | .473 | .435 | .396 | .358 | .319 | .282 | .245 | .211 | .179 | .150 |
| | | | | | | | | .123 | .100 | .080 | .063 | .049 | .037 | .028 | .021 | | |
| 3.600 | 3.565 | .000 | .035 | .98 | 3.17 | 1.0% | 19 | .473 | .435 | .396 | .358 | .319 | .282 | .245 | .211 | .179 | .150 |
| | | | | | | | | .123 | .100 | .080 | .063 | .049 | .037 | .028 | .021 | .015 | |
| 3.600 | 3.576 | .000 | .024 | .99 | 3.09 | .7% | 20 | .473 | .435 | .396 | .358 | .319 | .282 | .245 | .211 | .179 | .150 |
| | | | | | | | | .123 | .100 | .080 | .063 | .049 | .037 | .028 | .021 | .015 | .011 |
| 3.800 | 2.610 | .000 | 1.190 | .74 | 4.11 | 31.3% | 7 | .487 | .449 | .411 | .373 | .334 | .296 | .260 | | | |
| 3.800 | 2.835 | .000 | .965 | .78 | 4.07 | 25.4% | 8 | .487 | .449 | .411 | .373 | .334 | .296 | .260 | .225 | | |
| 3.800 | 3.026 | .000 | .774 | .81 | 4.02 | 20.4% | 9 | .487 | .449 | .411 | .373 | .334 | .296 | .260 | .225 | .192 | |
| 3.800 | 3.188 | .000 | .612 | .84 | 3.96 | 16.1% | 10 | .487 | .449 | .411 | .373 | .334 | .296 | .260 | .225 | .192 | .161 |
| 3.800 | 3.321 | .000 | .479 | .87 | 3.89 | 12.6% | 11 | .487 | .449 | .411 | .373 | .334 | .296 | .260 | .225 | .192 | .161 |
| | | | | | | | | .134 | | | | | | | | | |
| 3.800 | 3.431 | .000 | .369 | .89 | 3.81 | 9.7% | 12 | .487 | .449 | .411 | .373 | .334 | .296 | .260 | .225 | .192 | .161 |
| | | | | | | | | .134 | .109 | | | | | | | | |
| 3.800 | 3.519 | .000 | .281 | .91 | 3.73 | 7.4% | 13 | .487 | .449 | .411 | .373 | .334 | .296 | .260 | .225 | .192 | .161 |
| | | | | | | | | .134 | .109 | .088 | | | | | | | |
| 3.800 | 3.589 | .000 | .21 | .93 | 3.65 | 5.5% | 14 | .487 | .449 | .411 | .373 | .334 | .296 | .260 | .225 | .192 | .161 |
| | | | | | | | | .134 | .109 | .088 | .070 | | | | | | |
| 3.800 | 3.644 | .000 | .156 | .95 | 3.56 | 4.1% | 15 | .487 | .449 | .411 | .373 | .334 | .296 | .260 | .225 | .192 | .161 |
| | | | | | | | | .134 | .109 | .088 | .070 | .055 | | | | | |
| 3.800 | 3.686 | .000 | .114 | .96 | 3.47 | 3.0% | 16 | .487 | .449 | .411 | .373 | .334 | .296 | .260 | .225 | .192 | .161 |
| | | | | | | | | .134 | .109 | .088 | .070 | .055 | .042 | | | | |
| 3.800 | 3.718 | .000 | .082 | .97 | 3.38 | 2.2% | 17 | .487 | .449 | .411 | .373 | .334 | .296 | .260 | .225 | .192 | .161 |
| | | | | | | | | .134 | .109 | .088 | .070 | .055 | .042 | .032 | | | |
| 3.800 | 3.741 | .000 | .059 | .98 | 3.29 | 1.5% | 18 | .487 | .449 | .411 | .373 | .334 | .296 | .260 | .225 | .192 | .161 |
| | | | | | | | | .134 | .109 | .088 | .070 | .055 | .042 | .032 | .024 | | |
| 3.800 | 3.759 | .000 | .041 | .98 | 3.21 | 1.1% | 19 | .487 | .449 | .411 | .373 | .334 | .296 | .260 | .225 | .192 | .161 |

*SIMILAR TO A.T.&T. NEAL-WILKINSON TABLES(SEE SECTION 1)

*EQUIVALENT RANDOM THEORY: VMR(VARIANCE TO MEAN RATIO)– 4.0 - NO RETRIES (PEAKED TRAFFIC - ROUTE ADVANCEMENT OR DAY-TO-DAY VARIATION)

| HOURS FIRST ATTEMPT VMR=4.0 | HOURS CON-NECT | HOURS RECALL /RETRY | HOURS OVFL | CON-NECT VMR | OVFL VMR | PERCENT BLOCKED (GRADE) | LINE | HOURS CONNECTED TRAFFIC ON LINES 1 | 2 | 3 | 4 | 5 | 6 | 7 | 8 | 9 | 10 |
|---|---|---|---|---|---|---|---|---|---|---|---|---|---|---|---|---|---|
| | | | | | | | | .134 | .109 | .088 | .070 | .055 | .042 | .032 | .024 | .017 | |
| 3.800 | 3.772 | .000 | .028 | .99 | 3.13 | .7% | 20 | .487 | .449 | .411 | .373 | .334 | .296 | .260 | .225 | .192 | .161 |
| | | | | | | | | .134 | .109 | .088 | .070 | .055 | .042 | .032 | .024 | .017 | .013 |
| 3.800 | 3.781 | .000 | .019 | .99 | 3.05 | .5% | 21 | .487 | .449 | .411 | .373 | .334 | .296 | .260 | .225 | .192 | .161 |
| | | | | | | | | .134 | .109 | .088 | .070 | .055 | .042 | .032 | .024 | .017 | .013 |
| | | | | | | | | .009 | | | | | | | | | |
| 4.000 | 2.709 | .000 | 1.291 | .73 | 4.13 | 32.3% | 7 | .500 | .463 | .426 | .387 | .349 | .311 | .274 | | | |
| 4.000 | 2.947 | .000 | 1.053 | .76 | 4.10 | 26.3% | 8 | .500 | .463 | .426 | .387 | .349 | .311 | .274 | .238 | | |
| 4.000 | 3.152 | .000 | .848 | .80 | 4.05 | 21.2% | 9 | .500 | .463 | .426 | .387 | .349 | .311 | .274 | .238 | .204 | |
| 4.000 | 3.325 | .000 | .675 | .83 | 4.00 | 16.9% | 10 | .500 | .463 | .426 | .387 | .349 | .311 | .274 | .238 | .204 | .173 |
| 4.000 | 3.469 | .000 | .531 | .86 | 3.93 | 13.3% | 11 | .500 | .463 | .426 | .387 | .349 | .311 | .274 | .238 | .204 | .173 |
| | | | | | | | | .145 | | | | | | | | | |
| 4.000 | 3.588 | .000 | .412 | .88 | 3.86 | 10.3% | 12 | .500 | .463 | .426 | .387 | .349 | .311 | .274 | .238 | .204 | .173 |
| | | | | | | | | .145 | .119 | | | | | | | | |
| 4.000 | 3.685 | .000 | .315 | .90 | 3.78 | 7.9% | 13 | .500 | .463 | .426 | .387 | .349 | .311 | .274 | .238 | .204 | .173 |
| | | | | | | | | .145 | .119 | .096 | | | | | | | |
| 4.000 | 3.762 | .000 | .238 | .92 | 3.69 | 6.0% | 14 | .500 | .463 | .426 | .387 | .349 | .311 | .274 | .238 | .204 | .173 |
| | | | | | | | | .145 | .119 | .096 | .077 | | | | | | |
| 4.000 | 3.822 | .000 | .178 | .94 | 3.60 | 4.4% | 15 | .500 | .463 | .426 | .387 | .349 | .311 | .274 | .238 | .204 | .173 |
| | | | | | | | | .145 | .119 | .096 | .077 | .061 | | | | | |
| 4.000 | 3.869 | .000 | .131 | .95 | 3.52 | 3.3% | 16 | .500 | .463 | .426 | .387 | .349 | .311 | .274 | .238 | .204 | .173 |
| | | | | | | | | .145 | .119 | .096 | .077 | .061 | .047 | | | | |
| 4.000 | 3.905 | .000 | .095 | .96 | 3.43 | 2.4% | 17 | .500 | .463 | .426 | .387 | .349 | .311 | .274 | .238 | .204 | .173 |
| | | | | | | | | .145 | .119 | .096 | .077 | .061 | .047 | .036 | | | |
| 4.000 | 3.932 | .000 | .068 | .97 | 3.34 | 1.7% | 18 | .500 | .463 | .426 | .387 | .349 | .311 | .274 | .238 | .204 | .173 |
| | | | | | | | | .145 | .119 | .096 | .077 | .061 | .047 | .036 | .027 | | |
| 4.000 | 3.952 | .000 | .048 | .98 | 3.25 | 1.2% | 19 | .500 | .463 | .426 | .387 | .349 | .311 | .274 | .238 | .204 | .173 |
| | | | | | | | | .145 | .119 | .096 | .077 | .061 | .047 | .036 | .027 | .020 | |
| 4.000 | 3.967 | .000 | .033 | .99 | 3.17 | .8% | 20 | .500 | .463 | .426 | .387 | .349 | .311 | .274 | .238 | .204 | .173 |
| | | | | | | | | .145 | .119 | .096 | .077 | .061 | .047 | .036 | .027 | .020 | .015 |
| 4.000 | 3.977 | .000 | .023 | .99 | 3.09 | .6% | 21 | .500 | .463 | .426 | .387 | .349 | .311 | .274 | .238 | .204 | .173 |
| | | | | | | | | .145 | .119 | .096 | .077 | .061 | .047 | .036 | .027 | .020 | .015 |
| | | | | | | | | .010 | | | | | | | | | |
| 4.200 | 2.806 | .000 | 1.394 | .71 | 4.15 | 33.2% | 7 | .513 | .477 | .439 | .401 | .363 | .325 | .287 | | | |
| 4.200 | 3.057 | .000 | 1.143 | .75 | 4.13 | 27.2% | 8 | .513 | .477 | .439 | .401 | .363 | .325 | .287 | .251 | | |
| 4.200 | 3.274 | .000 | .926 | .78 | 4.08 | 22.0% | 9 | .513 | .477 | .439 | .401 | .363 | .325 | .287 | .251 | .217 | |
| 4.200 | 3.459 | .000 | .741 | .82 | 4.03 | 17.6% | 10 | .513 | .477 | .439 | .401 | .363 | .325 | .287 | .251 | .217 | .185 |
| 4.200 | 3.614 | .000 | .586 | .84 | 3.97 | 13.9% | 11 | .513 | .477 | .439 | .401 | .363 | .325 | .287 | .251 | .217 | .185 |
| | | | | | | | | .155 | | | | | | | | | |
| 4.200 | 3.743 | .000 | .457 | .87 | 3.90 | 10.9% | 12 | .513 | .477 | .439 | .401 | .363 | .325 | .287 | .251 | .217 | .185 |

*SIMILAR TO A.T.&T. NEAL-WILKINSON TABLES(SEE SECTION 1)

*EQUIVALENT RANDOM THEORY: VMR(VARIANCE TO MEAN RATIO)– 4.0 - NO RETRIES (PEAKED TRAFFIC - ROUTE ADVANCEMENT OR DAY-TO-DAY VARIATION)

| HOURS FIRST ATTEMPT VMR = 4.0 | HOURS CON-NECT | HOURS RECALL /RETRY | HOURS OVFL | CON-NECT VMR | OVFL VMR | PERCENT BLOCKED (GRADE) | LINE | HOURS CONNECTED TRAFFIC ON LINES 1 | 2 | 3 | 4 | 5 | 6 | 7 | 8 | 9 | 10 |
|---|---|---|---|---|---|---|---|---|---|---|---|---|---|---|---|---|---|
| | | | | | | | | .155 | .129 | | | | | | | | |
| 4.200 | 3.848 | .000 | .352 | .90 | 3.82 | 8.4% | 13 | .513 | .477 | .439 | .401 | .363 | .325 | .287 | .251 | .217 | .185 |
| | | | | | | | | .155 | .129 | .105 | | | | | | | |
| 4.200 | 3.932 | .000 | .268 | .92 | 3.74 | 6.4% | 14 | .513 | .477 | .439 | .401 | .363 | .325 | .287 | .251 | .217 | .185 |
| | | | | | | | | .155 | .129 | .105 | .084 | | | | | | |
| 4.200 | 3.999 | .000 | .201 | .93 | 3.65 | 4.8% | 15 | .513 | .477 | .439 | .401 | .363 | .325 | .287 | .251 | .217 | .185 |
| | | | | | | | | .155 | .129 | .105 | .084 | .067 | | | | | |
| 4.200 | 4.051 | .000 | .149 | .95 | 3.56 | 3.5% | 16 | .513 | .477 | .439 | .401 | .363 | .325 | .287 | .251 | .217 | .185 |
| | | | | | | | | .155 | .129 | .105 | .084 | .067 | .052 | | | | |
| 4.200 | 4.091 | .000 | .109 | .96 | 3.47 | 2.6% | 17 | .513 | .477 | .439 | .401 | .363 | .325 | .287 | .251 | .217 | .185 |
| | | | | | | | | .155 | .129 | .105 | .084 | .067 | .052 | .040 | | | |
| 4.200 | 4.122 | .000 | .078 | .97 | 3.38 | 1.9% | 18 | .513 | .477 | .439 | .401 | .363 | .325 | .287 | .251 | .217 | .185 |
| | | | | | | | | .155 | .129 | .105 | .084 | .067 | .052 | .040 | .030 | | |
| 4.200 | 4.144 | .000 | .056 | .98 | 3.30 | 1.3% | 19 | .513 | .477 | .439 | .401 | .363 | .325 | .287 | .251 | .217 | .185 |
| | | | | | | | | .155 | .129 | .105 | .084 | .067 | .052 | .040 | .030 | .023 | |
| 4.200 | 4.161 | .000 | .039 | .98 | 3.21 | .9% | 20 | .513 | .477 | .439 | .401 | .363 | .325 | .287 | .251 | .217 | .185 |
| | | | | | | | | .155 | .129 | .105 | .084 | .067 | .052 | .040 | .030 | .023 | .017 |
| 4.200 | 4.173 | .000 | .027 | .99 | 3.13 | .6% | 21 | .513 | .477 | .439 | .401 | .363 | .325 | .287 | .251 | .217 | .185 |
| | | | | | | | | .155 | .129 | .105 | .084 | .067 | .052 | .040 | .030 | .023 | .017 |
| | | | | | | | | .012 | | | | | | | | | |
| 4.400 | 2.890 | .000 | 1.510 | .70 | 4.20 | 34.3% | 7 | .523 | .488 | .451 | .414 | .376 | .338 | .301 | | | |
| 4.400 | 3.155 | .000 | 1.245 | .74 | 4.18 | 28.3% | 8 | .523 | .488 | .451 | .414 | .376 | .338 | .301 | .264 | | |
| 4.400 | 3.384 | .000 | 1.016 | .77 | 4.14 | 23.1% | 9 | .523 | .488 | .451 | .414 | .376 | .338 | .301 | .264 | .230 | |
| 4.400 | 3.581 | .000 | .819 | .80 | 4.09 | 18.6% | 10 | .523 | .488 | .451 | .414 | .376 | .338 | .301 | .264 | .230 | .197 |
| 4.400 | 3.748 | .000 | .652 | .83 | 4.03 | 14.8% | 11 | .523 | .488 | .451 | .414 | .376 | .338 | .301 | .264 | .230 | .197 |
| | | | | | | | | .167 | | | | | | | | | |
| 4.400 | 3.887 | .000 | .513 | .86 | 3.97 | 11.7% | 12 | .523 | .488 | .451 | .414 | .376 | .338 | .301 | .264 | .230 | .197 |
| | | | | | | | | .167 | .139 | | | | | | | | |
| 4.400 | 4.002 | .000 | .398 | .89 | 3.89 | 9.1% | 13 | .523 | .488 | .451 | .414 | .376 | .338 | .301 | .264 | .230 | .197 |
| | | | | | | | | .167 | .139 | .115 | | | | | | | |
| 4.400 | 4.094 | .000 | .306 | .91 | 3.81 | 6.9% | 14 | .523 | .488 | .451 | .414 | .376 | .338 | .301 | .264 | .230 | .197 |
| | | | | | | | | .167 | .139 | .115 | .093 | | | | | | |
| 4.400 | 4.169 | .000 | .231 | .93 | 3.72 | 5.3% | 15 | .523 | .488 | .451 | .414 | .376 | .338 | .301 | .264 | .230 | .197 |
| | | | | | | | | .167 | .139 | .115 | .093 | .074 | | | | | |
| 4.400 | 4.227 | .000 | .173 | .94 | 3.63 | 3.9% | 16 | .523 | .488 | .451 | .414 | .376 | .338 | .301 | .264 | .230 | .197 |
| | | | | | | | | .167 | .139 | .115 | .093 | .074 | .058 | | | | |
| 4.400 | 4.273 | .000 | .127 | .95 | 3.55 | 2.9% | 17 | .523 | .488 | .451 | .414 | .376 | .338 | .301 | .264 | .230 | .197 |
| | | | | | | | | .167 | .139 | .115 | .093 | .074 | .058 | .045 | | | |
| 4.400 | 4.307 | .000 | .093 | .97 | 3.46 | 2.1% | 18 | .523 | .488 | .451 | .414 | .376 | .338 | .301 | .264 | .230 | .197 |
| | | | | | | | | .167 | .139 | .115 | .093 | .074 | .058 | .045 | .035 | | |

*SIMILAR TO A.T.&T. NEAL-WILKINSON TABLES(SEE SECTION 1)

*EQUIVALENT RANDOM THEORY: VMR(VARIANCE TO MEAN RATIO)– 4.0 - NO RETRIES (PEAKED TRAFFIC - ROUTE ADVANCEMENT OR DAY-TO-DAY VARIATION)

| HOURS FIRST ATTEMPT VMR = 4.0 | HOURS CON- NECT | HOURS RECALL /RETRY | HOURS OVFL | CON- NECT VMR | OVFL VMR | PERCENT BLOCKED (GRADE) | LINE | HOURS CONNECTED TRAFFIC ON LINES 1 | 2 | 3 | 4 | 5 | 6 | 7 | 8 | 9 | 10 |
|---|---|---|---|---|---|---|---|---|---|---|---|---|---|---|---|---|---|
| 4.400 | 4.333 | .000 | .067 | .97 | 3.37 | 1.5% | 19 | .523 | .488 | .451 | .414 | .376 | .338 | .301 | .264 | .230 | .197 |
| | | | | | | | | .167 | .139 | .115 | .093 | .074 | .058 | .045 | .035 | .026 | |
| 4.400 | 4.353 | .000 | .047 | .98 | 3.28 | 1.1% | 20 | .523 | .488 | .451 | .414 | .376 | .338 | .301 | .264 | .230 | .197 |
| | | | | | | | | .167 | .139 | .115 | .093 | .074 | .058 | .045 | .035 | .026 | .019 |
| 4.400 | 4.367 | .000 | .033 | .99 | 3.20 | .8% | 21 | .523 | .488 | .451 | .414 | .376 | .338 | .301 | .264 | .230 | .197 |
| | | | | | | | | .167 | .139 | .115 | .093 | .074 | .058 | .045 | .035 | .026 | .019 |
| | | | | | | | | .014 | | | | | | | | | |
| 4.400 | 4.377 | .000 | .023 | .99 | 3.12 | .5% | 22 | .523 | .488 | .451 | .414 | .376 | .338 | .301 | .264 | .230 | .197 |
| | | | | | | | | .167 | .139 | .115 | .093 | .074 | .058 | .045 | .035 | .026 | .019 |
| | | | | | | | | .014 | .010 | | | | | | | | |
| 4.600 | 3.258 | .000 | 1.342 | .72 | 4.20 | 29.2% | 8 | .535 | .500 | .464 | .427 | .389 | .351 | .314 | .277 | | |
| 4.600 | 3.501 | .000 | 1.099 | .76 | 4.17 | 23.9% | 9 | .535 | .500 | .464 | .427 | .389 | .351 | .314 | .277 | .242 | |
| 4.600 | 3.709 | .000 | .891 | .79 | 4.12 | 19.4% | 10 | .535 | .500 | .464 | .427 | .389 | .351 | .314 | .277 | .242 | .209 |
| 4.600 | 3.887 | .000 | .713 | .82 | 4.07 | 15.5% | 11 | .535 | .500 | .464 | .427 | .389 | .351 | .314 | .277 | .242 | .209 |
| | | | | | | | | .178 | | | | | | | | | |
| 4.600 | 4.036 | .000 | .564 | .85 | 4.00 | 12.3% | 12 | .535 | .500 | .464 | .427 | .389 | .351 | .314 | .277 | .242 | .209 |
| | | | | | | | | .178 | .149 | | | | | | | | |
| 4.600 | 4.160 | .000 | .440 | .88 | 3.93 | 9.6% | 13 | .535 | .500 | .464 | .427 | .389 | .351 | .314 | .277 | .242 | .209 |
| | | | | | | | | .178 | .149 | .123 | | | | | | | |
| 4.600 | 4.260 | .000 | .340 | .90 | 3.85 | 7.4% | 14 | .535 | .500 | .464 | .427 | .389 | .351 | .314 | .277 | .242 | .209 |
| | | | | | | | | .178 | .149 | .123 | .101 | | | | | | |
| 4.600 | 4.341 | .000 | .259 | .92 | 3.77 | 5.6% | 15 | .535 | .500 | .464 | .427 | .389 | .351 | .314 | .277 | .242 | .209 |
| | | | | | | | | .178 | .149 | .123 | .101 | .081 | | | | | |
| 4.600 | 4.406 | .000 | .194 | .94 | 3.68 | 4.2% | 16 | .535 | .500 | .464 | .427 | .389 | .351 | .314 | .277 | .242 | .209 |
| | | | | | | | | .178 | .149 | .123 | .101 | .081 | .064 | | | | |
| 4.600 | 4.450 | .000 | .144 | .95 | 3.59 | 3.1% | 17 | .535 | .500 | .464 | .427 | .389 | .351 | .314 | .277 | .242 | .209 |
| | | | | | | | | .178 | .149 | .123 | .101 | .081 | .064 | .050 | | | |
| 4.600 | 4.494 | .000 | .106 | .96 | 3.50 | 2.3% | 18 | .535 | .500 | .464 | .427 | .389 | .351 | .314 | .277 | .242 | .209 |
| | | | | | | | | .178 | .149 | .123 | .101 | .081 | .064 | .050 | .039 | | |
| 4.600 | 4.524 | .000 | .076 | .97 | 3.41 | 1.7% | 19 | .535 | .500 | .464 | .427 | .389 | .351 | .314 | .277 | .242 | .209 |
| | | | | | | | | .178 | .149 | .123 | .101 | .081 | .064 | .050 | .039 | .029 | |
| 4.600 | 4.546 | .000 | .054 | .98 | 3.33 | 1.2% | 20 | .535 | .500 | .464 | .427 | .389 | .351 | .314 | .277 | .242 | .209 |
| | | | | | | | | .178 | .149 | .123 | .101 | .081 | .064 | .050 | .039 | .029 | .022 |
| 4.600 | 4.562 | .000 | .038 | .98 | 3.24 | .8% | 21 | .535 | .500 | .464 | .427 | .389 | .351 | .314 | .277 | .242 | .209 |
| | | | | | | | | .178 | .149 | .123 | .101 | .081 | .064 | .050 | .039 | .029 | .022 |
| | | | | | | | | .016 | | | | | | | | | |
| 4.600 | 4.573 | .000 | .027 | .99 | 3.16 | .6% | 22 | .535 | .500 | .464 | .427 | .389 | .351 | .314 | .277 | .242 | .209 |
| | | | | | | | | .178 | .149 | .123 | .101 | .001 | .064 | .050 | .039 | .029 | .022 |
| | | | | | | | | .016 | .012 | | | | | | | | |
| 4.800 | 3.349 | .000 | 1.451 | .71 | 4.25 | 30.2% | 8 | .544 | .510 | .475 | .438 | .401 | .364 | .326 | .290 | | |

*SIMILAR TO A.T.&T. NEAL-WILKINSON TABLES(SEE SECTION 1)

*EQUIVALENT RANDOM THEORY: VMR(VARIANCE TO MEAN RATIO)– 4.0 - NO RETRIES (PEAKED TRAFFIC - ROUTE ADVANCEMENT OR DAY-TO-DAY VARIATION)

| HOURS FIRST ATTEMPT VMR=4.0 | HOURS CONNECT | HOURS RECALL /RETRY | HOURS OVFL | CONNECT VMR | OVFL VMR | PERCENT BLOCKED (GRADE) | LINE | HOURS CONNECTED TRAFFIC ON LINES 1 | 2 | 3 | 4 | 5 | 6 | 7 | 8 | 9 | 10 |
|---|---|---|---|---|---|---|---|---|---|---|---|---|---|---|---|---|---|
| 4.800 | 3.604 | .000 | 1.196 | .75 | 4.22 | 24.9% | 9 | .544 | .510 | .475 | .438 | .401 | .364 | .326 | .290 | .255 | |
| 4.800 | 3.825 | .000 | .975 | .78 | 4.18 | 20.3% | 10 | .544 | .510 | .475 | .438 | .401 | .364 | .326 | .290 | .255 | .221 |
| 4.800 | 4.014 | .000 | .786 | .81 | 4.13 | 16.4% | 11 | .544 | .510 | .475 | .438 | .401 | .364 | .326 | .290 | .255 | .221 |
| | | | | | | | | .189 | | | | | | | | | |
| 4.800 | 4.174 | .000 | .626 | .84 | 4.07 | 13.0% | 12 | .544 | .510 | .475 | .438 | .401 | .364 | .326 | .290 | .255 | .221 |
| | | | | | | | | .189 | .160 | | | | | | | | |
| 4.800 | 4.307 | .000 | .493 | .87 | 4.00 | 10.3% | 13 | .544 | .510 | .475 | .438 | .401 | .364 | .326 | .290 | .255 | .221 |
| | | | | | | | | .189 | .160 | .133 | | | | | | | |
| 4.800 | 4.417 | .000 | .383 | .89 | 3.92 | 8.0% | 14 | .544 | .510 | .475 | .438 | .401 | .364 | .326 | .290 | .255 | .221 |
| | | | | | | | | .189 | .160 | .133 | .110 | | | | | | |
| 4.800 | 4.506 | .000 | .294 | .91 | 3.84 | 6.1% | 15 | .544 | .510 | .475 | .438 | .401 | .364 | .326 | .290 | .255 | .221 |
| | | | | | | | | .189 | .160 | .133 | .110 | .089 | | | | | |
| 4.800 | 4.577 | .000 | .223 | .93 | 3.75 | 4.6% | 16 | .544 | .510 | .475 | .438 | .401 | .364 | .326 | .290 | .255 | .221 |
| | | | | | | | | .189 | .160 | .133 | .110 | .089 | .071 | | | | |
| 4.800 | 4.633 | .000 | .167 | .94 | 3.66 | 3.5% | 17 | .544 | .510 | .475 | .438 | .401 | .364 | .326 | .290 | .255 | .221 |
| | | | | | | | | .189 | .160 | .133 | .110 | .089 | .071 | .056 | | | |
| 4.800 | 4.677 | .000 | .123 | .96 | 3.57 | 2.6% | 18 | .544 | .510 | .475 | .438 | .401 | .364 | .326 | .290 | .255 | .221 |
| | | | | | | | | .189 | .160 | .133 | .110 | .089 | .071 | .056 | .044 | | |
| 4.800 | 4.710 | .000 | .090 | .97 | 3.48 | 1.9% | 19 | .544 | .510 | .475 | .438 | .401 | .364 | .326 | .290 | .255 | .221 |
| | | | | | | | | .189 | .160 | .133 | .110 | .089 | .071 | .056 | .044 | .033 | |
| 4.800 | 4.735 | .000 | .065 | .97 | 3.40 | 1.3% | 20 | .544 | .510 | .475 | .438 | .401 | .364 | .326 | .290 | .255 | .221 |
| | | | | | | | | .189 | .160 | .133 | .110 | .089 | .071 | .056 | .044 | .033 | .025 |
| 4.800 | 4.754 | .000 | .046 | .98 | 3.31 | 1.0% | 21 | .544 | .510 | .475 | .438 | .401 | .364 | .326 | .290 | .255 | .221 |
| | | | | | | | | .189 | .160 | .133 | .110 | .089 | .071 | .056 | .044 | .033 | .025 |
| | | | | | | | | .019 | | | | | | | | | |
| 4.800 | 4.768 | .000 | .032 | .99 | 3.23 | .7% | 22 | .544 | .510 | .475 | .438 | .401 | .364 | .326 | .290 | .255 | .221 |
| | | | | | | | | .189 | .160 | .133 | .110 | .089 | .071 | .056 | .044 | .033 | .025 |
| | | | | | | | | .019 | .014 | | | | | | | | |
| 5.000 | 3.448 | .000 | 1.552 | .70 | 4.27 | 31.0% | 8 | .555 | .522 | .487 | .451 | .414 | .377 | .339 | .303 | | |
| 5.000 | 3.714 | .000 | 1.286 | .73 | 4.24 | 25.7% | 9 | .555 | .522 | .487 | .451 | .414 | .377 | .339 | .303 | .267 | |
| 5.000 | 3.947 | .000 | 1.053 | .77 | 4.21 | 21.1% | 10 | .555 | .522 | .487 | .451 | .414 | .377 | .339 | .303 | .267 | .233 |
| 5.000 | 4.147 | .000 | .853 | .80 | 4.16 | 17.1% | 11 | .555 | .522 | .487 | .451 | .414 | .377 | .339 | .303 | .267 | .233 |
| | | | | | | | | .200 | | | | | | | | | |
| 5.000 | 4.317 | .000 | .683 | .83 | 4.10 | 13.7% | 12 | .555 | .522 | .487 | .451 | .414 | .377 | .339 | .303 | .267 | .233 |
| | | | | | | | | .200 | .170 | | | | | | | | |
| 5.000 | 4.460 | .000 | .540 | .86 | 4.03 | 10.8% | 13 | .555 | .522 | .487 | .451 | .414 | .377 | .339 | .303 | .267 | .233 |
| | | | | | | | | .200 | .170 | .143 | | | | | | | |
| 5.000 | 4.578 | .000 | .422 | .88 | 3.96 | 8.4% | 14 | .555 | .522 | .487 | .451 | .414 | .377 | .339 | .303 | .267 | .233 |
| | | | | | | | | .200 | .170 | .143 | .118 | | | | | | |
| 5.000 | 4.675 | .000 | .325 | .90 | 3.88 | 6.5% | 15 | .555 | .522 | .487 | .451 | .414 | .377 | .339 | .303 | .267 | .233 |

*SIMILAR TO A.T.&T. NEAL-WILKINSON TABLES(SEE SECTION 1)

*EQUIVALENT RANDOM THEORY: VMR(VARIANCE TO MEAN RATIO)– 4.0 - NO RETRIES (PEAKED TRAFFIC - ROUTE ADVANCEMENT OR DAY-TO-DAY VARIATION)

| HOURS FIRST ATTEMPT VMR = 4.0 | HOURS CON-NECT | HOURS RECALL /RETRY | HOURS OVFL | CON-NECT VMR | OVFL VMR | PERCENT BLOCKED (GRADE) | LINE | HOURS CONNECTED TRAFFIC ON LINES 1 | 2 | 3 | 4 | 5 | 6 | 7 | 8 | 9 | 10 |
|---|---|---|---|---|---|---|---|---|---|---|---|---|---|---|---|---|---|
| | | | | | | | | .200 | .170 | .143 | .118 | .096 | | | | | |
| 5.000 | 4.752 | .000 | .248 | .92 | 3.79 | 5.0% | 16 | .555 | .522 | .487 | .451 | .414 | .377 | .339 | .303 | .267 | .233 |
| | | | | | | | | .200 | .170 | .143 | .118 | .096 | .077 | | | | |
| 5.000 | 4.813 | .000 | .187 | .94 | 3.70 | 3.7% | 17 | .555 | .522 | .487 | .451 | .414 | .377 | .339 | .303 | .267 | .233 |
| | | | | | | | | .200 | .170 | .143 | .118 | .096 | .077 | .061 | | | |
| 5.000 | 4.861 | .000 | .139 | .95 | 3.61 | 2.8% | 18 | .555 | .522 | .487 | .451 | .414 | .377 | .339 | .303 | .267 | .233 |
| | | | | | | | | .200 | .170 | .143 | .118 | .096 | .077 | .061 | .048 | | |
| 5.000 | 4.898 | .000 | .102 | .96 | 3.52 | 2.0% | 19 | .555 | .522 | .487 | .451 | .414 | .377 | .339 | .303 | .267 | .233 |
| | | | | | | | | .200 | .170 | .143 | .118 | .096 | .077 | .061 | .048 | .037 | |
| 5.000 | 4.926 | .000 | .074 | .97 | 3.44 | 1.5% | 20 | .555 | .522 | .487 | .451 | .414 | .377 | .339 | .303 | .267 | .233 |
| | | | | | | | | .200 | .170 | .143 | .118 | .096 | .077 | .061 | .048 | .037 | .028 |
| 5.000 | 4.947 | .000 | .053 | .98 | 3.35 | 1.1% | 21 | .555 | .522 | .487 | .451 | .414 | .377 | .339 | .303 | .267 | .233 |
| | | | | | | | | .200 | .170 | .143 | .118 | .096 | .077 | .061 | .048 | .037 | .028 |
| | | | | | | | | .021 | | | | | | | | | |
| 5.000 | 4.963 | .000 | .037 | .98 | 3.27 | .7% | 22 | .555 | .522 | .487 | .451 | .414 | .377 | .339 | .303 | .267 | .233 |
| | | | | | | | | .200 | .170 | .143 | .118 | .096 | .077 | .061 | .048 | .037 | .028 |
| | | | | | | | | .021 | .015 | | | | | | | | |
| 5.000 | 4.974 | .000 | .026 | .99 | 3.19 | .5% | 23 | .555 | .522 | .487 | .451 | .414 | .377 | .339 | .303 | .267 | .233 |
| | | | | | | | | .200 | .170 | .143 | .118 | .096 | .077 | .061 | .048 | .037 | .028 |
| | | | | | | | | .021 | .015 | .011 | | | | | | | |
| 5.500 | 3.660 | .000 | 1.840 | .67 | 4.36 | 33.5% | 8 | .577 | .545 | .512 | .477 | .442 | .406 | .369 | .333 | | |
| 5.500 | 3.957 | .000 | 1.543 | .70 | 4.35 | 28.1% | 9 | .577 | .545 | .512 | .477 | .442 | .406 | .369 | .333 | .297 | |
| 5.500 | 4.219 | .000 | 1.281 | .74 | 4.32 | 23.3% | 10 | .577 | .545 | .512 | .477 | .442 | .406 | .369 | .333 | .297 | .262 |
| 5.500 | 4.448 | .000 | 1.052 | .77 | 4.28 | 19.1% | 11 | .577 | .545 | .512 | .477 | .442 | .406 | .369 | .333 | .297 | .262 |
| | | | | | | | | .229 | | | | | | | | | |
| 5.500 | 4.645 | .000 | .855 | .80 | 4.23 | 15.5% | 12 | .577 | .545 | .512 | .477 | .442 | .406 | .369 | .333 | .297 | .262 |
| | | | | | | | | .229 | .197 | | | | | | | | |
| 5.500 | 4.813 | .000 | .687 | .83 | 4.17 | 12.5% | 13 | .577 | .545 | .512 | .477 | .442 | .406 | .369 | .333 | .297 | .262 |
| | | | | | | | | .229 | .197 | .168 | | | | | | | |
| 5.500 | 4.954 | .000 | .546 | .86 | 4.11 | 9.9% | 14 | .577 | .545 | .512 | .477 | .442 | .406 | .369 | .333 | .297 | .262 |
| | | | | | | | | .229 | .197 | .168 | .141 | | | | | | |
| 5.500 | 5.072 | .000 | .428 | .88 | 4.03 | 7.8% | 15 | .577 | .545 | .512 | .477 | .442 | .406 | .369 | .333 | .297 | .262 |
| | | | | | | | | .229 | .197 | .168 | .141 | .117 | | | | | |
| 5.500 | 5.168 | .000 | .332 | .90 | 3.95 | 6.0% | 16 | .577 | .545 | .512 | .477 | .442 | .406 | .369 | .333 | .297 | .262 |
| | | | | | | | | .229 | .197 | .168 | .141 | .117 | .096 | | | | |
| 5.500 | 5.245 | .000 | .255 | .92 | 3.86 | 4.6% | 17 | .577 | .545 | .512 | .477 | .442 | .406 | .369 | .333 | .297 | .262 |
| | | | | | | | | .229 | .197 | .168 | .141 | .117 | .096 | .078 | | | |
| 5.500 | 5.307 | .000 | .193 | .94 | 3.77 | 3.5% | 18 | .577 | .545 | .512 | .477 | .442 | .406 | .369 | .333 | .297 | .262 |
| | | | | | | | | .229 | .197 | .168 | .141 | .117 | .096 | .078 | .062 | | |
| 5.500 | 5.356 | .000 | .144 | .95 | 3.68 | 2.6% | 19 | .577 | .545 | .512 | .477 | .442 | .406 | .369 | .333 | .297 | .262 |

*SIMILAR TO A.T.&T. NEAL-WILKINSON TABLES(SEE SECTION 1)

*EQUIVALENT RANDOM THEORY: VMR(VARIANCE TO MEAN RATIO)– 4.0 - NO RETRIES (PEAKED TRAFFIC - ROUTE ADVANCEMENT OR DAY-TO-DAY VARIATION)

| HOURS FIRST ATTEMPT VMR=4.0 | HOURS CON-NECT | HOURS RECALL /RETRY | HOURS OVFL | CON-NECT VMR | OVFL VMR | PERCENT BLOCKED (GRADE) | LINE | HOURS CONNECTED TRAFFIC ON LINES 1 | 2 | 3 | 4 | 5 | 6 | 7 | 8 | 9 | 10 |
|---|---|---|---|---|---|---|---|---|---|---|---|---|---|---|---|---|---|
| | | | | | | | | .229 | .197 | .168 | .141 | .117 | .096 | .078 | .062 | .049 | |
| 5.500 | 5.393 | .000 | .107 | .96 | 3.59 | 1.9% | 20 | .577 | .545 | .512 | .477 | .442 | .406 | .369 | .333 | .297 | .262 |
| | | | | | | | | .229 | .197 | .168 | .141 | .117 | .096 | .078 | .062 | .049 | .038 |
| 5.500 | 5.422 | .000 | .078 | .97 | 3.51 | 1.4% | 21 | .577 | .545 | .512 | .477 | .442 | .406 | .369 | .333 | .297 | .262 |
| | | | | | | | | .229 | .197 | .168 | .141 | .117 | .096 | .078 | .062 | .049 | .038 |
| | | | | | | | | .029 | | | | | | | | | |
| 5.500 | 5.444 | .000 | .056 | .98 | 3.42 | 1.0% | 22 | .577 | .545 | .512 | .477 | .442 | .406 | .369 | .333 | .297 | .262 |
| | | | | | | | | .229 | .197 | .168 | .141 | .117 | .096 | .078 | .062 | .049 | .038 |
| | | | | | | | | .029 | .022 | | | | | | | | |
| 5.500 | 5.460 | .000 | .040 | .98 | 3.34 | .7% | 23 | .577 | .545 | .512 | .477 | .442 | .406 | .369 | .333 | .297 | .262 |
| | | | | | | | | .229 | .197 | .168 | .141 | .117 | .096 | .078 | .062 | .049 | .038 |
| | | | | | | | | .029 | .022 | .016 | | | | | | | |
| 5.500 | 5.472 | .000 | .028 | .99 | 3.25 | .5% | 24 | .577 | .545 | .512 | .477 | .442 | .406 | .369 | .333 | .297 | .262 |
| | | | | | | | | .229 | .197 | .168 | .141 | .117 | .096 | .078 | .062 | .049 | .038 |
| | | | | | | | | .029 | .022 | .016 | .012 | | | | | | |
| 6.000 | 4.193 | .000 | 1.807 | .67 | 4.41 | 30.1% | 9 | .598 | .568 | .536 | .503 | .469 | .434 | .398 | .362 | .326 | |
| 6.000 | 4.484 | .000 | 1.516 | .71 | 4.40 | 25.3% | 10 | .598 | .568 | .536 | .503 | .469 | .434 | .398 | .362 | .326 | .291 |
| 6.000 | 4.740 | .000 | 1.260 | .74 | 4.37 | 21.0% | 11 | .598 | .568 | .536 | .503 | .469 | .434 | .398 | .362 | .326 | .291 |
| | | | | | | | | .256 | | | | | | | | | |
| 6.000 | 4.964 | .000 | 1.036 | .78 | 4.33 | 17.3% | 12 | .598 | .568 | .536 | .503 | .469 | .434 | .398 | .362 | .326 | .291 |
| | | | | | | | | .256 | .224 | | | | | | | | |
| 6.000 | 5.157 | .000 | .843 | .81 | 4.28 | 14.0% | 13 | .598 | .568 | .536 | .503 | .469 | .434 | .398 | .362 | .326 | .291 |
| | | | | | | | | .256 | .224 | .193 | | | | | | | |
| 6.000 | 5.322 | .000 | .678 | .84 | 4.22 | 11.3% | 14 | .598 | .568 | .536 | .503 | .469 | .434 | .398 | .362 | .326 | .291 |
| | | | | | | | | .256 | .224 | .193 | .165 | | | | | | |
| 6.000 | 5.460 | .000 | .540 | .86 | 4.15 | 9.0% | 15 | .598 | .568 | .536 | .503 | .469 | .434 | .398 | .362 | .326 | .291 |
| | | | | | | | | .256 | .224 | .193 | .165 | .139 | | | | | |
| 6.000 | 5.576 | .000 | .424 | .88 | 4.07 | 7.1% | 16 | .598 | .568 | .536 | .503 | .469 | .434 | .398 | .362 | .326 | .291 |
| | | | | | | | | .256 | .224 | .193 | .165 | .139 | .115 | | | | |
| 6.000 | 5.670 | .000 | .330 | .91 | 3.99 | 5.5% | 17 | .598 | .568 | .536 | .503 | .469 | .434 | .398 | .362 | .326 | .291 |
| | | | | | | | | .256 | .224 | .193 | .165 | .139 | .115 | .094 | | | |
| 6.000 | 5.746 | .000 | .254 | .92 | 3.90 | 4.2% | 18 | .598 | .568 | .536 | .503 | .469 | .434 | .398 | .362 | .326 | .291 |
| | | | | | | | | .256 | .224 | .193 | .165 | .139 | .115 | .094 | .076 | | |
| 6.000 | 5.807 | .000 | .193 | .94 | 3.81 | 3.2% | 19 | .598 | .568 | .536 | .503 | .469 | .434 | .398 | .362 | .326 | .291 |
| | | | | | | | | .256 | .224 | .193 | .165 | .139 | .115 | .094 | .076 | .061 | |
| 6.000 | 5.856 | .000 | .144 | .95 | 3.72 | 2.4% | 20 | .598 | .568 | .536 | .503 | .469 | .434 | .398 | .362 | .326 | .291 |
| | | | | | | | | .256 | .224 | .193 | .165 | .139 | .115 | .094 | .076 | .061 | .048 |
| 6.000 | 5.893 | .000 | .107 | .96 | 3.63 | 1.8% | 21 | .598 | .568 | .536 | .503 | .469 | .434 | .398 | .362 | .326 | .291 |
| | | | | | | | | .256 | .224 | .193 | .165 | .139 | .115 | .094 | .076 | .061 | .048 |
| | | | | | | | | .037 | | | | | | | | | |

*SIMILAR TO A.T.&T. NEAL-WILKINSON TABLES(SEE SECTION 1)

*EQUIVALENT RANDOM THEORY: VMR(VARIANCE TO MEAN RATIO)– 4.0 - NO RETRIES (PEAKED TRAFFIC - ROUTE ADVANCEMENT OR DAY-TO-DAY VARIATION)

| HOURS FIRST ATTEMPT VMR = 4.0 | HOURS CON-NECT | HOURS RECALL /RETRY | HOURS OVFL | CON-NECT VMR | OVFL VMR | PERCENT BLOCKED (GRADE) | LINE | HOURS CONNECTED TRAFFIC ON LINES 1 | 2 | 3 | 4 | 5 | 6 | 7 | 8 | 9 | 10 |
|---|---|---|---|---|---|---|---|---|---|---|---|---|---|---|---|---|---|
| 6.000 | 5.922 | .000 | .078 | .97 | 3.54 | 1.3% | 22 | .598 | .568 | .536 | .503 | .469 | .434 | .398 | .362 | .326 | .291 |
| | | | | | | | | .256 | .224 | .193 | .165 | .139 | .115 | .094 | .076 | .061 | .048 |
| | | | | | | | | .037 | .029 | | | | | | | | |
| 6.000 | 5.943 | .000 | .057 | .98 | 3.46 | .9% | 23 | .598 | .568 | .536 | .503 | .469 | .434 | .398 | .362 | .326 | .291 |
| | | | | | | | | .256 | .224 | .193 | .165 | .139 | .115 | .094 | .076 | .061 | .048 |
| | | | | | | | | .037 | .029 | .022 | | | | | | | |
| 6.000 | 5.959 | .000 | .041 | .98 | 3.37 | .7% | 24 | .598 | .568 | .536 | .503 | .469 | .434 | .398 | .362 | .326 | .291 |
| | | | | | | | | .256 | .224 | .193 | .165 | .139 | .115 | .094 | .076 | .061 | .048 |
| | | | | | | | | .037 | .029 | .022 | .016 | | | | | | |
| 6.500 | 4.414 | .000 | 2.086 | .65 | 4.47 | 32.1% | 9 | .618 | .589 | .559 | .527 | .494 | .460 | .425 | .389 | .354 | |
| 6.500 | 4.732 | .000 | 1.768 | .68 | 4.46 | 27.2% | 10 | .618 | .589 | .559 | .527 | .494 | .460 | .425 | .389 | .354 | .318 |
| 6.500 | 5.016 | .000 | 1.484 | .72 | 4.44 | 22.8% | 11 | .618 | .589 | .559 | .527 | .494 | .460 | .425 | .389 | .354 | .318 |
| | | | | | | | | .284 | | | | | | | | | |
| 6.500 | 5.266 | .000 | 1.234 | .75 | 4.41 | 19.0% | 12 | .618 | .589 | .559 | .527 | .494 | .460 | .425 | .389 | .354 | .318 |
| | | | | | | | | .284 | .250 | | | | | | | | |
| 6.500 | 5.485 | .000 | 1.015 | .78 | 4.37 | 15.6% | 13 | .618 | .589 | .559 | .527 | .494 | .460 | .425 | .389 | .354 | .318 |
| | | | | | | | | .284 | .250 | .219 | | | | | | | |
| 6.500 | 5.673 | .000 | .827 | .81 | 4.32 | 12.7% | 14 | .618 | .589 | .559 | .527 | .494 | .460 | .425 | .389 | .354 | .318 |
| | | | | | | | | .284 | .250 | .219 | .189 | | | | | | |
| 6.500 | 5.834 | .000 | .666 | .84 | 4.25 | 10.2% | 15 | .618 | .589 | .559 | .527 | .494 | .460 | .425 | .389 | .354 | .318 |
| | | | | | | | | .284 | .250 | .219 | .189 | .161 | | | | | |
| 6.500 | 5.969 | .000 | .531 | .86 | 4.18 | 8.2% | 16 | .618 | .589 | .559 | .527 | .494 | .460 | .425 | .389 | .354 | .318 |
| | | | | | | | | .284 | .250 | .219 | .189 | .161 | .135 | | | | |
| 6.500 | 6.082 | .000 | .418 | .89 | 4.11 | 6.4% | 17 | .618 | .589 | .559 | .527 | .494 | .460 | .425 | .389 | .354 | .318 |
| | | | | | | | | .284 | .250 | .219 | .189 | .161 | .135 | .113 | | | |
| 6.500 | 6.174 | .000 | .326 | .91 | 4.02 | 5.0% | 18 | .618 | .580 | .559 | .527 | .494 | .460 | .425 | .389 | .354 | .318 |
| | | | | | | | | .284 | .250 | .219 | .189 | .161 | .135 | .113 | .092 | | |
| 6.500 | 6.249 | .000 | .251 | .93 | 3.94 | 3.9% | 19 | .618 | .589 | .559 | .527 | .494 | .460 | .425 | .389 | .354 | .318 |
| | | | | | | | | .284 | .250 | .219 | .189 | .161 | .135 | .113 | .092 | .075 | |
| 6.500 | 6.309 | .000 | .191 | .94 | 3.85 | 2.9% | 20 | .618 | .589 | .559 | .527 | .494 | .460 | .425 | .389 | .354 | .318 |
| | | | | | | | | .284 | .250 | .219 | .189 | .161 | .135 | .113 | .092 | .075 | .060 |
| 6.500 | 6.356 | .000 | .144 | .95 | 3.76 | 2.2% | 21 | .618 | .589 | .559 | .527 | .494 | .460 | .425 | .309 | .354 | .318 |
| | | | | | | | | .284 | .250 | .219 | .189 | .161 | .135 | .113 | .092 | .075 | .060 |
| | | | | | | | | .047 | | | | | | | | | |
| 6.500 | 6.393 | .000 | .107 | .96 | 3.67 | 1.6% | 22 | .618 | .509 | .559 | .527 | .494 | .460 | .425 | .389 | .354 | .318 |
| | | | | | | | | .284 | .250 | .219 | .189 | .161 | .135 | .113 | .092 | .075 | .060 |
| | | | | | | | | .047 | .037 | | | | | | | | |
| 6.500 | 6.422 | .000 | .078 | .97 | 3.58 | 1.2% | 23 | .618 | .589 | .559 | .527 | .494 | .460 | .425 | .389 | .354 | .318 |
| | | | | | | | | .284 | .250 | .219 | .189 | .161 | .135 | .113 | .092 | .075 | .060 |
| | | | | | | | | .047 | .037 | .028 | | | | | | | |

*SIMILAR TO A.T.&T. NEAL-WILKINSON TABLES(SEE SECTION 1)

*EQUIVALENT RANDOM THEORY: VMR(VARIANCE TO MEAN RATIO)– 4.0 - NO RETRIES (PEAKED TRAFFIC - ROUTE ADVANCEMENT OR DAY-TO-DAY VARIATION)

| HOURS FIRST ATTEMPT VMR=4.0 | HOURS CON-NECT | HOURS RECALL /RETRY | HOURS OVFL | CON-NECT VMR | OVFL VMR | PERCENT BLOCKED (GRADE) | LINE | HOURS CONNECTED TRAFFIC ON LINES 1 | 2 | 3 | 4 | 5 | 6 | 7 | 8 | 9 | 10 |
|---|---|---|---|---|---|---|---|---|---|---|---|---|---|---|---|---|---|
| 6.500 | 6.443 | .000 | .057 | .98 | 3.49 | .9% | 24 | .618 | .589 | .559 | .527 | .494 | .460 | .425 | .389 | .354 | .318 |
| | | | | | | | | .284 | .250 | .219 | .189 | .161 | .135 | .113 | .092 | .075 | .060 |
| | | | | | | | | .047 | .037 | .028 | .021 | | | | | | |
| 6.500 | 6.459 | .000 | .041 | .98 | 3.41 | .6% | 25 | .618 | .589 | .559 | .527 | .494 | .460 | .425 | .389 | .354 | .318 |
| | | | | | | | | .284 | .250 | .219 | .189 | .161 | .135 | .113 | .092 | .075 | .060 |
| | | | | | | | | .047 | .037 | .028 | .021 | .016 | | | | | |
| 7.000 | 4.608 | .000 | 2.392 | .62 | 4.54 | 34.2% | 9 | .634 | .607 | .578 | .547 | .516 | .483 | .449 | .415 | .380 | |
| 7.000 | 4.953 | .000 | 2.047 | .65 | 4.54 | 29.2% | 10 | .634 | .607 | .578 | .547 | .516 | .483 | .449 | .415 | .380 | .345 |
| 7.000 | 5.263 | .000 | 1.737 | .69 | 4.54 | 24.8% | 11 | .634 | .607 | .578 | .547 | .516 | .483 | .449 | .415 | .380 | .345 |
| | | | | | | | | .311 | | | | | | | | | |
| 7.000 | 5.540 | .000 | 1.460 | .72 | 4.51 | 20.9% | 12 | .634 | .607 | .578 | .547 | .516 | .483 | .449 | .415 | .380 | .345 |
| | | | | | | | | .311 | .277 | | | | | | | | |
| 7.000 | 5.785 | .000 | 1.215 | .76 | 4.48 | 17.4% | 13 | .634 | .607 | .578 | .547 | .516 | .483 | .449 | .415 | .380 | .345 |
| | | | | | | | | .311 | .277 | .244 | | | | | | | |
| 7.000 | 5.998 | .000 | 1.002 | .79 | 4.44 | 14.3% | 14 | .634 | .607 | .578 | .547 | .516 | .483 | .449 | .415 | .380 | .345 |
| | | | | | | | | .311 | .277 | .244 | .213 | | | | | | |
| 7.000 | 6.182 | .000 | .818 | .82 | 4.38 | 11.7% | 15 | .634 | .607 | .578 | .547 | .516 | .483 | .449 | .415 | .380 | .345 |
| | | | | | | | | .311 | .277 | .244 | .213 | .184 | | | | | |
| 7.000 | 6.339 | .000 | .661 | .84 | 4.32 | 9.4% | 16 | .634 | .607 | .578 | .547 | .516 | .483 | .449 | .415 | .380 | .345 |
| | | | | | | | | .311 | .277 | .244 | .213 | .184 | .157 | | | | |
| 7.000 | 6.472 | .000 | .528 | .87 | 4.25 | 7.5% | 17 | .634 | .607 | .578 | .547 | .516 | .483 | .449 | .415 | .380 | .345 |
| | | | | | | | | .311 | .277 | .244 | .213 | .184 | .157 | .133 | | | |
| 7.000 | 6.583 | .000 | .417 | .89 | 4.17 | 6.0% | 18 | .634 | .607 | .578 | .547 | .516 | .483 | .449 | .415 | .380 | .345 |
| | | | | | | | | .311 | .277 | .244 | .213 | .184 | .157 | .133 | .111 | | |
| 7.000 | 6.674 | .000 | .326 | .91 | 4.08 | 4.7% | 19 | .634 | .607 | .578 | .547 | .516 | .483 | .449 | .415 | .380 | .345 |
| | | | | | | | | .311 | .277 | .244 | .213 | .184 | .157 | .133 | .111 | .091 | |
| 7.000 | 6.748 | .000 | .252 | .93 | 4.00 | 3.6% | 20 | .634 | .607 | .578 | .547 | .516 | .483 | .449 | .415 | .380 | .345 |
| | | | | | | | | .311 | .277 | .244 | .213 | .184 | .157 | .133 | .111 | .091 | .074 |
| 7.000 | 6.807 | .000 | .193 | .94 | 3.91 | 2.8% | 21 | .634 | .607 | .578 | .547 | .516 | .483 | .449 | .415 | .380 | .345 |
| | | | | | | | | .311 | .277 | .244 | .213 | .184 | .157 | .133 | .111 | .091 | .074 |
| | | | | | | | | .059 | | | | | | | | | |
| 7.000 | 6.854 | .000 | .146 | .95 | 3.82 | 2.1% | 22 | .634 | .607 | .578 | .547 | .516 | .483 | .449 | .415 | .380 | .345 |
| | | | | | | | | .311 | .277 | .244 | .213 | .184 | .157 | .133 | .111 | .091 | .074 |
| | | | | | | | | .059 | .047 | | | | | | | | |
| 7.000 | 6.891 | .000 | .109 | .96 | 3.73 | 1.6% | 23 | .634 | .607 | .578 | .547 | .516 | .483 | .449 | .415 | .380 | .345 |
| | | | | | | | | .311 | .277 | .244 | .213 | .184 | .157 | .133 | .111 | .091 | .074 |
| | | | | | | | | .059 | .047 | .037 | | | | | | | |
| 7.000 | 6.919 | .000 | .081 | .97 | 3.64 | 1.2% | 24 | .634 | .607 | .578 | .547 | .516 | .483 | .449 | .415 | .380 | .345 |
| | | | | | | | | .311 | .277 | .244 | .213 | .184 | .157 | .133 | .111 | .091 | .074 |
| | | | | | | | | .059 | .047 | .037 | .028 | | | | | | |

*SIMILAR TO A.T.&T. NEAL-WILKINSON TABLES(SEE SECTION 1)

*EQUIVALENT RANDOM THEORY: VMR(VARIANCE TO MEAN RATIO)– 4.0 - NO RETRIES (PEAKED TRAFFIC - ROUTE ADVANCEMENT OR DAY-TO-DAY VARIATION)

| HOURS FIRST ATTEMPT VMR = 4.0 | HOURS CON-NECT | HOURS RECALL /RETRY | HOURS OVFL | CON-NECT VMR | OVFL VMR | PERCENT BLOCKED (GRADE) | LINE | HOURS CONNECTED TRAFFIC ON LINES 1 | 2 | 3 | 4 | 5 | 6 | 7 | 8 | 9 | 10 |
|---|---|---|---|---|---|---|---|---|---|---|---|---|---|---|---|---|---|
| 7.000 | 6.941 | .000 | .059 | .98 | 3.55 | .8% | 25 | .634 | .607 | .578 | .547 | .516 | .483 | .449 | .415 | .380 | .345 |
| | | | | | | | | .311 | .277 | .244 | .213 | .184 | .157 | .133 | .111 | .091 | .074 |
| | | | | | | | | .059 | .047 | .037 | .028 | .022 | | | | | |
| 7.000 | 6.957 | .000 | .043 | .98 | 3.47 | .6% | 26 | .634 | .607 | .578 | .547 | .516 | .483 | .449 | .415 | .380 | .345 |
| | | | | | | | | .311 | .277 | .244 | .213 | .184 | .157 | .133 | .111 | .091 | .074 |
| | | | | | | | | .059 | .047 | .037 | .028 | .022 | .016 | | | | |
| 7.500 | 5.171 | .000 | 2.329 | .63 | 4.59 | 31.1% | 10 | .650 | .624 | .597 | .568 | .537 | .506 | .473 | .439 | .405 | .371 |
| 7.500 | 5.508 | .000 | 1.992 | .66 | 4.59 | 26.6% | 11 | .650 | .624 | .597 | .568 | .537 | .506 | .473 | .439 | .405 | .371 |
| | | | | | | | | .336 | | | | | | | | | |
| 7.500 | 5.810 | .000 | 1.690 | .70 | 4.58 | 22.5% | 12 | .650 | .624 | .597 | .568 | .537 | .506 | .473 | .439 | .405 | .371 |
| | | | | | | | | .336 | .303 | | | | | | | | |
| 7.500 | 6.080 | .000 | 1.420 | .73 | 4.56 | 18.9% | 13 | .650 | .624 | .597 | .568 | .537 | .506 | .473 | .439 | .405 | .371 |
| | | | | | | | | .336 | .303 | .269 | | | | | | | |
| 7.500 | 6.317 | .000 | 1.183 | .76 | 4.52 | 15.8% | 14 | .650 | .624 | .597 | .568 | .537 | .506 | .473 | .439 | .405 | .371 |
| | | | | | | | | .336 | .303 | .269 | .238 | | | | | | |
| 7.500 | 6.525 | .000 | .975 | .79 | 4.47 | 13.0% | 15 | .650 | .624 | .597 | .568 | .537 | .506 | .473 | .439 | .405 | .371 |
| | | | | | | | | .336 | .303 | .269 | .238 | .207 | | | | | |
| 7.500 | 6.703 | .000 | .797 | .82 | 4.42 | 10.6% | 16 | .650 | .624 | .597 | .568 | .537 | .506 | .473 | .439 | .405 | .371 |
| | | | | | | | | .336 | .303 | .269 | .238 | .207 | .179 | | | | |
| 7.500 | 6.856 | .000 | .644 | .85 | 4.35 | 8.6% | 17 | .650 | .624 | .597 | .568 | .537 | .506 | .473 | .439 | .405 | .371 |
| | | | | | | | | .336 | .303 | .269 | .238 | .207 | .179 | .153 | | | |
| 7.500 | 6.985 | .000 | .515 | .87 | 4.28 | 6.9% | 18 | .650 | .624 | .597 | .568 | .537 | .506 | .473 | .439 | .405 | .371 |
| | | | | | | | | .336 | .303 | .269 | .238 | .207 | .179 | .153 | .129 | | |
| 7.500 | 7.093 | .000 | .407 | .89 | 4.20 | 5.4% | 19 | .650 | .624 | .597 | .568 | .537 | .506 | .473 | .439 | .405 | .371 |
| | | | | | | | | .336 | .303 | .269 | .238 | .207 | .179 | .153 | .129 | .107 | |
| 7.500 | 7.181 | .000 | .319 | .91 | 4.11 | 4.3% | 20 | .650 | .624 | .597 | .568 | .537 | .506 | .473 | .439 | .405 | .371 |
| | | | | | | | | .336 | .303 | .269 | .238 | .207 | .179 | .153 | .129 | .107 | .088 |
| 7.500 | 7.253 | .000 | .247 | .93 | 4.03 | 3.3% | 21 | .650 | .624 | .597 | .568 | .537 | .506 | .473 | .439 | .405 | .371 |
| | | | | | | | | .336 | .303 | .269 | .238 | .207 | .179 | .153 | .129 | .107 | .088 |
| | | | | | | | | .072 | | | | | | | | | |
| 7.500 | 7.311 | .000 | .189 | .94 | 3.94 | 2.5% | 22 | .650 | .624 | .597 | .568 | .537 | .506 | .473 | .439 | .405 | .371 |
| | | | | | | | | .336 | .303 | .269 | .238 | .207 | .179 | .153 | .129 | .107 | .088 |
| | | | | | | | | .072 | .058 | | | | | | | | |
| 7.500 | 7.357 | .000 | .143 | .95 | 3.85 | 1.9% | 23 | .650 | .624 | .597 | .568 | .537 | .506 | .473 | .430 | .405 | .371 |
| | | | | | | | | .336 | .303 | .269 | .238 | .207 | .179 | .153 | .129 | .107 | .088 |
| | | | | | | | | .072 | .058 | .046 | | | | | | | |
| 7.500 | 7.393 | .000 | .107 | .96 | 3.76 | 1.4% | 24 | .650 | .624 | .597 | .568 | .537 | .506 | .473 | .439 | .405 | .371 |
| | | | | | | | | .336 | .303 | .269 | .238 | .207 | .179 | .153 | .129 | .107 | .088 |
| | | | | | | | | .072 | .058 | .046 | .036 | | | | | | |
| 7.500 | 7.420 | .000 | .080 | .97 | 3.67 | 1.1% | 25 | .650 | .624 | .597 | .568 | .537 | .506 | .473 | .439 | .405 | .371 |

*SIMILAR TO A.T.&T. NEAL-WILKINSON TABLES(SEE SECTION 1)

*EQUIVALENT RANDOM THEORY: VMR(VARIANCE TO MEAN RATIO)– 4.0 - NO RETRIES (PEAKED TRAFFIC - ROUTE ADVANCEMENT OR DAY-TO-DAY VARIATION)

| HOURS FIRST ATTEMPT VMR = 4.0 | HOURS CON-NECT | HOURS RECALL /RETRY | HOURS OVFL | CON-NECT VMR | OVFL VMR | PERCENT BLOCKED (GRADE) | LINE | HOURS CONNECTED TRAFFIC ON LINES 1 | 2 | 3 | 4 | 5 | 6 | 7 | 8 | 9 | 10 |
|---|---|---|---|---|---|---|---|---|---|---|---|---|---|---|---|---|---|
| | | | | | | | | .336 | .303 | .269 | .238 | .207 | .179 | .153 | .129 | .107 | .088 |
| | | | | | | | | .072 | .058 | .046 | .036 | .028 | | | | | |
| 7.500 | 7.442 | .000 | .058 | .98 | 3.58 | .8% | 26 | .650 | .624 | .597 | .568 | .537 | .506 | .473 | .439 | .405 | .371 |
| | | | | | | | | .336 | .303 | .269 | .238 | .207 | .179 | .153 | .129 | .107 | .088 |
| | | | | | | | | .072 | .058 | .046 | .036 | .028 | .021 | | | | |
| 7.500 | 7.458 | .000 | .042 | .98 | 3.49 | .6% | 27 | .650 | .624 | .597 | .568 | .537 | .506 | .473 | .439 | .405 | .371 |
| | | | | | | | | .336 | .303 | .269 | .238 | .207 | .179 | .153 | .129 | .107 | .088 |
| | | | | | | | | .072 | .058 | .046 | .036 | .028 | .021 | .016 | | | |
| 8.000 | 5.363 | .000 | 2.637 | .60 | 4.66 | 33.0% | 10 | .664 | .639 | .613 | .585 | .556 | .526 | .494 | .462 | .429 | .395 |
| 8.000 | 5.725 | .000 | 2.275 | .64 | 4.67 | 28.4% | 11 | .664 | .639 | .613 | .585 | .556 | .526 | .494 | .462 | .429 | .395 |
| | | | | | | | | .361 | | | | | | | | | |
| 8.000 | 6.052 | .000 | 1.948 | .67 | 4.67 | 24.3% | 12 | .664 | .639 | .613 | .585 | .556 | .526 | .494 | .462 | .429 | .395 |
| | | | | | | | | .361 | .328 | | | | | | | | |
| 8.000 | 6.347 | .000 | 1.653 | .71 | 4.65 | 20.7% | 13 | .664 | .639 | .613 | .585 | .556 | .526 | .494 | .462 | .429 | .395 |
| | | | | | | | | .361 | .328 | .294 | | | | | | | |
| 8.000 | 6.609 | .000 | 1.391 | .74 | 4.62 | 17.4% | 14 | .664 | .639 | .613 | .585 | .556 | .526 | .494 | .462 | .429 | .395 |
| | | | | | | | | .361 | .328 | .294 | .262 | | | | | | |
| 8.000 | 6.840 | .000 | 1.160 | .77 | 4.59 | 14.5% | 15 | .664 | .639 | .613 | .585 | .556 | .526 | .494 | .462 | .429 | .395 |
| | | | | | | | | .361 | .328 | .294 | .262 | .231 | | | | | |
| 8.000 | 7.042 | .000 | .958 | .80 | 4.54 | 12.0% | 16 | .664 | .639 | .613 | .585 | .556 | .526 | .494 | .462 | .429 | .395 |
| | | | | | | | | .361 | .328 | .294 | .262 | .231 | .202 | | | | |
| 8.000 | 7.216 | .000 | .784 | .83 | 4.48 | 9.8% | 17 | .664 | .639 | .613 | .585 | .556 | .526 | .494 | .462 | .429 | .395 |
| | | | | | | | | .361 | .328 | .294 | .262 | .231 | .202 | .174 | | | |
| 8.000 | 7.365 | .000 | .635 | .85 | 4.41 | 7.9% | 18 | .664 | .639 | .613 | .585 | .556 | .526 | .494 | .462 | .429 | .395 |
| | | | | | | | | .361 | .328 | .294 | .262 | .231 | .202 | .174 | .149 | | |
| 8.000 | 7.491 | .000 | .509 | .87 | 4.34 | 6.4% | 19 | .664 | .639 | .613 | .585 | .556 | .526 | .494 | .462 | .429 | .395 |
| | | | | | | | | .361 | .328 | .294 | .262 | .231 | .202 | .174 | .149 | .126 | |
| 8.000 | 7.596 | .000 | .404 | .89 | 4.26 | 5.0% | 20 | .664 | .639 | .613 | .585 | .556 | .526 | .494 | .462 | .429 | .395 |
| | | | | | | | | .361 | .328 | .294 | .262 | .231 | .202 | .174 | .149 | .126 | .105 |
| 8.000 | 7.683 | .000 | .317 | .91 | 4.17 | 4.0% | 21 | .664 | .639 | .613 | .585 | .556 | .526 | .494 | .462 | .429 | .395 |
| | | | | | | | | .361 | .328 | .294 | .262 | .231 | .202 | .174 | .149 | .126 | .105 |
| | | | | | | | | .087 | | | | | | | | | |
| 8.000 | 7.753 | .000 | .247 | .93 | 4.08 | 3.1% | 22 | .664 | .639 | .613 | .585 | .556 | .526 | .494 | .462 | .429 | .395 |
| | | | | | | | | .361 | .328 | .294 | .262 | .231 | .202 | .174 | .149 | .126 | .105 |
| | | | | | | | | .087 | .071 | | | | | | | | |
| 8.000 | 7.810 | .000 | .190 | .94 | 3.99 | 2.4% | 23 | .664 | .639 | .613 | .585 | .556 | .526 | .494 | .462 | .429 | .395 |
| | | | | | | | | .361 | .328 | .294 | .262 | .231 | .202 | .174 | .149 | .126 | .105 |
| | | | | | | | | .087 | .071 | .057 | | | | | | | |
| 8.000 | 7.856 | .000 | .144 | .95 | 3.90 | 1.8% | 24 | .664 | .639 | .613 | .585 | .556 | .526 | .494 | .462 | .429 | .395 |
| | | | | | | | | .361 | .328 | .294 | .262 | .231 | .202 | .174 | .149 | .126 | .105 |

*SIMILAR TO A.T.&T. NEAL-WILKINSON TABLES(SEE SECTION 1)

*EQUIVALENT RANDOM THEORY: VMR(VARIANCE TO MEAN RATIO)– 4.0 - NO RETRIES (PEAKED TRAFFIC - ROUTE ADVANCEMENT OR DAY-TO-DAY VARIATION)

| HOURS FIRST ATTEMPT VMR=4.0 | HOURS CON-NECT | HOURS RECALL /RETRY | HOURS OVFL | CON-NECT VMR | OVFL VMR | PERCENT BLOCKED (GRADE) | LINE | HOURS CONNECTED TRAFFIC ON LINES 1 | 2 | 3 | 4 | 5 | 6 | 7 | 8 | 9 | 10 |
|---|---|---|---|---|---|---|---|---|---|---|---|---|---|---|---|---|---|
| | | | | | | | | .087 | .071 | .057 | .045 | | | | | | |
| 8.000 | 7.891 | .000 | .109 | .96 | 3.81 | 1.4% | 25 | .664 | .639 | .613 | .585 | .556 | .526 | .494 | .462 | .429 | .395 |
| | | | | | | | | .361 | .328 | .294 | .262 | .231 | .202 | .174 | .149 | .126 | .105 |
| | | | | | | | | .087 | .071 | .057 | .045 | .036 | | | | | |
| 8.000 | 7.919 | .000 | .081 | .97 | 3.72 | 1.0% | 26 | .664 | .639 | .613 | .585 | .556 | .526 | .494 | .462 | .429 | .395 |
| | | | | | | | | .361 | .328 | .294 | .262 | .231 | .202 | .174 | .149 | .126 | .105 |
| | | | | | | | | .087 | .071 | .057 | .045 | .036 | .028 | | | | |
| 8.000 | 7.940 | .000 | .060 | .98 | 3.63 | .7% | 27 | .664 | .639 | .613 | .585 | .556 | .526 | .494 | .462 | .429 | .395 |
| | | | | | | | | .361 | .328 | .294 | .262 | .231 | .202 | .174 | .149 | .126 | .105 |
| | | | | | | | | .087 | .071 | .057 | .045 | .036 | .028 | .021 | | | |
| 8.000 | 7.957 | .000 | .043 | .98 | 3.55 | .5% | 28 | .664 | .639 | .613 | .585 | .556 | .526 | .494 | .462 | .429 | .395 |
| | | | | | | | | .361 | .328 | .294 | .262 | .231 | .202 | .174 | .149 | .126 | .105 |
| | | | | | | | | .087 | .071 | .057 | .045 | .036 | .028 | .021 | .016 | | |
| 8.500 | 5.555 | .000 | 2.945 | .58 | 4.70 | 34.6% | 10 | .678 | .654 | .629 | .603 | .575 | .546 | .516 | .484 | .452 | .419 |
| 8.500 | 5.941 | .000 | 2.559 | .61 | 4.71 | 30.1% | 11 | .678 | .654 | .629 | .603 | .575 | .546 | .516 | .484 | .452 | .419 |
| | | | | | | | | .385 | | | | | | | | | |
| 8.500 | 6.293 | .000 | 2.207 | .65 | 4.72 | 26.0% | 12 | .678 | .654 | .629 | .603 | .575 | .546 | .516 | .484 | .452 | .419 |
| | | | | | | | | .385 | .352 | | | | | | | | |
| 8.500 | 6.611 | .000 | 1.889 | .68 | 4.71 | 22.2% | 13 | .678 | .654 | .629 | .603 | .575 | .546 | .516 | .484 | .452 | .419 |
| | | | | | | | | .385 | .352 | .319 | | | | | | | |
| 8.500 | 6.897 | .000 | 1.603 | .71 | 4.69 | 18.9% | 14 | .678 | .654 | .629 | .603 | .575 | .546 | .516 | .484 | .452 | .419 |
| | | | | | | | | .385 | .352 | .319 | .286 | | | | | | |
| 8.500 | 7.152 | .000 | 1.348 | .75 | 4.66 | 15.9% | 15 | .678 | .654 | .629 | .603 | .575 | .546 | .516 | .484 | .452 | .419 |
| | | | | | | | | .385 | .352 | .319 | .286 | .255 | | | | | |
| 8.500 | 7.376 | .000 | 1.124 | .78 | 4.62 | 13.2% | 16 | .678 | .654 | .629 | .603 | .575 | .540 | .510 | .484 | .452 | .419 |
| | | | | | | | | .385 | .352 | .319 | .286 | .255 | .224 | | | | |
| 8.500 | 7.572 | .000 | .928 | .80 | 4.57 | 10.9% | 17 | .678 | .654 | .629 | .603 | .575 | .546 | .516 | .484 | .452 | .419 |
| | | | | | | | | .385 | .352 | .319 | .286 | .255 | .224 | .196 | | | |
| 8.500 | 7.741 | .000 | .759 | .83 | 4.51 | 8.9% | 18 | .678 | .654 | .629 | .603 | .575 | .546 | .516 | .484 | .452 | .419 |
| | | | | | | | | .385 | .352 | .319 | .286 | .255 | .224 | .196 | .169 | | |
| 8.500 | 7.885 | .000 | .615 | .86 | 4.44 | 7.2% | 19 | .678 | .654 | .629 | .603 | .575 | .546 | .516 | .484 | .452 | .419 |
| | | | | | | | | .385 | .352 | .319 | .286 | .255 | .224 | .196 | .109 | .144 | |
| 8.500 | 8.006 | .000 | .494 | .88 | 4.36 | 5.8% | 20 | .678 | .654 | .629 | .603 | .575 | .546 | .516 | .484 | .452 | .419 |
| | | | | | | | | .385 | .352 | .310 | .286 | .255 | .224 | .196 | .169 | .144 | .122 |
| 8.500 | 8.108 | .000 | .392 | .90 | 4.28 | 4.6% | 21 | .678 | .654 | .629 | .603 | .575 | .546 | .516 | .484 | .452 | .419 |
| | | | | | | | | .385 | .352 | .319 | .286 | .255 | .224 | .196 | .169 | .144 | .122 |
| | | | | | | | | .102 | | | | | | | | | |
| 8.500 | 8.192 | .000 | .308 | .92 | 4.20 | 3.6% | 22 | .678 | .654 | .629 | .603 | .575 | .546 | .516 | .484 | .452 | .419 |
| | | | | | | | | .385 | .352 | .319 | .286 | .255 | .224 | .196 | .109 | .144 | .122 |
| | | | | | | | | .102 | .084 | | | | | | | | |

*SIMILAR TO A.T.&T. NEAL-WILKINSON TABLES(SEE SECTION 1)

*EQUIVALENT RANDOM THEORY: VMR(VARIANCE TO MEAN RATIO)– 4.0 - NO RETRIES (PEAKED TRAFFIC - ROUTE ADVANCEMENT OR DAY-TO-DAY VARIATION)

| HOURS FIRST ATTEMPT VMR = 4.0 | HOURS CON- NECT | HOURS RECALL /RETRY | HOURS OVFL | CON- NECT VMR | OVFL VMR | PERCENT BLOCKED (GRADE) | LINE | HOURS CONNECTED TRAFFIC ON LINES 1 | 2 | 3 | 4 | 5 | 6 | 7 | 8 | 9 | 10 |
|---|---|---|---|---|---|---|---|---|---|---|---|---|---|---|---|---|---|
| 8.500 | 8.260 | .000 | .240 | .93 | 4.11 | 2.8% | 23 | .678 | .654 | .629 | .603 | .575 | .546 | .516 | .484 | .452 | .419 |
| | | | | | | | | .385 | .352 | .319 | .286 | .255 | .224 | .196 | .169 | .144 | .122 |
| | | | | | | | | .102 | .084 | .068 | | | | | | | |
| 8.500 | 8.315 | .000 | .185 | .94 | 4.02 | 2.2% | 24 | .678 | .654 | .629 | .603 | .575 | .546 | .516 | .484 | .452 | .419 |
| | | | | | | | | .385 | .352 | .319 | .286 | .255 | .224 | .196 | .169 | .144 | .122 |
| | | | | | | | | .102 | .084 | .068 | .055 | | | | | | |
| 8.500 | 8.359 | .000 | .141 | .96 | 3.93 | 1.7% | 25 | .678 | .654 | .629 | .603 | .575 | .546 | .516 | .484 | .452 | .419 |
| | | | | | | | | .385 | .352 | .319 | .286 | .255 | .224 | .196 | .169 | .144 | .122 |
| | | | | | | | | .102 | .084 | .068 | .055 | .044 | | | | | |
| 8.500 | 8.394 | .000 | .106 | .97 | 3.84 | 1.2% | 26 | .678 | .654 | .629 | .603 | .575 | .546 | .516 | .484 | .452 | .419 |
| | | | | | | | | .385 | .352 | .319 | .286 | .255 | .224 | .196 | .169 | .144 | .122 |
| | | | | | | | | .102 | .084 | .068 | .055 | .044 | .035 | | | | |
| 8.500 | 8.421 | .000 | .079 | .97 | 3.75 | .9% | 27 | .678 | .654 | .629 | .603 | .575 | .546 | .516 | .484 | .452 | .419 |
| | | | | | | | | .385 | .352 | .319 | .286 | .255 | .224 | .196 | .169 | .144 | .122 |
| | | | | | | | | .102 | .084 | .068 | .055 | .044 | .035 | .027 | | | |
| 8.500 | 8.442 | .000 | .058 | .98 | 3.66 | .7% | 28 | .678 | .654 | .629 | .603 | .575 | .546 | .516 | .484 | .452 | .419 |
| | | | | | | | | .385 | .352 | .319 | .286 | .255 | .224 | .196 | .169 | .144 | .122 |
| | | | | | | | | .102 | .084 | .068 | .055 | .044 | .035 | .027 | .021 | | |
| 8.500 | 8.457 | .000 | .043 | .98 | 3.57 | .5% | 29 | .678 | .654 | .629 | .603 | .575 | .546 | .516 | .484 | .452 | .419 |
| | | | | | | | | .385 | .352 | .319 | .286 | .255 | .224 | .196 | .169 | .144 | .122 |
| | | | | | | | | .102 | .084 | .068 | .055 | .044 | .035 | .027 | .021 | .016 | |
| 9.000 | 6.144 | .000 | 2.856 | .59 | 4.75 | 31.7% | 11 | .691 | .668 | .645 | .619 | .593 | .565 | .536 | .505 | .474 | .441 |
| | | | | | | | | .409 | | | | | | | | | |
| 9.000 | 6.520 | .000 | 2.480 | .62 | 4.76 | 27.6% | 12 | .691 | .668 | .645 | .619 | .593 | .565 | .536 | .505 | .474 | .441 |
| | | | | | | | | .409 | .375 | | | | | | | | |
| 9.000 | 6.862 | .000 | 2.138 | .66 | 4.76 | 23.8% | 13 | .691 | .668 | .645 | .619 | .593 | .565 | .536 | .505 | .474 | .441 |
| | | | | | | | | .409 | .375 | .342 | | | | | | | |
| 9.000 | 7.172 | .000 | 1.828 | .69 | 4.75 | 20.3% | 14 | .691 | .668 | .645 | .619 | .593 | .565 | .536 | .505 | .474 | .441 |
| | | | | | | | | .409 | .375 | .342 | .310 | | | | | | |
| 9.000 | 7.450 | .000 | 1.550 | .72 | 4.73 | 17.2% | 15 | .691 | .668 | .645 | .619 | .593 | .565 | .536 | .505 | .474 | .441 |
| | | | | | | | | .409 | .375 | .342 | .310 | .278 | | | | | |
| 9.000 | 7.696 | .000 | 1.304 | .75 | 4.70 | 14.5% | 16 | .691 | .668 | .645 | .619 | .593 | .565 | .536 | .505 | .474 | .441 |
| | | | | | | | | .409 | .375 | .342 | .310 | .278 | .247 | | | | |
| 9.000 | 7.914 | .000 | 1.086 | .78 | 4.66 | 12.1% | 17 | .691 | .668 | .645 | .619 | .593 | .565 | .536 | .505 | .474 | .441 |
| | | | | | | | | .409 | .375 | .342 | .310 | .278 | .247 | .217 | | | |
| 9.000 | 8.103 | .000 | .897 | .81 | 4.60 | 10.0% | 18 | .691 | .668 | .645 | .619 | .593 | .565 | .536 | .505 | .474 | .441 |
| | | | | | | | | .409 | .375 | .342 | .310 | .278 | .247 | .217 | .189 | | |
| 9.000 | 8.266 | .000 | .734 | .84 | 4.54 | 8.2% | 19 | .691 | .668 | .645 | .619 | .593 | .565 | .536 | .505 | .474 | .441 |
| | | | | | | | | .409 | .375 | .342 | .310 | .278 | .247 | .217 | .189 | .163 | |
| 9.000 | 8.405 | .000 | .595 | .86 | 4.47 | 6.6% | 20 | .691 | .668 | .645 | .619 | .593 | .565 | .536 | .505 | .474 | .441 |

*SIMILAR TO A.T.&T. NEAL-WILKINSON TABLES(SEE SECTION 1)

*EQUIVALENT RANDOM THEORY: VMR(VARIANCE TO MEAN RATIO)– 4.0 - NO RETRIES (PEAKED TRAFFIC - ROUTE ADVANCEMENT OR DAY-TO-DAY VARIATION)

| HOURS FIRST ATTEMPT VMR = 4.0 | HOURS CON-NECT | HOURS RECALL /RETRY | HOURS OVFL | CON-NECT VMR | OVFL VMR | PERCENT BLOCKED (GRADE) | LINE | HOURS CONNECTED TRAFFIC ON LINES 1 | 2 | 3 | 4 | 5 | 6 | 7 | 8 | 9 | 10 |
|---|---|---|---|---|---|---|---|---|---|---|---|---|---|---|---|---|---|
| | | | | | | | | .409 | .375 | .342 | .310 | .278 | .247 | .217 | .189 | .163 | .139 |
| 9.000 | 8.523 | .000 | .477 | .88 | 4.39 | 5.3% | 21 | .691 | .668 | .645 | .619 | .593 | .565 | .536 | .505 | .474 | .441 |
| | | | | | | | | .409 | .375 | .342 | .310 | .278 | .247 | .217 | .189 | .163 | .139 |
| | | | | | | | | .117 | | | | | | | | | |
| 9.000 | 8.621 | .000 | .379 | .90 | 4.31 | 4.2% | 22 | .691 | .668 | .645 | .619 | .593 | .565 | .536 | .505 | .474 | .441 |
| | | | | | | | | .409 | .375 | .342 | .310 | .278 | .247 | .217 | .189 | .163 | .139 |
| | | | | | | | | .117 | .098 | | | | | | | | |
| 9.000 | 8.702 | .000 | .298 | .92 | 4.22 | 3.3% | 23 | .691 | .668 | .645 | .619 | .593 | .565 | .536 | .505 | .474 | .441 |
| | | | | | | | | .409 | .375 | .342 | .310 | .278 | .247 | .217 | .189 | .163 | .139 |
| | | | | | | | | .117 | .098 | .081 | | | | | | | |
| 9.000 | 8.768 | .000 | .232 | .93 | 4.13 | 2.6% | 24 | .691 | .668 | .645 | .619 | .593 | .565 | .536 | .505 | .474 | .441 |
| | | | | | | | | .409 | .375 | .342 | .310 | .278 | .247 | .217 | .189 | .163 | .139 |
| | | | | | | | | .117 | .098 | .081 | .066 | | | | | | |
| 9.000 | 8.821 | .000 | .179 | .95 | 4.04 | 2.0% | 25 | .691 | .668 | .645 | .619 | .593 | .565 | .536 | .505 | .474 | .441 |
| | | | | | | | | .409 | .375 | .342 | .310 | .278 | .247 | .217 | .189 | .163 | .139 |
| | | | | | | | | .117 | .098 | .081 | .066 | .053 | | | | | |
| 9.000 | 8.863 | .000 | .137 | .96 | 3.95 | 1.5% | 26 | .691 | .668 | .645 | .619 | .593 | .565 | .536 | .505 | .474 | .441 |
| | | | | | | | | .409 | .375 | .342 | .310 | .278 | .247 | .217 | .189 | .163 | .139 |
| | | | | | | | | .117 | .098 | .081 | .066 | .053 | .042 | | | | |
| 9.000 | 8.897 | .000 | .103 | .97 | 3.86 | 1.1% | 27 | .691 | .668 | .645 | .619 | .593 | .565 | .536 | .505 | .474 | .441 |
| | | | | | | | | .409 | .375 | .342 | .310 | .278 | .247 | .217 | .189 | .163 | .139 |
| | | | | | | | | .117 | .098 | .081 | .066 | .053 | .042 | .033 | | | |
| 9.000 | 8.923 | .000 | .077 | .97 | 3.77 | .9% | 28 | .691 | .668 | .645 | .619 | .593 | .565 | .536 | .505 | .474 | .441 |
| | | | | | | | | .409 | .375 | .342 | .310 | .278 | .247 | .217 | .189 | .163 | .139 |
| | | | | | | | | .117 | .098 | .081 | .066 | .053 | .042 | .033 | .026 | | |
| 9.000 | 8.943 | .000 | .057 | .98 | 3.68 | .6% | 29 | .691 | .668 | .645 | .619 | .593 | .565 | .536 | .505 | .474 | .441 |
| | | | | | | | | .409 | .375 | .342 | .310 | .278 | .247 | .217 | .189 | .163 | .139 |
| | | | | | | | | .117 | .098 | .081 | .066 | .053 | .042 | .033 | .026 | .020 | |
| 9.500 | 6.323 | .000 | 3.177 | .57 | 4.81 | 33.4% | 11 | .701 | .680 | .657 | .633 | .608 | .581 | .553 | .524 | .493 | .462 |
| | | | | | | | | .430 | | | | | | | | | |
| 9.500 | 6.721 | .000 | 2.779 | .60 | 4.83 | 29.3% | 12 | .701 | .680 | .657 | .633 | .608 | .581 | .553 | .524 | .493 | .462 |
| | | | | | | | | .430 | .398 | | | | | | | | |
| 9.500 | 7.086 | .000 | 2.414 | .63 | 4.84 | 25.4% | 13 | .701 | .680 | .657 | .633 | .608 | .581 | .553 | .524 | .493 | .462 |
| | | | | | | | | .430 | .398 | .365 | | | | | | | |
| 9.500 | 7.419 | .000 | 2.081 | .67 | 4.84 | 21.9% | 14 | .701 | .680 | .657 | .633 | .608 | .581 | .553 | .524 | .493 | .462 |
| | | | | | | | | .430 | .398 | .365 | .333 | | | | | | |
| 9.500 | 7.720 | .000 | 1.780 | .70 | 4.82 | 18.7% | 15 | .701 | .680 | .657 | .633 | .608 | .581 | .553 | .524 | .493 | .462 |
| | | | | | | | | .430 | .398 | .365 | .333 | .301 | | | | | |
| 9.500 | 7.990 | .000 | 1.510 | .73 | 4.80 | 15.9% | 16 | .701 | .680 | .657 | .633 | .608 | .581 | .553 | .524 | .493 | .462 |
| | | | | | | | | .430 | .398 | .365 | .333 | .301 | .270 | | | | |

*SIMILAR TO A.T.&T. NEAL-WILKINSON TABLES(SEE SECTION 1)

*EQUIVALENT RANDOM THEORY: VMR(VARIANCE TO MEAN RATIO)– 4.0 - NO RETRIES (PEAKED TRAFFIC - ROUTE ADVANCEMENT OR DAY-TO-DAY VARIATION)

| HOURS FIRST ATTEMPT VMR = 4.0 | HOURS CON- NECT | HOURS RECALL /RETRY | HOURS OVFL | CON- NECT VMR | OVFL VMR | PERCENT BLOCKED (GRADE) | LINE | HOURS CONNECTED TRAFFIC ON LINES 1 | 2 | 3 | 4 | 5 | 6 | 7 | 8 | 9 | 10 |
|---|---|---|---|---|---|---|---|---|---|---|---|---|---|---|---|---|---|
| 9.500 | 8.229 | .000 | 1.271 | .76 | 4.76 | 13.4% | 17 | .701 | .680 | .657 | .633 | .608 | .581 | .553 | .524 | .493 | .462 |
| | | | | | | | | .430 | .398 | .365 | .333 | .301 | .270 | .239 | | | |
| 9.500 | 8.440 | .000 | 1.060 | .79 | 4.71 | 11.2% | 18 | .701 | .680 | .657 | .633 | .608 | .581 | .553 | .524 | .493 | .462 |
| | | | | | | | | .430 | .398 | .365 | .333 | .301 | .270 | .239 | .211 | | |
| 9.500 | 8.624 | .000 | .876 | .82 | 4.66 | 9.2% | 19 | .701 | .680 | .657 | .633 | .608 | .581 | .553 | .524 | .493 | .462 |
| | | | | | | | | .430 | .398 | .365 | .333 | .301 | .270 | .239 | .211 | .184 | |
| 9.500 | 8.782 | .000 | .718 | .84 | 4.59 | 7.6% | 20 | .701 | .680 | .657 | .633 | .608 | .581 | .553 | .524 | .493 | .462 |
| | | | | | | | | .430 | .398 | .365 | .333 | .301 | .270 | .239 | .211 | .184 | .158 |
| 9.500 | 8.917 | .000 | .583 | .86 | 4.52 | 6.1% | 21 | .701 | .680 | .657 | .633 | .608 | .581 | .553 | .524 | .493 | .462 |
| | | | | | | | | .430 | .398 | .365 | .333 | .301 | .270 | .239 | .211 | .184 | .158 |
| | | | | | | | | .135 | | | | | | | | | |
| 9.500 | 9.031 | .000 | .469 | .89 | 4.44 | 4.9% | 22 | .701 | .680 | .657 | .633 | .608 | .581 | .553 | .524 | .493 | .462 |
| | | | | | | | | .430 | .398 | .365 | .333 | .301 | .270 | .239 | .211 | .184 | .158 |
| | | | | | | | | .135 | .114 | | | | | | | | |
| 9.500 | 9.127 | .000 | .373 | .90 | 4.36 | 3.9% | 23 | .701 | .680 | .657 | .633 | .608 | .581 | .553 | .524 | .493 | .462 |
| | | | | | | | | .430 | .398 | .365 | .333 | .301 | .270 | .239 | .211 | .184 | .158 |
| | | | | | | | | .135 | .114 | .095 | | | | | | | |
| 9.500 | 9.206 | .000 | .294 | .92 | 4.27 | 3.1% | 24 | .701 | .680 | .657 | .633 | .608 | .581 | .553 | .524 | .493 | .462 |
| | | | | | | | | .430 | .398 | .365 | .333 | .301 | .270 | .239 | .211 | .184 | .158 |
| | | | | | | | | .135 | .114 | .095 | .079 | | | | | | |
| 9.500 | 9.270 | .000 | .230 | .94 | 4.18 | 2.4% | 25 | .701 | .680 | .657 | .633 | .608 | .581 | .553 | .524 | .493 | .462 |
| | | | | | | | | .430 | .398 | .365 | .333 | .301 | .270 | .239 | .211 | .184 | .158 |
| | | | | | | | | .135 | .114 | .095 | .079 | .064 | | | | | |
| 9.500 | 9.322 | .000 | .178 | .95 | 4.09 | 1.9% | 26 | .701 | .680 | .657 | .633 | .608 | .581 | .553 | .524 | .493 | .462 |
| | | | | | | | | .430 | .398 | .365 | .333 | .301 | .270 | .239 | .211 | .184 | .158 |
| | | | | | | | | .135 | .114 | .095 | .079 | .064 | .052 | | | | |
| 9.500 | 9.364 | .000 | .136 | .96 | 4.00 | 1.4% | 27 | .701 | .680 | .657 | .633 | .608 | .581 | .553 | .524 | .493 | .462 |
| | | | | | | | | .430 | .398 | .365 | .333 | .301 | .270 | .239 | .211 | .184 | .158 |
| | | | | | | | | .135 | .114 | .095 | .079 | .064 | .052 | .042 | | | |
| 9.500 | 9.397 | .000 | .103 | .97 | 3.91 | 1.1% | 28 | .701 | .680 | .657 | .633 | .608 | .581 | .553 | .524 | .493 | .462 |
| | | | | | | | | .430 | .398 | .365 | .333 | .301 | .270 | .239 | .211 | .184 | .158 |
| | | | | | | | | .135 | .114 | .095 | .079 | .064 | .052 | .042 | .033 | | |
| 9.500 | 9.423 | .000 | .077 | .97 | 3.82 | .8% | 29 | .701 | .680 | .657 | .633 | .608 | .581 | .553 | .524 | .493 | .462 |
| | | | | | | | | .430 | .398 | .365 | .333 | .301 | .270 | .239 | .211 | .184 | .158 |
| | | | | | | | | .135 | .114 | .095 | .079 | .064 | .052 | .042 | .033 | .026 | |
| 9.500 | 9.442 | .000 | .058 | .98 | 3.73 | .6% | 30 | .701 | .680 | .657 | .633 | .608 | .581 | .553 | .524 | .493 | .462 |
| | | | | | | | | .430 | .398 | .365 | .333 | .301 | .270 | .239 | .211 | .184 | .158 |
| | | | | | | | | .135 | .114 | .095 | .079 | .064 | .052 | .042 | .033 | .026 | .020 |
| 10.000 | 6.504 | .000 | 3.496 | .55 | 4.83 | 35.0% | 11 | .712 | .692 | .671 | .648 | .623 | .598 | .571 | .542 | .513 | .483 |
| | | | | | | | | .452 | | | | | | | | | |

*SIMILAR TO A.T.&T. NEAL-WILKINSON TABLES(SEE SECTION 1)

***EQUIVALENT RANDOM THEORY: VMR(VARIANCE TO MEAN RATIO)– 4.0 - NO RETRIES (PEAKED TRAFFIC - ROUTE ADVANCEMENT OR DAY-TO-DAY VARIATION)**

| HOURS FIRST ATTEMPT VMR = 4.0 | HOURS CON-NECT | HOURS RECALL /RETRY | HOURS OVFL | CON-NECT VMR | OVFL VMR | PERCENT BLOCKED (GRADE) | LINE | HOURS CONNECTED TRAFFIC ON LINES 1 | 2 | 3 | 4 | 5 | 6 | 7 | 8 | 9 | 10 |
|---|---|---|---|---|---|---|---|---|---|---|---|---|---|---|---|---|---|
| 10.000 | 6.924 | .000 | 3.076 | .58 | 4.86 | 30.8% | 12 | .712 | .692 | .671 | .648 | .623 | .598 | .571 | .542 | .513 | .483 |
| | | | | | | | | .452 | .420 | | | | | | | | |
| 10.000 | 7.311 | .000 | 2.689 | .61 | 4.88 | 26.9% | 13 | .712 | .692 | .671 | .648 | .623 | .598 | .571 | .542 | .513 | .483 |
| | | | | | | | | .452 | .420 | .388 | | | | | | | |
| 10.000 | 7.667 | .000 | 2.333 | .64 | 4.88 | 23.3% | 14 | .712 | .692 | .671 | .648 | .623 | .598 | .571 | .542 | .513 | .483 |
| | | | | | | | | .452 | .420 | .388 | .355 | | | | | | |
| 10.000 | 7.990 | .000 | 2.010 | .68 | 4.88 | 20.1% | 15 | .712 | .692 | .671 | .648 | .623 | .598 | .571 | .542 | .513 | .483 |
| | | | | | | | | .452 | .420 | .388 | .355 | .323 | | | | | |
| 10.000 | 8.282 | .000 | 1.718 | .71 | 4.86 | 17.2% | 16 | .712 | .692 | .671 | .648 | .623 | .598 | .571 | .542 | .513 | .483 |
| | | | | | | | | .452 | .420 | .388 | .355 | .323 | .292 | | | | |
| 10.000 | 8.543 | .000 | 1.457 | .74 | 4.83 | 14.6% | 17 | .712 | .692 | .671 | .648 | .623 | .598 | .571 | .542 | .513 | .483 |
| | | | | | | | | .452 | .420 | .388 | .355 | .323 | .292 | .261 | | | |
| 10.000 | 8.775 | .000 | 1.225 | .77 | 4.79 | 12.3% | 18 | .712 | .692 | .671 | .648 | .623 | .598 | .571 | .542 | .513 | .483 |
| | | | | | | | | .452 | .420 | .388 | .355 | .323 | .292 | .261 | .232 | | |
| 10.000 | 8.978 | .000 | 1.022 | .80 | 4.74 | 10.2% | 19 | .712 | .692 | .671 | .648 | .623 | .598 | .571 | .542 | .513 | .483 |
| | | | | | | | | .452 | .420 | .388 | .355 | .323 | .292 | .261 | .232 | .204 | |
| 10.000 | 9.156 | .000 | .844 | .82 | 4.68 | 8.4% | 20 | .712 | .692 | .671 | .648 | .623 | .598 | .571 | .542 | .513 | .483 |
| | | | | | | | | .452 | .420 | .388 | .355 | .323 | .292 | .261 | .232 | .204 | .177 |
| 10.000 | 9.308 | .000 | .692 | .85 | 4.62 | 6.9% | 21 | .712 | .692 | .671 | .648 | .623 | .598 | .571 | .542 | .513 | .483 |
| | | | | | | | | .452 | .420 | .388 | .355 | .323 | .292 | .261 | .232 | .204 | .177 |
| | | | | | | | | .153 | | | | | | | | | |
| 10.000 | 9.439 | .000 | .561 | .87 | 4.54 | 5.6% | 22 | .712 | .692 | .671 | .648 | .623 | .598 | .571 | .542 | .513 | .483 |
| | | | | | | | | .452 | .420 | .388 | .355 | .323 | .292 | .261 | .232 | .204 | .177 |
| | | | | | | | | .153 | .130 | | | | | | | | |
| 10.000 | 9.549 | .000 | .451 | .89 | 4.46 | 4.5% | 23 | .712 | .692 | .671 | .648 | .623 | .598 | .571 | .542 | .513 | .483 |
| | | | | | | | | .452 | .420 | .388 | .355 | .323 | .292 | .261 | .232 | .204 | .177 |
| | | | | | | | | .153 | .130 | .110 | | | | | | | |
| 10.000 | 9.640 | .000 | .360 | .91 | 4.38 | 3.6% | 24 | .712 | .692 | .671 | .648 | .623 | .598 | .571 | .542 | .513 | .483 |
| | | | | | | | | .452 | .420 | .388 | .355 | .323 | .292 | .261 | .232 | .204 | .177 |
| | | | | | | | | .153 | .130 | .110 | .092 | | | | | | |
| 10.000 | 9.716 | .000 | .284 | .92 | 4.29 | 2.8% | 25 | .712 | .692 | .671 | .648 | .623 | .598 | .571 | .542 | .513 | .483 |
| | | | | | | | | .452 | .420 | .388 | .355 | .323 | .292 | .261 | .232 | .204 | .177 |
| | | | | | | | | .153 | .130 | .110 | .092 | .076 | | | | | |
| 10.000 | 9.778 | .000 | .222 | .94 | 4.20 | 2.2% | 26 | .712 | .692 | .671 | .648 | .623 | .598 | .571 | .542 | .513 | .483 |
| | | | | | | | | .452 | .420 | .388 | .355 | .323 | .292 | .261 | .232 | .204 | .177 |
| | | | | | | | | .153 | .130 | .110 | .092 | .076 | .062 | | | | |
| 10.000 | 9.829 | .000 | .171 | .95 | 4.11 | 1.7% | 27 | .712 | .692 | .671 | .648 | .623 | .598 | .571 | .542 | .513 | .483 |
| | | | | | | | | .452 | .420 | .388 | .355 | .323 | .292 | .261 | .232 | .204 | .177 |
| | | | | | | | | .153 | .130 | .110 | .092 | .076 | .062 | .050 | | | |
| 10.000 | 9.860 | .000 | .131 | .96 | 4.02 | 1.3% | 28 | .712 | .692 | .671 | .648 | .623 | .598 | .571 | .542 | .513 | .483 |

*SIMILAR TO A.T.&T. NEAL-WILKINSON TABLES(SEE SECTION 1)

## *EQUIVALENT RANDOM THEORY: VMR(VARIANCE TO MEAN RATIO)– 4.0 - NO RETRIES (PEAKED TRAFFIC - ROUTE ADVANCEMENT OR DAY-TO-DAY VARIATION)

| HOURS FIRST ATTEMPT VMR = 4.0 | HOURS CON-NECT | HOURS RECALL /RETRY | HOURS OVFL | CON-NECT VMR | OVFL VMR | PERCENT BLOCKED (GRADE) | LINE | HOURS CONNECTED TRAFFIC ON LINES 1 | 2 | 3 | 4 | 5 | 6 | 7 | 8 | 9 | 10 |
|---|---|---|---|---|---|---|---|---|---|---|---|---|---|---|---|---|---|
| | | | | | | | | .452 | .420 | .388 | .355 | .323 | .292 | .261 | .232 | .204 | .177 |
| | | | | | | | | .153 | .130 | .110 | .092 | .076 | .062 | .050 | .040 | | |
| 10.000 | 9.900 | .000 | .100 | .97 | 3.93 | 1.0% | 29 | .712 | .692 | .671 | .648 | .623 | .598 | .571 | .542 | .513 | .483 |
| | | | | | | | | .452 | .420 | .388 | .355 | .323 | .292 | .261 | .232 | .204 | .177 |
| | | | | | | | | .153 | .130 | .110 | .092 | .076 | .062 | .050 | .040 | .032 | |
| 10.000 | 9.925 | .000 | .075 | .98 | 3.84 | .7% | 30 | .712 | .692 | .671 | .648 | .623 | .598 | .571 | .542 | .513 | .483 |
| | | | | | | | | .452 | .420 | .388 | .355 | .323 | .292 | .261 | .232 | .204 | .177 |
| | | | | | | | | .153 | .130 | .110 | .092 | .076 | .062 | .050 | .040 | .032 | .025 |
| 10.000 | 9.944 | .000 | .056 | .98 | 3.75 | .6% | 31 | .712 | .692 | .671 | .648 | .623 | .598 | .571 | .542 | .513 | .483 |
| | | | | | | | | .452 | .420 | .388 | .355 | .323 | .292 | .261 | .232 | .204 | .177 |
| | | | | | | | | .153 | .130 | .110 | .092 | .076 | .062 | .050 | .040 | .032 | .025 |
| | | | | | | | | .019 | | | | | | | | | |
| 11.000 | 7.283 | .000 | 3.717 | .54 | 4.94 | 33.8% | 12 | .731 | .713 | .693 | .672 | .650 | .627 | .602 | .576 | .548 | .520 |
| | | | | | | | | .491 | .460 | | | | | | | | |
| 11.000 | 7.712 | .000 | 3.288 | .57 | 4.97 | 29.9% | 13 | .731 | .713 | .693 | .672 | .650 | .627 | .602 | .576 | .548 | .520 |
| | | | | | | | | .491 | .460 | .430 | | | | | | | |
| 11.000 | 8.111 | .000 | 2.889 | .60 | 4.99 | 26.3% | 14 | .731 | .713 | .693 | .672 | .650 | .627 | .602 | .576 | .548 | .520 |
| | | | | | | | | .491 | .460 | .430 | .398 | | | | | | |
| 11.000 | 8.477 | .000 | 2.523 | .63 | 5.00 | 22.9% | 15 | .731 | .713 | .693 | .672 | .650 | .627 | .602 | .576 | .548 | .520 |
| | | | | | | | | .491 | .460 | .430 | .398 | .367 | | | | | |
| 11.000 | 8.813 | .000 | 2.187 | .66 | 4.99 | 19.9% | 16 | .731 | .713 | .693 | .672 | .650 | .627 | .602 | .576 | .548 | .520 |
| | | | | | | | | .491 | .460 | .430 | .398 | .367 | .336 | | | | |
| 11.000 | 9.117 | .000 | 1.883 | .70 | 4.98 | 17.1% | 17 | .731 | .713 | .693 | .672 | .650 | .627 | .602 | .576 | .548 | .520 |
| | | | | | | | | .491 | .460 | .430 | .398 | .367 | .336 | .305 | | | |
| 11.000 | 9.392 | .000 | 1.608 | .73 | 4.96 | 14.6% | 18 | .731 | .713 | .693 | .672 | .650 | .627 | .602 | .576 | .548 | .520 |
| | | | | | | | | .491 | .460 | .430 | .398 | .367 | .336 | .305 | .274 | | |
| 11.000 | 9.637 | .000 | 1.363 | .76 | 4.92 | 12.4% | 19 | .731 | .713 | .693 | .672 | .650 | .627 | .602 | .576 | .548 | .520 |
| | | | | | | | | .491 | .460 | .430 | .398 | .367 | .336 | .305 | .274 | .245 | |
| 11.000 | 9.854 | .000 | 1.146 | .78 | 4.88 | 10.4% | 20 | .731 | .713 | .693 | .672 | .650 | .627 | .602 | .576 | .548 | .520 |
| | | | | | | | | .491 | .460 | .430 | .398 | .367 | .336 | .305 | .274 | .245 | .217 |
| 11.000 | 10.044 | .000 | .956 | .81 | 4.82 | 8.7% | 21 | .731 | .713 | .693 | .672 | .650 | .627 | .002 | .576 | .548 | .520 |
| | | | | | | | | .491 | .460 | .430 | .398 | .367 | .336 | .305 | .274 | .245 | .217 |
| | | | | | | | | .190 | | | | | | | | | |
| 11.000 | 10.209 | .000 | .791 | .83 | 4.76 | 7.2% | 22 | .731 | .713 | .693 | .672 | .650 | .627 | .602 | .576 | .548 | .520 |
| | | | | | | | | .491 | .460 | .430 | .398 | .367 | .336 | .305 | .274 | .245 | .217 |
| | | | | | | | | .190 | .165 | | | | | | | | |
| 11.000 | 10.352 | .000 | .648 | .86 | 4.69 | 5.9% | 23 | .731 | .713 | .693 | .672 | .650 | .627 | .602 | .576 | .548 | .520 |
| | | | | | | | | .491 | .460 | .430 | .398 | .367 | .336 | .305 | .274 | .245 | .217 |
| | | | | | | | | .190 | .165 | .142 | | | | | | | |
| 11.000 | 10.473 | .000 | .527 | .88 | 4.61 | 4.8% | 24 | .731 | .713 | .693 | .672 | .650 | .627 | .602 | .576 | .548 | .520 |

*SIMILAR TO A.T.&T. NEAL-WILKINSON TABLES(SEE SECTION 1)

*EQUIVALENT RANDOM THEORY: VMR(VARIANCE TO MEAN RATIO)– 4.0 - NO RETRIES (PEAKED TRAFFIC - ROUTE ADVANCEMENT OR DAY-TO-DAY VARIATION)

| HOURS FIRST ATTEMPT VMR = 4.0 | HOURS CON-NECT | HOURS RECALL /RETRY | HOURS OVFL | CON-NECT VMR | OVFL VMR | PERCENT BLOCKED (GRADE) | LINE | HOURS CONNECTED TRAFFIC ON LINES 1 | 2 | 3 | 4 | 5 | 6 | 7 | 8 | 9 | 10 |
|---|---|---|---|---|---|---|---|---|---|---|---|---|---|---|---|---|---|
| | | | | | | | | .491 | .460 | .430 | .398 | .367 | .336 | .305 | .274 | .245 | .217 |
| | | | | | | | | .190 | .165 | .142 | .121 | | | | | | |
| 11.000 | 10.576 | .000 | .424 | .90 | 4.53 | 3.9% | 25 | .731 | .713 | .693 | .672 | .650 | .627 | .602 | .576 | .548 | .520 |
| | | | | | | | | .491 | .460 | .430 | .398 | .367 | .336 | .305 | .274 | .245 | .217 |
| | | | | | | | | .190 | .165 | .142 | .121 | .103 | | | | | |
| 11.000 | 10.661 | .000 | .339 | .91 | 4.44 | 3.1% | 26 | .731 | .713 | .693 | .672 | .650 | .627 | .602 | .576 | .548 | .520 |
| | | | | | | | | .491 | .460 | .430 | .398 | .367 | .336 | .305 | .274 | .245 | .217 |
| | | | | | | | | .190 | .165 | .142 | .121 | .103 | .086 | | | | |
| 11.000 | 10.732 | .000 | .268 | .93 | 4.35 | 2.4% | 27 | .731 | .713 | .693 | .672 | .650 | .627 | .602 | .576 | .548 | .520 |
| | | | | | | | | .491 | .460 | .430 | .398 | .367 | .336 | .305 | .274 | .245 | .217 |
| | | | | | | | | .190 | .165 | .142 | .121 | .103 | .086 | .071 | | | |
| 11.000 | 10.790 | .000 | .210 | .94 | 4.26 | 1.9% | 28 | .731 | .713 | .693 | .672 | .650 | .627 | .602 | .576 | .548 | .520 |
| | | | | | | | | .491 | .460 | .430 | .398 | .367 | .336 | .305 | .274 | .245 | .217 |
| | | | | | | | | .190 | .165 | .142 | .121 | .103 | .086 | .071 | .058 | | |
| 11.000 | 10.837 | .000 | .163 | .95 | 4.17 | 1.5% | 29 | .731 | .713 | .693 | .672 | .650 | .627 | .602 | .576 | .548 | .520 |
| | | | | | | | | .491 | .460 | .430 | .398 | .367 | .336 | .305 | .274 | .245 | .217 |
| | | | | | | | | .190 | .165 | .142 | .121 | .103 | .086 | .071 | .058 | .047 | |
| 11.000 | 10.875 | .000 | .125 | .96 | 4.08 | 1.1% | 30 | .731 | .713 | .693 | .672 | .650 | .627 | .602 | .576 | .548 | .520 |
| | | | | | | | | .491 | .460 | .430 | .398 | .367 | .336 | .305 | .274 | .245 | .217 |
| | | | | | | | | .190 | .165 | .142 | .121 | .103 | .086 | .071 | .058 | .047 | .038 |
| 11.000 | 10.905 | .000 | .095 | .97 | 3.99 | .9% | 31 | .731 | .713 | .693 | .672 | .650 | .627 | .602 | .576 | .548 | .520 |
| | | | | | | | | .491 | .460 | .430 | .398 | .367 | .336 | .305 | .274 | .245 | .217 |
| | | | | | | | | .190 | .165 | .142 | .121 | .103 | .086 | .071 | .058 | .047 | .038 |
| | | | | | | | | .030 | | | | | | | | | |
| 11.000 | 10.928 | .000 | .072 | .98 | 3.90 | .7% | 32 | .731 | .713 | .693 | .672 | .650 | .627 | .602 | .576 | .548 | .520 |
| | | | | | | | | .491 | .460 | .430 | .398 | .367 | .336 | .305 | .274 | .245 | .217 |
| | | | | | | | | .190 | .165 | .142 | .121 | .103 | .086 | .071 | .058 | .047 | .038 |
| | | | | | | | | .030 | .023 | | | | | | | | |
| 12.000 | 8.071 | .000 | 3.929 | .53 | 5.04 | 32.7% | 13 | .748 | .731 | .713 | .694 | .674 | .652 | .629 | .605 | .580 | .553 |
| | | | | | | | | .526 | .497 | .468 | | | | | | | |
| 12.000 | 8.509 | .000 | 3.491 | .56 | 5.07 | 29.1% | 14 | .748 | .731 | .713 | .694 | .674 | .652 | .629 | .605 | .580 | .553 |
| | | | | | | | | .526 | .497 | .468 | .438 | | | | | | |
| 12.000 | 8.917 | .000 | 3.083 | .59 | 5.09 | 25.7% | 15 | .748 | .731 | .713 | .694 | .674 | .652 | .629 | .605 | .580 | .553 |
| | | | | | | | | .526 | .497 | .468 | .438 | .408 | | | | | |
| 12.000 | 9.294 | .000 | 2.706 | .62 | 5.11 | 22.6% | 16 | .748 | .731 | .713 | .694 | .674 | .652 | .629 | .605 | .580 | .553 |
| | | | | | | | | .526 | .497 | .468 | .438 | .408 | .377 | | | | |
| 12.000 | 9.640 | .000 | 2.360 | .65 | 5.11 | 19.7% | 17 | .748 | .731 | .713 | .694 | .674 | .652 | .629 | .605 | .580 | .553 |
| | | | | | | | | .526 | .497 | .468 | .438 | .408 | .377 | .346 | | | |
| 12.000 | 9.956 | .000 | 2.044 | .68 | 5.10 | 17.0% | 18 | .748 | .731 | .713 | .694 | .674 | .652 | .629 | .605 | .580 | .553 |
| | | | | | | | | .526 | .497 | .468 | .438 | .408 | .377 | .346 | .316 | | |

*SIMILAR TO A.T.&T. NEAL-WILKINSON TABLES(SEE SECTION 1)

*EQUIVALENT RANDOM THEORY: VMR(VARIANCE TO MEAN RATIO)– 4.0 - NO RETRIES (PEAKED TRAFFIC - ROUTE ADVANCEMENT OR DAY-TO-DAY VARIATION)

| HOURS FIRST ATTEMPT VMR = 4.0 | HOURS CON-NECT | HOURS RECALL /RETRY | HOURS OVFL | CON-NECT VMR | OVFL VMR | PERCENT BLOCKED (GRADE) | LINE | HOURS CONNECTED TRAFFIC ON LINES 1 | 2 | 3 | 4 | 5 | 6 | 7 | 8 | 9 | 10 |
|---|---|---|---|---|---|---|---|---|---|---|---|---|---|---|---|---|---|
| 12.000 | 10.243 | .000 | 1.757 | .71 | 5.08 | 14.6% | 19 | .748 | .731 | .713 | .694 | .674 | .652 | .629 | .605 | .580 | .553 |
| | | | | | | | | .526 | .497 | .468 | .438 | .408 | .377 | .346 | .316 | .286 | |
| 12.000 | 10.500 | .000 | 1.500 | .74 | 5.05 | 12.5% | 20 | .748 | .731 | .713 | .694 | .674 | .652 | .629 | .605 | .580 | .553 |
| | | | | | | | | .526 | .497 | .468 | .438 | .408 | .377 | .346 | .316 | .286 | .257 |
| 12.000 | 10.729 | .000 | 1.271 | .77 | 5.00 | 10.6% | 21 | .748 | .731 | .713 | .694 | .674 | .652 | .629 | .605 | .580 | .553 |
| | | | | | | | | .526 | .497 | .468 | .438 | .408 | .377 | .346 | .316 | .286 | .257 |
| | | | | | | | | .229 | | | | | | | | | |
| 12.000 | 10.931 | .000 | 1.069 | .80 | 4.95 | 8.9% | 22 | .748 | .731 | .713 | .694 | .674 | .652 | .629 | .605 | .580 | .553 |
| | | | | | | | | .526 | .497 | .468 | .438 | .408 | .377 | .346 | .316 | .286 | .257 |
| | | | | | | | | .229 | .203 | | | | | | | | |
| 12.000 | 11.109 | .000 | .891 | .82 | 4.89 | 7.4% | 23 | .748 | .731 | .713 | .694 | .674 | .652 | .629 | .605 | .580 | .553 |
| | | | | | | | | .526 | .497 | .468 | .438 | .408 | .377 | .346 | .316 | .286 | .257 |
| | | | | | | | | .229 | .203 | .177 | | | | | | | |
| 12.000 | 11.263 | .000 | .737 | .85 | 4.83 | 6.1% | 24 | .748 | .731 | .713 | .694 | .674 | .652 | .629 | .605 | .580 | .553 |
| | | | | | | | | .526 | .497 | .468 | .438 | .408 | .377 | .346 | .316 | .286 | .257 |
| | | | | | | | | .229 | .203 | .177 | .154 | | | | | | |
| 12.000 | 11.395 | .000 | .605 | .87 | 4.75 | 5.0% | 25 | .748 | .731 | .713 | .694 | .674 | .652 | .629 | .605 | .580 | .553 |
| | | | | | | | | .526 | .497 | .468 | .438 | .408 | .377 | .346 | .316 | .286 | .257 |
| | | | | | | | | .229 | .203 | .177 | .154 | .132 | | | | | |
| 12.000 | 11.508 | .000 | .492 | .89 | 4.67 | 4.1% | 26 | .748 | .731 | .713 | .694 | .674 | .652 | .629 | .605 | .580 | .553 |
| | | | | | | | | .526 | .497 | .468 | .438 | .408 | .377 | .346 | .316 | .286 | .257 |
| | | | | | | | | .229 | .203 | .177 | .154 | .132 | .113 | | | | |
| 12.000 | 11.603 | .000 | .397 | .90 | 4.59 | 3.3% | 27 | .748 | .731 | .713 | .694 | .674 | .652 | .629 | .605 | .580 | .553 |
| | | | | | | | | .526 | .497 | .468 | .438 | .408 | .377 | .346 | .316 | .286 | .257 |
| | | | | | | | | .229 | .203 | .177 | .154 | .132 | .113 | .095 | | | |
| 12.000 | 11.683 | .000 | .317 | .92 | 4.50 | 2.6% | 28 | .748 | .731 | .713 | .694 | .674 | .652 | .629 | .605 | .580 | .553 |
| | | | | | | | | .526 | .497 | .468 | .438 | .408 | .377 | .346 | .316 | .286 | .257 |
| | | | | | | | | .229 | .203 | .177 | .154 | .132 | .113 | .095 | .080 | | |
| 12.000 | 11.749 | .000 | .251 | .93 | 4.41 | 2.1% | 29 | .748 | .731 | .713 | .694 | .674 | .652 | .629 | .605 | .580 | .553 |
| | | | | | | | | .526 | .497 | .468 | .438 | .408 | .377 | .346 | .316 | .286 | .257 |
| | | | | | | | | .229 | .203 | .177 | .154 | .132 | .113 | .095 | .080 | .066 | |
| 12.000 | 11.803 | .000 | .197 | .95 | 4.32 | 1.6% | 30 | .748 | .731 | .713 | .094 | .674 | .652 | .629 | 605 | .580 | .553 |
| | | | | | | | | .526 | .497 | .468 | .438 | .408 | .377 | .346 | .316 | .286 | .257 |
| | | | | | | | | .229 | .203 | .177 | .154 | .132 | .113 | .095 | .080 | .066 | .054 |
| 12.000 | 11.846 | .000 | .154 | .96 | 4.23 | 1.3% | 31 | .748 | .731 | .713 | .694 | .674 | .652 | .629 | .605 | .580 | .553 |
| | | | | | | | | .526 | .497 | .468 | .438 | .408 | .377 | .346 | .316 | .286 | .257 |
| | | | | | | | | .229 | .203 | .177 | .154 | .132 | .113 | .095 | .080 | .066 | .054 |
| | | | | | | | | .044 | | | | | | | | | |
| 12.000 | 11.881 | .000 | .119 | .96 | 4.13 | 1.0% | 32 | .748 | .731 | .713 | .694 | .674 | .652 | .629 | .605 | .580 | .553 |
| | | | | | | | | .526 | .497 | .468 | .438 | .408 | .377 | .346 | .316 | .286 | .257 |

*SIMILAR TO A.T.&T. NEAL-WILKINSON TABLES(SEE SECTION 1)

*EQUIVALENT RANDOM THEORY: VMR(VARIANCE TO MEAN RATIO)– 4.0 - NO RETRIES (PEAKED TRAFFIC - ROUTE ADVANCEMENT OR DAY-TO-DAY VARIATION)

| HOURS FIRST ATTEMPT VMR = 4.0 | HOURS CON-NECT | HOURS RECALL /RETRY | HOURS OVFL | CON-NECT VMR | OVFL VMR | PERCENT BLOCKED (GRADE) | LINE | HOURS CONNECTED TRAFFIC ON LINES 1 | 2 | 3 | 4 | 5 | 6 | 7 | 8 | 9 | 10 |
|---|---|---|---|---|---|---|---|---|---|---|---|---|---|---|---|---|---|
| | | | | | | | | .229 | .203 | .177 | .154 | .132 | .113 | .095 | .080 | .066 | .054 |
| | | | | | | | | .044 | .035 | | | | | | | | |
| 12.000 | 11.909 | .000 | .091 | .97 | 4.04 | .8% | 33 | .748 | .731 | .713 | .694 | .674 | .652 | .629 | .605 | .580 | .553 |
| | | | | | | | | .526 | .497 | .468 | .438 | .408 | .377 | .346 | .316 | .286 | .257 |
| | | | | | | | | .229 | .203 | .177 | .154 | .132 | .113 | .095 | .080 | .066 | .054 |
| | | | | | | | | .044 | .035 | .028 | | | | | | | |
| 12.000 | 11.931 | .000 | .069 | .98 | 3.95 | .6% | 34 | .748 | .731 | .713 | .694 | .674 | .652 | .629 | .605 | .580 | .553 |
| | | | | | | | | .526 | .497 | .468 | .438 | .408 | .377 | .346 | .316 | .286 | .257 |
| | | | | | | | | .229 | .203 | .177 | .154 | .132 | .113 | .095 | .080 | .066 | .054 |
| | | | | | | | | .044 | .035 | .028 | .022 | | | | | | |
| 13.000 | 8.867 | .000 | 4.133 | .52 | 5.14 | 31.8% | 14 | .763 | .747 | .731 | .713 | .695 | .675 | .654 | .631 | .608 | .584 |
| | | | | | | | | .558 | .531 | .504 | .475 | | | | | | |
| 13.000 | 9.313 | .000 | 3.687 | .55 | 5.17 | 28.4% | 15 | .763 | .747 | .731 | .713 | .695 | .675 | .654 | .631 | .608 | .584 |
| | | | | | | | | .558 | .531 | .504 | .475 | .446 | | | | | |
| 13.000 | 9.729 | .000 | 3.271 | .58 | 5.20 | 25.2% | 16 | .763 | .747 | .731 | .713 | .695 | .675 | .654 | .631 | .608 | .584 |
| | | | | | | | | .558 | .531 | .504 | .475 | .446 | .416 | | | | |
| 13.000 | 10.116 | .000 | 2.884 | .61 | 5.21 | 22.2% | 17 | .763 | .747 | .731 | .713 | .695 | .675 | .654 | .631 | .608 | .584 |
| | | | | | | | | .558 | .531 | .504 | .475 | .446 | .416 | .386 | | | |
| 13.000 | 10.472 | .000 | 2.528 | .64 | 5.22 | 19.4% | 18 | .763 | .747 | .731 | .713 | .695 | .675 | .654 | .631 | .608 | .584 |
| | | | | | | | | .558 | .531 | .504 | .475 | .446 | .416 | .386 | .356 | | |
| 13.000 | 10.798 | .000 | 2.202 | .67 | 5.21 | 16.9% | 19 | .763 | .747 | .731 | .713 | .695 | .675 | .654 | .631 | .608 | .584 |
| | | | | | | | | .558 | .531 | .504 | .475 | .446 | .416 | .386 | .356 | .326 | |
| 13.000 | 11.096 | .000 | 1.904 | .70 | 5.19 | 14.6% | 20 | .763 | .747 | .731 | .713 | .695 | .675 | .654 | .631 | .608 | .584 |
| | | | | | | | | .558 | .531 | .504 | .475 | .446 | .416 | .386 | .356 | .326 | .297 |
| 13.000 | 11.364 | .000 | 1.636 | .73 | 5.17 | 12.6% | 21 | .763 | .747 | .731 | .713 | .695 | .675 | .654 | .631 | .608 | .584 |
| | | | | | | | | .558 | .531 | .504 | .475 | .446 | .416 | .386 | .356 | .326 | .297 |
| | | | | | | | | .268 | | | | | | | | | |
| 13.000 | 11.604 | .000 | 1.396 | .76 | 5.13 | 10.7% | 22 | .763 | .747 | .731 | .713 | .695 | .675 | .654 | .631 | .608 | .584 |
| | | | | | | | | .558 | .531 | .504 | .475 | .446 | .416 | .386 | .356 | .326 | .297 |
| | | | | | | | | .268 | .241 | | | | | | | | |
| 13.000 | 11.818 | .000 | 1.182 | .79 | 5.08 | 9.1% | 23 | .763 | .747 | .731 | .713 | .695 | .675 | .654 | .631 | .608 | .584 |
| | | | | | | | | .558 | .531 | .504 | .475 | .446 | .416 | .386 | .356 | .020 | .297 |
| | | | | | | | | .268 | .241 | .214 | | | | | | | |
| 13.000 | 12.007 | .000 | .993 | .81 | 5.02 | 7.6% | 24 | .763 | .747 | .731 | .713 | .695 | .675 | .654 | .631 | .608 | .584 |
| | | | | | | | | .558 | .531 | .504 | .475 | .446 | .416 | .386 | .356 | .326 | .297 |
| | | | | | | | | .268 | .241 | .214 | .189 | | | | | | |
| 13.000 | 12.172 | .000 | .828 | .83 | 4.96 | 6.4% | 25 | .763 | .747 | .731 | .713 | .695 | .675 | .654 | .631 | .608 | .584 |
| | | | | | | | | .558 | .531 | .504 | .475 | .446 | .416 | .386 | .050 | .326 | .297 |
| | | | | | | | | .268 | .241 | .214 | .189 | .165 | | | | | |
| 13.000 | 12.315 | .000 | .685 | .86 | 4.89 | 5.3% | 26 | .763 | .747 | .731 | .713 | .695 | .675 | .654 | .631 | .608 | .584 |

*SIMILAR TO A.T.&T. NEAL-WILKINSON TABLES(SEE SECTION 1)

*EQUIVALENT RANDOM THEORY: VMR(VARIANCE TO MEAN RATIO)– 4.0 - NO RETRIES (PEAKED TRAFFIC - ROUTE ADVANCEMENT OR DAY-TO-DAY VARIATION)

| HOURS FIRST ATTEMPT VMR=4.0 | HOURS CON-NECT | HOURS RECALL /RETRY | HOURS OVFL | CON-NECT VMR | OVFL VMR | PERCENT BLOCKED (GRADE) | LINE | HOURS CONNECTED TRAFFIC ON LINES 1 | 2 | 3 | 4 | 5 | 6 | 7 | 8 | 9 | 10 |
|---|---|---|---|---|---|---|---|---|---|---|---|---|---|---|---|---|---|
| | | | | | | | | .558 | .531 | .504 | .475 | .446 | .416 | .386 | .356 | .326 | .297 |
| | | | | | | | | .268 | .241 | .214 | .189 | .165 | .143 | | | | |
| 13.000 | 12.438 | .000 | .562 | .88 | 4.81 | 4.3% | 27 | .763 | .747 | .731 | .713 | .695 | .675 | .654 | .631 | .608 | .584 |
| | | | | | | | | .558 | .531 | .504 | .475 | .446 | .416 | .386 | .356 | .326 | .297 |
| | | | | | | | | .268 | .241 | .214 | .189 | .165 | .143 | .123 | | | |
| 13.000 | 12.543 | .000 | .457 | .90 | 4.73 | 3.5% | 28 | .763 | .747 | .731 | .713 | .695 | .675 | .654 | .631 | .608 | .584 |
| | | | | | | | | .558 | .531 | .504 | .475 | .446 | .416 | .386 | .356 | .326 | .297 |
| | | | | | | | | .268 | .241 | .214 | .189 | .165 | .143 | .123 | .105 | | |
| 13.000 | 12.631 | .000 | .369 | .91 | 4.64 | 2.8% | 29 | .763 | .747 | .731 | .713 | .695 | .675 | .654 | .631 | .608 | .584 |
| | | | | | | | | .558 | .531 | .504 | .475 | .446 | .416 | .386 | .356 | .326 | .297 |
| | | | | | | | | .268 | .241 | .214 | .189 | .165 | .143 | .123 | .105 | .088 | |
| 13.000 | 12.704 | .000 | .296 | .93 | 4.55 | 2.3% | 30 | .763 | .747 | .731 | .713 | .695 | .675 | .654 | .631 | .608 | .584 |
| | | | | | | | | .558 | .531 | .504 | .475 | .446 | .416 | .386 | .356 | .326 | .297 |
| | | | | | | | | .268 | .241 | .214 | .189 | .165 | .143 | .123 | .105 | .088 | .074 |
| 13.000 | 12.765 | .000 | .235 | .94 | 4.46 | 1.8% | 31 | .763 | .747 | .731 | .713 | .695 | .675 | .654 | .631 | .608 | .584 |
| | | | | | | | | .558 | .531 | .504 | .475 | .446 | .416 | .386 | .356 | .326 | .297 |
| | | | | | | | | .268 | .241 | .214 | .189 | .165 | .143 | .123 | .105 | .088 | .074 |
| | | | | | | | | .061 | | | | | | | | | |
| 13.000 | 12.815 | .000 | .185 | .95 | 4.37 | 1.4% | 32 | .763 | .747 | .731 | .713 | .695 | .675 | .654 | .631 | .608 | .584 |
| | | | | | | | | .558 | .531 | .504 | .475 | .446 | .416 | .386 | .356 | .326 | .297 |
| | | | | | | | | .268 | .241 | .214 | .189 | .165 | .143 | .123 | .105 | .088 | .074 |
| | | | | | | | | .061 | .050 | | | | | | | | |
| 13.000 | 12.856 | .000 | .144 | .96 | 4.28 | 1.1% | 33 | .763 | .747 | .731 | .713 | .695 | .675 | .654 | .631 | .608 | .584 |
| | | | | | | | | .558 | .531 | .504 | .475 | .446 | .416 | .386 | .356 | .326 | .297 |
| | | | | | | | | .268 | .241 | .214 | .189 | .165 | .143 | .123 | .105 | .088 | .074 |
| | | | | | | | | .061 | .050 | .041 | | | | | | | |
| 13.000 | 12.889 | .000 | .111 | .97 | 4.18 | .9% | 34 | .763 | .747 | .731 | .713 | .695 | .675 | .654 | .631 | .608 | .584 |
| | | | | | | | | .558 | .531 | .504 | .475 | .446 | .416 | .386 | .356 | .326 | .297 |
| | | | | | | | | .268 | .241 | .214 | .189 | .165 | .143 | .123 | .105 | .088 | .074 |
| | | | | | | | | .061 | .050 | .041 | .033 | | | | | | |
| 13.000 | 12.914 | .000 | .086 | .97 | 4.09 | .7% | 35 | .763 | .747 | .731 | .713 | .695 | .675 | .654 | .631 | .608 | .584 |
| | | | | | | | | .558 | .531 | .504 | .475 | .446 | .416 | .386 | .356 | .326 | .297 |
| | | | | | | | | .268 | .241 | .214 | .189 | .165 | .143 | .123 | .105 | .088 | .074 |
| | | | | | | | | .061 | .050 | .041 | .033 | .026 | | | | | |
| 13.000 | 12.935 | .000 | .065 | .98 | 4.00 | .5% | 36 | .763 | .747 | .731 | .713 | .695 | .675 | .654 | .631 | .608 | .584 |
| | | | | | | | | .558 | .531 | .504 | .475 | .446 | .416 | .386 | .356 | .326 | .297 |
| | | | | | | | | .268 | .241 | .214 | .189 | .165 | .143 | .123 | .105 | .088 | .074 |
| | | | | | | | | .061 | .050 | .041 | .033 | .026 | .020 | | | | |
| 14.000 | 9.189 | .000 | 4.811 | .49 | 5.19 | 34.4% | 14 | .776 | .761 | .746 | .730 | .713 | .695 | .676 | .655 | .633 | .611 |
| | | | | | | | | .587 | .562 | .536 | .509 | | | | | | |

*SIMILAR TO A.T.&T. NEAL-WILKINSON TABLES(SEE SECTION 1)

*EQUIVALENT RANDOM THEORY: VMR(VARIANCE TO MEAN RATIO)– 4.0 - NO RETRIES (PEAKED TRAFFIC - ROUTE ADVANCEMENT OR DAY-TO-DAY VARIATION)

| HOURS FIRST ATTEMPT VMR = 4.0 | HOURS CON-NECT | HOURS RECALL /RETRY | HOURS OVFL | CON-NECT VMR | OVFL VMR | PERCENT BLOCKED (GRADE) | LINE | HOURS CONNECTED TRAFFIC ON LINES 1 | 2 | 3 | 4 | 5 | 6 | 7 | 8 | 9 | 10 |
|---|---|---|---|---|---|---|---|---|---|---|---|---|---|---|---|---|---|
| 14.000 | 9.671 | .000 | 4.329 | .52 | 5.23 | 30.9% | 15 | .776 | .761 | .746 | .730 | .713 | .695 | .676 | .655 | .633 | .611 |
| | | | | | | | | .587 | .562 | .536 | .509 | .481 | | | | | |
| 14.000 | 10.124 | .000 | 3.876 | .55 | 5.27 | 27.7% | 16 | .776 | .761 | .746 | .730 | .713 | .695 | .676 | .655 | .633 | .611 |
| | | | | | | | | .587 | .562 | .536 | .509 | .481 | .453 | | | | |
| 14.000 | 10.547 | .000 | 3.453 | .58 | 5.29 | 24.7% | 17 | .776 | .761 | .746 | .730 | .713 | .695 | .676 | .655 | .633 | .611 |
| | | | | | | | | .587 | .562 | .536 | .509 | .481 | .453 | .424 | | | |
| 14.000 | 10.942 | .000 | 3.058 | .61 | 5.31 | 21.8% | 18 | .776 | .761 | .746 | .730 | .713 | .695 | .676 | .655 | .633 | .611 |
| | | | | | | | | .587 | .562 | .536 | .509 | .481 | .453 | .424 | .395 | | |
| 14.000 | 11.307 | .000 | 2.693 | .63 | 5.32 | 19.2% | 19 | .776 | .761 | .746 | .730 | .713 | .695 | .676 | .655 | .633 | .611 |
| | | | | | | | | .587 | .562 | .536 | .509 | .481 | .453 | .424 | .395 | .365 | |
| 14.000 | 11.643 | .000 | 2.357 | .66 | 5.32 | 16.8% | 20 | .776 | .761 | .746 | .730 | .713 | .695 | .676 | .655 | .633 | .611 |
| | | | | | | | | .587 | .562 | .536 | .509 | .481 | .453 | .424 | .395 | .365 | .336 |
| 14.000 | 11.951 | .000 | 2.049 | .69 | 5.30 | 14.6% | 21 | .776 | .761 | .746 | .730 | .713 | .695 | .676 | .655 | .633 | .611 |
| | | | | | | | | .587 | .562 | .536 | .509 | .481 | .453 | .424 | .395 | .365 | .336 |
| | | | | | | | | .307 | | | | | | | | | |
| 14.000 | 12.229 | .000 | 1.771 | .72 | 5.28 | 12.6% | 22 | .776 | .761 | .746 | .730 | .713 | .695 | .676 | .655 | .633 | .611 |
| | | | | | | | | .587 | .562 | .536 | .509 | .481 | .453 | .424 | .395 | .365 | .336 |
| | | | | | | | | .307 | .279 | | | | | | | | |
| 14.000 | 12.480 | .000 | 1.520 | .75 | 5.25 | 10.9% | 23 | .776 | .761 | .746 | .730 | .713 | .695 | .676 | .655 | .633 | .611 |
| | | | | | | | | .587 | .562 | .536 | .509 | .481 | .453 | .424 | .395 | .365 | .336 |
| | | | | | | | | .307 | .279 | .251 | | | | | | | |
| 14.000 | 12.705 | .000 | 1.295 | .78 | 5.20 | 9.2% | 24 | .776 | .761 | .746 | .730 | .713 | .695 | .676 | .655 | .633 | .611 |
| | | | | | | | | .587 | .562 | .536 | .509 | .481 | .453 | .424 | .395 | .365 | .336 |
| | | | | | | | | .307 | .279 | .251 | .225 | | | | | | |
| 14.000 | 12.904 | .000 | 1.096 | .80 | 5.15 | 7.8% | 25 | .776 | .761 | .746 | .730 | .713 | .695 | .676 | .655 | .633 | .611 |
| | | | | | | | | .587 | .562 | .536 | .509 | .481 | .453 | .424 | .395 | .365 | .336 |
| | | | | | | | | .307 | .279 | .251 | .225 | .199 | | | | | |
| 14.000 | 13.080 | .000 | .920 | .82 | 5.09 | 6.6% | 26 | .776 | .761 | .746 | .730 | .713 | .695 | .676 | .655 | .633 | .611 |
| | | | | | | | | .587 | .562 | .536 | .509 | .481 | .453 | .424 | .395 | .365 | .336 |
| | | | | | | | | .307 | .279 | .251 | .225 | .199 | .175 | | | | |
| 14.000 | 13.233 | .000 | .767 | .85 | 5.02 | 5.5% | 27 | .776 | .761 | .746 | .730 | .713 | .695 | .676 | .655 | .633 | .611 |
| | | | | | | | | .587 | .562 | .536 | .509 | .481 | .453 | .424 | .395 | .365 | .336 |
| | | | | | | | | .307 | .279 | .251 | .225 | .199 | .175 | .153 | | | |
| 14.000 | 13.366 | .000 | .634 | .87 | 4.94 | 4.5% | 28 | .776 | .761 | .746 | .730 | .713 | .695 | .676 | .655 | .633 | .611 |
| | | | | | | | | .587 | .562 | .536 | .509 | .481 | .453 | .424 | .395 | .365 | .336 |
| | | | | | | | | .307 | .279 | .251 | .225 | .199 | .175 | .153 | .133 | | |
| 14.000 | 13.479 | .000 | .521 | .89 | 4.86 | 3.7% | 29 | .776 | .761 | .746 | .730 | .713 | .695 | .676 | .655 | .633 | .611 |
| | | | | | | | | .587 | .562 | .536 | .509 | .481 | .453 | .424 | .395 | .365 | .336 |
| | | | | | | | | .307 | .279 | .251 | .225 | .199 | .175 | .153 | .133 | .114 | |
| 14.000 | 13.576 | .000 | .424 | .90 | 4.78 | 3.0% | 30 | .776 | .761 | .746 | .730 | .713 | .695 | .676 | .655 | .633 | .611 |

*SIMILAR TO A.T.&T. NEAL-WILKINSON TABLES(SEE SECTION 1)

*EQUIVALENT RANDOM THEORY: VMR(VARIANCE TO MEAN RATIO)– 4.0 - NO RETRIES (PEAKED TRAFFIC - ROUTE ADVANCEMENT OR DAY-TO-DAY VARIATION)

| HOURS FIRST ATTEMPT VMR = 4.0 | HOURS CON-NECT | HOURS RECALL /RETRY | HOURS OVFL | CON-NECT VMR | OVFL VMR | PERCENT BLOCKED (GRADE) | LINE | HOURS CONNECTED TRAFFIC ON LINES 1 | 2 | 3 | 4 | 5 | 6 | 7 | 8 | 9 | 10 |
|---|---|---|---|---|---|---|---|---|---|---|---|---|---|---|---|---|---|
| | | | | | | | | .587 | .562 | .536 | .509 | .481 | .453 | .424 | .395 | .365 | .336 |
| | | | | | | | | .307 | .279 | .251 | .225 | .199 | .175 | .153 | .133 | .114 | .097 |
| 14.000 | 13.657 | .000 | .343 | .92 | 4.69 | 2.4% | 31 | .776 | .761 | .746 | .730 | .713 | .695 | .676 | .655 | .633 | .611 |
| | | | | | | | | .587 | .562 | .536 | .509 | .481 | .453 | .424 | .395 | .365 | .336 |
| | | | | | | | | .307 | .279 | .251 | .225 | .199 | .175 | .153 | .133 | .114 | .097 |
| | | | | | | | | .081 | | | | | | | | | |
| 14.000 | 13.725 | .000 | .275 | .93 | 4.60 | 2.0% | 32 | .776 | .761 | .746 | .730 | .713 | .695 | .676 | .655 | .633 | .611 |
| | | | | | | | | .587 | .562 | .536 | .509 | .481 | .453 | .424 | .395 | .365 | .336 |
| | | | | | | | | .307 | .279 | .251 | .225 | .199 | .175 | .153 | .133 | .114 | .097 |
| | | | | | | | | .081 | .068 | | | | | | | | |
| 14.000 | 13.782 | .000 | .218 | .94 | 4.51 | 1.6% | 33 | .776 | .761 | .746 | .730 | .713 | .695 | .676 | .655 | .633 | .611 |
| | | | | | | | | .587 | .562 | .536 | .509 | .481 | .453 | .424 | .395 | .365 | .336 |
| | | | | | | | | .307 | .279 | .251 | .225 | .199 | .175 | .153 | .133 | .114 | .097 |
| | | | | | | | | .081 | .068 | .056 | | | | | | | |
| 14.000 | 13.828 | .000 | .172 | .95 | 4.41 | 1.2% | 34 | .776 | .761 | .746 | .730 | .713 | .695 | .676 | .655 | .633 | .611 |
| | | | | | | | | .587 | .562 | .536 | .509 | .481 | .453 | .424 | .395 | .365 | .336 |
| | | | | | | | | .307 | .279 | .251 | .225 | .199 | .175 | .153 | .133 | .114 | .097 |
| | | | | | | | | .081 | .068 | .056 | .046 | | | | | | |
| 14.000 | 13.865 | .000 | .135 | .96 | 4.32 | 1.0% | 35 | .776 | .761 | .746 | .730 | .713 | .695 | .676 | .655 | .633 | .611 |
| | | | | | | | | .587 | .562 | .536 | .509 | .481 | .453 | .424 | .395 | .365 | .336 |
| | | | | | | | | .307 | .279 | .251 | .225 | .199 | .175 | .153 | .133 | .114 | .097 |
| | | | | | | | | .081 | .068 | .056 | .046 | .038 | | | | | |
| 14.000 | 13.896 | .000 | .104 | .97 | 4.23 | .7% | 36 | .776 | .761 | .746 | .730 | .713 | .695 | .676 | .655 | .633 | .611 |
| | | | | | | | | .587 | .562 | .536 | .509 | .481 | .453 | .424 | .395 | .365 | .336 |
| | | | | | | | | .307 | .279 | .251 | .225 | .199 | .175 | .153 | .133 | .114 | .097 |
| | | | | | | | | .081 | .068 | .056 | .046 | .038 | .030 | | | | |
| 14.000 | 13.920 | .000 | .080 | .98 | 4.14 | .6% | 37 | .776 | .761 | .746 | .730 | .713 | .695 | .676 | .655 | .633 | .611 |
| | | | | | | | | .587 | .562 | .536 | .509 | .481 | .453 | .424 | .395 | .365 | .336 |
| | | | | | | | | .307 | .279 | .251 | .225 | .199 | .175 | .153 | .133 | .114 | .097 |
| | | | | | | | | .081 | .068 | .056 | .046 | .038 | .030 | .024 | | | |
| 15.000 | 9.994 | .000 | 5.006 | .49 | 5.28 | 33.4% | 15 | .787 | .774 | .760 | .746 | .730 | .713 | .695 | .676 | .656 | .635 |
| | | | | | | | | .613 | .590 | .565 | .540 | .514 | | | | | |
| 15.000 | 10.481 | .000 | 4.519 | .51 | 5.32 | 30.1% | 16 | .787 | .774 | .760 | .746 | .730 | .713 | .695 | .676 | .656 | .635 |
| | | | | | | | | .613 | .590 | .565 | .540 | .514 | .487 | | | | |
| 15.000 | 10.940 | .000 | 4.060 | .54 | 5.36 | 27.1% | 17 | .787 | .774 | .760 | .746 | .730 | .713 | .695 | .676 | .656 | .635 |
| | | | | | | | | .613 | .590 | .565 | .540 | .514 | .487 | .459 | | | |
| 15.000 | 11.371 | .000 | 3.629 | .57 | 5.39 | 24.2% | 18 | .787 | .774 | .760 | .746 | .730 | .713 | .695 | .676 | .656 | .635 |
| | | | | | | | | .613 | .590 | .565 | .540 | .514 | .487 | .459 | .431 | | |
| 15.000 | 11.773 | .000 | 3.227 | .60 | 5.41 | 21.5% | 19 | .787 | .774 | .760 | .746 | .730 | .713 | .695 | .676 | .656 | .635 |
| | | | | | | | | .613 | .590 | .565 | .540 | .514 | .487 | .459 | .431 | .402 | |

*SIMILAR TO A.T.&T. NEAL-WILKINSON TABLES(SEE SECTION 1)

*EQUIVALENT RANDOM THEORY: VMR(VARIANCE TO MEAN RATIO)– 4.0 - NO RETRIES (PEAKED TRAFFIC - ROUTE ADVANCEMENT OR DAY-TO-DAY VARIATION)

| HOURS FIRST ATTEMPT VMR=4.0 | HOURS CON-NECT | HOURS RECALL /RETRY | HOURS OVFL | CON-NECT VMR | OVFL VMR | PERCENT BLOCKED (GRADE) | LINE | HOURS CONNECTED TRAFFIC ON LINES 1 | 2 | 3 | 4 | 5 | 6 | 7 | 8 | 9 | 10 |
|---|---|---|---|---|---|---|---|---|---|---|---|---|---|---|---|---|---|
| 15.000 | 12.147 | .000 | 2.853 | .63 | 5.42 | 19.0% | 20 | .787 | .774 | .760 | .746 | .730 | .713 | .695 | .676 | .656 | .635 |
| | | | | | | | | .613 | .590 | .565 | .540 | .514 | .487 | .459 | .431 | .402 | .374 |
| 15.000 | 12.491 | .000 | 2.509 | .66 | 5.42 | 16.7% | 21 | .787 | .774 | .760 | .746 | .730 | .713 | .695 | .676 | .656 | .635 |
| | | | | | | | | .613 | .590 | .565 | .540 | .514 | .487 | .459 | .431 | .402 | .374 |
| | | | | | | | | .345 | | | | | | | | | |
| 15.000 | 12.808 | .000 | 2.192 | .68 | 5.41 | 14.6% | 22 | .787 | .774 | .760 | .746 | .730 | .713 | .695 | .676 | .656 | .635 |
| | | | | | | | | .613 | .590 | .565 | .540 | .514 | .487 | .459 | .431 | .402 | .374 |
| | | | | | | | | .345 | .316 | | | | | | | | |
| 15.000 | 13.096 | .000 | 1.904 | .71 | 5.39 | 12.7% | 23 | .787 | .774 | .760 | .746 | .730 | .713 | .695 | .676 | .656 | .635 |
| | | | | | | | | .613 | .590 | .565 | .540 | .514 | .487 | .459 | .431 | .402 | .374 |
| | | | | | | | | .345 | .316 | .288 | | | | | | | |
| 15.000 | 13.357 | .000 | 1.643 | .74 | 5.36 | 11.0% | 24 | .787 | .774 | .760 | .746 | .730 | .713 | .695 | .676 | .656 | .635 |
| | | | | | | | | .613 | .590 | .565 | .540 | .514 | .487 | .459 | .431 | .402 | .374 |
| | | | | | | | | .345 | .316 | .288 | .261 | | | | | | |
| 15.000 | 13.592 | .000 | 1.408 | .77 | 5.32 | 9.4% | 25 | .787 | .774 | .760 | .746 | .730 | .713 | .695 | .676 | .656 | .635 |
| | | | | | | | | .613 | .590 | .565 | .540 | .514 | .487 | .459 | .431 | .402 | .374 |
| | | | | | | | | .345 | .316 | .288 | .261 | .235 | | | | | |
| 15.000 | 13.801 | .000 | 1.199 | .79 | 5.27 | 8.0% | 26 | .787 | .774 | .760 | .746 | .730 | .713 | .695 | .676 | .656 | .635 |
| | | | | | | | | .613 | .590 | .565 | .540 | .514 | .487 | .459 | .431 | .402 | .374 |
| | | | | | | | | .345 | .316 | .288 | .261 | .235 | .209 | | | | |
| 15.000 | 13.986 | .000 | 1.014 | .81 | 5.21 | 6.8% | 27 | .787 | .774 | .760 | .746 | .730 | .713 | .695 | .676 | .656 | .635 |
| | | | | | | | | .613 | .590 | .565 | .540 | .514 | .487 | .459 | .431 | .402 | .374 |
| | | | | | | | | .345 | .316 | .288 | .261 | .235 | .209 | .185 | | | |
| 15.000 | 14.149 | .000 | .851 | .84 | 5.15 | 5.7% | 28 | .787 | .774 | .760 | .746 | .730 | .713 | .695 | .676 | .656 | .635 |
| | | | | | | | | .613 | .590 | .565 | .540 | .514 | .487 | .459 | .431 | .402 | .374 |
| | | | | | | | | .345 | .316 | .288 | .261 | .235 | .209 | .185 | .163 | | |
| 15.000 | 14.291 | .000 | .709 | .86 | 5.07 | 4.7% | 29 | .787 | .774 | .760 | .746 | .730 | .713 | .695 | .676 | .656 | .635 |
| | | | | | | | | .613 | .590 | .565 | .540 | .514 | .487 | .459 | .431 | .402 | .374 |
| | | | | | | | | .345 | .316 | .288 | .261 | .235 | .209 | .185 | .163 | .142 | |
| 15.000 | 14.414 | .000 | .586 | .88 | 5.00 | 3.9% | 30 | .787 | .774 | .760 | .746 | .730 | .713 | .695 | .676 | .656 | .635 |
| | | | | | | | | .613 | .590 | .565 | .540 | .514 | .487 | .459 | .431 | .402 | .374 |
| | | | | | | | | .345 | .316 | .288 | .261 | .235 | .209 | .185 | .163 | .142 | .123 |
| 15.000 | 14.519 | .000 | .481 | .89 | 4.91 | 3.2% | 31 | .787 | .774 | .760 | .746 | .730 | .713 | .695 | .676 | .656 | .635 |
| | | | | | | | | .613 | .590 | .565 | .540 | .514 | .487 | .459 | .431 | .402 | .374 |
| | | | | | | | | .345 | .316 | .288 | .261 | .235 | .209 | .185 | .163 | .142 | .123 |
| | | | | | | | | .105 | | | | | | | | | |
| 15.000 | 14.608 | .000 | .392 | .91 | 4.82 | 2.6% | 32 | .787 | .774 | .760 | .746 | .730 | .713 | .695 | .676 | .656 | .635 |
| | | | | | | | | .613 | .590 | .565 | .540 | .514 | .487 | .459 | .431 | .402 | .374 |
| | | | | | | | | .345 | .316 | .288 | .261 | .235 | .209 | .185 | .163 | .142 | .123 |
| | | | | | | | | .105 | .089 | | | | | | | | |

*SIMILAR TO A.T.&T. NEAL-WILKINSON TABLES(SEE SECTION 1)

*EQUIVALENT RANDOM THEORY: VMR(VARIANCE TO MEAN RATIO)– 4.0 - NO RETRIES (PEAKED TRAFFIC - ROUTE ADVANCEMENT OR DAY-TO-DAY VARIATION)

| HOURS FIRST ATTEMPT VMR=4.0 | HOURS CON- NECT | HOURS RECALL /RETRY | HOURS OVFL | CON- NECT VMR | OVFL VMR | PERCENT BLOCKED (GRADE) | LINE | HOURS CONNECTED TRAFFIC ON LINES 1 | 2 | 3 | 4 | 5 | 6 | 7 | 8 | 9 | 10 |
|---|---|---|---|---|---|---|---|---|---|---|---|---|---|---|---|---|---|
| 15.000 | 14.683 | .000 | .317 | .92 | 4.73 | 2.1% | 33 | .787 | .774 | .760 | .746 | .730 | .713 | .695 | .676 | .656 | .635 |
| | | | | | | | | .613 | .590 | .565 | .540 | .514 | .487 | .459 | .431 | .402 | .374 |
| | | | | | | | | .345 | .316 | .288 | .261 | .235 | .209 | .185 | .163 | .142 | .123 |
| | | | | | | | | .105 | .089 | .075 | | | | | | | |
| 15.000 | 14.746 | .000 | .254 | .94 | 4.64 | 1.7% | 34 | .787 | .774 | .760 | .746 | .730 | .713 | .695 | .676 | .656 | .635 |
| | | | | | | | | .613 | .590 | .565 | .540 | .514 | .487 | .459 | .431 | .402 | .374 |
| | | | | | | | | .345 | .316 | .288 | .261 | .235 | .209 | .185 | .163 | .142 | .123 |
| | | | | | | | | .105 | .089 | .075 | .063 | | | | | | |
| 15.000 | 14.798 | .000 | .202 | .95 | 4.55 | 1.3% | 35 | .787 | .774 | .760 | .746 | .730 | .713 | .695 | .676 | .656 | .635 |
| | | | | | | | | .613 | .590 | .565 | .540 | .514 | .487 | .459 | .431 | .402 | .374 |
| | | | | | | | | .345 | .316 | .288 | .261 | .235 | .209 | .185 | .163 | .142 | .123 |
| | | | | | | | | .105 | .089 | .075 | .063 | .052 | | | | | |
| 15.000 | 14.840 | .000 | .160 | .96 | 4.46 | 1.1% | 36 | .787 | .774 | .760 | .746 | .730 | .713 | .695 | .676 | .656 | .635 |
| | | | | | | | | .613 | .590 | .565 | .540 | .514 | .487 | .459 | .431 | .402 | .374 |
| | | | | | | | | .345 | .316 | .288 | .261 | .235 | .209 | .185 | .163 | .142 | .123 |
| | | | | | | | | .105 | .089 | .075 | .063 | .052 | .043 | | | | |
| 15.000 | 14.875 | .000 | .125 | .97 | 4.36 | .8% | 37 | .787 | .774 | .760 | .746 | .730 | .713 | .695 | .676 | .656 | .635 |
| | | | | | | | | .613 | .590 | .565 | .540 | .514 | .487 | .459 | .431 | .402 | .374 |
| | | | | | | | | .345 | .316 | .288 | .261 | .235 | .209 | .185 | .163 | .142 | .123 |
| | | | | | | | | .105 | .089 | .075 | .063 | .052 | .043 | .035 | | | |
| 15.000 | 14.903 | .000 | .097 | .97 | 4.27 | .6% | 38 | .787 | .774 | .760 | .746 | .730 | .713 | .695 | .676 | .656 | .635 |
| | | | | | | | | .613 | .590 | .565 | .540 | .514 | .487 | .459 | .431 | .402 | .374 |
| | | | | | | | | .345 | .316 | .288 | .261 | .235 | .209 | .185 | .163 | .142 | .123 |
| | | | | | | | | .105 | .089 | .075 | .063 | .052 | .043 | .035 | .028 | | |
| 16.000 | 10.805 | .000 | 5.195 | .48 | 5.37 | 32.5% | 16 | .798 | .786 | .773 | .759 | .745 | .729 | .713 | .695 | .677 | .657 |
| | | | | | | | | .636 | .615 | .592 | .568 | .544 | .518 | | | | |
| 16.000 | 11.297 | .000 | 4.703 | .51 | 5.42 | 29.4% | 17 | .798 | .786 | .773 | .759 | .745 | .729 | .713 | .695 | .677 | .657 |
| | | | | | | | | .636 | .615 | .592 | .568 | .544 | .518 | .492 | | | |
| 16.000 | 11.762 | .000 | 4.238 | .54 | 5.46 | 26.5% | 18 | .798 | .786 | .773 | .759 | .745 | .729 | .713 | .695 | .677 | .657 |
| | | | | | | | | .636 | .615 | .592 | .568 | .544 | .518 | .492 | .465 | | |
| 16.000 | 12.199 | .000 | 3.801 | .56 | 5.49 | 23.8% | 19 | .798 | .786 | .773 | .759 | .745 | .729 | .713 | .695 | .677 | .657 |
| | | | | | | | | .636 | .615 | .592 | .568 | .544 | .518 | .492 | .465 | .437 | |
| 16.000 | 12.608 | .000 | 3.392 | .59 | 5.51 | 21.2% | 20 | .798 | .786 | .773 | .759 | .745 | .729 | .713 | .695 | .677 | .657 |
| | | | | | | | | .636 | .615 | .592 | .568 | .544 | .518 | .492 | .465 | .437 | .409 |
| 16.000 | 12.989 | .000 | 3.011 | .62 | 5.52 | 18.8% | 21 | .798 | .786 | .773 | .759 | .745 | .729 | .713 | .695 | .677 | .657 |
| | | | | | | | | .636 | .615 | .592 | .568 | .544 | .518 | .492 | .465 | .437 | .409 |
| | | | | | | | | .381 | | | | | | | | | |
| 16.000 | 13.342 | .000 | 2.658 | .65 | 5.52 | 16.6% | 22 | .798 | .786 | .773 | .759 | .745 | .729 | .713 | .695 | .677 | .657 |
| | | | | | | | | .636 | .615 | .592 | .568 | .544 | .518 | .492 | .465 | .437 | .409 |
| | | | | | | | | .381 | .353 | | | | | | | | |

*SIMILAR TO A.T.&T. NEAL-WILKINSON TABLES(SEE SECTION 1)

*EQUIVALENT RANDOM THEORY: VMR(VARIANCE TO MEAN RATIO)– 4.0 - NO RETRIES (PEAKED TRAFFIC - ROUTE ADVANCEMENT OR DAY-TO-DAY VARIATION)

| HOURS FIRST ATTEMPT VMR = 4.0 | HOURS CON-NECT | HOURS RECALL /RETRY | HOURS OVFL | CON-NECT VMR | OVFL VMR | PERCENT BLOCKED (GRADE) | LINE | HOURS CONNECTED TRAFFIC ON LINES 1 | 2 | 3 | 4 | 5 | 6 | 7 | 8 | 9 | 10 |
|---|---|---|---|---|---|---|---|---|---|---|---|---|---|---|---|---|---|
| 16.000 | 13.667 | .000 | 2.333 | .68 | 5.51 | 14.6% | 23 | .798 | .786 | .773 | .759 | .745 | .729 | .713 | .695 | .677 | .657 |
| | | | | | | | | .636 | .615 | .592 | .568 | .544 | .518 | .492 | .465 | .437 | .409 |
| | | | | | | | | .381 | .353 | .325 | | | | | | | |
| 16.000 | 13.964 | .000 | 2.036 | .70 | 5.50 | 12.7% | 24 | .798 | .786 | .773 | .759 | .745 | .729 | .713 | .695 | .677 | .657 |
| | | | | | | | | .636 | .615 | .592 | .568 | .544 | .518 | .492 | .465 | .437 | .409 |
| | | | | | | | | .381 | .353 | .325 | .297 | | | | | | |
| 16.000 | 14.235 | .000 | 1.765 | .73 | 5.47 | 11.0% | 25 | .798 | .786 | .773 | .759 | .745 | .729 | .713 | .695 | .677 | .657 |
| | | | | | | | | .636 | .615 | .592 | .568 | .544 | .518 | .492 | .465 | .437 | .409 |
| | | | | | | | | .381 | .353 | .325 | .297 | .270 | | | | | |
| 16.000 | 14.479 | .000 | 1.521 | .76 | 5.43 | 9.5% | 26 | .798 | .786 | .773 | .759 | .745 | .729 | .713 | .695 | .677 | .657 |
| | | | | | | | | .636 | .615 | .592 | .568 | .544 | .518 | .492 | .465 | .437 | .409 |
| | | | | | | | | .381 | .353 | .325 | .297 | .270 | .244 | | | | |
| 16.000 | 14.697 | .000 | 1.303 | .78 | 5.39 | 8.1% | 27 | .798 | .786 | .773 | .759 | .745 | .729 | .713 | .695 | .677 | .657 |
| | | | | | | | | .636 | .615 | .592 | .568 | .544 | .518 | .492 | .465 | .437 | .409 |
| | | | | | | | | .381 | .353 | .325 | .297 | .270 | .244 | .219 | | | |
| 16.000 | 14.892 | .000 | 1.108 | .81 | 5.33 | 6.9% | 28 | .798 | .786 | .773 | .759 | .745 | .729 | .713 | .695 | .677 | .657 |
| | | | | | | | | .636 | .615 | .592 | .568 | .544 | .518 | .492 | .465 | .437 | .409 |
| | | | | | | | | .381 | .353 | .325 | .297 | .270 | .244 | .219 | .195 | | |
| 16.000 | 15.064 | .000 | .936 | .83 | 5.27 | 5.8% | 29 | .798 | .786 | .773 | .759 | .745 | .729 | .713 | .695 | .677 | .657 |
| | | | | | | | | .636 | .615 | .592 | .568 | .544 | .518 | .492 | .465 | .437 | .409 |
| | | | | | | | | .381 | .353 | .325 | .297 | .270 | .244 | .219 | .195 | .172 | |
| 16.000 | 15.215 | .000 | .785 | .85 | 5.20 | 4.9% | 30 | .798 | .786 | .773 | .759 | .745 | .729 | .713 | .695 | .677 | .657 |
| | | | | | | | | .636 | .615 | .592 | .568 | .544 | .518 | .492 | .465 | .437 | .409 |
| | | | | | | | | .381 | .353 | .325 | .297 | .270 | .244 | .219 | .195 | .172 | .151 |
| 16.000 | 15.346 | .000 | .654 | .87 | 5.12 | 4.1% | 31 | .798 | .786 | .773 | .759 | .745 | .729 | .713 | .695 | .677 | .657 |
| | | | | | | | | .636 | .615 | .592 | .568 | .544 | .518 | .492 | .465 | .437 | .409 |
| | | | | | | | | .381 | .353 | .325 | .297 | .270 | .244 | .219 | .195 | .172 | .151 |
| | | | | | | | | .131 | | | | | | | | | |
| 16.000 | 15.459 | .000 | .541 | .89 | 5.04 | 3.4% | 32 | .798 | .786 | .773 | .759 | .745 | .729 | .713 | .695 | .677 | .657 |
| | | | | | | | | .636 | .615 | .592 | .568 | .544 | .518 | .492 | .465 | .437 | .409 |
| | | | | | | | | .381 | .353 | .325 | .297 | .270 | .244 | .219 | .195 | .172 | .151 |
| | | | | | | | | .131 | .113 | | | | | | | | |
| 16.000 | 15.556 | .000 | .444 | .90 | 4.95 | 2.8% | 33 | .798 | .786 | .773 | .759 | .745 | .729 | .713 | .695 | .677 | .657 |
| | | | | | | | | .636 | .615 | .592 | .568 | .544 | .518 | .492 | .465 | .437 | .409 |
| | | | | | | | | .381 | .353 | .325 | .297 | .270 | .244 | .219 | .195 | .172 | .151 |
| | | | | | | | | .131 | .113 | .097 | | | | | | | |
| 16.000 | 15.639 | .000 | .361 | .92 | 4.87 | 2.3% | 34 | .798 | .786 | .773 | .759 | .745 | .729 | .713 | .695 | .677 | .657 |
| | | | | | | | | .636 | .615 | .592 | .568 | .544 | .518 | .492 | .465 | .437 | .409 |
| | | | | | | | | .381 | .353 | .325 | .297 | .270 | .244 | .219 | .195 | .172 | .151 |
| | | | | | | | | .131 | .113 | .097 | .082 | | | | | | |

*SIMILAR TO A.T.&T. NEAL-WILKINSON TABLES(SEE SECTION 1)

*EQUIVALENT RANDOM THEORY: VMR(VARIANCE TO MEAN RATIO)– 4.0 - NO RETRIES (PEAKED TRAFFIC - ROUTE ADVANCEMENT OR DAY-TO-DAY VARIATION)

| HOURS FIRST ATTEMPT VMR=4.0 | HOURS CON-NECT | HOURS RECALL /RETRY | HOURS OVFL | CON-NECT VMR | OVFL VMR | PERCENT BLOCKED (GRADE) | LINE | HOURS CONNECTED TRAFFIC ON LINES 1 | 2 | 3 | 4 | 5 | 6 | 7 | 8 | 9 | 10 |
|---|---|---|---|---|---|---|---|---|---|---|---|---|---|---|---|---|---|
| 16.000 | 15.708 | .000 | .292 | .93 | 4.77 | 1.8% | 35 | .798 | .786 | .773 | .759 | .745 | .729 | .713 | .695 | .677 | .657 |
| | | | | | | | | .636 | .615 | .592 | .568 | .544 | .518 | .492 | .465 | .437 | .409 |
| | | | | | | | | .381 | .353 | .325 | .297 | .270 | .244 | .219 | .195 | .172 | .151 |
| | | | | | | | | .131 | .113 | .097 | .082 | .069 | | | | | |
| 16.000 | 15.765 | .000 | .235 | .94 | 4.68 | 1.5% | 36 | .798 | .786 | .773 | .759 | .745 | .729 | .713 | .695 | .677 | .657 |
| | | | | | | | | .636 | .615 | .592 | .568 | .544 | .518 | .492 | .465 | .437 | .409 |
| | | | | | | | | .381 | .353 | .325 | .297 | .270 | .244 | .219 | .195 | .172 | .151 |
| | | | | | | | | .131 | .113 | .097 | .082 | .069 | .058 | | | | |
| 16.000 | 15.813 | .000 | .187 | .95 | 4.59 | 1.2% | 37 | .798 | .786 | .773 | .759 | .745 | .729 | .713 | .695 | .677 | .657 |
| | | | | | | | | .636 | .615 | .592 | .568 | .544 | .518 | .492 | .465 | .437 | .409 |
| | | | | | | | | .381 | .353 | .325 | .297 | .270 | .244 | .219 | .195 | .172 | .151 |
| | | | | | | | | .131 | .113 | .097 | .082 | .069 | .058 | .048 | | | |
| 16.000 | 15.852 | .000 | .148 | .96 | 4.50 | .9% | 38 | .798 | .786 | .773 | .759 | .745 | .729 | .713 | .695 | .677 | .657 |
| | | | | | | | | .636 | .615 | .592 | .568 | .544 | .518 | .492 | .465 | .437 | .409 |
| | | | | | | | | .381 | .353 | .325 | .297 | .270 | .244 | .219 | .195 | .172 | .151 |
| | | | | | | | | .131 | .113 | .097 | .082 | .069 | .058 | .048 | .039 | | |
| 16.000 | 15.884 | .000 | .116 | .97 | 4.40 | .7% | 39 | .798 | .786 | .773 | .759 | .745 | .729 | .713 | .695 | .677 | .657 |
| | | | | | | | | .636 | .615 | .592 | .568 | .544 | .518 | .492 | .465 | .437 | .409 |
| | | | | | | | | .381 | .353 | .325 | .297 | .270 | .244 | .219 | .195 | .172 | .151 |
| | | | | | | | | .131 | .113 | .097 | .082 | .069 | .058 | .048 | .039 | .032 | |
| 16.000 | 15.910 | .000 | .090 | .97 | 4.31 | .6% | 40 | .798 | .786 | .773 | .759 | .745 | .729 | .713 | .695 | .677 | .657 |
| | | | | | | | | .636 | .615 | .592 | .568 | .544 | .518 | .492 | .465 | .437 | .409 |
| | | | | | | | | .381 | .353 | .325 | .297 | .270 | .244 | .219 | .195 | .172 | .151 |
| | | | | | | | | .131 | .113 | .097 | .082 | .069 | .058 | .048 | .039 | .032 | .026 |
| 17.000 | 11.100 | .000 | 5.900 | .45 | 5.40 | 34.7% | 16 | .808 | .796 | .784 | .772 | .758 | .744 | .729 | .712 | .695 | .677 |
| | | | | | | | | .658 | .638 | .616 | .594 | .571 | .547 | | | | |
| 17.000 | 11.622 | .000 | 5.378 | .48 | 5.46 | 31.6% | 17 | .808 | .796 | .784 | .772 | .758 | .744 | .729 | .712 | .695 | .677 |
| | | | | | | | | .658 | .638 | .616 | .594 | .571 | .547 | .522 | | | |
| 17.000 | 12.118 | .000 | 4.882 | .50 | 5.51 | 28.7% | 18 | .808 | .796 | .784 | .772 | .758 | .744 | .729 | .712 | .695 | .677 |
| | | | | | | | | .658 | .638 | .616 | .594 | .571 | .547 | .522 | .496 | | |
| 17.000 | 12.588 | .000 | 4.412 | .53 | 5.55 | 26.0% | 19 | .808 | .796 | .784 | .772 | .758 | .744 | .729 | .712 | .695 | .677 |
| | | | | | | | | .658 | .638 | .616 | .594 | .571 | .547 | .522 | .496 | .470 | |
| 17.000 | 13.031 | .000 | 3.969 | .56 | 5.58 | 23.3% | 20 | .808 | .796 | .784 | .772 | .758 | .744 | .729 | .712 | .695 | .677 |
| | | | | | | | | .658 | .638 | .616 | .594 | .571 | .547 | .522 | .496 | .470 | .443 |
| 17.000 | 13.447 | .000 | 3.553 | .58 | 5.60 | 20.9% | 21 | .808 | .796 | .784 | .772 | .758 | .744 | .729 | .712 | .695 | .677 |
| | | | | | | | | .658 | .638 | .616 | .594 | .571 | .547 | .522 | .496 | .470 | .443 |
| | | | | | | | | .416 | | | | | | | | | |
| 17.000 | 13.835 | .000 | 3.165 | .61 | 5.62 | 18.6% | 22 | .808 | .796 | .784 | .772 | .758 | .744 | .729 | .712 | .695 | .677 |
| | | | | | | | | .658 | .638 | .616 | .594 | .571 | .547 | .522 | .496 | .470 | .443 |
| | | | | | | | | .416 | .388 | | | | | | | | |

*SIMILAR TO A.T.&T. NEAL-WILKINSON TABLES(SEE SECTION 1)

*EQUIVALENT RANDOM THEORY: VMR(VARIANCE TO MEAN RATIO)– 4.0 - NO RETRIES (PEAKED TRAFFIC - ROUTE ADVANCEMENT OR DAY-TO-DAY VARIATION)

| HOURS FIRST ATTEMPT VMR = 4.0 | HOURS CON-NECT | HOURS RECALL /RETRY | HOURS OVFL | CON-NECT VMR | OVFL VMR | PERCENT BLOCKED (GRADE) | LINE | HOURS CONNECTED TRAFFIC ON LINES 1 | 2 | 3 | 4 | 5 | 6 | 7 | 8 | 9 | 10 |
|---|---|---|---|---|---|---|---|---|---|---|---|---|---|---|---|---|---|
| 17.000 | 14.195 | .000 | 2.805 | .64 | 5.62 | 16.5% | 23 | .808 | .796 | .784 | .772 | .758 | .744 | .729 | .712 | .695 | .677 |
| | | | | | | | | .658 | .638 | .616 | .594 | .571 | .547 | .522 | .496 | .470 | .443 |
| | | | | | | | | .416 | .388 | .360 | | | | | | | |
| 17.000 | 14.528 | .000 | 2.472 | .67 | 5.62 | 14.5% | 24 | .808 | .796 | .784 | .772 | .758 | .744 | .729 | .712 | .695 | .677 |
| | | | | | | | | .658 | .638 | .616 | .594 | .571 | .547 | .522 | .496 | .470 | .443 |
| | | | | | | | | .416 | .388 | .360 | .333 | | | | | | |
| 17.000 | 14.834 | .000 | 2.166 | .69 | 5.60 | 12.7% | 25 | .808 | .796 | .784 | .772 | .758 | .744 | .729 | .712 | .695 | .677 |
| | | | | | | | | .658 | .638 | .616 | .594 | .571 | .547 | .522 | .496 | .470 | .443 |
| | | | | | | | | .416 | .388 | .360 | .333 | .306 | | | | | |
| 17.000 | 15.113 | .000 | 1.887 | .72 | 5.58 | 11.1% | 26 | .808 | .796 | .784 | .772 | .758 | .744 | .729 | .712 | .695 | .677 |
| | | | | | | | | .658 | .638 | .616 | .594 | .571 | .547 | .522 | .496 | .470 | .443 |
| | | | | | | | | .416 | .388 | .360 | .333 | .306 | .279 | | | | |
| 17.000 | 15.366 | .000 | 1.634 | .75 | 5.54 | 9.6% | 27 | .808 | .796 | .784 | .772 | .758 | .744 | .729 | .712 | .695 | .677 |
| | | | | | | | | .658 | .638 | .616 | .594 | .571 | .547 | .522 | .496 | .470 | .443 |
| | | | | | | | | .416 | .388 | .360 | .333 | .306 | .279 | .253 | | | |
| 17.000 | 15.593 | .000 | 1.407 | .77 | 5.50 | 8.3% | 28 | .808 | .796 | .784 | .772 | .758 | .744 | .729 | .712 | .695 | .677 |
| | | | | | | | | .658 | .638 | .616 | .594 | .571 | .547 | .522 | .496 | .470 | .443 |
| | | | | | | | | .416 | .388 | .360 | .333 | .306 | .279 | .253 | .228 | | |
| 17.000 | 15.797 | .000 | 1.203 | .80 | 5.45 | 7.1% | 29 | .808 | .796 | .784 | .772 | .758 | .744 | .729 | .712 | .695 | .677 |
| | | | | | | | | .658 | .638 | .616 | .594 | .571 | .547 | .522 | .496 | .470 | .443 |
| | | | | | | | | .416 | .388 | .360 | .333 | .306 | .279 | .253 | .228 | .204 | |
| 17.000 | 15.978 | .000 | 1.022 | .82 | 5.39 | 6.0% | 30 | .808 | .796 | .784 | .772 | .758 | .744 | .729 | .712 | .695 | .677 |
| | | | | | | | | .658 | .638 | .616 | .594 | .571 | .547 | .522 | .496 | .470 | .443 |
| | | | | | | | | .416 | .388 | .360 | .333 | .306 | .279 | .253 | .228 | .204 | .181 |
| 17.000 | 16.138 | .000 | .862 | .84 | 5.32 | 5.1% | 31 | .808 | .796 | .784 | .772 | .758 | .744 | .729 | .712 | .695 | .677 |
| | | | | | | | | .658 | .638 | .616 | .594 | .571 | .547 | .522 | .490 | .470 | .443 |
| | | | | | | | | .416 | .388 | .000 | .333 | .306 | .279 | .253 | .228 | .204 | .181 |
| | | | | | | | | .159 | | | | | | | | | |
| 17.000 | 16.277 | .000 | .723 | .86 | 5.24 | 4.3% | 32 | .808 | .796 | .784 | .772 | .758 | .744 | .729 | .712 | .695 | .677 |
| | | | | | | | | .658 | .638 | .616 | .594 | .571 | .547 | .522 | .496 | .470 | .443 |
| | | | | | | | | .416 | .388 | .360 | .333 | .306 | .279 | .253 | .228 | .204 | .181 |
| | | | | | | | | .159 | .140 | | | | | | | | |
| 17.000 | 16.398 | .000 | .602 | .88 | 5.17 | 3.5% | 33 | .808 | .796 | .784 | .772 | .758 | .744 | .729 | .712 | .695 | .677 |
| | | | | | | | | .658 | .638 | .010 | .594 | .571 | .547 | .522 | .496 | .470 | .443 |
| | | | | | | | | .416 | .388 | .360 | .333 | .306 | .279 | .253 | .228 | .204 | .181 |
| | | | | | | | | .159 | .140 | .121 | | | | | | | |
| 17.000 | 16.503 | .000 | .497 | .90 | 5.08 | 2.9% | 34 | .808 | .796 | .784 | .772 | .758 | .744 | .729 | .712 | .695 | .677 |
| | | | | | | | | .658 | .638 | .616 | .594 | .571 | .547 | .522 | .496 | .470 | .443 |
| | | | | | | | | .416 | .388 | .360 | .333 | .306 | .279 | .253 | .228 | .204 | .181 |
| | | | | | | | | .159 | .140 | .121 | .104 | | | | | | |

*SIMILAR TO A.T.&T. NEAL-WILKINSON TABLES(SEE SECTION 1)

*EQUIVALENT RANDOM THEORY: VMR(VARIANCE TO MEAN RATIO)– 4.0 - NO RETRIES (PEAKED TRAFFIC - ROUTE ADVANCEMENT OR DAY-TO-DAY VARIATION)

| HOURS FIRST ATTEMPT VMR = 4.0 | HOURS CON-NECT | HOURS RECALL /RETRY | HOURS OVFL | CON-NECT VMR | OVFL VMR | PERCENT BLOCKED (GRADE) | LINE | HOURS CONNECTED TRAFFIC ON LINES 1 | 2 | 3 | 4 | 5 | 6 | 7 | 8 | 9 | 10 |
|---|---|---|---|---|---|---|---|---|---|---|---|---|---|---|---|---|---|
| 17.000 | 16.592 | .000 | .408 | .91 | 4.99 | 2.4% | 35 | .808 | .796 | .784 | .772 | .758 | .744 | .729 | .712 | .695 | .677 |
| | | | | | | | | .658 | .638 | .616 | .594 | .571 | .547 | .522 | .496 | .470 | .443 |
| | | | | | | | | .416 | .388 | .360 | .333 | .306 | .279 | .253 | .228 | .204 | .181 |
| | | | | | | | | .159 | .140 | .121 | .104 | .089 | | | | | |
| 17.000 | 16.667 | .000 | .333 | .92 | 4.90 | 2.0% | 36 | .808 | .796 | .784 | .772 | .758 | .744 | .729 | .712 | .695 | .677 |
| | | | | | | | | .658 | .638 | .616 | .594 | .571 | .547 | .522 | .496 | .470 | .443 |
| | | | | | | | | .416 | .388 | .360 | .333 | .306 | .279 | .253 | .228 | .204 | .181 |
| | | | | | | | | .159 | .140 | .121 | .104 | .089 | .076 | | | | |
| 17.000 | 16.731 | .000 | .269 | .94 | 4.81 | 1.6% | 37 | .808 | .796 | .784 | .772 | .758 | .744 | .729 | .712 | .695 | .677 |
| | | | | | | | | .658 | .638 | .616 | .594 | .571 | .547 | .522 | .496 | .470 | .443 |
| | | | | | | | | .416 | .388 | .360 | .333 | .306 | .279 | .253 | .228 | .204 | .181 |
| | | | | | | | | .159 | .140 | .121 | .104 | .089 | .076 | .064 | | | |
| 17.000 | 16.784 | .000 | .216 | .95 | 4.72 | 1.3% | 38 | .808 | .796 | .784 | .772 | .758 | .744 | .729 | .712 | .695 | .677 |
| | | | | | | | | .658 | .638 | .616 | .594 | .571 | .547 | .522 | .496 | .470 | .443 |
| | | | | | | | | .416 | .388 | .360 | .333 | .306 | .279 | .253 | .228 | .204 | .181 |
| | | | | | | | | .159 | .140 | .121 | .104 | .089 | .076 | .064 | .053 | | |
| 17.000 | 16.828 | .000 | .172 | .96 | 4.62 | 1.0% | 39 | .808 | .796 | .784 | .772 | .758 | .744 | .729 | .712 | .695 | .677 |
| | | | | | | | | .658 | .638 | .616 | .594 | .571 | .547 | .522 | .496 | .470 | .443 |
| | | | | | | | | .416 | .388 | .360 | .333 | .306 | .279 | .253 | .228 | .204 | .181 |
| | | | | | | | | .159 | .140 | .121 | .104 | .089 | .076 | .064 | .053 | .044 | |
| 17.000 | 16.864 | .000 | .136 | .96 | 4.53 | .8% | 40 | .808 | .796 | .784 | .772 | .758 | .744 | .729 | .712 | .695 | .677 |
| | | | | | | | | .658 | .638 | .616 | .594 | .571 | .547 | .522 | .496 | .470 | .443 |
| | | | | | | | | .416 | .388 | .360 | .333 | .306 | .279 | .253 | .228 | .204 | .181 |
| | | | | | | | | .159 | .140 | .121 | .104 | .089 | .076 | .064 | .053 | .044 | .036 |
| 17.000 | 16.893 | .000 | .107 | .97 | 4.44 | .6% | 41 | .808 | .796 | .784 | .772 | .758 | .744 | .729 | .712 | .695 | .677 |
| | | | | | | | | .658 | .638 | .616 | .594 | .571 | .547 | .522 | .496 | .470 | .443 |
| | | | | | | | | .416 | .388 | .360 | .333 | .306 | .279 | .253 | .228 | .204 | .181 |
| | | | | | | | | .159 | .140 | .121 | .104 | .089 | .076 | .064 | .053 | .044 | .036 |
| | | | | | | | | .029 | | | | | | | | | |
| 18.000 | 11.918 | .000 | 6.082 | .45 | 5.49 | 33.8% | 17 | .816 | .806 | .795 | .783 | .770 | .757 | .743 | .728 | .712 | .695 |
| | | | | | | | | .677 | .658 | .639 | .618 | .596 | .574 | .550 | | | |
| 18.000 | 12.444 | .000 | 5.556 | .47 | 5.54 | 30.9% | 18 | .816 | .806 | .795 | .783 | .770 | .757 | .743 | .728 | .712 | .695 |
| | | | | | | | | .677 | .658 | .639 | .618 | .596 | .574 | .550 | .526 | | |
| 18.000 | 12.944 | .000 | 5.056 | .50 | 5.59 | 28.1% | 19 | .816 | .806 | .795 | .783 | .770 | .757 | .743 | .728 | .712 | .695 |
| | | | | | | | | .677 | .658 | .639 | .618 | .596 | .574 | .550 | .526 | .501 | |
| 18.000 | 13.419 | .000 | 4.581 | .53 | 5.63 | 25.4% | 20 | .816 | .806 | .795 | .783 | .770 | .757 | .743 | .728 | .712 | .695 |
| | | | | | | | | .677 | .658 | .639 | .618 | .596 | .574 | .550 | .526 | .501 | .475 |
| 18.000 | 13.868 | .000 | 4.132 | .55 | 5.67 | 23.0% | 21 | .816 | .806 | .795 | .783 | .770 | .757 | .743 | .728 | .712 | .695 |
| | | | | | | | | .677 | .658 | .639 | .618 | .596 | .574 | .550 | .526 | .501 | .475 |
| | | | | | | | | .448 | | | | | | | | | |

*SIMILAR TO A.T.&T. NEAL-WILKINSON TABLES(SEE SECTION 1)

*EQUIVALENT RANDOM THEORY: VMR(VARIANCE TO MEAN RATIO)– 4.0 - NO RETRIES (PEAKED TRAFFIC - ROUTE ADVANCEMENT OR DAY-TO-DAY VARIATION)

| HOURS FIRST ATTEMPT VMR = 4.0 | HOURS CON-NECT | HOURS RECALL /RETRY | HOURS OVFL | CON-NECT VMR | OVFL VMR | PERCENT BLOCKED (GRADE) | LINE | HOURS CONNECTED TRAFFIC ON LINES 1 | 2 | 3 | 4 | 5 | 6 | 7 | 8 | 9 | 10 |
|---|---|---|---|---|---|---|---|---|---|---|---|---|---|---|---|---|---|
| 18.000 | 14.289 | .000 | 3.711 | .58 | 5.69 | 20.6% | 22 | .816 | .806 | .795 | .783 | .770 | .757 | .743 | .728 | .712 | .695 |
| | | | | | | | | .677 | .658 | .639 | .618 | .596 | .574 | .550 | .526 | .501 | .475 |
| | | | | | | | | .448 | .422 | | | | | | | | |
| 18.000 | 14.684 | .000 | 3.316 | .61 | 5.71 | 18.4% | 23 | .816 | .806 | .795 | .783 | .770 | .757 | .743 | .728 | .712 | .695 |
| | | | | | | | | .677 | .658 | .639 | .618 | .596 | .574 | .550 | .526 | .501 | .475 |
| | | | | | | | | .448 | .422 | .395 | | | | | | | |
| 18.000 | 15.051 | .000 | 2.949 | .63 | 5.72 | 16.4% | 24 | .816 | .806 | .795 | .783 | .770 | .757 | .743 | .728 | .712 | .695 |
| | | | | | | | | .677 | .658 | .639 | .618 | .596 | .574 | .550 | .526 | .501 | .475 |
| | | | | | | | | .448 | .422 | .395 | .367 | | | | | | |
| 18.000 | 15.391 | .000 | 2.609 | .66 | 5.71 | 14.5% | 25 | .816 | .806 | .795 | .783 | .770 | .757 | .743 | .728 | .712 | .695 |
| | | | | | | | | .677 | .658 | .639 | .618 | .596 | .574 | .550 | .526 | .501 | .475 |
| | | | | | | | | .448 | .422 | .395 | .367 | .340 | | | | | |
| 18.000 | 15.705 | .000 | 2.295 | .69 | 5.70 | 12.8% | 26 | .816 | .806 | .795 | .783 | .770 | .757 | .743 | .728 | .712 | .695 |
| | | | | | | | | .677 | .658 | .639 | .618 | .596 | .574 | .550 | .526 | .501 | .475 |
| | | | | | | | | .448 | .422 | .395 | .367 | .340 | .313 | | | | |
| 18.000 | 15.992 | .000 | 2.008 | .71 | 5.68 | 11.2% | 27 | .816 | .806 | .795 | .783 | .770 | .757 | .743 | .728 | .712 | .695 |
| | | | | | | | | .677 | .658 | .639 | .618 | .596 | .574 | .550 | .526 | .501 | .475 |
| | | | | | | | | .448 | .422 | .395 | .367 | .340 | .313 | .287 | | | |
| 18.000 | 16.253 | .000 | 1.747 | .74 | 5.65 | 9.7% | 28 | .816 | .806 | .795 | .783 | .770 | .757 | .743 | .728 | .712 | .695 |
| | | | | | | | | .677 | .658 | .639 | .618 | .596 | .574 | .550 | .526 | .501 | .475 |
| | | | | | | | | .448 | .422 | .395 | .367 | .340 | .313 | .287 | .261 | | |
| 18.000 | 16.489 | .000 | 1.511 | .76 | 5.61 | 8.4% | 29 | .816 | .806 | .795 | .783 | .770 | .757 | .743 | .728 | .712 | .695 |
| | | | | | | | | .677 | .658 | .639 | .618 | .596 | .574 | .550 | .526 | .501 | .475 |
| | | | | | | | | .448 | .422 | .395 | .367 | .340 | .313 | .287 | .261 | .236 | |
| 18.000 | 16.702 | .000 | 1.298 | .79 | 5.56 | 7.2% | 30 | .816 | .806 | .795 | .783 | .770 | .757 | .743 | .728 | .712 | .695 |
| | | | | | | | | .677 | .658 | .639 | .618 | .596 | .574 | .550 | .526 | .501 | .475 |
| | | | | | | | | .448 | .422 | .395 | .367 | .340 | .313 | .287 | .261 | .236 | .212 |
| 18.000 | 16.891 | .000 | 1.109 | .81 | 5.50 | 6.2% | 31 | .816 | .806 | .795 | .783 | .770 | .757 | .743 | .728 | .712 | .695 |
| | | | | | | | | .677 | .658 | .639 | .618 | .596 | .574 | .550 | .526 | .501 | .475 |
| | | | | | | | | .448 | .422 | .395 | .367 | .340 | .313 | .287 | .261 | .236 | .212 |
| | | | | | | | | .189 | | | | | | | | | |
| 18.000 | 17.059 | .000 | .941 | .83 | 5.43 | 5.2% | 32 | .816 | .806 | .795 | .783 | .770 | .757 | .743 | .728 | .712 | .695 |
| | | | | | | | | .677 | .658 | .639 | .618 | .596 | .574 | .550 | .526 | .501 | .475 |
| | | | | | | | | .448 | .422 | .395 | .367 | .340 | .313 | .287 | .261 | .236 | .212 |
| | | | | | | | | .189 | .168 | | | | | | | | |
| 18.000 | 17.206 | .000 | .794 | .85 | 5.36 | 4.4% | 33 | .816 | .806 | .795 | .783 | .770 | .757 | .743 | .728 | .712 | .695 |
| | | | | | | | | .677 | .658 | .639 | .618 | .596 | .574 | .550 | .526 | .501 | .475 |
| | | | | | | | | .448 | .422 | .395 | .367 | .340 | .313 | .287 | .261 | .236 | .212 |
| | | | | | | | | .180 | .168 | .140 | | | | | | | |
| 18.000 | 17.335 | .000 | .665 | .87 | 5.29 | 3.7% | 34 | .816 | .806 | .795 | .783 | .770 | .757 | .743 | .728 | .712 | .695 |

*SIMILAR TO A.T.&T. NEAL-WILKINSON TABLES(SEE SECTION 1)

*EQUIVALENT RANDOM THEORY: VMR(VARIANCE TO MEAN RATIO)– 4.0 - NO RETRIES (PEAKED TRAFFIC - ROUTE ADVANCEMENT OR DAY-TO-DAY VARIATION)

| HOURS FIRST ATTEMPT VMR=4.0 | HOURS CON- NECT | HOURS RECALL /RETRY | HOURS OVFL | CON- NECT VMR | OVFL VMR | PERCENT BLOCKED (GRADE) | LINE | HOURS CONNECTED TRAFFIC ON LINES 1 | 2 | 3 | 4 | 5 | 6 | 7 | 8 | 9 | 10 |
|---|---|---|---|---|---|---|---|---|---|---|---|---|---|---|---|---|---|
| | | | | | | | | .677 | .658 | .639 | .618 | .596 | .574 | .550 | .526 | .501 | .475 |
| | | | | | | | | .448 | .422 | .395 | .367 | .340 | .313 | .287 | .261 | .236 | .212 |
| | | | | | | | | .189 | .168 | .148 | .129 | | | | | | |
| 18.000 | 17.447 | .000 | .553 | .89 | 5.20 | 3.1% | 35 | .816 | .806 | .795 | .783 | .770 | .757 | .743 | .728 | .712 | .695 |
| | | | | | | | | .677 | .658 | .639 | .618 | .596 | .574 | .550 | .526 | .501 | .475 |
| | | | | | | | | .448 | .422 | .395 | .367 | .340 | .313 | .287 | .261 | .236 | .212 |
| | | | | | | | | .189 | .168 | .148 | .129 | .112 | | | | | |
| 18.000 | 17.543 | .000 | .457 | .90 | 5.12 | 2.5% | 36 | .816 | .806 | .795 | .783 | .770 | .757 | .743 | .728 | .712 | .695 |
| | | | | | | | | .677 | .658 | .639 | .618 | .596 | .574 | .550 | .526 | .501 | .475 |
| | | | | | | | | .448 | .422 | .395 | .367 | .340 | .313 | .287 | .261 | .236 | .212 |
| | | | | | | | | .189 | .168 | .148 | .129 | .112 | .096 | | | | |
| 18.000 | 17.625 | .000 | .375 | .92 | 5.03 | 2.1% | 37 | .816 | .806 | .795 | .783 | .770 | .757 | .743 | .728 | .712 | .695 |
| | | | | | | | | .677 | .658 | .639 | .618 | .596 | .574 | .550 | .526 | .501 | .475 |
| | | | | | | | | .448 | .422 | .395 | .367 | .340 | .313 | .287 | .261 | .236 | .212 |
| | | | | | | | | .189 | .168 | .148 | .129 | .112 | .096 | .082 | | | |
| 18.000 | 17.694 | .000 | .306 | .93 | 4.94 | 1.7% | 38 | .816 | .806 | .795 | .783 | .770 | .757 | .743 | .728 | .712 | .695 |
| | | | | | | | | .677 | .658 | .639 | .618 | .596 | .574 | .550 | .526 | .501 | .475 |
| | | | | | | | | .448 | .422 | .395 | .367 | .340 | .313 | .287 | .261 | .236 | .212 |
| | | | | | | | | .189 | .168 | .148 | .129 | .112 | .096 | .082 | .069 | | |
| 18.000 | 17.753 | .000 | .247 | .94 | 4.84 | 1.4% | 39 | .816 | .806 | .795 | .783 | .770 | .757 | .743 | .728 | .712 | .695 |
| | | | | | | | | .677 | .658 | .639 | .618 | .596 | .574 | .550 | .526 | .501 | .475 |
| | | | | | | | | .448 | .422 | .395 | .367 | .340 | .313 | .287 | .261 | .236 | .212 |
| | | | | | | | | .189 | .168 | .148 | .129 | .112 | .096 | .082 | .069 | .058 | |
| 18.000 | 17.801 | .000 | .199 | .95 | 4.75 | 1.1% | 40 | .816 | .806 | .795 | .783 | .770 | .757 | .743 | .728 | .712 | .695 |
| | | | | | | | | .677 | .658 | .639 | .618 | .596 | .574 | .550 | .526 | .501 | .475 |
| | | | | | | | | .448 | .422 | .395 | .367 | .340 | .313 | .287 | .261 | .236 | .212 |
| | | | | | | | | .189 | .168 | .148 | .129 | .112 | .096 | .082 | .069 | .058 | .049 |
| 18.000 | 17.841 | .000 | .159 | .96 | 4.66 | .9% | 41 | .816 | .806 | .795 | .783 | .770 | .757 | .743 | .728 | .712 | .695 |
| | | | | | | | | .677 | .658 | .639 | .618 | .596 | .574 | .550 | .526 | .501 | .475 |
| | | | | | | | | .448 | .422 | .395 | .367 | .340 | .313 | .287 | .261 | .236 | .212 |
| | | | | | | | | .189 | .168 | .148 | .129 | .112 | .096 | .082 | .069 | .058 | .049 |
| | | | | | | | | .040 | | | | | | | | | |
| 18.000 | 17.874 | .000 | .126 | .97 | 4.56 | .7% | 42 | .816 | .806 | .795 | .783 | .770 | .757 | .743 | .728 | .712 | .695 |
| | | | | | | | | .677 | .658 | .639 | .618 | .596 | .574 | .550 | .526 | .501 | .475 |
| | | | | | | | | .448 | .422 | .395 | .367 | .340 | .313 | .287 | .261 | .236 | .212 |
| | | | | | | | | .189 | .168 | .148 | .129 | .112 | .096 | .082 | .069 | .058 | .049 |
| | | | | | | | | .040 | .033 | | | | | | | | |
| 18.000 | 17.901 | .000 | .099 | .97 | 4.47 | .5% | 43 | .816 | .806 | .795 | .783 | .770 | .757 | .743 | .728 | .712 | .695 |
| | | | | | | | | .677 | .658 | .639 | .618 | .596 | .574 | .550 | .526 | .501 | .475 |
| | | | | | | | | .448 | .422 | .395 | .367 | .340 | .313 | .287 | .261 | .236 | .212 |

*SIMILAR TO A.T.&T. NEAL-WILKINSON TABLES(SEE SECTION 1)

*EQUIVALENT RANDOM THEORY: VMR(VARIANCE TO MEAN RATIO)– 4.0 - NO RETRIES (PEAKED TRAFFIC - ROUTE ADVANCEMENT OR DAY-TO-DAY VARIATION)

| HOURS FIRST ATTEMPT VMR=4.0 | HOURS CON-NECT | HOURS RECALL /RETRY | HOURS OVFL | CON-NECT VMR | OVFL VMR | PERCENT BLOCKED (GRADE) | LINE | HOURS CONNECTED TRAFFIC ON LINES 1 | 2 | 3 | 4 | 5 | 6 | 7 | 8 | 9 | 10 |
|---|---|---|---|---|---|---|---|---|---|---|---|---|---|---|---|---|---|
| | | | | | | | | .189 | .168 | .148 | .129 | .112 | .096 | .082 | .069 | .058 | .049 |
| | | | | | | | | .040 | .033 | .027 | | | | | | | |
| 19.000 | 12.742 | .000 | 6.258 | .45 | 5.57 | 32.9% | 18 | .824 | .815 | .804 | .793 | .782 | .769 | .756 | .742 | .727 | .712 |
| | | | | | | | | .695 | .677 | .659 | .640 | .619 | .598 | .576 | .553 | | |
| 19.000 | 13.271 | .000 | 5.729 | .47 | 5.63 | 30.2% | 19 | .824 | .815 | .804 | .793 | .782 | .769 | .756 | .742 | .727 | .712 |
| | | | | | | | | .695 | .677 | .659 | .640 | .619 | .598 | .576 | .553 | .529 | |
| 19.000 | 13.775 | .000 | 5.225 | .50 | 5.68 | 27.5% | 20 | .824 | .815 | .804 | .793 | .782 | .769 | .756 | .742 | .727 | .712 |
| | | | | | | | | .695 | .677 | .659 | .640 | .619 | .598 | .576 | .553 | .529 | .504 |
| 19.000 | 14.254 | .000 | 4.746 | .52 | 5.72 | 25.0% | 21 | .824 | .815 | .804 | .793 | .782 | .769 | .756 | .742 | .727 | .712 |
| | | | | | | | | .695 | .677 | .659 | .640 | .619 | .598 | .576 | .553 | .529 | .504 |
| | | | | | | | | .479 | | | | | | | | | |
| 19.000 | 14.708 | .000 | 4.292 | .55 | 5.76 | 22.6% | 22 | .824 | .815 | .804 | .793 | .782 | .769 | .756 | .742 | .727 | .712 |
| | | | | | | | | .695 | .677 | .659 | .640 | .619 | .598 | .576 | .553 | .529 | .504 |
| | | | | | | | | .479 | .453 | | | | | | | | |
| 19.000 | 15.135 | .000 | 3.865 | .57 | 5.78 | 20.3% | 23 | .824 | .815 | .804 | .793 | .782 | .769 | .756 | .742 | .727 | .712 |
| | | | | | | | | .695 | .677 | .659 | .640 | .619 | .598 | .576 | .553 | .529 | .504 |
| | | | | | | | | .479 | .453 | .427 | | | | | | | |
| 19.000 | 15.535 | .000 | 3.465 | .60 | 5.80 | 18.2% | 24 | .824 | .815 | .804 | .793 | .782 | .769 | .756 | .742 | .727 | .712 |
| | | | | | | | | .695 | .677 | .659 | .640 | .619 | .598 | .576 | .553 | .529 | .504 |
| | | | | | | | | .479 | .453 | .427 | .401 | | | | | | |
| 19.000 | 15.909 | .000 | 3.091 | .63 | 5.81 | 16.3% | 25 | .824 | .815 | .804 | .793 | .782 | .769 | .756 | .742 | .727 | .712 |
| | | | | | | | | .695 | .677 | .659 | .640 | .619 | .598 | .576 | .553 | .529 | .504 |
| | | | | | | | | .479 | .453 | .427 | .401 | .374 | | | | | |
| 19.000 | 16.256 | .000 | 2.744 | .65 | 5.81 | 14.4% | 26 | .824 | .815 | .804 | .793 | .782 | .769 | .756 | .742 | .727 | .712 |
| | | | | | | | | .695 | .677 | .659 | .640 | .619 | .598 | .570 | .553 | .529 | .504 |
| | | | | | | | | .479 | .453 | .427 | .401 | .374 | .347 | | | | |
| 19.000 | 16.577 | .000 | 2.423 | .68 | 5.80 | 12.8% | 27 | .824 | .815 | .804 | .793 | .782 | .769 | .756 | .742 | .727 | .712 |
| | | | | | | | | .695 | .677 | .659 | .640 | .619 | .598 | .576 | .553 | .529 | .504 |
| | | | | | | | | .479 | .453 | .427 | .401 | .374 | .347 | .321 | | | |
| 19.000 | 16.872 | .000 | 2.128 | .71 | 5.78 | 11.2% | 28 | .824 | .815 | .804 | .793 | .782 | .769 | .756 | .742 | .727 | .712 |
| | | | | | | | | .695 | .677 | .659 | .640 | .619 | .598 | .576 | .553 | .529 | .504 |
| | | | | | | | | .479 | .453 | .427 | .401 | .374 | .347 | .321 | .295 | | |
| 19.000 | 17.141 | .000 | 1.859 | .73 | 5.75 | 9.8% | 29 | .824 | .815 | .804 | .793 | .782 | .769 | .756 | .742 | .727 | .712 |
| | | | | | | | | .695 | .677 | .659 | .640 | .619 | .598 | .576 | .553 | .529 | .504 |
| | | | | | | | | .479 | .453 | .427 | .401 | .374 | .347 | .321 | .295 | .269 | |
| 19.000 | 17.385 | .000 | 1.615 | .76 | 5.71 | 8.5% | 30 | .824 | .815 | .804 | .793 | .782 | .769 | .756 | .742 | .727 | .712 |
| | | | | | | | | .695 | .677 | .659 | .640 | .619 | .598 | .576 | .553 | .529 | .504 |
| | | | | | | | | .479 | .453 | .427 | .401 | .374 | .347 | .321 | .295 | .269 | .244 |
| 19.000 | 17.606 | .000 | 1.394 | .78 | 5.67 | 7.3% | 31 | .824 | .815 | .804 | .793 | .782 | .769 | .756 | .742 | .727 | .712 |
| | | | | | | | | .695 | .677 | .659 | .640 | .619 | .598 | .576 | .553 | .529 | .504 |

*SIMILAR TO A.T.&T. NEAL-WILKINSON TABLES(SEE SECTION 1)

*EQUIVALENT RANDOM THEORY: VMR(VARIANCE TO MEAN RATIO)– 4.0 - NO RETRIES (PEAKED TRAFFIC - ROUTE ADVANCEMENT OR DAY-TO-DAY VARIATION)

| HOURS FIRST ATTEMPT VMR = 4.0 | HOURS CON-NECT | HOURS RECALL /RETRY | HOURS OVFL | CON-NECT VMR | OVFL VMR | PERCENT BLOCKED (GRADE) | LINE | HOURS CONNECTED TRAFFIC ON LINES 1 | 2 | 3 | 4 | 5 | 6 | 7 | 8 | 9 | 10 |
|---|---|---|---|---|---|---|---|---|---|---|---|---|---|---|---|---|---|
| | | | | | | | | .479 | .453 | .427 | .401 | .374 | .347 | .321 | .295 | .269 | .244 |
| | | | | | | | | .220 | | | | | | | | | |
| 19.000 | 17.803 | .000 | 1.197 | .80 | 5.61 | 6.3% | 32 | .824 | .815 | .804 | .793 | .782 | .769 | .756 | .742 | .727 | .712 |
| | | | | | | | | .695 | .677 | .659 | .640 | .619 | .598 | .576 | .553 | .529 | .504 |
| | | | | | | | | .479 | .453 | .427 | .401 | .374 | .347 | .321 | .295 | .269 | .244 |
| | | | | | | | | .220 | .197 | | | | | | | | |
| 19.000 | 17.979 | .000 | 1.021 | .82 | 5.55 | 5.4% | 33 | .824 | .815 | .804 | .793 | .782 | .769 | .756 | .742 | .727 | .712 |
| | | | | | | | | .695 | .677 | .659 | .640 | .619 | .598 | .576 | .553 | .529 | .504 |
| | | | | | | | | .479 | .453 | .427 | .401 | .374 | .347 | .321 | .295 | .269 | .244 |
| | | | | | | | | .220 | .197 | .176 | | | | | | | |
| 19.000 | 18.134 | .000 | .866 | .84 | 5.48 | 4.6% | 34 | .824 | .815 | .804 | .793 | .782 | .769 | .756 | .742 | .727 | .712 |
| | | | | | | | | .695 | .677 | .659 | .640 | .619 | .598 | .576 | .553 | .529 | .504 |
| | | | | | | | | .479 | .453 | .427 | .401 | .374 | .347 | .321 | .295 | .269 | .244 |
| | | | | | | | | .220 | .197 | .176 | .155 | | | | | | |
| 19.000 | 18.271 | .000 | .729 | .86 | 5.40 | 3.8% | 35 | .824 | .815 | .804 | .793 | .782 | .769 | .756 | .742 | .727 | .712 |
| | | | | | | | | .695 | .677 | .659 | .640 | .619 | .598 | .576 | .553 | .529 | .504 |
| | | | | | | | | .479 | .453 | .427 | .401 | .374 | .347 | .321 | .295 | .269 | .244 |
| | | | | | | | | .220 | .197 | .176 | .155 | .136 | | | | | |
| 19.000 | 18.390 | .000 | .610 | .88 | 5.32 | 3.2% | 36 | .824 | .815 | .804 | .793 | .782 | .769 | .756 | .742 | .727 | .712 |
| | | | | | | | | .695 | .677 | .659 | .640 | .619 | .598 | .576 | .553 | .529 | .504 |
| | | | | | | | | .479 | .453 | .427 | .401 | .374 | .347 | .321 | .295 | .269 | .244 |
| | | | | | | | | .220 | .197 | .176 | .155 | .136 | .119 | | | | |
| 19.000 | 18.493 | .000 | .507 | .90 | 5.24 | 2.7% | 37 | .824 | .815 | .804 | .793 | .782 | .769 | .756 | .742 | .727 | .712 |
| | | | | | | | | .695 | .677 | .659 | .640 | .619 | .598 | .576 | .553 | .529 | .504 |
| | | | | | | | | .479 | .453 | .427 | .401 | .374 | .347 | .321 | .295 | .269 | .244 |
| | | | | | | | | .220 | .197 | .176 | .155 | .136 | .119 | .103 | | | |
| 19.000 | 18.581 | .000 | .419 | .91 | 5.15 | 2.2% | 38 | .824 | .815 | .804 | .793 | .782 | .769 | .756 | .742 | .727 | .712 |
| | | | | | | | | .695 | .677 | .659 | .640 | .619 | .598 | .576 | .553 | .529 | .504 |
| | | | | | | | | .479 | .453 | .427 | .401 | .374 | .347 | .321 | .295 | .269 | .244 |
| | | | | | | | | .220 | .197 | .176 | .155 | .136 | .119 | .103 | .088 | | |
| 19.000 | 18.656 | .000 | .344 | .92 | 5.06 | 1.8% | 39 | .824 | .815 | .804 | .793 | .782 | .769 | .756 | .742 | .727 | .712 |
| | | | | | | | | .695 | .677 | .659 | .640 | .619 | .598 | .576 | .553 | .529 | .504 |
| | | | | | | | | .479 | .453 | .427 | .401 | .374 | .347 | .321 | .206 | .269 | .244 |
| | | | | | | | | .220 | .197 | .176 | .155 | .136 | .119 | .103 | .088 | .075 | |
| 19.000 | 18.720 | .000 | .280 | .94 | 4.97 | 1.5% | 40 | .824 | .815 | .804 | .793 | .782 | .769 | .756 | .742 | .727 | .712 |
| | | | | | | | | .695 | .677 | .659 | .640 | .619 | .598 | .576 | .553 | .529 | .504 |
| | | | | | | | | .479 | .453 | .427 | .401 | .374 | .347 | .321 | .295 | .269 | .244 |
| | | | | | | | | .220 | .197 | .176 | .155 | .136 | .119 | .103 | .088 | .075 | .064 |
| 19.000 | 18.773 | .000 | .227 | .95 | 4.87 | 1.2% | 41 | .824 | .815 | .804 | .793 | .782 | .769 | .756 | .742 | .727 | .712 |
| | | | | | | | | .695 | .677 | .659 | .640 | .619 | .598 | .576 | .553 | .529 | .504 |

*SIMILAR TO A.T.&T. NEAL-WILKINSON TABLES(SEE SECTION 1)

*EQUIVALENT RANDOM THEORY: VMR(VARIANCE TO MEAN RATIO)– 4.0 - NO RETRIES (PEAKED TRAFFIC - ROUTE ADVANCEMENT OR DAY-TO-DAY VARIATION)

| HOURS FIRST ATTEMPT VMR=4.0 | HOURS CON-NECT | HOURS RECALL /RETRY | HOURS OVFL | CON-NECT VMR | OVFL VMR | PERCENT BLOCKED (GRADE) | LINE | HOURS CONNECTED TRAFFIC ON LINES 1 | 2 | 3 | 4 | 5 | 6 | 7 | 8 | 9 | 10 |
|---|---|---|---|---|---|---|---|---|---|---|---|---|---|---|---|---|---|
| | | | | | | | | .479 | .453 | .427 | .401 | .374 | .347 | .321 | .295 | .269 | .244 |
| | | | | | | | | .220 | .197 | .176 | .155 | .136 | .119 | .103 | .088 | .075 | .064 |
| | | | | | | | | .053 | | | | | | | | | |
| 19.000 | 18.818 | .000 | .182 | .96 | 4.78 | 1.0% | 42 | .824 | .815 | .804 | .793 | .782 | .769 | .756 | .742 | .727 | .712 |
| | | | | | | | | .695 | .677 | .659 | .640 | .619 | .598 | .576 | .553 | .529 | .504 |
| | | | | | | | | .479 | .453 | .427 | .401 | .374 | .347 | .321 | .295 | .269 | .244 |
| | | | | | | | | .220 | .197 | .176 | .155 | .136 | .119 | .103 | .088 | .075 | .064 |
| | | | | | | | | .053 | .044 | | | | | | | | |
| 19.000 | 18.854 | .000 | .146 | .96 | 4.68 | .8% | 43 | .824 | .815 | .804 | .793 | .782 | .769 | .756 | .742 | .727 | .712 |
| | | | | | | | | .695 | .677 | .659 | .640 | .619 | .598 | .576 | .553 | .529 | .504 |
| | | | | | | | | .479 | .453 | .427 | .401 | .374 | .347 | .321 | .295 | .269 | .244 |
| | | | | | | | | .220 | .197 | .176 | .155 | .136 | .119 | .103 | .088 | .075 | .064 |
| | | | | | | | | .053 | .044 | .037 | | | | | | | |
| 19.000 | 18.885 | .000 | .115 | .97 | 4.59 | .6% | 44 | .824 | .815 | .804 | .793 | .782 | .769 | .756 | .742 | .727 | .712 |
| | | | | | | | | .695 | .677 | .659 | .640 | .619 | .598 | .576 | .553 | .529 | .504 |
| | | | | | | | | .479 | .453 | .427 | .401 | .374 | .347 | .321 | .295 | .269 | .244 |
| | | | | | | | | .220 | .197 | .176 | .155 | .136 | .119 | .103 | .088 | .075 | .064 |
| | | | | | | | | .053 | .044 | .037 | .030 | | | | | | |
| 20.000 | 13.014 | .000 | 6.986 | .42 | 5.59 | 34.9% | 18 | .832 | .823 | .813 | .803 | .792 | .780 | .768 | .755 | .741 | .727 |
| | | | | | | | | .711 | .695 | .678 | .659 | .640 | .621 | .600 | .578 | | |
| 20.000 | 13.570 | .000 | 6.430 | .44 | 5.66 | 32.2% | 19 | .832 | .823 | .813 | .803 | .792 | .780 | .768 | .755 | .741 | .727 |
| | | | | | | | | .711 | .695 | .678 | .659 | .640 | .621 | .600 | .578 | .555 | |
| 20.000 | 14.102 | .000 | 5.898 | .47 | 5.71 | 29.5% | 20 | .832 | .823 | .813 | .803 | .792 | .780 | .768 | .755 | .741 | .727 |
| | | | | | | | | .711 | .695 | .678 | .659 | .640 | .621 | .600 | .578 | .555 | .532 |
| 20.000 | 14.610 | .000 | 5.390 | .49 | 5.76 | 27.0% | 21 | .832 | .823 | .813 | .803 | .792 | .780 | .768 | .755 | .741 | .727 |
| | | | | | | | | .711 | .695 | .678 | .659 | .640 | .621 | .600 | .578 | .555 | .532 |
| | | | | | | | | .508 | | | | | | | | | |
| 20.000 | 15.093 | .000 | 4.907 | .52 | 5.81 | 24.5% | 22 | .832 | .823 | .813 | .803 | .792 | .780 | .768 | .755 | .741 | .727 |
| | | | | | | | | .711 | .695 | .678 | .659 | .640 | .621 | .600 | .578 | .555 | .532 |
| | | | | | | | | .508 | .483 | | | | | | | | |
| 20.000 | 15.551 | .000 | 4.449 | .54 | 5.84 | 22.2% | 23 | .832 | .823 | .813 | .803 | .792 | .780 | .768 | .755 | .741 | .727 |
| | | | | | | | | .711 | .695 | .678 | .659 | .640 | .621 | .600 | .578 | .555 | .532 |
| | | | | | | | | .508 | .483 | .458 | | | | | | | |
| 20.000 | 15.983 | .000 | 4.017 | .57 | 5.87 | 20.1% | 24 | .832 | .823 | .813 | .803 | .792 | .780 | .768 | .755 | .741 | .727 |
| | | | | | | | | .711 | .695 | .678 | .659 | .640 | .621 | .600 | .578 | .555 | .532 |
| | | | | | | | | .508 | .483 | .458 | .432 | | | | | | |
| 20.000 | 16.390 | .000 | 3.610 | .59 | 5.89 | 18.1% | 25 | .832 | .823 | .813 | .803 | .792 | .780 | .768 | .755 | .741 | .727 |
| | | | | | | | | .711 | .695 | .678 | .659 | .640 | .621 | .600 | .578 | .555 | .532 |
| | | | | | | | | .508 | .483 | .458 | .432 | .406 | | | | | |
| 20.000 | 16.770 | .000 | 3.230 | .62 | 5.90 | 16.2% | 26 | .832 | .823 | .813 | .803 | .792 | .780 | .768 | .755 | .741 | .727 |

*SIMILAR TO A.T.&T. NEAL-WILKINSON TABLES(SEE SECTION 1)

*EQUIVALENT RANDOM THEORY: VMR(VARIANCE TO MEAN RATIO)– 4.0 - NO RETRIES (PEAKED TRAFFIC - ROUTE ADVANCEMENT OR DAY-TO-DAY VARIATION)

| HOURS FIRST ATTEMPT VMR=4.0 | HOURS CON-NECT | HOURS RECALL /RETRY | HOURS OVFL | CON-NECT VMR | OVFL VMR | PERCENT BLOCKED (GRADE) | LINE | HOURS CONNECTED TRAFFIC ON LINES 1 | 2 | 3 | 4 | 5 | 6 | 7 | 8 | 9 | 10 |
|---|---|---|---|---|---|---|---|---|---|---|---|---|---|---|---|---|---|
| | | | | | | | | .711 | .695 | .678 | .659 | .640 | .621 | .600 | .578 | .555 | .532 |
| | | | | | | | | .508 | .483 | .458 | .432 | .406 | .380 | | | | |
| 20.000 | 17.123 | .000 | 2.877 | .65 | 5.90 | 14.4% | 27 | .832 | .823 | .813 | .803 | .792 | .780 | .768 | .755 | .741 | .727 |
| | | | | | | | | .711 | .695 | .678 | .659 | .640 | .621 | .600 | .578 | .555 | .532 |
| | | | | | | | | .508 | .483 | .458 | .432 | .406 | .380 | .354 | | | |
| 20.000 | 17.451 | .000 | 2.549 | .67 | 5.89 | 12.7% | 28 | .832 | .823 | .813 | .803 | .792 | .780 | .768 | .755 | .741 | .727 |
| | | | | | | | | .711 | .695 | .678 | .659 | .640 | .621 | .600 | .578 | .555 | .532 |
| | | | | | | | | .508 | .483 | .458 | .432 | .406 | .380 | .354 | .328 | | |
| 20.000 | 17.753 | .000 | 2.247 | .70 | 5.88 | 11.2% | 29 | .832 | .823 | .813 | .803 | .792 | .780 | .768 | .755 | .741 | .727 |
| | | | | | | | | .711 | .695 | .678 | .659 | .640 | .621 | .600 | .578 | .555 | .532 |
| | | | | | | | | .508 | .483 | .458 | .432 | .406 | .380 | .354 | .328 | .302 | |
| 20.000 | 18.030 | .000 | 1.970 | .72 | 5.85 | 9.9% | 30 | .832 | .823 | .813 | .803 | .792 | .780 | .768 | .755 | .741 | .727 |
| | | | | | | | | .711 | .695 | .678 | .659 | .640 | .621 | .600 | .578 | .555 | .532 |
| | | | | | | | | .508 | .483 | .458 | .432 | .406 | .380 | .354 | .328 | .302 | .277 |
| 20.000 | 18.281 | .000 | 1.719 | .75 | 5.81 | 8.6% | 31 | .832 | .823 | .813 | .803 | .792 | .780 | .768 | .755 | .741 | .727 |
| | | | | | | | | .711 | .695 | .678 | .659 | .640 | .621 | .600 | .578 | .555 | .532 |
| | | | | | | | | .508 | .483 | .458 | .432 | .406 | .380 | .354 | .328 | .302 | .277 |
| | | | | | | | | .252 | | | | | | | | | |
| 20.000 | 18.510 | .000 | 1.490 | .77 | 5.77 | 7.5% | 32 | .832 | .823 | .813 | .803 | .792 | .780 | .768 | .755 | .741 | .727 |
| | | | | | | | | .711 | .695 | .678 | .659 | .640 | .621 | .600 | .578 | .555 | .532 |
| | | | | | | | | .508 | .483 | .458 | .432 | .406 | .380 | .354 | .328 | .302 | .277 |
| | | | | | | | | .252 | .228 | | | | | | | | |
| 20.000 | 18.715 | .000 | 1.285 | .79 | 5.72 | 6.4% | 33 | .832 | .823 | .813 | .803 | .792 | .780 | .768 | .755 | .741 | .727 |
| | | | | | | | | .711 | .695 | .678 | .659 | .640 | .621 | .600 | .578 | .555 | .532 |
| | | | | | | | | .508 | .483 | .458 | .432 | .406 | .380 | .354 | .328 | .302 | .277 |
| | | | | | | | | .252 | .228 | .205 | | | | | | | |
| 20.000 | 18.898 | .000 | 1.102 | .82 | 5.66 | 5.5% | 34 | .832 | .823 | .813 | .803 | .792 | .780 | .768 | .755 | .741 | .727 |
| | | | | | | | | .711 | .695 | .678 | .659 | .640 | .621 | .600 | .578 | .555 | .532 |
| | | | | | | | | .508 | .483 | .458 | .432 | .406 | .380 | .354 | .328 | .302 | .277 |
| | | | | | | | | .252 | .228 | .205 | .184 | | | | | | |
| 20.000 | 19.061 | .000 | .939 | .84 | 5.59 | 4.7% | 35 | .832 | .823 | .813 | .803 | .792 | .780 | .768 | .755 | .741 | .727 |
| | | | | | | | | .711 | .695 | .678 | .659 | .640 | .621 | .600 | .578 | .555 | .532 |
| | | | | | | | | .508 | .483 | .458 | .432 | .406 | .380 | .354 | .328 | .302 | .277 |
| | | | | | | | | .252 | .228 | .205 | .184 | .163 | | | | | |
| 20.000 | 19.205 | .000 | .795 | .86 | 5.52 | 4.0% | 36 | .832 | .823 | .813 | .803 | .792 | .780 | .768 | .755 | .741 | .727 |
| | | | | | | | | .711 | .695 | .678 | .659 | .640 | .621 | .600 | .578 | .555 | .532 |
| | | | | | | | | .508 | .483 | .458 | .432 | .406 | .380 | .354 | .328 | .302 | .277 |
| | | | | | | | | .252 | .228 | .205 | .184 | .163 | .144 | | | | |
| 20.000 | 19.331 | .000 | .669 | .87 | 5.44 | 3.3% | 37 | .832 | .823 | .813 | .803 | .792 | .780 | .768 | .755 | .741 | .727 |
| | | | | | | | | .711 | .695 | .678 | .659 | .640 | .621 | .600 | .578 | .555 | .532 |

*SIMILAR TO A.T.&T. NEAL-WILKINSON TABLES(SEE SECTION 1)

*EQUIVALENT RANDOM THEORY: VMR(VARIANCE TO MEAN RATIO)– 4.0 - NO RETRIES (PEAKED TRAFFIC - ROUTE ADVANCEMENT OR DAY-TO-DAY VARIATION)

| HOURS FIRST ATTEMPT VMR = 4.0 | HOURS CON-NECT | HOURS RECALL /RETRY | HOURS OVFL | CON-NECT VMR | OVFL VMR | PERCENT BLOCKED (GRADE) | LINE | HOURS CONNECTED TRAFFIC ON LINES 1 | 2 | 3 | 4 | 5 | 6 | 7 | 8 | 9 | 10 |
|---|---|---|---|---|---|---|---|---|---|---|---|---|---|---|---|---|---|
| | | | | | | | | .508 | .483 | .458 | .432 | .406 | .380 | .354 | .328 | .302 | .277 |
| | | | | | | | | .252 | .228 | .205 | .184 | .163 | .144 | .126 | | | |
| 20.000 | 19.440 | .000 | .560 | .89 | 5.36 | 2.8% | 38 | .832 | .823 | .813 | .803 | .792 | .780 | .768 | .755 | .741 | .727 |
| | | | | | | | | .711 | .695 | .678 | .659 | .640 | .621 | .600 | .578 | .555 | .532 |
| | | | | | | | | .508 | .483 | .458 | .432 | .406 | .380 | .354 | .328 | .302 | .277 |
| | | | | | | | | .252 | .228 | .205 | .184 | .163 | .144 | .126 | .110 | | |
| 20.000 | 19.535 | .000 | .465 | .91 | 5.27 | 2.3% | 39 | .832 | .823 | .813 | .803 | .792 | .780 | .768 | .755 | .741 | .727 |
| | | | | | | | | .711 | .695 | .678 | .659 | .640 | .621 | .600 | .578 | .555 | .532 |
| | | | | | | | | .508 | .483 | .458 | .432 | .406 | .380 | .354 | .328 | .302 | .277 |
| | | | | | | | | .252 | .228 | .205 | .184 | .163 | .144 | .126 | .110 | .095 | |
| 20.000 | 19.616 | .000 | .384 | .92 | 5.18 | 1.9% | 40 | .832 | .823 | .813 | .803 | .792 | .780 | .768 | .755 | .741 | .727 |
| | | | | | | | | .711 | .695 | .678 | .659 | .640 | .621 | .600 | .578 | .555 | .532 |
| | | | | | | | | .508 | .483 | .458 | .432 | .406 | .380 | .354 | .328 | .302 | .277 |
| | | | | | | | | .252 | .228 | .205 | .184 | .163 | .144 | .126 | .110 | .095 | .081 |
| 20.000 | 19.685 | .000 | .315 | .93 | 5.09 | 1.6% | 41 | .832 | .823 | .813 | .803 | .792 | .780 | .768 | .755 | .741 | .727 |
| | | | | | | | | .711 | .695 | .678 | .659 | .640 | .621 | .600 | .578 | .555 | .532 |
| | | | | | | | | .508 | .483 | .458 | .432 | .406 | .380 | .354 | .328 | .302 | .277 |
| | | | | | | | | .252 | .228 | .205 | .184 | .163 | .144 | .126 | .110 | .095 | .081 |
| | | | | | | | | .069 | | | | | | | | | |
| 20.000 | 19.743 | .000 | .257 | .94 | 4.99 | 1.3% | 42 | .832 | .823 | .813 | .803 | .792 | .780 | .768 | .755 | .741 | .727 |
| | | | | | | | | .711 | .695 | .678 | .659 | .640 | .621 | .600 | .578 | .555 | .532 |
| | | | | | | | | .508 | .483 | .458 | .432 | .406 | .380 | .354 | .328 | .302 | .277 |
| | | | | | | | | .252 | .228 | .205 | .184 | .163 | .144 | .126 | .110 | .095 | .081 |
| | | | | | | | | .069 | .058 | | | | | | | | |
| 20.000 | 19.792 | .000 | .208 | .95 | 4.90 | 1.0% | 43 | .832 | .823 | .813 | .803 | .792 | .780 | .768 | .755 | .741 | .727 |
| | | | | | | | | .711 | .695 | .678 | .659 | .640 | .621 | .600 | .578 | .555 | .532 |
| | | | | | | | | .508 | .483 | .458 | .432 | .406 | .380 | .354 | .328 | .302 | .277 |
| | | | | | | | | .252 | .228 | .205 | .184 | .163 | .144 | .126 | .110 | .095 | .081 |
| | | | | | | | | .069 | .058 | .049 | | | | | | | |
| 20.000 | 19.833 | .000 | .167 | .96 | 4.80 | .8% | 44 | .832 | .823 | .813 | .803 | .792 | .780 | .768 | .755 | .741 | .727 |
| | | | | | | | | .711 | .695 | .678 | .659 | .640 | .621 | .600 | .578 | .555 | .532 |
| | | | | | | | | .508 | .483 | .458 | .432 | .406 | .380 | .354 | .328 | .302 | .277 |
| | | | | | | | | .252 | .228 | .205 | .184 | .163 | .144 | .126 | .110 | .095 | .081 |
| | | | | | | | | .069 | .058 | .049 | .041 | | | | | | |
| 20.000 | 19.867 | .000 | .133 | .97 | 4.71 | .7% | 45 | .832 | .823 | .813 | .803 | .792 | .780 | .768 | .755 | .741 | .727 |
| | | | | | | | | .711 | .695 | .678 | .659 | .640 | .621 | .600 | .578 | .555 | .532 |
| | | | | | | | | .508 | .483 | .458 | .432 | .406 | .380 | .354 | .328 | .302 | .277 |
| | | | | | | | | .252 | .228 | .205 | .184 | .163 | .144 | .126 | .110 | .095 | .081 |
| | | | | | | | | .069 | .058 | .049 | .041 | .034 | | | | | |
| 20.000 | 19.894 | .000 | .106 | .97 | 4.62 | .5% | 46 | .832 | .823 | .813 | .803 | .792 | .780 | .768 | .755 | .741 | .727 |

*SIMILAR TO A.T.&T. NEAL-WILKINSON TABLES(SEE SECTION 1)

*EQUIVALENT RANDOM THEORY: VMR(VARIANCE TO MEAN RATIO)– 4.0 - NO RETRIES (PEAKED TRAFFIC - ROUTE ADVANCEMENT OR DAY-TO-DAY VARIATION)

| HOURS FIRST ATTEMPT VMR=4.0 | HOURS CON-NECT | HOURS RECALL /RETRY | HOURS OVFL | CON-NECT VMR | OVFL VMR | PERCENT BLOCKED (GRADE) | LINE | HOURS CONNECTED TRAFFIC ON LINES 1 | 2 | 3 | 4 | 5 | 6 | 7 | 8 | 9 | 10 |
|---|---|---|---|---|---|---|---|---|---|---|---|---|---|---|---|---|---|
| | | | | | | | | .711 | .695 | .678 | .659 | .640 | .621 | .600 | .578 | .555 | .532 |
| | | | | | | | | .508 | .483 | .458 | .432 | .406 | .380 | .354 | .328 | .302 | .277 |
| | | | | | | | | .252 | .228 | .205 | .184 | .163 | .144 | .126 | .110 | .095 | .081 |
| | | | | | | | | .069 | .058 | .049 | .041 | .034 | .028 | | | | |
| 22.000 | 14.694 | .000 | 7.306 | .42 | 5.73 | 33.2% | 20 | .845 | .837 | .829 | .820 | .811 | .800 | .790 | .778 | .766 | .754 |
| | | | | | | | | .740 | .726 | .711 | .695 | .678 | .661 | .642 | .623 | .603 | .582 |
| 22.000 | 15.254 | .000 | 6.746 | .44 | 5.80 | 30.7% | 21 | .845 | .837 | .829 | .820 | .811 | .800 | .790 | .778 | .766 | .754 |
| | | | | | | | | .740 | .726 | .711 | .695 | .678 | .661 | .642 | .623 | .603 | .582 |
| | | | | | | | | .560 | | | | | | | | | |
| 22.000 | 15.792 | .000 | 6.208 | .46 | 5.85 | 28.2% | 22 | .845 | .837 | .829 | .820 | .811 | .800 | .790 | .778 | .766 | .754 |
| | | | | | | | | .740 | .726 | .711 | .695 | .678 | .661 | .642 | .623 | .603 | .582 |
| | | | | | | | | .560 | .538 | | | | | | | | |
| 22.000 | 16.307 | .000 | 5.693 | .49 | 5.91 | 25.9% | 23 | .845 | .837 | .829 | .820 | .811 | .800 | .790 | .778 | .766 | .754 |
| | | | | | | | | .740 | .726 | .711 | .695 | .678 | .661 | .642 | .623 | .603 | .582 |
| | | | | | | | | .560 | .538 | .515 | | | | | | | |
| 22.000 | 16.797 | .000 | 5.203 | .51 | 5.95 | 23.6% | 24 | .845 | .837 | .829 | .820 | .811 | .800 | .790 | .778 | .766 | .754 |
| | | | | | | | | .740 | .726 | .711 | .695 | .678 | .661 | .642 | .623 | .603 | .582 |
| | | | | | | | | .560 | .538 | .515 | .491 | | | | | | |
| 22.000 | 17.264 | .000 | 4.736 | .53 | 5.99 | 21.5% | 25 | .845 | .837 | .829 | .820 | .811 | .800 | .790 | .778 | .766 | .754 |
| | | | | | | | | .740 | .726 | .711 | .695 | .678 | .661 | .642 | .623 | .603 | .582 |
| | | | | | | | | .560 | .538 | .515 | .491 | .466 | | | | | |
| 22.000 | 17.705 | .000 | 4.295 | .56 | 6.02 | 19.5% | 26 | .845 | .837 | .829 | .820 | .811 | .800 | .790 | .778 | .766 | .754 |
| | | | | | | | | .740 | .726 | .711 | .695 | .678 | .661 | .642 | .623 | .603 | .582 |
| | | | | | | | | .560 | .538 | .515 | .491 | .466 | .442 | | | | |
| 22.000 | 18.122 | .000 | 3.878 | .58 | 6.04 | 17.6% | 27 | .845 | .837 | .829 | .820 | .811 | .800 | .790 | .778 | .766 | .754 |
| | | | | | | | | .740 | .726 | .711 | .695 | .678 | .661 | .642 | .623 | .603 | .582 |
| | | | | | | | | .560 | .538 | .515 | .491 | .466 | .442 | .416 | | | |
| 22.000 | 18.513 | .000 | 3.487 | .61 | 6.06 | 15.9% | 28 | .845 | .837 | .829 | .820 | .811 | .800 | .790 | .778 | .766 | .754 |
| | | | | | | | | .740 | .726 | .711 | .695 | .678 | .661 | .642 | .623 | .603 | .582 |
| | | | | | | | | .560 | .538 | .515 | .491 | .466 | .442 | .416 | .391 | | |
| 22.000 | 18.878 | .000 | 3.122 | .63 | 6.06 | 14.2% | 29 | .845 | .837 | .829 | .820 | .811 | .800 | .790 | .778 | .766 | .754 |
| | | | | | | | | .740 | .726 | .711 | .695 | .678 | .661 | .642 | .623 | .603 | .582 |
| | | | | | | | | .560 | .538 | .515 | .491 | .466 | .442 | .416 | .391 | .365 | |
| 22.000 | 19.218 | .000 | 2.782 | .66 | 6.06 | 12.6% | 30 | .845 | .837 | .829 | .820 | .811 | .800 | .790 | .778 | .766 | .754 |
| | | | | | | | | .740 | .726 | .711 | .695 | .678 | .661 | .642 | .623 | .603 | .582 |
| | | | | | | | | .560 | .538 | .515 | .491 | .466 | .442 | .416 | .391 | .365 | .340 |
| 22.000 | 19.533 | .000 | 2.467 | .69 | 6.04 | 11.2% | 31 | .845 | .837 | .829 | .820 | .811 | .800 | .790 | .778 | .766 | .754 |
| | | | | | | | | .740 | .726 | .711 | .695 | .678 | .661 | .642 | .623 | .603 | .582 |
| | | | | | | | | .560 | .538 | .515 | .491 | .466 | .442 | .416 | .391 | .365 | .340 |
| | | | | | | | | .315 | | | | | | | | | |

*SIMILAR TO A.T.&T. NEAL-WILKINSON TABLES(SEE SECTION 1)

*EQUIVALENT RANDOM THEORY: VMR(VARIANCE TO MEAN RATIO)– 4.0 - NO RETRIES (PEAKED TRAFFIC - ROUTE ADVANCEMENT OR DAY-TO-DAY VARIATION)

| HOURS FIRST ATTEMPT VMR = 4.0 | HOURS CONNECT | HOURS RECALL /RETRY | HOURS OVFL | CONNECT VMR | OVFL VMR | PERCENT BLOCKED (GRADE) | LINE | HOURS CONNECTED TRAFFIC ON LINES 1 | 2 | 3 | 4 | 5 | 6 | 7 | 8 | 9 | 10 |
|---|---|---|---|---|---|---|---|---|---|---|---|---|---|---|---|---|---|
| 22.000 | 19.823 | .000 | 2.177 | .71 | 6.02 | 9.9% | 32 | .845 | .837 | .829 | .820 | .811 | .800 | .790 | .778 | .766 | .754 |
| | | | | | | | | .740 | .726 | .711 | .695 | .678 | .661 | .642 | .623 | .603 | .582 |
| | | | | | | | | .560 | .538 | .515 | .491 | .466 | .442 | .416 | .391 | .365 | .340 |
| | | | | | | | | .315 | .290 | | | | | | | | |
| 22.000 | 20.088 | .000 | 1.912 | .73 | 5.99 | 8.7% | 33 | .845 | .837 | .829 | .820 | .811 | .800 | .790 | .778 | .766 | .754 |
| | | | | | | | | .740 | .726 | .711 | .695 | .678 | .661 | .642 | .623 | .603 | .582 |
| | | | | | | | | .560 | .538 | .515 | .491 | .466 | .442 | .416 | .391 | .365 | .340 |
| | | | | | | | | .315 | .290 | .266 | | | | | | | |
| 22.000 | 20.330 | .000 | 1.670 | .76 | 5.95 | 7.6% | 34 | .845 | .837 | .829 | .820 | .811 | .800 | .790 | .778 | .766 | .754 |
| | | | | | | | | .740 | .726 | .711 | .695 | .678 | .661 | .642 | .623 | .603 | .582 |
| | | | | | | | | .560 | .538 | .515 | .491 | .466 | .442 | .416 | .391 | .365 | .340 |
| | | | | | | | | .315 | .290 | .266 | .242 | | | | | | |
| 22.000 | 20.549 | .000 | 1.451 | .78 | 5.90 | 6.6% | 35 | .845 | .837 | .829 | .820 | .811 | .800 | .790 | .778 | .766 | .754 |
| | | | | | | | | .740 | .726 | .711 | .695 | .678 | .661 | .642 | .623 | .603 | .582 |
| | | | | | | | | .560 | .538 | .515 | .491 | .466 | .442 | .416 | .391 | .365 | .340 |
| | | | | | | | | .315 | .290 | .266 | .242 | .219 | | | | | |
| 22.000 | 20.746 | .000 | 1.254 | .80 | 5.85 | 5.7% | 36 | .845 | .837 | .829 | .820 | .811 | .800 | .790 | .778 | .766 | .754 |
| | | | | | | | | .740 | .726 | .711 | .695 | .678 | .661 | .642 | .623 | .603 | .582 |
| | | | | | | | | .560 | .538 | .515 | .491 | .466 | .442 | .416 | .391 | .365 | .340 |
| | | | | | | | | .315 | .290 | .266 | .242 | .219 | .197 | | | | |
| 22.000 | 20.923 | .000 | 1.077 | .82 | 5.79 | 4.9% | 37 | .845 | .837 | .829 | .820 | .811 | .800 | .790 | .778 | .766 | .754 |
| | | | | | | | | .740 | .726 | .711 | .695 | .678 | .661 | .642 | .623 | .603 | .582 |
| | | | | | | | | .560 | .538 | .515 | .491 | .466 | .442 | .416 | .391 | .365 | .340 |
| | | | | | | | | .315 | .290 | .266 | .242 | .219 | .197 | .177 | | | |
| 22.000 | 21.080 | .000 | .920 | .84 | 5.72 | 4.2% | 38 | .845 | .837 | .829 | .820 | .811 | .800 | .790 | .778 | .766 | .754 |
| | | | | | | | | .740 | .726 | .711 | .695 | .678 | .661 | .642 | .623 | .603 | .582 |
| | | | | | | | | .560 | .538 | .515 | .491 | .466 | .442 | .416 | .391 | .365 | .340 |
| | | | | | | | | .315 | .290 | .266 | .242 | .219 | .197 | .177 | .157 | | |
| 22.000 | 21.219 | .000 | .781 | .86 | 5.64 | 3.6% | 39 | .845 | .837 | .829 | .820 | .811 | .800 | .790 | .778 | .766 | .754 |
| | | | | | | | | .740 | .726 | .711 | .695 | .678 | .661 | .642 | .623 | .603 | .582 |
| | | | | | | | | .560 | .538 | .515 | .491 | .466 | .442 | .416 | .391 | .365 | .340 |
| | | | | | | | | .315 | .290 | .266 | .242 | .219 | .197 | .177 | .157 | .139 | |
| 22.000 | 21.340 | .000 | .660 | .88 | 5.56 | 3.0% | 40 | .845 | .837 | .829 | .820 | .811 | .800 | .790 | .778 | .766 | .754 |
| | | | | | | | | .740 | .726 | .711 | .695 | .678 | .661 | .642 | .623 | .603 | .582 |
| | | | | | | | | .560 | .538 | .515 | .491 | .466 | .442 | .416 | .391 | .365 | .340 |
| | | | | | | | | .315 | .290 | .266 | .242 | .219 | .197 | .177 | .157 | .139 | .122 |
| 22.000 | 21.446 | .000 | .554 | .89 | 5.48 | 2.5% | 41 | .845 | .837 | .829 | .820 | .811 | .800 | .790 | .778 | .766 | .754 |
| | | | | | | | | .740 | .726 | .711 | .695 | .678 | .661 | .642 | .623 | .603 | .582 |
| | | | | | | | | .560 | .538 | .515 | .491 | .466 | .442 | .416 | .391 | .365 | .340 |
| | | | | | | | | .315 | .290 | .266 | .242 | .219 | .197 | .177 | .157 | .139 | .122 |

*SIMILAR TO A.T.&T. NEAL-WILKINSON TABLES(SEE SECTION 1)

*EQUIVALENT RANDOM THEORY: VMR(VARIANCE TO MEAN RATIO)– 4.0 - NO RETRIES (PEAKED TRAFFIC - ROUTE ADVANCEMENT OR DAY-TO-DAY VARIATION)

| HOURS FIRST ATTEMPT VMR=4.0 | HOURS CON-NECT | HOURS RECALL /RETRY | HOURS OVFL | CON-NECT VMR | OVFL VMR | PERCENT BLOCKED (GRADE) | LINE | HOURS CONNECTED TRAFFIC ON LINES 1 | 2 | 3 | 4 | 5 | 6 | 7 | 8 | 9 | 10 |
|---|---|---|---|---|---|---|---|---|---|---|---|---|---|---|---|---|---|
| | | | | | | | | .106 | | | | | | | | | |
| 22.000 | 21.538 | .000 | .462 | .91 | 5.39 | 2.1% | 42 | .845 | .837 | .829 | .820 | .811 | .800 | .790 | .778 | .766 | .754 |
| | | | | | | | | .740 | .726 | .711 | .695 | .678 | .661 | .642 | .623 | .603 | .582 |
| | | | | | | | | .560 | .538 | .515 | .491 | .466 | .442 | .416 | .391 | .365 | .340 |
| | | | | | | | | .315 | .290 | .266 | .242 | .219 | .197 | .177 | .157 | .139 | .122 |
| | | | | | | | | .106 | .092 | | | | | | | | |
| 22.000 | 21.617 | .000 | .383 | .92 | 5.30 | 1.7% | 43 | .845 | .837 | .829 | .820 | .811 | .800 | .790 | .778 | .766 | .754 |
| | | | | | | | | .740 | .726 | .711 | .695 | .678 | .661 | .642 | .623 | .603 | .582 |
| | | | | | | | | .560 | .538 | .515 | .491 | .466 | .442 | .416 | .391 | .365 | .340 |
| | | | | | | | | .315 | .290 | .266 | .242 | .219 | .197 | .177 | .157 | .139 | .122 |
| | | | | | | | | .106 | .092 | .079 | | | | | | | |
| 22.000 | 21.685 | .000 | .315 | .93 | 5.20 | 1.4% | 44 | .845 | .837 | .829 | .820 | .811 | .800 | .790 | .778 | .766 | .754 |
| | | | | | | | | .740 | .726 | .711 | .695 | .678 | .661 | .642 | .623 | .603 | .582 |
| | | | | | | | | .560 | .538 | .515 | .491 | .466 | .442 | .416 | .391 | .365 | .340 |
| | | | | | | | | .315 | .290 | .266 | .242 | .219 | .197 | .177 | .157 | .139 | .122 |
| | | | | | | | | .106 | .092 | .079 | .067 | | | | | | |
| 22.000 | 21.742 | .000 | .258 | .94 | 5.11 | 1.2% | 45 | .845 | .837 | .829 | .820 | .811 | .800 | .790 | .778 | .766 | .754 |
| | | | | | | | | .740 | .726 | .711 | .695 | .678 | .661 | .642 | .623 | .603 | .582 |
| | | | | | | | | .560 | .538 | .515 | .491 | .466 | .442 | .416 | .391 | .365 | .340 |
| | | | | | | | | .315 | .290 | .266 | .242 | .219 | .197 | .177 | .157 | .139 | .122 |
| | | | | | | | | .106 | .092 | .079 | .067 | .057 | | | | | |
| 22.000 | 21.790 | .000 | .210 | .95 | 5.02 | 1.0% | 46 | .845 | .837 | .829 | .820 | .811 | .800 | .790 | .778 | .766 | .754 |
| | | | | | | | | .740 | .726 | .711 | .695 | .678 | .661 | .642 | .623 | .603 | .582 |
| | | | | | | | | .560 | .538 | .515 | .491 | .466 | .442 | .416 | .391 | .365 | .340 |
| | | | | | | | | .315 | .290 | .266 | .242 | .219 | .197 | .177 | .157 | .139 | .122 |
| | | | | | | | | .106 | .092 | .079 | .067 | .057 | .048 | | | | |
| 22.000 | 21.830 | .000 | .170 | .96 | 4.92 | .8% | 47 | .845 | .837 | .829 | .820 | .811 | .800 | .790 | .778 | .766 | .754 |
| | | | | | | | | .740 | .726 | .711 | .695 | .678 | .661 | .642 | .623 | .603 | .582 |
| | | | | | | | | .560 | .538 | .515 | .491 | .466 | .442 | .416 | .391 | .365 | .340 |
| | | | | | | | | .315 | .290 | .266 | .242 | .219 | .197 | .177 | .157 | .139 | .122 |
| | | | | | | | | .106 | .092 | .079 | .067 | .057 | .048 | .040 | | | |
| 22.000 | 21.863 | .000 | .137 | .97 | 4.83 | .6% | 48 | .845 | .837 | .829 | .820 | .811 | .800 | .790 | .778 | .766 | .754 |
| | | | | | | | | .740 | .726 | .711 | .695 | .678 | .661 | .642 | .623 | .603 | .582 |
| | | | | | | | | .560 | .538 | .515 | .491 | .466 | .442 | .416 | .391 | .365 | .340 |
| | | | | | | | | .315 | .290 | .266 | .242 | .219 | .197 | .177 | .157 | .139 | .122 |
| | | | | | | | | .106 | .092 | .079 | .067 | .057 | .048 | .040 | .033 | | |
| 24.000 | 15.789 | .000 | 8.211 | .39 | 5.82 | 34.2% | 21 | .856 | .849 | .842 | .834 | .826 | .817 | .807 | .798 | .787 | .776 |
| | | | | | | | | .764 | .752 | .738 | .725 | .710 | .694 | .678 | .661 | .643 | .625 |
| | | | | | | | | .605 | | | | | | | | | |
| 24.000 | 16.374 | .000 | 7.626 | .41 | 5.89 | 31.8% | 22 | .856 | .849 | .842 | .834 | .826 | .817 | .807 | .798 | .787 | .776 |

*SIMILAR TO A.T.&T. NEAL-WILKINSON TABLES(SEE SECTION 1)

*EQUIVALENT RANDOM THEORY: VMR(VARIANCE TO MEAN RATIO)– 4.0 - NO RETRIES (PEAKED TRAFFIC - ROUTE ADVANCEMENT OR DAY-TO-DAY VARIATION)

| HOURS FIRST ATTEMPT VMR = 4.0 | HOURS CON-NECT | HOURS RECALL /RETRY | HOURS OVFL | CON-NECT VMR | OVFL VMR | PERCENT BLOCKED (GRADE) | LINE | HOURS CONNECTED TRAFFIC ON LINES 1 | 2 | 3 | 4 | 5 | 6 | 7 | 8 | 9 | 10 |
|---|---|---|---|---|---|---|---|---|---|---|---|---|---|---|---|---|---|
| | | | | | | | | .764 | .752 | .738 | .725 | .710 | .694 | .678 | .661 | .643 | .625 |
| | | | | | | | | .605 | .585 | | | | | | | | |
| 24.000 | 16.938 | .000 | 7.062 | .44 | 5.95 | 29.4% | 23 | .856 | .849 | .842 | .834 | .826 | .817 | .807 | .798 | .787 | .776 |
| | | | | | | | | .764 | .752 | .738 | .725 | .710 | .694 | .678 | .661 | .643 | .625 |
| | | | | | | | | .605 | .585 | .564 | | | | | | | |
| 24.000 | 17.481 | .000 | 6.519 | .46 | 6.01 | 27.2% | 24 | .856 | .849 | .842 | .834 | .826 | .817 | .807 | .798 | .787 | .776 |
| | | | | | | | | .764 | .752 | .738 | .725 | .710 | .694 | .678 | .661 | .643 | .625 |
| | | | | | | | | .605 | .585 | .564 | .543 | | | | | | |
| 24.000 | 18.001 | .000 | 5.999 | .48 | 6.07 | 25.0% | 25 | .856 | .849 | .842 | .834 | .826 | .817 | .807 | .798 | .787 | .776 |
| | | | | | | | | .764 | .752 | .738 | .725 | .710 | .694 | .678 | .661 | .643 | .625 |
| | | | | | | | | .605 | .585 | .564 | .543 | .520 | | | | | |
| 24.000 | 18.498 | .000 | 5.502 | .50 | 6.11 | 22.9% | 26 | .856 | .849 | .842 | .834 | .826 | .817 | .807 | .798 | .787 | .776 |
| | | | | | | | | .764 | .752 | .738 | .725 | .710 | .694 | .678 | .661 | .643 | .625 |
| | | | | | | | | .605 | .585 | .564 | .543 | .520 | .497 | | | | |
| 24.000 | 18.972 | .000 | 5.028 | .53 | 6.15 | 21.0% | 27 | .856 | .849 | .842 | .834 | .826 | .817 | .807 | .798 | .787 | .776 |
| | | | | | | | | .764 | .752 | .738 | .725 | .710 | .694 | .678 | .661 | .643 | .625 |
| | | | | | | | | .605 | .585 | .564 | .543 | .520 | .497 | .474 | | | |
| 24.000 | 19.421 | .000 | 4.579 | .55 | 6.19 | 19.1% | 28 | .856 | .849 | .842 | .834 | .826 | .817 | .807 | .798 | .787 | .776 |
| | | | | | | | | .764 | .752 | .738 | .725 | .710 | .694 | .678 | .661 | .643 | .625 |
| | | | | | | | | .605 | .585 | .564 | .543 | .520 | .497 | .474 | .450 | | |
| 24.000 | 19.847 | .000 | 4.153 | .57 | 6.21 | 17.3% | 29 | .856 | .849 | .842 | .834 | .826 | .817 | .807 | .798 | .787 | .776 |
| | | | | | | | | .764 | .752 | .738 | .725 | .710 | .694 | .678 | .661 | .643 | .625 |
| | | | | | | | | .605 | .585 | .564 | .543 | .520 | .497 | .474 | .450 | .425 | |
| 24.000 | 20.248 | .000 | 3.752 | .60 | 6.23 | 15.6% | 30 | .856 | .849 | .842 | .834 | .826 | .817 | .807 | .798 | .787 | .776 |
| | | | | | | | | .764 | .752 | .738 | .725 | .710 | .694 | .678 | .661 | .643 | .625 |
| | | | | | | | | .605 | .585 | .564 | .543 | .520 | .497 | .474 | .450 | .425 | .401 |
| 24.000 | 20.624 | .000 | 3.376 | .62 | 6.24 | 14.1% | 31 | .856 | .849 | .842 | .834 | .826 | .817 | .807 | .798 | .787 | .776 |
| | | | | | | | | .764 | .752 | .738 | .725 | .710 | .694 | .678 | .661 | .643 | .625 |
| | | | | | | | | .605 | .585 | .564 | .543 | .520 | .497 | .474 | .450 | .425 | .401 |
| | | | | | | | | .376 | | | | | | | | | |
| 24.000 | 20.975 | .000 | 3.025 | .65 | 6.23 | 12.6% | 32 | .856 | .849 | .842 | .834 | .826 | .817 | .807 | .798 | .787 | .776 |
| | | | | | | | | .764 | .752 | .738 | .725 | .710 | .694 | .678 | .661 | .643 | .625 |
| | | | | | | | | .605 | .585 | .564 | .543 | .520 | .497 | .474 | .450 | .425 | .401 |
| | | | | | | | | .376 | .351 | | | | | | | | |
| 24.000 | 21.302 | .000 | 2.698 | .67 | 6.23 | 11.2% | 33 | .856 | .849 | .842 | .834 | .826 | .817 | .807 | .798 | .787 | .776 |
| | | | | | | | | .764 | .752 | .738 | .725 | .710 | .694 | .678 | .661 | .643 | .625 |
| | | | | | | | | .605 | .585 | .564 | .543 | .520 | .497 | .474 | .450 | .425 | .401 |
| | | | | | | | | .376 | .351 | .327 | | | | | | | |
| 24.000 | 21.604 | .000 | 2.396 | .70 | 6.21 | 10.0% | 34 | .856 | .849 | .842 | .834 | .826 | .817 | .807 | .798 | .787 | .776 |
| | | | | | | | | .764 | .752 | .738 | .725 | .710 | .694 | .678 | .661 | .643 | .625 |

*SIMILAR TO A.T.&T. NEAL-WILKINSON TABLES(SEE SECTION 1)

*EQUIVALENT RANDOM THEORY: VMR(VARIANCE TO MEAN RATIO)– 4.0 - NO RETRIES (PEAKED TRAFFIC - ROUTE ADVANCEMENT OR DAY-TO-DAY VARIATION)

| HOURS FIRST ATTEMPT VMR = 4.0 | HOURS CON-NECT | HOURS RECALL /RETRY | HOURS OVFL | CON-NECT VMR | OVFL VMR | PERCENT BLOCKED (GRADE) | LINE | HOURS CONNECTED TRAFFIC ON LINES 1 | 2 | 3 | 4 | 5 | 6 | 7 | 8 | 9 | 10 |
|---|---|---|---|---|---|---|---|---|---|---|---|---|---|---|---|---|---|
| | | | | | | | | .605 | .585 | .564 | .543 | .520 | .497 | .474 | .450 | .425 | .401 |
| | | | | | | | | .376 | .351 | .327 | .302 | | | | | | |
| 24.000 | 21.882 | .000 | 2.118 | .72 | 6.18 | 8.8% | 35 | .856 | .849 | .842 | .834 | .826 | .817 | .807 | .798 | .787 | .776 |
| | | | | | | | | .764 | .752 | .738 | .725 | .710 | .694 | .678 | .661 | .643 | .625 |
| | | | | | | | | .605 | .585 | .564 | .543 | .520 | .497 | .474 | .450 | .425 | .401 |
| | | | | | | | | .376 | .351 | .327 | .302 | .279 | | | | | |
| 24.000 | 22.138 | .000 | 1.862 | .74 | 6.15 | 7.8% | 36 | .856 | .849 | .842 | .834 | .826 | .817 | .807 | .798 | .787 | .776 |
| | | | | | | | | .764 | .752 | .738 | .725 | .710 | .694 | .678 | .661 | .643 | .625 |
| | | | | | | | | .605 | .585 | .564 | .543 | .520 | .497 | .474 | .450 | .425 | .401 |
| | | | | | | | | .376 | .351 | .327 | .302 | .279 | .255 | | | | |
| 24.000 | 22.370 | .000 | 1.630 | .77 | 6.10 | 6.8% | 37 | .856 | .849 | .842 | .834 | .826 | .817 | .807 | .798 | .787 | .776 |
| | | | | | | | | .764 | .752 | .738 | .725 | .710 | .694 | .678 | .661 | .643 | .625 |
| | | | | | | | | .605 | .585 | .564 | .543 | .520 | .497 | .474 | .450 | .425 | .401 |
| | | | | | | | | .376 | .351 | .327 | .302 | .279 | .255 | .233 | | | |
| 24.000 | 22.581 | .000 | 1.419 | .79 | 6.05 | 5.9% | 38 | .856 | .849 | .842 | .834 | .826 | .817 | .807 | .798 | .787 | .776 |
| | | | | | | | | .764 | .752 | .738 | .725 | .710 | .694 | .678 | .661 | .643 | .625 |
| | | | | | | | | .605 | .585 | .564 | .543 | .520 | .497 | .474 | .450 | .425 | .401 |
| | | | | | | | | .376 | .351 | .327 | .302 | .279 | .255 | .233 | .211 | | |
| 24.000 | 22.771 | .000 | 1.229 | .81 | 5.99 | 5.1% | 39 | .856 | .849 | .842 | .834 | .826 | .817 | .807 | .798 | .787 | .776 |
| | | | | | | | | .764 | .752 | .738 | .725 | .710 | .694 | .678 | .661 | .643 | .625 |
| | | | | | | | | .605 | .585 | .564 | .543 | .520 | .497 | .474 | .450 | .425 | .401 |
| | | | | | | | | .376 | .351 | .327 | .302 | .279 | .255 | .233 | .211 | .190 | |
| 24.000 | 22.941 | .000 | 1.059 | .83 | 5.93 | 4.4% | 40 | .856 | .849 | .842 | .834 | .826 | .817 | .807 | .798 | .787 | .776 |
| | | | | | | | | .764 | .752 | .738 | .725 | .710 | .694 | .678 | .661 | .643 | .625 |
| | | | | | | | | .605 | .585 | .564 | .543 | .520 | .497 | .474 | .450 | .425 | .401 |
| | | | | | | | | .376 | .351 | .327 | .302 | .279 | .255 | .233 | .211 | .190 | .170 |
| 24.000 | 23.093 | .000 | .907 | .85 | 5.85 | 3.8% | 41 | .856 | .849 | .842 | .834 | .826 | .817 | .807 | .798 | .787 | .776 |
| | | | | | | | | .764 | .752 | .738 | .725 | .710 | .694 | .678 | .661 | .643 | .625 |
| | | | | | | | | .605 | .585 | .564 | .543 | .520 | .497 | .474 | .450 | .425 | .401 |
| | | | | | | | | .376 | .351 | .327 | .302 | .279 | .255 | .233 | .211 | .190 | .170 |
| | | | | | | | | .152 | | | | | | | | | |
| 24.000 | 23.227 | .000 | .773 | .87 | 5.78 | 3.2% | 42 | .856 | .849 | .842 | .834 | .826 | .817 | .807 | .798 | .787 | .776 |
| | | | | | | | | .764 | .752 | .738 | .725 | .710 | .694 | .678 | .661 | .643 | .625 |
| | | | | | | | | .605 | .585 | .564 | .543 | .520 | .497 | .474 | .450 | .425 | .401 |
| | | | | | | | | .376 | .351 | .327 | .302 | .279 | .255 | .233 | .211 | .190 | .170 |
| | | | | | | | | .152 | .134 | | | | | | | | |
| 24.000 | 23.345 | .000 | .655 | .88 | 5.70 | 2.7% | 43 | .856 | .849 | .842 | .834 | .826 | .817 | .807 | .798 | .787 | .776 |
| | | | | | | | | .764 | .752 | .738 | .725 | .710 | .694 | .678 | .661 | .643 | .625 |
| | | | | | | | | .605 | .585 | .564 | .543 | .520 | .497 | .474 | .450 | .425 | .401 |
| | | | | | | | | .376 | .351 | .327 | .302 | .279 | .255 | .233 | .211 | .190 | .170 |

*SIMILAR TO A.T.&T. NEAL-WILKINSON TABLES(SEE SECTION 1)

*EQUIVALENT RANDOM THEORY: VMR(VARIANCE TO MEAN RATIO)– 4.0 - NO RETRIES (PEAKED TRAFFIC - ROUTE ADVANCEMENT OR DAY-TO-DAY VARIATION)

| HOURS FIRST ATTEMPT VMR=4.0 | HOURS CON-NECT | HOURS RECALL /RETRY | HOURS OVFL | CON-NECT VMR | OVFL VMR | PERCENT BLOCKED (GRADE) | LINE | HOURS CONNECTED TRAFFIC ON LINES 1 | 2 | 3 | 4 | 5 | 6 | 7 | 8 | 9 | 10 |
|---|---|---|---|---|---|---|---|---|---|---|---|---|---|---|---|---|---|
| | | | | | | | | .152 | .134 | .118 | | | | | | | |
| 24.000 | 23.448 | .000 | .552 | .90 | 5.61 | 2.3% | 44 | .856 | .849 | .842 | .834 | .826 | .817 | .807 | .798 | .787 | .776 |
| | | | | | | | | .764 | .752 | .738 | .725 | .710 | .694 | .678 | .661 | .643 | .625 |
| | | | | | | | | .605 | .585 | .564 | .543 | .520 | .497 | .474 | .450 | .425 | .401 |
| | | | | | | | | .376 | .351 | .327 | .302 | .279 | .255 | .233 | .211 | .190 | .170 |
| | | | | | | | | .152 | .134 | .118 | .103 | | | | | | |
| 24.000 | 23.538 | .000 | .462 | .91 | 5.52 | 1.9% | 45 | .856 | .849 | .842 | .834 | .826 | .817 | .807 | .798 | .787 | .776 |
| | | | | | | | | .764 | .752 | .738 | .725 | .710 | .694 | .678 | .661 | .643 | .625 |
| | | | | | | | | .605 | .585 | .564 | .543 | .520 | .497 | .474 | .450 | .425 | .401 |
| | | | | | | | | .376 | .351 | .327 | .302 | .279 | .255 | .233 | .211 | .190 | .170 |
| | | | | | | | | .152 | .134 | .118 | .103 | .090 | | | | | |
| 24.000 | 23.615 | .000 | .385 | .92 | 5.43 | 1.6% | 46 | .856 | .849 | .842 | .834 | .826 | .817 | .807 | .798 | .787 | .776 |
| | | | | | | | | .764 | .752 | .738 | .725 | .710 | .694 | .678 | .661 | .643 | .625 |
| | | | | | | | | .605 | .585 | .564 | .543 | .520 | .497 | .474 | .450 | .425 | .401 |
| | | | | | | | | .376 | .351 | .327 | .302 | .279 | .255 | .233 | .211 | .190 | .170 |
| | | | | | | | | .152 | .134 | .118 | .103 | .090 | .077 | | | | |
| 24.000 | 23.681 | .000 | .319 | .93 | 5.34 | 1.3% | 47 | .856 | .849 | .842 | .834 | .826 | .817 | .807 | .798 | .787 | .776 |
| | | | | | | | | .764 | .752 | .738 | .725 | .710 | .694 | .678 | .661 | .643 | .625 |
| | | | | | | | | .605 | .585 | .564 | .543 | .520 | .497 | .474 | .450 | .425 | .401 |
| | | | | | | | | .376 | .351 | .327 | .302 | .279 | .255 | .233 | .211 | .190 | .170 |
| | | | | | | | | .152 | .134 | .118 | .103 | .090 | .077 | .066 | | | |
| 24.000 | 23.738 | .000 | .262 | .94 | 5.24 | 1.1% | 48 | .856 | .849 | .842 | .834 | .826 | .817 | .807 | .798 | .787 | .776 |
| | | | | | | | | .764 | .752 | .738 | .725 | .710 | .694 | .678 | .661 | .643 | .625 |
| | | | | | | | | .605 | .585 | .564 | .543 | .520 | .497 | .474 | .450 | .425 | .401 |
| | | | | | | | | .376 | .351 | .327 | .302 | .279 | .255 | .233 | .211 | .190 | .170 |
| | | | | | | | | .152 | .134 | .118 | .103 | .090 | .077 | .066 | .056 | | |
| 24.000 | 23.785 | .000 | .215 | .95 | 5.15 | .9% | 49 | .856 | .849 | .842 | .834 | .826 | .817 | .807 | .798 | .787 | .776 |
| | | | | | | | | .764 | .752 | .738 | .725 | .710 | .694 | .678 | .661 | .643 | .625 |
| | | | | | | | | .605 | .585 | .564 | .543 | .520 | .497 | .474 | .450 | .425 | .401 |
| | | | | | | | | .376 | .351 | .327 | .302 | .279 | .255 | .233 | .211 | .190 | .170 |
| | | | | | | | | .152 | .134 | .118 | .103 | .090 | .077 | .066 | .056 | .048 | |
| 24.000 | 23.825 | .000 | .175 | .96 | 5.05 | .7% | 50 | .856 | .849 | .842 | .834 | .826 | .817 | .807 | .798 | .787 | .776 |
| | | | | | | | | .764 | .752 | .738 | .725 | .710 | .694 | .678 | .661 | .643 | .625 |
| | | | | | | | | .605 | .585 | .564 | .543 | .520 | .497 | .474 | .450 | .425 | .401 |
| | | | | | | | | .376 | .351 | .327 | .302 | .279 | .255 | .233 | .211 | .190 | .170 |
| | | | | | | | | .152 | .134 | .118 | .103 | .090 | .077 | .066 | .056 | .048 | .040 |
| 24.000 | 23.859 | .000 | .141 | .97 | 4.96 | .6% | 51 | .856 | .849 | .842 | .834 | .826 | .817 | .807 | .798 | .787 | .776 |
| | | | | | | | | .764 | .752 | .738 | .725 | .710 | .694 | .678 | .661 | .643 | .625 |
| | | | | | | | | .605 | .585 | .564 | .543 | .520 | .497 | .474 | .450 | .425 | .401 |
| | | | | | | | | .376 | .351 | .327 | .302 | .279 | .255 | .233 | .211 | .190 | .170 |

*SIMILAR TO A.T.&T. NEAL-WILKINSON TABLES(SEE SECTION 1)

## *EQUIVALENT RANDOM THEORY: VMR(VARIANCE TO MEAN RATIO)– 4.0 - NO RETRIES (PEAKED TRAFFIC - ROUTE ADVANCEMENT OR DAY-TO-DAY VARIATION)

| HOURS FIRST ATTEMPT VMR = 4.0 | HOURS CON-NECT | HOURS RECALL /RETRY | HOURS OVFL | CON-NECT VMR | OVFL VMR | PERCENT BLOCKED (GRADE) | LINE | HOURS CONNECTED TRAFFIC ON LINES 1 | 2 | 3 | 4 | 5 | 6 | 7 | 8 | 9 | 10 |
|---|---|---|---|---|---|---|---|---|---|---|---|---|---|---|---|---|---|
| | | | | | | | | .152 | .134 | .118 | .103 | .090 | .077 | .066 | .056 | .048 | .040 |
| | | | | | | | | .033 | | | | | | | | | |
| 26.000 | 17.479 | .000 | 8.521 | .39 | 5.97 | 32.8% | 23 | .866 | .860 | .853 | .846 | .839 | .831 | .823 | .814 | .805 | .795 |
| | | | | | | | | .784 | .773 | .762 | .750 | .737 | .723 | .709 | .694 | .678 | .662 |
| | | | | | | | | .644 | .626 | .607 | | | | | | | |
| 26.000 | 18.067 | .000 | 7.933 | .41 | 6.04 | 30.5% | 24 | .866 | .860 | .853 | .846 | .839 | .831 | .823 | .814 | .805 | .795 |
| | | | | | | | | .784 | .773 | .762 | .750 | .737 | .723 | .709 | .694 | .678 | .662 |
| | | | | | | | | .644 | .626 | .607 | .588 | | | | | | |
| 26.000 | 18.635 | .000 | 7.365 | .43 | 6.11 | 28.3% | 25 | .866 | .860 | .853 | .846 | .839 | .831 | .823 | .814 | .805 | .795 |
| | | | | | | | | .784 | .773 | .762 | .750 | .737 | .723 | .709 | .694 | .678 | .662 |
| | | | | | | | | .644 | .626 | .607 | .588 | .568 | | | | | |
| 26.000 | 19.182 | .000 | 6.818 | .45 | 6.17 | 26.2% | 26 | .866 | .860 | .853 | .846 | .839 | .831 | .823 | .814 | .805 | .795 |
| | | | | | | | | .784 | .773 | .762 | .750 | .737 | .723 | .709 | .694 | .678 | .662 |
| | | | | | | | | .644 | .626 | .607 | .588 | .568 | .547 | | | | |
| 26.000 | 19.707 | .000 | 6.293 | .47 | 6.22 | 24.2% | 27 | .866 | .860 | .853 | .846 | .839 | .831 | .823 | .814 | .805 | .795 |
| | | | | | | | | .784 | .773 | .762 | .750 | .737 | .723 | .709 | .694 | .678 | .662 |
| | | | | | | | | .644 | .626 | .607 | .588 | .568 | .547 | .525 | | | |
| 26.000 | 20.210 | .000 | 5.790 | .50 | 6.27 | 22.3% | 28 | .866 | .860 | .853 | .846 | .839 | .831 | .823 | .814 | .805 | .795 |
| | | | | | | | | .784 | .773 | .762 | .750 | .737 | .723 | .709 | .694 | .678 | .662 |
| | | | | | | | | .644 | .626 | .607 | .588 | .568 | .547 | .525 | .503 | | |
| 26.000 | 20.690 | .000 | 5.310 | .52 | 6.31 | 20.4% | 29 | .866 | .860 | .853 | .846 | .839 | .831 | .823 | .814 | .805 | .795 |
| | | | | | | | | .784 | .773 | .762 | .750 | .737 | .723 | .709 | .694 | .678 | .662 |
| | | | | | | | | .644 | .626 | .607 | .588 | .568 | .547 | .525 | .503 | .480 | |
| 26.000 | 21.147 | .000 | 4.853 | .54 | 6.35 | 18.7% | 30 | .866 | .860 | .853 | .846 | .839 | .831 | .823 | .814 | .805 | .795 |
| | | | | | | | | .784 | .773 | .762 | .750 | .737 | .723 | .709 | .694 | .678 | .662 |
| | | | | | | | | .644 | .626 | .607 | .588 | .568 | .547 | .525 | .503 | .480 | .457 |
| 26.000 | 21.580 | .000 | 4.420 | .57 | 6.37 | 17.0% | 31 | .866 | .860 | .853 | .846 | .839 | .831 | .823 | .814 | .805 | .795 |
| | | | | | | | | .784 | .773 | .762 | .750 | .737 | .723 | .709 | .694 | .678 | .662 |
| | | | | | | | | .644 | .626 | .607 | .588 | .568 | .547 | .525 | .503 | .480 | .457 |
| | | | | | | | | .433 | | | | | | | | | |
| 26.000 | 21.990 | .000 | 4.010 | .59 | 6.39 | 15.4% | 32 | .066 | .860 | 853 | .846 | .839 | .831 | .823 | .814 | .805 | .795 |
| | | | | | | | | .784 | .773 | .762 | .750 | .737 | .723 | .709 | .694 | .070 | .662 |
| | | | | | | | | .644 | .626 | .607 | .588 | .568 | .547 | .525 | .503 | .480 | .457 |
| | | | | | | | | .433 | .410 | | | | | | | | |
| 26.000 | 22.375 | .000 | 3.625 | .61 | 6.40 | 13.9% | 33 | .866 | .860 | .853 | .846 | .839 | .831 | .823 | .814 | .805 | .795 |
| | | | | | | | | .784 | .773 | .762 | .750 | .737 | .723 | .709 | .694 | .678 | .662 |
| | | | | | | | | .644 | .626 | .607 | .588 | .568 | .547 | .525 | .503 | .480 | .457 |
| | | | | | | | | .433 | .410 | .386 | | | | | | | |
| 26.000 | 22.737 | .000 | 3.263 | .64 | 6.40 | 12.6% | 34 | .866 | .860 | .853 | .846 | .839 | .831 | .823 | .814 | .805 | .795 |
| | | | | | | | | .784 | .773 | .762 | .750 | .737 | .723 | .709 | .694 | .678 | .662 |

*SIMILAR TO A.T.&T. NEAL-WILKINSON TABLES(SEE SECTION 1)

*EQUIVALENT RANDOM THEORY: VMR(VARIANCE TO MEAN RATIO)– 4.0 - NO RETRIES (PEAKED TRAFFIC - ROUTE ADVANCEMENT OR DAY-TO-DAY VARIATION)

| HOURS FIRST ATTEMPT VMR=4.0 | HOURS CON-NECT | HOURS RECALL /RETRY | HOURS OVFL | CON-NECT VMR | OVFL VMR | PERCENT BLOCKED (GRADE) | LINE | HOURS CONNECTED TRAFFIC ON LINES 1 | 2 | 3 | 4 | 5 | 6 | 7 | 8 | 9 | 10 |
|---|---|---|---|---|---|---|---|---|---|---|---|---|---|---|---|---|---|
| | | | | | | | | .644 | .626 | .607 | .588 | .568 | .547 | .525 | .503 | .480 | .457 |
| | | | | | | | | .433 | .410 | .386 | .362 | | | | | | |
| 26.000 | 23.074 | .000 | 2.926 | .66 | 6.40 | 11.3% | 35 | .866 | .860 | .853 | .846 | .839 | .831 | .823 | .814 | .805 | .795 |
| | | | | | | | | .784 | .773 | .762 | .750 | .737 | .723 | .709 | .694 | .678 | .662 |
| | | | | | | | | .644 | .626 | .607 | .588 | .568 | .547 | .525 | .503 | .480 | .457 |
| | | | | | | | | .433 | .410 | .386 | .362 | .338 | | | | | |
| 26.000 | 23.388 | .000 | 2.612 | .69 | 6.38 | 10.0% | 36 | .866 | .860 | .853 | .846 | .839 | .831 | .823 | .814 | .805 | .795 |
| | | | | | | | | .784 | .773 | .762 | .750 | .737 | .723 | .709 | .694 | .678 | .662 |
| | | | | | | | | .644 | .626 | .607 | .588 | .568 | .547 | .525 | .503 | .480 | .457 |
| | | | | | | | | .433 | .410 | .386 | .362 | .338 | .314 | | | | |
| 26.000 | 23.678 | .000 | 2.322 | .71 | 6.36 | 8.9% | 37 | .866 | .860 | .853 | .846 | .839 | .831 | .823 | .814 | .805 | .795 |
| | | | | | | | | .784 | .773 | .762 | .750 | .737 | .723 | .709 | .694 | .678 | .662 |
| | | | | | | | | .644 | .626 | .607 | .588 | .568 | .547 | .525 | .503 | .480 | .457 |
| | | | | | | | | .433 | .410 | .386 | .362 | .338 | .314 | .290 | | | |
| 26.000 | 23.946 | .000 | 2.054 | .73 | 6.33 | 7.9% | 38 | .866 | .860 | .853 | .846 | .839 | .831 | .823 | .814 | .805 | .795 |
| | | | | | | | | .784 | .773 | .762 | .750 | .737 | .723 | .709 | .694 | .678 | .662 |
| | | | | | | | | .644 | .626 | .607 | .588 | .568 | .547 | .525 | .503 | .480 | .457 |
| | | | | | | | | .433 | .410 | .386 | .362 | .338 | .314 | .290 | .267 | | |
| 26.000 | 24.191 | .000 | 1.809 | .75 | 6.29 | 7.0% | 39 | .866 | .860 | .853 | .846 | .839 | .831 | .823 | .814 | .805 | .795 |
| | | | | | | | | .784 | .773 | .762 | .750 | .737 | .723 | .709 | .694 | .678 | .662 |
| | | | | | | | | .644 | .626 | .607 | .588 | .568 | .547 | .525 | .503 | .480 | .457 |
| | | | | | | | | .433 | .410 | .386 | .362 | .338 | .314 | .290 | .267 | .245 | |
| 26.000 | 24.414 | .000 | 1.586 | .78 | 6.24 | 6.1% | 40 | .866 | .860 | .853 | .846 | .839 | .831 | .823 | .814 | .805 | .795 |
| | | | | | | | | .784 | .773 | .762 | .750 | .737 | .723 | .709 | .694 | .678 | .662 |
| | | | | | | | | .644 | .626 | .607 | .588 | .568 | .547 | .525 | .503 | .480 | .457 |
| | | | | | | | | .433 | .410 | .386 | .362 | .338 | .314 | .290 | .267 | .245 | .223 |
| 26.000 | 24.617 | .000 | 1.383 | .80 | 6.19 | 5.3% | 41 | .000 | .860 | .853 | .846 | .839 | .831 | .823 | .814 | .805 | .795 |
| | | | | | | | | .784 | .773 | .762 | .750 | .737 | .723 | .709 | .694 | .678 | .662 |
| | | | | | | | | .644 | .626 | .607 | .588 | .568 | .547 | .525 | .503 | .480 | .457 |
| | | | | | | | | .433 | .410 | .386 | .362 | .338 | .314 | .290 | .267 | .245 | .223 |
| | | | | | | | | .203 | | | | | | | | | |
| 26.000 | 24.799 | .000 | 1.201 | .82 | 6.13 | 4.6% | 42 | .866 | .860 | .853 | .846 | .839 | .831 | .823 | .814 | .805 | .795 |
| | | | | | | | | .784 | .773 | .762 | .750 | .737 | .723 | .709 | .694 | .678 | .662 |
| | | | | | | | | .644 | .626 | .607 | .588 | .568 | .547 | .525 | .503 | .480 | .457 |
| | | | | | | | | .433 | .410 | .386 | .362 | .338 | .314 | .290 | .267 | .245 | .223 |
| | | | | | | | | .203 | .183 | | | | | | | | |
| 26.000 | 24.963 | .000 | 1.037 | .84 | 6.06 | 4.0% | 43 | .866 | .860 | .853 | .846 | .839 | .831 | .823 | .814 | .805 | .795 |
| | | | | | | | | .784 | .773 | .762 | .750 | .737 | .723 | .709 | .694 | .678 | .662 |
| | | | | | | | | .644 | .626 | .607 | .588 | .568 | .547 | .525 | .503 | .480 | .457 |
| | | | | | | | | .433 | .410 | .386 | .362 | .338 | .314 | .290 | .267 | .245 | .223 |

*SIMILAR TO A.T.&T. NEAL-WILKINSON TABLES(SEE SECTION 1)

*EQUIVALENT RANDOM THEORY: VMR(VARIANCE TO MEAN RATIO)– 4.0 - NO RETRIES (PEAKED TRAFFIC - ROUTE ADVANCEMENT OR DAY-TO-DAY VARIATION)

| HOURS FIRST ATTEMPT VMR = 4.0 | HOURS CON-NECT | HOURS RECALL /RETRY | HOURS OVFL | CON-NECT VMR | OVFL VMR | PERCENT BLOCKED (GRADE) | LINE | HOURS CONNECTED TRAFFIC ON LINES 1 | 2 | 3 | 4 | 5 | 6 | 7 | 8 | 9 | 10 |
|---|---|---|---|---|---|---|---|---|---|---|---|---|---|---|---|---|---|
| | | | | | | | | .203 | .183 | .164 | | | | | | | |
| 26.000 | 25.109 | .000 | .891 | .85 | 5.99 | 3.4% | 44 | .866 | .860 | .853 | .846 | .839 | .831 | .823 | .814 | .805 | .795 |
| | | | | | | | | .784 | .773 | .762 | .750 | .737 | .723 | .709 | .694 | .678 | .662 |
| | | | | | | | | .644 | .626 | .607 | .588 | .568 | .547 | .525 | .503 | .480 | .457 |
| | | | | | | | | .433 | .410 | .386 | .362 | .338 | .314 | .290 | .267 | .245 | .223 |
| | | | | | | | | .203 | .183 | .164 | .146 | | | | | | |
| 26.000 | 25.239 | .000 | .761 | .87 | 5.91 | 2.9% | 45 | .866 | .860 | .853 | .846 | .839 | .831 | .823 | .814 | .805 | .795 |
| | | | | | | | | .784 | .773 | .762 | .750 | .737 | .723 | .709 | .694 | .678 | .662 |
| | | | | | | | | .644 | .626 | .607 | .588 | .568 | .547 | .525 | .503 | .480 | .457 |
| | | | | | | | | .433 | .410 | .386 | .362 | .338 | .314 | .290 | .267 | .245 | .223 |
| | | | | | | | | .203 | .183 | .164 | .146 | .130 | | | | | |
| 26.000 | 25.353 | .000 | .647 | .89 | 5.82 | 2.5% | 46 | .866 | .860 | .853 | .846 | .839 | .831 | .823 | .814 | .805 | .795 |
| | | | | | | | | .784 | .773 | .762 | .750 | .737 | .723 | .709 | .694 | .678 | .662 |
| | | | | | | | | .644 | .626 | .607 | .588 | .568 | .547 | .525 | .503 | .480 | .457 |
| | | | | | | | | .433 | .410 | .386 | .362 | .338 | .314 | .290 | .267 | .245 | .223 |
| | | | | | | | | .203 | .183 | .164 | .146 | .130 | .114 | | | | |
| 26.000 | 25.453 | .000 | .547 | .90 | 5.74 | 2.1% | 47 | .866 | .860 | .853 | .846 | .839 | .831 | .823 | .814 | .805 | .795 |
| | | | | | | | | .784 | .773 | .762 | .750 | .737 | .723 | .709 | .694 | .678 | .662 |
| | | | | | | | | .644 | .626 | .607 | .588 | .568 | .547 | .525 | .503 | .480 | .457 |
| | | | | | | | | .433 | .410 | .386 | .362 | .338 | .314 | .290 | .267 | .245 | .223 |
| | | | | | | | | .203 | .183 | .164 | .146 | .130 | .114 | .100 | | | |
| 26.000 | 25.540 | .000 | .460 | .91 | 5.65 | 1.8% | 48 | .866 | .860 | .853 | .846 | .839 | .831 | .823 | .814 | .805 | .795 |
| | | | | | | | | .784 | .773 | .762 | .750 | .737 | .723 | .709 | .694 | .678 | .662 |
| | | | | | | | | .644 | .626 | .607 | .588 | .568 | .547 | .525 | .503 | .480 | .457 |
| | | | | | | | | .433 | .410 | .386 | .362 | .338 | .314 | .290 | .267 | .245 | .223 |
| | | | | | | | | .203 | .183 | .164 | .146 | .130 | .114 | .100 | .087 | | |
| 26.000 | 25.615 | .000 | .385 | .92 | 5.55 | 1.5% | 49 | .866 | .860 | .853 | .846 | .839 | .831 | .823 | .814 | .805 | .795 |
| | | | | | | | | .784 | .773 | .762 | .750 | .737 | .723 | .709 | .694 | .678 | .662 |
| | | | | | | | | .644 | .626 | .607 | .588 | .568 | .547 | .525 | .503 | .480 | .457 |
| | | | | | | | | .433 | .410 | .386 | .362 | .338 | .314 | .290 | .267 | .245 | .223 |
| | | | | | | | | .203 | .183 | .164 | .146 | .130 | .114 | .100 | .087 | .075 | |
| 26.000 | 25.680 | .000 | .320 | .94 | 5.46 | 1.2% | 50 | .866 | .860 | .853 | .046 | .839 | .831 | .823 | .814 | .805 | .795 |
| | | | | | | | | .784 | .773 | .762 | .750 | .737 | .723 | .709 | .694 | .678 | .662 |
| | | | | | | | | .644 | .626 | .607 | .588 | .568 | .547 | .525 | .503 | .480 | .457 |
| | | | | | | | | .433 | .410 | .386 | .362 | .338 | .314 | .290 | .267 | .245 | .223 |
| | | | | | | | | .203 | .183 | .164 | .146 | .130 | .114 | .100 | .087 | .075 | .065 |
| 26.000 | 25.735 | .000 | .265 | .94 | 5.37 | 1.0% | 51 | .866 | .860 | .853 | .846 | .839 | .831 | .823 | .814 | .805 | .795 |
| | | | | | | | | .784 | .773 | .762 | .750 | .737 | .723 | .709 | .694 | .678 | .662 |
| | | | | | | | | .644 | .626 | .607 | .588 | .568 | .547 | .525 | .503 | .480 | .457 |
| | | | | | | | | .433 | .410 | .386 | .362 | .338 | .314 | .290 | .267 | .245 | .223 |

*SIMILAR TO A.T.&T. NEAL-WILKINSON TABLES(SEE SECTION 1)

*EQUIVALENT RANDOM THEORY: VMR(VARIANCE TO MEAN RATIO)– 4.0 - NO RETRIES (PEAKED TRAFFIC - ROUTE ADVANCEMENT OR DAY-TO-DAY VARIATION)

| HOURS FIRST ATTEMPT VMR = 4.0 | HOURS CON-NECT | HOURS RECALL /RETRY | HOURS OVFL | CON-NECT VMR | OVFL VMR | PERCENT BLOCKED (GRADE) | LINE | HOURS CONNECTED TRAFFIC ON LINES 1 | 2 | 3 | 4 | 5 | 6 | 7 | 8 | 9 | 10 |
|---|---|---|---|---|---|---|---|---|---|---|---|---|---|---|---|---|---|
| | | | | | | | | .203 | .183 | .164 | .146 | .130 | .114 | .100 | .087 | .075 | .065 |
| | | | | | | | | .055 | | | | | | | | | |
| 26.000 | 25.782 | .000 | .218 | .95 | 5.27 | .8% | 52 | .866 | .860 | .853 | .846 | .839 | .831 | .823 | .814 | .805 | .795 |
| | | | | | | | | .784 | .773 | .762 | .750 | .737 | .723 | .709 | .694 | .678 | .662 |
| | | | | | | | | .644 | .626 | .607 | .588 | .568 | .547 | .525 | .503 | .480 | .457 |
| | | | | | | | | .433 | .410 | .386 | .362 | .338 | .314 | .290 | .267 | .245 | .223 |
| | | | | | | | | .203 | .183 | .164 | .146 | .130 | .114 | .100 | .087 | .075 | .065 |
| | | | | | | | | .055 | .047 | | | | | | | | |
| 26.000 | 25.822 | .000 | .178 | .96 | 5.18 | .7% | 53 | .866 | .860 | .853 | .846 | .839 | .831 | .823 | .814 | .805 | .795 |
| | | | | | | | | .784 | .773 | .762 | .750 | .737 | .723 | .709 | .694 | .678 | .662 |
| | | | | | | | | .644 | .626 | .607 | .588 | .568 | .547 | .525 | .503 | .480 | .457 |
| | | | | | | | | .433 | .410 | .386 | .362 | .338 | .314 | .290 | .267 | .245 | .223 |
| | | | | | | | | .203 | .183 | .164 | .146 | .130 | .114 | .100 | .087 | .075 | .065 |
| | | | | | | | | .055 | .047 | .040 | | | | | | | |
| 26.000 | 25.855 | .000 | .145 | .97 | 5.08 | .6% | 54 | .866 | .860 | .853 | .846 | .839 | .831 | .823 | .814 | .805 | .795 |
| | | | | | | | | .784 | .773 | .762 | .750 | .737 | .723 | .709 | .694 | .678 | .662 |
| | | | | | | | | .644 | .626 | .607 | .588 | .568 | .547 | .525 | .503 | .480 | .457 |
| | | | | | | | | .433 | .410 | .386 | .362 | .338 | .314 | .290 | .267 | .245 | .223 |
| | | | | | | | | .203 | .183 | .164 | .146 | .130 | .114 | .100 | .087 | .075 | .065 |
| | | | | | | | | .055 | .047 | .040 | .033 | | | | | | |
| 28.000 | 18.573 | .000 | 9.427 | .37 | 6.05 | 33.7% | 24 | .874 | .869 | .863 | .857 | .850 | .843 | .836 | .828 | .820 | .811 |
| | | | | | | | | .802 | .792 | .782 | .771 | .760 | .748 | .735 | .722 | .708 | .693 |
| | | | | | | | | .678 | .662 | .645 | .627 | | | | | | |
| 28.000 | 19.182 | .000 | 8.818 | .39 | 6.12 | 31.5% | 25 | .874 | .869 | .863 | .857 | .850 | .843 | .836 | .828 | .820 | .811 |
| | | | | | | | | .802 | .792 | .782 | .771 | .760 | .748 | .735 | .722 | .708 | .693 |
| | | | | | | | | .678 | .662 | .645 | .627 | .609 | | | | | |
| 28.000 | 19.773 | .000 | 8.227 | .41 | 6.19 | 29.4% | 26 | .874 | .869 | .863 | .857 | .850 | .843 | .836 | .828 | .820 | .811 |
| | | | | | | | | .802 | .792 | .782 | .771 | .760 | .748 | .735 | .722 | .708 | .693 |
| | | | | | | | | .678 | .662 | .645 | .627 | .609 | .590 | | | | |
| 28.000 | 20.343 | .000 | 7.657 | .43 | 6.26 | 27.3% | 27 | .874 | .869 | .863 | .857 | .850 | .843 | .836 | .828 | .820 | .811 |
| | | | | | | | | .802 | .792 | .782 | .771 | .760 | .748 | .735 | .722 | .708 | .693 |
| | | | | | | | | .678 | .662 | .645 | .627 | .609 | .590 | .571 | | | |
| 28.000 | 20.894 | .000 | 7.100 | .45 | 6.32 | 25.4% | 28 | .874 | .869 | .863 | .857 | .850 | .843 | .836 | .828 | .820 | .811 |
| | | | | | | | | .802 | .792 | .782 | .771 | .760 | .748 | .735 | .722 | .708 | .690 |
| | | | | | | | | .678 | .662 | .645 | .627 | .609 | .590 | .571 | .550 | | |
| 28.000 | 21.423 | .000 | 6.577 | .47 | 6.37 | 23.5% | 29 | .874 | .869 | .863 | .857 | .850 | .843 | .836 | .028 | .820 | .811 |
| | | | | | | | | .802 | .792 | .782 | .771 | .760 | .748 | .735 | .722 | .708 | .693 |
| | | | | | | | | .678 | .662 | .645 | .027 | .609 | .590 | .571 | .550 | .529 | |
| 28.000 | 21.931 | .000 | 6.069 | .49 | 6.42 | 21.7% | 30 | .874 | .869 | .863 | .857 | .850 | .843 | .836 | .828 | .820 | .811 |
| | | | | | | | | .802 | .792 | .782 | .771 | .760 | .748 | .735 | .722 | .708 | .693 |

*SIMILAR TO A.T.&T. NEAL-WILKINSON TABLES(SEE SECTION 1)

*EQUIVALENT RANDOM THEORY: VMR(VARIANCE TO MEAN RATIO)– 4.0 - NO RETRIES (PEAKED TRAFFIC - ROUTE ADVANCEMENT OR DAY-TO-DAY VARIATION)

| HOURS FIRST ATTEMPT VMR=4.0 | HOURS CON-NECT | HOURS RECALL /RETRY | HOURS OVFL | CON-NECT VMR | OVFL VMR | PERCENT BLOCKED (GRADE) | LINE | HOURS CONNECTED TRAFFIC ON LINES 1 | 2 | 3 | 4 | 5 | 6 | 7 | 8 | 9 | 10 |
|---|---|---|---|---|---|---|---|---|---|---|---|---|---|---|---|---|---|
| | | | | | | | | .678 | .662 | .645 | .627 | .609 | .590 | .571 | .550 | .529 | .508 |
| 28.000 | 22.417 | .000 | 5.583 | .51 | 6.46 | 19.9% | 31 | .874 | .869 | .863 | .857 | .850 | .843 | .836 | .828 | .820 | .811 |
| | | | | | | | | .802 | .792 | .782 | .771 | .760 | .748 | .735 | .722 | .708 | .693 |
| | | | | | | | | .678 | .662 | .645 | .627 | .609 | .590 | .571 | .550 | .529 | .508 |
| | | | | | | | | .486 | | | | | | | | | |
| 28.000 | 22.880 | .000 | 5.120 | .54 | 6.50 | 18.3% | 32 | .874 | .869 | .863 | .857 | .850 | .843 | .836 | .828 | .820 | .811 |
| | | | | | | | | .802 | .792 | .782 | .771 | .760 | .748 | .735 | .722 | .708 | .693 |
| | | | | | | | | .678 | .662 | .645 | .627 | .609 | .590 | .571 | .550 | .529 | .508 |
| | | | | | | | | .486 | .463 | | | | | | | | |
| 28.000 | 23.321 | .000 | 4.679 | .56 | 6.53 | 16.7% | 33 | .874 | .869 | .863 | .857 | .850 | .843 | .836 | .828 | .820 | .811 |
| | | | | | | | | .802 | .792 | .782 | .771 | .760 | .748 | .735 | .722 | .708 | .693 |
| | | | | | | | | .678 | .662 | .645 | .627 | .609 | .590 | .571 | .550 | .529 | .508 |
| | | | | | | | | .486 | .463 | .441 | | | | | | | |
| 28.000 | 23.738 | .000 | 4.262 | .58 | 6.55 | 15.2% | 34 | .874 | .869 | .863 | .857 | .850 | .843 | .836 | .828 | .820 | .811 |
| | | | | | | | | .802 | .792 | .782 | .771 | .760 | .748 | .735 | .722 | .708 | .693 |
| | | | | | | | | .678 | .662 | .645 | .627 | .609 | .590 | .571 | .550 | .529 | .508 |
| | | | | | | | | .486 | .463 | .441 | .417 | | | | | | |
| 28.000 | 24.133 | .000 | 3.867 | .61 | 6.56 | 13.8% | 35 | .874 | .869 | .863 | .857 | .850 | .843 | .836 | .828 | .820 | .811 |
| | | | | | | | | .802 | .792 | .782 | .771 | .760 | .748 | .735 | .722 | .708 | .693 |
| | | | | | | | | .678 | .662 | .645 | .627 | .609 | .590 | .571 | .550 | .529 | .508 |
| | | | | | | | | .486 | .463 | .441 | .417 | .394 | | | | | |
| 28.000 | 24.503 | .000 | 3.497 | .63 | 6.57 | 12.5% | 36 | .874 | .869 | .863 | .857 | .850 | .843 | .836 | .828 | .820 | .811 |
| | | | | | | | | .802 | .792 | .782 | .771 | .760 | .748 | .735 | .722 | .708 | .693 |
| | | | | | | | | .678 | .662 | .645 | .627 | .609 | .590 | .571 | .550 | .529 | .508 |
| | | | | | | | | .486 | .463 | .441 | .417 | .394 | .371 | | | | |
| 28.000 | 24.851 | .000 | 3.149 | .65 | 6.57 | 11.2% | 37 | .874 | .869 | .863 | .857 | .850 | .843 | .836 | .828 | .820 | .811 |
| | | | | | | | | .802 | .792 | .782 | .771 | .760 | .748 | .735 | .722 | .708 | .693 |
| | | | | | | | | .678 | .662 | .645 | .627 | .609 | .590 | .571 | .550 | .529 | .508 |
| | | | | | | | | .486 | .463 | .441 | .417 | .394 | .371 | .347 | | | |
| 28.000 | 25.175 | .000 | 2.825 | .68 | 6.56 | 10.1% | 38 | .874 | .869 | .863 | .857 | .850 | .843 | .836 | .828 | .820 | .811 |
| | | | | | | | | .802 | .792 | .782 | .771 | .760 | .748 | .735 | .722 | .708 | .693 |
| | | | | | | | | .678 | .662 | .645 | .627 | .609 | .590 | .571 | .550 | .529 | .508 |
| | | | | | | | | .486 | .463 | .441 | .417 | .394 | .371 | .347 | .324 | | |
| 28.000 | 25.476 | .000 | 2.524 | .70 | 6.54 | 9.0% | 39 | .874 | .869 | .863 | .857 | .850 | .843 | .836 | .828 | .820 | .811 |
| | | | | | | | | .802 | .792 | .782 | .771 | .760 | .748 | .735 | .722 | .708 | .693 |
| | | | | | | | | .678 | .662 | .645 | .627 | .609 | .590 | .571 | .550 | .529 | .508 |
| | | | | | | | | .486 | .463 | .441 | .417 | .394 | .371 | .347 | .324 | .301 | |
| 28.000 | 25.755 | .000 | 2.245 | .72 | 6.51 | 8.0% | 40 | .874 | .869 | .863 | .857 | .850 | .843 | .836 | .828 | .820 | .811 |
| | | | | | | | | .802 | .792 | .782 | .771 | .760 | .748 | .735 | .722 | .708 | .693 |
| | | | | | | | | .678 | .662 | .645 | .627 | .609 | .590 | .571 | .550 | .529 | .508 |

*SIMILAR TO A.T.&T. NEAL-WILKINSON TABLES(SEE SECTION 1)

## *EQUIVALENT RANDOM THEORY: VMR(VARIANCE TO MEAN RATIO)– 4.0 - NO RETRIES (PEAKED TRAFFIC - ROUTE ADVANCEMENT OR DAY-TO-DAY VARIATION)

| HOURS FIRST ATTEMPT VMR=4.0 | HOURS CON-NECT | HOURS RECALL /RETRY | HOURS OVFL | CON-NECT VMR | OVFL VMR | PERCENT BLOCKED (GRADE) | LINE | HOURS CONNECTED TRAFFIC ON LINES 1 | 2 | 3 | 4 | 5 | 6 | 7 | 8 | 9 | 10 |
|---|---|---|---|---|---|---|---|---|---|---|---|---|---|---|---|---|---|
| | | | | | | | | .486 | .463 | .441 | .417 | .394 | .371 | .347 | .324 | .301 | .279 |
| 28.000 | 26.011 | .000 | 1.989 | .74 | 6.47 | 7.1% | 41 | .874 | .869 | .863 | .857 | .850 | .843 | .836 | .828 | .820 | .811 |
| | | | | | | | | .802 | .792 | .782 | .771 | .760 | .748 | .735 | .722 | .708 | .693 |
| | | | | | | | | .678 | .662 | .645 | .627 | .609 | .590 | .571 | .550 | .529 | .508 |
| | | | | | | | | .486 | .463 | .441 | .417 | .394 | .371 | .347 | .324 | .301 | .279 |
| | | | | | | | | .257 | | | | | | | | | |
| 28.000 | 26.246 | .000 | 1.754 | .76 | 6.43 | 6.3% | 42 | .874 | .869 | .863 | .857 | .850 | .843 | .836 | .828 | .820 | .811 |
| | | | | | | | | .802 | .792 | .782 | .771 | .760 | .748 | .735 | .722 | .708 | .693 |
| | | | | | | | | .678 | .662 | .645 | .627 | .609 | .590 | .571 | .550 | .529 | .508 |
| | | | | | | | | .486 | .463 | .441 | .417 | .394 | .371 | .347 | .324 | .301 | .279 |
| | | | | | | | | .257 | .235 | | | | | | | | |
| 28.000 | 26.461 | .000 | 1.539 | .79 | 6.38 | 5.5% | 43 | .874 | .869 | .863 | .857 | .850 | .843 | .836 | .828 | .820 | .811 |
| | | | | | | | | .802 | .792 | .782 | .771 | .760 | .748 | .735 | .722 | .708 | .693 |
| | | | | | | | | .678 | .662 | .645 | .627 | .609 | .590 | .571 | .550 | .529 | .508 |
| | | | | | | | | .486 | .463 | .441 | .417 | .394 | .371 | .347 | .324 | .301 | .279 |
| | | | | | | | | .257 | .235 | .214 | | | | | | | |
| 28.000 | 26.655 | .000 | 1.345 | .81 | 6.32 | 4.8% | 44 | .874 | .869 | .863 | .857 | .850 | .843 | .836 | .828 | .820 | .811 |
| | | | | | | | | .802 | .792 | .782 | .771 | .760 | .748 | .735 | .722 | .708 | .693 |
| | | | | | | | | .678 | .662 | .645 | .627 | .609 | .590 | .571 | .550 | .529 | .508 |
| | | | | | | | | .486 | .463 | .441 | .417 | .394 | .371 | .347 | .324 | .301 | .279 |
| | | | | | | | | .257 | .235 | .214 | .194 | | | | | | |
| 28.000 | 26.830 | .000 | 1.170 | .82 | 6.26 | 4.2% | 45 | .874 | .869 | .863 | .857 | .850 | .843 | .836 | .828 | .820 | .811 |
| | | | | | | | | .802 | .792 | .782 | .771 | .760 | .748 | .735 | .722 | .708 | .693 |
| | | | | | | | | .678 | .662 | .645 | .627 | .609 | .590 | .571 | .550 | .529 | .508 |
| | | | | | | | | .486 | .463 | .441 | .417 | .394 | .371 | .347 | .324 | .301 | .279 |
| | | | | | | | | .257 | .235 | .214 | .194 | .175 | | | | | |
| 28.000 | 26.988 | .000 | 1.012 | .84 | 6.19 | 3.6% | 46 | .874 | .869 | .863 | .857 | .850 | .843 | .836 | .828 | .820 | .811 |
| | | | | | | | | .802 | .792 | .782 | .771 | .760 | .748 | .735 | .722 | .708 | .693 |
| | | | | | | | | .678 | .662 | .645 | .627 | .609 | .590 | .571 | .550 | .529 | .508 |
| | | | | | | | | .486 | .463 | .441 | .417 | .394 | .371 | .347 | .324 | .301 | .279 |
| | | | | | | | | .257 | .235 | .214 | .194 | .175 | .158 | | | | |
| 28.000 | 27.129 | .000 | .871 | .86 | 6.11 | 3.1% | 47 | .874 | .869 | .863 | .857 | .850 | .843 | .836 | .828 | .820 | .811 |
| | | | | | | | | .802 | .792 | .782 | .771 | .760 | .748 | .735 | .722 | .708 | .693 |
| | | | | | | | | .678 | .662 | .645 | .627 | .609 | .590 | .571 | .550 | .529 | .508 |
| | | | | | | | | .486 | .463 | .441 | .417 | .394 | .371 | .347 | .324 | .301 | .279 |
| | | | | | | | | .257 | .235 | .214 | .194 | .175 | .158 | .141 | | | |
| 28.000 | 27.253 | .000 | .747 | .88 | 6.03 | 2.7% | 48 | .874 | .869 | .863 | .857 | .850 | .843 | .836 | .828 | .820 | .811 |
| | | | | | | | | .802 | .792 | .782 | .771 | .760 | .748 | .735 | .722 | .708 | .693 |
| | | | | | | | | .678 | .662 | .645 | .627 | .609 | .590 | .571 | .550 | .529 | .508 |
| | | | | | | | | .486 | .463 | .441 | .417 | .394 | .371 | .347 | .324 | .301 | .279 |

*SIMILAR TO A.T.&T. NEAL-WILKINSON TABLES(SEE SECTION 1)

*EQUIVALENT RANDOM THEORY: VMR(VARIANCE TO MEAN RATIO)– 4.0 - NO RETRIES (PEAKED TRAFFIC - ROUTE ADVANCEMENT OR DAY-TO-DAY VARIATION)

| HOURS FIRST ATTEMPT VMR = 4.0 | HOURS CON-NECT | HOURS RECALL /RETRY | HOURS OVFL | CON-NECT VMR | OVFL VMR | PERCENT BLOCKED (GRADE) | LINE | HOURS CONNECTED TRAFFIC ON LINES 1 | 2 | 3 | 4 | 5 | 6 | 7 | 8 | 9 | 10 |
|---|---|---|---|---|---|---|---|---|---|---|---|---|---|---|---|---|---|
| | | | | | | | | .257 | .235 | .214 | .194 | .175 | .158 | .141 | .125 | | |
| 28.000 | 27.364 | .000 | .636 | .89 | 5.94 | 2.3% | 49 | .874 | .869 | .863 | .857 | .850 | .843 | .836 | .828 | .820 | .811 |
| | | | | | | | | .802 | .792 | .782 | .771 | .760 | .748 | .735 | .722 | .708 | .693 |
| | | | | | | | | .678 | .662 | .645 | .627 | .609 | .590 | .571 | .550 | .529 | .508 |
| | | | | | | | | .486 | .463 | .441 | .417 | .394 | .371 | .347 | .324 | .301 | .279 |
| | | | | | | | | .257 | .235 | .214 | .194 | .175 | .158 | .141 | .125 | .110 | |
| 28.000 | 27.460 | .000 | .540 | .90 | 5.86 | 1.9% | 50 | .874 | .869 | .863 | .857 | .850 | .843 | .836 | .828 | .820 | .811 |
| | | | | | | | | .802 | .792 | .782 | .771 | .760 | .748 | .735 | .722 | .708 | .693 |
| | | | | | | | | .678 | .662 | .645 | .627 | .609 | .590 | .571 | .550 | .529 | .508 |
| | | | | | | | | .486 | .463 | .441 | .417 | .394 | .371 | .347 | .324 | .301 | .279 |
| | | | | | | | | .257 | .235 | .214 | .194 | .175 | .158 | .141 | .125 | .110 | .097 |
| 28.000 | 27.545 | .000 | .455 | .92 | 5.77 | 1.6% | 51 | .874 | .869 | .863 | .857 | .850 | .843 | .836 | .828 | .820 | .811 |
| | | | | | | | | .802 | .792 | .782 | .771 | .760 | .748 | .735 | .722 | .708 | .693 |
| | | | | | | | | .678 | .662 | .645 | .627 | .609 | .590 | .571 | .550 | .529 | .508 |
| | | | | | | | | .486 | .463 | .441 | .417 | .394 | .371 | .347 | .324 | .301 | .279 |
| | | | | | | | | .257 | .235 | .214 | .194 | .175 | .158 | .141 | .125 | .110 | .097 |
| | | | | | | | | .084 | | | | | | | | | |
| 28.000 | 27.618 | .000 | .382 | .93 | 5.67 | 1.4% | 52 | .874 | .869 | .863 | .857 | .850 | .843 | .836 | .828 | .820 | .811 |
| | | | | | | | | .802 | .792 | .782 | .771 | .760 | .748 | .735 | .722 | .708 | .693 |
| | | | | | | | | .678 | .662 | .645 | .627 | .609 | .590 | .571 | .550 | .529 | .508 |
| | | | | | | | | .486 | .463 | .441 | .417 | .394 | .371 | .347 | .324 | .301 | .279 |
| | | | | | | | | .257 | .235 | .214 | .194 | .175 | .158 | .141 | .125 | .110 | .097 |
| | | | | | | | | .084 | .073 | | | | | | | | |
| 28.000 | 27.681 | .000 | .319 | .94 | 5.58 | 1.1% | 53 | .874 | .869 | .863 | .857 | .850 | .843 | .836 | .828 | .820 | .811 |
| | | | | | | | | .802 | .792 | .782 | .771 | .760 | .748 | .735 | .722 | .708 | .693 |
| | | | | | | | | .678 | .662 | .645 | .627 | .609 | .590 | .571 | .550 | .529 | .508 |
| | | | | | | | | .486 | .463 | .441 | .417 | .394 | .371 | .347 | .324 | .301 | .279 |
| | | | | | | | | .257 | .235 | .214 | .194 | .175 | .158 | .141 | .125 | .110 | .097 |
| | | | | | | | | .084 | .073 | .063 | | | | | | | |
| 28.000 | 27.735 | .000 | .265 | .95 | 5.48 | .9% | 54 | .874 | .869 | .863 | .857 | .850 | .843 | .836 | .828 | .820 | .811 |
| | | | | | | | | .802 | .792 | .782 | .771 | .760 | .748 | .735 | .722 | .708 | .693 |
| | | | | | | | | .678 | .662 | .645 | .627 | .609 | .590 | .571 | .550 | .529 | .508 |
| | | | | | | | | .486 | .463 | .441 | .417 | .394 | .371 | .347 | .324 | .301 | .279 |
| | | | | | | | | .257 | .235 | .214 | .194 | .175 | .158 | .141 | .125 | .110 | .097 |
| | | | | | | | | .084 | .073 | .063 | .054 | | | | | | |
| 28.000 | 27.781 | .000 | .219 | .95 | 5.39 | .8% | 55 | .874 | .869 | .863 | .857 | .850 | .843 | .836 | .828 | .820 | .811 |
| | | | | | | | | .802 | .792 | .782 | .771 | .760 | .748 | .735 | .722 | .708 | .693 |
| | | | | | | | | .678 | .662 | .645 | .627 | .609 | .590 | .571 | .550 | .529 | .508 |
| | | | | | | | | .486 | .463 | .441 | .417 | .394 | .371 | .347 | .324 | .301 | .279 |
| | | | | | | | | .257 | .235 | .214 | .194 | .175 | .158 | .141 | .125 | .110 | .097 |

*SIMILAR TO A.T.&T. NEAL-WILKINSON TABLES(SEE SECTION 1)

*EQUIVALENT RANDOM THEORY: VMR(VARIANCE TO MEAN RATIO)– 4.0 - NO RETRIES (PEAKED TRAFFIC - ROUTE ADVANCEMENT OR DAY-TO-DAY VARIATION)

| HOURS FIRST ATTEMPT VMR = 4.0 | HOURS CON-NECT | HOURS RECALL /RETRY | HOURS OVFL | CON-NECT VMR | OVFL VMR | PERCENT BLOCKED (GRADE) | LINE | HOURS CONNECTED TRAFFIC ON LINES 1 | 2 | 3 | 4 | 5 | 6 | 7 | 8 | 9 | 10 |
|---|---|---|---|---|---|---|---|---|---|---|---|---|---|---|---|---|---|
| | | | | | | | | .084 | .073 | .063 | .054 | .046 | | | | | |
| 28.000 | 27.820 | .000 | .180 | .96 | 5.29 | .6% | 56 | .874 | .869 | .863 | .857 | .850 | .843 | .836 | .828 | .820 | .811 |
| | | | | | | | | .802 | .792 | .782 | .771 | .760 | .748 | .735 | .722 | .708 | .693 |
| | | | | | | | | .678 | .662 | .645 | .627 | .609 | .590 | .571 | .550 | .529 | .508 |
| | | | | | | | | .486 | .463 | .441 | .417 | .394 | .371 | .347 | .324 | .301 | .279 |
| | | | | | | | | .257 | .235 | .214 | .194 | .175 | .158 | .141 | .125 | .110 | .097 |
| | | | | | | | | .084 | .073 | .063 | .054 | .046 | .039 | | | | |
| 28.000 | 27.853 | .000 | .147 | .97 | 5.20 | .5% | 57 | .874 | .869 | .863 | .857 | .850 | .843 | .836 | .828 | .820 | .811 |
| | | | | | | | | .802 | .792 | .782 | .771 | .760 | .748 | .735 | .722 | .708 | .693 |
| | | | | | | | | .678 | .662 | .645 | .627 | .609 | .590 | .571 | .550 | .529 | .508 |
| | | | | | | | | .486 | .463 | .441 | .417 | .394 | .371 | .347 | .324 | .301 | .279 |
| | | | | | | | | .257 | .235 | .214 | .194 | .175 | .158 | .141 | .125 | .110 | .097 |
| | | | | | | | | .084 | .073 | .063 | .054 | .046 | .039 | .033 | | | |
| 30.000 | 19.657 | .000 | 10.343 | .35 | 6.12 | 34.5% | 25 | .882 | .877 | .871 | .866 | .860 | .854 | .847 | .840 | .833 | .825 |
| | | | | | | | | .817 | .808 | .799 | .790 | .779 | .769 | .758 | .746 | .733 | .720 |
| | | | | | | | | .707 | .692 | .677 | .662 | .645 | | | | | |
| 30.000 | 20.285 | .000 | 9.715 | .37 | 6.19 | 32.4% | 26 | .882 | .877 | .871 | .866 | .860 | .854 | .847 | .840 | .833 | .825 |
| | | | | | | | | .817 | .808 | .799 | .790 | .779 | .769 | .758 | .746 | .733 | .720 |
| | | | | | | | | .707 | .692 | .677 | .662 | .645 | .628 | | | | |
| 30.000 | 20.896 | .000 | 9.104 | .39 | 6.27 | 30.3% | 27 | .882 | .877 | .871 | .866 | .860 | .854 | .847 | .840 | .833 | .825 |
| | | | | | | | | .817 | .808 | .799 | .790 | .779 | .769 | .758 | .746 | .733 | .720 |
| | | | | | | | | .707 | .692 | .677 | .662 | .645 | .628 | .611 | | | |
| 30.000 | 21.488 | .000 | 8.512 | .41 | 6.34 | 28.4% | 28 | .882 | .877 | .871 | .866 | .860 | .854 | .847 | .840 | .833 | .825 |
| | | | | | | | | .817 | .808 | .799 | .790 | .779 | .769 | .758 | .746 | .733 | .720 |
| | | | | | | | | .707 | .692 | .677 | .662 | .645 | .628 | .611 | .592 | | |
| 30.000 | 22.062 | .000 | 7.938 | .43 | 6.40 | 26.5% | 29 | .882 | .877 | .871 | .866 | .860 | .854 | .847 | .840 | .833 | .825 |
| | | | | | | | | .817 | .808 | .799 | .790 | .779 | .769 | .758 | .746 | .733 | .720 |
| | | | | | | | | .707 | .692 | .677 | .662 | .645 | .628 | .611 | .592 | .573 | |
| 30.000 | 22.615 | .000 | 7.385 | .45 | 6.46 | 24.6% | 30 | .882 | .877 | .871 | .866 | .860 | .854 | .847 | .840 | .833 | .825 |
| | | | | | | | | .817 | .808 | .799 | .790 | .779 | .769 | .758 | .746 | .733 | .720 |
| | | | | | | | | .707 | .692 | .677 | .662 | .645 | .628 | .611 | .592 | .573 | .554 |
| 30.000 | 23.149 | .000 | 6.851 | .47 | 6.52 | 22.8% | 31 | .882 | .877 | .871 | .866 | .860 | .854 | .847 | .840 | .833 | .825 |
| | | | | | | | | .817 | .808 | .799 | .790 | .779 | .769 | .758 | .746 | .733 | .720 |
| | | | | | | | | .707 | .692 | .677 | .662 | .645 | .628 | .611 | .592 | .573 | .554 |
| | | | | | | | | .533 | | | | | | | | | |
| 30.000 | 23.661 | .000 | 6.339 | .49 | 6.57 | 21.1% | 32 | .882 | .877 | .871 | .866 | .860 | .854 | .847 | .840 | .833 | .825 |
| | | | | | | | | .817 | .808 | .799 | .790 | .779 | .769 | .758 | .746 | .733 | .720 |
| | | | | | | | | .707 | .692 | .677 | .662 | .645 | .628 | .611 | .592 | .573 | .554 |
| | | | | | | | | .533 | .512 | | | | | | | | |
| 30.000 | 24.152 | .000 | 5.848 | .51 | 6.61 | 19.5% | 33 | .882 | .877 | .871 | .866 | .860 | .854 | .847 | .840 | .833 | .825 |

*SIMILAR TO A.T.&T. NEAL-WILKINSON TABLES(SEE SECTION 1)

*EQUIVALENT RANDOM THEORY: VMR(VARIANCE TO MEAN RATIO)– 4.0 - NO RETRIES (PEAKED TRAFFIC - ROUTE ADVANCEMENT OR DAY-TO-DAY VARIATION)

| HOURS FIRST ATTEMPT VMR = 4.0 | HOURS CON-NECT | HOURS RECALL /RETRY | HOURS OVFL | CON-NECT VMR | OVFL VMR | PERCENT BLOCKED (GRADE) | LINE | HOURS CONNECTED TRAFFIC ON LINES 1 | 2 | 3 | 4 | 5 | 6 | 7 | 8 | 9 | 10 |
|---|---|---|---|---|---|---|---|---|---|---|---|---|---|---|---|---|---|
| | | | | | | | | .817 | .808 | .799 | .790 | .779 | .769 | .758 | .746 | .733 | .720 |
| | | | | | | | | .707 | .692 | .677 | .662 | .645 | .628 | .611 | .592 | .573 | .554 |
| | | | | | | | | .533 | .512 | .491 | | | | | | | |
| 30.000 | 24.621 | .000 | 5.379 | .53 | 6.65 | 17.9% | 34 | .882 | .877 | .871 | .866 | .860 | .854 | .847 | .840 | .833 | .825 |
| | | | | | | | | .817 | .808 | .799 | .790 | .779 | .769 | .758 | .746 | .733 | .720 |
| | | | | | | | | .707 | .692 | .677 | .662 | .645 | .628 | .611 | .592 | .573 | .554 |
| | | | | | | | | .533 | .512 | .491 | .469 | | | | | | |
| 30.000 | 25.069 | .000 | 4.931 | .55 | 6.68 | 16.4% | 35 | .882 | .877 | .871 | .866 | .860 | .854 | .847 | .840 | .833 | .825 |
| | | | | | | | | .817 | .808 | .799 | .790 | .779 | .769 | .758 | .746 | .733 | .720 |
| | | | | | | | | .707 | .692 | .677 | .662 | .645 | .628 | .611 | .592 | .573 | .554 |
| | | | | | | | | .533 | .512 | .491 | .469 | .447 | | | | | |
| 30.000 | 25.493 | .000 | 4.507 | .58 | 6.70 | 15.0% | 36 | .882 | .877 | .871 | .866 | .860 | .854 | .847 | .840 | .833 | .825 |
| | | | | | | | | .817 | .808 | .799 | .790 | .779 | .769 | .758 | .746 | .733 | .720 |
| | | | | | | | | .707 | .692 | .677 | .662 | .645 | .628 | .611 | .592 | .573 | .554 |
| | | | | | | | | .533 | .512 | .491 | .469 | .447 | .425 | | | | |
| 30.000 | 25.895 | .000 | 4.105 | .60 | 6.72 | 13.7% | 37 | .882 | .877 | .871 | .866 | .860 | .854 | .847 | .840 | .833 | .825 |
| | | | | | | | | .817 | .808 | .799 | .790 | .779 | .769 | .758 | .746 | .733 | .720 |
| | | | | | | | | .707 | .692 | .677 | .662 | .645 | .628 | .611 | .592 | .573 | .554 |
| | | | | | | | | .533 | .512 | .491 | .469 | .447 | .425 | .402 | | | |
| 30.000 | 26.274 | .000 | 3.726 | .62 | 6.73 | 12.4% | 38 | .882 | .877 | .871 | .866 | .860 | .854 | .847 | .840 | .833 | .825 |
| | | | | | | | | .817 | .808 | .799 | .790 | .779 | .769 | .758 | .746 | .733 | .720 |
| | | | | | | | | .707 | .692 | .677 | .662 | .645 | .628 | .611 | .592 | .573 | .554 |
| | | | | | | | | .533 | .512 | .491 | .469 | .447 | .425 | .402 | .379 | | |
| 30.000 | 26.631 | .000 | 3.369 | .64 | 6.73 | 11.2% | 39 | .882 | .877 | .871 | .866 | .860 | .854 | .847 | .840 | .833 | .825 |
| | | | | | | | | .817 | .808 | .799 | .790 | .779 | .769 | .758 | .746 | .733 | .720 |
| | | | | | | | | .707 | .692 | .677 | .662 | .645 | .628 | .611 | .592 | .573 | .554 |
| | | | | | | | | .533 | .512 | .491 | .469 | .447 | .425 | .402 | .379 | .356 | |
| 30.000 | 26.965 | .000 | 3.035 | .67 | 6.72 | 10.1% | 40 | .882 | .877 | .871 | .866 | .860 | .854 | .847 | .840 | .833 | .825 |
| | | | | | | | | .817 | .808 | .799 | .790 | .779 | .769 | .758 | .746 | .733 | .720 |
| | | | | | | | | .707 | .692 | .677 | .662 | .645 | .628 | .611 | .592 | .573 | .554 |
| | | | | | | | | .533 | .512 | .491 | .469 | .447 | .425 | .402 | .379 | .356 | .334 |
| 30.000 | 27.276 | .000 | 2.724 | .69 | 6.70 | 9.1% | 41 | .882 | .877 | .871 | .866 | .860 | .854 | .847 | .840 | .833 | .825 |
| | | | | | | | | .817 | .808 | .799 | .790 | .779 | .769 | .758 | .746 | .733 | .720 |
| | | | | | | | | .707 | .692 | .677 | .662 | .645 | .628 | .611 | .592 | .573 | .554 |
| | | | | | | | | .533 | .512 | .491 | .469 | .447 | .425 | .402 | .379 | .356 | .334 |
| | | | | | | | | .311 | | | | | | | | | |
| 30.000 | 27.565 | .000 | 2.435 | .71 | 6.68 | 8.1% | 42 | .882 | .877 | .871 | .866 | .860 | .854 | .847 | .840 | .833 | .825 |
| | | | | | | | | .817 | .808 | .799 | .790 | .779 | .769 | .758 | .746 | .733 | .720 |
| | | | | | | | | .707 | .692 | .677 | .662 | .645 | .628 | .611 | .592 | .573 | .554 |
| | | | | | | | | .533 | .512 | .491 | .469 | .447 | .425 | .402 | .379 | .356 | .334 |

*SIMILAR TO A.T.&T. NEAL-WILKINSON TABLES(SEE SECTION 1)

# *EQUIVALENT RANDOM THEORY: VMR(VARIANCE TO MEAN RATIO)– 4.0 - NO RETRIES (PEAKED TRAFFIC - ROUTE ADVANCEMENT OR DAY-TO-DAY VARIATION)

| HOURS FIRST ATTEMPT VMR = 4.0 | HOURS CON-NECT | HOURS RECALL /RETRY | HOURS OVFL | CON-NECT VMR | OVFL VMR | PERCENT BLOCKED (GRADE) | LINE | HOURS CONNECTED TRAFFIC ON LINES 1 | 2 | 3 | 4 | 5 | 6 | 7 | 8 | 9 | 10 |
|---|---|---|---|---|---|---|---|---|---|---|---|---|---|---|---|---|---|
| | | | | | | | | .311 | .289 | | | | | | | | |
| 30.000 | 27.832 | .000 | 2.168 | .73 | 6.65 | 7.2% | 43 | .882 | .877 | .871 | .866 | .860 | .854 | .847 | .840 | .833 | .825 |
| | | | | | | | | .817 | .808 | .799 | .790 | .779 | .769 | .758 | .746 | .733 | .720 |
| | | | | | | | | .707 | .692 | .677 | .662 | .645 | .628 | .611 | .592 | .573 | .554 |
| | | | | | | | | .533 | .512 | .491 | .469 | .447 | .425 | .402 | .379 | .356 | .334 |
| | | | | | | | | .311 | .289 | .267 | | | | | | | |
| 30.000 | 28.078 | .000 | 1.922 | .75 | 6.61 | 6.4% | 44 | .882 | .877 | .871 | .866 | .860 | .854 | .847 | .840 | .833 | .825 |
| | | | | | | | | .817 | .808 | .799 | .790 | .779 | .769 | .758 | .746 | .733 | .720 |
| | | | | | | | | .707 | .692 | .677 | .662 | .645 | .628 | .611 | .592 | .573 | .554 |
| | | | | | | | | .533 | .512 | .491 | .469 | .447 | .425 | .402 | .379 | .356 | .334 |
| | | | | | | | | .311 | .289 | .267 | .246 | | | | | | |
| 30.000 | 28.303 | .000 | 1.697 | .77 | 6.56 | 5.7% | 45 | .882 | .877 | .871 | .866 | .860 | .854 | .847 | .840 | .833 | .825 |
| | | | | | | | | .817 | .808 | .799 | .790 | .779 | .769 | .758 | .746 | .733 | .720 |
| | | | | | | | | .707 | .692 | .677 | .662 | .645 | .628 | .611 | .592 | .573 | .554 |
| | | | | | | | | .533 | .512 | .491 | .469 | .447 | .425 | .402 | .379 | .356 | .334 |
| | | | | | | | | .311 | .289 | .267 | .246 | .225 | | | | | |
| 30.000 | 28.509 | .000 | 1.491 | .79 | 6.51 | 5.0% | 46 | .882 | .877 | .871 | .866 | .860 | .854 | .847 | .840 | .833 | .825 |
| | | | | | | | | .817 | .808 | .799 | .790 | .779 | .769 | .758 | .746 | .733 | .720 |
| | | | | | | | | .707 | .692 | .677 | .662 | .645 | .628 | .611 | .592 | .573 | .554 |
| | | | | | | | | .533 | .512 | .491 | .469 | .447 | .425 | .402 | .379 | .356 | .334 |
| | | | | | | | | .311 | .289 | .267 | .246 | .225 | .205 | | | | |
| 30.000 | 28.695 | .000 | 1.305 | .81 | 6.44 | 4.3% | 47 | .882 | .877 | .871 | .866 | .860 | .854 | .847 | .840 | .833 | .825 |
| | | | | | | | | .817 | .808 | .799 | .790 | .779 | .769 | .758 | .746 | .733 | .720 |
| | | | | | | | | .707 | .692 | .677 | .662 | .645 | .628 | .611 | .592 | .573 | .554 |
| | | | | | | | | .533 | .512 | .491 | .469 | .447 | .425 | .402 | .379 | .356 | .334 |
| | | | | | | | | .311 | .289 | .267 | .246 | .225 | .205 | .186 | | | |
| 30.000 | 28.864 | .000 | 1.136 | .83 | 6.38 | 3.8% | 48 | .882 | .877 | .871 | .866 | .860 | .854 | .847 | .840 | .833 | .825 |
| | | | | | | | | .817 | .808 | .799 | .790 | .779 | .769 | .758 | .746 | .733 | .720 |
| | | | | | | | | .707 | .692 | .677 | .662 | .645 | .628 | .611 | .592 | .573 | .554 |
| | | | | | | | | .533 | .512 | .491 | .469 | .447 | .425 | .402 | .379 | .356 | .334 |
| | | | | | | | | .311 | .289 | .267 | .246 | .225 | .205 | .186 | .168 | | |
| 30.000 | 29.015 | .000 | .985 | .85 | 6.31 | 3.3% | 49 | .882 | .877 | .871 | .866 | .860 | .854 | .847 | .840 | .833 | .825 |
| | | | | | | | | .817 | .808 | .799 | .790 | .779 | .769 | .758 | .746 | .733 | .720 |
| | | | | | | | | .707 | .692 | .677 | .662 | .645 | .628 | .611 | .592 | .573 | .554 |
| | | | | | | | | .533 | .512 | .491 | .469 | .447 | .425 | .402 | .379 | .356 | .334 |
| | | | | | | | | .311 | .289 | .267 | .246 | .225 | .205 | .186 | .168 | .151 | |
| 30.000 | 29.150 | .000 | .850 | .86 | 6.23 | 2.8% | 50 | .882 | .877 | .871 | .866 | .860 | .854 | .847 | .840 | .833 | .825 |
| | | | | | | | | .817 | .808 | .799 | .790 | .779 | .769 | .758 | .746 | .733 | .720 |
| | | | | | | | | .707 | .692 | .677 | .662 | .645 | .628 | .611 | .592 | .573 | .554 |
| | | | | | | | | .533 | .512 | .491 | .469 | .447 | .425 | .402 | .379 | .356 | .334 |

*SIMILAR TO A.T.&T. NEAL-WILKINSON TABLES(SEE SECTION 1)

*EQUIVALENT RANDOM THEORY: VMR(VARIANCE TO MEAN RATIO)– 4.0 - NO RETRIES (PEAKED TRAFFIC - ROUTE ADVANCEMENT OR DAY-TO-DAY VARIATION)

| HOURS FIRST ATTEMPT VMR=4.0 | HOURS CON-NECT | HOURS RECALL /RETRY | HOURS OVFL | CON-NECT VMR | OVFL VMR | PERCENT BLOCKED (GRADE) | LINE | HOURS CONNECTED TRAFFIC ON LINES 1 | 2 | 3 | 4 | 5 | 6 | 7 | 8 | 9 | 10 |
|---|---|---|---|---|---|---|---|---|---|---|---|---|---|---|---|---|---|
| | | | | | | | | .311 | .289 | .267 | .246 | .225 | .205 | .186 | .168 | .151 | .135 |
| 30.000 | 29.270 | .000 | .730 | .88 | 6.15 | 2.4% | 51 | .882 | .877 | .871 | .866 | .860 | .854 | .847 | .840 | .833 | .825 |
| | | | | | | | | .817 | .808 | .799 | .790 | .779 | .769 | .758 | .746 | .733 | .720 |
| | | | | | | | | .707 | .692 | .677 | .662 | .645 | .628 | .611 | .592 | .573 | .554 |
| | | | | | | | | .533 | .512 | .491 | .469 | .447 | .425 | .402 | .379 | .356 | .334 |
| | | | | | | | | .311 | .289 | .267 | .246 | .225 | .205 | .186 | .168 | .151 | .135 |
| | | | | | | | | .120 | | | | | | | | | |
| 30.000 | 29.376 | .000 | .624 | .89 | 6.06 | 2.1% | 52 | .882 | .877 | .871 | .866 | .860 | .854 | .847 | .840 | .833 | .825 |
| | | | | | | | | .817 | .808 | .799 | .790 | .779 | .769 | .758 | .746 | .733 | .720 |
| | | | | | | | | .707 | .692 | .677 | .662 | .645 | .628 | .611 | .592 | .573 | .554 |
| | | | | | | | | .533 | .512 | .491 | .469 | .447 | .425 | .402 | .379 | .356 | .334 |
| | | | | | | | | .311 | .289 | .267 | .246 | .225 | .205 | .186 | .168 | .151 | .135 |
| | | | | | | | | .120 | .106 | | | | | | | | |
| 30.000 | 29.470 | .000 | .530 | .91 | 5.97 | 1.8% | 53 | .882 | .877 | .871 | .866 | .860 | .854 | .847 | .840 | .833 | .825 |
| | | | | | | | | .817 | .808 | .799 | .790 | .779 | .769 | .758 | .746 | .733 | .720 |
| | | | | | | | | .707 | .692 | .677 | .662 | .645 | .628 | .611 | .592 | .573 | .554 |
| | | | | | | | | .533 | .512 | .491 | .469 | .447 | .425 | .402 | .379 | .356 | .334 |
| | | | | | | | | .311 | .289 | .267 | .246 | .225 | .205 | .186 | .168 | .151 | .135 |
| | | | | | | | | .120 | .106 | .093 | | | | | | | |
| 30.000 | 29.551 | .000 | .449 | .92 | 5.88 | 1.5% | 54 | .882 | .877 | .871 | .866 | .860 | .854 | .847 | .840 | .833 | .825 |
| | | | | | | | | .817 | .808 | .799 | .790 | .779 | .769 | .758 | .746 | .733 | .720 |
| | | | | | | | | .707 | .692 | .677 | .662 | .645 | .628 | .611 | .592 | .573 | .554 |
| | | | | | | | | .533 | .512 | .491 | .469 | .447 | .425 | .402 | .379 | .356 | .334 |
| | | | | | | | | .311 | .289 | .267 | .246 | .225 | .205 | .186 | .168 | .151 | .135 |
| | | | | | | | | .120 | .106 | .093 | .082 | | | | | | |
| 30.000 | 29.622 | .000 | .378 | .93 | 5.78 | 1.3% | 55 | .882 | .877 | .871 | .866 | .860 | .854 | .847 | .840 | .833 | .825 |
| | | | | | | | | .817 | .808 | .799 | .790 | .779 | .769 | .758 | .746 | .733 | .720 |
| | | | | | | | | .707 | .692 | .677 | .662 | .645 | .628 | .611 | .592 | .573 | .554 |
| | | | | | | | | .533 | .512 | .491 | .469 | .447 | .425 | .402 | .379 | .356 | .334 |
| | | | | | | | | .311 | .289 | .267 | .246 | .225 | .205 | .186 | .168 | .151 | .135 |
| | | | | | | | | .120 | .106 | .093 | .082 | .071 | | | | | |
| 30.000 | 29.683 | .000 | .317 | .94 | 5.69 | 1.1% | 56 | .882 | .877 | .871 | .866 | .860 | .854 | .847 | .840 | .833 | .825 |
| | | | | | | | | .817 | .808 | .799 | .790 | .779 | .769 | .758 | .746 | .733 | .720 |
| | | | | | | | | .707 | .692 | .677 | .662 | .645 | .628 | .611 | .592 | .573 | .554 |
| | | | | | | | | .533 | .512 | .491 | .469 | .447 | .425 | .402 | .379 | .356 | .334 |
| | | | | | | | | .311 | .289 | .267 | .246 | .225 | .205 | .186 | .168 | .151 | .135 |
| | | | | | | | | .120 | .106 | .093 | .082 | .071 | .061 | | | | |
| 30.000 | 29.736 | .000 | .264 | .95 | 5.59 | .9% | 57 | .882 | .877 | .871 | .866 | .860 | .854 | .847 | .840 | .833 | .825 |
| | | | | | | | | .817 | .808 | .799 | .790 | .779 | .769 | .758 | .746 | .733 | .720 |
| | | | | | | | | .707 | .692 | .677 | .662 | .645 | .628 | .611 | .592 | .573 | .554 |

*SIMILAR TO A.T.&T. NEAL-WILKINSON TABLES(SEE SECTION 1)

*EQUIVALENT RANDOM THEORY: VMR(VARIANCE TO MEAN RATIO)– 4.0 - NO RETRIES (PEAKED TRAFFIC - ROUTE ADVANCEMENT OR DAY-TO-DAY VARIATION)

| HOURS FIRST ATTEMPT VMR=4.0 | HOURS CON-NECT | HOURS RECALL /RETRY | HOURS OVFL | CON-NECT VMR | OVFL VMR | PERCENT BLOCKED (GRADE) | LINE | HOURS CONNECTED TRAFFIC ON LINES 1 | 2 | 3 | 4 | 5 | 6 | 7 | 8 | 9 | 10 |
|---|---|---|---|---|---|---|---|---|---|---|---|---|---|---|---|---|---|
| | | | | | | | | .533 | .512 | .491 | .469 | .447 | .425 | .402 | .379 | .356 | .334 |
| | | | | | | | | .311 | .289 | .267 | .246 | .225 | .205 | .186 | .168 | .151 | .135 |
| | | | | | | | | .120 | .106 | .093 | .082 | .071 | .061 | .053 | | | |
| 30.000 | 29.781 | .000 | .219 | .95 | 5.50 | .7% | 58 | .882 | .877 | .871 | .866 | .860 | .854 | .847 | .840 | .833 | .825 |
| | | | | | | | | .817 | .808 | .799 | .790 | .779 | .769 | .758 | .746 | .733 | .720 |
| | | | | | | | | .707 | .692 | .677 | .662 | .645 | .628 | .611 | .592 | .573 | .554 |
| | | | | | | | | .533 | .512 | .491 | .469 | .447 | .425 | .402 | .379 | .356 | .334 |
| | | | | | | | | .311 | .289 | .267 | .246 | .225 | .205 | .186 | .168 | .151 | .135 |
| | | | | | | | | .120 | .106 | .093 | .082 | .071 | .061 | .053 | .045 | | |
| 30.000 | 29.819 | .000 | .181 | .96 | 5.40 | .6% | 59 | .882 | .877 | .871 | .866 | .860 | .854 | .847 | .840 | .833 | .825 |
| | | | | | | | | .817 | .808 | .799 | .790 | .779 | .769 | .758 | .746 | .733 | .720 |
| | | | | | | | | .707 | .692 | .677 | .662 | .645 | .628 | .611 | .592 | .573 | .554 |
| | | | | | | | | .533 | .512 | .491 | .469 | .447 | .425 | .402 | .379 | .356 | .334 |
| | | | | | | | | .311 | .289 | .267 | .246 | .225 | .205 | .186 | .168 | .151 | .135 |
| | | | | | | | | .120 | .106 | .093 | .082 | .071 | .061 | .053 | .045 | .038 | |

*SIMILAR TO A.T.&T. NEAL-WILKINSON TABLES(SEE SECTION 1)